Basic
Fluid Mechanics

by

David C. Wilcox

DCW Industries, Inc.
La Cañada, California

Basic Fluid Mechanics

Copyright © 1997 by DCW Industries, Inc. All rights reserved.

First Printing: April, 1997

DCW Industries, Inc.
5354 Palm Drive, La Cañada, California 91011
818/790-3844 (FAX) 818/952-1272
e-mail: dcwilcox@ix.netcom.com
World Wide Web: http://webknx.com/dcw

This book was prepared with LaTeX as implemented by Personal TeX, Inc. of Mill Valley, California. It was printed in the United States of America by KNI, Inc., Anaheim, California and bound by Stauffer Edition Binding Company, Inc. of Monterey Park, California..

Library of Congress Cataloging in Publication Data

Wilcox, David C.
    Basic Fluid Mechanics/David C. Wilcox—1st ed.
    Includes bibliography, index and $3\frac{1}{2}$ inch floppy disk.
    1. Fluid Mechanics.
    2. Computational Fluid Dynamics.
Catalog Card Number 97-91556

ISBN   0-9636051-4-3

Dedicated to Rosie and Sabby
for having such a wonderful daughter

Made in the USA

About the Author

Dr. David C. Wilcox, was born in Wilmington, Delaware. He did his undergraduate studies from 1963 to 1966 at the Massachusetts Institute of Technology, graduating with a Bachelor of Science degree in Aeronautics and Astronautics. From 1966 to 1967, he was employed by the McDonnell Douglas Aircraft Division in Long Beach, California, and began his professional career under the guidance of A. M. O. Smith. His experience with McDonnell Douglas focused on subsonic and transonic flow calculations. From 1967 to 1970, he attended the California Institute of Technology, graduating with a Ph.D. in Aeronautics. In 1970 he joined TRW Systems, Inc. in Redondo Beach, California, where he performed studies of both high- and low-speed fluid-mechanical and heat-transfer problems, such as turbulent hypersonic flow and thermal radiation from a flame. From 1972 to 1973, he was a staff scientist for Applied Theory, Inc., in Los Angeles, California. He participated in many research efforts involving numerical computation and analysis of fluid flows such as separated turbulent flow, transitional flow and hypersonic plume-body interaction. In 1973, he founded DCW Industries, Inc., a La Cañada, California firm engaged in engineering research, software development and publishing, for which he is currently the President. He has taught several fluid mechanics and applied mathematics courses at the University of Southern California and at the University of California, Los Angeles.

Dr. Wilcox has numerous publications on turbulence modeling, computational fluid dynamics, boundary-layer separation, boundary-layer transition, thermal radiation, and rapidly rotating fluids. His publications include two graduate-level texts, *Turbulence Modeling for CFD* and *Perturbation Methods in the Computer Age*. He is an Associate Fellow of the American Institute of Aeronautics and Astronautics (AIAA) and has served as an Associate Editor for the AIAA Journal.

Contents

Index of Home Experiments

The book contains several straightforward experiments that can be performed without special laboratory equipment. The experiments can be found on the following pages.

Notation

This section includes the most commonly used notation in this book. In order to avoid departing too much from conventions normally used in literature on general fluid mechanics, a few symbols denote more than one quantity.

English Symbols

Symbol	Definition
a	Speed of sound
$\mathbf{a}$	Acceleration vector
A	Area; constant in Sutherland's law
b	Open-channel width; impeller-blade width
$B(t)$	Extensive variable for a system
c	Celerity
c_f	Skin friction, $\tau_w/(\frac{1}{2}\rho u_e^2)$
c_p, c_v	Specific heat at constant pressure, volume
$C(x)$	Airfoil camber line, $\frac{1}{2}\left[Y_u(x) + Y_l(x)\right]$
C_D	Drag coefficient, $D/(\frac{1}{2}\rho U^2 A)$
C_H	Head coefficient
C_L	Lift coefficient, $L/(\frac{1}{2}\rho U^2 A)$
C_p	Pressure coefficient
C_P	Power coefficient
C_Q	Capacity coefficient; suction coefficient
C^+	Additive constant in the law of the wall
d, D	Diameter
dS	Differential surface area
dV	Differential volume
D, D_f, D_p	Drag, friction drag, pressure drag
D_h	Hydraulic diameter
$\mathcal{D}$	Doublet strength
$\mathcal{D}_a$	Numerical dispersion coefficient
e	Specific internal energy
E	Total energy; open-channel specific energy
$\mathcal{E}$	Specific total energy
f	Friction factor, $\tau_w/(\frac{1}{8}\rho\bar{u}^2)$; frequency
$\mathbf{f}$	Body force vector per unit mass
$\mathbf{f}_v$	Viscous force vector per unit area
F	Force vector magnitude, $|\mathbf{F}|$

F_{buoy}	Buoyancy force
F_D	Drag force
F_H, F_V	Horizontal force, vertical force
Fr_L	Froude number based on length L
F_x, F_y, F_z	Force components in x, y, z directions
F	Force vector
F$_b$, **F**$_s$	Body-force vector, surface-force vector
F$_D$	Drag force vector
g, **g**	Gravitational acceleration magnitude, vector
G	Amplitude factor
h	Specific enthalpy; height
h_L	Head loss
h_p	Head supplied by a pump
h_t	Head given up to a turbine
H	Total enthalpy; channel height; shape factor, δ^*/θ
$\mathcal{H}$	Specific total enthalpy
$\mathcal{H}(t)$	Heaviside step function
H	Angular-momentum vector
i, **j**, **k**	Unit vectors in x, y, z directions
I	Moment of inertia
k	Boltzmann's constant; thermal conductivity; turbulence energy
k_s	Surface roughness height
K	Loss coefficient
Kn	Knudsen number
ℓ_e	Entrance length in a pipe
ℓ_{mfp}	Mean free path
ℓ_{mix}	Mixing length
L	Lift force; characteristic length scale
L_e	Equivalent length
L	Lift vector
m	Mass
$\dot{m}$	Mass-flow rate
M	Mach number; total mass
M	Moment vector
n	Number density; Manning roughness coefficient
n	Outer unit normal vector
N_{CFL}	Courant-Friedrichs-Lewy (CFL) number
N_s	Specific speed
p	Static pressure
$\hat{p}$	Reduced pressure, $p + \rho g z$
p_o	Atmospheric pressure
P	Perimeter
Pr, Pr_t	Prandtl number, turbulent Prandtl number
P	Momentum vector
q	Specific heat transfer; heat added per unit mass
q_t	Turbulent heat flux
$q(\xi)$	Source-sheet strength, $dQ/d\xi$
q	Heat-flux vector
Q	Volume-flow rate; heat transfer; source strength

$\dot{Q}$	Heat transfer rate
r	Recovery factor
r, θ, z	Cylindrical polar coordinates
$\mathbf{r}$	Position vector
$\mathbf{r}_{cp}$	Center of pressure vector
R	Perfect gas constant; radius
R_h	Hydraulic radius
R, θ, ϕ	Spherical coordinates
$\mathcal{R}$	Radius of curvature
$\mathbf{R}$	Reaction force vector
Re_L, Re_x	Reynolds number based on length L, x
Re_θ	Reynolds number based on momentum thickness
s	Distance along a streamline; specific entropy
S	Surface area; constant in Sutherland's law
S_o	Bottom slope in an open channel
St	Strouhal number
S_{xx}, S_{yy}, S_{zz}	Normal strain rates
S_{xy}, S_{yz}, S_{zx}	Shear strain rates
$[\mathbf{S}]$	Strain-rate tensor
t	Time
$\mathbf{t}$	Unit vector tangent to a surface or curve
T	Temperature; thrust; time
T_{aw}	Adiabatic-wall temperature
$T(x)$	Airfoil half-thickness, $\frac{1}{2}\left[Y_u(x) - Y_l(x)\right]$
u, v, w	Cartesian velocity components in x, y, z directions
u_r, u_θ, w	Cylindrical velocity components in r, θ, z directions
u_R, u_θ, u_ϕ	Spherical velocity components in R, θ, ϕ directions
u_n, u_t	Velocity components in normal, tangential directions
$\mathbf{u}$	Velocity vector
u_τ	Friction velocity, $\sqrt{\tau_w/\rho}$
U, V, W	Velocity components in x, y, z directions
U_j	Jet velocity
U_m	Maximum or centerline velocity
$\mathbf{U}_s$	Velocity vector of a moving surface
v_{mix}	Mixing velocity
v_{th}	Thermal velocity
v_w	Surface injection velocity
V	Volume
$\mathbf{V}, \mathbf{w}$	Absolute, relative velocity
$\mathcal{V}$	Body-force potential
w_e	Exit velocity relative to control volume
W	Weight; work
We	Weber number
W_R	Rocket velocity
$\dot{W}$	Rate at which work is done (power)
$\dot{W}_m$	Mechanical power
$\dot{W}_p$	Hydraulic power (power supplied by a pump)
$\dot{W}_s$	Shaft work, $\dot{W}_t - \dot{W}_p$

$\dot{W}_t$	Power given up to a turbine
x, y, z	Rectangular Cartesian coordinates
x_{cp}, z_{cp}	Center of pressure components
y	Open-channel depth
y_c	Critical depth
y^+	Dimensionless, sublayer-scaled, distance, $u_\tau y/\nu$
$Y_l(x), Y_u(x)$	Airfoil lower, upper surface
$\bar{z}$	Centroid

Greek Symbols

Symbol	Definition
α	Lapse rate; kinetic-energy correction factor; angle
β	Intensive variable; Falkner-Skan parameter; shock angle; blade angle
β_T	Equilibrium parameter, $(\delta^*/\tau_w)d\overline{p}/dx$
γ	Specific-heat ratio, c_p/c_v
$\gamma(\xi)$	Vortex-sheet strength, $d\Gamma_a/d\xi$
Γ	Circulation
Γ_a	Aerodynamic circulation, $-\Gamma$
δ	Boundary-layer thickness; shock stand-off distance
δ_m, δ_s	Rayleigh, Stokes layer thickness
δ^*	Displacement thickness, $\int_0^\delta \left(1 - \frac{u}{u_e}\right) dy$
$[\delta]$	Identity matrix (tensor)
Δ	Defect-layer length scale
$\Delta x, \Delta y, \Delta z$	Finite-difference cell size in x, y, z direction
ϵ	Small parameter; dissipation rate
ζ	Second viscosity coefficient
η	Fluid particle dimension; similarity variable
η_p, η_t	Pump, turbine efficiency
θ	Momentum thickness, $\int_0^\delta \frac{u}{u_e}\left(1 - \frac{u}{u_e}\right) dy$; angle
κ	Kármán constant; wave number
μ	Molecular viscosity
$\mu(M)$	Mach angle, $\sin^{-1}(1/M)$
μ_t	Eddy viscosity
ν	Kinematic viscosity, μ/ρ
$\nu(M)$	Prandtl-Meyer function
ν_a	Artificial viscosity
ν_t	Kinematic eddy viscosity, μ_t/ρ
ξ	Fluid particle dimension
$\tilde{\Pi}$	Coles' wake-strength parameter
ρ	Mass density
σ	Surface tension
$[\sigma]$	Stress tensor
τ	Shear stress; compressibility; time
τ_t	Reynolds shear stress, $-\rho\overline{u'v'}$
τ_w	Surface shear stress
$\tau_{xx}, \tau_{yy}, \tau_{zz}$	Viscous normal stresses

$\tau_{xy}, \tau_{yz}, \tau_{zx}$	Viscous shear stresses
$\boldsymbol{\tau}$	Torque vector
$[\boldsymbol{\tau}]$	Viscous stress tensor
υ	Specific volume, $1/\rho$
ϕ	Velocity potential; angle
ψ	Streamfunction
ω	Vorticity magnitude; specific dissipation rate
$\boldsymbol{\omega}$	Vorticity vector
Ω	Angular velocity magnitude
$\boldsymbol{\Omega}$	Angular velocity vector

Subscripts

Symbol	Definition
1	Ahead of shock
2	Behind shock
abs	Absolute
avg	Average
$crit$	Critical
cv	Control volume
e	Boundary-layer edge; exit plane
inc	Incompressible
$irrev$	Irreversible
m	Model; maximum
max	Maximum
o	Centerline; initial value
p	Prototype
rev	Reversible
t	Total (stagnation, reservoir)
w	Wall (surface)
∞	Freestream

Superscripts

Symbol	Definition
$+$	Sublayer-scaled value
$*$	Throat (sonic-point); best-efficiency point

Other

Symbol	Definition
$<f>$	Statistical average of f
$\overline{f}$	Statistical average of f
$\tilde{f}$	Mass average of f
f'	Fluctuating part of f
Im	Imaginary part

Preface

This book has been developed from the author's lecture notes used in presenting fluid mechanics courses at the University of California, Los Angeles and the University of Southern California. Many elementary fluid mechanics texts have been written during the past two decades, and most have a common denominator of minimal mathematical complexity. This is done primarily to emphasize the physics of fluid motion without losing the student in the equations. While this is certainly a desirable goal, it is often achieved at the expense of rigor. The most glaring example of this compromise is the unsatisfactory treatment many elementary fluid mechanics texts give in deriving the Reynolds Transport Theorem, where, for example, one-dimensional methods are often used to make a "derivation" in a two-dimensional geometry.

Avoidance of all but the most fundamental concepts of calculus precludes many elementary texts from including important concepts such as demonstration of Galilean invariance of the Euler equation and the classical Helmholtz Theorem. However, with judicious use of just a handful of common vector calculus concepts, the mean value theorem of calculus and the chain rule for differentiation, a great deal of rigor can be included in the presentation of a first course in fluid mechanics. These are mathematical tools that are normally taught in the freshman and sophomore years at most engineering schools.

Thus, the primary goal of this book is to provide a rigorous and understandable introduction to the fascinating field of fluid mechanics with a classical point of view. While maintaining a commitment to mathematical rigor throughout, the text continually emphasizes the physics of fluid motion. Mathematical results are repeatedly reinforced and verified by appealing to physical arguments. To avoid making derivations for simplified (non-general) geometries, the text makes extensive use of basic vector calculus, most notably the divergence theorem. The text accommodates the reader who needs a review of vector calculus by providing all that is needed to follow the text in Appendix C.

The all-important Reynolds Transport Theorem is derived rigorously in a one-dimensional geometry and heuristically generalized for three-dimensional flows. The integral forms of the basic conservation laws are developed for a control volume in a straightforward manner using the Reynolds Transport Theorem. Differential forms are then deduced, and simple properties of the Euler and Bernoulli equations are discussed. Presenting the derivations in this order stresses the central role played by the integral conservation forms. The first 9 chapters focus on the integral conservation laws. Subsequent chapters turn to the differential forms.

This text has been written with a full recognition of the increasingly important role that Computational Fluid Dynamics (CFD) plays. The modern fluid dynamicist has three research tools, namely, analytical methods, experimentation and CFD. The rapid emergence of CFD since the 1960's has eliminated the usefulness of many approximate, ad hoc techniques. That is not to say that all such methods should be abandoned, as some analytical solutions are needed to check the accuracy and reasonableness of computer programs. They can also help identify key design parameters that would require large numbers of computer runs to discover.

Only the most useful of such theories are included, however, to make room for an introduction to CFD as an integrated part of fluid mechanics.

By stressing the role of CFD, the text attempts to reflect the modern state of fluid mechanics. The emergence of CFD as an important tool mandates a shift in emphasis from elaboration of approximate design theories to a presentation promoting understanding of basic concepts, problem formulation and the ability to judge the correctness of numerical results. In this spirit, each of the final five chapters concludes with rudimentary concepts of CFD. If this book had been written 20 or 30 years ago, the concluding sections might have illustrated an approximate analytical technique.

Another way the text differs from many others is in its total integration of examples into the explanation of concepts. While appearing as part of the narrative, the examples still serve the important purpose for the reader who can "learn by doing" — but not overdoing! There is a tendency in some fluid mechanics and mathematics texts to overwhelm the reader with examples. This can have the undesired effect of steering the reader away from the fundamentals in favor of mundane problem solving. To assist the reader in developing the ability to apply the concepts, many of the most complex homework problems at the end of each chapter have multiple parts that lead the reader through a logical sequence of steps to develop the solution. In this sense, some of the homework problems can be viewed as pseudo-examples.

Numerous homework problems require only an algebraic result. More often than not, the answers are even rigged to involve only rational numbers to help simplify the algebra. This is a reflection of a desire to stress the importance of understanding the physics, rather than improving the student's ability to use a calculator. It also permits emphasizing the importance of good engineering practices such as checking mathematical results for dimensional consistency and examining limiting cases for which properties of the solution are evident. Of course, some feel for the magnitudes of quantities of interest in practical fluid flows is needed, and problems have been included throughout the text that require a numerical answer.

The material presented in this book is appropriate for a two-term, junior or senior level undergraduate series of courses, or as an introductory text for graduate students with minimal prior knowledge of fluid mechanics. The first 10 chapters provide sufficient material for an introductory course, while the balance of the text can be presented in a subsequent course. Successful study of this material requires an understanding of basic calculus and elementary mechanics. Some exposure to thermodynamics is helpful but not required. Chapter 7 provides an overview of the most important concepts needed from the science of thermodynamics. A degree of proficiency in solving ordinary differential equations is also needed to master Chapters 1 through 10. Exposure to the calculus of several variables is needed for the more-advanced material beginning in Chapter 11. A knowledge of the Fortran language will permit the reader to take full advantage of the software described in Appendices E and F.

The first 11 chapters contain more than enough material for a one-quarter course at most universities. This permits the instructor to choose the topics most pertinent to the curriculum. All students will benefit from the material presented in Chapters 1, 2, 4 through 7 and 10. These chapters provide the foundation for learning the classical control volume method, and the invaluable tool of dimensional analysis. This exposes the student to the primary techniques that have been used so successfully in developing both design methods and general theoretical tools. Practical applications of the control volume method are found in Chapter 7, viz., pipe flow and open-channel flow. With its focus on turbomachinery, Chapter 9 provides a quintessential example of how effective the combination of dimensional analysis and the control volume method is in developing an understanding of complex fluid flows.

Similarly, Chapters 11 through 15 include more than enough for a one-quarter (and probably a one-semester) course. If the material on CFD is omitted, most of the material can be covered in a standard course. However, that would defeat part of the intended purpose of this text. Since a personal computer now has the power of a Cray I super computer, it is imperative that today's students learn the elements of CFD, and the sooner the better. Most of the CFD homework problems can be done by changing just a few lines of the supplied computer programs. While including the CFD material will require alteration of the curriculum, it is a change that is inevitable. Many universities have already opted to include CFD at the undergraduate level, and this text will help promote that objective.

As a final comment on the design of the book, there are no photographs either within this book or on the cover. This has been a conscious decision based on the notion that photos add little quantitative understanding, and are included in many books to provide a flashy appearance. Some argue that flashy books stimulate a student's interest. *I firmly believe it is the teacher's responsibility to provide such simulation.* Several books with beautiful color photos of fluid flows are available — most notably *An Album of Fluid Motion* by M. D. Van Dyke — and such a book can be recommended by a teacher who feels such visual displays are helpful.

I extend my thanks to several friends and colleagues for their invaluable help as I prepared the manuscript, most notably Drs. P. Bradshaw, A. Lavine, C. G. Speziale, P. J. Roache and F. K. Browand. As usual, my dear friend Peter Bradshaw read the entire manuscript, taking me to task whenever I failed to explain things clearly or split an infinitive. The book includes a countless number of improvements for which Peter is responsible.

I owe special thanks to two of my best graduate students at USC, Chris Landry and Patrick Yee, for thoughtful reviews that led to important improvements of the book. Patrick and Chris also reviewed the complete manuscript, and gave me a point of view from the people the book is intended for. Chris even worked most of the book's 1,000 homework problems while reviewing the solution manual — a truly remarkable feat!

I thank the hundreds of undergraduate students at UCLA and USC who have had their introduction to fluid mechanics in my classes, especially for their many thought-provoking questions that have helped me improve my own understanding of fluid mechanics. It is largely at the urging of those students that I have written this text.

Finally, I owe yet another debt to my wonderful wife Barbara for helping me survive the transition from strictly aerospace research activities to all these other things I do well.

David C. Wilcox

Chapter 1

Introduction

The twentieth century ushered in the era of flight. The century had barely begun when, in 1903, the Wright brothers created the first heavier-than-air craft to fly under its own power. Since that historic flight, aviation has brought us faster and more efficient vehicles that permit mankind to "soar with the eagles." Just 44 years later, Captain Charles (Chuck) Yeager demonstrated that it is possible to fly at speeds in excess of the speed of sound, an issue that was very much in doubt at the time. Captain Yeager's rocket-powered flight is often cited as the beginning of the space age.

The development of rockets has taken us beyond the confines of Earth's gravitational field. Those of us old enough to be tuned in, *on color television broadcasting live from the Moon* (an achievement in its own right), still remember the chills up and down our spine on July 20, 1969, when Neil Armstrong said, as he became the first man to set foot on the Moon, **"That's one small step for man, one giant leap for mankind."** Today, we see routine flights of the Space Shuttle performing missions in Earth orbit.

All of these achievements have been accomplished partly because of the vast accumulation of knowledge regarding the motion of fluids. The lifting and propulsive forces generated by the Wright Flyer have been thoroughly understood, and dramatically improved designs have evolved for modern aircraft. Theoretical studies of flows at supersonic speeds were conducted by Prandtl in the early part of the twentieth century, long before supersonic flight was even dreamed of. Those studies provided part of the basis for Captain Yeager's historic flight. Intense theoretical and experimental analysis in the 1950's and 1960's provided accurate predictions of the heating that a space vehicle would have to be protected against to permit safe reentry into the Earth's atmosphere. The emergence of Computational Fluid Dynamics (CFD) has put the computer at our disposal to aid in the design of advanced vehicles such as the Space Shuttle.

But, while evidence of its importance and relevance to real-world problems is most obvious in the achievements with aircraft and rockets, fluid mechanics is not limited solely to such problems. The aerodynamic drag of automobiles in the 1990's is less than half the drag of models designed in the 1930's. The flow of gases through the engine of an automobile and even its air-conditioning system are routinely computed by today's auto builders.

A challenging area of fluid-mechanics research is the flow of blood through arteries and, ultimately, through the human heart. Developing a thorough understanding of blood flow through the body will surely help control and/or eliminate one of the most serious health hazards in today's world, namely, arteriosclerosis and heart disease.

Applications involving fluid motion extend to all branches of engineering. Aeronautical and mechanical engineers' interests range from basic studies of fluid motion, most notably turbulence, to everyday problems involving power generation, heating and ventilation, computer disk-drive design, etc. Civil engineers focus on interaction of aerodynamic forces with structures such as bridges and piers. Electrical engineers seek reduced costs in forming microchips, where acids must flow in a controlled manner to create desired patterns on silicon and other semiconductor materials. Chemical engineers must accurately determine reaction rates, a particularly acute problem when the velocity is high enough for the flow to be turbulent.

This book has been written with the object of introducing the reader to the exciting field of fluid mechanics. Relative to many popular texts on the topic, the approach taken is perhaps a bit more from the point of view of a physicist who really wants to use his knowledge to build everyday practical devices. Another way of saying this is the point of view is that of an engineer who shares Prof. Keith Stewartson's disdain for things that are "unrigorous." Either way, the book strives for mathematical rigor throughout, without concealing the exciting physical concepts involved. So, read on with the understanding that the intent is to challenge your intellect, and to encourage you to hone your skills in mathematical and physical reasoning. Aiming high is a good strategy to follow if you truly want to reach for the stars!

To begin our study, we first note that the field we call **fluid mechanics** is the branch of physics concerned with fluid motion. Most substances found in nature can be thought of as either a solid or a fluid, where the two most common fluids are gases and liquids. The goals of this book are twofold. Our most important goal is to explain how to apply Newton's laws of motion to fluids. Our second goal is to present a brief overview of the general field of fluid mechanics. The first thing we need to do is define what we mean by a fluid.

1.1   Simplistic Definition of a Fluid

Anything that flows. Liquids and gases are the most obvious examples of substances that will flow. Traffic and glass are more subtle examples that can be treated as fluids, in the sense that their motion can be described with equations derived from fluid-mechanics principles.

The flow of traffic has been modeled as a fluid whose density (number of automobiles per unit distance) varies with speed. A particularly interesting application of traffic-flow theory is to the timing of traffic lights. In the 1960's and 1970's, many cities posted signs indicating the speed required to avoid catching a red light on a given route. The speed and timing of the lights were determined from observations of normal traffic flow and a theory developed from the mass and momentum conservation laws of fluid mechanics.

Glass flow is an even more subtle application of fluid mechanics. Although not obvious, glass is a fluid at room temperature, albeit with a very large coefficient of friction (viscosity). This is evident from windows in old structures such as the twin-tower Cathedral in Cologne, Germany. Some windows in the church are more than 200 years old, and show evidence that gravity has caused the glass to flow toward the bottom of each pane. Although it has been a painfully slow process, gravity is gradually overcoming the resisting frictional forces.

1.2   Rigorous Definition of a Fluid

A substance that cannot be in static equilibrium under the action of oblique stresses. On the one hand, if only normal forces such as pressure act, a fluid will adjust to the applied pressure with a change in volume. After adjusting, it remains at rest as illustrated in Figure 1.1(a).

On the other hand, if a tangential stress[1] such as shear due to friction is present, the fluid deforms and continues moving. Such a flow has an oblique stress because the resultant force from the tangential shear and normal pressure acts at an oblique angle.

For example, consider flow in a channel as shown in Figure 1.1(b). If we apply different pressures $p_1 > p_2$ at the ends of the channel, frictional forces develop at the channel walls to balance the pressure difference. As a result, the initially rectangular section of fluid moves at different velocities across the channel, and thus undergoes significant distortion. By contrast, a solid will resist a shearing force until it yields.

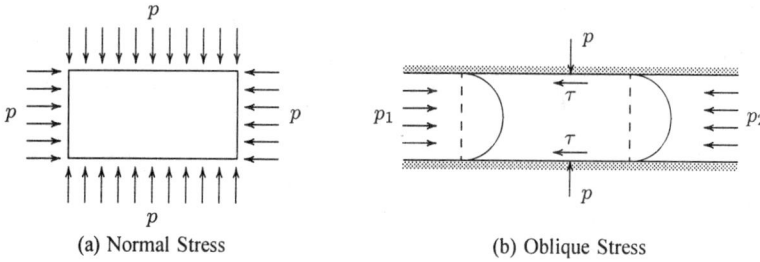

(a) Normal Stress   (b) Oblique Stress

Figure 1.1: *Fluid response to normal and oblique stresses. The forces are not drawn to scale in (b); shear stresses are generally much smaller than the total pressure acting.*

1.3   Continuum Approximation

In reality, a fluid is made up of molecules. It is not a continuous substance and, in principle, we must apply the appropriate laws of motion to each molecule to describe the overall motion. This would be an extremely difficult task given the large number of molecules present for any practical flow. For example, there are $3.34 \cdot 10^{19}$ molecules in a 1 mm^3 drop of water and $2.69 \cdot 10^{19}$ molecules in 1 cm^3 of air at **Standard Temperature and Pressure (STP)** — 0° C and 1 atm. To simplify our task, we introduce the **continuum approximation.**

We regard a fluid as being a **continuous** substance as opposed to being made up of discrete particles, i.e., molecules. We pretend that the basic properties (e.g., pressure, density, viscosity) of the smallest subdivisions of a fluid approach a unique limiting value as we reduce the size of the subdivision. Our justification for making this approximation is as follows. We can average random thermal motion of molecules provided the number of molecules in a **fluid particle** is very large, a fluid particle being defined as a volume of fluid whose size is extremely small compared to the characteristic dimension of the flow under consideration.

Figure 1.2 illustrates the limiting process we use to define a fluid property such as mass density, ρ, where mass density is mass per unit volume at each point in a given flow. In terms of a fluid particle, if its mass is Δm and its volume is ΔV, the mass density is given by

$$\rho = \lim_{\Delta V \to \Delta V_\ell} \frac{\Delta m}{\Delta V} \tag{1.1}$$

where ΔV_ℓ is a limiting volume below which an insufficient number of molecules can be found to define a meaningful statistical average. As shown in the figure, for larger values of ΔV, say $\Delta V > \Delta V_u$, the ratio of Δm to ΔV will vary if the fluid has nontrivial spatial

---

[1] A stress, by definition, is a force per unit area.

variations in its properties. In the continuum limit, we postulate that a wide separation of scales exists, i.e., that

$$\Delta V_\ell \ll \Delta V_u \tag{1.2}$$

When this is true, the limiting process defined in Equation (1.1) is well defined, with $\Delta m / \Delta V$ approaching a constant limiting value at each point in the fluid.

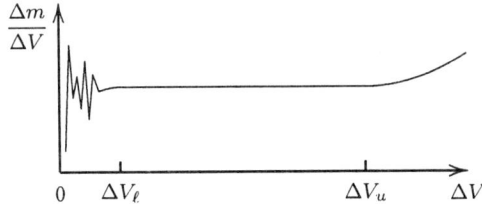

Figure 1.2: *Continuum definition of mass density at a point in a fluid.*

For example, consider flow past a sphere of one centimeter diameter in a standard wind tunnel. If we consider a fluid particle to occupy a volume the size of a dust particle (diameter 1 micron = 10^{-4} cm), for air at STP we find approximately 14 million molecules. Clearly, we can define meaningful statistical averages if we have such a large number of molecules in our sample. Thus, the continuum limit applies.

Now let's move our sphere 200 miles above the Earth's surface. At this altitude, our dust-particle sized volume contains 10^{-8} molecules, i.e., to find a single molecule we need a volume a hundred million times larger than that of a dust particle. The continuum limit is meaningless for this flow.

We have a continuum provided we can define a fluid particle in which a sufficient number of molecules exist to define statistical averages of molecular motion. The vast majority of practical fluid-flow applications satisfy this constraint.

1.4   Microscopic and Macroscopic Views

With the continuum approximation, we view the fluid from the **macroscopic** level. By contrast, in viewing the fluid on a scale of the order of a molecule, we observe things on the **microscopic** level. Hence, we refer to continuum fluid properties like mass density as macroscopic properties. In this section, we will explore the manner in which density, pressure, temperature and other thermodynamic variables relate to what actually happens at the microscopic level.

Density. Mass density, usually called simply density, is denoted by ρ and has dimensions of mass per unit volume. Thus,

$$\rho = mn \tag{1.3}$$

where m is mass of a molecule and n is the number density, i.e., the number of molecules per unit volume. Values of ρ for air and water at atmospheric pressure and a temperature of 20° C (68° F) are as follows.

$$\rho = \begin{cases} 1.20 & \text{kg/m}^3 \quad (.00234 \text{ slug/ft}^3), \quad \text{Air} \\ 998 & \text{kg/m}^3 \quad (1.94 \text{ slug/ft}^3), \quad \text{Water} \end{cases} \tag{1.4}$$

Table 1.1: *Densities of Common Liquids*

Liquid	T (°C)	ρ (kg/m^3)	T (°F)	ρ (slug/ft^3)
Ethyl alcohol	20	789	68	1.53
Carbon tetrachlorode	20	1,590	68	3.09
Glycerin	20	1,260	68	2.44
Kerosene	20	814	68	1.58
Mercury	20	13,550	68	26.30
Oil: SAE 10W	38	870	100	1.69
Oil: SAE 10W-30	38	880	100	1.71
Oil: SAE 30	38	880	100	1.71
Seawater	16	1,030	60	1.99

Table 1.2: *Density of Water as a Function of Temperature*

T (°C)	ρ (kg/m^3)	T (°F)	ρ (slug/ft^3)
0	1,000	32	1.94
10	1,000	50	1.94
20	998	68	1.94
30	996	86	1.93
40	992	104	1.93
50	988	122	1.92
60	983	140	1.91
70	978	158	1.90
80	972	176	1.89
90	965	194	1.87
100	958	212	1.86

The density of a liquid is a weak function of pressure and depends mainly on temperature. Table 1.1 includes densities of common liquids, while Table 1.2 includes the variation of density with temperature for water.

The density of a gas varies with both pressure and temperature. The perfect gas law, which is given below [see Equation (1.9)], can be used to compute the density of a given gas.

Temperature. The kinetic theory [cf. Jeans (1962)] definition of temperature, T, is

$$\tfrac{3}{2}kT = \left\langle \tfrac{1}{2}m\mathbf{u} \cdot \mathbf{u} \right\rangle \tag{1.5}$$

where k is Boltzmann's constant, m is molecular mass, $\mathbf{u}$ is molecular velocity and $<>$ denotes statistical average. That is, the temperature of a fluid is directly proportional to the average kinetic energy of the molecular motion. Standard temperature is as follows.

$$T_{standard} = 0° \text{ Centigrade} = 32° \text{ Fahrenheit} \tag{1.6}$$

In terms of **absolute temperature**, we have

$$T_{standard} = 273.16 \text{ Kelvins} = 491.67° \text{ Rankine} \tag{1.7}$$

Pressure. For a fluid at rest we observe that the fluid exerts a pressure on the walls of its container. In fact, the only force exerted by the fluid is everywhere normal to the surface of

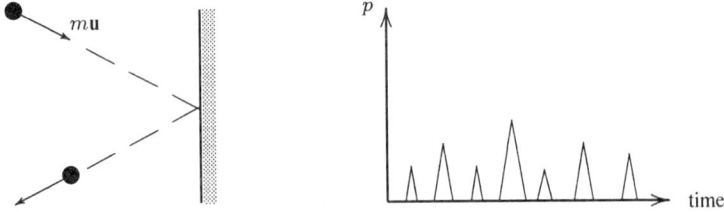

Figure 1.3: *Impulses from molecules striking a container wall in the limiting case of number density, $n \ll 1$.*

the container. We also observe that if the fluid is moving, it exerts an oblique force on the walls of the container. We define the pressure, p, as the normal force per unit area exerted by a given fluid particle on its immediate neighbors, whether or not the fluid is moving.

Pressure is a scalar quantity and thus acts equally in all directions at a given point in the fluid. It can change from point to point in a given flow as required to balance local accelerations. The force attending pressure is formed as the product of the pressure, the area of the surface on which it acts and the unit normal to the surface. We refer to pressure as a normal stress. We will take up the notion of oblique stresses when we discuss effects of friction in Section 1.8.

At the microscopic level, fluid molecules are continuously moving in a random manner. Collisions with a container wall give rise to an instantaneous force (impulse) on the wall. If the number density is small, the resulting "pressure" appears as a series of spikes as illustrated in Figure 1.3. By contrast, for a very large number of collisions, we see a more uniform variation of this force, similar to the variation depicted in Figure 1.4. This is the continuum limit ($n \gg 1$), and we use a statistical average, $<p>$, to define the pressure. Pressure is commonly expressed in many different units. For example, the standard value of p in the atmosphere is

$$p = \begin{cases} 1 & \text{atm} & \text{(atmospheres)} \\ 14.7 & \text{psi} & \text{(pounds per square inch)} \\ 101 & \text{kPa} & \text{(kiloPascals)} \\ 760 & \text{mmHg} & \text{(millimeters of mercury)} \\ 2116.8 & \text{psf} & \text{(pounds per square foot)} \end{cases} \tag{1.8}$$

where one Pascal = one Newton per square meter. Pressure is often quoted in terms of **absolute** (psia) and **gage** (psig) values, with the latter relative to the atmospheric pressure. Thus, $p = 14.9$ psi is an absolute pressure of 14.9 psia and a gage pressure of 0.2 psig.

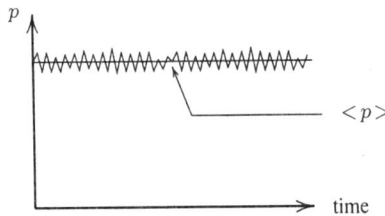

Figure 1.4: *Pressure from molecules striking a container wall in the limiting case of number density, $n \gg 1$.*

1.5 Thermodynamic Properties of Gases

When our fluid is a gas, there are some additional useful thermodynamic relations and properties of interest in the study of fluid mechanics. For most applications, we treat the gas as though it is a **perfect gas**. Also, when we are interested in heat transfer, there are several energy-related properties of interest to us.

Equation of State for a Perfect Gas. If the intermolecular forces in a gas are negligible and intermolecular collisions are perfectly elastic, the pressure, density and temperature are related by the perfect gas equation of state,

$$p = \rho R T \tag{1.9}$$

where R is the **perfect-gas constant**. Table 1.3 includes values of R for common gases.

Table 1.3: *Thermodynamic Properties of Common Gases*

Gas	γ	J/(kg·K)		ft·lb/(slug·°R)	
		R	c_p	R	c_p
Air	1.40	287	1,004	1,716	6,003
Carbon dioxide	1.30	189	841	1,130	5,028
Helium	1.66	2,077	5,225	12,419	31,240
Hydrogen	1.40	4,124	14,180	24,677	84,783
Methane	1.31	518	2,208	3,098	13,783
Nitrogen	1.40	297	1,039	1,776	6,212
Oxygen	1.40	260	910	1,555	5,440

Other Useful Quantities and Notation. The heat contained by a fluid is quantified in terms of the **internal energy**, e. In general, the basic postulates of thermodynamics tell us e is a function of any two independent thermodynamic state variables, a point we will address in Chapter 7. The internal energy is usually written as a function of temperature and **specific volume**, $v = 1/\rho$. For a wide range of practical conditions, we find that $e = e(T)$. The gas is said to be **thermally perfect** when e is a function only of T. Additionally, **enthalpy**, h, is

$$h = e + p/\rho \tag{1.10}$$

By contrast to internal energy, h is usually written as a function of temperature and pressure. As with e, $h = h(T)$ for a thermally-perfect gas. Enthalpy and internal energy have units of J/kg (ft·lb/slug).

There are two **specific-heat coefficients** denoted by c_v and c_p. By definition,

$$c_v = \left(\frac{\partial e}{\partial T}\right)_v \qquad \text{and} \qquad c_p = \left(\frac{\partial h}{\partial T}\right)_p \tag{1.11}$$

where c_v is appropriate to a constant-volume process and c_p to a constant-pressure process. Both of the specific-heat coefficients have units of J/(kg·K) [ft·lb/(slug·°R)] and, in general, depend upon T. Except at very high temperatures, c_v and c_p are constant for air. When c_v and c_p are constant, we say the gas is **calorically perfect**. It follows that, for a calorically-perfect gas, $e = c_v T$ and $h = c_p T$. Finally, the ratio of the specific heats is denoted by γ (the symbol introduced by Rankine) and referred to as the **specific-heat ratio**. Thus, we write

$$\gamma = \frac{c_p}{c_v} \tag{1.12}$$

Table 1.3 includes values of γ and c_p for common gases.

1.6   Compressibility

If we hold temperature constant and apply a pressure to a container filled with gas, its density will change according to

$$\rho = \frac{p}{RT} \tag{1.13}$$

so that density changes, $\Delta\rho$, are given by

$$\Delta\rho = \frac{\Delta p}{RT} \qquad (\text{constant } T) \tag{1.14}$$

In so doing, we **compress** the gas. The formal definition of the **compressibility**, τ, is

$$\tau = \frac{1}{\rho}\frac{\partial\rho}{\partial p} \tag{1.15}$$

Some authors prefer to work with the **bulk modulus**, which is the reciprocal of the compressibility. Hence, for our isothermal perfect gas, we have $\partial\rho/\partial p = 1/(RT)$ so that

$$\tau = \frac{1}{\rho}\cdot\frac{1}{RT} = \frac{1}{p} \tag{1.16}$$

Compressibility is a property of the fluid. The compressibility for gases is much larger than corresponding values for liquids. For example, at atmospheric pressure and constant temperature,

$$\tau = \begin{cases} 1.00\cdot10^{-5} & \text{m}^2/\text{N} \;\; (4.79\cdot10^{-4}\ \text{ft}^2/\text{lb}), & \text{Air} \\ 4.65\cdot10^{-10} & \text{m}^2/\text{N} \;\; (2.23\cdot10^{-8}\ \text{ft}^2/\text{lb}), & \text{Water} \end{cases} \tag{1.17}$$

Thus, since Equation (1.15) implies $\Delta\rho/\rho = \tau\Delta p$, a very large Δp for water relative to that required for air (21,500 times greater) must be applied to obtain even a small fractional change in ρ. Table 1.4 lists τ for several common fluids.

Table 1.4: *Compressibility and Sound Speed of Common Fluids for a Pressure of 1 atm*

Fluid	T (°C)	τ (m²/N)	a (m/sec)	T (°F)	τ (ft²/lb)	a (ft/sec)
Air	15.6	$1.00\cdot10^{-5}$	341	60.0	$4.79\cdot10^{-4}$	1119
Ether	15.0	$1.63\cdot10^{-9}$	1032	59.0	$7.82\cdot10^{-8}$	3386
Ethyl alcohol	20.0	$9.43\cdot10^{-10}$	1213	68.0	$4.52\cdot10^{-8}$	3980
Glycerin	15.6	$2.21\cdot10^{-10}$	1860	60.0	$1.06\cdot10^{-8}$	6102
Helium	15.6	$1.00\cdot10^{-5}$	998	60.0	$4.79\cdot10^{-4}$	3274
Mercury	15.6	$0.35\cdot10^{-10}$	1450	60.0	$0.17\cdot10^{-8}$	4757
Water	15.6	$4.65\cdot10^{-10}$	1481	60.0	$2.23\cdot10^{-8}$	4859

Fluids that require very large changes in pressure to cause even a minor change in density are termed **incompressible**. In a mathematical sense, we can say incompressible flow occurs when $\tau \to 0$, so that $\Delta\rho = \rho\tau\Delta p \to 0$. Hence, for our immediate purposes, we will consider an incompressible flow to be constant-density flow. As we will see in Chapter 8, this is not the conventional definition. It would imply that motion in a "stratified" medium such as the ocean whose density varies with depth should be classified "compressible." This is incorrect. When we study compressibility effects we will find that a better definition of incompressible flow is **very low Mach number flow**, where Mach number is the ratio of flow speed to the speed of sound. This is a far better indicator of incompressibility, and is satisfactory for most practical applications.[2] Table 1.4 lists the speed of sound for several common fluids and shows, for example, that the speed of sound in water is more than four times the speed in air.

Until we specifically address effects of compressibility, most of our applications will be for incompressible flows. This greatly simplifies the equations we have to deal with and actually precludes the need to include the energy-conservation principle for most problems. This means that incompressible flow problems can be solved using only mass and momentum conservation concepts. This by no means limits our study to a small class of fluid-flow problems. Incompressible flow covers the flow of liquids and "low-speed" gas flows, where low speed means the Mach number is less than about 0.3. Examples are the motion of a torpedo, an automobile and a Piper Cub with typical speeds of 40 mph, 65 mph and 200 mph, respectively. Based on these speeds and the sound speeds listed in Table 1.4, the Mach numbers would be 0.012, 0.085 and 0.262.

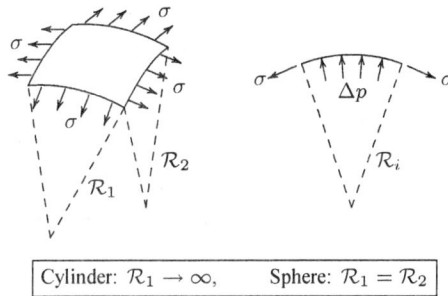

Figure 1.5: *Stretched-membrane analogy for surface tension.*

1.7 Surface Tension

There is an interesting effect that occurs near liquid-gas and immiscible liquid interfaces, viz., surface tension. At such an interface, we find the surface to be like a stretched membrane. A spider can walk on water, a steel needle placed gently in a pan of water floats, water "beads up" on your freshly waxed car. These are all commonly observed examples of surface tension.

Surface tension is actually a very complex phenomenon, involving concepts from physical chemistry [cf. Adamson (1960)]. A simplified explanation that captures the dominant physics is as follows. On the one hand, molecules within the main body of the fluid are surrounded by molecules with attractive forces equal in all directions. On the other hand, molecules at the surface are attracted only by molecules from within the fluid and thus exhibit a stronger attraction for other surface molecules to achieve an overall balance of molecular forces. This modified "force field" results in the surface behaving like a stretched membrane. We can thus describe surface tension mathematically by making an analogy to a membrane, in which a uniform tensile force, σ, acts tangent to the surface. However, there is no physical membrane present — only a distribution of molecular forces that acts like the tension in a membrane.

For example, any curvilinear surface can be described in terms of two principal radii of curvature, $\mathcal{R}_1$ and $\mathcal{R}_2$ (Figure 1.5). In analogy to a stretched membrane, the pressure

---

[2]Even this definition has its limitations. For example, flow in an internal-combustion engine occurs at low Mach number, yet has large density changes caused by the very high pressure within the combustion chamber.

difference, Δp, that can be supported by a surface tension, σ, on a general surface is given by **Laplace's formula** [see Landau and Lifshitz (1966)]:

$$\frac{\sigma}{\mathcal{R}_1} + \frac{\sigma}{\mathcal{R}_2} = \Delta p \tag{1.18}$$

As an example of how we can use this formula, consider the pressure within a 1 mm diameter spherical droplet of water. We would like to determine the pressure relative to the atmospheric pressure outside. To apply Laplace's formula, we make an imaginary cut through the center of the droplet as shown in Figure 1.6. Since the principal radii of curvature are both equal to the sphere radius, $R = \frac{1}{2} \cdot 10^{-3}$ m, and the surface tension for water is $7.3 \cdot 10^{-2}$ N/m, Equation (1.18) tells us

$$\Delta p = \frac{2\sigma}{R} = \frac{2 \cdot 7.3 \cdot 10^{-2} \text{ N/m}}{\frac{1}{2} \cdot 10^{-3} \text{ m}} = 292 \ \frac{\text{N}}{\text{m}^2} \tag{1.19}$$

Alternatively, we can balance forces and arrive at the same result with some added insight into the physics involved. The net contribution from the atmosphere is p_o times the projection of the hemisphere onto a vertical plane. If this point is unclear, consider the limiting case $\Delta p = 0$ and $\sigma = 0$ in which the atmospheric pressure on the two surfaces must be in balance. Thus,

$$\sum F = (p_o + \Delta p)\, \pi R^2 - p_o \pi R^2 - 2\pi R \sigma = 0 \tag{1.20}$$

Simplifying, we find

$$\Delta p \pi R^2 = 2\pi R \sigma \qquad \Longrightarrow \qquad \Delta p = \frac{2\sigma}{R} \tag{1.21}$$

which is identical to what we obtained above using Laplace's formula.

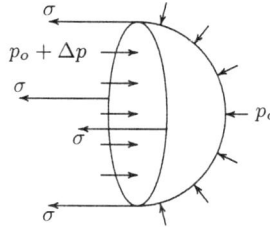

Figure 1.6: *Force diagram for a droplet.*

As a second interesting example of the effect of surface tension, consider the phenomenon known as **capillary action**. A tube of diameter d is immersed in a container filled with a liquid of density ρ as shown in Figure 1.7. The fluid is drawn up into the tube a distance Δh. Surface tension, σ, acts around the perimeter of the tube and supports the weight of the column of fluid. The surface tension is tangent to the surface as shown, making an angle ϕ with the tube surface. The vertical component of the surface tension balances the weight of the column of fluid so that

$$\pi d \sigma \cos\phi = \rho g \Delta h (\pi d^2 / 4) \tag{1.22}$$

Thus, the average height to which the fluid is drawn up the tube is given by

$$\Delta h = \frac{4\sigma}{\rho g d} \cos\phi \tag{1.23}$$

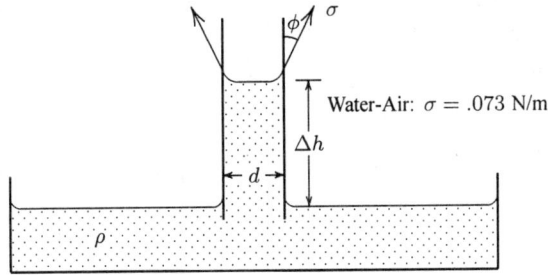

Figure 1.7: *Capillary action – cross-sectional view. Note that σ has dimensions of force per unit length. It acts along an arc, not on a surface.*

The height is an average since our computation approximates the column of fluid as being exactly rectangular. In reality, its shape is curved as shown in Figure 1.7, corresponding to the commonly observed **meniscus**.

In general, surface tension depends upon the surface material as well as the nature of the two fluids involved, and it is a function of pressure and temperature. It is also considerably affected by dirt on the surface. Table 1.5 lists surface tension for several common liquids. The angle ϕ is called the **wetting angle** or the **contact angle**. For water interfacing air in a glass tube, we find that $\phi \approx 0°$. By contrast, for mercury interfacing air in a glass tube, the surface tension acts downward with $\phi \approx 129°$. Similarly, for water, air and paraffin wax, the wetting angle is $105°$ [cf. Sabersky et al. (1989)]. As a consequence of having $\phi > 90°$, the column of fluid is drawn downward by capillary action in these two cases.

So, in the example above, assume the tube is made of glass with diameter $d = 1$ mm. Also, assume the liquid is water, and the tube is open to the atmosphere so that $\rho = 998$ kg/m^3, $\sigma = .073$ N/m and $\phi = 0°$. Then, the height of the column of water drawn up the tube is

$$\Delta h = \frac{(4)(.073 \text{ N/m})(1)}{\left(998 \text{ kg/m}^3\right)\left(9.807 \text{ m/sec}^2\right)(.001 \text{ m})} = .0298 \text{ m} \approx 30 \text{ mm} \qquad (1.24)$$

Thus, due to capillary action, the water is drawn 30 tube diameters above the container surface.

Table 1.5: *Surface Tension of Common Fluids*

Fluid$_1$	Fluid$_2$	T (° C)	σ (N/m)	T (° F)	σ (lb/ft)
Water	Air	0	.076	32	.0052
Water	Air	20	.073	68	.0050
Water	Air	100	.059	212	.0040
Mercury	Air	20	.466	68	.0319
Mercury	Water	20	.375	68	.0257
Carbon tetrachloride	Air	20	.027	68	.0018
Carbon tetrachloride	Water	20	.045	68	.0030
Ethyl alcohol	Air	20	.023	68	.0016
Gasoline	Air	16	.022	60	.0015
Glycerin	Air	20	.063	68	.0043
SAE 10 Oil	Air	16	.036	60	.0025
SAE 30 Oil	Air	16	.035	60	.0024
Seawater	Air	16	.073	60	.0050

By contrast, if we change the fluid to mercury leaving all other conditions the same, we have $\rho = 13550$ kg/m^3, $\sigma = .466$ N/m and $\phi = 129°$. For this case, capillary action draws the fluid 9 tube diameters below the container surface:

$$\Delta h = \frac{(4)(.466 \text{ N/m})(-.629)}{\left(13550 \text{ kg/m}^3\right)\left(9.807 \text{ m/sec}^2\right)(.001 \text{ m})} = -.0088 \text{ m} \approx -9 \text{ mm} \qquad (1.25)$$

1.8   Viscosity

Fluids such as honey or molasses tend to flow very slowly down an inclined plane. Water goes much faster. The former fluids are more **viscous**. Recall that for a container filled with a fluid at rest we observe only normal stresses. If the fluid is moving, we also observe oblique stresses on the walls of the container. In fact, these oblique stresses act everywhere in the fluid if the fluid is moving.

To understand the origin of viscosity, we need to take another look at what goes on at the molecular level. The motion at this level is random in both magnitude and direction. Referring to Figure 1.8, consider an imaginary plane in a fluid that, on the average, is moving with velocity U_1 for $y > 0$ and U_2 for $y < 0$, and assume $U_1 > U_2$. Molecules migrating across $y = 0$ are **typical of where they come from**. That is, molecules moving up bring a momentum deficit and vice versa. This gives rise to a stress on a surface whose unit normal lies in the y direction and that acts in the x direction. We define this type of stress as a shear stress.

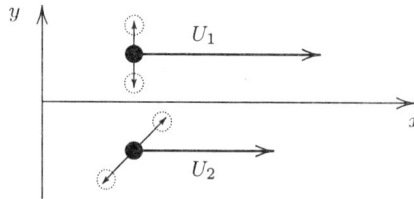

Figure 1.8: *Close-up view of molecular motion.*

When you step off a conveyer belt (moving walkway), you experience this phenomenon. You are typical of where you came from, namely a region where you had nonzero forward momentum with respect to the stationary floor. As your feet touch the floor, you arrive in a region where you have an excess amount of momentum and your body will lurch forward until you regain your balance. Note that, assuming you step directly to your side to exit the belt, the "force" on your body is normal to the direction you have chosen to move.

In a viscous fluid, the velocity in the x direction, u, varies continuously in the y direction. For a majority of fluids encountered in engineering applications, we find that the shear stress, τ, is proportional to the slope of the velocity profile, du/dy. Thus, we can say that

$$\tau = \mu \frac{du}{dy} \qquad (1.26)$$

where μ is the viscosity coefficient with units kg/(m·sec) [slug/(ft·sec)]. When τ is a linear function of du/dy, we have what is referred to as a **Newtonian fluid**. For some fluids such as long-chain polymers, we have a nonlinear relation between τ and du/dy. Such a fluid is called a **non-Newtonian fluid**.

Table 1.6: *Kinematic Viscosities of Common Liquids and Gases*

Liquid	T (° C)	$10^6 \nu$ (m²/sec)	T (° F)	$10^5 \nu$ (ft²/sec)
Ethyl alcohol	20	1.51	68	1.62
Carbon tetrachlorode	20	0.60	68	0.65
Gasoline	16	0.46	60	0.49
Glycerin	20	1190.00	68	1280.00
Kerosene	20	2.37	68	2.55
Mercury	20	0.12	68	0.13
Oil: SAE 10W	38	41.00	100	44.00
Oil: SAE 10W-30	38	76.00	100	82.00
Oil: SAE 30	38	110.00	100	118.00
Air	15	14.60	59	15.71
Carbon dioxide	15	7.84	59	8.44
Helium	15	114.00	59	123.00
Hydrogen	15	101.00	59	109.00
Methane	15	15.90	59	17.10
Nitrogen	15	14.50	59	15.60
Oxygen	15	15.00	59	16.10

Table 1.7: *Kinematic Viscosity of Water as a Function of Temperature*

T (° C)	$10^6 \nu$ (m²/sec)	T (° F)	$10^5 \nu$ (ft²/sec)
0	1.79	32	1.93
10	1.31	50	1.41
20	1.00	68	1.08
30	0.80	86	0.86
40	0.66	104	0.71
50	0.55	122	0.59
60	0.47	140	0.51
70	0.41	158	0.44
80	0.36	176	0.39
90	0.33	194	0.36
100	0.29	212	0.31

Often in fluid mechanics, especially for incompressible flows, we find ourselves dealing with the ratio μ/ρ. This happens so frequently we choose to dignify the ratio with its own symbol and name. We define the **kinematic viscosity**, ν, as follows.

$$\nu = \frac{\mu}{\rho} \tag{1.27}$$

Kinematic viscosity has units of m²/sec (ft²/sec). Usually $d\mu/dT$ and $d\nu/dT$ are positive for a gas while $d\mu/dT$ and $d\nu/dT$ are negative for a liquid. Table 1.6 includes kinematic viscosities of common liquids and gases, while Table 1.7 includes the variation of ν with temperature for water. For air and water at atmospheric pressure and 20° C (68° F), the kinematic viscosity is

$$\nu = \begin{cases} 1.51 \cdot 10^{-5} \text{ m}^2/\text{sec} \quad (1.62 \cdot 10^{-4} \text{ ft}^2/\text{sec}), & \text{Air} \\ 1.00 \cdot 10^{-6} \text{ m}^2/\text{sec} \quad (1.08 \cdot 10^{-5} \text{ ft}^2/\text{sec}), & \text{Water} \end{cases} \tag{1.28}$$

The viscosity of air is well approximated by **Sutherland's Law**, which is an empirical equation that is quite accurate for a wide range of temperatures. The formula is

$$\mu = \frac{AT^{3/2}}{T + S} \tag{1.29}$$

where A and S are empirical constants. Note that T is the absolute temperature, and is thus given either in Kelvins (K) or degrees Rankine ($^\circ$R). The values of A and S in USCS units are

$$A = 2.27 \cdot 10^{-8} \, \frac{\text{slug}}{\text{ft} \cdot \text{sec} \cdot (^\circ\text{R})^{1/2}}, \qquad S = 198.6 \, ^\circ\text{R} \tag{1.30}$$

The values for SI units are

$$A = 1.46 \cdot 10^{-6} \, \frac{\text{kg}}{\text{m} \cdot \text{sec} \cdot \text{K}^{1/2}}, \qquad S = 110.3 \, \text{K} \tag{1.31}$$

1.9 Examples of Viscosity Dominated Flows

There are some flows that are completely dominated by viscous effects. In this section, we consider two such flows, i.e., two-dimensional **Couette flow** and axisymmetric **Hagen-Poiseuille flow**. Our discussion here is, of necessity, heuristic and limited to examining properties of the solutions without benefit of a rigorous derivation. This is true because we will not develop the viscous-flow equations of motion until Chapter 12. We examine the rigorous derivation of the solutions for both of these flows in Chapter 13.

1.9.1 Couette Flow

Couette flow occurs when we have a fluid between two infinite parallel planes as shown in Figure 1.9. The lower plane is at rest while the upper plane moves with constant velocity U, and the pressure is constant. This flow can be realized in a laboratory by having a conveyer belt close to a stationary surface. If the distance h is very small compared to the length of the belt, we approximate the idealized Couette-flow geometry provided we are not too close to the ends of the belt.

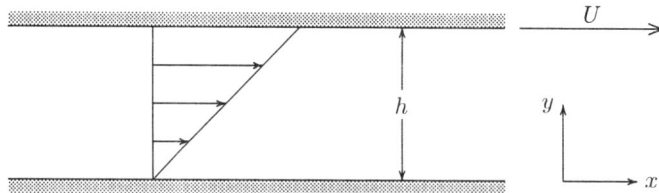

Figure 1.9: *Couette flow.*

We expect the solution to depend only upon distance between the plates. If we look close to the solid boundaries, we observe that the fluid "sticks" to the boundaries. This is known as the **no-slip** surface boundary condition on the velocity. This boundary condition will be discussed in more detail in Section 10.3. Thus, our solution must satisfy $u(0) = 0$ and $u(h) = U$, where $u(y)$ is the velocity component in the x direction. Indeed, an exact solution to the fluid mechanics equations of motion exists in which flow properties are functions only of y (Section 13.1). The velocity varies linearly with distance across the channel, viz.,

$$u(y) = U\frac{y}{h} \tag{1.32}$$

The corresponding shear stress is

$$\tau = \mu\frac{du}{dy} = \frac{\mu U}{h} \tag{1.33}$$

Hence, the shear stress is not confined to the walls of the channel. Rather, it is constant across the channel and acts throughout. We can rewrite Equation (1.33) in terms of nondimensional parameters as follows.

$$C_f = \frac{\tau}{\frac{1}{2}\rho U^2} = \frac{2}{(\rho U h/\mu)} = \frac{2}{Re} \tag{1.34}$$

The dimensionless grouping on the left-hand side of Equation (1.34) is called the **skin-friction coefficient**, and is usually denoted by c_f. The dimensionless grouping in the denominator of the right-hand side of Equation (1.34) is called the **Reynolds number**, and is usually denoted by Re. The linear velocity profile, constant-shear-stress solution is observed experimentally provided

$$\frac{\rho U h}{\mu} \lesssim 1500 \tag{1.35}$$

This is known as the **laminar** flow solution. At larger values of Reynolds number, this solution is unstable to small disturbances, and the flow becomes **turbulent**. When the flow is turbulent, the velocity and pressure vary rapidly with time and the effective shear stress is much larger than the value given above. Chapter 14 includes an introduction to turbulence.

The Couette-flow solution is often used as a "building-block" solution. Any application that has flow through a small gap can be described in terms of the linear velocity variation in the gap. For example, consider a cube of mass M and side H sliding on a thin lubricating film (see Figure 1.10). The lubricant has viscosity μ and thickness h. If we further assume that $h \ll H$, we can use Equation (1.33) to find the shear stress on the surface of the cube exposed to the lubricant. If there is a force, F, pushing the cube, then application of Newton's law tells us the cube's motion is governed by

$$M\frac{dU}{dt} = F - \frac{\mu H^2}{h}U \tag{1.36}$$

where $U(t)$ is the instantaneous velocity of the cube. Practical applications involving flow in a small gap include flow between a computer disk and read/write head, and an air-hockey game.

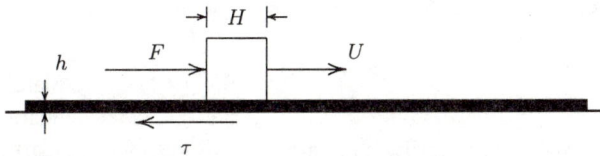

Figure 1.10: *Cube sliding on a thin lubricating film.*

1.9.2   Hagen-Poiseuille Flow

Hagen-Poiseuille Flow is a second simple example of a flow completely dominated by viscosity. We consider flow through an infinitely long pipe with circular cross section (see Figure 1.11). As with Couette flow, this is an exact solution to the fluid mechanics equations of motion (Section 13.3).

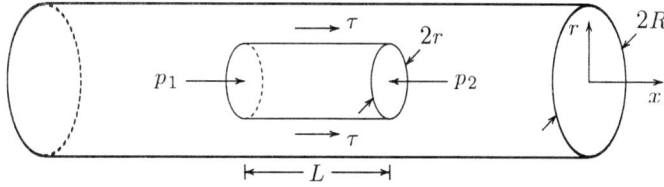

Figure 1.11: *Hagen-Poiseuille (pipe) flow.*

Consider the flow in a cylindrical section of the pipe of radius r and length L as shown in Figure 1.11. Assume the pressure on the left side of the section is p_1 and the pressure on the right side of the section is p_2. We choose our sign convention so that positive shear stress on the cylindrical section is in the positive x direction as shown.

The pressure difference is balanced by the shear stress acting on the circumference of the cylindrical section. Hence, we have

$$p_1 \pi r^2 - p_2 \pi r^2 + 2\pi r L \tau = 0 \tag{1.37}$$

Thus, the shear stress is given by

$$\tau = -\frac{(p_1 - p_2)}{L}\frac{r}{2} \tag{1.38}$$

So, when $p_1 > p_2$, the shear stress is negative and therefore resists the motion. We can determine the velocity profile, $u(r)$, by integrating Equation (1.26) with r replacing y. The integration is done subject to the no-slip boundary condition that tells us $u(R) = 0$. The resulting expression for the velocity then becomes

$$u(r) = \frac{(p_1 - p_2)}{4\mu L}(R^2 - r^2) \tag{1.39}$$

The maximum velocity, u_m, occurs on the centerline ($r = 0$) and is given by

$$u_m = \frac{(p_1 - p_2)}{4\mu L}R^2 \tag{1.40}$$

We observe this (laminar-flow) solution experimentally provided the Reynolds number based on u_m is as follows.

$$\frac{\rho\, u_m R}{\mu} \lesssim 2300 \tag{1.41}$$

If great care is taken to prevent transition to turbulence by keeping disturbances to a minimum, laminar flow can be achieved for pipe flow at a Reynolds number as high as 40,000 [Schlichting (1979)]. We will study pipe flow in greater detail in Chapters 7, 13 and 15.

1.10   Fluid-Flow Regimes

Figure 1.12 outlines the major regimes of classical continuum fluid mechanics. It also serves as a "road map" for this text, indicating the chapters in which the various regimes are discussed.

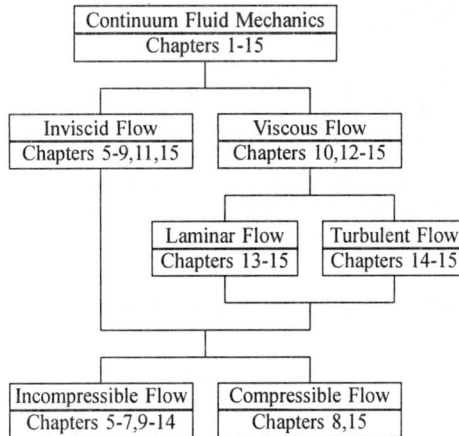

Figure 1.12: *Major regimes of fluid mechanics.*

The first major subdivision of the discipline is between the flow of inviscid and viscous fluids. An inviscid fluid is also referred to as an **ideal fluid**. We define an ideal fluid as one in which the stresses acting are everywhere normal to the element of surface on which they are measured, whether or not the fluid is moving. This is sometimes referred to as a **perfect fluid** or a **frictionless fluid**. This is a drastic simplification as compared to a viscous fluid, which is often referred to as a **real fluid**.

For a moving fluid, the approximation of treating it as ideal cannot hold everywhere. For fluids of small viscosity such as water and air, we observe two important facts. On the one hand, since viscous stresses are proportional to velocity gradients [cf. Equation (1.26)], neglect of viscous stresses breaks down where velocity gradients are large. On the other hand, the flow around a streamlined body behaves almost as if the fluid were frictionless except in very thin layers next to the surface and in a thin wake downstream. These near-surface layers were discovered by Prandtl in 1904 who named them **boundary layers**. Velocity gradients are very large in boundary layers. We will find out how thin boundary layers are in Chapter 14.

The concept of an Ideal Fluid is very useful, and much of our analysis in the first 11 chapters assumes effects of friction are unimportant. There are many interesting problems we can solve ignoring effects of friction, and the attendant mathematical simplification permits us to focus on the physical behavior of fluids without getting lost in the algebra.

However, there are problems for which an inviscid solution is completely unsatisfactory. Inviscid flow past a sphere or cylinder, for example, bears little similarity to the flow observed in nature. The inviscid solution predicts that there is no net force on the object which is contrary to physical observations. Even when we include viscosity, the flow past solid objects falls into two different categories.

On the one hand, at very low flow speeds, fluid motion is very smooth, and flow patterns remain unchanged from one instant to the next. This type of motion is termed **laminar flow**. We have seen examples of exact laminar-flow solutions in Section 1.9, and several more are presented in Chapter 13. On the other hand, at higher flow speeds, laminar flow becomes unstable and undergoes transition to what is known as **turbulent flow**. Turbulence is a state in which all flow properties vary rapidly with time, with a wide range of excited frequencies. All laminar flows ultimately become turbulent, with 'higher flow speeds' corresponding to the mildest of motions such as a small puff of wind. Although extensive research efforts have focused on the topic, to paraphrase Prof. Richard Feynman, turbulence remains one of the most noteworthy unsolved scientific problems of the twentieth century. Chapter 14 includes a brief introduction to turbulence.

The final major subdivision addressed in this book is between compressible and incompressible flows. As discussed in Section 1.6, there are many incompressible flow applications in everyday life. It is equally true that compressible flow abounds in practical engineering design and application. Transmission of gas in pipelines at high pressure, flow past a commercial airliner, and flow beneath an Indy 500 racer all involve compressible flow. Chapters 8 and 15 focus explicitly on compressible flow.

1.11   Brief History of Fluid Mechanics

The first part of this text focuses almost entirely on inviscid fluids. The second part turns to the more complex analysis of viscous fluids. Approaching fluid mechanics in this manner, to some extent, reflects the chronological development of the subject. The quantitative approach to fluid mechanics began in earnest with Isaac Newton's seventeenth century formulation of the laws of classical mechanics, including a postulate for how effects of friction should appear in a fluid. Eighteenth century mathematicians, most notably Bernoulli, d'Alembert, Euler, Lagrange and Laplace, developed the theory of frictionless flow. One unsettling result of this early work was a famous corollary, known as d'Alembert's Paradox, proving that no forces can develop for inviscid flow past an object in an unbounded fluid. This is, of course, at variance with physical reality.

Eighteenth and nineteenth century attempts at applying Newton's law of friction to resolve d'Alembert's Paradox proved unsatisfactory, raising serious doubts about theoretical fluid mechanics. Largely rejecting theoretical methods, the field of hydraulics was formulated by researchers such as Chézy, Weber, Hagen, Poiseuille, Darcy and Weisbach. Their work was based almost entirely on empiricism and experimentation, with little relation to basic physics.

It wasn't until the end of the nineteenth century that a unified theory of fluid mechanics including both theoretical and experimental methods took hold. The first important developments that helped unify the theory (with key contributions by Froude, Rayleigh and Reynolds) were dimensional analysis and dynamic similitude. These techniques, although adding minimal physical understanding, helped correlate measurements and establish parallels between experimental observations and the differential equations of motion.

Prandtl's discovery of the boundary layer in 1904 resolved d'Alembert's Paradox, and showed that the early nineteenth century work of Navier and Stokes, which generalized Newton's law of friction, was indeed correct. The viscous theory of fluid motion has advanced dramatically during the twentieth century with profound contributions from pioneers such as Kolmogorov, Prandtl, Taylor and von Kármán.

During the second half of the twentieth century, a third branch of the unified theory of fluid mechanics has evolved, viz., Computational Fluid Dynamics, or CFD. Using digital computers, we can now solve very complex fluid-flow problems. Early researchers such as Richardson, Courant, Friedrichs, Lewy, Lax and von Neumann helped build the foundation of what has become an important tool for both basic and applied research activities. The list of active CFD researchers has expanded dramatically since the 1960's, and many of the most significant contributors in this rapidly developing field are currently expanding its horizons.

Problems

1.1 Suppose we had a very fast electronic counter that can count at a rate of one million molecules per second. How long would it take to count the molecules in 1 mm^3 of air at STP?

1.2 Suppose we had a very fast electronic counter that can count at a rate of one million molecules per second. How long would it take to count the molecules in a cubic micron of water? Note that 1 micron is 10^{-4} cm.

1.3 A typical amoeba measures about .01 mm in size. You can assume its shape is a perfect sphere and that it is swimming in water.

   (a) How many water molecules are there in an amoeba-sized volume?

   (b) If an amoeba considers a fluid particle to be a spherical volume with 10^{-4} of its own diameter, how many molecules are there in such a fluid particle?

1.4 Turbulent flow consists of a wide range of swirling motions referred to as *eddies*. For flow near the windshield of an automobile moving at highway speed, the smallest eddy has a diameter, η, of about 0.0001 inch. Approximately how many air molecules are contained in a cube with side η?

1.5 The average distance between air molecules 100 km up in the atmosphere is approximately $3 \cdot 10^{-5}$ cm. How many molecules are contained in a spherical volume of diameter 10^{-4} cm?

1.6 The *mean free path* of a gas, ℓ_{mfp}, is the average distance traveled by a molecule between collisions. The continuum limit is valid if the characteristic dimension of the flow is very large compared to ℓ_{mfp}. From the kinetic theory of gases, we know that $\ell_{mfp} = 1.5\nu/\sqrt{\gamma RT}$. Compute the ratio of the diameter of Imperial Standard 50 Gauge wire, $d = 0.001$ in, to the mean free path of air at 68° F. Does flow of air at this temperature past this type of wire fall within the continuum limit?

1.7 The *mean free path* of a gas, ℓ_{mfp}, is the average distance traveled by a molecule between collisions. The continuum limit is valid if the characteristic dimension of the flow is very large compared to ℓ_{mfp}. From the kinetic theory of gases, we know that $\ell_{mfp} = 1.5\nu/\sqrt{\gamma RT}$. Compute the ratio of the diameter of Imperial Standard 45 Gauge wire, $d = 0.0071$ cm, to the mean free path of air at 20° C. Does flow of air at this temperature past this type of wire fall within the continuum limit?

1.8 Experience shows that a gas fails to behave as a continuum when it has fewer than 10^{12} molecules per mm^3. If the temperature of air is 20° C, what is the lower bound on the pressure, in Pascals, for the continuum limit to be valid? **HINT:** Write the perfect gas law as $p = mnRT$, where m is mass of a molecule and n is number density.

1.9 For a temperature of 70° F, what pressure (in psi) is required to compress air to a point where its density is a tenth that of water?

1.10 For a pressure of $p = 20$ psi and temperature $T = 86°$ F, determine the density for Air, CO_2, He, H_2, CH_4, N_2 and O_2.

1.11 At a temperature of 10° C and a pressure of 103 kN/m^2, what is the ratio of the density of water to the density of air, ρ_w/ρ_a?

1.12 Compute the weights, in Newtons, of 1 m^3 of methane and of oxygen pressurized to 500 kPa and held at a temperature of 20° C.

1.13 Compute the weights, in pounds, of 1 ft^3 of carbon dioxide and of helium pressurized to 100 psi and held at a temperature of 68° F.

1.14 The *specific weight* of a fluid is simply ρg. Determine the specific weights of air and water, in lb/ft^3, at 68° F and 1 atm.

1.15 The *specific weight* of a fluid is simply ρg. Determine the specific weights of air and water, in N/m^3, at $20°$ C and 1 atm.

1.16 The *specific gravity*, SG, of a gas is the ratio of its density to the density of air at $20°$ C and 1 atm.

(a) Verify that, at $20°$ C and 1 atm, if the gas constant is R, then $SG = R_{air}/R$.

(b) Compute the specific gravity of CO_2, He, H_2, CH_4, N_2 and O_2.

1.17 The *specific gravity*, SG, of a liquid is the ratio of its density to the density of water at $68°$ F and 1 atm. Compute the specific gravity of ethyl alcohol. carbon tetrachloride, mercury and seawater at $68°$ F and 1 atm.

1.18 Assuming water vapor can be treated as a perfect gas, what is its gas constant, R, if, for a pressure of 400 kPa and a temperature of $15°$ C, the density is 3 kg/m^3?

1.19 Assuming water vapor can be treated as a perfect gas, what is its gas constant, R, if, for a pressure of 50 psi and a temperature of $50°$ F, the density is $5.1 \cdot 10^{-3}$ $slug/ft^3$?

1.20 The perfect-gas constant, R, is given by $R = \mathcal{R}/\mathcal{M}$, where $\mathcal{M}$ is the molecular weight of the gas and $\mathcal{R}$ = 49,700 ft·lb/(slug·°R) is the *universal gas constant*. What is the effective molecular weight of air?

1.21 The perfect-gas constant, R, is given by $R = \mathcal{R}/\mathcal{M}$, where $\mathcal{M}$ is the molecular weight of the gas and $\mathcal{R}$ = 8,310 J/(kg·K) is the *universal gas constant*. Using the value of R in Table 1.3 for carbon dioxide, what is its molecular weight? Does this agree with what you learned in chemistry class? **HINT:** Check the periodic tables in your chemistry book.

1.22 For a *calorically-perfect* gas, the specific-heat coefficients, c_p and c_v, and hence the specific-heat ratio, $\gamma = c_p/c_v$, are all constant. For such a gas, derive expressions for c_p and c_v in terms of R and γ.

1.23 At high pressure and low temperature, the perfect gas law is inaccurate. A better approximation is provided by the *van der Waal's* formula, which is

$$(p + a\rho^2)(1 - b\rho) = \rho RT$$

where $a = 27(RT_c/8)^2/p_c$ and $b = (RT_c/8)/p_c$. Determine the density, ρ_c, corresponding to p_c and T_c. **HINT:** Assume $p_c = N\rho_c RT_c$ and solve for N, noting that the resulting cubic equation for N has a triple root.

1.24 What is the fractional change in density, $\Delta\rho/\rho$, for air and water caused by a change in pressure, $\Delta p = 0.5$ atm, starting from $p = 1$ atm? Assume constant temperature.

1.25 What is the fractional change in density, $\Delta\rho/\rho$, for air and water caused by a change in pressure, $\Delta p = 5$ psi, starting from $p = 14.7$ psi? Assume constant temperature.

1.26 What pressure change, Δp, starting from atmospheric pressure, is required to cause a 1% change in density for helium, mercury and water? Assume the process is isothermal, and express your answers in MPa (megaPascals).

1.27 What pressure change, Δp, starting from atmospheric pressure, is required to cause a 1% change in density for air, ether and glycerin? Assume the process is isothermal, and express your answers in psi.

1.28 An empirical formula relating pressure and density for seawater with temperature held constant is

$$p/p_o \approx (\alpha + 1)(\rho/\rho_o)^7 - \alpha$$

where p_o and ρ_o are conditions at the surface and the dimensionless constant α is approximately 3000. The pressure at the deepest part of the Pacific Ocean is about 16,170 psi. What is the density of seawater at this depth in $slug/ft^3$? What is the percentage difference from the value at the surface?

1.29 An empirical formula relating pressure and density for water with temperature held constant is

$$p/p_o \approx (\alpha + 1)(\rho/\rho_o)^7 - \alpha$$

where p_o and ρ_o are conditions at the surface and α is a dimensionless constant.

(a) According to this formula, what is the compressibility, τ, of water as a function of p, p_o and α?

(b) Appealing to Equation (1.4), determine the numerical value of the constant α.

1.30 If a gas is compressed through a frictionless process with no heat exchange, the pressure is related to density by the isentropic relation, $p = A\rho^\gamma$, where A is a constant and γ is the specific-heat ratio.

(a) Compute the compressibility, τ_s, for such a process. Express your answer in terms of γ and p.

(b) Compare your result with the isothermal τ for a gas when $p = 1$ atm.

1.31 A spherical soap bubble has an inside radius, R, film thickness, $t \ll R$, and surface tension σ. Find the difference between internal and ambient pressure. Note that a bubble has two liquid-air interfaces.

1.32 A small droplet of water surrounded by air at an unspecified temperature has a diameter of $D = 0.02$ in. The pressure within the droplet is $\Delta p = 0.075$ psi above ambient pressure. What is the surface tension?

1.33 A small droplet of water surrounded by air at an unspecified temperature has a diameter of $D = 5$ mm. The pressure within the droplet is $\Delta p = 50$ Pa above ambient pressure. What is the surface tension?

1.34 Compute the ratio of the diameter of a small droplet of mercury immersed in air with the diameter it would have when immersed in water. Assume the pressure within the droplet exceeds the ambient value by the same amount in both cases.

1.35 Compute the ratio of the diameter of a small droplet of carbon tetrachloride immersed in air with the diameter it would have when immersed in water. Assume the pressure within the droplet exceeds the ambient value by the same amount in both cases.

1.36 To what height will glycerin rise in a cylindrical capillary tube of diameter, $d = 1/16$ in? Assume the wetting angle is $\phi = 0°$.

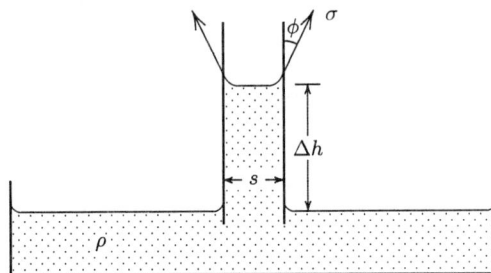

Problems 1.36, 1.37, 1.38, 1.39, 1.40

1.37 To what height will ethyl alcohol rise in a cylindrical capillary tube of diameter, $d = 1.7$ mm? Assume the wetting angle is $\phi = 0°$.

1.38 To what height will mercury rise in a cylindrical capillary tube of diameter, $d = 0.5$ in? Assume the wetting angle is $\phi = 129°$.

1.39 Suppose a square tube of width s is immersed in a large container. If the fluid density is ρ, the surface tension is σ and the wetting angle is ϕ, to what height, Δh, will the fluid rise?

1.40 Suppose a rectangular tube s wide by $2s$ thick (out of the page) is immersed in a large container. If the fluid density is ρ, the surface tension is σ and the wetting angle is ϕ, to what height, Δh, will the fluid rise?

1.41 An open tube of diameter d is inserted into a pan of ethyl alcohol. A second tube of diameter $\frac{5}{2}d$ is inserted into a pan of water. Due to capillary action, the fluids rise to heights of Δh_1 and Δh_2 for the alcohol and water, respectively. Compute the ratio $\Delta h_2/\Delta h_1$. Assume the wetting angle is the same in both tubes and $T = 20°$ C.

1.42 A round capillary tube of diameter d is inserted into a pan of water. A second tube, also of diameter d, is inserted into a pan of liquid whose density is $\frac{3}{2}\rho_w$, where ρ_w is the density of water. If the liquid rises to a height Δh in water and $0.2\Delta h$ in the unknown liquid, what is the surface tension of the unknown liquid? Assume the wetting angle is the same in both tubes and $T = 68°$ F.

1.43 Consider a liquid-air interface next to a plane wall. As shown in the figure, surface tension causes the liquid to assume the shape shown. The surface tension is σ, the fluid density is ρ and the wetting angle is ϕ. You can assume the pressure difference across the interface and the radius of curvature of the interface are

$$\Delta p \approx \rho g \eta \quad \text{and} \quad \frac{1}{\mathcal{R}} \approx \frac{d^2\eta}{dx^2}$$

Determine the shape of the interface and the maximum height, h.

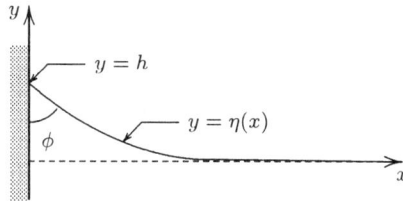

Problem 1.43

1.44 The viscosity of water as a function of temperature can be represented by the following empirical relationship known as *Andrade's equation.*

$$\mu = \mu_r e^{S/T}$$

The empirical *Sutherland equation*, Equation (1.29), is often used to compute the viscosity of air. In both cases, the quantities A, S and μ_r are empirical coefficients that can be determined from measurements — all three are positive numbers. Using these formulas and assuming constant pressure, determine the sign of $d\nu/dT$ for water and for air.

1.45 The viscosity of air can be approximated by a power law according to

$$\mu \approx CT^{0.7}$$

where T is absolute temperature and C is a constant.

  (a) Determine C by insisting that the value of μ at $T = 68°$ F matches the value given by *Sutherland's equation.*

  (b) Make a graph of μ according to the power law developed in Part (a) and *Sutherland's equation* for $-250°$ F $\leq T \leq 2000°$ F.

1.46 The viscosity of water can be approximated by *Andrade's equation*, viz.,

$$\mu \approx \mu_r e^{S/T}$$

where T is absolute temperature. The quantities μ_r and S are empirical constants.

  (a) Using values from Tables 1.2 and 1.7 for $T = 10°$ C and $T = 90°$ C, determine the values of μ_r and S.

  (b) Using the values of μ_r and S determined in Part (a), make a graph comparing the tabulated values of μ with *Andrade's equation* for $0°$ C $\leq T \leq 100°$ C.

1.47 Heavy cream is a non-Newtonian fluid whose shear stress can be approximated by

$$\tau \approx 0.12 \left(\frac{du}{dy}\right)^{2/3}$$

where all quantities are expressed in SI units.

(a) Determine the effective viscosity, μ_e, such that $\tau = \mu_e du/dy$.

(b) Compute the ratio of μ_e to the viscosity of water, μ_w, at $20°$ C.

(c) Make a graph of μ_e/μ_w for values of du/dy between 0.01 sec^{-1} and 10 sec^{-1}.

1.48 The velocity distribution, $u(y)$, for *Plane Poiseuille* flow between two parallel walls is given by $u(y) = u_m(y/h)(1 - y/h)$, where u_m is constant, y is measured from the lower wall and the space between the walls is h. Plot the velocity profile as u/u_m vs. y/h, and determine the shear stress at the walls as a function of μ, u_m and h.

1.49 Compute the average velocity for Couette flow, $\bar{u}$, defined by

$$\bar{u} \equiv \frac{1}{h}\int_0^h u(y)dy$$

1.50 Consider Couette flow with a plate-separation distance, $h = 0.3$ mm, and velocity, $U = 2$ m/sec. Measurements show that the shear stress is $\tau = 8$ Pa.

(a) What is the viscosity of the fluid between the plates?

(b) What is the maximum density for which the answer of Part (a) is valid?

1.51 Consider Couette flow with a plate-separation distance, $h = 1$ in, and velocity, $U = 12$ ft/sec. Measurements show that the shear stress is $\tau = 0.2$ lb/ft^2.

(a) What is the viscosity of the fluid between the plates?

(b) What is the maximum density for which the answer of Part (a) is valid?

1.52 A *rotating-cylinder viscometer* can be used to measure the viscosity of a liquid. This device consists of a cylinder of radius R_i and length L rotating at angular velocity Ω inside a concentric cylinder of radius R_o. The device is designed to have $R_o - R_i \ll R_i$ and $R_o - R_i \ll L$. Determine the torque, T, on the rotating cylinder as a function of R_i, R_o, L, Ω and μ. Neglect any torque on the bottom of the inner cylinder.

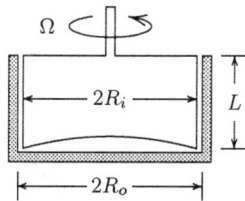

Problem 1.52

1.53 A plate moving with velocity U in its own plane lies between two stationary parallel plates as shown. The viscosity of the fluid below the moving plate is μ, while the fluid above has viscosity 4μ.

(a) Make a graph of the velocity, u/U, versus y/h for $0 \le y/h \le 3$.

(b) Compute the shear stress on the upper and lower walls.

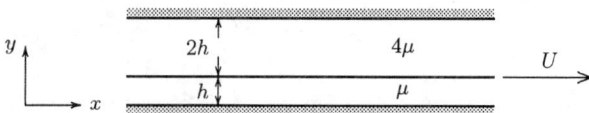

Problem 1.53

1.54 A cylinder of mass M slides horizontally in a lubricated, but otherwise empty, pipe. The radial clearance between cylinder and pipe is d. The diameter and length of the cylinder are D and L, respectively. Assuming the cylinder's deceleration is $-A$ when the speed is U, derive a formula for the viscosity of the fluid. If the weight of the cylinder is 20 lb, $d = .001$ in, $A = 2$ ft/sec^2, $U = 20$ ft/sec, $D = 6$ in and $L = 5$ in, determine the value of μ in slug/(ft·sec).

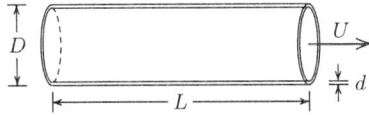

Problems 1.54, 1.55

1.55 A cylinder of mass M slides horizontally in a lubricated, but otherwise empty, pipe. The radial clearance between cylinder and pipe is d. The diameter and length of the cylinder are D and L, respectively. Assuming the cylinder's deceleration is $-A$ when the speed is U, derive a formula for the viscosity of the fluid. If the weight of the cylinder is 100 N, $d = .025$ mm, $A = 1$ m/sec^2, $U = 5$ m/sec, $D = 15$ cm and $L = 10$ cm, determine the value of μ in kg/(m·sec).

1.56 An air hockey game has a cylindrical puck of diameter D, height h, and density ρ_p. The air film under the puck has a thickness $\frac{1}{12}h \ll D$. The puck will slow down due to the viscous force of the air on the bottom of the puck. The air viscosity is μ. What time is required for the puck to decelerate to e^{-2} of its initial speed? Express your answer as a function of ρ_p, h and μ.

1.57 In manufacturing recording tape, the tape is coated with a lubricant. This is done by pulling it through a narrow gap filled with the lubricant. Consider tape whose width and thickness are w and h, respectively. The maximum tensile force the tape can withstand is T. The tape is centered in a gap of height $3h$, the lubricant has viscosity μ, and the length of tape within the gap is ℓ. What is the maximum velocity at which the tape can be pulled through the gap?

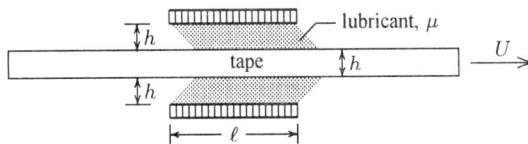

Problem 1.57

1.58 A solid circular cylinder of radius r and length ℓ slides inside a vertical smooth pipe having an inside radius R. The small space between the cylinder and pipe is lubricated with an oil film that has a viscosity μ. Derive a formula for the rate of descent of the cylinder in the vertical pipe. Assume that the cylinder has a weight W and is concentric with the pipe as it falls. Use the formula to find the rate of descent of a cylinder of 100 mm diameter that slides inside a 100.5 mm pipe. The cylinder is 200 mm long and weighs 20 N. The lubricant has viscosity $\mu = 0.35$ kg/(m·sec).

1.59 A block of mass M slides on a thin film of oil whose viscosity is μ. The film thickness is h and the area of the block exposed to the oil is A. Another block of mass m is attached with a cord as shown. When released, the mass m exerts tension on the cord, causing the blocks to accelerate. You may neglect friction in the pulley and air resistance. Assuming the instantaneous speed of the block is $U(t)$, derive a differential equation for U as a function of time. Obtain an algebraic expression for the maximum speed of the block. **HINT:** Draw free-body diagrams for each mass with the tension force acting on each.

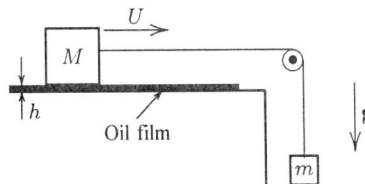

Problem 1.59

1.60 A cube of density ρ and side s slides down an inclined plane coated with a thin lubricating film of thickness h and viscosity μ. Derive a differential equation for the motion of the cube. Solve the differential equation, assuming the cube starts from rest at time $t = 0$. What is the terminal velocity for $\rho = 800$ kg/m^3, $s = .05$ m, $h = .001$ m, $\mu = 0.04$ kg/(m·sec) and $\alpha = 30°$?

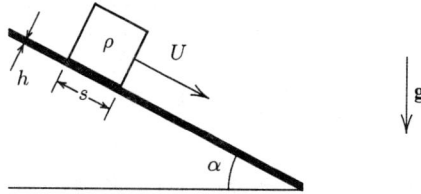

Problems 1.60, 1.61, 1.62

1.61 A cube of density ρ and side s slides down an inclined plane coated with a thin lubricating film of thickness h and viscosity μ. Derive a differential equation for the motion of the cube. Solve the differential equation, assuming the cube starts from rest at time $t = 0$. What is the terminal velocity for $\rho = 1.5$ slug/ft^3, $s = 2$ in, $h = .02$ in, $\mu = 0.0023$ slug/(ft·sec) and $\alpha = 30°$?

1.62 A cube of density ρ and side s slides down an inclined plane coated with a thin lubricating film of thickness h and viscosity μ. Its terminal velocity is U.

(a) Develop a formula for the thickness of the lubricating film.

(b) Compute h in mm for $\rho = 800$ kg/m^3, $s = .05$ m, $\mu = 0.067$ kg/(m·sec), $\alpha = 30°$ and $U = 5$ m/sec.

(c) What is the maximum density of the fluid in the lubricating film for which the solution is valid?

1.63 Develop an equation for the *skin friction*, c_f, as a function of *Reynolds number*, Re_R, for pipe flow, where

$$c_f \equiv \frac{-\tau(R)}{\frac{1}{2}\rho u_m^2} \quad \text{and} \quad Re_R \equiv \frac{\rho u_m R}{\mu}$$

1.64 Consider the flow in a pipe of radius R.

(a) Compute the average velocity for pipe flow, $\bar{u}$, defined by

$$\bar{u} \equiv \frac{1}{\pi R^2} \int_0^{2\pi} \int_0^R u(r) r \, dr \, d\theta$$

(b) Develop an equation for the *friction factor*, f, as a function of the *Reynolds number*, Re_D, according to the following definitions.

$$f \equiv \frac{-\tau(R)}{\frac{1}{8}\rho \bar{u}^2} \quad \text{and} \quad Re_D \equiv \frac{\rho \bar{u} D}{\mu}$$

1.65 Consider laminar flow of blood in the aorta, which can be analyzed using the pipe-flow solution.

(a) Show that, for a given pressure difference, the volume flow rate, Q, increases as R^4, where

$$Q \equiv 2\pi \int_0^R u(r) r \, dr$$

(b) Now, write your answer in terms of the maximum velocity, u_m, and the radius, R. The maximum velocity in the aorta is $u_m \approx 1$ ft/sec and the radius is $R \approx 0.6$ in. How long does it take for 1 in^3 of blood to flow through the aorta?

1.66 Consider laminar flow of blood in the aorta, which can be analyzed using the pipe-flow solution.

  (a) Show that, for a given pressure difference, the volume flow rate, Q, increases as R^4, where

$$Q \equiv 2\pi \int_0^R u(r)r\,dr$$

  (b) Now, write your answer in terms of the maximum velocity, u_m, and the radius, R. The maximum velocity in the aorta is $u_m \approx 30$ cm/sec and the radius is $R \approx 1.5$ cm. How long does it take for 100 cm^3 of blood to flow through the aorta?

1.67 The density and viscosity of human blood are $\rho = 1058$ kg/m^3 and $\mu = 3.3 \cdot 10^{-3}$ kg/(m·sec). The maximum speed of blood in the aorta of a large man is roughly $u_m \approx 0.30$ m/sec. If the diameter of the aorta is $d \approx 3$ cm, what is the *Reynolds number*, $\rho u_m d / \mu$? Will the flow be laminar or turbulent?

1.68 The density and viscosity of human blood are $\rho = 2.06$ slug/ft^3 and $\mu = 6.9 \cdot 10^{-5}$ slug/(ft·sec). The maximum speed of blood in the aorta of a large man is roughly $u_m \approx 1.0$ ft/sec. If the diameter of the aorta is $d \approx 1.2$ in, what is the *Reynolds number*, $\rho u_m d / \mu$? Will the flow be laminar or turbulent?

1.69 A pump moves water at 50° F through a small tube of diameter $D = 0.2$ in and length $L = 8$ in. Because of a malfunction, the pump creates a suction-pressure difference of only $p_1 - p_2 = 0.01$ psi.

  (a) What is the speed at the center of the tube cross section?

  (b) What is the *Reynolds number*, $Re_R \equiv u_m R / \nu$?

  (c) What is the total viscous force acting on the tube in pounds?

1.70 A pump moves water at 10° C through a small tube of diameter $D = 5$ mm and length $L = 20$ cm. Because of a malfunction, the pump creates a suction-pressure difference of only $p_1 - p_2 = 70$ Pa.

  (a) What is the speed at the center of the tube cross section?

  (b) What is the *Reynolds number*, $Re_R \equiv u_m R / \nu$?

  (c) What is the total viscous force acting on the tube in Newtons?

Chapter 2

Dimensional Analysis

Dimensional analysis is a particularly valuable tool whose applicability extends to all branches of physics. It is used extensively in fluid mechanics, especially in correlating experimental data. On the theoretical side, dimensional analysis is indispensable in checking consistency of algebraic relationships derived from basic principles. It is especially helpful in determining the relevant scales in a given problem, correlating experimental data and extrapolating measurements on small-scale models to large-scale objects. Some of the most important insight into the complicated phenomenon of turbulence has come from dimensional analysis [cf. Kolmogorov (1941), Tennekes and Lumley (1983), Landahl and Mollo-Christensen (1992), Bradshaw (1972) or Wilcox (1993)]. This chapter presents the methodology involved in dimensional analysis.

For many engineering students, these glowing remarks about dimensional concepts may appear surprising. Often, deriving an algebraic formula proves far less tedious than the frustrating exercise that follows in substituting actual numbers into the formula. The frustration is invariably associated with the units of dimensional quantities. For example, if a length is given in inches, it has to be divided by 12 to convert to feet in order to be consistent with a velocity given in feet per second. And that is a relatively minor issue compared to figuring out how many foot-pounds there are in a British thermal unit or foot-pounds per second in a horsepower! When the problem employs metric units, even an experienced engineer sometimes has to look up the definition of some of the more obscure units such as the Pascal, the Joule and the Watt.

Herein lies an important distinction, namely, the difference between the separate concepts of **dimensions** and **units**. By convention, we refer to the dimensions of a physical quantity in a generic sense. For example, we say that the dimensions of velocity are length per unit time, without reference to any specific measure. When we use the term units, we make specific reference to a consistent set of measures that characterize the dimensions. Hence, we say the units of velocity are feet per second, meters per second or perhaps even (the legendary and rarely used) furlongs per fortnight. For most engineering students (and even those of us with years of experience), the problems encountered in computing numerical answers are usually associated with units.

With dimensional analysis, we deal with dimensions, so that our primary operations are unaffected by any particular choice of units. However, as we will discover, a solid grasp of the principles of dimensional analysis will even simplify the chore of cranking out a number from the elegant and glorious equations we work so hard to derive!

27

2.1 Basic Premises

Clearly, the laws of physics are independent of the system of units we choose. In fluid mechanics, the two most commonly used sets of units are Standard International (SI) units and the U. S. Customary System (USCS).[1] In these two systems, we express mass, length, time and temperature as follows.

System	Mass, M	Length, L	Time, T	Temperature, Θ
SI	kilogram (kg)	meter (m)	second (sec)	Kelvin (K)
USCS	slug (slug)	foot (ft)	second (sec)	° Rankine (° R)

In dimensional analysis, we introduce a symbolic representation of the basic units. Specifically, we let M, L, T and Θ denote mass, length, time and temperature, respectively. Additionally, the dimensions of a given quantity are indicated by enclosing the quantity in square brackets. Thus, a typical dimensional-analysis equation for the viscosity of a fluid is

$$[\mu] = \frac{M}{LT} \tag{2.1}$$

In words, this equation says the dimensions of viscosity are mass per unit length per unit time. If we are using SI units, the viscosity thus has units kg/(m·sec). Similarly, in USCS units, its units are slug/(ft·sec).

The four dimensions listed above are defined to be **independent dimensions**. All other dimensions of interest can be expressed in terms of the independent dimensions. For mechanical systems, M, L, T and Θ are sufficient to define all dimensional quantities. If electrical properties are included, we gain additional independent dimensions such as charge or resistance.

As an example of how other dimensions can be expressed in terms of our independent dimensions, consider pressure. It is normally expressed as a force per unit area, such as Newtons per square meter or pounds per square foot. As a mnemonic device, we can take advantage of our knowledge of Newton's second law of motion to conclude that force is simply mass times acceleration. Hence,

$$[p] = \frac{\text{Force}}{L^2} = \frac{ML/T^2}{L^2} = \frac{M}{LT^2} \tag{2.2}$$

Thus, for the purposes of dimensional analysis, the units of pressure in SI and USCS units are kg/(m·sec^2) and slug/(ft·sec^2), respectively.

For the sake of maximum utility, any meaningful equation relating physical quantities should be valid regardless of the system of units employed. Not all formulas quoted in engineering literature satisfy this condition, and this limits their value. For example, most sailing enthusiasts are familiar with the formula for the **theoretical hull speed**[2] given by

$$U_{hull} = 1.34\sqrt{L_w} \tag{2.3}$$

---

[1] Appendix A discusses physical dimensions and units for the benefit of readers who require a review of alternative systems of units.

[2] Attempting to sail beyond this speed will cause the boat to be literally sucked under the water by the low pressure that develops on the bottom of the hull. This phenomenon, often referred to as *sailing under*, may account for many of the ships lost in the Bermuda Triangle.

where U_{hull} is the sailboat speed in knots (nautical miles per hour) and L_w is the length of the hull at the waterline in feet. This tells us that for $L_w = 20$ feet the theoretical hull speed is 6 knots. Our knowledge of dimensions also tells us that the coefficient 1.34 is not a pure number, but rather has the dimensions of $L^{1/2}/T$, which can be seen from the following dimensional equivalent of Equation (2.3).

$$\underbrace{[U_{hull}]}_{L/T} = \underbrace{[1.34]}_{L^{1/2}/T} \underbrace{[\sqrt{L_w}]}_{L^{1/2}} \tag{2.4}$$

Substituting a value for L_w in any units other than feet thus yields a meaningless result for the hull speed. The coefficient 1.34 would have to be replaced with an alternative value to permit specifying L_w in meters, for example.

For reasons that will become clear later in this chapter when we define the Froude number, we should expect to see the acceleration due to gravity, g, under the square root in Equation (2.3). So let's revise Equation (2.3) in this manner, using the information that a 20 foot hull (at the waterline) yields a hull speed of 6 knots, which reference to Table A.4 of Appendix A shows is 10.128 ft/sec. Hence, since the acceleration of gravity is $g = 32.174$ ft/sec^2, we find $U_{hull}/\sqrt{gL_w} = 0.40$. Therefore, we propose the following alternative to Equation (2.3).

$$U_{hull} = 0.40\sqrt{gL_w} \tag{2.5}$$

In this revised equation, the number 0.40 is dimensionless because $\sqrt{gL_w}$ has the same dimensions as velocity. Substituting a value for L_w of 20 feet yields a hull speed of 10.128 ft/sec, or 6 knots. However, unlike Equation (2.3), if we quote L_w in meters, we obtain the correct value of the hull speed in meters per second, viz., 3.089 m/sec. Reference to Table A.4 shows that this is also 6 knots.

One of the most important facts that serves as the foundation of dimensional analysis is the following. All equations developed from the basic laws of physics are **dimensionally homogeneous**. That is, the dimensions of all terms appearing in such an equation are the same. Equation (2.5) is an example.

As a second example, consider the momentum equation for a real fluid, i.e., the famous **Navier-Stokes equation**. We will derive and study this equation when we introduce viscous effects in Chapter 12. For one-dimensional motion, the Navier-Stokes equation assumes the following form.

$$\rho\frac{\partial u}{\partial t} + \rho u\frac{\partial u}{\partial x} = -\frac{\partial p}{\partial x} + \mu\frac{\partial^2 u}{\partial x^2} \tag{2.6}$$

The quantity ρ is density (M/L^3), u is velocity (L/T), x is distance (L), t is time (T), p is pressure (M/LT^2), and μ is viscosity (M/LT).

Despite the fact that we may be completely unfamiliar with this equation, we can still assess its dimensional consistency with the information already at our disposal. This is an important point about dimensional analysis. Most of our focus is on the relationships amongst the dimensions of physical quantities, and not on the equations they happen to satisfy. This is especially useful when we begin our study of a new field, where we may be unfamiliar with the governing equations. Using dimensional analysis, we can quickly determine key parameters governing the basic physics of a given application, *independent of the detailed equations involved.*

Proceeding term by term with this somewhat mysterious equation, we have the following.

$$\left.\begin{aligned}
\left[\rho\frac{\partial u}{\partial t}\right] &= \frac{M}{L^3}\frac{L/T}{T} = \frac{M}{L^2T^2} \\[2mm]
\left[\rho u\frac{\partial u}{\partial x}\right] &= \frac{M}{L^3}\frac{L}{T}\frac{L/T}{L} = \frac{M}{L^2T^2} \\[2mm]
\left[\frac{\partial p}{\partial x}\right] &= \frac{M/LT^2}{L} = \frac{M}{L^2T^2} \\[2mm]
\left[\mu\frac{\partial^2 u}{\partial x^2}\right] &= \frac{M}{LT}\frac{L/T}{L^2} = \frac{M}{L^2T^2}
\end{aligned}\right\} \qquad (2.7)$$

Thus, as claimed, the Navier-Stokes equation is dimensionally homogeneous.

Obviously, one purpose dimensional analysis serves is to check for dimensional consistency of algebraic results. This is especially helpful when the result has been obtained from a lengthy algebraic exercise. If this were the only end served by dimensional analysis, it would hardly deserve an entire chapter of discussion. However, as we will see in the following sections, the method has even more important uses. There are important conclusions we can draw regarding a given problem where the only information we use is the dimensions of relevant physical quantities. Also, the method provides the proper scaling laws that permit extrapolation of measurements made on small-scale models to full-scale objects.

There are two key facts to keep in mind as we study dimensional analysis. On the one hand, as already mentioned, we have no need to solve any differential equations. In fact, *we don't even have to know what the governing equations are.* The method would appear to be very powerful from this point of view. The most noteworthy example is the pioneering work of Kolmogorov (1941), which laid the theoretical foundation for much of our present-day understanding of turbulence. Many important aspects of his research were based on dimensional analysis. On the other hand, dimensional analysis is no substitute for understanding of the physics, and its use as an analytical tool should not be overestimated. Of greatest importance, be aware that *we establish the relevant physical processes when we select the dimensional quantities in our analysis.* Thus, in Kolmogorov's case, he combined great physical insight into the nature of turbulence with dimensional analysis to discover some of the phenomenon's mysteries.

2.2   Buckingham Π Theorem

Buckingham (1915) proved that for any given physical problem, the number of independent **dimensionless groupings** needed to correlate is n less[3] than the total number of variables, where n is the number of independent dimensions. By definition, a dimensionless grouping is an algebraic combination of the variables that has no dimensions. Equation (1.34), for example, involves the dimensionless groupings known as skin friction, $\tau/(\frac{1}{2}\rho U^2)$, and Reynolds number, $\rho U h/\mu$. Also, our discussion of a sailboat's theoretical hull speed noted that $U_{hull}/\sqrt{gL_w}$ is a dimensionless grouping. This general result regarding the number of dimensionless groupings is known as the **Buckingham Π Theorem**.

---

[3]There are exceptions, however, and the number of dimensionless groupings can be larger. It will be obvious from the dimensional analysis when this somewhat pathological case arises.

If a problem involves only mass, length and time as independent dimensions, then n is 3. If the dimensional quantities also involve temperature, then n increases to 4, provided mass, length and time are still relevant. This section first introduces the classical **indicial method** developed by Buckingham. After illustrating how we interpret the results of a dimensional analysis, the section gives an alternative method for finding dimensionless groupings developed by E. S. Taylor.

2.2.1 Finding Dimensionless Groupings

To illustrate the ramifications of this theorem, consider viscous flow past a sphere as illustrated in Figure 2.1. Our object is to correlate the force that develops on the sphere, F, in terms of the dimensionless parameters in the problem. Assuming the sphere is not heated or cooled and that gravitational effects (or those of any other external force) are of no consequence, there are a total of five dimensional quantities in the problem, viz.,

F = force on the sphere (ML/T^2)

μ = viscosity of the fluid (M/LT)

ρ = density of the fluid (M/L^3)

U = velocity of the fluid (L/T)

D = diameter of the sphere (L)

Note that in selecting our variables, we have made two assumptions regarding the relevant physical processes. We ignore temperature of the fluid and the sphere as dimensional variables because of our assumption about heating or cooling. We exclude gravitational acceleration, g, because of our assumption about external forces. This is the manner in which our physical understanding establishes the relevant variables.

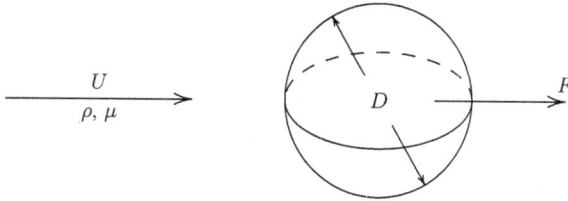

Figure 2.1: *Viscous flow past a sphere.*

Appealing to the Buckingham Π theorem, we have a total of 5 variables and 3 independent dimensions, i.e., M, L and T. Therefore, there are 2 dimensionless groupings. In order to find the two groupings, we postulate a dimensional equation relating the dimensions of F to the other four variables, viz.,

$$[F] = [\mu]^{a_1} [\rho]^{a_2} [U]^{a_3} [D]^{a_4} \tag{2.8}$$

where a_1, a_2, a_3 and a_4 are unknown exponents. Our goal is to rearrange this equation so that it assumes a special form. In general, using dimensional analysis, we can show that the force on the sphere is the product of factors as follows.

$$[F] = [F_0] \prod_{i=1}^{N-1} [P_i]^{b_i} \tag{2.9}$$

where $N = 2$ is the number of dimensionless groupings, F_0 has dimensions of force, and the exponents b_i are functions of the a_i. The quantities P_i, known as **pi factors**, are dimensionless. To understand the motivation for the terminology "Π Theorem," note that by definition,

$$\prod_{i=1}^{N-1} [P_i]^{b_i} = [P_1]^{b_1} [P_2]^{b_2} [P_3]^{b_3} \cdots [P_{N-1}]^{b_{N-1}} \qquad (2.10)$$

The operator $\prod$ is known as the **product pi**, and is analogous to the standard summation operator, $\sum$. In the present case, there will be one dimensionless grouping, P_1, on the right-hand side of our equation. The second dimensionless grouping predicted by the Buckingham Π Theorem, P_0, is the ratio of F to F_0. In other words, we are trying to arrive at a result of the form

$$\frac{F}{\text{Quantity with dimensions of force}} = \mathcal{F}(\text{dimensionless grouping}) \qquad (2.11)$$

We use the following 4-step procedure, known as the **indicial method**, to arrive at the desired dimensionless groupings.

Step 1. To begin the process, we first identify the dimensions of all the variables in the problem. While we have already done this above, we now state the results in algebraic form suitable for substitution into Equation (2.8). Thus, we have:

$$[F] = \frac{ML}{T^2}, \quad [\mu] = \frac{M}{LT}, \quad [\rho] = \frac{M}{L^3}, \quad [U] = \frac{L}{T}, \quad [D] = L \qquad (2.12)$$

Step 2. Next, we substitute from Equation (2.12) into Equation (2.8). This yields an equation relating the independent dimensions, M, L and T, i.e.,

$$\begin{aligned} MLT^{-2} &= M^{a_1} L^{-a_1} T^{-a_1} M^{a_2} L^{-3a_2} L^{a_3} T^{-a_3} L^{a_4} \\ &= M^{a_1+a_2} L^{-a_1-3a_2+a_3+a_4} T^{-a_1-a_3} \end{aligned} \qquad (2.13)$$

Since, from an algebraic point of view, M, L and T are independent variables, the only way this equation can be satisfied is to have identical exponents on both sides of the equation. Thus, we arrive at the following three coupled, linear algebraic equations, known as the **indicial equations**.

$$\left. \begin{aligned} a_1 + a_2 &= 1 \\ -a_1 - 3a_2 + a_3 + a_4 &= 1 \\ -a_1 - a_3 &= -2 \end{aligned} \right\} \qquad (2.14)$$

Observe that we have only 3 equations for 4 unknowns. In general, the number of unknowns will exceed the number of equations by one less than the number of dimensionless groupings predicted by the Buckingham Π Theorem.

While this will be the case in most applications, there is an exception. Specifically, the number of dimensionless groupings increases by one for each linearly-dependent equation present. An equation is **linearly dependent** if it can be derived as the sum of constant multiples of the other equations. For example, if the third of a set of three linear equations is the same as 2 times the first equation plus 3 times the second equation, then it is linearly dependent. None of Equations (2.14) can be derived as a linear combination of the others, and we thus describe them as being **linearly independent**. The indicial method almost always yields linearly-independent equations.

Step 3. We now solve the coupled set of equations for the exponents. This is usually straightforward since there is often weak coupling amongst the equations. In the present case, for example, the first and third of Equations (2.14) involve only two of the exponents. Thus, we find immediately that

$$a_1 = 2 - a_3 \tag{2.15}$$

while a_2 is given by

$$a_2 = 1 - a_1 = 1 - (2 - a_3) = a_3 - 1 \tag{2.16}$$

Thus, substituting these results into the second of Equations (2.14) permits solving for a_4 as a function of a_3, viz.,

$$-(2 - a_3) - 3(a_3 - 1) + a_3 + a_4 = 1 \quad \Longrightarrow \quad a_4 = a_3 \tag{2.17}$$

Step 4. We collect our solution for the exponents and substitute into Equation (2.8), wherefore

$$[F] = [\mu]^{2-a_3} [\rho]^{a_3-1} [U]^{a_3} [D]^{a_3} \tag{2.18}$$

Regrouping terms, we arrive at the final result, viz.,

$$[F] = \left[\frac{\mu^2}{\rho} \right] \left[\frac{\rho U D}{\mu} \right]^{a_3} \tag{2.19}$$

Therefore, the two dimensionless groupings are

$$P_0 = \frac{\rho F}{\mu^2} \quad \text{and} \quad P_1 = \frac{\rho U D}{\mu} \tag{2.20}$$

Before proceeding to interpretation of this result, it is worthwhile to pause and discuss two subtle points. First, as an alternative to Equation (2.8), we could begin with the following.

$$1 = [F]^{a_0} [\mu]^{a_1} [\rho]^{a_2} [U]^{a_3} [D]^{a_4} \tag{2.21}$$

This equation corresponds to stating that products of all of the variables raised to appropriate powers are dimensionless. Using the procedure above, we would arrive at an equation with two dimensionless groupings and two undetermined exponents, viz.,

$$1 = \left[\frac{\rho F}{\mu^2} \right]^{a_0} \left[\frac{\rho U D}{\mu} \right]^{a_3} \tag{2.22}$$

While the results are equivalent, the algebra is more tedious as we have 5 unknowns in place of 4. Thus, Equation (2.8) is the preferred starting point since it leads to fewer algebraic operations.

Second, suppose we had included a variable such as the temperature of the freestream fluid, T_∞, and no other temperature. Upon completion of the dimensional-analysis algebra, we would find that

$$[F] = \left[\frac{\mu^2}{\rho} \right] \left[\frac{\rho U D}{\mu} \right]^{a_3} [T_\infty]^0 \tag{2.23}$$

i.e., the exponent of T_∞ is zero. This means either the temperature is irrelevant or we have inadvertently left out some other pertinent quantity involving temperature. The converse is untrue, however. Had we included an irrelevant quantity such as surface tension, we would arrive at a perfectly valid dimensionless grouping, which would be of no importance to the application.

2.2.2 Interpretation of Results

It is important to understand that all we have shown is existence of the two dimensionless groupings in Equation (2.20). Because the number of unknown exponents exceeds the number of equations by one, the exponent a_3 is undetermined. Hence, while the formalism of the method led to Equation (2.19), we have not shown that $\rho F/\mu^2$ is proportional to some power of the dimensionless grouping known as **Reynolds number**, Re, defined by

$$Re = \frac{\rho U D}{\mu} \qquad (2.24)$$

Rather, our analysis has shown that the dimensionless force, $\rho F/\mu^2$, is some function of Re. At this point, we have extracted as much analytical information as we can from dimensional analysis. However, this information is extremely useful for several purposes, including:

- Establishing the order of magnitude of flow properties;

- Correlating experimental data;

- Developing scaling laws to extrapolate measurements on small models to larger models.

For example, we can reasonably expect that μ^2/ρ provides the order of magnitude of the force on the sphere for some range of Reynolds numbers. As we will see below, this is indeed true at relatively small Reynolds numbers. However, as we will also find, flow past a sphere is a bit of a pathological case as the dimensionless force can assume a wide range of values as we vary Reynolds number over many orders of magnitude.

This reflects the fact that flow past a sphere undergoes dramatic changes as Reynolds number increases from very small to very large values. To appreciate the complexity of this flow, the curious reader might want to refer to the Van Dyke (1982) collection of fluid-motion photographs. The book provides a spectacular group of photographs depicting the many different Reynolds-number regimes for flow past a sphere.

One of the most powerful purposes of dimensional analysis is to aid in correlating measurements. If our object is to correlate the force on the sphere as a function of the relevant variables, without dimensional analysis we are dealing with a function of four variables, viz.,

$$F = \mathcal{G}(\rho, U, D, \mu) \qquad (2.25)$$

By contrast, dimensional analysis shows that the dimensionless force is a function of a single variable, because our analysis tells us that

$$F = \frac{\mu^2}{\rho}\mathcal{F}(Re) \qquad (2.26)$$

Obviously, Equation (2.26) is far easier to correlate than Equation (2.25). Using a single graph, all data from various experiments will fall on a single curve if we create the plot with the dimensionless groupings on the ordinate and abscissa.

Perhaps the most remarkable thing we have shown is that our measurements fall on the same curve even if one set of experiments is done in water and another in air. That is, dimensional analysis tells us that regardless of the fluid used, the (appropriately scaled) force depends only upon Reynolds number. This is a remarkable fact, and certainly not an intuitive one, whose discovery would require a large number of experiments varying ρ, U, D and μ independently. Nevertheless, we have discovered this fact with no knowledge of the complexities involved in viscous flow past a sphere, of which there are many.

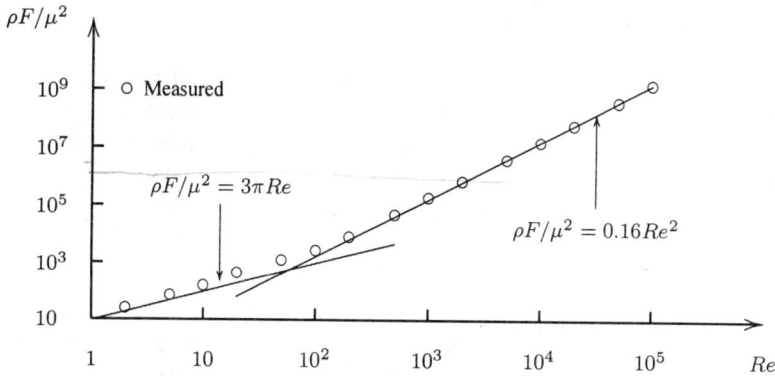

Figure 2.2: *Correlation of measured force on a sphere using $\rho F/\mu^2$.*

Figure 2.2 presents graphical results of hypothetical experiments for flow past a sphere. As shown, if we make a log-log plot of $\rho F/\mu^2$ versus Re for a wide range of Reynolds numbers, there are two clear limiting cases. On the one hand, in the limit $Re \to 0$, the measured values asymptotically approach linear variation of $\rho F/\mu^2$ with Re. If our measurements were very accurate, our graph would indicate that

$$\frac{\rho F}{\mu^2} \to 3\pi Re \quad \text{as} \quad Re \to 0 \qquad (2.27)$$

This result can be shown to be valid for small Reynolds numbers by solving the exact equations of motion for flow past a sphere [see, for example, Panton (1996)]. This is the theoretical limiting case known as **Stokes flow**, or **creeping flow**.

On the other hand, for large Reynolds number, the measured values of $\rho F/\mu^2$ asymptotically approach quadratic variation with Re, viz.,

$$\frac{\rho F}{\mu^2} \to 0.16 Re^2 \quad \text{as} \quad Re \to \infty \qquad (2.28)$$

Although this behavior, corresponding to turbulent fluid motion, has not been predicted from first principles, the coefficient 0.16 has been found experimentally by many researchers [cf. Schlichting (1979)].

While we have made an important simplification in correlating the experimental data, we can extract additional useful information by manipulating the results of our dimensional analysis further. We might want to do this, for example, to focus on a specific range of Reynolds numbers where a discernible trend is evident.

Imagine that you are interested in low Reynolds number applications. Your experimental data clearly indicate a linear variation of $\rho F/\mu^2$ with Re. You can further verify this trend, including the theoretical coefficient 3π, by making measurements at even smaller Reynolds numbers. Furthermore, it would be advantageous to plot the ratio of $\rho F/\mu^2$ to Re versus Re, since the ratio asymptotes to a constant value as $Re \to 0$. This is permissible as the product of two dimensionless groupings remains dimensionless. Hence, the dimensionless force becomes

$$\frac{\rho F}{\mu^2} Re^{-1} = \frac{\rho F}{\mu^2} \frac{\mu}{\rho \, UD} = \frac{F}{\mu UD} \qquad (2.29)$$

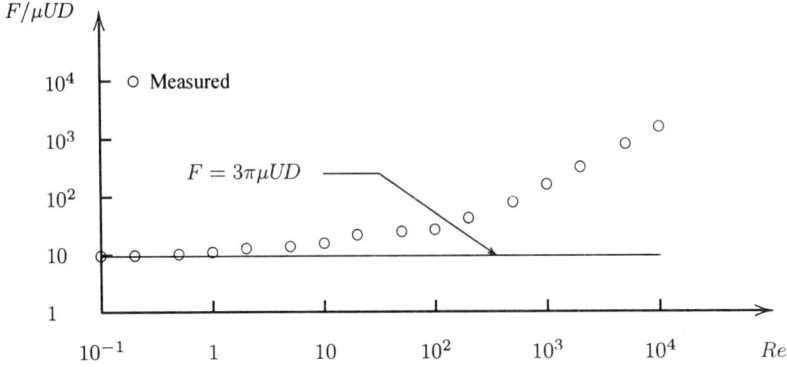

Figure 2.3: *Correlation of measured force on a sphere using $F/\mu UD$.*

Figure 2.3 shows the data plotted in terms of $F/\mu UD$. Clearly, the data tend toward the theoretical limit, thus providing confidence in the measured values. If you had no knowledge of the theoretical limiting value for $Re \rightarrow 0$ (which is normally the case for general applications), your measurements might provide a correlation that could be used for design purposes.

Finally, let's turn our attention to high Reynolds number applications, such as the flow past a tennis ball, a golf ball or a baseball. In this limit, your experimental data suggest that there is a quadratic variation of $\rho F/\mu^2$ with Re. Thus, plotting the ratio of $\rho F/\mu^2$ to Re^2 versus Re would appear to be appropriate. On a log-log plot, rescaling in this manner magnifies details of the shape of the curve in the limit $Re \rightarrow \infty$, just as the rescaling of Equation (2.29) magnifies details at low Reynolds number. Therefore, forming the desired ratio, the dimensionless force now becomes

$$\frac{\rho F}{\mu^2} Re^{-2} = \frac{\rho F}{\mu^2} \frac{\mu^2}{\rho^2 U^2 D^2} = \frac{F}{\rho U^2 D^2} \qquad (2.30)$$

Figure 2.4 presents a log-log plot of $F/\rho U^2 D^2$ as a function of Re. As shown in the figure, the force data tend toward a constant value of 0.16 as the Reynolds number increases.

As a concluding remark, there is often an element of nonuniqueness in dimensional analysis similar to what we have found for viscous flow past a sphere. In our analysis of the sphere, for example, all of the proposed dimensionless forms of the force are valid. With a bit of reflection, this should not come as a surprise. Recall that we had more unknowns than equations when we formulated the problem, so that we can hardly expect to find a unique solution. Although there usually is an optimum choice of the dimensionless groupings, we don't know which form is best until we actually correlate our data. This is, of course, taking the view that we know little about the problem at the start.

2.2.3   E. S. Taylor's Method

There is a useful alternative to the indicial method for determining dimensionless groupings developed by E. S. Taylor (1974). Briefly stated, it is a method of successive elimination of dimensions. It proceeds in several easy stages, with one important advantage over the indicial method. That is, an experienced person with knowledge of common dimensionless groupings can accelerate and guide the process. For example, an experienced fluid-flow engineer would expect Reynolds number, Re, to be one of the dimensionless groupings. If we had solved

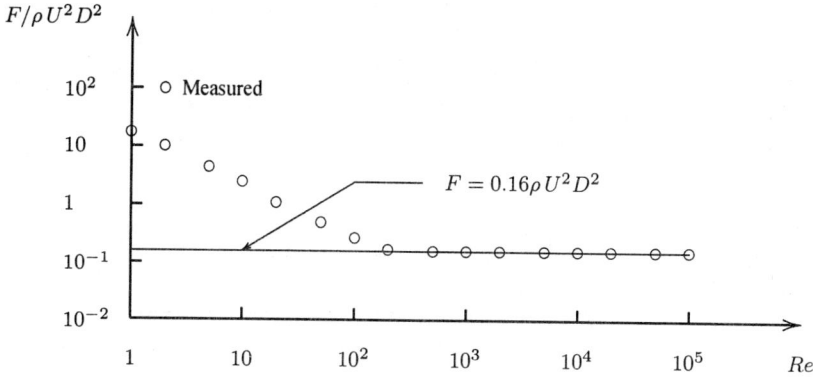

Figure 2.4: *Correlation of measured force on a sphere using* $F/\rho U^2 D^2$.

Equations (2.14) in a different order, we could arrive at $1/Re$ or $\sqrt{Re}$. This can be avoided completely with the Taylor method.

After using the Buckingham Π Theorem to determine the number of dimensionless group-ings, the first thing we do is write down a matrix whose rows each represent the dimensions of one of the quantities. In our example for flow past a sphere, we have

	M	L	T
μ	1	-1	-1
ρ	1	-3	0
U	0	1	-1
D	0	1	0
F	1	1	-2

The columns are the coefficients of a_1, a_2, a_3 and a_4 and the right hand sides in Equa-tions (2.14).

Now, we choose the simplest-looking column, in this problem the first, and divide all but one of the mass-containing quantities by that one. Division is accomplished by subtracting all elements of one row from those of another. Selecting ρ as the divisor, we rearrange our matrix by subtracting the second row from the first and the last. The rearranged matrix is

	M	L	T
μ/ρ	0	2	-1
ρ	1	-3	0
U	0	1	-1
D	0	1	0
F/ρ	0	4	-2

Note that ρ is now the only quantity that has mass as one of its dimensions, and therefore cannot appear in any dimensionless groups except as μ/ρ or F/ρ. Hence, we can delete it along with the resulting empty mass column to arrive at

	L	T
μ/ρ	2	-1
U	1	-1
D	1	0
F/ρ	4	-2

Inspection of the time column suggests dividing U by μ/ρ, which is clearly required as one step in forming the Reynolds number, $\rho U D/\mu$. Thus, we now eliminate the time dimension in all quantities except μ/ρ. To do this, we subtract the first row from the second row, and we subtract twice the first row from the last. The matrix simplifies to:

	L	T
μ/ρ	2	-1
$\rho U/\mu$	-1	0
D	1	0
$\rho F/\mu^2$	0	0

At this point, we eliminate μ/ρ from further consideration and delete the empty time column. All that remains at this point is a single column.

	L
$\rho U/\mu$	-1
D	1
$\rho F/\mu^2$	0

The 0 in the column next to $\rho F/\mu^2$ means this quantity is one of the dimensionless groupings. This is the same dimensionless form of the force we arrived at using the indicial method. Also, further inspection of the nonzero entries in the length column shows that multiplying $\rho U/\mu$ by D yields the second dimensionless grouping. We do this by adding the first and second rows. Thus, we conclude that the two dimensionless groupings are

$$P_0 = \frac{\rho F}{\mu^2} \quad \text{and} \quad P_1 = \frac{\rho U D}{\mu} \tag{2.31}$$

This is identical to what we obtained [see Equation (2.20)] using the indicial method.

2.3 Common Dimensionless Groupings

There are many dimensionless groupings that occur in fluid-mechanical applications. One of the customs that has evolved is to name these groupings after famous pioneers in the field. Some of the most frequently encountered follow.

- **Reynolds Number:** This grouping, named after Osborne Reynolds (1842-1912), provides a measure of the effects of viscous stresses on a flow. For most practical fluid flows, the Reynolds number is very large. It often appears with a subscript denoting the length that has been used. If fluid density is ρ, velocity is U, length is L and viscosity is μ, the Reynolds number is

$$Re_L = \frac{\rho U L}{\mu} \tag{2.32}$$

- **Mach Number:** This famous parameter honors the important contributions of Ernst Mach (1838-1916), a pioneer in compressible fluid dynamics. The Mach number is the ratio of flow velocity, U, to the speed of sound, a, viz.,

$$M = \frac{U}{a} \tag{2.33}$$

- **Froude Number:** Named after William Froude (1810-1879), this number provides a measure of the importance of gravity waves in a fluid. Normally used for liquids, it is analogous to the Mach number. As with the Reynolds number, it is customary to include the length used as a subscript. If flow velocity is U, gravitational acceleration is g and length is L, the Froude number is

$$Fr_L = \frac{U}{\sqrt{gL}} \tag{2.34}$$

Inspection of Equation (2.5) shows that a sailboat's theoretical hull speed corresponds to a Froude number of 0.40.

- **Prandtl Number:** When heat transfer is important, the number named in honor of Ludwig Prandtl (1875-1953) appears. If the specific heat at constant pressure of the fluid is c_p, viscosity is μ and **thermal conductivity** is k, then the Prandtl number is

$$Pr = \frac{c_p\mu}{k} \tag{2.35}$$

Unlike the other dimensionless parameters in our list, the Prandtl number is a property of the fluid alone, independent of the flow.

- **Strouhal Number:** This dimensionless grouping, named after Vincenz Strouhal (1850-1922), pertains to flows that vary with time. If the flow involves a periodic variation with frequency ω, for example, and if the flow velocity and length scale are U and L, respectively, the Strouhal number is

$$St = \frac{\omega L}{U} \tag{2.36}$$

- **Weber Number:** This number honors the research of Moritz Weber (1871-1951), who investigated flows in which surface tension, σ, is important. If the fluid density is ρ, flow velocity is U, and characteristic length is L, the Weber number is defined by

$$We = \frac{\rho U^2 L}{\sigma} \tag{2.37}$$

2.4 Dynamic Similitude

As mentioned in Subsection 2.2.2, dimensional analysis has a third important use that we have not yet discussed. We have seen how the method provides estimates of the appropriate scale and order of magnitude of flow properties. We have also demonstrated how powerful dimensional analysis is for correlating data. This section focuses on how we can use the results of dimensional analysis to extrapolate measurements made on small **models** to full-size objects, or **prototypes**.

To understand how we can accomplish such scaling, we must pause and consider how dimensional analysis relates to the physical equations governing a given problem. As an example, we use the differential equation discussed at the beginning of this chapter, namely, the Navier-Stokes equation. Recall that for one-dimensional motion of a viscous fluid, the equation is

$$\rho\frac{\partial u}{\partial t} + \rho u\frac{\partial u}{\partial x} = -\frac{\partial p}{\partial x} + \mu\frac{\partial^2 u}{\partial x^2} \tag{2.38}$$

where we assume the temperature is constant so that μ is constant. Suppose a small one-dimensional sound source of size L moving at a constant velocity U emits a tone with frequency ω. For simplicity, we assume the motion can be represented by Equation (2.38). We can rewrite this equation in terms of dimensionless variables if we define the following quantities:

$$\bar{u} = \frac{u}{U}, \quad \bar{t} = \omega t, \quad \bar{x} = \frac{x}{L}, \quad \bar{p} = \frac{p}{\rho_\infty U^2}, \quad \bar{\rho} = \frac{\rho}{\rho_\infty} \tag{2.39}$$

where ρ_∞ is the density of the fluid far from the sound source. Upon transforming, we arrive at

$$St\,\bar{\rho}\frac{\partial \bar{u}}{\partial \bar{t}} + \bar{\rho}\,\bar{u}\frac{\partial \bar{u}}{\partial \bar{x}} = -\frac{\partial \bar{p}}{\partial \bar{x}} + \frac{1}{Re}\frac{\partial^2 \bar{u}}{\partial \bar{x}^2} \tag{2.40}$$

where $St = \omega L/U$ and $Re = \rho_\infty U L/\mu$ are the Strouhal and Reynolds numbers, respectively. The equation, in this dimensionless form, applies to any fluid and frequency. It is ideally suited for numerical solution on a computer, and the only problem-specific input required for a given geometry would be St and Re, regardless of the individual values of the dimensional quantities in the problem.

Although we could have deduced these two dimensionless groupings with no knowledge of Equation (2.38), understanding the connection with the governing equations provides additional insight that has a direct bearing on the concept known as **dynamic similitude**. This simple example shows that when we perform a dimensional analysis, we are finding the dimensionless groupings that appear when the governing equations are cast in nondimensional form. Since flow past both a small-scale model and the prototype is governed by the same physical laws and associated equations, matching the dimensionless groupings produces exactly the same solution in terms of dimensionless variables. To obtain the actual flowfield for both cases requires using equations such as Equation (2.39) to arrive at dimensional properties. When all relevant dimensionless groupings match for a model and a prototype, the flows are said to be **dynamically similar**, or to stand in **dynamic similitude**.

To see how we establish dynamic similitude between a model and a prototype, we focus on two general classes of applications, viz., airplanes and ships. For the airplane, which is an **aerodynamic** vehicle, we must match the Reynolds number, Re, and the Mach number, M. As we will see in the following example, we come pretty close to exact dynamic similitude in modern **wind tunnels**. By contrast, for the ship, which is a **hydrodynamic** vehicle, we must match the Reynolds number and, in place of the Mach number, we must match the Froude number, Fr. We'll see that simultaneously matching both Re and Fr cannot be done on Earth primarily because we cannot change g. Thus, we never achieve exact similitude for a ship.

2.4.1 Similitude for an Airplane

First consider an airplane. The relevant dimensional quantities are fluid density, ρ, viscosity, μ, speed of sound, a, velocity, U, and length, L. We would like to extrapolate measured lift or drag force, F, for the model to the prototype. Performing dimensional analysis shows that there are three dimensionless groupings, viz.,

$$\frac{F}{\rho\,U^2 L^2}, \quad M = \frac{U}{a}, \quad Re = \frac{\rho\,U L}{\mu}$$

Hence, we conclude that the force behaves according to

$$F = \rho\,U^2 L^2 \mathcal{F}(M, Re) \tag{2.41}$$

We thus achieve dynamic similitude by matching M and Re, and use Equation (2.41) to extrapolate the measured force.

Letting subscript m denote the model and subscript p the prototype, matching Mach and Reynolds numbers means we require

$$\frac{U_p}{a_p} = \frac{U_m}{a_m} \quad \text{and} \quad \frac{\rho_p U_p L_p}{\mu_p} = \frac{\rho_m U_m L_m}{\mu_m} \tag{2.42}$$

where L is the total length of the airplane (or any other convenient distance). Although we can always use a different gas, we will assume our experiments are done in air. For air, it is a fact that over a wide range of temperatures, the viscosity and speed of sound vary with temperature according to[4]

$$\mu \propto T^{0.7} \quad \text{and} \quad a = \sqrt{\gamma R T} \propto T^{0.5} \tag{2.43}$$

So, matching Mach number means that the ratio of the wind-tunnel flow velocity to the prototype velocity is given by

$$\frac{U_m}{U_p} = \frac{a_m}{a_p} = \sqrt{\frac{T_m}{T_p}} \tag{2.44}$$

Also, matching Reynolds number means the ratio of the model length to the prototype length is

$$\frac{L_m}{L_p} = \frac{\rho_p}{\rho_m} \frac{U_p}{U_m} \frac{\mu_m}{\mu_p} = \frac{\rho_p}{\rho_m} \sqrt{\frac{T_p}{T_m}} \left(\frac{T_m}{T_p}\right)^{0.7} = \frac{\rho_p}{\rho_m} \left(\frac{T_m}{T_p}\right)^{0.2} \tag{2.45}$$

Therefore, we achieve dynamic similitude provided:

$$\frac{U_m}{U_p} = \sqrt{\frac{T_m}{T_p}} \quad \text{and} \quad \frac{L_m}{L_p} = \frac{\rho_p}{\rho_m} \left(\frac{T_m}{T_p}\right)^{0.2} \tag{2.46}$$

The weak dependence of L_m/L_p on temperature ratio means that we can almost independently control the model velocity and length by adjusting the air temperature and density, respectively. We have great flexibility with both quantities through heating or cooling the air supply and adjusting its pressure. For example, assume the prototype will fly at an altitude of 20 kilometers where the temperature is 218 Kelvins and the air density is 0.336 kg/m^3. If the air in the wind tunnel is at room temperature (293 Kelvins) and atmospheric pressure so that the density is 1.20 kg/m^3, we find

$$\frac{U_m}{U_p} = 1.16 \quad \text{and} \quad \frac{L_m}{L_p} = 1.06 \frac{\rho_p}{\rho_m} = 0.30 \tag{2.47}$$

Thus, maintaining atmospheric conditions in the tunnel, we can achieve exact dynamic similitude with a 30% scale model provided the air in the wind tunnel moves 16% faster than the prototype's velocity. In order to have a model one-tenth the size of the prototype, we can use a combination of pressurization and cooling of the tunnel's air supply to increase the density by about a factor of 3.

Because it is relatively easy to achieve dynamic similitude for an airplane, we can confidently use dimensional-analysis results to predict the forces and moments on prototypes based on measurements for a wind-tunnel model. This capability plays an important role in the fact

---

[4]The fact that $a = \sqrt{\gamma R T}$ is an exact theoretical result that we will derive in Chapter 8, while $\mu \propto T^{0.7}$ is purely empirical.

that the airplane is one of the most efficient and reliable vehicles ever built. In the present example, combining Equations (2.41) and (2.46) shows that the force scales according to

$$\frac{F_p}{F_m} = \frac{\rho_p}{\rho_m}\frac{U_p^2}{U_m^2}\frac{L_p^2}{L_m^2} = \frac{\rho_m}{\rho_p}\left(\frac{T_p}{T_m}\right)^{1.4} = 2.36 \tag{2.48}$$

2.4.2   Similitude for a Ship

For a ship, the situation is quite different. Consider a model to be tested in a **towing tank**. Dimensional analysis tells us the wave drag, W, for example, scales according to

$$W = \rho g L^3 \mathcal{F}(Fr, Re) \tag{2.49}$$

where Fr and Re are Froude and Reynolds numbers, respectively. Matching Froude numbers means

$$\frac{U_m}{\sqrt{gL_m}} = \frac{U_p}{\sqrt{gL_p}} \quad\Longrightarrow\quad \frac{U_m}{U_p} = \sqrt{\frac{L_m}{L_p}} \tag{2.50}$$

so that the ratio of model to prototype velocity decreases as the ratio of model to prototype size decreases.

   If we use water in the towing tank with no heating or cooling of the water in the tank (which would be very expensive), matching Reynolds numbers yields

$$\frac{\rho_m U_m L_m}{\mu_m} = \frac{\rho_p U_p L_p}{\mu_p} \quad\Longrightarrow\quad \frac{U_m}{U_p} = \frac{L_p}{L_m} \tag{2.51}$$

where we make use of the fact that the water density and viscosity are identical for the model and prototype. This tells us the ratio of model to prototype velocity increases as the ratio of model to prototype size decreases. Thus, the condition needed to match Reynolds number is the opposite of what is required to match Froude number. That is, matching both dimensionless groupings means

$$\sqrt{\frac{L_m}{L_p}} = \frac{L_p}{L_m} \quad\Longrightarrow\quad \left(\frac{L_p}{L_m}\right)^{3/2} = 1 \tag{2.52}$$

which is possible only if we have

$$L_p = L_m \tag{2.53}$$

Hence, using water at the same temperature as that of the prototype's environment, we can achieve dynamic similitude only with a full-scale model! We cannot practicably modify density or viscosity because inexpensive and safe liquids with a density greater than water are unavailable, and changing temperature of the water is expensive and difficult to control. If we could change g, we would have the same flexibility afforded in a wind tunnel where a different gas can be used. Unfortunately, the laws of gravitation prevent this. In practice, designers match Froude number only. This is not too bad if the model is large, say half size.

Problems

2.1 The *Dean number*, De, is pertinent for flow in curved pipes. It is defined by

$$De \equiv \frac{UR^{3/2}}{\nu \mathcal{R}^{1/2}}$$

where U is velocity, R is pipe radius, ν is kinematic viscosity and $\mathcal{R}$ is radius of curvature of the pipe. Verify that De is dimensionless.

2.2 The *Eckert number*, Ec, is relevant for flows with heat transfer, and is defined by

$$Ec \equiv \frac{U^2}{c_p \Delta T}$$

where U is velocity, c_p is specific-heat coefficient, and ΔT is temperature difference. Verify that Ec is dimensionless.

2.3 The gradient *Richardson number* occurs in density-stratified flows. It is defined by

$$Ri \equiv \frac{-g d\rho/dz}{\rho \left(dU/dz\right)^2}$$

where g is gravitational acceleration, ρ is density, U is velocity and z is vertical distance. Verify that Ri is dimensionless.

2.4 The *Grashof number*, Gr, is a key dimensionless grouping for buoyancy-driven flows defined by

$$Gr \equiv \frac{\beta g \ell^3 \Delta T}{\nu^2}$$

The quantities β, g, ℓ, ΔT and ν are the thermal-expansion coefficient, gravitational acceleration, length, temperature difference and kinematic viscosity, respectively. What are the dimensions of β?

2.5 The *Stanton number*, St, is a dimensionless grouping defined by

$$St \equiv \frac{\dot{h}}{\rho U c_p}$$

where $\dot{h}$ is the heat-transfer coefficient, ρ is density, U is velocity and c_p is specific-heat coefficient. What are the dimensions of $\dot{h}$?

2.6 The *Nusselt number*, Nu, is a dimensionless grouping defined by

$$Nu \equiv Pr \frac{s\dot{h}}{\mu c_p}$$

where Pr is the Prandtl number, $\dot{h}$ is the heat-transfer coefficient, s is arc length along a surface, μ is viscosity and c_p is specific-heat coefficient. What are the dimensions of $\dot{h}$?

2.7 What would the constant 1.34 in Equation (2.3) have to be replaced by to be valid for U_{hull} in knots and L_w in meters?

2.8 The shape of a hanging drop of liquid satisfies the following empirical equation.

$$\frac{\left(\rho - \rho_o\right) g d^3}{C} = \sigma$$

The quantities ρ and ρ_o are the density of the liquid and the surrounding air. Also, g is gravitational acceleration, d is the drop's diameter at its equator, σ is surface tension, and C is an empirical constant. What are the dimensions of C?

2.9 Many formulas from the hydraulics literature involve empirical coefficients determined from experiments with a single fluid. An example is the *Hazen-Williams* formula for volume-flow rate, Q, (dimensions L^3/T) in a pipe, as a function of pipe diameter, D, and pressure gradient, dp/dx:

$$Q = 61.9D^{2.63} \left(\frac{dp}{dx} \right)^{0.54}$$

(a) What are the dimensions of the constant 61.9?

(b) Suppose we rewrite this equation in terms of a revised empirical coefficient $C = 61.9\rho^{0.54}$, where ρ is the density of water. Show that C is only weakly dependent on dimensions, i.e., it is almost a dimensionless coefficient.

2.10 The classical *Stokes-Oseen equation* for the drag, F, in low Reynolds number flow past a sphere is

$$F = 3\pi\mu UD + \frac{9}{16}\pi\rho U^2 D^2$$

where ρ, μ, U and D are density, viscosity, velocity and sphere diameter, respectively.

(a) Is this equation dimensionally homogeneous?

(b) Rewrite this equation in terms of the Reynolds number, $Re = \rho UD/\mu$, and the drag coefficient, $C_D \equiv F/(\frac{1}{2}\rho U^2 A)$, where $A = \frac{\pi}{4}D^2$.

2.11 The following empirical differential equation can be used to solve for the velocity, u, in turbulent channel flow.

$$\nu \frac{d^2u}{dy^2} + \frac{d}{dy} \left(\ell_{mix}^2 \left| \frac{du}{dy} \right| \frac{du}{dy} \right) = \frac{1}{\rho} \frac{dp}{dx}$$

The quantities ν, ρ and p denote kinematic viscosity, density and pressure, respectively. Also, x and y are distance parallel to and normal to the channel walls. For this equation to be dimensionally homogeneous, what must the dimensions of ℓ_{mix} be?

2.12 The following empirical differential equation determines a quantity ω known as the specific dissipation rate in a one-dimensional, time-dependent flow.

$$\frac{\partial\omega}{\partial t} + u\frac{\partial\omega}{\partial x} = \alpha \left(\frac{\partial u}{\partial x} \right)^2 - \beta\omega^2 + \frac{\partial}{\partial x} \left(\nu \frac{\partial\omega}{\partial x} \right)$$

The quantity x is spatial distance and t is time. Also, u denotes velocity, ν is kinematic viscosity and β is a dimensionless coefficient. For this equation to be dimensionally homogeneous, what must the dimensions of ω and α be?

2.13 For a perfect gas, the speed of sound, a, is known to be a function of pressure, p, and density, ρ. Verify that the Buckingham Π Theorem implies there are no dimensionless groupings. Show that the indicial equations are linearly dependent, and verify that

$$a = \text{constant} \cdot \sqrt{p/\rho}$$

2.14 For turbulent flow close to a solid boundary, the fluctuating velocities scale with the *friction velocity*, u_τ. The friction velocity is a function of the surface shear stress, τ_w, and density, ρ. Verify that the Buckingham Π Theorem implies there are no dimensionless groupings. Show that the indicial equations are linearly dependent, and (choosing the constant of proportionality to be 1) verify that

$$u_\tau = \sqrt{\tau_w/\rho}$$

2.15 For a calorically-perfect gas, the specific internal energy, e, whose dimensions are L^2/T^2, is known to be a function of the perfect-gas constant, R, and temperature, T. Verify that the Buckingham Π Theorem implies there are no dimensionless groupings. Show that the indicial equations are linearly dependent, and verify that

$$e = \text{constant} \cdot RT$$

2.16 The energy, E, of an electron in orbit about the nucleus of an atom is a function of its rest mass, m_o, its charge, q, the permittivity of free space, ϵ_o, and Planck's constant, h. The dimensions of h and ϵ_o are

$$[h] = \frac{ML^2}{T} \quad \text{and} \quad [\epsilon_o] = \frac{Q^2T^2}{ML^3}$$

where Q denotes the independent dimension of electrical charge. Using dimensional analysis, deduce a formula for E as a function of m_o, q, ϵ_o and h.

2.17 For flow through a horizontal tube, the axial pressure gradient, dp/dx, is a function of the viscosity, μ, mean velocity, U_m, and tube diameter, D. Using dimensional analysis, deduce a formula for dp/dx as a function of μ, U_m and D.

2.18 For flow in a rotating tank of liquid, the radial pressure gradient, dp/dr, is a function of the density, ρ, angular velocity, ω, and radial distance from the center of the tank, r. Using dimensional analysis, deduce a formula for dp/dr as a function of ρ, ω and r.

2.19 Waves of very short wavelength at the interface between a liquid and a gas are known as *capillary waves* or *ripples*. This type of wave is produced when a small object is dropped into a pool of water. The propagation speed, c, is a function of the liquid density, ρ, the surface tension, σ, and the wavelength, λ. Deduce a formula for c as a function of ρ, σ and λ.

2.20 Many relatively simple turbulent flows can be represented by introducing a kinematic *eddy viscosity*, ν_t, in analogy to laminar flow. One empirical model for the eddy viscosity assumes it is a function of the specific turbulence kinetic energy, k, (dimensions L^2/T^2) and the rate of dissipation of energy, ϵ (dimensions L^2/T^3). Develop a formula for ν_t as a function of k and ϵ.

2.21 Many relatively simple turbulent flows can be represented by introducing a kinematic *eddy viscosity*, ν_t, in analogy to laminar flow. One empirical model for the eddy viscosity assumes it is a function of the specific turbulence kinetic energy, k, (dimensions L^2/T^2) and the specific rate of dissipation of energy, ω (dimensions $1/T$). Develop a formula for ν_t as a function of k and ω.

2.22 Kolmogorov argued that for high Reynolds number, in terms of Fourier analysis, the turbulence energy spectrum contains an *inertial subrange*. He further argued that in the inertial subrange, the turbulence-energy spectrum function, $E(\kappa)$, (dimensions L^3/T^2) depends only upon the dissipation rate, ϵ, (dimensions L^2/T^3) and the wave number, κ (dimensions $1/L$). Develop a formula for $E(\kappa)$ as a function of ϵ and κ.

2.23 When a large explosion occurs in the atmosphere, it creates a very strong spherical wave known as a *blast wave*. If the energy released is E (dimensions ML^2/T^2) and the ambient density is ρ_o, develop an equation for the radius of the blast wave, $R(t)$, as a function of E, ρ_o and time, t.

2.24 The centerline velocity in a turbulent *plane jet* of fluid blowing into a stationary fluid from a small orifice is $u_m(x)$. Very far from the orifice, u_m is a function only of the specific momentum flux, J, (dimensions L^3/T^2) and distance from the origin, x. Deduce a formula for $u_m(x)$ as a function of J and x.

2.25 In *magnetohydrodynamic (MHD)* flows (i.e., flows in which the fluid is electrically conducting), fluid-mechanics equations must be coupled with Maxwell's equations of electricity and magnetism. For sufficiently low frequencies, we introduce Ampere's law and Ohm's law in which the magnetic permeability of free space, μ_o, and electrical conductivity, σ, appear, where

$$[\mu_o] = \frac{\Omega T}{L} \quad \text{and} \quad [\sigma] = \frac{1}{\Omega L}$$

The quantity Ω denotes the independent dimension of electrical resistance. For an incompressible, inviscid MHD flow with density ρ, use dimensional analysis to deduce the pertinent dimensionless parameters. Assume U and L are the appropriate velocity and length scales, and that the force on a body is F. How many independent dimensionless groupings are there? What are the dimensionless groupings?

2.26 We have studied Couette flow of a viscous fluid when one wall moves with a constant velocity, U, parallel to a stationary wall a distance h away. Consider the unsteady case where the upper wall starts moving from rest at time $t = 0$, with a constant velocity U. Assuming the velocity can be written as $u/U = f(y, \mu, t, h, \rho)$, determine the relevant dimensionless groupings.

2.27 Under certain conditions, liquid droplets in a moving gas will break up. The relevant dimensional quantities for droplet break up are the gas density, ρ, velocity, U, droplet diameter, D, gas viscosity, μ, and surface tension, σ. How many dimensionless groupings are there? What are the dimensionless groupings?

2.28 At very high temperatures, *quantum-mechanical* effects become important and the internal energy, e, of a gas becomes a function of the perfect-gas constant, R, temperature, T, Planck's constant, h, Boltzmann's constant, k, and the fundamental vibration frequency of the molecule, ν. In SI units, $h = 6.625 \cdot 10^{-34}$ J · sec and $k = 1.380 \cdot 10^{-13}$ J/K. How many independent dimensionless groupings are there? What are the dimensionless groupings?

2.29 For natural convection above a mildly heated, infinite flat plate, the difference between the plate temperature and ambient temperature, ΔT, is thought to depend upon the specific-heat coefficient, c_p, thermal conductivity, k, viscosity, μ, fluid density, ρ, and gravitational acceleration, g. The units of c_p are ft²/(sec²·°R), while the units of k are slug·ft/(sec³·°R). How many independent dimensionless groupings are there? What are the dimensionless groupings?

2.30 For flow through a smooth straight pipe, the pressure gradient, dp/dx, is a function of viscosity, μ, velocity, U, pipe diameter, D, and density, ρ. How many independent dimensionless groupings are there? What are the dimensionless groupings? (This is one of the original problems discussed by Buckingham.)

2.31 For flow in a curved pipe of constant radius, R, the critical dimensional quantities are velocity, U, kinematic viscosity, ν, and radius of curvature, $\mathcal{R}$. How many independent dimensionless groupings are there? What are the dimensionless groupings? Manipulate your answer to show that one of the dimensionless groupings is the Reynolds number.

2.32 An air-hockey game uses a porous surface with air injected at a rate $\dot{m}$ (dimensions M/T). A puck of mass M moves at velocity U while floating at a distance h above the surface on a cushion of air. The kinematic viscosity of air is ν. How many independent dimensionless groupings are there? What are the dimensionless groupings? Manipulate your answer to show that one of the dimensionless groupings is the Reynolds number based on h.

2.33 The pressure rise, Δp, through a centrifugal pump is a function of pump diameter, D, angular speed, Ω, (dimensions $1/T$) volume-flow rate, Q, (dimensions L^3/T) and fluid density, ρ. Using dimensional analysis, develop an equation for Δp as the product of a quantity with the same dimensions as Δp and a function of all relevant dimensionless groupings.

2.34 The torque, $\mathcal{T}$, on an axial-flow turbine is a function of fluid density, ρ, rotor diameter, D, angular rotation rate, Ω, (dimensions $1/T$) and volume-flow rate, Q (dimensions L^3/T). Using dimensional analysis, develop an equation for $\mathcal{T}$ as the product of a quantity with the same dimensions as $\mathcal{T}$ and a function of all relevant dimensionless groupings.

2.35 For flow of a viscous fluid close to a solid boundary rotating with angular velocity Ω (dimensions $1/T$), a region known as the *Ekman layer* develops. The thickness of the Ekman layer, δ, depends upon Ω, the freestream velocity, U, and the kinematic viscosity of the fluid, ν. Using dimensional analysis, develop an equation for δ as the product of a quantity independent of U with the same dimensions as δ and a function of all relevant dimensionless groupings.

2.36 Because of a phenomenon known as *vortex shedding*, a flagpole will oscillate at a frequency ω when the wind blows at velocity U. The diameter of the flagpole is D and the kinematic viscosity of air is ν. Using dimensional analysis, develop an equation for ω as the product of a quantity independent of ν with the same dimensions as ω and a function of all relevant dimensionless groupings.

2.37 The relevant dimensional quantities for flow between the read-write head of a floppy-disk drive and the disk surface are the kinematic viscosity of air, ν, the disk rotation rate, Ω, (dimensions $1/T$) the distance between the head and the disk, h, and the average height of disk surface irregularities, k_s. What are the dimensionless parameters most appropriate for developing an empirical correlation for the flow velocity, U, between the head and the disk?

2.38 We wish to correlate the drag, D, on a body of characteristic length L moving at velocity U. It is moving through a viscous fluid with viscosity μ and density ρ. The problem is further complicated by occurring in a coordinate frame rotating with angular velocity Ω (dimensions $1/T$). How many independent dimensionless groupings are there? What are the dimensionless groupings?

2.39 For *inviscid* flow past a body in a rotating, stratified fluid, the significant dimensional quantities are density, ρ, gravitational acceleration, g, coordinate-system rotation rate, Ω, (dimensions $1/T$) flow velocity, U, and length of the body, L. How many independent dimensionless groupings are there? What are the dimensionless groupings?

2.40 For *inviscid* flow past an oscillating body in a rotating fluid, the significant dimensional quantities are density, ρ, oscillation frequency, ω, coordinate system rotation rate, Ω, (dimensions $1/T$) flow velocity, U, and length of the body, L. How many independent dimensionless groupings are there? What are the dimensionless groupings?

2.41 The vorticity, ω, (dimensions $1/T$) at a point in an axisymmetric flowfield depends on initial circulation, Γ_o, (dimensions L^2/T) radius, r, time, t, and kinematic viscosity of the fluid, ν. We wish to find a set of dimensionless parameters for organizing experimental data. How many dimensionless groupings are there? What are the dimensionless groupings?

2.42 In turbulent flow over a rough surface with mass injection, the velocity, U, is thought to be a function of the surface shear stress, τ_w, density, ρ, surface roughness height, k_s, surface-injection velocity, v_w, and viscosity, μ. How many independent dimensionless groupings are there? What are the dimensionless groupings? Express your answer in terms of $u_\tau \equiv \sqrt{\tau_w/\rho}$.

2.43 The load-carrying capacity, W, (a force) of a journal bearing is known to depend on three lengths, viz., bearing diameter, D, length, ℓ, and clearance, c. Additionally, W depends upon angular speed, ω, (dimensions $1/T$) and viscosity of the lubricant, μ. How many dimensionless groupings are there? What are the dimensionless groupings?

2.44 We wish to correlate the drag force, F, on a rough sphere of diameter D. The freestream velocity is U, density is ρ, viscosity is μ, and the height of a roughness element on the sphere is k_s. How many dimensionless groupings are there? What are the dimensionless groupings?

2.45 For natural convection above a strongly heated plate, the difference between plate temperature and ambient temperature, ΔT, depends upon ambient temperature, T_∞, specific-heat coefficient, c_p, thermal conductivity, k, viscosity, μ, fluid density, ρ, and gravitational acceleration, g. The units of k are slug·ft/(sec^3·°R). How many independent dimensionless groupings are there? What are the dimensionless groupings? Manipulate your answer so that one of the groupings is the Prandtl number and so that only the Prandtl number involves k.

2.46 For flow above a heated plate, the difference between plate temperature and ambient temperature, ΔT, depends upon ambient temperature, T_∞, specific-heat coefficient, c_p, thermal conductivity, k, viscosity, μ, fluid density, ρ, and freestream velocity, U. The units of k are kg·m/(sec^3·K). How many independent dimensionless groupings are there? What are the dimensionless groupings? Manipulate your answer so that one of the groupings is the Prandtl number and so that only the Prandtl number involves k.

2.47 For inviscid flow approaching a stagnation point, the surface temperature, T_w, depends upon freestream temperature, T_∞, the specific-heat coefficients, c_p and c_v, freestream velocity, U, and speed of sound, a. Show that the indicial equations are linearly dependent and that the surface temperature can be written as $T_w = T_\infty F(\gamma, M, a/\sqrt{c_p T_\infty})$, where $\gamma = c_p/c_v$ and $M = U/a$.

2.48 The head, Δh_d, (dimensions L) developed by a turbomachine depends upon the rotor diameter, d, rotational speed, N, (dimensions $1/T$) volume-flow rate, Q, (dimensions L^3/T) kinematic viscosity, ν, and the acceleration due to gravity, g. How many independent dimensionless groupings are there? What are the dimensionless groupings?

2.49 A thin layer of spherical particles rests on the bottom of a horizontal tube. When an incompressible fluid flows through the tube, we observe that at some critical velocity, U_c, the particles will rise and be transported along the tube. Assume the critical velocity to be a function of pipe diameter, D, particle diameter, d, fluid density, ρ, fluid viscosity, μ, particle density, ρ_p, and the acceleration of gravity, g. How many independent dimensionless groupings are there? What are the dimensionless groupings?

2.50 The power required to drive a propeller, P, is known to depend upon the freestream velocity, U, fluid density, ρ, speed of sound, a, angular velocity of the propeller, ω, (dimensions $1/T$) diameter of the propeller, D, and the viscosity of the fluid, μ. How many independent dimensionless groupings are there? What are the dimensionless groupings? Manipulate your answer to show that three of the dimensionless groupings are the Mach, Strouhal and Reynolds numbers.

2.51 The thrust, T, (a force) of a screw propeller is known to be a function of its diameter, D, rate of revolution, n, (dimensions $1/T$) speed of advance, U, fluid density, ρ, kinematic viscosity, ν, and gravitational acceleration, g. How many independent dimensionless groupings are there? What are the dimensionless groupings? Manipulate your answer to show that three of the dimensionless groupings are the Strouhal, Reynolds and Froude numbers. (This is one of the original problems discussed by Buckingham.)

2.52 Modern helicopters use tip injection to enhance power and control. The thrust, T, (a force) of such a helicopter blade is a function of span, ℓ, (a length), rotation rate, Ω, (dimensions $1/T$), translation speed, U, the speed of sound, a, fluid density, ρ, and tip injection velocity, V_t. How many independent dimensionless groupings are there? What are the dimensionless groupings? Manipulate your answer to show that two of the dimensionless groupings are the Strouhal and Mach numbers.

2.53 We have a 1:20 scale model of an aircraft that will fly at very low Mach number, so low that we can consider the flow to be incompressible. For wind-tunnel testing, we cool the air to the point where its kinematic viscosity, ν, is half the value of normal atmospheric air at the intended altitude where the aircraft will fly. To guarantee dynamic similitude, what is the required ratio of the velocity in the wind tunnel to the velocity of the full-scale aircraft?

2.54 A 1:16 scale model of a submarine is tested in a highly pressurized wind tunnel with $p = 20$ atm and $T = 100^\circ$ F. The prototype submarine will move at $U_p = 12$ mph in seawater, whose kinematic viscosity is $\nu_p = 1.12 \cdot 10^{-5}$ ft^2/sec.

(a) Assuming dynamic similitude can be achieved by matching just the Reynolds number, what must the wind-tunnel flow speed be?

(b) If the speed of sound of air at 100° F is 791 mph, what is the wind-tunnel Mach number?

2.55 An airplane wing with a chord length of $c_p = 1.5$ m, and a span of $S_p = 9$ m has been designed for a cruise speed of $U_p = 70$ m/sec. A 1:10 scale model will be tested in a water channel.

(a) If the model fluid and the prototype fluid both have a temperature of 20° C, what must the water-channel speed, U_m, be to match model and prototype Reynolds numbers?

(b) For the conditions of Part (a), what are the Mach numbers for the prototype and the model?

2.56 To determine the drag of a blimp, tests will be done on a 1:24 scale model in a water channel. The full-scale blimp will move at 20 ft/sec in 68° F air. The water in the channel will also be at 68° F.

(a) What must the water-channel velocity be to establish dynamic similitude? Ignore Mach and Froude numbers in arriving at your answer.

(b) If the measured drag is 500 lb, what is the drag on the full-scale blimp?

(c) How much power, in horsepower, is required to move the blimp at this speed?

2.57 Wind-tunnel tests show that the drag force on a 1:4 scale model of an automobile is 750 N. The prototype will move at 60 km/hr in air at the same air conditions.

(a) What must the wind-tunnel velocity be to achieve dynamic similitude? Ignore Mach number in arriving at your answer.

(b) What will the drag force on the prototype be?

(c) How much power, in horsepower, is required to move the automobile at this speed?

2.58 A 1:10 scale model of a tractor-trailer is tested in a pressurized wind tunnel with air at a density $\rho_m = .00585$ slug/ft^3 and a temperature of 68° F. The height of the model is $H_m = 1.2$ ft and the flow speed is $U_m = 240$ ft/sec. The measured drag is $F_m = 72$ lb.

(a) Assuming that the Mach number is unimportant and that the prototype moves in air at 68° F and 1 atm, verify that to achieve dynamic similitude, the condition $F = \rho U^2 H^2 f(Re_H)$ implies $\rho_p F_p = \rho_m F_m$.

(b) Determine the prototype velocity in mph and the prototype drag force in lb.

(c) How much power, in horsepower, is required to move the tractor trailer at this speed?

2.59 An airplane is designed to fly at a speed of 600 mph at an elevation of 30,000 ft, where the temperature is -56° F and the pressure is 4 psi. A 1:10 scale wind-tunnel model will be tested at a temperature of 70° F. The viscosity of air at 30,000 ft is approximately $3 \cdot 10^{-7}$ slug/(ft·sec). It is $3.8 \cdot 10^{-7}$ slug/(ft·sec) in the wind tunnel. Noting that the speed of sound of air varies as $a \propto \sqrt{T}$, find the wind-tunnel velocity (in mph) and pressure (in atm) required to match both Mach and Reynolds numbers of the prototype.

2.60 Wind-tunnel tests are proposed for a full-scale golf ball in a tunnel whose maximum flow speed is $U_m = 180$ ft/sec. The temperature of the tunnel air supply can be adjusted to change the viscosity, μ_m. A professional golfer can hit a ball at $U_p = 240$ ft/sec with an angular spin rate of $\omega_p = 24$ sec^{-1}. Dimensional analysis shows that dynamic similitude can be achieved by matching Strouhal and Reynolds numbers,

$$St \equiv \omega D/U, \qquad Re \equiv \rho UD/\mu$$

where ρ is fluid density and $D = 1.68$ inches is the diameter of a golf ball.

(a) If the air density for model and prototype is the same, what must ω_m and μ_m be?

(b) Assuming $\mu \propto T^{0.7}$ and $T_p = 80°$ F, what must the wind-tunnel temperature be?

2.61 A 225:1 scale model of a harbor breakwater is constructed for testing. The object of the tests is to determine effects of tides and waves generated by a storm. Typical storm waves have an amplitude, h_p, of 5 ft and travel with a velocity, U_p, of 30 ft/sec. Dimensional analysis shows that dynamic similitude can be achieved by matching Froude number and h/H, where H is breakwater height.

(a) For identical model and prototype fluid properties, what must the amplitude and speed of the model waves be?

(b) To simulate effects of tides, we must match Strouhal number defined by

$$St \equiv \frac{H}{U\tau}$$

where τ is the tidal period. If the time between tides is $\tau_p = 12$ hours, what must the model period, τ_m, be?

2.62 An experiment is being conducted on Earth to help design a water-channel that will be constructed on the moon, where the gravitational acceleration is one-fifth that on Earth. The experiments match both Reynolds and Froude numbers using water at the prototype temperature. The prototype channel will be $L_p = 2$ ft wide and will have a flow speed of $U_p = 5$ ft/sec. Determine the width and flow speed of the model channel.

2.63 A 1:9 scale model of a ship's screw propeller will be tested in water at the same temperature that will be encountered in practice. The flow speed past the model screw is $U_m = 2$ m/sec and its rotation rate is $N_m = 540$ rpm. Its diameter is $D_m = 30$ cm. Dimensional analysis indicates the Froude number, Fr, and Strouhal number, St, must be matched to achieve dynamic similitude, where

$$Fr \equiv U/\sqrt{gD} \qquad \text{and} \qquad St \equiv ND/U$$

(a) What are the corresponding forward velocity and rotation rate of the prototype screw?

(b) For the conditions of Part (a) and $T = 20°\,$C, what are the Reynolds numbers based on D for the model and the prototype?

2.64 A graduate student performs wind-tunnel studies of a 1:25 scale submarine model. The goal of the experiments is to simulate conditions when the submarine moves at 10 knots in seawater at 40° F, for which the kinematic viscosity is $\nu_p = 1.66 \cdot 10^{-5}$ ft²/sec. The wind tunnel operates at 68° F and 1 atm.

(a) The graduate student's experiments match Reynolds numbers for model and prototype. What is the wind tunnel flow speed, U_m?

(b) Noting that the speed of sound in the wind tunnel is $a_m = \sqrt{\gamma R T_m}$ (T_m is absolute temperature, γ is specific-heat ratio and R is the perfect-gas constant), compute the Mach number in the wind tunnel. Can you explain why the student's thesis adviser is beside himself?

2.65 An inexperienced recent graduate has conducted a wind-tunnel experiment on a 1:20 scale model of a missile designed to fly at 400 mph. The wind-tunnel temperature and pressure were the same as those for the low-flying model, and the tunnel flow speed was adjusted to match model and prototype Reynolds numbers.

(a) What was the wind-tunnel flow speed?

(b) The wind-tunnel temperature, T_m, was 60° F so that the speed of sound in the tunnel was $a_m = 1{,}119$ ft/sec. Noting that the peak temperature on the model was given by

$$T_{max} = T_m \left[1 + \frac{\gamma - 1}{2} r M_m^2 \right]$$

where T_m is absolute temperature, M_m is Mach number, $\gamma \approx 1.2$ and $r \approx 0.85$, explain why the aluminum model vaporized during the first test. **HINT:** Aluminum boils at 2057° C.

2.66 Experiments are planned for a small one-man submarine that uses surface heating to maintain laminar flow, and thus permit high enough velocity to outrun torpedoes due to its very low drag force. Pertinent flow parameters are the drag force, F, water density, ρ, velocity, U, submarine volume, V, freestream viscosity, μ_e, and viscosity corresponding to the surface temperature, μ_w.

(a) Using dimensional analysis, verify that the drag force can be written as

$$F = \rho U^2 V^{2/3} f(Re, \mu_w/\mu_e), \qquad Re \equiv \rho U V^{1/3}/\mu_e$$

(b) Experiments will be done on a 1:3 scale model with freestream and surface temperatures the same as will be encountered by the prototype. The prototype has a volume of $V_p = 600$ ft³, and will move in fresh water at 25 knots. If the experiments are also done in fresh water, what must the model velocity, U_m, be?

2.67 A model of a centrifugal pump that will be used in a nuclear-reactor cooling system is 5 times larger than the prototype. The prototype coolant is liquid sodium. For the intended operating conditions, the kinematic viscosity of liquid sodium is $\nu_p = 3.16 \cdot 10^{-7}$ m²/sec. The prototype pump has an angular rotation speed of $N_p = 1800$ rpm and a volume-flow rate of $Q_p = 0.0278$ m³/sec. The model fluid is water at 20° C.

(a) Assuming the pertinent dimensional quantities are ν, Q, N and impeller diameter, D, determine the appropriate dimensionless groupings required for dynamic similitude.

(b) What are the model volume-flow rate, Q_m, and angular rotation speed, N_m?

2.68 Ignoring viscosity, the wave resistance on a ship, R, is known to be a function of density, ρ, gravitational acceleration, g, velocity, U, and length, L.

(a) Determine the appropriate dimensionless groupings.

(b) The wave resistance of a model of a ship at 1:36 scale is 2 lb at a model speed of 4 ft/sec. What are the corresponding velocity and wave resistance of the prototype?

2.69 Consider the experimental determination of the force acting on a plate that is being dragged through a fluid. The force on the plate per unit width (into the page), F, is known to depend upon fluid density, ρ, fluid viscosity, μ, plate length, L, plate thickness, t, and the plate velocity, U.

(a) How many dimensionless groupings are there? Determine the dimensionless groupings. **NOTE:** The quantity F has dimensions of force/length.

(b) The prototype and model fluids have the following properties.

Property	Prototype	Model
$\rho \left(\frac{kg}{m^3}\right)$	800	1000
$\mu \left(\frac{kg}{m \cdot sec}\right)$	$3.0 \cdot 10^{-2}$	$1.5 \cdot 10^{-3}$

We use the same plate in the model experiment as we are interested in for the prototype fluid. If the prototype velocity of interest is 25 m/sec, what should the model velocity, U_m, be to have similitude between the prototype and the model?

(c) The experiment is performed with the model fluid at velocity U_m determined in Part (b), and the force per unit width is found to be $F_m = 10$ N/m. What is the corresponding force per unit width, F_p, for the plate moving in the prototype fluid at the prototype velocity?

2.70 An oil-drilling platform is planned for a part of the ocean where average currents are $U = 0.5$ ft/sec. Typical waves have a height of 8 ft and a frequency of $\omega = 0.5$ sec^{-1}. Experiments will be conducted on a 1:16 scale model.

(a) If the width of the platform is L, verify that the pertinent dimensionless groupings are:

$$St = \frac{\omega h}{U}, \quad Fr = \frac{U}{\sqrt{gh}}, \quad \frac{h}{L}$$

(b) What current speed, wave height and wave frequency must be used in the experiments in order to establish dynamic similitude?

Chapter 3

Effects of Gravity on Pressure

Our first topic as we begin our detailed exploration of fluid mechanics concerns fluids that are at rest, or in static equilibrium. The only force acting in a static fluid is the pressure, p. Even in the absence of fluid motion, the presence of a gravitational field gives rise to some interesting phenomena. On the one hand, we learned in Chapter 1 that pressure acts equally in all directions at a given point in a fluid. On the other hand, the gravitational field causes a **pressure gradient**, and we find that pressure increases with depth in a fluid. There is no contradiction as the former observation applies to a single point, while gravitational effects are felt over a finite distance. This chapter focuses on fluid-pressure variation in a gravitational field for a static fluid. For the obvious reason, this is often referred to as **fluid statics**.

One of the most interesting formulas we will develop is the **hydrostatic relation** for pressure. This relation shows why pressure increases with depth in the ocean and decreases with altitude in the atmosphere. It can also be used to measure the pressure at any point in a stationary fluid indirectly by simply measuring the height of a column of fluid.

We will develop a method for computing forces and moments on both planar and curved surfaces. With a clever balance of forces, we will replace the problem of integrating a varying pressure over a general curvilinear surface with a simple set of algebraic operations. We will accomplish this by deriving formulas for a point force and moment arm that give the mechanical equivalent of the force and moment due to the distributed pressure. This method is useful for determining the loads on a structure such as a dam or a flow-control valve.

We conclude the chapter with a discussion of buoyancy and a famous result known as **Archimedes' Principle**. This principle says the buoyancy force on a floating or submerged body is equal to the weight of the displaced fluid. The section on buoyancy includes a simple experiment you can perform at home.

3.1 The Hydrostatic Relation

For a fluid at rest in a gravitational field, the only forces acting are gravity and the pressure within the fluid. In the absence of gravity, the pressure in the fluid would obviously be the same at all points. Thus, the presence of a gravitational field must give rise to a variation in pressure. In this section, we will determine how the pressure varies using two methods. In the first derivation, we will use a simple geometrical shape to help illuminate the physics, albeit in a contrived configuration. Then, we will derive the result for a completely general geometry.

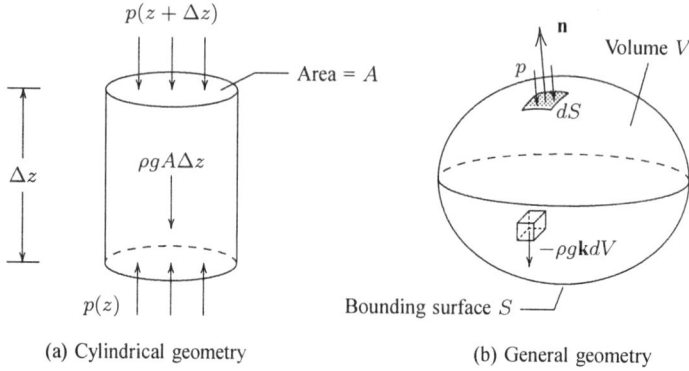

Figure 3.1: *Fluid volumes in a gravitational field.*

Consider the cylinder of fluid shown in Figure 3.1(a). The forces acting vertically are the pressure on the top and bottom of the cylinder and the weight of the fluid within. If ρ is fluid density and g is acceleration of gravity (9.807 m/sec^2, 32.174 ft/sec^2), the weight of the fluid is $\rho g A \Delta z$, where z is measured upwards. Balancing forces, we have

$$p(z + \Delta z)A + \rho g A \Delta z - p(z)A = 0 \tag{3.1}$$

which can be rewritten as

$$\frac{p(z + \Delta z) - p(z)}{\Delta z} = -\rho g \tag{3.2}$$

Thus, taking the limit $\Delta z \to 0$,

$$\frac{dp}{dz} = -\rho g \tag{3.3}$$

Although we considered a special geometry, this result is completely general. To see this, consider the arbitrary volume V bounded by a closed surface S shown in Figure 3.1(b). If $\mathbf{n}$ is the outer unit normal to the differential surface element dS, the net pressure force exerted *by the surroundings on the surface* is the closed surface integral of $-p\,\mathbf{n}$ (the minus sign makes it an inward directed force). That is, the net pressure force acting on the volume is

$$\textbf{Pressure Force} = - \oiint_S p\, \mathbf{n}\, dS \tag{3.4}$$

If the density of the fluid contained in the volume is ρ and the gravitational acceleration vector is given by $-g\mathbf{k}$ where $\mathbf{k}$ is a unit vector in the z direction, the weight of the fluid is the volume integral of $-\rho g \mathbf{k}$, i.e.,

$$\textbf{Weight} = - \iiint_V \rho g \mathbf{k}\, dV \tag{3.5}$$

Balancing forces we have

$$\oiint_S p\, \mathbf{n}\, dS + \iiint_V \rho g \mathbf{k}\, dV = \mathbf{0} \tag{3.6}$$

Using the divergence theorem (see Appendix C), we can convert the pressure integral to a volume integral. Combining the two resulting volume integrals yields

$$\iiint_V (\nabla p + \rho g \mathbf{k}) dV = \mathbf{0} \tag{3.7}$$

Since the integral vanishes for arbitrary volume, necessarily the integrand must vanish. Hence, we conclude that

$$\nabla p = -\rho g \mathbf{k} \tag{3.8}$$

Thus, in component form, we have shown that for an arbitrary volume,

$$\frac{\partial p}{\partial x} = \frac{\partial p}{\partial y} = 0 \quad \text{and} \quad \frac{\partial p}{\partial z} = -\rho g \tag{3.9}$$

which verifies Equation (3.3). We can integrate immediately for fluids such as liquids in which density is constant. The resulting pressure variation is

$$p + \rho g z = \text{constant} \tag{3.10}$$

We call this the **hydrostatic relation**. Equation (3.10) tells us that pressure increases linearly with depth (remember that negative values of z correspond to increasing depth). An example of this is water in a swimming pool. At a depth of 10 feet the absolute pressure is $p = p_o - \rho g z$, where p_o is atmospheric pressure, so that

$$p = 14.7 \text{ psi} - \frac{\left(1.94 \text{ slug/ft}^3\right)\left(32.174 \text{ ft/sec}^2\right)(-10 \text{ ft})}{144 \text{ in}^2/\text{ft}^2} = 19.0 \text{ psi} \tag{3.11}$$

A similar calculation shows that the pressure at 130 feet in the ocean is 72.8 psi (note that the density of seawater is 2.0 slugs/ft^3). This high a pressure is sufficient to cause nitrogen to dissolve in a diver's bloodstream, leading to impaired judgment and a painful condition known as the **bends** if the diver returns to the surface too rapidly. Impaired judgment is very noticeable at a depth of 130 feet, which is the established maximum depth for sport diving.[1]

3.2 Atmospheric Pressure Variation

The pressure in the Earth's atmosphere varies in a more complicated manner than the simple linear relation in Equation (3.10). The reason for this is the following. Atmospheric air behaves like a perfect gas so that its density is given by

$$\rho = \frac{p}{RT} \tag{3.12}$$

Hence, Equation (3.3) assumes the following form.

$$\frac{dp}{dz} = -\frac{g}{RT}p \tag{3.13}$$

If we knew the variation of temperature T with altitude, integration of Equation (3.13) would be straightforward. There is a simple model for the atmosphere over the United States known as the U. S. Standard Atmosphere [U. S. Government Printing Office (1976)]]. This model represents average conditions in the United States at 40° N latitude (e.g., New York City). In the U. S. Standard Atmosphere, the region from the Earth's surface ($z = 0$) up

---

[1]Using no breathing equipment, divers have exceeded this depth and risen to the surface rapidly with no ill effects. However, their pulse rates have gone so low at these great depths as to be in a state close to death – this practice is not recommended!

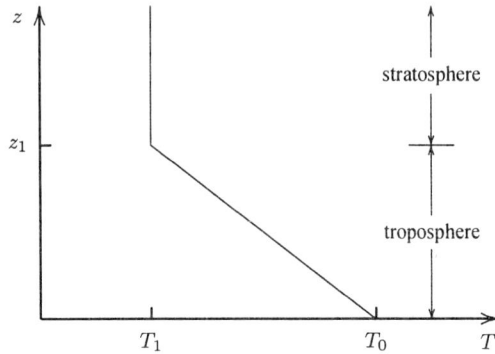

Figure 3.2: *Temperature in the U. S. Standard Atmosphere; z_1 = 11.0 km.*

to z = 11.0 km (6.84 miles) is called the **troposphere** and the temperature decreases linearly with altitude according to

$$T = T_0 - \alpha z, \qquad 0 \le z \le 11.0 \text{ km} \tag{3.14}$$

where the coefficient α is the **lapse rate**, and T_0 is the average surface temperature. The region from z = 11.0 km (6.84 mi) to z = 20.1 km (12.5 miles) is called the **stratosphere**. The temperature in this idealized model is constant in the stratosphere, and denoted by T_1. Figure 3.2 shows the temperature variation in the troposphere and the stratosphere. Table 3.1 lists values of α, T_0 and T_1 in SI and USCS units.[2]

Table 3.1: *Properties of the U. S. Standard Atmosphere*

Property	SI Units	USCS Units
α	6.50 K/km	18.85° R/mi
T_0	288 K (15° C)	518.4° R (59° F)
T_1	218 K (-55° C)	392.4° R (-67° F)

Tropospheric Pressure Variation. Hence, in the troposphere, combining Equations (3.13) and (3.14) yields

$$\frac{dp}{p} = -\frac{g \, dz}{R(T_0 - \alpha z)} \tag{3.15}$$

Integrating, we find that the pressure varies according to

$$p = p_0 \left[1 - \frac{\alpha z}{T_0} \right]^{g/(\alpha R)}, \qquad 0 \le z \le 11.0 \text{ km} \tag{3.16}$$

where p_0 = 101 kPa (14.7 psi) is the pressure at sea level. The exponent $g/(\alpha R)$ in Equation (3.16) is approximately 5.26. Note that the pressure falls to 22.5 kPa (3.28 psi) at the upper boundary of the troposphere.

---

[2]Temperature in the U. S. Standard Atmosphere increases (in a nontrivial manner) above the stratosphere.

Stratospheric Pressure Variation. The integration is even easier in the stratosphere since temperature is constant. The pressure varies as follows.

$$p = p_1 \exp\left[-\frac{g(z - z_1)}{RT_1}\right], \qquad 11.0 \text{ km} \leq z \leq 20.1 \text{ km} \tag{3.17}$$

The pressure $p_1 = 22.5$ kPa (3.28 psi) follows from insisting that Equations (3.16) and (3.17) yield the same pressure at the interface between the troposphere and stratosphere. Thus, the pressure (and density) fall off exponentially in the stratosphere, and we sometimes refer to this as an **exponential atmosphere**.

The cabin pressure in a modern airliner at cruise altitude is typically about 9.5 psi. As the airplane descends to land, low pressure air is trapped inside your ears. This is what causes the "popping" sensation you experience when you yawn and allow the pressure to equilibrate. We can use Equation (3.16) to compute the altitude in the U. S. Standard Atmosphere at which this pressure prevails. Solving for z, we find

$$z = \frac{T_0}{\alpha}\left[1 - \left(\frac{p}{p_0}\right)^{\alpha R/g}\right] = \frac{288 \text{ K}}{6.50 \text{ K/km}}\left[1 - \left(\frac{9.5}{14.7}\right)^{1/5.26}\right] = 3.53 \text{ km} \tag{3.18}$$

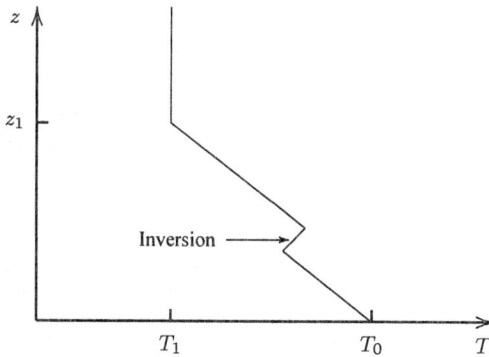

Figure 3.3: *Temperature variation over Los Angeles.*

The atmospheric temperature variation near Los Angeles (and many other large cities) deviates from the U. S. Standard Atmosphere in a significant manner. Specifically, there is a region in the troposphere, typically at an altitude of about 0.5 km (0.3 mi), known as the **inversion layer**, in which temperature increases with increasing altitude (Figure 3.3). This layer is present because the Los Angeles area is almost completely enclosed by high mountains. As air descends from the mountains, the sun heats it and creates a warm layer that rises above cooler air blowing in from the Pacific Ocean. This is the primary mechanism that creates the temperature inversion, with the heavier cool air trapped near the surface. The inversion layer puts a "lid" on the area that traps surface emissions responsible for smog. The prevailing winds in the Los Angeles area are unable to relieve the pollution problem because of this lid. Rather, they merely move the smoggy air from one part of the region to another.

3.3 Pressure Measurement

We can use the hydrostatic relation to measure pressure indirectly by making a much simpler measurement of the height of a column of fluid. This technique is called manometry. As an

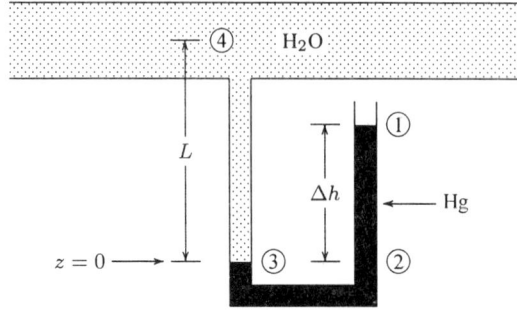

Figure 3.4: *U-Tube manometer.*

example, consider the **U-Tube Manometer** shown in Figure 3.4. The hydrostatic relation tells us that the pressure at point 2 is equal to the pressure at point 1 plus the weight of the column of mercury above. Hence,

$$p_1 + \rho_{Hg} g \Delta h = p_2 \tag{3.19}$$

Similarly, the pressure at point 3 is equal to the pressure at point 4 plus the weight of the column of water above, i.e.,

$$p_4 + \rho_{H_2O} g L = p_3 \tag{3.20}$$

Finally, since $z_3 = z_2$, the hydrostatic relation also tells us that $p_3 = p_2$. Consequently, we can equate the left hand sides of Equations (3.19) and (3.20), which yields the following.

$$p_4 - p_1 = (\rho_{Hg} \Delta h - \rho_{H_2O} L) g \tag{3.21}$$

The implication of Equation (3.21) is clear. If the U-Tube is open to the atmosphere, then p_1 is the pressure of the atmosphere. The density of mercury and water are known quantities. Thus, by simply measuring L and the height of the mercury column, Δh, we can infer the pressure in the water channel.

As a more complicated example, consider the arrangement shown in Figure 3.5 designed to monitor the pressure at the interface of the water and oil in the tank. To make the problem a little more interesting, we assume some air has been trapped in the attached manometer. We solve this problem by repeated use of the hydrostatic relation, with one important word of caution. Since Equation (3.10) follows from Equation (3.9) only when density is constant, we

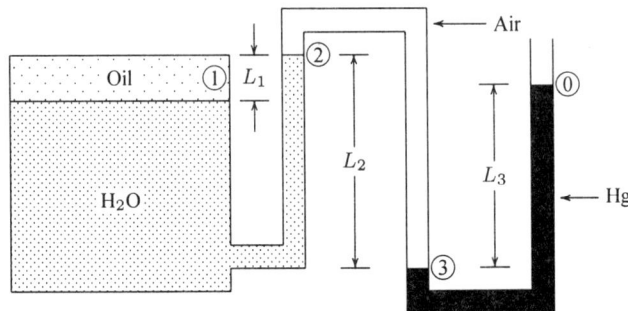

Figure 3.5: *More complicated manometry example.*

must stay within the same fluid when applying Equation (3.10). Working first in the column of mercury, we have

$$p_3 = p_0 + \rho_{Hg} g L_3 \qquad (3.22)$$

Next, applying the hydrostatic relation in the column of air yields

$$p_3 = p_2 + \rho_{Air} g L_2 \qquad (3.23)$$

Finally, for the column of water, we have

$$p_1 = p_2 + \rho_{H_2O} g L_1 \qquad (3.24)$$

We thus have three equations for the three unknown pressures, p_1, p_2 and p_3. A little straightforward algebra shows that the pressure at the oil-water interface is as follows.

$$p_1 = p_0 + (\rho_{Hg} L_3 + \rho_{H_2O} L_1 - \rho_{Air} L_2) g \qquad (3.25)$$

Finally, if L_1, L_2 and L_3 are all of the same order of magnitude, we can neglect the contribution from the trapped air because $\rho_{Air} \ll \rho_{Hg}, \rho_{H_2O}$. Therefore,

$$p_1 \approx p_0 + (\rho_{Hg} L_3 + \rho_{H_2O} L_1) g \qquad (3.26)$$

Suppose, for example, the measured heights are $L_1 = 100$ mm, $L_2 = 450$ mm and $L_3 = 400$ mm. Using densities corresponding to a temperature of 20° C, we find a pressure difference of $p_1 - p_0 = 54.133$ kPa if we ignore the air. Including the contribution from the air, the pressure difference is $p_1 - p_0 = 54.138$ kPa — a difference of .01%.

3.4 Hydrostatic Forces on Plane Surfaces

Because pressure in a non-moving fluid varies only with depth, the force on a submerged horizontal surface of area A is simply $F = pA$, where p is the pressure at the current depth. When the surface is inclined to the horizontal we must work a bit harder to evaluate the force. Specifically, as illustrated in Figure 3.6, we must integrate the varying pressure over the area of interest.

As we will see in this section, because p varies linearly with depth, we can compute its integral over a surface inclined to the horizontal in terms of simple geometrical properties.

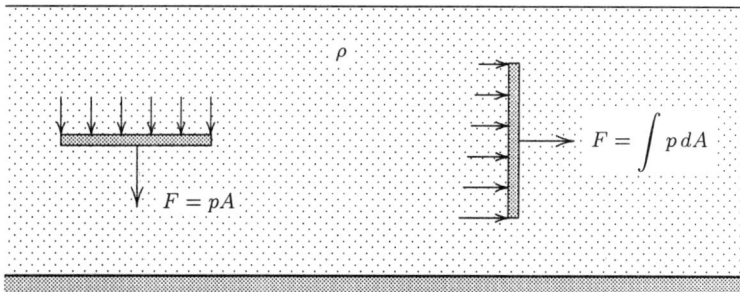

Figure 3.6: *Submerged surfaces and pressure distributions for horizontal and vertical orientations.*

Specifically, if we know the area and the centroid of the submerged surface, we can compute not only the force, but also the moment, on the surface. To see how this is done, we consider a plane surface inclined to the horizontal at an angle α as shown in Figure 3.7. It is submerged in a liquid of density ρ that has an air-liquid interface, often referred to as a **free surface**, at $z = 0$. Note that the coordinates are chosen so that z is positive in the direction of the gravitational acceleration vector,[3] x is parallel to the free surface, and y is out of the page. The figure includes a second set of Cartesian coordinates $(\xi\eta\zeta)$ aligned with the submerged surface that we will make use of below.

It is important to pause at this point and stress that we are dealing with a *three-dimensional geometry*. Figure 3.7 shows a side view of the inclined surface, i.e., the xz plane. Although its thickness is exaggerated in the figure, the surface appears as a straight line from this view. The force, **F**, is normal to the surface and lies in the xz plane. As shown, the angle α determines the direction of the force. The figure also shows a cross-sectional view of the surface in the $\eta\zeta$ plane. The geometry of the surface determines the magnitude of the force vector.

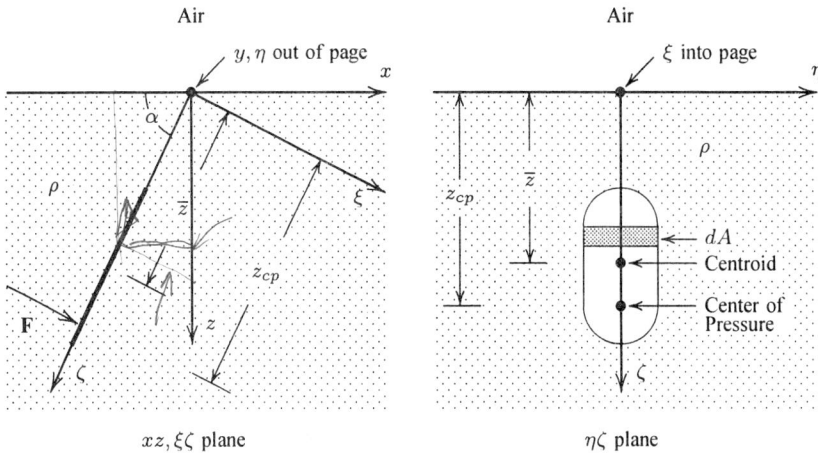

$xz, \xi\zeta$ plane $\qquad\qquad\qquad$ $\eta\zeta$ plane

Figure 3.7: *Submerged surface inclined to the horizontal.*

The hydrostatic relation tells us the absolute pressure, p_{abs}, increases linearly with depth. That is, changing the sign of z in Equation (3.3) yields

$$\frac{dp_{\text{abs}}}{dz} = \rho g \tag{3.27}$$

If the atmospheric pressure is p_o, the absolute pressure is given by

$$p_{\text{abs}} = p_o + \rho g z \tag{3.28}$$

To simplify our analysis, we will work with the gage pressure, p, i.e.,

$$p = p_{\text{abs}} - p_o \tag{3.29}$$

and note that $z = \zeta \sin\alpha$. Hence, we have

$$p = \rho g \zeta \sin\alpha \tag{3.30}$$

---

[3]In considering free-surface problems, we often define positive z downward for convenience. This obviates carrying minus signs throughout the analysis to follow.

Now, on a surface element dA, the differential force exerted by the fluid to the left of the surface is given by $dF = p\,dA$ so that the force on the entire surface is given by the following integral.

$$F = \iint_A p\,dA = \rho g \sin\alpha \iint_A \zeta\,d\eta d\zeta \tag{3.31}$$

By definition, the centroid, $\bar{z}$, of the planar surface A is

$$\bar{z} \equiv \frac{1}{A} \iint_A \zeta\,d\eta d\zeta \tag{3.32}$$

Hence, the magnitude of the force on the inclined surface, $F = |\mathbf{F}|$, can be expressed in terms of the centroid of the planar surface as follows.

$$F = \rho g \bar{z} A \sin\alpha \tag{3.33}$$

Since the force is normal to the submerged surface, in vector form we have

$$\mathbf{F} = \rho g \bar{z} A \sin\alpha\,(\mathbf{i}\cos\alpha + \mathbf{k}\sin\alpha) \tag{3.34}$$

What we have computed is a *single point force* equivalent to the distributed pressure acting on the surface. If we know the depth of submergence of the planar surface's centroid and its area, we can compute the force from Equation (3.33), thus eliminating the need to do an integral.

When forces are applied to an object, they cause the object's "center of gravity" to accelerate. The center of gravity and centroid are coincident for a constant density object. Because pressure increases with depth, it should be obvious that the equivalent force, $\mathbf{F}$, must act through a point a bit deeper than the centroid. This will give rise to a moment about the centroid. We will now determine an effective lever arm to go with $\mathbf{F}$ that yields the same moment as the distributed pressure. The point corresponding to this lever arm is called the **center of pressure**. We denote the center of pressure by z_{cp} (see Figure 3.7).

Taking moments about the η axis, we have $\mathbf{M} = \mathbf{r}_{cp} \times \mathbf{F} = z_{cp}F\mathbf{j}$. Hence,

$$z_{cp}F = \int \zeta\,dF \tag{3.35}$$

Then, since $dF = p\,dA$, we have

$$z_{cp}F = \iint_A \zeta p\,dA = \rho g \sin\alpha \iint_A \zeta^2\,d\eta d\zeta \tag{3.36}$$

The integral of ζ^2 appearing in Equation (3.36) is the familiar **moment of inertia** of the planar surface A relative to the η axis. We denote the moment of inertia by I_o, so that[4]

$$I_o = \iint_A \zeta^2\,d\eta d\zeta \tag{3.37}$$

For our purposes, it is more convenient to reference the moment of inertia to the centroid. We accomplish this by shifting the coordinate axes from the free surface to the centroid. To

---

[4]This moment of inertia is usually denoted by I_{zz} in classical mechanics, and it is one of 9 components of a 3 x 3 tensor (matrix). Since we are using only one component of the inertia tensor, we omit the customary subscripts.

accomplish this, we simply replace ζ by $\zeta - \bar{z}$. Thus, the moment of inertia relative to the centroid, I, is given by

$$
\begin{aligned}
I &\equiv \iint_A (\zeta - \bar{z})^2 dA \\
&= \iint_A (\zeta^2 - 2\bar{z}\zeta + \bar{z}^2) dA \\
&= I_o - 2\bar{z} \iint_A \zeta \, dA + \bar{z}^2 \iint_A dA \\
&= I_o - 2\bar{z}^2 A + \bar{z}^2 A \\
&= I_o - \bar{z}^2 A
\end{aligned}
\tag{3.38}
$$

Therefore, we can express the moment of inertia relative to the surface, I_o, as a function of I, A and $\bar{z}$, viz.,

$$
I_o = I + \bar{z}^2 A \tag{3.39}
$$

Equation (3.39) is known as the **parallel-axis theorem**. Substituting Equations (3.37) and (3.39) into Equation (3.36) yields

$$
z_{cp} F = \rho g \sin \alpha \left(I + \bar{z}^2 A \right) \tag{3.40}
$$

Then, using Equation (3.33), we can eliminate F and solve for z_{cp}. That is, substituting for F, we have

$$
z_{cp} \rho g A \bar{z} \sin \alpha = \rho g \sin \alpha \left(I + \bar{z}^2 A \right) \tag{3.41}
$$

Dividing through by $\rho g A \bar{z} \sin \alpha$, the final result is as follows.

$$
z_{cp} = \bar{z} + \frac{I}{\bar{z} A} \tag{3.42}
$$

Equation (3.42) is especially useful because I and A depend only upon the shape of the planar surface in the $\eta\zeta$ plane. It is completely independent of the fluid, and the only way in which depth of submergence affects z_{cp} is through the dependence upon $\bar{z}$. The same is true of the force, which is proportional to the area, A, and $\bar{z}$. Since the moment of inertia, area and centroid are all positive, necessarily z_{cp} lies below the centroid. Also, note that $z_{cp} \to \bar{z}$ as $\bar{z} \to \infty$.

Because our formulas for the force and center of pressure can be easily determined once A, I and $\bar{z}$ are known, it is worthwhile to tabulate their values for common geometries. Figure 3.8 includes area, moment of inertia and centroid for a triangle, a rectangle, a circle and a semicircle.

To illustrate how we can use Equations (3.33) and (3.42) to analyze fluid-statics problems, consider the forces acting on the triangular gate in Figure 3.9. From the left, the fluid develops a force of magnitude F that is balanced by the sum of the reaction force at the hinge, R_H, and the reaction force at the top of the gate, R_T, i.e.,

$$
F = R_T + R_H \tag{3.43}
$$

There is actually a fourth force acting, viz., the atmospheric pressure on the right-hand side of the gate. However, we implicitly account for this force when we use Equation (3.33). That is, since we reference all pressures to the atmospheric level [see Equation (3.29)], F is the difference between the integral of the absolute fluid pressure from the left, $p_{abs}(z)$, and the

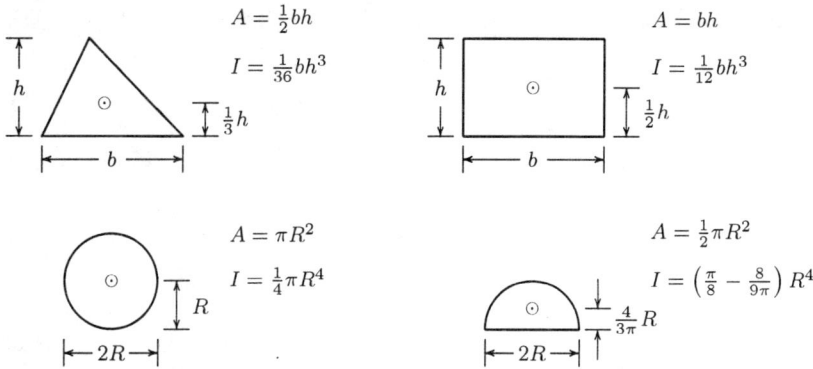

Figure 3.8: *Areas, moments of inertia and centroids (indicated by a ⊙) for common geometries; in all cases, the moment of inertia is relative to a horizontal axis passing through the centroid.*

atmospheric pressure from the right, p_o. Equivalently, we can regard the adjusted pressure, $p = p_{abs} - p_o$, as being zero in the atmosphere, so that the net force is zero. We will return to this point in greater detail below.

The essence of the problem is as follows. There are three unknown forces to be determined, i.e., F, R_H and R_T. We thus need three equations to determine the solution. Equation (3.43) is the first of the three, while Equation (3.33) provides a second. To arrive at a third equation, we must balance moments about a suitable axis. As we will see below, computing the lever arm for the hydrostatic force involves the center of pressure.

Turning first to the force F, it is given by Equation (3.33). Hence, we must determine the area, A, and the location of the centroid, $\bar{z}$. As indicated in Figure 3.9, the base of the triangular gate is w and its altitude is $h/\sin \alpha$. Thus, the area of the gate is

$$A = \frac{1}{2} \frac{wh}{\sin \alpha} \tag{3.44}$$

Measuring along the ζ axis, which is aligned with the gate, the top of the gate is located a distance $h/\sin \alpha$ below the surface. The centroid is an additional $\frac{2}{3}h/\sin \alpha$ below the top of

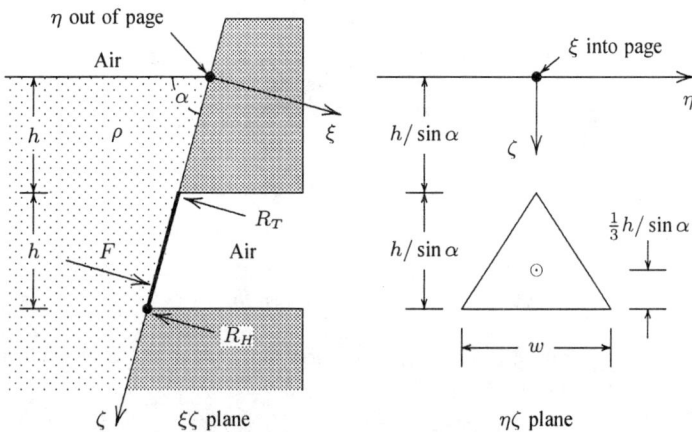

Figure 3.9: *Forces acting on a triangular gate.*

the gate. Adding these two contributions, we arrive at the following.

$$\overline{z} = \frac{h}{\sin\alpha} + \frac{2}{3}\frac{h}{\sin\alpha} = \frac{5}{3}\frac{h}{\sin\alpha} \tag{3.45}$$

Therefore, substituting Equations (3.44) and (3.45) into Equation (3.33) yields the hydrostatic force normal to the gate, viz.,

$$F = \rho g \left(\frac{1}{2}\frac{wh}{\sin\alpha}\right)\left(\frac{5}{3}\frac{h}{\sin\alpha}\right)\sin\alpha = \frac{5}{6}\frac{\rho g w h^2}{\sin\alpha} \tag{3.46}$$

At this point, we have two of the three required equations for determining the complete solution. As noted earlier, this is as much information as we can extract from balancing forces. The third equation required to close our system of equations follows from balancing moments. As we will see, the moment due to the hydrostatic force depends upon the center of pressure.

Hence, turning to the moment calculation, it is convenient to take moments about the axis perpendicular to the page passing through the hinge. As shown in Figure 3.10, the lever arms for the three forces are

$$\text{Lever Arm} = \begin{cases} 0, & R_H \\ \dfrac{h}{\sin\alpha}, & R_T \\ 2\dfrac{h}{\sin\alpha} - z_{cp}, & F \end{cases} \tag{3.47}$$

Hence, balancing moments yields:

$$F\left(2\frac{h}{\sin\alpha} - z_{cp}\right) = R_T\frac{h}{\sin\alpha} + R_H \cdot 0 \quad\Longrightarrow\quad \frac{R_T}{F} = 2 - \frac{z_{cp}}{h}\sin\alpha \tag{3.48}$$

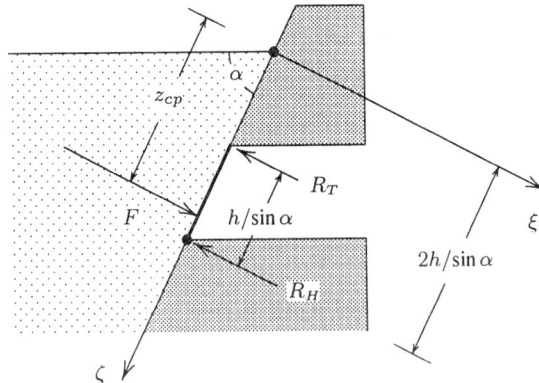

Figure 3.10: *Lever arms for the forces.*

Now, the center of pressure is given by Equation (3.42). For the triangular gate, reference to Figure 3.8 shows that the moment of inertia is

$$I = \frac{1}{36}w\left(\frac{h}{\sin\alpha}\right)^3 \tag{3.49}$$

Thus, the center of pressure is located at

$$
z_{cp} = \overline{z} + \frac{I}{\overline{z}A} = \frac{5}{3}\frac{h}{\sin\alpha} + \frac{\frac{1}{36}w\left(\dfrac{h}{\sin\alpha}\right)^3}{\left(\dfrac{5}{3}\dfrac{h}{\sin\alpha}\right)\left(\dfrac{1}{2}\dfrac{wh}{\sin\alpha}\right)} = \frac{17}{10}\frac{h}{\sin\alpha} \tag{3.50}
$$

Hence, we have

$$
\frac{R_T}{F} = 2 - \frac{17}{10} = \frac{3}{10} \quad \Longrightarrow \quad \frac{R_H}{F} = \frac{7}{10} \tag{3.51}
$$

Thus, the reaction forces acting on the gate are

$$
R_H = \frac{7}{12}\frac{\rho g w h^2}{\sin\alpha} \quad \text{and} \quad R_T = \frac{1}{4}\frac{\rho g w h^2}{\sin\alpha} \tag{3.52}
$$

As a concluding comment on the calculation of forces and moments on submerged surfaces, consider the atmospheric pressure, p_o. Recall from Equation (3.29) that we have been dealing with what is referred to as **gage pressure**, defined as the difference between the **absolute pressure**, p_{abs}, and p_o. Pressure is similar to potential energy in the sense that changes in pressure give rise to forces and motion, and a reference value can usually be chosen arbitrarily. Atmospheric pressure is a natural reference since it is the value at the free surface, $z = 0$.

For many applications, the net force and moment on an object follow from integrands proportional to $(p_{abs} - p_o)$. This will be true, for example, when we have a liquid on one side of a barrier and air on the other side. Figure 3.11(a) shows such a configuration, including the pressure forces acting on both sides of the barrier. If the area and centroid of the submerged part of the barrier are A and $\overline{z}$, respectively, and the total area of the barrier is A_{tot}, then the net force is

$$
F = \underbrace{\rho g \overline{z} A + p_o A_{tot}}_{\substack{Force\ from \\ the\ left}} - \underbrace{p_o A_{tot}}_{\substack{Force\ from \\ the\ right}} = \rho g \overline{z} A \tag{3.53}
$$

Thus, the contribution from the atmospheric pressure cancels exactly, and all that remains is the hydrostatic contribution to the force. The same is true for the net moment.

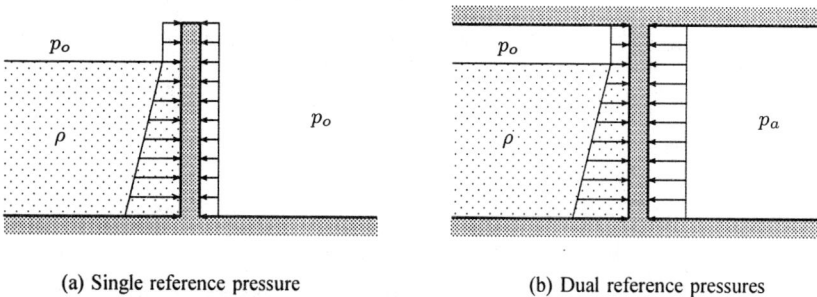

(a) Single reference pressure   (b) Dual reference pressures

Figure 3.11: *Pressure forces on a barrier including reference pressures.*

By contrast, if different reference pressures occur in a problem, we must account for them in computing forces and moments. For example, consider Figure 3.11(b), which corresponds to a closed chamber pressurized on one side. In the left chamber, the air is at atmospheric pressure, p_o. In the right chamber, the pressure is $p_a > p_o$. The net force on the barrier in this case is

$$F = \underbrace{\rho g \bar{z} A + p_o A_{tot}}_{\substack{Force\ from \\ the\ left}} - \underbrace{p_a A_{tot}}_{\substack{Force\ from \\ the\ right}} = \rho g \bar{z} A - (p_a - p_o) A_{tot} \qquad (3.54)$$

Because there is a difference in reference pressure, we see that the net force is not equal to the hydrostatic force. The difference is exactly equal to the difference between the reference pressure forces acting on each side of the barrier.

3.5 Hydrostatic Forces on Curved Surfaces

In principle, computing the hydrostatic force on a curved surface is more difficult than on a plane surface. We must now do an integral on a potentially complex contour, which can prove to be a nontrivial task. However, with a little cleverness, we can reduce this more-complex problem to one we have already solved. To understand how we can accomplish this end, consider the arc AB shown in Figure 3.12. We proceed as follows.

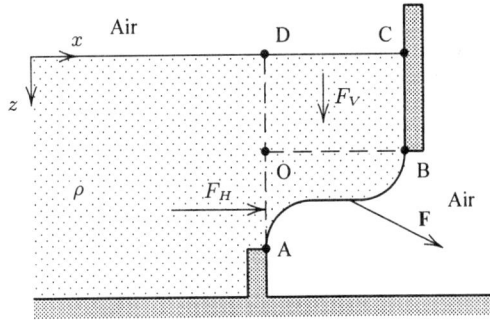

Figure 3.12: *Force on a curved surface. OA is the projection of the arc AB on a vertical plane.*

1. Resolve the force vector into its x and z components, i.e., $\mathbf{F} = F_x \mathbf{i} + F_z \mathbf{k}$, where $\mathbf{i}$ and $\mathbf{k}$ are unit vectors in the x and z directions, respectively.

2. Compute F_H, the force on surface OA, using plane surface methods, i.e., $F_H = \rho g \bar{z} A_{OA}$.

3. Compute F_V, the weight of the column of fluid above AB from the geometry, i.e., $F_V = \rho g V_{ABCD}$.

4. Balance forces in the x and z directions to determine F_x and F_z in terms of F_H and F_V. Hence, in the x direction,

$$F_x + F_{CB} = F_H + F_{OD}$$

But, $F_{CB} = F_{OD}$ (same pressure, opposite unit normal), so that

$$F_x = F_H$$

Similarly, balancing forces in the z direction shows that

$$F_z = F_V$$

As an example, consider the rectangular gate of unit width out of the page aligned with arc AB in Figure 3.13 separating a fluid of density ρ and air. We will use this problem as a first illustration of the curved-arc methodology. Because the arc is a straight line segment oriented at $45°$ to the horizontal, and since $\mathbf{F}$ is normal to the arc, necessarily $F_x = F_z$. We can use this fact as a check on our computations.

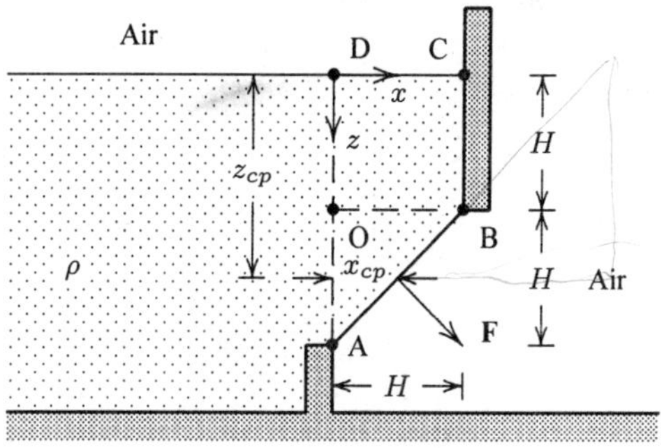

Figure 3.13: *Fluid-statics problem; the gate has unit width out of the page.*

The x component of $\mathbf{F}$ is equal to the hydrostatic force on the projection of the gate surface onto the vertical plane. The projection is a surface of unit width out of the page aligned with arc OA. Because the projected surface is a rectangle, its centroid is $\frac{1}{2}H$ below point O, which is $\frac{3}{2}H$ below the surface. Therefore, the centroid location is $\overline{z} = \frac{3}{2}H$ and the area of the surface is $A = H \cdot 1 = H$ (the 1 represents unit width). Thus, we can use Equation (3.33) with $\alpha = 90°$ wherefore

$$F_x = \frac{3}{2}\rho g H^2 \tag{3.55}$$

The weight of the column of fluid above AB is the sum of the weight of the fluid in OBCD and the fluid in OAB. The volumes of these columns are $V_{OBCD} = H^2 \cdot 1 = H^2$ and $V_{OAB} = \frac{1}{2}H^2 \cdot 1 = \frac{1}{2}H^2$. Thus, the z component of $\mathbf{F}$ is as follows.

$$F_z = \rho g H^2 + \frac{1}{2}\rho g H^2 = \frac{3}{2}\rho g H^2 \tag{3.56}$$

Hence, as expected, we have shown that $F_x = F_z$.

We can compute the center of pressure using Equation (3.42) where $\overline{z}$, A and I are for the projection of AB onto the vertical plane. For this gate geometry, i.e., a rectangle of height $h = H$ and width $b = 1$, reference to Figure 3.8 shows that $I = \frac{1}{12}H^3$. Thus, the z component of the center of pressure is given by

$$z_{cp} = \frac{3}{2}H + \frac{\frac{1}{12}H^3}{\frac{3}{2}H \cdot H} = \frac{14}{9}H \tag{3.57}$$

For planar surfaces, the center of pressure lies on the surface. Hence, for this problem, relative to point D, the equation of arc AB is $x = 2H - z$ so that

$$x_{cp} = 2H - \frac{14}{9}H = \frac{4}{9}H \tag{3.58}$$

We cannot assume that $\mathbf{r}_{cp}$ lies on the surface for curvilinear arcs. After all, it is an equivalent lever arm for an equivalent force. To find the x component of the center of pressure, x_{cp}, for general curvilinear arcs, we again take moments about the y axis. Assuming, for simplicity, that our surface is symmetric about $y = 0$, the moment is

$$\mathbf{M} = \mathbf{r}_{cp} \times \mathbf{F} = (z_{cp}F_x - x_{cp}F_z)\mathbf{j} \tag{3.59}$$

In general, to determine z_{cp}, we simply use Equation (3.42) on OA, the projection of AB on the vertical plane. To determine x_{cp}, we must satisfy

$$x_{cp}F_z = \int x\, dF \tag{3.60}$$

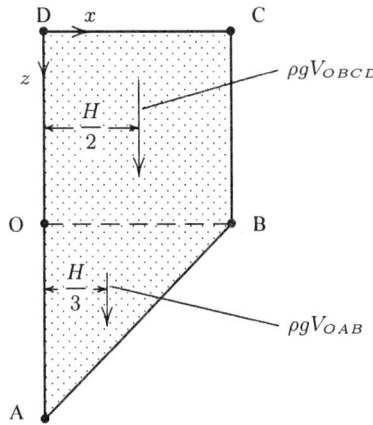

Figure 3.14: *Moment arms for computing* x_{cp}.

As an example of a straightforward way to evaluate the integral in Equation (3.60), consider the sample problem. The contribution to the moment from the combined arc ABC relative to the z axis is the sum of the moments from the weight of OBCD and the weight of OAB. As shown in Figure 3.14, the lever arm for OBCD (a square) is $\frac{1}{2}H$ relative to the z axis, while the lever arm for OAB (a triangle) is $\frac{1}{3}H$. These are distances of the centroids of these volumes from the z axis in the xz plane. (Note that we have previously considered the surface geometry in the yz plane.) The moment on ABC must exactly balance the moments on OBCD and OAB. Thus,

$$x_{cp}F_z = \int x\, dF = \frac{1}{2}H \cdot \rho g \underbrace{V_{OBCD}}_{H^2} + \frac{1}{3}H \cdot \rho g \underbrace{V_{OAB}}_{\frac{1}{2}H^2} = \frac{2}{3}\rho g H^3 \tag{3.61}$$

Combining Equations (3.56), (3.60) and (3.61) yields

$$\frac{3}{2}\rho g H^2 x_{cp} = \frac{2}{3}\rho g H^3 \qquad \Longrightarrow \qquad x_{cp} = \frac{4}{9}H \tag{3.62}$$

This matches the result quoted in Equation (3.58).

3.6   Equivalent Pressure Fields and Superposition

Because pressure varies only with depth, the magnitudes of the forces acting on the arcs shown in Figure 3.15 are identical. Hence, if we compute the force vector, $\mathbf{F}_1$, with fluid of density

ρ to the left of the arc, clearly the force is $-\mathbf{F}_1$ when the same fluid lies to the right of the arc. Mathematically, we can say

$$\mathbf{F}_2 = -\mathbf{F}_1 \tag{3.63}$$

This is a useful fact since the method described in Section 3.5 is ideally suited for problems with fluid above the arc, but requires modification when the fluid is below the arc.

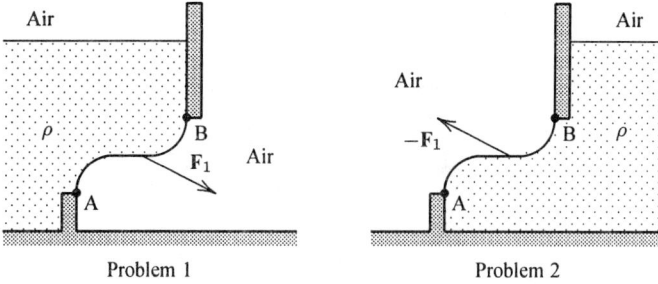

Figure 3.15: *Equivalent static pressure fields.*

Another useful technique for solving fluid-statics problems is to replace a given problem with a series of simpler problems. We can do this by using the principle of superposition. Again, we seek replacement problems whose sum or difference has an equivalent pressure field. As an example, consider an arc separating two fluids of different densities as shown in Figure 3.16.

To solve this problem, we use superposition. That is, we first replace the problem in Figure 3.16 by the sum of Problem 1 of Figure 3.15 with $\rho = \rho_1$ and Problem 2 with $\rho = \rho_2$. The force from Problem 1 points downward while the force from Problem 2 points upward. Hence, the total force will be

$$\mathbf{F} = \mathbf{F}_1 + \mathbf{F}_2 \tag{3.64}$$

Another way of viewing this problem is to say the solution is the difference between Problem 1 with $\rho = \rho_1$ and another Problem 1 with $\rho = \rho_2$. But, this is identical to solving Problem 1 with $\rho = \rho_1 - \rho_2$. The reason this works is because the pressure depends only upon depth, and pressure is proportional to the density. Consequently, at a given depth z, the pressure at any point on the upper part of the arc is $\rho_1 g z$ pushing to the right while the contribution from the lower part of the arc is $\rho_2 g z$ pushing in the opposite direction. This means the net pressure acting on the arc at a depth z is $(\rho_1 - \rho_2)g z$, which is exactly the pressure exerted by a single fluid above the arc of density $(\rho_1 - \rho_2)$.

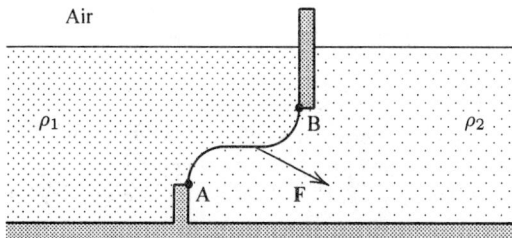

Figure 3.16: *Superposition example.*

3.7   Buoyancy

By definition, the **buoyancy force** on an object, $\mathbf{F}_{buoy}$, is the net pressure force exerted by the fluid on a given object submerged in a non-moving fluid. If the object is more dense than the surrounding fluid, its weight exceeds the buoyancy force and it sinks. If the object is less dense than the fluid, it will float. We express the buoyancy force mathematically as

$$\mathbf{F}_{buoy} = - \oiint_S p\, \mathbf{n}\, dS \qquad (3.65)$$

where we integrate about the surface bounding the object. Then, using the divergence theorem, we can convert the closed surface integral to a volume integral:

$$\mathbf{F}_{buoy} = - \iiint_V \nabla p\, dV \qquad (3.66)$$

with the integration extending over the volume of fluid displaced by the object. But, we know from the hydrostatic relation [Equation (3.8)] that $\nabla p = -\rho g \mathbf{k}$ in the fluid that has been displaced. Thus, we conclude finally that

$$\mathbf{F}_{buoy} = g\mathbf{k} \iiint_V \rho\, dV = M g \mathbf{k} \qquad (3.67)$$

where M is the **mass of the displaced fluid**. Note that $\mathbf{F}_{buoy}$ acts vertically upward. This is known as **Archimedes' Principle**. It is valid for both floating and submerged bodies.

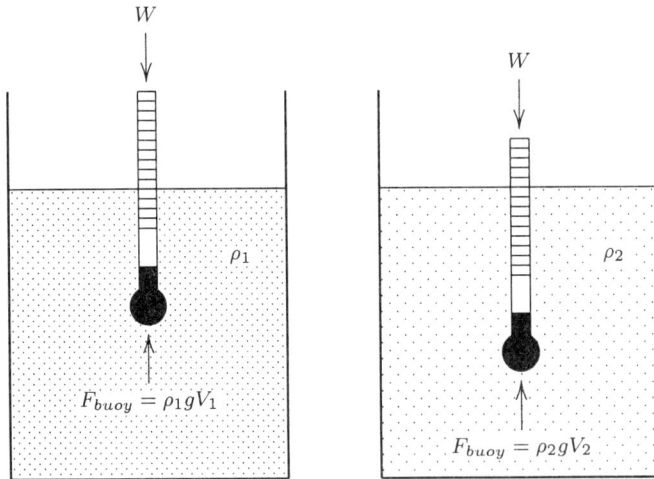

Figure 3.17: *Mass Hydrometer;* $\rho_1 > \rho_2$.

Figure 3.17 illustrates a practical application of Archimedes' Principle. The figure shows a **hydrometer**, which is a bottle of known weight, W, with a graded scale to indicate depth of submergence. The hydrometer sinks to a greater depth in the lighter fluid because more fluid must be displaced in order to generate a buoyancy force that balances W. In the figure, $\rho_1 > \rho_2$. If the submerged volumes are V_1 and V_2, then we have

$$\rho_1 g V_1 = W = \rho_2 g V_2 \qquad (3.68)$$

Consequently, the densities stand in the following ratio.

$$\rho_1/\rho_2 = V_2/V_1 \qquad (3.69)$$

We can thus calibrate our hydrometer so that by reading a simple graded scale we know the density of a given fluid relative to a known reference density of a fluid such as water.

AN EXPERIMENT YOU CAN DO AT HOME

To discover for yourself a subtle effect of buoyancy in the comfort of your home, try the following simple experiment. Obtain a soda bottle, a wine bottle, or any container with a top small enough to be covered by your thumb. Fill it nearly to the top with water or any other readily available liquid. Insert a piece of cork in the bottle or container so that it floats. Now cover the top completely with your thumb. Press down firmly and, if you have done everything correctly, the cork sinks. Remove the pressure, and it returns to the surface.

The explanation for what happens is as follows. When you press down, you increase the pressure we have called p_o. This, in turn, increases the absolute pressure, $p_{abs} = p_o + \rho g z$ [see Equation (3.28)]. The cork, being a porous material, experiences two changes. First it is compressed, which reduces its volume and hence the buoyancy force. Second, it absorbs some of the water because of this increased pressure, and becomes more massive. Pressing hard enough causes the cork to undergo compression and to absorb a sufficient amount of water that its density becomes greater than that of water. As a result, it sinks. Removing the increased pressure causes the cork to expand, release the absorbed water and return to its original density.

Problems

3.1 The pressure at one point in a container filled with a liquid of unknown density is 15.0 psi. At another point 6 in deeper in the container the pressure is 15.2 psi. What is the density of the liquid?

3.2 The pressure at one point in a container filled with a liquid of unknown density is 110 kPa. At another point 6.1 cm deeper in the container the pressure is 118.1 kPa. What is the density of the liquid?

3.3 James Bond has just thrown Blofeld into a vat of mercury. If Blofeld, chained to a piece of Plutonium, sinks to a depth of 12 ft, will he have to worry about the bends as he floats back to the surface?

3.4 Consider the tank shown with three layers of unmixed fluids. The temperature is 20° C and the pressure at the bottom of the tank is 259 kPa. Also, the tank is open to the atmosphere at the top. What is the density of the unknown fluid, ρ_u, if h = 1 meter?

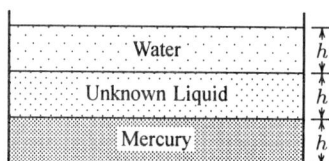

Problem 3.4

3.5 Consider the tank shown with three layers of unmixed fluids. The temperature is 68° F and the pressure at the bottom of the tank is 16.4 psi. Also, the tank is open to the atmosphere at the top. What is the density of the unknown fluid, ρ_u, if h = 3 inches?

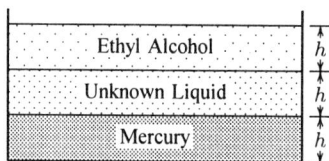

Problem 3.5

3.6 Consider the two layers of fluid shown. What is the pressure a distance h from the bottom?

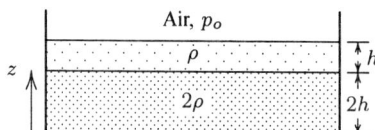

Problem 3.6

3.7 In 1644, E. Torricelli invented the *mercury barometer*. It consists of a glass tube closed at one end with the open end immersed in a container filled with mercury. The tube is initially filled with mercury and is turned upside down when inserted in the container, creating a near vacuum in the top of the tube.

(a) Assuming the temperature is 20° C, what is the height, h, (expressed in millimeters) of the mercury in the tube if the atmospheric pressure, p_o, is 103 kPa?

(b) If the barometer used water instead of mercury, what would h be? Express your answer in meters.

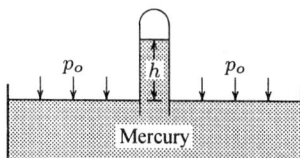

Problem 3.7

3.8 Consider a density-stratified fluid in which the density varies with depth, z, as

$$\rho = \rho_o \left(1 + z/z_o\right)$$

where z is measured downward from the surface, ρ_o is the density at the surface and z_o is a constant. Determine the pressure as a function of depth.

3.9 A small aircraft is cruising at an altitude of one mile. Assuming atmospheric conditions are accurately represented by the U. S. Standard Atmosphere, determine the pressure and temperature outside the aircraft.

3.10 An airliner is cruising at an altitude of 35,000 ft. Assuming atmospheric conditions are accurately represented by the U. S. Standard Atmosphere, determine the pressure and temperature outside the airliner.

3.11 A high-altitude aircraft is cruising at an altitude of 15 km. Assuming atmospheric conditions are accurately represented by the U. S. Standard Atmosphere, determine the pressure and temperature outside the aircraft.

3.12 The Washington Monument in the District of Columbia is 550 feet high. Assuming atmospheric conditions are as given by the U. S. Standard Atmosphere, determine the decrease in pressure and temperature relative to surface values at the top of the Monument.

3.13 The power output of an automobile engine is proportional to the mass-flow rate of the air supplied to the carburetor. The mass-flow rate, in turn, is proportional to the atmospheric air density. If a Ford Mustang GT delivers 220 horsepower at sea level, what is the corresponding power delivered in Denver, Colorado, which is approximately 1.6 km above sea level. Assume the U. S. Standard Atmosphere accurately describes conditions in Denver and at sea level.

3.14 Consider an atmosphere with an inversion layer for which the temperature varies according to:

$$T = \begin{cases} T_o - \alpha z, & 0 \leq z \leq \frac{1}{4}T_o/\alpha \\ \frac{3}{4}T_o + \alpha(z - \frac{1}{4}T_o/\alpha), & \frac{1}{4}T_o/\alpha \leq z \leq \frac{3}{10}T_o/\alpha \\ \frac{4}{5}T_o, & z \geq \frac{3}{10}T_o/\alpha \end{cases}$$

(a) Make a graph of the temperature profile.

(b) Assuming the atmosphere is a perfect gas, determine the pressure in the inversion layer as a function of altitude. Express your answer in terms of the pressure at sea level, p_o, the perfect-gas constant, R, gravitational acceleration, g, as well as T_o, α and z.

3.15 In an *adiabatic atmosphere*, the pressure varies with density according to

$$p = A\rho^\gamma$$

where γ is the specific-heat ratio and A is a constant. Determine the adiabatic lapse rate, α, in such an atmosphere assuming the temperature at the surface is T_o. Express your answer in terms of γ, g, and the perfect-gas constant, R.

3.16 Compute the pressure at Point A in terms of atmospheric pressure, p_o, density of water, ρ, gravitational acceleration, g, and the distance h. Use the fact that $\rho_{H_2O} = \rho$, $\rho_{Oil} = 0.80\rho$ and $\rho_{Hg} = 13.55\rho$.

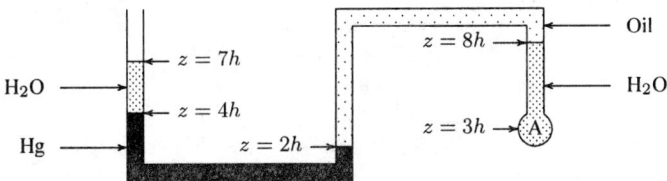

Problem 3.16

3.17 For the setup shown, what is the density $\tilde{\rho}$ if the pressure at Point A is $p_o + 3.3\rho g H$, where p_o is atmospheric pressure?

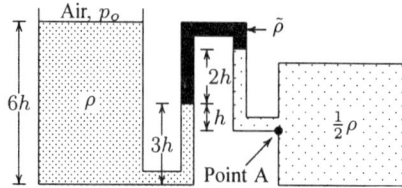

Problems 3.17, 3.18

3.18 For the setup shown, what is the pressure at Point A if $\tilde{\rho} = \frac{1}{4}\rho$?

3.19 An open-ended mercury manometer is attached to an oil tank. Water has entered the tank and lies at the bottom in a layer of thickness h. The air trapped in the tank is also pressurized at a pressure, p_a. Compute the differential difference, Δz, in the manometer's mercury column, assuming $N = 6$, $n = 1$, $h = 1$ m and $p_a = 2p_o$, where $p_o = 101$ kPa is atmospheric pressure. Also, assume the oil has a density, $\rho_{Oil} = 850$ kg/m^3.

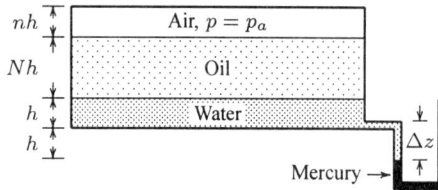

Problems 3.19, 3.20

3.20 An open-ended mercury manometer is attached to an oil tank. Water has entered the tank and lies at the bottom in a layer of thickness h. Some of the air trapped in the tank has been evacuated so that $p_a < p_o$, where $p_o = 14.7$ psi is atmospheric pressure. Find p_a if the differential difference, Δz, in the manometer's mercury column is 1 ft, assuming $N = 10$, $n = 3$ and $h = 2$ ft. Also, assume the oil has a density, $\rho_{Oil} = 1.71$ slug/ft^3.

3.21 An object rests on top of a piston in a cylindrical tank filled with a liquid of density $\rho = 900$ kg/m^3. The attached pressure gage indicates a pressure, p_g, of 50 kPa. How much does the object weigh if the gage is $\Delta z = 0.25$ m above the piston as shown? The tank diameter, D, is 0.5 m.

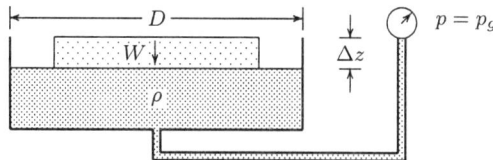

Problems 3.21, 3.22, 3.23

3.22 An object rests on top of a piston in a cylindrical tank filled with a liquid of density $\rho = 1.75$ slug/ft^3. The object weighs 1 ton and the diameter of the tank, D, is 1.5 ft. If the attached pressure gage indicates a pressure, p_g, of 7.47 psi, how far above the piston, Δz, is the gage located?

3.23 An object with a weight, W, of 100 kN rests on top of a piston in a cylindrical tank filled with a liquid of density ρ. The diameter of the tank, D, is 1 m. If the attached pressure gage, located a distance $\Delta z = 2$ m above the piston, indicates a pressure, p_g, of 100 kPa, what is the liquid density, ρ?

3.24 A *differential manometer* is a device connected between two tanks as shown. It is designed to measure the pressure difference $p_1 - p_2$.

(a) Determine $p_1 - p_2$ between the two tanks as a function of ρ_1, ρ_2, g, h_1, h_2, λ and ρ_{Hg}.

(b) Suppose $h_1 = h_2 = 1$ m, $\lambda = 10$ cm, fluid 1 is kerosene, fluid 2 is glycerin and the temperature is $20°$ C. What is $p_1 - p_2$?

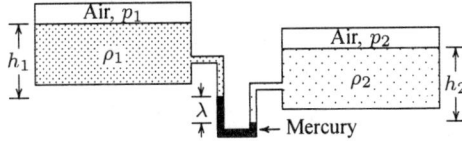

Problems 3.24, 3.25

3.25 A *differential manometer* is a device connected between two tanks as shown. It is designed to measure the pressure difference $p_1 - p_2$.

(a) Determine the distance λ as a function of ρ_1, ρ_2, g, h_1, h_2, p_1, p_2 and ρ_{Hg}.

(b) Suppose $h_1 = 1$ ft, $h_2 = 0.75$ ft, $p_1 = p_2$, fluid 1 is ethyl alcohol, fluid 2 is carbon tetrachloride and the temperature is $68°$ F. What is λ?

3.26 An *inclined manometer* is a large spherical container with an inclined tube of small diameter attached. The tube is inclined at an angle α to the horizontal and has a scale with markings separated by 1 cm. If the liquid in the manometer is carbon tetrachloride, what must the angle α be if each marking corresponds to a change in pressure, Δp_a, of 100 Pa? Assume the temperature is $20°$ C.

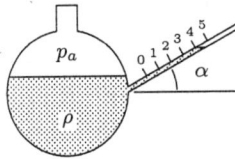

Problems 3.26, 3.27

3.27 An *inclined manometer* is a large spherical container with an inclined tube of small diameter attached. The tube is inclined at an angle α to the horizontal and has a scale with markings separated by 1 in. We would like each marking to correspond to a change in pressure, Δp_a, of 1 lb/ft^2? We are considering several fluids including ethyl alcohol, water, glycerin and mercury. Assuming the temperature is $68°$ F, compute the angle α required for all 4 of the candidate fluids.

3.28 A *micromanometer* can be used to measure very small pressure differences. The device uses two fluids of slightly different densities, ρ and $\rho + \Delta\rho$, where $\Delta\rho$ is small compared to ρ. The cylindrical reservoirs are very large ($D \gg d$) so that the reservoir levels are nearly equal when fluid moves from the reservoir to the tube and vice versa.

(a) Develop a formula for h as a function of p_1, p_2, ρ, $\Delta\rho$ and g.

(b) Assume the pressure difference is $p_1 - p_2 = 10$ Pa and that the lighter fluid is ethyl alcohol. Contrast the height, h, when the heavier fluid is kerosene with the value when the heavier fluid is carbon tetrachloride. Assume the temperature is $20°$ C, and express your answers in millimeters.

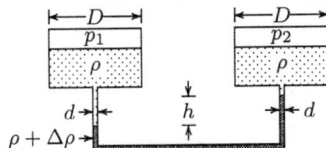

Problems 3.28, 3.29

3.29 A *micromanometer* can be used to measure very small pressure differences. The device uses two fluids of slightly different densities, ρ and $\rho + \Delta\rho$, where $\Delta\rho$ is small compared to ρ. Assume the fluid levels in the cylindrical reservoirs are equal when $p_1 = p_2$. Develop a formula for h as a function of p_1, p_2, ρ, $\Delta\rho$, D, d and g.

3.30 A *hydraulic jack* uses a fluid of density ρ and a lever of length h to lift an object of weight W. If the piston areas are A and $5A$, what force, F, must be applied to support an object of weight W?

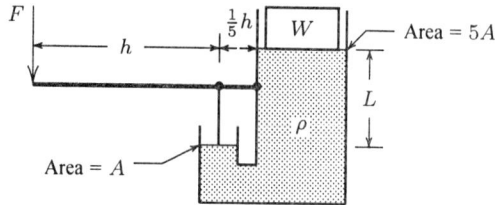

Problem 3.30

3.31 Use the parallel-axis theorem to verify that the moments for a semicircle and a circle quoted in Figure 3.8 are consistent.

3.32 Use the parallel-axis theorem to determine the moment of inertia for the double triangular geometry shown.

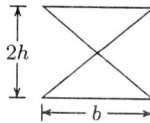

Problem 3.32

3.33 Compute the centroid location, $\bar{z}$, area, A, and moment of inertia, I, for the diamond shaped geometry shown.

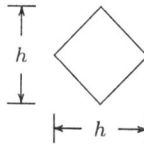

Problem 3.33

3.34 Compute the centroid location, $\bar{z}$, area, A, and moment of inertia, I, for the elliptical geometry shown. Verify that your results match those of a circle when $a = b$.

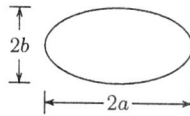

Problem 3.34

3.35 Determine the magnitude of the force, to the nearest kN, and the center of pressure on the dam if the water is $h = 10$ m deep, the dam is $3h$ wide and $\alpha = 60°$.

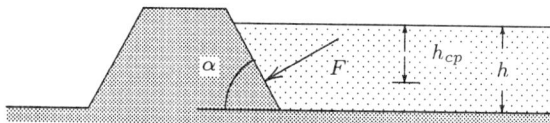

Problems 3.35, 3.36

3.36 Determine the magnitude of the force, to the nearest ton, and the center of pressure on the dam if the water is $h = 10$ ft deep, the dam is $3h$ wide and $\alpha = 60°$.

3.37 A square gate of side h can rotate about a pivot as shown. At what depth of the gate's top, H, will the gate open if $h_p = \frac{2}{5}h$?

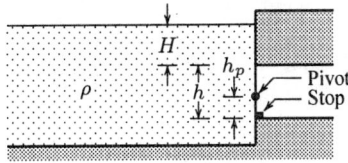

Problems 3.37, 3.38, 3.39, 3.40

3.38 A circular gate of diameter h can rotate about a pivot as shown. At what depth of the gate's top, H, will the gate open if $h_p = \frac{2}{5}h$?

3.39 Determine the pivot location, h_p, for a rectangular gate of width $2h$ (out of the page) such that it will open when $H = h$.

3.40 Determine the pivot location, h_p, for a triangular gate of base $2h$ (out of the page) such that it will open when $H = h$. The base lies below the pivot.

3.41 For the rectangular gate shown, the stop can resist a force of up to 5 tons. The gate is $1.5h$ wide (out of the page). For what depth H will the stop fail if $h = 3$ ft?

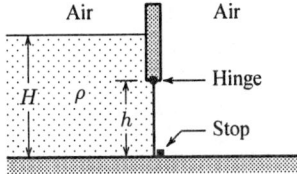

Problem 3.41

3.42 A gate of width $\frac{1}{2}H$ (out of the page) is designed to open and release fresh water when the tide goes out. The density of fresh water is ρ and the density of ocean water is $(1 + \epsilon)\rho$. Neglecting the weight of the gate, verify that the ocean level, h, at which the gate will open can be determined from the following cubic equation.

$$(1 + \epsilon)\left(\frac{h}{H}\right)^2 \left(7 - 2\frac{h}{H}\right) = 5$$

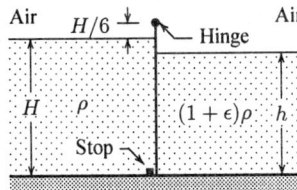

Problem 3.42

3.43 A hinged, triangular access port is provided in the side of a form containing liquid concrete (density ρ). Determine the hydrodynamic force exerted on the access port by the concrete. What force is needed at the upper tip of the access port to prevent it from opening?

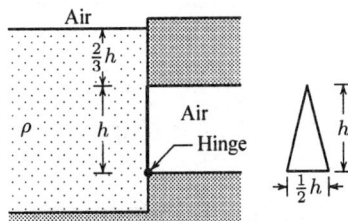

Problem 3.43

3.44 Consider the rectangular gate shown below. The width of the gate (normal to the page) is $2L$, where L is the distance (measured parallel to the gate) from the bottom to the fulcrum. You may neglect the gate's weight.

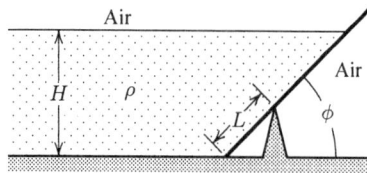

Problem 3.44

(a) Compute the height H to which a fluid of density ρ must rise in order to tip over the gate. Express your answer in terms of L and ϕ.

(b) What is the force normal to the gate when the gate is just about to tip over? Express your answer in terms of ρ, g, L and ϕ.

(c) Indicate why your answer is sensible for the limiting cases $\phi \to 0$ and $\phi \to \pi/2$?

3.45 The gate shown is hinged at point A. It is square so that its width (normal to the page) is $\sqrt{2}H$. Compute the force at point B required to hold the gate closed. You may ignore the gate's weight.

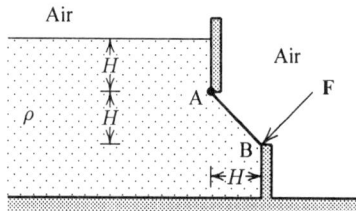

Problem 3.45

3.46 Consider the hinged gate with a weight connected through a pulley arrangement as shown. The gate is square with side H. The weight is a cube with side s and density 2ρ. You may ignore the gate's weight and any friction that might develop in the pulley. At what depth, h, will the fluid cause the gate to rotate in the clockwise direction?

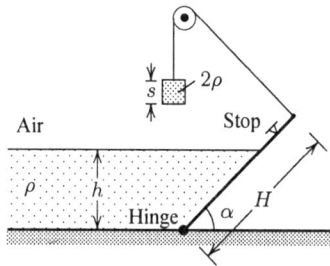

Problem 3.46

3.47 Consider the L-shaped gate, ABC, shown below. The gate has width H out of the page and is free to pivot about the hinge at B. You may neglect the weight of the gate. At what depth h will the gate automatically open?

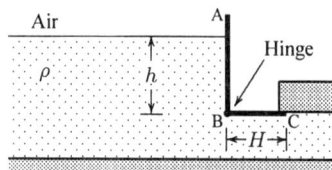

Problem 3.47

3.48 Determine the force due to hydrostatic pressure acting on the hinge of the gate for:

(a) a square gate.

(b) a triangular gate with base width L (the base is at the stop).

To simplify your analysis, first develop a general expression for the force on the hinge and then proceed to the specific gate geometry.

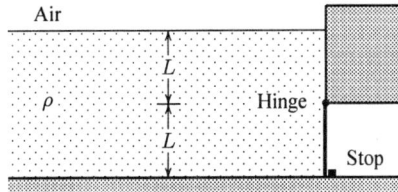

Problem 3.48

3.49 A circular gate of radius R is held in place by pivots about a horizontal axis through its centroid, and a gate stop. What force must be applied to the bottom of the gate by the stop to hold it closed?

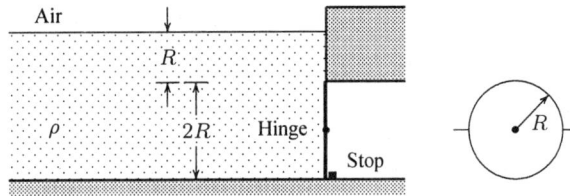

Problem 3.49

3.50 Determine the force due to hydrostatic pressure acting on the hinge of the gate shown. The fluid density is ρ and the gate width (out of the page) is $6H$.

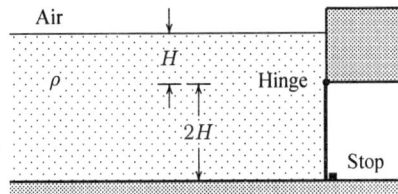

Problem 3.50

3.51 The force on the gate's hinge, F_h, is 2/5 times the hydrodynamic force acting on the gate, F. The gate is rectangular and has unit width out of the page. Determine the depth of the hinge, h.

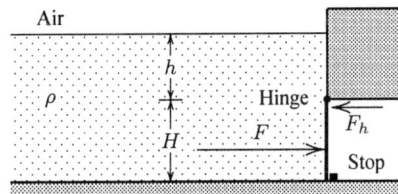

Problems 3.51, 3.52

3.52 The force on the gate's hinge, F_h, is 7/12 times the hydrodynamic force acting on the gate, F. The gate is triangular with base H at the hinge. Determine the depth of the hinge, h.

3.53 The gate shown is a quarter circle of radius H, and is $4H$ wide (out of the page). Determine the applied horizontal force, F, required to hold the gate closed. You may ignore the weight of the gate.

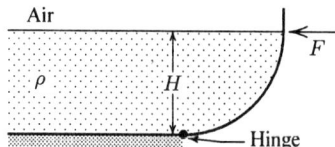

HINT: Use the fact that, relative to the hinge:

$$x_{cp} = \frac{4H}{3\pi}$$

Problem 3.53

3.54 A dam is made up of three linear sections as shown. The dam is $4h$ wide (out of the page) and $\alpha = \tan^{-1}(1/2) = 26.565°$.

(a) Determine the force, **F**, on the dam.

(b) Determine the z coordinate of the center of pressure, z_{cp}.

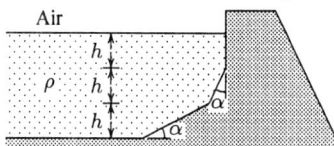

Problem 3.54

3.55 A dam has a parabolic shape defined by $z = 2h(1 - x/h)^2$ for $z > 0$. The dam is $6h$ wide (out of the page).

(a) Determine the force, **F**, on the dam.

(b) Determine the z coordinate of the center of pressure, z_{cp}.

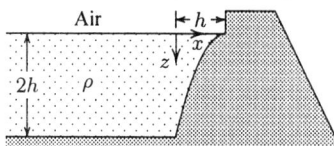

Problem 3.55

3.56 A dam has a parabolic shape defined by $z = h\sqrt{1 - x/h}$ for $0 < x < h$. The dam is $4h$ wide (out of the page).

(a) Determine the force, **F**, on the dam.

(b) Determine the z coordinate of the center of pressure, z_{cp}.

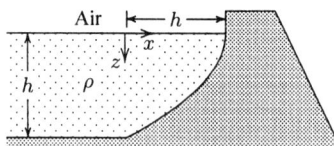

Problem 3.56

3.57 A swimming pool has a set of n steps at one end. Each step has a horizontal and vertical length of h/n, where h is the total depth. The width of the steps (out of the page) is also h. Compute the hydrostatic force on the set of steps.

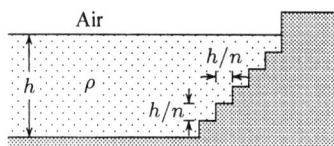

HINT: Use the fact that

$$\sum_{i=1}^{n} i = \frac{n(n+1)}{2}$$

Problem 3.57

3.58 A dam has a quarter-circle shape. The dam is $12h$ wide (out of the page).

(a) Determine the force, **F**, on the dam.

(b) Determine the z coordinate of the center of pressure, z_{cp}.

(c) Determine the x coordinate of the center of pressure, x_{cp}.

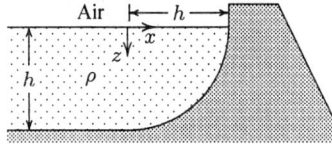

Problem 3.58

3.59 Consider the arc AB separating two liquids of density ρ and 2ρ as shown. Determine the depth h as a function of H such that the net force on the arc in the x direction vanishes. For this value of h, what is the vertical force on the arc assuming its weight is negligible. The area of OAB is $\frac{2}{3}H^2$ and the arc has unit width out of the page.

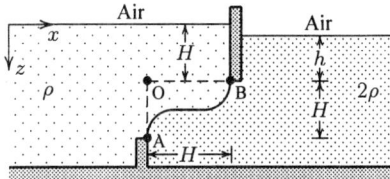

Problem 3.59

3.60 Determine the force on the gate AB if the upper layer of fluid has density $\rho_1 = \rho$ and the lower layer has density $\rho_2 = 2\rho$. The gate is rectangular and has unit width (out of the page).

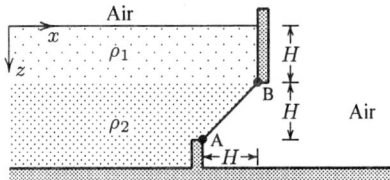

Problem 3.60, 3.61

3.61 Determine the force on the gate AB if the upper layer of fluid has density $\rho_1 = \rho$ and the lower layer has density $\rho_2 = 3\rho$. The gate is rectangular and has width $3H$ (out of the page).

3.62 Arc AB separates fluids of density ρ_1 and ρ_2. The arc has unit width normal to the page. The area of OAB is $\frac{2}{3}H^2$, and the centroid of OAB lies at $x = \frac{1}{4}H$.

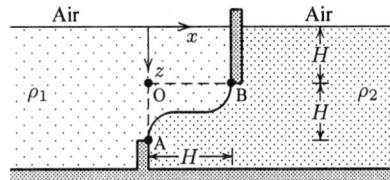

Problem 3.62

(a) Compute the force components, F_x and F_z, acting on the arc AB. Is your answer sensible for the limiting case $\rho_1 = \rho_2$?

(b) Compute the vertical location of the center of pressure, z_{cp}.

(c) Compute the horizontal location of the center of pressure, x_{cp}. (Take moments relative to OA.)

3.63 Compute the net force on the arc AB, assuming that the area of OAB is $\frac{1}{2}H^2$. Also, assume unit width out of the page.

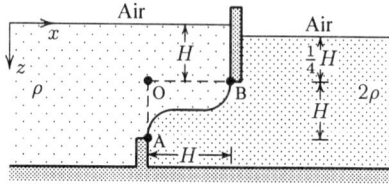

Problem 3.63

3.64 Consider a gate AB of constant width, H, out of the page separating two liquids of density ρ_1 and ρ_2. The net vertical force on the gate, F_z, is zero. Assuming the area of OAB is $H^2/3$, determine the density ratio, ρ_2/ρ_1, and the horizontal force, F_x.

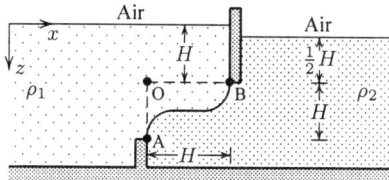

Problem 3.64

3.65 Consider the circular-arc gate shown. You may assume the gate has unit width out of the page. Express your answers in terms of ρ, g and ℓ.

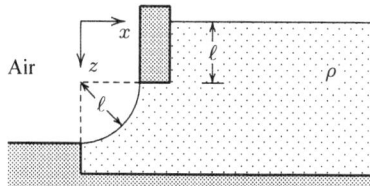

Problem 3.65

(a) Compute the hydrostatic force components, F_x and F_z, acting on the arc AB.

(b) Compute the vertical location of the center of pressure, z_{cp}.

(c) Compute the horizontal location of the center of pressure, x_{cp}.

3.66 Consider a container of water (density = ρ) with the shape shown below. The container has width h in the direction normal to the page. Determine the hydrostatic force, **F**, and center of pressure, z_{cp}, on the circular arc AB.

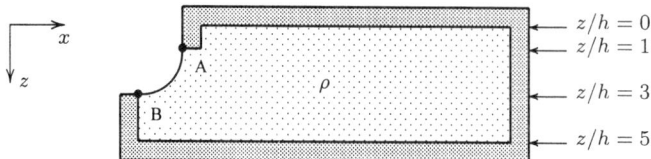

Problem 3.66

3.67 One fourth of an object of density ρ_o and volume V is submerged in a fluid of density ρ. What is the magnitude of the buoyancy force on the object?

3.68 A cement slab of irregular shape weighs 180 lb in air. When it is submerged in a tank filled with water at 68° F, a force of 117.6 lb is required to lift it from the bottom of the tank. What is the volume of the slab? What is its density?

3.69 An irregularly-shaped object made of copper weighs 2.160 kN in air. When it is submerged in a tank filled with water at $10°$ C, a force of 1.915 kN is required to lift it from the bottom of the tank. What is the volume of the object? What is its density?

3.70 King Hiro ordered a new crown to be made of pure gold. When he received the crown, he suspected that other metals had been used in its construction. Archimedes found that, when immersed in water (density ρ), the crown's weight was 7/8 of its actual weight. Assuming the crown was made of a mixture of steel (density 7.8ρ) and gold (density 19.3ρ), what percentage, by weight, was gold?

3.71 A cube of balsa wood (density, $\rho_b = 0.24$ slug/ft^3) measures 4 inches on its sides. What force is required to hold it in place completely submerged in water at $68°$ F?

3.72 A spherical balloon of diameter 12.5 m is filled with helium and is pressurized to 122 kPa. Determine the tension in the mooring line if the surrounding air is at 1 atm and $20°$ C. You can ignore the weight of the material from which the balloon is constructed.

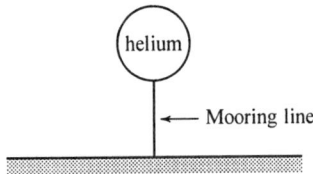

Problem 3.72

3.73 Consider a coal-carrying barge loaded with 100 tons of coal making its way down the Mississippi River. The empty barge weighs 25 tons. If the barge is 15 ft wide, 40 ft long and 8 ft high, what is its depth below the surface of the water?

3.74 To what depth, D, will the cube shown below float in the two-liquid reservoir? The density of the cube is $\frac{5}{4}\rho$.

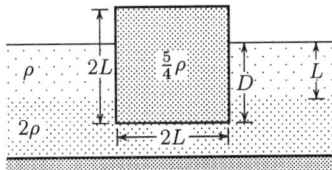

Problem 3.74

3.75 An object of density 4ρ is immersed in a two-layer fluid as shown. If the object has constant cross-sectional area, A (out of the page), how much of the object's volume lies in the lower layer?

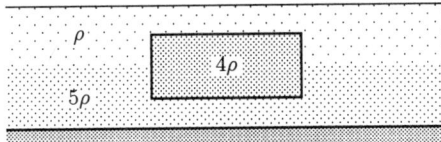

Problem 3.75

3.76 To what depth, D, will the block shown below float in the two-liquid reservoir? The density of the block is $\frac{4}{5}\rho$ and it has unit width out of the page.

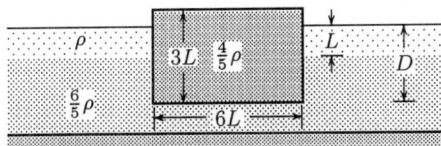

Problem 3.76

3.77 A cube-shaped "balloon" filled with helium (density ρ_{He}) rises through the stratosphere for which the density is $\rho = \rho_o \exp(-gz/RT)$. The "balloon" is designed so that ρ_{He} and h remain constant for all z.

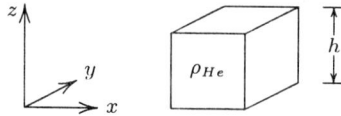

Problem 3.77

(a) Compute the buoyancy force on the balloon. The lower face is located at z and the upper face at $(z + h)$.

(b) Now, assuming $gh/RT \ll 1$, to what altitude will the balloon rise?

3.78 The fully enclosed tank shown below is divided into two independent chambers by a thin wall of negligible thickness. One chamber contains a fluid of density ρ while the other contains a fluid of density 3ρ. A prism of height H out of the page is attached to the wall as shown. If the density of the prism is $\frac{3}{2}\rho$, compute the force required to hold the prism in place.

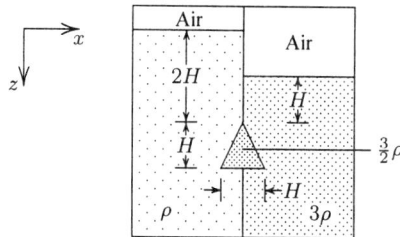

Problem 3.78

3.79 A wooden block with square cross section is in equilibrium when immersed in a fluid of density ρ as shown. The width of the block is H out of the page, and the side of the cross section is L. Ignoring any effects of friction in the hinge, determine the density of the block, ρ_b.

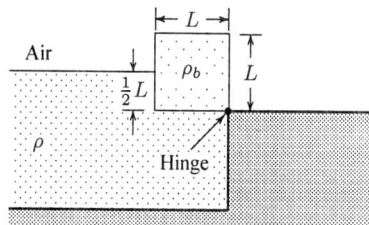

Problem 3.79

3.80 An enclosed storage tank is designed to take water in from a reservoir when the water depth becomes too large by means of a gate that opens. The gate is hinged at the top, is $6L$ high, and L wide (out of the page). A compressor is used to regulate the air pressure in the tank. Find the minimum pressure differential, $p_a - p_o$, required to keep the gate closed when water with depth $2L$ depth already exists in the tank.

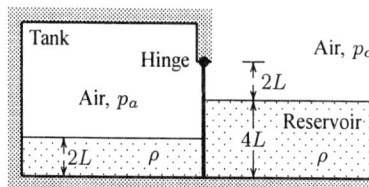

Problem 3.80

3.81 Compute the force, F, and center of pressure, z_{cp}, on the vertical wall below, assuming unit width out of the page. **NOTE:** The formulas developed for constant density do not apply to this problem.

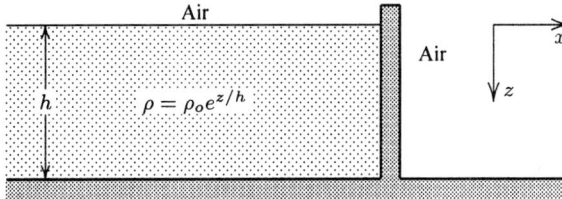

Problem 3.81

3.82 An object of constant cross-sectional area, A, height, h, and density, ρ_m, is floating in a fluid whose density varies with depth as $\rho = \rho_o(1 + 6z/h)$.

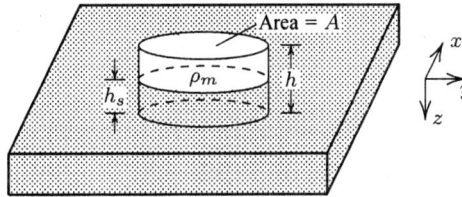

Problem 3.82

(a) If $\rho_m = 2\rho_o$, what fraction of the object, h_s/h, is submerged?

(b) What is the minimum ρ_m/ρ_o for which the object is totally submerged?

3.83 Compute the force on the gate AB for a density-stratified fluid in which the density is given by $\rho(z) = \rho_o(1 + \alpha z/H)$, where α is a constant. The projection of gate on OA is rectangular and has unit width out of the page. Also, the weight of the fluid in OAB is $\frac{1}{4}\rho_o g H^2(1 + \alpha/2)$. Does your answer make sense in the limit $\alpha \to 0$?

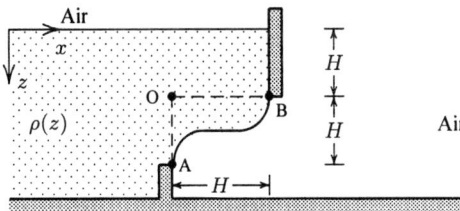

Problem 3.83

3.84 A cube-shaped weight of density ρ_o is connected to a rectangular gate of width $2h$ out of the page as shown. The gate is designed to open when the water level, H, drops below $\frac{5}{2}h$. Determine the required value for ρ_o.

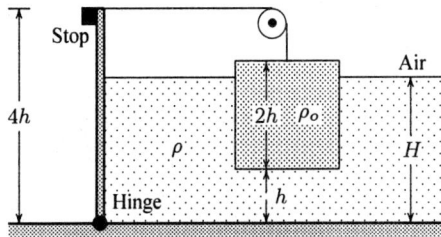

Problem 3.84

3.85 Determine the minimum volume of steel-reinforced concrete needed to keep the rectangular gate shown in a closed position. The width of the gate out of the page is w, the density of water is ρ and the density of the concrete block is 3ρ.

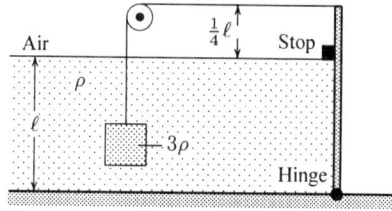

Problem 3.85

3.86 A rectangular gate of width $4L$ out of the page is hinged at point A and holds back a fluid of density ρ as shown. A cube of density 9ρ and side $\frac{1}{2}L$ is connected to the gate through its centroid with a pulley arrangement so that the weight exerts a force normal to the gate. Neglecting the weight of the gate, determine the minimum force, P, required to hold the gate closed.

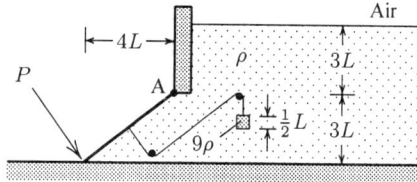

Problem 3.86

3.87 A liquid of density $N\rho$ lies below a liquid of density ρ. A rectangular piston of width H out of the page is attached to a totally submerged object of density $\frac{1}{3}\rho$ and volume $4\sqrt{2}H^3$, with a pulley attachment as shown. The pulley cable makes a right angle with the piston, which is always at a $45°$ angle to the horizontal, and the pressure below the piston is atmospheric.

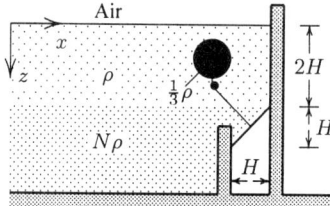

Problem 3.87

(a) Without writing any equations, explain why the x and z components of the hydrostatic force on the piston must be equal.

(b) If there is zero net force on the piston, what is the value of N?

3.88 A hemispherical dome is located one dome radius, R, below the surface of a liquid of density, ρ. Determine the force components needed to hold the dome in place, and the vertical coordinate of the center of pressure, z_{cp}.

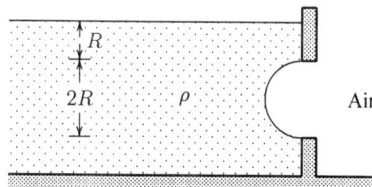

Problem 3.88

3.89 A cylindrical barrier holds water back as shown. If the length of the cylinder is L and its radius is R, determine its weight and the force exerted against the wall. Let ρ denote the density of water.

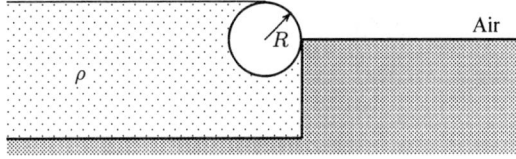

Problem 3.89

3.90 The figure below depicts a cylindrical *weir*, or low dam, of diameter D. The weir separates two reservoirs containing a fluid of density ρ as shown, and has a width w (out of the page). Determine the hydrostatic force acting on the weir.

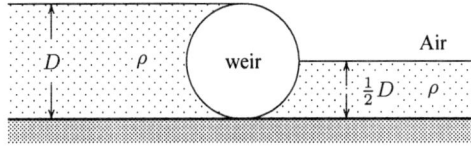

Problem 3.90

Chapter 4

Kinematics

Before we can embark on a study of fluids in motion, we must develop the equations governing conservation of mass, momentum and energy. While this point is obvious, there is a more subtle point we must address. Specifically, a central issue in developing the equations is the way we choose to describe spatial coordinates and velocities. The choice is not arbitrary, and the one we make for most fluid-mechanical applications differs from the choice made for elementary applications of Newton's laws of motion, i.e., for classical mechanics.

In this chapter, we address this fundamental difference. It is far more convenient in fluid mechanics to focus upon a fixed point or volume rather than to track individual fluid particles. The mathematical description used in fluid mechanics is known as the **Eulerian** description as opposed to the **Lagrangian** description used in classical mechanics. This is a key element of the topic known as **kinematics**. We take the term kinematics to mean basic relationships amongst coordinates and velocities required to describe a given problem, as opposed to developing and solving the dynamical equations of motion.

We begin by showing the difference between the Eulerian and Lagrangian descriptions, including a simple example. Then, we introduce two interrelated flow properties known as **vorticity** and **circulation**. Although our discussion is necessarily brief at this point, the central importance of vorticity in understanding the physics of fluid motion warrants special attention at all levels of study. The chapter also discusses **streamlines**, **streaklines** and **pathlines**. All are useful for visualizing details of a given flowfield, especially in experimental investigations.

After introducing these basic kinematical concepts, we derive a key integral relationship expressing the rate of change of properties in a fixed volume that contains different fluid particles at each instant. This is the famous continuum mechanics theorem known as the **Reynolds Transport Theorem**. We will make use of this theorem in Chapters 5 and 7 in deriving the basic conservation laws for a fluid.

4.1   Eulerian Versus Lagrangian Description

Regardless of which kinematical description we choose, we must express position and velocity as vectors. Throughout this text, we will use the same notation, so it is worthwhile to identify our notation at this point. Most of our work will be done in rectangular Cartesian coordinates for which the position vector, $\mathbf{r}$, and velocity vector, $\mathbf{u}$, are

$$\mathbf{r} = x\,\mathbf{i} + y\,\mathbf{j} + z\,\mathbf{k} \qquad \text{and} \qquad \mathbf{u} = u\,\mathbf{i} + v\,\mathbf{j} + w\,\mathbf{k} \qquad (4.1)$$

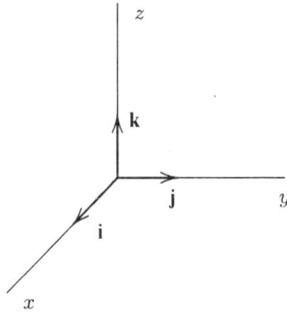

Velocity vector:
$$\mathbf{u} = u\,\mathbf{i} + v\,\mathbf{j} + w\,\mathbf{k}$$

Position vector:
$$\mathbf{r} = x\,\mathbf{i} + y\,\mathbf{j} + z\,\mathbf{k}$$

Figure 4.1: *Rectangular Cartesian coordinates.*

As shown in Figure 4.1, the quantities (x, y, z) are rectangular Cartesian coordinates, while (u, v, w) and $(\mathbf{i}, \mathbf{j}, \mathbf{k})$ are the velocity components and unit vectors, respectively, in the (x, y, z) directions.

In applying Newton's second law of motion, we compute the motion of an object by setting its mass times its acceleration equal to the sum of all forces acting. Inherent in the approach taken in classical mechanics is the way we describe the object's location, velocity and acceleration. Specifically, classical mechanics uses the **Lagrangian** description in which we track a specified object for all time so that its position vector, $\mathbf{r}$, is given by

$$\mathbf{r} = \mathbf{r}(\mathbf{r}_o, t) \quad \text{where} \quad \mathbf{r} = \mathbf{r}_o \quad \text{at} \quad t = t_o \tag{4.2}$$

Equation (4.2) states the essence of the Lagrangian description. The position of a given fluid particle at a time t is a function of time and its initial position. The velocity vector, $\mathbf{u}$, is the rate of change of position vector, $\mathbf{r}$, with respect to time:

$$\mathbf{u} = \left(\frac{\partial \mathbf{r}}{\partial t}\right)_{\mathbf{r}_o} \tag{4.3}$$

where the subscript $\mathbf{r}_o$ indicates that the particle's initial position is held constant in computing the derivative. This is a consequence of the fact that we are following the same particle around as it moves through space. If the sum of all external forces acting on an object of mass m is $\mathbf{F}$, its motion according to the laws of classical mechanics is governed by the *linear* partial differential equation:

$$\mathbf{F} = m\mathbf{a} = m\left(\frac{\partial^2 \mathbf{r}}{\partial t^2}\right)_{\mathbf{r}_o} \tag{4.4}$$

where $\mathbf{a}$ denotes acceleration.

The primary advantage of using the Lagrangian description is the linearity of the momentum equation, i.e., Equation (4.4). Nevertheless, the Lagrangian description is inconvenient in fluid mechanics for two key reasons. First, to implement this description we would have to keep track of a very large number of fluid particles to adequately resolve our continuum field. Second, the relative positions of fluid particles change by nontrivial amounts, so that viscous stresses are very complicated functions in the Lagrangian description. Its primary applications are in rarefied (very low density) gasdynamics, free-surface (water-air interface) flows and some combustion applications.

The vast majority of fluid mechanics problems are solved using the **Eulerian** description. In this alternative kinematical formulation, we focus upon a fixed point in space and observe fluid particles as they pass by. The velocity is expressed as a function of time and of the position in space at which we make our observations, viz.,

$$\mathbf{u} = \mathbf{u}(\mathbf{r}, t) = \mathbf{u}(x, y, z, t) \tag{4.5}$$

Clearly, we cannot compute the acceleration of a fluid particle by taking the partial derivative with respect to time. If we did this, we would, in effect, be taking the difference between the velocities of two different fluid particles in forming the derivative. Instead, we must follow the fluid particle for a differential time increment to determine the differential change in its velocity. Using this information we can compute the acceleration. In other words, we still must differentiate the velocity with respect to time, holding the fluid particle's initial Lagrangian coordinates constant.

To illustrate why $\partial \mathbf{u}/\partial t$ is not the true acceleration in the Eulerian description, consider the following example. Suppose you are focusing on a fixed point on a highway. At time $t = 0$, Granny B passes the point you are observing moving at 60 mph = 88 ft/sec. Because a Highway Patrol car is just ahead, all of the traffic is also moving at 60 mph, so that Granny B maintains a constant speed. Four seconds later, Leadfoot D passes your observation point moving at 90 mph = 132 ft/sec. Having noticed the "slow-moving" traffic, Leadfoot D has been rapidly decelerating to avoid colliding with Granny B and/or receiving a speeding ticket from the Highway Patrolman. Using your observations at the fixed point on the highway, you would estimate that traffic is *accelerating* at your observation point according to

$$a \approx \frac{\Delta u}{\Delta t} = \frac{132 \text{ ft/sec} - 88 \text{ ft/sec}}{4 \text{ sec}} = 11 \text{ ft/sec}^2 \tag{4.6}$$

In reality, Granny B has zero acceleration and Leadfoot D is *decelerating*, so that your computation is obviously incorrect.

Although the required process for computing the acceleration is conceptually straightforward, the mathematical details are not. The problem we face is the following. On the one hand, we need information about the same fluid particle's velocity at two different times in order to compute its acceleration. In terms of the Lagrangian description, all we have to do is form the following limit:

$$\mathbf{a} = \left(\frac{\partial \mathbf{u}}{\partial t}\right)_{\mathbf{r}_o} = \lim_{\Delta t \to 0} \frac{\mathbf{u}(\mathbf{r}_o, t + \Delta t) - \mathbf{u}(\mathbf{r}_o, t)}{\Delta t} \tag{4.7}$$

The operation is simple because $\mathbf{r}_o$ and t are independent of each other, and following the same fluid particle means holding $\mathbf{r}_o$ constant.

On the other hand, our choice of the Eulerian description means we have information about the velocity throughout the flowfield as a function of time and position relative to a fixed set of coordinates, (x, y, z). However, the Eulerian description includes no explicit information about where an individual fluid particle has been in the past nor where it will be at a future time. To obtain the information we need, we must regard the position vector as being an implicit function of time, i.e., $\mathbf{r} = \mathbf{r}(t)$, so that

$$\mathbf{a} = \lim_{\Delta t \to 0} \frac{\mathbf{u}(\mathbf{r} + \Delta \mathbf{r}, t + \Delta t) - \mathbf{u}(\mathbf{r}, t)}{\Delta t} \tag{4.8}$$

If we consider an infinitesimally small time increment, Δt, we can approximate the velocity as being $\mathbf{u}(\mathbf{r}, t)$ for the entire time it moves.[1] Hence, the distance traveled by the fluid particle

---

[1] The error in $\Delta \mathbf{r} \approx \mathbf{u}\Delta t$ due to this approximation goes to zero as $(\Delta t)^2$ because the velocity approaches its value at time t in the limit $\Delta t \to 0$.

is approximately $\mathbf{u}\Delta t$. Thus, as illustrated in Figure 4.2, a fluid particle located at position $\mathbf{r}(t)$ at time t, will move to

$$\mathbf{r} + \Delta\mathbf{r} \approx \mathbf{r}(t) + \mathbf{u}(\mathbf{r}, t)\Delta t \tag{4.9}$$

Then, in terms of Cartesian coordinates ($\mathbf{r} = x\mathbf{i} + y\mathbf{j} + z\mathbf{k}$ and $\mathbf{u} = u\mathbf{i} + v\mathbf{j} + w\mathbf{k}$), we can say that

$$\Delta\mathbf{r} = \mathbf{u}\Delta t \quad\Longrightarrow\quad \Delta x = u\Delta t, \quad \Delta y = v\Delta t, \quad \Delta z = w\Delta t \tag{4.10}$$

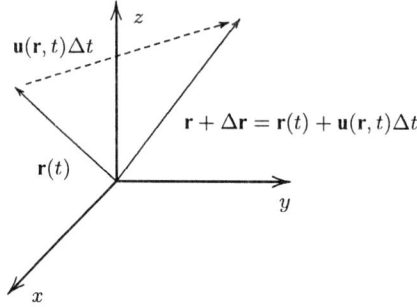

Figure 4.2: *Trajectory of a fluid particle.*

Now, expanding the velocity at time $t + \Delta t$ in Taylor series, and retaining only terms linear in Δt, we find

$$
\begin{aligned}
\mathbf{u}(\mathbf{r} + \Delta\mathbf{r}, t + \Delta t) &= \mathbf{u}(x + \Delta x, y + \Delta y, z + \Delta z, t + \Delta t) \\
&\approx \mathbf{u}(x, y, z, t) + \frac{\partial\mathbf{u}}{\partial x}\Delta x + \frac{\partial\mathbf{u}}{\partial y}\Delta y + \frac{\partial\mathbf{u}}{\partial z}\Delta z + \frac{\partial\mathbf{u}}{\partial t}\Delta t
\end{aligned}
\tag{4.11}
$$

Therefore, combining Equations (4.10) and (4.11) and returning to the vector shorthand $\mathbf{u}(\mathbf{r}, t)$ for $\mathbf{u}(x, y, z, t)$, the change in velocity of the fluid particle between times $t + \Delta t$ and t is

$$\mathbf{u}(\mathbf{r} + \Delta\mathbf{r}, t + \Delta t) - \mathbf{u}(\mathbf{r}, t) \approx u\frac{\partial\mathbf{u}}{\partial x}\Delta t + v\frac{\partial\mathbf{u}}{\partial y}\Delta t + w\frac{\partial\mathbf{u}}{\partial z}\Delta t + \frac{\partial\mathbf{u}}{\partial t}\Delta t \tag{4.12}$$

With a little rearrangement of terms, we can rewrite Equation (4.12) as

$$\lim_{\Delta t \to 0}\frac{\mathbf{u}(\mathbf{r} + \Delta\mathbf{r}, t + \Delta t) - \mathbf{u}(\mathbf{r}, t)}{\Delta t} = \frac{\partial\mathbf{u}}{\partial t} + u\frac{\partial\mathbf{u}}{\partial x} + v\frac{\partial\mathbf{u}}{\partial y} + w\frac{\partial\mathbf{u}}{\partial z} \tag{4.13}$$

Consequently, denoting the acceleration by $d\mathbf{u}/dt$, Equation (4.13) tells us

$$\frac{d\mathbf{u}}{dt} = \frac{\partial\mathbf{u}}{\partial t} + u\frac{\partial\mathbf{u}}{\partial x} + v\frac{\partial\mathbf{u}}{\partial y} + w\frac{\partial\mathbf{u}}{\partial z} \tag{4.14}$$

Finally, we can rewrite Equation (4.14) in more compact vector notation by introducing the ∇ operator. Hence,

$$\frac{d\mathbf{u}}{dt} = \frac{\partial\mathbf{u}}{\partial t} + \left(u\frac{\partial}{\partial x} + v\frac{\partial}{\partial y} + w\frac{\partial}{\partial x}\right)\mathbf{u} = \frac{\partial\mathbf{u}}{\partial t} + \mathbf{u}\cdot\nabla\mathbf{u} \tag{4.15}$$

where

$$\nabla = \mathbf{i}\frac{\partial}{\partial x} + \mathbf{j}\frac{\partial}{\partial y} + \mathbf{k}\frac{\partial}{\partial z} \tag{4.16}$$

Equation (4.15) is known as the **Eulerian derivative** of the velocity. In general, we regard the differential operator

$$\frac{d}{dt} = \frac{\partial}{\partial t} + \mathbf{u} \cdot \nabla \qquad (4.17)$$

as the *rate of change following a fluid particle*. Note that in Equation (4.17), the partial derivative with respect to time is the **unsteady** contribution, while the second term on the right-hand side is known as **convection** (some authors call it **advection**).

For example, consider the following velocity vector,

$$\mathbf{u} = Axt\,\mathbf{i} + By\,\mathbf{j} \qquad (4.18)$$

where A and B are constants. Since the individual velocity components are $u = Axt$ and $v = By$, the acceleration corresponding to this velocity is

$$
\begin{aligned}
\frac{d\mathbf{u}}{dt} &= \frac{\partial}{\partial t}\underbrace{(Axt\,\mathbf{i} + By\,\mathbf{j})}_{\mathbf{u}} + \underbrace{Axt\frac{\partial}{\partial x}}_{u\partial/\partial x}\underbrace{(Axt\,\mathbf{i} + By\,\mathbf{j})}_{\mathbf{u}} + \underbrace{By\frac{\partial}{\partial y}}_{v\partial/\partial y}\underbrace{(Axt\,\mathbf{i} + By\,\mathbf{j})}_{\mathbf{u}} \\
&= Ax\,\mathbf{i} + Axt(At\,\mathbf{i}) + By(B\,\mathbf{j}) = Ax\left(1 + At^2\right)\mathbf{i} + B^2 y\,\mathbf{j} \qquad (4.19)
\end{aligned}
$$

The Eulerian derivative effectively bridges the gap between the Lagrangian and Eulerian descriptions. To help illustrate why this is true, consider the convective part of the Eulerian derivative. We can form a unit vector, $\mathbf{n}$, parallel to the velocity vector by writing

$$\mathbf{n} = \frac{\mathbf{u}}{|\mathbf{u}|} \qquad (4.20)$$

Then, the convective part of the Eulerian derivative can be rewritten as

$$\mathbf{u} \cdot \nabla = |\mathbf{u}|\,\mathbf{n} \cdot \nabla = |\mathbf{u}|\frac{\partial}{\partial n} \qquad (4.21)$$

where $\partial/\partial n$ is the derivative in the direction of $\mathbf{n}$. Hence, the convective part of the Eulerian derivative represents the rate of change in the direction of motion, i.e., following the fluid particle.

The advantages of using the Eulerian description are that we have no need to track a large number of fluid particles, and stresses are very simple to express. However, there is one important disadvantage to using the Eulerian description that we can regard as an example of the applied mathematician's **conservation of difficulty** principle. Specifically, we find that the equations of motion are no longer linear. Rather, because of the presence of products of velocities and their derivatives in terms such as $u\partial u/\partial x$, the momentum equation is **quasi-linear**.[2] We will discuss this point further in Chapter 5. As a final comment, note that some authors refer to the Eulerian derivative as the **material derivative** and some denote it as D/Dt.

For an illustration of differences between the Lagrangian and Eulerian descriptions, consider the following. Imagine that in a one-dimensional flow, each fluid particle begins motion in the x direction at time $t = 0$ with velocity $u = Kx_o$ from point $x = x_o$, and continues at that speed. Since each particle moves at constant velocity, its position, x, after a time t has elapsed is given by

$$x = x_o + ut = x_o + Kx_o t \qquad (4.22)$$

---

[2]A differential equation is termed **nonlinear** only if the highest derivative appears nonlinearly.

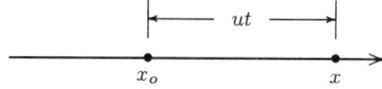

Figure 4.3: *One-dimensional motion.*

Figure 4.3 depicts this type of motion. The velocity and acceleration follow by taking the first and second partial derivatives, respectively, with respect to time. Thus, we find that

$$u = \frac{\partial x}{\partial t} = K x_o \quad \text{and} \quad a = \frac{\partial u}{\partial t} = 0 \tag{4.23}$$

Because the description of the motion is stated in Lagrangian terms, i.e., in terms of the motion of individual particles, we are able to write the expressions above almost by inspection. Summarizing, the Lagrangian description of this motion is as follows.

$$\left. \begin{array}{l} x(x_o, t) = x_o(1 + Kt) \\ u(x_o, t) = K x_o \\ a(x_o, t) = 0 \end{array} \right\} \quad \text{Lagrangian} \tag{4.24}$$

In the Eulerian description, we focus upon a fixed value of x. Since each fluid particle travels with constant velocity $u = K x_o$, after a time t it has traveled a distance $K x_o t$. By definition, the distance traveled is also $x - x_o$. So, for this motion we know that

$$K x_o t = x - x_o \tag{4.25}$$

wherefore

$$x_o = \frac{x}{1 + Kt} \tag{4.26}$$

That is, a particle with velocity $K x_o$ came from this point. Now, since $u = K x_o$, we can express the velocity as a function of x and t as

$$u = K x_o = \frac{K x}{1 + Kt} \tag{4.27}$$

Finally, we compute the acceleration by taking the Eulerian derivative of the velocity. There follows

$$a = \frac{du}{dt} = \underbrace{\frac{\partial u}{\partial t}}_{\partial u/\partial t} + \underbrace{u}_{u} \underbrace{\frac{\partial u}{\partial x}}_{\partial u/\partial x} = -\underbrace{\frac{K^2 x}{(1 + Kt)^2}}_{\partial u/\partial t} + \underbrace{\frac{K x}{(1 + Kt)}}_{u} \underbrace{\frac{K}{(1 + Kt)}}_{\partial u/\partial x} = 0 \tag{4.28}$$

Hence, we again find that the acceleration is zero, as it must be. Summarizing, the velocity and acceleration in the Eulerian description are:

$$\left. \begin{array}{l} u(x, t) = \dfrac{K x}{1 + Kt} \\[2mm] a(x, t) = 0 \end{array} \right\} \quad \text{Eulerian} \tag{4.29}$$

Note that in the Eulerian description we observe the motion at a specified point, and where a fluid particle came from is of no interest. Hence, we require only velocity and acceleration. According to Equation (4.29), the velocity at a given point x decreases monotonically with time. This reflects the fact that as time goes by, we see a fluid particle that has traveled from an increasingly distant point, i.e., closer to $x = 0$. Clearly, the closer the particle's initial location is to the origin, the smaller its velocity. Thus, the velocity must indeed decrease as time passes.

4.2 Steady and Unsteady Flows

To further illustrate ramifications of the Eulerian description, consider the flow past an airplane. If you are seated on the airplane and observe a point above one of the wings, you see the velocity relative to the vehicle. If the plane is cruising at constant velocity, and if you imagine a vector aligned with the fluid velocity, the vector has an unchanging magnitude and direction. As illustrated in Figure 4.4, you observe a different velocity vector if you focus on another point above the wing. However, provided you are not looking at a point too close to the surface, the velocity will be independent of time at all points. If the local velocity is **u**, we state what you see in this situation as follows.

$$\frac{\partial \mathbf{u}}{\partial t} = \mathbf{0} \tag{4.30}$$

Any flow satisfying Equation (4.30) is called a **steady flow**. By contrast, when $\partial \mathbf{u}/\partial t$ is nonzero, the flow is said to be **unsteady**.

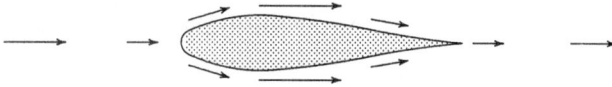

Figure 4.4: *Velocity vectors as seen by a wing-fixed observer.*

Steady flows are clearly easier to analyze both experimentally and analytically because the number of independent variables is reduced by one (time). Usually, a flow is steady if the flow geometry is constant in time, although this is not a guarantee that the flow will be steady, i.e., it is a necessary but not a sufficient condition. Many practical fluid-flow problems can be treated as steady, and many computational methods have been developed specifically for steady flows.

There are many important unsteady flows such as flow past an advancing helicopter blade, a rapidly maneuvering airplane, flow in a reciprocating engine, etc. Even flow past a stationary cylinder can be unsteady, depending upon flow conditions. While many of these examples involve **periodic flow** (e.g., varying sinusoidally in time), there are also flows that involve **random motions** (e.g., varying in a non-periodic manner with time). For example, if we move our observation point very close to the airplane wing discussed above, we will see high-frequency variations in both magnitude and direction of the velocity vector. This apparently random motion is known as **turbulence** (Chapter 14), and is present in most practical flows. However, we often approximate turbulent motion as being steady in the mean, so that the problem can still be treated by extensions of steady-flow methods.

As discussed above, for steady flow the properties at a given point in space are constant. However, unless the flow is **uniform**, i.e., the same at all spatial locations, the properties of a fluid particle will change with time as it makes its way from one point in the flow to another. As shown in Figure 4.4, a fluid particle decelerates as it approaches the leading edge of a wing, accelerates to a maximum value larger than the freestream velocity at some point over the wing, and eventually decelerates back toward the freestream value. At each point along the wing we see steady velocities, although the velocity varies with position. There is no contradiction here as we simply observe different fluid particles at each instant in the Eulerian description.

4.3   Vorticity and Circulation

One of the most important quantities in fluid mechanics is known as **vorticity**. The vorticity, denoted by ω, is defined as the curl of the velocity.

$$\omega = \nabla \times \mathbf{u} \tag{4.31}$$

Although we defer the proof to Chapter 12, the vorticity is twice the local rate of rotation of a fluid particle as it moves through the flow. It is a bit like angular velocity, and gives rise to gyroscopic-like effects in a fluid flow. Flows that have zero vorticity are said to be **irrotational**. When the vorticity is nonzero, the flow is **rotational**.

Most importantly, as we will see in Chapter 10, vorticity plays a central role in the development of forces on an object such as the lifting force that develops on an airplane wing. Vorticity is so important that entire books have been written about it [e.g. Saffman (1993)].

While vorticity plays a critical role in theoretical and computational fluid dynamics, it has one important drawback for the experimentalist. It cannot be measured directly, and has to be obtained from derivatives of measured velocity data. Unless measurements are made at closely-spaced locations (preferably with non-intrusive techniques), differentiating measured quantities is a notoriously inaccurate process, so that reliable experimental data for vorticity are very difficult to obtain.

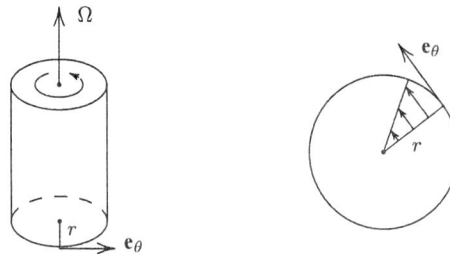

Figure 4.5: *Rigid-body rotation.*

For an example of a flow with vorticity, consider the flow inside a rotating cylinder (see Figure 4.5). If the cylinder has been rotating for a long time, all of the fluid inside will be rotating as though the container and its contents were a rigid body. We thus refer to this situation as **rigid-body rotation**. The velocity vector is given by

$$\mathbf{u} = \Omega r \mathbf{e}_\theta \tag{4.32}$$

where Ω is the angular velocity of the container, r is radial distance from the center of the cylinder and $\mathbf{e}_\theta$ is a unit vector in the circumferential direction. Taking the curl of this velocity, we find that the vorticity vector is (see Section D.2 in Appendix D for the curl in cylindrical coordinates):

$$\omega = \frac{1}{r}\frac{\partial}{\partial r}\left(r u_\theta\right)\mathbf{k} = \frac{1}{r}\frac{d}{dr}\left(\Omega r^2\right)\mathbf{k} = 2\Omega\,\mathbf{k} \tag{4.33}$$

Thus, we see that rigid-body rotation is a rotational flow and its vorticity is twice the angular velocity.

Now consider the fluid motion associated with a hurricane. Near the center of a hurricane, usually referred to as the eye, the air is essentially in rigid-body rotation and its velocity is given by Equation (4.32). Hence, very close to the center, velocities are low, which is the reason the eye is found to be relatively calm. Moving outward from the center, the velocity achieves a maximum and then decreases with radius. Beyond the point where it reaches a maximum, the velocity can be approximated by

$$\mathbf{u} = \frac{K}{r}\mathbf{e}_\theta \tag{4.34}$$

where K is a constant (see Figure 4.6). Taking the curl, we find

$$\boldsymbol{\omega} = \frac{1}{r}\frac{\partial}{\partial r}(ru_\theta)\,\mathbf{k} = \frac{1}{r}\frac{dK}{dr}\mathbf{k} = \mathbf{0} \tag{4.35}$$

Hence, the flow in the outer part of a hurricane is an example of irrotational flow. Interestingly, the flow defined by the velocity in Equation (4.34) is known as the **potential vortex**.

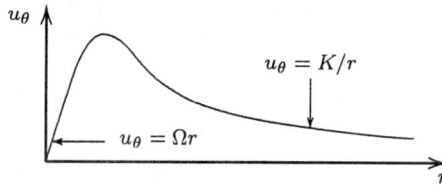

Figure 4.6: *Idealized variation of velocity in a hurricane.*

At first glance, it might appear odd to call a flow that has zero vorticity a "potential vortex." To understand the nomenclature, we must introduce an important vorticity-related quantity known as **circulation**, Γ. Circulation is defined by

$$\Gamma = \oint_C \mathbf{u} \cdot d\mathbf{s} \tag{4.36}$$

where C is a closed contour and $d\mathbf{s}$ is a differential vector tangent to the contour. Equation (4.36) assumes integration on contour C proceeds in a counterclockwise direction, which is the classical definition of a contour integral. In aerodynamic applications, the convention is to integrate in the clockwise direction, and this yields the same magnitude but the opposite sign. We distinguish this difference by defining the **aerodynamic circulation**, Γ_a, according to

$$\Gamma_a = -\Gamma = -\oint_C \mathbf{u} \cdot d\mathbf{s} \tag{4.37}$$

Using Stokes' Theorem (Appendix C), we find

$$\oint_C \mathbf{u} \cdot d\mathbf{s} = \iint_A (\nabla \times \mathbf{u}) \cdot \mathbf{n}\, dA \tag{4.38}$$

where $\mathbf{n}$ is a unit normal to the plane containing the contour C. From this, we conclude that the circulation is the integral of the vorticity component normal to the area bounded by the contour, i.e.,

$$\Gamma = \iint_A \boldsymbol{\omega} \cdot \mathbf{n}\, dA \tag{4.39}$$

Thus, the circulation provides a measure of the strength of the vorticity contained within the bounding contour.

Now consider a potential vortex. Using Equation (4.36) with a circular contour of radius r, the differential vector tangent to the contour is $r\,d\theta\mathbf{e}_\theta$. Hence, since velocity is given by Equation (4.34), the circulation for a potential vortex is

$$\Gamma = \int_0^{2\pi} \left(\frac{K}{r}\right) r\,d\theta = 2\pi K \tag{4.40}$$

This result can be shown to hold on any contour that bounds $r = 0$. Hence, the potential vortex is a flow with constant circulation. This would appear to be inconsistent with Equation (4.39) as the vorticity is zero. However, the velocity is singular at $r = 0$, so that its derivative with respect to r is indeterminate. Since the circulation is the same regardless of how small a circular contour we use, we conclude that this flow is irrotational everywhere except at the origin. That is, a potential vortex has all of its vorticity concentrated at a single point.

As a concluding remark, although very useful in inviscid flow theory, the potential vortex is a mathematical idealization and cannot be realized in nature all the way to the origin. Physical flowfields have no singular points, and vorticity is always distributed over a finite region. It is nevertheless a useful approximation in high Reynolds number flows, as the size of the finite region over which vorticity is typically distributed is often very small on the overall scale of the flow.

4.4   Streamlines, Streaklines and Pathlines

One of the most commonly used concepts for describing a given flowfield is the **streamline**. By definition, a streamline is a curve in the flowfield to which the local velocity vector is tangent. We can derive a differential equation for a streamline by noting that the cross product of two parallel vectors is zero. So, if $d\mathbf{r} = dx\mathbf{i} + dy\mathbf{j} + dz\mathbf{k}$ is a differential element on the streamline, then

$$\mathbf{u} \times d\mathbf{r} = \mathbf{i}(v\,dz - w\,dy) + \mathbf{j}(w\,dx - u\,dz) + \mathbf{k}(u\,dy - v\,dx) = \mathbf{0} \tag{4.41}$$

Hence, the differential equation of a streamline in Cartesian coordinates is:

$$\frac{dx}{u} = \frac{dy}{v} = \frac{dz}{w} \tag{4.42}$$

In two dimensions, we can rewrite this in a more illuminating form,

$$\frac{dy}{dx} = \frac{v}{u} \tag{4.43}$$

Equation (4.43) says the slope of the streamline, dy/dx, is equal to the angle the velocity vector makes with the horizontal, i.e., the velocity is tangent to the streamline. Figure 4.7 shows the streamlines for a **converging flow** and for the potential vortex. In converging flow the streamlines are straight lines, while the streamlines for vortex flow are concentric circles. Note that, by definition, there is no flow across (normal to) a streamline. Hence, solid boundaries, across which there can be no flow, are streamlines in any flow.

For steady flows, streamlines can be visualized experimentally by injecting dye, smoke or some other tracer substance. For unsteady flows, however, there is no practical way to visualize streamlines. When we inject tracer material continuously into an unsteady flow, we observe the path traveled by all particles that have passed through the injection point.

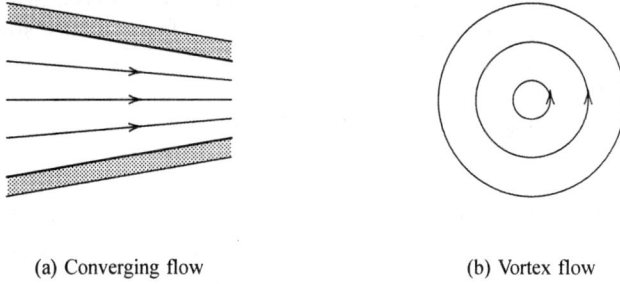

(a) Converging flow   (b) Vortex flow

Figure 4.7: *Examples of streamlines.*

However, there is no reason for this path to be a streamline. We call such a path a **streakline**. Streaklines are primarily an experimenter's tool, and are rarely used in analytical work.

Experimenters have another flow visualization method that is based on the Lagrangian description. Specifically, by taking a time exposure photograph of a marked fluid particle, we can trace the fluid particle's **pathline**. The pathline follows the Lagrangian coordinates of a single fluid particle.

To illustrate these concepts, consider the following idealized unsteady flow. For time $t \leq \tau$, the velocity is uniform with magnitude U in the x direction. At $t = \tau$, the velocity is instantaneously rotated through an angle ϕ, and held constant thereafter. That is, the velocity vector is

$$\mathbf{u} = \begin{cases} U\mathbf{i}\,, & t \leq \tau \\ U\,(\mathbf{i}\cos\phi + \mathbf{j}\sin\phi)\,, & t > \tau \end{cases} \tag{4.44}$$

So, for $t \leq \tau$, we have

$$\frac{dy}{dx} = \frac{v}{u} = 0 \quad \Longrightarrow \quad y = \text{constant} \tag{4.45}$$

while, for $t > \tau$,

$$\frac{dy}{dx} = \frac{v}{u} = \tan\phi \quad \Longrightarrow \quad y = x\tan\phi + \text{constant} \tag{4.46}$$

Thus, the streamlines for this flow are given by families of straight lines, i.e.,

$$y = \begin{cases} \text{constant}\,, & t \leq \tau \\ x\tan\phi + \text{constant}\,, & t > \tau \end{cases} \tag{4.47}$$

If we inject dye at the origin, the pathline is defined by the solution to the following initial-value problem.

$$\frac{d\mathbf{r}}{dt} = \mathbf{u}, \qquad \mathbf{r}(0) = \mathbf{0} \tag{4.48}$$

where $\mathbf{r} = x\mathbf{i} + y\mathbf{j}$ is the (two-dimensional) position vector. In component form, we have the following initial value problem for $t \leq \tau$.

$$\frac{dx}{dt} = U, \quad \frac{dy}{dt} = 0, \quad x(0) = 0, \quad y(0) = 0 \tag{4.49}$$

Integrating, we find

$$x(t) = Ut \quad \text{and} \quad y(t) = 0, \qquad t \leq \tau \tag{4.50}$$

To solve for $t > \tau$, we must follow the same fluid particle. The solution just obtained tells us that at $t = \tau$, the particle's position is $x = U\tau$ and $y = 0$. Thus, we must solve the following problem for $t > \tau$.

$$\frac{dx}{dt} = U\cos\phi, \quad \frac{dy}{dt} = U\sin\phi, \quad x(\tau) = U\tau, \quad y(\tau) = 0 \tag{4.51}$$

The solution is

$$x(t) = U\tau + U(t - \tau)\cos\phi \quad \text{and} \quad y(t) = U(t - \tau)\sin\phi \tag{4.52}$$

Hence, summarizing, the complete solution is

$$x\mathbf{i} + y\mathbf{j} = \begin{cases} Ut\mathbf{i}, & t \le \tau \\ U\tau\mathbf{i} + U(t - \tau)(\mathbf{i}\cos\phi + \mathbf{j}\sin\phi), & t > \tau \end{cases} \tag{4.53}$$

so that the pathline is given by

$$y = \begin{cases} 0, & t \le \tau \\ (x - U\tau)\tan\phi, & t > \tau \end{cases} \tag{4.54}$$

Finally, we can determine the streaklines by a simple graphical construction. First, we note that injected dye travels along the instantaneous pathlines. Hence, for $t \le \tau$, the streakline is defined by $y = 0$. At $t = \tau$, the streakline is the line segment between $x = 0$ and $x = U\tau$. For $t > \tau$, the dye injected at the origin travels along the line $y = x\tan\phi$. The line of dye along the x axis at $t = \tau$ propagates up and to the right at an angle ϕ. Figure 4.8 illustrates the streamlines, a streakline and a pathline for this idealized flow. As shown, all three are different. However, until the velocity changes at $t = \tau$, the streamlines, streaklines and pathlines are all coincident. This is true in general for steady flow.

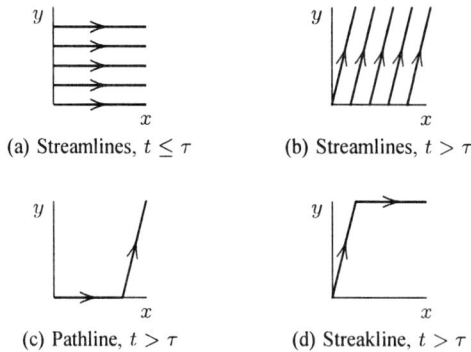

(a) Streamlines, $t \le \tau$ (b) Streamlines, $t > \tau$

(c) Pathline, $t > \tau$ (d) Streakline, $t > \tau$

Figure 4.8: *Streamlines, streaklines and pathlines for the example.*

Normally, the streamlines and pathlines are different for unsteady flows. However, if $\mathbf{u}$ is parallel to $\partial\mathbf{u}/\partial t$, so that their cross product vanishes, they are coincident [Eringen (1980)]. Thus, we can say

$$\mathbf{u} \times \frac{\partial\mathbf{u}}{\partial t} = \mathbf{0} \quad \Longrightarrow \quad \text{Streamlines} \parallel \text{Pathlines} \tag{4.55}$$

A straightforward example of such a flow is one for which the velocity varies sinusoidally with time such as $\mathbf{u} = \mathbf{u}_o(x, y, z)\sin(\omega t)$.

4.5 Extensive and Intensive Properties

We turn our attention now to the derivation of an important kinematical result known as the **Reynolds Transport Theorem**. We will use this theorem in Chapters 5 and 7 to formulate the basic laws for conservation of mass, momentum and energy for a fluid. However, before proceeding to the formal derivation in Section 4.7, we must first introduce some preliminary concepts. That is the purpose of this and the next section.

In deriving the Reynolds Transport Theorem and the basic conservation laws, we must define three important concepts. Specifically, we must introduce the **system**, **extensive** properties and **intensive** properties. They are defined as follows.

- A **system** is the same collection of fluid particles for all time. A system thus has constant mass.

- An **extensive property** is dependent on the amount of fluid. Examples are mass (M), momentum (**P**), total energy (E), potential energy (Mgz) and kinetic energy ($\frac{1}{2}M\mathbf{u}\cdot\mathbf{u}$).

- An **intensive property** is one that is independent of the amount of fluid. Examples are velocity (**u**), internal energy (e), **specific** potential energy (gz) and specific kinetic energy ($\frac{1}{2}\mathbf{u}\cdot\mathbf{u}$). Note that each intensive property example cited here corresponds to an extensive variable per unit mass, which is usually referred to by adding the leading word "specific".

Extensive and intensive properties are related through a volume integral. In general, if the extensive variable is B and the corresponding intensive variable is β, we have

$$B = \iiint_V \rho\,\beta\,dV \qquad (4.56)$$

For example, the mass ($B = M$, $\beta = 1$) and momentum ($B = \mathbf{P}$, $\beta = \mathbf{u}$) of a system are given by

$$M = \iiint_V \rho\,dV \quad \text{and} \quad \mathbf{P} = \iiint_V \rho\,\mathbf{u}\,dV \qquad (4.57)$$

4.6 Surface Fluxes

Now we consider the rate at which fluid flows across a plane of area A. Imagine that the density, ρ, and the velocity in the x direction, u, are constant. As indicated in Figure 4.9, the volume of fluid crossing the plane in a time Δt is $\Delta V = Au\Delta t$. Hence, the mass crossing the plane is

$$\Delta m = \rho u A \Delta t \qquad (4.58)$$

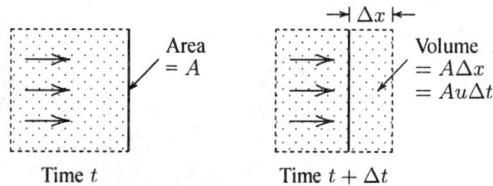

Figure 4.9: *Mass flow across a plane.*

Thus, in the limit $\Delta t \to 0$, the instantaneous mass flux, $\dot{m}$, is given by

$$\dot{m} \equiv \frac{dm}{dt} = \rho u A \tag{4.59}$$

We can easily generalize this result for the case where ρ and u vary over the plane. That is, we apply the analysis above to a differential area dA to conclude that $d\dot{m} = \rho u\, dA$. Then, integrating over the plane, we obtain

$$\dot{m} = \iint_A \rho u\, dA \tag{4.60}$$

We can also generalize the mass-flux integral to non-planar surfaces that are at an oblique angle to the velocity vector. In terms of a general velocity vector, $\mathbf{u}$, only the normal component carries fluid across the surface (Figure 4.10) so that the differential mass flux is given by $d\dot{m} = \rho\, \mathbf{u} \cdot \mathbf{n}\, dA$, where $\mathbf{n}$ is a unit normal to the surface. Thus, the most general result for the mass flux across a surface is given by

$$\dot{m} = \iint_A \rho\, \mathbf{u} \cdot \mathbf{n}\, dA \tag{4.61}$$

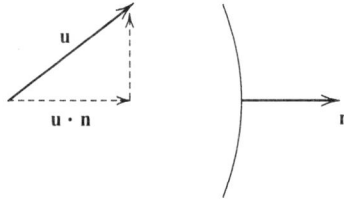

Figure 4.10: *Mass flow across a curvilinear surface.*

Finally, the arguments leading to Equation (4.61) focus specifically upon mass flux. Clearly, if we are interested in the flux of any property such as momentum or energy, all that is needed is to replace ρ by the appropriate property per unit volume. In other words, the differential flux of extensive property B is given by $dB = \rho\beta\, \mathbf{u} \cdot \mathbf{n}\, dA$, where β is the corresponding intensive property. Hence, the flux of extensive property B across a general curvilinear surface is given by

$$\dot{B} = \iint_A \rho\beta\, \mathbf{u} \cdot \mathbf{n}\, dA \tag{4.62}$$

4.7   Reynolds Transport Theorem

We now have sufficient background information to derive a famous theorem of continuum mechanics known as the **Reynolds Transport Theorem**, whose origin traces from the pioneering work of Reynolds (1895). Modern use of the theorem in fluid mechanics began in the 1950's. The theorem establishes a mathematical link between a Lagrangian or **system-oriented** approach in which we track the same collection of fluid particles, and an Eulerian or **control-volume** approach in which we focus on a fixed volume that contains different fluid particles at different times. This linkage is necessary for developing conservation laws in fluid mechanics where the latter approach is the most natural. By contrast, Newton's second law and the laws of thermodynamics, upon which conservation of momentum and energy are based, are formulated from a system-oriented point of view.

Once we establish the theorem, it becomes a simple task to derive the conservation principles for fluid motion. To understand why we need this theorem, consider what we have discussed in this chapter thus far. We have seen that using the Eulerian description permits us to focus on a fixed point in space and to observe the flow as time passes. Because we see different fluid particles at each instant, we have learned that special care must be taken in computing the rate of change of fluid properties. This led to the Eulerian derivative defined in Equation (4.17). We have also learned that one of the primary reasons we use the Eulerian description rather than the Lagrangian description in fluid mechanics is to obviate the need to follow a large number of fluid particles through the flowfield. Thus far, we have only partially accomplished this end. That is, we know how to focus our attention on a fixed point in space. However, space consists of an infinite number of points in a continuum. It would be far more convenient if we could expand our focus to a finite volume, or at least to a differential volume element. We refer to such a region as a **control volume**.

Hence, we pose the following problem. Given a fixed control volume, how do we compute the rate of change of its mass, momentum and energy? Just as with observing the flow at a single point, we face the problem that our control volume contains different fluid particles at each instant. This stands in distinct contrast to a **system**, which always contains the same fluid particles.

Our strategy is the same as that applied in establishing the Eulerian derivative, which we aptly describe as the *rate of change following a fluid particle*. By analogy, we derive the Reynolds Transport Theorem by computing the *rate of change following a system* that is coincident with our control volume at a specified time, t. This is depicted in a one-dimensional geometry in Figure 4.11.

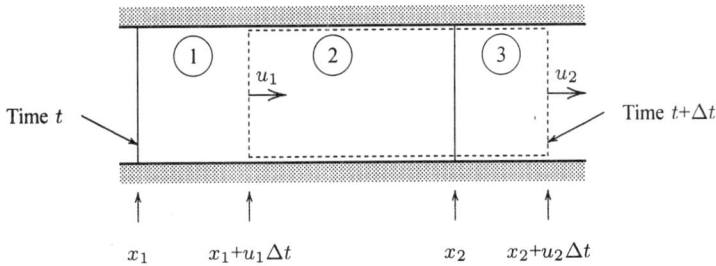

Figure 4.11: *A control volume and a system coincident at time t. The control volume is regions 1 and 2. At time $t + \Delta t$, the system lies in regions 2 and 3.*

Our goal in the following derivation is to determine the rate of change of a given extensive property, B, for the *stationary* control volume. The control volume is the region between $x = x_1$ and $x = x_2$. In an infinitesimally small time increment, Δt, the fluid in the control volume at the initial time, t, moves as indicated in Figure 4.11. Because Δt is infinitesimal, we can approximate the velocity at each end of the system coincident with the control volume at time t as being constant during the motion. Any error introduced by this approximation will vanish when we take the limit $\Delta t \to 0$. Hence, at time $t + \Delta t$, the *moving* system lies between $x_1 + u_1 \Delta t$ and $x_2 + u_2 \Delta t$, where u_1 and u_2 denote the velocities of the left and right boundaries of the system, respectively. Thus, at time t, the value of $B(t)$ is given by

$$B(t) = \iiint_V \rho \beta \, dV = A \int_{x_1}^{x_2} \rho(x, t) \beta(x, t) dx \qquad (4.63)$$

where β is the intensive variable corresponding to B, A is the (constant) cross-sectional area and the volume, V, is just $(x_2 - x_1)A$. Because this is a one-dimensional geometry, A can assume any value we wish. At time $t + \Delta t$,

$$B(t + \Delta t) = A \int_{x_1 + u_1 \Delta t}^{x_2 + u_2 \Delta t} \rho(x, t + \Delta t)\beta(x, t + \Delta t)dx \qquad (4.64)$$

Thus, we find that

$$\frac{B(t + \Delta t) - B(t)}{\Delta t} = \underbrace{A \int_{x_1}^{x_2} \frac{\rho(x, t + \Delta t)\beta(x, t + \Delta t) - \rho(x, t)\beta(x, t)}{\Delta t} dx}_{\text{Regions 1 and 2}}$$

$$+ \underbrace{\frac{A}{\Delta t} \int_{x_2}^{x_2 + u_2 \Delta t} \rho(x, t + \Delta t)\beta(x, t + \Delta t)\, dx}_{\text{Region 3}}$$

$$- \underbrace{\frac{A}{\Delta t} \int_{x_1}^{x_1 + u_1 \Delta t} \rho(x, t + \Delta t)\beta(x, t + \Delta t)\, dx}_{\text{Region 1}} \qquad (4.65)$$

Before we can take the limit $\Delta t \to 0$, we must pay special attention to the last two integrals in Equation (4.65). To determine their limiting values, we make use of the **mean value theorem of calculus** (see Appendix C). For the first of the two integrals, for example, we have

$$\int_{x_2}^{x_2 + u_2 \Delta t} \rho(x, t + \Delta t)\beta(x, t + \Delta t)dx$$

$$= \rho(x_2 + \lambda u_2 \Delta t, t + \Delta t)\beta(x_2 + \lambda u_2 \Delta t, t + \Delta t)u_2 \Delta t \qquad (4.66)$$

where the quantity λ lies between 0 and 1. Consequently,

$$\lim_{\Delta t \to 0} \frac{A}{\Delta t} \int_{x_2}^{x_2 + u_2 \Delta t} \rho(x, t + \Delta t)\beta(x, t + \Delta t)dx = A(\rho u \beta)_2 \qquad (4.67)$$

with subscript 2 denoting values at $x = x_2$ and time t. A similar result holds for the last integral in Equation (4.65). Thus, in the limit $\Delta t \to 0$, we arrive at

$$\frac{dB}{dt} = A \int_{x_1}^{x_2} \frac{\partial}{\partial t}(\rho\beta)dx + A\left[(\rho u \beta)_2 - (\rho u \beta)_1\right] \qquad (4.68)$$

Because the cross-sectional area of the control volume is constant, we can rewrite this equation as follows.

$$\frac{dB}{dt} = \iiint_V \frac{\partial}{\partial t}(\rho\beta)dV + \underbrace{\iint_A (\rho u \beta)_2 dA}_{\dot{B}_{\text{out}}} - \underbrace{\iint_A (\rho u \beta)_1 dA}_{\dot{B}_{\text{in}}} \qquad (4.69)$$

For this simple one-dimensional geometry, we know that the **outer unit normal** at the left boundary is $\mathbf{n} = -\mathbf{i}$ and the velocity vector is $\mathbf{u} = u_1\mathbf{i}$, wherefore $\mathbf{u} \cdot \mathbf{n} = -u_1$. At the right boundary, we have $\mathbf{n} = \mathbf{i}$ and $\mathbf{u} = u_2 \mathbf{i}$ so that $\mathbf{u} \cdot \mathbf{n} = u_2$. Hence, we can rewrite Equation (4.69) as

$$\frac{dB}{dt} = \iiint_V \frac{\partial}{\partial t} (\rho\beta) \, dV + \iint_A (\rho\beta \, \mathbf{u} \cdot \mathbf{n})_2 \, dA + \iint_A (\rho\beta \, \mathbf{u} \cdot \mathbf{n})_1 \, dA \qquad (4.70)$$

Finally, for this admittedly simple geometry, the surface bounding the control volume consists of the left and right cross sections. All other boundaries are of no consequence because the motion is strictly one dimensional. Hence, for this simple geometry, the sum of the two area integrals in Equation (4.70) can be replaced by a single area integral over the "closed" surface bounding the control volume, viz.,

$$\frac{dB}{dt} = \iiint_V \frac{\partial}{\partial t} (\rho\beta) \, dV + \oiint_S \rho\beta \, \mathbf{u} \cdot \mathbf{n} \, dS \qquad (4.71)$$

In words, Equation (4.71) says the rate of change of a specified variable B for the system coincident with the control volume at time t is equal to the sum of the instantaneous rate of change of the property within the control volume and the net flux of the property out of the control volume.

For some applications, it is helpful to take advantage of the fact that, for a stationary control volume, the $\partial/\partial t$ and volume integral operations commute. That is, the integration extends over spatial coordinates, which are independent of time for a non-moving control volume. These operations also commute in the case where the control volume moves without changing size or shape. In either case, we can thus rewrite Equation (4.71) as follows.

$$\frac{dB}{dt} = \frac{\partial}{\partial t} \iiint_V \rho\beta \, dV + \oiint_S \rho\beta \, \mathbf{u} \cdot \mathbf{n} \, dS \qquad (4.72)$$

Equation (4.71) is the Reynolds Transport Theorem in its most basic form, and it applies to general three-dimensional geometries such as the one shown in Figure 4.12. Although we have not proven so here, the theorem is valid for general three-dimensional motion and even for a moving control volume, with the understanding that $\mathbf{u} \cdot \mathbf{n}$ is relative to the moving control volume's boundary. A three-dimensional derivation requires more advanced mathematical concepts, such as the multi-dimensional analog of the mean value theorem of calculus, and will not be presented here.

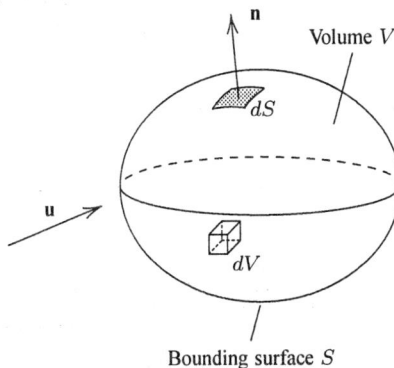

Figure 4.12: *A general three-dimensional geometry.*

Equation (4.71) is a special case of **Leibnitz's Theorem**, which applies to an arbitrary number of dimensions. In the form quoted, it applies to three-dimensional space, and we will use it for cases where β will be either a scalar or a vector. It is most recognizable in its one-dimensional form, which is quoted in some elementary calculus books (and which we derived above), viz.,

$$\frac{d}{dt}\int_{x_1(t)}^{x_2(t)} f(x,t)\,dx = \int_{x_1}^{x_2} \frac{\partial f}{\partial t}\,dx + f(x_2,t)\frac{dx_2}{dt} - f(x_1,t)\frac{dx_1}{dt} \qquad (4.73)$$

In the one-dimensional case, this shows that differentiating an integral with variable limits consists of two contributions. The first contribution accounts for the time variation of the integrand. The second accounts for the fact that the limits of the integral are changing. The analog in three dimensions should now be clear. Differentiating a volume integral consists of two contributions also, the first being a volume integral of the derivative of the integrand with respect to t. The second results from the changing limits of integration of the volume integral. The result is the closed surface integral (one integral less than a volume integral), with $\mathbf{u} \cdot \mathbf{n}$ representing the analog of dx_2/dt and $-dx_1/dt$ in Equation (4.73).

As we will see in Chapters 5 and 7, using the Reynolds Transport Theorem permits expressing the classical laws for conservation of mass, momentum and energy in integral form. The most important point to note about Equations (4.71) and (4.72) is the following. In either equation, both integrals pertain to the control volume rather than to the system. This is important because the physical principles we will apply to deduce conservation laws are most naturally stated for a moving system. As discussed at the beginning of this section, the Reynolds Transport Theorem bridges the gap between the Eulerian-description-oriented control volume and the Lagrangian-description-oriented system.

4.8   Vector Calculus and Multiple Integrals

This book makes extensive use of vector calculus and associated theorems involving multiple (volume and surface) integrals. At first glance, this might appear a bit formidable for the beginning engineering student. So why use these concepts?

The answer is simple. Force is a *vector*. Velocity is a *vector*. The direction of the force due to pressure acting on a surface is in the direction of the *vector* normal to the surface on which the pressure acts. Everywhere we look, we find a vector! The only kind of motion that can be represented strictly with scalar quantities is one dimensional. There are few real applications involving any kind of motion that can be described as one dimensional. Hence, of necessity, we must deal with vectors in order to tackle any but the most trivial problems.

Although we could still avoid vectors by working with individual components of forces and velocities, for example, the compactness of vector notation saves a great deal of writing (and reading). To fully appreciate this aspect of vector notation, simply consider the curl of the velocity, i.e., the vorticity vector. Calling the vorticity ω, in vector notation we say

$$\boldsymbol{\omega} = \nabla \times \mathbf{u} \qquad (4.74)$$

To avoid vector notation, we must deal with its three components, ω_x, ω_y and ω_z:

$$\omega_x = \frac{\partial w}{\partial y} - \frac{\partial v}{\partial z}, \qquad \omega_y = \frac{\partial u}{\partial z} - \frac{\partial w}{\partial x}, \qquad \omega_z = \frac{\partial v}{\partial x} - \frac{\partial u}{\partial y} \qquad (4.75)$$

The compactness of vector notation is an attractive aspect motivating its use.

Concerning the use of multiple integrals, most of the integrands we will deal with in this text will vary in, at worst, just one direction. Often, we will even assume the integrands are constant so that the integrals can be replaced by a summation involving products of integrands and areas or volumes. So, their use is more symbolic than anything else for the highly idealized (for the sake of algebraic simplicity) classroom examples. However, in practical applications, the integrands will not be so simple, and evaluating the integrals will be important. There is no point in developing bad habits at this fundamental a level, so we retain the full integrals.

The greatest advantage we will have in working with volume and bounding-surface integrals will occur when we derive basic laws of fluid motion. All of our derivations will be valid for arbitrary geometries, as opposed to the standard approach of using a simple geometry with special symmetries that permit easy evaluation of integrals. Without using the integral approach, we generally confine our derivations to rectangular Cartesian coordinates, which is an unnecessary limitation. Recall, for example, our two derivations of the hydrostatic relation in Chapter 3. We did a simple scalar development using Figure 3.1(a). The full vector form of the hydrostatic relation followed from using volume and surface integrals to analyze the arbitrary shape of Figure 3.1(b). Once you feel comfortable with the divergence theorem, which is discussed in Appendix C, you should be comfortable with either derivation.

The reason, then, for using volume and surface integrals along with the useful integral theorems of Appendix C is simple. We can avoid the tedium of developing differential, and other, relationships for contrived geometries, with no guarantee that the results are valid for any but the special geometry chosen. In taking advantage of the integral theorems, we minimize the amount of tedious math required to derive basic principles, *thus facilitating a greater focus on the physics of fluid motion.* This, above all, is the primary rationale for using vector calculus, multiple integrals and the integral-transform theorems.

Now that you are hopefully sold on the approach, take the time to review the material in Appendix C if you have not already done so. We will be using the various theorems throughout the rest of the text.

Problems

4.1 Compute the acceleration vector for the following velocity vectors.

(a) $\mathbf{u} = A(t)x\,\mathbf{i} - A(t)y\,\mathbf{j}$

(b) $\mathbf{u} = A(t)x^2\mathbf{i} + A(t)y^2\mathbf{j}$

4.2 Compute the acceleration vector for the following velocity vectors, where U, δ and a are constant.

(a) $\mathbf{u} = U\left(1 - e^{-y/\delta}\right)\mathbf{i}$

(b) $\mathbf{u} = U\sin(x - at)\mathbf{i}$

(c) $\mathbf{u} = f(y)\mathbf{i} + g(x)\mathbf{j}$

4.3 Compute the acceleration vector for the following velocity vectors, where U and H are constant.

(a) $\mathbf{u} = \dfrac{U}{H}\left(x\,\mathbf{i} + y\,\mathbf{j} - 2z\,\mathbf{k}\right)$

(b) $\mathbf{u} = \dfrac{U}{H^2}\left(x^2\,\mathbf{i} + xy\,\mathbf{j} - 3xz\,\mathbf{k}\right)$

4.4 The velocity for incompressible flow in a converging nozzle can be approximated by

$$\mathbf{u} = U\left(1 + 2\frac{x}{L}\right)\mathbf{i} - 2U\frac{y}{L}\mathbf{j}$$

where U and L are characteristic velocity and length scales, respectively. Compute the acceleration vector.

4.5 Consider the incompressible flow whose velocity components are given by $u = Ux/L$, $v = -Uy/L$, where U and L are characteristic velocity and length scales, respectively. Determine the acceleration vector.

4.6 The axial velocity in a conical nozzle is

$$\mathbf{u} = U_o\left(1 - x/L\right)^{-2}\mathbf{i}$$

where U_o and L are constant velocity and length scales, respectively.

(a) If we treat the flow as being one dimensional, what is the acceleration, $\mathbf{a}$?

(b) If the magnitude of the velocity at $x = x_2$ is 1/4 of its value at $x = x_1$, what is the magnitude of the acceleration at $x = x_2$ relative to its value at $x = x_1$?

4.7 The velocity for *Couette flow* is

$$\mathbf{u} = U\frac{y}{h}\,\mathbf{i}$$

where U is the velocity of the moving plate and h is the distance between the plates. What is the acceleration, $\mathbf{a}$?

4.8 Consider the one-dimensional flow whose velocity is given by

$$\mathbf{u} = \frac{x}{t}\,\mathbf{i}$$

(a) Determine the unsteady, convective and total accelerations.

(b) Repeat your computations for $u = -x/t$.

4.9 Consider a one-dimensional flow whose velocity is

$$\mathbf{u} = \left(U_o - \frac{x}{\tau}\right)e^{-t/\tau}\,\mathbf{i}$$

where U_o and τ are constant velocity and time scales. Compute the unsteady, convective and total accelerations.

4.10 For flow near the *stagnation point* of an accelerating cylinder, the velocity is

$$\mathbf{u} = \frac{2U(t)}{R}(x\,\mathbf{i} - y\,\mathbf{j})$$

where R is the cylinder's radius and $U(t)$ is the cylinder's speed.

(a) Compute the unsteady, convective and total accelerations.

(b) Determine $U(t)$ and a_x if $a_y = 0$. Assume $U(0) = U_o$.

4.11 For cylindrical coordinates, the acceleration in two-dimensional flow is

$$a_r = \frac{\partial u_r}{\partial t} + u_r\frac{\partial u_r}{\partial r} + \frac{u_\theta}{r}\frac{\partial u_r}{\partial \theta} - \frac{u_\theta^2}{r}$$

$$a_\theta = \frac{\partial u_\theta}{\partial t} + u_r\frac{\partial u_\theta}{\partial r} + \frac{u_\theta}{r}\frac{\partial u_\theta}{\partial \theta} + \frac{u_r u_\theta}{r}$$

Find the acceleration vector for the following flows.

(a) Rigid-body rotation, $\mathbf{u} = \Omega r\,\mathbf{e}_\theta$ (Ω = constant)

(b) Potential vortex, $\mathbf{u} = \dfrac{\Gamma}{2\pi r}\,\mathbf{e}_\theta$ (Γ = constant)

4.12 For cylindrical coordinates, the acceleration in two-dimensional flow is

$$a_r = \frac{\partial u_r}{\partial t} + u_r\frac{\partial u_r}{\partial r} + \frac{u_\theta}{r}\frac{\partial u_r}{\partial \theta} - \frac{u_\theta^2}{r}$$

$$a_\theta = \frac{\partial u_\theta}{\partial t} + u_r\frac{\partial u_\theta}{\partial r} + \frac{u_\theta}{r}\frac{\partial u_\theta}{\partial \theta} + \frac{u_r u_\theta}{r}$$

Find the acceleration vector for the following flows.

(a) Flow toward a sink, $\mathbf{u} = -\dfrac{Q}{2\pi r}\,\mathbf{e}_r$ (Q = constant)

(b) Doublet flow, $\mathbf{u} = -\dfrac{\mathcal{D}}{2\pi r^2}(\mathbf{e}_r\cos\theta + \mathbf{e}_\theta\sin\theta)$ ($\mathcal{D}$ = constant)

4.13 Consider the flowfield for which the velocity vector is

$$\mathbf{u} = Ax\mathbf{i} - Ay\mathbf{j}$$

where A is a constant of dimensions 1/time. Determine the Lagrangian description of the fluid-particle position vector, $\mathbf{r} = x\mathbf{i} + y\mathbf{j}$, in terms of A, t and initial coordinates, x_o and y_o.

4.14 Consider a flow for which the velocity vector is

$$\mathbf{u} = Ax^2\mathbf{i} + Ay^2\mathbf{j}$$

where A is a constant of dimensions 1/(length·time). Determine the Lagrangian description of the fluid-particle position vector, $\mathbf{r} = x\,\mathbf{i} + y\,\mathbf{j}$, in terms of A, t and initial coordinates, x_o and y_o.

4.15 The velocity for incompressible flow above an infinite flat wall at $y = 0$ oscillating with velocity $u(0,t) = U\cos(\Omega t)$ is given by

$$u(y,t) = Ue^{-ky}\cos(\Omega t - ky), \quad v(y,t) = w(y,t) = 0$$

where t is time, Ω is frequency, y is distance normal to the surface and k is a constant of dimensions 1/length. Determine the Lagrangian description of the fluid-particle coordinates, x and y, in terms of U, Ω, k, t and initial coordinates, x_o and y_o.

4.16 Consider incompressible flow through a converging nozzle, for which the velocity can be approximated by

$$\mathbf{u} = U \left(1 + 2\frac{x}{L}\right) \mathbf{i} - 2U\frac{y}{L}\mathbf{j}$$

where U and L are characteristic velocity and length scales, respectively. Determine the Lagrangian description of the fluid-particle coordinates, x and y, in terms of U, L, t and initial coordinates, x_o and y_o.

4.17 In cylindrical coordinates, the rate of change of the position vector, $\mathbf{r}$, is given by

$$\frac{d\mathbf{r}}{dt} = \frac{d}{dt}(r\mathbf{e}_r) = \frac{dr}{dt}\mathbf{e}_r + r\frac{d\theta}{dt}\mathbf{e}_\theta$$

Determine the Lagrangian description of the fluid-particle coordinates, r and θ, for *rigid-body rotation* in terms of Ω, t and initial coordinates, r_o and θ_o.

4.18 In cylindrical coordinates, the rate of change of the position vector, $\mathbf{r}$, is given by

$$\frac{d\mathbf{r}}{dt} = \frac{d}{dt}(r\mathbf{e}_r) = \frac{dr}{dt}\mathbf{e}_r + r\frac{d\theta}{dt}\mathbf{e}_\theta$$

Determine the Lagrangian description of the fluid-particle coordinates, r and θ, for a *potential vortex* in terms of K, t and initial coordinates, r_o and θ_o.

4.19 In cylindrical coordinates, the rate of change of the position vector, $\mathbf{r}$, is given by

$$\frac{d\mathbf{r}}{dt} = \frac{d}{dt}(r\mathbf{e}_r) = \frac{dr}{dt}\mathbf{e}_r + r\frac{d\theta}{dt}\mathbf{e}_\theta$$

Determine the Lagrangian description of the fluid-particle coordinates, r and θ, for a *point source*, whose velocity vector is

$$\mathbf{u} = \frac{Q}{2\pi r}\mathbf{e}_r$$

Express your answer in terms of Q, t and initial coordinates, r_o and θ_o.

4.20 Compute the vorticity for the following velocity vectors, where U, L and δ are constant.

(a) $\mathbf{u} = U\left[(x/L)\mathbf{i} - (y/L)\mathbf{j}\right]$

(b) $\mathbf{u} = U\left(1 - e^{-y/\delta}\right)\mathbf{i}$

(c) $\mathbf{u} = f(y)\mathbf{i} + g(x)\mathbf{j}$

4.21 Compute the vorticity for the following velocity vectors, where U, L and δ are constant.

(a) $\mathbf{u} = Ax\,\mathbf{i} + Ay\,\mathbf{j} - 2Az\,\mathbf{k}$

(b) $\mathbf{u} = Ay\,\mathbf{i} + Ax\,\mathbf{j}$

(c) $\mathbf{u} = U\left(1 - e^{-z/\delta}\right)(\mathbf{i} + \mathbf{j})$

4.22 A velocity field is given in terms of dimensionless quantities by

$$\mathbf{u} = Ax\,\mathbf{i} + z\,\mathbf{j} + (y - z)\mathbf{k}$$

Is there any value of the constant A for which the flow is irrotational?

4.23 A velocity field is given in terms of dimensionless quantities by

$$\mathbf{u} = Ay^3 t\,\mathbf{i} + 3xy^2 t\,\mathbf{j}$$

Is there any value of the constant A for which the flow is irrotational?

4.24 In cylindrical coordinates for two-dimensional flow, the vorticity is given by

$$\boldsymbol{\omega} = \frac{1}{r}\left[\frac{\partial}{\partial r}(ru_\theta) - \frac{\partial u_r}{\partial \theta}\right]\mathbf{k}$$

Compute the vorticity for the following velocity vectors, where U and R are constant.

(a) $\mathbf{u} = 4U\dfrac{r}{R}\cosh\theta\,\mathbf{e}_r + 2U\dfrac{r}{R}\sinh\theta\,\mathbf{e}_\theta$

(b) $\mathbf{u} = U\left(\dfrac{r}{R}\right)^2\mathbf{e}_r - 2U\left(\dfrac{r}{R}\right)^2\theta\,\mathbf{e}_\theta$

(c) $\mathbf{u} = 2U\dfrac{r}{R}\cos\theta\,\mathbf{e}_r - U\dfrac{r}{R}\sin\theta\,\mathbf{e}_\theta$

4.25 In cylindrical coordinates for two-dimensional flow, the vorticity is given by

$$\boldsymbol{\omega} = \frac{1}{r}\left[\frac{\partial}{\partial r}(ru_\theta) - \frac{\partial u_r}{\partial \theta}\right]\mathbf{k}$$

Compute the vorticity for the following velocity vectors, where U and R are constant.

(a) $\mathbf{u} = U\left[1 - (R/r)^2\right]\cos\theta\,\mathbf{e}_r - U\left[1 + (R/r)^2\right]\sin\theta\,\mathbf{e}_\theta$

(b) $\mathbf{u} = U(r/R)^2\sin 2\theta\,\mathbf{e}_r + U\left[R\theta/r + 2(r/R)^2\cos 2\theta\right]\mathbf{e}_\theta$

(c) $\mathbf{u} = (UR/r)(\mathbf{e}_r + 3\mathbf{e}_\theta)$

4.26 For cylindrical coordinates, the vorticity is given by

$$\boldsymbol{\omega} = \begin{vmatrix} \dfrac{1}{r}\mathbf{e}_r & \mathbf{e}_\theta & \dfrac{1}{r}\mathbf{k} \\[2mm] \dfrac{\partial}{\partial r} & \dfrac{\partial}{\partial \theta} & \dfrac{\partial}{\partial z} \\[2mm] u_r & ru_\theta & w \end{vmatrix}$$

Compute the vorticity of the following flowfield for which all quantities are dimensionless.

$$\mathbf{u} = \frac{2\cos\theta}{r^2}\mathbf{e}_r + \frac{\sin\theta}{r^2}\mathbf{e}_\theta + \frac{z\cos\theta}{r^3}\mathbf{k}$$

4.27 The velocity for incompressible flow above an infinite flat wall at $y = 0$ oscillating with velocity $u(0,t) = U\cos(\Omega t)$ is given by

$$u(y,t) = Ue^{-ky}\cos(\Omega t - ky), \quad v(y,t) = w(y,t) = 0$$

where t is time, Ω is frequency, y is distance normal to the surface and k is a constant.

(a) Compute the acceleration vector.

(b) Compute the vorticity vector.

4.28 The velocity for incompressible flow above an infinite flat wall at $y = 0$ impulsively set in motion with constant velocity U at $t = 0$ is given by

$$u(y,t) = U\left[1 - \operatorname{erf}\left(\frac{y}{2\sqrt{\nu t}}\right)\right], \quad v(y,t) = w(y,t) = 0$$

where t is time, y is distance normal to the surface, ν is kinematic viscosity and $\operatorname{erf}(\eta)$ is the *error function* defined by

$$\operatorname{erf}(\eta) = \frac{2}{\sqrt{\pi}}\int_0^\eta e^{-\hat{\eta}^2}\,d\hat{\eta}$$

(a) Compute the acceleration vector.

(b) Compute the vorticity vector.

4.29 Consider the flow whose velocity is given by

$$\mathbf{u} = U\frac{y}{H}\mathbf{i} - U\frac{x}{H}\mathbf{j}$$

where U and H are constants. Compute the circulation, $\Gamma = \oint_C \mathbf{u} \cdot d\mathbf{s}$, on the rectangular contour shown. Verify that your result is consistent with Equation (4.39).

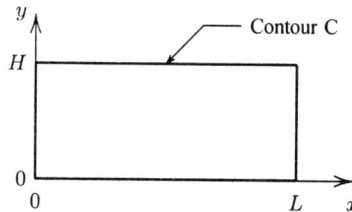

Problems 4.29, 4.30, 4.31

4.30 Consider the flow whose velocity is given by

$$\mathbf{u} = Ax\,\mathbf{i} - Ay\,\mathbf{j}$$

where A is a constant. Compute the circulation, $\Gamma = \oint_C \mathbf{u} \cdot d\mathbf{s}$, on the rectangular contour shown. Verify that your result is consistent with Equation (4.39).

4.31 Consider the flow whose velocity is given by

$$\mathbf{u} = 2U\frac{xy}{H^2}\mathbf{i} - U\frac{y^2}{H^2}\mathbf{j}$$

where U and H are constants. Compute the circulation, $\Gamma = \oint_C \mathbf{u} \cdot d\mathbf{s}$, on the rectangular contour shown. Verify that your result is consistent with Equation (4.39).

4.32 Verify that the circulation of a *potential vortex*, for which

$$\mathbf{u} = \frac{K}{r}\,\mathbf{e}_\theta$$

is $\Gamma = 2\pi K$ for a rectangular contour centered about the origin. **HINT:** Make use of the trigonometric identity

$$\tan^{-1}\left(\frac{H}{L}\right) + \tan^{-1}\left(\frac{L}{H}\right) = \frac{\pi}{2}$$

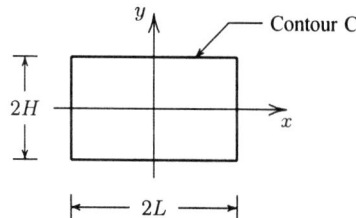

Problem 4.32

4.33 Consider the flowfield whose velocity is given by

$$\mathbf{u} = Ax\,\mathbf{i} - Ay\,\mathbf{j}$$

where A is a constant. Derive an equation defining the streamlines for this flow. Sketch a few streamlines for the upper half plane, i.e., for $y > 0$. Indicate flow direction on the streamlines for $A > 0$.

4.34 Consider the flowfield whose velocity is given by

$$\mathbf{u} = \frac{Ax}{x^2 + y^2}\,\mathbf{i} + \frac{Ay}{x^2 + y^2}\,\mathbf{j}$$

where A is a constant. Derive an equation defining the streamlines for this flow. Sketch a few streamlines for the entire xy plane. Indicate flow direction on the streamlines for $A > 0$.

4.35 Consider the flowfield whose velocity is given by

$$\mathbf{u} = \frac{Ay}{x^2 + y^2}\,\mathbf{i} - \frac{Ax}{x^2 + y^2}\,\mathbf{j}$$

where A is a constant. Derive an equation defining the streamlines for this flow. Sketch a few streamlines for the entire xy plane. Indicate flow direction on the streamlines for $A > 0$.

4.36 Consider the flowfield whose velocity is given by

$$\mathbf{u} = Ay\,\mathbf{i} + Ax\,\mathbf{j}$$

where A is a constant. Derive an equation defining the streamlines for this flow. Sketch a few streamlines for the region $y > |x|$. Indicate flow direction on the streamlines for $A > 0$.

4.37 Consider the flowfield whose velocity is given in terms of dimensionless quantities by

$$\mathbf{u} = x\,\mathbf{i} + x^2 y\,\mathbf{j}$$

Derive an equation defining the streamlines for this flow. Sketch a few streamlines for the upper half plane, i.e., for $y > 0$. Indicate flow direction on the streamlines.

4.38 Suppose the velocity components for a two-dimensional, incompressible flowfield are given by

$$u = f(y) \quad \text{and} \quad v = g(x)$$

where $f(y)$ and $g(x)$ are arbitrary functions.

(a) What must these two functions be in order to have irrotational flow?

(b) Determine an equation for the streamlines, using your result of Part (a).

4.39 Consider the flowfield whose velocity is given by

$$\mathbf{u} = \frac{A\cos\theta}{r^2}\,\mathbf{e}_r + \frac{A\sin\theta}{r^2}\,\mathbf{e}_\theta$$

where A is a constant. The differential equation (in two dimensions) for streamlines in cylindrical coordinates is

$$\frac{dr}{u_r} = \frac{r\,d\theta}{u_\theta}$$

Derive an equation defining the streamlines for this flow. Sketch a few streamlines. Indicate flow direction on the streamlines for $A > 0$. **HINT:** After developing the equation for the streamlines, convert to Cartesian coordinates.

4.40 The velocity for ideal flow past a cylinder is

$$\mathbf{u} = U \left[1 - \frac{R^2}{r^2} \right] \cos \theta \, \mathbf{e}_r - U \left[1 + \frac{R^2}{r^2} \right] \sin \theta \, \mathbf{e}_\theta$$

where U is freestream velocity and R is cylinder radius. The differential equation (in two dimensions) for streamlines in cylindrical coordinates is

$$\frac{dr}{u_r} = \frac{r \, d\theta}{u_\theta}$$

Derive an equation defining the streamlines for this flow. **HINT:** Use the fact that expanding in partial fractions,

$$\frac{1}{(r^2 - R^2) \, r} = \frac{1}{2R^2 (r - R)} + \frac{1}{2R^2 (r + R)} - \frac{1}{R^2 r}$$

4.41 Consider the unsteady flow with velocity given by

$$\mathbf{u} = U \, \mathbf{i} + U \sin \omega t \, \mathbf{j}$$

where U and ω are constant velocity and frequency, respectively.

  (a) Derive an equation for the instantaneous streamlines. Sketch a few streamlines for $\omega t = 0$, $\pi/2$, π and $3\pi/2$.

  (b) Solve for the pathline of a fluid particle passing through $x = y = 0$ at $t = 0$. Eliminate t from your answer and sketch the pathline.

4.42 Consider the unsteady flow with velocity given by

$$\mathbf{u} = U \, \mathbf{i} + U \cos[k(x - Ut)] \, \mathbf{j}$$

where U and k are constant velocity and wave number, respectively.

  (a) Derive an equation for the instantaneous streamlines. Sketch a few streamlines for a fixed time.

  (b) Solve for the pathline of a fluid particle passing through $x = x_o$ and $y = 0$ at $t = 0$. Eliminate t from your answer and sketch the pathline.

4.43 For flow near the *stagnation point* of an accelerating body, the flow is given by

$$\mathbf{u} = A(t) \left[x \, \mathbf{i} - y \, \mathbf{j} \right]$$

  (a) Derive an equation for the instantaneous streamlines.

  (b) Solve for the pathline of a fluid particle passing through $x = x_o$ and $y = y_o$ at $t = 0$. Eliminate t from your answer.

  (c) Compute the cross product of $\mathbf{u}$ and $\partial \mathbf{u}/\partial t$. Are your answers of Parts (a) and (b) consistent with this result?

4.44 Show that the differential equation of a streamline in cylindrical polar coordinates is

$$\frac{dr}{u_r} = \frac{r \, d\theta}{u_\theta} = \frac{dz}{w}$$

HINT: As part of your derivation, use the fact that

$$d\mathbf{e}_r = \frac{d\mathbf{e}_r}{d\theta} d\theta = \mathbf{e}_\theta \, d\theta$$

4.45 Show that the differential equation of a streamline in spherical coordinates is

$$\frac{dR}{u_R} = \frac{R\sin\phi\, d\theta}{u_\theta} = \frac{R\, d\phi}{u_\phi}$$

HINT: As part of your derivation, use the fact that

$$d\mathbf{e}_R = \frac{\partial \mathbf{e}_R}{\partial\theta} d\theta + \frac{\partial \mathbf{e}_R}{\partial\phi} d\phi = \mathbf{e}_\theta \sin\phi\, d\theta + \mathbf{e}_\phi d\phi$$

4.46 An empirical equation for the velocity distribution in a horizontal *open channel* of height h is

$$u(y) = U_{max}(y/h)^{1/n}$$

where U_{max} is maximum velocity, y is normal distance and n lies between 6 and 8.

(a) Compute the mass-flow rate as a function of U_{max}, h and n. Assume unit width out of the page.

(b) Couette flow is the limiting case $n = 1$. Compute the ratio of mass flux for $n = 7$ to that for Couette flow.

4.47 Consider a uniform flow with $\mathbf{u} = U\mathbf{i}$. Compute the mass-flow rate per unit width (out of the page) through a 30° circular arc of radius R.

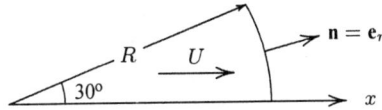

Problem 4.47

4.48 Consider flow near a *stagnation point* for which the velocity is

$$\mathbf{u} = Ax\,\mathbf{i} - Ay\,\mathbf{j}$$

where A is a positive constant. Compute the mass flux across Plane 1 and Plane 2.

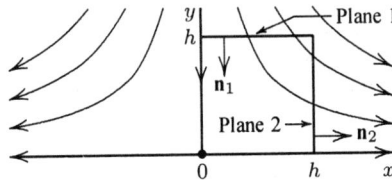

Problem 4.48

4.49 Consider flow near a *stagnation point* for which the velocity is

$$\mathbf{u} = Ar\cos 2\theta\, \mathbf{e}_r - Ar\sin 2\theta\, \mathbf{e}_\theta$$

where A is a positive constant. Compute the mass flux, $\dot{M}$, and momentum flux, $\dot{\mathbf{P}}$, across the quarter-circle arc of radius R in the first quadrant.

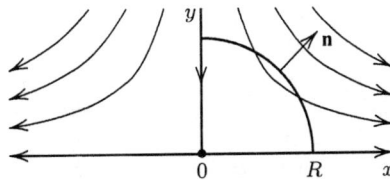

Problem 4.49

4.50 An incompressible flow has a velocity vector given by

$$\mathbf{u} = U(1 + x/h)\,\mathbf{i} - Uy/h\,\mathbf{j} - U\,\mathbf{k}$$

where U and h are constant reference velocity and length scales, respectively. Compute the flux of x momentum across the square planar area parallel to the xz plane at $y = h$.

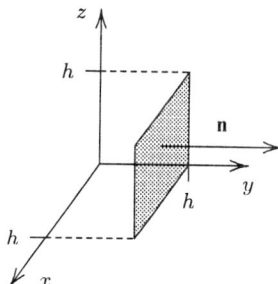

Problems 4.50, 4.51

4.51 An incompressible flow has a velocity vector given by

$$\mathbf{u} = U\frac{xz}{h^2}\,\mathbf{i} - U\frac{yz}{h^2}\,\mathbf{j} + U\frac{xy}{h^2}\,\mathbf{k}$$

where U and h are constant reference velocity and length scales, respectively. Compute the flux of z momentum across the square planar area parallel to the xz plane at $y = h$.

4.52 The velocity for *Couette flow* is

$$u(y) = U\frac{y}{h}$$

where U is the velocity of the upper wall and h is channel height. Compute the mass flux, $\dot{M}$, momentum flux, $\dot{P}$, and kinetic energy flux, $\dot{K}$, through a channel cross section.

4.53 Compute the net flux of fluid out of a sphere of radius R if density is constant and the velocity is

$$u/U = 2(y/R)e^{-x/R} + J_{1/2}(y/R), \qquad v/U = (y/R)^2 e^{-x/R} + Y_{1/2}(x/R)$$

where U is constant. Also, $J_{1/2}(\xi)$ and $Y_{1/2}(\xi)$ are Bessel functions of order $1/2$. **HINT:** This problem simplifies greatly if you use one of the theorems in Appendix C.

4.54 Compute the net flux of fluid out of a sphere of radius R if density is constant and the velocity is

$$u/U = 2(y/R)e^{-x/R} + \text{cn}(y/R), \qquad v/U = (y/R)^2 e^{-x/R} + \text{sn}(x/R)$$

where U is constant. Also, $\text{cn}(\xi)$ and $\text{sn}(\xi)$ are Jacobian elliptic functions. **HINT:** This problem simplifies greatly if you use one of the theorems in Appendix C.

4.55 A group of people is leaving a room in single file through a doorway that is 3 ft by 7 ft. Allowing for space between people, each person occupies 36 ft^3 as he or she approaches the doorway. The line of people is moving at a rate of 2 in/sec. Using the Reynolds Transport Theorem (with number density replacing mass density), determine the rate of change of the number of people in the room.

4.56 Helium-filled balloons are escaping through a hole of area A in the top of a circus tent. The number of balloons per unit volume just below the hole is n, and their velocity as they escape is U. Because of ragged edges at the hole, some of the balloons burst and fall back into the tent at a rate of $\dot{N}_{pop}$. Using the Reynolds Transport Theorem (with number density replacing mass density), determine the rate of change of the number of inflated balloons in the circus tent.

4.57 Ants, frightened by the presence of an anteater, are vacating their underground nest. The number of ants per unit volume near the exit is n, the exit area is A and the ants are moving at a velocity U as they exit. The anteater is consuming ants within the nest at a rate of $\dot{N}_{zap}$. Using the Reynolds Transport Theorem (with number density replacing mass density), determine the rate of change of the number of live ants in the nest.

4.58 Beginning with the Reynolds Transport Theorem for an extensive/intensive variable pair, B and β, show that

$$\dot{B}(t) = \iiint_V \left[\frac{d}{dt}(\rho\beta) + \rho\beta\nabla\cdot\mathbf{u} \right] dV$$

where d/dt is the Eulerian derivative.

4.59 The two-dimensional *Laplacian* of the velocity vector, $\mathbf{u}$, is defined by

$$\nabla^2\mathbf{u} = \frac{\partial^2\mathbf{u}}{\partial x^2} + \frac{\partial^2\mathbf{u}}{\partial x^2}$$

By direct substitution, verify the following identity in Cartesian coordinates for two-dimensional flows.

$$\nabla^2\mathbf{u} = \nabla(\nabla\cdot\mathbf{u}) - \nabla\times(\nabla\times\mathbf{u})$$

4.60 A useful vector identity for theoretical fluid mechanics is

$$\mathbf{u}\cdot\nabla\mathbf{u} = \nabla\left(\frac{1}{2}\mathbf{u}\cdot\mathbf{u}\right) - \mathbf{u}\times(\nabla\times\mathbf{u})$$

By direct substitution, verify this identity in Cartesian coordinates for two-dimensional flow.

Chapter 5

Conservation of Mass and Momentum

This chapter derives the basic conservation of mass and momentum laws for a fluid in both integral and differential form. Because of their simplicity and common occurrence in nature, our primary focus will be on incompressible flows. Mass- and momentum-conservation laws provide a sufficient number of equations to solve incompressible-flow problems for which temperature and heat transfer are of no interest. If thermal considerations are required or if viscous losses are important, we require an additional equation based on conservation of energy. We defer development of the energy-conservation law to Chapter 7.

After presenting the conservation forms, we examine a few properties of fluid motion, including one of the most famous results of fluid mechanics known as **Bernoulli's equation**. This result, derived from the momentum equation, serves as a conservation of mechanical energy law, valid for incompressible flows under a set of commonly-observed constraints. It relates pressure, velocity and potential energy in a moving fluid, and simplifies to the hydrostatic relation when the velocity is zero.

Like the hydrostatic relation, Bernoulli's equation serves as the principle upon which important measurement devices are based. We will learn of two such devices known as the **Pitot tube** and the **Pitot-static tube**. The chapter includes a simple home experiment demonstrating Bernoulli's equation.

We also demonstrate **Galilean invariance** of the equations of motion. This is a nontrivial consideration for two key reasons. First, a Galilean transformation is a linear operation while the Eulerian description makes the momentum equation quasi-linear. Thus, it is not obvious that the fluid mechanics equations of motion are invariant under a Galilean transformation. Second, if our equations are not invariant, we cannot use measurements in a wind tunnel for a stationary model to infer forces on a prototype moving into a fluid that is at rest. This would greatly complicate the job of the experimenter who would have to design models that are in motion in a wind tunnel.

We derived the Reynolds Transport Theorem in Chapter 4 to bridge the gap between the Lagrangian and Eulerian descriptions. It is now possible to apply the familiar conservation laws of classical physics to a volume that contains different fluid particles at each instant. To derive equations expressing conservation of mass, momentum and energy, we appeal to the definition of a system, Newton's second law of motion and the first law of thermodynamics, respectively.

In terms of the nomenclature introduced for the Reynolds Transport Theorem development, we work with corresponding extensive (B) and intensive (β) variable pairs listed in Table 5.1. Note that in the case of the energy conservation principle, which we have deferred to Chapter 7, the intensive variable is the sum of the internal energy, e, kinetic energy, $\frac{1}{2}\mathbf{u} \cdot \mathbf{u}$, and the body-force potential, $\mathcal{V}$ — all per unit mass.

Table 5.1: *Extensive and Intensive Variable Pairs in Conservation Principles*

Property	B	β	Foundation
Mass	M = Mass	1	Definition of a System
Momentum	$\mathbf{P}$ = Momentum	$\mathbf{u}$	Newton's Second Law of Motion
Energy	E = Total Energy	$e + \frac{1}{2}\mathbf{u} \cdot \mathbf{u} + \mathcal{V}$	First Law of Thermodynamics

5.1   Conservation of Mass

Consider a general control volume, V, bounded by a closed surface S. Figure 5.1 illustrates such a volume, including a differential volume element, dV, a differential surface element, dS, and an **outer unit normal** vector, $\mathbf{n}$. Let $B = M$, i.e., the mass of the control volume, so that $\beta = 1$. Then, since

$$B = \iiint_V \rho\beta \, dV \tag{5.1}$$

there follows

$$M = \iiint_V \rho \, dV \tag{5.2}$$

By definition, a *system* always contains the same collection of fluid particles. Consequently, its mass is constant for all time, which means

$$\frac{dM}{dt} = 0 \tag{5.3}$$

Therefore, invoking the Reynolds Transport Theorem, we arrive at the integral form of the mass-conservation principle for a *control volume*.

$$\iiint_V \frac{\partial \rho}{\partial t} \, dV + \oiint_S \rho \, \mathbf{u} \cdot \mathbf{n} \, dS = 0 \tag{5.4}$$

The first term on the left-hand side of Equation (5.4) represents the *instantaneous rate of change of mass in the control volume*. The second term represents the *net flux of mass out of the control volume*.

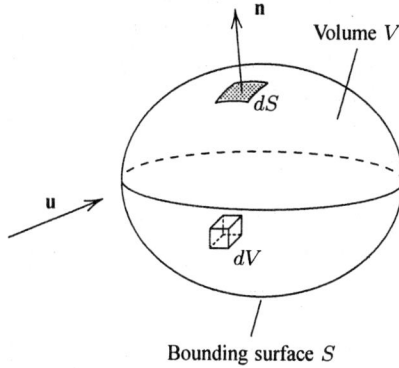

Figure 5.1: *A general control volume for mass conservation.*

We can deduce a differential equation for mass conservation by using the divergence theorem (see Appendix C) to convert the closed surface integral to a volume integral.

$$\iiint_V \frac{\partial \rho}{\partial t}\, dV + \iiint_V \nabla \cdot (\rho\, \mathbf{u})\, dV = 0 \qquad (5.5)$$

Then, since integration is a linear operation,

$$\iiint_V \left[\frac{\partial \rho}{\partial t} + \nabla \cdot (\rho\, \mathbf{u}) \right] dV = 0 \qquad (5.6)$$

Because our result holds for an arbitrary volume, we conclude that the integrand in Equation (5.6) must vanish. We thus deduce the following differential equation governing mass conservation *at every point within the control volume.*

$$\frac{\partial \rho}{\partial t} + \nabla \cdot (\rho\, \mathbf{u}) = 0 \qquad (5.7)$$

This equation is often referred to as the **continuity equation**. Equation (5.7) is in what is known as **conservation form**. By definition, this means the differential equation consists of the sum of the time derivative of one quantity and the divergence of another. We can expand $\nabla \cdot (\rho\, \mathbf{u})$ and rewrite Equation (5.7) as

$$\frac{\partial \rho}{\partial t} + \mathbf{u} \cdot \nabla \rho + \rho \nabla \cdot \mathbf{u} = 0 \qquad (5.8)$$

The sum of the first two terms on the left-hand side of Equation (5.8) is the Eulerian derivative of ρ. Thus, we arrive at the continuity equation in **primitive-variable form**.

$$\frac{d\rho}{dt} + \rho \nabla \cdot \mathbf{u} = 0 \qquad (5.9)$$

As a final comment on the differential form of mass conservation, note that for incompressible flow, where we regard ρ as a constant, Equation (5.9) simplifies to

$$\nabla \cdot \mathbf{u} = 0 \qquad \text{(Incompressible)} \qquad (5.10)$$

whether the flow is steady or unsteady.

As an example, consider two-dimensional flow approaching a special point known as a stagnation point. In the immediate vicinity of the stagnation point, the velocity vector is given by $\mathbf{u} = A(x\,\mathbf{i} - y\,\mathbf{j})$, where x and y are tangent to and normal to the surface, respectively. The quantity A is a constant. Taking the divergence of this vector, we find

$$\nabla \cdot \mathbf{u} = \left(\mathbf{i}\frac{\partial}{\partial x} + \mathbf{j}\frac{\partial}{\partial y}\right) \cdot (\mathrm{A}x\,\mathbf{i} - Ay\,\mathbf{j}) = A - A = 0 \tag{5.11}$$

Thus, we conclude that this flow is incompressible.

5.2   Conservation of Momentum

For simplicity, we consider an inviscid, or perfect, fluid so that only pressure, with no viscous force, acts on any surface. We will address viscous effects briefly in Chapter 10, and in more complete detail in Chapters 13 and 14. Letting $\mathbf{P}$ denote the momentum vector, the momentum of the control volume shown in Figure 5.2 is

$$\mathbf{P} = \iiint_V \rho\,\mathbf{u}\,dV \tag{5.12}$$

so that our extensive variable is $\mathbf{P}$, while the intensive variable is $\mathbf{u}$.

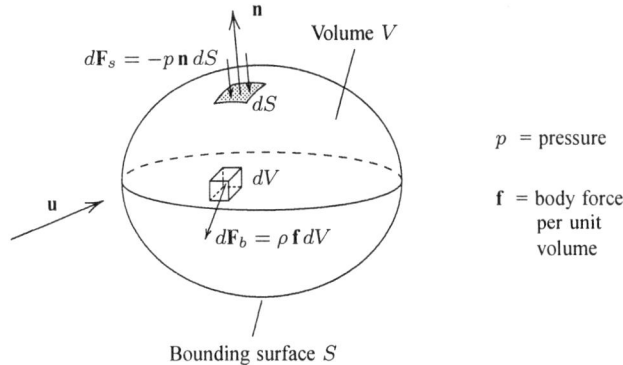

Figure 5.2: *A general control volume for momentum conservation.*

Now, we know from Newton's second law of motion applied to the system coincident with the control volume at an instant in time that

$$\frac{d\mathbf{P}}{dt} = \mathbf{F}_s + \mathbf{F}_b \tag{5.13}$$

where $\mathbf{F}_s$ and $\mathbf{F}_b$ denote **surface force** and **body force** exerted by the surroundings on the system, respectively. We define a surface force as one that is transmitted across the surface bounding the system. By contrast, a body force acts at a distance, the most common examples being gravitational, electrical and magnetic forces.

Because we have confined our focus to a perfect fluid, the only surface force acting is the fluid pressure. Since $\mathbf{n}$ is an outer unit normal, the pressure force exerted by the surroundings on a differential surface element dS is $-p\,\mathbf{n}\,dS$. Hence, the net surface force due to pressure imposed by the surroundings on the system is

$$\mathbf{F}_s = - \oiint_S p\,\mathbf{n}\,dS \tag{5.14}$$

Turning to the body force, we introduce the **specific body force vector, f**, whose dimensions are force per unit mass. The net body force on the system is given by the following volume integral.

$$\mathbf{F}_b = \iiint_V \rho\,\mathbf{f}\,dV \tag{5.15}$$

As an example, for gravity we would say $\mathbf{f} = \mathbf{g} = -g\mathbf{k}$, where $g = 32.174$ ft/sec^2 (9.807 m/sec^2) is the acceleration due to gravity. In this case, the force $\mathbf{F}_b$ would be the control volume's weight.

Thus, invoking the Reynolds Transport Theorem, we arrive at the conservation of momentum principle for a control volume, viz.,

$$\iiint_V \frac{\partial}{\partial t}(\rho\,\mathbf{u})\,dV + \oiint_S \rho\,\mathbf{u}\,(\mathbf{u}\cdot\mathbf{n})\,dS = - \oiint_S p\,\mathbf{n}\,dS + \iiint_V \rho\,\mathbf{f}\,dV \tag{5.16}$$

The first term on the left-hand side of Equation (5.16) represents the *instantaneous rate of change of momentum in the control volume*. The second term represents the *net flux of momentum out of the control volume*. The two terms on the right-hand side are the *net pressure force* and *net body force* exerted by the surroundings on the control volume.

As with mass conservation, we can deduce a differential equation for momentum conservation by using the divergence theorem to convert the closed surface integrals to volume integrals. For the sake of clarity, we will work with the z component of Equation (5.16), i.e.,

$$\iiint_V \frac{\partial}{\partial t}(\rho w)\,dV + \oiint_S \rho w\,(\mathbf{u}\cdot\mathbf{n})\,dS = -\mathbf{k}\cdot \oiint_S p\,\mathbf{n}\,dS + \iiint_V \rho f_z\,dV \tag{5.17}$$

where we isolate the z component of the pressure integral by taking a dot product with the unit normal in the z direction, $\mathbf{k}$. Rearranging terms a little, we have

$$\iiint_V \frac{\partial}{\partial t}(\rho w)\,dV + \oiint_S (\rho w\mathbf{u})\cdot\mathbf{n}\,dS + \oiint_S p\,\mathbf{k}\cdot\mathbf{n}\,dS - \iiint_V \rho f_z\,dV = 0 \tag{5.18}$$

Using the divergence theorem, there follows

$$\iiint_V \left[\frac{\partial}{\partial t}(\rho w) + \nabla\cdot(\rho w\mathbf{u}) + \nabla\cdot(p\,\mathbf{k}) - \rho f_z \right] dV = 0 \tag{5.19}$$

Because the volume is arbitrary, the integrand must vanish at all points. Hence, we obtain the following differential equation (in conservation form) governing the variation of z momentum.

$$\frac{\partial}{\partial t}(\rho w) + \nabla\cdot(\rho\,\mathbf{u}\,w) = -\nabla\cdot(p\,\mathbf{k}) + \rho f_z \tag{5.20}$$

This result can be simplified by expanding the individual terms. Using the chain rule, the first term on the left-hand side becomes

$$\frac{\partial}{\partial t}(\rho w) = w\frac{\partial \rho}{\partial t} + \rho\frac{\partial w}{\partial t} \tag{5.21}$$

Using the standard rules governing the ∇ operator (see Appendix C), the second term on the left-hand side expands according to

$$\nabla \cdot (\rho\, \mathbf{u}\, w) = w \nabla \cdot (\rho\, \mathbf{u}) + \rho\, \mathbf{u} \cdot \nabla w \qquad (5.22)$$

Finally, the term involving the pressure is

$$\nabla \cdot (p\, \mathbf{k}) = \frac{\partial p}{\partial z} \qquad (5.23)$$

Collecting all this, and rearranging a little, Equation (5.20) can be rewritten as

$$w \left[\frac{\partial \rho}{\partial t} + \nabla \cdot (\rho\, \mathbf{u}) \right] + \rho \left[\frac{\partial w}{\partial t} + \mathbf{u} \cdot \nabla w \right] = -\frac{\partial p}{\partial z} + \rho f_z \qquad (5.24)$$

Inspection of Equation (5.7) shows that the first term in brackets vanishes. Also, the second term in brackets is the Eulerian derivative of w. Hence, we arrive at the final form of the equation for conservation of z momentum.

$$\rho \frac{dw}{dt} = -\frac{\partial p}{\partial z} + \rho f_z \qquad (5.25)$$

In a similar way, it is a straightforward matter to show that the equations for momentum conservation in the x and y directions are

$$\rho \frac{du}{dt} = -\frac{\partial p}{\partial x} + \rho f_x \qquad (5.26)$$

$$\rho \frac{dv}{dt} = -\frac{\partial p}{\partial y} + \rho f_y \qquad (5.27)$$

Combining Equations (5.25) through (5.27) into a single vector equation leads to

$$\rho \frac{d\mathbf{u}}{dt} = -\nabla p + \rho\, \mathbf{f} \qquad (5.28)$$

This equation, valid for a perfect fluid, is known as **Euler's equation**. It is in primitive-variable form, and can be used to compute the details of general fluid motion at every point in a flow. Note that, in words, this equation says that *mass per unit volume times acceleration equals the sum of forces per unit volume*, i.e., it is Newton's second law of motion per unit volume.

5.3 Bernoulli's Equation

In order to compute flow of an inviscid, or perfect, fluid about a specified object we must solve the continuity equation [Equation (5.7)] and Euler's equation [Equation (5.28)], subject to appropriate boundary conditions. Although closed-form solutions exist for a few simple geometries, general flowfield solutions require carefully formulated computer programs. However, under special conditions, we can integrate the Euler equation to yield a famous result known as **Bernoulli's equation**.

To derive this result, we note first that the convective acceleration, $\mathbf{u} \cdot \nabla \mathbf{u}$, that appears in Euler's equation, can be rewritten as follows (see Appendix C).

$$\mathbf{u} \cdot \nabla \mathbf{u} = \nabla \left(\frac{1}{2} \mathbf{u} \cdot \mathbf{u} \right) - \mathbf{u} \times (\nabla \times \mathbf{u}) \qquad (5.29)$$

Hence, since total acceleration is the sum of instantaneous and convective contributions, i.e., $d\mathbf{u}/dt = \partial\mathbf{u}/\partial t + \mathbf{u} \cdot \nabla\mathbf{u}$, in terms of vorticity, $\omega = \nabla \times \mathbf{u}$, Euler's equation becomes

$$\rho\frac{\partial\mathbf{u}}{\partial t} + \rho\nabla\left(\frac{1}{2}\mathbf{u}\cdot\mathbf{u}\right) - \rho\,\mathbf{u}\times\omega = -\nabla p + \rho\,\mathbf{f} \qquad (5.30)$$

Now, consider flows that are:

1. Steady ($\partial\mathbf{u}/\partial t = \mathbf{0}$);

2. Incompressible (ρ = constant);

3. Irrotational ($\omega = \mathbf{0}$);

4. Subject to a conservative body force ($\mathbf{f} = -\nabla\mathcal{V}$).

The quantity $\mathcal{V}$ is the **body-force potential**. Under these four conditions, Equation (5.30) simplifies to

$$\nabla\left(p + \frac{1}{2}\rho\,\mathbf{u}\cdot\mathbf{u} + \rho\mathcal{V}\right) = \mathbf{0} \qquad (5.31)$$

Since this holds for all points in the flow, necessarily the quantity in parentheses is constant. Therefore, we conclude that

$$p + \frac{1}{2}\rho\,\mathbf{u}\cdot\mathbf{u} + \rho\mathcal{V} = \text{constant} \qquad (5.32)$$

Equation (5.32) is known as **Bernoulli's equation**. Although we arrived at this result by integrating the momentum equation, it is actually an equation for *conservation of mechanical energy*. This is especially clear when the body force is gravity for which the body-force potential is $\mathcal{V} = gz$. Note that in this spirit, we can regard pressure as the pressure-force potential. Equation (5.32) says the sum of the pressure, p, kinetic energy per unit volume (also known as the **dynamic pressure** or **dynamic head**), $\frac{1}{2}\rho\,\mathbf{u}\cdot\mathbf{u}$, and potential energy per unit volume, $\rho\mathcal{V}$, is constant. There is no contradiction here. A constant-density fluid flow is specified completely by the momentum and continuity equations, so any other property can be deduced from them.

As a final comment on the conditions required for Bernoulli's equation to hold, note that we can actually relax the irrotationality condition. When we do this, Equation (5.32) holds along a streamline, although the constant is different on each streamline. To see this, note first that for steady, incompressible flow with a conservative body force, $\mathbf{f} = -\nabla\mathcal{V}$, Euler's equation simplifies to

$$\mathbf{u}\cdot\nabla\mathbf{u} + \nabla\left(\frac{p}{\rho} + \mathcal{V}\right) = \mathbf{0} \qquad (5.33)$$

Using natural coordinates (see Appendix D), we can write the component of Equation (5.33) along a streamline as

$$u\frac{\partial u}{\partial s} + \frac{\partial}{\partial s}\left(\frac{p}{\rho} + \mathcal{V}\right) = 0 \qquad (5.34)$$

Then, noting that $u\partial u/\partial s = \partial(\frac{1}{2}u^2)/\partial s$, we have

$$\frac{\partial}{\partial s}\left(\frac{p}{\rho} + \frac{1}{2}u^2 + \mathcal{V}\right) = 0 \qquad (5.35)$$

This shows that the quantity in parentheses is constant only along a streamline as opposed to being constant everywhere. By contrast, we found above that when the flow is irrotational, the gradient of this quantity [cf. Equation (5.31)] vanishes, which is a much stronger condition. Nevertheless, we conclude that even when the flow has nonzero vorticity,

$$p + \frac{1}{2}\rho\, \mathbf{u} \cdot \mathbf{u} + \rho\mathcal{V} = \text{constant on streamlines} \tag{5.36}$$

where we note that $u^2 = \mathbf{u} \cdot \mathbf{u}$. This result is interesting from a conceptual point of view, but is not very helpful in general practice. That is, we don't know where the streamlines are until we have solved the equations of motion. Hence, this form of Bernoulli's equation is more limited than the form for irrotational flow.

AN EXPERIMENT YOU CAN DO AT HOME
Contributed by Prof. Peter Bradshaw, Stanford University

To observe one of the implications of Bernoulli's equation, you will need a paper napkin, a thin sheet of paper such as toilet paper, or even a sheet of notebook paper. Grasp the paper in each hand and hold it up to your mouth. Blow gently, being careful that you blow only on the top side. It may take a couple practice tries, but when done correctly, the paper will rise to a horizontal orientation. You might even want to try all of the types of paper noted above to observe that you must blow harder for the more massive sheets.

From Bernoulli's equation, it should be clear that by creating a moving stream of air over the top of the paper, the pressure decreases. Thus, you create a suction on the top surface that causes the paper to rise.

As an example of the use of Bernoulli's equation, consider a large tank with a small hole a distance h below the surface (Figure 5.3). A jet of fluid issues from the hole with velocity U. The relevant body force is gravity so that Bernoulli's equation becomes

$$p + \frac{1}{2}\rho\, \mathbf{u} \cdot \mathbf{u} + \rho g z = \text{constant} \tag{5.37}$$

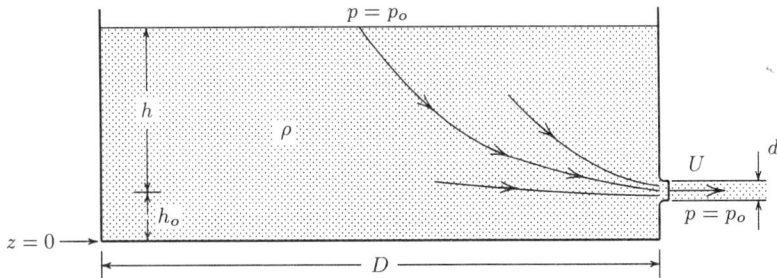

Figure 5.3: *A leaky cylindrical tank; diameter = D.*

As is generally true for thin jets, the surrounding air impresses atmospheric pressure, p_o, throughout the jet. The tank is open to the atmosphere so that the pressure at the top of the tank is also p_o. If the tank diameter, D, is very large compared to the diameter of the small hole, d, we can ignore the velocity of the fluid at the top of the tank. Hence, we can use Bernoulli's equation to relate a point at the top of the tank to a point in the jet to show that

$$p_o + \rho g(h + h_o) = p_o + \frac{1}{2}\rho U^2 + \rho g h_o \tag{5.38}$$

Thus, the velocity in the jet is given by

$$U = \sqrt{2gh} \tag{5.39}$$

Note that we can set the origin, $z = 0$, anywhere we wish. This is true since the potential energy appears on both sides of Equation (5.38), so that only the difference in potential energy between the surface and the jet, $\rho g h$, matters.

In the example above, we used Bernoulli's equation to relate conditions at two specified points in the flow. Often it is helpful to determine the "constant" in Bernoulli's equation by evaluating each term at a point in the flow where all terms are known. Typically, we seek a point that lies very far from solid boundaries where the flow is uniform. To illustrate this, consider water flowing with uniform velocity, U_1, as shown in Figure 5.4. The water enters a uniform-diameter tube at some point below the surface. We can determine the velocity of the water leaving the tube at a height z_2 above the surface as follows.

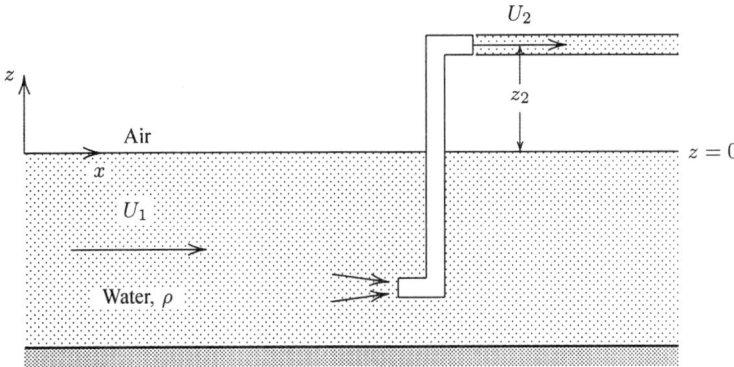

Figure 5.4: *Water flowing past a vertical tube.*

Clearly, one point where we know everything about the flow is at the free surface very far upstream. Because of the interface with the air, the pressure must be atmospheric, p_o. Letting the free surface lie at $z = 0$, the potential energy is zero. Since the flow is uniform far upstream, we also know that the velocity is U_1. Thus, we have

$$p + \frac{1}{2}\rho \, \mathbf{u} \cdot \mathbf{u} + \rho g z = p_o + \frac{1}{2}\rho U_1^2 \tag{5.40}$$

The most important point about using Bernoulli's equation in this way is that Equation (5.40) holds at *all points in the flow*. Hence, since $p = p_o$ in the jet of fluid issuing from the tube, we find

$$p_o + \frac{1}{2}\rho U_2^2 + \rho g z_2 = p_o + \frac{1}{2}\rho U_1^2 \tag{5.41}$$

wherefore

$$U_2 = \sqrt{U_1^2 - 2gz_2} \tag{5.42}$$

It is also interesting to determine the pressure in the primary stream of fluid. Provided we are not too close to either the tube or the bottom, we expect the flow to be uniform with velocity $\mathbf{u} = U_1\,\mathbf{i}$. Then, Bernoulli's equation becomes

$$p + \frac{1}{2}\rho U_1^2 + \rho gz = p_o + \frac{1}{2}\rho U_1^2 \quad \Longrightarrow \quad p = p_o - \rho gz \tag{5.43}$$

Thus, in the moving stream, the pressure satisfies the hydrostatic relation. Close to the tube, the velocity will deviate from the freestream value and, correspondingly, the pressure will depart from the hydrostatic relation. In a real fluid, viscous effects would result in nonuniform velocity near the bottom, which would also cause the pressure to differ from the hydrostatic value.

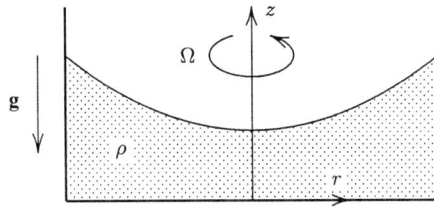

Figure 5.5: *Rotating tank of incompressible fluid.*

Our final example is a flow for which Bernoulli's equation does not hold, viz., incompressible flow in a rotating cylindrical tank (Figure 5.5). We assume the tank has been rotating for a long time, so that the fluid all moves with constant angular velocity, $\boldsymbol{\Omega} = \Omega\mathbf{k}$. That is, the fluid is in a state of **rigid-body rotation** (recall our discussion of flow in a rotating cylinder in Section 4.3) so that the velocity at any point in the fluid is

$$\mathbf{u} = \boldsymbol{\Omega} \times \mathbf{r} = \Omega r\mathbf{e}_\theta \tag{5.44}$$

where r is radial distance from the center of the tank, and $\mathbf{e}_\theta$ is a unit vector in the circumferential direction. Hence, as shown in Section 4.3, the vorticity is

$$\boldsymbol{\omega} = \nabla \times \mathbf{u} = 2\boldsymbol{\Omega} \tag{5.45}$$

Because the vorticity is nonvanishing, we cannot use Bernoulli's equation to help analyze this flow. Rather, we must solve Euler's equation if we wish to determine, for example, the pressure in the fluid.

Since the geometry is symmetric about the z axis, we use the axisymmetric form of the Euler equation. From Appendix D, we find that the three components are as follows.

$$\left.\begin{aligned}
\rho u_r \frac{\partial u_r}{\partial r} + \rho \frac{u_\theta}{r}\frac{\partial u_r}{\partial \theta} + \rho w \frac{\partial u_r}{\partial z} - \rho \frac{u_\theta^2}{r} &= -\frac{\partial p}{\partial r} \\[2mm]
\rho u_r \frac{\partial u_\theta}{\partial r} + \rho \frac{u_\theta}{r}\frac{\partial u_\theta}{\partial \theta} + \rho w \frac{\partial u_\theta}{\partial z} + \rho \frac{u_r u_\theta}{r} &= -\frac{1}{r}\frac{\partial p}{\partial \theta} \\[2mm]
\rho u_r \frac{\partial w}{\partial r} + \rho \frac{u_\theta}{r}\frac{\partial w}{\partial \theta} + \rho w \frac{\partial w}{\partial z} &= -\frac{\partial p}{\partial z} - \rho g
\end{aligned}\right\} \tag{5.46}$$

Now, for rigid-body rotation, we know that the radial (u_r) and axial (w) velocity components vanish and the circumferential component is given by $u_\theta = \Omega r$. Hence, Equations (5.46) simplify to

$$\left.\begin{aligned} -\rho\Omega^2 r &= -\frac{\partial p}{\partial r} \\[2mm] 0 &= -\frac{\partial p}{\partial \theta} \\[2mm] 0 &= -\frac{\partial p}{\partial z} - \rho g \end{aligned}\right\} \tag{5.47}$$

To solve this coupled set of equations, we begin by integrating the first of the three with respect to r. Note that when we perform an integration of a function of r, θ and z with respect to r, we introduce a *function of integration*, $f(\theta, z)$. This is the analog of the *constant of integration* that appears when we integrate a function of a single variable. Therefore, we have

$$p(r, \theta, z) = \frac{1}{2}\rho\Omega^2 r^2 + f(\theta, z) \tag{5.48}$$

Next, we differentiate Equation (5.48) with respect to θ and substitute into the second of Equations (5.47), viz.,

$$\frac{\partial p}{\partial \theta} = \frac{\partial f}{\partial \theta} = 0 \quad \implies \quad f(\theta, z) = F(z) \tag{5.49}$$

That is, we have shown that at most our function of integration is a function only of z. This is consistent with the axial symmetry of the flow. So, the pressure is now given by

$$p(r, \theta, z) = \frac{1}{2}\rho\Omega^2 r^2 + F(z) \tag{5.50}$$

Finally, to determine the function $F(z)$, we differentiate Equation (5.50) with respect to z and substitute into the last of Equations (5.47). This yields

$$\frac{\partial p}{\partial z} = \frac{dF}{dz} = -\rho g \quad \implies \quad F(z) = -\rho g z + \text{constant} \tag{5.51}$$

wherefore the pressure is given by

$$p(r, \theta, z) = \frac{1}{2}\rho\Omega^2 r^2 - \rho g z + \text{constant} \tag{5.52}$$

Thus, we conclude that the pressure in the rotating tank satisfies the following equation.

$$p + \rho g z - \frac{1}{2}\rho\Omega^2 r^2 = \text{constant} \tag{5.53}$$

Although similar in form, this is not Bernoulli's equation. The kinetic energy term appears with a minus sign, which is a result of the rotational nature of this particular flow. Rather, it is a solution to Euler's equation. *This example illustrates that Euler's equation applies to all flows while Bernoulli's equation does not.*

As a final observation, it is always a major simplification if we can use Bernoulli's equation to relate pressure and velocity, rather than having to solve Euler's equation. The reason is obvious, i.e., Bernoulli's equation is a solution to Euler's equation so that no additional computation is needed. With the exception of certain idealized flows, such as the example above (where the velocity was given), solutions to Euler's equation are difficult to obtain and usually require a numerical solution. For this reason, Bernoulli's equation, whenever valid, provides a major simplification for determining flow properties.

5.4   Velocity Measurement Techniques

We can use Bernoulli's equation to infer velocity from a pressure measurement. This is useful because pressure is fundamentally easier to measure than velocity. To understand how this is done, we must first introduce the notion of a **stagnation point**. Then, we discuss two measurement devices known as the **Pitot tube** and the **Pitot-static tube**.

5.4.1   Stagnation Points

Figure 5.6 illustrates stagnation points in ideal two-dimensional flow past a cylinder. The streamline coincident with the x axis upstream and downstream of the cylinder is the **dividing streamline**. This streamline splits at the front of the cylinder so that half of the fluid moves over the cylinder and half moves below. The streamlines rejoin at the back of the cylinder. We have a very special situation at these two points. Specifically, we have the intersection of two perpendicular streamlines, namely, the dividing streamline and the cylinder surface. Now, since streamlines are parallel to the velocity, necessarily $v = 0$ approaching the cylinder. Also, flow very close to the front of the cylinder must have $u = 0$ (see Figure 5.6) to be tangent to the surface. Thus, at the point where the streamlines collide, we must have

$$\mathbf{u} = 0 \qquad \text{(Stagnation Point)} \qquad (5.54)$$

So, we have two points on the cylinder where the velocity vanishes, and we refer to these points as **stagnation points**.

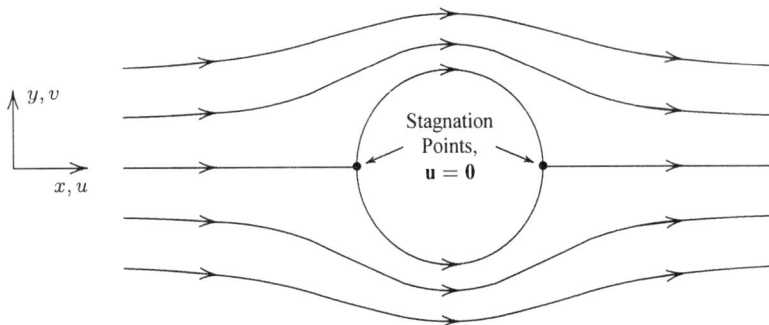

Figure 5.6: *Ideal flow past a cylinder.*

5.4.2   Pitot Tube

The **Pitot tube** is one of the simplest devices based on Bernoulli's equation that provides an indirect measurement of velocity. Figure 5.7 illustrates flow in the immediate vicinity of a Pitot tube placed a distance d below the surface of a flowing stream of water. The tube typically has a very small diameter such as that characteristic of a hypodermic needle. As shown, the water fills the Pitot tube up to a point a distance h above the free surface. Because the fluid in the tube cannot move, there must be a stagnation point at the tube's entrance below the surface. As we will now show, this device permits determining the velocity by a simple measurement of the distance the fluid rises above the surface. It is thus the analog, for moving fluids, of the U-Tube manometer discussed in Section 3.3.

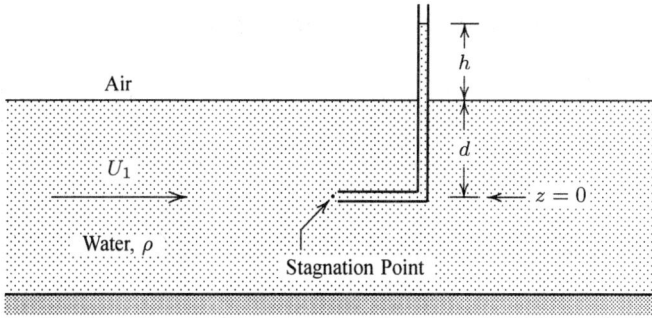

Figure 5.7: *Pitot tube.*

Now, Bernoulli's equation tells us that for this flow we have

$$p + \frac{1}{2}\rho\, \mathbf{u} \cdot \mathbf{u} + \rho g z = \text{constant} \tag{5.55}$$

We can evaluate the constant by selecting a point in the flowfield where the values of pressure, velocity and z are all known. As with the vertical-tube example considered earlier (see Figure 5.4), we select a point far upstream of the Pitot tube at the free surface. At this point, the pressure is equal to the atmospheric pressure, p_o, the velocity assumes its freestream value, U_1, and $z = d$ (note that we are choosing the origin to be coincident with the lower portion of the Pitot tube). Hence,

$$\text{constant} = p_o + \frac{1}{2}\rho\, U_1^2 + \rho g d \tag{5.56}$$

Now, at the top of the tube, which is open to the atmosphere, we know the pressure is p_o, the velocity is zero and $z = d + h$. Hence, applying Bernoulli's equation, we have

$$p_o + \rho g(d + h) = p_o + \frac{1}{2}\rho\, U_1^2 + \rho g d \tag{5.57}$$

Simplifying, we arrive at the following straightforward relation between the flow velocity and the height of the column of fluid in the Pitot tube.

$$U_1 = \sqrt{2gh} \tag{5.58}$$

While the Pitot tube permits a correlation between the height of a column of fluid and the fluid velocity, it is clearly limited to flows with uniform velocity, and requires a point where velocity and pressure are both known. The device has no provision for flows in which the velocity varies with z, e.g., near the bottom of the channel shown in Figure 5.7 where viscous effects are important.

5.4.3 Pitot-Static Tube

The **Pitot-static tube** is another measuring device based on Bernoulli's equation that can be used for more general velocity distributions. This device makes two separate pressure measurements. The first measurement is done with a standard Pitot tube, which measures the pressure at the tip of the probe. Because this is a stagnation point, the Pitot tube measures the **stagnation pressure**. The second measurement is at a point downstream of the probe tip sufficiently distant (typically 8 tube diameters) that the flow has returned to its freestream

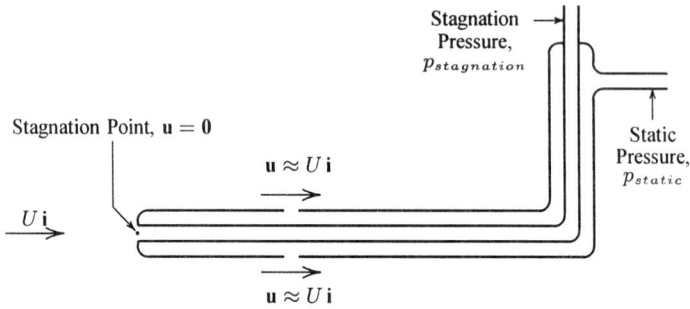

Figure 5.8: *Pitot-static tube.*

value, U. The pressure at this point is the freestream pressure, also referred to as the **static pressure**. Figure 5.8 schematically depicts a Pitot-static tube.

Assuming the tube is extremely thin (as it must be to avoid changing the flow), we can ignore the difference in depth of the stagnation point and the static pressure tap. Hence, from Bernoulli's equation, we have

$$p_{stagnation} = p_{static} + \frac{1}{2}\rho U^2 \tag{5.59}$$

That is, the stagnation pressure is the sum of the static pressure and the **dynamic pressure**, $\frac{1}{2}\rho U^2$. Therefore the local velocity is given by

$$U = \sqrt{\frac{2\left(p_{stagnation} - p_{static}\right)}{\rho}} \tag{5.60}$$

Clearly, the Pitot-static tube is not limited to uniform velocity distributions. Furthermore, the device is essentially self calibrating in the sense that no reference pressure or velocity is needed. Although somewhat sensitive to misalignment with flow direction, it is one of the most useful tools in experimental fluid mechanics.

5.5   Galilean Invariance of Euler's Equation

Suppose we have a body advancing into a quiescent fluid with constant velocity $\mathbf{u} = -U\mathbf{i}$ as illustrated in Figure 5.9(a). Clearly this flow is unsteady for an observer in the main body of fluid since the geometry looks different at each instant. Thus, we cannot use Bernoulli's equation to relate pressure and velocity, even if the flow is incompressible and irrotational, with conservative body forces.

Now, suppose we observe the motion in a coordinate frame translating with the same velocity as the body [Figure 5.9(b)]. In this coordinate frame, the flow geometry does not change with time, and the motion is in fact steady. This is a dramatic improvement from an analytical point of view. In addition to reducing the number of independent variables by one (time is no longer of any consequence), we can use the powerful Bernoulli equation provided, of course, that the flow is also inviscid, irrotational, incompressible and subject only to conservative body forces.

This is a smooth move provided the Euler equation is invariant under this transformation. What we have done is made the classical **Galilean transformation**. When we say the Euler

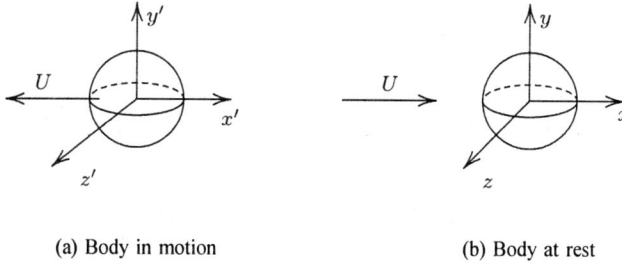

(a) Body in motion   (b) Body at rest

Figure 5.9: *Motion of a body in different coordinate frames.*

equation is invariant under such a transformation, we mean the equation holds when we write the equation in terms of all transformed velocities and pertinent flow properties. The physical meaning of invariance is straightforward. From elementary physics we know that Newton's laws of motion are Galilean invariant for motion of discrete particles, and there is no reason for this to change for fluids. However, there is some cause for concern as the Galilean transformation is a linear operation and the Eulerian description introduces nonlinear terms in the time derivative. Hence, the purpose of this section is to demonstrate Galilean invariance of the Euler equation.

Before proceeding to the proof, it is worthwhile to pause and discuss the reason why Galilean invariance matters. Very simply, if the equations of fluid mechanics were not Galilean invariant, any measurements made in a wind tunnel with fluid moving past a stationary model could not be used to predict forces on a full-scale model moving through a fluid at rest. Rather, the model would have to move through the wind tunnel to simulate the full-scale object's motion, with all the difficulties attending acceleration from rest, attainment of steady flow, and deceleration to rest prior to crashing into the end of the tunnel. Clearly, it is preferable to have a stationary model, and Galilean invariance of the equations of motion guarantees applicability of the measurements to a moving object.

Letting primed quantities denote conditions in the frame where the fluid is at rest and the body moves, the Euler equation is given by

$$\frac{\partial \mathbf{u}'}{\partial t'} + \mathbf{u}' \cdot \nabla' \mathbf{u}' = -\frac{\nabla' p}{\rho} + \mathbf{f} \tag{5.61}$$

Clearly, since p and ρ are thermodynamic properties of the fluid, they cannot depend upon the coordinate frame from which we make our observations. Likewise, the body force, $\mathbf{f}$, will be independent of the coordinate frame provided it doesn't depend explicitly upon velocity or position. Body forces that are coordinate-system dependent do exist, e.g., Coriolis force (see Appendix D, Section D.4), but their presence signifies a noninertial frame for which Galilean invariance does not hold.

By definition, in a Galilean transformation, the coordinates and time transform according to (see Figure 5.9):

$$x = x' + Ut, \quad y = y', \quad z = z', \quad t = t' \tag{5.62}$$

where unprimed quantities correspond to the frame in which the body is at rest. Also, the velocity vectors in the two coordinate frames are related by

$$\mathbf{u} = \mathbf{u}' + U\mathbf{i} \tag{5.63}$$

Clearly, spatial differentiation is unaffected by a Galilean transformation, so that

$$\frac{\partial}{\partial x} = \frac{\partial}{\partial x'}, \quad \frac{\partial}{\partial y} = \frac{\partial}{\partial y'}, \quad \frac{\partial}{\partial z} = \frac{\partial}{\partial z'} \quad \implies \quad \nabla = \nabla' \tag{5.64}$$

By contrast, temporal differentiation is affected. From the chain rule, the time derivative transforms as follows.

$$\left(\frac{\partial}{\partial t'}\right)_{x'} = \left(\frac{\partial t}{\partial t'}\right)_{x'} \left(\frac{\partial}{\partial t}\right)_{x} + \left(\frac{\partial x}{\partial t'}\right)_{x'} \left(\frac{\partial}{\partial x}\right)_{t} = \frac{\partial}{\partial t} + U \frac{\partial}{\partial x} \tag{5.65}$$

Note that we have omitted y and z from Equation (5.65) for the sake of brevity as all attending derivatives vanish (only x depends upon t). So, transforming the unsteady term first,

$$\frac{\partial \mathbf{u}'}{\partial t'} = \left(\frac{\partial}{\partial t} + U \frac{\partial}{\partial x}\right)(\mathbf{u} - U\mathbf{i}) = \frac{\partial \mathbf{u}}{\partial t} + U \frac{\partial \mathbf{u}}{\partial x} \tag{5.66}$$

Thus, the unsteady term is most certainly not invariant under a Galilean transformation.[1] Now consider the convective acceleration. We have

$$\mathbf{u}' \cdot \nabla' \mathbf{u}' = (\mathbf{u} - U\mathbf{i}) \cdot \nabla (\mathbf{u} - U\mathbf{i}) = \mathbf{u} \cdot \nabla \mathbf{u} - U \frac{\partial \mathbf{u}}{\partial x} \tag{5.67}$$

which shows that the convective acceleration is not Galilean invariant either. However, when we sum the unsteady and convective acceleration terms, we find

$$\frac{\partial \mathbf{u}'}{\partial t'} + \mathbf{u}' \cdot \nabla' \mathbf{u}' = \frac{\partial \mathbf{u}}{\partial t} + U \frac{\partial \mathbf{u}}{\partial x} + \mathbf{u} \cdot \nabla \mathbf{u} - U \frac{\partial \mathbf{u}}{\partial x} = \frac{\partial \mathbf{u}}{\partial t} + \mathbf{u} \cdot \nabla \mathbf{u} \tag{5.68}$$

In other words, the sum of the unsteady and convective acceleration terms, which is the Eulerian derivative, is invariant under a Galilean transformation. Therefore, the Euler equation in the transformed coordinate frame is

$$\frac{\partial \mathbf{u}}{\partial t} + \mathbf{u} \cdot \nabla \mathbf{u} = -\frac{\nabla p}{\rho} + \mathbf{f} \tag{5.69}$$

which is identical to Equation (5.61) with all primes omitted. Thus, the Euler equation is invariant under a Galilean transformation.

5.6 Summary of Conservation Equations

It is worthwhile to summarize the equations developed in this chapter. Considering first the integral conservation forms, the equations of motion for an inviscid fluid are as follows.

Mass Conservation:

$$\iiint_V \frac{\partial \rho}{\partial t} \, dV + \oiint_S \rho \, \mathbf{u} \cdot \mathbf{n} \, dS = 0 \tag{5.70}$$

Momentum Conservation:

$$\iiint_V \frac{\partial}{\partial t}(\rho \, \mathbf{u}) \, dV + \oiint_S \rho \, \mathbf{u}(\mathbf{u} \cdot \mathbf{n}) \, dS = -\oiint_S p \, \mathbf{n} \, dS + \iiint_V \rho \, \mathbf{f} \, dV \tag{5.71}$$

---

[1] As discussed at the beginning of this section, a term is invariant under the transformation if and only if transforming the term is equivalent to dropping primes.

Equation of State:

$$\rho = \begin{cases} \dfrac{p}{RT}, & \text{gases} \\[2mm] \text{constant}, & \text{liquids} \end{cases} \tag{5.72}$$

Turning to the differential forms of the conservation laws, we have deduced the continuity and Euler equations that govern conservation of mass and momentum, respectively. The equations are:

Continuity:

$$\frac{\partial \rho}{\partial t} + \nabla \cdot (\rho\, \mathbf{u}) = 0 \tag{5.73}$$

Euler's Equation:

$$\rho \frac{d\mathbf{u}}{dt} = -\nabla p + \rho\, \mathbf{f} \tag{5.74}$$

An excellent exercise in any branch of mathematical physics, or more generally for any mathematics problem, is to count unknowns and equations. Considering liquids first, the unknowns are the density, ρ, pressure, p, and the three velocity components, (u, v, w). Thus, we have a total of five unknowns. Conservation of mass and the equation of state are both scalar equations, while momentum is a vector equation with three components. Thus, we have five equations to solve for five unknowns. Our mathematical system is said to be closed as we have a sufficient number of equations to solve for the unknowns.

Turning to gases, note that the equation of state introduces the temperature as an additional unknown. We thus have six unknowns for a gas. However, conservation of mass, momentum and the state equation still account for only five equations. Our system is not closed as we lack a sufficient number of equations to solve for all of the unknowns.

Actually, we don't have enough equations for a liquid either if the temperature is required, as it would be for a flow with heat transfer. In both cases we must also consider energy conservation in order to completely specify all properties in a given fluid flow. Nevertheless, there are a wide range of problems we can solve without considering energy conservation. Specifically, as long as we confine our attention to incompressible flow without heat transfer, we can treat ρ as a constant. For such flows, we have five equations and five unknowns for both liquids and gases. As we will learn in Chapter 8, variations in the density of a gas are negligible for low-speed flows, i.e., for flows with Mach number less than about 0.3. We will consider energy conservation in Chapter 7.

Problems

5.1 Consider the following velocity field (in which all quantities are dimensionless):

$$\mathbf{u} = 6x^2 y\,\mathbf{i} + 2x^3\,\mathbf{j} + 10\,\mathbf{k}$$

Is the flow incompressible? Is the flow irrotational?

5.2 Consider the following velocity field (in which all quantities are dimensionless):

$$\mathbf{u} = x^2 y\,\mathbf{i} + xy^2\,\mathbf{j} - 4xyz\,\mathbf{k}$$

Is the flow incompressible? Is the flow irrotational?

5.3 The velocity for a two-dimensional flow is

$$u = \frac{C(y^2 - x^2)}{(y^2 + x^2)^2}, \qquad v = -\frac{2Cxy}{(y^2 + x^2)^2}, \qquad C = \text{constant}$$

Does this velocity field satisfy continuity? Is the flow irrotational?

5.4 The velocity for a two-dimensional flow is

$$u = \frac{Cy}{(y^2 + x^2)^{3/2}}, \qquad v = -\frac{Cx}{(y^2 + x^2)^{3/2}}, \qquad C = \text{constant}$$

Does this velocity field satisfy continuity? Is the flow irrotational?

5.5 The velocity for a two-dimensional flow is

$$\mathbf{u} = Ur\cos\theta\,\mathbf{e}_r - 2Ur\sin\theta\,\mathbf{e}_\theta, \qquad U = \text{constant}$$

Does this velocity field satisfy continuity? Is the flow irrotational?

5.6 The velocity for a two-dimensional flow is

$$u_r = U\left(1 - \frac{R^2}{r^2}\right)\cos\theta, \qquad u_\theta = U\frac{R}{r} - U\left(1 + \frac{R^2}{r^2}\right)\sin\theta$$

where U and R are constants. Does this velocity field satisfy continuity? Is the flow irrotational?

5.7 Consider the following velocity field:

$$u = xt + y, \qquad v = Ax + Byt, \qquad w = 0$$

where A and B are constants and all quantities are dimensionless. The flow is incompressible and irrotational. Determine the values of A and B necessary to guarantee these conditions. Compute the total acceleration vector as a function of x, y and t.

5.8 Consider a two-dimensional, incompressible, irrotational flow for which the x component of the velocity vector is

$$u = Axy, \qquad A = \text{constant}$$

What must the y component, v, be?

5.9 Consider a two-dimensional, incompressible, irrotational flow for which the radial component of the velocity vector is

$$u_r = \frac{A}{r^2}\cos\theta, \qquad A = \text{constant}$$

What must the circumferential component, u_θ, be?

5.10 Consider a two-dimensional, incompressible, irrotational flow for which the circumferential component of the velocity vector is

$$u_\theta = -3Ar^2 \sin 3\theta, \qquad A = \text{constant}$$

What must the radial component, u_r, be?

5.11 For steady flow of a gas through a nozzle, we can approximate the velocity as

$$\mathbf{u} = U_o(1 + x/x_o)\,\mathbf{i}$$

where U_o and x_o are constant reference velocity and length, respectively. What is the density, ρ, if its value at $x = 0$ is ρ_o? At what point does the density fall to 80% of ρ_o?

5.12 Consider a one-dimensional, compressible flow in which the density exponentially decreases from ρ_o to ρ_∞, i.e.,

$$\rho = \rho_o - (\rho_o - \rho_\infty)\,e^{-t/\tau}$$

where τ is a constant of dimensions time. If the velocity at $x = 0$ is $u(0, t) = u_o$, where u_o is constant, determine $u(x, t)$ as a function of u_o, ρ_o, ρ, x, t and τ.

5.13 The divergence of the velocity in cylindrical and spherical coordinates is given in Appendix D. What is the most general form of the velocity for incompressible flow if the following conditions hold?

(a) Axially-symmetric flow with $u_\theta = w = 0$.

(b) Spherically-symmetric flow with $u_\theta = u_\phi = 0$.

5.14 Beginning with the continuity equation, verify that the divergence of the velocity is equal to the rate of change of specific volume following a fluid particle per unit specific volume, i.e., that

$$\nabla \cdot \mathbf{u} = \frac{1}{v}\frac{dv}{dt}$$

where the specific volume is defined by $v \equiv 1/\rho$.

5.15 Beginning with the Reynolds Transport Theorem, show for a given extensive, intensive variable pair, B and β, that

$$\dot{B}(t) = \iiint_V \rho\,\frac{d\beta}{dt}\,dV$$

where d/dt is the Eulerian derivative and the integration is carried out over the control volume.

5.16 For inviscid flow past an airfoil, the streamlines are as shown. If the freestream velocity, U_0, is 50 m/sec, what is the pressure difference, $p_2 - p_1$, at points where the velocities are $U_1 = 95$ m/sec and $U_2 = 75$ m/sec? Assume that the fluid density, ρ, is 1.1 kg/m^3 and that Bernoulli's equation applies. Express your answer in kPa.

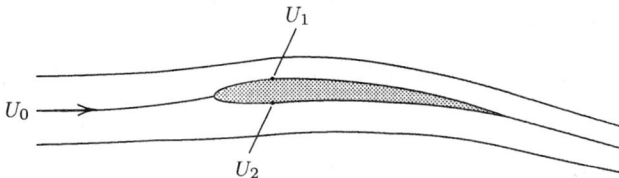

Problems 5.16, 5.17

5.17 For inviscid flow past an airfoil, the streamlines are as shown. If the freestream velocity, U_0, is 160 ft/sec, what is the pressure difference, $p_2 - p_1$, at points where the velocities are $U_1 = 300$ ft/sec and $U_2 = 250$ ft/sec? Assume that the fluid density, ρ, is .002 slug/ft^3 and that Bernoulli's equation applies. Express your answer in psi.

5.18 The velocity in the outlet pipe from a large reservoir of depth $h = 20$ ft is $U = 25$ ft/sec. Due to the rounded entrance to the pipe, the flow can be assumed to be irrotational. Also, the reservoir is so large that the flow is essentially steady. With these conditions, what is the pressure at point A as a function of the fluid density, ρ, gravitational acceleration, g, atmospheric pressure, p_o, as well as U and h? Determine the value of $p - p_o$ in psi for water whose density is $\rho = 1.94$ slug/ft^3.

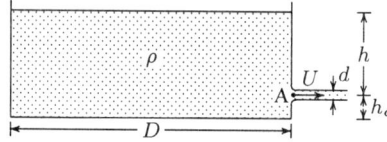

Problems 5.18, 5.19

5.19 The velocity in the outlet pipe from a large reservoir of depth $h = 10$ m is $U = 15$ m/sec. Due to the rounded entrance to the pipe, the flow can be assumed to be irrotational. Also, the reservoir is so large that the flow is essentially steady. With these conditions, what is the pressure at point A as a function of the fluid density, ρ, gravitational acceleration, g, atmospheric pressure, p_o, as well as U and h? Determine the value of $p - p_o$ in kPa for water whose density is $\rho = 1000$ kg/m^3.

5.20 We wish to determine the maximum pressure on your hand when you hold it out the window of your automobile. Assume the ambient pressure is 1 atm and the temperature is 68° F. Assuming the conditions required for Bernoulli's equation to hold are satisfied, compute the maximum pressure (in atm) when you are in the following two situations.

(a) Cruising along a highway at 65 mph.

(b) Driving your Indy 500 racer at 195 mph.

5.21 An eager student performs the home experiment described in this chapter. The student blows a stream of air over one side of an $8\frac{1}{2}$ inch by 11 inch sheet of paper weighing $W = 0.01$ lb. The density of air is $\rho = .00234$ slug/ft^3. Assuming the stream blows over the entire surface, what velocity, U, is required to support the weight of the sheet of paper in a horizontal position? What is the pressure difference between the upper and lower surfaces of the paper?

Sheet of paper

Problems 5.21, 5.22

5.22 An eager student performs the home experiment described in this chapter. The student blows a stream of air over one side of a 21 cm by 29.7 cm sheet of paper. The density of air is $\rho = 1.20$ kg/m^3. Assuming the stream blows over the entire surface at a velocity, $U = 1.10$ m/sec, and the paper is in a horizontal position, what is the weight of the paper in Newtons? What is the pressure difference between the upper and lower surfaces of the paper?

5.23 Consider incompressible flow through a pipe and nozzle that emits a vertical jet. The flow is steady, irrotational, has density ρ and the only body force is gravity. What is the velocity of the jet at the nozzle exit, U_j? To what height, z_{max}, will the jet of fluid rise? Express your answers in terms of ρ, g, h, U_i and $p_i - p_o$. Determine the numerical values of U_j and z_{max} if $p_i - p_o = 55.3$ kPa, $h = 1$ m, $U_i = 3$ m/sec and $\rho = 1000$ kg/m^3.

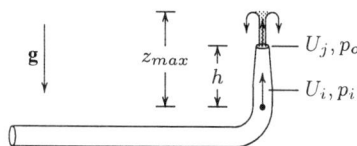

Problems 5.23, 5.24

5.24 Consider incompressible flow through a pipe and nozzle that emits a vertical jet. The flow is steady, irrotational, has density ρ and the only body force is gravity. If the jet of fluid rises to a height $z_{max} = 2h$, what is the value of $p_i - p_o$ as a function of ρ, g, h and U_i? Determine the numerical value of $p_i - p_o$ in psi if $\rho = 2.01$ slug/ft^3, $h = 3$ ft and $U_i = 10$ ft/sec.

5.25 A downward-facing round nozzle is attached to a hose through which an incompressible fluid of density ρ flows. The hose diameter is D and the mass flux is $\dot{m} = \frac{\pi}{4}\rho|w|d^2$, where d is nozzle diameter and w is vertical velocity. Treating the flow as one-dimensional, determine how the nozzle diameter must vary with z in order to have atmospheric pressure, p_o, throughout the nozzle. Assume the flow is steady, irrotational and the only body force acting is gravity.

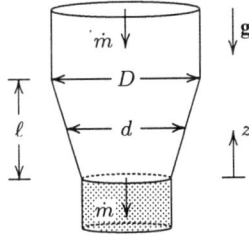

Problems 5.25, 5.26

5.26 A downward-facing round nozzle is attached to a hose through which an incompressible fluid of density ρ flows. The hose diameter is D and the mass flux is $\dot{m} = \frac{\pi}{4}\rho|w|d^2$, where d is nozzle diameter and w is vertical velocity. Treating the flow as one-dimensional, determine how the nozzle diameter must vary with z in order to have

$$\frac{dp}{dz} = -\frac{1}{2}\rho g$$

throughout the nozzle. The quantity g is gravitational acceleration. Assume the flow is quasi-steady, irrotational and the only body force acting is gravity.

5.27 Consider a large tank with a siphon tube attached. The tube has constant diameter, d. You can assume the flow is quasi-steady, incompressible, irrotational and that gravity is the only body force.

(a) Determine the outlet velocity, U, and the minimum pressure in the siphon tube, p_{min}, as a function of gravitational acceleration, g, fluid density, ρ, atmospheric pressure, p_o, and the distances ℓ, h_1 and h_2. **HINT:** Make use of the fact that the mass flux through the tube, $\dot{m} = \frac{\pi}{4}\rho|\mathbf{u}|d^2$, is constant.

(b) Calculate U and p_{min} for $h_1 = 2$ m, $h_2 = 1$ m, $\ell = 1$ m, $\rho = 1000$ kg/m^3 and $p_o = 1$ atm.

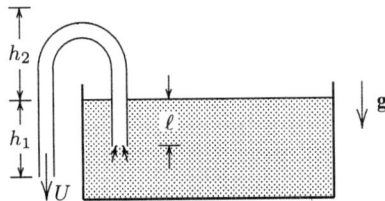

Problems 5.27, 5.28

5.28 Consider a large tank with a siphon tube attached. The tube has constant diameter, d. You can assume the flow is quasi-steady, incompressible, irrotational and that gravity is the only body force.

(a) Determine the outlet velocity, U, and the minimum pressure in the siphon tube, p_{min}, as a function of gravitational acceleration, g, fluid density, ρ, atmospheric pressure, p_o, and the distances ℓ, h_1 and h_2. **HINT:** Make use of the fact that the mass flux through the tube, $\dot{m} = \frac{\pi}{4}\rho|\mathbf{u}|d^2$, is constant.

(b) Calculate U and p_{min} for $h_1 = 5$ ft, $h_2 = 4$ ft, $\ell = 2$ ft, $\rho = 1.94$ slug/ft^3 and $p_o = 1$ atm.

5.29 A large closed tank is pressurized as shown. A jet of fluid issues from a small hole. Assume the flow is incompressible, irrotational, quasi-steady and the only body force acting is gravity.

(a) Determine the pressure, p_a, required to double the jet velocity relative to the value realized for atmospheric pressure, p_o, in the upper chamber.

(b) Compute the value of p_a in atm when $h = 5$ m and $\rho = 998$ kg/m^3. What is the jet velocity, U, for this pressure?

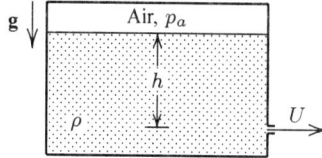

Problems 5.29, 5.30

5.30 A large closed tank is pressurized as shown. A jet of fluid issues from a small hole. Assume the flow is incompressible, irrotational, quasi-steady and the only body force acting is gravity.

(a) Determine the pressure, p_a, required to triple the jet velocity relative to the value realized for atmospheric pressure, p_o, in the upper chamber.

(b) Compute the value of p_a in atm when $h = 20$ ft and $\rho = 1.94$ slug/ft^3. What is the jet velocity, U, for this pressure?

5.31 Consider an incompressible, two-dimensional flow with velocity vector

$$\mathbf{u} = Ax\,\mathbf{i} - Ay\,\mathbf{j}$$

where A is a constant. You can ignore body forces.

(a) Does this velocity field satisfy the continuity equation?

(b) Is the flow irrotational?

(c) If the flow is inviscid, what is the pressure, $p(x,y)$, if $p(0,0) = p_t$?

5.32 The average velocity of water in a nozzle increases from $u_1 = 10$ ft/sec to $u_2 = 60$ ft/sec. Assuming the average velocity varies linearly with distance along the nozzle, x, and that the length of the nozzle is $\ell = 1$ ft, estimate the pressure gradient dp/dx at a point midway through the nozzle. The density of water is $\rho = 1.94$ slug/ft^3. You may assume the flow can be approximated as one dimensional.

5.33 Consider an unsteady flow in an incompressible fluid in which gravitational effects are important. The velocity and gravitational vectors are

$$\mathbf{u} = U\,\mathbf{i} + U\cos[\kappa(x - Ut)]\mathbf{k}, \qquad \mathbf{g} = -g\,\mathbf{k}$$

where U and κ are constants and g is gravitational acceleration. Determine the pressure, $p(x,y,z,t)$ for this flow.

5.34 The U-tube shown rotates about axis O-O at angular velocity Ω. Determine the new positions of the water surfaces, ℓ_1 and ℓ_2. Assume the diameter of the thickest part of the U-tube, $2d$, is very small compared to L_3.

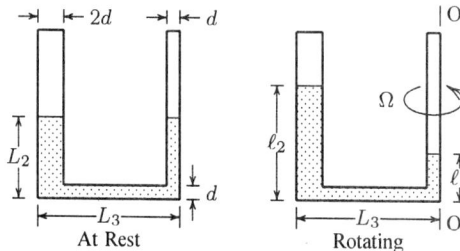

Problem 5.34

5.35 The constant-diameter U-tube shown rotates about axis O-O at angular velocity Ω.

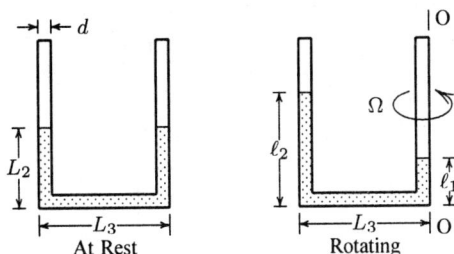

At Rest   Rotating

Problem 5.35

(a) Determine the new positions of the water surfaces, ℓ_1 and ℓ_2. Neglect the diameter of the U-tube in your computations.

(b) What does your answer for Part (a) predict for rotation rates in excess of the critical value defined by $\Omega_{crit} = 2\sqrt{gL_2}/L_3$? Explain how to reformulate the problem with a diagram of the fluid in the U-tube when $\Omega > \Omega_{crit}$.

5.36 Fluid in a cylindrical tank of radius R rotates about the z axis with angular velocity Ω. The fluid has been rotating for a time sufficient to establish rigid-body rotation. The initial fluid level (indicated by the dashed line) is h, the fluid density is ρ and atmospheric pressure is p_o.

(a) Find the equation of the free surface, $z = \zeta(r)$, when the tank rotates, noting that $p = p_o$ at the free surface.

(b) Compute h_{max} and h_{min} as functions of h, Ω, R and g. **NOTE:** The volume of fluid in the tank is given by

$$\text{Volume} = 2\pi \int_0^R \int_0^{\zeta(r)} dz\, r\, dr$$

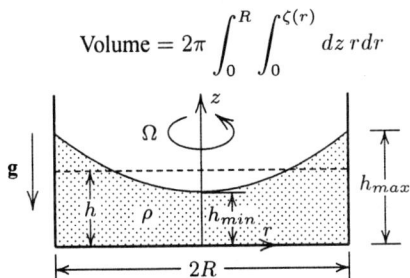

Problems 5.36, 5.37

5.37 Fluid in a cylindrical tank of radius R rotates about the z axis with angular velocity Ω. The fluid has been rotating for a time sufficient to establish rigid-body rotation. The initial fluid level (indicated by the dashed line) is h, the fluid density is ρ and atmospheric pressure is p_o.

(a) Find the equation of the free surface, $z = \zeta(r)$, when the tank rotates, noting that $p = p_o$ at the free surface.

(b) Find the rotation rate for which the center of the container just becomes exposed, i.e., $h_{min} = 0$. **NOTE:** The volume of fluid in the tank is given by

$$\text{Volume} = 2\pi \int_0^R \int_0^{\zeta(r)} dz\, r\, dr$$

5.38 The cross-sectional area of a pipe is $A(x) = A_o F(x)$, where A_o is the area at $x = 0$, $F(0) = 1$ and $F(x)$ is a strange function you've never heard of. Assume the flow is inviscid, incompressible, can be approximated as one dimensional and that body forces are negligible. Using the fact that mass flux, $\dot{m} = \rho u A$, is constant and $p(0) = p_o$, compute the pressure throughout the pipe. How does $p(x)$ vary with increasing area? How does it vary with decreasing area? Explain your results.

5.39 A truck carries a tank of water that is open at the top with length ℓ, width w and depth h. Assuming the driver will not accelerate the truck at a rate greater than a, what is the maximum depth, h_o, to which the tank may be filled to prevent spilling any water? Assume constant acceleration, and note that the pressure is constant and equal to its atmospheric value at the free surface.

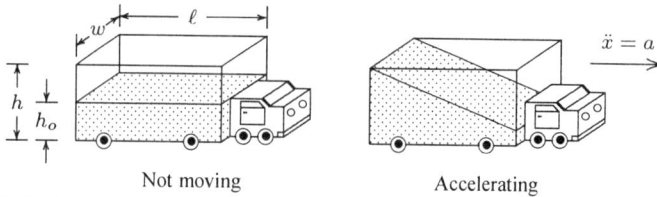

Not moving   Accelerating

Problems 5.39, 5.40

5.40 A truck carries a tank of water that is open at the top with length ℓ, width w and depth h. If the truck is $\frac{2}{3}$ full and $\ell = 3h$, what is the maximum acceleration, a, that can be sustained without spilling any water? Assume constant acceleration, and note that the pressure is constant and equal to its atmospheric value at the free surface.

5.41 Imagine you are rushing to the university to avoid being late for your fluid-mechanics class. Your coffee cup is resting next to you in your car. In your haste to get to class, you accelerate at λ g's, i.e., your acceleration is $a = \lambda g$ where λ is a constant. What is the maximum height, h_o, to which the cup can be filled to avoid spilling any coffee? Assume the cup is a cylinder of height $h = 3$ inches and diameter $d = 3$ inches. **HINT:** To conserve mass in this geometry, necessarily $h_o = \frac{1}{2}(h + h_{min})$. As a percentage, determine how full the cup can be if you are driving your Volkswagen Bug ($\lambda = 1/6$) or your Corvette ($\lambda = 2/5$).

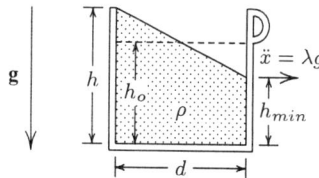

Problems 5.41, 5.42

5.42 Imagine you are rushing to the university to avoid being late for your fluid-mechanics class. Your coffee cup is resting next to you in your car. In your haste to get to class, you accelerate at λ g's, i.e., your acceleration is $a = \lambda g$ where λ is a constant. What is the maximum value of λ possible to avoid spilling if the cup is initially 85% full? Assume the cup is a cylinder of height $h = 8$ cm and diameter $d = 10$ cm. **HINT:** To conserve mass in this geometry, necessarily $h_o = \frac{1}{2}(h + h_{min})$.

5.43 A small car containing an incompressible fluid of density ρ is rolling down an inclined plane. Show that the free surface is planar and determine the angle it makes with the horizontal, β. Ignore any friction in the wheels. **HINT:** Do your work in a coordinate system for which x and z are parallel to and normal to the inclined plane, respectively. Also, make use of the trigonometric identity

$$\tan(\beta \pm \alpha) = \frac{\tan\beta \pm \tan\alpha}{1 \mp \tan\beta \tan\alpha}$$

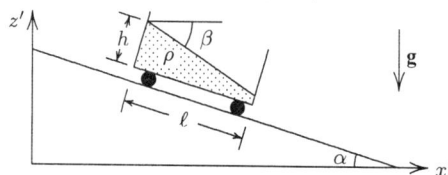

Problems 5.43, 5.44

5.44 A small car containing an incompressible fluid of density ρ is rolling down an inclined plane. Determine the location and value of the maximum pressure in the car. Ignore any friction in the wheels. **HINT:** Use a coordinate system with x and z parallel to and normal to the inclined plane, respectively.

5.45 A Pitot-static tube is placed in a flow of air with $\rho = 1.20$ kg/m^3. The stagnation- and static-pressure taps read 102.5 kPa and 101 kPa, respectively. What is the velocity of the air? If the velocity changes to 80 m/sec and the static pressure is unchanged, what is the corresponding stagnation pressure?

5.46 A Pitot-static tube is placed in a flow of helium with $\rho = 3.2 \cdot 10^{-4}$ slug/ft^3. The static-pressure tap reads 12.0 psi and the flow velocity is 300 ft/sec. What is the stagnation pressure? If the stagnation pressure changes to 12.2 psi and the static pressure is unchanged, what is the corresponding velocity?

5.47 Consider a poorly designed Pitot-static tube with a single static-pressure hole at the top of the tube as shown. If the tube radius is r, develop a formula for the true velocity, U_{true}, as a function of gravitational acceleration, g, radius, r, and the velocity U inferred from Equation (5.60). If $r = 0.01$ inch, determine the percentage error in velocity for an indicated velocity of $U = 1$, 10 and 100 ft/sec.

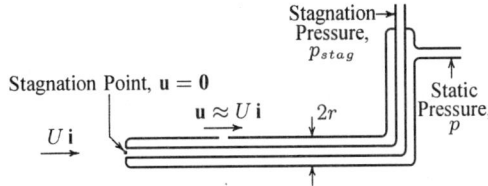

Problem 5.47

5.48 A *Venturi meter* is a device used to measure fluid velocities and flow rates for incompressible, steady flow. As shown, the pressure is measured at two sections of a pipe with different cross-sectional areas. You may assume the flow is irrotational and that effects of body forces can be ignored. Noting that for steady flow the mass-flow rate, $\dot{m}$, is constant and equal to $\rho U A$, where ρ is density, U is average velocity and A is cross-sectional area, verify that $\dot{m}$ is

$$\dot{m} = \rho A_2 \sqrt{\frac{2(p_1 - p_2)}{\rho(1 - A_2^2/A_1^2)}}$$

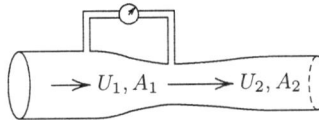

Problems 5.48, 5.49, 5.50

5.49 A *Venturi meter* is a device used to measure fluid velocities and flow rates for incompressible, steady flow. As shown, the pressure is measured at two sections of a pipe with different cross-sectional areas. A straightforward derivation shows that the volume-flow rate, Q, is

$$Q = A_2 \sqrt{\frac{2(p_1 - p_2)}{\rho(1 - A_2^2/A_1^2)}}$$

where p is pressure, ρ is density and A is cross-sectional area. Consider a Venturi meter that has $A_1 = 12$ ft^2 and $A_2 = 10$ ft^2. If the volume flow rate is $Q = 100$ ft^3/sec, what is the pressure difference, $p_1 - p_2$, if the fluid flowing is air, water or mercury?

5.50 A *Venturi meter* is a device used to measure fluid velocities and flow rates for incompressible, steady flow. As shown, the pressure is measured at two sections of a pipe with different cross-sectional areas. A straightforward derivation shows that the volume-flow rate, Q, is

$$Q = A_2 \sqrt{\frac{2(p_1 - p_2)}{\rho(1 - A_2^2/A_1^2)}}$$

where p is pressure, ρ is density and A is cross-sectional area. Consider a Venturi meter that has $A_1 = 1$ m^2 and $A_2 = 0.7$ m^2. The attached pressure gage can accurately measure pressures no smaller than 0.1 Pa. If air of density $\rho = 1.20$ kg/m^3 is flowing, what is the minimum flow rate that can be accurately measured?

5.51 A body moves at constant velocity $U = 10$ m/sec through water ($\rho = 1000$ kg/m^3). The difference between the pressure at the front stagnation point on the body and at Point A is $p_{stag} - p_A = 32$ kPa. What is the velocity at Point A?

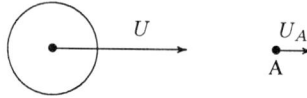

Problems 5.51, 5.52

5.52 A body moves at constant velocity $U = 30$ ft/sec through water ($\rho = 1.94$ slug/ft^3). The difference between the pressure at the front stagnation point on the body and at Point A is $p_{stag} - p_A = 4.91$ psi. What is the velocity at Point A?

5.53 Body A travels through water at a constant speed of $U_A = 12$ m/sec. Velocities at points B and C are induced by the moving body and have magnitudes of $U_B = 4$ m/sec and $U_C = 2$ m/sec. If the density of water is 1000 kg/m^3 and effects of gravity can be ignored, what is $p_B - p_C$?

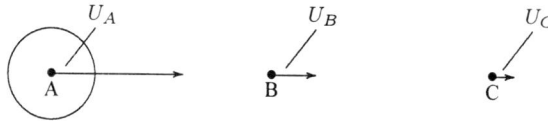

Problems 5.53, 5.54

5.54 Body A travels through water at a constant speed of $U_A = 20$ ft/sec. Velocities at points B and C are induced by the moving body and have magnitudes of $U_B = 8$ ft/sec and $U_C = 4$ ft/sec. If the density of water is 1.94 slug/ft^3 and effects of gravity can be ignored, what is $p_B - p_C$ (in psi)?

5.55 Reference to Appendix D, Section D.4 shows that for inviscid flow in a coordinate system rotating about the z axis with angular velocity $\boldsymbol{\Omega} = \Omega \mathbf{k}$, the momentum equation assumes the following form.

$$\frac{\partial \mathbf{u}'}{\partial t'} + \mathbf{u}' \cdot \nabla' \mathbf{u}' = -\frac{\nabla' p}{\rho} - \boldsymbol{\Omega} \times \boldsymbol{\Omega} \times \mathbf{r}' - 2\boldsymbol{\Omega} \times \mathbf{u}'$$

The last two terms on the right-hand side of this equation are the centrifugal and Coriolis forces, respectively. Taking advantage of the results developed in Section 5.5, determine the form this equation assumes under a Galilean transformation. Is this equation Galilean invariant?

5.56 In terms of natural coordinates (see Appendix D, Section D.5), the s and n components of Euler's equation are

$$\rho u \frac{\partial u}{\partial s} = -\frac{\partial p}{\partial s} - \rho \frac{\partial \mathcal{V}}{\partial s} \quad \text{and} \quad \rho \frac{u^2}{\mathcal{R}} = -\frac{\partial p}{\partial n} - \rho \frac{\partial \mathcal{V}}{\partial n}$$

As discussed in Section 5.3, integrating the streamwise, or s, component yields

$$p + \frac{1}{2}\rho u^2 + \rho \mathcal{V} = F(n)$$

where $F(n)$ is constant along a streamline (because $n = $ constant on a streamline). Verify that, with $\omega = u/\mathcal{R} - \partial u/\partial n$ denoting the vorticity,

$$F'(n) = -\rho \omega u$$

5.57 For incompressible, irrotational flows, even when the flow is unsteady, the velocity vector, $\mathbf{u}$, can be written as $\mathbf{u} = \nabla \phi$, where $\phi(x, y, z, t)$ is known as the *velocity potential*. Beginning with the vector identity

$$\mathbf{u} \cdot \nabla \mathbf{u} = \nabla \left(\frac{1}{2}\mathbf{u} \cdot \mathbf{u}\right) - \mathbf{u} \times (\nabla \times \mathbf{u})$$

derive an unsteady-flow replacement for Bernoulli's equation. Assume a conservative body force is present so that $\mathbf{f} = -\nabla \mathcal{V}$.

5.58 A *barotropic fluid* is one for which the pressure is a function only of density, i.e., $p = p(\rho)$. For a steady, irrotational flow of a barotropic fluid with a conservative body force, $\mathbf{f} = -\nabla \mathcal{V}$, derive the following replacement for Bernoulli's equation.

$$\int \frac{dp}{\rho} + \frac{1}{2} \mathbf{u} \cdot \mathbf{u} + \mathcal{V} = \text{constant}$$

HINT: Introduce a function $F(\rho)$ defined by

$$\frac{\nabla p}{\rho} = \frac{1}{\rho} \frac{dp}{d\rho} \nabla \rho = F'(\rho) \nabla \rho$$

5.59 Beginning with the two-dimensional component form of the continuity and Euler equations, Verify that they can be written as follows.

$$\frac{\partial \mathbf{U}}{\partial t} + \frac{\partial \mathbf{F}}{\partial x} + \frac{\partial \mathbf{G}}{\partial y} = 0$$

where $\mathbf{U}$ is the column vector defined by

$$\mathbf{U} = \left\{ \begin{array}{c} \rho \\ \rho u \\ \rho v \end{array} \right\}$$

while $\mathbf{F}$ and $\mathbf{G}$ are column vectors that you must determine. The equations of fluid mechanics are often written in this form for computational studies.

5.60 For hurricanes, the Coriolis acceleration is much larger than the convective acceleration. Reference to Appendix D, Section D.4 shows that the Euler equation simplifies to

$$2\rho \, \mathbf{\Omega} \times \mathbf{u} = -\nabla \hat{p}$$

where $\hat{p}$ is a reduced pressure that includes the centrifugal acceleration and $\mathbf{\Omega}$ is Earth's angular velocity. Based on this equation, explain why hurricanes rotate counterclockwise in the northern hemisphere and clockwise in the southern hemisphere. **HINT:** To simplify your explanation, consider hurricanes centered at the north pole and the south pole.

Chapter 6

Control Volume Method

This chapter emphasizes use of the integral conservation laws for a carefully chosen **control volume** on which we can indirectly compute forces arising from complex pressure fields within. Before launching on our study of this powerful and useful method, it is worthwhile to pause and discuss the method's generality. As we will see, one way of using the **control volume method** is to provide a *global view* as opposed to a *detailed view* of the fluid motion through a finite-sized volume. When used in this way, the method does not discern precise details of how fluid velocity, pressure, etc. vary throughout a flow. Rather, it provides integrated values such as the total mass flux across a surface or the net pressure force on a solid surface.

The control volume method can thus be used as a consistency check of measured or computed properties. For example, if we measure or compute density and velocity at several points across the inlets to and outlets from a specified volume, we can compute the mass flux at the inlets and outlets by numerically integrating the experimental values. These integrated values can then be compared to results of a control-volume analysis.

The control volume method is, of course, entirely consistent with the differential equations for mass and momentum conservation developed in Chapter 5. This is obvious since we deduced the differential equations from the very integral-conservation laws upon which the control volume method is based. As a consequence, if we subdivide a finite control volume into a number of smaller volumes, we can observe more flow details. In the limit of infinitesimal subdivisions, our solution will approach the continuum solution of the differential equations.

We implement the control volume method for a collection of small sub-volumes in the field of **Computational Fluid Dynamics (CFD)**. In order to replace the differential equations governing motion of a fluid by algebraic equations suitable for computer analysis, the standard starting point is to first establish a computational grid [cf. Knupp and Steinberg (1993)]. This involves dividing physical space into a collection of contiguous cells as illustrated in Figure 6.1.

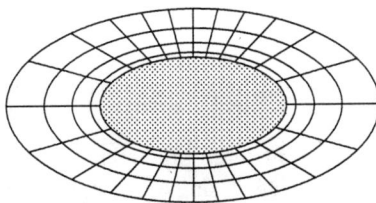

Figure 6.1: *Computational grid.*

The cells are close enough to be considered infinitesimal on the overall scale of the problem. Flow properties are then defined at the vertices or perhaps at the centroid of each cell. To derive the algebraic equations that will be solved by a computer, the integral conservation equations are applied to each cell [cf. Anderson (1995)]. This is done to insure that properties are conserved on both a global and a detailed level. This is so important for compressible flows that older numerical procedures that are not based on the exact conservation principles have been almost universally rejected by modern CFD researchers.

We begin this chapter by outlining proper implementation of the control volume method. Because of their relative simplicity, applications involving the mass-conservation principle are considered first. More complex problems that require combined use of mass and momentum conservation follow, including applications for which forces are computed indirectly in terms of mass and momentum flowing through the control volume. In contrast to Bernoulli's equation, we will see that the momentum-conservation integral always applies, even when the flow is unsteady, rotational and subject to non-conservative body forces. We will also find that the control volume can move, and even accelerate.

6.1   Preliminaries

Before proceeding to applications of the control volume method, it is worthwhile at this point to provide a framework and some useful preliminary information. To do so, this section discusses the following three key issues:

- Guidelines for selecting control-volume boundaries;

- A useful theorem that simplifies handling the pressure integral;

- A quick test of the method for a problem we solved earlier, viz., the fluid-statics problem.

6.1.1   Overview of the Method

The basic objective of the control volume method is to use the conservation laws in their integral form to analyze global properties of a given fluid-flow problem. Because the integral laws involve a volume integral and integrals over the surface bounding the volume, our first requirement is to select a suitable volume for evaluating the integrals. There are two primary principles that influence selection of the control volume.

- Portions of the control surface across which fluid flows should have their unit normal as nearly as possible parallel to the fluid velocity vector.

- To the greatest extent possible, the pressure should be either known or superfluous on portions of the control surface that are not coincident with a solid boundary.

Although these principles are helpful as a starting point in establishing control-volume boundaries, they can be mutually exclusive. Hence, a bit of thought should always precede selection of the control volume.

Figure 6.2 illustrates application of these principles for simple geometries. Flow A has fluid leaving at an angle to the horizontal. Using a rectangular control volume would introduce a sinusoidal function in evaluating the flow rate, which is proportional to $(\mathbf{u} \cdot \mathbf{n})$. However, selecting a control volume identical to the shape of the object obviates the need to introduce

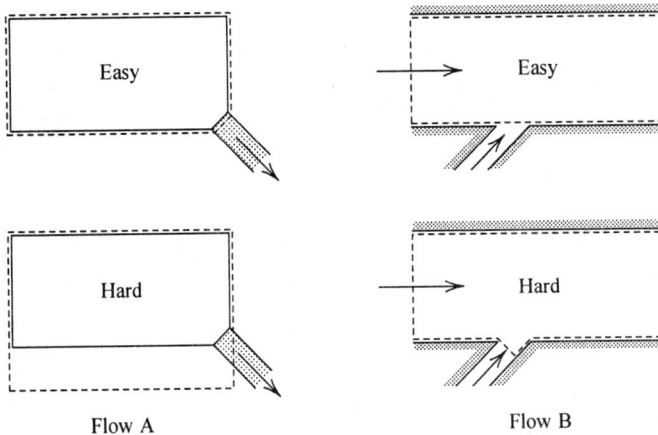

Figure 6.2: *Selecting a control volume (denoted by dashed lines).*

trigonometric functions in the flow rate. Such functions will be needed to properly address momentum conservation, but the overall computations are simplified by using the non-rectangular control volume. In this case, there is no obvious reason to consider the pressure, so only the first principle noted above affects the choice of the control-volume boundaries.

By contrast, Flow B is an example in which the two principles cannot both be satisfied, and we are forced to choose one over the other. The channel shown has fluid injected into the main channel from an inlet below in which the pressure is unknown. If we are interested only in mass and horizontal-momentum conservation, a rectangular control volume is most appropriate. That is, the unit normal at the inlet is vertical so that the horizontal component of the pressure integral vanishes, regardless of the pressure in the inlet. By contrast, the alternative control volume shown, which is consistent with having the boundaries everywhere perpendicular to **u**, would require a rather detailed computation to show that the pressure integral is zero.

There are several additional steps that should be taken to properly implement the control volume method. In general, once the control volume is selected, the sequence of computations should be as follows.

1. **Apply Conservation of Mass:** This is the simplest of the conservation principles as it is a scalar equation and involves, at most, one volume integral and one closed-surface integral. In addition to establishing one equation relating flow properties, this step yields $(\mathbf{u} \cdot \mathbf{n})$ at all points on the control-volume boundary, including sign. This factor appears in all of the conservation principles and can be re-used in subsequent steps.

2. **Apply Conservation of Momentum:** Use as many components of this vector equation as are appropriate for the problem at hand. Most importantly, when you evaluate the momentum-flux integral, viz.,

$$\oiint_S \rho\, \mathbf{u}(\mathbf{u} \cdot \mathbf{n})\, dS$$

treat the integrand as the product of two separate factors. The first is the *quantity being carried across the control surface*, i.e., the momentum per unit volume, $\rho\, \mathbf{u}$. The second is the *rate at which it is being carried across the surface*, $(\mathbf{u} \cdot \mathbf{n})\, dS$. If you make this distinction, you will minimize the number of sign errors you make.

3. **Count Equations and Unknowns:** Check to see if the number of unknowns and the number of equations you have at this point are equal to determine if you have a **closed** system of equations. By definition, a system of equations is closed if the number of equations matches the number of unknowns.

4. **Appeal to Energy Conservation or Bernoulli's Equation if Necessary:** If there are more unknowns than equations, in the most general case, you must use the energy-conservation principle (see Chapter 7) to determine an additional equation. However, if the flow is steady, incompressible, irrotational and involves conservative body forces, then Bernoulli's equation can be used — it is an expression of the conservation of mechanical energy.

5. **Check the Physics:** Throughout the process, pause and use physical reasoning to determine the validity of signs and limiting cases. Also check for dimensional consistency of your results.

To reinforce the importance of the final step, note that the control volume method provides an excellent bookkeeping tool for analyzing fluid-flow problems. It provides a methodology for keeping track of mass and momentum fluxes and forces on a control volume. However, it is a means to an end, not the end itself. That is, it is a mathematical tool whose purpose is to aid in understanding the physics of the problem at hand. *You should not lose sight of the fact that understanding the physics through application of the conservation principles to the problem at hand is the desired end of the control volume method.*

6.1.2 Useful Control Volume Theorem

In evaluating the force and moment on a control volume caused by the pressure field, we can replace p by $(p - p_o)$ where p_o is a constant. In applications, we will often select p_o as atmospheric pressure. The proof is simple and requires use of the divergence theorem for the force and Gauss' theorem for the moment (see Appendix C for details on these theorems). We focus on the force here, leaving the proof for the moment to the Problems section.

Because integration is a linear operation, we can say

$$\oiint_S (p - p_o)\mathbf{n}\, dS = \oiint_S p\, \mathbf{n}\, dS - \oiint_S p_o \mathbf{n}\, dS \tag{6.1}$$

Using the divergence theorem on the integral of p_o, there follows:

$$\oiint_S p_o \mathbf{n}\, dS = \iiint_V \nabla p_o\, dV \tag{6.2}$$

But, since p_o is a constant, necessarily $\nabla p_o = \mathbf{0}$ so that this integral vanishes. Therefore,

$$\oiint_S (p - p_o)\mathbf{n}\, dS = \oiint_S p\, \mathbf{n}\, dS \tag{6.3}$$

This should come as no great surprise as it is pressure differences that cause fluid motion. Simply pumping up the ambient value has no effect on a closed volume. Nevertheless, failing to reference pressure to the ambient value, p_o, is a common source of error in the control volume method, and is thus worthy of mention.

6.1.3 Fluid Statics Revisited

In this section we will employ a useful engineering technique to begin learning how to use the conservation principles in their integral form. Specifically, we will apply the conservation forms in a special limiting case for which we already know the answer. In particular, we will use the integral forms to determine the forces in a non-moving, or static, fluid. We can compare our results to those obtained in Chapter 3.

Consider the arc AB illustrated in Figure 6.3. We determined the force on the arc, **F**, in Section 3.5. Our computations showed that the vertical force is equal to the weight of the column of fluid above the arc, while the horizontal force is equal to the force on arc AB's projection on a vertical plane.

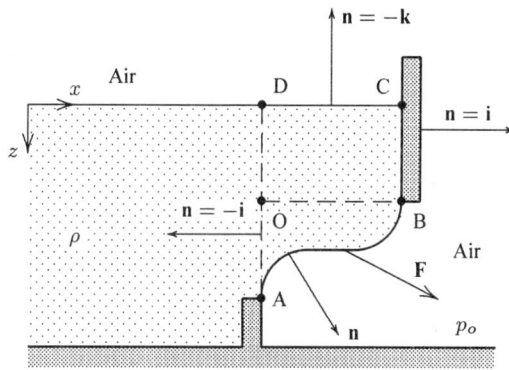

Figure 6.3: *Force on a curved surface for a stationary fluid. OA is the projection of the arc AB on a vertical plane.*

To obtain the force on arc AB, we begin with the momentum-conservation principle with gravity as the body force ($\mathbf{f} = \mathbf{g} = g\mathbf{k}$):

$$\iiint_V \frac{\partial}{\partial t}(\rho\,\mathbf{u})\,dV + \oiint_S \rho\,\mathbf{u}(\mathbf{u}\cdot\mathbf{n})\,dS = -\oiint_S p\,\mathbf{n}\,dS + \iiint_V \rho\,\mathbf{g}\,dV \tag{6.4}$$

For fluid statics, the velocity vector vanishes. Substituting $\mathbf{u} = \mathbf{0}$ into Equation (6.4) yields a balance between the net pressure force on the control volume and its weight, i.e.,

$$\oiint_S p\,\mathbf{n}\,dS = \iiint_V \rho\,\mathbf{g}\,dV \tag{6.5}$$

Also, with zero velocity, Bernoulli's equation simplifies to

$$p + \rho g z = \text{constant} = p_o \tag{6.6}$$

which is simply the **hydrostatic relation**, Equation (3.10).

Now, as shown in the preceding section, we can replace p by $(p - p_o)$ in the pressure integral appearing in Equation (6.5). Defining our control volume as the region bounded by the closed contour OABCD, the closed surface integral is the sum of integrals on arcs AB, BC, CD, DO and OA. Since our geometry is actually three dimensional, we also have integrals on the end planes whose unit normals point into and out of the page. For the sake of brevity, we omit these contributions in the following equation. We will explain why they sum to zero below. Thus, Equation (6.5) becomes:

$$\iint_{S_{AB}} (p - p_o)\mathbf{n}\, dS + \iint_{S_{BC}} (p - p_o)\mathbf{i}\, dS + \iint_{S_{CD}} (p - p_o)(-\mathbf{k})\, dS$$
$$+ \iint_{S_{DO}} (p - p_o)(-\mathbf{i})\, dS + \iint_{S_{OA}} (p - p_o)(-\mathbf{i})\, dS = \iiint_{V_{OABCD}} \rho g \mathbf{k}\, dV \qquad (6.7)$$

Note that when we evaluate a closed surface integral, it is a sum of integrals with integration always in the positive direction. Thus, for example, the z integration on surfaces BC and DO begins at the free surface so that the lower limit of integration in both integrals is 0. *The sign of the integrals is determined solely by the unit normal.*

Hence, the sum of the integrals on arcs BC and DO cancel. This is true because p depends only upon z, and both arcs include the same range of values of z. Thus, the force on both arcs has the same magnitude, but acts in opposite directions. The sum of the forces on the end planes vanishes for precisely the same reason, so that their omission in Equation (6.7) is justified.

Also, because $p = p_o$ on the free surface (arc CD), the integral is zero. Finally, the force on the arc is the difference between the integral along arc AB of the hydrostatic pressure, p, (from above) and the integral along arc AB of the atmospheric pressure, p_o, (from below) i.e.,

$$\mathbf{F} = \iint_{S_{AB}} (p - p_o)\mathbf{n}\, dS \qquad (6.8)$$

Therefore, since the integrals on S_{BC}, S_{CD} and S_{DO} sum to zero, Equation (6.7) simplifies to:

$$\mathbf{F} = \mathbf{i} \iint_{S_{OA}} (p - p_o)\, dS + \mathbf{k} \iiint_{V_{OABCD}} \rho g\, dV \qquad (6.9)$$

In words, we have shown that, consistent with our analysis of fluid statics in Chapter 3, the vertical force on arc AB is equal to the weight of the column of fluid above the arc, while the horizontal force is equal to the force on arc AB's projection on a vertical plane.

6.2 Conservation of Mass

There are some problems that can be analyzed using only the mass-conservation integral. In this section, we discuss three such problems. Each brings out important aspects of the control volume method and the integral law for mass conservation.

6.2.1 Steady Flow in a Pipe

Consider steady flow in a pipe of varying cross section as shown in Figure 6.4. If the pipe is curved as illustrated in the figure, the flow within is very complicated. In addition to the primary motion along the axis of the pipe, cross-sectional regions of swirling flow appear. The swirling patterns are known as **secondary motions**, and were first observed in curved pipe flow (with constant cross-sectional area) by Prandtl. The precise details aren't really important for our present purposes. We simply mention secondary motions to stress that an innocent looking geometry can have a very complicated flow — and we can tackle it fearlessly with the control volume method! Theoretical computations that predict details of secondary motions are very difficult to perform and are often very inaccurate. Nevertheless, the control

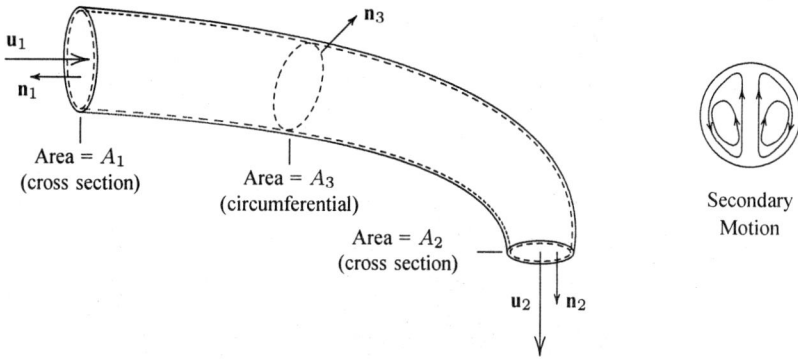

Figure 6.4: *Flow in a pipe; the secondary-motion offset shows cross-sectional streamlines in flow through a curved pipe first observed by Prandtl.*

volume method is very easy to apply to this flow and results obtained are exact, albeit lacking subtle details.

We would like to develop a relation between flow properties at the inlet (subscript 1) and at the outlet (subscript 2). To analyze this flow, we begin with the integral form of mass conservation, viz.,

$$\iiint_V \frac{\partial \rho}{\partial t} \, dV + \oiint_S \rho \, \mathbf{u} \cdot \mathbf{n} \, dS = 0 \tag{6.10}$$

Because the flow is steady, the first integral vanishes, and all that remains is

$$\oiint_S \rho \, \mathbf{u} \cdot \mathbf{n} \, dS = 0 \tag{6.11}$$

In words, this equation says that *the net flux of fluid out of the control volume is zero*. It is a *net flux* because we are evaluating a closed surface integral. It is a flux *out of the control volume* because we are using an *outer unit normal*.

To evaluate the closed surface integral, we must first select a suitable control volume. The boundaries of the control volume are indicated with dashed lines in Figure 6.4. The control volume is bounded by three distinct surfaces as follows.

1. A cross section at the inlet of area A_1 and unit normal $\mathbf{n}_1$

2. A cross section at the outlet of area A_2 and unit normal $\mathbf{n}_2$

3. The circumferential area coincident with the pipe between inlet and outlet of area A_3 and unit normal $\mathbf{n}_3$

Thus, the closed-surface integral expands to the following.

$$\iint_{A_1} \rho_1 \mathbf{u}_1 \cdot \mathbf{n}_1 \, dA + \iint_{A_2} \rho_2 \mathbf{u}_2 \cdot \mathbf{n}_2 \, dA + \iint_{A_3} \rho_3 \mathbf{u}_3 \cdot \mathbf{n}_3 \, dA = 0 \tag{6.12}$$

Now, because there is no flow through the pipe walls, we know that $\mathbf{u}_3 \cdot \mathbf{n}_3 = 0$. Thus, the integral along the circumference (i.e., on A_3) is exactly zero. Also, because we are working with an outer unit normal, necessarily

$$\mathbf{u}_1 \cdot \mathbf{n}_1 = -u_1 \quad \text{and} \quad \mathbf{u}_2 \cdot \mathbf{n}_2 = u_2 \tag{6.13}$$

where u_1 and u_2 denote the magnitudes of the velocity vectors $\mathbf{u}_1$ and $\mathbf{u}_2$, respectively. Therefore, Equation (6.12) simplifies to:

$$\iint_{A_1} \rho_1 u_1 \, dA = \iint_{A_2} \rho_2 u_2 \, dA \tag{6.14}$$

In words, this equation says *the mass flux into the control volume equals the mass flux out.*

This expression illuminates several points. We can use it, for example, to verify a numerical solution that provides the actual values of density and velocity at the inlet and outlet (and, presumably, on all cross sections). By verifying that Equation (6.14) is satisfied, we check that mass is conserved in a global sense. If it is not, the numerical solution is defective. In a similar way, we can determine the accuracy and error bounds for measured density and velocity profiles.

In many of the applications to follow, we will simplify the analysis by treating flow properties as though they are uniform on cross sections. For a real fluid this is never true as the no-slip boundary condition always gives rise to non-uniform velocity profiles. What we will be doing is working with averaged flow properties. For example, the average mass flux per unit area, $\overline{\rho u}$, on a given cross section is

$$\overline{\rho u} \equiv \frac{1}{A} \iint_{A} \rho u \, dA \tag{6.15}$$

Thus, in terms of averaged mass flux per unit area, Equation (6.14) simplifies to

$$\overline{\rho_1 u_1} A_1 = \overline{\rho_2 u_2} A_2 \quad \Longrightarrow \quad \overline{\rho u} A = \text{constant} \tag{6.16}$$

As a final comment, since A_2 can be any cross section in the pipe, we conclude that the mass-flow rate is constant through all cross sections.

6.2.2   Sphere Falling in a Cylinder

As a second example of a problem that can be solved using only the mass-conservation law, consider a sphere of radius R that is falling in a closed cylinder filled with an incompressible fluid of density, ρ [Figure 6.5(a)]. We assume that the sphere falls axially, i.e., it remains centered within the cylinder, whose radius is, say, twice that of the sphere, $2R$. Finally, we

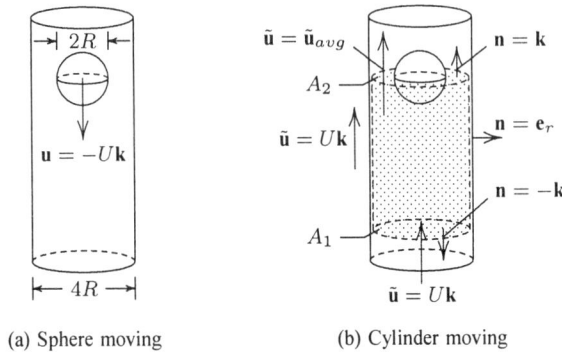

(a) Sphere moving   (b) Cylinder moving

Figure 6.5: *Sphere falling in a cylinder; the control volume is the shaded region.*

observe that the sphere falls at velocity U. We would like to determine the average velocity of the surrounding fluid at the midsection of the sphere.

This problem differs from the application of the preceding section in an important way. Specifically, the fluid motion in this problem is unsteady, while the pipe-flow problem was postulated to be steady. We can use either of the following strategies for solving this problem.

1. If the sphere falls at a constant velocity, we can use a Galilean transformation to recast the problem as a steady-flow problem. This permits using a stationary control volume.

2. We can select a control volume that moves with the sphere. This method is more general as it applies even if the sphere's velocity varies with time.

Considering the constant-velocity case first, we use Strategy 1 and change coordinate frames with a Galilean transformation as follows. In the transformed frame, the sphere is at rest and the cylinder moves upward with velocity vector $\tilde{\mathbf{u}} = U\mathbf{k}$ as indicated in Figure 6.5(b). Because the cylinder wall is always vertical, the geometry for an *infinitely-long cylinder* is unchanged as time passes under this transformation. That is, unless we paint stripes on the cylinder walls, for example, an observer riding on the sphere cannot see any difference in the geometry as the cylinder passes. Even with such stripes, the fluid motion would experience no change, of course, so we clearly replace the original unsteady problem with an equivalent steady problem.

The control volume is the shaded region bounded by the dashed lines in Figure 6.5(b). The upper boundary is the horizontal surface between the cylinder and the sphere and the hemispherical surface below the midsection. The lower boundary is a horizontal surface far below the sphere. The boundary is presumed to be sufficiently distant that the incoming flow is uniform, i.e., it is unaffected by the presence of the sphere. To close the surface bounding the control volume, we select a cylindrical surface coincident with the portion of the cylinder between the upper and lower control-volume boundaries. The outer unit normal, $\mathbf{n}$, is shown on various parts of the bounding surface. Note that, although this is not shown, in order to be an outer unit normal on the surface of the sphere, $\mathbf{n}$ must point into the sphere.

Because the flow is steady in this reference frame, conservation of mass tells us that the net flux of fluid out of the control volume is zero, viz.,

$$\oiint_S \rho\, \tilde{\mathbf{u}} \cdot \mathbf{n}\, dS = 0 \tag{6.17}$$

where $\tilde{\mathbf{u}}$ denotes the Galilean-transformed velocity defined by

$$\tilde{\mathbf{u}} = \mathbf{u} - \mathbf{u}_{sphere} = \mathbf{u} + U\mathbf{k} \tag{6.18}$$

Noting that $\tilde{\mathbf{u}} \cdot \mathbf{n} = 0$ on the sphere and the cylinder, our mass-conservation integral assumes the following form:

$$\iint_{A_1} \rho\, U\mathbf{k} \cdot (-\mathbf{k})\, dA + \iint_{A_2} \rho\, \tilde{\mathbf{u}} \cdot \mathbf{k}\, dA = 0 \tag{6.19}$$

where A_1 is the area of the lower boundary, which is simply the area of the cylinder, i.e., $4\pi R^2$. The area A_2 is the part of the control-volume boundary between the sphere and the cylinder. It is the difference between the cylinder area and a circle of radius R, which is $3\pi R^2$. Thus, treating $\tilde{\mathbf{u}}$ as a constant on the upper control-volume boundary (which, as we saw in the preceding section, is the equivalent of using the average velocity), we can perform the indicated integrals. Thus, we have

$$-4\pi R^2 \rho\, U + 3\pi R^2 \rho\, \tilde{u}_{avg} = 0 \quad \Longrightarrow \quad \tilde{u}_{avg} = \frac{4}{3}U \tag{6.20}$$

We can pause at this point to check for physical consistency. That is, note that the first term on the left-hand side of Equation (6.20) has a minus sign and the second term has a plus sign. The first term corresponds to fluid entering the control volume, which is a negative flux out. Conversely, the second term corresponds to fluid leaving the control volume, which is a positive flux out. Thus, the signs make sense from a physical point of view.

Finally, in terms of the original coordinate frame in which the cylinder is stationary and the sphere is moving, the average velocity at the midsection of the sphere, u_{avg}, is

$$u_{avg} = \frac{1}{3}U \tag{6.21}$$

In general, because the Galilean transformation formally applies only to a coordinate frame translating at a constant velocity, we cannot use the approach just described if the velocity varies with time. Rather, we must implement Strategy 2 and use a moving control volume. Before we can do this, however, we must re-examine the mass-conservation principle. For general (time-dependent) fluid motion, we know that

$$\iiint_V \frac{\partial \rho}{\partial t}\, dV + \oiint_S \rho\, \mathbf{u} \cdot \mathbf{n}\, dS = 0 \tag{6.22}$$

When we use a moving control volume, as discussed at the end of Section 4.7, this equation holds with the understanding that the velocity **u** corresponds to the fluid velocity *relative to the control-volume boundary*. Thus, if $\mathbf{u}_{abs}$ denotes the **absolute velocity of the fluid**, and $\mathbf{u}_{cv}$ the velocity of the control volume, we can rewrite Equation (6.22) as follows.

$$\iiint_V \frac{\partial \rho}{\partial t}\, dV + \oiint_S \rho\, (\mathbf{u}_{abs} - \mathbf{u}_{cv}) \cdot \mathbf{n}\, dS = 0 \tag{6.23}$$

Since the flow is incompressible, we regard the density as a constant so that the volume integral of $\partial \rho / \partial t$ vanishes. We again select a control volume extending from the midsection of the sphere to a distance far below the sphere. In contrast to the formulation above, the control volume now moves downward with a velocity

$$\mathbf{u}_{cv} = -U\,\mathbf{k} \tag{6.24}$$

Because the lower surface of the sphere forms part of the control-volume boundary, clearly there is no mass flux across this part of the boundary (i.e., $\mathbf{u}_{abs} - \mathbf{u}_{cv} = \mathbf{0}$ on the sphere). Since the unit normal lies at a right angle to the control-volume velocity vector on its cylindrical boundary, the mass flux vanishes there also. Hence, conservation of mass tells us that

$$\iint_{A_1} \rho\, [\mathbf{0} - (-U\,\mathbf{k})] \cdot (-\mathbf{k})\, dA + \iint_{A_2} \rho\, [u_{avg}\mathbf{k} - (-U\,\mathbf{k})] \cdot \mathbf{k}\, dA = 0 \tag{6.25}$$

Simplifying, we find

$$-4\pi R^2 \rho\, U + 3\pi R^2 \rho\, (u_{avg} + U) = 0 \quad \Longrightarrow \quad u_{avg} = \frac{1}{3}U \tag{6.26}$$

and the result is identical to Equation (6.21).

6.2.3 Deforming Control Volume

Consider a cylindrical plunger that moves downward into a conical receptacle filled with an incompressible fluid of density ρ as shown in Figure 6.6. The plunger moves downward with

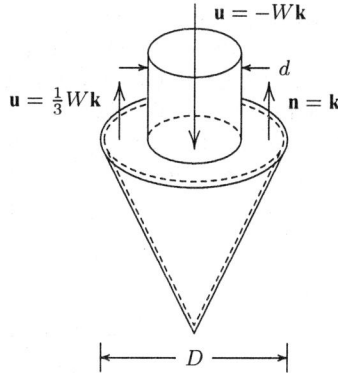

Figure 6.6: *Cylindrical plunger moving into a conical receptacle.*

a velocity $\mathbf{u} = -W\mathbf{k}$, and forces fluid to move upward through the area between the plunger and cone walls. Because the area changes and the fluid is incompressible, the average upward velocity of the fluid escaping across the plane the plunger is crossing must vary as the plunger moves deeper into the cone. By average, we mean the average value over the cross section. The chosen geometry causes complicated motion that is clearly not one dimensional. To make the problem specific, we want to determine the point at which the average upward flow speed is 1/3 that of the plunger. For present purposes, it will suffice to determine the cone diameter at which this condition holds.

Clearly, because the geometry of the plunger and the cone are dissimilar, we cannot use a Galilean transformation to recast this problem as a steady flow. Rather, we have two choices of control volumes, both of which require evaluation of the unsteady term in the mass-conservation principle, Equation (6.22). The two possibilities are as follows.

1. Select a stationary control volume whose upper boundary lies at the plane where the average upward flow speed is $\frac{1}{3}W$.

2. Select a control volume whose upper boundary is the plane parallel to the bottom of the plunger for all time.

In both cases, the control volume deforms as time passes, most notably, in a way that the total volume, V, changes.

Consider first a stationary control volume, situated such that the plunger has just arrived at the plane where the upward flow speed is $\frac{1}{3}W$. Conservation of mass tells us that

$$\iiint_V \frac{\partial \rho}{\partial t}\, dV + \oiint_S \rho\, \mathbf{u} \cdot \mathbf{n}\, dS = 0 \tag{6.27}$$

Now, although the flow is incompressible, we cannot say that $\partial \rho / \partial t = 0$ *everywhere* in the control volume. That is, as the plunger crosses the plane, it displaces fluid. Within the region where fluid is displaced, the density changes from ρ to zero, so that $\partial \rho / \partial t \neq 0$ within this region. The point is, the volume integral in Equation (6.27) represents the rate of change of mass in the control volume, which is changing with time. This is more obvious if we interchange the partial derivative with the volume integral, viz.,

$$\frac{\partial}{\partial t} \iiint_V \rho\, dV + \oiint_S \rho\, \mathbf{u} \cdot \mathbf{n}\, dS = 0 \tag{6.28}$$

Although the rate of change of ρ can be quantified, doing so requires introducing advanced mathematical concepts (i.e., step functions and delta functions). Dealing directly with the integral is straightforward, and is the preferred approach.

The total mass of fluid in the control volume decreases in direct proportion to the rate at which the plunger displaces the fluid. As illustrated in Figure 6.7, in time Δt, the control volume decreases in size by an amount

$$\Delta V = -\frac{\pi}{4} W d^2 \Delta t \qquad \Longrightarrow \qquad \frac{dV}{dt} = -\frac{\pi}{4} W d^2 \tag{6.29}$$

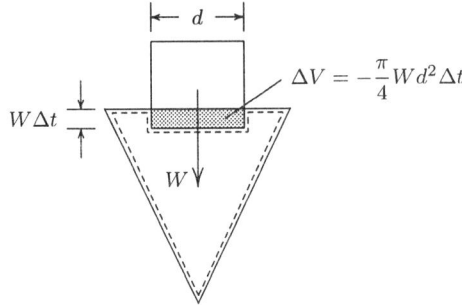

Figure 6.7: *Volume change as the plunger moves into the cone — stationary control volume; the shaded region is the incremental volume change.*

We can use this information to compute the unsteady term. Because the density is constant, we have

$$\frac{\partial}{\partial t} \iiint_V \rho\, dV = \rho \frac{\partial}{\partial t} \iiint_V dV = \rho \frac{dV}{dt} \tag{6.30}$$

Therefore, conservation of mass for the stationary control volume simplifies to

$$\rho \frac{dV}{dt} + \oiint_S \rho\, \mathbf{u} \cdot \mathbf{n}\, dS = 0 \tag{6.31}$$

Finally, the only part of the control-volume boundary across which fluid passes is the horizontal plane between the plunger and the cone. Therefore, we can evaluate the net mass-flux (surface) integral as follows.

$$\oiint_S \rho\, \mathbf{u} \cdot \mathbf{n}\, dS = \rho\, \frac{1}{3} W \mathbf{k} \cdot \mathbf{k} \frac{\pi}{4} \left(D^2 - d^2 \right) = \frac{\pi}{12} \rho\, W \left(D^2 - d^2 \right) \tag{6.32}$$

So, substituting Equations (6.29) and (6.32) into Equation (6.31), we have

$$-\frac{\pi}{4} \rho\, W d^2 + \frac{\pi}{12} \rho\, W \left(D^2 - d^2 \right) = 0 \tag{6.33}$$

Solving for D is straightforward, and the final answer is

$$D = 2d \tag{6.34}$$

Note that, as with the pipe-flow and falling-sphere problems of the two preceding subsections, the detailed flow for this problem is quite complicated and certainly is not one dimensional. Nevertheless, the control volume method yields an exact answer for the average vertical velocity. It is exact because the method is based on the integral form of the mass-conservation principle, which holds regardless of how complicated the flowfield happens to be. While the method may conceal the complexity of the detailed flowfield, it provides an accurate global view.

Now consider the control volume whose upper boundary moves with the plunger. For this case, we must use the mass-conservation principle in the following form.

$$\frac{\partial}{\partial t} \iiint_V \rho \, dV + \oiint_S \rho \, (\mathbf{u}_{abs} - \mathbf{u}_{cv}) \cdot \mathbf{n} \, dS = 0 \tag{6.35}$$

The control volume is now decreasing in size by

$$\Delta V = -\frac{\pi}{4} W D^2 \Delta t \tag{6.36}$$

where D is the current cross-sectional area of the cone (see Figure 6.8). Thus,

$$\frac{dV}{dt} = -\frac{\pi}{4} W D^2 \tag{6.37}$$

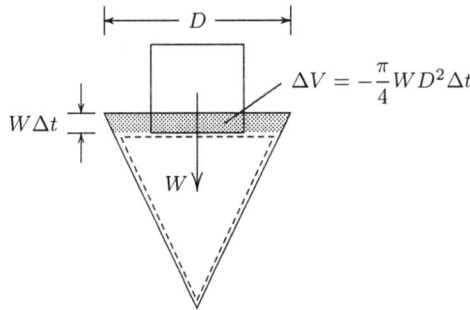

Figure 6.8: *Volume change as the plunger moves into the cone — moving control volume; the shaded region is the incremental volume change.*

Turning to the mass-flux integral, the only mass flowing across the control-volume surface is again on the plane between the plunger and the cone. Thus,

$$\oiint_S \rho \, (\mathbf{u}_{abs} - \mathbf{u}_{cv}) \cdot \mathbf{n} \, dS = \rho \left[\frac{1}{3} W \mathbf{k} - (-W \mathbf{k}) \right] \cdot \mathbf{k} \frac{\pi}{4} \left(D^2 - d^2 \right)$$

$$= \frac{\pi}{3} \rho W \left(D^2 - d^2 \right) \tag{6.38}$$

So, using Equations (6.37) and (6.38), we can evaluate the terms in the mass-conservation Equation (6.35) to arrive at

$$-\frac{\pi}{4} \rho W D^2 + \frac{\pi}{3} \rho W \left(D^2 - d^2 \right) = 0 \tag{6.39}$$

which again yields $D = 2d$.

As a concluding remark, both control volumes used in this section deform as time passes. The boundary of the first control volume moves only where the relative velocity is zero, so the computation appears to be the same as that for a stationary control volume. By contrast, there is fluid crossing the boundary of the second control volume, and we must account for the control-volume movement in Equation (6.35). Conceptually, it is usually easiest to employ the first type of control volume. Nevertheless, both control volumes yield the same answer, and the choice of which to use is a matter of taste.

6.3   Conservation of Mass and Momentum

Only a limited class of problems can be solved using the mass-conservation principle alone. In order to analyze more general problems, we must appeal to momentum conservation, including as many components as the geometry or other considerations dictate. For full three-dimensional applications, the momentum-conservation principle introduces three additional equations. This section focuses on problems that require combined use of the mass- and momentum-conservation principles.

6.3.1   Channel Flow With Suction

Consider two-dimensional flow between two parallel surfaces as shown in Figure 6.9. We refer to such a region as a channel, and the flow within is known as **channel flow**. We also have mass removal, or **suction**, at the upper wall, with the vertical velocity given by

$$v(x, H) = C_Q\, u, \qquad C_Q = C_Q(x) \tag{6.40}$$

The upper wall is made of a porous material, and the dimensionless coefficient, C_Q, is known as the **suction coefficient**. In the analysis to follow, we will neglect body forces and friction. Also, we assume the flow is incompressible and steady. Finally, for simplicity, we assume p and u are uniform across the channel, so that we say

$$p = p(x) \qquad \text{and} \qquad u = u(x) \tag{6.41}$$

The goal of our analysis in this section is to illustrate how the combined mass- and momentum-conservation principles can be used to derive simplified one-dimensional equations governing the averaged velocity and pressure profiles in the channel. First, we consider mass conservation. Because the flow is steady, we know that

$$\oiint_S \rho\, \mathbf{u} \cdot \mathbf{n}\, dS = 0 \tag{6.42}$$

Since there is no flow through the bottom of the channel, the closed-surface integral is the sum of three integrals. Specifically, it is the sum of the integrals over the inlet, the outlet and the upper surface. Expanding the integral here is trivial since we assume constant velocities at the inlet and outlet so that

$$-\rho\, U_1 H + \rho\, U_2 H + \int_0^L \rho v\, dx = 0 \tag{6.43}$$

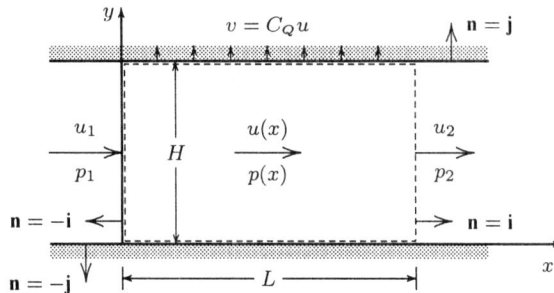

Figure 6.9: *Flow in a channel with mass removal from the porous upper wall.*

or, substituting from Equation (6.40) for v on the upper surface,

$$\int_0^L \rho \, C_Q \, u \, dx = \rho \, (U_1 - U_2) \, H \tag{6.44}$$

In words, this equation tells us the total mass removed from the upper wall [the integral on the left-hand side of Equation (6.44)] equals the difference between the mass that flows into the channel at $x = 0$ and the mass that flows out at $x = L$.

Turning now to momentum conservation, present purposes allow us to consider only the x direction. Thus, we have

$$\oiint_S \rho u (\mathbf{u} \cdot \mathbf{n}) \, dS = -\mathbf{i} \cdot \oiint_S p \, \mathbf{n} \, dS \tag{6.45}$$

Note that, as discussed in Subsection 6.1.1, we regard the integrand in the momentum-flux integral, i.e., the integral on the left-hand side of Equation (6.45), as consisting of the product of two quantities. The first is the property being carried across the surface, ρu. The second is the rate at which it is being carried across the boundary, $(\mathbf{u} \cdot \mathbf{n}) dS$. Hence, noting the orientation of the unit normals on all bounding surfaces, we expand the integral as follows.

$$
\begin{aligned}
\oiint_S \rho u \, (\mathbf{u} \cdot \mathbf{n}) \, dS &= (\rho \, U_1)(-U_1 H) + (\rho \, U_2)(U_2 H) + \int_0^L (\rho u) v \, dx \\
&= -\rho \, U_1^2 H + \rho \, U_2^2 H + \int_0^L \rho C_Q u^2 \, dx
\end{aligned}
\tag{6.46}
$$

The integral on the right-hand side of Equation (6.46) arises because x momentum is being carried out of the control volume at $y = H$ by the flux in the y direction.

Turning to the pressure integral, we have:

$$\oiint_S p \, \mathbf{n} \, dS = p_1(-\mathbf{i})H + p_2(\mathbf{i})H + \int_0^L p(x)(\mathbf{j}) dx + \int_0^L p(x)(-\mathbf{j}) dx \tag{6.47}$$

Now, we know that the unit vectors satisfy the following relations.

$$\mathbf{i} \cdot \mathbf{i} = 1 \quad \text{and} \quad \mathbf{i} \cdot \mathbf{j} = 0 \tag{6.48}$$

Thus,

$$-\mathbf{i} \cdot \oiint_S p \, \mathbf{n} \, dS = (p_1 - p_2) \, H \tag{6.49}$$

Observe that the unknown pressure on the channel walls has no effect on momentum conservation in the x direction because it exerts a force only in the y direction. As a check on the physics, note that the term on the left-hand side of Equation (6.49), including the minus sign, is the net force on the control volume. If the pressure at the inlet exceeds that at the outlet, i.e., if $p_1 > p_2$, the net force must be positive. Inspection of Equation (6.49) shows that this is true. Hence, x-momentum conservation for the channel simplifies to

$$\rho \left(U_2^2 - U_1^2 \right) H + \int_0^L \rho C_Q u^2 \, dx = (p_1 - p_2) \, H \tag{6.50}$$

It is helpful to pause and discuss what we have accomplished in our analysis of the channel-flow problem considered in this section. The most important point is the appearance of terms involving C_Q in the mass-conservation Equation (6.44) and in the momentum Equation (6.50). While it is obvious that mass removal should affect mass conservation, it is a bit more subtle why it appears in the momentum equation. The origin of the term with C_Q in the momentum equation is the momentum-flux integral on the left-hand side of Equation (6.45), and our assumption[1] that u is a function only of x. Its presence could be easily overlooked if we are not careful in separating what is being carried across the control-volume surface, ρu, and the rate at which it is carried across, $v\,dx = (\mathbf{u} \cdot \mathbf{n})dx$. Thus, the most important lesson to be gleaned from this application is the importance of separating the terms in the momentum-flux integral, and how helpful the control volume method is in delineating the proper separation.

6.3.2   Indirect Force Computation — Rotational Flow

An interesting use of the control volume method is for the indirect computation of forces. When we have flow about or through a solid object, the fluid exerts forces on the object at its surface. If we know the pressure (and viscous stresses for a viscous fluid), we can integrate over the surface of the object to find the net force. We do this in analyzing fluid-statics problems where we know the pressure from the hydrostatic relation.

Alternatively, if we know something about the flow such as velocities at the surface bounding a specified volume, we can compute the net force on the object indirectly. We have already shown this for the non-moving fluid case (Subsection 6.1.3). The following sections show how it can be done in the more complex case where the fluid is moving.

Consider the channel shown in Figure 6.10 with two branches exhausting to the atmosphere. When the lower branch develops a leak, we find that the velocity in the upper branch is 4/5 that of the lower branch and a jet of velocity U_j is emitted. Additionally, there is no net vertical force on the channel. Because the channel has rectangular cross sections, the flow is very complicated internally and is strongly rotational. Our objectives are to compute V and U_j and to determine the horizontal force, F_x, on the channel, ignoring body forces.

As indicated in the figure, we use a control volume entirely inside the channel. Since we assume the flow is steady, mass conservation tells us that

$$\oiint_S \rho\,\mathbf{u} \cdot \mathbf{n}\,dS = 0 \tag{6.51}$$

Hence, integrating about the closed surface, we find

$$\rho\left(-UA\right) + \rho\left(\frac{4}{5}V\frac{A}{4}\right) + \rho\left(V\frac{A}{4}\right) + \rho\left(U_j\frac{A}{36}\right) = 0 \tag{6.52}$$

Therefore,

$$\frac{9}{20}V + \frac{1}{36}U_j = U \tag{6.53}$$

Because the vertical force is known to be zero, it is most convenient in this problem to consider y momentum before considering x momentum. Replacing p by $(p - p_o)$ where p_o is atmospheric pressure and ignoring body forces, the exact equation is

$$\oiint_S \rho\,v(\mathbf{u} \cdot \mathbf{n})dS = -\mathbf{j} \cdot \oiint_S (p - p_o)\mathbf{n}\,dS \tag{6.54}$$

where we take the dot product of the pressure integral and $\mathbf{j}$ to extract the y component. Remembering to segregate the integrand into the product of what's being carried across the

---

[1]In a viscous flow, $u = 0$ at the wall so that no horizontal momentum would be carried out of the control volume.

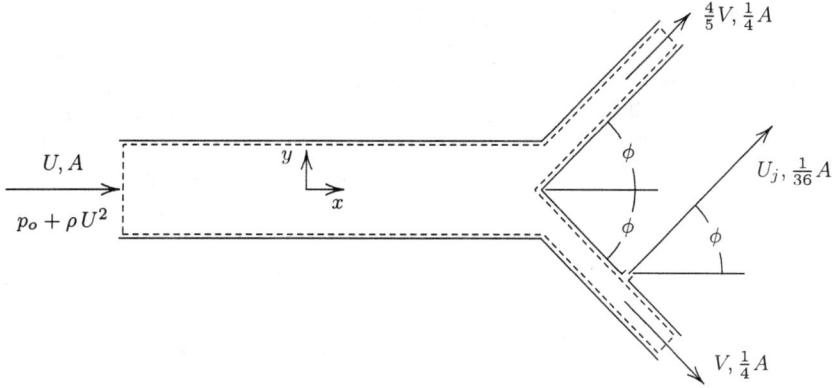

Figure 6.10: *Flow in a channel with a leak; the flow yields zero normal force ($F_y = 0$) on the channel.*

surface and the rate at which it's being carried across, conservation of y momentum simplifies to the following.

$$\left(\frac{4}{5}\rho V \sin\phi\right)\left(\frac{4}{5}V\frac{A}{4}\right) + (\rho U_j \sin\phi)\left(U_j\frac{A}{36}\right) + (-\rho V \sin\phi)\left(V\frac{A}{4}\right) = 0 \qquad (6.55)$$

where we take advantage of the fact that $F_y = 0$. Since the factor $\rho A \sin\phi$ appears in all terms, the equation simplifies immediately to

$$\frac{4}{25}V^2 + \frac{1}{36}U_j^2 - \frac{1}{4}V^2 = 0 \qquad \Longrightarrow \qquad U_j = \frac{9}{5}V \qquad (6.56)$$

At this point, we can solve directly for V and U_j by substituting into Equation (6.53), viz.,

$$\frac{9}{20}V + \frac{1}{36}\cdot\frac{9}{5}V = U \qquad \Longrightarrow \qquad \frac{1}{2}V = U \qquad (6.57)$$

Therefore, the velocities are

$$V = 2U \qquad \text{and} \qquad U_j = \frac{18}{5}U \qquad (6.58)$$

In a similar way, the equation for conservation of x momentum is

$$\oiint_S \rho\, u(\mathbf{u}\cdot\mathbf{n})dS = -\mathbf{i}\cdot\oiint_S (p - p_o)\mathbf{n}\, dS \qquad (6.59)$$

Again being careful to segregate the integrand's two factors, we find

$$(\rho U)(-UA) + \left(\frac{4}{5}\rho V \cos\phi\right)\left(\frac{4}{5}V\frac{A}{4}\right) + (\rho U_j \cos\phi)\left(U_j\frac{A}{36}\right)$$
$$+ (\rho V \cos\phi)\left(V\frac{A}{4}\right) = -F_x + \rho U^2 A \qquad (6.60)$$

Note that in evaluating the pressure integral, the x component of the force *exerted by the fluid on the pipe*, F_x, is the integral of $(p - p_o)\mathbf{n}$ over all solid parts of the channel. This is true because **n** points outward, so that we are computing the force exerted by the control volume on the surroundings. The complete integration domain also includes the three outlets and the inlet. Since the outlets all exhaust to the atmosphere, $p = p_o$, so that they contribute nothing to the integral. By contrast, $p = p_o + \rho U^2$ at the inlet, which yields a contribution of $\rho U^2 A$. Simplifying, there follows:

$$-\rho U^2 + \frac{4}{25}\rho V^2 \cos\phi + \frac{1}{36}\rho U_j^2 \cos\phi + \frac{1}{4}\rho V^2 \cos\phi = -\frac{F_x}{A} + \rho U^2 \qquad (6.61)$$

Finally, solving for the force we have

$$F_x = 2(1 - \cos\phi)\rho U^2 A \qquad (6.62)$$

At this point, we can pause and check the appropriateness of our result by considering two limiting cases. First, when $\phi \to 0°$, there are no vertical channel surfaces for the pressure to act on so that we expect the horizontal force to vanish. Second, when $\phi \to 90°$, we have positive horizontal momentum entering the control volume and zero horizontal momentum exiting. This would be similar to aiming a fire hose at a vertical surface. Thus, we expect the net force to be to the right. Our result is consistent with both limiting cases.

6.3.3   Indirect Force Computation — Irrotational Flow

Sometimes mass and momentum conservation fail to yield enough equations for the unknown properties in the problem. When this happens, we must in general appeal to conservation of energy to close the set of equations. For steady, incompressible flow problems with conservative body forces, we can appeal to Bernoulli's equation to obtain the required additional equations.

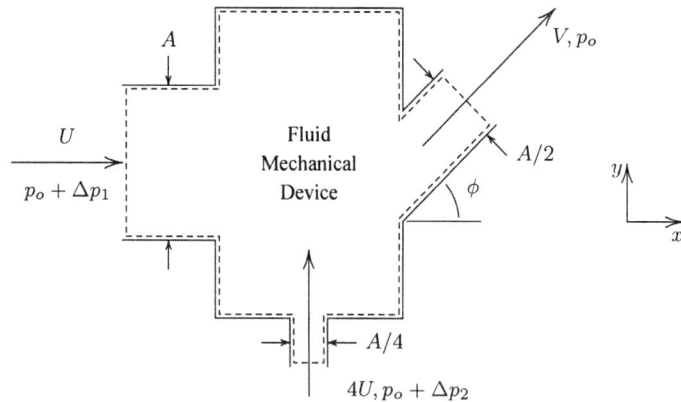

Figure 6.11: *Irrotational flow through a fluid mechanical device; A denotes area.*

As an example of such a problem, consider the "fluid mechanical device" shown in Figure 6.11. An incompressible fluid flows *steadily* and *irrotationally* through the device and exhausts to the atmosphere. We would like to determine the angle ϕ that causes the vertical component of the force on the device to vanish. Also, we would like to know the horizontal component of the force for this angle. We assume properties are constant across inlets to and the outlet from the device and that body forces can be ignored.

As in the example of the preceding section, we select a control volume that lies entirely within the device. Since the flow is assumed steady, we know that mass conservation is

$$\oiint_S \rho \mathbf{u} \cdot \mathbf{n} \, dS = 0 \tag{6.63}$$

Using the assumption that velocities are constant across inlets and the outlet, the closed-surface integral can be trivially evaluated, i.e.,

$$\rho(-UA) + \rho\left(-4U\frac{A}{4}\right) + \rho\left(V\frac{A}{2}\right) = 0 \tag{6.64}$$

We can solve for V immediately, wherefore

$$V = 4U \tag{6.65}$$

Since the flow is steady and no body forces are present, the x component of the momentum-conservation equation is

$$\oiint_S \rho u(\mathbf{u} \cdot \mathbf{n})dS = -\mathbf{i} \cdot \oiint_S (p - p_o)\mathbf{n} \, dS \tag{6.66}$$

where we again replace the pressure p by $(p - p_o)$. Evaluating in the usual manner yields:

$$(\rho U)(-UA) + (\rho V \cos \phi)\left(V\frac{A}{2}\right) = \Delta p_1 A - F_x \tag{6.67}$$

where the x component of the force *exerted by the fluid on the device* is denoted by F_x. Taking advantage of Equation (6.65) to eliminate V, the force becomes

$$F_x = \Delta p_1 A - (8 \cos \phi - 1)\rho U^2 A \tag{6.68}$$

Similarly, the y component of the momentum-conservation equation is

$$\oiint_S \rho v(\mathbf{u} \cdot \mathbf{n})dS = -\mathbf{j} \cdot \oiint_S (p - p_o)\mathbf{n} \, dS \tag{6.69}$$

Performing the integration around the closed surface yields:

$$(4\rho U)\left(-4U\frac{A}{4}\right) + (\rho V \sin \phi)\left(V\frac{A}{2}\right) = \Delta p_2 \frac{A}{4} - F_y \tag{6.70}$$

so that

$$F_y = \frac{1}{4}\Delta p_2 A - (8 \sin \phi - 4)\rho U^2 A \tag{6.71}$$

At this point, we have insufficient information to solve for the stated unknown properties in the problem. That is, there are *six unknowns* in the problem, viz., V, ϕ, F_x, F_y, Δp_1 and Δp_2. However, we have only *four equations*, three of which are Equations (6.65), (6.68), (6.71). The fourth equation is the given fact that F_y vanishes, i.e.,

$$F_y = 0 \tag{6.72}$$

the latter being part of the statement of the problem. Because the problem is cast as one involving conditions under which Bernoulli's equation applies, we thus use it to obtain the missing equations.

Since Bernoulli's equation holds throughout the flow, we can use it as many times as we need to arrive at a closed system of equations. First, we apply it between Inlet 1 and the outlet. Hence,

$$p_o + \Delta p_1 + \frac{1}{2}\rho U^2 = p_o + \frac{1}{2}\rho V^2 = p_o + 8\rho U^2 \tag{6.73}$$

where we again make use of Equation (6.65). Therefore, solving for the pressure differential,

$$\Delta p_1 = \frac{15}{2}\rho U^2 \tag{6.74}$$

Now, applying Bernoulli's equation between Inlet 2 and the outlet, we find:

$$p_o + \Delta p_2 + \frac{1}{2}\rho(4U)^2 = p_o + 8\rho U^2 \tag{6.75}$$

Solving for Δp_2,

$$\Delta p_2 = 0 \tag{6.76}$$

Hence, in order to have vanishing vertical force, substituting Equation (6.76) into Equation (6.71) shows that

$$(8\sin\phi - 4)\rho U^2 = 0 \tag{6.77}$$

Because ρ and U are nonzero, we conclude that

$$\sin\phi = \frac{1}{2} \quad \Longrightarrow \quad \phi = 30^\circ \tag{6.78}$$

Finally, combining Equations (6.68) and (6.74) yields the horizontal force, i.e.,

$$F_x = \frac{15}{2}\rho U^2 A - (8\cos\phi - 1)\rho U^2 A = \left(\frac{17}{2} - 8\cos\phi\right)\rho U^2 A \tag{6.79}$$

Using the fact that ϕ is 30°, the x component of the force is

$$F_x = 1.57\rho U^2 A \tag{6.80}$$

6.3.4   Reaction Forces

In all of the examples presented thus far, we have selected a control volume that contains only fluid. In order to compute the force on an object, we have chosen a control volume for which part of the bounding surface is coincident with the object's surface. Then, the force is given by the integral of pressure (and viscous stress if present) over the surface of the object. By contrast, for some applications, it is more straightforward to select a control volume that contains the object under consideration. The question then arises about how we might determine *the force exerted by the fluid on the object*. The answer is, we introduce a *reaction force*, which is *the force required to hold the control volume in place*. As we will see in this section, the reaction force is the negative of the force on the object.

To illustrate these concepts, consider a jet of fluid impinging at a right angle on a disk as shown in Figure 6.12. The diameter of the jet is D and its average velocity is U. For steady flow, momentum conservation tells us that

$$\iint_S \rho\,\mathbf{u}(\mathbf{u}\cdot\mathbf{n})\,dS = -\iint_S (p - p_o)\mathbf{n}\,dS \tag{6.81}$$

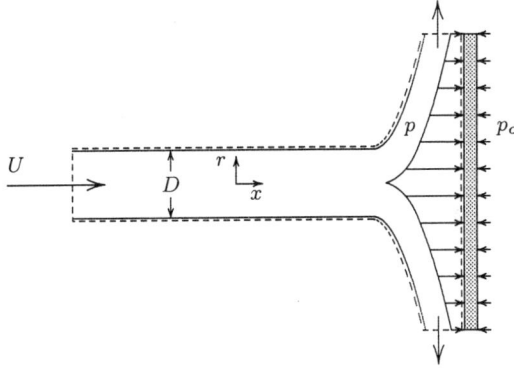

Figure 6.12: *Jet impinging on a disk with the disk outside the control volume.*

By symmetry, there can be no net force or momentum flux in the radial direction so that the momentum flux and the force must lie in the x direction. The only surface across which x momentum moves is the jet cross section upstream of the disk. Hence, the momentum-flux integral is

$$\oiint_S \rho\, \mathbf{u}(\mathbf{u}\cdot\mathbf{n})\, dS = -\frac{\pi}{4}\rho\, U^2 D^2\, \mathbf{i} \tag{6.82}$$

The force on the disk, $\mathbf{F}_{disk}$, is the difference between (a) the integral of the pressure on the disk surface exposed to the jet, and (b) the force exerted by the atmosphere on the opposite side of the disk. That is, the force on the disk is

$$\mathbf{F}_{disk} = \mathbf{i}\iint_{disk} (p - p_o)\, dS \tag{6.83}$$

Finally, provided the disk diameter is large compared to the jet diameter, the pressure is approximately p_o on the disk far from its center. Also, the surrounding fluid impresses atmospheric pressure both on the edge of and throughout the jet. Thus, the integral of $(p - p_o)$ on the control volume's bounding surface is zero except on the disk. Hence,

$$-\oiint_S (p - p_o)\mathbf{n}\, dS = -\mathbf{i}\iint_{disk} (p - p_o)\, dS = -\mathbf{F}_{disk} \tag{6.84}$$

Therefore, conservation of momentum tells us that

$$\mathbf{F}_{disk} = \frac{\pi}{4}\rho\, U^2 D^2\, \mathbf{i} \tag{6.85}$$

Now, consider the control volume shown in Figure 6.13. The disk is included within the control volume. An immediate advantage of this control volume is the fact that $p = p_o$ at all points on the bounding surface. Hence, the pressure integral vanishes. We must revise the momentum conservation equation for this control volume by adding a **reaction force, R**, viz.,

$$\oiint_S \rho\, \mathbf{u}(\mathbf{u}\cdot\mathbf{n})\, dS = -\oiint_S (p - p_o)\mathbf{n}\, dS + \mathbf{R} \tag{6.86}$$

The physical meaning of the reaction force is as stated above — it is the force required to hold the control volume in place.

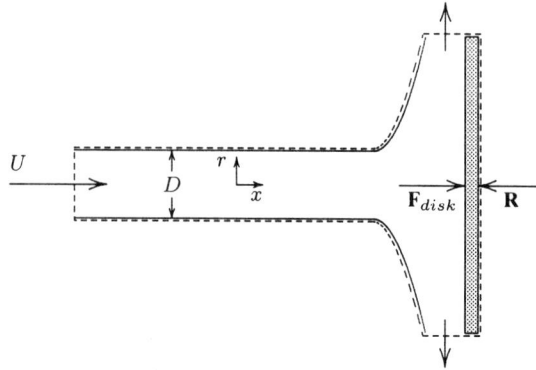

Figure 6.13: *Jet impinging on a disk with the disk inside the control volume.*

Now, the momentum flux is still given by Equation (6.82). Since the pressure integral is zero, we conclude immediately that the reaction force is given by

$$\mathbf{R} = -\frac{\pi}{4}\rho\, U^2 D^2\,\mathbf{i} \tag{6.87}$$

Clearly, the force required to hold the disk in place is just the negative of the force exerted by the fluid on the disk, i.e.,

$$\mathbf{R} + \mathbf{F}_{disk} = \mathbf{0} \tag{6.88}$$

wherefore the force is again given by Equation (6.85). Since both types of control volumes yield the same results, either can be used successfully. It is a matter of convenience as to which type will be used in a given problem.

As a final comment, we can check the reasonableness of the sign of our computed force for this problem. Clearly the force must be to the right, else we would be predicting that a jet will draw a disk toward its source! Clearly, Equation (6.85) is in the correct direction.

6.3.5   Nonuniform Cross-Sectional Velocity Profiles

In every control volume example of the preceding sections, we have assumed the velocity is constant on cross sections. This is untrue for the flow of a real fluid, for example, because viscous effects give rise to nonuniform velocity profiles. Since all fluids in nature are viscous, and since other common effects such as body forces can lead to nonuniform velocity profiles, we must assess the importance of assuming uniform cross-sectional velocity. We saw in Subsection 6.2.1 that using uniform profiles is equivalent to working with average properties on a given cross section. Although this is convenient and exact for mass conservation, it is an approximation for momentum conservation. This is because the velocity appears linearly in the mass-conservation equation, while it appears quadratically in the momentum-conservation equation.

To illustrate why this is a problem, consider Couette flow (see Section 1.9) for which the plates are separated by a distance h. We know that the velocity varies linearly with distance from the lower plate, i.e., the horizontal velocity is

$$u(y) = U\frac{y}{h} \tag{6.89}$$

Then, the average velocity is given by

$$\overline{U} = \frac{1}{h} \int_0^h U \frac{y}{h} \, dy = \frac{1}{2} U \tag{6.90}$$

Using this average velocity, if the channel width is w, we would then conclude that the cross-sectional mass flux is

$$\iint_A \rho \, u \, dA = \rho \overline{U} hw = \frac{1}{2} \rho U hw \tag{6.91}$$

Similarly, using the average velocity, the momentum flux through the same cross section would be

$$\iint_A \rho \, u^2 \, dA \approx \rho \overline{U}^2 hw = \frac{1}{4} \rho U^2 hw \tag{6.92}$$

Although the mass flux is correct, the momentum flux is not. If we substitute the exact velocity into the integrand, we find

$$\iint_A \rho u^2 \, dA = \rho U^2 w \int_0^h \left(\frac{y}{h}\right)^2 \, dy = \frac{1}{3} \rho U^2 hw \tag{6.93}$$

Thus, using the average velocity yields a 25% error in the momentum flux! Mathematically speaking, our error results from the fact that the square of the mean is not equal to the mean of the square, i.e.,

$$\overline{U}^2 \neq \overline{U^2} \tag{6.94}$$

Couette flow is an extreme case in this respect. In most practical viscous-flow applications, the error in the momentum flux will be less than 10%. This is true because most flows of engineering interest are turbulent, and the attending velocity profiles are not too far from uniform in pipes and channels, for example. We will explore this in more detail when we analyze pipe flow in Chapter 7. Nevertheless, for the sake of rigor and completeness, we should know how to account for nonuniform velocity profiles.

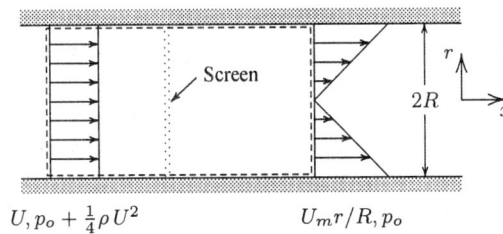

Figure 6.14: *Axisymmetric flow through a screen — cross-sectional view.*

To illustrate how the control volume method is implemented when nonuniform velocity profiles are accounted for, consider a variable mesh screen in a wind tunnel. The mesh produces a linear, axisymmetric velocity profile[2] as indicated in Figure 6.14 in the air flow through a circular-cross-section duct of radius R. The measured pressures upstream and downstream of the screen are as indicated and are uniform over the cross section, which is an excellent approximation for this type of flow. Neglecting the viscous force exerted by the

---

[2]Note that the velocity profile shown is chosen to simplify the mathematics of this problem. In any physical flow, the centerline velocity would be nonvanishing at any finite distance from the screen, regardless of its porosity.

duct wall on the air flow, we want to compute the screen drag force. The flow is assumed to be steady and incompressible, with no body forces acting.

We select a control volume that extends from the uniform velocity profile upstream of the screen to the linear profile downstream. Turning first to mass conservation, we begin with the general integral form, viz.,

$$\oiint_S \rho\, \mathbf{u} \cdot \mathbf{n}\, dS = 0 \tag{6.95}$$

Substituting the appropriate velocity profiles into the integral yields

$$- \rho\, U \pi R^2 + \int_0^{2\pi} \int_0^R \underbrace{(\rho)}_{\rho} \underbrace{(U_m r / R)}_{(\mathbf{u} \cdot \mathbf{n})} \underbrace{(r\, dr\, d\theta)}_{dS} = 0 \tag{6.96}$$

Performing the indicated integral, a little straightforward algebra shows that the maximum velocity, U_m, is

$$U_m = \frac{3}{2} U \tag{6.97}$$

Focusing now on momentum conservation, since the screen lies within the control volume, we must add a reaction force. Noting that the reaction force is minus the drag force on the screen, F_x, the x component of the momentum conservation equation is

$$\oiint_S \rho\, u(\mathbf{u} \cdot \mathbf{n})\, dS = -\mathbf{i} \cdot \oiint_S (p - p_o)\, \mathbf{n}\, dS - F_x \tag{6.98}$$

The sign chosen corresponds to a positive value for F_x. That is, the screen will tend to resist the flow, and should thus tend to reduce the net momentum flux, i.e., it will decelerate the flow. We can use this observation to check our result after we complete the computation. Substituting the various profiles into the integrals yields the following equation.

$$(\rho\, U)\left(-\pi R^2 U\right) + \int_0^{2\pi} \int_0^R \underbrace{(\rho\, U_m r / R)}_{\rho\, u} \underbrace{(U_m r / R)}_{(\mathbf{u} \cdot \mathbf{n})} \underbrace{(r\, dr\, d\theta)}_{dS} = \frac{1}{4} \rho\, U^2 \pi R^2 - F_x \tag{6.99}$$

Solving for the force, we have

$$
\begin{aligned}
F_x &= \frac{1}{4} \pi\, \rho\, U^2 R^2 + \pi\, \rho\, U^2 R^2 - 2\pi\, \rho\, U_m^2 R^{-2} \int_0^R r^3\, dr \\
&= \frac{5}{4} \pi\, \rho\, U^2 R^2 - \frac{1}{2} \pi\, \rho\, U_m^2 R^2
\end{aligned}
\tag{6.100}
$$

Finally, observing that U_m is given in terms of U by Equation (6.97), a bit more algebra shows that the drag force on the screen is:

$$F_x = \frac{1}{8} \pi\, \rho\, U^2 R^2 \tag{6.101}$$

As expected, F_x is positive so that the computed force is in the correct direction.

6.4   Accelerating Control Volume

In Subsection 6.2.2 we saw that using a moving control volume is advantageous for some problems. However, the falling-sphere problem that we analyzed in that section required generalizing only the mass-conservation principle for a moving control volume. Thus, we made no attempt to generalize the momentum-conservation equation. In this section, we will first establish the appropriate form of the momentum-conservation principle in a moving control volume. Then, we will use it to compute the forces on an accelerating rocket.

6.4.1   Inertial and Noninertial Coordinate Frames

Before we can proceed, we must discuss the concept of an **inertial frame**. Newton's laws of motion are valid in what we call an inertial frame of reference. The ultimate inertial frame lies at the "center of the universe," which is completely at rest. Any coordinate system that moves at constant velocity relative to the center of the universe is also an inertial frame, i.e., all inertial frames are related by a **Galilean transformation.** Since the laws of motion ultimately involve *changes* in mass, momentum and energy, we have no need to know where either the center of the universe is or what the velocity of our coordinate system is relative to it. This is true for momentum, for example, because the constant velocity of our coordinate frame cancels when we take differences between velocities observed in our moving frame.

However, we don't always work in an inertial frame. For example, because of its motion about the sun, or more generally about the center of the universe, a coordinate system fixed on the Earth is not an inertial system. To be precise in our calculations, we would have to account for the centrifugal and Coriolis accelerations attending the Earth's motion. Nevertheless, it would be exceedingly pedantic to account for such effects in typical Earth-bound applications. Engineering fluid-flow problems occur on such a small scale relative to galactic distances that the Earth's motion relative to the center of the universe can be ignored. We are all familiar with such a contrast between observations on vastly differing scales. Consider, for example, the flat appearance of the horizon as observed from a beach or a low-flying aircraft. Hence, an Earth-fixed coordinate frame may be regarded as an inertial frame for many engineering applications.

In the problems we analyze in general engineering practice, it is sometimes convenient to work in a **noninertial frame** relative to the Earth.[3] When we do this, we must account for the fact that *momentum relative to the inertial frame is conserved.* With this understanding, we can write the equations for mass and momentum conservation as follows:

$$\iiint_V \frac{\partial \rho}{\partial t} dV + \oiint_S \rho\, \mathbf{u} \cdot \mathbf{n}\, dS = 0 \tag{6.102}$$

$$\iiint_V \frac{\partial}{\partial t} \left(\rho\, \mathbf{u}_{abs} \right) dV + \oiint_S \rho\, \mathbf{u}_{abs} \left(\mathbf{u} \cdot \mathbf{n} \right) dS = \mathbf{F} \tag{6.103}$$

where $\mathbf{u}_{abs}$ is the velocity relative to the inertial frame, $\mathbf{u}$ is the velocity relative to the control volume, and $\mathbf{F}$ is the sum of all external forces acting on the control volume. As discussed in Subsection 6.2.2, the rate at which fluid crosses the control-volume surface is proportional to the velocity relative to the moving control volume, regardless of the control-volume velocity. Also, the absolute velocity, $\mathbf{u}_{abs}$, is related to the control-volume velocity, $\mathbf{u}_{cv}$, as follows.

$$\mathbf{u}_{abs} = \mathbf{u}_{cv} + \mathbf{u} \tag{6.104}$$

Before tackling a non-trivial problem, let's check our generalized conservation principles on a problem for which we know the answer. Specifically, we can examine the limiting case

---

[3]Hurricanes are an example. Because of their large extent (hundreds of miles), they are strongly affected by the Earth's rotation.

of a control volume translating at constant velocity. If Equations (6.102) and (6.103) are correct, then we should be able to replace the absolute velocity by the relative velocity in both equations. Since Equation (6.102) involves only $\mathbf{u}$, all we actually have to show is that the momentum equation reduces to the proper form. So, we assume that

$$\mathbf{u}_{cv} = \text{constant} \tag{6.105}$$

Then, we can rewrite Equation (6.103) as follows.

$$\iiint_V \frac{\partial}{\partial t}(\rho\mathbf{u})\,dV + \oiint_S \rho\mathbf{u}\,(\mathbf{u}\cdot\mathbf{n})\,dS + \mathbf{u}_{cv}\left[\iiint_V \frac{\partial\rho}{\partial t}dV + \oiint_S \rho\mathbf{u}\cdot\mathbf{n}\,dS\right] = \mathbf{F} \tag{6.106}$$

But, the sum of the terms in brackets multiplied by the constant control-volume velocity vanishes by virtue of the mass-conservation principle. This shows that, under a Galilean transformation, the equations of motion for a moving control volume can be rewritten in terms of the velocity observed in the moving frame. Hence, our generalized control-volume conservation principles are consistent with Galilean invariance, as they must be.

6.4.2   Ascending Rocket

To illustrate how we use the generalized conservation principles, consider a rocket as it ascends through the atmosphere. Referring to Figure 6.15, we select a control volume whose bounding surface is coincident with the outer surface of the rocket. The speed of the rocket relative to the Earth is W_R. Hence, the velocity of the control volume for this application is

$$\mathbf{u}_{cv} = W_R\,\mathbf{k} \tag{6.107}$$

We assume the rocket remains oriented vertically as it rises so that the velocity of the exhausting rocket gases is given by

$$\mathbf{u} = -w_e\,\mathbf{k} \qquad \text{at the nozzle exit plane} \tag{6.108}$$

where $\mathbf{u}$ is the velocity relative to the rocket, i.e., the velocity observed from within the rocket. The area of the nozzle exit plane is A_e, and the density of the rocket exhaust gases is ρ_e. Finally, the forces acting on the rocket are aerodynamic drag, D, and its weight, Mg, where M is the instantaneous mass and g is acceleration due to gravity. Our goal is to derive equations of motion for the rocket in terms of its instantaneous mass and velocity.

To begin our analysis, we consider mass conservation. The mass within the control volume includes the rocket and the fuel contained within the rocket. Because exhaust products are being ejected from the rocket nozzle, the mass of the control volume changes with time. Since we have chosen a control volume whose boundaries are coincident with the rocket, clearly its volume is constant. Therefore, we can interchange the partial derivative with respect to t and the volume integral in Equation (6.102). Thus,

$$\frac{d}{dt}\iiint_V \rho\,dV + \oiint_S \rho\,\mathbf{u}\cdot\mathbf{n}\,dS = 0 \tag{6.109}$$

Note that we can replace the partial derivative by the total derivative since doing the indicated integration over volume removes all spatial dependence. Now, by definition, the mass of the control volume is the volume integral of the density of all material within the rocket, i.e.,

$$M = \iiint_V \rho\,dV \tag{6.110}$$

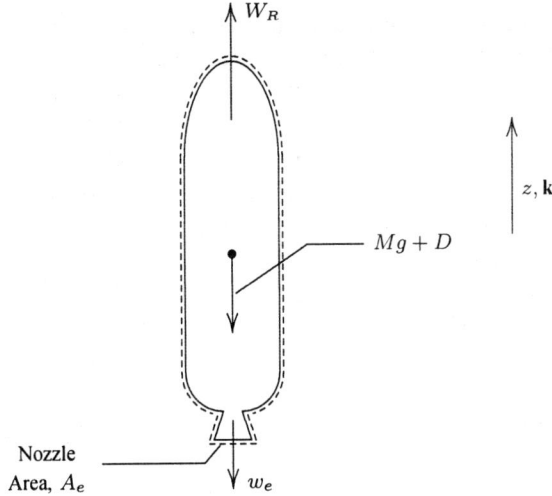

Figure 6.15: *Motion of an ascending rocket.*

This integral includes the rocket hardware as well as any fluids within the rocket. Also, the only part of the control volume across which mass passes is at the nozzle exit plane. Since the unit normal at the exit plane is $\mathbf{n} = -\mathbf{k}$, we can do the net-mass-flux integral to obtain

$$\frac{dM}{dt} + \rho_e w_e A_e = 0 \tag{6.111}$$

It's worthwhile to pause and check the sign of the predicted rate of change of the rocket's mass. Because the rocket is constantly ejecting fuel-exhaust products, the combined mass of the rocket and its fuel must decrease. Since ρ_e, w_e and A_e are all positive, clearly Equation (6.111) implies that $dM/dt < 0$. Thus, we have the correct sign.

Turning now to momentum conservation, assuming the rocket is axially symmetric, we can focus specifically on the z component of the conservation equation. Letting w denote the z component of $\mathbf{u}$, the conservation principle becomes

$$\iiint_V \frac{\partial}{\partial t} \left[\rho \left(w + W_R \right) \right] dV + \oiint_S \rho \left(w + W_R \right) \left(\mathbf{u} \cdot \mathbf{n} \right) dS$$

$$= -\mathbf{k} \cdot \oiint_S p \, \mathbf{n} \, dS - \iiint_V \rho \, g \, dV - D_f \tag{6.112}$$

First, let's consider the forces acting on the rocket. As shown, there are three types of forces, viz.,

- The net pressure force: $D_p = -\mathbf{k} \cdot \oiint_S p \, \mathbf{n} \, dS$

- The weight of the rocket: $\iiint_V \rho \, g \, dV$

- The friction drag force: D_f

We include the **friction drag**, D_f, which is one part of the aerodynamic drag. The other contribution that we regard as part of the total aerodynamic drag is contained within the pressure integral, and we call it **pressure drag**, D_p. By definition, the total aerodynamic drag, D, is the sum of the friction and pressure drag, i.e.,

$$D = D_f + D_p \tag{6.113}$$

The drag for the rocket can be obtained, for example, from wind-tunnel measurements. However, because of the complications attending simulation of the rocket exhaust, typical tests would determine D_f and D_p in the absence of the rocket exhaust. To account for this difference, we make a subtle distinction. Specifically, the primary difference between measured and actual pressures occurs at the nozzle exhaust plane, where the actual pressure, p_e, is much greater than the exhaust-free value. Hence, the total pressure drag is the value measured under no-exhaust conditions, plus the contribution due to the pressure increase caused by the exhaust. Denoting the pressure on the rocket without the exhausting fuel by p_a, we can write the net pressure force as

$$-\mathbf{k} \cdot \oiint_S p\, \mathbf{n}\, dS = -\mathbf{k} \cdot \oiint_S p_a\, \mathbf{n}\, dS + (p_e - p_a)\, A_e \equiv -D_p + (p_e - p_a)\, A_e \tag{6.114}$$

Finally, the weight of the rocket and fuel contained within the control volume is given by

$$\iiint_V \rho\, g\, dV = M g \tag{6.115}$$

Substituting these results into Equation (6.112), the equation for the rocket's motion simplifies to:

$$\iiint_V \frac{\partial}{\partial t}\left[\rho\left(w + W_R\right)\right] dV + \oiint_S \rho\left(w + W_R\right)(\mathbf{u} \cdot \mathbf{n})\, dS$$
$$= -D - M g + (p_e - p_a)\, A_e \tag{6.116}$$

Turning now to the unsteady term, we assume the flow within the rocket engine, as observed from the engine room, is steady. This assumption means the instantaneous change of relative momentum is zero, i.e.,

$$\iiint_V \frac{\partial}{\partial t}\left(\rho\, w\right) dV = 0 \tag{6.117}$$

Again noting that the chosen control volume remains constant in size, we can interchange $\partial/\partial t$ with the volume integral, wherefore

$$\iiint_V \frac{\partial}{\partial t}\left(\rho\, W_R\right) dV = \frac{d}{dt}\iiint_V \rho\, W_R\, dV = \frac{d}{dt}\left(M W_R\right) \tag{6.118}$$

Because the nozzle exit plane is the only part of the surface with fluid crossing, the net momentum-flux integral can be evaluated as follows:

$$\oiint_S \rho\left(w + W_R\right)(\mathbf{u} \cdot \mathbf{n})\, dS = \iint_{A_e} \rho_e\left(-w_e + W_R\right) w_e\, dA$$
$$= \rho_e\left(-w_e + W_R\right) w_e A_e \tag{6.119}$$

Substituting Equations (6.118) and (6.119) into Equation (6.116), the momentum equation becomes

$$\frac{d}{dt}\left(M W_R\right) + \rho_e w_e\left(W_R - w_e\right) A_e = -D - M g + (p_e - p_a)\, A_e \tag{6.120}$$

We can make an additional simplification by expanding the first term and regrouping as follows.

$$M \frac{dW_R}{dt} + W_R \left[\frac{dM}{dt} + \rho_e w_e A_e \right] - \rho_e w_e^2 A_e = -D - Mg + (p_e - p_a) A_e \qquad (6.121)$$

But, the term in brackets multiplied by W_R vanishes by virtue of mass conservation [Equation (6.111)]. Therefore, the final equation for the rocket's motion is

$$M \frac{dW_R}{dt} = T - D - Mg, \qquad T \equiv \left[(p_e - p_a) + \rho_e w_e^2 \right] A_e \qquad (6.122)$$

The quantity T is known as the **thrust**, and is used to rate the performance of a rocket engine. Note that in words, Equation (6.122) says

(Mass)·(Acceleration) = (Thrust) - (Drag) - (Weight)

which is consistent with what our knowledge of basic mechanics would lead us to expect. Hence, application of the control volume method to an accelerating vehicle produces an equation that is completely consistent with our experience in elementary physics.

Problems

6.1 Consider the control volume indicated by the dashed lines in the inclined channel whose width (out of the page) is $10h$. The velocity is constant throughout the control volume. Compute the area, A, outer unit normal, $\mathbf{n}$, normal velocity, $\mathbf{u} \cdot \mathbf{n}$, and volume flux, $\mathbf{u} \cdot \mathbf{n}A$, at the inlet and the outlet.

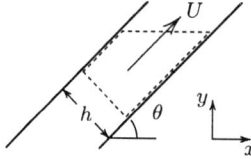

Problem 6.1

6.2 Consider the pipe segment with one face slanted to the horizontal at an angle α. The velocity is constant throughout the pipe segment. Compute the area, A, outer unit normal, $\mathbf{n}$, normal velocity, $\mathbf{u} \cdot \mathbf{n}$, and volume flux, $\mathbf{u} \cdot \mathbf{n}A$, at the inlet and the outlet. **HINT:** The area of an ellipse with semimajor axis a and semiminor axis b is πab.

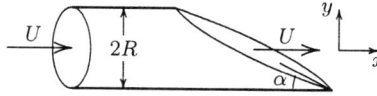

Problem 6.2

6.3 Consider the control volume indicated by the dashed lines in the inclined channel whose width (out of the page) is $3H$. Compute the area, A, and outer unit normal, $\mathbf{n}$, for all three faces shown. Verify from your results that, if the pressure, p, is constant throughout the control volume,

$$p\,\mathbf{n}_1 A_1 + p\,\mathbf{n}_2 A_2 + p\,\mathbf{n}_3 A_3 = 0$$

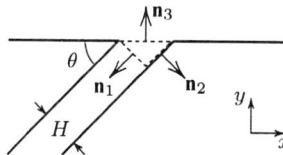

Problem 6.3

6.4 What is the net pressure force exerted by the surroundings on the pipe shown? The control volume (indicated by the dashed contour) lies entirely outside of the pipe, where the pressure is equal to its atmospheric value, p_o, everywhere except at the inlet where it is p_i. Assume pressure is constant on all pipe cross sections.

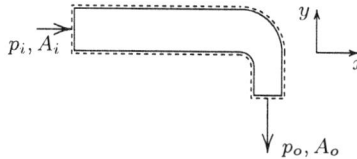

Problem 6.4

6.5 What is the net pressure force exerted by the surroundings on the channel shown? The control volume (indicated by the dashed contour) lies entirely outside of the channel, where the pressure is atmospheric ($p = p_o$) everywhere except at the inlets. Assume pressure is constant on all pipe cross sections.

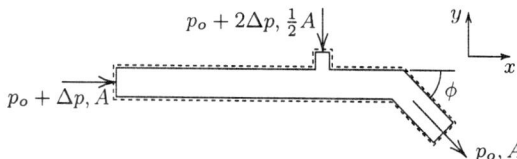

Problem 6.5

6.6 What is the net pressure force exerted by the surroundings on the duct section shown? The duct width (out of the page) is $2H$. The control volume (indicated by the dashed contour) lies entirely outside of the duct, where the pressure is atmospheric ($p = p_o$) everywhere except on duct cross sections. Assume pressure is constant on all duct cross sections.

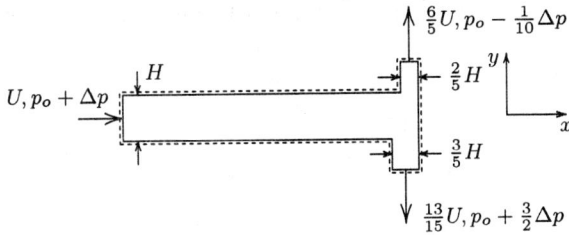

Problems 6.6, 6.7

6.7 For the duct section shown, you can assume the velocity is constant on all cross sections and that the flow is incompressible. Also, the duct width (out of the page) is $5H$. At the inlet and outlets, determine $\mathbf{n}$, $(\mathbf{u} \cdot \mathbf{n})$, $\iint \rho u(\mathbf{u} \cdot \mathbf{n})dA$ and $\iint \rho v(\mathbf{u} \cdot \mathbf{n})dA$.

6.8 Consider incompressible flow into a 180° bend. The pipe cross-sectional area and velocity at the inlet and the outlet are the same and equal to A and U, respectively. Assume the velocity is constant on all cross sections. At the inlet and outlet, determine $\mathbf{n}$, $(\mathbf{u} \cdot \mathbf{n})$, and $\iint \rho u(\mathbf{u} \cdot \mathbf{n})dA$.

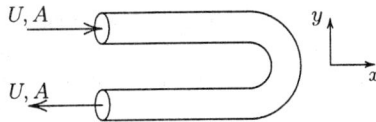

Problem 6.8

6.9 Consider the two control volumes shown. The velocity vector at the outlet is

$$\mathbf{u} = U\left[\mathbf{i}\cos\phi - \mathbf{j}\sin\phi\right]$$

The jet cross-sectional area is A and the flow is incompressible. For the part of the control-volume surface passing through the jet, determine $\mathbf{n}$, $(\mathbf{u} \cdot \mathbf{n})$, $\iint \rho u(\mathbf{u} \cdot \mathbf{n})dA$ and $\iint \rho v(\mathbf{u} \cdot \mathbf{n})dA$.

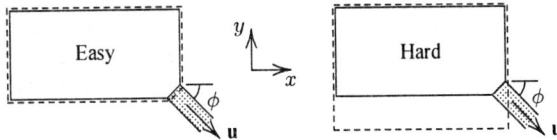

Problems 6.9

6.10 Verify that in computing the moment on a control volume due to the pressure, p can be replaced by $(p - p_o)$, i.e., that

$$\oiint_S \mathbf{r} \times (p - p_o)\mathbf{n}\,dS = \oiint_S \mathbf{r} \times p\mathbf{n}\,dS$$

HINT: Use Gauss' Theorem (Appendix C, Section C.4).

6.11 Water at 10° C flows steadily with a mass flow rate $\dot{m} = 62.8$ kg/sec through the nozzle shown. What are the average velocities U and u if the diameters are $D = 20$ cm and $d = 5$ cm?

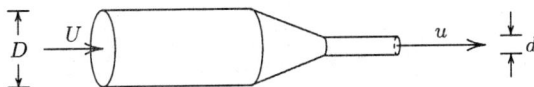

Problems 6.11, 6.12

6.12 Water at 68° F flows steadily with a mass flow rate $\dot{m} = 0.857$ slug/sec through the nozzle shown. What are the average velocities U and u if the diameters are $D = 9$ in and $d = 3$ in?

6.13 For the cylindrical tank with attached cylindrical pipes shown, compute the rate at which the water level is changing for $\alpha = \frac{1}{2}$ and $\beta = \frac{2}{5}$. Indicate whether the tank is filling or emptying. Express your answer for dh/dt as a function of U, d and D.

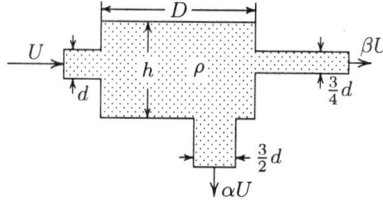

Problems 6.13, 6.14

6.14 For the cylindrical tank with attached cylindrical pipes shown, the constant β is 1. Determine the value of the constant α for which the water level in the tank is constant.

6.15 A wind tunnel has a porous wall in its test section to help reduce viscous effects. The diameter of the cylindrical test section is D, its length is L and the inlet velocity is U_i. The test-section surface has N holes per unit area of diameter d, and the suction velocity is v_w. Assuming the flow is steady and incompressible with density ρ, determine the outlet velocity, U_o. If $U_i = 50$ m/sec, $D = 1$ m, $L = 5$ m and $v_w = 10$ m/sec, compute the value of U_o for $d = 5$ mm and $N = 510$ holes/m^2 and for $d = 7$ mm and $N = 1040$ holes/m^2.

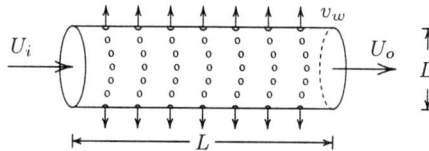

Problem 6.15

6.16 The figure illustrates a *jet pump*. At Section 1, a high-speed jet of fluid is injected into a uniform flow of velocity U_1 in a duct of area A. The fluid mixes and, at Section 2, returns to nominally uniform flow with velocity U_2. If the jet velocity is $U_j = 10U_1$ and the jet area is $A_j = \frac{1}{5}A$, what is the velocity at Section 2? Assume the flow is steady and incompressible with density ρ.

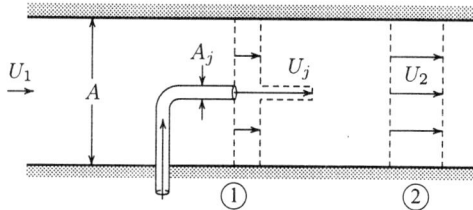

Problem 6.16

6.17 A cylindrical tank of diameter D is supplied with an incompressible fluid of density ρ by a pipe of diameter $\frac{1}{10}D$ and velocity U. Fluid leaves the tank through another horizontal pipe of diameter $\frac{1}{10}D$ and a vertical pipe of diameter $\frac{1}{5}D$. If the water level does not change with time and the velocity in the horizontal pipe is $U_h = \frac{1}{4}U$, what is the velocity in the vertical pipe, U_v?

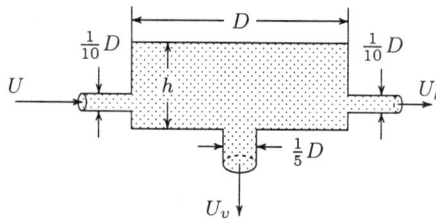

Problems 6.17, 6.18

6.18 A cylindrical tank of diameter D is supplied with an incompressible fluid of density ρ by a pipe of diameter $\frac{1}{10}D$ and velocity U. Fluid leaves the tank through another horizontal pipe of diameter $\frac{1}{10}D$ and a vertical pipe of diameter $\frac{1}{5}D$. If the velocity in the horizontal pipe is $U_h = \frac{1}{5}U$ and the velocity in the vertical pipe is $U_v = \frac{9}{20}U$, at what rate, dh/dt, is the level changing in the tank?

6.19 A subway train has a cross-sectional area, $A_{cs} = \frac{1}{10}A$, where A is the area of the tunnel through which it moves. The train is traveling at a constant velocity U. What is the average velocity, u, between the train and the tunnel walls in the indicated direction?

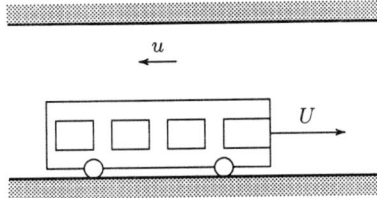

Problems 6.19, 6.20

6.20 A subway train moves at a constant velocity U through a tunnel of area A. If the average velocity between the train and the tunnel walls is $u = \frac{1}{4}U$ in the direction shown, what is the cross-sectional area of the train, A_{cs}?

6.21 A sphere of radius R falls axially through an infinitely long cylinder of radius $\frac{3}{2}R$ at a speed U. What is the mean upward velocity of the surrounding (incompressible) fluid at the sphere midsection?

6.22 Two parallel disks of diameter D are brought together, each having a speed V as shown. The fluid between the disks is incompressible with density ρ. You may assume uniform velocity across any cylindrical section, i.e., that radial velocity, u, depends only upon radial distance from the origin. Use a stationary control volume to solve.

(a) Using a cylindrical control volume of radius $r < D/2$, determine u as a function of r, h and V.

(b) Show that the unsteady and convective accelerations at $r = D/2$ are given by $A_{\text{unsteady}} = V^2 D/h^2$ and $A_{\text{convective}} = \frac{1}{2}V^2 D/h^2$.

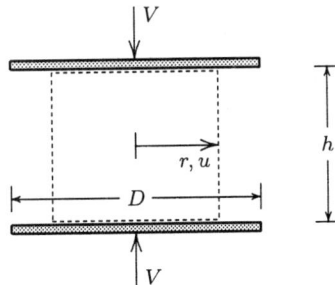

Problems 6.22, 6.23

6.23 Two parallel disks of diameter D are brought together, each having a speed V as shown. The fluid between the disks is incompressible with density ρ. You may assume uniform velocity across any cylindrical section, i.e., that radial velocity, u, depends only upon radial distance from the origin. Use a deforming control volume to solve.

(a) Using a cylindrical control volume of radius $r < D/2$, determine u as a function of r, h and V.

(b) Show that the unsteady and convective accelerations at $r = D/2$ are given by $A_{\text{unsteady}} = V^2 D/h^2$ and $A_{\text{convective}} = \frac{1}{2}V^2 D/h^2$.

6.24 A plunger moves downward in a conical receptacle as shown. The receptacle is filled with an incompressible fluid of density ρ. At what level (z as a function of d) above the bottom of the receptacle will the mean upward velocity of the fluid between the bottom tip of the plunger and the receptacle wall match the downward velocity of the plunger? Use a stationary control volume to solve.

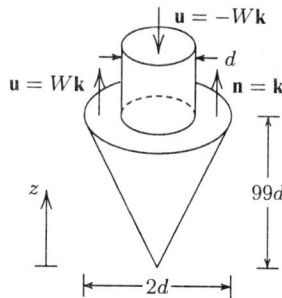

Problems 6.24, 6.25

6.25 A plunger moves downward in a conical receptacle as shown. The receptacle is filled with an incompressible fluid of density ρ. At what level (z as a function of d) above the bottom of the receptacle will the mean upward velocity of the fluid between the bottom tip of the plunger and the receptacle wall match the downward velocity of the plunger? Use a deforming control volume to solve.

6.26 The velocity in the outlet pipe from the reservoir is U. Assuming the diameter of the reservoir, D, is 25 times the diameter, d, at point A, what is the rate at which the surface recedes, dh/dt? Express your result in terms of U. Use a stationary control volume to solve.

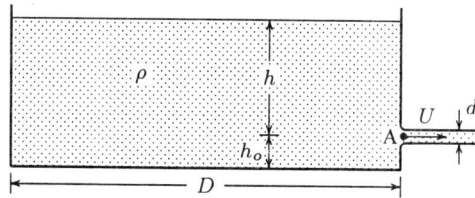

Problem 6.26, 6.27

6.27 The velocity in the outlet pipe from the reservoir is U. Assuming the diameter of the reservoir, D, is 25 times the diameter, d, at point A, what is the rate at which the surface recedes, dh/dt? Express your result in terms of U. Use a deforming control volume to solve.

6.28 The mass flow rate through a nozzle is $\dot{m} = 0.65 p_c A_t / \sqrt{RT_c}$, where p_c and T_c are pressure and temperature in the rocket chamber and R is the gas constant of the gases in the chamber. The constant propellant burning rate (surface regression rate) is $\dot{r} = a p_c^n$, where a and n are constants. As a result, gas of density ρ_p enters the chamber with velocity $u_p = -\dot{r}$. You can ignore the motion of the propellant surface.

(a) Denoting the propellant surface burning area by A_p, show that

$$p_c = (a\rho_p/0.65)^{1/(1-n)} (A_p/A_t)^{1/(1-n)} (RT_c)^{1/[2(1-n)]}$$

(b) If the operating chamber pressure of a rocket motor is 4 megaPascals (MPa) and the exponent $n = 0.3$, how much will the chamber pressure increase if a crack develops in the grain, increasing the burning area by 17%? You may assume all variables except p_c and A_p remain constant.

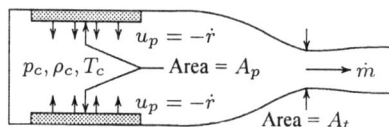

Problems 6.28, 6.29

6.29 The mass flow rate through a nozzle is $\dot{m} = 0.65 p_c A_t / \sqrt{RT_c}$, where p_c and T_c are pressure and temperature in the rocket chamber and R is the gas constant of the gases in the chamber. The propellant burning rate (surface regression rate) can be expressed as $\dot{r} = a p_c^n$, where a and n are constants. As a result, gas of density ρ_p enters the chamber with velocity $u_p = -\dot{r}$. You can ignore the motion of the propellant surface.

(a) Denoting the propellant surface burning area by A_p, show that

$$p_c = (a\rho_p/0.65)^{1/(1-n)} (A_p/A_t)^{1/(1-n)} (RT_c)^{1/[2(1-n)]}$$

(b) If the operating chamber pressure of a rocket motor is 500 psi and the exponent $n = 0.3$, how much will the chamber pressure increase if a crack develops in the grain, increasing the burning area by 13.5%? You may assume all variables except p_c and A_p remain constant.

6.30 The flow in the inlet between two parallel plates develops from uniform flow, $u_1(y) = U_o$, to a parabolic distribution downstream given by $u_2(y) = Ay(h - y)$, where A is a constant and h is distance between the plates. If the flow is incompressible and steady, what is the maximum velocity at the downstream location? What is the average velocity downstream?

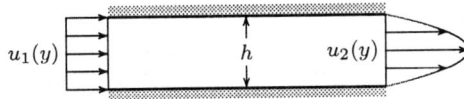

Problem 6.30

6.31 The vessel shown has height h normal to the page. There is no flow into the vessel, but a chemical reaction that occurs inside generates gas that leaves through the four openings, each $\frac{1}{2}r_o$ by h in cross-sectional area as shown. The velocity relative to the vessel, V, varies with radius as $V = 2V_o(7/4 - r/r_o)$. The flow is incompressible with density ρ_o. Find the rate of change of mass in the vessel, dM/dt. Express your answer in terms of ρ_o, V_o, r_o and h.

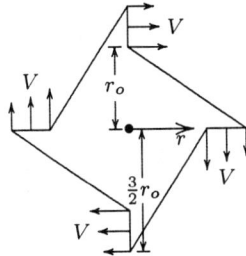

Problems 6.31, 6.32

6.32 The vessel shown has height h normal to the page. There is no flow into the vessel, but a chemical reaction that occurs inside generates gas that leaves through the four openings, each $\frac{1}{2}r_o$ by h in cross-sectional area as shown. The velocity relative to the vessel, V, and the density of the gas, ρ, leaving vary with radius as $V = V_o(r_o/r)$ and $\rho = \rho_o(1 + r/r_o)$. Find the rate of change of mass in the vessel, dM/dt. Express your answer in terms of ρ_o, V_o, r_o and h.

6.33 An incompressible fluid of density ρ flows in a duct with velocity U and area A as shown. Fluid of the same density is injected with velocity $2U$ from the lower duct wall. Assume the flow is steady, and neglect effects of gravity and viscosity. If all flow properties are constant on cross sections, compute the pressure difference between points 1 and 2. **NOTE:** U_2 is not given, you must solve for it.

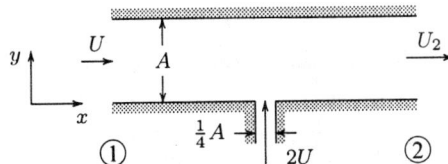

Problem 6.33

6.34 Two incompressible plane jets of the same velocity, V, and density, ρ, meet head on and the flow shown results. Ignoring body forces, derive an expression for $\cos\phi$ as a function of h_1 and h_2. Does your answer make sense for $\phi = 90°$?

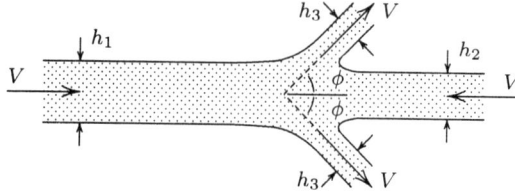

Problem 6.34

6.35 A horizontal pipe (i.e., neglect body forces) carries water to three jets discharging into the atmosphere from the end of the pipe as shown. The velocity of all three jets is V. We observe that there is no net force on the pipe. What is the angle ϕ and what is the overpressure, Δp (i.e., the excess over atmospheric) at an upstream section in the pipe? Express your answer for Δp in terms of water density, ρ, and the jet velocity, V. **NOTE:** U is not given, you must solve for it.

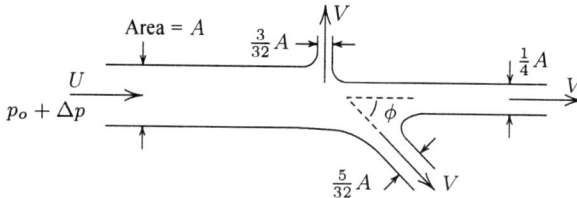

Problem 6.35

6.36 A two-dimensional channel of width H has a slot of width h as shown. Fluid is injected through the slot at an angle ϕ to the horizontal. The fluid is incompressible with density ρ and body forces can be neglected. Velocity and pressure can be assumed constant across the channel.

(a) Derive an equation stating mass conservation for the channel.

(b) Derive an equation stating conservation of x momentum for the channel.

(c) Now, assume $\phi = 60°$ so that $\cos\phi = 1/2$. Combine your results of Parts (a) and (b) with Bernoulli's equation to determine U_2 as a function of U_1, H and h.

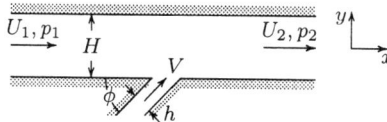

Problem 6.36

6.37 A two-dimensional channel of width H has two slots of width h as shown. Fluid is injected through the lower slot at an angle ϕ to the horizontal and normal to the flow direction from the upper slot. The fluid is incompressible with density ρ and body forces can be neglected. Velocity and pressure can be assumed constant across the channel.

(a) Derive an equation stating mass conservation for the channel.

(b) Derive an equation stating conservation of x momentum for the channel.

(c) Now, assume $\phi = 60°$ so that $\cos\phi = 1/2$. Combine your results of Parts (a) and (b) with Bernoulli's equation to determine U_2 as a function of U_1, H and h.

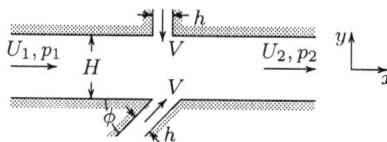

Problem 6.37

6.38 Incompressible fluid of density ρ flows through a container as shown. We wish to design the container appendages so that for a given velocity U, the horizontal force on the container vanishes. Pressure is atmospheric at all three openings, the flow is steady and body forces can be neglected. What is the angle ϕ?

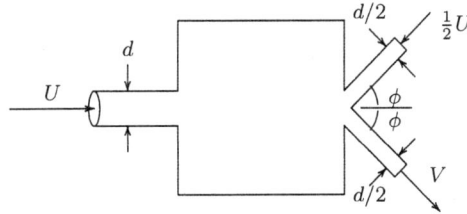

Problem 6.38

6.39 A vane deflects a jet of incompressible fluid with density ρ through $180°$. The incident and reflected jets are circular with diameter D. You may ignore effects of gravity and the weight of the vane. What is the net force on the vane if the incident (absolute) jet velocity is U_i and the vane is moving at (absolute) velocity $U_v = \frac{1}{2}U_i$? The velocity U_o is not given, you must solve for it.

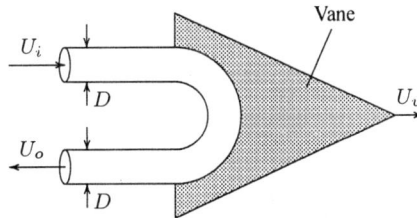

Problems 6.39, 6.40

6.40 A vane deflects a jet of incompressible fluid with density ρ through $180°$. The incident and reflected jets are circular with diameter D. You may ignore effects of gravity and the weight of the vane. What is the net force on the vane if the incident (absolute) jet velocity is U_i and the vane is moving at (absolute) velocity $U_v = \frac{1}{4}U_i$? The velocity U_o is not given, you must solve for it.

6.41 A water jet of velocity $U_j = 10$ ft/sec impinges on a small cart, causing it to move at a constant velocity $U = 2$ ft/sec. If the jet diameter is $d = 0.2$ ft and the water density is $\rho = 1.94$ slug/ft^3, determine the force on the cart.

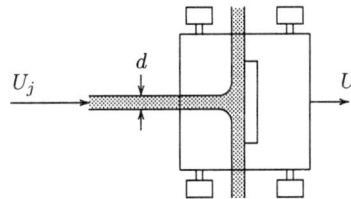

Problems 6.41, 6.42

6.42 A water jet of cross-sectional area A with velocity U_j and density ρ causes a cart to move at a constant velocity $U = \frac{1}{5}U_j$.

(a) What is the rolling resistance of the cart?

(b) The jet velocity is changed to $\hat{U}_j$ and we observe that the cart velocity doubles, i.e., $U = \frac{2}{5}U_j$, where U_j is the original jet velocity. What must the incident jet velocity be? **NOTE:** The rolling resistance is independent of cart velocity.

6.43 An incompressible jet of density ρ and velocity U is deflected by a cone of half angle $\theta = 60°$. The cone moves toward the jet at speed $U_c = \frac{1}{2}U$. Determine the thickness of the radially spreading jet as it leaves the cone, τ, and the force required to move the cone. You may assume the flow is irrotational and that body forces can be neglected.

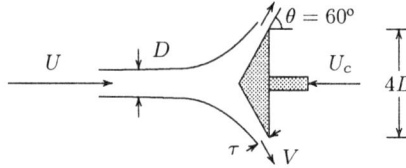

Problems 6.43, 6.44

6.44 An incompressible jet of density ρ and velocity U is deflected by a cone of half angle $\theta = 60°$. The cone moves toward the jet at speed $U_c = U$. Determine the thickness of the radially spreading jet as it leaves the cone, τ, and the force required to move the cone. You may assume the flow is irrotational and that body forces can be neglected.

6.45 The figure illustrates a *jet pump*. Compute the pressure coefficient,

$$C_p = (p_2 - p_1)/\rho\, U_o^2$$

assuming the flow is steady, incompressible, body forces can be neglected and that velocity and pressure are uniform across Sections 0, 1, and 2. In the jet, the flow is *rotational*. In computing the velocity at Station 1, be sure to account for the blockage due to the pump, whose area is $A_j = A_o/100$. Also, $U_j = 10U_o$. **HINT:** Use a control volume between Sections 0 and 1 to determine U_1, and another between Sections 1 and 2 to complete the solution.

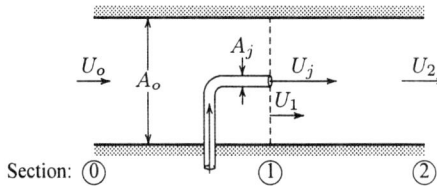

Problems 6.45, 6.46

6.46 The figure illustrates a *jet pump*. We want to design the pump so that the pressure coefficient is

$$C_p = (p_2 - p_1)/\rho\, U_o^2 = 0.75$$

Assume the flow is steady, incompressible, body forces can be neglected and that velocity and pressure are uniform across Sections 0, 1, and 2. In the jet, the flow is *rotational*. In computing the velocity at Station 1, be sure to account for the blockage due to the pump, whose area is $A_j = A_o/10$. What is the jet velocity, U_j. **HINT:** Use a control volume between Sections 0 and 1 to determine U_1, and another between Sections 1 and 2 to complete the solution.

6.47 Water flows in a pipe that lies in a horizontal plane (so that gravitational effects can be ignored). The flow is *rotational* and $\Delta p = \frac{1}{4}\rho U^2$. What are the velocities, V_1 and V_2, and what must the angle ϕ be in order to have no net force on the pipe?

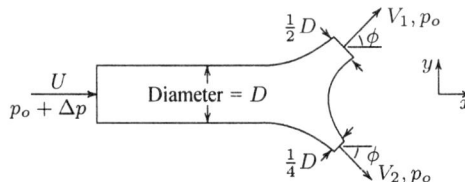

Problem 6.47

6.48 A very cheap faucet made of tin springs a leak so that an extra horizontal spray makes its use practical only when the user is wearing a raincoat. Assuming the flow to be *rotational*, what is the vertical force required to hold the faucet in place if (thanks to the leak) the net horizontal force is zero? Express your answer in terms of water density, ρ, inlet velocity, U_i, and the constant cross-sectional area, A. Neglect body forces and note that neither U_f nor U_j are given, you must solve for them.

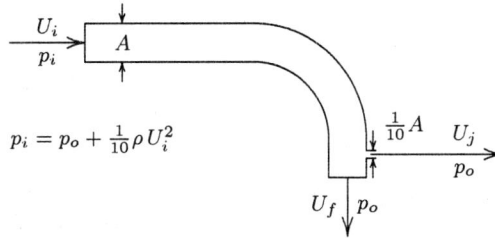

Problem 6.48

6.49 Consider steady, incompressible flow in a pipe that exhausts to the atmosphere as shown below. The flow is asymmetric because of the sharp bend that causes flow separation at the lower corner. The flow is *strongly rotational*. We observe that the force required to hold the pipe in place is in the direction shown. What is the overpressure, Δp, at the inlet? **NOTE:** V is not given, you must solve for it. You may neglect body forces and assume all properties are constant across inlet and outlets.

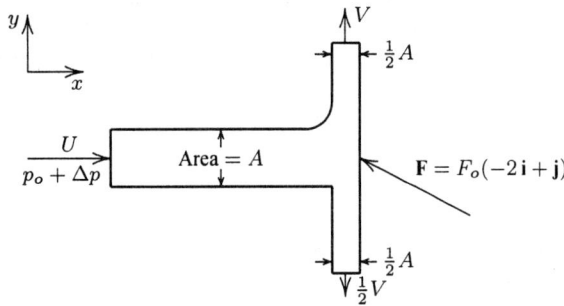

Problem 6.49

6.50 A *splitter plate* inserted part way into a two-dimensional incompressible stream produces the split stream shown. You may ignore viscous effects and body forces, and assume the flow is steady. The fluid density is ρ.

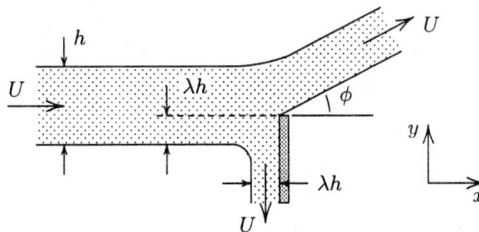

Problem 6.50

(a) Determine the width of the upper stream in terms of h and λ.

(b) Determine $\sin \phi$ as a function of λ.

(c) Compute the horizontal force, F_x, exerted by the fluid on the splitter plate as a function of ρ, U, h and λ. Explain why your answer for F_x makes sense for the limiting case $\lambda \to 0$.

6.51 Consider steady, incompressible flow through the device shown. The fluid density is ρ and the inlet velocity is U. The fluid at both exits exhausts to the atmosphere. Ignoring body forces, determine the force exerted by the fluid on the device as a function of ρ, U and diameter D if the inlet pressure is $p_i = p_o + \frac{1}{16}\rho U^2$.

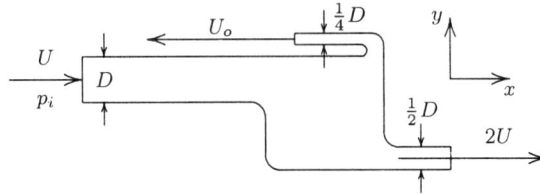

Problems 6.51, 6.52

6.52 Consider steady, incompressible flow through the device shown. The fluid density is ρ and the inlet velocity is U. The fluid at both exits exhausts to the atmosphere. Ignoring body forces, determine the inlet pressure, p_i, as a function of ρ, U and atmospheric pressure, p_o, if the force exerted by the fluid on the device is $\mathbf{F} = \frac{17}{16}\rho U^2 D^2 \,\mathbf{i}$.

6.53 A fixed vane turns an incompressible jet of density ρ through an angle α. The flow is steady, body forces can be ignored and velocity can be assumed constant on cross sections. Determine the force required to hold the vane in place as a function of U_1, U_2, α and the mass flux in the jet, $\dot{m}$.

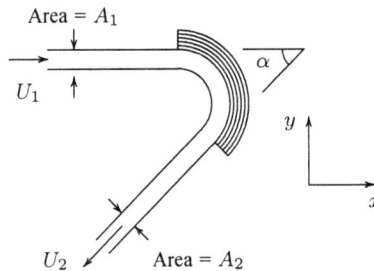

Problem 6.53

6.54 Sand of density ρ is being loaded onto a barge as shown. The velocity of the sand is U, the cross-sectional area of the pipe is A, and the mass-flow rate through the pipe is $\dot{m} = \rho U A$.

(a) If the mass of the barge is M_o, what is the total mass of sand and barge a time t after the pipe starts emitting sand?

(b) Ignoring the instantaneous rate of change of momentum in your control volume, determine the tension, T, in the mooring line as a function of $\dot{m}$, U and α.

(c) Find the buoyancy force on the barge as a function of M_o, g, $\dot{m}$, t, U and α.

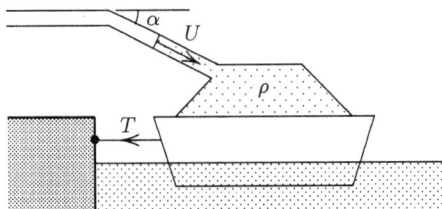

Problem 6.54

6.55 A channel with three branches has no net force in the x direction. The flow is steady, incompressible and is not subject to any body forces. Because of the sharp corners, however, the flow is strongly rotational. The inlet velocity, U_i, channel area, A, and the angle $\phi = 60°$ are given. Also, the pressure at the inlet and central outlet branch are $p_i = p_\infty + \Delta p$ and $p_o = p_\infty + \frac{11}{2}\Delta p$, where p_∞ is atmospheric pressure and the overpressure is $\Delta p = \frac{1}{8}\rho U_i^2$. The upper and lower branches exhaust to the atmosphere at velocity V. Determine the velocities V and U_o, and the vertical force, F_y, required to hold the channel in place. **NOTE:** There are two possible solutions for this problem, and you should find both.

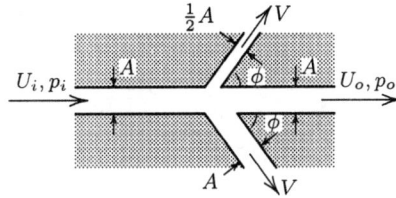

Problem 6.55

6.56 A steady stream of incompressible, inviscid fluid flowing in the xy plane is deflected by two vanes as shown below. The deflection angle for each vane is θ, and the inlet and outlet streams are a distance Δy apart. Gravity acts normal to the xy plane. The mass flow rate is $\dot{m}$, and the speed at all points in the stream is U. Note that this stream is a free stream, i.e., it is not under pressure in a pipe. Compute the force exerted by the fluid on the top vane, and the total force exerted by the fluid on both vanes.

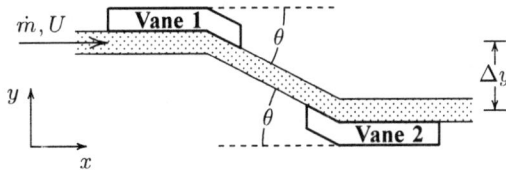

Problem 6.56

6.57 Consider the incompressible, steady, strongly rotational flow from a pipe of diameter D into a spherical container as shown. The mass of the sphere, including the fluid inside is M. Fluid exits the sphere through two jets as shown. Both jets have velocity U_j and diameter $\frac{1}{2}D$. The pressure in the pipe is $p = p_o + \frac{1}{5}\rho W^2$, where ρ is the fluid density, p_o is atmospheric pressure and W is the inlet velocity. If the vertical component of the force required to hold the sphere in place is zero, what is the horizontal component, F_x? Express your answer in terms of M, g and the angle ϕ.

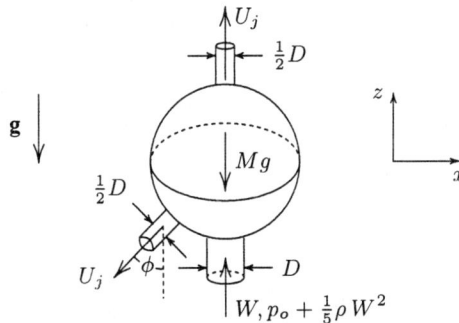

Problems 6.57, 6.58

6.58 Consider the incompressible, steady, strongly rotational flow from a pipe of diameter D into a spherical container as shown. Fluid exits the sphere through two jets as shown. Both jets have velocity U_j and diameter $\frac{1}{2}D$. The pressure in the pipe is $p = p_o + \frac{1}{5}\rho W^2$, where ρ is the fluid density, p_o is atmospheric pressure and W is the inlet velocity. The weight of the sphere, including the fluid inside is $Mg = \frac{\pi}{20}\rho W^2 D^2$. If the vertical component of the force required to hold the sphere in place, F_z, is twice the horizontal component, F_x, what is the angle ϕ?

6.59 An incompressible fluid of density ρ flows through the multiple nozzle shown and discharges to the atmosphere at all three outlets. You may assume the flow is steady, strongly rotational and that properties are uniform on all cross sections. Also, the nozzle lies in a horizontal plane so that effects of gravity can be neglected. We observe that the components of the net force required to hold the nozzle in place are such that $R_y = 8R_x$. Determine the pressure differential at the inlet, Δp.

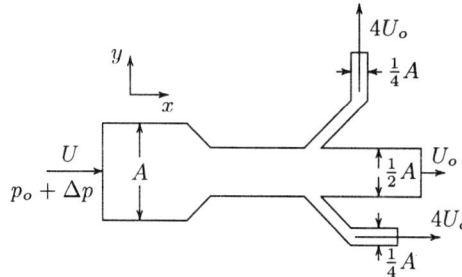

Problem 6.59

6.60 A pipe carries water to two jets that discharge into the atmosphere from the pipe as shown. The exit velocity for both jets is V and the jet diameters are half the inlet pipe diameter. What force is required to hold the pipe in place? Express your answer in terms of fluid density, ρ, jet velocity, V, and pipe diameter, D. You may assume the flow to be steady, irrotational and that body forces can be ignored.

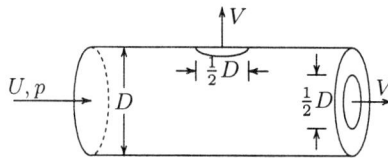

Problem 6.60

6.61 An incompressible fluid of density ρ flows through the nozzle shown and discharges to the atmosphere. Determine the force required to hold the nozzle in place. You may assume the flow is irrotational and that all flow properties are uniform across the nozzle. Express your answer as a function of ρ, U and D.

Problem 6.61

6.62 Water flows *steadily* and *irrotationally* up the vertical pipe of diameter D as shown, and flows radially outwards between two circular, horizontal end plates. The diameter of the end plates is $\frac{5}{2}D$.

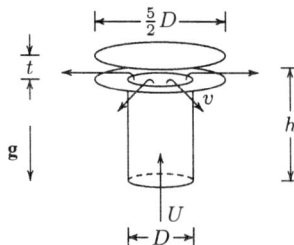

Problem 6.62

(a) Assuming the distance between the plates is $t = D/20$, determine the radial velocity, v, in terms of the inlet velocity, U.

(b) Assuming $t \ll h$ and the pressure at the inlet is $p_o + 4\rho gh$ (p_o is atmospheric pressure), determine the inlet velocity, U, in terms of g and h.

6.63 The nozzle shown below discharges water into the atmosphere through two outlets. What force through the mounting bolts is required to hold the nozzle in place? You may assume the flow is steady, irrotational and you may neglect effects of gravity. The jet velocities, V, (for which you must solve) are equal. Express your answer in terms of fluid density, ρ, inlet velocity, U, cross-sectional area, A, and the angle ϕ.

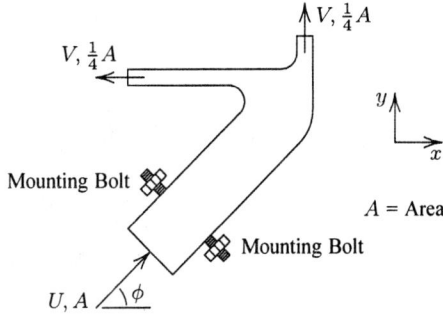

Problem 6.63

6.64 Consider the cone shown with a hole of diameter d at its center. A round jet of diameter D, density ρ, and speed U flows toward the cone concentrically. What is the force required to hold the cone in place if the jet issuing from the hole also has speed U? You may assume the flow is incompressible, steady and body forces can be neglected.

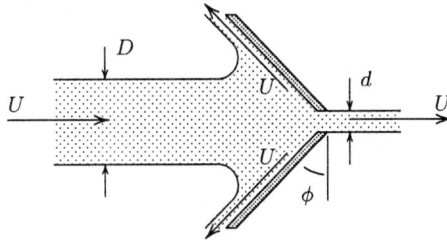

Problem 6.64

6.65 Consider the large reservoir of liquid of density ρ as shown. At depth h there is a small hole of diameter d through which a jet of velocity V issues. The jet strikes a cube with side $10d$ made of a material whose density is $\pi\rho$. The coefficient of sliding friction between cube and surface is $\frac{1}{2}$. Assume the flow is steady, incompressible, irrotational, and neglect effects of gravity *in the jet*.

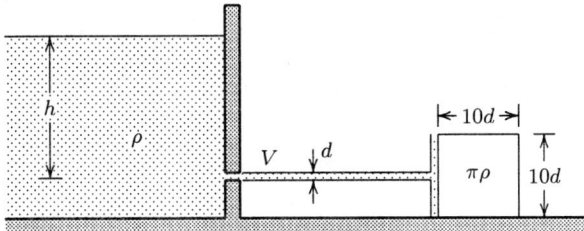

Problem 6.65

(a) What is the jet velocity?

(b) What is the force on the cube from the jet? **NOTE:** Ignore any horizontal momentum in the fluid as it moves beyond the edges of the cube.

(c) What is the maximum value of h for which the cube will not move?

6.66 A two-dimensional jet of velocity U and width H impinges on a wall at an angle ϕ. The fluid is incompressible and has density ρ. Suppose that, far from the impingement point, the widths of the *wall jets* are equal so that $h_1 = h_2 = h$. Neglect body forces and assume the flow is steady, inviscid, and that velocities are uniform across the jets.

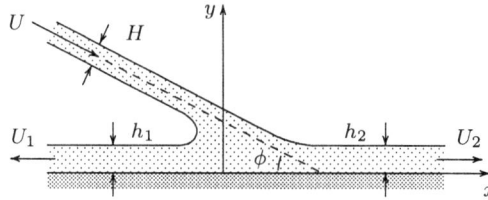

Problems 6.66, 6.67

(a) What is the force on the wall? *Be sure to indicate direction.*

(b) What are the velocities U_1 and U_2 as functions of H, h, U and ϕ?

(c) Check the limiting case $\phi = 90°$ and explain why your answer to Part (b) is sensible.

6.67 A two-dimensional jet of velocity U and width H impinges on a wall at an angle ϕ. The fluid is incompressible and has density ρ. Suppose that, far from the impingement point, the velocities of the *wall jets* are equal so that $U_1 = U_2 = U$. Neglect body forces and assume the flow is steady, inviscid, and that velocities are uniform across the jets. Determine h_1 and h_2 as functions of H and ϕ.

6.68 Consider steady, incompressible flow into a 180° bend. The pipe diameter is constant so that $D_1 = D_2 = D$. The pressure at both Sections 1 and 2 is $p_o + \Delta p$, where p_o is atmospheric pressure. Assume the velocity and pressure are constant on all cross sections and that body forces can be ignored. What is the force exerted by the fluid on the portion of the pipe between Sections 1 and 2? Do conditions at Sections 1 and 2 satisfy Bernoulli's equation?

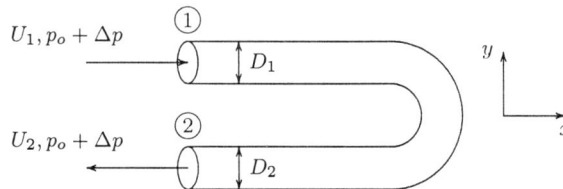

Problems 6.68, 6.69

6.69 Consider steady, incompressible flow into a 180° bend. The pipe diameters at Sections 1 and 2 are $D_1 = D$ and $D_2 = \frac{11}{10}D$, while the pressure at both Sections 1 and 2 is $p_o + \Delta p$, where p_o is atmospheric pressure. Assume the velocity and pressure are constant on all cross sections and that body forces can be ignored. If the horizontal force exerted by the fluid on the portion of the pipe between Sections 1 and 2 is $F_x = \frac{5}{8}\pi\rho U_1^2 D^2$, what is the overpressure, Δp? Do conditions at Sections 1 and 2 satisfy Bernoulli's equation?

6.70 An incompressible jet of fluid of density ρ, velocity W and diameter d_o issues vertically from a wall and supports a disk of mass M and diameter D. We want to determine the height h to which the disk will rise. Solve this problem using two control volumes, CV_1 and CV_2, as shown. You can assume the flow is steady, the flow in CV_1 is irrotational, and velocities are constant across all cross sections.

(a) Derive the basic equations required to solve the problem for the three unknown quantities w, d and h. You can regard ρ, W, d_o, M and τ as known quantities.

(b) Now, simplify the equations derived in Part (a) assuming that $\tau \ll h$ and that $\rho D^2 \tau \ll M$.

(c) Solve for the height h. Express your answer in terms of the dimensionless grouping $2gh/W^2$.

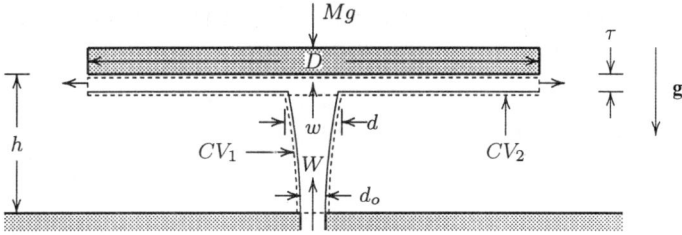

Problem 6.70

6.71 An incompressible fluid flows steadily and irrotationally through the two-dimensional channel as shown and exhausts to the atmosphere. What must the angle ϕ be in order for the vertical component of the force on the channel walls to vanish, and what is the horizontal component of the force for this angle? Assume properties are constant across the channel and body forces can be ignored. Express your answers in terms of ρ, U, and A. **NOTE:** The quantities V, Δp_1 and Δp_2 are not given, you must solve for them.

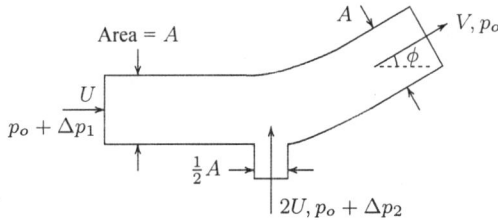

Problem 6.71

6.72 A water jet (density ρ) with an initial velocity W and diameter d_o issues vertically from a wall and supports a hemispherical shell at a height h as shown. You can assume the flow is steady, irrotational and that velocities are constant across all cross sections. Also, you can ignore the weight of the fluid in the control volume. **NOTE:** Neither the downflow velocity, w, nor the area of the downflowing fluid are given, you must solve for them.

(a) Using the indicated control volume, determine the weight of the hemispherical shell, Mg, where g is the acceleration of gravity.

(b) Compute Mg for $\rho = 1000$ kg/m^3, $W = 30$ m/sec, $d_o = 5$ cm and $h = 1$ m.

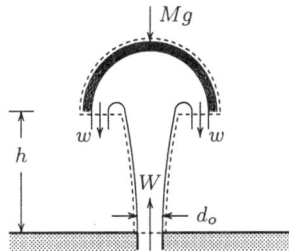

Problems 6.72, 6.73

6.73 A water jet (density ρ) with an initial velocity W and diameter d_o issues vertically from a wall and supports a hemispherical shell at a height h as shown. You can assume the flow is steady, irrotational and that velocities are constant across all cross sections. Also, assume the volume of the fluid contained in the indicated control volume is $V = \frac{\pi}{2}d_o^2 h$. **NOTE:** Neither the downflow velocity, w, nor the area of the downflowing fluid are given, you must solve for them.

(a) Using the indicated control volume, determine the weight of the hemispherical shell, Mg, where g is the acceleration of gravity.

(b) Compute Mg for $\rho = 1.94$ slug/ft^3, $W = 25$ ft/sec, $d_o = 1$ in and $h = 6$ in.

6.74 A water jet (density ρ) with an initial velocity W and diameter d_o issues vertically from a wall and supports a conical object that rises to a height h as shown. You can assume the flow is steady, irrotational and that velocities are constant across all cross sections. Also, you can ignore the weight of the fluid in the control volume. **NOTE:** Neither the upflow velocity, w, nor the area of the upflowing fluid at the base of the cone are given, you must solve for them.

  (a) Using the indicated control volume, determine the weight of the cone, Mg, where g is the acceleration of gravity.

  (b) Compute Mg for $\rho = 1.94$ slug/ft^3, $W = 22$ ft/sec, $d_o = 2$ in, $h = 1$ ft, $\phi = 30°$.

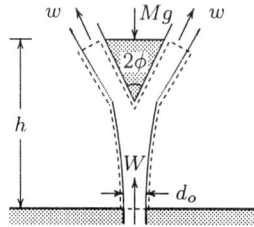

Problems 6.74, 6.75

6.75 A water jet (density ρ) with an initial velocity W and diameter d_o issues vertically from a wall and supports a conical object that rises to a height h as shown. You can assume the flow is steady, irrotational and that velocities are constant across all cross sections. Also, assume the volume of the fluid contained in the indicated control volume is $V = \frac{\pi}{3}d_o^2 h$. **NOTE:** Neither the upflow velocity, w, nor the area of the upflowing fluid at the base of the cone are given, you must solve for them.

  (a) Using the indicated control volume, determine the weight of the cone, Mg, where g is the acceleration of gravity.

  (b) Compute Mg for $\rho = 998$ kg/m^3, $W = 24$ m/sec, $d_o = 10$ cm, $h = 1$ m, $\phi = 57°$.

6.76 For laminar viscous flow near the entrance to a pipe, the velocity distribution changes from uniform to parabolic as shown. At the fully-developed outlet section, the velocity varies according to

$$u_2 = U_{\max}\left[1 - \left(\frac{r}{R}\right)^2\right]$$

Derive a formula for the net resisting frictional force, F_τ, on the pipe section as a function of U, ρ, p_1, p_2, and $D = 2R$ (the pipe diameter).

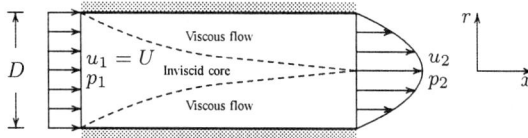

Problem 6.76

6.77 An incompressible fluid of density ρ enters a square duct of constant height h with uniform velocity U. The duct makes a 90° bend that distorts the flow to produce the linear velocity profile shown at the exit, with $v_{max} = 2v_{min}$. You may assume the flow is steady and that body forces are negligible. Determine v_{min} as a function of U. Also, assuming Bernoulli's equation holds between the inlet and the high-speed side of the exit, determine the inlet pressure coefficient, $C_p = (p - p_o)/(\frac{1}{2}\rho U^2)$.

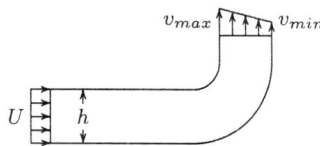

Problem 6.77

6.78 A circular cylinder of diameter D and length L (out of page) is mounted in a *rectangular* duct with height $4D$. Far upstream, the flow is uniform with velocity U_∞ and pressure p_1. Far downstream, the streamlines again become parallel, and the shape of the velocity profile is as shown while the pressure is p_2. Using the control volume indicated, the symmetry of the geometry, and the fact that the drag coefficient, C_D, and outlet velocity, u_2, are given by $C_D = F_D/(\rho U_\infty^2 DL) = 4$ and $u_2 = U(y/2D)^4$, determine the pressure difference $(p_1 - p_2)$. You can ignore effects of friction on the duct walls and assume pressure is independent of y at the duct inlet and outlet. **NOTE:** The peak velocity, U, is not given, you must solve for it.

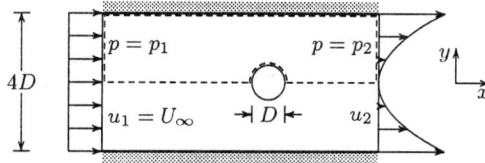

Problems 6.78, 6.79

6.79 A circular cylinder of diameter D and length L (out of page) is mounted in a *rectangular* duct with height $4D$. Far upstream, the flow is uniform with velocity U_∞ and pressure p_1. Far downstream, the streamlines again become parallel, and the shape of the velocity profile is as shown while the pressure is p_2. Using the control volume indicated, the symmetry of the geometry, and the fact that the drag coefficient, C_D, and outlet velocity, u_2, are given by $C_D = F_D/(\rho U_\infty^2 DL) = 3$ and $u_2 = U(y/2D)^2$, determine the pressure difference $(p_1 - p_2)$. You can ignore effects of friction on the duct walls and assume pressure is independent of y at the duct inlet and outlet. **NOTE:** The peak velocity, U, is not given, you must solve for it.

6.80 Consider the two-dimensional *reducing bend* shown below. At the inlet, the velocity varies linearly with distance across the channel and the pressure is unknown. The flow is uniform at both outlets, is incompressible with density ρ and the flow exhausts to the atmosphere. The flow is steady and body forces can be neglected. Determine V as a function of U, and compute the vertical force exerted by the fluid on the bend as a function of ρ, U, h and ϕ.

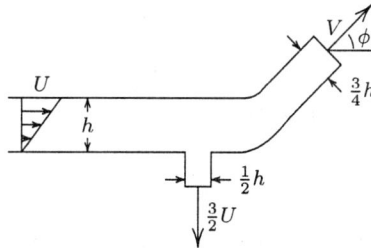

Problem 6.80

6.81 An incompressible fluid of density ρ flows steadily through the nozzle shown and discharges to the atmosphere. The pressure at the inlet is $p = p_o + \frac{1}{4}\rho U^2$. Assume the pressure is uniform on all cross sections, body forces are unimportant and that velocity is uniform on the inlet plane.

(a) Determine the force required to hold the nozzle in place assuming u is uniform $[u(r) = U_e]$ on the exit plane.

(b) Determine the force if $u(r) = U_e(1 - 16r^2/D^2)$ on the exit plane, where r is radial distance from the centerline and D is the initial nozzle diameter. How does the force differ from the value determined in Part (a)?

Problem 6.81

6.82 An incompressible fluid of density ρ flows steadily through the nozzle shown and discharges to the atmosphere. The pressure at the inlet is $p = p_o + \frac{1}{8}\rho U^2$. Assume the pressure is uniform on all cross sections, body forces are unimportant and that velocity is uniform on the inlet plane.

  (a) Determine the force required to hold the nozzle in place assuming u is uniform $[u(r) = U_e]$ on the exit plane.

  (b) Determine the force if $u(r) = U_e(1 - 4r/D)^{1/7}$ on the exit plane, where r is radial distance from the centerline and D is the initial nozzle diameter. How does the force differ from the value determined in Part (a)? **HINT:** To make evaluation of the integrals on the exit plane easier, use the change of integration variable defined by $\eta = 1 - 4r/D$.

Problem 6.82

6.83 Consider steady, incompressible flow of a fluid with density ρ in a pipe of constant radius R with two outlets as shown. Although the configuration is symmetric in every other respect, the inner surface of the upper outlet is rough while the lower outlet is smooth. As a result, the velocities are

$$V_u(r) = 1.155V(1 - r/R)^{1/10} \quad \text{and} \quad V_\ell(r) = 1.224V(1 - r/R)^{1/7}$$

Body forces are unimportant and both outlets exhaust to the atmosphere.

  (a) Verify that the average velocity at both outlets is V. Ignoring the variation of the velocities on cross sections, what is the vertical force required to hold the pipe in place?

  (b) Taking account of the variable velocities, determine the force as a function of ρ, V and R. **HINT:** To make evaluation of the integrals on the exit plane easier, use the change of integration variable defined by $\eta = 1 - 4r/D$.

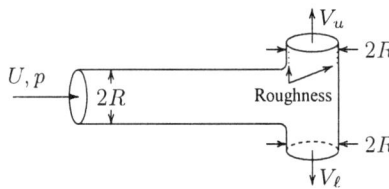

Problem 6.83

6.84 Consider steady, incompressible, viscous flow above a flat plate. The velocity above the plate is given by $u(x, y) = U_\infty f(y/\delta)$, where U_∞ is the constant freestream velocity and $\delta(x)$ is the thickness of the viscous region. Pressure is constant and equal to p_∞, and you can assume that $u = U_\infty$ for $y \geq \delta$.

  (a) Show that, with ρ denoting fluid density, the drag force per unit width (out of the page) is

$$F_D = \int_0^\delta \rho \left(U_\infty - u\right) u \, dy$$

  (b) For turbulent flow, a good approximation for very high Reynolds numbers is $u = U_\infty(y/\delta)^{1/7}$. Compute D as a function of ρ, U_∞ and δ.

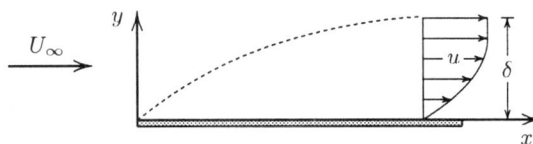

Problem 6.84

6.85 Consider steady, incompressible, viscous flow above a flat plate of length L with surface mass removal, which is referred to as *suction*. The velocity above the plate is given by $u(x, y) = U_\infty f(y/\delta)$, where U_∞ is the constant freestream velocity and $\delta(x)$ is the thickness of the viscous region. Pressure is constant and equal to p_∞, and you can assume that $u = U_\infty$ for $y \geq \delta$. The surface velocity is given by $\mathbf{u} = -C_Q U_\infty \, \mathbf{j}$, where C_Q is the constant suction coefficient. Show that, with ρ denoting fluid density and $\dot{m} = \rho U_\infty C_Q L$, the drag force per unit width (out of the page) is

$$F_D = \dot{m} U_\infty + \int_0^\delta \rho \left(U_\infty - u \right) u \, dy$$

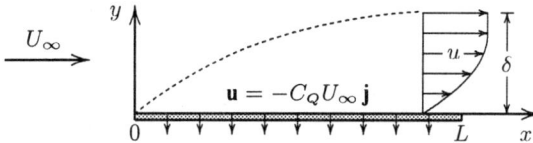

Problem 6.85

6.86 Captain Kirk is returning to the Enterprise in a shuttle when Klingons turn on their *tractor beam*. To counter the beam's effect, i.e., to continue moving on the same course at a constant velocity, V, Kirk fires a rocket as shown. The tractor-beam force is

$$\mathbf{F} = -B^2 M^4 \, \mathbf{i}$$

where B is a constant, and M is the total mass of the shuttle (fuel plus vehicle). Assuming fuel density, ρ_e, velocity, u_e, and exit area, A_e, are constant at the rocket exit plane, and that pressure at the rocket exit plane is negligibly small, determine the total mass of Kirk's shuttle as a function of time. The initial mass (at time $t = 0$) is M_o.

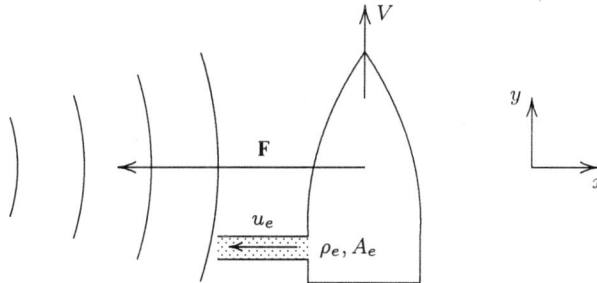

Problems 6.86, 6.87

6.87 Captain Picard is returning to the Enterprise in a shuttle when Klingons turn on their *tractor beam*. To counter the beam's effect, i.e., to continue moving on the same course at a constant velocity, V, Picard fires a rocket as shown. The tractor-beam force is

$$\mathbf{F} = -T^2 M^2 \, \mathbf{i}$$

where T is a constant, and M is the total mass of the shuttle (fuel plus vehicle). Assuming fuel density, ρ_e, velocity, u_e, and exit area, A_e, are constant at the rocket exit plane, and that pressure at the rocket exit plane is negligibly small, determine the total mass of Picard's shuttle as a function of time. The initial mass (at time $t = 0$) is M_o.

6.88 A small boat is powered by an air jet of diameter d, density ρ_e and velocity u_e relative to the boat. The density of water is ρ and the hydrodynamic drag on the boat's hull is

$$F_D = \rho U^2 A C_D$$

where U is the speed of the boat, A is an effective cross-sectional area and C_D is the drag coefficient. You can neglect the aerodynamic drag on the boat.

(a) Derive appropriate equations for conservation of mass and momentum.

(b) Solve for the steady-state velocity, U, in terms of u_e, ρ_e, ρ, d, A and C_D.

(c) Compute the steady-state speed in knots for $\rho A C_D = 0.72$ slug/ft, $d = 1$ in, $\rho_e = .00234$ slug/ft^3 and $u_e = 1002$ ft/sec.

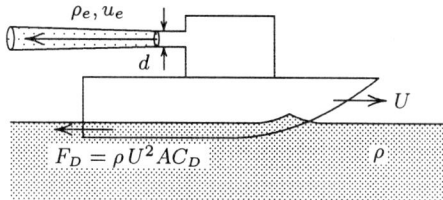

Problems 6.88, 6.89

6.89 A small boat is powered by an air jet of diameter d, density ρ_e and velocity u_e relative to the boat. The density of water is ρ and the hydrodynamic drag on the boat's hull is

$$F_D = \rho U^2 A C_D$$

where U is the speed of the boat, A is an effective cross-sectional area and C_D is the drag coefficient. You can neglect any other form of drag on the boat including the net pressure force.

(a) Derive appropriate equations for conservation of mass and momentum.

(b) Show that the instantaneous velocity, $U(t)$, is

$$U(t) = U_f \left[\frac{1 - (M/M_o)^{2u_e/U_f}}{1 + (M/M_o)^{2u_e/U_f}} \right]$$

where $M(t)$ is the instantaneous mass of the boat and compressed air, $M_o = M(0)$ and U_f is a velocity to be determined as part of the solution.

(c) Compute the steady-state speed, U_f, in knots for $\rho A C_D = 30$ kg/m, $d = 3$ cm, $\rho_e = 1.2$ kg/m^3 and $u_e = 290$ m/sec.

6.90 A tank of water sits on a sled. High pressure in the tank is maintained by a compressor so that the water leaving the tank through the orifice does so at a constant speed, u_e, relative to the tank. The orifice area is A. The instantaneous mass of the sled, water, tank and compressor is M, the water density is ρ and the coefficient of sliding friction between sled and ice is μ. The sled starts from rest with an initial mass, M_o. Ignore the aerodynamic drag on the sled.

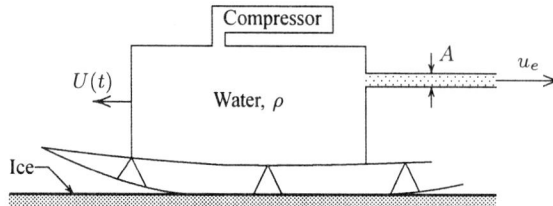

Problem 6.90

(a) Derive appropriate equations for conservation of mass and momentum.

(b) Solve for the instantaneous velocity, $U(t)$, in terms of u_e, M, M_o, μ, g and t (g = gravitational acceleration, t = time).

Chapter 7

Conservation of Energy

At the end of Chapter 5, by comparing unknowns and equations we showed that, in general, we need an additional equation to have a complete description of any flowfield, whether or not it is incompressible. Our count showed that, *for gases*, we have only five equations for six unknowns. We are also short one equation *for liquids* because, although we have five equations and five unknowns, our equations yield no information about temperature.

An important distinction must be made at this point. By dealing strictly with inviscid, incompressible flows, we have been able to use Bernoulli's equation as an energy-conservation principle. However, in many practical flows, thermal energy cannot be ignored in the overall energy balance. Common examples are incompressible flows with heat transfer and compressible flows. Thus, we must consider conservation of energy, explicitly including thermal energy, to complete our analysis.

This chapter first provides a brief overview of thermodynamics. We then derive the integral conservation law for energy in a control volume. The differential form of the energy equation follows and is expressed in alternative forms. The differential form is restricted to inviscid flows with no heat transfer. This is necessary as including friction and heat transfer requires developing a description of these effects, a step we have not yet taken. Chapter 15 presents the complete viscous-flow version of the energy-conservation principle. In this chapter, we develop a simplified form of the energy equation that can be used in place of Bernoulli's equation when heat transfer and frictional losses are important.

The chapter concludes with a section on pipe flow and a section on open-channel flow. These topics are ideally suited for control volume analysis and illustrate the type of approach to fluid-mechanics problems characteristic of the nineteenth-century era of hydraulics.

7.1 Thermodynamics

Before we can formulate an equation for conservation of energy, we require some familiarity with a few basic concepts of **thermodynamics**. The science of thermodynamics deals with relations between heat and work. Its foundation consists of two general laws of nature, i.e., the first and second laws of thermodynamics. This section presents a brief overview of thermodynamics, focusing on those concepts that will be needed in the following sections. It is not intended as a thorough exposition of the topic. For a more complete introduction to thermodynamics, see an introductory text such as Lee and Sears (1963), Reynolds and Perkins (1977), Van Wylen and Sonntag (1986), Cengel and Boles (1994) or Wark (1966).

197

7.1.1 Fundamental Concepts

There are four fundamental concepts from thermodynamics central to formulation of an energy-conservation principle. The first concerns basic definitions, most of which we have already discussed in Section 4.5 and Chapter 6. The second is the concept of thermodynamic equilibrium. The third is the type of process, while the fourth is that of the equation of state and state variables.

- **Basic Definitions.** We have already used some of the nomenclature of thermodynamics in formulating the control volume method. The concepts of a system, a control volume, extensive properties and intensive properties that we introduced earlier all find their origins in classical thermodynamics.

  The primary focus in thermodynamics is usually on a *system*, which refers to "a definite quantity of matter bounded by some closed surface which is impervious to the flow of matter" [cf. Lee and Sears (1963)]. The surface can be either solid or a streamsurface, i.e., a surface parallel to streamlines. This is entirely consistent with the definition presented in Section 4.5, which, in the context of a fluid, is *the same collection of fluid particles for all time.*

  Just as in fluid mechanics, thermodynamicists find it convenient to introduce the concept of the control volume, which is sometimes referred to as an *open system*. The control volume is bounded by a control surface, across which matter may flow. There is no difference between thermodynamic control volumes and those analyzed and discussed in Chapter 6.

  Everything outside the boundary of a system is referred to as the *surroundings*. As we will see, distinguishing between a system (or control volume) and the surroundings is important in determining the transfer direction of quantities such as heat.

  Extensive and intensive properties are usually defined for a thermodynamic system as follows. We imagine that the system is divided into several parts. A property of the system whose value, for the entire system, is equal to the sum of its values for the separate parts of the system, is referred to as an extensive property. Extensive properties are a function of the quantity of matter (fluid) present, while intensive properties are not. Volume, mass and total energy are all examples of extensive properties. Properties that are independent of the size of the system (supposing them to be the same in all parts of the system) are known as intensive properties. Examples are temperature, pressure and density.

  As noted in Section 4.5, a property expressed per unit mass is called a *specific* property. Thus, if M is the total mass of a volume of fluid moving with constant velocity V, then $\frac{1}{2}MV^2$ is the total kinetic energy, an extensive property. By contrast, $\frac{1}{2}V^2$ is the specific kinetic energy.

- **Thermodynamic Equilibrium.** In general, when a system with nonuniform properties is isolated from interaction with its surroundings, those properties will generally change as time passes. If the pressure varies, for example, parts of the system may move, perhaps expanding or contracting. Ultimately, such motion stops, and when this occurs, we say the system has reached a state of *mechanical equilibrium*.

  In thermodynamics, we consider more properties than those associated with mechanical issues. If the temperature varies through the system, for example, we observe that it varies in a manner that may or may not be directly coupled with the mechanical motion. In fact, the temperature can vary on a time scale quite different from that of the mechanical time scale, and can even occur in the absence of all motion. When all changes in temperature cease, the system is in *thermal equilibrium*.

Finally, the most general system can contain substances that undergo chemical reactions. The reactions will occur on a third time scale dictated by the nature of the reactants. The chemical time scale is often much shorter than the mechanical and thermal time scales, but in any case is distinct from the former two. When all chemical-reaction activity ceases, a state of *chemical equilibrium* is reached.

A system that is simultaneously in mechanical, thermal and chemical equilibrium is said to be in *thermodynamic equilibrium*.

- **Processes.** The strict definition of thermodynamic equilibrium requires a totally static end state for a system. A natural question to pose is, of what value is the concept of thermodynamic equilibrium in the study of a fluid in motion? The answer is contained in the classical approach to thermodynamics which addresses systems changing in such a manner that the departure from equilibrium is infinitesimally small. That is, a system that changes in time is viewed as taking a succession of small steps from one equilibrium state to another. This approach is often described as a *quasi-static* or a *quasi-steady* solution. We used the quasi-steady approach to simplify the analysis of the leaky tank in Section 5.3 (see Figure 5.3).

Many processes are characterized by having some property of the system held constant during the process. We generally append the prefix "iso" to name the process. For example, a constant volume process is called *isovolumetric*, while a constant energy process is called *isoenergetic*. Somewhat more subtle variations sometimes occur in naming such processes. That is, a constant temperature process is called *isothermal*, while a process with constant pressure is referred to as *isobaric* (from the Greek word *baros*, which means heavy — an isobar on a weather map is a line of constant pressure).

We call a process *reversible* if, upon completion of the process, the initial state of *both the surroundings and the system* can be restored. This will be possible only if the process remains in thermodynamic equilibrium at all times, and has no losses due, for example, to friction. If the initial states of the system and its surroundings cannot be simultaneously returned to their initial state, the process is *irreversible*. We will discuss reversible and irreversible processes in more detail below in the context of entropy and the second law of thermodynamics.

Finally, when a process involves no transfer of heat between a system and its surroundings, we call the process *adiabatic*.

- **Equation of State and State Variables.** Experimentation has shown that, for a pure substance, there is a lower bound on the number of properties of the substance that can be given arbitrary values. That is, there is a distinct number of independent variables (properties) sufficient to define the thermodynamic state of a system. The functional form of the relationship amongst the properties is called the *equation of state*.

Even more significantly, when the properties of a system are expressed exclusively in terms of intensive properties, the equation of state of a substance can be expressed in terms of *just three* such properties. The perfect-gas law is an example — pressure, p, density, ρ, and temperature, T, are related according to $p = \rho R T$, where R is the perfect-gas constant for the gas under consideration. In general, we can write

$$F(p, \rho, T) = 0 \qquad (7.1)$$

where F is the appropriate functional relationship for the substance.[1] The interrelated properties appearing in the state equation are referred to as *state variables*, and are understood to correspond to a given equilibrium state. The practical implication of having such a relationship is the following: Equation (7.1) provides an implicit relationship defining any one of the state variables as a function of the other two, i.e., any two state variables define the state uniquely.

7.1.2 The First Law of Thermodynamics

The first law of thermodynamics is the basic energy-conservation principle relating all modes of energy transfer. In words, the first law says that the change in a system's energy, dE, equals the heat added to the system, δQ, minus the work done by the system on its surroundings, δW. Thus, we can say

$$dE = \delta Q - \delta W \qquad (7.2)$$

In writing this equation, we make a distinction between **exact** and **inexact differentials**. Because the energy is a state variable, it depends only on the initial and final states. Thus, a differential change is independent of the process, and we represent this as a perfect differential, dE. By contrast, both the heat added and the work done are very much dependent on the process. For example, the heat added might occur as a result of radiation or thermal conduction, which are quite different processes. Similarly, work might be done under isothermal conditions or perhaps at constant pressure, again quite different processes. Simply specifying the end states in the limits of an integral is insufficient as the integration path affects the final result. We distinguish this difference by denoting the differential changes as inexact differentials, δQ and δW.

A useful result that holds for a reversible process is as follows. If we let V denote volume, the work done in a reversible process (sometimes called **piston work**) is $\delta W = pdV$. Thus, we can rewrite the first law as

$$dE = \delta Q - pdV \qquad \text{(reversible process)} \qquad (7.3)$$

As a corollary result, we can also write the first law of thermodynamics in terms of specific variables. When we do this, we introduce the specific internal energy, e, heat added per unit mass, q, and specific volume, $v = 1/\rho$. The first law becomes:

$$de = \delta q - pd(1/\rho) \qquad \text{(First Law per unit mass)} \qquad (7.4)$$

7.1.3 The Second Law of Thermodynamics

The second law of thermodynamics makes a statement about the directionality of thermodynamic processes. To understand why the first law is insufficient as it stands, consider a simple experiment. Place an ice cube in contact with a hot object. Experience and common sense tells us the ice cube will heat up (and melt) while the object will have its temperature reduced. However, the first law does not prevent the reverse from occurring, so long as energy is conserved.

The second law of thermodynamics says, in words, that the **entropy** of a closed system and its surroundings must always increase. By definition, the entropy, s, is a state variable defined by

$$ds = \frac{\delta q_{rev}}{T} \qquad (7.5)$$

---

[1] Any three state variables can appear as variables in the equation of state — it is not restricted to p, ρ and T.

where δq_{rev} denotes an incremental amount of heat added reversibly to the system, and T is the system temperature. Note that δq_{rev} is not necessarily the actual heat added to the system. It is the effective heat that would have to be added reversibly to achieve the differential change in entropy from one equilibrium state to another according to Equation (7.5). If the actual heat added to the system is δq, we can rewrite this equation as

$$ds = \frac{\delta q}{T} + ds_{irrev} \tag{7.6}$$

where ds_{irrev} is the change in entropy due to irreversible processes such as viscous dissipation, heat conduction and mass diffusion within the system. Nature dictates that these processes always cause the entropy to increase, i.e., that

$$ds_{irrev} \geq 0 \tag{7.7}$$

We have equality in the special case of a reversible process. Since δq is positive by definition, the second law states that the **Clausius inequality** holds, viz.,

$$ds \geq \frac{\delta q}{T} \geq 0 \tag{7.8}$$

7.1.4 Combined First and Second Laws

One of the most useful equations of thermodynamics is known as **Gibbs' equation**. It is the result of combining the first and second laws, and permits a quantitative measure of the entropy of a system. To derive this equation, consider a reversible process. Then, the first law of thermodynamics is given by Equation (7.4) so that, per unit mass,

$$de = \delta q - pd(1/\rho) \tag{7.9}$$

Now, since the process is reversible, necessarily we have $ds_{irrev} = 0$ in Equation (7.6), wherefore

$$\delta q = T\,ds \tag{7.10}$$

Thus, combining Equations (7.9) and (7.10), we can rewrite the first law as

$$T\,ds = de + pd(1/\rho) \tag{7.11}$$

Equation (7.11) is the famous **Gibbs' equation**. Because it involves only state variables (and hence perfect differentials), it relates conditions between different equilibrium states, and is thus independent of the process. That is, Gibbs' equation holds for both reversible and irreversible processes.

As an example of how Gibbs' relation permits developing a quantitative expression for entropy, consider a perfect gas that is also calorically perfect. Because we have a perfect gas, the equation of state is $p = \rho R T$. By definition (see Section 1.5), a calorically-perfect gas has constant specific-heat coefficients, c_p and c_v, so that the internal energy and enthalpy are given by $e = c_v T$ and $h = c_p T$. Thus, from the definition of enthalpy,

$$h = e + \frac{p}{\rho} = c_v T + RT = (c_v + R)T \tag{7.12}$$

Now, since $h = c_p T$,

$$c_p T = (c_v + R)T \quad \Longrightarrow \quad c_p = c_v + R \tag{7.13}$$

Additionally, if we make use of the fact that the specific-heat ratio is defined by $\gamma = c_p/c_v$, then

$$\gamma = \frac{c_p}{c_v} = \frac{c_v + R}{c_v} \quad \Longrightarrow \quad c_v = \frac{R}{\gamma - 1}, \quad c_p = \frac{\gamma R}{\gamma - 1} \qquad (7.14)$$

So, for a calorically-perfect gas, $de = c_v dT$, wherefore

$$T ds = c_v dT - \frac{p}{\rho^2} d\rho \qquad (7.15)$$

The differential change in entropy is

$$ds = c_v \frac{dT}{T} - \frac{p}{\rho T} \frac{d\rho}{\rho} \qquad (7.16)$$

Then, using the perfect-gas law and noting from Equation (7.14) that the perfect-gas constant can be written as $R = (\gamma - 1)c_v$, we arrive at:

$$ds = c_v \frac{dT}{T} - R \frac{d\rho}{\rho} = c_v \frac{dT}{T} - (\gamma - 1)c_v \frac{d\rho}{\rho} \qquad (7.17)$$

Hence, integration yields

$$s = c_v \ln \left(\frac{T}{\rho^{\gamma-1}} \right) + \text{constant} \qquad (7.18)$$

But, we can rewrite the argument of the natural logarithm as follows.

$$\frac{T}{\rho^{\gamma-1}} = \frac{\rho T}{\rho^\gamma} = \frac{p}{R \rho^\gamma} \qquad (7.19)$$

Thus, absorbing R in the constant, the entropy for a perfect gas is given by

$$s = c_v \ln \left(\frac{p}{\rho^\gamma} \right) + \text{constant} \qquad (7.20)$$

Of particular interest is the special case of **isentropic** flow, i.e., flow with constant entropy. Clearly, the argument of the logarithm is constant so that

$$p = A\rho^\gamma, \quad A = \text{constant} \qquad (7.21)$$

7.1.5 The First Law for a Moving Fluid

The basis of energy conservation for a fluid is the first law of thermodynamics. As noted above, the first law in its most elementary form says that, for a system:

$$dE = \delta Q - \delta W \qquad (7.22)$$

where

$$
\begin{array}{rcl}
E & = & \text{Total energy of the system} \\
Q & = & \text{Heat transferred to the system} \\
W & = & \text{Work done by the system on the surroundings}
\end{array}
$$

Equivalently, assuming the flow remains close to thermodynamic equilibrium, we can regard the flow as being quasi-steady (in the thermodynamic sense) and rewrite Equation (7.22) as a differential equation, viz,

$$\frac{dE}{dt} = \dot{Q} - \dot{W} \qquad (7.23)$$

where $\dot{Q} \equiv \delta Q/dt$ and where $\dot{W} \equiv \delta W/dt$. In the following sections, we will evaluate the terms in Equation (7.23) as functions of properties in a control volume. This will provide the final equation needed to analyze general fluid-flow problems.

7.2 Integral Form of the Energy Equation

Consider a system in which the only body forces acting are conservative, i.e., represented by $\mathbf{f} = -\nabla \mathcal{V}$, where $\mathcal{V}$ is the force potential. We also assume the force potential is independent of time. For general fluid motion, the total energy consists of the sum of internal energy, kinetic energy, and potential energy. Then, the appropriate extensive/intensive variable pair is

$$B = E \quad \text{and} \quad \beta = e + \frac{1}{2}\mathbf{u} \cdot \mathbf{u} + \mathcal{V} \tag{7.24}$$

Alternatively, we could exclude potential energy and include the body force in the work term — we will discuss this alternative below. The three terms in the intensive variable, β, are internal, kinetic, and potential energy per unit mass, respectively. Thus, the Reynolds Transport Theorem tells us that

$$\frac{dE}{dt} = \iiint_V \frac{\partial}{\partial t}\left[\rho\left(e + \frac{1}{2}\mathbf{u} \cdot \mathbf{u} + \mathcal{V}\right)\right]dV + \iint_S \rho\left(e + \frac{1}{2}\mathbf{u} \cdot \mathbf{u} + \mathcal{V}\right)(\mathbf{u} \cdot \mathbf{n})\,dS \tag{7.25}$$

In general, the work done by a force, $\mathbf{F}$, in moving a distance $d\mathbf{r}$ is defined by the dot product of $\mathbf{F}$ and $d\mathbf{r}$, i.e.,

$$dW = \mathbf{F} \cdot d\mathbf{r} = \mathbf{F} \cdot \frac{d\mathbf{r}}{dt}dt = \mathbf{F} \cdot \mathbf{u}\,dt \tag{7.26}$$

Now, a system will do work on its surroundings through a surface force, $\mathbf{F}_s$, so that

$$\dot{W} = \iint_S \mathbf{F}_s \cdot \mathbf{u}\,dS \tag{7.27}$$

The surface force exerted by a system on its surroundings is minus the force exerted by the surroundings on the system. With no loss of generality, we can say for a viscous fluid that the surface force acting on a system is $-p\,\mathbf{n} + \mathbf{f}_v$, where the vector $\mathbf{f}_v$ denotes the surface force due to viscosity (see Figure 7.1). Although we defer formulation of the viscous force to Chapter 12, we can proceed with the development of the energy equation using this symbolic representation of the friction force. So, the surface force exerted by the system on its surroundings is

$$\mathbf{F}_s = p\,\mathbf{n} - \mathbf{f}_v \tag{7.28}$$

Thus, the rate at which the system does work on its surroundings becomes

$$\dot{W} = \underbrace{\iint_S p\,(\mathbf{u} \cdot \mathbf{n})\,dS}_{Pressure\ Work} - \underbrace{\iint_S \mathbf{f}_v \cdot \mathbf{u}\,dS}_{Friction\ Work} \tag{7.29}$$

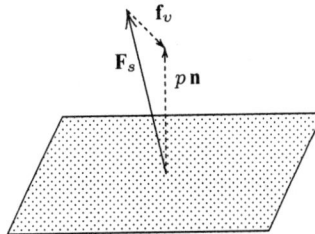

Figure 7.1: *Pressure and viscous forces on a surface element; the viscous force, $\mathbf{f}_s$, can have both normal and tangential components.*

Substituting Equations (7.25) and (7.29) into the first law of thermodynamics [Equation (7.23)], we arrive at the following.

$$\iiint_V \frac{\partial}{\partial t}\left[\rho\left(e + \frac{1}{2}\mathbf{u}\cdot\mathbf{u} + \mathcal{V}\right)\right]dV + \oiint_S \rho\left(e + \frac{1}{2}\mathbf{u}\cdot\mathbf{u} + \mathcal{V}\right)(\mathbf{u}\cdot\mathbf{n})\,dS$$
$$= \dot{Q} - \oiint_S p\,(\mathbf{u}\cdot\mathbf{n})\,dS + \oiint_S \mathbf{f}_v\cdot\mathbf{u}\,dS \qquad (7.30)$$

Note that since we have not specified the type of heat transfer, we represent the net rate of heat transfer by the original symbol, $\dot{Q}$.

We can move the pressure work integral from the right-hand to the left-hand side of this equation by first observing that

$$\oiint_S p\,(\mathbf{u}\cdot\mathbf{n})\,dS = \oiint_S \rho\left(\frac{p}{\rho}\right)(\mathbf{u}\cdot\mathbf{n})\,dS \qquad (7.31)$$

We then combine the pressure-work integral with the energy-flux integral [i.e., the closed surface integral on the left-hand side of Equation (7.30)]. To help accomplish this end, recall that the enthalpy, h, is defined by

$$h = e + p/\rho \qquad (7.32)$$

This yields a slightly more compact energy-conservation equation, viz.,

$$\iiint_V \frac{\partial}{\partial t}\left[\rho\left(e + \frac{1}{2}\mathbf{u}\cdot\mathbf{u} + \mathcal{V}\right)\right]dV + \oiint_S \rho\left(h + \frac{1}{2}\mathbf{u}\cdot\mathbf{u} + \mathcal{V}\right)(\mathbf{u}\cdot\mathbf{n})\,dS$$
$$= \dot{Q} + \oiint_S \mathbf{f}_v\cdot\mathbf{u}\,dS \qquad (7.33)$$

In words, this equation says that *the instantaneous rate of change of total energy in the control volume plus the net flux of total enthalpy equals the sum of the heat transfer rate to the control volume and the friction work.* For purposes of using the integral form, all of the applications in this chapter involve the energy equation in the following form.

$$\iiint_V \frac{\partial}{\partial t}\left[\rho\left(e + \frac{1}{2}\mathbf{u}\cdot\mathbf{u} + \mathcal{V}\right)\right]dV + \oiint_S \rho\left(h + \frac{1}{2}\mathbf{u}\cdot\mathbf{u} + \mathcal{V}\right)(\mathbf{u}\cdot\mathbf{n})\,dS$$
$$= \dot{Q} - \dot{W}_s \qquad (\mathbf{f} = -\nabla\mathcal{V}) \qquad (7.34)$$

where we define $\dot{W}_s$ as

$$\dot{W}_s = -\oiint_S \mathbf{f}_v\cdot\mathbf{u}\,dS \qquad (7.35)$$

Either the quantities $\dot{Q}$ and $\dot{W}_s$ are given or sufficient information is provided to permit evaluation of the two integrals, with $\dot{Q}$ and $\dot{W}_s$ being part of the solution. The negative of the friction work, $\dot{W}_s$, is generally called the **shaft work**. It represents the useful work that is done by the fluid-mechanical device contained in the control volume. The name, shaft work, originates from devices, such as a turbine or a pump, that operate with a shaft.[2]

---

[2]Strictly speaking, shaft work is the integral over the area enclosing the shaft. All other contributions to the integral are sometimes called shear work, which is non-zero only if the surface velocity is non-zero. Shear work is rarely significant compared to shaft work, so that the entire closed-surface integral is very nearly equal to the shaft work.

As a final comment, if a nonconservative body force is acting, we cannot define a force potential, $\mathcal{V}$. To develop the energy-conservation principle in this case, we must exclude potential energy from B and β in Equation (7.24). That is, we replace Equation (7.24) by

$$B = E \quad \text{and} \quad \beta = e + \frac{1}{2}\mathbf{u} \cdot \mathbf{u} \qquad (\mathbf{f} \neq -\nabla \mathcal{V}) \qquad (7.36)$$

Then, we must develop an expression for the work done by the body force in a manner similar to the way we handled the surface forces. The work done, per unit volume, by the body force is $\rho \mathbf{f} \cdot \mathbf{u}$. Since a body force is exerted by the surroundings on the control volume, the work done by the control volume on the surroundings, per unit volume, is $-\rho \mathbf{f} \cdot \mathbf{u}$. Hence, the rate at which the body force effectively does work on the surroundings is $d\dot{W} = -\rho \mathbf{f} \cdot \mathbf{u} \, dV$. Consequently, the work term defined in Equation (7.27) must be replaced by

$$\dot{W} = \oiint_S \mathbf{F}_s \cdot \mathbf{u} \, dS - \iiint_V \rho \mathbf{f} \cdot \mathbf{u} \, dV \qquad (\mathbf{f} \neq -\nabla \mathcal{V}) \qquad (7.37)$$

Aside from these two adjustments, all other steps in the derivation are the same as above. Thus, the resulting energy-conservation principle for a general body force is

$$\iiint_V \frac{\partial}{\partial t}\left[\rho\left(e + \frac{1}{2}\mathbf{u} \cdot \mathbf{u}\right)\right] dV + \oiint_S \rho\left(h + \frac{1}{2}\mathbf{u} \cdot \mathbf{u}\right)(\mathbf{u} \cdot \mathbf{n}) \, dS$$

$$= \dot{Q} - \dot{W}_s + \iiint_V \rho \mathbf{f} \cdot \mathbf{u} \, dV \qquad (\mathbf{f} \neq -\nabla \mathcal{V}) \qquad (7.38)$$

As an example of a straightforward application of the energy-conservation principle, consider a simple turbine as illustrated in Figure 7.2. We assume the flow is steady, and that the mass flux through the turbine is $\dot{m}$. The velocity and enthalpy are as indicated at inlet and outlet, and the heat transfer, $\dot{Q}$, is given. Properties at the inlet are denoted by subscript 1 and those at the outlet by subscript 2. Both are assumed uniform over the cross section. Ignoring effects of body forces, we want to compute the shaft work as a function of the given flow properties, including the heat transfer rate from the surroundings to the turbine, $\dot{Q}$. We begin by selecting a control volume whose surface (indicated by the dashed lines) coincides with the turbine walls as shown in the figure.

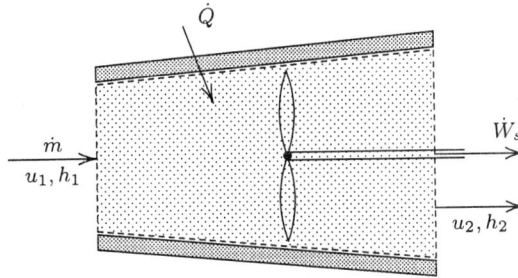

Figure 7.2: *A simple turbine.*

Turning first to mass conservation, the fact that the flow is steady tells us the net mass flux out of the control volume is zero, i.e.,

$$\oiint_S \rho \mathbf{u} \cdot \mathbf{n} \, dS = 0 \qquad (7.39)$$

Thus, denoting inlet and outlet areas by A_1 and A_2, respectively, we have

$$-\rho_1 u_1 A_1 + \rho_2 u_2 A_2 = 0 \tag{7.40}$$

Therefore, the mass flux is

$$\dot{m} = \rho_1 u_1 A_1 = \rho_2 u_2 A_2 \tag{7.41}$$

We learn nothing from momentum conservation because, for each component introduced, we introduce an unknown reaction force. Thus, we turn to the energy equation. Because the flow is steady and we are ignoring effects of body forces for simplicity, we have

$$\oiint_S \rho \left(h + \frac{1}{2} \mathbf{u} \cdot \mathbf{u} \right) (\mathbf{u} \cdot \mathbf{n}) \, dS = \dot{Q} - \dot{W}_s \tag{7.42}$$

wherefore the shaft work is given by

$$\begin{aligned}
\dot{W}_s &= \dot{Q} - \oiint_S \rho \left(h + \frac{1}{2} \mathbf{u} \cdot \mathbf{u} \right) (\mathbf{u} \cdot \mathbf{n}) \, dS \\
&= \dot{Q} - \left[\rho_1 \left(h_1 + \frac{1}{2} u_1^2 \right)(-u_1 A_1) + \rho_2 \left(h_2 + \frac{1}{2} u_2^2 \right)(u_2 A_2) \right] \\
&= \dot{Q} + \rho_1 u_1 A_1 \left(h_1 + \frac{1}{2} u_1^2 \right) - \rho_2 u_2 A_2 \left(h_2 + \frac{1}{2} u_2^2 \right)
\end{aligned} \tag{7.43}$$

Finally, using the result obtained from mass conservation, we can rewrite the expression for $\dot{W}_s$ in terms of $\dot{m}$, viz.,

$$\dot{W}_s = \dot{Q} + \dot{m} \left[(h_1 - h_2) + \frac{1}{2} \left(u_1^2 - u_2^2 \right) \right] \tag{7.44}$$

7.3 Differential Form for Adiabatic, Inviscid Flow

Because details of the rate of heat transfer, $\dot{Q}$, and the shaft work, $\dot{W}_s$, have not been specified, it is impossible at this point to write them as either a surface or a volume integral. We thus confine our attention to a differential form of the energy conservation equation in the special case of adiabatic flow (for which $\dot{Q} = 0$) of a frictionless fluid (for which $\dot{W}_s = 0$). To simplify our analysis, it is convenient to define the **specific total energy**, $\mathcal{E}$, and **specific total enthalpy**, $\mathcal{H}$, as follows.

$$\mathcal{E} = e + \frac{1}{2} \mathbf{u} \cdot \mathbf{u} + \mathcal{V} \quad \text{and} \quad \mathcal{H} = h + \frac{1}{2} \mathbf{u} \cdot \mathbf{u} + \mathcal{V} \tag{7.45}$$

In terms of these variables, the integral energy-conservation principle for an adiabatic, inviscid flow becomes

$$\iiint_V \frac{\partial}{\partial t} (\rho \mathcal{E}) \, dV + \oiint_S \rho \mathcal{H} (\mathbf{u} \cdot \mathbf{n}) \, dS = 0 \tag{7.46}$$

Next, we make use of the divergence theorem to convert the surface integral to a volume integral. There follows:

$$\oiint_S \rho \mathcal{H} (\mathbf{u} \cdot \mathbf{n}) \, dS = \iiint_V \nabla \cdot (\rho \mathbf{u} \mathcal{H}) \, dV \tag{7.47}$$

Thus, we can group the two terms in the energy equation under a single volume integral:

$$\iiint_V \left[\frac{\partial}{\partial t} (\rho \mathcal{E}) + \nabla \cdot (\rho \mathbf{u} \mathcal{H}) \right] dV = 0 \qquad (7.48)$$

wherefore the conservation form of the differential equation governing energy conservation is

$$\frac{\partial}{\partial t} (\rho \mathcal{E}) + \nabla \cdot (\rho \, \mathbf{u} \mathcal{H}) = 0 \qquad (7.49)$$

It is interesting to recast this equation in primitive-variable form as we have done for the mass- and momentum-conservation principles. To do this, we first note that total enthalpy and total energy are related by

$$\mathcal{H} = \mathcal{E} + \frac{p}{\rho} \qquad \Longrightarrow \qquad \rho \mathcal{H} = \rho \mathcal{E} + p \qquad (7.50)$$

So,

$$\frac{\partial}{\partial t} (\rho \mathcal{E}) = \frac{\partial}{\partial t} (\rho \mathcal{H}) - \frac{\partial p}{\partial t} \qquad (7.51)$$

By combining Equations (7.49) and (7.51), the energy equation becomes

$$\frac{\partial}{\partial t} (\rho \mathcal{H}) + \nabla \cdot (\rho \, \mathbf{u} \mathcal{H}) = \frac{\partial p}{\partial t} \qquad (7.52)$$

Using the chain rule to expand terms on the left-hand side, and regrouping yields

$$\rho \left[\frac{\partial \mathcal{H}}{\partial t} + \mathbf{u} \cdot \nabla \mathcal{H} \right] + \mathcal{H} \left[\frac{\partial \rho}{\partial t} + \nabla \cdot (\rho \, \mathbf{u}) \right] = \frac{\partial p}{\partial t} \qquad (7.53)$$

The first term in brackets is the Eulerian derivative of the total enthalpy, $d\mathcal{H}/dt$, while the second term in brackets vanishes by virtue of mass conservation [Equation (5.7)]. Therefore, the primitive-variable form of the energy equation is:

$$\rho \frac{d\mathcal{H}}{dt} = \frac{\partial p}{\partial t} \qquad (7.54)$$

where the total enthalpy, $\mathcal{H}$, is defined in Equation (7.45). We can make an immediate observation that follows from Equation (7.54). On the one hand, if a flow is steady, the total enthalpy will remain constant along streamlines. On the other hand, if the flow is unsteady, the $\partial p / \partial t$ term causes the total enthalpy of a fluid particle to vary as it moves through the flowfield. That is, because kinetic energy is not Galilean invariant, neither is $\mathcal{H}$.

7.4 Entropy Generation

As noted earlier in this chapter (Subsection 7.1.4), we can appeal to the combined first and second laws of thermodynamics, i.e., to Gibbs' equation, to determine the entropy for a given system or control volume. It is instructive to develop the equation for entropy in an adiabatic, inviscid medium, subject to conservative body forces implied by the energy conservation law developed in the preceding section. First, note that

$$
\begin{aligned}
T ds &= de + p \, d(1/\rho) \\
&= d(e + p/\rho) - \frac{dp}{\rho} \\
&= dh - \frac{dp}{\rho} \qquad (7.55)
\end{aligned}
$$

Hence, the entropy must satisfy the following differential equation.

$$\rho T \frac{ds}{dt} = \rho \frac{dh}{dt} - \frac{dp}{dt} \tag{7.56}$$

Now, the momentum equation for an inviscid fluid is simply Euler's equation, which is

$$\rho \frac{d\mathbf{u}}{dt} = -\nabla p + \rho \mathbf{f} = -\nabla p - \rho \nabla \mathcal{V} \tag{7.57}$$

where we note that the body force is assumed to be conservative. Taking the dot product of Euler's equation with the velocity and noting from the chain rule that $\mathbf{u} \cdot d\mathbf{u}/dt = d(\frac{1}{2}\mathbf{u} \cdot \mathbf{u})/dt$, we have

$$\rho \frac{d}{dt} \left(\frac{1}{2} \mathbf{u} \cdot \mathbf{u} \right) = -\mathbf{u} \cdot \nabla p - \rho \, \mathbf{u} \cdot \nabla \mathcal{V} \tag{7.58}$$

Hence, since we have also assumed that the body force (and thus its potential function, $\mathcal{V}$) is independent of time, then

$$\frac{\partial \mathcal{V}}{\partial t} = 0 \quad \Longrightarrow \quad \mathbf{u} \cdot \nabla \mathcal{V} = \frac{d\mathcal{V}}{dt} \tag{7.59}$$

wherefore Equation (7.58) can be rearranged to read

$$0 = \rho \frac{d}{dt} \left(\frac{1}{2} \mathbf{u} \cdot \mathbf{u} + \mathcal{V} \right) + \mathbf{u} \cdot \nabla p \tag{7.60}$$

This equation represents conservation of mechanical energy. Now, adding the respective sides of Equations (7.56) and (7.60) yields

$$\rho T \frac{ds}{dt} = \underbrace{\rho \frac{d}{dt} \left(h + \frac{1}{2} \mathbf{u} \cdot \mathbf{u} + \mathcal{V} \right)}_{\rho \, d\mathcal{H}/dt} - \underbrace{\left(\frac{dp}{dt} - \mathbf{u} \cdot \nabla p \right)}_{\partial p/\partial t} \tag{7.61}$$

Finally, reference to the differential form of the energy-conservation law [Equation (7.54)] tells us the right-hand side of Equation (7.61) vanishes. Thus, we arrive at the important result that for inviscid, adiabatic flow with conservative body forces, the rate of change of entropy following a fluid particle is zero:

$$\rho T \frac{ds}{dt} = 0 \tag{7.62}$$

Consistent with our naming conventions discussed in Subsection 7.1.1, we refer to a constant-entropy flow (or process) as **isentropic**.

7.5   Relation to Bernoulli's Equation

We can use the results of the two preceding sections to illustrate the connection between exact energy conservation and Bernoulli's equation. Recall that we derived the equation in Section 5.3 by manipulating the momentum equation. Hence, it represents conservation of mechanical energy, and excludes thermal effects. We escaped the need to consider any but mechanical forms of energy with the postulate that the flow is incompressible. In this section, we will see exactly what the limits of Bernoulli's equation are regarding effects of compressibility.

For *steady flow* of a perfect fluid with no heat transfer, Equation (7.54) simplifies to

$$\frac{d\mathcal{H}}{dt} = 0 \tag{7.63}$$

where we note that $d/dt = \mathbf{u} \cdot \nabla$ for steady flow. Therefore, under these conditions, we conclude that

$$h + \frac{1}{2}\mathbf{u} \cdot \mathbf{u} + \mathcal{V} = \text{constant on a streamline} \tag{7.64}$$

We call this condition **particle isenthalpic**.

In a liquid, ρ is constant. Entropy is constant in the absence of heat transfer and viscous effects, and we infer from Equation (7.55) that

$$dh = dp/\rho \quad \Longrightarrow \quad h = p/\rho + \text{constant} \quad \text{(for isentropic flow of a liquid)} \tag{7.65}$$

Therefore the energy-conservation principle for a steady, adiabatic, inviscid flow simplifies to Bernoulli's equation, viz.,

$$\frac{p}{\rho} + \frac{1}{2}\mathbf{u} \cdot \mathbf{u} + \mathcal{V} = \text{constant} \tag{7.66}$$

In a gas, we cannot neglect density changes. Thus, demonstrating that the energy-conservation principle is consistent with Bernoulli's equation under incompressible, or nearly incompressible, conditions is a bit more subtle. We begin by noting that, if we select a reference point at infinity (i.e., very far from the point of interest), then the energy-conservation principle tells us

$$h + \frac{1}{2}\mathbf{u} \cdot \mathbf{u} + \mathcal{V} = h_\infty + \frac{1}{2}U_\infty^2 + \mathcal{V}_\infty \tag{7.67}$$

Now, noting that adiabatic, inviscid flows with conservative body forces are isentropic, one thermodynamic state variable (entropy) is fixed. Thus, all other thermodynamic state variables can be expressed as a function of a single state variable. So, if we regard enthalpy as a function of pressure, we can expand in Taylor series according to

$$h = h_\infty + \left(\frac{dh}{dp}\right)_\infty (p - p_\infty) + \frac{1}{2}\left(\frac{d^2h}{dp^2}\right)_\infty (p - p_\infty)^2 + \cdots \tag{7.68}$$

For a perfect, calorically-perfect gas in an isentropic flow, we know from Equation (7.21) that

$$p = A\rho^\gamma, \quad A = \text{constant} \quad \Longrightarrow \quad \rho = (p/A)^{1/\gamma} \tag{7.69}$$

Also, the enthalpy can be written as

$$h = c_p T = \frac{c_p}{R}\frac{p}{\rho} = \frac{\gamma}{\gamma - 1}\frac{p}{\rho} \tag{7.70}$$

where we make use of Equation (7.14) to replace the ratio of c_p to R by the factor $\gamma/(\gamma - 1)$. Then, combining Equations (7.69) and (7.70), the enthalpy for a perfect, calorically-perfect gas in an isentropic flow is given by

$$h = \frac{\gamma}{\gamma - 1}A^{1/\gamma}p^{(\gamma-1)/\gamma} \tag{7.71}$$

We can differentiate this equation to show that

$$\left(\frac{dh}{dp}\right)_\infty = \frac{1}{\rho_\infty}, \qquad \left(\frac{d^2h}{dp^2}\right)_\infty = -\frac{1}{\rho_\infty^2 a_\infty^2} \tag{7.72}$$

where a_∞ is the speed of sound.[3] Therefore, the enthalpy is given by the Taylor series:

$$
\begin{aligned}
h &= h_\infty + \frac{(p - p_\infty)}{\rho_\infty} - \frac{1}{2}\frac{(p - p_\infty)^2}{\rho_\infty^2 a_\infty^2} + \cdots \\
&= h_\infty + \frac{(p - p_\infty)}{\rho_\infty} - \frac{1}{2}U_\infty^2 M_\infty^2 \left(\frac{p - p_\infty}{\rho_\infty U_\infty^2}\right)^2 + \cdots
\end{aligned}
\tag{7.73}
$$

where $M_\infty = U_\infty / a_\infty$ is the Mach number. Hence, the energy-conservation Equation (7.67) can be approximated by

$$
h_\infty + \frac{(p - p_\infty)}{\rho_\infty} - \frac{1}{2}U_\infty^2 M_\infty^2 \left(\frac{p - p_\infty}{\rho_\infty U_\infty^2}\right)^2 + \frac{1}{2}\mathbf{u}\cdot\mathbf{u} + \mathcal{V} \approx h_\infty + \frac{1}{2}U_\infty^2 + \mathcal{V}_\infty
\tag{7.74}
$$

Canceling h_∞ and regrouping terms, we have

$$
\frac{p}{\rho_\infty} + \frac{1}{2}\mathbf{u}\cdot\mathbf{u} + \mathcal{V} \approx \frac{p_\infty}{\rho_\infty} + \frac{1}{2}U_\infty^2 \left[1 + M_\infty^2 \left(\frac{p - p_\infty}{\rho_\infty U_\infty^2}\right)^2\right] + \mathcal{V}_\infty
\tag{7.75}
$$

Thus, in the limit of small Mach number, we can say (with ρ_∞ replaced by ρ)

$$
\frac{p}{\rho} + \frac{1}{2}\mathbf{u}\cdot\mathbf{u} + \mathcal{V} \approx \frac{p_\infty}{\rho} + \frac{1}{2}U_\infty^2 + \mathcal{V}_\infty
\tag{7.76}
$$

which is again Bernoulli's equation. Therefore, as with a liquid, we see that the limiting form of the exact energy conservation law (as Mach number approaches zero) is Bernoulli's equation. This also suggests that the proper definition of incompressible flow is the limiting case $M_\infty \to 0$. We will take this matter up in detail in Chapter 8.

7.6   Approximate Form of the Energy Equation

One important application that requires use of the energy-conservation principle, including heat transfer and viscous work terms, is flow in pipes. Delivery of gas and oil from an oil well to a refinery, for example, requires pipes that can be hundreds of miles long. Viscous losses are nontrivial for flow through such pipe systems, especially at tees or corners and at junctions between pipes of differing areas. In this section, we develop an approximate energy conservation equation that accounts for viscous and thermal losses and that can be used as a replacement for Bernoulli's equation in such flows.

We confine our attention to steady, incompressible flow and assume that the body force is gravity so that $\mathcal{V} = gz$, where g is the acceleration due to gravity and z is vertical distance. For steady flow, the integral form of the energy-conservation principle simplifies to

$$
\oiint_S \rho\left(h + \frac{1}{2}\mathbf{u}\cdot\mathbf{u} + gz\right)(\mathbf{u}\cdot\mathbf{n})\,dS = \dot{Q} - \dot{W}_s
\tag{7.77}
$$

Considering a control volume that is coincident with the pipe walls, the only boundaries of the control volume across which fluid passes are the inlet and outlet. Thus, referring to Figure 7.3,

---

[3]We have made use of the fact that, as we will show in Section 8.3, the speed of sound for a perfect gas is given by $a = \sqrt{\gamma p/\rho}$.

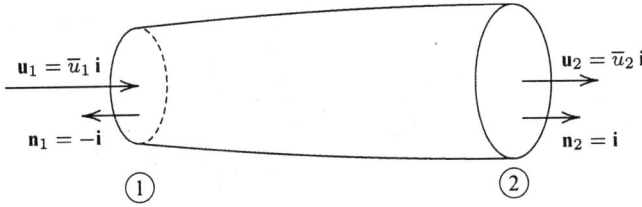

Figure 7.3: *Incompressible pipe flow.*

the net flux of total enthalpy is

$$\oiint_S \rho \left(h + \frac{1}{2} \mathbf{u} \cdot \mathbf{u} + gz \right) (\mathbf{u} \cdot \mathbf{n}) \, dS = - \iint_{A_1} \rho \left(h_1 + \frac{1}{2} u_1^2 + gz_1 \right) u_1 \, dA$$
$$+ \iint_{A_2} \rho \left(h_2 + \frac{1}{2} u_2^2 + gz_2 \right) u_2 \, dA \qquad (7.78)$$

Note that u_1 and u_2 are the speeds at the inlet and outlet, respectively. In general, they vary over the cross sections. At this point, we have made no approximation to the exact conservation principle. To make further headway, we assume that, on each cross section of the pipe, the enthalpy and elevation are nearly constant across the section so that

$$\iint_{A_i} \rho \left(h_i + gz_i \right) u_i \, dA \approx \rho \left(h_i + gz_i \right) \iint_{A_i} u_i \, dA$$
$$= \left(h_i + gz_i \right) \rho \overline{u}_i A_i = \left(h_i + gz_i \right) \dot{m} \qquad (7.79)$$

where $\overline{u}_i$ is the average velocity on cross section i defined by

$$\overline{u}_i = \frac{1}{A_i} \iint_{A_i} u_i \, dA \qquad (7.80)$$

We exercise a bit of caution in computing the kinetic-energy flux. That is, as discussed in Subsection 6.3.5, we must take account of the fact that the average of u_i^3 is not equal to the cube of the average of u_i (the former is always larger). To account for the difference, we introduce a factor α_i, i.e.,

$$\iint_{A_i} \rho \left(\frac{1}{2} u_i^2 \right) u_i \, dA = \frac{1}{2} \rho \iint_{A_i} u_i^3 \, dA = \frac{1}{2} \rho \overline{u}_i^3 \iint_{A_i} \left(\frac{u_i}{\overline{u}_i} \right)^3 dA \qquad (7.81)$$

where $\overline{u}_i$ is again the average value of u_i. We now define the **kinetic-energy correction factor**, α_i, as

$$\alpha_i \equiv \frac{1}{A_i} \iint_{A_i} \left(\frac{u_i}{\overline{u}_i} \right)^3 dA \qquad (7.82)$$

Therefore, the kinetic-energy flux integral is

$$\iint_{A_i} \rho \left(\frac{1}{2} u_i^2 \right) u_i \, dA = \frac{1}{2} \rho \alpha_i \overline{u}_i^3 A_i = \frac{1}{2} \alpha_i \overline{u}_i^2 \dot{m} \qquad (7.83)$$

So, substituting these averaged values into the exact energy-conservation equation yields:

$$\dot{Q} - \dot{W}_s = - \left(h_1 + \frac{1}{2} \alpha_1 \overline{u}_1^2 + gz_1 \right) \dot{m} + \left(h_2 + \frac{1}{2} \alpha_2 \overline{u}_2^2 + gz_2 \right) \dot{m} \qquad (7.84)$$

or,

$$\frac{\dot{Q} - \dot{W}_s}{\dot{m}g} + \frac{h_1}{g} + \alpha_1 \frac{\overline{u}_1^2}{2g} + z_1 = \frac{h_2}{g} + \alpha_2 \frac{\overline{u}_2^2}{2g} + z_2 \qquad (7.85)$$

Next, we rewrite the shaft work, $\dot{W}_s$, as the difference between the power given up to a device such as a turbine, $\dot{W}_t$, and the power supplied by a device such as a pump, $\dot{W}_p$, i.e.,

$$\dot{W}_s = \dot{W}_t - \dot{W}_p \qquad (7.86)$$

The idea here is that we will apply this approximate equation [cf. Equation (7.79) for one of the approximations] to pipe systems including components that add and/or remove energy. A pump adds energy to the system, while a turbine extracts energy from the system. Then, rewriting the enthalpy as $h = e + p/\rho$, the equation for energy conservation becomes

$$\frac{\dot{W}_p}{\dot{m}g} + \frac{p_1}{\rho g} + \alpha_1 \frac{\overline{u}_1^2}{2g} + z_1 = \frac{\dot{W}_t}{\dot{m}g} + \frac{p_2}{\rho g} + \alpha_2 \frac{\overline{u}_2^2}{2g} + z_2 + \left[\frac{(e_2 - e_1)}{g} - \frac{\dot{Q}}{\dot{m}g} \right] \qquad (7.87)$$

Finally, we define the **head supplied by a pump**, h_p, the **head given up to a turbine**, h_t, and the **head loss**, h_L, by the following equations.

$$h_p \equiv \frac{\dot{W}_p}{\dot{m}g}, \quad h_t \equiv \frac{\dot{W}_t}{\dot{m}g}, \quad h_L \equiv \left[\frac{(e_2 - e_1)}{g} - \frac{\dot{Q}}{\dot{m}g} \right] \qquad (7.88)$$

Clearly, all terms in Equations (7.87) and (7.88) have dimensions of length.

There is always a head loss in a viscous fluid. This is a process in which mechanical energy is converted to thermal energy through heat transfer and dissipative processes. It appears as a loss in the overall energy balance. In terms of these parameters, the final form of our approximate equation for incompressible pipe flow is:

$$\frac{p_1}{\rho g} + \alpha_1 \frac{\overline{u}_1^2}{2g} + z_1 + h_p = \frac{p_2}{\rho g} + \alpha_2 \frac{\overline{u}_2^2}{2g} + z_2 + h_t + h_L \qquad (7.89)$$

This equation can be used as a replacement for Bernoulli's equation when viscous and heat transfer effects are important. Keep in mind that this relation is approximate, and is intended mainly for application to steady, incompressible flow through pipe systems.

For example, consider a pipe system that pumps water from one elevation to another as shown in Figure 7.4. We want to know how powerful a pump is needed to have an outlet velocity that is twice the inlet velocity. We first determine the head supplied by the pump, h_p, by using Equation (7.89). Ignoring any head loss in the pipe system, we thus have:

$$\frac{p_1}{\rho g} + \alpha_1 \frac{\overline{u}_1^2}{2g} + z_1 + h_p = \frac{p_2}{\rho g} + \alpha_2 \frac{\overline{u}_2^2}{2g} + z_2 \qquad (7.90)$$

Suppose further that the pressure at is the same at inlet and outlet so that $p_1 = p_2$. Suppose also that the kinetic-energy correction factors are equal and given by $\alpha_1 = \alpha_2 = 1.06$. Then,

$$0.53 \frac{U^2}{g} + h_p = 0.53 \frac{(2U)^2}{g} + \Delta z \qquad (7.91)$$

Therefore, the head supplied by the pump must be

$$h_p = 1.59 \frac{U^2}{g} + \Delta z \qquad (7.92)$$

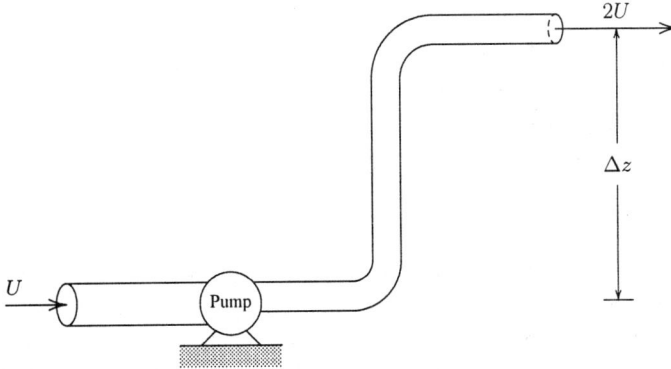

Figure 7.4: *Pipe system with a pump.*

To compute the power supplied by the pump, we appeal to the definition of head [Equation (7.88)], to conclude that

$$\dot{W}_p = \dot{m}gh_p = 1.59\dot{m}U^2 + \dot{m}g\Delta z \tag{7.93}$$

Suppose, for example, we are given that the flow rate is $\dot{m} = 0.5$ kg/sec, the inlet velocity is $U = 30$ m/sec, and the change in elevation is $\Delta z = 100$ m. Then, the power required of the pump is

$$\begin{aligned} \dot{W}_p &= (1.59)(.5)(900) + (.5)(9.807)(100) \text{ kg} \cdot \text{m}^2/\text{sec}^3 \\ &= 1206 \text{ Joules/sec} = 1.206 \text{ kWatts} \end{aligned} \tag{7.94}$$

It is worthwhile at this point to pause and discuss the kinetic-energy correction factor, α, including an example of how it can be computed. The first point is that, in practical pipe-flow applications, its value is a bit larger than unity — typically 1.05. Thus, you might be tempted to ignore it and simply approximate $\alpha = 1$. Some of the homework problems include this approximation for the sake of algebraic simplicity. However, you should be aware that this is inconsistent with the spirit of Equation (7.89), and should not be done for practical applications. For example, the head loss, h_L, is often a small fraction of the kinetic energy — sometimes less than 10%. Thus, if you approximate $\alpha \approx 1$, then all head losses should be ignored as well. This would, of course, defeat much of the purpose of using Equation (7.89) as a replacement for Bernoulli's equation.

Turning to computation of α, consider the laminar pipe-flow velocity profile discussed in Subsection 1.9.2. We found that the velocity is

$$u(r) = u_m \left(1 - r^2/R^2\right) \tag{7.95}$$

where u_m is the maximum velocity, r is distance from the centerline and R is the pipe radius. We are, of course, assuming the pipe has circular cross section. First, we must compute the average velocity, $\bar{u}$. Because the pipe has circular cross section, it is most convenient to work in cylindrical coordinates. Thus, since the cross-sectional area is $A = \pi R^2$ and the differential area is $dA = r\,dr\,d\theta$, we have

$$\begin{aligned}
\bar{u} &= \frac{1}{\pi R^2} \int_0^{2\pi} \int_0^R u(r)\, r\, dr\, d\theta \\
&= \frac{2\pi u_m}{\pi R^2} \int_0^R \left(1 - \frac{r^2}{R^2}\right) r\, dr \\
&= 2u_m \int_0^1 \left(1 - \frac{r^2}{R^2}\right) \frac{r}{R}\, d\left(\frac{r}{R}\right) \\
&= 2u_m \int_0^1 \left(1 - \xi^2\right) \xi\, d\xi = 2u_m \int_0^1 \left(\xi - \xi^3\right) d\xi \qquad (\xi \equiv r/R) \\
&= 2u_m \left(\frac{1}{2}\xi^2 - \frac{1}{4}\xi^4\right)\Bigg|_{\xi=0}^{\xi=1} = \frac{1}{2}u_m \qquad\qquad (7.96)
\end{aligned}$$

Then, from the definition of α given in Equation (7.82),

$$\begin{aligned}
\alpha &= \frac{1}{\pi R^2} \int_0^{2\pi} \int_0^R \left[\frac{u_m\left(1 - r^2/R^2\right)}{u_m/2}\right]^3 r\, dr\, d\theta \\
&= \frac{8 \cdot (2\pi)}{\pi R^2} \int_0^R \left(1 - \frac{r^2}{R^2}\right)^3 r\, dr \\
&= 16 \int_0^1 \left(1 - \frac{r^2}{R^2}\right)^3 \frac{r}{R}\, d\left(\frac{r}{R}\right) \\
&= 16 \int_0^1 \left(1 - \xi^2\right)^3 \xi\, d\xi = 16 \int_0^1 \left(1 - 3\xi^2 + 3\xi^4 - \xi^6\right) \xi\, d\xi \\
&= 16 \left(\frac{1}{2}\xi^2 - \frac{3}{4}\xi^4 + \frac{1}{2}\xi^6 - \frac{1}{8}\xi^8\right)\Bigg|_{\xi=0}^{\xi=1} = 2 \qquad\qquad (7.97)
\end{aligned}$$

Although laminar pipe flow is uncommon in engineering applications, this computation underscores the caution we exercised in evaluating the kinetic-energy flux integral. In this special case, assuming $\bar{u}^3 = \overline{u^3}$ introduces a factor-of-two error!

7.7   Flow in Pipes

Recall that in Subsection 1.9.2, we analyzed incompressible flow in an infinitely long pipe. We found that while the velocity varies across the pipe cross section, it is independent of distance along the pipe, x. This independence of x is called **fully-developed** flow. The solution we obtained is observed in finite-length pipes provided the pipe's overall length, L, is very large compared to its diameter, D, and provided we make our observations at a sufficient distance from the ends of the pipe. Figure 7.5 schematically depicts the way in which pipe flow develops from a rounded inlet to the fully-developed region.

As shown, near the inlet viscous effects are confined to the region close to the pipe wall, known as the boundary layer (Chapter 14 discusses boundary layers in great detail). Eventually, these boundary layers grow in thickness until they merge on the centerline. At a distance, ℓ_e, from the inlet, measurements show that the flow becomes fully developed, where

$$\frac{\ell_e}{D} = \begin{cases} 0.06 Re_D, & \text{Laminar} \\ 4.4 Re_D^{1/6}, & \text{Turbulent} \end{cases} \qquad\qquad (7.98)$$

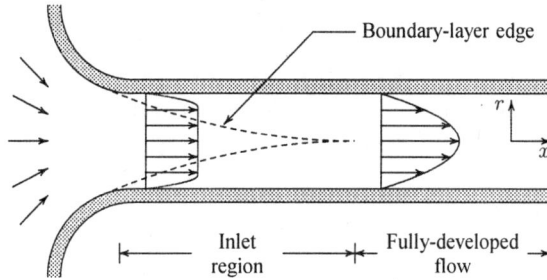

Figure 7.5: *Fully-developed pipe flow with the radial scale magnified. The length of the inlet region depends upon Reynolds number.*

Thus, for example, the entrance length for laminar pipe flow with Re_D = 1000 is 60 pipe diameters. While we have not yet developed the viscous-flow equations required to analyze viscous pipe flow, we can nevertheless make use of analytical and empirical results regarding head loss for pipes and pipe systems. Formal justification for some of the relationships discussed in the following subsections is provided in Chapters 13 and 14.

7.7.1 Friction and Head Loss in Straight Pipes

As discussed in Subsection 1.9.2, laminar flow is possible only at relatively small Reynolds numbers. Unless great care is exercised to minimize flow disturbances (and to thus permit larger Reynolds numbers), the laminar velocity profile of Equation (7.95) is observed only if the Reynolds number, Re_D, satisfies[4]

$$Re_D = \frac{\overline{u}D}{\nu} \lesssim 2300 \qquad \text{(Laminar flow)} \qquad (7.99)$$

At greater values of the Reynolds number, the flow undergoes transition to turbulence, characterized by time-varying fluctuations in all flow properties (we will discuss turbulence in Chapter 14). Again, in the absence of any attempts to minimize flow disturbances, the flow in a pipe will be completely turbulent when

$$Re_D = \frac{\overline{u}D}{\nu} \gtrsim 3000 \qquad \text{(Turbulent flow)} \qquad (7.100)$$

For intermediate Reynolds numbers, the flow is described as being **transitional**, i.e., in a state of transition from laminar to turbulent flow.

The head loss in a pipe, h_L, is a function of the friction at the pipe walls, the average kinetic energy of the flow and the length to diameter ratio. It is expressed by the **Darcy-Weisbach equation** as follows.

$$h_L = f \frac{\overline{u}^2}{2g} \frac{L}{D} \qquad (7.101)$$

In Equation (7.101), L is the length of the pipe and f is the dimensionless **Darcy friction factor** defined in terms of the shear stress at the pipe wall, τ_w, by

$$f \equiv \frac{\tau_w}{\frac{1}{8}\rho\overline{u}^2} \qquad (7.102)$$

---

[4]This criterion is consistent with Equation (1.41) because $\overline{u} = u_m/2$ and $D = 2R$.

It is worthwhile to pause and discuss how the friction factor is actually determined for pipe flow. While instruments are available to measure it directly, the special nature of pipe flow permits an indirect measurement. In Subsection 1.9.2, we showed that the shear stress varies linearly with distance across a pipe [Equation (1.38)]. This result holds for both laminar and turbulent flow. The shear stress at the wall is[5]

$$\tau_w = \frac{R}{2}\left(\frac{p_1 - p_2}{L}\right) \quad \Longrightarrow \quad f = \frac{D}{\frac{1}{2}\rho\,\overline{u}^2}\left(\frac{p_1 - p_2}{L}\right) \tag{7.103}$$

Thus, the friction factor can be inferred from the imposed pressure gradient and the measured average velocity.

For **perfectly-smooth pipes**, the friction factor depends upon whether or not the flow is turbulent. Its value is

$$\left.\begin{array}{ll} f = \dfrac{64}{Re_D}, & \text{Laminar} \\[3mm] 1/\sqrt{f} = -2\log_{10}\left(\dfrac{2.51}{Re_D\sqrt{f}}\right), & \text{Turbulent} \end{array}\right\} \tag{7.104}$$

The laminar value follows from the exact solution, while the turbulent-flow formula has been justified by asymptotic analysis [cf. Wilcox (1995)] and confirmed by correlation of measurements.

For **rough pipes**, if we have uniformly distributed, or **sand-grain**, roughness elements of average height k_s, the Colebrook (1939) formula provides a value accurate to within about 15% over a wide range of Reynolds numbers. The **Colebrook formula** is as follows.

$$1/\sqrt{f} = -2\log_{10}\left(\frac{k_s/D}{3.7} + \frac{2.51}{Re_D\sqrt{f}}\right) \tag{7.105}$$

Alternatively, we can use Figure 7.6, which is known as the **Moody diagram** [Moody (1944)], to find the friction factor for laminar and turbulent flow, including effects of roughness. The figure consists of two families of curves.

The first family is friction factor, f, as a function of Reynolds number, Re_D, for several values of the dimensionless roughness height, k_s/D. The curves are graphical representations of Equation (7.105). The family consists of the nearly horizontal curves whose axes are at the left and bottom of the plot. As shown in the figure, for rough surfaces the friction factor tends toward a constant value at high Reynolds number. We can use the friction factor curves to determine head loss when the velocity (or flow rate) and pipe dimensions are known.

For example, consider a 100 foot length of steel pipe of diameter 1 inch. Assume the fluid in the pipe is water at $68°$ F flowing at an average speed, $\overline{u}$, of 10 ft/sec. Reference to Table 1.7 shows that the kinematic viscosity, ν, is $1.08 \cdot 10^{-5}$ ft^2/sec. First, we compute the Reynolds number.

$$Re_D = \frac{\overline{u}D}{\nu} = \frac{(10 \text{ ft/sec}) \cdot \left(\frac{1}{12} \text{ ft}\right)}{1.08 \cdot 10^{-5} \text{ ft}^2/\text{sec}} = 7.716 \cdot 10^4 \tag{7.106}$$

Next, from the insert in Figure 7.6, the roughness height is $k_s = 1.5 \cdot 10^{-4}$ ft. Thus,

$$\frac{k_s}{D} = \frac{1.5 \cdot 10^{-4} \text{ ft}}{1/12 \text{ ft}} = .0018 \tag{7.107}$$

---

[5]For reasons that will become clear when we introduce the stress tensor in Chapter 12, positive shear at the pipe surface corresponds to a change of sign in Equation (1.38).

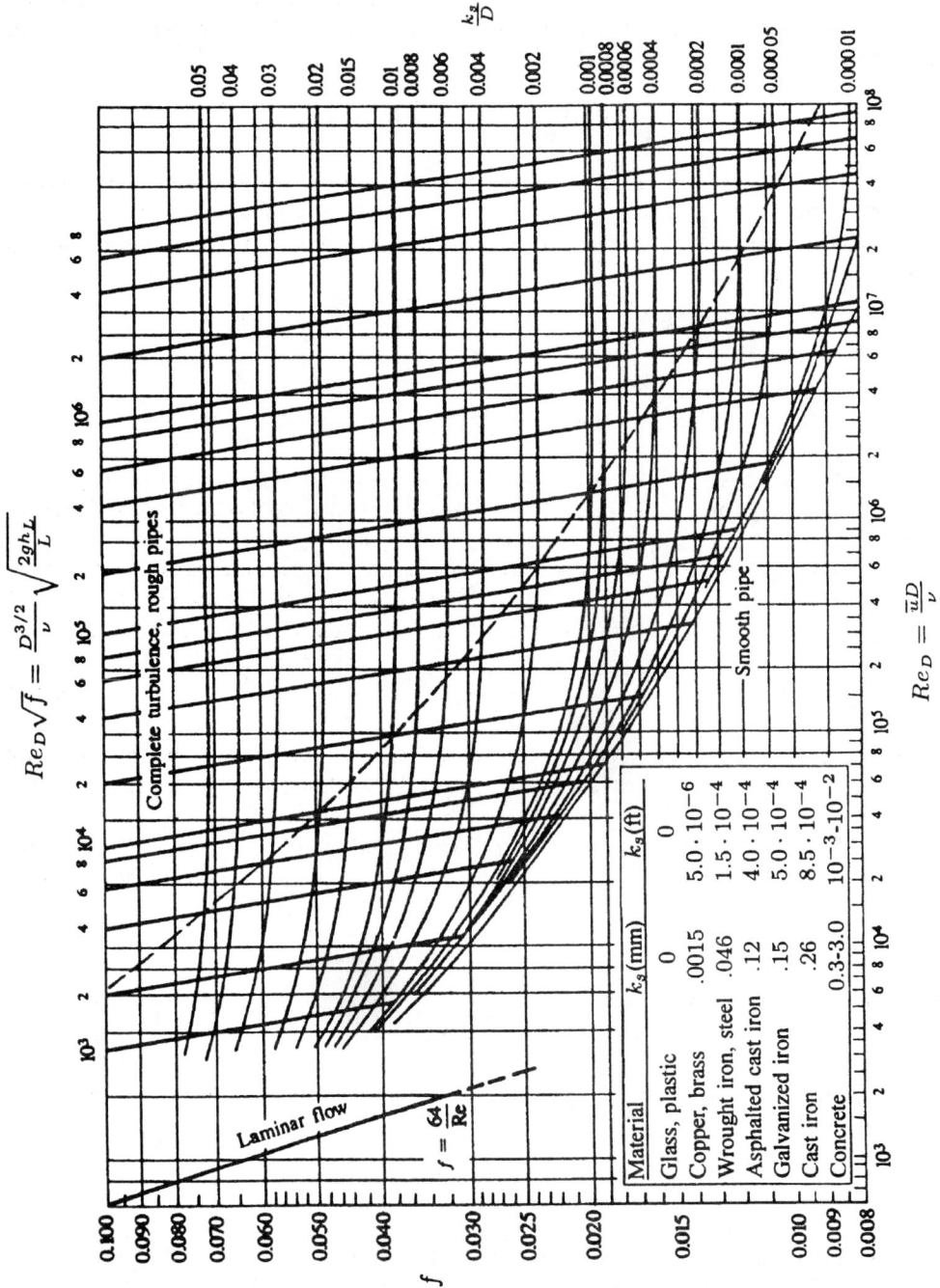

Figure 7.6: *The Moody diagram [From Moody (1944) — used with permission of the ASME].*

From the Moody diagram, the friction factor, f, is

$$f = .025 \tag{7.108}$$

Therefore, the head loss is

$$h_L = f \frac{\bar{u}^2}{2g} \frac{L}{D} = .025 \left(\frac{100 \text{ ft}^2/\text{sec}^2}{2 \cdot 32.174 \text{ ft/sec}^2} \right) \left(\frac{100 \text{ ft}}{1/12 \text{ ft}} \right) = 46.6 \text{ ft} \tag{7.109}$$

The second family of curves represent the loci of constant head loss, which can be determined from the product of Re_D and $\sqrt{f}$. That is, from Equation (7.101),

$$f\bar{u}^2 = D \frac{2gh_L}{L} \quad \Longrightarrow \quad Re_D \sqrt{f} = \frac{D^{3/2}}{\nu} \sqrt{\frac{2gh_L}{L}} \tag{7.110}$$

This family of curves consists of straight lines. The constant value of $Re_D \sqrt{f}$ on each line is the value at the intersection with the upper horizontal axis. We can use this family to solve for the flow rate when the h_L and D are known.

For example, suppose the head loss per kilometer of 20 cm diameter, asphalted cast-iron pipe is 12.2 meters. The fluid is again water, and the temperature is 20° C. We would like to determine the flow velocity. Reference to Table 1.7 shows that the kinematic viscosity is $\nu = 1.0 \cdot 10^{-6}$ m^2/sec. First, we compute the dimensionless head-loss parameter defined in Equation (7.110), viz.,

$$\frac{D^{3/2}}{\nu} \sqrt{\frac{2gh_L}{L}} = \frac{(0.2 \text{ m})^{3/2}}{1.0 \cdot 10^{-6} \text{ m}^2/\text{sec}} \sqrt{\frac{2 \cdot (9.807 \text{ m/sec}^2) \cdot (12.2 \text{ m})}{1000 \text{ m}}} = 4.38 \cdot 10^4 \tag{7.111}$$

The inset in Figure 7.6 tells us the roughness height is .12 mm = .00012 m. Thus, the dimensionless roughness height for this pipe is

$$\frac{k_s}{D} = \frac{.00012 \text{ m}}{.2 \text{ m}} = .0006 \tag{7.112}$$

Reference to the Moody diagram shows that the head-loss line with head-loss parameter equal to $4.38 \cdot 10^4$ crosses the friction curve with $k_s/D = .0006$ when

$$f = .019 \tag{7.113}$$

Finally, we can solve for the velocity using Equation (7.101), i.e.,

$$\bar{u} = \sqrt{\frac{2gh_L D}{fL}} = \sqrt{\frac{2 \cdot (9.807 \text{ m/sec}^2) \cdot (12.2 \text{ m}) \cdot (.2 \text{ m})}{.019 \cdot (1000 \text{ m})}} = 1.59 \frac{\text{m}}{\text{sec}} \tag{7.114}$$

7.7.2   Non-Circular Cross Sections and Minor Losses

Up to this point, we have considered head loss in a straight pipe of constant cross-sectional area. In real piping systems, we have several complicating factors such as inlets, outlets, bends, and abrupt changes in area, that create additional head loss. Losses in straight sections discussed in the preceding subsection are usually referred to as **major losses**. By contrast, losses attending inlets, outlets, etc. are termed **minor losses**, and are usually of secondary importance. There are applications where minor losses dominate, however, typically when the straight sections are short.

Empirical correlations have been developed to describe minor head losses. They are usually expressed in terms of a dimensionless **loss coefficient**, K, viz.,

$$h_L = K \frac{\bar{u}^2}{2g} \tag{7.115}$$

Head loss is also expressed in terms of an **equivalent length** of straight pipe, L_e, so that the head loss becomes

$$h_L = f \frac{\bar{u}^2}{2g} \frac{L_e}{D} \tag{7.116}$$

Measurements have been made to establish loss coefficients and equivalent lengths for pipes with circular cross sections. They can be used as approximations for noncircular geometries as well. To do this, we simply introduce the **hydraulic diameter**, D_h. If A is cross-sectional area and P is perimeter, D_h is defined by

$$D_h = \frac{4A}{P} \tag{7.117}$$

Note that the perimeter P, includes all surfaces exposed to the fluid, i.e., the so-called **wetted area**. If we have flow between two concentric cylinders, for example, P is the sum of the perimeters of both cylinders. As can be easily demonstrated, the hydraulic diameter is equal to the physical diameter for a circular cross section. This approximation yields satisfactory results provided cross sections do not depart too much from circular shape. In the case of rectangular geometries, for example, the height to width ratio should not exceed 4.

Consider a rectangular cross section of width w and height h (Figure 7.7). Such cross sections are typical of air-conditioning, heating and ventilating ducts. The area is $A = hw$, while the perimeter is $P = 2(w + h)$. Thus, the hydraulic diameter is

$$D_h = \frac{4A}{P} = \frac{4hw}{2(w + h)} = \frac{2h}{1 + h/w} \tag{7.118}$$

In the limiting case of a square, for which $w = h$, D_h is equal to h.

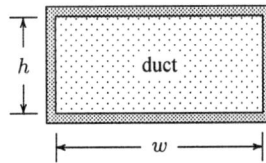

Figure 7.7: *Rectangular cross section typical of an air-conditioning duct.*

It might seem questionable that any reasonable degree of accuracy can be attained by what appears to be a completely ad hoc approximation, i.e., using the hydraulic diameter to adapt results for a circular cross section to general shapes. However, the approximation is far less ad hoc than it appears for two key reasons.

First, as long as we deal with fully-developed flow so that properties do not change in the streamwise direction, flow in any pipe or duct involves a balance between the pressure force and the viscous force [cf. Equation (1.37)]. The control volume approach tells us that the pressure force is proportional to the cross-sectional area, A, while the viscous force is proportional to the perimeter, P, of the wetted area. Hence, the viscous force adjusts according to A and P independent of the detailed geometry.

Second, for high Reynolds number flow characteristic of many applications, the most significant viscous effects contributing to head loss are confined mainly to a very thin region near the surface. Thus, regardless of the actual shape, the walls of the duct or pipe look planar on the scale of the thin viscous region, and are more-or-less independent of the precise geometry. This is a quintessential example of how powerful the control volume method can be in developing useful engineering methods.

All of the following correlations are limited to incompressible flows, and are generally reliable only for steady flow. Because the correlations rest on an empirical foundation, there is no guarantee that they can be used far beyond their established data base. This was a symptomatic weakness of the hydraulics era in fluid mechanics characterized by the work of Chézy, Weber, Hagen, Poiseuille, Darcy and Weisbach during the nineteenth century. That is not to say that empiricism is entirely without merit, as it is often the only approach practicable with the analytical techniques available. Much of twentieth century work on predicting turbulence, for example, is based on little more than experimentation, dimensional analysis and empirical relationships [cf. Wilcox (1993)].

Inlet. Figure 7.8 depicts an inlet, sometimes called an entrance, with a bend radius, $\mathcal{R}$. The inset includes values of K for various ratios of $\mathcal{R}$ to pipe diameter, D. Note that the limiting case $\mathcal{R}/D \to 0$ corresponds to a square inlet.

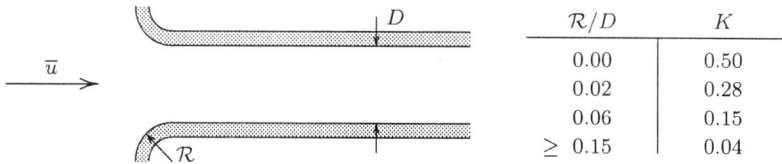

$\mathcal{R}/D$	K
0.00	0.50
0.02	0.28
0.06	0.15
$\geq$ 0.15	0.04

Figure 7.8: *Head loss coefficient at an inlet.*

Figure 7.9 schematically depicts flow at a square inlet. The primary reason for head loss is the inertia of the fluid, which prevents it from following the exact shape of the inlet. Rather, the streamlines are curved as shown and a small region of reverse, or separated, flow exists near the inlet. The fluid swirls around in the reverse-flow region, producing nontrivial viscous losses.

Because of the obstruction presented by the **separation bubble**, the primary flow in the core of the pipe has a reduced forward-flow area, an effect that is referred to as **vena contracta**. The size of the separated region and the vena contracta can be reduced significantly by rounding the inlet. As noted in the inset in Figure 7.8, the loss coefficient can be reduced to an insignificant value of 0.04 for a radius of curvature, $\mathcal{R}$, as small as $0.15D$.

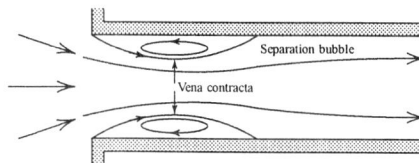

Figure 7.9: *Schematic of flow near a square inlet.*

Contraction. In a contraction, the diameter of the pipe changes from an initial value of D_1 to a smaller value of D_2. Figure 7.10 illustrates such a change in diameter. The inset includes the loss coefficient for a gradual (60°) contraction and a sudden (180°) contraction. The head loss is based on the velocity in the smaller section of the pipe.

As with inlet flow, the primary mechanism for head loss in a contraction is flow separation (at both corners) and the vena contracta. The head loss can be reduced by rounding the corners to permit more gradual turning of the flow. For a sudden contraction, correlation of measurements shows that

$$K \approx \frac{1}{2}\left[1 - \left(\frac{D_2}{D_1}\right)^2\right] \qquad \text{(sudden contraction)} \qquad (7.119)$$

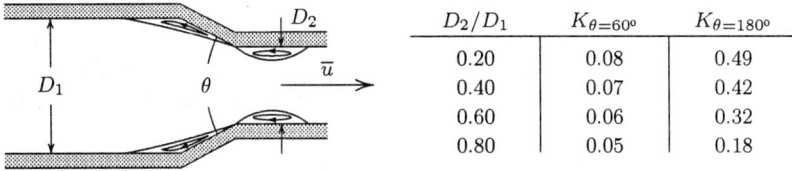

D_2/D_1	$K_{\theta=60°}$	$K_{\theta=180°}$
0.20	0.08	0.49
0.40	0.07	0.42
0.60	0.06	0.32
0.80	0.05	0.18

Figure 7.10: *Head loss coefficient at a contraction.*

Expansion. In an expansion, the diameter of the pipe changes from an initial value of D_1 to a larger value of D_2. Figure 7.11 illustrates such a change in diameter. The inset includes the loss coefficient for a gradual (10°) expansion and a sudden (180°) expansion. The head loss is based on the velocity in the smaller section of the pipe.

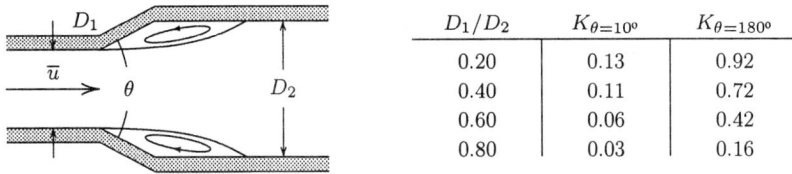

D_1/D_2	$K_{\theta=10°}$	$K_{\theta=180°}$
0.20	0.13	0.92
0.40	0.11	0.72
0.60	0.06	0.42
0.80	0.03	0.16

Figure 7.11: *Head loss coefficient at an expansion.*

As with inlet and contraction flow, the fluid cannot negotiate the sudden turn in an expansion. Its inertia carries it from the first corner along the trajectory sketched in Figure 7.11, known as the **dividing streamline**, which joins with the pipe surface downstream of the second corner as shown. The flow recirculates below the dividing streamline with significant head loss due to viscous effects.

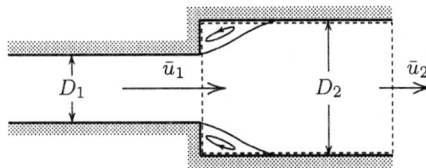

Figure 7.12: *Flow into a sudden expansion.*

We can derive a remarkably accurate expression for the head loss in a sudden expansion using the control volume method. The only approximations we make are that the pressure on the vertical face within the separation bubble is equal to the inlet pressure and that the kinetic-energy correction factor, α, is 1. For steady flow, conservation of mass, momentum and energy for the control volume indicated by the dashed contour in Figure 7.12 yield:

$$\bar{u}_1 D_1^2 = \bar{u}_2 D_2^2 \tag{7.120}$$

$$\rho \left(\bar{u}_2^2 D_2^2 - \bar{u}_1^2 D_1^2 \right) = p_1 D_1^2 - p_2 D_2^2 \approx (p_1 - p_2) D_2^2 \tag{7.121}$$

$$\frac{p_1}{\rho g} + \frac{\bar{u}_1^2}{2g} = \frac{p_2}{\rho g} + \frac{\bar{u}_2^2}{2g} + h_L \tag{7.122}$$

Noting the definition of the loss coefficient, K, given in Equation (7.115), a straightforward algebraic exercise shows that

$$K \approx \left[1 - \left(\frac{D_1}{D_2} \right)^2 \right]^2 \qquad \text{(sudden expansion)} \tag{7.123}$$

90° Smooth Bend. In a smooth bend, the loss coefficient varies with the radius of the bend, $\mathcal{R}$. Figure 7.13 illustrates a 90° bend in a pipe of constant diameter, D. The inset includes values of K for various ratios of $\mathcal{R}$ to pipe diameter, D. As $\mathcal{R}/D$ increases, K first falls because "secondary flow" (see cross-section inset) losses decrease but then rises because the "bend" contains a greater length of pipe. For very large $\mathcal{R}/D$, the equivalent length, L_e, [see Equation (7.116)] is just $\pi \mathcal{R}/2$.

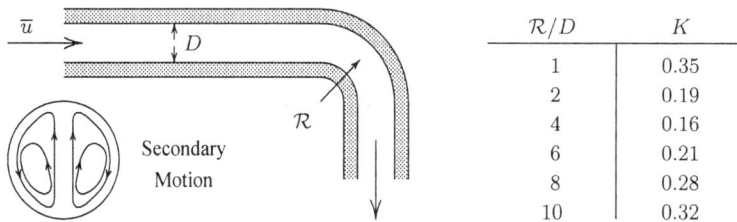

$\mathcal{R}/D$	K
1	0.35
2	0.19
4	0.16
6	0.21
8	0.28
10	0.32

Figure 7.13: *Head loss coefficient at a 90° smooth bend.*

Flow separation also plays a role in the head loss for a bend. Interestingly, the type of secondary motion depicted in Figure 7.13 occurs only for turbulent flow. Secondary motion also occurs for turbulent flow in ducts with noncircular cross sections, even when the duct is straight.

These are a sampling of the types of geometries for which loss coefficients have been empirically determined. Extensive measurements have been made to account for effects of valves, elbows of varying angles, and bends with internal vanes, just to mention a few. Useful summaries of head losses for other features of piping systems can be found in many of the current introductory fluid mechanics texts such as Fay (1994), Fox and McDonald (1992), Munson et al. (1990), Roberson and Crowe (1990), Sabersky et al. (1989), Shames (1992), and White (1994). If accurate head losses are needed, manufacturer's data should be used.

As an example of how important these effects can be, consider the following situation. For the sake of aesthetics, a plumber installs a 90° bend between two points in the pipe system for a swimming pool. The same functionality can be accomplished with a straight pipe as shown in Figure 7.14. It would be interesting to see what penalty is paid, as measured by head loss and the attendant added burden on the pool's pump, to satisfy the plumber's desire

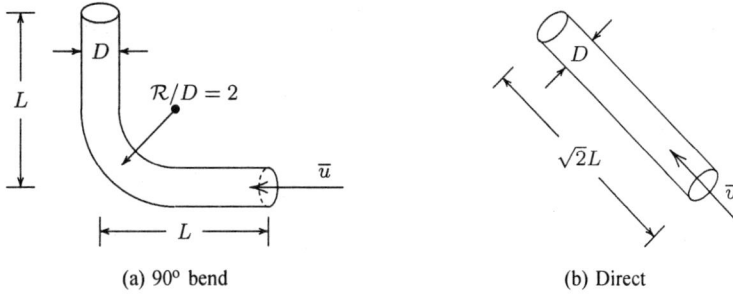

(a) 90° bend　　　　　　　　　　(b) Direct

Figure 7.14: *Swimming pool piping alternatives.*

for aesthetics. To make the problem more precise, we know that the pipe is smooth and the flow Reynolds number is $Re_D = 1.75 \cdot 10^6$. The flow velocity is observed to remain constant with or without the bend, and $L = 8D$. Using either Equation (7.104) or the Moody diagram (Figure 7.6), the friction factor is

$$f = .0106 \qquad (7.124)$$

Without the bend, Figure 7.14(b) shows that the length of the pipe connecting the two points is $\sqrt{2}L$. Hence the head loss is

$$h_L = f \frac{\bar{u}^2}{2g} \frac{\sqrt{2}L}{D} = \left(\sqrt{2}f \frac{L}{D} \right) \frac{\bar{u}^2}{2g} \qquad (7.125)$$

Then, in terms of the given properties, we have

$$h_L = \left(\sqrt{2} \cdot .0106 \cdot 8 \right) \frac{\bar{u}^2}{2g} = .12 \frac{\bar{u}^2}{2g} \qquad (7.126)$$

By contrast, with the bend [Figure 7.14(a)], reference to the inset in Figure 7.13 shows that the loss factor for $\mathcal{R}/D = 2$ is

$$K = .19 \qquad (7.127)$$

Thus, the total head loss is

$$h_L = f \frac{\bar{u}^2}{2g} \frac{2L}{D} + K \frac{\bar{u}^2}{2g} = \left(2f \frac{L}{D} + K \right) \frac{\bar{u}^2}{2g} \qquad (7.128)$$

Again, substituting the given values we find

$$h_L = (2 \cdot .0106 \cdot 8 + .19) \frac{\bar{u}^2}{2g} = .36 \frac{\bar{u}^2}{2g} \qquad (7.129)$$

Therefore, for the sake of a pretty pipe arrangement, our plumber has tripled the head loss through this particular section of the pool's piping system!

7.7.3　Multiple-Pipe Systems

All of the discussion above focuses on a single pipe. However, there are many important applications that involve two or more pipes. Examples are the pipes in the water distribution system for a housing development and the complex series of bronchial and other tubes in a human lung. The way we analyze complex pipe networks is similar to the method used by electrical engineers to analyze electrical circuits.

To understand the similarity, recall that we can express head loss, h_L, in terms of either the loss coefficient, K, or the equivalent length, L_e, defined in Equations (7.115) and (7.116). Hence, we can say

$$h_L = \overline{u}^2 \hat{R}, \qquad \hat{R} = \frac{K}{2g} \text{ or } \hat{R} = \frac{f}{2g}\frac{L_e}{D} \qquad (7.130)$$

where $\hat{R}$ is an effective resistance. Now, Ohm's law for electrical circuits tells us $V = iR$, where V is voltage, i is current and R is resistance. Thus, head loss is analogous to voltage drop, $\hat{R}$ is analogous to resistance and $\overline{u}^2$ is analogous to current. Note that we don't have a formal mathematical analogy since velocity does not appear linearly in the energy equation. Thus, some, but not all, of the standard circuit theory methods can be applied to pipe systems.

The foundation of our approach is conservation of mass and energy, which is analogous to conservation of charge and voltage. To illustrate how pipe systems can be analyzed, we consider pipes attached in series and in parallel. Figure 7.15 shows typical arrangements of both types.

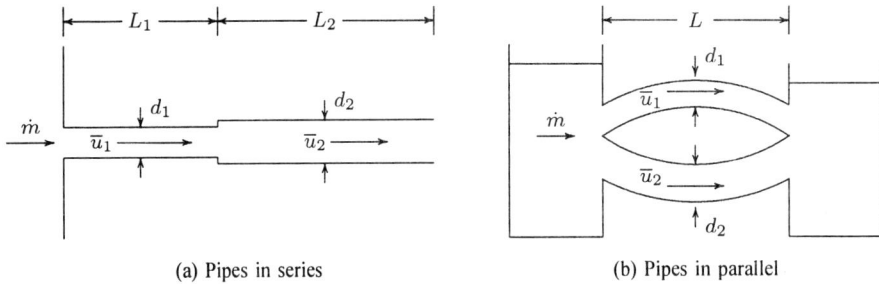

(a) Pipes in series   (b) Pipes in parallel

Figure 7.15: *Pipes in series and in parallel.*

Pipes in Series. Figure 7.15(a) shows two pipes of different diameters, i.e., $d_1 < d_2$, connected in series. Conservation of mass and energy tell us that the mass flow is constant, while the overall head loss is the sum of the head losses in each pipe. Thus, we have

$$\dot{m} = \dot{m}_1 = \dot{m}_2 \qquad (7.131)$$

$$h_L = h_{L1} + h_{L2} \qquad (7.132)$$

Now, for the first pipe, we have a minor loss at the entrance and a major loss proportional to the friction factor, f_1, pipe length, L_1, etc. Pipe 2 has a similar major loss and a minor loss due to sudden expansion. Thus, using Equation (7.123) to approximate the loss at a sudden expansion, we have

$$h_{L1} = \frac{1}{2}\frac{\overline{u}_1^2}{2g} + f_1\frac{L_1}{d_1}\frac{\overline{u}_1^2}{2g}, \qquad h_{L2} = \left[1 - \left(\frac{d_1}{d_2}\right)^2\right]^2\frac{\overline{u}_2^2}{2g} + f_2\frac{L_2}{d_2}\frac{\overline{u}_2^2}{2g} \qquad (7.133)$$

Also, using the fact that $\dot{m} = \frac{\pi}{4}\rho\overline{u}d^2$, we find from Equation (7.131)

$$\overline{u}_2 = \overline{u}_1\left(\frac{d_1}{d_2}\right)^2 \qquad (7.134)$$

To be specific, suppose the ratio of length to diameter ratio, L_i/d_i, is 10 for both pipes, $d_1 = d$, $d_2 = \sqrt{2}d$ and $\overline{u}_1 = U$. Then, Equation (7.134) shows that

$$\overline{u}_1 = U, \qquad \overline{u}_2 = \frac{1}{2}U \tag{7.135}$$

Substituting these values into (and combining) Equations (7.132) and (7.133) yields

$$h_L = \left[\frac{9}{16} + 10\left(f_1 + \frac{1}{4}f_2\right)\right]\frac{U^2}{2g} \tag{7.136}$$

To complete the solution, all that remains is to compute the friction factor in each pipe, which, in turn, depends upon the Reynolds number. Since we have not specified enough information to compute Reynolds number, this is as far as we can take the solution here.

Note that because mass flow is linear in $\overline{u}$ while head loss is quadratic in $\overline{u}$, we do not have the analogy to electrical circuit theory that would say the total resistance is equal to the sum of the resistances in the two pipes. This becomes very clear if we substitute Equations (7.134) and (7.135) into Equation (7.130), wherefore

$$h_L = \left[\hat{R}_1 + \hat{R}_2\left(\frac{d_1}{d_2}\right)^4\right]\frac{\overline{u}_1^2}{2g} = \left[\hat{R}_1 + \frac{1}{4}\hat{R}_2\right]\frac{U^2}{2g} \tag{7.137}$$

showing that head loss in the first pipe is the most significant. This is true because the velocity is twice as large in the first pipe, and the loss varies as $\overline{u}^2$.

Pipes in Parallel. Figure 7.15(b) shows two pipes of different diameters connected in parallel. As above, we assume $d_1 < d_2$. Conservation of mass and energy tell us that the mass flow is equal to the sum of the mass flows in each pipe, while the overall head loss is the same for each pipe. Therefore, we say

$$\dot{m} = \dot{m}_1 + \dot{m}_2 \tag{7.138}$$

$$h_L = h_{L1} = h_{L2} \tag{7.139}$$

To simplify the analysis, assume that both pipes are such that the product of the equivalent length, L_e, and friction factor, f, is invariant. Then,

$$h_{L1} = \left(\frac{fL_e}{2g}\right)\frac{\overline{u}_1^2}{d_1}, \quad h_{L2} = \left(\frac{fL_e}{2g}\right)\frac{\overline{u}_2^2}{d_2} \quad \Longrightarrow \quad \overline{u}_2 = \overline{u}_1\sqrt{\frac{d_2}{d_1}} \tag{7.140}$$

Also, using the fact that $\dot{m} = \frac{\pi}{4}\rho\overline{u}d^2$, we find

$$\overline{u}_1 d_1^2 + \overline{u}_2 d_2^2 = \frac{4\dot{m}}{\pi\rho} \tag{7.141}$$

Substituting Equation (7.140) into Equation (7.141),

$$\overline{u}_1 d_1^2 + \overline{u}_1\sqrt{\frac{d_2}{d_1}}d_2^2 = \frac{4\dot{m}}{\pi\rho} \quad \Longrightarrow \quad \overline{u}_1\left[1 + \left(\frac{d_2}{d_1}\right)^{5/2}\right] = \frac{4\dot{m}}{\pi\rho d_1^2} \tag{7.142}$$

Therefore, the velocities in the pipes are

$$\left.\begin{aligned}
\overline{u}_1 &= \frac{4\dot{m}}{\pi\rho d_1^2}\left[1 + \left(\frac{d_2}{d_1}\right)^{5/2}\right]^{-1} \\[2em]
\overline{u}_2 &= \frac{4\dot{m}}{\pi\rho d_1^2}\left[1 + \left(\frac{d_2}{d_1}\right)^{5/2}\right]^{-1}\sqrt{\frac{d_2}{d_1}}
\end{aligned}\right\} \tag{7.143}$$

As a final comment, these ideas can be extended to a network of pipes in order to analyze the water distribution system for a housing development or even an entire city. A segment of a pipe network would appear as in Figure 7.16. There are two basic rules that must be applied to solve such a network, and they are identical to the famous **Kirchhoff's rules** for electric circuits. Specifically:

1. The sum of the mass-flow rates entering any junction point is equal to the sum of the mass-flow rates leaving this point.

2. The algebraic sum of the head loss around any closed loop of the network equals zero.

Rule 1 simply means that whatever flows in to a given point must also flow out. Rule 2 means the total enthalpy can assume only one value at a given point. Unlike classical electric circuits, pipe-network equations are nonlinear and require iterative solutions. Such solutions are readily obtained with a computer. Jepson (1976), for example, describes numerical techniques used in solving pipe-network problems.

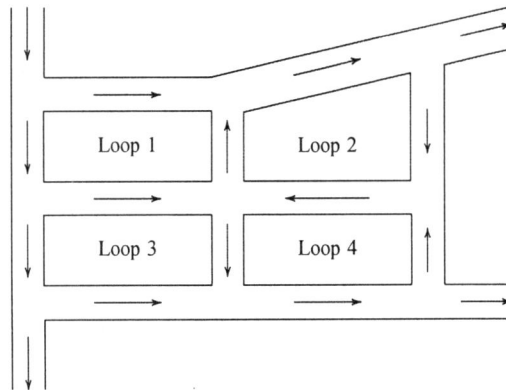

Figure 7.16: *Pipe network.*

7.8   Open-Channel Flow

Our final topic of this chapter is the flow of a liquid (usually water) that occurs in channels with a surface that interfaces with a gas (usually air), i.e., when the flow has a **free surface**. Open-channel flows occur naturally, with typical examples being streams and rivers. Man-made open channels are very common, some of the most familiar being irrigation canals, sewers, aqueducts, drainage ditches and spillways. More subtle examples are the rain gutters on your home and the water running down a driveway when you wash your car.

Open-channel flow, while similar in some ways to pipe flow, includes the extra complication of the free surface. The boundary condition appropriate for a free surface is quite different from that imposed at a solid boundary, and gives rise to interesting phenomena that do not appear in pipe flow. We will see that the gravitational force, rather than the pressure gradient, drives open-channel flows and that the Froude number plays a significant role. Because there is a free surface, we must address the question of the role that surface tension might play. The latter point is easily addressed. For virtually all applications of interest, the Weber number is so large that surface tension can be ignored.

The primary equations used to describe flow in an open channel are conservation of mass and energy. The momentum conservation principle plays a secondary role. Focusing on steady, incompressible flow, we treat the flow in a **quasi-one-dimensional** sense. That is, we use the control volume method to relate conditions on cross-sectional planes. Consider Figure 7.17, which is an idealized section of a typical open channel.

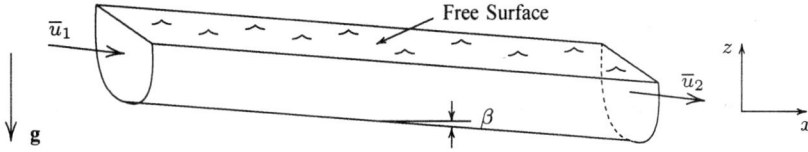

Figure 7.17: *A section of a typical open channel.*

The reason the momentum equation is of secondary importance is because we address only flows for which changes in the streamwise direction occur gradually. This will be true if cross-sectional areas and depth do not change rapidly with streamwise distance and if the inclination of the channel to the horizontal direction is small, i.e., $\beta \ll 1$. When these conditions are satisfied, the vertical velocity is very small compared to the horizontal velocity. Now, the pressure at a free surface is equal to the atmospheric pressure, p_o, even when the fluid at the surface is moving, i.e.,

$$p = p_o \qquad \text{(at a free surface)} \qquad (7.144)$$

Thus, a streamwise pressure gradient cannot occur at the free surface. In fact, straight open-channel flows have no significant streamwise pressure gradient at any part of the cross sections. The primary balance of forces is between the gravitational force and the viscous force on the solid boundaries of the channel. Thus, for steady flow, the vertical (z) component of the momentum equation is as follows.

$$\underbrace{\oiint_S \rho w \left(\mathbf{u} \cdot \mathbf{n}\right) dS}_{\text{Negligible}} = -\mathbf{k} \cdot \oiint_S (p - p_o)\,\mathbf{n}\,dS - \iiint_V \rho g\,dV + \mathbf{k} \cdot \underbrace{\oiint_S \mathbf{f}_v\,dS}_{\text{Negligible}} \qquad (7.145)$$

Note that the vector $\mathbf{f}_v$ is the viscous force vector introduced in Equation (7.28). The most important conclusion we can draw from the momentum equation is that, after applying the divergence theorem to the pressure integral in Equation (7.145), we have

$$0 \approx -\iiint_V \frac{\partial p}{\partial z}\,dV - \iiint_V \rho g\,dV \qquad \Longrightarrow \qquad \frac{\partial p}{\partial z} \approx -\rho g \qquad (7.146)$$

Hence, we see that the pressure satisfies the hydrostatic relation. To use the streamwise component of the momentum equation, we require some information about the viscous force. Because the flow is incompressible, we can use either the mass and momentum equations or the energy equation. As with pipe flows, it is more convenient to use the energy equation for open-channel flows.

Since we assume the flow is steady, the mass-flow rate, $\dot{m}$, across any cross section is constant. Thus, we can say

$$Q = \frac{\dot{m}}{\rho} = \bar{u}A = \text{constant} \qquad (7.147)$$

where Q is the **volume-flow rate**, $\bar{u}$ is the average streamwise flow velocity and A is the cross-sectional area.

Turning to the energy-conservation equation, we can use the approximate equation derived earlier for pipe-flow applications, viz., Equation (7.89). We make the following three simplifications to the basic conservation equation.

1. Since none are present for the intended applications, we drop turbine and pump terms, h_t and h_p.

2. Because of the large uncertainties associated with the irregular geometries of rivers, streams, etc., we assume the kinetic-energy correction factor, α, is unity throughout. Any errors attending this approximation will be minor relative to typical assumptions made to idealize a real application.

3. Since open channels have negligible streamwise pressure gradient, necessarily $p_1 \approx p_2$.

Thus, Equation (7.89) simplifies to a balance amongst kinetic energy, potential energy, and viscous effects as represented by a head loss, i.e.,

$$\frac{\bar{u}_1^2}{2g} + z_1 = \frac{\bar{u}_2^2}{2g} + z_2 + h_L \tag{7.148}$$

If the depth of the channel is denoted by y, then for a channel segment of length ℓ, reference to Figure 7.18 shows that depth and altitude are related as follows.

$$z_1 = y_1 + \ell \tan\beta \qquad \text{and} \qquad z_2 = y_2 \tag{7.149}$$

So, denoting the slope of the bottom by

$$S_o \equiv \tan\beta \tag{7.150}$$

the energy-conservation equation becomes

$$y_1 + \frac{\bar{u}_1^2}{2g} + S_o\ell = y_2 + \frac{\bar{u}_2^2}{2g} + h_L \tag{7.151}$$

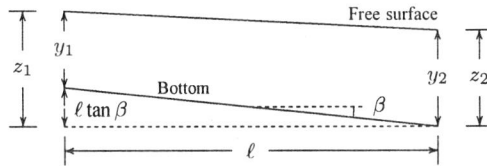

Figure 7.18: *Open-channel geometry — side view.*

To make further progress with these equations, we must determine the head loss. Experience has shown that we can use the results for pipe flow by introducing the **hydraulic radius**, R_h. By definition, the hydraulic radius is

$$R_h = \frac{A}{P} \tag{7.152}$$

where A is cross-sectional area and P is the perimeter of the wetted, solid surface. R_h differs from the hydraulic diameter used for pipe-flow analysis to the extent that it excludes the free surface, while D_h is based on the perimeter bounding the cross-sectional area. This reflects

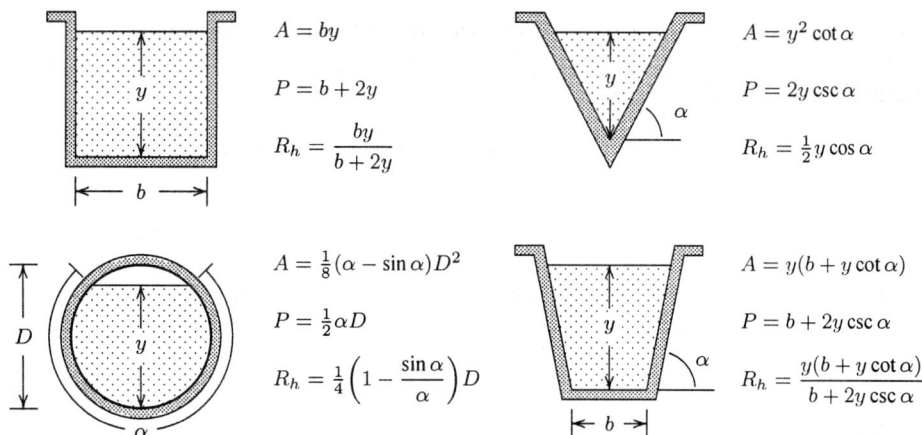

Figure 7.19: *Areas, perimeters and hydraulic radii for common geometries.*

the fact that the free surface cannot support a shear stress. Figure 7.19 includes A, P and R_h for several simple geometries.

An infinitely wide rectangular channel, which might approximate a thin sheet of water flowing down an incline, is an interesting limiting case. For such a "channel," which is simply a two-dimensional flow, the hydraulic radius is equal to the depth, viz.,

$$R_h = \lim_{b \to \infty} \left(\frac{by}{b + 2y} \right) = y \tag{7.153}$$

In the case of a circular cross section, the information in Figure 7.19 shows that if the channel is full ($\alpha = 2\pi$), then $R_h = \frac{1}{4}D$. This reflects the factor of 4 difference between the definition of hydraulic diameter for pipes [Equation (7.117)] and hydraulic radius for open channels [Equation (7.152)]. We can use the Moody diagram and the correlation formulas for friction factor, f, and head loss, h_L, provided we compute the hydraulic diameter according to

$$D_h = 4R_h \tag{7.154}$$

7.8.1 Uniform Flow

As an example of how we can use the formulas developed above to solve a typical open-channel flow problem, consider a straight channel with a very small inclination to the horizontal. Many rivers and canals have slopes of order $S_o \approx .0001$, and this is the value we will use. For this small a slope, the flow will maintain constant depth provided cross-sectional shape remains unchanged. This is referred to as **uniform flow** or **normal flow**.

We will analyze a triangular cross section like the one depicted in Figure 7.19 with $\alpha = 45°$. The volume-flow rate is $Q = 5$ m³/sec and the channel is constructed of brass for which the roughness height is $k_s = 0.6$ mm. Using a value for kinematic viscosity of $\nu = 10^{-6}$ m²/sec, we wish to determine the normal depth of the water in the channel, y.

The governing equations are mass and energy conservation. From mass conservation applied to the specified triangular cross section, we know that

$$Q = \bar{u}A, \qquad A = y^2 \qquad \Longrightarrow \qquad \bar{u} = \frac{Q}{y^2} \tag{7.155}$$

Since the volume flux and the cross-sectional area are constant, so are the average flow velocity and the depth. That is, at any two streamwise sections, we have $\bar{u}_1 = \bar{u}_2$ and $y_1 = y_2$. Hence, Equation (7.151) simplifies to

$$S_o \ell = h_L = f \frac{\ell}{4R_h} \frac{\bar{u}^2}{2g} \quad \Longrightarrow \quad \bar{u}^2 = \frac{8g S_o R_h}{f} \tag{7.156}$$

Observe that we have used the Darcy-Weisbach Equation (7.101) with $D = 4R_h$ to evaluate the head loss.

Now, reference to Figure 7.19 shows that the hydraulic radius for our triangular cross section is

$$R_h = \frac{\sqrt{2}}{4} y \tag{7.157}$$

Substituting Equations (7.155) and (7.157) into Equation (7.156) yields

$$\frac{Q^2}{y^4} = \frac{8g S_o}{f} \frac{\sqrt{2}}{4} y \quad \Longrightarrow \quad y^5 = \frac{Q^2 f}{2\sqrt{2}\, g S_o} \tag{7.158}$$

Therefore, the depth in the channel is

$$y = \left[\frac{Q^2}{2\sqrt{2}\, g S_o} \right]^{1/5} f^{1/5} \tag{7.159}$$

We cannot solve directly for y because the friction coefficient, f, is not yet known. It depends upon Reynolds number and the ratio of k_s to D_h which, in turn, depend upon flow velocity and depth. Thus, we will have to solve with an iterative procedure. As we will see, however, our solution will converge very rapidly, mainly because most open-channel flows occur at high Reynolds number and involve very rough wetted surfaces, so the variation of f with Reynolds number is very slow.

For the given values, the equation relating y and f becomes

$$y = \left[\frac{\left(5 \text{ m}^3/\text{sec}\right)^2}{2\sqrt{2} \left(9.807 \text{ m/sec}^2\right)(.0001)} \right]^{1/5} f^{1/5} = 6.18 f^{1/5} \tag{7.160}$$

Omitting the detailed arithmetic calculations for the sake of brevity, the Reynolds number, Re_{D_h}, and dimensionless roughness-height, k_s/D_h, are

$$Re_{D_h} = \frac{\bar{u} D_h}{\nu} = \frac{7.07 \cdot 10^6}{y} \quad \text{and} \quad \frac{k_s}{D_h} = \frac{4.25 \cdot 10^{-4}}{y} \tag{7.161}$$

To solve, we first make a guess for the friction factor and then use Equations (7.160) and (7.161) to determine Reynolds number and the roughness parameter. Reference to the Moody diagram (Figure 7.6) provides a new value for the friction factor. When the guessed and computed values of f agree to within some desired margin of error, the solution has converged.

Table 7.1 shows a two-step iteration that yields an acceptable solution for f. The initial guess for the friction factor is $f = .0100$. As shown, using the corresponding values of Re_{D_h} and k_s/D_h implies $f \approx .0135$. Using $f = .0135$ as the guess in the second iteration again yields a computed value of $f = .0135$. Substitution of the values for the second iteration into

Table 7.1: *Computational Details for a Triangular Open Channel*

f_{guess}	y (m)	Re_{D_h}	k_s/D_h	$f_{computed}$
.0100	2.46	$2.87 \cdot 10^6$	$1.73 \cdot 10^{-4}$	.0135
.0135	2.61	$2.71 \cdot 10^6$	$1.63 \cdot 10^{-4}$	.0135

the Colebrook formula [Equation (7.105)] shows that it is satisfied to within 0.5%. Thus, we conclude that the depth of the fluid in the triangular channel is

$$y = 2.61 \text{ m} \qquad (7.162)$$

As mentioned above, the iterative procedure converges very rapidly because typical open-channel flows occur at high Reynolds number, and the wetted surface is generally quite rough. Inspection of the Moody diagram shows that, under such conditions, the friction factor is independent of Reynolds number. Correspondingly, the Colebrook formula simplifies to

$$\frac{1}{\sqrt{f}} \approx -2 \log_{10} \left(\frac{k_s/D_h}{3.7} \right) \qquad \text{(Completely-rough flow)} \qquad (7.163)$$

This equation defines a very slow variation of f with k_s/D_h. For example, increasing k_s/D_h by a factor of 1000 from 10^{-4} to 10^{-1} corresponds to f increasing by less than a factor of 10 (from .012 to .102). Additionally, Equation (7.160) shows that the depth is a slowly varying function of f. Hence, any sensible first guess for f will lead to a close approximation to the actual value of f after a single iteration.

7.8.2 Chézy-Manning Equations

Perhaps the most famous formulation for uniform open-channel flow was developed by French engineer Antoine Chézy (1718-1798) in experimental studies of the River Seine and the Courpalet Canal. We can solve for average channel velocity, $\bar{u}$, from Equation (7.156), wherefore

$$\bar{u} = \sqrt{\frac{8g}{f}} \sqrt{R_h S_o} \qquad (7.164)$$

When the channel shape and roughness are constant, so is the factor $\sqrt{8g/f}$. Calling this constant C, we have the famous **Chézy equation**, viz.,

$$\bar{u} = C \sqrt{R_h S_o} \qquad (7.165)$$

The coefficient C is known as the **Chézy coefficient**. It is a dimensional quantity, having dimensions $L^{1/2}/T$, and will thus be different for each new application. In his experiments, Chézy determined that it varies from 60 ft$^{1/2}$/sec (33 m$^{1/2}$/sec) for small rough channels to about 160 ft$^{1/2}$/sec (88 m$^{1/2}$/sec) for large smooth channels. The use of such dimensional, empirically determined, coefficients limits the usefulness of these formulas to the single fluid for which they have been determined, namely, water.

Through the late nineteenth and early twentieth centuries, a great deal of research focused on developing correlations of the Chézy coefficient with roughness, slope and channel shape. Chow (1959), for example, gives a lucid discussion of many correlations devised during this period. By far, the most popular correlation is attributed to Robert Manning (1816-1897). The **Manning correlation** is

$$C = \frac{\chi}{n} R_h^{1/6}, \qquad \chi = 1.49 \text{ ft}^{1/3}/\text{sec} = 1.00 \text{ m}^{1/3}/\text{sec} \qquad (7.166)$$

where the dimensionless coefficient n is the **Manning roughness coefficient**. We can arrive at this formula by noting that for completely-rough pipes and channels, the limiting form of the Colebrook formula, Equation (7.163), is closely approximated by

$$f \approx .113 \left(\frac{k_s}{R_h}\right)^{1/3}, \qquad .001 < \frac{k_s}{R_h} < .05 \qquad (7.167)$$

Substituting this formula for f into $C = \sqrt{8g/f}$ yields

$$C \approx \left(8.4g^{1/2}/k_s^{1/6}\right) R_h^{1/6} \qquad (7.168)$$

which reproduces the proportionality between C and $R_h^{1/6}$ deduced empirically by Manning. Comparison with Equation (7.166) implies that, for a range of roughness heights, the Manning roughness coefficient is given by

$$n \approx \chi \left(\frac{k_s^{1/6}}{8.4g^{1/2}}\right), \qquad .001 < \frac{k_s}{R_h} < .05 \qquad (7.169)$$

Table 7.2 lists values of n determined by Manning for a variety of open-channels with roughness height varying from 0.3 mm to 5 m. Combining Equations (7.165) and (7.166), we arrive at the famous **Chézy-Manning equation**, i.e.,

$$\bar{u} = \frac{\chi}{n} S_o^{1/2} R_h^{2/3}, \qquad \chi = 1.49 \text{ ft}^{1/3}/\text{sec} = 1.00 \text{ m}^{1/3}/\text{sec} \qquad (7.170)$$

All of the values for n in Table 7.2 include an error band. This reflects the fact that this empirical parameter depends upon Reynolds number, channel geometry, and other factors that have been glossed over in this rather simplistic formulation. The advantage of the Chézy-Manning equation is the ease with which it can be applied.

To understand the last point, we revisit the triangular channel discussed in the preceding subsection. Recall from Equations (7.155) and (7.157) that the velocity and hydraulic radius are given by

$$\bar{u} = \frac{Q}{y^2}, \qquad R_h = \frac{\sqrt{2}}{4} y = 2^{-3/2} y \qquad (7.171)$$

where $Q = 5$ m^3/sec is the volume-flow rate and y is depth. So, since the slope is $S_o = .0001$, the Chézy-Manning Equation (7.170) becomes

$$\frac{Q}{y^2} = \frac{1}{n}(.0001)^{1/2} \left(2^{-3/2} y\right)^{2/3} = \frac{y^{2/3}}{200n} \qquad (7.172)$$

Thus, solving for y in terms of Q and n, we find

$$y = (200Qn)^{3/8} \qquad (7.173)$$

Since the channel walls are made of brass, reference to Table 7.2 shows that the Manning friction coefficient is $n = .011 \pm .002$, i.e.,

$$.009 < n < .013 \qquad (7.174)$$

Table 7.2: *Experimental Values of the Manning Roughness Coefficient*

Type of Channel	n	k_s (ft)	k_s (mm)
Artificially lined channels:			
Glass	$.010 \pm .002$	0.0011	0.3
Brass	$.011 \pm .002$	0.0019	0.6
Smooth steel	$.012 \pm .002$	0.0032	1.0
Cast iron	$.013 \pm .003$	0.0051	1.6
Brickwork	$.015 \pm .002$	0.0120	3.7
Asphalt	$.016 \pm .003$	0.0180	5.4
Corrugated metal	$.022 \pm .005$	0.1200	37.0
Rubble masonry	$.025 \pm .005$	0.2600	80.0
Excavated earth channels:			
Clean	$.022 \pm .004$	0.12	37
Gravelly	$.025 \pm .005$	0.26	80
Weedy	$.030 \pm .005$	0.80	240
Stony, large cobbles	$.035 \pm .010$	1.50	500
Natural channels:			
Clean and straight	$.030 \pm .005$	0.80	240
Major rivers	$.035 \pm .010$	1.50	500
Sluggish, deep pools	$.040 \pm .010$	3.00	900
Floodplains:			
Pasture, farmland	$.035 \pm .010$	1.5	500
Light brush	$.050 \pm .020$	6.0	2000
Heavy brush	$.075 \pm .025$	15.0	5000

Therefore, according to the Chézy-Manning formula, the depth of the water in the channel lies in the range

$$2.28 \text{ m} < y < 2.62 \text{ m} \tag{7.175}$$

Recall that, using the more-accurate Moody diagram/Colebrook formula values for the friction factor, we found that $y = 2.61$ m. Had we used the nominal value of $n = .011$, our predicted depth would have been 2.46 m, which is about 6% lower than 2.61 m. Thus, the Chézy-Manning formula is reasonably accurate for this example.

In general, the Manning approximation is accurate for a limited range of roughness heights. Equation (7.169) provides an indication as to why its range of applicability might be limited in this manner. In practice, it yields unrealistically low friction, and thus high flow rate, for both deep smooth channels (small k_s/R_h) and shallow rough channels (large k_s/R_h).

7.8.3 Surface Wave Speed and Flow Classification

While uniform flow in open channels is of interest for practical design of irrigation systems, flood control studies, etc., there are other flow regimes that are of practical interest. There are two primary ways for classifying open-channel flow, viz., in terms of depth variation and in terms of wave-propagation properties.

Focusing first on depth variation, we have already examined the simplest case of uniform flow in which depth is constant. If the flow encounters an obstacle or if the cross-sectional area changes, then the depth will vary. When this occurs, we say the flow is **varied**. If the flow changes gradually so that it can still be treated analytically in a quasi-one-dimensional manner, we say the flow is **gradually varied**. By contrast, if depth (and other) variations occur so rapidly that a more-complicated multidimensional analysis is required, we say the flow is **rapidly varied**. Summarizing, there are three classes of open-channel flows based on depth variation as follows.

1. *Uniform Flow:* Constant depth and slope, which can be analyzed using simple algebraic equations.

2. *Gradually-Varied Flow:* Depth and/or slope varying sufficiently slowly that the flow can be described with a straightforward one-dimensional, ordinary differential equation [see Equation (7.201) below].

3. *Rapidly-Varied Flow:* Depth and/or slope varying so fast that a two- or three-dimensional analysis is required.

Turning to the second way of classifying open-channel flows, we note that one of the most interesting consequences of having a free surface is the presence of waves that propagate both along and below the surface. Waves on the surface are most easily observed, as anyone who has visited a beach or taken a ride on a boat can attest. An interesting question to pose is, what is the speed of a small wave traveling along a free surface? We might also want to know if the speed depends upon the depth, and if so, what the dependence is. We refer to such waves as **gravity waves**. The speed of propagation, which we denote as c, is often called the **wave celerity**.

We can compute c by analyzing a weak wave traveling along a free surface of an infinitely-wide open channel of undisturbed depth y. As shown in Figure 7.20(a), after the wave passes, it imparts a velocity, Δu, to the fluid. For a weak wave, we assume the changes in flow properties are very small, i.e.,

$$|\Delta y| \ll y, \qquad |\Delta u| \ll c \qquad (7.176)$$

We can replace this unsteady problem with an equivalent steady problem by making a Galilean transformation so that the wave is stationary and the flow is from left to right, as shown in Figure 7.20(b). In this coordinate system, the velocity to the left of the wave is c, while the velocity to the right of the wave is $c - \Delta u$.

To analyze the wave, we select the control volume indicated by the dashed contour in Figure 7.20(b). Because we are using a one-dimensional geometry, our analysis can be viewed as applying to a volume of unit width out of the page. Since the flow is steady, conservation of mass tells us the net flux of fluid out of the control volume vanishes. Thus,

$$\iint_S \rho\, \mathbf{u} \cdot \mathbf{n}\, dS = 0 \qquad (7.177)$$

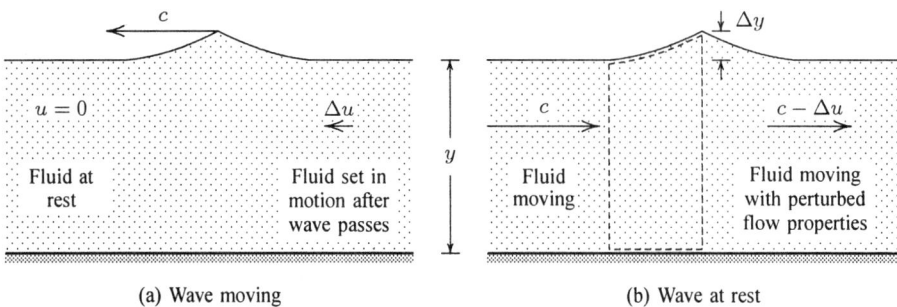

(a) Wave moving   (b) Wave at rest

Figure 7.20: *Analysis of a gravity wave in an open channel; a Galilean transformation produces a steady flow with fluid passing through the stationary wave.*

Hence, since **n** is the *outer* unit normal, evaluating the closed surface integral gives

$$-\rho c y + \rho(c - \Delta u)(y + \Delta y) = 0 \tag{7.178}$$

Expanding the terms in parentheses, there follows

$$-\rho c y + \rho c y + \rho c \Delta y - \rho y \Delta u - \rho \Delta y \Delta u = 0 \tag{7.179}$$

wherefore

$$y \Delta u = c \Delta y - \Delta u \Delta y \tag{7.180}$$

Since we are assuming the surface wave causes an infinitesimally small change in all flow properties (in particular, since $|\Delta u| \ll c$), we can drop the last term in Equation (7.180) as it is quadratic in changes while the other two terms are linear. Hence, we conclude from mass conservation that the changes in velocity and depth are related as follows.

$$\Delta u \approx \frac{c}{y} \Delta y \tag{7.181}$$

If we ignore viscous effects, the x component of the momentum-conservation equation for steady flow is

$$\oiint_S \rho u (\mathbf{u} \cdot \mathbf{n}) \, dS = -\mathbf{i} \cdot \oiint_S (p - p_o) \mathbf{n} \, dS + \mathbf{i} \cdot \iiint_V \rho g \, \mathbf{k} \, dV \tag{7.182}$$

where we take the dot product of **i** with the net pressure and gravity integrals to extract their x components. Now, the gravity force is in the vertical direction and has no effect on x momentum. Also, as argued above, the pressure is hydrostatic [see Equation (7.146)] so that

$$p - p_o = -\rho g z \tag{7.183}$$

Expanding the various integrals, with $z = 0$ at the free surface, we have

$$\rho c(-c y) + \rho(c - \Delta u)[(c - \Delta u)(y + \Delta y)] = -\int_{-y}^{0} \rho g z \, dz + \int_{-y-\Delta y}^{0} \rho g z \, dz \tag{7.184}$$

We can simplify the momentum equation by combining the two pressure integrals and changing integration variables according to $\zeta = -z$, so that

$$\rho c^2 y = [\rho(c - \Delta u)(y + \Delta y)](c - \Delta u) + \int_{y}^{y + \Delta y} \rho g \zeta \, d\zeta \tag{7.185}$$

Then, using Equation (7.178), and evaluating the pressure integral, we can simplify this result further, viz.,

$$\begin{aligned} \rho c^2 y &= \rho c y(c - \Delta u) + \frac{1}{2}\rho g \left[(y + \Delta y)^2 - y^2 \right] \\ &= \rho c^2 y - \rho c y \Delta u + \frac{1}{2}\rho g (2y + \Delta y)\Delta y \end{aligned} \tag{7.186}$$

Therefore, dropping the term quadratic in Δy, conservation of momentum yields the following relation between the changes in velocity and depth.

$$\rho c y \Delta u \approx \rho g y \Delta y \qquad \Longrightarrow \qquad \Delta u \approx \frac{g}{c} \Delta y \tag{7.187}$$

Finally, we can combine Equations (7.181) and (7.187) to arrive at

$$c^2 \approx gy \quad \implies \quad c = \sqrt{gy} \tag{7.188}$$

We can now reveal the second way of classifying open-channel flows. If we compute the ratio of the average flow velocity, $\overline{u}$, to the celerity, c, we have the Froude number, Fr, i.e.,

$$Fr = \frac{\overline{u}}{\sqrt{gy}} \tag{7.189}$$

In terms of the Froude number, there are three types of flow:

1. $Fr < 1$...*Subcritical Flow*;

2. $Fr = 1$...*Critical Flow*;

3. $Fr > 1$...*Supercritical Flow*.

The difference between subcritical and supercritical flow is in the way the flow is affected by an obstruction or a cross-sectional change. In subcritical flow, waves are created by a change in flow conditions. The waves move faster than the oncoming flow, and can propagate upstream at a speed $\tilde{u} \approx \sqrt{gy} - \overline{u}$. Thus, the oncoming flow has advance warning that flow conditions are changing, and it can adjust accordingly in a gradual manner. In supercritical flow, the oncoming stream moves faster than surface waves, which are swept downstream. Thus, the flow has no advance warning that conditions are changing, and the transition caused by an obstacle, for example, is much more abrupt.

It would be interesting to pause and compute the Froude number for the triangular channel considered in the two preceding subsections. Based on results obtained, the depth is $y = 2.61$ m and the average velocity is

$$\overline{u} = \frac{Q}{y^2} = \frac{5 \text{ m}^3/\text{sec}}{(2.61 \text{ m/sec})^2} = 0.734 \ \frac{\text{m}}{\text{sec}} \tag{7.190}$$

Also, the wave speed is

$$c = \sqrt{gy} = \sqrt{\left(9.807 \ \frac{\text{m}}{\text{sec}^2}\right)(2.61 \text{ m})} = 5.06 \ \frac{\text{m}}{\text{sec}} \tag{7.191}$$

Therefore, for our triangular open channel, the Froude number is

$$Fr = \frac{\overline{u}}{\sqrt{gy}} = 0.145 \tag{7.192}$$

so that the flow is subcritical.

The Froude number appears explicitly in the equation of motion for *gradually-varied* flow, i.e., for flow in which depth and/or cross sections change gradually. To see the variation, recall that for a segment of an open channel of length ℓ, the energy-conservation principle [see Equation (7.151) and Figure 7.18] is

$$y_1 + \frac{\overline{u}_1^2}{2g} + S_o \ell = y_2 + \frac{\overline{u}_2^2}{2g} + h_L \tag{7.193}$$

We can derive a differential equation by considering a channel segment of differential length $\ell = \Delta x$. Then, the various quantities appearing in Equation (7.193) can be written in terms of differential changes, viz.,

$$\left.\begin{array}{l} \overline{u}_1 = \overline{u}, \qquad \overline{u}_2 = \overline{u} + \Delta\overline{u} \\[2mm] y_1 = y, \qquad y_2 = y + \Delta y \\[2mm] h_L = f\dfrac{\Delta x}{D_h}\dfrac{\overline{u}^2}{2g} \end{array}\right\} \tag{7.194}$$

Substituting Equation (7.194) into Equation (7.193), we obtain

$$y + \frac{\overline{u}^2}{2g} + S_o \Delta x = y + \Delta y + \frac{\overline{u}^2}{2g} + \frac{\overline{u}}{g}\Delta\overline{u} + \frac{(\Delta\overline{u})^2}{2g} + f\frac{\Delta x}{D_h}\frac{\overline{u}^2}{2g} \tag{7.195}$$

Canceling like terms, dropping the term quadratic in $\Delta\overline{u}$ and rearranging a little, there follows

$$S_o \approx \frac{\Delta y}{\Delta x} + \frac{\overline{u}}{g}\frac{\Delta\overline{u}}{\Delta x} + \frac{f}{D_h}\frac{\overline{u}^2}{2g} \tag{7.196}$$

Taking the limit $\Delta x \to 0$, the differential equation governing gradually-varied flow is

$$S_o = \frac{dy}{dx} + \frac{\overline{u}}{g}\frac{d\overline{u}}{dx} + \frac{f}{D_h}\frac{\overline{u}^2}{2g} \tag{7.197}$$

We can simplify this equation further by appealing to mass conservation, which permits us to eliminate $d\overline{u}/dx$. For the sake of simplicity, assume the channel has rectangular cross section with width b. Then, mass conservation is $\overline{u}by = $ constant, so that

$$by\frac{d\overline{u}}{dx} + b\overline{u}\frac{dy}{dx} = 0 \qquad \Longrightarrow \qquad \frac{d\overline{u}}{dx} = -\frac{\overline{u}}{y}\frac{dy}{dx} \tag{7.198}$$

Finally, it is customary to denote the head-loss term by S, i.e.,

$$S \equiv \frac{f}{D_h}\frac{\overline{u}^2}{2g} \tag{7.199}$$

Note that S is the slope corresponding to uniform flow with the same volume-flow rate, Q [cf. Equation (7.156)]. Substituting Equations (7.198) and (7.199) into Equation (7.197) gives

$$\left[1 - \frac{\overline{u}^2}{gy}\right]\frac{dy}{dx} = S_o - S \tag{7.200}$$

Recognizing the fact that the quantity in brackets is simply $[1 - Fr^2]$, the final form of the differential equation for gradually-varied flow in a rectangular open channel is

$$\frac{dy}{dx} = \frac{S_o - S}{1 - Fr^2} \tag{7.201}$$

As discussed by White (1994), for example, there are actually 12 different sub-cases of gradually-varied open-channel flow that can be discerned from Equation (7.201), depending upon the Froude number and upon $(S_o - S)$. As an example, consider a bottom slope, S_o, which exceeds the slope appropriate for uniform flow, S, so that $S_o - S > 0$. If the flow is subcritical, i.e., if $Fr < 1$, then the depth increases with distance, and the velocity decreases. By contrast, the depth decreases and the velocity increases when the flow is supercritical.

7.8.4   Specific Energy

We can gain some qualitative feel for the dynamics of open-channel flows by introducing the **specific energy**, E, defined by

$$E = y + \frac{\overline{u}^2}{2g}$$

(7.202)

Focusing on a rectangular channel of width b for simplicity, we know from mass conservation that the volume-flow rate is $Q = \overline{u}by$. Thus, we can rewrite the specific energy as

$$E = y + \frac{Q^2}{2gb^2y^2}$$

(7.203)

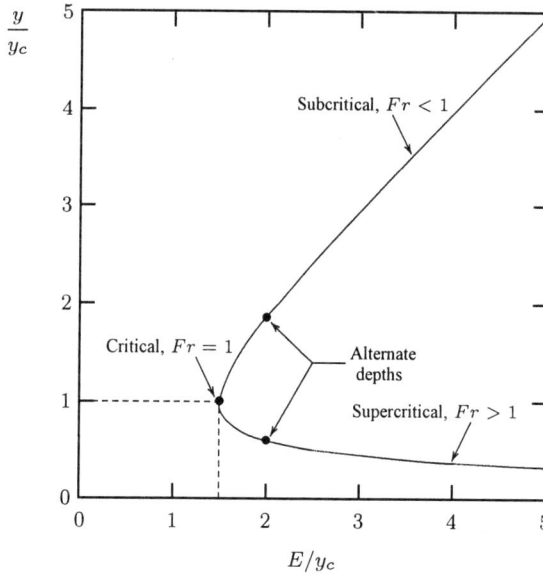

Figure 7.21: *Specific energy for flow in an open channel;* y_c *is critical depth.*

Figure 7.21 graphically displays specific energy for a fixed value of $Q^2/(gb^2)$. Inspection of the graph shows that E has a minimum value. We can locate this minimum by setting dE/dy equal to zero, i.e.,

$$\frac{dE}{dy} = 1 - \frac{Q^2}{gb^2y^3} = 0 \quad \Longrightarrow \quad y = \left(\frac{Q^2}{gb^2}\right)^{1/3} \equiv y_c$$

(7.204)

For reasons that will become obvious below, the quantity y_c is known as the **critical depth**. Dividing Equation (7.203) through by y_c, there follows:

$$\frac{E}{y_c} = \frac{y}{y_c} + \frac{1}{2}\left(\frac{y_c}{y}\right)^2$$

(7.205)

Hence, the minimum value of E is

$$E_{min} = E(y_c) = \frac{3}{2}y_c$$

(7.206)

Using this value in the original definition of E, Equation (7.202), we can solve for the velocity at the point where E achieves its minimum value, viz.,

$$y_c + \frac{\overline{u}_c^2}{2g} = \frac{3}{2}y_c \quad \Longrightarrow \quad \overline{u}_c = \sqrt{gy_c} \tag{7.207}$$

The specific-energy curve is the locus of all states possible for a given flow rate and channel width [so that $Q^2/(2gb^2)$ is constant]. Now, since the volume-flow rate is constant on the curve, we know that

$$Q = \overline{u}_c b y_c = \overline{u}by \quad \Longrightarrow \quad \overline{u} = \overline{u}_c \frac{y_c}{y} \tag{7.208}$$

Hence, the Froude number at any point on the curve is

$$Fr = \frac{\overline{u}}{\sqrt{gy}} = \frac{\overline{u}_c y_c}{\sqrt{gy}\,y} = \frac{\sqrt{gy_c}\,y_c}{\sqrt{gy}\,y} = \left(\frac{y_c}{y}\right)^{3/2} \tag{7.209}$$

Therefore, the Froude number at the critical depth is 1, corresponding to critical flow, which is the reason we call y_c the *critical depth*. This equation also shows that the Froude number is less than 1 on the upper branch of the curve, wherefore the flow is subcritical. By contrast, it is greater than 1 on the lower branch, corresponding to supercritical flow.

As shown in Figure 7.21, for a given volume-flow rate and specific energy, there are two possible depths. These are referred to as **alternate depths**. Since the total energy is the same, the subcritical branch corresponds to a state with high potential energy and low kinetic energy, while the distribution of energy is the opposite on the supercritical branch.

Flow under a **sluice gate** provides an example of flow in which the alternate depths occur for a fixed value of E. A sluice gate, depicted in Figure 7.22, is a control structure used to regulate flow rate. As shown, the subcritical fluid upstream of the gate has high potential energy and low kinetic energy. After passing through the gate, the volume-flow rate is unchanged and, in the absence of head losses, so is the specific energy. The flow becomes supercritical with low potential energy and high kinetic energy. For fixed values of E and Q, the depth upstream of the gate is alternate to the depth downstream.

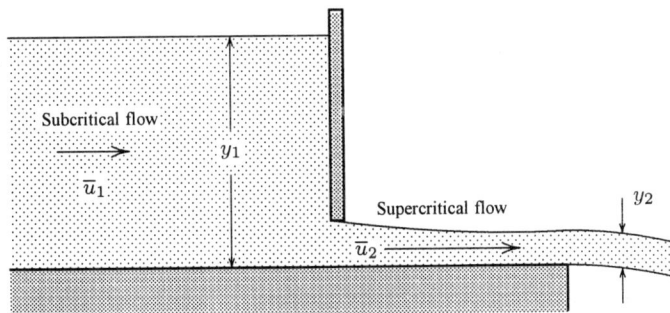

Figure 7.22: *Flow through a sluice gate.*

For example, consider flow in a rectangular open channel that is regulated by a sluice gate. The channel has width $b = 10$ ft, depth $y_1 = 5$ ft and volume-flow rate $Q = 100$ ft^3/sec. Conservation of mass tells us the average flow velocity is $\overline{u} = Q/(by_1) = 2$ ft/sec. Also, using Equation (7.204), a short calculation shows that the critical depth is $y_c = 1.46$ ft, wherefore $y_1/y_c = 3.42$. From Figure 7.21, the alternate depth is $y_2/y_c \approx 0.40$. Therefore, the depth downstream of the sluice gate is $y_2 = 0.584$ ft = 7 inches.

7.8.5 Hydraulic Jump

A sluice gate is an example of a flow that changes from subcritical to supercritical, with a negligibly small head loss. When the opposite change occurs with the upstream flow being supercritical and the downstream flow being subcritical, we have what is known as a **hydraulic jump**. This situation is sometimes imposed as a design feature, by forcing an abrupt change in depth, or may occur naturally because of the prevailing depth downstream. The hydraulic jump is analogous to a shock wave in a compressible fluid, which we will discuss in Chapter 8.

As shown in Figure 7.23, the flow upstream of the jump is fast and shallow. After the jump, the flow becomes slow and deep, and there is a substantial head loss. Because they are very effective in dissipating energy and in mixing, hydraulic jumps are commonly used in spillways and in sewage-treatment plants.

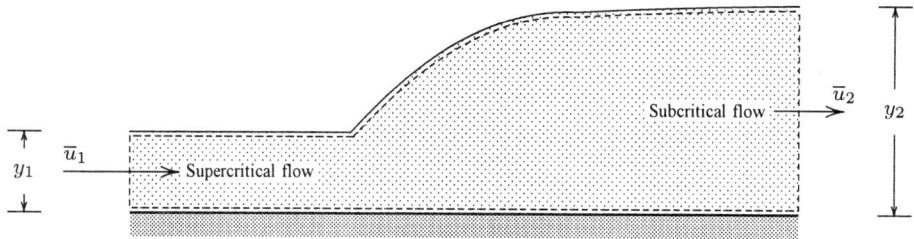

Figure 7.23: *Hydraulic jump.*

We can analyze a hydraulic jump using the control volume indicated by the dashed contour in Figure 7.23. For simplicity, we consider a horizontal bottom and an infinitely-wide channel. We also neglect the viscous force in the momentum equation. Our approach is the same as that used to compute the velocity of a surface wave in Subsection 7.8.3, with the exception that we do not assume changes in velocity and depth are small. Referring to Equations (7.178) and (7.186), we replace c by $\overline{u}_1$, $c - \Delta u$ by $\overline{u}_2$, y by y_1 and $y + \Delta y$ by y_2, wherefore

$$\rho \overline{u}_1 y_1 = \rho \overline{u}_2 y_2 \tag{7.210}$$

$$\rho \overline{u}_1^2 y_1 = \rho \overline{u}_2^2 y_2 + \frac{1}{2} \rho g \left(y_2^2 - y_1^2 \right) \tag{7.211}$$

$$y_1 + \frac{\overline{u}_1^2}{2g} = y_2 + \frac{\overline{u}_2^2}{2g} + h_L \tag{7.212}$$

We include the energy Equation (7.212) to permit solving for the head loss across the jump.

Omitting details of the algebra for the sake of brevity, we can solve for the depth ratio, y_2/y_1, velocity ratio, $\overline{u}_2/\overline{u}_1$, Froude number after the jump, Fr_2, and head loss, h_L, as a function of the Froude number ahead of the jump, Fr_1. The results are as follows.

$$Fr_2 = Fr_1 \left[\frac{2}{\sqrt{1 + 8Fr_1^2} - 1} \right]^{3/2} \tag{7.213}$$

$$\frac{\overline{u}_2}{\overline{u}_1} = \frac{2}{\sqrt{1 + 8Fr_1^2} - 1} \tag{7.214}$$

$$\frac{y_2}{y_1} = \frac{1}{2} \left[\sqrt{1 + 8Fr_1^2} - 1 \right] \tag{7.215}$$

$$h_L = \frac{(y_2 - y_1)^3}{4y_1 y_2} \tag{7.216}$$

Since, by definition, head loss is always positive, necessarily a hydraulic jump causes depth to increase. Inspection of Equation (7.215) shows that we have $y_2 \geq y_1$ when $Fr_1 \geq 1$, while $y_2 < y_1$ when $Fr_1 < 1$. *Thus, a hydraulic jump can only cause a transition from supercritical to subcritical flow.*

As an example, consider flow in a wide channel with an incident Froude number of $Fr_2 = 2$ and a volume-flow rate per unit width of $Q = 11.4$ ft²/sec, in which a hydraulic jump occurs. We would like to determine flow properties in the channel. First, we determine the incident depth and velocity by noting that

$$\overline{u}_1 = \frac{Q}{y_1} \quad \text{and} \quad \overline{u}_1 = Fr_1\sqrt{gy_1} \quad \Longrightarrow \quad y_1 = \left(\frac{Q^2}{4g}\right)^{1/3} \tag{7.217}$$

For the given numerical values, we find

$$y_1 = 1.00 \text{ ft} \quad \text{and} \quad \overline{u}_1 = 11.4 \frac{\text{ft}}{\text{sec}} \tag{7.218}$$

Using Equations (7.213) through (7.216), a straightforward arithmetic exercise shows that:

$$Fr_2 = 0.55, \quad \frac{\overline{u}_2}{\overline{u}_1} = 0.42, \quad \frac{y_2}{y_1} = 2.37, \quad \frac{h_L}{y_1} = 0.27 \tag{7.219}$$

Therefore, downstream of the hydraulic jump, we have

$$y_2 = 2.37 \text{ ft} \quad \text{and} \quad \overline{u}_2 = 4.8 \frac{\text{ft}}{\text{sec}} \tag{7.220}$$

Finally, note that the specific energy in the flow upstream of the hydraulic jump is given by $E_1 = y_1 + \overline{u}_1^2/(2g) = 3.0$ ft, so that the relative loss of energy is

$$\frac{h_L}{E_1} = .09 \tag{7.221}$$

For this low a Froude number, a hydraulic jump is usually thought of as being relatively weak. Even so, 9% of the energy in the flow is dissipated into heat. For Froude numbers in excess of 9, the head loss can be as high as 85% of the incident specific energy.

Problems

7.1 Find the compressibility, τ_s, for isentropic flow of a perfect gas defined by

$$\tau_s \equiv \frac{1}{\rho}\left(\frac{\partial \rho}{\partial p}\right)_s$$

where subscript s means entropy is held constant. Compare your result with the isothermal compressibility discussed in Section 1.6.

7.2 Beginning with Gibbs' equation, viz., $Tds = de + pd(1/\rho)$, first introduce the enthalpy, h, and show that $Tds = dh - dp/\rho$. Then, prove that for a perfect gas that is also calorically perfect, the equation of state can be written as

$$\frac{p}{p_o} = \left(\frac{T}{T_o}\right)^{\gamma/(\gamma-1)} \exp\left[\frac{(s_o - s)}{R}\right]$$

where subscript 'o' denotes reference state conditions.

7.3 Beginning with Gibbs' equation, viz., $Tds = de + pd(1/\rho)$, determine how entropy varies for an *isothermal* flow of a perfect, calorically-perfect gas.

7.4 Beginning with Gibbs' equation, viz., $Tds = de + pd(1/\rho)$, determine how entropy varies for an *isobaric* flow of a perfect, calorically-perfect gas.

7.5 A container filled with 4 kg of nitrogen at $p_1 = 0.2$ MPa and $T_1 = 0°$ C is compressed isentropically to a pressure $p_2 = 0.5$ MPa. Determine the final temperature, T_2, and the work done, W_c, in compressing the gas. **NOTE:** Table 1.3 includes relevant thermodynamic properties of nitrogen.

7.6 The internal energy of a perfect, *thermally-perfect* gas is

$$e(T) = e_o + 8R\sqrt{T_o T}$$

where e_o and T_o are constants. Noting that $de = c_v dT$ and $c_p = c_v + R$ for a thermally-perfect gas, determine c_v and c_p. What is the specific-heat ratio, γ, when $T = 4T_o$?

7.7 For a *thermally-perfect* gas, by definition $e = e(T)$ and $h = h(T)$. Noting that

$$c_p = \left(\frac{\partial h}{\partial T}\right)_p \quad \text{and} \quad c_v = \left(\frac{\partial e}{\partial T}\right)_v$$

verify that $c_p - c_v = R$ for a thermally-perfect gas.

7.8 Show that for a perfect gas, the specific-heat coefficients satisfy $c_p - c_v = R$, even in the most general case where $h = h(p, T)$ and $e = e(v, T)$. **HINT:** First show that

$$c_p - c_v = R\left[1 + \frac{1}{p}\left(\frac{\partial e}{\partial v}\right)_T\right]$$

where $v = 1/\rho$ is specific volume. Then use Gibbs' equation, $Tds = de + pdv$, to determine $(\partial e/\partial v)_T$, noting that $[\partial/\partial v(\partial s/\partial T)_v]_T = [\partial/\partial T(\partial s/\partial v)_T]_v$.

7.9 In a *polytropic* process, the pressure and density of a perfect gas are related by $p = p_1(\rho/\rho_1)^n$, where n is a constant and subscript '1' denotes initial condition.

(a) If 2 slugs of air undergo a reversible, polytropic process in which the air is initially at $p_1 = 14.7$ psi and $T_1 = 60°$ F and changes to $p_2 = 25$ psi and a volume of $V_2 = 541$ ft^3, what is the polytropic exponent, n?

(b) Determine the work done by the surroundings on the gas, W, where $W \equiv \int_{V_1}^{V_2} p\, dV$. Express your answer in foot pounds.

(c) Determine the heat transferred from the surroundings to the gas, Q, in Btu's.

7.10 In a *polytropic* process, the pressure and density of a perfect gas are related by $p = p_1(\rho/\rho_1)^n$, where n is a constant and subscript '1' denotes initial condition.

(a) If 50 kg of air undergo a reversible, polytropic process in which the air is initially at $p_1 = 100$ kPa and $T_1 = 20°$ C and changes to $p_2 = 200$ kPa and $T_2 = 63.6°$ C, what is the polytropic exponent, n?

(b) Determine the work done by the surroundings on the gas, W, where $W \equiv \int_{V_1}^{V_2} p\, dV$. Express your answer in Joules.

(c) Determine the heat transferred from the surroundings to the gas, Q, in Joules.

7.11 Air flows steadily in a pipe of diameter $d = 8$ cm at a rate of $\dot{m} = 0.5$ kg/sec. The pressure and temperature at the upstream end of the pipe are $p_1 = 150$ kPa and $T_1 = 20°$ C. If we ignore heat transfer, viscous effects and body forces, what are the velocity, u_2, and temperature, T_2, in the jet of air at the outlet as it exhausts into the atmosphere? Assume that atmospheric pressure is $p_2 = 100$ kPa. Also, assume that air is a perfect, calorically-perfect gas and that the specific heat at constant pressure and perfect-gas constant are $c_p = 1004$ m²/(sec²·K) and $R = 287$ m²/(sec²·K). Develop a quadratic equation for T_2/T_1 before determining numerical values for T_2 and u_2.

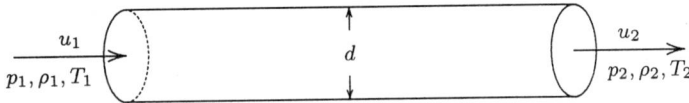

Problem 7.11

7.12 The pump shown supplies energy to the flow such that the upstream absolute pressure is p and the downstream pressure is $3p$. The steady mass-flow rate is $\dot{m}$ and the pipe diameters are as shown. What is the power, P, delivered by the pump to the flow? The temperature increases by ΔT. Also, you may neglect effects of gravity. Express your answer for the power in terms of $\dot{m}$, p, d, water density, ρ, and specific-heat coefficient, c_v.

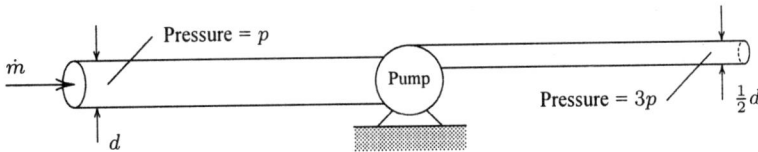

Problem 7.12

7.13 The mass-flow rate of water through a turbine is $\dot{m}$ and the pressure drop through the turbine is $p_2 - p_1 = -\Delta p$. Heat-transfer effects can be neglected and the flow is steady.

(a) Determine the power, P, delivered to the turbine from the water. Neglect internal-energy changes.

(b) Compute P for $\dot{m} = 200$ kg/sec, $\Delta p = 200$ kPa, $\rho = 1000$ kg/m³, $d = 0.2$ m and $\Delta z = 2$m. Express your answer in kilowatts and in horsepower.

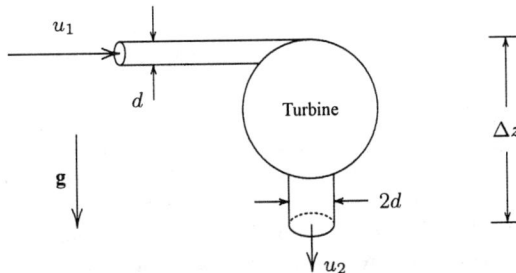

Problem 7.13

7.14 Consider a centrifugal water pump for which the mass-flow rate is $\dot{m}$. The flow is steady with water entering normal to the xy plane at Point 1 and exiting at Point 2 as shown. Gravitational effects can be neglected, the water exhausts to the atmosphere and $\Delta h = h_1 - h_2 = 8.5\mathbf{u}_1 \cdot \mathbf{u}_1$. If the power required to drive the pump is

$$\dot{W}_s = \frac{6\dot{m}^3}{5\pi^2\rho^2 d^4}$$

what is the heat transfer rate, $\dot{Q}$?

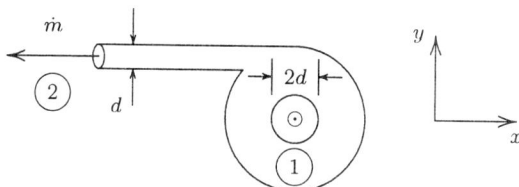

Problems 7.14, 7.15

7.15 Consider a centrifugal water pump that is sufficiently insulated to prevent any heat transfer from the surroundings. The power required to drive the pump is $\dot{W}_s$, and the mass-flow rate through the pump is $\dot{m}$. The flow is steady with water entering normal to the xy plane at Point 1 and exiting to the atmosphere at Point 2. Gravitational effects can be neglected, and $\Delta h = h_1 - h_2 = 8.5\mathbf{u}_1 \cdot \mathbf{u}_1$.

(a) Compute the force in the xy plane required to hold the pump in place as a function of $\dot{m}$ and $\dot{W}_s$. No other variables should appear in your answer.

(b) What is the force if $\dot{m} = 11$ slug/sec and $\dot{W}_s = 50$ hp.

7.16 Verify that Equation (7.38) reduces to Equation (7.34) for a conservative body force whose potential function, $\mathcal{V}$, is independent of time.

7.17 A power plant uses a river to discharge waste heat. Heat is transferred from the plant to the river through a heat exchanger at a rate $\dot{Q} = 60$ MW. The volume-flow rate in the river is $\dot{V} = 3$ m³/sec and the river temperature upstream of the power plant is $T_i = 20°$ C. The flow is steady and the river has constant cross-sectional area. Also, you can neglect frictional work amongst the river, ground and atmosphere. What is the temperature downstream of the power plant, T_f? Note that $c_p = 4200$ J/(kg·K) and $\rho = 998$ kg/m³ for the river water.

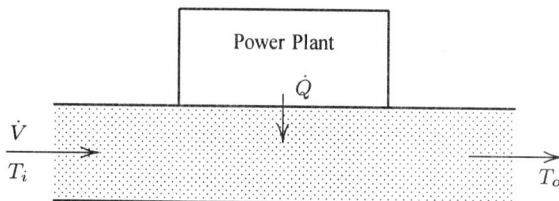

Problems 7.17, 7.18

7.18 A power plant uses a river to discharge waste heat. Heat is transferred from the plant to the river through a heat exchanger at a rate $\dot{Q}$. The volume-flow rate in the river is $\dot{V} = 50$ ft³/sec and difference between the temperature downstream and upstream of the power plant is $\Delta T = T_f - T_i = 5°$ F. The flow is steady and the river has constant cross-sectional area. Also, you can neglect frictional work amongst the river, ground and atmosphere. What is the heat-transfer rate from the power plant to the river? Note that $\rho c_p = 62.6$ Btu/(ft³·°R) for the river water.

7.19 A laser is used to energize the flow of air through a channel of height H and width $5H$ (out of the page). The design objective is to increase the velocity in the channel by 10%. Ignoring viscous losses and effects of gravity, what heat-transfer rate, $\dot{Q}$, must be supplied by the laser if $u_1 = 200$ ft/sec and $T_1 = -150°$ F? Assume the flow is steady, pressure is constant and equal to 1 atm throughout the channel and $H = 1$ ft. If an affordable laser delivers 10-20 Btu/sec, can the design objective be realized?

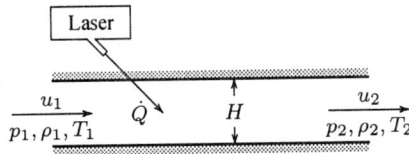

Problem 7.19

7.20 Beginning with the one-dimensional, adiabatic, inviscid differential form of the energy equation,

$$\rho \frac{d\mathcal{H}}{dt} = \rho \frac{\partial \mathcal{H}}{\partial t} + \rho u \frac{\partial \mathcal{H}}{\partial x} = \frac{\partial p}{\partial t}$$

and making use of the corresponding one-dimensional form of Euler's equation, verify that specific enthalpy, h, satisfies the following equation.

$$\rho \frac{dh}{dt} = \frac{dp}{dt}$$

7.21 For turbulent channel flow, a useful approximation for the velocity profile between the channel wall and centerline is

$$\frac{u}{u_m} = \left(\frac{y}{H} \right)^n, \qquad 0 \leq y \leq H$$

where u_m is centerline velocity, y is distance from the channel wall and H is channel half height. Also, n is a number between 1/6 and 1/8. Determine the kinetic-energy correction factor, α, as a function of n. Evaluate α for $n = 1/6$, $n = 1/7$ and $n = 1/8$.

7.22 For turbulent pipe flow, a useful approximation for the velocity profile is

$$\frac{u}{u_m} = \left(\frac{y}{r_o} \right)^n, \qquad y = r_o - r$$

where u_m is centerline velocity, y is distance from the surface of the pipe, r_o is pipe radius and r is distance from the centerline. Also, n is a number between 1/6 and 1/8. Show that the kinetic-energy correction factor, α, is given by:

$$\alpha = \frac{(n+1)^3 (n+2)^3}{4(1+3n)(2+3n)}$$

Evaluate α for $n = 1/6$, $n = 1/7$ and $n = 1/8$.

7.23 Consider the section of pipe shown below. The radius of the pipe is constant and equal to R. At the inlet, the velocity is constant and equal to U. At the outlet, the velocity is given by $u = u_o(r)$, where r is distance from the pipe centerline. The difference between inlet pressure, p_i, and outlet pressure, p_o, is $p_i - p_o = \rho U^2$, where ρ is the density of the (incompressible) fluid in the pipe. The pipe lies in a horizontal plane so that potential energy is constant.

(a) Determine U_{max} as a function of U.

(b) Verify that $\alpha_1 = 1$ and $\alpha_2 = 2.7$.

(c) Compute the head loss, h_L, as a function of U and the acceleration of gravity, g.

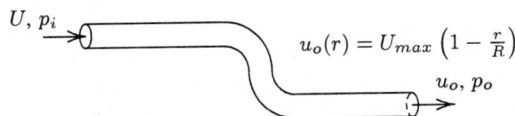

Problem 7.23

7.24 Consider flow through a constant diameter pipe that carries water from an elevation z_1 to another elevation z_2. The average velocity at elevation z_1 is $\overline{u}_1 = U$ and the overall head loss through the pipe is $h_L = .03U^2/(2g)$. Curiously, in terms of the averaged velocities $\overline{u}_1$ and $\overline{u}_2$, Bernoulli's equation is satisfied between elevations z_1 and z_2. Determine the difference between the kinetic-energy correction factors, $\alpha_1 - \alpha_2$.

7.25 Water is pumped through a pipe with a bend lying in a horizontal plane in such a way that pressure, p, and average velocity, U, remain constant throughout. The kinetic-energy correction factor upstream of the bend, α_1, exceeds the downstream value by 0.05, i.e., $\alpha_1 = \alpha_2 + 0.05$. If the head loss through the bend is $h_L = 0.15U^2/(2g)$, what head, h_p, must the pump deliver?

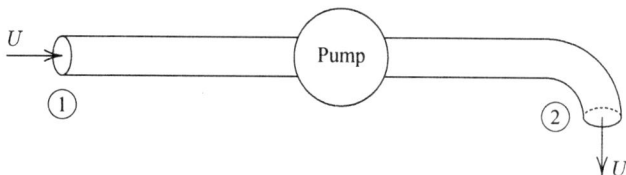

Problem 7.25

7.26 An incompressible fluid of kinematic viscosity $\nu = 10^{-4}$ m^2/sec flows through a pipe of diameter $D = 10$ cm.

   (a) If the average flow velocity, $\overline{u}$, is 1 m/sec, is the flow laminar or turbulent? What is the approximate value of the friction coefficient, f?

   (b) If the pipe is perfectly smooth and $\overline{u}$ increases to 10 m/sec, will the flow most likely be laminar or turbulent? What is the friction coefficient, f?

   (c) Suppose, however, that the pipe surface is very rough. Suppose further that increasing $\overline{u}$ to 100 m/sec yields no change in the value of f determined in Part (a) above. What must the relative roughness, k_s/D, be?

7.27 Consider a pipe of diameter $D = \frac{1}{2}$ m and length $L = 100$ m. A liquid of kinematic viscosity $\nu = 5 \cdot 10^{-7}$ m^2/sec flows through the pipe with average velocity $\overline{u} = 10$ m/sec. Determine the friction factor, f, and head loss, h_L, for pipes made of glass, copper and cast iron.

7.28 Consider a pipe of diameter $D = \frac{3}{2}$ in and length $L = 300$ ft. A liquid of kinematic viscosity $\nu = 1.25 \cdot 10^{-6}$ ft^2/sec flows through the pipe with average velocity $\overline{u} = 5$ ft/sec. Determine the friction factor, f, and head loss, h_L, for pipes made of plastic, steel and galvanized iron.

7.29 The head loss for a 1 mile long, perfectly-smooth pipe is $h_L = 48.64$ ft. Water with kinematic viscosity $\nu = 10^{-5}$ ft^2/sec flows through the pipe, whose diameter, D, is 9 inches. What is the average flow velocity, $\overline{u}$?

7.30 The head loss for a 2 km long, galvanized-iron pipe is $h_L = 96.68$ m. Water with kinematic viscosity $\nu = 10^{-6}$ m^2/sec flows through the pipe, whose diameter, D, is 7.5 cm. What is the average flow velocity, $\overline{u}$?

7.31 For plastic piping, what diameter, D, is needed to convey $\dot{V} = .03$ m^3/sec of fuel oil if the system can support a head loss of $h_L = 7$ m over a length $L = 300$ m? The kinematic viscosity of the fuel oil is $\nu = 2 \cdot 10^{-4}$ m^2/sec. Assume and verify that the flow is laminar.

7.32 With great care, laminar pipe flow can be realized for Reynolds number, Re_D, as high as 50000. For water ($\nu = 10^{-5}$ ft^2/sec) flowing in a 2 in diameter, 100 ft long pipe, compute the head loss, in inches, for laminar and turbulent flow when $Re_D = 50000$.

7.33 A liquid of density ρ is pumped from a tank to the atmosphere. The diameter of the pipe connecting the tank and pump is $2D$. After leaving the pump the pipe diameter is D. If the head supplied by the pump is $h_p = 2.4\Delta z$, where Δz is as indicated in the figure, what is the dimensionless head loss, $2gh_L/U^2$? Express your answer in terms of Δz, U and g. You may assume that throughout this flow the kinetic-energy correction factor, α, is 1.05. What is $2gh_L/U^2$ if $\Delta z = 5$ ft and $U = 15$ ft/sec?

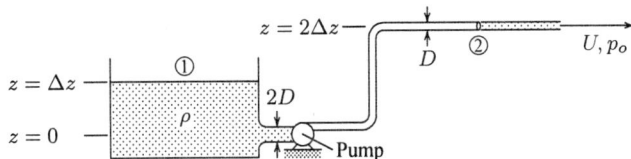

Problems 7.33, 7.34

7.34 A liquid of density ρ is pumped from a tank to the atmosphere. The diameter of the pipe connecting the tank and pump is $2D$. After leaving the pump the pipe diameter is D. If the head loss is assumed to be $h_L = \frac{3}{2}U^2/(2g)$, where U is the velocity in the smaller pipe, what head must the pump supply? Express your answer in terms of Δz, U and g. You may assume that throughout this flow the kinetic-energy correction factor, α, is 1.0. What is h_p if $\Delta z = 2$ m and $U = 10$ m/sec?

7.35 Consider the hydroelectric plant shown. The head loss between Points 1 and 2 is $h_L = \frac{1}{5}H$. If the mass-flow rate through the turbine is $\dot{m}$, how much power is transferred from the water to the turbine? In developing your answer, assume the kinetic-energy correction factor, α, is 1.06 throughout the flow, and that the water emitted by the turbine exhausts to the atmosphere.

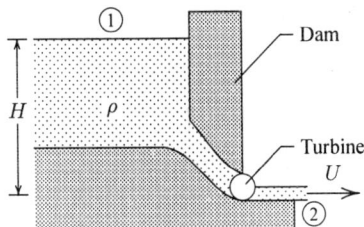

Problems 7.35, 7.36

7.36 Consider the hydroelectric plant shown. The head loss between Points 1 and 2 is $h_L = \frac{1}{20}H$. If the mass-flow rate through the turbine is $\dot{m}$, how much power is transferred from the water to the turbine? In developing your answer, assume the kinetic-energy correction factor, α, is 1.04 throughout the flow, and that the water emitted by the turbine exhausts to the atmosphere. Also, the Froude number (based on H) of the water emitted by the turbine is 0.31.

7.37 What head, h_p, must be supplied by the pump in order to pump water at a rate $\dot{m}$ from the lower to the upper reservoir? Assume $\alpha_1 = \alpha_2 = 1$ and that the head loss in a pipe of length L and diameter D is given by $h_L = .0234\frac{U^2}{2g}\frac{L}{D}$. For the two pipes, we have $D_1 = D$, $D_2 = D/2$, $L_1/D_1 = 100$ and $L_2/D_2 = 9900/16$. Express your solution for h_p in terms of ρ, Δz, $\dot{m}$, g and D.

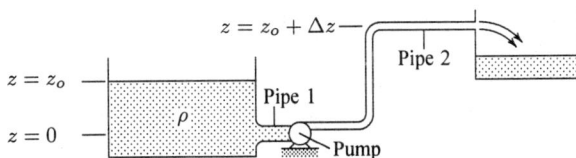

Problems 7.37, 7.38

7.38 What head, h_p, must be supplied by the pump in order to pump water at a rate $\dot{m}$ from the lower to the upper reservoir? Assume $\alpha_1 = \alpha_2 = 1$ and that the head loss in a pipe of length L and diameter D is given by $h_L = .028\frac{U^2}{2g}\frac{L}{D}$. For the two pipes, we have $D_1 = D$, $D_2 = D/2$, $L_1/D_1 = 20$ and $L_2/D_2 = 300$. Express your solution for h_p in terms of ρ, Δz, $\dot{m}$, g and D.

7.39 Consider incompressible, steady flow in a pipe that undergoes a sudden expansion as indicated below. The pipe lies in a horizontal plane so that potential energy changes can be ignored. The kinetic-energy correction factor, α, is 1.04 throughout the pipe, and the pressure change is found from measurements to be given by

$$C_p \equiv \frac{p_2 - p_1}{\frac{1}{2}\rho \bar{u}_1^2} = \frac{1}{2}$$

The head loss for the sudden expansion is known from measurements to be of the form:

$$h_L = K \frac{\bar{u}_1^2}{2g}$$

Determine the velocity $\bar{u}_2$ as a function of $\bar{u}_1$ and the coefficient K.

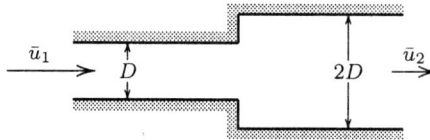

Problem 7.39

7.40 The volume-flow rate through a hydroturbine operating between two large bodies of water of density ρ is $\dot{V}$, and the head loss between Points 1 and 2 is $h_L = \frac{3}{20}H$. If the reference points lie at the free surfaces far from the turbine, what power, $\dot{W}_t$, is generated? Express your answer in terms of ρ, $\dot{m}$, H and gravitational acceleration, g. Compute $\dot{W}_t$ in kW for $\dot{V} = 113.5$ m³/sec, $\rho = 1000$ kg/m³ and $H = 10$ m.

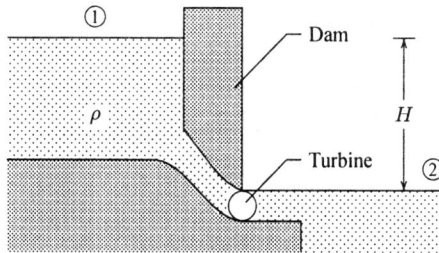

Problem 7.40

7.41 Compute the hydraulic diameter for a concentric annulus. Note that the perimeter is that of the *wetted area*, i.e., it is the sum of the perimeters of both circles. Does your answer make sense in the limiting cases $\lambda = 0$ and $\lambda = 1$?

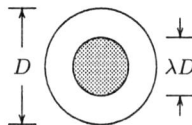

Problem 7.41

7.42 A square duct of side h has a cylindrical obstruction of diameter d centered as shown. If the hydraulic diameter is $D_h = d$, what is h? Note that the perimeter is that of the *wetted area*, i.e., it is the sum of the duct and obstruction perimeters.

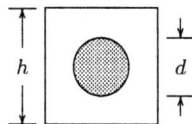

Problem 7.42

7.43 Consider a straight duct required to carry a given mass flux, $\dot{m}$, of a fluid of density ρ over a distance L. The flow is laminar and the duct cross-sectional area is A. We wish to determine the head loss for several alternative cross sections. Show that the head loss, h_L, is given by

$$h_L = \frac{P^2}{4\pi A} h_{L_o}$$

where P is perimeter and h_{L_o} is the head loss for a circular cross section. Compute h_L/h_{L_o} for the following:

(a) A square of side h;

(b) A rectangle of sides $2h \times h$;

(c) An equilateral triangle of side h.

7.44 Water is extracted from a reservoir through a pipe of diameter $D_1 = D$ and length $L_1 = L$ connected to a second pipe of diameter $D_2 = \frac{1}{2}D$ and length $L_2 = \frac{1}{2}L$. The pipe is a distance Δz below the reservoir surface. The friction factor in the large pipe is $f_1 = .020$, while that of the small pipe is $f_2 = .024$.

(a) Ignoring minor losses and assuming $\alpha = 1$, what is the exit velocity, U? Express your answer as a function of D, L, Δz and gravitational acceleration, g.

(b) Determine U for $D = 10$ cm, $L = 200$ m and $\Delta z = 20$ m.

(c) Now, include the inlet and sudden contraction losses and repeat Parts (a) and (b).

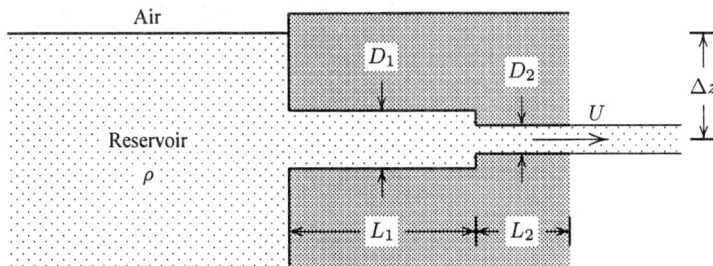

Problems 7.44, 7.45, 7.46

7.45 Water is extracted from a reservoir through a pipe of diameter $D_1 = D$ and length $L_1 = L$ connected to a second pipe of diameter $D_2 = \frac{1}{2}D$ and length $L_2 = \frac{1}{2}L$. The pipe is a distance Δz below the reservoir surface. The friction factor in the large pipe is $f_1 = .040$, while that of the small pipe is $f_2 = .048$. The desired pipe exit velocity is U.

(a) Ignoring minor losses and assuming $\alpha = 1$, what is Δz? Express your answer as a function of D, L, U and gravitational acceleration, g.

(b) Determine Δz for $D = 6$ in, $L = 100$ ft and $U = 10$ ft/sec.

(c) Now, include the inlet and sudden contraction losses and repeat Parts (a) and (b).

7.46 Water is extracted from a reservoir through a pipe of diameter $D_1 = D$ and length $L_1 = L$ connected to a second pipe of diameter $D_2 = \frac{3}{5}D$ and length $L_2 = \frac{2}{5}L$. The pipe is a distance Δz below the reservoir surface. The friction factor in the large pipe is $f_1 = .0120$, while that of the small pipe is $f_2 = .0135$. The pipe exit velocity is U.

(a) Ignoring minor losses and assuming $\alpha = 1$, what is D? Express your answer as a function of L, Δz, U and gravitational acceleration, g.

(b) Determine D for $L = 410$ m, $U = 4$ m/sec and $\Delta z = 40$ m.

(c) Now, include the inlet and sudden contraction losses and repeat Parts (a) and (b).

7.47 Consider a circular pipe with a $60°$ gradual contraction followed by a $10°$ gradual expansion as shown. Assuming $L = 100D$, $k_s = .006D$ and $Re_D > 10^7$, determine the dimensionless head loss, gh_L/U^2 for d/D = 0.2, 0.4, 0.6 and 0.8.

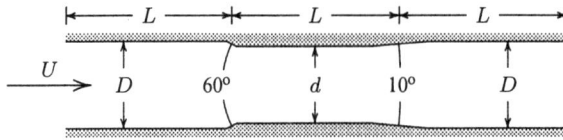

Problem 7.47

7.48 Water flows from a large reservoir through a pipe of constant diameter D as shown. The inlet to the pipe is square and the radius of curvature of each $90°$ bend is $\mathcal{R} = 4D$. The kinetic-energy correction factor is $\alpha = 1.06$ throughout.

(a) If the pipe exit velocity is U and $h/D = 100$, develop an equation for the friction factor, f, as a function of D, U and gravitational acceleration, g.

(b) If U = 7.13 m/sec, D = 11.2 cm and kinematic viscosity, $\nu = 10^{-6}$ m²/sec, determine the roughness height, k_s, of the pipe.

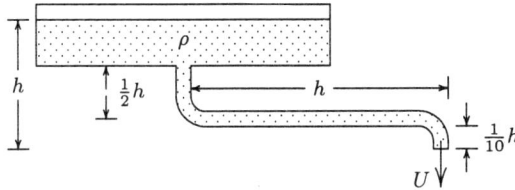

Problems 7.48, 7.49

7.49 Water flows from a large reservoir through a pipe of constant diameter D as shown. The inlet to the pipe is square and the radius of curvature of each $90°$ bend is $\mathcal{R} = 2D$. The kinetic-energy correction factor is $\alpha = 1.05$ throughout.

(a) If $h/D = 100$, develop an equation for the pipe exit velocity, U, as a function of friction factor, f, h and gravitational acceleration, g.

(b) If the pipe is perfectly smooth, h = 10 ft and kinematic viscosity, $\nu = 10^{-5}$ ft²/sec, determine U. **HINT:** You can solve this problem iteratively as follows: (1) make a guess for U; (2) compute Re_D; (3) determine f from the Moody diagram; (4) compute U from the formula derived in Part (a); (5) return to step 1 and repeat until U is determined to within 0.1 ft/sec.

7.50 Consider a reservoir with a pipe and nozzle attached as shown. The inlet to the pipe is not rounded, nor is the entrance to the final pipe section leading into the nozzle. The radius of curvature of both $90°$ bends is $\mathcal{R} = .06D$. The only loss in the nozzle is due to the $60°$ contraction. The friction factor in the pipe is $f = .0255$ when the diameter is D, and $f = .0200$ when the diameter is $\frac{5}{2}D$. Determine the nozzle exit velocity, U, as a function of L and gravitational acceleration, g. Assume $\alpha = 1$ throughout.

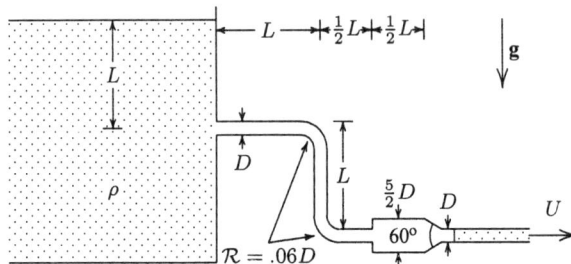

Problem 7.50

7.51 An incompressible fluid of density ρ flows through the multiple nozzle shown and discharges to the atmosphere as shown. You may assume the flow is steady and that properties are uniform across the nozzle. Also, the nozzle lies in a horizontal plane so that potential energy effects are unimportant. If the head loss between the inlet and the largest outlet pipe (the one with diameter $D/2$) is $h_L = \frac{9}{25}U^2/(2g)$ and the kinetic-energy correction factor is $\alpha = 1$ throughout, determine the force required to hold the pipe in place. Express your answer in terms of ρ, U and D.

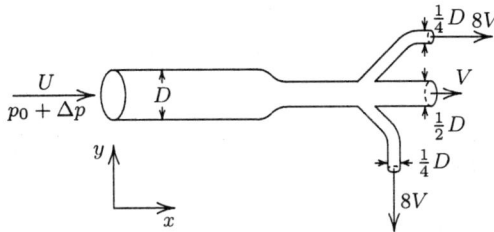

Problem 7.51

7.52 Water flows steadily through a *Black Box* as shown. The pressure at the inlet pipe a distance L from the box is $(p_o + \Delta p)$, where p_o denotes atmospheric pressure. The two identical outlet pipes, also of length L, exhaust to the atmosphere. The entire geometry lies in a horizontal plane, so that gravitational effects can be neglected. Also, assume there is no net reaction force from the *Black Box*.

(a) Use conservation of mass and momentum to determine U_o as a function of U_i, and to derive a relation between U_i, $\Delta p/\rho$, L, D, and the friction coefficients f_i and f_o for the inlet and outlet pipes, respectively.

(b) Assuming the total head loss as the fluid flows from the inlet pipe, through the Black Box and through each of the outlet pipes is h_L, express $\Delta p/\rho$ as a function of U_i, g and h_L. Assume $\alpha = 1$ throughout.

(c) Combining results from Parts (a) and (b), show that if $h_L = 0.1U_i^2/(2g)$ and $L/D = 11$, the friction factors are related by $f_i + 4f_o = 0.1$.

(d) Assuming flow throughout the system occurs at Reynolds number based on pipe diameter in excess of 10^7 and also that the roughness height in the inlet pipe is given by $k_s/D = .001$, determine the approximate value of k_s in the outlet pipes. Express your answer in terms of D, the diameter of the inlet pipe.

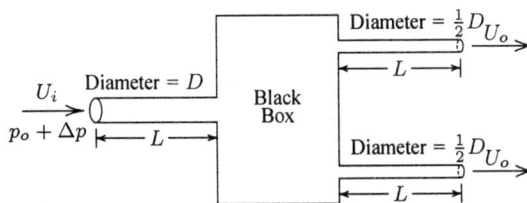

Problem 7.52

7.53 For laminar flow through three pipes in series as shown, the equivalent length is $L_e = 3L$. If $L/D = 10$, what is the Reynolds number, Re_D?

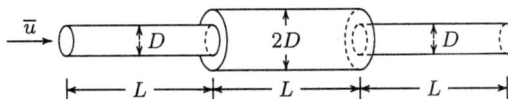

Problems 7.53, 7.54

7.54 For turbulent flow through three pipes in series as shown, the equivalent length is $L_e = 3L$. If $L/D = 100$, what is the Reynolds number, Re_D? Use the *Blasius formula* for the friction factor, $f \approx 0.3164Re_D^{-1/4}$.

7.55 Water flows at 68° F between two reservoirs with a parallel piping system as shown. All pipes have circular cross section and are made of galvanized iron.

(a) Ignoring minor losses, compute the flow velocity in each pipe as a function of g, d, h, L and the three friction factors, f_1, f_2 and f_3.

(b) Assuming Reynolds numbers are high enough to achieve completely-rough flow so that the friction factors approach their high-Reynolds-number asymptotic values, determine the flow velocities for $D = 2$ in, $h = 100$ ft and $L = 1000$ ft. Use the Moody diagram to estimate the friction factors.

(c) Using your results from Part (b), compute the Reynolds number in each pipe. Compare the left- and right-hand sides of the Colebrook equation for the friction factors used in Part (b).

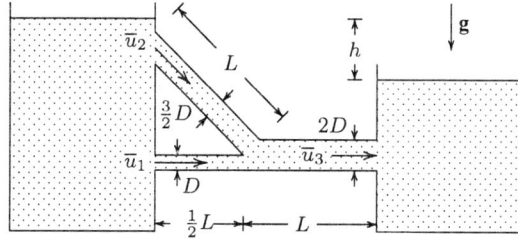

Problems 7.55, 7.56

7.56 Water flows at 20° C between two reservoirs with a parallel piping system as shown. All pipes have circular cross section and are made of cast iron.

(a) Ignoring minor losses, compute the flow velocity in each pipe as a function of g, d, h, L and the three friction factors, f_1, f_2 and f_3.

(b) Assuming Reynolds numbers are high enough to achieve completely-rough flow so that the friction factors approach their high-Reynolds-number asymptotic values, determine the flow velocities for $D = 1.3$ cm, $h = 200$ m and $L = 400$ m. Use the Moody diagram to estimate the friction factors.

(c) Using your results from Part (b), compute the Reynolds number in each pipe. Compare the left- and right-hand sides of the Colebrook equation for the friction factors used in Part (b).

7.57 SAE30 oil with $\rho = 1.71$ slug/ft^3 and $\nu = 1.18 \cdot 10^{-3}$ ft^2/sec flows through a parallel pipe system. The pressure difference is $p_1 - p_2 = 3.9$ psi, while the pipe diameters and lengths are $D_a = 1$ in, $D_b = 2$ in, $L_a = 10$ ft and $L_b = 20$ ft. The diameter, altitude, z, and α are the same at Points 1 and 2. Ignoring minor losses, compute the volume-flow rate, Q. **HINT:** Assume the flow is laminar. Confirm this assumption after solving.

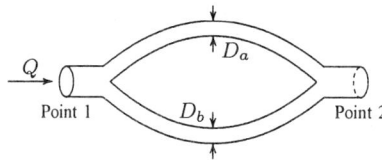

Problems 7.57, 7.58

7.58 Glycerin with $\rho = 1260$ kg/m^3 and $\nu = 1.19 \cdot 10^{-3}$ m^2/sec flows through a parallel pipe system. The volume-flow rate is $Q = 0.25$ m^3/sec, while the pipe diameters and lengths are $D_a = D_b = 10$ cm, $L_a = 20$ m and $L_b = 10$ m. The diameter, altitude, z, and α are the same at Points 1 and 2. Ignoring minor losses, compute the pressure drop, $p_1 - p_2$. **HINT:** Assume the flow is laminar. Confirm this assumption after solving.

7.59 A wide, 1 cm sheet of water at 20° C runs down a 5° glass surface with $k_s = 0.3$ mm. Compute the average flow velocity, $\bar{u}$, first by using the Chézy-Manning equation and then by using the Moody diagram and the Colebrook formula.

7.60 A wide, 1 inch sheet of water at $68°$ F runs down a $10°$ asphalt surface with $k_s = .018$ ft. Compute the average flow velocity, $\bar{u}$, first by using the Chézy-Manning equation and then by using the Moody diagram and the Colebrook formula.

7.61 Water flows in a canal with trapezoidal cross section as shown. The bottom slope is $S_o = .0016$, while the canal dimensions are $b = 4$ m, $y = 2$ m, $\alpha = 30°$. Using the Chézy-Manning approach, determine the volume-flow rate, Q, and the Froude number, Fr, if the canal is lined with brickwork and also if it is weedy.

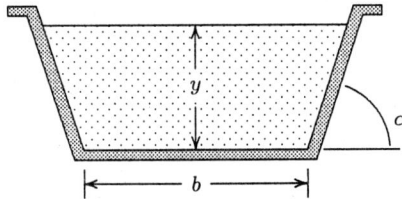

Problems 7.61, 7.62, 7.63

7.62 Water flows in a canal with trapezoidal cross section as shown. The bottom slope is $S_o = .0009$, while the canal dimensions are $b = 10$ ft, $y = 7$ ft, $\alpha = 50°$. Using the Chézy-Manning approach, determine the volume-flow rate, Q, the Froude number, Fr, and the Reynolds number Re_{D_h}, if the canal is lined with smooth steel and also if it has a stony surface. Assume $\nu = 10^{-5}$ ft²/sec.

7.63 A natural channel can be approximated as having trapezoidal cross section with $b = 1000$ ft, $y = 300$ ft and $\alpha = 15°$. The bottom slope is $S_o = .0001$ and the Froude number is $Fr = 0.112$. What is the Manning roughness coefficient, n?

7.64 We wish to maximize the volume-flow rate in an open channel of rectangular cross section under uniform-flow conditions. The optimum shape is called the *best hydraulic cross section*. Our goal is to find the *aspect ratio*, b/y, (b = width, y = depth) that corresponds to the best hydraulic cross section.

(a) Verify that the area of the channel, A, satisfies $A^{5/2}y = \lambda\left(A + 2y^2\right)$ where λ is a constant depending on the bottom slope, S_o, the volume-flow rate, Q, and the Manning friction coefficient, n.

(b) Find the minimum value of $A(y)$ and the corresponding aspect ratio. **HINT:** After setting $dA/dy = 0$, eliminate λ by using the equation derived in Part (a).

7.65 We wish to maximize the volume-flow rate in an open channel of triangular cross section under uniform-flow conditions. The optimum shape is called the *best hydraulic cross section*. Our goal is to find the angle, α, that corresponds to the best hydraulic cross section.

(a) Verify that the area of the channel, A, satisfies $A^5 y^2 = 4\lambda\left(A^2 + y^4\right)$ where λ is a constant depending on the bottom slope, S_o, the volume-flow rate, Q, and the Manning friction coefficient, n.

(b) Find the minimum value of $A(y)$ and the corresponding angle α. **HINT:** After setting $dA/dy = 0$, eliminate λ by using the equation derived in Part (a).

7.66 A *flume* must be designed to deliver water from a mountain lake to a small hydroelectric power plant. The required volume-flow rate is $Q = 2.5$ m³/sec, the bottom slope is $S_o = 0.0016$ and the Manning friction coefficient is $n = 0.016$. Determine flume size for a rectangular cross section with $b = 3y$, a semi-circular cross section with $\alpha = \pi$ and a triangular cross section with $\alpha = \frac{1}{3}\pi$ (see Figure 7.19 for geometric details). Make a table summarizing area, A, perimeter, P, hydraulic radius, R_h, and Froude number for all three cases.

7.67 At two different times, a pebble is dropped into an open channel that is very wide. Each pebble creates a circular ripple that is convected downstream as shown. The channel has constant depth y and the channel flow speed is U.

(a) Noting that the wave speed for weak radial waves is the same as for planar waves, verify that U is given by

$$U = \frac{\ell \sqrt{gy}}{r_1 - r_2}$$

(b) If the upstream wavefronts are coincident, what is the Froude number, Fr?

(c) Compute U and Fr for $y = 50$ cm, $r_1 = 10$ m, $r_2 = 5$ m and $\ell = 18$ m.

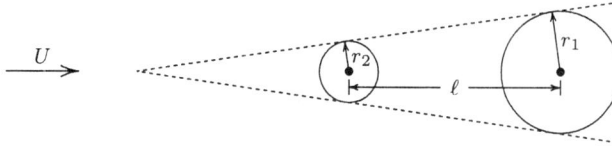

Problem 7.67

7.68 A small stone is dropped into a brook. After one second, the ripple created by the stone has traveled 4 m downstream. Assuming the brook has a constant depth $y = 75$ cm, what is the flow speed? Also, how far upstream will the ripple travel in one second?

7.69 A small stone is dropped into a brook. After one second, the ripple created by the stone has traveled 3 ft upstream and 5 ft downstream. Assuming the brook has constant depth, what are the flow speed and the depth?

7.70 A small surface disturbance is created in a uniform open-channel flow in a very wide channel. The flow velocity is U and the channel depth is y. There are two observers, one a distance ℓ upstream of the point of disturbance and the other a distance ℓ downstream. How much time, Δt, passes after the wave reaches the downstream observer before the upstream observer senses the wave? Express your answer as a function of ℓ, U and Froude number, Fr. Make a graph of the dimensionless time, $U\Delta t/\ell$, versus Fr for $0 \leq Fr < 1$ and discuss what happens when $Fr > 1$.

7.71 Consider a gravelly rectangular channel of width $b = 11$ ft that has a constant volume-flow rate of $Q = 500$ ft^3/sec. Determine the critical depth, y_c, and the *critical slope*, S_c, i.e., the slope given by the Chézy-Manning equation with critical depth and critical velocity. What angle, θ_c, corresponds to this bottom slope?

7.72 A flood occurs over farmland and the volume-flow rate per unit width, Q/b, is 9 m^2/sec. If the depth is $y = 1$ m, is the flow subcritical or supercritical? What is the ratio of the specific energy to its minimum value?

7.73 Consider a sluice gate in a rectangular channel of width b. The volume-flow rate is Q, and head loss can be neglected.

(a) Determine Q as a function of b, y_1, g and y_2/y_1.

(b) If $y_2 = \frac{1}{2}y_c$, where y_c is the critical depth, what is y_2/y_1?

(c) For the conditions of Part (b), what are the Froude numbers upstream and downstream of the sluice gate?

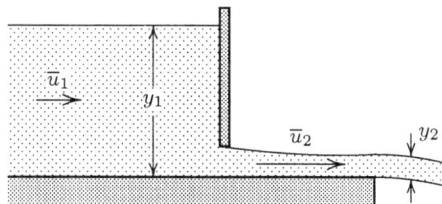

Problems 7.73, 7.74

7.74 The Froude number downstream of a sluice gate shown in the figure is $Fr = 8$. What is the Froude number upstream of the gate? **HINT:** Show that y_1/y_c satisfies a cubic equation of the form

$$(y_1/y_c)^3 - N(y_1/y_c)^2 + \frac{1}{2} = 0$$

for which the desired root is $y_1/y_c \approx N$.

7.75 A hydraulic jump occurs in a wide rectangular flume. The volume-flow rate per unit width is $Q/b = 20$ ft^2/sec and the depth before the jump is $y_1 = 11$ in. Determine the depth after the jump, y_2, and the head loss, h_L.

7.76 Water flows down a steep hill in a wide rectangular open channel as shown. A hydraulic jump occurs, abruptly increasing the channel depth from y_1 to y_2. The volume-flow rate is $Q = 750$ ft^3/sec, the channel width is $b = 15$ ft and the depth before the jump is $y_1 = 2.05$ ft. Compute the depth after the jump, and the Froude numbers and velocities upstream and downstream of the jump.

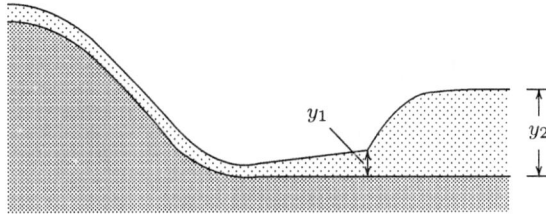

Problems 7.76, 7.77

7.77 Water flows down a steep hill in a wide rectangular open channel as shown. A weak hydraulic jump occurs, increasing the channel depth from y_1 to y_2. The volume-flow rate is $Q = 56$ m^3/sec, the channel width is $b = 20$ m and the depth after the jump is $y_2 = 1$m. Compute the depth before the jump, and the Froude numbers and velocities upstream and downstream of the jump. **HINT:** Use the fact that, for a weak hydraulic jump, $Fr_2 \approx 2 - Fr_1$.

7.78 Water flows in a wide rectangular open channel, without head loss, over a bump. Because of the prevailing depth downstream of the bump, a hydraulic jump occurs.

(a) Determine $\bar{u}_0$, $\bar{u}_1$ and Fr_1 as functions of y_0, y_1 and gravitational acceleration, g.

(b) Compute Fr_1 for $y_0 = 3$ m and $y_1 = 1.4$ m.

(c) Compute y_2 for the Froude number of Part (b).

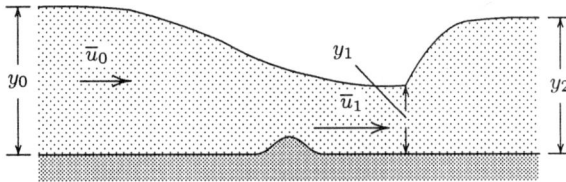

Problem 7.78

7.79 Show for a weak hydraulic jump ($Fr_1 \approx 1$) that the Froude number downstream of the jump is approximated by

$$Fr_2 \approx 2 - Fr_1$$

Compare values for Fr_2 obtained using this equation with values for Fr_2 from the exact solution for $Fr_1 = 1$, 1.05, 1.1 and 1.2. **HINT:** Let $Fr_1 = 1 + \epsilon$ where $\epsilon \ll 1$ and expand the exact relationship between Fr_2 and Fr_1.

7.80 Verify that the ratio of the head loss across a hydraulic jump to the initial specific energy is

$$\frac{h_L}{E_1} = \frac{\left(\sqrt{1+8Fr_1^2} - 3\right)^3}{8\left(\sqrt{1+8Fr_1^2} - 1\right)(2 + Fr_1^2)}$$

Compute this ratio, as a percentage, for $Fr_1 = 2, 5, 10$ and 20.

Chapter 8

One-Dimensional Compressible Flow

In this chapter, we take up the study of flows in which density variations cannot be ignored. Thus far, most of the fluid-flow applications we have dealt with have been for incompressible fluids. We have tentatively chosen to characterize such flows as having constant density, albeit with a warning in Section 1.6 that an incompressible flow is formally the limiting case of vanishing compressibility, τ. We will show that, for a gas, this occurs in the limit of very small Mach number. The first section provides a basis for classifying the primary compressible-flow regimes in terms of Mach number.

Compressible flow is a fascinating branch of fluid mechanics that introduces us to interesting new physical phenomena. In contrast to incompressible flow, where the energy equation is uncoupled from the equations for mass and momentum conservation, we must now include all three conservation principles to arrive at a closed set of equations. We will also need the equation of state to relate the various thermodynamic properties.

Despite the increased number of equations we have to deal with, we will find the mathematics to be far simpler than in most other branches of fluid mechanics. The latter point is true because a great deal of insight can be gained by analyzing steady, inviscid, one-dimensional flows, most notably **isentropic flow** and **normal shock waves**. Such applications are the primary focus of this chapter. We take up viscous, compressible flows and two-dimensional applications in Chapter 15.

Examples of compressible flows abound in everyday life. Common compressible flow applications are an aircraft cruising faster than three-tenths the speed of sound (i.e., approximately 200 mph at sea level) and flow within the standard combustion reciprocating engine. The flow of traffic on a large highway behaves much like flow in a compressible medium. Equally important for the modern comforts we enjoy is flow through rocket engines (to launch communication satellites), gas- or steam-turbine engines (to generate electricity), pipelines at high pressure (to deliver natural gas for heating and cooking) and supersonic wind tunnels (to help design and improve performance of these devices).

There is even an interesting analogy between compressible flows and open-channel flows (Section 7.8). Flow through a sluice gate and a hydraulic jump are analogs of isentropic flow and a shock wave, respectively. Although details are beyond the scope of this chapter, the differential equations governing wave propagation in open channels are identical to those for a compressible fluid with a specific-heat ratio, γ, of 2.

8.1   Classification

With regard to compressibility, flows are classified according to the Mach number, M, defined as follows.

$$M = \frac{U}{a} = \frac{\text{Flow Speed}}{\text{Sound Speed}} \tag{8.1}$$

There are five different Mach-number regimes, each of which is summarized in Table 8.1. As we will see later in this chapter, most analytical expressions for compressible flows involve the square of the Mach number. Thus, for example, incompressible flow extends all the way to a Mach number of about 0.3. This is true because, to a first approximation, the density in the flow of a gas, ρ, differs from its incompressible value, ρ_t, by a fraction equal to $\frac{1}{2}M^2$ [see Equation (8.61) below]:

$$\rho \approx \rho_t \left(1 - \frac{1}{2}M^2 \right) \tag{8.2}$$

When the Mach number is 0.3, this corresponds to a difference of less than 5%. Because the speed of sound is generally very large relative to typical flow velocities, virtually all liquid flows are incompressible. This is not true for gases, however, as we saw in Section 7.5 when we developed Bernoulli's equation from the energy-conservation law. In the sections to follow, we will deal mostly with air.

Table 8.1: *Compressible-Flow Regimes*

Mach Number Range	Type of Flow	M^2
0.0 - 0.3	Incompressible	0.00 - 0.09
0.3 - 0.8	Subsonic	0.09 - 0.64
0.8 - 1.2	Transonic	0.64 - 1.44
1.2 - 5.0	Supersonic	1.44 - 25
>5	Hypersonic	>25

The primary difference amongst the various Mach number regimes is in the way waves propagate. Briefly, for incompressible and subsonic flow, waves propagate at a speed much greater than characteristic flow velocities. Consequently, any disturbance in the flow causes waves to move in both the upstream and downstream directions. By contrast, waves move at speeds less than flow velocities in supersonic and hypersonic flow, and flow disturbances can only propagate downstream. The transonic-flow regime is mathematically complicated because it includes regions of both subsonic and supersonic flow. We will illuminate some of the differences amongst the Mach-number regimes in Section 8.5.

Note that the Mach number ranges are not intended to mark precise boundaries, but are simply ranges consistent with common nomenclature. Some engineers, for example, prefer to consider a Mach 4 flow to be in the low hypersonic range, and many of the approximations that have been developed from hypersonic flow theory are quite accurate for Mach numbers as small as 3. At the lower end of the Mach number spectrum, many Computational Fluid Dynamicists often consider Mach 0.1 to be the upper bound for an accurate incompressible-flow computation.

8.2 Thermodynamic Properties of Air

We have discussed thermodynamics in some detail in Section 7.1. We also summarized some common thermodynamic properties of gases in Section 1.5. Because we will be dealing primarily with air in this chapter, it is worthwhile to list its thermodynamic properties here for convenience.

- We assume the thermodynamic properties of air are given by the perfect-gas law, which relates pressure, p, density, ρ, and temperature, T, as follows:

$$p = \rho R T \tag{8.3}$$

where R = 1716 ft·lb/(slug·°R) [287 J/(kg·K)] is the perfect-gas constant.

- For temperatures up to about 3600° R (2000 K), air is well approximated as being calorically perfect. Thus, its specific-heat coefficients, c_p and c_v, are constant. The internal energy, e and enthalpy, h, are hence given by

$$e = c_v T \quad \text{and} \quad h = c_p T \tag{8.4}$$

- As shown in Subsection 7.1.4, the specific-heat coefficients can be expressed in terms of γ and R as follows.

$$c_p = \frac{\gamma}{\gamma - 1} R \quad \text{and} \quad c_v = \frac{1}{\gamma - 1} R \tag{8.5}$$

- Because c_p and c_v are constant, so is the specific-heat ratio, γ. Its value for air is

$$\gamma = \frac{c_p}{c_v} = 1.4 \tag{8.6}$$

- Using the combined first and second laws of thermodynamics (Subsection 7.1.4), we have shown that the entropy, s, for a perfect, calorically-perfect gas is

$$s = c_v \, \ell n \left(\frac{p}{\rho^\gamma} \right) + \text{constant} \tag{8.7}$$

which, for isentropic flow, tells us that

$$p = A\rho^\gamma, \quad A = \text{constant} \qquad \text{(Isentropic flow)} \tag{8.8}$$

Another useful isentropic relation follows from Equation (7.18), which tells us $T/\rho^{\gamma-1}$ is constant for isentropic flow. Thus, $\rho/T^{1/(\gamma-1)} = \text{constant}$, whence

$$p = B T^{\gamma/(\gamma-1)}, \quad B = \text{constant} \qquad \text{(Isentropic flow)} \tag{8.9}$$

8.3 Speed of Sound

We can compute the speed of sound for a perfect gas by analyzing an **acoustic wave**, i.e., a weak sound wave traveling through the gas. We denote the sound speed by a, while the pressure and density of the undisturbed medium are p and ρ. As shown in Figure 8.1(a), after the wave passes, the pressure and density are $p + \Delta p$ and $\rho + \Delta \rho$. Also, the wave imparts a velocity, Δu, to the fluid. For an acoustic wave, we assume the changes in flow properties are very small, i.e.,

$$|\Delta p| \ll p, \quad |\Delta \rho| \ll \rho, \quad |\Delta u| \ll a \tag{8.10}$$

We can simplify our analysis a bit by making a Galilean transformation so that the wave is stationary and the flow is from left to right, as shown in Figure 8.1(b). In this coordinate frame, the velocity upstream (to the left of the wave) is a. Similarly, the velocity downstream (to the right of the wave) is $a - \Delta u$.

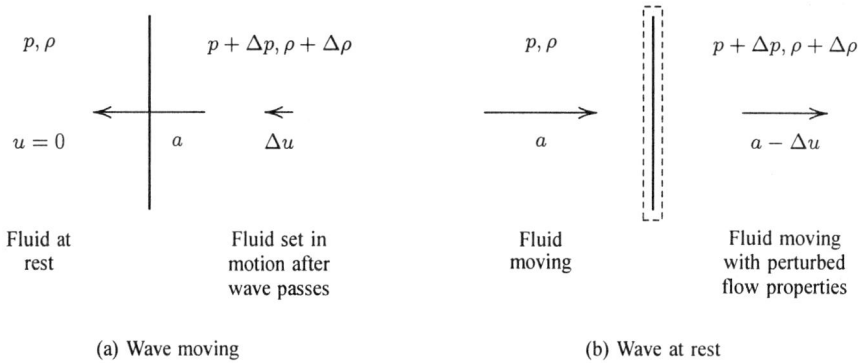

| p, ρ | $p + \Delta p, \rho + \Delta \rho$ | p, ρ | $p + \Delta p, \rho + \Delta \rho$ |

| $u = 0$ | a | Δu | a | $a - \Delta u$ |

| Fluid at rest | Fluid set in motion after wave passes | Fluid moving | Fluid moving with perturbed flow properties |

(a) Wave moving   (b) Wave at rest

Figure 8.1: *Analysis of an acoustic wave; a Galilean transformation produces a steady flow with fluid passing through the stationary wave.*

To make further progress, we can implement our tried and proven control volume method. We select a control volume whose boundaries are just upstream and downstream of the wave. Because we are using a one-dimensional geometry, we can select unit area. Since the flow is steady, conservation of mass tells us the net flux of fluid out of the control volume vanishes. Thus,

$$\oiint_S \rho \, \mathbf{u} \cdot \mathbf{n} \, dS = 0 \tag{8.11}$$

Hence, remembering that $\mathbf{n}$ is always an *outer* unit normal, doing the integrals yields

$$-\rho a + (\rho + \Delta \rho)(a - \Delta u) = 0 \tag{8.12}$$

Now, expanding the terms in parentheses, we have

$$-\rho a + \rho a - \rho \Delta u + a \Delta \rho - \Delta \rho \Delta u = 0 \tag{8.13}$$

so that

$$\rho \Delta u = a \Delta \rho - \Delta \rho \Delta u \tag{8.14}$$

Since we have assumed the wave causes an infinitesimally small change in all flow properties (in particular, since $|\Delta u| \ll a$), we can ignore the last term in Equation (8.14) as it is quadratic in changes while the other two terms are linear. Thus, we conclude from conservation of mass that the changes in velocity and density are related as follows.

$$\rho \Delta u \approx a \Delta \rho \qquad (8.15)$$

If we ignore body forces and viscous effects,[1] momentum conservation simplifies to

$$\oiint_S \rho u(\mathbf{u} \cdot \mathbf{n}) \, dS = -\mathbf{i} \cdot \oiint_S p \, \mathbf{n} \, dS \qquad (8.16)$$

where we take the dot product of $\mathbf{i}$ with the net pressure integral to extract its x component. Therefore, expanding the various integrals, we have

$$-\rho a^2 + (\rho + \Delta \rho)(a - \Delta u)^2 = p - (p + \Delta p) \qquad (8.17)$$

We can simplify the momentum equation immediately by canceling p, wherefore

$$\rho a^2 = \Delta p + (\rho + \Delta \rho)(a - \Delta u)^2 \qquad (8.18)$$

Then, using Equation (8.12), we can simplify this result further, viz.,

$$\rho a^2 = \Delta p + \rho a(a - \Delta u) = \Delta p + \rho a^2 - \rho a \Delta u \qquad (8.19)$$

Therefore, conservation of momentum yields the following (exact) relation between the changes in velocity and pressure.

$$\rho a \Delta u = \Delta p \qquad (8.20)$$

Finally, we can combine Equations (8.15) and (8.20) to arrive at

$$a^2 \Delta \rho \approx \Delta p \qquad (8.21)$$

If we take the limit $\Delta \rho \to 0$, neglecting the higher order term in Equation (8.14) is now seen to be of no consequence. Thus, we can say

$$a^2 = \lim_{\Delta \rho \to 0} \frac{\Delta p}{\Delta \rho} \qquad (8.22)$$

The indicated operation must be a partial derivative since all thermodynamic state variables such as sound speed depend upon two other state variables. So, we have to specify what thermodynamic variable is held constant when we differentiate.

Recall that the energy-conservation equation for adiabatic, inviscid flow can be written as [see Equation (7.62)]:

$$\rho T \frac{ds}{dt} = 0 \qquad (8.23)$$

We will find in Chapter 15 that both heat transfer and viscous effects (dissipation) cause entropy changes. Nevertheless, even when we include heat transfer and viscous effects, they are negligible because acoustic signals travel slow enough that large gradients in T and u cannot arise [cf. Liepmann and Roshko (1963)]. This means that entropy changes caused by heat transfer and viscous dissipation, which depend on temperature and velocity gradients, are negligible for an acoustic wave. This is not true for stronger waves, but we are assuming that

---

[1] The latter are in fact zero even in a viscous fluid provided the flow is uniform ahead of and behind the wave.

we have an infinitesimal disturbance. Thus, Equation (8.23) is a good approximation for an acoustic wave, i.e., the wave is isentropic, wherefore

$$a^2 = \left(\frac{\partial p}{\partial \rho}\right)_s \tag{8.24}$$

where subscript s means the partial derivative is computed with entropy held constant. Equation (8.24) provides the general definition of the speed of sound in a gas.

We can use the isentropic relations for a gas to compute the speed of sound in terms of thermodynamic state variables. Recalling Equation (8.8), differentiation yields

$$a^2 = \gamma A \rho^{\gamma-1} = \gamma \frac{p}{\rho} = \gamma R T \tag{8.25}$$

Therefore, the speed of sound for a perfect gas is

$$a = \sqrt{\gamma R T} \tag{8.26}$$

For example, on a hot summer day when the temperature is 95° F, and in the middle of winter in Massachusetts when the temperature dips to 5° F, Equation (8.26) predicts that the speed of sound in air is

$$a = \begin{cases} 1154 \text{ ft/sec,} & \text{Summer}: \quad (T = 554.67°\text{R}) \\ 1057 \text{ ft/sec,} & \text{Winter}: \quad (T = 464.67°\text{R}) \end{cases} \tag{8.27}$$

The temperature appearing in Equation (8.26) is absolute temperature (°R in this example).

As a final comment, recall from Section 1.6 that the compressibility of a fluid, τ, is defined as

$$\tau = \frac{1}{\rho}\frac{\partial \rho}{\partial p} \tag{8.28}$$

Comparison with Equation (8.24) shows that, for isentropic flow, the compressibility is

$$\tau = \frac{1}{\rho a^2} \tag{8.29}$$

Thus, since incompressible flow corresponds to the limit $\tau \to 0$, we have shown that incompressible flow also corresponds to the limiting case of very large sound speed. That is, for finite flow velocity, U, the Mach number approaches 0 for $a \gg U$, which is consistent with the discussion of the preceding subsection.

8.4   Subsonic Versus Supersonic Flow

There is a fundamental difference between the flow attending motion below and above the speed of sound. Consider a small body moving with velocity U. As the body moves, it collides with molecules that are then set into motion. The molecules transmit information to their neighboring molecules acoustically so that information that the object is coming is transmitted in all directions at the speed of sound. Thus, if the motion is **subsonic** so that $U < a$, the motion appears as shown in Figure 8.2.

Each circle (which is the planar projection of a spherical wave) corresponds to the acoustic wave front of a signal emitted by the object as it moves. The outermost circle is the wave

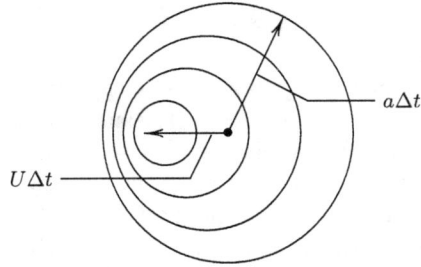

Figure 8.2: *Subsonic motion of a small object.*

front formed at the beginning of our observations, so that if we have followed the motion for a time Δt, its radius is $a\Delta t$. The inner three circles denote the wave fronts formed at times $\frac{1}{4}\Delta t$, $\frac{1}{2}\Delta t$ and $\frac{3}{4}\Delta t$ with corresponding radii of $\frac{3}{4}a\Delta t$, $\frac{1}{2}a\Delta t$ and $\frac{1}{4}a\Delta t$, respectively. Clearly, because the acoustic wave fronts move faster than the object, the fluid upstream has advance warning that the object is coming. As a result, the fluid particles move aside to allow smooth passage of the object.

Now consider what happens when the velocity of the body is **supersonic** so that $U > a$. As shown in Figure 8.3, when the body collides with a molecule, there can be no advance warning. This is true because the body is moving faster than the velocity at which molecules transmit information. Only molecules within a conical region behind the body have any knowledge that the body is passing through the fluid.

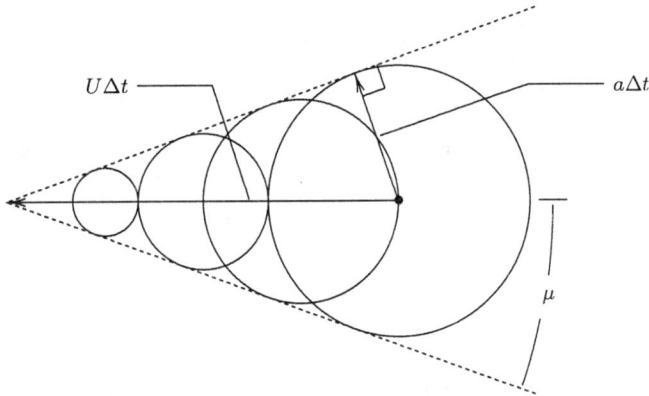

Figure 8.3: *Supersonic motion of a small object.*

The sides of the cone, whose planar projection is a wedge, are tangent to the family of spherical wave fronts. They correspond to a propagating wave front known as a **Mach wave**. The region downstream of the conical wave front is referred to as the **Mach cone**. The half angle of the Mach cone, μ, known as the **Mach angle**, is readily computed from its geometry. Inspection of Figure 8.3 shows that the sine of μ is the ratio of distance traveled by a molecule, $a\Delta t$, to the distance traveled by the object, $U\Delta t$. Therefore,

$$\mu = \sin^{-1}\left(\frac{a}{U}\right) = \sin^{-1}\left(\frac{1}{M}\right) \tag{8.30}$$

8.5   Analysis of a Streamtube

We can discover some of the interesting differences amongst the various Mach-number ranges by analyzing adiabatic, inviscid flow of a compressible fluid in a **streamtube** of varying area. By definition, a streamtube is a tube whose sides are streamlines (Figure 8.4). Note that because we have assumed the flow is inviscid, the sides of the tube could even be solid boundaries. The object of the following control-volume analysis is to express differential changes in velocity, du, and density, $d\rho$, in terms of the Mach number, M, and the change in area, dA. When we have developed the desired relations, which are called the **streamtube equations**, we will be in a position to draw some useful conclusions, and to illuminate some of the interesting properties and idiosyncrasies of compressible flows.

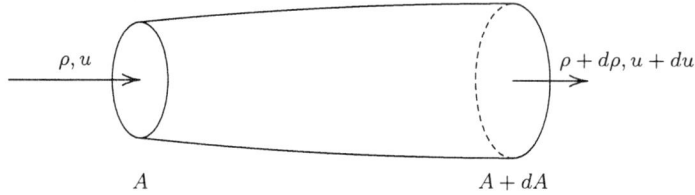

Figure 8.4: *Schematic of a streamtube.*

For steady flow, we know from continuity that the net mass flux out of the streamtube vanishes, i.e.,

$$\oiint_S \rho\, \mathbf{u} \cdot \mathbf{n}\, dS = 0 \qquad (8.31)$$

Expanding the closed surface integral into its three contributions at the inlet, the outlet and on the perimeter, we have

$$\iint_A \rho\, \mathbf{u} \cdot \mathbf{n}\, dS + \iint_{A+dA} \rho\, \mathbf{u} \cdot \mathbf{n}\, dS + \iint_{Perimeter} \rho\, \mathbf{u} \cdot \mathbf{n}\, dS = 0 \qquad (8.32)$$

By definition of a streamline, no fluid flows across the perimeter. Also, since $\mathbf{n}$ is an outer unit normal, $\mathbf{u} \cdot \mathbf{n} = -u$ at the inlet while $\mathbf{u} \cdot \mathbf{n} = u$ at the outlet. Thus, we conclude that

$$\iint_A \rho\, u\, dS = \iint_{A+dA} \rho\, u\, dS \qquad (8.33)$$

Finally, we treat flow through our streamtube as a **quasi-one-dimensional** flow, i.e., we regard all flow properties as being constant on cross sections. In practice, this requires A to vary only slowly with x. Hence, Equation (8.33) tells us

$$\rho u A = (\rho + d\rho)(u + du)(A + dA) \qquad (8.34)$$

Expanding the right-hand side yields

$$\rho u A = \rho u A + u A\, d\rho + \rho A\, du + \rho u\, dA + (\text{higher order terms}) \qquad (8.35)$$

where the "higher order terms" are quadratic and cubic in differential changes. Dropping higher order terms, canceling $\rho u A$ and dividing through by $\rho u A$ gives

$$\frac{d\rho}{\rho} + \frac{du}{u} + \frac{dA}{A} = 0 \qquad (8.36)$$

Now, consider momentum conservation. We have a balance between the net flux of x momentum and the x component of the net pressure integral so that

$$\oiint_S \rho u (\mathbf{u} \cdot \mathbf{n}) \, dS = -\mathbf{i} \cdot \oiint_S p \, \mathbf{n} \, dS \tag{8.37}$$

Hence, expanding the closed surface integrals and noting the proper orientation of outer unit normals, we have

$$-\iint_A \rho \, u^2 \, dS + \iint_{A+dA} \rho \, u^2 \, dS + \iint_{Perimeter} \rho \, u (\mathbf{u} \cdot \mathbf{n}) \, dS$$
$$= \iint_A p \, dS - \iint_{A+dA} p \, dS - \iint_{Perimeter} p \, (\mathbf{i} \cdot \mathbf{n}) \, dS \tag{8.38}$$

Hence, again regarding the flow as quasi-one-dimensional, we have

$$-\rho u^2 A + (\rho + d\rho)(u + du)^2 (A + dA) = pA - (p + dp)(A + dA)$$
$$- \iint_{Perimeter} p \, (\mathbf{i} \cdot \mathbf{n}) \, dS \tag{8.39}$$

In order to compute the integral of $p \, (\mathbf{i} \cdot \mathbf{n})$ on the perimeter, it is instructive to examine a side view of the streamtube as shown in Figure 8.5. Inspection of the geometry shows that the quantity $-(\mathbf{i} \cdot \mathbf{n}) dS$ is just the projection of the streamline on the vertical plane. However, this projection is also the differential area change, so that

$$dA = -(\mathbf{i} \cdot \mathbf{n}) dS \tag{8.40}$$

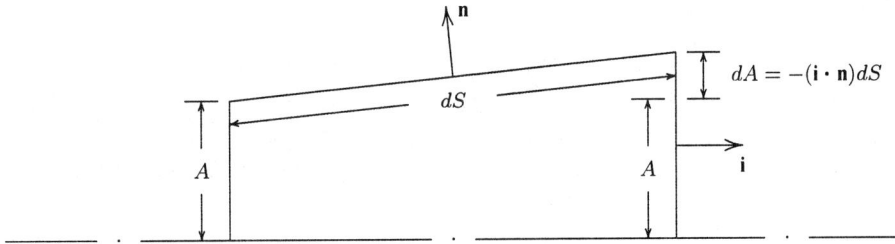

Figure 8.5: *Side view of the streamtube.*

Thus, noting Equation (8.40), we can evaluate the integral on the right-hand side of Equation (8.39). There follows

$$-\iint_{Perimeter} p \, (\mathbf{i} \cdot \mathbf{n}) \, dS = p \, dA \tag{8.41}$$

Also, note that by making use of Equation (8.34), we have

$$(\rho + d\rho)(u + du)^2 (A + dA) = \rho u A (u + du) \tag{8.42}$$

Substituting Equations (8.41) and (8.42) into Equation (8.39), momentum conservation tells us

$$-\rho u^2 A + \rho u A (u + du) = pA - (p + dp)(A + dA) + p \, dA \tag{8.43}$$

Expanding all terms in parentheses and canceling like terms yields

$$\rho u A\, du = -A\, dp - dp\, dA \tag{8.44}$$

Finally, dropping the $dp\, dA$ term, which is of higher order, and dividing through by A, we arrive at the desired differential form of the momentum equation, viz.,

$$\rho u\, du = -dp \tag{8.45}$$

We now have sufficient information to derive equations for the differential changes in density and velocity. First, we must eliminate the pressure. From momentum conservation [Equation (8.45)], we have

$$u\, du = -\frac{dp}{\rho} = -\frac{dp}{d\rho}\frac{d\rho}{\rho} = -a^2\frac{d\rho}{\rho} \tag{8.46}$$

Note that the final equality in Equation (8.46) holds because the flow is adiabatic and inviscid, and must therefore be isentropic. Hence, $dp/d\rho = (\partial p/\partial \rho)_s = a^2$. We can thus express $d\rho/\rho$ as a function of du/u. Using $M^2 = u^2/a^2$, we have

$$\frac{d\rho}{\rho} = -\frac{u}{a^2}du = -M^2\frac{du}{u} \tag{8.47}$$

Substituting this result into Equation (8.36) then yields

$$-M^2\frac{du}{u} + \frac{du}{u} + \frac{dA}{A} = 0 \tag{8.48}$$

Hence, using Equations (8.47) and (8.48), we arrive at the desired **streamtube equations**, viz.,

$$\frac{du}{u} = -\frac{1}{1-M^2}\frac{dA}{A} \quad \text{and} \quad \frac{d\rho}{\rho} = \frac{M^2}{1-M^2}\frac{dA}{A} \tag{8.49}$$

We can use these differential relations to analyze the various compressible-flow regimes.

Because there is a fundamental difference between subsonic and supersonic flow, we will examine behavior predicted by the streamtube equations in these two cases. Additionally, we will analyze the limiting behavior for incompressible flow ($M \to 0$) and for sonic conditions ($M \to 1$).

Incompressible Flow, $M \to 0$: When the speed of sound is so large that the Mach number becomes very small, we have incompressible flow. Then, Equation (8.49) tells us the relation between changes in velocity and area is

$$\frac{du}{u} = -\frac{dA}{A} \tag{8.50}$$

Because the differential changes in u and A differ in sign, a decrease in area yields an increase in velocity, and vice versa. Clearly, this must occur in order to have constant mass flow through each cross section.

Subsonic Flow, $M < 1$: Inspection of Equation (8.49) shows that du still follows $-dA$, although the change in u is greater than in the incompressible case because $1/(1-M^2) > 1$. The velocity must change more than it does for $M \to 0$ as the density changes also. For example, when the area decreases, the density decreases. Thus, to conserve mass, the additional increase in velocity (relative to the incompressible case) accounts for the reduction in fluid density. A similar argument holds for increasing area.

Supersonic Flow, $M > 1$:   When the Mach number exceeds 1, it is illuminating to rewrite Equation (8.49) as follows.

$$\frac{du}{u} = \frac{1}{M^2 - 1} \frac{dA}{A} \quad \text{and} \quad \frac{d\rho}{\rho} = -\frac{M^2}{M^2 - 1} \frac{dA}{A} \tag{8.51}$$

Now we find that du has the same sign as dA so that a decrease in area yields a decrease in velocity, while an increase in area corresponds to an increase in velocity. This is quite different from what occurs for $M < 1$. On the one hand, for decreasing area, ρ rises so rapidly that u must decrease in order to conserve mass. This is very similar to our everyday observations. Consider what happens when an accident causes the loss of a lane on a highway, resulting in "bumper-to-bumper" traffic. The density of automobiles increases so much that the same flux of automobiles can occur only at reduced speed. On the other hand, for increasing area, ρ drops so rapidly that u must increase. This is similar to what happens once the automobiles pass beyond the accident and the lost lane is regained.

The Sonic Point, M = 1:   We are left with an interesting question of what happens when the Mach number is exactly 1. Both of Equations (8.49) have $1 - M^2$ in the denominator. Hence, since singular behavior does not occur in real fluid flows, the numerators must also vanish. Thus, we can only have $M = 1$ at a point where $dA = 0$, i.e., a point of minimum or maximum area. When it's a minimum, we call such a point a **throat**. However, the inverse is not true. We do not necessarily have $M = 1$ at a throat. If $M \neq 1$ at a throat, then $du = 0$, so that u attains either a minimum or a maximum value at the throat.

8.6   Total Conditions

Again, we consider a steady flow of a frictionless gas with no heat transfer. In the absence of body forces, appeal to Equation (7.62) tells us the flow is isentropic. Also, we know the total enthalpy is constant [see Equation (7.64)], wherefore

$$h + \frac{1}{2}\mathbf{u} \cdot \mathbf{u} = \text{constant on a streamline} \tag{8.52}$$

Assuming the gas is calorically perfect, we know that

$$h = c_p T \quad \text{and} \quad c_p = \text{constant} \tag{8.53}$$

Therefore, dividing Equation (8.52) through by c_p, we have:

$$T + \frac{\mathbf{u} \cdot \mathbf{u}}{2c_p} = \text{constant} \tag{8.54}$$

We can simplify further by recalling Equations (8.5) and (8.26), so that

$$c_p = \frac{\gamma R}{\gamma - 1} = \frac{\gamma R T}{(\gamma - 1)T} = \frac{a^2}{(\gamma - 1)T} \tag{8.55}$$

So, our energy-conservation equation becomes

$$T + \frac{\gamma - 1}{2}\frac{\mathbf{u} \cdot \mathbf{u}}{a^2}T = \text{constant} \tag{8.56}$$

Thus, introducing the Mach number, we conclude finally that

$$T \left[1 + \frac{\gamma - 1}{2} M^2 \right] = \text{constant} \tag{8.57}$$

The constant appearing in Equation (8.57) is so important in compressible-flow theory that we give it the special name **total temperature**, and denote it by T_t. Therefore, we define the total temperature by

$$T_t = T \left[1 + \frac{\gamma - 1}{2} M^2 \right] \tag{8.58}$$

Total temperature is also referred to as **stagnation temperature** and **reservoir temperature**. It can be thought of as the stagnation temperature in the sense that the fluid temperature would be T_t if it were brought to rest isentropically. This is not true at a stagnation point in a real fluid however, as viscous effects cause losses that reduce the temperature somewhat. Hence, calling T_t stagnation temperature is a bit of a misnomer. It can be thought of as the reservoir temperature in a typical supersonic wind tunnel (Figure 8.6). Gas is pressurized and maintained at a temperature T_t in a large reservoir. The flow then moves through a small nozzle and accelerates from rest to the wind-tunnel design Mach number.

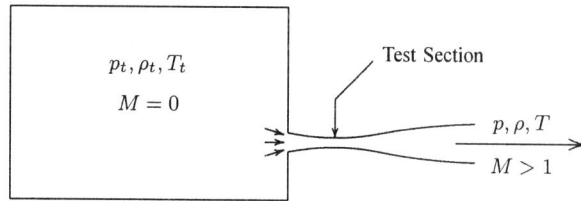

Figure 8.6: *Schematic of a supersonic wind tunnel.*

We can also define **total pressure**, p_t, and **total density**, ρ_t, from the isentropic relations [see Equations (8.8) and (8.9)]. That is,

$$\frac{p_t}{p} = \left(\frac{\rho_t}{\rho} \right)^{\gamma} = \left(\frac{T_t}{T} \right)^{\gamma/(\gamma - 1)} \tag{8.59}$$

Hence, substituting Equation (8.58) into Equation (8.59), there follows

$$p_t = p \left[1 + \frac{\gamma - 1}{2} M^2 \right]^{\gamma/(\gamma - 1)} \tag{8.60}$$

$$\rho_t = \rho \left[1 + \frac{\gamma - 1}{2} M^2 \right]^{1/(\gamma - 1)} \tag{8.61}$$

Like the total temperature, the total pressure and total density are constant in isentropic flow. Table B.1 in Appendix B includes p/p_t, ρ/ρ_t and T/T_t as a function of Mach number for $\gamma = 1.4$.

As an example, suppose air is flowing at a Mach number of 0.7, and the freestream temperature is 20° C = 293.16 K. We would like to know the temperature the fluid would have if we brought it to rest isentropically. From the Isentropic Flow Table B.1, we find $T/T_t = 0.9107$. Thus, the total temperature is 321.9 K = 48.7° C.

8.7 Normal Shock Waves

In Section 8.3 we analyzed an acoustic wave, which is very weak and produces infinitesimally small changes in fluid properties. Not all waves in a compressible medium are found to be weak. The **sonic boom** produced by supersonic aircraft, for example, is caused by a strong pressure wave known as a **shock wave**. Shock waves produce finite changes in flow properties and, as we will discuss in this section, the changes occur over an extremely thin region.

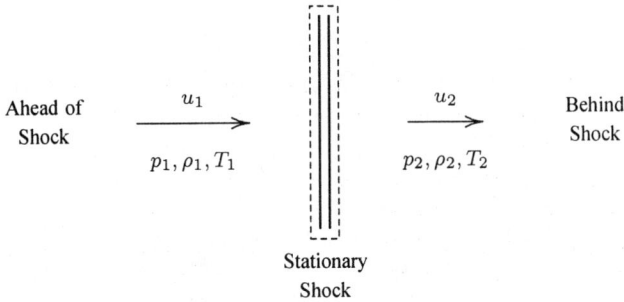

Figure 8.7: *Shock wave and associated terminology.*

Consider a one-dimensional wave propagating into an undisturbed fluid as illustrated in Figure 8.7. The pressure, density and temperature of the undisturbed fluid are p_1, ρ_1 and T_1, respectively. If the wave propagates to the left with a wave speed of u_1, a Galilean transformation puts the wave at rest, and the fluid **ahead of the shock** (i.e., the undisturbed fluid) moves to the right with speed u_1. After the fluid passes through the wave, the flow properties are p_2, ρ_2, T_2 and u_2. We refer to the latter region as lying **behind the shock**.

For an inviscid fluid with no heat transfer or body forces, because the flow is steady, conservation of mass, momentum and energy for the indicated control volume simplify to the following.

$$\rho_1 u_1 = \rho_2 u_2 \tag{8.62}$$

$$\rho_1 u_1^2 + p_1 = \rho_2 u_2^2 + p_2 \tag{8.63}$$

$$h_1 + \frac{1}{2}u_1^2 = h_2 + \frac{1}{2}u_2^2 \tag{8.64}$$

We can solve these equations with a little algebra and we find two solutions. The first solution has $p_2 = p_1$, $\rho_2 = \rho_1$, $T_2 = T_1$ and $u_2 = u_1$, which corresponds to no wave, whereby the solution is continuous in a mathematical sense. By contrast, the second solution has all flow properties changing from Region 1 to Region 2. This is a rather profound conclusion because we have specified nothing regarding the width of our control volume or the wave. That is, the second solution allows for solution discontinuities in all flow properties. This corresponds to a shock wave, and is consistent with measurements.

The width of a shock wave is a few **mean free paths**, where the mean free path is the average distance a molecule travels before suffering a collision with another molecule. Since the continuum approximation regards molecular dimensions as negligibly small, a shock wave has zero width from a continuum point of view. The second solution is most conveniently written in terms of the Mach number ahead of the shock, M_1. The solution gives what are known as the **normal-shock relations**, viz.,

$$M_2^2 = \frac{1 + \dfrac{\gamma - 1}{2} M_1^2}{\gamma M_1^2 - \dfrac{\gamma - 1}{2}} \tag{8.65}$$

$$\frac{\rho_2}{\rho_1} = \frac{u_1}{u_2} = \frac{(\gamma + 1) M_1^2}{2 + (\gamma - 1) M_1^2} \tag{8.66}$$

$$\frac{p_2}{p_1} = 1 + \frac{2\gamma}{\gamma + 1} \left(M_1^2 - 1 \right) \tag{8.67}$$

$$\frac{T_2}{T_1} = 1 + \frac{2(\gamma - 1)}{(\gamma + 1)^2} \left(\frac{1 + \gamma M_1^2}{M_1^2} \right) \left(M_1^2 - 1 \right) \tag{8.68}$$

Table B.1 in Appendix B includes the normal-shock relations as a function of Mach number.

Shock waves occur in nature whenever we have supersonic flow past objects. They also occur for supersonic flow within practical fluid mechanical devices such as jet engines and rocket nozzles. As a final comment, note that the dimensionless grouping

$$\frac{\Delta p}{p_1} \equiv \frac{p_2 - p_1}{p_1} = \frac{2\gamma}{\gamma + 1} \left(M_1^2 - 1 \right) \tag{8.69}$$

is known as the **shock strength**. Clearly, the strength of a shock wave increases as the Mach number increases.

For gases with a γ of 1.4, we can use the Shock Tables of Section B.1 in Appendix B to determine conditions behind a shock wave. For example, suppose we have a normal shock wave in a Mach 3 flow with an ambient pressure and of 1 atm and an ambient temperature of 50° F $= 509.67^\circ$ R. Reference to the shock tables shows that for $M_1 = 3.00$, conditions behind the shock are $M_2 = 0.4752$, $p_2/p_1 = 10.333$ and $T_2/T_1 = 2.679$. Therefore, behind the shock wave, we have (with a little rounding):

$$M_2 = 0.475, \quad p_2 = 10.3 \text{ atm}, \quad T_2 = 1365^\circ \text{R} \qquad \text{(Air)} \tag{8.70}$$

Now suppose the gas we are interested in is carbon dioxide, which has a specific-heat ratio of $\gamma = 1.3$. All ambient conditions and the Mach number remain unchanged. Because γ differs from 1.4, we cannot use the tables of Appendix B. Rather, we must use the normal-shock relations given in Equations (8.65) through (8.68). A little arithmetic shows that flow conditions behind the shock wave are $M_2 = 0.4511$, $p_2/p_1 = 10.043$ and $T_2/T_1 = 2.280$. Therefore,

$$M_2 = 0.451, \quad p_2 = 10.0 \text{ atm}, \quad T_2 = 1162^\circ \text{R} \qquad \text{(CO}_2) \tag{8.71}$$

8.8 Directionality

Equation (8.65) has an interesting property. Specifically, a simple algebraic exercise shows that it is still valid if we interchange indices 1 and 2. This suggests that if we have a solution that causes the Mach number, M_1, to change to M_2, then there is also a solution with M_2 ahead of the shock and M_1 behind the shock. Figure 8.8 illustrates the variation of M_2 with M_1. When the flow ahead of the shock is supersonic ($M_1 > 1$), the flow behind the shock is subsonic ($M_2 < 1$), and vice versa.

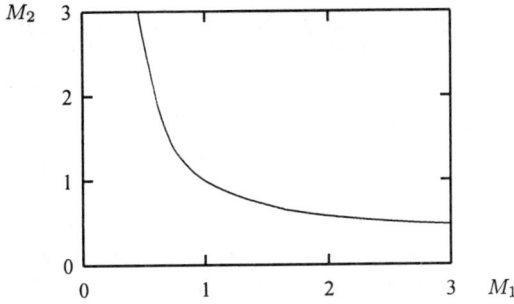

Figure 8.8: *Variation of M_2 with M_1 for $\gamma = 1.4$.*

Thus, there are two possibilities regarding the nature of the shock wave. The first possibility is a transition from supersonic flow ahead of the shock to subsonic flow behind the shock. Since the shock relations show that the density increases for $M_1 > 1$, we call this a **compression shock**. The second possibility is a transition from subsonic flow ahead of the shock to supersonic flow behind the shock so that $M_1 < 1$. In this case, the density decreases and we call it an **expansion shock**.

The second case, the expansion shock, does not occur in nature because it would result in a decrease in entropy. This is, of course, in conflict with the second law of thermodynamics, which tells us that, in a closed system, entropy is always increasing. Substituting Equations (8.66) and (8.67) into Equation (8.7) yields the entropy change across a shock, viz.,

$$\frac{s_2 - s_1}{c_v} = \ell n \left[1 + \frac{2\gamma}{\gamma + 1} \left(M_1^2 - 1 \right) \right] - \gamma \ell n \left[\frac{(\gamma + 1) M_1^2}{2 + (\gamma - 1) M_1^2} \right] \tag{8.72}$$

Examination of Equation (8.72) shows that $s_2 = s_1$ when $M_1 = 1$ and $s_2 > s_1$ for large values of M_1. However, because of its algebraic complexity, it is not obvious how the entropy changes when $M_1 < 1$. We can make further progress by examining a graph of entropy change across a shock as a function of Mach number.

Figure 8.9 shows the dimensionless change in entropy as a function of Mach number ahead of the shock, M_1. Clearly, $M_1 < 1$ yields $s_2 - s_1 < 0$, which is forbidden by the second law of thermodynamics. By contrast, $s_2 - s_1 > 0$ for $M_1 > 1$, which is consistent with the second law.

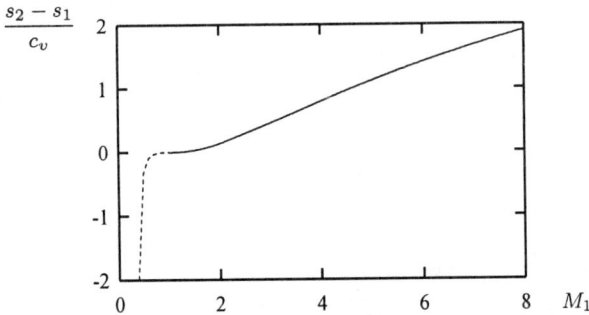

Figure 8.9: *Entropy change across a shock wave.*

These arguments establish the **directionality** of a shock wave. Obviously it always causes a transition from supersonic flow ahead of the shock to subsonic flow behind the shock.

Knowing that shocks always have $M_1 > 1$, we can now determine how all flow properties change across a shock:

- p, ρ and T increase

- T_t is constant

- u, p_t and ρ_t decrease

Because total enthalpy is conserved across a shock, necessarily the total temperature is constant. However, the shock converts some of the flow's kinetic energy into thermal energy in such a way that the total pressure and total density decrease.

8.9   Laval Nozzle

Our final topic is the **Laval nozzle**, which is the basis of the supersonic wind tunnel. Figure 8.10 illustrates a nozzle with subsonic flow upstream of the throat. Our goal is to arrive at a nozzle that achieves sonic flow ($M = 1$) at the throat and permits acceleration to supersonic flow. Since expansion shocks do not exist, we must find a way to accelerate the flow isentropically.

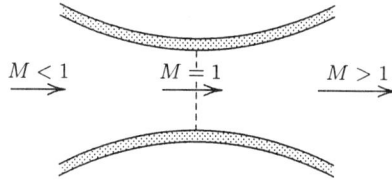

Figure 8.10: *Laval nozzle.*

We can analyze this flow using control volume methods, just as we did in analyzing streamtube flow in Section 8.5. However, we have no need to deal with differential forms. Rather, we deal directly with mass conservation in its integrated form, which tells us that

$$\rho u A = \rho^* u^* A^* \tag{8.73}$$

where superscript * denotes conditions at the throat. So, the ratio of the nozzle area to the throat area is

$$\frac{A}{A^*} = \frac{\rho^* u^*}{\rho u} = \frac{\rho^*}{\rho} \frac{a^* M^*}{a M} \tag{8.74}$$

Then, noting that $a = \sqrt{\gamma R T}$ for a perfect gas [Equation (8.26)], we have

$$\frac{A}{A^*} = \frac{\rho^*}{\rho} \sqrt{\frac{T^*}{T}} \frac{M^*}{M} \tag{8.75}$$

If our nozzle achieves sonic conditions at the throat, necessarily

$$M^* = 1 \qquad \text{(Sonic conditions at throat)} \tag{8.76}$$

Therefore, the area ratio is given by

$$\frac{A}{A^*} = \frac{\rho^*}{\rho} \sqrt{\frac{T^*}{T}} \frac{1}{M} \tag{8.77}$$

At this point, our formulation is specific to the case of a nozzle that indeed achieves sonic conditions at the throat. However, Equation (8.77) still applies even if the nozzle is operating at off-design conditions (i.e., for $M^* < 1$). In this case we cannot say that A^* is the physical area of the throat. Rather, it is a **fictitious area** whose value represents the area that would exist in the nozzle if the flow were accelerated isentropically to Mach 1.

Now, since the flow is isentropic, the total temperature and total density are both constant, wherefore

$$\left.\begin{array}{rcl} T^*\left[1+\dfrac{\gamma-1}{2}M^{*2}\right] &=& T\left[1+\dfrac{\gamma-1}{2}M^2\right] \\[3mm] \rho^*\left[1+\dfrac{\gamma-1}{2}M^{*2}\right]^{\frac{1}{\gamma-1}} &=& \rho\left[1+\dfrac{\gamma-1}{2}M^2\right]^{\frac{1}{\gamma-1}} \end{array}\right\} \tag{8.78}$$

Then, using the fact that, by hypothesis, $M^* = 1$, we have

$$\left.\begin{array}{rcl} \dfrac{T^*}{T} &=& \dfrac{2}{\gamma+1}\left[1+\dfrac{\gamma-1}{2}M^2\right] \\[3mm] \dfrac{\rho^*}{\rho} &=& \left\{\dfrac{2}{\gamma+1}\left[1+\dfrac{\gamma-1}{2}M^2\right]\right\}^{\frac{1}{\gamma-1}} \end{array}\right\} \tag{8.79}$$

Substituting Equations (8.79) into Equation (8.77) yields the following equation for the area ratio.

$$\frac{A}{A^*} = \frac{1}{M}\left\{\frac{2}{\gamma+1}\left[1+\frac{\gamma-1}{2}M^2\right]\right\}^{\frac{\gamma+1}{2(\gamma-1)}} \tag{8.80}$$

This equation yields the area ratio as a function of Mach number for *isentropic* flow. As with the isentropic relations for p/p_t, ρ/ρ_t and T/T_t, Table B.1 in Appendix B includes A/A^* as a function of Mach number. Inspection of Equation (8.80) reveals some interesting limiting values of the reciprocal of A/A^*.

$$A^*/A = 0 \text{ when } M = 0, \quad A^*/A = 1 \text{ when } M = 1, \quad A^*/A \to 0 \text{ as } M \to \infty \tag{8.81}$$

Figure 8.11 presents the variation of A^*/A with M for a specific-heat ratio, $\gamma = 1.4$. As shown, the curve is double valued, featuring a subsonic and a supersonic branch.

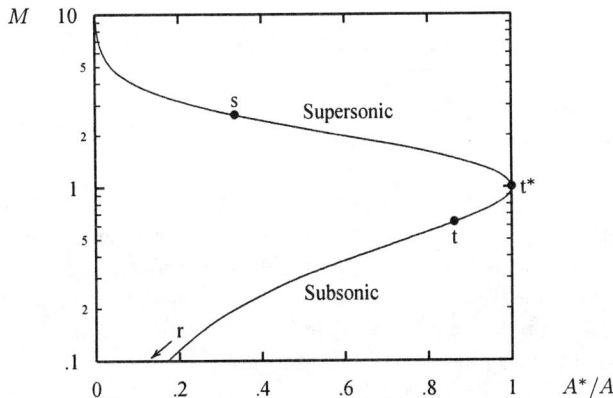

Figure 8.11: *Area-ratio variation with Mach number for* $\gamma = 1.4$.

For the following discussion, it is instructive to express the area ratio as a function of pressure. Because the flow is isentropic, we know that

$$\frac{p_t}{p} = \left[1 + \frac{\gamma - 1}{2}M^2\right]^{\frac{\gamma}{\gamma-1}} \tag{8.82}$$

With a little algebra, there follows

$$\frac{A^*}{A} = \frac{\left(\dfrac{p}{p_t}\right)^{\frac{1}{\gamma}}\left[1 - \left(\dfrac{p}{p_t}\right)^{\frac{\gamma-1}{\gamma}}\right]^{\frac{1}{2}}}{\left(\dfrac{\gamma-1}{2}\right)^{\frac{1}{2}}\left(\dfrac{2}{\gamma+1}\right)^{\frac{\gamma+1}{2(\gamma-1)}}} \tag{8.83}$$

Subsonic Case. Let Point r denote reservoir conditions and Point t the throat. If the pressure at the nozzle exit plane is too high relative to the reservoir (total) pressure, sonic conditions will not be attained at the throat. Flow will remain subsonic throughout the nozzle. With reference to Figure 8.11, we move along the subsonic branch from r up to t and return along the same branch. Cases a and b in Figure 8.12 correspond to the subsonic case. Note that the Mach number at the throat increases as the exit pressure decreases.

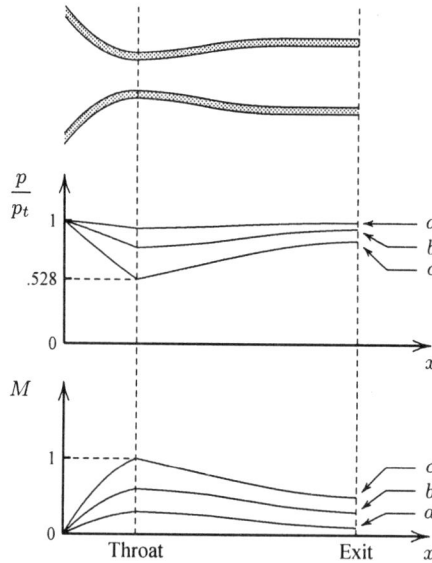

Figure 8.12: *Pressure and Mach number for subsonic flow; $\gamma = 1.4$.*

Clearly, the flow throughout the nozzle is controlled by the exit pressure. Also, the mass flow through the nozzle, $\dot{m} = \rho u A$, increases in a monotone fashion with decreasing exit pressure. As we continue to reduce the pressure at the nozzle exit, we eventually reach a value, p_c, for which the flow is sonic at the throat. In this case, we reach Point t^* in Figure 8.11. We can compute the pressure at the throat when sonic conditions are realized by substituting $M = 1$ into Equation (8.82). We thus have

$$\frac{p^*}{p_t} = \left(\frac{2}{\gamma+1}\right)^{\frac{\gamma}{\gamma-1}} = 0.528 \qquad (\gamma = 1.4) \tag{8.84}$$

When the flow at the throat of a Laval nozzle is sonic, we have a condition referred to as **choked flow**. Once the flow is choked, any further decrease in the exit pressure yields no change in the mass-flow rate, $\dot{m}$, through the nozzle!

Sonic Case. For the case where the throat just goes sonic, a second solution exists, albeit with a much lower exit pressure, p_e. Figure 8.13(a) includes the two solutions. As shown, for the supersonic solution, the flow continues expanding after the throat, and we follow the supersonic branch to Point s in Figure 8.11. This is known as the **design case**, and the flow is completely shock free.

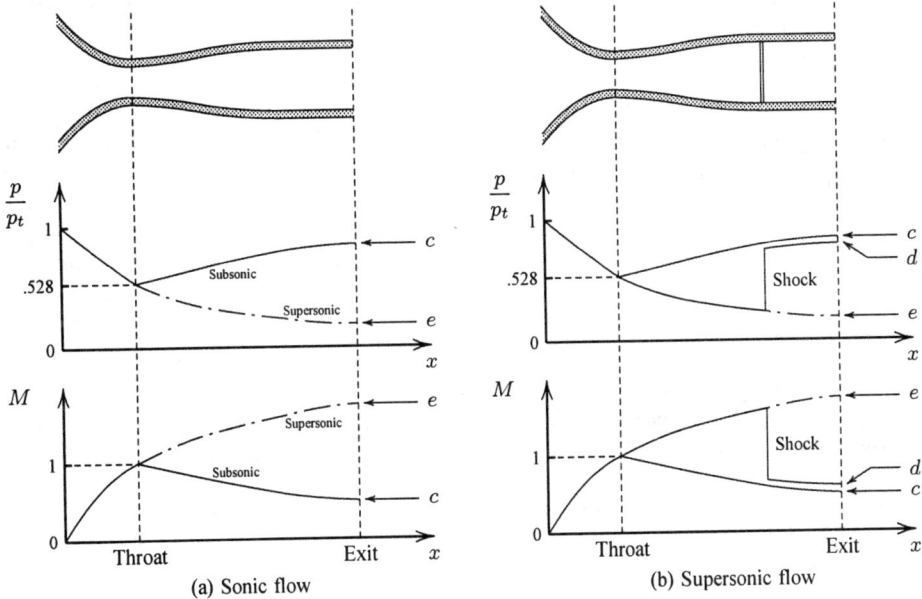

Figure 8.13: *Pressure and Mach number for sonic and supersonic flow;* $\gamma = 1.4$.

Supersonic Case. For any other exit pressures below p_c and above p_e, no isentropic solutions exist. Rather, as illustrated in Figure 8.13(b), either a normal shock wave appears at some point in the nozzle (Case d) or an oblique shock appears outside the nozzle (see Chapter 15 for a discussion of oblique shocks). This case is referred to as **overexpanded**. Finally, if the pressure is less than p_e, a series of nonisentropic expansion waves occur at the nozzle exit, and the flow is referred to as **underexpanded**. For further details, see any of the excellent texts on compressible flow such as Shapiro (1953), Liepmann and Roshko (1963) or Anderson (1990).

As an example, consider a supersonic wind tunnel designed to have Mach 2 flow in the test section with conditions corresponding to the U. S. Standard Atmosphere at sea level, i.e., $p_{exit} = 1$ atm and $T_{exit} = 288$ K. We would like to determine the exit-area ratio and the reservoir conditions needed to achieve this design. From the Isentropic Flow and Shock Tables in Section B.1 of Appendix B, for Mach 2 we find $A_{exit}/A^* = 1.6875$, $p_{exit}/p_t = 0.1278$ and $T_{exit}/T_t = 0.5556$. Thus, rounding to three significant figures:

$$A_{exit} = 1.69 A^*, \quad p_t = 7.82 \text{ atm}, \quad T_t = 518 \text{ K} \qquad (8.85)$$

Problems

8.1 Determine the Mach numbers and corresponding flow classification for the following. Use Table 1.3 to determine appropriate gas properties.

(a) Methane at 26° C flowing with a velocity of 414.5 m/sec.

(b) Helium at 26° C flowing with a velocity of 610 m/sec.

(c) Air at 0° C flowing with a velocity of 994 m/sec.

(d) Air at -49.16° C flowing with a velocity of 300 m/sec.

8.2 Determine the Mach numbers and corresponding flow classification for the following. Use Table 1.3 to determine appropriate gas properties.

(a) Carbon dioxide at 50° F flowing with a velocity of 1000 ft/sec.

(b) Hydrogen at 50° F flowing with a velocity of 1000 ft/sec.

(c) Air at 0° F flowing with a velocity of 1000 ft/sec.

(d) Helium at -300° F flowing with a velocity of 9500 ft/sec.

8.3 For density variations to become important, the Mach number must be at least 0.3. What flow speed, in mph, must be attained to reach Mach 0.3 in air and in water? See Table 1.4 for speed of sound values.

8.4 Compute the velocity required to reach the transonic flow regime ($M \approx 0.8$) in air and water. Express your answers in km/hr. See Table 1.4 for speed of sound values.

8.5 According to the U. S. Standard Atmosphere, the temperature in the atmosphere is $T = T_o - \alpha z$, where α = 6.50 K/km, z is altitude and T_o is surface temperature. On a day when T_o = 10° C, a missile flies at Mach 0.6. Determine its speed in m/sec if it is flying at z = 25 m and at z = 2 km.

8.6 According to the U. S. Standard Atmosphere, the temperature in the atmosphere is $T = T_o - \alpha z$, where α = 18.85° R/mi, z is altitude and T_o is surface temperature. At an altitude of 4 miles, a rocket is flying at 1300 mph, which corresponds to Mach 2. What is the surface temperature to the nearest ° F?

8.7 An empirical formula relating pressure and density for seawater is

$$p/p_o \approx (\alpha + 1)(\rho/\rho_o)^7 - \alpha$$

where p_o and ρ_o are surface values and α = 3000. If p_o = 1 atm and ρ_o = 1030 kg/m^3, what is the speed of sound at a depth where ρ = 1040 kg/m^3? Compare with the fresh-water value of 1481 m/sec.

8.8 Consider a liquid of density ρ_ℓ that has gas bubbles uniformly distributed. A fraction λ of the total volume, V, is occupied by the air. The fluid mixture is compressed isentropically and $\lambda \ll 1$.

(a) Show that differential pressure and volume changes are related by

$$\frac{dp}{p} = -\frac{\gamma}{\lambda}\frac{dV}{V}$$

where γ is the specific-heat ratio of the gas. Neglect any volume change for the liquid.

(b) Explain why $d\rho/\rho = -dV/V$ and $\rho \approx (1 - \lambda)\rho_\ell$.

(c) Combine results of Parts (a) and (b) to show that the speed of sound, a, is

$$a = \sqrt{\frac{\gamma}{\lambda(1 - \lambda)}\frac{p}{\rho_\ell}}$$

(d) Suppose the gas is air and the liquid is water with ρ_ℓ = 1000 kg/m^3. Compute a/a_o for the mixture, where a_o = 1481 m/sec is the sound speed in pure water, when λ = 0.01, 0.02 and 0.03.

8.9 A classical result of the *kinetic theory of gases* tells us that the internal energy for each degree of freedom is $\frac{1}{2}RT$, where R is the perfect-gas constant and T is absolute temperature. A molecule of a monatomic gas is viewed as a sphere and thus has 3 translational degrees of freedom. A diatomic molecule appears as a 'barbell' with 5 degrees of freedom, 3 translational and 2 rotational (the rotational mode about the axis can be ignored). At very high temperatures, two additional vibrational modes appear for diatomic molecules.

(a) According to this theory, what is γ for a perfect, calorically-perfect gas with n degrees of freedom?

(b) Compare with measured values (Table 1.3) for helium ($n = 3$), air ($n = 5$), carbon dioxide at room temperature ($n = 5$) and carbon dioxide at very high temperature ($n = 7$).

8.10 If we account for *quantum-mechanical* effects, we find that the specific heat for carbon dioxide is

$$c_v = \frac{5}{2}R + \frac{(\Theta_{vib}/T)^2 \, e^{\Theta_{vib}/T}}{(e^{\Theta_{vib}/T} - 1)^2} R, \qquad \Theta_{vib} = 954 \text{ K}$$

(a) Using the fact that $c_p - c_v = R$, compute the specific-heat ratio, γ, for $T = 10°$ C, $100°$ C, $1000°$ C and $5000°$ C.

(b) What temperature does the nominal value, $\gamma = 1.30$, correspond to? **HINT:** Solve by trial and error — it lies between $300°$ C and $400°$ C.

8.11 The Mach numbers at two points in an isentropic flow of air are $M_a = 0.3$ and $M_b = 0.6$. Determine p_a/p_b and T_a/T_b. Also, compute T_{ta}/T_{tb} for this flow.

8.12 The Mach numbers at two points in an isentropic flow of air are $M_a = 3$ and $M_b = 6$. Determine p_a/p_b and T_a/T_b. Also, compute T_{ta}/T_{tb} for this flow.

8.13 Wind-tunnel tests are being planned for Mach 1.6 airflow past a scale model of an airplane. The reservoir temperature will be $20°$ C and the pressure in the wind-tunnel test section must be 1 atm. Assuming the flow will be isentropic, determine the required reservoir pressure and the flow velocity in the test section.

8.14 Wind-tunnel tests are being planned for Mach 2.5 airflow past a scale model of a cruise missile. The reservoir pressure will be 250 psi and the temperature in the wind-tunnel test section must be $40°$ F. Assuming the flow will be isentropic, determine the required reservoir temperature and the density in the test section.

8.15 Hypersonic experiments are being done in a wind tunnel with a reservoir temperature of $150°$ C. If the test-section Mach number is 10, determine the temperature in the test section if the working gas is air ($\gamma = 1.40$) and if it is helium ($\gamma = 1.66$).

8.16 Air undergoes an isentropic expansion through a nozzle from a stagnation temperature of $65°$ C. What is the Mach number at the point where the flow velocity is 180 m/sec?

8.17 Helium ($\gamma = 5/3$) experiences an isentropic expansion from a stagnation temperature of $200°$ F. What is the Mach number at points where the flow velocity is 100 ft/sec, 500 ft/sec and 1000 ft/sec?

8.18 Gas flows with velocity, $U = 1000$ m/sec, pressure, $p = 150$ kPa, and temperature, $T = 250°$ C. We want to determine the pressure and temperature the gas would reach if it were brought to rest isentropically. Do your computations for air and for methane. See Table 1.3 for properties of methane.

8.19 Gas flows with velocity, $U = 1500$ ft/sec, pressure, $p = 20$ psi, and temperature, $T = 94°$ F. We want to determine the pressure and temperature the gas would reach if it were brought to rest isentropically. Do your computations for air and for hydrogen. See Table 1.3 for properties of hydrogen.

8.20 Helium ($\gamma = 5/3$) is brought to rest isentropically. If its pressure increases from 8 kPa to 256 kPa, at what Mach number is the helium flowing?

8.21 Consider subsonic flow of air approaching a Pitot-static tube. The flow is essentially isentropic. Determine the Mach number when the measured total and static pressures are as follows.

(a) $p = 101$ kPa, $p_t = 126$ kPa

(b) $p = 14.7$ psi, $p_t = 26.5$ psi

(c) $p = 1.06$ atm, $p_t = 1.11$ atm

8.22 Consider subsonic flow of helium ($\gamma = 5/3$) approaching a Pitot-static tube. The flow is essentially isentropic. Determine the Mach number when the measured total and static pressures are as follows.

(a) $p = 101$ kPa, $p_t = 126$ kPa

(b) $p = 14.7$ psi, $p_t = 26.5$ psi

(c) $p = 1.06$ atm, $p_t = 1.11$ atm

8.23 Consider subsonic flow past an airplane wing. The flow speed far ahead of the wing is U_∞. The flow accelerates over the wing, reaching a maximum speed of $\frac{7}{5}U_\infty$. The freestream density and temperature are $\rho_\infty = 1.20$ kg/m^3 and $T_\infty = 290$ K, respectively. Compute the Mach number and the percentage change in temperature at the point where the maximum velocity is achieved for the following.

(a) $U_\infty = 100$ m/sec

(b) $U_\infty = 200$ m/sec

8.24 Consider subsonic flow past an airplane wing. The flow speed far ahead of the wing is U_∞. The flow accelerates over the wing, reaching a maximum speed of $\frac{3}{2}U_\infty$. The freestream density and temperature are $\rho_\infty = .00234$ slug/ft^3 and $T_\infty = 519°$R, respectively. Compute the Mach number and the percentage change in density at the point where the maximum velocity is achieved for the following.

(a) $U_\infty = 100$ mph

(b) $U_\infty = 450$ mph

8.25 In Chapter 5 we learned that, for incompressible flow, the flow speed, U_p, can be obtained from a Pitot-static tube according to

$$U_p = \sqrt{\frac{2\,(p_t - p)}{\rho}}$$

where ρ is density, p_t is total pressure and p is static pressure.

(a) Assuming isentropic, subsonic flow, show that U_p/U is the following function of Mach number, M, where U is the true flow speed:

$$\frac{U_p}{U} = \frac{1}{M}\sqrt{\frac{2}{\gamma}\left[\left(1 + \frac{\gamma-1}{2}M^2\right)^{\gamma/(\gamma-1)} - 1\right]}$$

(b) Verify that $U_p \rightarrow U$ as $M \rightarrow 0$.

(c) Compute U_p/U for $M = 0.2, 0.4, 0.6, 0.8$ and 1.0. Below what Mach number is the difference between U_p and U less than 2%?

8.26 Superman is traveling faster than a speeding bullet. In fact, he is traveling at 10 times the speed of sound in the ambient atmosphere (temperature 60°F at 14.7 psi). Estimate the maximum temperature on his head (stagnation temperature). Also, determine the maximum pressure on his head, taking into consideration the normal shock wave standing in front of him.

8.27 Just ahead of a normal shock, conditions are given by $M_1 = 3$, $T_1 = 600°$R and $p_1 = 30$ psi. Determine the values of M_2, T_{t2} and p_2 (i.e., conditions behind the shock) for air ($\gamma = 7/5$) and helium ($\gamma = 5/3$). Use the shock tables for the parts of this problem to which they apply.

8.28 Just ahead of a normal shock, conditions are given by $M_1 = 4$, $T_1 = 400$ K and $p_1 = 10$ kPa. Determine the values of M_2, T_{t2} and p_2 (i.e., conditions behind the shock) for air ($\gamma = 7/5$) and carbon dioxide ($\gamma = 9/7$). Use the shock tables for the parts of this problem to which they apply.

8.29 Just behind a normal shock, the Mach number is $M_2 = 0.7$ and the temperature is $T_2 = 123°$ C. Determine the values of M_1 and T_1 (i.e., conditions ahead of the shock) for air ($\gamma = 7/5$) and for helium ($\gamma = 5/3$). Use the shock tables for the parts of this problem to which they apply.

8.30 The static pressure increases from 100 kPa to 600.5 kPa across a shock wave in air. If the static temperature of the fluid ahead of the shock is 100 K, what is the temperature behind the shock?

8.31 Wind-tunnel experiments are being planned to simulate Mach 3 flow past a blunt-nosed object. We require the static pressure just behind the normal shock standing in front of the object to be 1 atm. What must the pressure be in the wind-tunnel reservoir for air and for helium ($\gamma = 5/3$)?

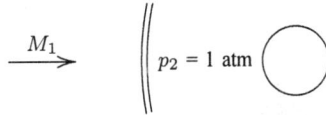

$$\xrightarrow{M_1} \quad \left\| \; p_2 = 1 \text{ atm} \quad \bigcirc \right.$$

Problems 8.31, 8.32

8.32 Wind-tunnel experiments are being planned to simulate Mach 2 flow past a blunt-nosed object. We require the static pressure just behind the normal shock standing in front of the object to be 1 atm. What must the pressure be in the wind-tunnel reservoir for air and for carbon dioxide ($\gamma = 1.30$)?

8.33 A Pitot-static tube is inserted in the test section of a supersonic wind tunnel with a reservoir pressure of 9 atm. The stagnation-pressure tap indicates $p_t = 1.17$ atm. Accounting for the *detached shock* standing ahead of the tube, what is the test-section Mach number if the working gas is air?

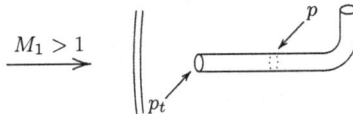

$$\xrightarrow{M_1 > 1} \quad \left\| \right.$$

Problems 8.33, 8.34, 8.35

8.34 A Pitot-static tube is inserted in the test section of a supersonic wind tunnel with a reservoir pressure of 12 atm. The test-section Mach number is 3 and the working gas is helium ($\gamma = 5/3$). Accounting for the *detached shock* upstream of the tube, what pressure, p_t, does the stagnation-pressure tap indicate?

8.35 A Pitot-static tube in a Mach 3 air flow indicates a total pressure, $p_t = 101$ kPa. Taking account of the *detached shock* standing upstream of the tube, determine the static pressure measured by the tube.

8.36 You have no calculator available, but you do have your shock tables (see Section B.1 of Appendix B). You also know that

$$M_2^2 = \frac{1 + \frac{1}{2}(\gamma - 1)M_1^2}{\gamma M_1^2 - \frac{1}{2}(\gamma - 1)}$$

In your shock tube you have air with $M_1 = 2.65$. You wish to change your gas to helium ($\gamma = 5/3$), but your application requires having the same value for M_2 that you have with air. What must M_1 be when you use helium? **NOTE:** This problem is intended to encourage you to make rational engineering approximations. In this spirit, any interpolation you do in the shock tables should be done to two significant figures, and no more.

8.37 A shock wave is moving to the left with a velocity W as shown. It is advancing into a uniform freestream with velocity U_∞ and Mach number M_∞. The specific-heat ratio of the gas is γ. If the shock velocity is γU_∞, what is M_∞?

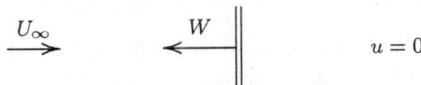

$$\xrightarrow{U_\infty} \qquad \xleftarrow{W} \; \left\| \qquad u = 0 \right.$$

Problem 8.37

8.38 Show that the lowest Mach number and the largest density ratio possible downstream of a normal shock wave are

$$M_2 = \sqrt{\frac{(\gamma - 1)}{2\gamma}} \quad \text{and} \quad \frac{\rho_2}{\rho_1} = \frac{(\gamma + 1)}{(\gamma - 1)}$$

What are the limiting values of M_2 and ρ_2/ρ_1 for air?

8.39 Show for a weak shock wave ($M \approx 1$) that the Mach number downstream of the shock is approximated by

$$M_2^2 = 2 - M_1^2$$

Compare values for M_2 obtained using this equation with values for M_2 from the shock tables, with $M_1 = 1$, 1.05, 1.1 and 1.2. **HINT:** Let $M_1^2 = 1 + \epsilon$, where $\epsilon \ll 1$, and expand the exact relationship between M_1 and M_2.

8.40 Verify that for weak shocks, the entropy change across a shock wave is

$$\frac{s_2 - s_1}{R} \approx \frac{2\gamma}{(\gamma + 1)^2} \frac{\left(M_1^2 - 1\right)^3}{3} = \frac{\gamma + 1}{12\gamma^2} \left(\frac{\Delta p}{p_1}\right)^3$$

HINT: Substitute $\epsilon \equiv M_1^2 - 1 \ll 1$ into Equation (8.72) and expand in Taylor series, using the fact that

$$ln(1 + a\epsilon) = a\epsilon - \frac{1}{2}a^2\epsilon^2 + \frac{1}{3}a^3\epsilon^3 + \cdots$$

8.41 Beginning with the normal-shock relation for Mach number, viz.,

$$M_2^2 = \frac{1 + \frac{1}{2}(\gamma - 1)M_1^2}{\gamma M_1^2 - \frac{1}{2}(\gamma - 1)}$$

solve for M_1^2 as a function of M_2^2. Compare your result with the equation you are starting with.

8.42 The normal-shock relations give flow properties as a function of Mach number ahead of the shock. Alternatively, we can express properties as functions of the pressure ratio, p_2/p_1. In this form, they are called the *Rankine-Hugoniot relations*. Verify that the Rankine-Hugoniot relations for density and temperature are

$$\frac{\rho_2}{\rho_1} = \frac{1 + \dfrac{\gamma + 1}{\gamma - 1}\dfrac{p_2}{p_1}}{\dfrac{p_2}{p_1} + \dfrac{\gamma + 1}{\gamma - 1}} \quad \text{and} \quad \frac{T_2}{T_1} = 1 + \frac{\left(\dfrac{p_2}{p_1}\right)^2 - 1}{\dfrac{\gamma + 1}{\gamma - 1}\dfrac{p_2}{p_1} + 1}$$

8.43 What is the pressure, p^*/p_t, at the throat of a Laval nozzle if the working gas is helium ($\gamma = 5/3$)?

8.44 What is the pressure, p^*/p_t, at the throat of a Laval nozzle if the working gas is methane ($\gamma = 1.31$)?

8.45 What is the pressure, p^*/p_t, at the throat of a Laval nozzle if the working gas is carbon dioxide ($\gamma = 1.3$)?

8.46 A wind tunnel is designed to have a Mach number of 2.5, a static pressure of 1.5 psi, and a static temperature of -10° F in the test section. Determine the area ratio of the nozzle required and the reservoir conditions that must be maintained if air is to be used.

8.47 Consider a supersonic wind tunnel designed by an eager young research assistant.

(a) If the test-section Mach number is 5 and the test-section area is 50 cm^2, what is the throat area?

(b) If, as a result of poor design, a shock forms upstream of the test section and the test-section Mach number is only 0.504, what is the throat area?

8.48 If the test-section Mach number of a supersonic wind tunnel using air is 2.8 and the test-section area is 35 cm^2, what is the throat area?

8.49 If the test-section Mach number of a supersonic wind tunnel using helium is 3 and the test-section area is 36 cm^2, what is the throat area?

8.50 Verify that the mass-flow rate, $\dot{m}$, for flow of a perfect, calorically-perfect gas through a choked nozzle is

$$\dot{m} = \frac{p_t A^*}{\sqrt{T_t}} \sqrt{\frac{\gamma}{R} \left(\frac{2}{\gamma+1}\right)^{(\gamma+1)/(\gamma-1)}}$$

8.51 A perfect gas experiences an isentropic expansion through a Laval nozzle to a point where the area is 1.787 times the throat area. The Mach number at this point is 2. What is the specific-heat ratio, γ, of the gas? **HINT:** Assume $\gamma = 1 + 2/n$ where $n \geq 3$ is an integer.

8.52 Air flows through a Laval nozzle. The exit pressure is sufficiently high to cause a normal shock to form inside the nozzle at a point where the nozzle area is 30 cm^2. Determine the pressure just downstream of the shock if the reservoir pressure is 600 kPa and the throat area is 15 cm^2.

8.53 Air flows from a reservoir with total pressure 800 kPa through a supersonic nozzle. The ratio of the nozzle-exit area to its throat area is $A_e/A^* = 3.5$. If the exit pressure is $p_e = 101$ kPa, will a shock be present in the nozzle?

8.54 Air flows from a reservoir through a supersonic nozzle. The ratio of the nozzle-exit area to its throat area is $A_e/A^* = 3$. If the exit pressure is $p_e = 1$ atm, what must the reservoir pressure be for the *design case*, i.e., for a shock-free flow?

8.55 Transonic wind-tunnel tests are complicated by the fact that *model blockage* can cause significant changes in the flow. To see why, consider a tunnel with a test-section Mach number, $M = 1.10$, area, $A = 10$ ft^2, and reservoir temperature, $T_t = 68°$ F. The working gas is air.

(a) Compute the test-section flow speed, U, and the nozzle-throat area, A^*.

(b) If a model of cross-sectional area $A_m = .006A$ is placed in the tunnel, what is the new test-section Mach number?

(c) Compute the test-section flow speed, $\tilde{U}$, with the model of Part (b) present. What is the percentage difference from the flow speed for the unblocked tunnel?

8.56 The ratio of exit to throat area for a Laval nozzle is $A_e/A_t = 5$. A normal shock stands inside the nozzle at a point where $A/A_t = 3$. The total pressure ahead of the shock is 900 kPa. What is the nozzle-exit Mach number?

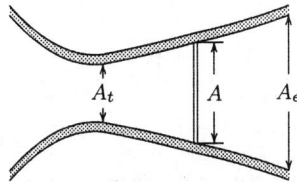

Problems 8.56, 8.57

8.57 The ratio of exit to throat area for a Laval nozzle is $A_e/A_t = 10$. A normal shock stands inside the nozzle at a point where $A/A_t = 2.5$. The total pressure ahead of the shock is 900 kPa. What is the nozzle-exit Mach number?

8.58 One problem in creating high Mach number flows is condensation of the oxygen component in air when the temperature drops below 50 K. Consider isentropic flow through a nozzle in which the reservoir temperature is 300 K.

(a) At what Mach number will condensation of oxygen occur?

(b) If the radius of the nozzle is given by

$$r = r_o\sqrt{1 + \frac{30x^2}{x^2 + 1}}$$

where the throat is located at $x = 0$, at what value of x will the oxygen first begin to condense?

8.59 Compute the area ratio, A/A^*, for isentropic flow at Mach 2 as a function of γ. Make a graph of your results. What are the limiting values of A/A^* for $\gamma \to 1$ and $\gamma \to \infty$?

8.60 A blow-down supersonic wind tunnel is supplied with air from a large reservoir as shown. The Mach number in the test section is $M_2 = 2$, and the pressure is below atmospheric so that a shock wave is formed just at the exit. The pressure at Point 3 immediately behind the shock is 14.7 psi.

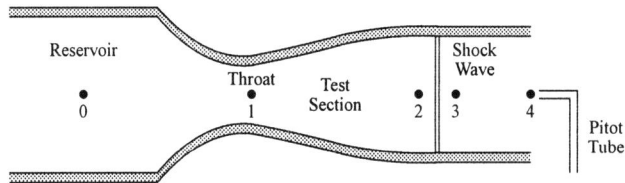

Problem 8.60

(a) Find p_0, p_1, p_2, and M_3.

(b) A Pitot tube is placed in the exit jet as shown. What is the pressure p_4? Why is $p_4 < p_0$? How does T_4 compare with T_0?

8.61 In 1985, Cadillac introduced a digital display of the temperature outside the automobile. Assume the temperature sensor actually measures stagnation temperature. If the temperature outside your Cadillac is 70° F and, thanks to the protection of your radar jammer, you are traveling at 76.9 mph (112.8 ft/sec), determine the percentage error in the indicated temperature, where

$$\text{Error} = 100 \cdot (T_{\text{actual}} - T_{\text{indicated}})/T_{\text{actual}}$$

Suppose the Supersonic Transport (SST) were to use the same device to measure outside temperature. If the SST moves at Mach 2, compute the percentage error, assuming a *detached normal shock wave* is present just upstream of the sensor.

8.62 Verify that the pressure coefficient, C_p, for a calorically-perfect gas can be written as

$$C_p = \frac{2}{\gamma M_\infty^2}\left(\frac{p}{p_\infty} - 1\right)$$

where subscript ∞ denotes freestream conditions.

8.63 The Newtonian approximation is often used in hypersonic flow theory. In this approximation we set the specific-heat ratio, $\gamma = 1$. What are the normal shock relations, i.e., M_2^2, p_2/p_1, ρ_2/ρ_1 and T_2/T_1 as functions of M_1, according to this approximation?

8.64 A Pitot-static tube mounted on an airplane indicates the static pressure is 95 kPa and the stagnation pressure is 130 kPa. Determine the two possible Mach numbers at which the plane might be moving.

8.65 Consider the flow of air through a duct of varying area. At the inlet (Station 1), the Mach number is $M_1 = 3$. A normal shock lies at Station 2. If the cross-sectional areas stand in the ratio $A_1 : A_2 : A_3 = 4 : 3 : 6$, what is the Mach number, M_3, at Station 3?

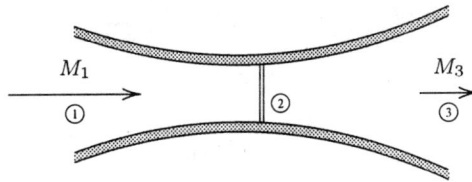

Problems 8.65, 8.66

8.66 Consider the flow of air through a duct of varying area. At the inlet (Station 1), the Mach number is $M_1 = 1.7$. A normal shock lies at Station 2. If the cross-sectional areas stand in the ratio $A_1 : A_2 : A_3 = 5 : 4 : 5$, what is the Mach number, M_3, at Station 3?

8.67 Air from a large reservoir flows steadily and subsonically through a nozzle. The subsonic jet impinges on a plate. The force required to hold the plate in place is F, the reservoir pressure is p_t and the nozzle-exit area is A_e.

(a) Letting p_o denote atmospheric pressure, verify that

$$F = \frac{2\gamma}{\gamma - 1} \left[\left(\frac{p_t}{p} \right)^{(\gamma-1)/\gamma} - 1 \right] p_o A_e$$

(b) If $F = 150$ N, $A_e = 30$ cm^2 and $p_o = 101$ kPa, determine p_t and the nozzle-exit Mach number.

(c) For the conditions of Part (b), what is the force if p_t increases by 25%? Verify that the flow will still be subsonic.

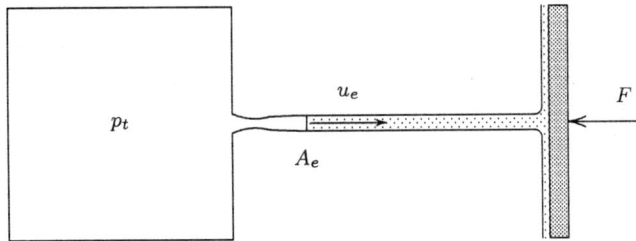

Problems 8.67, 8.68

8.68 Air from a large reservoir flows steadily and subsonically through a nozzle. The subsonic jet impinges on a plate. The force required to hold the plate in place is F, the reservoir pressure is p_t and the nozzle-exit area is A_e.

(a) Letting p_o denote atmospheric pressure, verify that

$$F = \frac{2\gamma}{\gamma - 1} \left[\left(\frac{p_t}{p} \right)^{(\gamma-1)/\gamma} - 1 \right] p_o A_e$$

(b) If $F = 50$ lb, $A_e = 5$ in^2 and $p_o = 14.7$ psi, determine p_t and the nozzle-exit Mach number.

(c) If the working gas is changed to helium at the same total pressure determined in Part (b), what is the force? What is the Mach number at the nozzle exit?

8.69 A shock wave propagates into a quiescent fluid in which the static pressure is p_1 and the speed of sound is a_1.

(a) If the pressure behind the shock is p_2, show that the shock moves at a speed U_s, where

$$U_s = a_1 \sqrt{\frac{\gamma - 1}{2\gamma} + \frac{\gamma + 1}{2\gamma} \frac{p_2}{p_1}}$$

(b) A nuclear explosion can generate a pressure of 5000 psi. Using the result of Part (a), compute the speed of the shock created by such an explosion on a day when $p = 14.7$ psi and $T_1 = 75°$ F.

8.70 Oxygen begins to dissociate at a temperature of 2000 K. When this happens, air cannot be approximated as a calorically-perfect gas.

(a) At what Mach number will this temperature be reached for flow past a blunt-nosed body if the freestream temperature is $T_\infty = 20°$ C?

(b) For an insulated body, the temperature at the stagnation point is actually a bit less than the total temperature. It is

$$T_{aw} = T_\infty \left[1 + \frac{\gamma - 1}{2} r M^2 \right]$$

where $r = 0.85$ is the *recovery factor*. How does this change your estimate of Part (a)?

(c) A common approximation for hypersonic flows is to treat air as a perfect, calorically-perfect gas with $\gamma = 1.2$. Repeat your estimate of Part (b) with $\gamma = 1.2$.

Chapter 9

Turbomachinery

A turbomachine is a device that can either add energy to a flowing fluid or extract energy from it. The origin of the prefix *turbo* is found in Latin, with the translation of "that which spins." In this chapter, we will investigate some of the features of flow through turbomachines that are commonly found in practical applications. Turbomachinery is utilized in our modern world in myriad contexts such as: (a) propulsion systems for airplanes, automobiles, ships and liquid rockets; (b) power-generating plants; (c) gas, oil and water pumping stations; and (d) numerous other settings ranging from heart-assist pumps to refrigeration facilities.

Flow through a turbomachine is generally three-dimensional, viscous and unsteady (actually periodic). Consequently, accurate analytical predictions can only be made with large-scale computations [see, for example, Lakshminarayana (1996)]. Nevertheless, useful insight can be obtained using the control volume method and reducing the analysis to essentially one-dimensional computations. That is the purpose of this chapter.

We begin by classifying turbomachines in terms of their geometry and how they affect the flow's energy and direction. Then, to aid in analysis of turbomachinery, we develop an equation for conservation of angular momentum. From the general integral-conservation laws, we derive the **Euler turbomachine equations**. We focus on incompressible flow, for simplicity, with emphasis on water pumps. The analysis applies equally to low-speed gas flows (e.g., ventilation blowers, wind-tunnel fans and windmills). The balance of the chapter applies these equations to turbomachines. Efficiency factors and performance curves are introduced to quantify how practical pumps and turbines behave. Above all, this chapter shows how the combination of the control volume method (Chapter 6) and dimensional analysis (Chapter 2) lays the foundation for developing an understanding of complex fluid flows.

9.1 Classification

There are several ways in which we can classify turbomachines based on their function and geometry. For our purposes, four ways will suffice, viz.,

1. Energy Considerations — how they affect the energy of a given flow

2. Geometrical Considerations — whether or not the turbomachine is enclosed

3. Exit-Flow Considerations — the direction of the outlet flow

4. Number of stages — single-stage and multistage machines

285

Energy Considerations: When a turbomachine adds energy to the flow, it is commonly referred to as a **pump** or a **compressor**. A turbomachine that extracts energy from the flow is called a **turbine**. We made this distinction in Chapter 7 when we developed an approximate form of the energy-conservation equation [see Equation (7.89)] containing changes in total head due to pumps and turbines.

Geometrical Considerations: Many of the most important turbomachines of engineering interest are enclosed by a permanent housing called the **shroud**. Examples of such turbomachines are jet-engine compressors, gas or steam turbines and centrifugal pumps. For the obvious reason, these machines are described as **enclosed turbomachines**. Not all turbomachines are enclosed, with the most important examples being airplane propellers, ship screws and wind turbines. These devices are called **extended turbomachines**.

Exit-Flow Considerations: Figure 9.1 illustrates the three types of flow exiting from common turbomachines. When the primary flow direction remains parallel to the axis as in Figure 9.1(a), we describe the machine as an **axial** turbomachine. By contrast, if the flow turns and exits the machine in the radial direction as depicted in Figure 9.1(b), we have a **radial** or **centrifugal** turbomachine. If the exit flow is partially axial and partially radial as in Figure 9.1(c), we call the device a **mixed-flow** turbomachine.

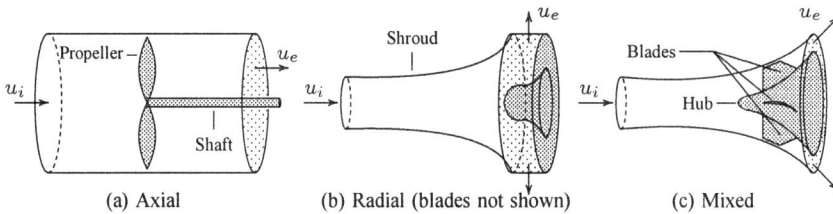

Figure 9.1: *Exit-flow directions for turbomachines.*

Number of Stages: The total-head (or total-pressure) change for a simple, **single-stage**, turbomachine is limited. To achieve larger changes, conceptually, we can put several single-stage machines in series. In practice, we call such machines **multistage** compressors (or pumps) and multistage turbines. Figure 9.2 depicts two stages in a section of a compressor. Each stage consists of a row of **rotor** blades attached to the hub, followed by a row of **stator** blades attached to the shroud or **casing**. There will be several rows of rotor and stator blades in a multistage arrangement. The function of the rotating rotor blades is to add energy to the flow. The stationary stator blades then remove the angular momentum induced by the torque on the rotor, with a corresponding increase in static pressure.

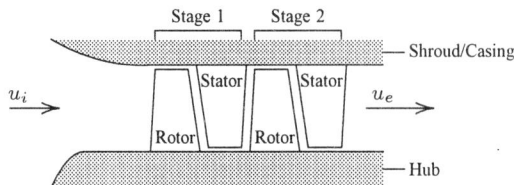

Figure 9.2: *Schematic of two stages in an axial-flow compressor.*

9.2   Conservation of Angular Momentum

The mass- and momentum-conservation principles are sufficient to solve incompressible flow problems. However, it is sometimes convenient to work with **angular momentum** rather than conventional **linear momentum**. This is especially true in the context of turbomachinery where, for example, we wish to compute the net torque as part of our analysis. In this section, we derive an integral-conservation law for angular momentum and apply the principle to a simple rotating, turbine-like device, viz., a lawn sprinkler.

9.2.1   Derivation

We can derive an integral conservation law for angular momentum in the same manner we have used to derive the principles discussed in Chapters 5 and 7. To motivate the form of the conservation principle, consider Newton's second law for a single particle, i.e.,

$$\mathbf{F} = \frac{d}{dt}(m\mathbf{u}) \tag{9.1}$$

where m is mass. Now, the torque, $\boldsymbol{\tau}$, and angular momentum, $\mathbf{H}$, relative to a fixed point are defined by

$$\boldsymbol{\tau} = \mathbf{r} \times \mathbf{F} \quad \text{and} \quad \mathbf{H} = m\mathbf{r} \times \mathbf{u} \tag{9.2}$$

where $\mathbf{r}$ is the position vector from the fixed point to the point at which $\mathbf{F}$ is applied. Differentiating the angular momentum vector with respect to time shows that

$$\frac{d\mathbf{H}}{dt} = \frac{d\mathbf{r}}{dt} \times (m\mathbf{u}) + \mathbf{r} \times \frac{d}{dt}(m\mathbf{u}) \tag{9.3}$$

Noting that $\mathbf{u} = d\mathbf{r}/dt$ and substituting from Equation (9.1), we can rewrite the rate of change of angular momentum as

$$\frac{d\mathbf{H}}{dt} = m(\mathbf{u} \times \mathbf{u}) + \mathbf{r} \times \mathbf{F} \tag{9.4}$$

Finally, the first term on the right-hand side of Equation (9.4) vanishes since the cross product of a vector with itself is identically zero. Noting the definition of torque, we conclude that the classical equation for angular-momentum conservation is

$$\frac{d\mathbf{H}}{dt} = \boldsymbol{\tau} \tag{9.5}$$

We can recast this equation in a form appropriate for a control volume by using the Reynolds Transport Theorem to express $d\mathbf{H}/dt$ in terms of the absolute[1] angular momentum per unit volume, $\rho\,\mathbf{r} \times \mathbf{u}_{abs}$. Letting $\mathbf{w}$ denote velocity relative to the rotating control volume to be consistent with classical turbomachine notation, we have the following integral law.

$$\iiint_V \frac{\partial}{\partial t}(\rho\,\mathbf{r} \times \mathbf{u}_{abs})\,dV + \oiint_S \rho\,\mathbf{r} \times \mathbf{u}_{abs}\,(\mathbf{w} \cdot \mathbf{n})\,dS = \boldsymbol{\tau} \tag{9.6}$$

Before proceeding to application of this principle, it is helpful to pause and discuss some of its subtleties. First, all of the forces acting (pressure, body forces, viscous stresses) contribute to the torque and should be included where appropriate. Second, this principle has been derived from the linear momentum law, and is therefore not an independent source of equations. Third, the corresponding differential form leads to the interesting conclusion that the viscous stress tensor (which we will define and discuss in Chapter 12) is symmetric, but otherwise yields no useful equations to describe the motion of the fluid.

---

[1]See Subsection 6.4.1 for a discussion of absolute and relative momentum.

9.2.2   Lawn Sprinkler

As an example of how we implement this principle, consider the lawn sprinkler depicted in Figure 9.3. The sprinkler rotates in the counterclockwise direction with angular velocity (measured in radians/second = $\sec^{-1}$)

$$\boldsymbol{\Omega} = \Omega\,\mathbf{k} \tag{9.7}$$

Water enters with velocity U from a vertical pipe through the center and exits with velocity w_e relative to the nozzle. The nozzle area is A_e and is located a distance ℓ from the center. We would like to compute the torque produced by the lawn sprinkler, assuming the flow is steady and incompressible. We also assume body forces can be ignored.

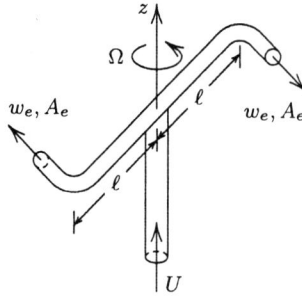

Figure 9.3: *A lawn sprinkler.*

By symmetry, the torque vector must act about the z axis. Thus, denoting the magnitude of the torque by τ_r, our integral conservation law simplifies to

$$-\tau_r = \mathbf{k} \cdot \oiint_S \rho\,\mathbf{r} \times \mathbf{u}_{abs}\,(\mathbf{w} \cdot \mathbf{n})\,dS \tag{9.8}$$

where our control volume is coincident with the sprinkler surface and rotates with the same angular velocity. We include a minus sign with τ_r because we have chosen a control volume that completely surrounds the sprinkler. Consequently, the torque appearing in our conservation law is the torque exerted by the surroundings on the sprinkler (i.e., it is a "reaction torque"). We take moments about the z axis so that

$$\mathbf{r} = r\mathbf{e}_r \tag{9.9}$$

where $\mathbf{e}_r$ is a unit vector in the radial direction and r is distance from the z axis. Clearly, the angular momentum at the inlet is zero since $r = 0$. The only other parts of the bounding surface that have nonzero angular-momentum flux are the two nozzles. Now, at the nozzles, we have

$$\mathbf{r} = \ell\mathbf{e}_r, \quad \mathbf{w} = -w_e\mathbf{e}_\theta, \quad \mathbf{u}_{abs} = (\Omega\ell - w_e)\,\mathbf{e}_\theta \tag{9.10}$$

where $\mathbf{e}_\theta$ is a unit vector in the circumferential direction. Thus, the absolute angular momentum per unit mass at the nozzles is

$$\mathbf{r} \times \mathbf{u}_{abs} = \ell\,(\Omega\ell - w_e)\,(\mathbf{e}_r \times \mathbf{e}_\theta) = \ell\,(\Omega\ell - w_e)\,\mathbf{k} \tag{9.11}$$

Thus, substituting Equation (9.11) into Equation (9.8), we have

$$-\tau_r = \mathbf{k} \cdot \oiint_S \underbrace{\rho\,\ell\,(\Omega\ell - w_e)\,\mathbf{k}}_{\rho\,\mathbf{r} \times \mathbf{u}_{abs}}\ \underbrace{w_e}_{(\mathbf{w} \cdot \mathbf{n})}\ dS = \rho\,\ell\,(\Omega\ell - w_e)\,w_e\,(2A_e) \tag{9.12}$$

Therefore, the reaction torque on the sprinkler is

$$\tau_r = 2\rho w_e A_e \ell \left(w_e - \Omega \ell \right) \tag{9.13}$$

We can rearrange our result in a more interesting form by first using conservation of mass. Clearly, the mass flux, $\dot{m}$, passing through the sprinkler is equal to the combined mass flux through the two exits, wherefore

$$\dot{m} = 2\rho w_e A_e \tag{9.14}$$

Then, the torque produced by the sprinkler, $\tau_{\text{sprinkler}} = -\tau_r$, is given by

$$\tau_{\text{sprinkler}} = \dot{m} \left(\ell u_\theta \right)_{abs} \tag{9.15}$$

Thus, the torque produced by the sprinkler is the product of the mass-flow rate and the absolute angular momentum per unit mass at the nozzles. Notably, details of the axial inlet flow do not appear explicitly in Equation (9.15). We will see in the next section that this result is true for a centrifugal pump and, in fact, is valid for turbomachines in general.

9.3 The Euler Turbomachine Equations

The geometry of turbomachines varies greatly, even for the same type of application. Regardless of the type of machine, there is always a rotating component whose purpose is to do work on the fluid. This is accomplished by a rotor (Figure 9.2) which, in the context of a pump, is called an **impeller** (Figure 9.4). In addition, there is a stationary member such as the stator of Figure 9.2 (or a **volute** for a centrifugal pump) whose purpose is to guide the flow after it passes through the rotor or impeller. There is sometimes an additional row of stationary blades known as **guide vanes** that help direct the flow prior to entering the rotor or impeller.

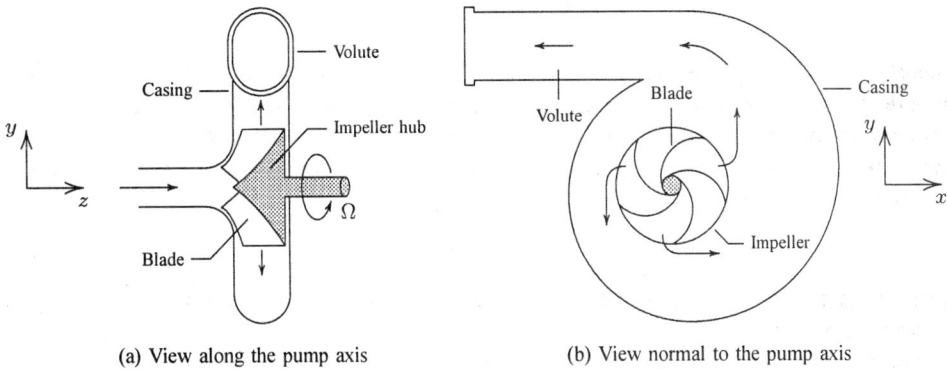

(a) View along the pump axis   (b) View normal to the pump axis

Figure 9.4: *A centrifugal pump; the clearance between the impeller and the casing is exaggerated – it is actually very small in practice.*

To make our discussion more specific, consider a centrifugal pump as sketched in Figure 9.4. The figure shows a side view (yz plane) and a view normal to the pump axis (xy plane). As shown, fluid enters parallel to the z axis, and flows toward the impeller. The impeller rotates at angular velocity Ω, imparting angular momentum to the flow and increasing its pressure.[2] The impeller depicted has what are called **backward-curved** blades. This is not a universal feature as **radial** (not-curved) and **forward-curved** blades are also used.

---

[2]The angular rotation rate is often denoted by n or N in turbomachinery literature with units of rpm or Hz.

The impeller causes the flow to rotate and directs it radially outward. The fluid eventually leaves the impeller and passes into the volute. Since the volute area increases as the flow advances, mass conservation dictates a decrease in velocity, accompanied by an additional increase in pressure.

In general, a detailed analysis requires a computer solution tailored to the specific turbomachine geometry. There are, nevertheless, important global properties of even the most complicated turbomachines that can be gleaned from our trustworthy control volume method. Since rotation is an integral part of turbomachine flows, we will use the angular-momentum conservation principle developed in the preceding section along with the mass- and energy-conservation principles developed in Chapters 5 and 7, respectively.

The most interesting part of the flow through a centrifugal pump is between the impeller blades. Thus, we select a control volume that lies between two successive blades. Figure 9.5(a) illustrates the impeller geometry from a view along the axis of rotation, which includes the blade shapes. Figure 9.5(b) is a side view in the yz plane, usually referred to as the **meridional plane**. The blade widths at the inlet and outlet are b_1 and b_2, respectively. Also, the radial distance from the axis to the hub at the inlet is r_1, while its value at the outlet is r_2. Our control volume is the region bounded by the hub, the casing [omitted in Figure 9.5(a) to avoid clutter] and two adjacent blades. Control-volume boundaries in the yz and xy planes are indicated by dashed lines in Figures 9.5(b) and 9.6, respectively.

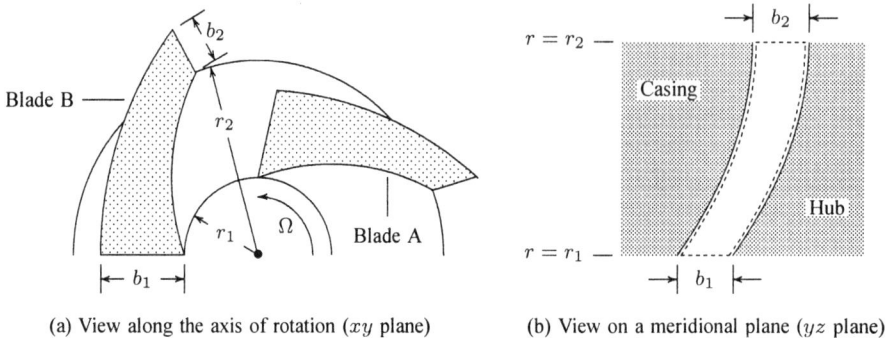

(a) View along the axis of rotation (xy plane)   (b) View on a meridional plane (yz plane)

Figure 9.5: *Closeup view of impeller geometry.*

Now, we shift our attention to Figure 9.6, which is a true axial view similar to Figure 9.5(a) with the blade shapes removed. In developing the **Euler turbomachine equations**, we assume the flow is steady and incompressible. We also assume that the blade width is small compared to the hub radius, i.e., that

$$b_1 \ll r_1 \quad \text{and} \quad b_2 \ll r_2 \tag{9.16}$$

When this is true, we can neglect variations in radial distance between the hub and the casing.

Fluid enters the impeller with velocity relative to the rotating blade of magnitude w_1. We assume this velocity is parallel to the blade at its leading edge, i.e., at $r = r_1$. The blade lies at an angle β_1 to the local tangent. Since the impeller velocity is Ωr_1 at the hub, the absolute velocity is the vector sum of these two velocities, and we denote its magnitude by V_1. Similarly, we assume the flow is parallel to the blade at its trailing edge ($r = r_2$), which makes an angle β_2 with the local tangent to the hub. The impeller and absolute velocity magnitude at $r = r_2$ are Ωr_2 and V_2, respectively.

As mentioned above, the control volume chosen is enclosed by the blades, hub and casing, which are solid boundaries, and by the surfaces $r = r_1$ and $r = r_2$, across which fluid can

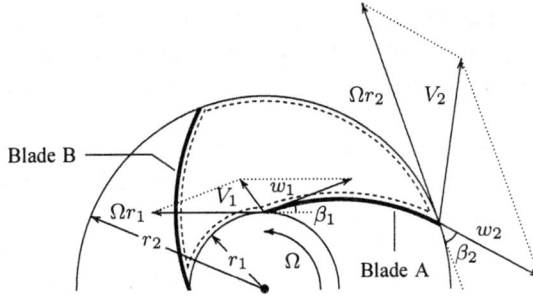

Figure 9.6: *Control volume between two impeller blades with flow directions (xy plane).*

flow. The radius vectors, absolute velocity vectors and unit normal vectors at the inlet and outlet portions of the control volume are as follows.

Inlet: $\quad \mathbf{r} = r_1 \mathbf{e}_r, \quad \mathbf{u}_{abs} = \mathbf{V}_1 = \mathbf{w}_1 + \Omega r_1 \mathbf{e}_\theta, \quad \mathbf{n} = -\mathbf{e}_r$ $\qquad$ (9.17)

Outlet: $\quad \mathbf{r} = r_2 \mathbf{e}_r, \quad \mathbf{u}_{abs} = \mathbf{V}_2 = \mathbf{w}_2 + \Omega r_2 \mathbf{e}_\theta, \quad \mathbf{n} = \mathbf{e}_r$ $\qquad$ (9.18)

The vectors $\mathbf{e}_r$ and $\mathbf{e}_\theta$ are the standard radial and circumferential unit vectors for a cylindrical coordinate system (see Appendix D, Section D.2). Note that the impeller rotates in the counter-clockwise direction, which corresponds to positive rotation in a mathematical sense. That is, the angular velocity vector, Ω, points in the positive z direction so that

$$\boldsymbol{\Omega} = \Omega \mathbf{k} \qquad (9.19)$$

This is consistent with the convention that the angular coordinate θ increases in the counter-clockwise direction.

Mass Conservation: Because the flow is steady, the integral form of mass conservation simplifies to

$$\oiint_S \rho\, \mathbf{w} \cdot \mathbf{n}\, dS = 0 \qquad (9.20)$$

Expanding the closed-surface integral, we find

$$
\begin{aligned}
0 &= \left\{ \iint_{r=r_1} + \iint_{r=r_2} + \iint_{\text{Blade A}} + \iint_{\text{Blade B}} + \iint_{\text{Hub}} + \iint_{\text{Casing}} \right\} \rho\, \mathbf{w} \cdot \mathbf{n}\, dS \\
&= \iint_{r=r_1} \rho\, \mathbf{w}_1 \cdot (-\mathbf{e}_r)\, dS + \iint_{r=r_2} \rho\, \mathbf{w}_2 \cdot (\mathbf{e}_r)\, dS = -\rho w_{r1} A_1 + \rho w_{r2} A_2 \qquad (9.21)
\end{aligned}
$$

We have made use of the fact that there is no flow through the blades, hub and casing in the second line of Equation (9.21), and assumed the inlet and outlet areas between the blades are A_1 and A_2, respectively. Hence, denoting the volume-flow rate by Q, conservation of mass tells us

$$\rho w_{r1} A_1 = \rho w_{r2} A_2 = \rho Q \qquad (9.22)$$

where w_{r1} and w_{r2} denote the radial components of $\mathbf{w}_1$ and $\mathbf{w}_2$, respectively. Since the flow between all blades is the same (unless, for some reason, the blade spacing is nonuniform), we can regard Q as either the volume-flow rate between two successive blades or the total volume-flow rate through the pump. In the latter case, the areas A_1 and A_2 are the total areas at the inlet and outlet of the impeller. If the blade widths (out of the page) at the inlet and outlet are b_1 and b_2, respectively [Figure 9.5(b)], then $A_1 \approx 2\pi r_1 b_1$ and $A_2 \approx 2\pi r_2 b_2$. Hence, the radial components of the inlet and outlet velocity vectors are

$$w_{r1} = \frac{Q}{2\pi r_1 b_1} \quad \text{and} \quad w_{r2} = \frac{Q}{2\pi r_2 b_2} \tag{9.23}$$

Angular-Momentum Conservation: Clearly, the torque that develops on the impeller is aligned with the z axis. So, calling the torque exerted by the fluid on the impeller τ, the z component of the angular-momentum conservation principle [Equation (9.6)] for steady flow simplifies to

$$\tau = \mathbf{k} \cdot \oiint_S \rho\, \mathbf{r} \times \mathbf{u}_{abs}\, (\mathbf{w} \cdot \mathbf{n})\, dS \tag{9.24}$$

Appealing to Equations (9.17) and (9.18) and again using the fact that there is no flow through the blades, hub and casing, the closed-surface integral expands as follows.

$$\begin{aligned}
\tau &= \mathbf{k} \cdot \iint_{r=r_1} \rho\, r_1 \mathbf{e}_r \times \mathbf{V}_1\, [\mathbf{w}_1 \cdot (-\mathbf{e}_r)]\, dS + \mathbf{k} \cdot \iint_{r=r_2} \rho\, r_2 \mathbf{e}_r \times \mathbf{V}_2\, [\mathbf{w}_2 \cdot (\mathbf{e}_r)]\, dS \\
&= \mathbf{k} \cdot [\rho r_1\, (V_{\theta 1}\mathbf{k})\, (-Q) + \rho r_2\, (V_{\theta 2}\mathbf{k})\, (Q)] = \mathbf{k} \cdot [-\rho r_1 V_{\theta 1} Q \mathbf{k} + \rho r_2 V_{\theta 2} Q \mathbf{k}] \quad (9.25)
\end{aligned}$$

where $V_{\theta 1}$ and $V_{\theta 2}$ are the circumferential components of the absolute velocity vectors, $\mathbf{V}_1$ and $\mathbf{V}_2$, respectively. Therefore, taking the dot product and regrouping terms, the torque is

$$\tau = \rho Q\, [r_2 V_{\theta 2} - r_1 V_{\theta 1}] \tag{9.26}$$

Finally, the rate at which work is done by this torque is $\dot{W}_p = \boldsymbol{\Omega} \cdot \boldsymbol{\tau}$. Since both Ω and τ point in the positive z direction, the power is

$$\dot{W}_p = \rho \Omega Q\, [r_2 V_{\theta 2} - r_1 V_{\theta 1}] \tag{9.27}$$

Equations (9.26) and (9.27) are known as the **Euler turbomachine equations**. They are based on averaged properties on cross sections, and thus represent a global view. One of the most interesting properties of these equations is the absence of axial inlet flow properties. As with the simple lawn-sprinkler example of Subsection 9.2.2, we see that the torque is the product of the mass-flow rate, $\dot{m} = \rho Q$, and the change in absolute angular momentum per unit mass of the fluid as it moves through the impeller.

While our derivation has focused on a pump, there is nothing that precludes application of the Euler turbomachine equations to devices that extract energy from the flow, such as turbines. The only difference between pump and turbine applications occurs in the sign of $\dot{W}_p$. That is, when Ω and τ are in the same direction, $\dot{W}_p$ is positive so that work is being done on the fluid. This corresponds to a pump or a compressor, which adds energy to the fluid. By contrast, if Ω and τ are in opposite directions, $\dot{W}_p$ is negative so that work is being done by the fluid on the rotor. This corresponds to a turbine.

Energy Conservation: Although we can obtain complete details of the solution from the Euler turbomachine equations, it is instructive to see how our analysis relates to the energy-conservation principle developed in Chapter 7. First, note that we can determine the head supplied by the pump, $h_p = \Delta p/(\rho g)$, by combining the definition of head in Equation (7.88) with the second of the Euler relations, Equation (9.27), wherefore

$$h_p \equiv \frac{\dot{W}_p}{\dot{m} g} = \frac{\Omega}{g}\, [r_2 V_{\theta 2} - r_1 V_{\theta 1}] \tag{9.28}$$

Inspection of Equations (9.17) and (9.18) shows that, in general, the absolute and relative velocities are given by

$$\mathbf{V} = \mathbf{w} + \Omega r \mathbf{e}_\theta \quad \Longrightarrow \quad V_r = w_r, \quad V_\theta = w_\theta + \Omega r \tag{9.29}$$

Thus, we can rewrite the head supplied by the pump as

$$h_p = \frac{\Omega}{g} \left[r_2 \left(w_{\theta 2} + \Omega r_2 \right) - r_1 \left(w_{\theta 1} + \Omega r_1 \right) \right] \tag{9.30}$$

Now, recall that for steady, incompressible flow, the approximate form of the energy-conservation principle is [cf. Equation (7.89)]:

$$\frac{p_1}{\rho g} + \alpha_1 \frac{\mathbf{V}_1 \cdot \mathbf{V}_1}{2g} + z_1 + h_p = \frac{p_2}{\rho g} + \alpha_2 \frac{\mathbf{V}_2 \cdot \mathbf{V}_2}{2g} + z_2 + h_t + h_L \tag{9.31}$$

where all notation is as in Section 7.6. Since our flow involves only a pump, obviously the head supplied to a turbine, h_t, is zero. Then, if we assume we have a perfect (inviscid) fluid, we can say the kinetic-energy correction factors, α_1 and α_2, are unity and the head loss, h_L, is zero. Therefore, energy conservation simplifies to

$$\frac{p_1}{\rho g} + \frac{V_{r1}^2 + V_{\theta 1}^2}{2g} + z_1 + h_p = \frac{p_2}{\rho g} + \frac{V_{r2}^2 + V_{\theta 2}^2}{2g} + z_2 \tag{9.32}$$

Combining Equations (9.29), (9.30) and (9.32), after a bit of straightforward algebra, we find

$$\frac{p_1}{\rho g} + \frac{w_{r1}^2 + w_{\theta 1}^2 - \Omega^2 r_1^2}{2g} + z_1 = \frac{p_2}{\rho g} + \frac{w_{r2}^2 + w_{\theta 2}^2 - \Omega^2 r_2^2}{2g} + z_2 \tag{9.33}$$

Finally, multiplying through by ρg, we conclude that

$$p + \frac{1}{2} \rho \, \mathbf{w} \cdot \mathbf{w} + \rho g z - \frac{1}{2} \rho \Omega^2 r^2 = \text{constant} \tag{9.34}$$

This is often referred to as Bernoulli's equation in a rotating coordinate system. Recall that we derived a similar equation in Subsection 5.3 as an exact solution to the Euler equation for a rigid-body rotation [cf. Equation (5.53)].

9.4 Application to a Centrifugal Pump

While the Euler turbomachine equations are relatively simple, their application is complicated somewhat by the three-dimensional geometry characteristic of virtually all turbomachines. In this section, we apply the equations to the centrifugal pump depicted in Figure 9.4. While their range of applicability extends to most turbomachines, it is convenient at this point to limit our scope to a centrifugal pump, which served as our focus in developing the theory. We will apply the equations to other turbomachines in Section 9.6.

9.4.1 Velocity Triangles

Inspection of the Euler turbomachine Equations (9.26) and (9.27) shows that the individual velocity components of the rotor and the blade at both the impeller inlet and outlet appear explicitly. Thus, to analyze a given pump, sufficient information must be available to define the velocities. To help clarify things, it is useful to discuss a concept known as the **velocity triangle**. This is simply a vector diagram based on the velocity vectors at the inlet and outlet, with sufficient information to determine all components.

Figure 9.7 shows a velocity triangle for the centrifugal pump of Figure 9.6. We assume flow relative to the rotor enters and leaves tangent to the blade profile. Recall that β is the blade angle relative to the impeller, and Figure 9.7 applies to any part of the blade from leading to trailing edge. Also, α is the angle the absolute velocity makes with the impeller.

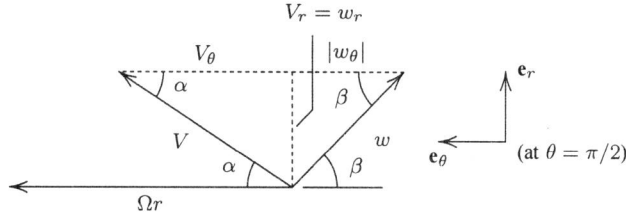

Figure 9.7: *Velocity triangle for a centrifugal pump (xy plane).*

In order to use the Euler turbomachine equations to determine the performance of a given machine, we require three types of input data. *First*, the geometry of the device must be specified. For our idealized formulation, the geometry is prescribed in terms of the impeller radii (r_1, r_2), the inlet and outlet blade widths (b_1, b_2) and the inlet and outlet blade angles (β_1, β_2). *Second*, we must know the angular rotation rate, Ω. *Third*, the inlet-flow direction is quantified in terms of the angle α_1. From this information, combined with conservation of mass, we can determine the volume-flow rate, Q, all velocity components of the relative and absolute velocities ($\mathbf{w}_1$, $\mathbf{V}_1$) and the angle α_2.

Volume-Flow Rate: From the velocity triangle, clearly

$$|w_\theta| = w_r \cot\beta \quad \text{and} \quad V_\theta = V_r \cot\alpha \tag{9.35}$$

Now, since $\mathbf{V} = \mathbf{w} + \Omega r\mathbf{e}_\theta$, while $\mathbf{V} = V_r\mathbf{e}_r + V_\theta\mathbf{e}_\theta$ and $\mathbf{w} = w_r\mathbf{e}_r - |w_\theta|\mathbf{e}_\theta$, necessarily the radial and circumferential components for this velocity triangle are related by

$$V_\theta + |w_\theta| = \Omega r \quad \text{and} \quad V_r = w_r \tag{9.36}$$

Note that there is no inconsistency with Equation (9.29), which made no explicit assumption about the sign of w_θ — here, we explicitly assume it is negative. Thus, at the inlet where all quantities have subscript 1, we find

$$w_{r1} \cot\alpha_1 + w_{r1} \cot\beta_1 = \Omega r_1 \tag{9.37}$$

which yields the following solution for w_{r1}.

$$w_{r1} = \frac{\Omega r_1}{\cot\alpha_1 + \cot\beta_1} \tag{9.38}$$

Finally, combining Equations (9.38) and (9.23), the volume-flow rate through the pump is

$$Q = \frac{2\pi b_1 \Omega r_1^2}{\cot\alpha_1 + \cot\beta_1} \tag{9.39}$$

As a final comment, the **design flow rate**, Q^*, is the maximum volume-flow rate through the pump. It usually occurs when the inlet flow is purely radial as it enters the impeller, i.e., for $\alpha_1 = \pi/2$.

Velocity Components: At the inlet, we have already shown that the radial and circumferential components of the absolute velocity are $V_{r1} = w_{r1}$ and $V_{\theta 1} = V_{r1} \cot \alpha_1$, respectively. Therefore, making use of Equation (9.38), the circumferential velocity at the inlet is

$$V_{\theta 1} = \frac{\Omega r_1 \cot \alpha_1}{\cot \alpha_1 + \cot \beta_1} \tag{9.40}$$

Conservation of mass, Equation (9.23), shows that the radial velocity components at the inlet and outlet [see Figure 9.5(b)] are related by

$$\frac{w_{r2}}{w_{r1}} = \left(\frac{r_1}{r_2}\right)\left(\frac{b_1}{b_2}\right) \tag{9.41}$$

Therefore, using the value of w_{r1} from Equation (9.38), the radial velocity at the outlet is

$$w_{r2} = \left(\frac{r_1}{r_2}\right)\left(\frac{b_1}{b_2}\right)\frac{\Omega r_1}{\cot \alpha_1 + \cot \beta_1} \tag{9.42}$$

Again referring to the velocity triangle in Figure 9.7, we see that

$$|w_{\theta 2}| = w_{r2} \cot \beta_2 \tag{9.43}$$

Thus, the circumferential component of the velocity relative to the blade at the outlet is

$$|w_{\theta 2}| = \left(\frac{r_1}{r_2}\right)\left(\frac{b_1}{b_2}\right)\frac{\Omega r_1 \cot \beta_2}{\cot \alpha_1 + \cot \beta_1} \tag{9.44}$$

Finally, as discussed above, $V_{\theta 2} + |w_{\theta 2}| = \Omega r_2$, wherefore the circumferential component of the absolute velocity is

$$
\begin{aligned}
V_{\theta 2} &= \Omega r_2 - |w_{\theta 2}| \\
&= \Omega r_2 - \left(\frac{r_1}{r_2}\right)\left(\frac{b_1}{b_2}\right)\frac{\Omega r_1 \cot \beta_2}{\cot \alpha_1 + \cot \beta_1} \\
&= \Omega r_2 \left[1 - \left(\frac{r_1}{r_2}\right)^2 \left(\frac{b_1}{b_2}\right)\frac{\cot \beta_2}{\cot \alpha_1 + \cot \beta_1}\right]
\end{aligned}
\tag{9.45}
$$

Outlet Flow Angle: The final quantity of interest is the angle between the absolute velocity and the impeller at the outlet, α_2. Again referring to the velocity triangle in Figure 9.7, the angle is given by

$$\tan \alpha_2 = \frac{V_{r2}}{V_{\theta 2}} = \frac{w_{r2}}{V_{\theta 2}} \tag{9.46}$$

Combining Equations (9.42), (9.45) and (9.46), after a short calculation we find

$$\alpha_2 = \tan^{-1}\left[\frac{\left(\dfrac{r_1}{r_2}\right)^2 \left(\dfrac{b_1}{b_2}\right)}{\cot \alpha_1 + \cot \beta_1 - \left(\dfrac{r_1}{r_2}\right)^2 \left(\dfrac{b_1}{b_2}\right)\cot \beta_2}\right] \tag{9.47}$$

In summary, the volume-flow rate, relative and absolute velocity vectors at the inlet and outlet, and the outlet flow angle for a centrifugal pump are as follows.

$$
\left.\begin{aligned}
Q &= \frac{2\pi b_1 \Omega r_1^2}{\cot\alpha_1 + \cot\beta_1} \\[2mm]
\mathbf{w}_1 &= \frac{\Omega r_1}{\cot\alpha_1 + \cot\beta_1}\left[\mathbf{e}_r - \cot\beta_1\,\mathbf{e}_\theta\right] \\[2mm]
\mathbf{V}_1 &= \frac{\Omega r_1}{\cot\alpha_1 + \cot\beta_1}\left[\mathbf{e}_r + \cot\alpha_1\,\mathbf{e}_\theta\right] \\[2mm]
\mathbf{w}_2 &= \frac{\Omega r_1}{\cot\alpha_1 + \cot\beta_1}\left(\frac{r_1}{r_2}\right)\left(\frac{b_1}{b_2}\right)\left[\mathbf{e}_r - \cot\beta_2\,\mathbf{e}_\theta\right] \\[2mm]
\mathbf{V}_2 &= \frac{\Omega r_1}{\cot\alpha_1 + \cot\beta_1}\left(\frac{r_1}{r_2}\right)\left(\frac{b_1}{b_2}\right)\mathbf{e}_r \\[2mm]
&+ \;\Omega r_2\left[1 - \left(\frac{r_1}{r_2}\right)^2\left(\frac{b_1}{b_2}\right)\frac{\cot\beta_2}{\cot\alpha_1 + \cot\beta_1}\right]\mathbf{e}_\theta \\[2mm]
\alpha_2 &= \tan^{-1}\left[\frac{\left(\dfrac{r_1}{r_2}\right)^2\left(\dfrac{b_1}{b_2}\right)}{\cot\alpha_1 + \cot\beta_1 - \left(\dfrac{r_1}{r_2}\right)^2\left(\dfrac{b_1}{b_2}\right)\cot\beta_2}\right]
\end{aligned}\right\} \qquad (9.48)
$$

This completes our formulation of an idealized turbomachine theory. Table 9.1 lists the most useful results of the theory, including the equation numbers for the various properties of interest. In practice, these equations provide estimates of pump performance that can differ from actual values by as much as 25%. There are a variety of reasons why our predictions are so far from the true performance of a real device, all of which stem from the complexity of turbomachine flows. Remember that we are dealing with a three-dimensional flowfield with complicated geometry. Thus, achieving this level of accuracy with a simple one-dimensional formulation is actually remarkable!

Some of the most significant complicating effects are leakage, mismatch, circulation loss and friction loss. **Leakage** is the movement of fluid between the impeller and the casing. Although the gap is small, any pressure difference between upper and lower blade surfaces causes flow in the gaps, with an attendant loss in flow rate and head. This is a three-dimensional effect that has been ignored in our simplified approach. **Mismatch** pertains to

Table 9.1: *Summary of the Euler Turbomachine Equations*

Equation(s)	Turbomachine Property
(9.26)	Torque (Euler turbomachine equation)
(9.27)	Power (Euler turbomachine equation)
(9.28)	Head change
(9.34)	Pressure (Bernoulli's-equation equivalent)
(9.48)	Volume-flow rate and flow velocities

the relation between the inlet flow and the blade passages. Ideally, the flow enters normal to the impeller and tangent to the blades. When these conditions are not met, the pump's performance is overestimated by our idealized model. Similarly, **circulation loss** occurs when the flow does not exit tangent to the blades. This can be caused by separation or stall, a condition where the flow actually reverses direction near the blade trailing edge. This condition reduces the increase in pressure achieved by the pump. There are two types of **friction losses**, viz., losses in the bearings supporting rotating metallic parts and losses caused by the fluid viscosity. Our theory ignores viscous effects, which can alter the flow, especially as the fluid approaches the blade trailing edge. Lakshminarayana (1996) addresses all of these, and other, issues in great detail.

9.4.2 Computing Pump Properties

To illustrate how the Euler turbomachine equations can be used to predict pump performance, consider the following. A typical commercial centrifugal water pump has the following characteristics.

- **Blade Geometry:** $r_1 = 10$ cm, $r_2 = 18$ cm, $b_1 = b_2 = 4.5$ cm, $\beta_1 = 30°$, $\beta_2 = 20°$

- **Angular-Rotation Rate:** $\Omega = 24$ Hz

- **Inlet-Flow Direction:** $\alpha_1 = 90°$

We would like to compute the volume-flow rate, Q, the power required to drive the pump, $\dot{W}_p$, and the increase in head as the water moves through the pump, h_p. Also, to assess how accurate our theory is, we can compare results of our computations with the actual pump properties as determined by the manufacturer, viz.,

$$Q = 0.22 \ \frac{m^3}{sec}, \quad \dot{W}_p = 98 \text{ kW}, \quad h_p = 46 \text{ m} \qquad \text{(Measured Pump Properties)} \quad (9.49)$$

The first thing we must do is convert Ω from cycles per second to radians per second. Since there are 2π radians in a complete cycle, we multiply by 2π. Therefore,

$$\Omega = \left(2\pi \ \frac{\text{radians}}{\text{cycle}} \right) \left(24 \ \frac{\text{cycles}}{\text{sec}} \right) = 150.8 \text{ sec}^{-1} \qquad (9.50)$$

Then, from the first of Equations (9.48), the volume-flow rate is

$$Q = \frac{2\pi b_1 \Omega r_1^2}{\cot \alpha_1 + \cot \beta_1} = \frac{2\pi (0.045 \text{ m}) \left(150.8 \text{ sec}^{-1} \right) (0.1 \text{ m})^2}{\cot 90° + \cot 30°} = 0.246 \ \frac{m^3}{sec} \qquad (9.51)$$

Comparing to the first of Equations (9.49), we see that the predicted value for Q is 12% percent higher than the measured value. While this difference is modest, differences between predicted and measured flow rate are generally larger for off-design conditions, i.e., for $\alpha_1 \neq \pi/2$.

Before we can compute the power and change in head, we must determine the velocities at the inlet and outlet. Using the rest of Equations (9.48), a little arithmetic shows that the velocities (in m/sec) are:

$$
\left.
\begin{aligned}
\mathbf{w}_1 &= \underbrace{8.71}_{w_{r1}}\,\mathbf{e}_r - \underbrace{15.08}_{|w_{\theta 1}|}\,\mathbf{e}_\theta \\[2mm]
\mathbf{V}_1 &= \underbrace{8.71}_{V_{r1}}\,\mathbf{e}_r \\[2mm]
\mathbf{w}_2 &= \underbrace{4.84}_{w_{r2}}\,\mathbf{e}_r - \underbrace{13.28}_{|w_{\theta 2}|}\,\mathbf{e}_\theta \\[2mm]
\mathbf{V}_2 &= \underbrace{4.84}_{V_{r2}}\,\mathbf{e}_r + \underbrace{13.86}_{V_{\theta 2}}\,\mathbf{e}_\theta
\end{aligned}
\right\}
\tag{9.52}
$$

Also, the angle α_2 follows from the last of Equations (9.48), which yields

$$
\alpha_2 = 19.2^\circ
\tag{9.53}
$$

Figure 9.8 shows the velocity vectors at the inlet and outlet for this pump.

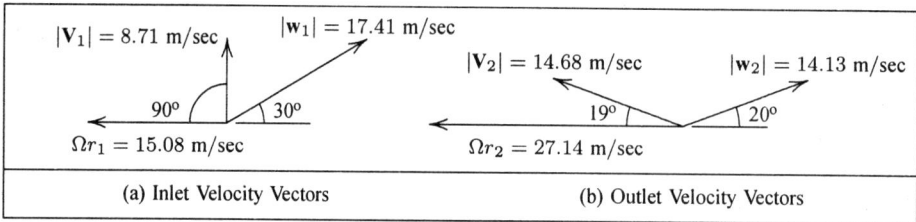

| (a) Inlet Velocity Vectors | (b) Outlet Velocity Vectors |

Figure 9.8: *Velocity vectors for a commercial centrifugal water pump.*

Next, we determine the power required to drive the pump from Equation (9.27). Thus, using the fact that the density of water is $\rho = 998$ kg/m^3 at 20° C,

$$
\begin{aligned}
\dot{W}_p &= \rho\Omega Q \left[r_2 V_{\theta 2} - r_1 V_{\theta 1} \right] \\[2mm]
&= \left(998\,\frac{\text{kg}}{\text{m}^3} \right) \left(150.8\,\text{sec}^{-1} \right) \left(0.246\,\frac{\text{m}^3}{\text{sec}} \right) \left[(0.18\,\text{m}) \left(13.86\,\frac{\text{m}}{\text{sec}} \right) - 0 \right] \\[2mm]
&= 92364\,\frac{\text{kg}\cdot\text{m}^2}{\text{sec}^3} = 92.4\,\text{kW}
\end{aligned}
\tag{9.54}
$$

Reference to the second of Equations (9.49) shows that our predicted value is just 6% lower than the measured power.

Finally, the increase in head that the fluid receives after passing through the pump is

$$
h_p = \frac{\dot{W}_p}{\dot{m}g} = \frac{\dot{W}_p}{\rho Q g}
\tag{9.55}
$$

From the values computed above, we have

$$
h_p = \frac{92364\,\text{kg}\cdot\text{m}^2/\text{sec}^3}{\left(998\,\text{kg/m}^3 \right) (0.246\,\text{m}^3/\text{sec}) \left(9.807\,\text{m/sec}^2 \right)} = 38.4\,\text{m}
\tag{9.56}
$$

The last of Equations (9.49) tells us the actual head increase is 46 m, so that our predicted gain is 17% less than measured.

9.4.3 Efficiency

The rate at which the blades do work on the fluid for a pump (and vice versa for a turbine), $\dot{W}_p$, is referred to as the **hydraulic power**. The power required to drive a pump (or power extracted from a turbine), $\Omega\tau$, is known as the **mechanical power**. According to the Euler turbomachine equations, the hydraulic power is equal to the mechanical power. In practice, they are not equal because of the various losses occurring in real flow through the machine.

In the case of a pump, the hydraulic power is less than the mechanical power. Conversely, a turbine absorbs more hydraulic power than it delivers in the form of mechanical power. To describe this mathematically, we introduce the **efficiency**. For a pump, the efficiency, η_p, is defined by

$$\eta_p \equiv \frac{\dot{W}_p}{\Omega\tau} = \frac{\rho Q g h_p}{\Omega\tau} \qquad \text{(Pump)} \qquad (9.57)$$

Similarly, we define the turbine efficiency, η_t, according to

$$\eta_t \equiv \frac{\Omega\tau}{\dot{W}_p} = \frac{\Omega\tau}{\rho Q g h_p} \qquad \text{(Turbine)} \qquad (9.58)$$

In either case, the efficiency is always less than one. The goal of a turbomachine designer is to achieve as large a value of either η_p or η_t over as wide a range of volume-flow rates as possible

For example, the power, $\dot{W}_p$, for the centrifugal pump discussed in the preceding subsection was quoted in Equation (9.49) as 98 kW. This is actually the power transferred from the pump to the fluid. The manufacturer states that the mechanical power required to drive the pump is $\Omega\tau = 115$ kW. Hence, for our example, the efficiency is

$$\eta_p = \frac{98 \text{ kW}}{115 \text{ kW}} = 0.85 \qquad (9.59)$$

9.4.4 Performance Curves

The ultimate indicator of a turbomachine's performance comes from extensive testing over a range of operating conditions. A standard format for quantifying turbomachine performance is a family of graphs, known as **performance curves**. Figure 9.9 shows typical centrifugal-pump performance curves. By convention, the curves are drawn for a constant rotation speed,

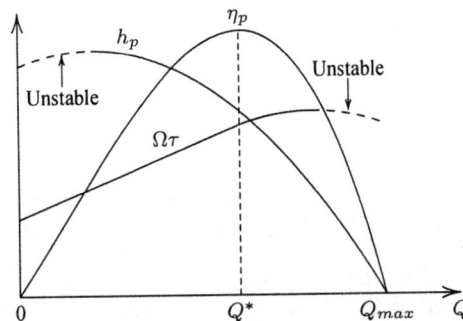

Figure 9.9: *Typical performance curves for a centrifugal pump.*

Ω, which is usually given in revolutions per minute (rpm). The three curves shown are the dimensionless efficiency, η_p, the head, h_p, in feet and the mechanical power, $\Omega\tau$, in horsepower. The latter curve is usually identified as the **brake horsepower**. Finally, the volume-flow rate, Q, is generally quoted in gallons per minute for liquids and cubic feet per minute for gases.

There are several interesting points about the performance curves worthy of mention. Focusing first on the head, it is nearly constant at small flow rates and falls to zero at a point where a maximum flow rate, Q_{max}, is reached. This is the maximum flow rate possible for the pump. The dashed part of the curve near $Q = 0$ labeled *unstable* involves a complete breakdown of the flow characterized by oscillations in pressure and mass flow. This breakdown is called **surge**, and can occur whenever h_p increases with Q. While it causes irregular operation for a liquid pump, it can lead to catastrophic failure of a gas compressor or even an entire jet engine.

The mechanical power increases slowly with flow rate and then drops off slightly as the flow rate approaches Q_{max}. The dashed part of the curve near $Q = Q_{max}$ marked *unstable* corresponds to another type of flow breakdown that is especially serious for axial-flow pumps. A key design objective is to minimize the range of flow rates over which unstable conditions can occur.

Finally, as the flow rate increases from zero, the efficiency increases to a peak value at the *design flow rate*, Q^*. Recall that assuming the inlet flow enters the impeller radially in our idealized turbomachine theory yields a reasonably accurate estimate of Q^*. At higher flow rates the efficiency drops off and ultimately falls to zero when the maximum flow rate, Q_{max}, is reached. The point where the efficiency achieves its maximum value is referred to as the **best-efficiency point**.

9.5 Dimensional Considerations

We saw in Subsection 9.4.4 that performance curves for common turbomachines are sometimes plotted in terms of dimensional quantities. While this practice is convenient for designers, it has one important disadvantage. In doing this, the information provided is valid *only for the fluid used in establishing the curves*. As we learned in Chapter 2, dimensional analysis provides a method for correlating experimental data in terms of dimensionless groupings that permits testing with one fluid and extrapolating the test results to a different fluid. It also provides scaling laws between machines of different sizes.

9.5.1 Primary Dimensionless Groupings

As we have seen, there are four key quantities commonly used to characterize centrifugal-pump (and other turbomachine) performance, viz., the change in head, h_p, the mechanical power, $\dot{W}_m = \Omega\tau$, the volume-flow rate, Q, and the efficiency, η_p (or η_t for a turbine). Of these four quantities, only the efficiency is dimensionless. An interesting question to pose is, what are the relevant dimensionless groupings for turbomachines?

To answer this question, we must first decide on the relevant dimensional quantities for flow through a turbomachine. Of course, the rotation speed, Ω, and geometry of the machine are significant. The latter is best characterized for a centrifugal pump by the impeller diameter, D. Also the key fluid properties that should be considered are density, ρ, and viscosity, μ.

Since gravitational acceleration, g, plays a role only when the working fluid is a liquid, it is customary to view the quantity gh_p, which is the hydraulic power per unit mass-flow rate, as a dimensional output quantity rather than h_p. Equivalently, we could work directly with the hydraulic power, $\dot{W}_p = \rho Q g h_p = Q \Delta p$, and arrive at the same results. Note that by dismissing g in this manner, we are making the judgment that buoyancy effects and gravity waves are unimportant in the dynamics of a turbomachine. We will also ignore the effects of surface roughness in our dimensional analysis. Although it will play a bit of a role, it is usually of secondary importance to the overall performance of a properly functioning turbomachine.

Thus, we postulate that $g h_p$ and $\dot{W}_m$ are each functions of five dimensional quantities:

$$g h_p = f_1(Q, D, \Omega, \rho, \mu) \quad \text{and} \quad \dot{W}_m = f_2(Q, D, \Omega, \rho, \mu) \qquad (9.60)$$

From dimensional analysis, we wish to correlate $g h_p$ and $\dot{W}_m$ with Q, D, Ω, ρ and μ. In both cases, using the techniques developed in Chapter 2, we find that there are six dimensional quantities and three independent dimensions. The Buckingham Π Theorem tells us there are three dimensionless groupings for each problem. Specifically, we find

$$g h_p = \Omega^2 D^2 \mathcal{F}_1 \left(\frac{Q}{\Omega D^3}, \frac{\rho \Omega D^2}{\mu} \right) \quad \text{and} \quad \dot{W}_m = \rho \Omega^3 D^5 \mathcal{F}_2 \left(\frac{Q}{\Omega D^3}, \frac{\rho \Omega D^2}{\mu} \right) \qquad (9.61)$$

Thus, we conclude that the change in the head and the mechanical power are both functions of the dimensionless flow rate, $Q/(\Omega D^3)$, and the Reynolds (or inverse Ekman) number, $Re = \rho \Omega D^2 / \mu$. In general, Reynolds numbers are very large for turbomachines — typically, $Re > 10^7$. At large Reynolds numbers, fluid flows tend to be relatively insensitive to its magnitude. Thus, it is customary to ignore the Reynolds number, which leaves the following three dimensionless groupings.

$$
\begin{aligned}
C_Q &= \frac{Q}{\Omega D^3} && \text{(Capacity coefficient)} \\[2mm]
C_H &= \frac{g h_p}{\Omega^2 D^2} && \text{(Head coefficient)} \\[2mm]
C_P &= \frac{\dot{W}_m}{\rho \Omega^3 D^5} && \text{(Power coefficient)}
\end{aligned}
\qquad (9.62)
$$

Finally, we can now determine the efficiency as a function of these dimensionless groupings by appealing to its definition in Equation (9.57), viz.,

$$\eta_p = \frac{\rho Q g h_p}{\dot{W}_m} = \frac{\rho \left(\Omega D^3 C_Q \right) \left(\Omega^2 D^2 C_H \right)}{\rho \Omega^3 D^5 C_P} = \frac{C_Q C_H}{C_P} \qquad (9.63)$$

Note that this explicitly demonstrates the interdependence of η, h_p, $\dot{W}_m$ and Q, which was obscured by presenting the performance curves in terms of dimensional quantities.

As an example of how the performance curves appear when plotted in terms of dimensionless groupings, we consider another popular centrifugal water pump. The pump comes in two sizes, one with an impeller diameter of 38 inches and a smaller model with an impeller diameter of 32 inches. All other geometrical parameters (casing, blade-to-casing clearance, etc.) are scaled in the same ratio. When conventional performance curves are used, two separate sets of curves must be provided, and each will be valid for just one rotation rate. However, using dimensionless groupings, the performance curves all collapse close to a single set, regardless of pump size and rotation rate!

Table 9.2: *Flow Properties for a Commercial Centrifugal Pump*

38-inch pump, 710 rpm				32-inch pump, 1170 rpm			
Q (gal/min)	h_p (ft)	$\dot{W}_m$ (hp)	η_p	Q (gal/min)	h_p (ft)	$\dot{W}_m$ (hp)	η_p
180	273	-	-	180	498	-	-
4215	267	-	-	4170	496	-	-
5560	266	607	.617	7310	491	1413	.643
6725	265	694	.650	10270	465	1611	.750
8430	263	805	.697	14440	431	1862	.846
12645	255	959	.851	18565	395	2095	.886
16860	236	1108	.909	22735	344	2265	.874
21075	210	1225	.914	26545	293	2360	.834
25290	176	1289	.874				
26635	169	1311	.869				

Table 9.2 lists Q, h_p, $\dot{W}_m$ and η_p for the 38-inch pump operating at 710 rpm and for the 32-inch pump operating at 1170 rpm. While the range of flow rates is the same, the smaller pump yields roughly twice the head of the larger pump, and is somewhat less efficient. Using Equations (9.62) and (9.63), we can recast the data of Table 9.2 in dimensionless form. Figure 9.10 shows the dimensionless performance curves for the two pumps, with the open circles corresponding to the larger pump and the closed circles to the smaller pump. The curves are least-squares fits to the data, which fall within 6% of the measurements for all three flow properties, η_p, h_p and $\dot{W}_m$. The minor departures from a single universal curve are attributable to differences in Reynolds number. Even so, the differences are sufficiently small to demonstrate the power of dimensional analysis in correlating measurements.

9.5.2 Specific Speed

There is another dimensionless grouping known as the **specific speed** commonly used in turbomachinery design, that is independent of pump size. It is called N_s and is based on properties at the *best-efficiency point*, Q^* and h_p^*. It can be formed as follows.

$$N_s = \frac{C_{Q^*}^{1/2}}{C_{H^*}^{3/4}} = \left(\frac{Q^*}{\Omega D^3}\right)^{1/2} \left(\frac{\Omega^2 D^2}{gh_p^*}\right)^{3/4} = \frac{\Omega (Q^*)^{1/2}}{(gh_p^*)^{3/4}} \tag{9.64}$$

As will be discussed in the next section, the specific speed is useful for selecting the type of turbomachine for a given application, especially for systems where volume-flow rate, head and rotation speed are dictated by the overall design requirements.

For the centrifugal pumps described in the performance curves of Figure 9.10, the least-squares curve indicates the best-efficiency point occurs when

$$C_{Q^*} = 0.0183 \quad \text{and} \quad C_{H^*} = 0.123 \tag{9.65}$$

Thus, for these pumps, the specific speed is

$$N_s = \frac{(0.0183)^{1/2}}{(0.123)^{3/4}} = 0.65 \tag{9.66}$$

In general, the specific speed for centrifugal pumps ranges from about 0.13 to 1.0.

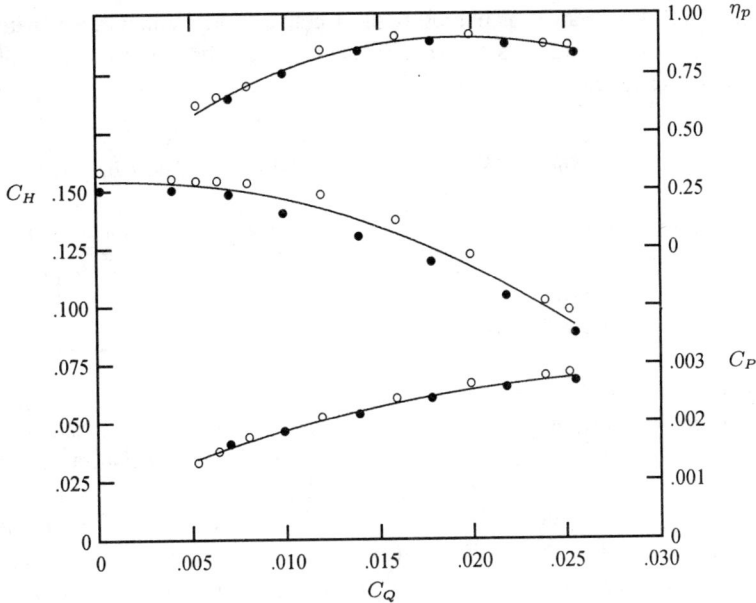

Figure 9.10: *Dimensionless performance curves for a centrifugal pump;* ○ *38-inch pump;* ● *32-inch pump;* —— *least-squares curve fit.*

9.6 Common Turbomachines

In this concluding section, we will briefly discuss common turbomachines found in practical applications. The discussion is, of necessity, mostly qualitative. While the Euler turbomachine equations reveal general properties of most pumps, compressors and turbines, they are far too simplistic to describe details of the complex flowfields involved. Rather, we must turn to computational methods if we wish to gain insight into detailed flow through the machine under investigation. Powerful computer programs are available to predict the performance of complex turbomachines, but they require intensive computational resources and are beyond the scope of this text.

Our use of the control volume method and dimensional analysis lays the foundation for understanding some aspects of turbomachinery. There are aspects we have omitted, such as the shapes of rotor and stator blades and their importance in a machine's performance. The interested reader can learn more about the physical and engineering aspects of turbomachines from the extensive literature on the topic. Some of the most noteworthy books are those by Shepherd (1956), Csanady (1964), Betz (1966), Dixon (1978), Logan (1981), Bathe (1984) and Lakshminarayana (1996). The text by Lakshminarayana cites the most up-to-date methods for analyzing turbomachines.

9.6.1 Pumps and Compressors

While most of our discussion thus far has focused on the centrifugal pump, there are other types of pumps that find practical application. Recall from Figure 9.1 that in addition to centrifugal pumps, which have purely radial outflow, there are also axial-flow pumps and mixed-flow pumps. The latter type of pump has both axial and radial velocity components at the outlet. The performance of pumps is most conveniently summarized in terms of the *specific speed*, N_s. Table 9.3 lists the specific-speed range for the three types of pumps as classified by outflow type.

Table 9.3: *Specific-Speed Ranges for Common Liquid Pumps*

Type of Pump	Specific-Speed Range
Radial-Flow (Centrifugal) Pumps	$0.13 < N_s < 1.00$
Mixed-Flow Pumps	$1.00 < N_s < 3.00$
Axial-Flow Pumps	$N_s > 3.00$

In general, centrifugal pumps are regarded as *high-head, low-flow* machines. They are very effective for pumping in many applications such as the water distribution system in a multi-story building. However, there are situations where an increase in head is of secondary importance relative to the volume-flow rate. A typical example would be pumping water between two reservoirs of slightly different depths. In such an application, we require a *low-head, high-flow* machine. By definition, the specific speed is small for the former type of pump and large for the latter. When we wish to pump large amounts of fluid and the increase in head is less important, mixed-flow and axial-flow pumps are more appropriate.

By convention, when the working fluid is a liquid, we categorize a turbomachine that adds energy to the flow as a "pump." When the working fluid is a gas, we use three different terms depending upon how large a pressure increase is imparted to the gas. By convention, when the pressure increase is 40 psi or more, we call the turbomachine a **compressor**. A machine that imparts a pressure rise of 1 psi or less is called a **fan**. We describe a machine with pressure increase between 1 and 40 psi as a **blower**. Table 9.4 summarizes typical specific-speed ranges for fans, blowers and compressors.

Table 9.4: *Specific-Speed Ranges for Common Gas "Pumps"*

Type of Pump	Specific-Speed Range
Fans	$0.50 < N_s < 2.00$
Blowers	$1.00 < N_s < 3.00$
Compressors	$N_s > 3.00$

While we have not discussed the shapes of the blades in the impeller of a pump, there is one interesting feature that can be discerned from our analysis. In particular, recall that the change in head for the fluid is given by [cf. Equation (9.28)]

$$h_p = \frac{\Omega}{g} \left[r_2 V_{\theta 2} - r_1 V_{\theta 1} \right] \tag{9.67}$$

Also, combining Equations (9.39) and (9.45), we have

$$V_{\theta 2} = \Omega r_2 - \frac{Q \cot \beta_2}{2 \pi r_2 b_2} \tag{9.68}$$

Now, ignoring the initial angular momentum, $r_1 V_{\theta 1}$, for simplicity, substituting Equation (9.68) into Equation (9.67) shows that we can write the head as

$$h_p = h_{po} + \frac{dh_p}{dQ} Q \qquad (9.69)$$

where

$$h_{po} \equiv \frac{\Omega^2 r_2^2}{g} \quad \text{and} \quad \frac{dh_p}{dQ} \equiv -\frac{\Omega \cot \beta_2}{2\pi b_2 g} \qquad (9.70)$$

This is a useful result for the following reason. The angle β_2 is the angle between the impeller and the blade at the trailing edge. When $\beta_2 < 90°$, we have *backward-curved* blades as depicted in Figures 9.4, 9.5 and 9.6. Since $\cot \beta_2 > 0$, necessarily $dh_p/dQ < 0$. By contrast, when $\beta_2 > 90°$, the blades are *forward curved*, and we have $dh_p/dQ > 0$. The special case of $\beta_2 = 90°$ corresponds to *radial* blades and $dh_p/dQ = 0$. Therefore, for small volume-flow rates, the h_p curve has negative slope for backward-curved blades, zero slope for radial blades, and positive slope for forward-curved blades. As discussed by Lakshminarayana (1996), a positive slope can lead to surge, so that backward-curved or radial blades are generally preferred, especially for compressors.

9.6.2 Reaction and Impulse Turbines

The purpose of a turbine is extract power from the fluid that can be used for things such as power generation and propulsion. There are two primary types of turbines known as the **reaction turbine** and the **impulse turbine**.

The reaction turbine is a *low-head, high-flow* turbomachine. It is similar to the pumps we have studied in the sense that fluid flows continuously through the passages between rotor (and stator) blades. As with pumps, one of the ways of classifying reaction turbines is in terms of the exit-flow direction. Thus, there are radial-flow, mixed-flow and axial-flow turbines, whose behavior is adequately described by the Euler turbomachine equations. Radial-flow and mixed-flow turbines are called **Francis turbines** in honor of James B. Francis, an early American pioneer in turbine design. Axial-flow turbines such as the one sketched in Figure 9.1(a) use a propeller and are appropriately called **propeller turbines**.

By contrast, an impulse turbine is a *high-head, low-flow* device. An example is the **Pelton wheel**, sketched in Figure 9.11, which uses a nozzle to create a high-speed jet that strikes a series of "buckets" at the same position as they pass by. Thus, the blade passages, i.e., the regions between the buckets, are not filled with fluid and the pressure is essentially constant as it flows past the blades. The pressure is constant since the pressure in a jet is always equal to the pressure of the surrounding fluid.

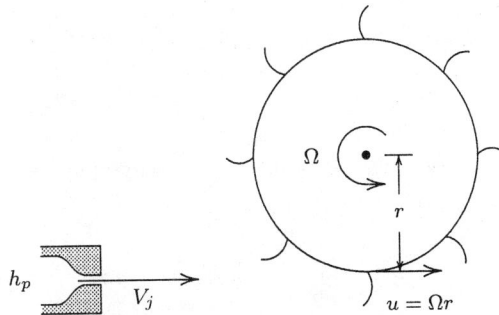

Figure 9.11: *Schematic of a Pelton wheel.*

The general principle upon which the Pelton wheel works is straightforward. Fluid from a reservoir with high potential energy flows through a nozzle. Its potential energy is thus transformed to kinetic energy in the jet issuing from the nozzle. The jet strikes the buckets at the wheel rim, and its kinetic energy is transmitted to the wheel to sustain the rotation. Designs exist that use more than one jet to enhance the power of the Pelton wheel. For simplicity, we will analyze a single-jet Pelton wheel.

We can formulate a simple theory by using a rotating control volume. We assume the buckets are sufficiently close that the jet always strikes one of the buckets. When this is true, we can approximate the motion as being steady. As shown in Figure 9.12, the fluid from the jet exits the bucket at an angle ϕ to the direction of the incident jet stream. We assume the area of the jet remains constant as it is turned by the bucket. We also assume the bucket width is very small compared to the wheel radius, so that the radial coordinate is approximately equal to the wheel radius, r, everywhere.

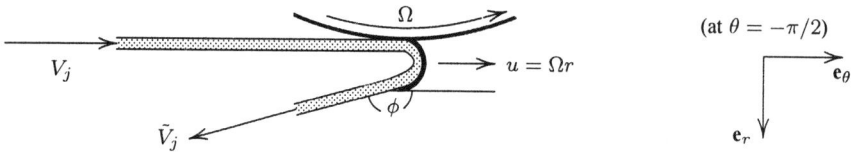

Figure 9.12: *Closeup view of a Pelton-wheel bucket.*

Since the flow is steady, conservation of mass becomes

$$\oiint_S \rho \left(\mathbf{w} \cdot \mathbf{n} \right) dS = 0 \qquad (9.71)$$

where $\mathbf{w}$ is the fluid velocity relative to the control volume. We select a control volume that is coincident with the jet boundaries. Hence, the only places where fluid crosses the bounding surface of the control volume is through the incident- and deflected-jet cross sections. Now, the incident flow in the jet has absolute velocity, $\mathbf{u}_{abs} = V_j \mathbf{e}_\theta$. Since the control volume moves with velocity $\mathbf{u}_{cv} = \Omega r \mathbf{e}_\theta$, we have the following at $\theta = -\pi/2$.

Incident Jet: $\mathbf{r} = r\mathbf{e}_r, \quad \mathbf{u}_{abs} = V_j\mathbf{e}_\theta, \quad \mathbf{w} = (V_j - \Omega r)\,\mathbf{e}_\theta, \quad \mathbf{n} = -\mathbf{e}_\theta$ (9.72)

Similarly, for the deflected jet, the vectors of interest are

Deflected Jet: $\mathbf{r} = r\mathbf{e}_r, \quad \mathbf{u}_{abs} = \tilde{V}_j\mathbf{n}, \quad \mathbf{w} = \tilde{V}_j\mathbf{n} - \Omega r\,\mathbf{e}_\theta, \quad \mathbf{n} = \mathbf{e}_\theta\cos\phi + \mathbf{e}_r\sin\phi$ (9.73)

Denoting the cross-sectional area of the jet by A, the mass-conservation integral simplifies to

$$\rho\left[-(V_j - \Omega r)\right]A + \rho\left[\tilde{V}_j - \Omega r\cos\phi\right]A = 0 \quad \Longrightarrow \quad \tilde{V}_j = V_j - \Omega r(1 - \cos\phi) \quad (9.74)$$

Since the flow is steady, conservation of angular momentum tells us the torque, τ_b, exerted by the fluid on the bucket is given by

$$\tau_b = \mathbf{k} \cdot \oiint_S \rho\,\mathbf{r} \times \mathbf{u}_{abs}\,(\mathbf{w} \cdot \mathbf{n})\,dS \qquad (9.75)$$

Therefore, using Equations (9.72) and (9.73) with subscripts *inc* and *def* denoting incident and deflected, respectively, we find

$$
\tau_b = \mathbf{k} \cdot \left\{ \underbrace{\rho\left(rV_j\mathbf{k}\right)}_{\rho(\mathbf{r}\,\times\,\mathbf{u}_{abs})_{inc}} \underbrace{\left[-\left(V_j - \Omega r\right) A\right]}_{(\mathbf{w}\,\cdot\,\mathbf{n})_{inc}A} + \rho \underbrace{\left(r\tilde{V}_j \cos\phi\mathbf{k}\right)}_{\rho(\mathbf{r}\,\times\,\mathbf{u}_{abs})_{def}} \underbrace{\left[\left(\tilde{V}_j - \Omega r \cos\phi\right) A\right]}_{(\mathbf{w}\,\cdot\,\mathbf{n})_{def}A} \right\} \tag{9.76}
$$

So, using the fact that the volume-flow rate is given by $Q = (\mathbf{w}\cdot\mathbf{n})A$, and introducing Equation (9.74) to eliminate $\tilde{V}_j$, the torque on the bucket is

$$
\begin{aligned}
\tau_b &= -\rho Q r \left(V_j - \tilde{V}_j \cos\phi \right) \\
&= -\rho Q r \left[V_j(1 - \cos\phi) + \Omega r \cos\phi(1 - \cos\phi) \right] \\
&= -\rho Q r (1 - \cos\phi)\left(V_j + \Omega r \cos\phi \right)
\end{aligned} \tag{9.77}
$$

The sum of the torque on the Pelton-wheel shaft, τ, and the torque on the bucket, τ_b, is zero so that $\tau = -\tau_b$. Thus,

$$
\tau = \rho Q r (1 - \cos\phi)\left(V_j + \Omega r \cos\phi \right) \tag{9.78}
$$

Finally, typical Pelton-wheel buckets are designed so that the angle ϕ is as close as possible to $180°$, with a value of $165°$ being common. Consequently, we can approximate $\cos\phi \approx -1$, so that the final expression for the torque is

$$
\tau \approx 2\rho Q r \left(V_j - \Omega r \right) \tag{9.79}
$$

The corresponding power output from the Pelton wheel is

$$
\Omega\tau \approx 2\rho Q \Omega r \left(V_j - \Omega r \right) \tag{9.80}
$$

At this point, we can derive an interesting result regarding the maximum power. Letting $U = \Omega r$, we can rewrite the power as

$$
\Omega\tau = 2\rho Q U \left(V_j - U \right) \qquad \Longrightarrow \qquad \frac{d\Omega\tau}{dU} = 2\rho r Q \left(V_j - 2U \right) \tag{9.81}
$$

Setting $d\Omega\tau/dU = 0$ and noting that $d^2\Omega\tau/dU^2 < 0$, we can locate the maximum value of $\Omega\tau$. It occurs when $U = V_j/2$, or, in terms of Ωr,

$$
\Omega r = \frac{1}{2}V_j \qquad \Longrightarrow \qquad \Omega\tau = 2\rho Q \Omega^2 r^2 = \text{maximum} \tag{9.82}
$$

Based on our idealized model of the Pelton wheel, we can develop an expression for the efficiency, η_t. First, however, we must determine how the jet velocity is related to the head in the reservoir, h_p. From energy conservation, we know that

$$
\frac{p_{res}}{\rho g} + \frac{\overline{u}_{res}^2}{2g} + h_p = \frac{p_{jet}}{\rho g} + \frac{V_j^2}{2g} \tag{9.83}
$$

Ideally, all of the total head in the reservoir is converted to kinetic energy as the fluid passes through the nozzle and forms the jet. Assuming the reservoir pressure, $p_{res} \approx p_{jet}$ and that the reservoir velocity, $\overline{u}_{res}$, is negligible, the jet velocity is given by

$$
V_j \approx \sqrt{2g h_p} \tag{9.84}
$$

We now have sufficient information to compute the efficiency of the Pelton wheel. Substituting Equations (9.80) and (9.84) into Equation (9.58), the efficiency is

$$\eta_t = \frac{\Omega\tau}{\rho Q g h_p} = 4\frac{\Omega r}{\sqrt{2gh_p}}\left(1 - \frac{\Omega r}{\sqrt{2gh_p}}\right) \qquad (9.85)$$

Figure 9.13 compares Equation 9.85 with measurements [White (1994)]. Consistent with Equation (9.82) and the fact that h_p = constant in our idealized theory, the Pelton wheel operates at maximum efficiency when $\Omega r/\sqrt{2gh_p} = \Omega r/V_j = \frac{1}{2}$. This corresponds to fluid exiting with no momentum or energy left and without vertical motion (in absolute terms). Since we have ignored all losses, the theoretical peak efficiency is 1. The measurements indicate a peak efficiency of 80% when $\Omega r/\sqrt{2gh_p} = 0.47$. In general, Pelton wheels are not as efficient as Francis or propeller turbines.

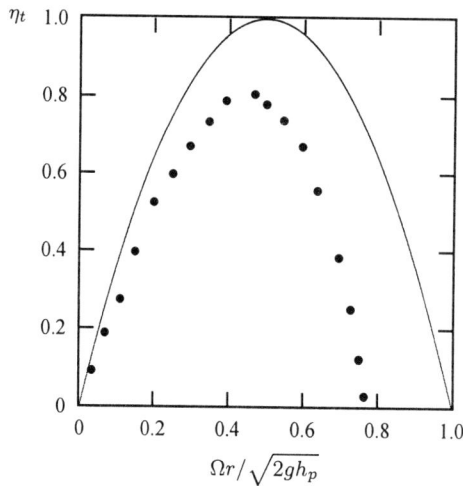

Figure 9.13: *Comparison of theoretical and measured efficiency for a Pelton wheel;* —— *Theoretical;* • *Measured [from White (1994)].*

We conclude our discussion of turbines by summarizing specific-speed ranges for the machines mentioned in this section. Table 9.5 lists the ranges for Pelton wheels, Francis turbines and propeller turbines. Note that a **Kaplan turbine** is a special type of propeller turbine.

Table 9.5: *Specific-Speed Ranges for Common Turbines*

Type of Turbine	Specific-Speed Range
Pelton wheel, single jet	$N_s < 0.13$
Pelton wheel, multijet	$0.10 < N_s < 0.25$
Francis turbines	$0.25 < N_s < 2.10$
Propeller turbines	$1.80 < N_s < 3.50$
Kaplan turbines	$3.00 < N_s < 5.00$

Problems

9.1 Consider a lawn sprinkler for which the water is ejected through two nozzles at an angle ϕ as shown.

(a) Determine the reaction torque on the sprinkler as a function of water density, ρ, ejection velocity, w_e, nozzle area, A_e, rotation rate, Ω, sprinkler-arm length, ℓ, and ϕ.

(b) Suppose the nozzle diameter is $\frac{1}{4}$ in, $\rho = 1.94$ slug/ft^3, $\ell = 3$ in and $\Omega = 300$ rpm. Also, assume the volume-flow rate is $Q = 0.05$ ft^3/sec. Compute the torque for $\phi = 0°$, $30°$ and $60°$.

(c) For the conditions of Part (b), at what angle does the torque vanish?

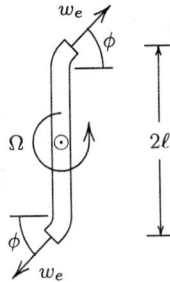

Problems 9.1, 9.2

9.2 The lawn sprinkler shown ejects water through two nozzles at an angle ϕ to the circumferential direction. The mass-flow rate through the device is $\dot{m} = 5$ kg/sec, and the area of each nozzle is $A_e = 1$ cm^2. The water density is $\rho = 1000$ kg/m^3, the sprinkler-arm length is $\ell = 10$ cm, while the sprinkler rotates at $\Omega = 600$ rpm.

(a) Compute the power, P, required to drive the sprinkler, neglecting all friction losses. Express your answer as a function of $\dot{m}$, ℓ, Ω, w_e and ϕ.

(b) Compute the value of w_e in m/sec.

(c) If the power is $P = 195$ W, what is the angle ϕ?

9.3 We would like to determine the conditions under which the outlet flow from the rotor or impeller of a turbomachine is purely radial. Using the solution developed in this chapter for a centrifugal pump, and assuming $b_1 = b_2$ while $\alpha_1 = \pi/2$, derive an expression for the exit blade angle, β_2, as a function of r_1, r_2 and β_1 that yields $\alpha = \pi/2$. Make a graph of your results for $\beta_1 = 30°$ and $0 \leq r_1/r_2 \leq 1$.

9.4 We want to determine the change in pressure across the impeller of a centrifugal pump, $\Delta p = p_2 - p_1$. As a first estimate, we can use the solution to the Euler turbomachine equations we developed in this chapter.

(a) Beginning with Bernoulli's equation in rotating coordinates [Equation (9.34)], determine Δp as a function of ρ, Ω, r_1, r_2, b_1, b_2, β_1 and β_2. Assume $\alpha_1 = \pi/2$.

(b) What is the pressure change for the pump discussed in Subsection 9.4.2?

9.5 A useful approximation for the head in a pump is given by

$$h_p = h_{po} - AQ^2$$

where h_{po} is the limiting value of the head as $Q \to 0$ and A is a constant of dimensions T^2/L^5. Using the solution to the Euler turbomachine equations we developed in this chapter, determine h_{po} and A. Assume the inlet flow is radial so that $\alpha_1 = \pi/2$.

9.6 Consider a centrifugal pump whose blade geometry is described by $r_1 = 4$ in, $r_2 = 8$ in, $b_1 = 4$ in, $b_2 = 4$ in, $\beta_1 = 30°$ and $\beta_2 = 90°$. If the angular-rotation rate is $\Omega = 1440$ rpm and $\alpha_1 = 90°$, determine the volume-flow rate and power required to drive the pump if the working fluid is kerosene at $68°$ F (see Table 1.1).

9.7 Consider a centrifugal pump whose blade geometry is described by $r_1 = 15$ cm, $r_2 = 25$ cm, $b_1 = 4$ cm, $b_2 = 6$ cm, $\beta_1 = 25°$ and $\beta_2 = 90°$. If the angular-rotation rate is $\Omega = 960$ rpm and $\alpha_1 = 90°$, determine the volume-flow rate and power required to drive the pump if the working fluid is ethyl alcohol at $20°$ C (see Table 1.1).

9.8 A small laboratory pump is used to move mercury ($\rho = 13550$ kg/m^3) between two tanks. It is a centrifugal pump with $r_1 = 3$ cm, $r_2 = 6$ cm, $b_1 = 3$ cm, $b_2 = 4$ cm, $\beta_1 = 35°$ and $\beta_2 = 45°$. If the angular-rotation rate is $\Omega = 600$ rpm and $\alpha_1 = 90°$, determine the volume-flow rate and power required to drive the pump. Express your answer for Q in liters per second (L/sec).

9.9 A commercial water pump is used to empty an overflowing septic tank. Due to suspended solid material, the density of the fluid is 2.10 slug/ft^3. The pump is of the centrifugal type with $r_1 = 4$ in, $r_2 = 12$ in, $b_1 = 2$ in, $b_2 = 1$ in, $\beta_1 = 25°$ and $\beta_2 = 15°$. If the angular-rotation rate is $\Omega = 1500$ rpm and $\alpha_1 = 90°$, determine the volume-flow rate (in gal/min) and power (in hp) required to drive the pump.

9.10 A radial fan is used to circulate fresh air ($\rho = .00234$ slug/ft^3) through a mine shaft. The fan operates at $\Omega = 6000$ rpm, has a volume-flow rate of 600 ft^3/min, and requires 1 hp to run. The blade geometry is given by $r_1 = 6$ in, $r_2 = 9$ in, $b_1 = 1$ in and $b_2 = 1$ in. Also, the inlet flow has $\alpha_1 = 90°$. Determine the inlet and outlet blade angles, β_1 and β_2.

9.11 A radial fan is used to circulate fresh air ($\rho = 1.20$ kg/m^3) through a warehouse. The fan operates at $\Omega = 4000$ rpm, has a volume-flow rate of 0.8 m^3/sec, and requires 2 kW to run. The blade geometry is given by $r_1 = 14$ cm, $r_2 = 20$ cm, $b_1 = 4$ cm and $b_2 = 5$ cm. Also, the inlet flow has $\alpha_1 = 90°$. Determine the inlet and outlet blade angles, β_1 and β_2.

9.12 A centrifugal fan is used to power a low-speed (incompressible-flow) wind tunnel. The impeller geometry is described by $r_1 = 24$ in, $r_2 = 36$ in, $b_1 = 2$ in, $b_2 = 2$ in, $\beta_1 = 30°$ and $\beta_2 = 16°$. The angular-rotation rate is 600 rpm and $\alpha_1 = 90°$. The working fluid is air with $\rho = .00234$ slug/ft^3.

(a) Compute the volume-flow rate and the power required to run the fan.

(b) Determine the velocity triangles at the inlet. Be sure to compute the angle α_2.

9.13 A centrifugal fan is used to power a low-speed (incompressible-flow) wind tunnel. The impeller geometry is described by $r_1 = 60$ cm, $r_2 = 90$ cm, $b_1 = 5$ cm, $b_2 = 4$ cm, $\beta_1 = 27°$ and $\beta_2 = 18°$. The angular-rotation rate is 600 rpm and $\alpha_1 = 80°$. The working fluid is air with $\rho = 1.20$ kg/m^3.

(a) Compute the volume-flow rate and the power required to run the fan.

(b) Determine the velocity triangles at the inlet. Be sure to compute the angle α_2.

9.14 Q is designing a miniature radial fan for Agent 007, who will use it to discharge a very potent sleep-inducing gas into the air-conditioning ducts at Spectre headquarters. The density of the gas is $\rho = 1.08$ kg/m^3 and the fan must deliver a volume-flow rate of 40 cm^3/sec with a power consumption of 150 μW (microWatts). The blade geometry must be such that $r_1 = 6$ mm, $b_1 = \frac{1}{2}$ mm, $b_2 = \frac{1}{2}$ mm, $\beta_1 = 30°$ and $\beta_2 = 20°$, while $\alpha_1 = 90°$.

(a) Determine the angular-rotation speed, Ω, and the outlet radius, r_2.

(b) Determine the velocity triangles at the inlet. Be sure to compute the angle α_2.

9.15 Develop an alternative to Equation (9.70) for dh_p/dQ, taking account of the initial angular momentum, $r_1 V_{\theta 1}$. If the inlet flow angle is $\alpha_1 = 80°$, determine the critical angle, β_{2c}, at which $dh_p/dQ = 0$ for $0 \le b_1/b_2 \le 2$. Make a graph of your results and discuss the meaning for the case $b_1/b_2 = 2$.

9.16 The maximum volume-flow rate, Q_{max}, for a centrifugal pump occurs when $h_p \to 0$. Verify that, according to the Euler turbomachine equations, if $\alpha_1 = 90°$ then

$$Q_{max} = 2\pi \Omega r_2^2 b_2 \tan \beta_2$$

9.17 For fans, blowers and compressors, performance curves are customarily presented in terms of pressure rise, Δp, expressed in "inches of H_2O." In terms of these units, the hydraulic power is given by $\dot{W}_p = \rho_w Q g \Delta p$ where ρ_w is the density of water (1.94 slug/ft^3), Q is volume-flow rate and g is gravitational acceleration. If the pressure rise is $\Delta p = 12$ inches of H_2O, what is the change in head, h_p, if the working fluid is water, air ($\rho = .00234$ slug/ft^3), or helium ($\rho = .000323$ slug/ft^3)? Express your answers in feet.

9.18 The head, h_p, and power, $\Omega\tau$, for an axial-flow pump vary with volume-flow rate according to

$$h_p = h_o\left[1 - Q^2/28\right] \qquad \text{and} \qquad \Omega\tau = P_o\left[1 + Q^2/14\right]$$

where h_o and P_o are constants and Q is given in ft^3/sec. Determine the volume-flow rate at the *best-efficiency point*, Q^*.

9.19 The performance curves for a centrifugal pump are shown in the figure. For the practical operating range of the pump, the head, h_p, and power, $\Omega\tau$, can be approximated by

$$h_p \approx 167 - \frac{3}{80}Q \qquad \text{and} \qquad \Omega\tau \approx 45 + \frac{1}{160}Q, \qquad (800 < Q < 2400)$$

with h_p in ft, $\Omega\tau$ in hp and Q in gal/min. Determine the volume-flow rate of water at the *best-efficiency point*, Q^*, and the corresponding efficiency, η_p.

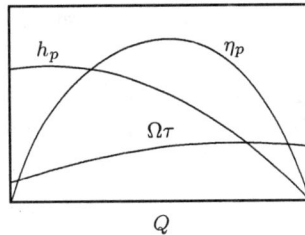

Problems 9.19, 9.20, 9.21

9.20 The performance curves for a centrifugal pump are shown in the figure. The head, h_p, and power, $\Omega\tau$, can be approximated by

$$h_p \approx 140\left[1 - \left(\frac{Q-400}{2800}\right)^2\right] \qquad \text{and} \qquad \Omega\tau \approx 60 - 40\left(\frac{Q-2400}{2400}\right)^2$$

with h_p in ft, $\Omega\tau$ in hp and Q in gal/min.

(a) Determine the volume-flow rate of water at the *best-efficiency point*, Q^*, and the corresponding efficiency, η_p. **HINT:** Solve by trial and error, noting that 1600 gal/min $< Q^* <$ 1800 gal/min.

(b) For what range of Q is this pump likely to operate without any type of instability?

9.21 The performance curves for a centrifugal water pump are shown in the figure. For small volume-flow rates, the head varies as

$$h_p \approx 137 + \frac{1}{70}Q$$

where h_p is given in ft and Q in gal/min.

(a) Verify that, according to the Euler turbomachine equations, if $\alpha_1 = \pi/2$ then

$$h_p = \frac{\Omega^2 r_2^2}{g} - \frac{\Omega Q \cot\beta_2}{2\pi g b_2}$$

(b) Now, assume $b_2 = \frac{1}{5}r_2$ and $\Omega = 1450$ rpm. What are the values of r_2, b_2 and β_2 for this pump?

9.22 The following data describe the measured performance of a pump operating at 1750 rpm. Complete the pump performance-curve tables by computing the power in hp. Above what volume-flow rate should we be concerned about possible unstable behavior?

Q (gal/min)	h_p (ft)	η_p (%)
0	124	0
400	120	49
800	112	64
1000	104	68
1200	96	71
1400	84	70
1500	76	69
1600	67	67
1800	41	50

Problem 9.22

9.23 It is common practice in the pump industry to compute a dimensional specific speed according to

$$\tilde{N}_s = \frac{\Omega \, (Q^*)^{1/2}}{(h_p^*)^{3/4}}$$

with Ω in rpm, Q^* in gal/min and h_p^* in ft. Determine the dimensions of $\tilde{N}_s$ and compute the ratio of $\tilde{N}_s$ to N_s.

9.24 A centrifugal water pump of diameter 14.62 in operating at 2134 rpm has the following performance figures: $Q^* = 6$ ft³/sec; $h_p^* = 330$ ft; $\dot{W}_m^* = 255$ hp. Compute C_{Q^*}, C_{H^*}, C_{P^*}, η_p and N_s.

9.25 A centrifugal water pump of diameter 40 cm operating at 2400 rpm has the following performance figures: $Q^* = 0.2$ m³/sec; $h_p^* = 100$ m; $\dot{W}_m^* = 215$ kW. Compute C_{Q^*}, C_{H^*}, C_{P^*}, η_p and N_s.

9.26 A blower of diameter $D = 24$ in operating at $\Omega = 2000$ rpm has the following performance figures: $Q^* = 335$ ft³/sec; $h_p^* = 430$ ft; $\dot{W}_m^* = 26$ hp. Compute C_{Q^*}, C_{H^*}, C_{P^*}, η_p and N_s for this blower if the working fluid is air with $\rho = .00234$ slug/ft³.

9.27 A compressor of diameter $D = 1$ m operating at $\Omega = 6000$ rpm has the following performance figures: $Q^* = 140$ m³/sec; $h_p^* = 1750$ m; $\dot{W}_m^* = 4.2$ MW. Compute C_{Q^*}, C_{H^*}, C_{P^*}, η_p and N_s for this compressor if the working fluid is air with $\rho = 1.20$ kg/m³.

9.28 A 36-inch pump operating at 720 rpm has capacity and head coefficients of $C_{Q^*} = 0.02$ and $C_{H^*} = 0.125$ at its *best-efficiency point*, where $\eta_p = 0.83$. We want to design a pump with the same dimensionless performance characteristics that will have a volume-flow rate of $Q^* = 60000$ gal/min at $\Omega = 1200$ rpm. What must the pump diameter be (to the nearest inch) and how much power will be needed to drive it? Assume the efficiency is the same for both pumps.

9.29 Measurements for a model mixed-flow pump with a discharge opening of diameter $D = 1.8$ m operating at $\Omega = 225$ rpm are as listed in the table. Determine the size and rotation speed of a pump with the same dimensionless performance parameters in order to have $Q^* = 6$ m³/sec and $h_p^* = 18$ m. Make a table of h_p as a function of Q for this pump.

Q (m³/sec)	h_p (m)	η_p (%)
5.66	18.3	69
7.25	16.8	80
8.58	15.2	86
9.77	13.7	88
11.21	11.4	84
13.00	7.6	64

Problem 9.29

9.30 From the manufacturer's specifications, we know that the minimum specific speed at which a centrifugal pump can operate is $(N_s)_{min} = 0.18$. The pump must deliver a volume-flow rate of $Q^* = 3000$ gal/min against a head of $h_p^* = 1002$ ft. Find the minimum rotation speed, Ω_{min}, in rpm.

9.31 Larger pumps and turbines are generally more efficient than smaller models of the same design. Moody developed the following empirical formula for estimating the effect:

$$\frac{1 - \eta_2}{1 - \eta_1} = \left(\frac{D_1}{D_2}\right)^{1/4}$$

where η is efficiency, D is diameter, and subscripts identify the two turbomachines. Consider a 70% efficient pump with a volume-flow rate of 1000 gal/min against a head of 81 ft. What is the efficiency of a larger, geometrically-similar pump delivering 5000 gal/min against a head of 25 ft?

9.32 Larger pumps and turbines are generally more efficient than smaller models of the same design. Moody developed the following empirical formula for estimating the effect:

$$\frac{1 - \eta_2}{1 - \eta_1} = \left(\frac{D_1}{D_2}\right)^{1/4}$$

where η is efficiency, D is diameter, and subscripts identify the two turbomachines. An 89% efficient Francis turbine has an output of 400 kW. A smaller, geometrically-similar turbine has half the volume-flow rate with twice the drop in pressure (head). How much smaller is the turbine and what is the power output, taking account of the reduced efficiency?

9.33 For turbines, a useful dimensionless measure of the output power relative to the available head is called the *power-specific speed*, N_{sp}. It increases linearly with rotation speed, Ω, and is independent of rotor diameter, D. Starting with the power coefficient, C_{P^*}, and the head coefficient, C_{H^*}, deduce N_{sp} as a function of Ω, $\dot{W}_m^*$, ρ, g and h_p^*.

9.34 We have shown from angular-momentum conservation that the torque on a Pelton wheel is given by Equation (9.78). Some authors use a less-rigorous derivation to conclude that the torque is

$$\tau \approx \rho Q r (1 - \cos\phi)(V_j - \Omega r)$$

For the case $\Omega r = \frac{1}{2}V_j$, we wish to determine the angle ϕ for which τ is a maximum.

   (a) Determine the angle and τ_{max} using Equation (9.78)

   (b) Determine the angle and τ_{max} using the equation that results from the less-rigorous derivation.

9.35 Nozzle losses in a Pelton wheel are accounted for through the empirical formula

$$V_j = C_v\sqrt{2gh_p}, \qquad 0.92 \leq C_v \leq 0.98$$

where C_v is called the *velocity coefficient*.

   (a) Assuming there are no other losses, determine the wheel velocity, Ωr, and maximum efficiency, $(\eta_t)_{max}$, as a function of C_v.

   (b) What value of C_v corresponds to $\Omega r / \sqrt{2gh_p} = 0.47$ at the point of maximum efficiency?

   (c) How small would C_v have to be to match the value of $(\eta_t)_{max} = 0.8$ indicated in Figure 9.13?

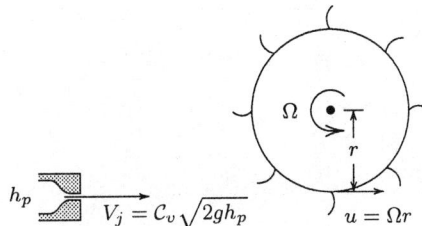

Problem 9.35

9.36 A 10 ft diameter Pelton wheel rotates at 200 rpm. The volume-flow rate of water is 120 ft^3/sec, the head is 775 ft and the power delivered is 8400 hp.

(a) Determine the specific speed, N_s and the efficiency, η_t.

(b) If the torque on the wheel is given by Equation (9.79), what is the jet velocity, V_j?

9.37 A Pelton wheel has a diameter D = 4.95 m and delivers 20 MW of power. It has a specific speed of N_s = 0.10 and operates at 82% efficiency. The diameter of the jet driving the wheel is d = 35 cm. You may assume that $\Omega r/V_j = \frac{1}{2}$, where Ω is the rotation rate, V_j is the jet velocity and $r = D/2$. Determine the rotation rate in rpm, assuming the fluid density is ρ = 1000 kg/m^3.

9.38 A power plant uses a 90% efficient turbine that develops 40 MW at 82 rpm under a head of 30 m. What type of turbine is it? Assume the working fluid is water with ρ = 1000 kg/m^3.

9.39 A secret research laboratory uses a 95% efficient turbine that develops 25000 hp at 500 rpm under a head of 130 ft. What type of turbine is it? Assume the working fluid is water with ρ = 2 slug/ft^3.

9.40 A family named Robinson uses a 75% efficient turbine that develops 13 kW at 18 rpm under a head of 6 m. What type of turbine is it? Assume the working fluid is water with ρ = 1000 kg/m^3.

Chapter 10

Vorticity and Viscosity

This chapter is an interlude. It can be viewed both as a post script to the material covered in Chapters 1 through 9 and as a preview of what will follow in Chapters 11 through 15. In the spirit of a post script to what we've already covered, we have gotten about as much out of inviscid control-volume theory as we can. Dealing with finite-sized control volumes provides important gross flow properties but no details of motion within the control volume.

Looking ahead, the chapters to follow examine fluid motion at a more fundamental level. To accomplish this we do two things. *First*, we turn our focus to the differential equations of motion. In doing this, we effectively decrease the size of our control volume to infinitesimal size. We have implicitly done this each time we have used the divergence theorem to group everything under a single volume integral to deduce a differential equation. If this statement is unclear, review Section C.6 of Appendix C, keeping in mind that the results hold for a differential-sized control volume. Beginning with Chapter 11, our primary focus shifts from finite control volumes to the differential forms of the conservation laws. *Second*, we include viscous effects in our analysis. Beginning with Chapter 12, we revise the conservation laws with friction and heat-transfer effects included.

Before moving on to our study of the differential forms of the conservation laws, it is worthwhile to pause and discuss the behavior of real, or viscous, fluids. One of the most important aspects of real-fluid behavior is the presence of vorticity, and the role viscosity plays in its creation. Vorticity is perhaps the most important of all fluid-flow properties. For example, without it we would have no lift on an airplane's wing or a sailboat's sail. The dynamics of the gulf stream and of hurricanes are characterized by the strength of the vorticity involved. On the one hand, the classical theory of fluid mechanics and aerodynamics deals with vorticity as an essentially inviscid-flow phenomenon. On the other hand, as we will see in this chapter, inviscid fluids in an unbounded medium have no physical mechanism for developing vorticity. We will also see the dilemma of eighteenth- and nineteenth-century mathematicians and physicists who concluded that viscous effects were too small to have a significant effect on fluid motion.

This dilemma is best characterized by two famous inviscid-flow results known as the **Helmholtz Theorem** and **d'Alembert's Paradox**. These two mathematical results, which we will discuss in this chapter, leave us with an apparent contradiction between theory and experiment. The theory says it is impossible to develop aerodynamic forces on objects such as an airplane wing, while experimental evidence shows that great forces do in fact develop.

It took Prandtl's pioneering work on viscous effects, first published in 1904, to resolve this dilemma, and to put theoretical fluid mechanics on a solid footing as a rational science.

Prandtl argued that, in fluids such as water and air, close to a solid boundary viscous effects are confined to very thin layers known as **boundary layers**. As a result, very large velocity gradients develop — so large that significant viscous forces are present. The large velocity gradients also create vorticity at solid boundaries, thus providing the missing physical mechanism for generating vorticity. This chapter is offered as a quick look at this historical road of discovery, which ultimately revolutionized fluid mechanics as a science. It is also intended to stress the close tie between viscosity and vorticity that is often overlooked in a first course on fluid mechanics. The chapter concludes with a fascinating home experiment that reveals a subtle phenomenon related to an interaction of vorticity and viscosity.

10.1 The Vortex Force

One way to understand the connection between vorticity and the force that develops when fluid flows past an object is to rearrange Euler's equation to reveal the presence of an apparent force known as the **vortex force**. As noted by Saffman (1993), the notion of the vortex force finds its origins in work by Prandtl in the early twentieth century, and was used extensively by von Kármán and associates in developing the theory of aerodynamics.

To see how the vortex force appears in the fluid-mechanical equations of motion, remember that Euler's equation can be written as [see Equation (5.30)]:

$$\rho \frac{\partial \mathbf{u}}{\partial t} + \rho \nabla \left(\frac{1}{2} \mathbf{u} \cdot \mathbf{u} \right) - \rho \mathbf{u} \times \omega = -\nabla p + \rho \mathbf{f} \tag{10.1}$$

where ω is the vorticity vector given by $\omega = \nabla \times \mathbf{u}$. The term proportional to the cross product of $\mathbf{u}$ and ω is similar to the Coriolis acceleration that appears in a rotating coordinate frame (see Section D.4 of Appendix D). Just as we often treat the Coriolis term as an apparent force, so we can rewrite Euler's equation as follows.

$$\rho \frac{\partial \mathbf{u}}{\partial t} + \rho \nabla \left(\frac{1}{2} \mathbf{u} \cdot \mathbf{u} \right) = \underbrace{\rho \mathbf{u} \times \omega}_{Vortex\ Force} -\nabla p + \rho \mathbf{f} \tag{10.2}$$

The vortex force acts in a direction normal to the flow direction. For example, in two dimensions, the vorticity vector is

$$\omega = \left(\frac{\partial v}{\partial x} - \frac{\partial u}{\partial y} \right) \mathbf{k} \tag{10.3}$$

Figure 10.1 illustrates the direction of the vortex force for a two-dimensional flowfield. As shown, if the flow has vorticity, a force will be present at 90° to the direction of flow. This is the situation that arises when we move an object such as an airplane wing through the air.

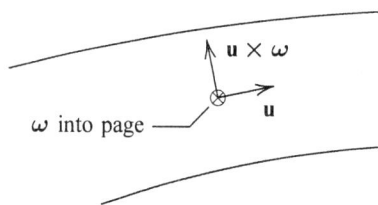

Figure 10.1: *The vortex force in a two-dimensional flow.*

10.2 Helmholtz's Theorem and d'Alembert's Paradox

Given the importance of vorticity in developing a force on an object in a fluid, it would be helpful to determine the conditions necessary to create vorticity. There is a profound result known as the **Helmholtz Theorem** that created a major stumbling block for mathematicians and scientists in developing the basic theoretical foundations of fluid mechanics. Specifically, the Helmholtz Theorem states the following.

> *If a fluid particle moving in an unbounded frictionless, incompressible fluid under conservative body forces has zero vorticity initially, it always has zero vorticity.*

Thus, for example, when a flow starts from rest this theorem tells us there is no physical mechanism in an unbounded frictionless, incompressible fluid to develop vorticity.

To prove this theorem, we begin with Euler's equation for an incompressible flow with a conservative body force given by $\mathbf{f} = -\nabla \mathcal{V}$, viz.,

$$\frac{\partial \mathbf{u}}{\partial t} + \nabla \left(\frac{1}{2} \mathbf{u} \cdot \mathbf{u} \right) - \mathbf{u} \times \omega = -\nabla \left(\frac{p}{\rho} + \mathcal{V} \right) \tag{10.4}$$

We can derive an equation for the vorticity by taking the curl of this differential equation. First, we note that the curl of the gradient of a scalar vanishes (see Appendix C). Hence,

$$\nabla \times \nabla \left(\frac{1}{2} \mathbf{u} \cdot \mathbf{u} \right) = \mathbf{0} \quad \text{and} \quad \nabla \times \nabla \left(\frac{p}{\rho} + \mathcal{V} \right) = \mathbf{0} \tag{10.5}$$

Since the curl commutes with time differentiation, the curl of the first term in Equation (10.4) becomes

$$\nabla \times \frac{\partial \mathbf{u}}{\partial t} = \frac{\partial}{\partial t} \left(\nabla \times \mathbf{u} \right) = \frac{\partial \omega}{\partial t} \tag{10.6}$$

Reference to Appendix C shows that the curl of the cross product of $\mathbf{u}$ and ω is

$$\nabla \times (\mathbf{u} \times \omega) = (\omega \cdot \nabla) \mathbf{u} + \mathbf{u} (\nabla \cdot \omega) - (\mathbf{u} \cdot \nabla) \omega - \omega (\nabla \cdot \mathbf{u}) \tag{10.7}$$

Since the divergence of the curl of any vector is zero, we can say

$$\nabla \cdot \omega = \nabla \cdot (\nabla \times \mathbf{u}) = 0 \tag{10.8}$$

Finally, since the flow is incompressible,

$$\nabla \cdot \mathbf{u} = 0 \tag{10.9}$$

Collecting these vector identities and substituting into Equation (10.4) yields the following differential equation for the vorticity vector.

$$\frac{\partial \omega}{\partial t} + \mathbf{u} \cdot \nabla \omega = \omega \cdot \nabla \mathbf{u} \tag{10.10}$$

Thus, since the sum of the two terms on the left-hand side of this equation is simply the Eulerian derivative of ω, the final form of the differential equation governing the vorticity is:

$$\frac{d\omega}{dt} = \omega \cdot \nabla \mathbf{u} \tag{10.11}$$

The term on the right-hand side of Equation (10.11) is the **vortex stretching** term. We will discuss vortex stretching when we revisit the kinematics of a fluid particle in Chapter 12.

We are now in a position to prove the Helmholtz Theorem. Because the proof is very simple for two-dimensional flow, we will confine our attention to two dimensions. Although we omit the proof here, the theorem applies to three-dimensional flows also. By definition, for two-dimensional flow in the xy plane, we know that

$$w = 0 \quad \text{and} \quad \frac{\partial}{\partial z} \to 0 \tag{10.12}$$

Noting that the vorticity has only a z component for two-dimensional flow, i.e.,

$$\boldsymbol{\omega} = \left(\frac{\partial v}{\partial x} - \frac{\partial u}{\partial y} \right) \mathbf{k} = \omega \mathbf{k} \tag{10.13}$$

necessarily the vortex-stretching term vanishes, viz.,

$$\boldsymbol{\omega} \cdot \nabla \mathbf{u} = \omega \frac{\partial \mathbf{u}}{\partial z} = \mathbf{0} \tag{10.14}$$

Therefore, in a two-dimensional flow, we have shown that

$$\frac{d\boldsymbol{\omega}}{dt} = \mathbf{0} \tag{10.15}$$

In words, this equation says *the rate of change of vorticity following a fluid particle is zero.* Hence, if $\boldsymbol{\omega}$ is initially zero, then it will remain zero for all time. Thus, we have proven the Helmholtz Theorem for two-dimensional flows. As mentioned above, the theorem holds for three-dimensional flows as well.

Consequently, if we have an object that is initially at rest and begins moving at $t = 0$, its initial vorticity is zero. At all subsequent times its vorticity remains zero, wherefore no force develops on the object. However, this is completely inconsistent with everyday observations. An airplane wing develops a tremendous lifting force, a sail causes a sailboat to be able to sail nearly into the wind, a hand extended out of an automobile's window feels a strong force. Practical experience tells us such objects do develop forces, indicating that the vorticity does not remain zero.

Eighteenth and nineteenth century scientists were able to solve Euler's equation for slender objects. They were also able to make crude wind-tunnel measurements late in the nineteenth century that confirmed their solutions for most of the flowfield, with one important exception. The wind-tunnel models developed forces that the theory failed to predict.

These apparently contradictory results are referred to as **d'Alembert's Paradox**. That is, measurements and theory were completely consistent in many details, providing evidence that the theory was, for the most part, correct. However, the theory failed to predict the most important detail — the force! This was indeed a paradox. Complete details regarding the Helmholtz Theorem and d'Alembert's Paradox can be found at an introductory level in the book by Anderson (1989), or in the advanced text by Saffman (1993).

As a final comment, the Helmholtz Theorem and d'Alembert's Paradox left practicing engineers with a sense that theoretical fluid mechanics was of little value for design purposes. Rather, the topic was viewed by many as an intellectual exercise for mathematicians, devoid of practical use. This point of view certainly had a lot to do with the empiricism and lack of rigor in much of the work of hydraulics engineers of the eighteenth and nineteenth centuries. The methods used by hydraulicists often did achieve their design objectives, but with little or no understanding of why any new innovation they stumbled upon might function as well as it did.

10.3   Boundary Conditions at a Solid Boundary

To help resolve d'Alembert's Paradox, we must understand what happens in the immediate vicinity of a solid surface past which a fluid is flowing. If the surface has unit normal $\mathbf{n}$, we can always write the velocity vector, $\mathbf{u}$, as

$$\mathbf{u} = u_n \mathbf{n} + u_t \mathbf{t} \tag{10.16}$$

where u_n is the magnitude of the component of $\mathbf{u}$ normal to the surface. The vector $u_t \mathbf{t} = \mathbf{u} - u_n \mathbf{n}$ is the contribution to the velocity parallel to the surface, where $\mathbf{t}$ is a unit vector tangent to the surface (see Figure 10.2). Since there can be no flow through the boundary, if the surface is stationary, we can say that

$$\mathbf{u} \cdot \mathbf{n} = u_n = 0 \qquad \text{(at the surface)} \tag{10.17}$$

This result is valid for any fluid, whether we regard the fluid as real or ideal. For an ideal fluid, i.e., for an inviscid or frictionless fluid, this is the only condition we can specify at a solid boundary. Thus, we can have a non-vanishing tangential velocity component in an ideal fluid.

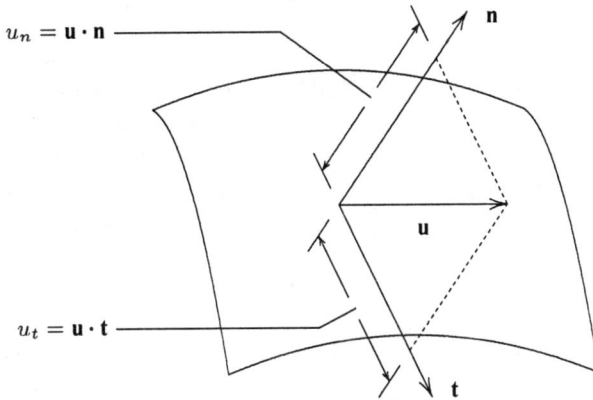

Figure 10.2: *Solid boundary and associated vectors.*

By contrast, for a real fluid, we find that fluid "sticks" to solid boundaries. That is, we find from observation of flow past stationary solid boundaries that $u_t = 0$. This is known as the **no-slip boundary condition**. As we will discover in Section 10.5, including viscosity in the momentum equation increases the order of the momentum equation, thus permitting (and requiring) additional boundary conditions.

If the surface moves with velocity $\mathbf{U}_s$, all of the comments above apply to the velocity relative to the surface. Thus, the boundary condition on the velocity appropriate at a solid boundary is as follows.

$$\left. \begin{array}{ll} (\mathbf{u} - \mathbf{U}_s) \cdot \mathbf{n} = 0, & \text{Ideal Fluid} \\[2mm] (\mathbf{u} - \mathbf{U}_s) = \mathbf{0}, & \text{Real Fluid} \end{array} \right\} \tag{10.18}$$

We will use this information in the next section to help explain how an object moving in a real fluid develops vorticity and, hence, a force.

10.4   Viscous Effects and Vorticity Generation

Consider flow past an *airfoil*, i.e., a two-dimensional cross section of a wing. Figure 10.3 contrasts the streamlines for a perfect fluid and a real fluid. As shown, for a perfect fluid, the streamlines have an unusual kink near the trailing edge of the airfoil, and there is a rear stagnation point on the upper surface upstream of the trailing edge. The streamlines shown are consistent with the Helmholtz Theorem, corresponding to irrotational flow. Assuming the motion started from rest, the irrotational flowfield that develops produces zero vorticity and zero force on the airfoil.

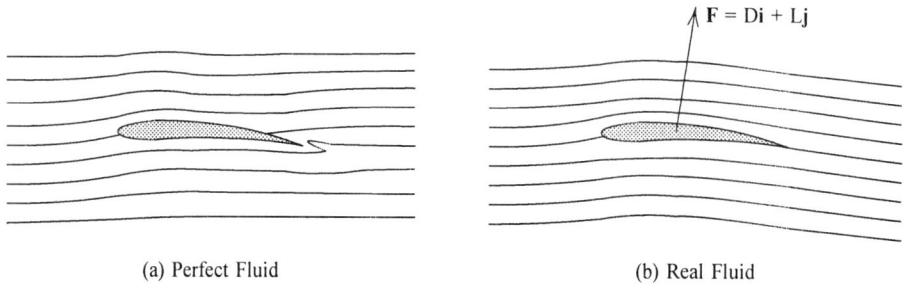

(a) Perfect Fluid   (b) Real Fluid

Figure 10.3: *Flow past an airfoil.*

By contrast, for a real fluid, the stagnation point lies precisely at the trailing edge. The streamlines leave the trailing edge smoothly without any kinks, and without the attending rapid local acceleration that would be required to negotiate the abrupt turning associated with ideal flow. This is a manifestation of the effect of friction on the flow. The primary purposes that friction serves in fluid flows are to drive the flow toward equilibrium and to smooth out discontinuities. Most importantly, the real flow develops vorticity even when the motion starts from rest, with an attendant force, **F**, that is usually written as

$$\mathbf{F} = D\mathbf{i} + L\mathbf{j} \tag{10.19}$$

where D is called **drag** and L is called **lift**. To further underscore the connection between vorticity and the force on the object, a reasonable approximation for the lift per unit width on the object is given by the **Kutta-Joukowski law** that we will derive in Chapter 11.

$$L = \rho U_\infty \Gamma_a \tag{10.20}$$

The quantity U_∞ is freestream velocity and Γ_a is the aerodynamic circulation [see Equation (4.37)].

The question remains about how the vorticity develops in the flow. For simplicity, consider a surface element parallel to the freestream flow direction, x. Recall from the preceding section that the no-slip boundary condition means both the tangential and normal velocity components, u and v, vanish at the surface. Since we assume for the purpose of our illustrative argument that the surface is parallel to the x axis, necessarily $\partial v/\partial x = 0$ as v vanishes at any two adjacent points on the surface. Hence, in a two-dimensional flow, the vorticity at the surface is

$$\omega = -\frac{\partial u}{\partial y}\mathbf{k} \tag{10.21}$$

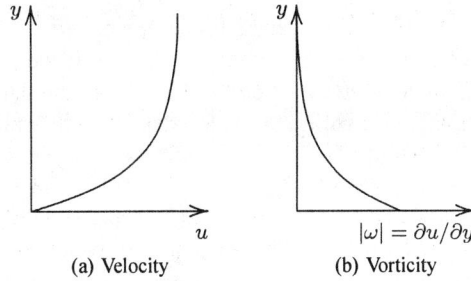

(a) Velocity   (b) Vorticity

Figure 10.4: *Velocity and vorticity close to a solid boundary.*

Now, as shown in Figure 10.4, as the distance above the surface, y, increases, u increases from zero to its freestream value. Thus, except at special points known as separation points,[1] the velocity gradient, $\partial u / \partial y$, is nonvanishing at the surface. Hence, in a real fluid, the kinematical constraint imposed by the no-slip boundary condition causes the surface value of the vorticity to be nonzero. Although we have demonstrated this phenomenon for a simplified planar geometry, it is true for arbitrary shapes.

The vorticity that is continually created at the surface diffuses into the flow under the action of the fluid's viscosity.[2] Then, it is swept into the main flow by the inertia of fluid particles passing near the surface. This is the process through which forces develop on an object immersed in a flowing fluid. It should not be surprising that friction plays a central role in controlling the force on an object. Consider the role friction plays, for example, in creating traction for an automobile's tires. Imagine what would happen when you press the accelerator pedal if your automobile were parked on a frictionless street. The idealization of a frictionless fluid is to the airfoil what that frictionless street is to your motionless automobile with rapidly spinning wheels!

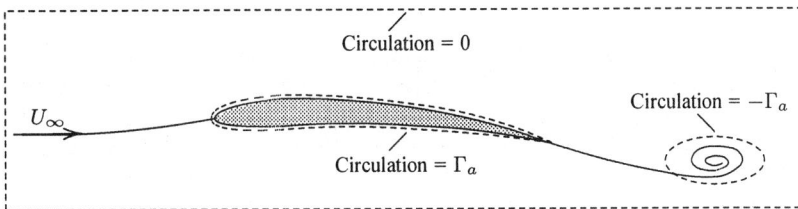

Figure 10.5: *An airfoil and a shed vortex.*

The streamline pattern shown in Figure 10.3(a) is observed at the beginning of the motion of an airfoil. Consistent with the discussion above, because of the action of viscous forces, the stagnation point moves toward the trailing edge. During this transient stage, vorticity is developing as a result of viscous diffusion. When the stagnation point reaches the trailing edge, a vortex is shed by the airfoil (Figure 10.5), and it propagates downstream. If we denote the strength of the vortex in terms of its circulation as $-\Gamma_a$, we discover two interesting facts. First, if we compute the circulation for a contour containing the airfoil and the shed vortex, the circulation about the contour is zero. Second, if we compute the circulation for a contour containing only the airfoil, we find that it is Γ_a.

---

[1] A separation point is a point where $\partial u / \partial y = 0$ — we will discuss separation in Chapter 14.

[2] The physical process of vorticity diffusion is the same as heat diffusing through a substance heated at a boundary, as we will see in Section 10.5.

The former fact provides a useful method for including lift in an inviscid computation. Noting that the Helmholtz Theorem tells us that $d\omega/dt = \mathbf{0}$, if we could somehow magically inject vorticity into a flow, it would never go away. Thus, it is completely consistent to imagine that we have a vortex of strength Γ_a within the airfoil that is permanently bound to that location. We refer to such a vortex as a **bound vortex**. Then, if we solve the Euler equation, subject to the condition that the flow leaves the trailing edge smoothly as in Figure 10.3(b), we have a solution that includes lift. It turns out that there is a unique value of Γ_a that causes the flow to leave the trailing edge in this manner. We refer to the boundary condition that makes the flow leave the trailing edge smoothly as the **Kutta condition**.

The latter fact is consistent with the Helmholtz Theorem. Although we haven't proven it here, a similar result known as **Kelvin's Circulation Theorem** is written in terms of circulation. It states that the circulation about any contour always containing the same fluid particles is constant for an inviscid, incompressible fluid in an unbounded domain, i.e., $d\Gamma_a/dt = 0$. Thus, the net circulation of the airfoil and the shed vortex must be zero. When we use the Kutta condition, we imagine that the shed vortex lies at infinity, and as noted above, the force on the airfoil is given by $\rho U_\infty \Gamma_a$. Conversely, the force on the shed vortex is $-\rho U_\infty \Gamma_a$, so that *the net force on the airfoil and shed vortex is zero*, as demanded by the Helmholtz Theorem.

As a final comment, we will develop this method for computing lift when we study potential-flow theory in Chapter 11. It yields a good approximation for a wide range of conditions, and only breaks down when the airfoil is inclined to the flow direction at a large enough angle to cause a major breakdown in the flow structure known as separation. However, the method fails to account for the drag force. Drag can be computed only by accounting for viscous effects.

10.5   Diffusion of Vorticity

When friction is included in the momentum equation, it appears as an additional surface force. Chapter 12 gives details of how the friction force is represented. As we will see, the Euler equation is replaced by the most fundamental equation governing the motion of a fluid, viz., the **Navier-Stokes** equation. For incompressible flow, the Navier-Stokes equation is

$$\rho \frac{d\mathbf{u}}{dt} = -\nabla p + \mu \nabla^2 \mathbf{u} + \rho \mathbf{f} \tag{10.22}$$

where μ is viscosity (assumed constant here) and the differential operator ∇^2 is the **Laplacian operator** (see Appendix D) defined by

$$\nabla^2 \equiv \frac{\partial^2}{\partial x^2} + \frac{\partial^2}{\partial y^2} + \frac{\partial^2}{\partial z^2} \tag{10.23}$$

Note first that the Navier-Stokes equation, because of the presence of the viscous term, $\mu \nabla^2 \mathbf{u}$, is a second-order differential equation. By contrast, the Euler equation is of first order. Thus, we must specify additional boundary conditions in general, such as the no-slip condition. Note also that taking the curl of this equation involves the same sequence of operations used for Euler's equation (see Section 10.2). The only difference is the viscous term, for which

$$\nabla \times \left(\mu \nabla^2 \mathbf{u} \right) = \mu \nabla^2 \boldsymbol{\omega} \tag{10.24}$$

Therefore, if we have an incompressible, viscous fluid subjected to conservative body forces, the equation for the vorticity becomes (with $\nu = \mu/\rho$):

$$\frac{d\boldsymbol{\omega}}{dt} = \boldsymbol{\omega} \cdot \nabla \mathbf{u} + \nu \nabla^2 \boldsymbol{\omega} \tag{10.25}$$

The last term on the right-hand side of this equation represents the physical process of diffusion. As discussed in the preceding section, this is the physical mechanism through which vorticity is created in flow past solid objects. If the "vortex-stretching" term, $\boldsymbol{\omega} \cdot \nabla \mathbf{u}$, is omitted and d/dt regarded as a pure time derivative, Equation (10.25) is identical with the unsteady heat-conduction equation in any number of spatial directions.

As noted in the introductory comments at the beginning of this chapter, mathematicians and scientists prior to the twentieth century were uncertain about the validity of the Newtonian (or Stokes) model for viscous stresses [Equation (1.26)]. This uncertainty was based on the observation that for the large Reynolds numbers encountered in practical flows,

$$\left| \mu \nabla^2 \mathbf{u} \right| \ll \left| \rho \, \mathbf{u} \cdot \nabla \mathbf{u} \right| \tag{10.26}$$

They were left with the dilemma that their best model for effects of friction seemed to provide an unsatisfactory explanation for the generation of vorticity in a real fluid.

However, Equation (10.26) implicitly assumes that derivatives of the velocity (velocity gradients) are of the order of U/L where U and L are characteristic flow velocity and length, e.g., L could be the length of the body under consideration, and the same characteristic length applies to derivatives in any direction. In 1904, Prandtl announced his discovery that in slightly viscous fluids (in the sense that $\mu \ll \rho U L$) such as air and water, viscous effects are confined to thin layers near a solid boundary. The thickness of this layer, say δ, and not the length of the body, is the characteristic length for y derivatives. Succinctly put, Prandtl argued that the streamwise length scale is $L_x = L$, while the normal length scale is $L_y = \delta$.

Furthermore, velocity gradients are so large in these layers, which he chose to call **boundary layers**, that $\mu \partial^2 \mathbf{u}/\partial y^2$ and $\rho \, \mathbf{u} \cdot \nabla \mathbf{u}$ are of comparable order of magnitude (y denotes distance normal to the surface). Prandtl further showed that outside of the boundary layer, the flow behaves as though it is inviscid, in complete agreement with the observations of his predecessors. His pioneering work proved that the Newtonian/Stokes model for representing the shear stress is correct. He also proved that the mechanism of vorticity diffusion is of sufficient magnitude to create the vorticity that leads to lift and drag. Prandtl's discovery of the boundary layer thus resolved d'Alembert's Paradox.

AN EXPERIMENT YOU CAN DO AT HOME
Excerpted from Granger (1988)

To observe an interesting interaction of vorticity and viscosity, you
must assemble a cup, some water, tea leaves and a spoon. The ideal
cup is as nearly cylindrical as possible and should be about 2/3 full of
water. Add a few tea leaves and vigorously stir with the spoon in a
circular motion for a few seconds. Now, remove the spoon and marvel
in wonder as the tea leaves settle in a nice little heap at the center of
the cup!

As discussed by Granger (1988), this is one of Prof. Nicholas Rott's
favorite experiments. It shows some of the interesting features of what
are known as the *spin-up* and *spin-down* processes. The stirring causes
the fluid to rotate in a nearly rigid-body state. The centrifugal force
causes the tea leaves to assemble along the cylindrical boundary of
the cup. Because of the no-slip condition at the cup boundaries, thin
boundary layers form along the bottom and sides of the cup. The fluid
in these boundary layers flows radially outward along the bottom, up the
sides and back into the main body of fluid. The circulation is completed
by a cylindrical front that advances from the cup boundary toward the
non-rotating core during spin up.

When the stirring ceases, a cylindrical front moves back toward the
cup boundaries, leaving the fluid near the center of the cup motionless.
During the spin-down process, the overall circulation reverses and the
tea leaves are deposited in the motionless core at the center of the cup.

While the motion is primarily inviscid, the flow is strongly influ-
enced by the boundary layer at the bottom of the cup, which is known
as an *Ekman layer*. As fluid moves radially outward, the layer draws
fluid downward in order to conserve mass. This so-called *Ekman suction* provides the mechanism for the unusual spin-up and spin-down
dynamics. Maxworthy (1968) describes this flow in a paper entitled "A
Storm in a Teacup."

Problems

10.1 Compute the rate at which work is done by the vortex force.

10.2 Are there any conditions under which the vortex force is conservative for inviscid, incompressible flow subjected to conservative body forces? **HINT:** A conservative force has zero curl.

10.3 *Beltrami flow* is an idealized type of flow sometimes used in turbomachinery analysis. In this type of flow, the velocity vector, **u**, is everywhere parallel to the vorticity vector, $\boldsymbol{\omega}$.

(a) What is the vortex force in Beltrami flow?

(b) If the body force is conservative with $\mathbf{f} = -\nabla \mathcal{V}$, determine the pressure, p, as a function of $\mathcal{V}$, **u** and density ρ for steady, incompressible Beltrami flow.

10.4 There are physical mechanisms for generating vorticity in an inviscid, compressible flow. To see this, we can develop *Crocco's equation* as follows.

(a) Beginning with Gibbs' equation [Equation (7.11)], verify that the entropy, s, is given by

$$T\nabla s = \nabla h - \frac{\nabla p}{\rho}$$

where p, ρ, T and h are pressure, density, temperature and enthalpy, respectively.

(b) For steady flow with no body forces, use Euler's equation to eliminate $\nabla p/\rho$ and verify that

$$\mathbf{u} \times \boldsymbol{\omega} = \nabla h_t - T\nabla s$$

where $h_t = h + \frac{1}{2}\mathbf{u} \cdot \mathbf{u}$ is the total enthalpy. This is *Crocco's equation*, which shows that variations in either h_t or s can lead to creation of vorticity.

(c) Based on this equation, if h_t is constant, what must be true of irrotational flows? What must be true of isentropic flows? The answers to these two equations constitute *Crocco's theorem*.

10.5 Derive the compressible-flow vorticity equation by starting with the momentum equation in the following form.

$$\frac{\partial \mathbf{u}}{\partial t} + \nabla\left(\frac{1}{2}\mathbf{u} \cdot \mathbf{u}\right) - \mathbf{u} \times \boldsymbol{\omega} = -\frac{\nabla p}{\rho} - \nabla \mathcal{V}$$

(a) Mimic the steps in the text to arrive at a differential equation for $\boldsymbol{\omega}$ in as simplified a form as possible. Use the continuity equation to replace $\nabla \cdot \mathbf{u}$ by a term proportional to $d\rho/dt$.

(b) Rewrite the equation derived in Part (a) as a differential equation for $\boldsymbol{\omega}/\rho$.

(c) How does your result change if the flow is *barotropic*, i.e., if the pressure depends only upon ρ?

10.6 A *Boeing 747* jumbo jet weighs 750,000 lb when fully fueled with 150 passengers aboard. To generate sufficient lift to take off at this weight, the 747 must achieve a runway speed of 140 mph. Assuming the circulation, Γ, increases linearly with speed (as it will be for unchanged flap settings and angle of attack), what runway speed is required to take off with 325 passengers? Assume an average passenger with luggage weighs 200 lb.

10.7 Compute the aerodynamic circulation for the following aircraft.

(a) The *Gossamer Condor*, the first human-powered aircraft, for which $L = 2.2$ lb/ft, $U_\infty = 10$ mph and $\rho = .00234$ slug/ft^3.

(b) A small, single-engine airplane for which $L = 100$ lb/ft, $U_\infty = 200$ mph and $\rho = .00216$ slug/ft^3.

(c) A *Boeing 747* for which $L = 4000$ lb/ft, $U_\infty = 570$ mph and $\rho = 7.0 \cdot 10^{-4}$ slug/ft^3.

10.8 Compute the aerodynamic circulation, Γ_a, for Couette flow, using the dashed rectangular contour shown. Recall from Chapter 1 that the velocity is

$$u(y) = Uy/h$$

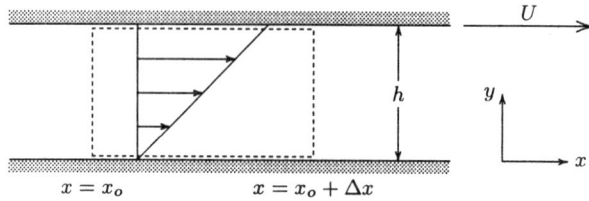

Problem 10.8

10.9 For flow past a flat plate, the boundary layer is as shown. The dotted line depicts the edge of the layer, and the flow is essentially uniform ($\mathbf{u} \approx U\,\mathbf{i}$) and inviscid above the edge. Also, the vertical velocity, v, in the boundary layer is very small compared to the horizontal component, i.e., $v \ll u$. The velocity at $x = L$ is $u(L, y) = Uf(y/\delta)$, where f is a function that has to be determined numerically. The point of this problem is to show that, regardless of the precise details of flow within the boundary layer, the circulation is a function only of the freestream velocity, U, and the plate length, L.

(a) Using the dashed rectangular contour shown, compute the circulation from its definition, $\Gamma = \oint \mathbf{u} \cdot d\mathbf{s}$. Neglect any contributions involving v.

(b) Now compute the circulation from $\Gamma = \int \int \boldsymbol{\omega} \cdot \mathbf{k}\, dA$, and verify that the answer matches that of Part (a).

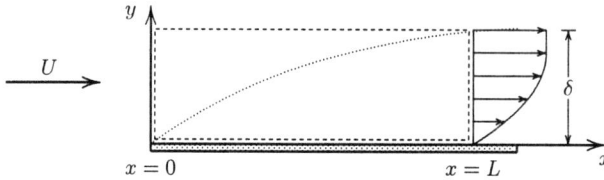

Problem 10.9

10.10 We want to analyze the aerodynamic circulation, Γ_a, for flow in a two-dimensional channel. The velocity profile for channel flow is

$$u(y) = u_m \left[1 - 4 \left(\frac{y}{h} \right)^2 \right]$$

where u_m is the maximum velocity.

(a) Using the dashed rectangular contour shown, verify that the circulation is zero.

(b) Now use a rectangular contour that extends from either of the channel walls to the centerline, and compute the nonzero circulation.

(c) Compute the vorticity and, noting that $\Gamma = \iint_A \omega\, dx dy$ where ω is the vorticity component normal to the xy plane, explain the results of Parts (a) and (b).

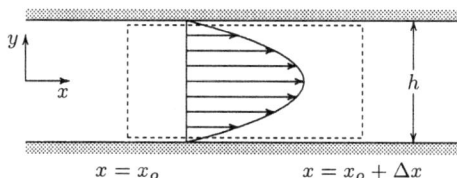

Problem 10.10

10.11 A simplified model of a hurricane approximates the flow as a rigid-body rotation in the inner core and a *potential vortex* outside of the core (cf. Section 4.3). That is, the circumferential velocity, u_θ, is

$$u_\theta \approx \begin{cases} \Omega r, & r \leq R \\ \dfrac{\Gamma}{2\pi r}, & r \geq R \end{cases}$$

where r is radial distance, R is core radius, Ω is the angular-rotation rate in the core, and Γ is the circulation. If a hurricane has peak winds of 100 mph and the core radius is 120 ft, what is the circulation? What is the angular-rotation rate (in rpm) in the core?

10.12 For laminar flow over a flat surface, the boundary-layer thickness, δ, is

$$\delta \approx 5.0x Re_x^{-1/2} \qquad \text{(Laminar)}$$

where x is distance along the surface and Re_x is Reynolds number based on x. Using this formula, estimate the thickness of the boundary layer near the trailing edge of a sailboat's sail where $x = 10$ ft. Assume the winds are very light so that sailboat is traveling at a frustrating speed of $U_\infty = 3$ knots. Also, assume the kinematic viscosity of air is $\nu = 1.62 \cdot 10^{-4}$ ft²/sec².

10.13 For laminar flow over a flat surface, the boundary-layer thickness, δ, is

$$\delta \approx 5.0x Re_x^{-1/2} \qquad \text{(Laminar)}$$

where x is distance along the surface and Re_x is Reynolds number based on x. Using this formula, estimate the thickness of the boundary layer on an 8-meter long submarine that uses surface heating to maintain laminar flow. The submarine is moving at $U_\infty = 30$ knots, and the kinematic viscosity of the water near the submarine surface is $\nu = 8 \cdot 10^{-7}$ m²/sec².

10.14 For turbulent flow over a flat surface, the boundary-layer thickness, δ, is

$$\delta \approx 0.37x Re_x^{-0.2} \qquad \text{(Turbulent)}$$

where x is distance along the surface and Re_x is Reynolds number based on x. Using this formula, estimate the thickness of the boundary layer near the trailing edge of the wing on a *Boeing 747* traveling at $U_\infty = 570$ mph with $\nu = 4.27 \cdot 10^{-4}$ ft²/sec². Assume the chord length is 25 ft. To appreciate how thin the boundary layer is, compute the angle $\tan^{-1}(\delta/x)$, which would be the angle measured from the leading edge if the wing were flat.

10.15 For turbulent flow over a flat surface, the boundary-layer thickness, δ, is

$$\delta \approx 0.37x Re_x^{-0.2} \qquad \text{(Turbulent)}$$

where x is distance along the surface and Re_x is Reynolds number based on x. Using this formula, estimate the thickness of the boundary layers above and below the water line near the stern of a 100-meter long ship traveling at $U_\infty = 15$ knots. The kinematic viscosity of air (at 20° C) is $\nu_a = 1.51 \cdot 10^{-5}$ m²/sec², while the value for water (at 10° C) is $\nu_w = 1.31 \cdot 10^{-6}$ m²/sec². To appreciate how thin these boundary layers are, compute the angle $\tan^{-1}(\delta/x)$, for both cases, which would be the angle measured from the leading edge if the side of the hull were flat.

10.16 The surface shear stress, τ_w, for viscous flow over a solid boundary is

$$\tau_w = \mu \left(\frac{\partial u}{\partial y} \right)_{y=0}$$

How is τ_w related to the vorticity at the surface, ω_w, in a two-dimensional flow? For simplicity, you may assume the surface is planar.

10.17 The velocity, **u**, in a two-dimensional, incompressible boundary layer can be approximated by

$$\mathbf{u} \approx U\frac{y}{\delta}\left(1 - \frac{y}{\delta}\right)\mathbf{i}, \qquad \delta \approx 5\sqrt{\frac{\nu x}{U}}$$

where U is freestream velocity, y is distance from the surface, δ is boundary-layer thickness, ν is kinematic viscosity and x is streamwise distance. Compute the dimensionless vorticity at the surface, $\nu\omega_w/U^2$, as a function of Reynolds number, $Re_x = Ux/\nu$.

10.18 To see how much larger variations in velocity normal to a surface are relative to streamwise variations, it is instructive to examine the boundary layer on a 30° wedge. The *Falkner-Skan* solution tells us the velocity in the boundary layer is

$$u(x,y) = u_e(x)f\left(\frac{y}{\delta}\right), \qquad u_e(x) = u_o\left(\frac{x}{L}\right)^{1/6}, \qquad \delta(x) \approx 6\sqrt{\frac{\nu L}{u_o}}\left(\frac{x}{L}\right)^{5/12}$$

where $u_e(x)$ is the velocity above the boundary layer, f is a function satisfying $f(0) = 0$ and $f(\eta) \to 1$ as $\eta \to \infty$, and $\delta(x)$ is the boundary-layer thickness. Also, u_o and L are characteristic velocity and length scales, while ν is kinematic viscosity.

(a) Develop an estimate of $\partial u/\partial x$ by ignoring the variation of $f(y/\delta)$, i.e., by computing du_e/dx.

(b) Develop an estimate of $\partial u/\partial y$ by approximating it as $\Delta u/\Delta y \approx u_e/\delta$.

(c) Determine the ratio of $\partial u/\partial y$ to $\partial u/\partial x$ at $x/L = 1$ for $Re \equiv u_oL/\nu = 10^4$, 10^5 and 10^6.

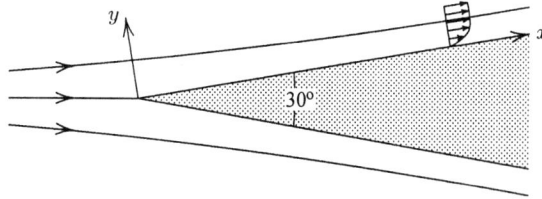

Problem 10.18

10.19 For flow over a porous flat plate with *suction*, or surface mass removal, the horizontal velocity, u, is a function only of distance from the surface, y, and is given by

$$u(y) = U_\infty\left[1 - \exp\left(-C_QU_\infty y/\nu\right)\right]$$

where C_Q is the suction coefficient, U_∞ is the freestream velocity and ν is kinematic viscosity.

(a) Compute the z component of the vorticity, ω, for constant U_∞, C_Q and ν.

(b) The boundary-layer thickness, δ, is usually defined as the value of y where $u = 0.99U_\infty$. Using this definition, determine δ for this flow.

(c) Let L denote the length of the plate and let $Re = U_\infty L/\nu$ denote the Reynolds number. Rewrite the velocity and vorticity in dimensionless form.

(d) Assuming $C_Q = 0.01$, compute δ/L for $Re = 10^3$ and 10^4.

(e) Make graphs of u/U_∞ and $|\nu\omega/U_\infty^2|$ versus y/L with $0 \le y/L \le 1$ for $C_Q = 0.01$. Make a single graph for each property including results on the same graph for the two Reynolds numbers, $Re = 10^3$ and 10^4.

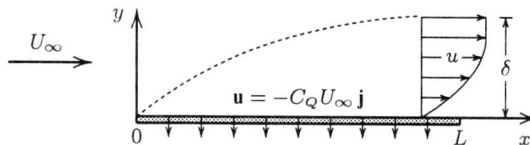

Problem 10.19

10.20 The phenomenon illustrated in the home experiment described in this chapter is relevant to missiles carrying liquid fuel that spin for directional stability. Detailed analysis shows that during *spin up*, the radial position, r, of the front as it moves from the missile's inner surface to its centerline is given by

$$r = Re^{-t\sqrt{\nu\Omega}/h}$$

where R and h are fuel-compartment radius and height, ν is kinematic viscosity and Ω is rotation rate. Determine the approximate spin-up time, i.e., the time to achieve nearly rigid-body rotation ($r/R \approx .05$), for the following.

(a) A missile with $R = 1$ m, $h = 2$ m, $\nu = 1.5 \cdot 10^{-6}$ m^2/sec and $\Omega = 10$ sec^{-1}.

(b) The home experiment with $R = 3$ in, $h = 3$ in, $\nu = 1.08 \cdot 10^{-5}$ ft^2/sec and $\Omega = 120$ rpm.

(c) Due to turbulence, the spin-up time for Part (b) is only $t_{true} = 3$ sec. Assuming the equation above is valid with an effective kinematic viscosity, ν_{eff}, replacing ν, what is ν_{eff}/ν?

Chapter 11

Potential Flow

Virtually all of the focus of the preceding chapters has been on implementation of the integral form of the conservation laws. As we have seen, this permits determining gross flow properties such as average flow rate, momentum flux and forces acting on an object. However, this approach yields no information about details of the fluid motion within the chosen control volume. That is, the integral conservation laws provide a *global view* of a given flow.

If we wish to know more about how the fluid moves within a control volume, we must take a *detailed view*. This requires solving, subject to appropriate boundary conditions, the differential equations that we inferred from the integral conservation laws. Beginning with this chapter, we shift our focus to the differential forms. This will permit determining fluid velocity, pressure and other flow properties throughout a given control volume. Note that, since we will actually be dropping the control-volume concept, in describing the flow domain, we will replace the term *control volume* with the commonly-used term of *flowfield*.

Because we have adopted the Eulerian description, we know that the differential equations describing fluid motion are *quasi-linear*. This is best exemplified by the acceleration, which is the Eulerian derivative of the velocity, i.e.,

$$\frac{d\mathbf{u}}{dt} = \frac{\partial \mathbf{u}}{\partial t} + \mathbf{u} \cdot \nabla \mathbf{u} = \frac{\partial \mathbf{u}}{\partial t} + u\frac{\partial \mathbf{u}}{\partial x} + v\frac{\partial \mathbf{u}}{\partial y} + w\frac{\partial \mathbf{u}}{\partial z} \qquad (11.1)$$

While $\partial \mathbf{u}/\partial t$ is a linear function of the velocity, the last three terms are not. The x component of the term $u\partial \mathbf{u}/\partial x$, for example, is $u\partial u/\partial x$, which is clearly not a linear function of the velocity component u.

Since the basic differential conservation laws are not linear equations, their solution can be very difficult, and must usually be carried out numerically. There is one specialized branch of fluid mechanics where the primary equations of motion, with no approximation, are linear. Specifically, if we confine our focus to inviscid, incompressible flows with no vorticity, there is a dramatic simplification of the fluid-flow equations. This branch of fluid mechanics is known as **potential-flow theory**.

Lest you think this might constitute a pathological case, recall what Prandtl found when he resolved d'Alembert's Paradox. In addition to finding that viscous effects are confined to thin boundary layers, he also found that the flow outside the boundary layers behaves just as the inviscid equations predict, at least for streamlined bodies. Thus, since viscous effects are confined to extremely thin layers, it is reasonable to ignore viscosity for the rest of the flowfield, which often accounts for most of the flow domain.

We know that incompressible flow is very common in nature, so only the assumption of irrotational flow might be a cause for concern. In practice, vorticity is often confined to small regions such as the boundary layers close to a solid surface. Although effects of vorticity can extend over great distances, the flow away from the concentrated vorticity is usually irrotational. For example, we will see that flow past a lifting airfoil turns out to be irrotational. The circulation required to achieve lift results from a fictitious vortex that we embed within the airfoil to balance the vortex shed by the airfoil at the beginning of its motion (see Figure 10.5). This fictitious vortex is separate from the physical flowfield. The chapter includes a home experiment that demonstrates the connection between vorticity and lift.

Some of the earliest work in Computational Fluid Dynamics focused on potential-flow theory. Under the guidance of A. M. O. Smith [see Hess and Smith (1966)] of the Douglas Aircraft Company, for example, potential-flow computations were routinely done for complete airplane fuselages in the early 1960's. A great deal of effort in CFD focused on potential-flow theory well into the 1980's, and some activity continues today.

11.1 Differential Equations of Motion

We begin with the differential form of the mass- and momentum-conservation principles. As summarized in Chapter 5, the appropriate equations are the continuity and Euler equations. For incompressible flow, conservation of mass simplifies to [see Equation (5.10)]

$$\nabla \cdot \mathbf{u} = 0 \qquad (11.2)$$

For a flow with a conservative body force given by $f = -\nabla \mathcal{V}$, Euler's equation [Equation (5.28)] is

$$\rho \frac{d\mathbf{u}}{dt} = -\nabla p - \rho \nabla \mathcal{V} \qquad (11.3)$$

Taking advantage of the vector identity for $\mathbf{u} \cdot \nabla \mathbf{u}$ given in Appendix C, we can rewrite the acceleration as

$$\frac{d\mathbf{u}}{dt} = \frac{\partial \mathbf{u}}{\partial t} + \nabla \left(\frac{1}{2} \mathbf{u} \cdot \mathbf{u} \right) - \mathbf{u} \times (\nabla \times \mathbf{u}) \qquad (11.4)$$

Then, for irrotational flow, because $\nabla \times \mathbf{u} = \mathbf{0}$, Euler's equation for incompressible, irrotational flow with a conservative body force is

$$\frac{\partial \mathbf{u}}{\partial t} + \nabla \left(\frac{1}{2} \mathbf{u} \cdot \mathbf{u} \right) = -\nabla \left(\frac{p}{\rho} \right) - \nabla \mathcal{V} \qquad (11.5)$$

Finally, we can move all terms to the left-hand side of this equation. The result is the desired form of Euler's equation that we will work with in this chapter, viz.,

$$\frac{\partial \mathbf{u}}{\partial t} + \nabla \left(\frac{p}{\rho} + \frac{1}{2} \mathbf{u} \cdot \mathbf{u} + \mathcal{V} \right) = \mathbf{0} \qquad (11.6)$$

11.2 Mathematical Foundation

To understand the origin of the name, *potential flow*, consider the following. For an incompressible, irrotational flow of a perfect (inviscid) fluid, we can make use of the fact that the velocity field is curl and divergence free. That is, by definition of irrotational flow, we have

$$\nabla \times \mathbf{u} = \mathbf{0} \qquad (11.7)$$

As discussed in the preceding section, conservation of mass for an incompressible fluid tells us

$$\nabla \cdot \mathbf{u} = 0 \tag{11.8}$$

First consider the irrotationality condition, Equation (11.7). As noted in Appendix C, for any scalar function ϕ,

$$curl\ grad\ \phi = \nabla \times \nabla\phi \equiv \mathbf{0} \tag{11.9}$$

Hence, if a function ϕ exists such that

$$\mathbf{u} = \nabla\phi \tag{11.10}$$

then Equation (11.7) is automatically satisfied. By definition, the quantity ϕ is the **velocity potential**, and this is the basis of the name potential-flow theory.

Second, consider the incompressibility condition, Equation (11.8). For any vector $\boldsymbol{\Psi}$, reference to Appendix C shows that

$$div\ curl\ \boldsymbol{\Psi} = \nabla \cdot (\nabla \times \boldsymbol{\Psi}) \equiv 0 \tag{11.11}$$

So, if we can find a vector $\boldsymbol{\Psi}$ such that

$$\mathbf{u} = \nabla \times \boldsymbol{\Psi} \tag{11.12}$$

then the incompressibility condition is automatically satisfied.

As we will see in the following subsections, by virtue of Equations (11.7) and (11.8), the functions ϕ and $\boldsymbol{\Psi}$ exist and, to within an additive constant, are unique. Although potential-flow theory is not confined to two-dimensional flow, we will limit our study of the topic to two dimensions to simplify the analysis. One advantage of specializing to two-dimensional flow is that the vector $\boldsymbol{\Psi}$ has only one component, which is normal to the plane of flow. Thus, $\boldsymbol{\Psi}$ can be written as

$$\boldsymbol{\Psi} = \psi\,\mathbf{k} \tag{11.13}$$

where ψ is a scalar and $\mathbf{k}$ is a unit vector normal to the xy plane. The quantity ψ is known as the **streamfunction**.

Use of the streamfunction is not confined to potential-flow theory, as the condition for its existence is simply that the flow be incompressible. As we will see later, it is frequently used in viscous-flow problems as well. By contrast, the velocity potential exists only if the flow is irrotational. Since viscous flows are inherently rotational, ϕ is confined to inviscid-flow applications.

11.2.1 Cylindrical Coordinates

Often the symmetry of a problem makes it convenient to work in cylindrical coordinates, r and θ, for potential-flow problems. In fact, we will make continuous use of both Cartesian and cylindrical coordinates in this chapter, often mixing both in the same problem. Thus, it is worthwhile to summarize key relations between the two systems, with regard to both coordinate axes and velocity components. Additional details can be found in Appendix D.

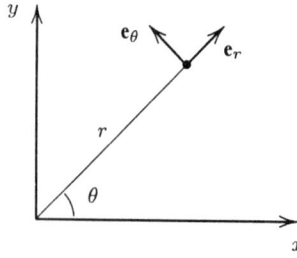

Figure 11.1: *Cylindrical coordinates.*

As illustrated in Figure 11.1, in two dimensions the cylindrical polar coordinate system is related to the rectangular Cartesian coordinate system by

$$x = r\cos\theta, \qquad y = r\sin\theta \tag{11.14}$$

where

$$r = \sqrt{x^2 + y^2}, \qquad \theta = \tan^{-1}\left(\frac{y}{x}\right) \tag{11.15}$$

Also, in cylindrical coordinates, we have unit vectors, $\mathbf{e}_r$ and $\mathbf{e}_\theta$, in the radial and circumferential directions, respectively. These unit vectors are related to those of the Cartesian system as follows.

$$\mathbf{e}_r = \mathbf{i}\cos\theta + \mathbf{j}\sin\theta, \qquad \mathbf{e}_\theta = -\mathbf{i}\sin\theta + \mathbf{j}\cos\theta \tag{11.16}$$

The velocity vector has a component u_r in the radial direction and a component u_θ in the circumferential direction, so that

$$\mathbf{u} = u_r\,\mathbf{e}_r + u_\theta\,\mathbf{e}_\theta \tag{11.17}$$

The Cartesian (u, v) and cylindrical (u_r, u_θ) velocity components are related to each other as follows (see Appendix D).

$$\left.\begin{aligned}
u = u_r\cos\theta - u_\theta\sin\theta \quad &\text{and} \quad v = u_r\sin\theta + u_\theta\cos\theta \\
u_r = u\cos\theta + v\sin\theta \quad &\text{and} \quad u_\theta = -u\sin\theta + v\cos\theta
\end{aligned}\right\} \tag{11.18}$$

11.2.2 Velocity-Potential Representation

For simplicity, it is convenient to begin our discussion in terms of rectangular Cartesian coordinates. In component form, the velocity potential, ϕ, is related to the velocity components by

$$u = \frac{\partial\phi}{\partial x} \quad \text{and} \quad v = \frac{\partial\phi}{\partial y} \tag{11.19}$$

Then, from the continuity Equation (11.8), we have

$$\frac{\partial u}{\partial x} + \frac{\partial v}{\partial y} = \frac{\partial^2\phi}{\partial x^2} + \frac{\partial^2\phi}{\partial y^2} = \nabla^2\phi \tag{11.20}$$

Hence, the velocity potential satisfies **Laplace's equation**, viz.,

$$\nabla^2\phi = 0 \tag{11.21}$$

As noted above, Laplace's equation is a linear partial differential equation about which a great deal is known [e.g., Kellogg (1953)]. We will generate numerous solutions to this equation in this chapter.

For problems in which cylindrical symmetry is present, we use cylindrical coordinates. Reference to Appendix D shows that the velocity potential is related to the velocity by

$$\nabla\phi = \mathbf{e}_r \frac{\partial\phi}{\partial r} + \mathbf{e}_\theta \frac{1}{r}\frac{\partial\phi}{\partial\theta} \qquad (11.22)$$

Therefore, the relation between the velocity components and the velocity potential becomes

$$u_r = \frac{\partial\phi}{\partial r} \quad \text{and} \quad u_\theta = \frac{1}{r}\frac{\partial\phi}{\partial\theta} \qquad (11.23)$$

Finally, ϕ still satisfies Laplace's equation, which transforms to the following.

$$r\frac{\partial}{\partial r}\left(r\frac{\partial\phi}{\partial r}\right) + \frac{\partial^2\phi}{\partial\theta^2} = 0 \qquad (11.24)$$

11.2.3 Streamfunction Representation

For two-dimensional flow, the streamfunction, ψ, is related to the rectangular Cartesian velocity components by

$$u = \frac{\partial\psi}{\partial y} \quad \text{and} \quad v = -\frac{\partial\psi}{\partial x} \qquad (11.25)$$

Then, from Equations (11.7), (11.12) and (11.13), there follows immediately

$$\nabla\times\mathbf{u} = \mathbf{k}\left(\frac{\partial v}{\partial x} - \frac{\partial u}{\partial y}\right) = -\mathbf{k}\left(\frac{\partial^2\psi}{\partial x^2} + \frac{\partial^2\psi}{\partial y^2}\right) = -\mathbf{k}\nabla^2\psi = \mathbf{0} \qquad (11.26)$$

Hence, the streamfunction also satisfies Laplace's equation, viz.,

$$\nabla^2\psi = 0 \qquad (11.27)$$

For cylindrical coordinates, we know from Appendix D that

$$\nabla\times(\psi\mathbf{k}) = \mathbf{e}_r\frac{1}{r}\frac{\partial\psi}{\partial\theta} - \mathbf{e}_\theta\frac{\partial\psi}{\partial r} \qquad (11.28)$$

so that the relation between the velocity components and the streamfunction is

$$u_r = \frac{1}{r}\frac{\partial\psi}{\partial\theta} \quad \text{and} \quad u_\theta = -\frac{\partial\psi}{\partial r} \qquad (11.29)$$

Finally, Laplace's equation for the streamfunction in cylindrical coordinates is also satisfied and is as follows.

$$r\frac{\partial}{\partial r}\left(r\frac{\partial\psi}{\partial r}\right) + \frac{\partial^2\psi}{\partial\theta^2} = 0 \qquad (11.30)$$

The velocity-potential and streamfunction representations form the basis of potential-flow theory. As noted above, one of the key advantages to this approach is the fact that Laplace's equation is linear. Thus, we can use features such as **superposition** and the various uniqueness and existence theorems for linear partial differential equations. As a convenience for future reference, Table 11.1 summarizes the velocity potential and streamfunction representations in both Cartesian and cylindrical coordinates.

Table 11.1: *Velocity-Potential and Streamfunction Representations*

	Velocity Potential	Streamfunction
General	$\mathbf{u} = \nabla \phi$ $\nabla^2 \phi = 0$ (from $\nabla \cdot \mathbf{u} = 0$)	$\mathbf{u} = \nabla \times \psi \mathbf{k}$ $\nabla^2 \psi = 0$ (from $\nabla \times \mathbf{u} = \mathbf{0}$)
2 Dimensional (Cartesian)	$u = \dfrac{\partial \phi}{\partial x}, \quad v = \dfrac{\partial \phi}{\partial y}$ $\dfrac{\partial^2 \phi}{\partial x^2} + \dfrac{\partial^2 \phi}{\partial y^2} = 0$	$u = \dfrac{\partial \psi}{\partial y}, \quad v = -\dfrac{\partial \psi}{\partial x}$ $\dfrac{\partial^2 \psi}{\partial x^2} + \dfrac{\partial^2 \psi}{\partial y^2} = 0$
Axisymmetric (Cylindrical)	$u_r = \dfrac{\partial \phi}{\partial r}, \quad u_\theta = \dfrac{1}{r}\dfrac{\partial \phi}{\partial \theta}$ $r\dfrac{\partial}{\partial r}\left(r\dfrac{\partial \phi}{\partial r}\right) + \dfrac{\partial^2 \phi}{\partial \theta^2} = 0$	$u_r = \dfrac{1}{r}\dfrac{\partial \psi}{\partial \theta}, \quad u_\theta = -\dfrac{\partial \psi}{\partial r}$ $r\dfrac{\partial}{\partial r}\left(r\dfrac{\partial \psi}{\partial r}\right) + \dfrac{\partial^2 \psi}{\partial \theta^2} = 0$

11.3   Streamlines and Equipotential Lines

The most useful feature of the streamfunction representation is its relation to streamlines. Consider the curve defined by

$$\psi(x, y) = \text{constant} \tag{11.31}$$

On this curve, an incremental change in ψ is given by:

$$d\psi = \frac{\partial \psi}{\partial x} dx + \frac{\partial \psi}{\partial y} dy = 0 \tag{11.32}$$

Then using Equation (11.25), we can rewrite this equation as

$$-v \, dx + u \, dy = 0 \tag{11.33}$$

or,

$$\frac{dx}{u} = \frac{dy}{v} \tag{11.34}$$

But, as discussed in Section 4.4, this is the equation of a streamline, i.e., a curve that is tangent to the local flow velocity. Therefore, $\psi(x, y) = \text{constant}$ defines a streamline.

Furthermore, the mass flux between two streamlines is proportional to the difference between the values of ψ on the two streamlines. To see this, consider flow across an arc ab with arc length Δs (see Figure 11.2) with the control volume shown. Clearly, the mass flowing across the arc, $\Delta \dot{m}_{ab}$, is

$$\Delta \dot{m}_{ab} = \rho \, \mathbf{u} \cdot \mathbf{n}_{ab} \Delta s \tag{11.35}$$

Now, for steady flow, we know that the net mass flux out of the control volume vanishes, wherefore

$$\oiint_S \rho \, \mathbf{u} \cdot \mathbf{n} \, dS = 0 \tag{11.36}$$

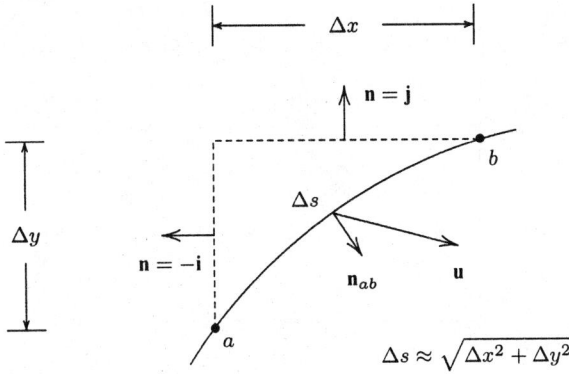

Figure 11.2: *Mass flux across an arc of length Δs.*

Hence, for the chosen control volume,

$$\rho\, \mathbf{u} \cdot \mathbf{n}_{ab}\, \Delta s + \rho\, \underbrace{[\mathbf{u} \cdot \mathbf{j}]}_{=v}\, \Delta x + \rho\, \underbrace{[\mathbf{u} \cdot (-\mathbf{i})]}_{=-u}\, \Delta y = 0 \tag{11.37}$$

or,

$$\Delta \dot{m}_{ab} + \rho v \Delta x - \rho u \Delta y = 0 \qquad \Longrightarrow \qquad \Delta \dot{m}_{ab} = \rho\, [u \Delta y - v \Delta x] \tag{11.38}$$

In differential form, we thus have

$$d\dot{m}_{ab} = \rho\, [u\, dy - v\, dx] \tag{11.39}$$

Now, using Equation (11.25), we can rewrite this equation in terms of the streamfunction, so that the differential mass flux across arc ab becomes

$$d\dot{m}_{ab} = \rho \left[\frac{\partial \psi}{\partial y} dy + \frac{\partial \psi}{\partial x} dx \right] = \rho\, d\psi \tag{11.40}$$

Therefore, $d\psi$ is the differential volume flux, viz.,

$$d\psi = \frac{1}{\rho} d\dot{m}_{ab} \tag{11.41}$$

In a similar way, we can compute a differential change in the velocity potential.

$$d\phi = \frac{\partial \phi}{\partial x} dx + \frac{\partial \phi}{\partial y} dy = u\, dx + v\, dy \tag{11.42}$$

So, on lines for which $\phi = $ constant, which are called the **equipotential lines**, there follows

$$\left(\frac{dy}{dx} \right)_{\phi=constant} = -\frac{u}{v} \tag{11.43}$$

As shown above, on lines for which $\psi = $ constant (the streamlines), we have

$$\left(\frac{dy}{dx} \right)_{\psi=constant} = \frac{v}{u} \tag{11.44}$$

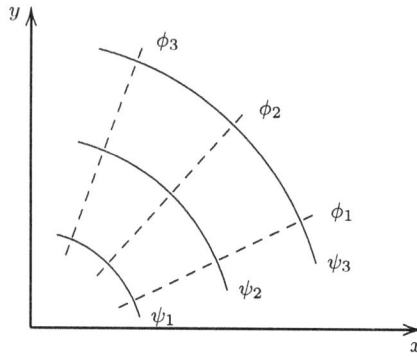

Figure 11.3: *Equipotential lines and streamlines.*

Thus, the slopes of the equipotential lines and the streamlines are negative reciprocals, which means these lines are mutually perpendicular. In general, we can construct families of curves as illustrated in Figure 11.3.

11.4   Bernoulli's Equation

Thus far, the equations we have discussed are the kinematical constraint of irrotationality, Equation (11.7), and mass conservation, Equation (11.8). However, aside from specializing Euler's equation for incompressible, irrotational flow [see Equation (11.6)], we have not discussed momentum conservation. The astute reader will observe that since the original mass-conservation equation is linear, it is no surprise that the governing equation for the velocity potential, $\nabla^2 \phi = 0$, is also linear. Similarly, the irrotationality condition, from which the streamfunction equation, $\nabla^2 \psi = 0$, is derived, is a linear equation. But, what about Euler's equation, which is where the nonlinearity associated with the Eulerian derivative lies? The answer to this question reflects the central role played by vorticity in determining the structure of an incompressible flow.

To explain this point, consider the following. If we introduce the velocity potential into Equation (11.6), we find

$$\nabla \left(\frac{\partial \phi}{\partial t} + \frac{p}{\rho} + \frac{1}{2} \mathbf{u} \cdot \mathbf{u} + \mathcal{V} \right) = \mathbf{0} \tag{11.45}$$

where we note that $\partial / \partial t$ and ∇ commute. Therefore, since its gradient, and thus all spatial derivatives are zero, the quantity within parentheses can be, at most, a function only of time, t. We thus have an unsteady-flow equivalent of Bernoulli's equation, viz.,

$$\frac{\partial \phi}{\partial t} + \frac{p}{\rho} + \frac{1}{2} \mathbf{u} \cdot \mathbf{u} + \mathcal{V} = F(t) \tag{11.46}$$

where $F(t)$ must be determined from initial and/or boundary conditions. This equation serves as a relation between the pressure and the velocity field for a given flow. Most importantly, if the velocity field is known, we can use Equation (11.46) to compute the pressure.

The question thus becomes, can we compute the velocity independent of the unsteady equivalent of Bernoulli's equation? Because the equation for ϕ is independent of pressure, we need only determine whether the initial and boundary conditions are independent of pressure. Regarding initial conditions, the strongest dependence on pressure they can have is upon the initial pressure. Hence, initial conditions are obviously independent of the instantaneous value of p. Regarding boundary conditions, the most common boundary condition imposed is at a solid boundary, through which there can be no flow. That is,

$$\mathbf{u} \cdot \mathbf{n} = 0 \qquad \text{(at solid boundaries)} \qquad (11.47)$$

In terms of the velocity potential, we therefore have

$$\frac{\partial \phi}{\partial n} = 0 \qquad \text{(at solid boundaries)} \qquad (11.48)$$

where n is the direction normal to the solid boundary. This is referred to as a *Neumann-type boundary condition*. We also impose farfield conditions, which are typically either undisturbed flow or uniform flow. Hence, the most common boundary conditions are completely independent of pressure.

Thus, in the context of potential-flow theory, we can pose a problem with an equation independent of the pressure, viz.,

$$\frac{\partial^2 \phi}{\partial x^2} + \frac{\partial^2 \phi}{\partial y^2} = 0 \qquad (11.49)$$

Further, it is usually solved subject to boundary conditions of Neumann type that are also independent of pressure. The extensive theory developed by mathematicians for Laplace's equation tells us that a unique solution to this problem exists which is, of necessity, independent of pressure. In other words, we can solve a potential-flow problem with no prior knowledge of the pressure. Once we have this solution in hand, Equation (11.46) provides the pressure.

Most importantly, this shows that all of the nonlinearity appears in the equation for pressure, which is not required to determine the velocity. That is, the mass-conservation principle uncouples from the momentum-conservation equation, thus permitting a solution for $\mathbf{u}$ independent of p.

This shows that, at least in a mathematical sense, the pressure is a passive quantity. In terms of the physics of fluid motion, this is consistent with the notion that the vorticity is the driving force in an incompressible flow. The condition of irrotationality places a strong kinematic constraint on the way fluid particles move that we will discuss in detail when we reconsider kinematics in Chapter 12. In general, we can envision the logical sequence of cause and effect as follows.

1. The vorticity field dictates the local acceleration in the flow.

2. The pressure adjusts to balance the acceleration.

The mathematical structure of the potential-flow problem is completely consistent with this chronology. When the vorticity is non-vanishing, the continuity and momentum equations are coupled and this interpretation is less obvious.

As a final comment, while time appears in Equation (11.46), it does not appear in Laplace's equation. Hence, we can generate the solution for an unsteady flow problem by simply reflecting the time dependence in the farfield and/or surface boundary conditions. This mirrors the physical fact that incompressible flow is the limiting case of infinite sound speed, so that changes in the flow are communicated throughout the flowfield instantaneously. Thus, even when flow properties are changing with time, we are effectively treating the flow in a quasi-steady manner.

11.5   Fundamental Solutions

Because Laplace's equation is linear, we can take advantage of the powerful concept of superposition. Specifically, if $\phi_1(x, y)$ and $\phi_2(x, y)$ are both solutions to Laplace's equation, then

$$\phi(x, y) = c_1\phi_1(x, y) + c_2\phi_2(x, y) \tag{11.50}$$

is also a solution, where c_1 and c_2 are arbitrary constants. Furthermore, we can even construct a solution of the form

$$\phi(x, y) = \int\int \left[c_1(\xi, \eta)\phi_1(x - \xi, y - \eta) + c_2(\xi, \eta)\phi_2(x - \xi, y - \eta) \right] d\xi d\eta \tag{11.51}$$

where $c_1(\xi, \eta)$ and $c_2(\xi, \eta)$ are arbitrary functions.

By using the superposition principle, we can construct solutions to complex problems based on some very simple solutions known as **fundamental solutions**. The fundamental solutions would correspond to ϕ_1 and ϕ_2. Generating a solution to a complex problem would then require solving for c_1 and c_2, for example. In this section, we will examine commonly used fundamental solutions to Laplace's equation, including both the velocity potential and the streamfunction. We will use fundamental solutions in the following sections to solve some interesting problems.

11.5.1   Uniform Flow

Description. The first fundamental solution we will consider is that of a uniform flow, i.e., a flow with constant velocity. Assuming the flow is parallel to the x axis, the constant velocity components are

$$u = U \quad \text{and} \quad v = 0 \tag{11.52}$$

Velocity Potential. To determine the velocity potential, $\phi(x, y)$, we note that

$$\frac{\partial\phi}{\partial x} = U \quad \text{and} \quad \frac{\partial\phi}{\partial y} = 0 \tag{11.53}$$

Integrating the first of Equations (11.53) over x, we find

$$\phi(x, y) = Ux + f(y) \tag{11.54}$$

where $f(y)$ is a function of integration. Now, we differentiate with respect to y and use the second of Equations (11.53), wherefore

$$\frac{\partial\phi}{\partial y} = \frac{df}{dy} = 0 \quad \Longrightarrow \quad f(y) = \text{constant} \tag{11.55}$$

Therefore, the velocity potential for uniform flow is

$$\phi(x, y) = Ux + \text{constant} \tag{11.56}$$

Clearly this potential function satisfies Laplace's equation since, by inspection, $\partial^2\phi/\partial x^2$ and $\partial^2\phi/\partial y^2$ both vanish.

Streamfunction. We follow a similar procedure to compute the streamfunction. First, by definition of the streamfunction, ψ, we know that

$$\frac{\partial\psi}{\partial y} = U \quad \text{and} \quad \frac{\partial\psi}{\partial x} = 0 \tag{11.57}$$

Integrating the first of Equations (11.57) over y, we find

$$\psi(x, y) = Uy + g(x) \tag{11.58}$$

where $g(x)$ is a function of integration. Differentiating with respect to x and using the second of Equations (11.57), we arrive at

$$\frac{\partial \psi}{\partial x} = \frac{dg}{dx} = 0 \quad \Longrightarrow \quad g(x) = \text{constant} \tag{11.59}$$

Thus, the streamfunction for uniform flow is

$$\psi(x, y) = Uy + \text{constant} \tag{11.60}$$

As with the potential function, the streamfunction obviously satisfies Laplace's equation since $\partial^2 \psi / \partial x^2$ and $\partial^2 \psi / \partial y^2$ both vanish.

Streamlines and Equipotential Lines. From Equation (11.60), the streamlines are simply a family of straight lines parallel to the x axis, i.e., lines along which y is constant. Similarly, from Equation (11.56), the equipotentials are a family of straight lines parallel to the y axis. Figure 11.4 shows the streamlines and equipotentials for uniform flow.

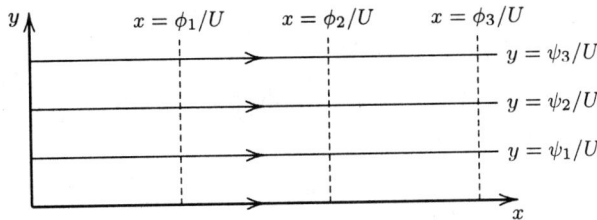

Figure 11.4: *Streamlines (——) and equipotential lines (- - -) for uniform flow.*

11.5.2 Potential Vortex

Description. We discussed the potential vortex in Section 4.3 when we introduced the concepts of vorticity and circulation. By definition, the velocity field for this flow is

$$u_r = 0 \quad \text{and} \quad u_\theta = \frac{\Gamma}{2\pi r} \tag{11.61}$$

where Γ is the circulation, and r is distance from the origin. Note that the circumferential velocity is singular at the origin, i.e., $u_\theta \to \infty$ as $r \to 0$. This is characteristic of all the fundamental solutions we will discuss, with the exception of uniform flow. To constitute part of the solution to a physical flowfield, such singular points must lie outside the solution domain — we will see how this is accomplished in subsequent sections. As a final comment, recall that the circulation for the potential vortex is the line integral of velocity on a contour enclosing the vortex. Selecting a circle of radius r, we have

$$\oint \mathbf{u} \cdot d\mathbf{s} = \int_0^{2\pi} (u_\theta \mathbf{e}_\theta) \cdot (r d\theta \mathbf{e}_\theta) = \int_0^{2\pi} \frac{\Gamma}{2\pi r} r d\theta = \Gamma \tag{11.62}$$

Velocity Potential. To determine the velocity potential, $\phi(r, \theta)$, we begin with

$$\frac{\partial \phi}{\partial r} = 0 \quad \text{and} \quad \frac{1}{r}\frac{\partial \phi}{\partial \theta} = \frac{\Gamma}{2\pi r} \tag{11.63}$$

Integrating the second of Equations (11.63) over θ, we find

$$\phi(r, \theta) = \frac{\Gamma}{2\pi}\theta + f(r) \tag{11.64}$$

where $f(r)$ is a function of integration. Now, we differentiate with respect to r and use the first of Equations (11.63), so that

$$\frac{\partial \phi}{\partial r} = \frac{df}{dr} = 0 \quad \implies \quad f(r) = \text{constant} \tag{11.65}$$

Therefore, the velocity potential for the potential vortex is

$$\phi(r, \theta) = \frac{\Gamma}{2\pi}\theta + \text{constant} \tag{11.66}$$

To verify that ϕ satisfies Laplace's equation, note that it is a function only of θ. Hence, because it is linear in θ, both terms in Laplace's equation vanish.

Streamfunction. In terms of cylindrical coordinates, the streamfunction, ψ, satisfies the following equations.

$$\frac{1}{r}\frac{\partial \psi}{\partial \theta} = 0 \quad \text{and} \quad \frac{\partial \psi}{\partial r} = -\frac{\Gamma}{2\pi r} \tag{11.67}$$

Integrating the second of Equations (11.67) over r, we find

$$\psi(r, \theta) = -\frac{\Gamma}{2\pi}\ell n r + g(\theta) \tag{11.68}$$

where $g(\theta)$ is a function of integration. Differentiating with respect to θ yields

$$\frac{1}{r}\frac{\partial \psi}{\partial \theta} = \frac{1}{r}\frac{dg}{d\theta} = 0 \quad \implies \quad g(\theta) = \text{constant} \tag{11.69}$$

Thus, the streamfunction for the potential vortex is

$$\psi(r, \theta) = -\frac{\Gamma}{2\pi}\ell n r + \text{constant} \tag{11.70}$$

To verify that the streamfunction satisfies Laplace's equation, note first that it is a function of r only. Also, $r\partial \psi/\partial r = -\Gamma/(2\pi) = \text{constant}$, so that Laplace's equation is trivially satisfied.

Streamlines and Equipotential Lines. From Equation (11.70), the streamlines are simply a family of concentric circles centered at the origin. The equipotentials are a family of radial lines radiating from the origin. Figure 11.5 shows the streamlines and equipotentials for the potential vortex.

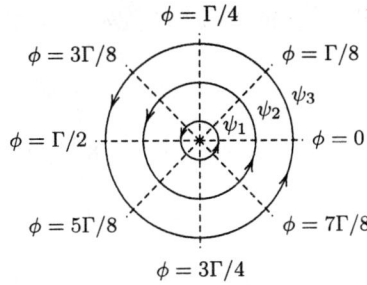

Figure 11.5: *Streamlines (——) and equipotential lines (- - -) for a potential vortex; the radii of the circular streamlines are $r_i = \exp(-2\pi\psi_i/\Gamma)$.*

11.5.3 Source and Sink

Description. The source is another singular solution. It represents a continuous source of fluid emanating from the origin. The source provides a constant volume flux of fluid, Q. When the value of Q is negative, we refer to this fundamental solution as a *sink*. The velocity components for a source are

$$u_r = \frac{Q}{2\pi r} \quad \text{and} \quad u_\theta = 0 \tag{11.71}$$

where Q is the source strength (volume flux), and r is distance from the origin. The mass flowing across any boundary enclosing a source is constant. For a circle of radius r centered at the origin, we have

$$\dot{m} = \oiint_S \rho\, \mathbf{u} \cdot \mathbf{n}\, dS = \int_0^{2\pi} \rho\, (u_r \mathbf{e}_r) \cdot \mathbf{e}_r\, r\, d\theta = \int_0^{2\pi} \rho\, \frac{Q}{2\pi r} r\, d\theta = \rho Q \tag{11.72}$$

Velocity Potential. The velocity potential, $\phi(r, \theta)$, satisfies

$$\frac{\partial \phi}{\partial r} = \frac{Q}{2\pi r} \quad \text{and} \quad \frac{1}{r}\frac{\partial \phi}{\partial \theta} = 0 \tag{11.73}$$

Integrating the first of Equations (11.73) over r yields

$$\phi(r, \theta) = \frac{Q}{2\pi} \ell n\, r + f(\theta) \tag{11.74}$$

where $f(\theta)$ is a function of integration. Hence, differentiating with respect to θ and using the second of Equations (11.73), there follows

$$\frac{\partial \phi}{\partial \theta} = \frac{df}{d\theta} = 0 \quad \Longrightarrow \quad f(\theta) = \text{constant} \tag{11.75}$$

Thus, the velocity potential is

$$\phi(r, \theta) = \frac{Q}{2\pi} \ell n\, r + \text{constant} \tag{11.76}$$

To verify that ϕ satisfies Laplace's equation, note that it is a function only of r, and that $r\partial\phi/\partial r = Q/(2\pi) = $ constant. Substitution into Laplace's equation thus shows that it is satisfied.

Streamfunction. Turning to the streamfunction, ψ, we know that

$$\frac{1}{r}\frac{\partial \psi}{\partial \theta} = \frac{Q}{2\pi r} \quad \text{and} \quad \frac{\partial \psi}{\partial r} = 0 \tag{11.77}$$

Integrating the first of Equations (11.77) over θ shows that

$$\psi(r, \theta) = \frac{Q}{2\pi}\theta + g(r) \tag{11.78}$$

where $g(r)$ is a function of integration. Differentiation with respect to r and using the second of Equations (11.77) gives

$$\frac{\partial \psi}{\partial r} = \frac{dg}{dr} = 0 \quad \Longrightarrow \quad g(r) = \text{constant} \tag{11.79}$$

Hence, the streamfunction for the source is

$$\psi(r, \theta) = \frac{Q}{2\pi}\theta + \text{constant} \tag{11.80}$$

Because $\psi(r, \theta)$ is a linear function of θ only, it obviously satisfies Laplace's equation.

Streamlines and Equipotential Lines. Equation (11.80) shows that the streamlines are a family of radial lines with origin at $r = 0$. Similarly, Equation (11.76) shows that the equipotential lines are a family of concentric circles centered at the origin. Figure 11.6 depicts the streamlines and equipotentials for a source and a sink.

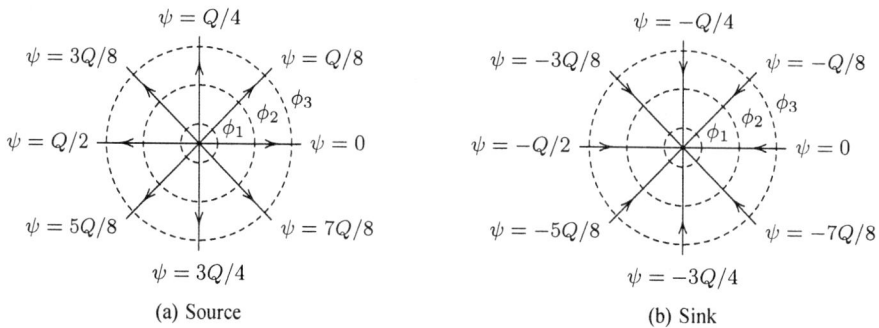

Figure 11.6: *Streamlines (——) and equipotential lines (- - -) for a source and a sink; the radii of the circular equipotential lines are $r_i = \exp(-2\pi\phi_i/Q)$.*

11.5.4 Doublet

Description. Consider a source of strength Q located on the x axis at $x = -a$, and a sink of strength $-Q$ located on the same axis at $x = a$. Figure 11.7(a) shows the geometry for such a source/sink superposition. A doublet is formed when the distance between the source and sink goes to zero in such a way that the product of the spacing and source strength is constant,

$$\lim_{a \to 0} 2aQ = \mathcal{D} \tag{11.81}$$

where $\mathcal{D}$ is a constant referred to as the doublet strength. The doublet is a useful fundamental solution that finds extensive use in potential-flow theory. There is no net flow out of a contour

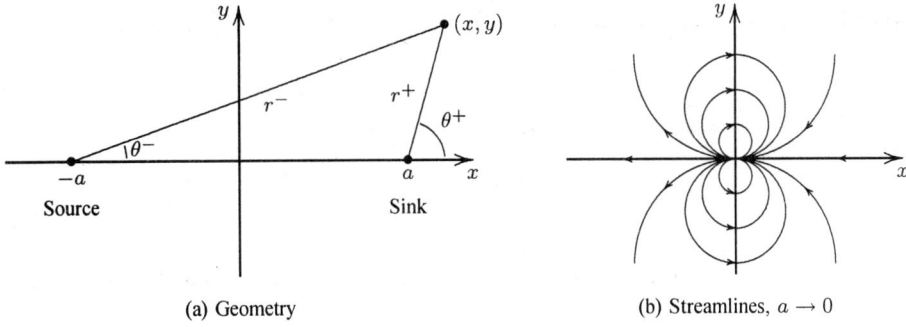

(a) Geometry   (b) Streamlines, $a \to 0$

Figure 11.7: *Source-sink geometry and streamlines.*

enclosing a doublet. Rather, fluid flows from the source to the sink with the streamlines approaching circular shapes tangent to the x axis as shown in Figure 11.7(b). The flow directions indicated on the streamlines correspond to positive doublet strength.

Velocity Potential. Using the superposition principle, we can construct the velocity potential as the sum of the potentials for a source and a sink. Hence, in terms of Cartesian coordinates,

$$\phi(x, y) = \lim_{a \to 0} \frac{Q}{2\pi} \left(\ell n r^- - \ell n r^+ \right) \tag{11.82}$$

The quantities r^+ and θ^+ are the cylindrical coordinates corresponding to coordinate axes coincident with the sink, while r^- and θ^- are cylindrical coordinates centered at the location of the source. They are defined as follows.

$$\left. \begin{aligned} \text{Sink}: \quad & r^+ = \sqrt{(x-a)^2 + y^2}, \quad \theta^+ = \tan^{-1}\left(\frac{y}{x-a}\right) \\ \text{Source}: \quad & r^- = \sqrt{(x+a)^2 + y^2}, \quad \theta^- = \tan^{-1}\left(\frac{y}{x+a}\right) \end{aligned} \right\} \tag{11.83}$$

Hence, we can rewrite the velocity potential as

$$\phi(x, y) = \lim_{a \to 0} \left(\frac{2aQ}{2\pi} \right) \frac{\ell n \sqrt{(x+a)^2 + y^2} - \ell n \sqrt{(x-a)^2 + y^2}}{2a} \tag{11.84}$$

Then, noting the definition of the doublet strength quoted in Equation (11.81), we can rearrange this equation as follows.

$$\phi(x, y) = \frac{D}{2\pi} \lim_{a \to 0} \frac{\ell n \sqrt{(x+a)^2 + y^2} - \ell n \sqrt{(x-a)^2 + y^2}}{2a} \tag{11.85}$$

But, by definition, the indicated limit is simply the partial derivative with respect to x of the function $\ell n \sqrt{x^2 + y^2} = \ell n r$. Therefore, we have

$$\phi(x, y) = \frac{D}{2\pi} \frac{\partial}{\partial x} \ell n r \tag{11.86}$$

We can simplify this further by taking the derivative, i.e.,

$$\frac{\partial}{\partial x} \ell n r = \frac{1}{r} \frac{\partial r}{\partial x} = \frac{1}{r} \frac{\partial}{\partial x} \sqrt{x^2 + y^2} = \frac{1}{r} \frac{x}{r} = \frac{x}{r^2} \tag{11.87}$$

Finally, since the Cartesian coordinate x is related to cylindrical coordinates by $x = r\cos\theta$, we conclude that

$$\frac{\partial}{\partial x}\ell nr = \frac{\cos\theta}{r} \tag{11.88}$$

Thus, substituting Equation (11.88) into Equation (11.86), we arrive at the velocity potential for a doublet, viz.,

$$\phi(r,\theta) = \frac{\mathcal{D}}{2\pi r}\cos\theta \tag{11.89}$$

Streamfunction. We can derive the streamfunction using the same procedure. Again using superposition, the streamfunction for the source-sink pair is

$$\psi(x,y) = \lim_{a\to0}\frac{Q}{2\pi}\left(\theta^- - \theta^+\right) = \lim_{a\to0}\left(\frac{2aQ}{2\pi}\right)\frac{\tan^{-1}\left(\dfrac{y}{x+a}\right) - \tan^{-1}\left(\dfrac{y}{x-a}\right)}{2a} \tag{11.90}$$

Taking the limit, the streamfunction thus becomes

$$\psi(x,y) = \frac{\mathcal{D}}{2\pi}\frac{\partial}{\partial x}\tan^{-1}\left(\frac{y}{x}\right) = -\frac{\mathcal{D}}{2\pi}\frac{y}{x^2+y^2} \tag{11.91}$$

Then noting that $y = r\sin\theta$ and $r^2 = x^2 + y^2$, the streamfunction for a doublet is

$$\psi(r,\theta) = -\frac{\mathcal{D}}{2\pi r}\sin\theta \tag{11.92}$$

Streamlines and Equipotential Lines. We can most conveniently determine the equations of the streamlines and equipotentials by using Cartesian coordinates. Considering the streamlines first, Equation (11.92) can be rewritten as

$$r^2 = -\frac{\mathcal{D}}{2\pi\psi}r\sin\theta \quad\Longrightarrow\quad x^2 + y^2 = -2\left(\frac{\mathcal{D}}{4\pi\psi}\right)y \tag{11.93}$$

or,

$$x^2 + \left(y + \frac{\mathcal{D}}{4\pi\psi}\right)^2 = \left(\frac{\mathcal{D}}{4\pi\psi}\right)^2 \tag{11.94}$$

Similarly, the equipotentials follow from Equation (11.89), viz.,

$$r^2 = \frac{\mathcal{D}}{2\pi\phi}r\cos\theta \quad\Longrightarrow\quad x^2 + y^2 = 2\left(\frac{\mathcal{D}}{4\pi\phi}\right)x \tag{11.95}$$

so that

$$\left(x - \frac{\mathcal{D}}{4\pi\phi}\right)^2 + y^2 = \left(\frac{\mathcal{D}}{4\pi\phi}\right)^2 \tag{11.96}$$

Therefore, the streamlines are a family of circles of radii $\mathcal{D}/(4\pi|\psi|)$ tangent to the x axis, and the equipotentials are a family of circles of radii $\mathcal{D}/(4\pi|\phi|)$ tangent to the y axis. Figure 11.7(b) shows the streamlines.

11.5.5   Comments on Use of Fundamental Solutions

In the following sections we will use the fundamental solutions described above to construct solutions to some interesting fluid-flow problems. Table 11.2 summarizes the velocity potentials and streamfunctions for the four fundamental solutions covered in this section. Note that we recast the uniform-flow results in cylindrical coordinates for the sake of consistency.[1]

Table 11.2: *Fundamental-Solution Velocity Potentials and Streamfunctions*

Flow	Velocity Potential, $\phi(r, \theta)$	Streamfunction, $\psi(r, \theta)$
Uniform Flow	$U r \cos\theta$	$U r \sin\theta$
Potential Vortex	$\dfrac{\Gamma}{2\pi}\theta$	$-\dfrac{\Gamma}{2\pi}\ell n\, r$
Source	$\dfrac{Q}{2\pi}\ell n\, r$	$\dfrac{Q}{2\pi}\theta$
Doublet	$\dfrac{\mathcal{D}}{2\pi r}\cos\theta$	$-\dfrac{\mathcal{D}}{2\pi r}\sin\theta$

Before proceeding, it is worthwhile to pause and summarize a few salient observations regarding potential-flow theory, and to describe what our solution strategy is for the balance of this chapter.

Simulating Solid Boundaries. Because a streamline is a contour to which the velocity is everywhere tangent, there can be no flow normal to a streamline. Thus, when we build a solution based on fundamental solutions that contains a closed streamline, we can always consider it to be the boundary of a solid object. We call such a boundary a **dividing streamline**. Closed bodies will result, for example, when the net source strength is zero, such as with the *doublet*. In general, we introduce *sources* and *sinks* to determine the shape of the body of interest. Any streamline, closed or not, can always be regarded as a solid boundary.

Suppressing Singular Behavior. As noted earlier, with one the exception of *uniform flow*, all of the fundamental solutions discussed in this section contain a singular point, i.e., a point where the velocity is infinite. Because such points do not occur in nature, the question naturally arises about how we can use fundamental solutions to construct a physically realistic flowfield solution. The answer is simple. The solutions we construct confine the singularities to a region external to the flowfield. In the case of a closed body, for example, the singularity lies within the interior of the dividing streamline. Note that the region within the dividing streamline is a separate flowfield according to potential-flow theory, although the singularities preclude using it as a description of a physically realizable flow.

---

[1]All of the singular potentials and streamfunctions in the Table assume the origin is located at $(x, y) = (0, 0)$. If we place the source, vortex or doublet at some other location, say $(x, y) = (\xi, \eta)$, the quantities r and θ are given by $r = \sqrt{(x - \xi)^2 + (y - \eta)^2}$ and $\theta = \tan^{-1}[(y - \eta)/(x - \xi)]$.

Simulating Forces on Objects. We can simulate a lifting force in our solution by adding potential vortices. Although the potential vortex is irrotational, it has constant circulation that corresponds to having all of the vorticity concentrated at the origin, where the solution is singular. Again, provided we locate the vortex within a dividing streamline, the solution will correspond to a physically realizable flow with lift. As discussed at the end of Chapter 10 (Section 10.4), this is entirely consistent with the Helmholtz Theorem. That is, any vorticity or circulation present in the flow at an initial instant will remain for all time.

Direct and Indirect Solutions. Laplace's equation is one of the basic equations of mathematical physics, and its properties and methods of solution are well established. Entire books have been written about Laplace's equation [cf. Kellogg (1953)] and its application in fluid mechanics [e.g., Robertson (1965) and Milne-Thompson (1960)]. In the following sections, we will construct what are known as **indirect solutions**, in which we distribute fundamental solutions and infer the shape generated by our selection of source, vortex, or doublet strengths. While this provides algebraically manageable solutions, we have virtually no control over the geometrical shapes that we obtain the solution for. To generate a solution for a given geometry, we require a **direct solution**. Such solutions are far more pertinent in general engineering practice. However, they usually require solution of an integral equation such as Equation (11.51), and fall beyond the scope of this text. Nevertheless, the reader should be aware that such methods exist [cf. Hess (1975)] and are generally implemented numerically. We touch on numerical solution methods for potential-flow problems in Section 11.9. There are also approximate theories that can be treated analytically, most notably, linear airfoil theory. We will discuss this theory briefly in Section 11.8.

Complex Variables. As a final comment, all of the fundamental solutions can be recast in terms of complex-variable theory [cf. Churchill and Brown (1990), Carrier, Krook and Pearson (1966), or Whittaker and Watson (1963)]. In this context, which is inherently limited to two-dimensional flows, there are many powerful advanced concepts from complex variables such as contour integration [Hildebrand (1976), Wilcox (1995)] and conformal mapping [Carrier, Krook and Pearson (1966)] that greatly enhance the range of problems that can be solved. Many texts describe the use of complex variables for potential-flow problems. Panton (1996) provides a particularly lucid discussion. The Problems section includes a few examples.

11.6 Flow Past a Cylinder

Constructing the solution for flow past a cylinder is especially illuminating as it illustrates many of the most important features of potential-flow theory. The solution involves superposition of three of the fundamental solutions of the preceding section, and shows many of the subtleties of the method. We begin by superposing a uniform flow and a doublet to generate the basic solution. Consistent with the Helmholtz Theorem, we will find that there is no force on the cylinder. Then, we add a vortex at the center of the cylinder so that its singularity is isolated from the main flow. We will find that the vortex gives rise to a lifting force.

11.6.1 The Basic Solution

Statement of the Problem. We consider a cylinder of radius R immersed in a flow that has velocity $\mathbf{u} = U\,\mathbf{i}$ far upstream (Figure 11.8). Hence, assuming the flow is steady, inviscid,

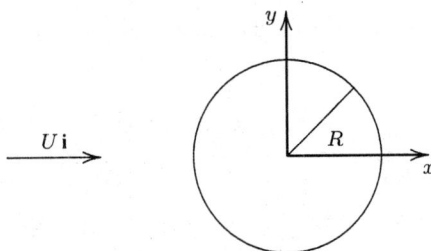

Figure 11.8: *Flow past a cylinder.*

incompressible and irrotational, the problem at hand can therefore be stated as

$$\mathbf{u} = \nabla \phi, \qquad \nabla^2 \phi = 0 \tag{11.97}$$

which must be solved subject to

$$\mathbf{u} \to U \mathbf{i} \quad \text{as} \quad r \to \infty, \qquad \mathbf{u} \cdot \mathbf{n} = 0 \quad \text{on} \quad r = R \tag{11.98}$$

The first of Equations (11.98) is the farfield condition that the flow be uniform far from the cylinder. The second is the condition that there be no flow across the cylinder surface. In terms of cylindrical coordinates, Equations (11.98) become

$$\left.\begin{aligned} u_r &\to U \cos\theta, & r &\to \infty \\ u_\theta &\to -U \sin\theta, & r &\to \infty \\ u_r &= 0, & r &= R \end{aligned}\right\} \tag{11.99}$$

There are ways to solve this linear problem directly. We could use a separation of variables, for example, i.e., assume $\phi(r,\theta) = R(r)\Theta(\theta)$, substitute into Laplace's equation and the boundary conditions, and solve. We will not pursue a direct solution however, as our purpose here is to examine the role of fundamental solutions in potential-flow theory. Thus, we will simply make a guess about what the solution is, with an adjustable parameter (the doublet strength), and verify that we have solved the correct problem with proper selection of the parameter. Alternatively, we can view what we are doing as follows. First, we construct a solution by superposition of fundamental solutions. Then, we determine the problem we have solved. From either point of view, we now proceed to the solution.

Velocity Potential and Streamfunction. If we superpose a uniform flow with freestream velocity U and a doublet of strength $\mathcal{D}$, reference to Table 11.2 shows that the velocity potential and streamfunction are as follows.

$$\left.\begin{aligned} \phi(r,\theta) &= Ur\cos\theta + \frac{\mathcal{D}}{2\pi r}\cos\theta \\ \psi(r,\theta) &= Ur\sin\theta - \frac{\mathcal{D}}{2\pi r}\sin\theta \end{aligned}\right\} \tag{11.100}$$

Note that with no loss of generality, we have set the constants that appear with ϕ and ψ equal to zero. Thus, the streamlines for this flow are given by

$$\left(Ur - \frac{\mathcal{D}}{2\pi r}\right)\sin\theta = \text{constant} \tag{11.101}$$

In the special case where the constant is zero, there are two solutions, viz.,

$$\left.\begin{array}{l} \sin\theta = 0 \qquad\qquad \Longrightarrow \quad y = r\sin\theta = 0 \\[2mm] Ur - \dfrac{\mathcal{D}}{2\pi r} = 0 \quad \Longrightarrow \quad r^2 = \dfrac{\mathcal{D}}{2\pi U} \end{array}\right\} \tag{11.102}$$

Therefore, if we define

$$R = \sqrt{\frac{\mathcal{D}}{2\pi U}} \tag{11.103}$$

then the last of Equations (11.102) shows that this streamline corresponds to flow about a cylinder of radius R. Also, the solution with $y = 0$ corresponds to the parts of the dividing streamline upstream and downstream of the cylinder. Eliminating $\mathcal{D}$, the velocity potential and streamfunction are

$$\left.\begin{array}{l} \phi(r,\theta) = U\left(r + \dfrac{R^2}{r}\right)\cos\theta \\[4mm] \psi(r,\theta) = U\left(r - \dfrac{R^2}{r}\right)\sin\theta \end{array}\right\} \tag{11.104}$$

Because both the uniform flow and the doublet satisfy Laplace's equation, so must their sum. Hence, we know that our solution satisfies the basic equations of motion. To be certain we have satisfied the boundary conditions, we should compute the velocity and verify that Equations (11.99) are satisfied.

Velocity Components. The corresponding velocity components are

$$\left.\begin{array}{lcl} u_r & = & \dfrac{\partial\phi}{\partial r} \quad = \quad U\left(1 - \dfrac{R^2}{r^2}\right)\cos\theta \\[4mm] u_\theta & = & \dfrac{1}{r}\dfrac{\partial\phi}{\partial\theta} \quad = \quad -U\left(1 + \dfrac{R^2}{r^2}\right)\sin\theta \end{array}\right\} \tag{11.105}$$

Clearly, the velocity approaches the freestream velocity as $r \to \infty$, where the contribution from the doublet approaches zero as $1/r^2$. Inspection of the solution for u_r shows that it vanishes on the surface of the cylinder. Therefore, the boundary conditions are satisfied. Since the solution satisfies Laplace's equation and the boundary conditions, we know that this is the (unique) solution.

Note that, because the doublet lies within the cylinder, there is no singular behavior for $r > R$. Figure 11.9 shows the streamlines for the flow both inside and outside the cylinder. The streamlines within the cylinder (shown as dashed lines) resemble those of pure doublet flow. For our purposes, we can ignore the flow for $r < R$ as the surface, $r = R$, is a closed streamline, so that there is no mixing of fluid from the two regions.

Stagnation Points. To locate any stagnation points that might be present in the flow, we seek points where *both* u_r and u_θ vanish. For the case at hand, appeal to Equations (11.105) yields

$$\left.\begin{array}{l} \left(1 - \dfrac{R^2}{r^2}\right)\cos\theta = 0 \\[4mm] \left(1 + \dfrac{R^2}{r^2}\right)\sin\theta = 0 \end{array}\right\} \tag{11.106}$$

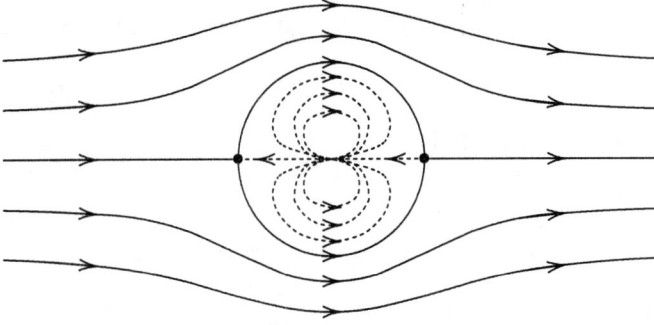

Figure 11.9: *Streamlines for flow past a cylinder;* • *denotes a stagnation point.*

Clearly, since $(1 + R^2/r^2)$ is nonzero at all points in the flowfield $(r > R)$, if any stagnation points are present, they must occur at points where $\sin \theta = 0$. Now, since $\cos \theta \neq 0$ at points where $\sin \theta = 0$, necessarily the factor $(1 - R^2/r^2) = 0$, so that $r = R$. Finally, $\sin \theta = 0$ when $\theta = 0$ and $\theta = \pi$. Therefore, the flow has two stagnation points located at

$$(r, \theta) = (R, \pi) \quad \text{and} \quad (r, \theta) = (R, 0) \tag{11.107}$$

corresponding to the points where the dividing streamline ($y = 0$ upstream of the cylinder) splits and rejoins (to $y = 0$ downstream of the cylinder), respectively. The stagnation points are shown in Figure 11.9.

Another interesting observation about the velocity field is the variation on the surface of the cylinder. When $r = R$, we have

$$u_r = 0, \quad u_\theta = -2U \sin \theta \tag{11.108}$$

where the minus sign reflects the fact that u_θ is positive in the counterclockwise direction. Thus, the magnitude of the velocity on the cylinder increases from zero at the leading stagnation point ($\theta = \pi$) to twice the freestream velocity at the midpoint ($\theta = \pi/2$), and drops back to zero at the trailing stagnation point. Because of the high velocity at the top of the cylinder, the spacing between the streamlines must decrease from the upstream spacing in order to conserve mass (see Figure 11.9).

11.6.2 Force on the Cylinder

We have sufficient information to compute the force on the cylinder. We can compute it by integrating the pressure over the cylinder surface. Letting **n** denote the outer unit normal to the cylinder, the net force exerted by the fluid on the cylinder is

$$\mathbf{F} = - \oiint_S p\, \mathbf{n}\, dS \tag{11.109}$$

Now, taking advantage of the fact that we can replace p by $(p - p_\infty)$ [see Subsection 6.1.2], and noting that

$$\mathbf{n} = \mathbf{e}_r = \mathbf{i} \cos \theta + \mathbf{j} \sin \theta, \quad dS = R\, d\theta \tag{11.110}$$

the force (per unit length) is thus given by

$$\mathbf{F} = - \int_0^{2\pi} (p - p_\infty)(\mathbf{i} \cos \theta + \mathbf{j} \sin \theta)\, R\, d\theta \tag{11.111}$$

For steady flow with no body force, Bernoulli's equation tells us the pressure is

$$p + \frac{1}{2}\rho\,\mathbf{u}\cdot\mathbf{u} = p_\infty + \frac{1}{2}\rho U^2 \tag{11.112}$$

On the surface of the cylinder, we have

$$\mathbf{u}\cdot\mathbf{u} = u_r^2(R,\theta) + u_\theta^2(R,\theta) = 4U^2\sin^2\theta \tag{11.113}$$

Therefore, the pressure difference for flow on the cylinder surface is

$$p - p_\infty = \frac{1}{2}\rho U^2\left(1 - 4\sin^2\theta\right) \tag{11.114}$$

Then,

$$\mathbf{F} = -\frac{1}{2}\rho U^2 R \int_0^{2\pi}\left(1 - 4\sin^2\theta\right)\left(\mathbf{i}\cos\theta + \mathbf{j}\sin\theta\right)d\theta \tag{11.115}$$

Finally, we define the lift, L, and drag, D, (per unit cylinder height) as follows.

$$\mathbf{F} = \mathbf{i}D + \mathbf{j}L \tag{11.116}$$

Hence, for our cylinder, the lift and drag according to the potential-flow solution are

$$\left.\begin{array}{l} L = -\dfrac{1}{2}\rho U^2 R \displaystyle\int_0^{2\pi}\left(1 - 4\sin^2\theta\right)\sin\theta\,d\theta = 0 \\[1.2em] D = -\dfrac{1}{2}\rho U^2 R \displaystyle\int_0^{2\pi}\left(1 - 4\sin^2\theta\right)\cos\theta\,d\theta = 0 \end{array}\right\} \tag{11.117}$$

Therefore, the net force on the cylinder is zero. This result is expected of course, as the flow, by construction of our solution, has zero vorticity. Examination of the pressure variation over the cylinder surface clearly illustrates why the net force is zero. Figure 11.10 shows the **pressure coefficient**, C_p, defined by

$$C_p \equiv \frac{p - p_\infty}{\frac{1}{2}\rho U^2} = 1 - 4\sin^2\theta \tag{11.118}$$

on the upper surface as a function of θ, which is measured from the positive x axis.

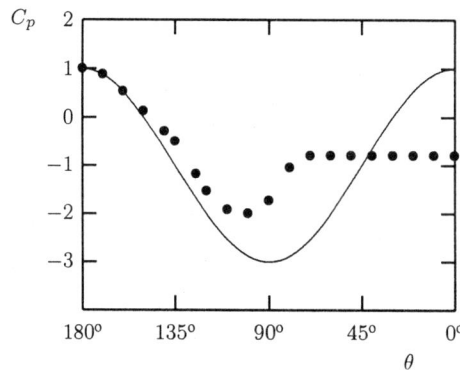

Figure 11.10: *Pressure coefficient for flow past a cylinder; —— Potential-flow theory; • Measured [Patel (1968)].*

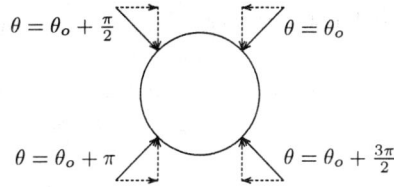

$\theta = \theta_o + \frac{\pi}{2}$ $\theta = \theta_o$

$\theta = \theta_o + \pi$ $\theta = \theta_o + \frac{3\pi}{2}$

Figure 11.11: *Pressure-force balance for flow past a cylinder.*

By symmetry, the pressure on the lower surface must be identical to that on the upper surface. Because the unit normal has opposite directions on the upper and lower surfaces, clearly there can be no lift. This conclusion would be true even for a viscous flow. It is less obvious for the drag, however. According to our potential-flow solution, the pressure is also symmetric about the midsection of the cylinder, i.e., about the y axis. Thus, for any pressure force exerted on the front face of the cylinder, there is an equal and opposite pressure force exerted on the back face. Figure 11.11 depicts the balance of pressure forces schematically for a given angle θ_o.

Figure 11.10 also includes measured C_p values [Patel (1968)]. As shown, except within about 75° of the leading stagnation point, the pressure on a cylinder in a real fluid is quite different from the potential-flow prediction. The large difference is caused by a phenomenon known as **separation**. In a real flow, the no-slip boundary condition (Section 10.3) tells us the velocity vanishes on the cylinder surface. This means the fluid very close to the surface, i.e., within the boundary layer, has low momentum. As we move beyond the top of the cylinder, the pressure begins to increase and decelerates the flow. The fluid in the boundary layer lacks sufficient momentum to overcome the increasing pressure, and leaves the surface before reaching the rear stagnation point predicted by our potential-flow solution. Rather, it enters a viscous region known as the **wake**. The flow in the wake near the cylinder has much lower pressure, and the asymmetric pressure distribution (Figure 11.10) yields a drag force.

This is illustrated schematically in Figure 11.12. Points where the flow leaves the surface are known as **separation points**. For very high Reynolds numbers, measurements show that separation occurs at about $\theta = 60°$ (120° from the forward stagnation point). For lower Reynolds numbers, it can occur as far forward as $\theta = 98°$. We will discuss separation in greater detail in Chapter 14.

As a final comment, a cylinder is not a thin, streamlined body. Thus, the fact that the predicted and measured pressure distributions differ so greatly is unsurprising, and should not be regarded as a failure of potential-flow theory. Thick, non-streamlined bodies are simply beyond the range of applicability of the theory. By contrast, for streamlined objects such as airplane wings, which are designed to pass with minimum resistance through the air, the theory provides an excellent prediction of most features of the flow.

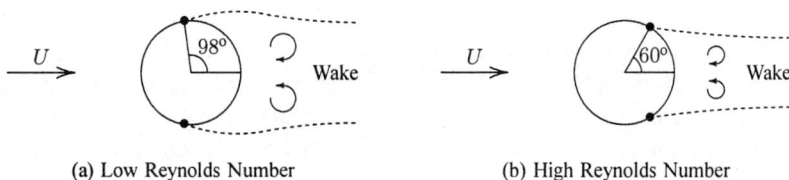

(a) Low Reynolds Number   (b) High Reynolds Number

Figure 11.12: *Flow separation for real flow past a cylinder;* • *denotes a separation point.*

However, if a streamlined object is inclined to the flow at too large an angle to the oncoming flow, separation occurs and, again, the potential-flow prediction will differ greatly from observations. In general, potential-flow methods fail to provide a physically realistic solution when increasing pressure is encountered near a solid boundary.

11.6.3   Adding a Vortex

To illustrate how we can add a lifting force to a potential-flow solution, we pose the following questions. If we add a potential vortex at the origin, what happens to our flow? Does the geometry of the dividing streamline change? What physical flow does the new solution correspond to? What is the magnitude and direction of the force predicted by the theoretical solution? To answer these questions, we will add a vortex and examine the resulting solution.

Velocity Potential and Streamfunction. To add a vortex, appeal to Table 11.2 and Equations (11.104) tells us the new velocity potential and streamfunction assume the following form:

$$\left. \begin{array}{l} \phi(r,\theta) = U\left(r + \dfrac{R^2}{r}\right)\cos\theta + \dfrac{\Gamma}{2\pi}\theta \quad + \text{(constant)}_\phi \\[4mm] \psi(r,\theta) = U\left(r - \dfrac{R^2}{r}\right)\sin\theta - \dfrac{\Gamma}{2\pi}\ell n\, r + \text{(constant)}_\psi \end{array} \right\} \qquad (11.119)$$

With no loss of generality, we can select $\text{(constant)}_\phi = 0$. However, it is advantageous to choose

$$\text{(constant)}_\psi = \frac{\Gamma}{2\pi}\ell n\, R \qquad (11.120)$$

for then, we still have $\psi(R,\theta) = 0$, i.e., the original cylinder remains on the streamline for which ψ vanishes. Thus, we propose the following forms for ϕ and ψ.

$$\left. \begin{array}{l} \phi(r,\theta) = U\left(r + \dfrac{R^2}{r}\right)\cos\theta + \dfrac{\Gamma}{2\pi}\theta \\[4mm] \psi(r,\theta) = U\left(r - \dfrac{R^2}{r}\right)\sin\theta - \dfrac{\Gamma}{2\pi}\ell n\left(\dfrac{r}{R}\right) \end{array} \right\} \qquad (11.121)$$

Velocity Components. Working with the streamfunction, the velocity components for this flow are

$$\left. \begin{array}{lclcl} u_r & = & \dfrac{1}{r}\dfrac{\partial\psi}{\partial\theta} & = & U\left(1 - \dfrac{R^2}{r^2}\right)\cos\theta \\[4mm] u_\theta & = & -\dfrac{\partial\psi}{\partial r} & = & -U\left(1 + \dfrac{R^2}{r^2}\right)\sin\theta + \dfrac{\Gamma}{2\pi r} \end{array} \right\} \qquad (11.122)$$

Inspection of the velocity components shows that our solution still satisfies the original boundary conditions of Equation (11.99). Since the $\psi = 0$ streamline still defines the original circular shape and the farfield and surface boundary conditions are satisfied, our solution still corresponds to flow past a cylinder. Regarding the velocity field, the circumferential velocity on the cylinder surface is

$$u_\theta(R,\theta) = -2U\sin\theta + \frac{\Gamma}{2\pi R} \qquad (11.123)$$

Nonrotating	Subcritical	Critical	Supercritical

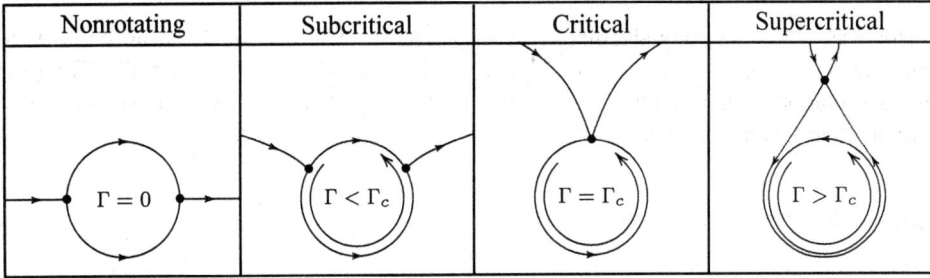

Figure 11.13: *Dividing streamlines for rotating-cylinder flow;* $\Gamma_c = 4\pi RU$ *is the critical circulation;* • *denotes a stagnation point.*

The second term on the right-hand side is the contribution from the potential vortex. Because it is a constant, it corresponds to the cylinder rotating at an angular velocity of $\Gamma/(2\pi R^2)$. Thus, the problem we have solved with our superposition of a uniform flow, a doublet and a potential vortex is flow past a rotating cylinder.

Stagnation Points. Interestingly, as illustrated in Figure 11.13, there are two possibilities for the location of the stagnation points in this flow, depending on the strength of the circulation, Γ. The first possibility has the stagnation points lying on the cylinder surface. Since $r = R$ guarantees that $u_r = 0$, we examine u_θ to determine their locations. Thus, we have

$$u_\theta(R, \theta) = 0 \quad \Longrightarrow \quad -2U \sin\theta + \frac{\Gamma}{2\pi R} = 0 \qquad (11.124)$$

Solving for θ, we find

$$r = R, \quad \theta = \sin^{-1}\left(\frac{\Gamma}{4\pi RU}\right) \qquad (\Gamma \le 4\pi RU) \qquad (11.125)$$

As long as $\Gamma < 4\pi RU$, there will be two stagnation points on the cylinder surface with the values of θ given by Equation (11.125). When $\Gamma = 4\pi RU$, there is a single stagnation point at $\theta = 90°$. For larger values of Γ, the stagnation point moves above the cylinder. To solve for the location of the stagnation point, note first that u_r always vanishes along the y axis where $\theta = \pi/2$, so that $\cos\theta = 0$. Hence, since $\sin\theta$ must equal 1, u_θ is zero when

$$0 = -U\left(1 + \frac{R^2}{r^2}\right) + \frac{\Gamma}{2\pi r} \qquad (11.126)$$

which can be rearranged to read

$$\left(\frac{r}{R}\right)^2 - 2\left(\frac{\Gamma}{4\pi RU}\right)\left(\frac{r}{R}\right) + 1 = 0 \qquad (11.127)$$

Solving, the location of the stagnation point is given by

$$r = \frac{\Gamma}{4\pi U}\left[1 + \sqrt{1 - \left(\frac{4\pi RU}{\Gamma}\right)^2}\right], \quad \theta = \frac{\pi}{2} \qquad (\Gamma > 4\pi RU) \qquad (11.128)$$

Figure 11.13 shows the dividing streamline for several values of circulation relative to the **critical circulation**, $\Gamma_c = 4\pi RU$. For the supercritical case, i.e., for $\Gamma > \Gamma_c$, the cylinder lies within the dividing streamline, so that fluid is trapped within the two closed streamlines. Alternatively, we could view our supercritical solution as flow past the tear-drop shaped geometry defined by the dividing streamline.

Observe that the dividing streamlines presented in Figure 11.13 are shown only in the immediate vicinity of the cylinder. Although not obvious from the figure, which is based on exact streamline coordinates, even in the supercritical case, the streamlines ultimately are parallel to the x axis far from the body. However, the effects of the vortex are much stronger than those of the doublet. This is evident by inspection of the velocity field. While the doublet contribution to the velocity falls off as $1/r^2$ far from the cylinder, the vortex contribution falls off as $1/r$. Hence, nontrivial effects of the vortex persist much farther than those of the doublet. This explains the strongly distorted streamline shapes near our rotating cylinder.

11.6.4   Force Computation Redone

Examination of the velocity components shows that, because the vortex contribution adds a counterclockwise circumferential component at the cylinder surface (provided $\Gamma > 0$), it increases the velocity at the bottom of the cylinder. Bernoulli's equation then tells us the pressure is reduced. Similarly, the velocity decreases at the top of the cylinder, corresponding to an increase in pressure. Hence, we should expect to find a downward, or negative lift, force on the cylinder. To compute the magnitude of the force, we again note that the force per unit length of the cylinder is given by Equation (11.111), viz.,

$$\mathbf{F} = -\int_0^{2\pi} (p - p_\infty)\,(\mathbf{i}\cos\theta + \mathbf{j}\sin\theta)\;Rd\theta \tag{11.129}$$

The pressure follows from Bernoulli's equation, which simplifies to

$$p - p_\infty = \frac{1}{2}\rho\left(U^2 - \mathbf{u}\cdot\mathbf{u}\right) = \frac{1}{2}\rho U^2\left[1 - \left(2\sin\theta - \frac{\Gamma}{2\pi RU}\right)^2\right] \tag{11.130}$$

where we use the fact that the radial velocity component, u_r, vanishes on the cylinder surface, so that $\mathbf{u}\cdot\mathbf{u} = u_\theta^2(R,\theta)$. Hence, expanding the quadratic term, we have

$$p - p_\infty = \frac{1}{2}\rho U^2\left\{\left[1 - \left(\frac{\Gamma}{2\pi RU}\right)^2\right] + \frac{2\Gamma}{\pi RU}\sin\theta - 4\sin^2\theta\right\} \tag{11.131}$$

Now, the first and third terms on the right-hand side of this expression for $p - p_\infty$ are even functions of θ. Hence, when multiplied by the unit normal, which is an odd function of θ, they must integrate to zero (integrating from 0 to 2π is the same as integrating from $-\pi$ to π). This means the force on the cylinder is simply

$$\begin{aligned}
\mathbf{F} &= -\frac{1}{2}\rho U^2\int_0^{2\pi}\frac{2\Gamma}{\pi RU}\sin\theta\,(\mathbf{i}\cos\theta + \mathbf{j}\sin\theta)\;Rd\theta \\
&= -\frac{\rho U\Gamma}{\pi}\int_0^{2\pi}\left(\mathbf{i}\sin\theta\cos\theta + \mathbf{j}\sin^2\theta\right)d\theta
\end{aligned} \tag{11.132}$$

Finally, we make use of the fact that, from elementary calculus,

$$\left.\begin{aligned}
\int_0^{2\pi}\sin\theta\cos\theta\,d\theta &= 0 \\[1em]
\int_0^{2\pi}\sin^2\theta\,d\theta &= \pi
\end{aligned}\right\} \tag{11.133}$$

Therefore the force on a counterclockwise rotating cylinder, according to potential-flow theory, is

$$\mathbf{F} = -\rho U \Gamma \mathbf{j} \qquad (11.134)$$

As expected, our solution predicts a negative lift force and zero drag. This force is generally referred to as the **Magnus force**, and has the interesting feature of being perpendicular to the freestream flow direction. Thus, the lift force per unit length of the cylinder, L, is

$$L = -\rho U \Gamma \qquad (11.135)$$

Equation (11.135) is the famous result known as the **Kutta-Joukowski law** or the **Kutta-Joukowski lift theorem**. This result was stated as follows by Kutta and Joukowski, two researchers who developed Equation (11.135) independent of each other's efforts in 1902 and 1906, respectively.

> *"According to inviscid theory, the lift per unit depth of any cylinder of any shape immersed in a uniform stream equals $\rho U \Gamma$, where Γ is the total net circulation contained within the body shape. The direction of the lift is 90° from the stream direction, rotating opposite to the circulation."*

Figure 11.14 compares computed and measured [Rouse (1946) and Hoerner and Borst (1975)] lift and drag coefficients, C_L and C_D, for flow past a rotating cylinder. By definition, these dimensionless coefficients are

$$C_L \equiv \frac{\tilde{L}}{\frac{1}{2}\rho U^2 (2Rb)}, \qquad C_D \equiv \frac{\tilde{D}}{\frac{1}{2}\rho U^2 (2Rb)} \qquad (11.136)$$

where b is cylinder height. The quantities $\tilde{L} = Lb$ and $\tilde{D} = Db$ are total lift and drag, respectively. In the experiments, the cylinder rotates in the clockwise direction with angular velocity ω, so that the circulation is

$$\Gamma = -2\pi \omega R^2 \qquad (11.137)$$

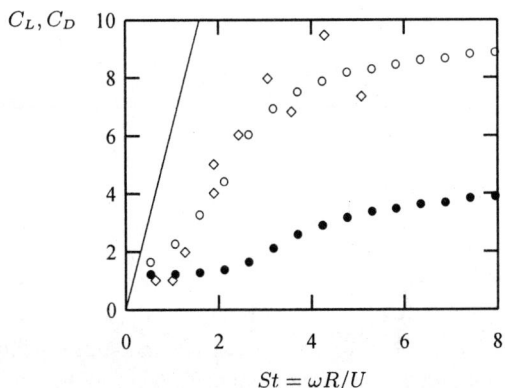

Figure 11.14: *Lift and drag coefficients for flow past a cylinder rotating with angular velocity ω; —— potential-flow theory; $\circ$, • Measured C_L, C_D [Rouse (1946)]; $\diamond$ Measured C_L [Hoerner and Borst (1975)].*

Results are plotted as a function of the Strouhal number, St, defined by

$$St = \frac{\omega R}{U} \qquad (11.138)$$

In terms of our potential-flow solution, the lift and drag coefficients are

$$C_L = 2\pi St, \qquad C_D = 0 \qquad \text{(Potential-flow theory)} \qquad (11.139)$$

As shown, measurements indicate that less than half the potential-flow value of C_L is realized, and there is a significant drag force. The discrepancies are caused by flow separation. Interestingly, the lift force is much larger than that on a typical airfoil of similar size. Hence, generating lift with a rotating cylinder can be a useful aerodynamic device in spite of flow separation. This is remarkable because separation is generally regarded as catastrophic failure for most aerodynamic devices, especially for an airfoil.

To make it easier to determine the direction of the Magnus force, we can rewrite the Kutta-Joukowski lift theorem in vector form as

$$\mathbf{F} = -\rho \, \mathbf{U} \times \boldsymbol{\Gamma} \qquad (11.140)$$

where the circulation vector, $\boldsymbol{\Gamma}$, is parallel to the direction of the bound vorticity. Using this generalized form of the Kutta-Joukowski law, we can apply what we have learned to help understand the motion of a baseball. Figure 11.15 illustrates three of the most common ingredients of a successful pitcher's arsenal, viz., a fastball, a sinker and a curveball. While the ball is a sphere rather than a cylinder, the Magnus force is present, and acts in the direction indicated by Equation (11.140).

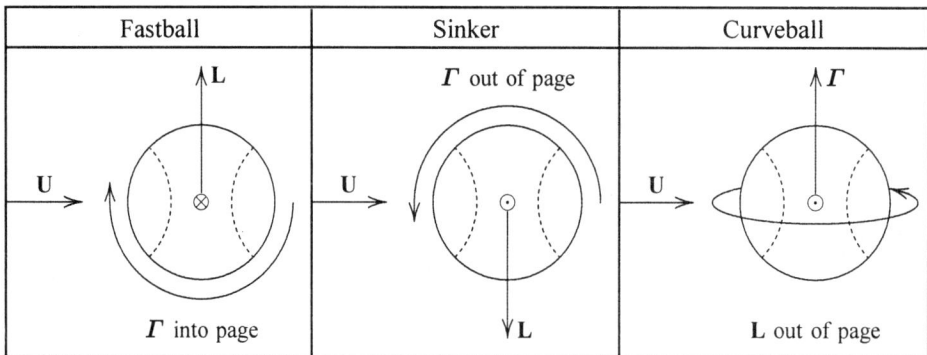

Figure 11.15: *Various motions of a baseball.*

As illustrated, a pitcher's fastball is thrown with "backspin" corresponding to clockwise circulation. This is done to counter the effect of gravity and to permit the ball to travel in a more nearly straight line. This permits a talented pitcher to more precisely control the point where the ball passes through the strike zone. Countering gravity is just as important for long throws, where the object is to get the ball to its desired destination as quickly as possible. An example is an outfielder attempting to throw out a runner trying to advance a base. Clearly, the more level the throw, the shorter the distance it must travel. Imparting backspin produces lift, which permits a flatter (and thus shorter) trajectory.

By contrast, some pitchers have mastered the so-called sinker. With this pitch, a counterclockwise spin is imparted to the ball as it is thrown, thus producing a downward force.

As the ball approaches the batter, the downward force causes the ball to drop. If the batter succeeds in making contact with the ball, it is usually hit downward, i.e., on the ground. This has given rise to the colorful description of the plight of batters facing a pitcher with an effective sinker, i.e., that they spend the afternoon "killing worms" (presumably living in the ground near home plate).

With the curveball, the pitcher provides spin in the vertical direction. This causes a sideways force that makes the ball move continually farther away from the batter as it approaches home plate, if he bats with the same hand as the pitcher throws. If the batter bats with the opposite hand, the ball will "curve" toward him as it approaches. In either case, the pitch is usually more difficult to hit than a pitch thrown in a straight line.

As it turns out, the most effective curveball is a combination sinker and pure curveball. That is, the orientation of the spin should be at an angle to the vertical so that it not only moves away from (or toward) the batter, but also drops. This is true because changing the elevation of the bat is physically more difficult than simply extending or pulling in the arms to compensate for a curveball. For the reader interested in exploring the fluid dynamics of a baseball further, the book by Adair (1990) provides an excellent starting point.

AN EXPERIMENT YOU CAN DO AT HOME
Excerpted from Shames (1992)

You can observe the Magnus force on a rotating cylinder by assembling the following materials:

- An $8\frac{1}{2}$ inch by $11\frac{1}{2}$ inch sheet of paper

- A pair of scissors or an exacto knife

- Some tape

- A three-foot long piece of string

To minimize weight, cut the sheet of paper down to an $8\frac{1}{2}$ inch by 4 inch segment. Now, roll this segment into a cylinder (of height $8\frac{1}{2}$ inches) and tape it in a way that conceals the seam, and permits it to roll freely. Allowing a little overlap to make handling easy while taping, you will have a cylinder that is roughly 1 inch in diameter. Tie the string around the center of the cylinder and wrap it around until about 4 inches are free. You are now ready to do the experiment.

Place the cylinder three feet from the edge of a table top with the free end of the string looping beneath the cylinder and lying flat on the surface. Jerk the string horizontally causing it to unwind and let go. When done properly, the translating cylinder will lift above the plane of the table, thus demonstrating the Magnus force.

11.7   Other Interesting Solutions

This section includes several interesting potential-flow solutions to problems that illustrate a few more subtle features of the theory. We will make use of some of these solutions when we include viscous effects in later chapters. First, we use potential-flow theory to solve for accelerating flow past a cylinder. Then, we examine flow past a classic body known as a Rankine oval, which we will analyze further in Chapter 14. Next, we analyze flow past a wedge, which will be of use when we discuss effects of pressure on a boundary layer in Chapter 14. Another application is flow near a stagnation point — we will revisit stagnation-point flow in Chapter 13. We conclude by showing how general solid boundaries can be simulated by suitable placement of fundamental solutions, viz., with the method of images.

11.7.1   Accelerating Cylinder

We noted at the end of Section 11.4 that Laplace's equation does not explicitly include time. This means we can solve an unsteady-flow problem in a quasi-steady sense, simply substituting time-dependent boundary conditions into what would otherwise be a steady-flow solution. The unsteady equivalent of Bernoulli's equation, Equation (11.46), shows that the pressure is directly affected by unsteadiness. This opens the possibility that forces might develop in unsteady potential flow. As we will see in the present example, this is indeed the case.

Basic Solution. We focus on potential flow past a cylinder of radius R where the freestream velocity, $U(t)$, varies with time, t. The preceding section tells us the velocity potential is

$$\phi(r, \theta, t) = U(t) \left(r + \frac{R^2}{r} \right) \cos \theta \qquad (11.141)$$

The unsteady Bernoulli's equation tells us that, in the absence of body forces,

$$\frac{\partial \phi}{\partial t} + \frac{p}{\rho} + \frac{1}{2} \mathbf{u} \cdot \mathbf{u} = F(t) \qquad (11.142)$$

where $F(t)$ is a function that must be determined from the boundary conditions.

Surface Pressure. Noting that

$$\frac{\partial \phi}{\partial t} = \frac{dU}{dt} \left(r + \frac{R^2}{r} \right) \cos \theta \qquad (11.143)$$

we see that because $\cos \theta = 0$ on the y axis, necessarily $\partial \phi / \partial t = 0$. Hence, we know every term on the left-hand side of Equation (11.142) on the y axis as $|y| \to \infty$. Therefore, we have

$$F(t) = \frac{p_\infty}{\rho} + \frac{1}{2} U^2 \qquad (11.144)$$

Substituting Equation (11.144) into Equation (11.142) yields the pressure:

$$p = p_\infty + \frac{1}{2} \rho \left(U^2 - u_r^2 - u_\theta^2 \right) - \rho \frac{\partial \phi}{\partial t} \qquad (11.145)$$

Force on the Cylinder. Our immediate goal in this exercise is to compute any force that might develop on the cylinder. So, noting that on the cylinder, $u_r = 0$ and $u_\theta = -2U \sin \theta$, the pressure simplifies to

$$p = p_\infty + \frac{1}{2}\rho U^2 \left(1 - 4\sin^2\theta\right) - \rho\frac{\partial\phi}{\partial t} \tag{11.146}$$

Also, reference to Equation (11.143) shows that on the surface of the cylinder,

$$\frac{\partial\phi}{\partial t} = 2R\frac{dU}{dt}\cos\theta \tag{11.147}$$

Therefore, the pressure on the surface of the cylinder is

$$p = p_\infty + \frac{1}{2}\rho U^2 \left(1 - 4\sin^2\theta\right) - 2\rho R\frac{dU}{dt}\cos\theta \tag{11.148}$$

As with steady-flow, the force on the cylinder is given by Equation (11.111). The only difference between the steady-flow pressure [Equation (11.114)] and the pressure for the present case is the term proportional to dU/dt. Since we know the other terms yield no force, we can omit all but this term and write

$$\begin{aligned}
\mathbf{F} &= -\int_0^{2\pi} \left(-2\rho R\frac{dU}{dt}\cos\theta\right)(\mathbf{i}\cos\theta + \mathbf{j}\sin\theta)\, R\,d\theta \\
&= 2\rho R^2\frac{dU}{dt}\int_0^{2\pi}\left(\mathbf{i}\cos^2\theta + \mathbf{j}\sin\theta\cos\theta\right)d\theta = 2\pi\rho R^2\frac{dU}{dt}\mathbf{i} \tag{11.149}
\end{aligned}$$

Thus, the cylinder develops a drag force proportional to the acceleration of the freestream flow. Consistent with the symmetry of the flow, the lift is zero.

Virtual Mass. When an object moves through an incompressible fluid, it pushes a finite mass of fluid out of its way. When the object accelerates, as in the present example, the surrounding fluid that is set in motion must also accelerate. This is the reason the cylinder experiences a drag force. A classical result of potential-flow theory is the following [cf. Milne-Thompson (1960) or Saffman (1993)]. For an accelerating body, the force that develops on the body is

$$\mathbf{F} = (m + m_h)\frac{d\mathbf{U}}{dt} \tag{11.150}$$

where m is the actual mass of the body and m_h is the **virtual mass**[2] defined in terms of the total relative kinetic energy of the fluid by

$$\frac{1}{2}m_h\mathbf{U}\cdot\mathbf{U} \equiv \iiint_V \frac{1}{2}\rho\,\mathbf{U}_{rel}\cdot\mathbf{U}_{rel}\,dV \tag{11.151}$$

where $\mathbf{U}_{rel}$ is the fluid velocity relative to the object, and the integration is performed for the entire volume surrounding the object.

For the case at hand, since the fluid within the dividing streamline has density ρ, the mass of the body (per unit length) is the product of ρ and the area of the cylinder, so that

$$m = \pi\rho R^2 \tag{11.152}$$

The relative velocity is simply the doublet contribution, so that

$$\mathbf{U}_{rel} = -U\frac{R^2}{r^2}\cos\theta\,\mathbf{e}_r - U\frac{R^2}{r^2}\sin\theta\,\mathbf{e}_\theta \quad\Longrightarrow\quad \mathbf{U}_{rel}\cdot\mathbf{U}_{rel} = U^2\frac{R^4}{r^4} \tag{11.153}$$

---

[2]Virtual mass is also referred to as **added mass** and **hydrodynamic mass**.

The virtual mass per unit length thus becomes

$$m_h = \int_0^{2\pi} \int_R^\infty \rho \left(\frac{R}{r}\right)^4 r\, dr\, d\theta = 2\pi \rho R^4 \int_R^\infty \frac{dr}{r^3} = \pi \rho R^2 \qquad (11.154)$$

which, coincidentally, is equal to the mass of fluid displaced by the cylinder. So, according to the virtual mass concept, the force on the cylinder should be

$$\mathbf{F} = \left(\pi \rho R^2 + \pi \rho R^2\right) \frac{dU}{dt}\mathbf{i} = 2\pi \rho R^2 \frac{dU}{dt}\mathbf{i} \qquad (11.155)$$

Comparison with Equation (11.149) shows that the force is identical to the result obtained by integrating the pressure directly.

11.7.2 Rankine Oval

Thus far, none of our applications has made direct use of the source or the sink. In this application, we superpose a uniform stream with a source and a sink of equal strengths to obtain a closed body known as the **Rankine oval** (Figure 11.16). In general, when we construct a potential-flow solution from fundamental solutions, the resulting body will be closed provided the sum of the strengths of all sources matches the sum of the strengths of all sinks.

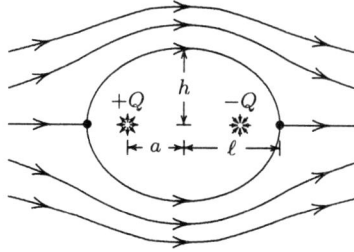

Figure 11.16: *Streamlines for flow past a Rankine oval; • denotes a stagnation point.*

We place a source of strength Q on the x axis at $x = -a$ and a sink of the same strength on the x axis at $x = a$. That is, we select the same geometry considered when we introduced the doublet (see Figure 11.7). However, the spacing between the source and the sink remains finite for the Rankine oval.

Streamfunction. As shown when we formulated the doublet relationships, the streamfunction for this flow is

$$\psi = Uy + \frac{Q}{2\pi}\left(\theta^- - \theta^+\right); \qquad \theta^- = \tan^{-1}\left(\frac{y}{x+a}\right), \quad \theta^+ = \tan^{-1}\left(\frac{y}{x-a}\right) \qquad (11.156)$$

We can simplify the algebra by making use of the trigonometric identity for the tangent of the difference between two angles, i.e.,

$$\tan\left(\theta^- - \theta^+\right) = \frac{\tan\theta^- - \tan\theta^+}{1 + \tan\theta^- \tan\theta^+} \qquad (11.157)$$

Thus, for the Rankine oval angles, θ^- and θ^+,

$$\tan\left(\theta^- - \theta^+\right) = \frac{\dfrac{y}{x+a} - \dfrac{y}{x-a}}{1 + \dfrac{y^2}{x^2 - a^2}} = -\frac{2ay}{x^2 + y^2 - a^2} \qquad (11.158)$$

Substituting into Equation (11.156), we arrive at a somewhat simplified form of the stream-function for flow past a Rankine oval.

$$\psi = Uy - \frac{Q}{2\pi} \tan^{-1} \left(\frac{2ay}{x^2 + y^2 - a^2} \right) \tag{11.159}$$

Determining Body Shape. Before proceeding with our analysis, it is convenient to recast the streamfunction and coordinates in nondimensional form. Using the half-spacing distance, a, as the appropriate length scale, we find

$$\overline{\psi} = \overline{y} - \overline{Q} \tan^{-1} \left(\frac{2\overline{y}}{\overline{x}^2 + \overline{y}^2 - 1} \right) \tag{11.160}$$

where

$$\overline{\psi} \equiv \frac{\psi}{Ua}, \qquad \overline{x} \equiv \frac{x}{a}, \qquad \overline{y} \equiv \frac{y}{a}, \qquad \overline{Q} \equiv \frac{Q}{2\pi Ua} \tag{11.161}$$

The body lies on the $\psi = 0$ streamline. Hence, the body shape is defined, in terms of dimensionless quantities, by the following equation.

$$0 = \overline{y} - \overline{Q} \tan^{-1} \left(\frac{2\overline{y}}{\overline{x}^2 + \overline{y}^2 - \overline{a}^2} \right) \quad \Longrightarrow \quad \tan \left(\frac{\overline{y}}{\overline{Q}} \right) = \frac{2\overline{y}}{\overline{x}^2 + \overline{y}^2 - 1} \tag{11.162}$$

Equation (11.162) is a transcendental equation that cannot be solved in closed form if we regard $\overline{y}$ as a function of $\overline{x}$. However, we can solve this equation for $\overline{x}$ as an explicit function of $\overline{y}$. The solution is

$$\overline{x} = \pm \sqrt{1 + 2\overline{y} \cot \left(\overline{y}/\overline{Q} \right) - \overline{y}^2} \tag{11.163}$$

The overall shape of the Rankine oval is best characterized by its height, $2h$, and length, 2ℓ. To determine the length, we must find the value of $\overline{x}$ that corresponds to $\overline{y} = 0$. Similarly, the height is the value of $\overline{y}$ corresponding to $\overline{x} = 0$. Considering the length first,

$$\overline{x} = \lim_{\overline{y} \to 0} \sqrt{1 + 2\overline{y} \cot \left(\overline{y}/\overline{Q} \right) - \overline{y}^2} = \sqrt{1 + 2\overline{Q}} \tag{11.164}$$

Returning to dimensional parameters, we have

$$\frac{\ell}{a} = \sqrt{1 + \frac{Q}{\pi Ua}} \tag{11.165}$$

Now, we turn to computation of the height. Setting $\overline{x} = 0$ in Equation (11.162), the value of $\overline{y}$ at the top of the oval satisfies

$$\tan \left(\frac{\overline{y}}{\overline{Q}} \right) = \frac{2\overline{y}}{\overline{y}^2 - 1} \tag{11.166}$$

We can simplify this equation by taking advantage of the trigonometric half-angle formula for the tangent, i.e.,

$$\tan \alpha = \frac{2 \tan \left(\frac{\alpha}{2} \right)}{1 - \tan^2 \left(\frac{\alpha}{2} \right)} = \frac{2 \cot \left(\frac{\alpha}{2} \right)}{\cot^2 \left(\frac{\alpha}{2} \right) - 1} \tag{11.167}$$

Hence, making the identification $\alpha = \overline{y}/\overline{Q}$, we have

$$\frac{2 \cot\left(\dfrac{\overline{y}}{2\overline{Q}}\right)}{\cot^2\left(\dfrac{\overline{y}}{2\overline{Q}}\right) - 1} = \frac{2\overline{y}}{\overline{y}^2 - 1} \tag{11.168}$$

By inspection, it must be the case that

$$\overline{y} = \cot\left(\frac{\overline{y}}{2\overline{Q}}\right) \tag{11.169}$$

Although this is still a transcendental equation that must be solved numerically, it is nevertheless simpler than the original equation. Rewriting the relationship in terms of dimensional quantities, we have

$$\frac{h}{a} = \cot\left(\frac{\pi U a}{Q}\frac{h}{a}\right) \tag{11.170}$$

Table 11.3 includes values for the height and width of the Rankine oval as a function of the dimensionless source strength. Note that as the source strength becomes very large, the spacing between the source and sink becomes less and less significant. Both $h/a \to \infty$ and $\ell/a \to \infty$ so that, in effect, the limit $Q/(\pi U a) \to \infty$ is the same as the limit $a \to 0$. In other words, as the source strength increases, our source/sink pair approaches a doublet. Hence, we should expect the solution to asymptotically approach the flow past a circular cylinder (for which $h = \ell$). This is borne out in Table 11.3.

Table 11.3: *Dimensions of a Rankine Oval*

$Q/(\pi U a)$	h/a	ℓ/a	h/ℓ
0	0.000	1.000	0.000
.01	0.016	1.005	0.015
.1	0.143	1.049	0.136
1	0.860	1.414	0.608
10	3.111	3.317	0.938
100	9.983	10.050	0.993
∞	∞	∞	1.000

11.7.3   Wedge

Any function that satisfies Laplace's equation can be regarded as a possible solution to some potential-flow problem. Hence, given such a function, it is interesting to see what the streamlines look like, determine any useful dividing streamlines, locate stagnation points, etc. For example, consider the function

$$\psi(r, \theta) = \psi_o r^n \sin n\theta \tag{11.171}$$

where ψ_o and n are constants. The exponent n is dimensionless and the coefficient ψ_o has dimensions $(\text{length})^{2-n}/(\text{time})$. This is indeed a solution to Laplace's equation, which can be verified by direct substitution.

Velocity Components. Before determining the shape of the streamlines implied by this stream-function, it is worthwhile to first compute the velocity components. This will permit us to determine flow direction on the streamlines and to locate any stagnation points that might exist. Hence, differentiating the streamfunction, we find

$$
\left.
\begin{aligned}
u_r &= \frac{1}{r}\frac{\partial \psi}{\partial \theta} = n\psi_o r^{n-1}\cos n\theta \\[2mm]
u_\theta &= -\frac{\partial \psi}{\partial r} = -n\psi_o r^{n-1}\sin n\theta
\end{aligned}
\right\}
\qquad (11.172)
$$

Determining Body Shape. Although there is no guarantee that the $\psi = 0$ streamline defines a meaningful dividing streamline [recall the rotating cylinder and Equation (11.120)], it usually serves as a good starting point. For the streamfunction at hand,

$$
\psi = 0 \quad \Longrightarrow \quad \theta = \frac{m\pi}{n}, \quad m = \text{integer} \qquad (11.173)
$$

If we confine our attention to the cases $m = 0$ and $m = 1$, this defines the wedge-shaped region between $\theta = 0$ and $\theta = \pi/n$. Along the sides of the wedge, the circumferential velocity, u_θ, is zero, while the radial velocity varies with radial distance according to

$$
u_r(r,0) = n\psi_o r^{n-1} \quad \text{and} \quad u_r(r,\pi/n) = -n\psi_o r^{n-1} \qquad (11.174)
$$

Inspection of Equations (11.172) shows that, as long as $n > 1$, there is a single stagnation point at $r = 0$. Figure 11.17 shows several wedge solutions for angles ranging from 45° to 225° ($\pi/4$ to $5\pi/4$).

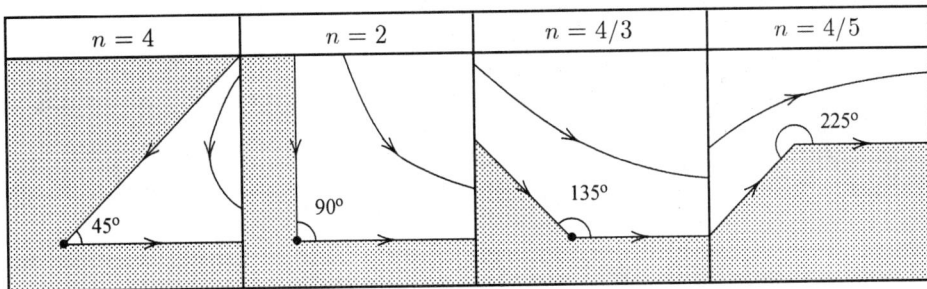

Figure 11.17: *Dividing streamlines for wedge flow; • denotes a stagnation point.*

The potential-flow wedge solution is valid only in the immediate vicinity of the origin when $n > 1$. This is true because the velocity becomes infinite as $r \to \infty$. Similarly, because the solution is infinite at the origin for $n < 1$, it is valid provided we exclude the immediate vicinity of the origin. This is true of all potential-flow solutions. That is, as can be shown from complex-variable theory, the only function that is completely bounded for the entire region $0 \le r < \infty$ is a constant. We see this with the fundamental solutions, for example, where the vortex, source and doublet velocities are singular as $r \to 0$, and otherwise well behaved.

If we continue to $m = 2, 3, \dots$ in Equation (11.173), we find additional wedges, each with an included angle of π/n. They may overlap, of course, depending upon the value of n.

We can also use more than a single wedge in a flowfield. For example, consider the case $n = 4/3$. If we use the three streamlines defined by setting $m = -1, 0, +1$, assuming $\psi_o > 0$, we would have flow past the trailing edge of a body with a sharp trailing edge as illustrated in

Figure 11.18(a). The primary flow direction is from left to right. The body is defined by the streamlines with $m = \pm 1$. The $m = 0$ streamline lies along the plane of symmetry between the upper and lower streams.

Similarly, if we have $\psi_o < 0$, the flow direction is reversed, and the solution corresponds to flow near the leading edge of a sharp-nosed body. This is illustrated in Figure 11.18(b). For this case, flow is from right to left. We will use this leading-edge solution in Chapter 14 when we discuss effects of varying pressure on a boundary layer.

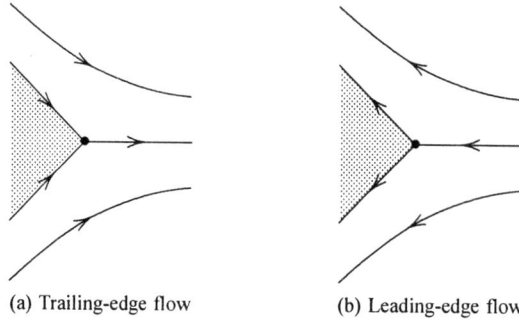

(a) Trailing-edge flow   (b) Leading-edge flow

Figure 11.18: *Dividing streamlines for flow near sharp leading and trailing edges;* • *denotes a stagnation point.*

11.7.4 Stagnation Point

One especially interesting case of wedge flow worthy of additional discussion occurs when we select $n = 2$. If we focus on a single wedge, Figure 11.17 shows that this corresponds to flow into a 90° corner. However, if we use the streamlines with $m = 0, 1, 2$ in Equation (11.173), then we can regard the x axis as the body surface with the y axis being an axis of symmetry for the flow. The resulting flow is shown in Figure 11.19.

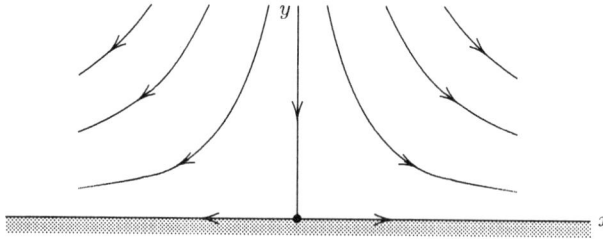

Figure 11.19: *Stagnation-point flow;* • *denotes the stagnation point.*

Because it assumes a simple form, it is worthwhile to recast the streamfunction in terms of Cartesian coordinates. That is,

$$\psi = \psi_o r^2 \sin 2\theta = 2\psi_o r^2 \sin\theta \cos\theta = 2\psi_o xy \tag{11.175}$$

Clearly, the streamlines are a family of hyperbolas given by $xy = $ constant. The velocity components are

$$u = \frac{\partial \psi}{\partial y} = 2\psi_o x \quad \text{and} \quad v = -\frac{\partial \psi}{\partial x} = -2\psi_o y \tag{11.176}$$

We noted in the preceding section that, because the velocity becomes infinite far from the stagnation point at $x = y = 0$, the solution can be valid only in the "immediate vicinity" of the origin. We can make a quantitative statement about what this phrase means in the special case of flow past a cylinder. For consistency in notation, consider flow past a cylinder with the freestream flow oriented along the y axis as shown in Figure 11.20.

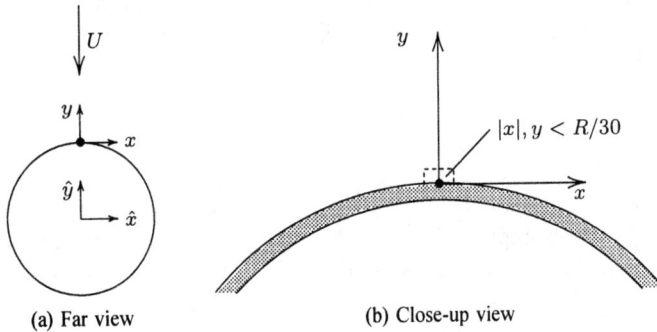

(a) Far view   (b) Close-up view

Figure 11.20: *Flow past a cylinder revisited.*

Since we have rotated our coordinate system by 90° relative to that used in Section 11.6, we simply replace the circumferential angle θ by $\theta + \pi/2$. Thus, the radial velocity component follows from the first of Equations (11.105), and becomes

$$u_r(r, \theta) = U\left(1 - \frac{R^2}{r^2}\right)\cos\left(\theta + \frac{\pi}{2}\right) = -U\left(1 - \frac{R^2}{r^2}\right)\sin\theta \tag{11.177}$$

Hence, along the positive y axis, which corresponds to $\theta = \pi/2$, we see that

$$u_r(r, \pi/2) = -U\left(1 - \frac{R^2}{\hat{y}^2}\right) \tag{11.178}$$

where $\hat{y}$ is measured from the center of the cylinder. Then, changing to the coordinate system whose origin lies at the forward stagnation point [Figure 11.20(a)], and noting that the Cartesian velocity component, v, is equal to u_r, we have

$$v(x, y) = -U\left[1 - \frac{R^2}{(R + y)^2}\right] = -U\left[1 - \frac{1}{(1 + y/R)^2}\right] \tag{11.179}$$

Finally, expanding in Taylor series, we know that

$$\frac{1}{(1 + y/R)^2} = 1 - 2\left(\frac{y}{R}\right) + 3\left(\frac{y}{R}\right)^2 + \cdots \tag{11.180}$$

Thus, substituting Equation (11.180) into Equation (11.179), the vertical velocity approaching the stagnation point is

$$v = -2U\left(\frac{y}{R}\right)\left[1 - \frac{3}{2}\left(\frac{y}{R}\right) + \cdots\right] \tag{11.181}$$

Hence, for $y \ll R$, the velocity is closely approximated by

$$v \approx -2\frac{U}{R}y \tag{11.182}$$

Comparison with the second of Equations (11.176) shows that our stagnation-point solution is valid with ψ_o chosen to be

$$\psi_o = \frac{U}{R} \qquad (11.183)$$

To quantify the terminology "immediate vicinity," we simply examine the Taylor series expansion for v. Clearly, our solution is accurate as long as the first term in the series that we neglect is small compared to the last term retained. Specifically, if we insist that our solution be valid to within 5%, then the immediate vicinity of the stagnation point means

$$\frac{3}{2}\frac{y}{R} < \frac{1}{20} \qquad \Longrightarrow \qquad y < \frac{R}{30} \qquad (11.184)$$

The dashed rectangle in Figure 11.20(b) shows the range of applicability of the stagnation-point solution. As an example, if we consider flow past a $2\frac{1}{2}$ foot circular cylinder, the stagnation-point solution will be valid in the region approximately 1 inch away from the stagnation point.

As a final comment, using Bernoulli's equation, we can solve for the pressure. Denoting the pressure at the stagnation point by p_o, then,

$$p + \frac{1}{2}\rho\left(2\psi_o\right)^2\left(x^2 + y^2\right) = p_o \qquad \Longrightarrow \qquad p = p_o - 2\rho\psi_o^2\left(x^2 + y^2\right) \qquad (11.185)$$

We will make use of this, and other features of the solution, when we discuss viscous effects on flow near a stagnation point in Chapter 13.

11.7.5   The Method of Images

All of the flows discussed thus far have occurred in a fluid that extends infinitely far from the object of interest. There are important applications that involve the presence of a solid boundary at a finite distance from an object. For example, we might want to compute flow past an object close to a plane such as an airfoil or an automobile-like shape to simulate motion on a runway or highway, respectively. We might also want to simulate the constraining walls of a wind tunnel to see how much difference there is between flow past an object in a finite and an infinite fluid, thus helping establish error bounds for measured flow properties.

A useful technique for simulating a planar boundary is known as the **method of images**. This method is best explained by example. Consider a source of strength Q. If we place the source at the point $(x, y) = (0, h)$ and another source of strength Q at $(x, y) = (0, -h)$, the streamfunction is

$$\psi(x, y) = \underbrace{\frac{Q}{2\pi}\tan^{-1}\left(\frac{y - h}{x}\right)}_{\text{Source at }(0,h)} + \underbrace{\frac{Q}{2\pi}\tan^{-1}\left(\frac{y + h}{x}\right)}_{\text{Source at }(0,-h)} \qquad (11.186)$$

Differentiation shows that the velocity components for this flow are

$$\left.\begin{aligned}
u &= \frac{\partial\psi}{\partial y} &&= \frac{Q}{2\pi}\left[\frac{x}{x^2 + (y - h)^2} + \frac{x}{x^2 + (y + h)^2}\right] \\[2mm]
v &= -\frac{\partial\psi}{\partial x} &&= \frac{Q}{2\pi}\left[\frac{y - h}{x^2 + (y - h)^2} + \frac{y + h}{x^2 + (y + h)^2}\right]
\end{aligned}\right\} \qquad (11.187)$$

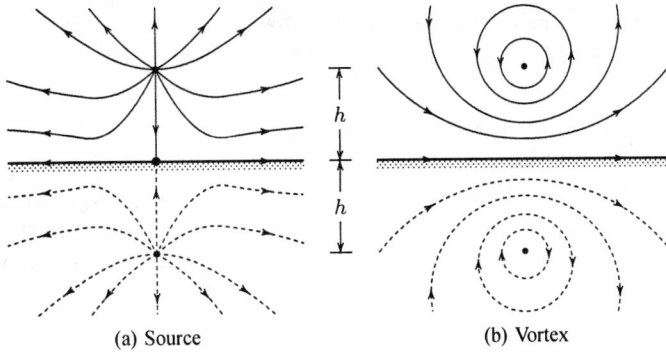

(a) Source   (b) Vortex

Figure 11.21: *Method of images for a source and a vortex.*

Hence, the velocity along the x axis is given by

$$u(x,0) = \frac{Qx}{\pi\left(x^2 + h^2\right)}, \qquad v(x,0) = 0 \qquad \Longrightarrow \qquad \mathbf{u} \cdot \mathbf{n} = 0 \ \text{ on } \ y = 0 \qquad (11.188)$$

where $\mathbf{n} = \mathbf{j}$ is the unit normal to the x axis. Therefore, we have a symmetry plane aligned with the x axis. The source below the axis is known as the **mirror image** of the source above the axis. Figure 11.21(a) shows a few streamlines for a source and its mirror image.

In a similar way, the mirror image of a vortex of strength Γ a distance h above the x axis is a vortex of *opposite strength*, $-\Gamma$, a distance h below the axis. The streamfunction for a vortex and its image is

$$\psi(x,y) = \underbrace{-\frac{\Gamma}{2\pi}\ell n\sqrt{x^2 + (y - h)^2}}_{\text{Counterclockwise Vortex at } (0,h)} + \underbrace{\frac{\Gamma}{2\pi}\ell n\sqrt{x^2 + (y + h)^2}}_{\text{Clockwise Vortex at } (0,-h)} \qquad (11.189)$$

while the velocity components are

$$\left. \begin{aligned} u &= \frac{\partial \psi}{\partial y} &&= \frac{\Gamma}{2\pi}\left[\frac{y + h}{x^2 + (y + h)^2} - \frac{y - h}{x^2 + (y - h)^2}\right] \\[2mm] v &= -\frac{\partial \psi}{\partial x} &&= \frac{\Gamma}{2\pi}\left[\frac{x}{x^2 + (y - h)^2} - \frac{x}{x^2 + (y + h)^2}\right] \end{aligned} \right\} \qquad (11.190)$$

Again, we see that the velocity along the x axis has only an x component, i.e.,

$$u(x,0) = \frac{\Gamma h}{\pi\left(x^2 + h^2\right)}, \qquad v(x,0) = 0 \qquad (11.191)$$

Figure 11.21(b) shows a few streamlines for a vortex and its mirror image.

Note that for both the source and the vortex, because the velocity on the x axis is nonuniform, Bernoulli's equation tells us the pressure is also nonuniform. Hence, it is reasonable to expect that there is a nonzero force on the axis caused by the presence of the source and vortex pairs. Equivalently, we can view this as the force exerted by the image source or vortex on the other. To determine the force, we observe first that along the x axis, the pressure is

$$p = p_\infty - \frac{1}{2}\rho u^2(x,0) \qquad (11.192)$$

where we note that the fluid velocity goes to zero far from the source or vortex. The force per unit depth exerted on the wall from the source or vortex above the x axis is given by

$$\mathbf{F}_w = \int_{-\infty}^{+\infty} (p - p_\infty)\,\mathbf{j}\,dx \tag{11.193}$$

Thus, appealing to Equations (11.188) and (11.191) for $u(x, 0)$, the force on the x axis is

$$\mathbf{F}_w = \begin{cases} -\mathbf{j}\dfrac{\rho Q^2}{2\pi^2} \displaystyle\int_{-\infty}^{+\infty} \dfrac{x^2 dx}{\left(x^2 + h^2\right)^2}, & \text{Source} \\[4mm] -\mathbf{j}\dfrac{\rho \Gamma^2}{2\pi^2} \displaystyle\int_{-\infty}^{+\infty} \dfrac{h^2 dx}{\left(x^2 + h^2\right)^2}, & \text{Vortex} \end{cases} \tag{11.194}$$

Finally, reference to a standard table of integrals (or through a substitution such as $x = h\tan\phi$) shows that

$$\int_{-\infty}^{+\infty} \frac{x^2 dx}{\left(x^2 + h^2\right)^2} = \frac{\pi}{2h} \quad \text{and} \quad \int_{-\infty}^{+\infty} \frac{h^2 dx}{\left(x^2 + h^2\right)^2} = \frac{\pi}{2h} \tag{11.195}$$

Therefore, the force exerted on the wall is

$$\mathbf{F}_w = \begin{cases} -\dfrac{\rho Q^2}{4\pi h}\mathbf{j}, & \text{Source} \\[4mm] -\dfrac{\rho \Gamma^2}{4\pi h}\mathbf{j}, & \text{Vortex} \end{cases} \tag{11.196}$$

In both cases, the force is directed downward, increases with the strength of the source or vortex, and increases as the distance from the x axis decreases.

Figure 11.21 shows that, in addition to producing a symmetry plane, the introduction of images induces changes throughout the flow. The streamlines are modified even in the immediate vicinity of the source and the vortex. Furthermore, the distortion increases as we move the source or vortex closer to the symmetry plane. Thus, for more complex flows, we can expect similar distortion. For example, consider flow past a cylinder of radius R. Suppose we wish to construct a solution for flow past a cylinder whose center is a distance nR above a plane surface, where n is a number greater than 1 (see Figure 11.22).

As a first approximation, we will superpose a uniform flow with a doublet of strength $\mathcal{D} = 2\pi R^2 U$ [see Equation (11.103)] at $(x, y) = (0, nR)$ and an image doublet of the same

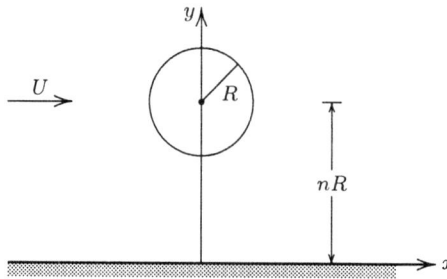

Figure 11.22: *Flow past a cylinder above a solid boundary.*

strength at $(x, y) = (0, -nR)$. We know that both the doublet and its image must have the same sign since the doublet is simply the limiting case of a source superposed with a sink. So, the streamfunction is

$$\psi = Uy - \underbrace{U \frac{R^2}{r^+} \sin \theta^+}_{\text{Doublet at } (0, nR)} - \underbrace{U \frac{R^2}{r^-} \sin \theta^-}_{\text{Doublet at } (0, -nR)} \tag{11.197}$$

where r^+ and r^- are the radial distances measured from the doublets at $(0, nR)$ and $(0, -nR)$, respectively, while θ^+ and θ^- are the angles measured from $y = nR$ and $y = -nR$, i.e.,

$$\left. \begin{array}{ll} r^+ = \sqrt{x^2 + (y - nR)^2}, & \theta^+ = \tan^{-1}\left(\dfrac{y - nR}{x}\right) \\[4mm] r^- = \sqrt{x^2 + (y + nR)^2}, & \theta^- = \tan^{-1}\left(\dfrac{y + nR}{x}\right) \end{array} \right\} \tag{11.198}$$

Because of the influence of the mirror-image doublet, the shape of the dividing streamline must differ from that given by a superposition of the original doublet and the uniform stream. Although noticeable, the effect is relatively small, even for $n = 2$. To illustrate the magnitude of the effect, Table 11.4 lists the velocities at the original stagnation-point locations for several values of n. As shown, for $n > 6$, deviations from the solution for flow past a cylinder are less than 0.2%.

Table 11.4: *Induced Velocities at Original Stagnation Points*

Stagnation Point Location	$n = 2$		$n = 3$		$n = 6$	
	u/U	v/U	u/U	v/U	u/U	v/U
Forward	0.052	-0.028	0.026	-0.009	0.002	-0.001
Rearward	0.052	0.028	0.026	0.009	0.002	0.001

The velocities in the table suggest that the upstream portion of the dividing streamline shifts upward slightly, and vice versa for the downstream portion. This turns out to be true for this flow. Also, the effective body shape is very nearly circular, although the center is shifted a bit to the right. The modified geometry is sketched in Figure 11.23. The distortion can be removed by adding so-called "corrective images" [cf. Robertson (1965)]. For general problems, this consists of adding an (often infinite) series of imaged sources, doublets, etc. to restore the original dividing streamline. Numerical solution, implementing an iterative procedure, is usually required for all but the simplest problems.

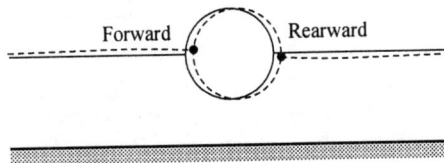

Figure 11.23: *Schematic showing the effect of the mirror doublet on the dividing streamline. The solid contour (——) is the original dividing streamline and the dashed line (- - -) is the shifted streamline; • denotes a stagnation point.*

11.8　Airfoil Flow

One of the most important applications for aeronautical engineers is that of flow past a wing. In the context of two-dimensional flow, we focus on a cross section of a wing, which is known as an **airfoil**. Study of the two-dimensional airfoil is relevant to airplane design as the **wingspan** (distance from the fuselage to the wing tip) of a typical wing is usually very large compared to the **chord length** (arc length from leading to trailing edge). Thus, the airfoil solution is accurate except close to the fuselage/wing junction and the wing tip. In this section, we briefly discuss some aspects of airfoil theory.

11.8.1　The Kutta Condition

In Section 10.4, we discussed the difference between inviscid and viscous flow about an airfoil. We saw that, as a direct consequence of the Helmholtz Theorem, if we could somehow create vorticity in an inviscid flow, it would remain for all time. In the context of an airfoil, this means that the circulation about the airfoil would remain constant for all time. Thus, if we embed a vortex, or series of vortices, within an airfoil, the flow will be irrotational, have constant circulation, and be completely consistent with both the Helmholtz Theorem and potential-flow theory.

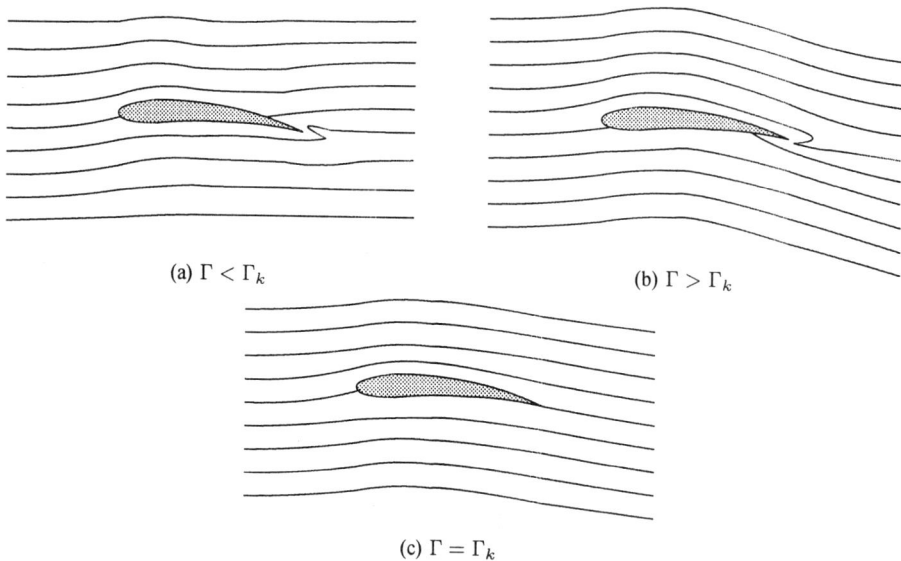

(a) $\Gamma < \Gamma_k$　　　　　　　　　　　　(b) $\Gamma > \Gamma_k$

(c) $\Gamma = \Gamma_k$

Figure 11.24: *Possible potential-flow solutions for flow past an airfoil.*

The question then arises as to what the strength of the bound vorticity within the airfoil should be. As it turns out, solutions to Laplace's equation for flow around a closed surface in an unbounded medium are not unique. Figure 11.24 illustrates the three possibilities for flow around an airfoil. As shown, when the circulation is below a certain value, which we will call Γ_k, the rear stagnation point lies on the upper surface [Figure 11.24(a)]. The special case of zero circulation falls into this category. Also, if the circulation exceeds Γ_k, the rear stagnation point lies on the lower surface of the airfoil [Figure 11.24(b)]. As discussed in Section 10.4, neither of these flows occurs in nature because of the infinite accelerations that would attend flow turning abruptly at the trailing edge. The viscosity of any real fluid prevents such turning.

As a result, the flow shown in Figure 11.24(c) develops in which the flow leaves the trailing edge smoothly. There is a unique circulation that corresponds to this solution, i.e., Γ_k.

Thus, we arrive at a unique solution by insisting that the circulation be such that the flow leaves the trailing edge of the airfoil smoothly. This is known as the **Kutta condition**.[3] The Kutta condition is based on the empirical observation that flow leaves an airfoil trailing edge smoothly, and is therefore not a provable result. It is, rather, a "rule of thumb" that works well, and settles the issue of uniqueness. The appropriate value of Γ_k varies with the freestream velocity, the *angle of attack* (i.e., the angle at which the airfoil is inclined to the freestream flow direction), and the precise shape of the airfoil.

Potential flow past general airfoil geometries can be routinely calculated numerically, and many computer programs are commercially available for this purpose. Using the theory of complex variables, it is possible to begin with the rotating cylinder solution of Subsection 11.6.3 and generate airfoil solutions analytically. In this approach, a technique known as conformal mapping [Carrier, Krook and Pearson (1966), Hildebrand (1976)] is used to transform the cylinder to an airfoil geometry. This approach generates a family of airfoils known as **Joukowski airfoils**. For more details, see Panton (1996) or Anderson (1989).

11.8.2   Source and Vortex Sheets

In principle, for a given airfoil shape, we can find a distribution of sources, sinks and vortices that produce the appropriate geometry and lift. Based on what we have done thus far, we could place these singular solutions to Laplace's equation within the airfoil, and achieve a physically realistic solution, free of singularities. As mentioned earlier, however, this approach generally requires solution of an integral equation. To see how such an equation arises, consider flow past a symmetric, nonlifting airfoil such as the one illustrated in Figure 11.25.

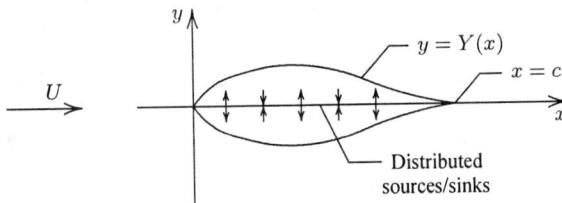

Figure 11.25: *Flow past a symmetric, nonlifting airfoil.*

Imagine that we have a continuous distribution of sources that we place on the x axis. This means we superpose a collection of sources with differential strength $dQ = q(\xi)d\xi$ located at $(x, y) = (\xi, 0)$. Then, the streamfunction for flow past the airfoil is

$$\psi(x, y) = Uy + \frac{1}{2\pi} \int_0^c q(\xi) \tan^{-1}\left(\frac{y}{x - \xi}\right) d\xi \qquad (11.199)$$

where U is the freestream velocity. If we denote the shape of the airfoil by $y = Y(x)$, then the dividing streamline, $\psi = 0$, becomes

$$UY(x) + \frac{1}{2\pi} \int_0^c q(\xi) \tan^{-1}\left(\frac{Y(x)}{x - \xi}\right) d\xi = 0 \qquad (11.200)$$

---

[3]Credit for this boundary condition also belongs to Joukowski and Chaplygin. Some authors even prefer to call it the **Kutta-Joukowski condition** — reference to Chaplygin is less common.

This is a classical integral equation of Fredholm type for $q(\xi)$ that can be solved for general shapes only by numerical methods.

For very thin bodies, i.e., for $Y(x) \ll c$, we can simplify this integral equation to a form more amenable to analytical solution. As we will see in this subsection, we can treat thickness and lift effects separately, provided the angle of attack, α, is also small. To understand how we implement these approximations in what is known as **linear airfoil theory**, we must first introduce two new concepts known as the **source sheet** and the **vortex sheet**.

Source Sheet. By definition, a source sheet is a planar area over which sources and sinks are continuously distributed. In the context of two-dimensional flow, the sheet is a curve. For the symmetric airfoil pictured in Figure 11.25, the source sheet is the part of the x axis between $x = 0$ and $x = c$. We have already introduced the idea of a differential source of strength $q(\xi)$. In terms of our previous notation for a source, $q(\xi) = dQ/d\xi$. The source sheet stands alone without a freestream. Thus, the streamfunction for a source sheet is

$$\psi(x, y) = \frac{1}{2\pi} \int_0^c q(\xi)\theta(x - \xi, y)\, d\xi, \qquad \theta(x - \xi, y) \equiv \tan^{-1}\left(\frac{y}{x - \xi}\right) \qquad (11.201)$$

As with a discrete distribution of sources and sinks, if the net integrated source strength vanishes, superposition with a uniform flow yields a closed body, viz.,

$$\int_0^c q(\xi)\, d\xi = 0 \qquad \Longrightarrow \qquad \text{Closed body} \qquad (11.202)$$

The vertical velocity component, $v(x, y)$, is

$$v(x, y) = -\frac{1}{2\pi} \int_0^c q(\xi)\frac{\partial\theta}{\partial x}d\xi = \frac{1}{2\pi} \int_0^c \frac{q(\xi)\, y\, d\xi}{(x - \xi)^2 + y^2} \qquad (11.203)$$

We can compute the velocity just above $[v(x, 0^+)]$ and below $[v(x, 0^-)]$ the axis by noting that

$$\tan\theta = \frac{y}{x - \xi} \qquad \Longrightarrow \qquad x - \xi = y\cot\theta, \quad d\xi = y\csc^2\theta\, d\theta \qquad (11.204)$$

Hence, the vertical velocity can be rewritten as

$$v(x, y) = \frac{1}{2\pi} \int_{\theta_o}^{\theta_c} \frac{q(x - y\cot\theta)\, y\, (y\csc^2\theta\, d\theta)}{y^2\, (1 + \cot^2\theta)} = \frac{1}{2\pi} \int_{\theta_o}^{\theta_c} q(x - y\cot\theta)\, d\theta \qquad (11.205)$$

where the limits of the integral are defined by

$$\theta_o \equiv \tan^{-1}\left(\frac{y}{x}\right) \qquad \text{and} \qquad \theta_c \equiv \tan^{-1}\left(\frac{y}{x - c}\right) \qquad (11.206)$$

Finally, for $y \to 0$, we conclude that $q(x - y\cot\theta) \approx q(x)$, wherefore

$$v(x, 0^{\pm}) \approx \frac{q(x)}{2\pi}(\theta_c - \theta_o) \qquad (11.207)$$

This limiting process illustrates, in effect, that contributions from the sources at $x = \xi$ are negligible except when x is close to ξ. This result is true for all geometries, and is not limited to slender bodies. As shown in Figure 11.26, the values for θ_o and θ_c depend on the location of the point of interest. As we should expect, the difference between θ_o and θ_c is nonzero

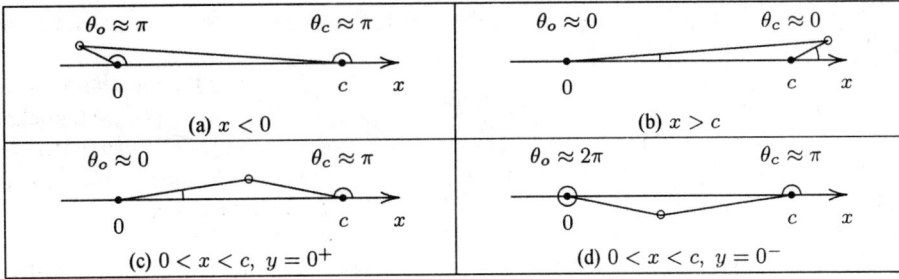

Figure 11.26: *Limiting angles for a source sheet; the sheet lies along the x axis between $x = 0$ and $x = c$; $\circ$ denotes a point just above or below the axis.*

only for points between $x = 0$ and $x = c$, i.e., directly above and below the source sheet. Inspection of Figure 11.26 shows that

$$v(x, 0^\pm) \approx \pm \frac{q(x)}{2} \tag{11.208}$$

Therefore, a source sheet produces a jump in the vertical velocity across the sheet. By contrast, the horizontal velocity is continuous across a source sheet as can be seen by direct computation. That is,

$$u(x, y) = \frac{1}{2\pi} \int_0^c q(\xi) \frac{\partial \theta}{\partial y} d\xi = \frac{1}{2\pi} \int_0^c \frac{q(\xi)(x - \xi) d\xi}{(x - \xi)^2 + y^2} \tag{11.209}$$

Substituting $y = 0$, we arrive at the following for a source sheet.

$$\left. \begin{array}{l} v(x, 0^+) - v(x, 0^-) = q(x) \\[2mm] u(x, 0) = \dfrac{1}{2\pi} \int_0^c \dfrac{q(\xi)\, d\xi}{x - \xi} \end{array} \right\} \quad \text{(Source sheet)} \tag{11.210}$$

Vortex Sheet. We define a vortex sheet as a planar area (a curve in two dimensions) over which vortices are distributed. Consistent with the aerodynamic convention, we will work with the aerodynamic circulation, Γ_a, which is positive for a clockwise line integral. We imagine a continuous distribution of vortices of differential strength given by $\gamma(\xi) = d\Gamma_a/d\xi$ (see Figure 11.27).

For reasons that will become obvious, it is most convenient to work with the velocity potential for the vortex sheet. The velocity potential is

$$\phi(x, y) = -\frac{1}{2\pi} \int_0^c \gamma(\xi)\theta(x - \xi, y)\, d\xi, \qquad \theta(x - \xi, y) \equiv \tan^{-1}\left(\frac{y}{x - \xi}\right) \tag{11.211}$$

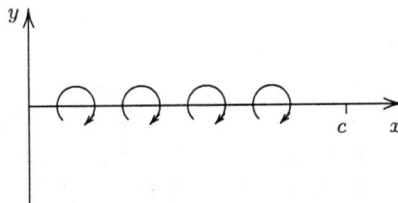

Figure 11.27: *Vortex sheet on the x axis between $x = 0$ and $x = c$.*

As a result of superposition, it should be clear that the total circulation about the vortex sheet is simply the integral of $\gamma(\xi)$, i.e.,

$$\Gamma_a = \int_0^c \gamma(\xi)\, d\xi \tag{11.212}$$

Differentiating the velocity potential, the velocity components are

$$\left. \begin{aligned} u(x, y) &= -\frac{1}{2\pi} \int_0^c \gamma(\xi) \frac{\partial \theta}{\partial x} d\xi = \frac{1}{2\pi} \int_0^c \frac{\gamma(\xi)\, y\, d\xi}{(x - \xi)^2 + y^2} \\[2mm] v(x, y) &= -\frac{1}{2\pi} \int_0^c \gamma(\xi) \frac{\partial \theta}{\partial y} d\xi = -\frac{1}{2\pi} \int_0^c \frac{\gamma(\xi)(x - \xi)\, d\xi}{(x - \xi)^2 + y^2} \end{aligned} \right\} \tag{11.213}$$

Proceeding in a manner entirely analogous to that used for the source sheet, we find that v is continuous across the sheet while

$$u(x, 0^{\pm}) = \pm \frac{\gamma(x)}{2} \tag{11.214}$$

Thus, the limiting forms of the velocity components as $y \to 0$ are as follows.

$$\left. \begin{aligned} u(x, 0^+) - u(x, 0^-) &= \gamma(x) \\[2mm] v(x, 0) &= -\frac{1}{2\pi} \int_0^c \frac{\gamma(\xi)\, d\xi}{x - \xi} \end{aligned} \right\} \qquad \text{(Vortex sheet)} \tag{11.215}$$

11.8.3 Linear Airfoil Theory

Airfoil Geometry. Figure 11.28 shows the conventional mathematical description of an airfoil. The segment of the x axis between $x = 0$ and $x = c$ passes through the leading and trailing edges of the airfoil, and is referred to as the **chord line**. The upper and lower surfaces are represented as $y = Y_u(x)$ and $y = Y_l(x)$, respectively. We define the **camber line**, $y = C(x)$, as the average of $Y_u(x)$ and $Y_l(x)$, while the **half-thickness**, $T(x)$, is half their difference. That is, we define

$$C(x) \equiv \frac{1}{2}\left[Y_u(x) + Y_l(x)\right] \qquad \text{and} \qquad T(x) \equiv \frac{1}{2}\left[Y_u(x) - Y_l(x)\right] \tag{11.216}$$

Now, let's represent the velocity, **u**, as the sum of the freestream velocity, **U**, defined by

$$\mathbf{U} = U\left(\mathbf{i}\cos\alpha + \mathbf{j}\sin\alpha\right) \tag{11.217}$$

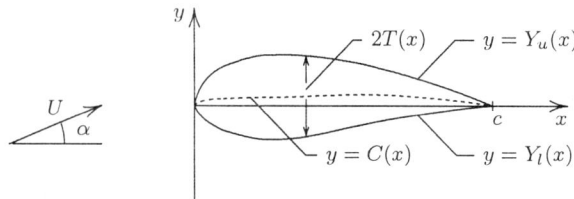

Figure 11.28: *Mathematical description of an airfoil; $y = C(x)$ is the camber line, $2T(x)$ is the thickness and α is the angle of attack; vertical scale exaggerated.*

and a perturbation, $\hat{\mathbf{u}}$, caused by the presence of the airfoil. If the airfoil is thin and the angle of attack small, the perturbation velocity is small compared to the freestream velocity. Thus, the complete velocity vector is given by

$$\mathbf{u} = \mathbf{U} + \hat{\mathbf{u}}, \qquad |\hat{\mathbf{u}}| \ll |\mathbf{U}| \tag{11.218}$$

Since the upper and lower surfaces of the airfoil are streamlines, $v/u = dy/dx$ so that the exact boundary conditions on the velocity are as follows:

$$\left.\begin{aligned}
\frac{U \sin\alpha + \hat{v}\left[x, Y_u(x)\right]}{U \cos\alpha + \hat{u}\left[x, Y_u(x)\right]} &= Y_u'(x) \\[2mm]
\frac{U \sin\alpha + \hat{v}\left[x, Y_l(x)\right]}{U \cos\alpha + \hat{u}\left[x, Y_l(x)\right]} &= Y_l'(x)
\end{aligned}\right\} \tag{11.219}$$

where prime denotes differentiation with respect to x. Then, noting that since $\hat{u} \ll U\cos\alpha$, we can linearize the boundary conditions to obtain

$$\left.\begin{aligned}
\hat{v}\left(x, 0^+\right) &\approx U\cos\alpha Y_u'(x) - U\sin\alpha \\[2mm]
\hat{v}\left(x, 0^-\right) &\approx U\cos\alpha Y_l'(x) - U\sin\alpha
\end{aligned}\right\} \tag{11.220}$$

Note that, in addition to dropping $\hat{u}$, we have made the approximation that the boundary condition is applied on the x axis rather than at the airfoil surface. Therefore, in terms of the camber, $C(x)$, and thickness, $T(x)$, we have the following approximate conditions for the vertical velocity just above and below the airfoil.

$$\left.\begin{aligned}
\hat{v}\left(x, 0^+\right) - \hat{v}\left(x, 0^-\right) &\approx 2U\cos\alpha \frac{dT}{dx} \\[2mm]
\hat{v}\left(x, 0^+\right) + \hat{v}\left(x, 0^-\right) &\approx 2U\cos\alpha \frac{dC}{dx} - 2U\sin\alpha
\end{aligned}\right\} \tag{11.221}$$

Lifting Force. The lifting force on the airfoil can be obtained by integrating the pressure over its surface. The pressure, in turn, follows from Bernoulli's equation. In terms of the velocity as expressed in Equation (11.218), we have

$$p + \frac{1}{2}\rho\left(\mathbf{U} + \hat{\mathbf{u}}\right)\cdot\left(\mathbf{U} + \hat{\mathbf{u}}\right) = p_\infty + \frac{1}{2}\rho\,\mathbf{U}\cdot\mathbf{U} \tag{11.222}$$

Thus, the difference between p and p_∞ is

$$\begin{aligned}
p - p_\infty &= \frac{1}{2}\rho\left[\mathbf{U}\cdot\mathbf{U} - \left(\mathbf{U}\cdot\mathbf{U} + 2\mathbf{U}\cdot\hat{\mathbf{u}} + \hat{\mathbf{u}}\cdot\hat{\mathbf{u}}\right)\right] \\[2mm]
&= -\rho\,\mathbf{U}\cdot\hat{\mathbf{u}} - \frac{1}{2}\rho\,\hat{\mathbf{u}}\cdot\hat{\mathbf{u}}
\end{aligned} \tag{11.223}$$

Because $|\hat{\mathbf{u}}| \ll |\mathbf{U}|$, we can neglect the last term, wherefore

$$p - p_\infty \approx -\rho\,\mathbf{U}\cdot\hat{\mathbf{u}} = -\rho\left(U\hat{u}\cos\alpha + U\hat{v}\sin\alpha\right) \tag{11.224}$$

Finally, for small angle of attack, $\hat{u}\cos\alpha \approx \hat{u}$ and $\hat{v}\sin\alpha \approx \hat{v}\alpha \ll \hat{v}$, which tells us we can also neglect the contribution from the vertical velocity. Thus, for thin airfoils at small angle of attack,

$$p - p_\infty \approx -\rho\,U\hat{u} \tag{11.225}$$

We are now prepared to compute the airfoil's lift. Integrating over the surface of the airfoil, we have

$$
\begin{aligned}
\mathbf{F} &= -\oiint_S (p - p_\infty)\,\mathbf{n}\,dS \approx \oiint_S \rho U \hat{u}\,\mathbf{n}\,dS \\
&\approx \rho U \int_0^c \left[\hat{u}(x, 0^+)(\mathbf{j}) + \hat{u}(x, 0^+)(-\mathbf{j}) \right] dx \\
&\approx \rho U \mathbf{j} \int_0^c \left[\hat{u}(x, 0^+) - \hat{u}(x, 0^-) \right] dx
\end{aligned}
\tag{11.226}
$$

where we note that for small angles of attack, $\mathbf{n} \approx \pm\mathbf{j}$ on the upper and lower surfaces, respectively. Thus, if we represent the airfoil by superposing a freestream with a source sheet and a vortex sheet, the jump in $\hat{u}$ comes entirely from the vortex sheet, and is given by Equation (11.215). Denoting the lift (per unit depth) by L so that $\mathbf{F} = L\,\mathbf{j}$, we conclude that

$$
L = \rho U \int_0^c \gamma(x)\,dx = \rho U \Gamma_a
\tag{11.227}
$$

Since, by definition $\Gamma_a = -\Gamma$, this result is completely consistent with the Kutta-Joukowski lift theorem, Equation (11.140).

The Thickness Problem. In linear airfoil theory, we distinguish between the thickness problem and the lift problem. We represent thickness with a source sheet and lift with a vortex sheet. The thickness problem can actually be solved directly if the object is symmetric and, therefore, nonlifting. This can be seen immediately by combining the boundary condition of Equation (11.221) with $\alpha = 0$ and the source-sheet jump condition of Equation (11.210). The result is

$$
q(x) = 2U \frac{dT}{dx}
\tag{11.228}
$$

Therefore, if we specify the shape of a symmetric airfoil, we know the source strength as an explicit function of its thickness. The rest of the flowfield is given by

$$
\psi(x, y) = Uy + \frac{1}{2\pi} \int_0^c q(\xi) \tan^{-1}\left(\frac{y}{x - \xi} \right) d\xi
\tag{11.229}
$$

Because of the form of the integrand, this integral can be difficult to evaluate in closed form, even for simple geometries. It can be evaluated numerically with little difficulty, however. Nevertheless, since no forces are directly involved, the thickness problem is relatively uninteresting in the overall theory.

The Lift Problem. Computing the lift for a given airfoil is of much greater interest. If we ignore the thickness of the airfoil, we only need to specify the camber line, $C(x)$. The solution is given by superposition of a uniform flow and a vortex sheet. Unlike the thickness problem, however, we cannot solve directly. Rather, we observe first that for a vortex sheet, Equation (11.215) tells us

$$
\hat{v}(x, 0^+) + \hat{v}(x, 0^-) \approx -\frac{1}{\pi} \int_0^c \frac{\gamma(\xi)\,d\xi}{x - \xi}
\tag{11.230}
$$

Then, substituting this result into the second of Equations (11.221), we have

$$
2U \cos\alpha \frac{dC}{dx} - 2U \sin\alpha = -\frac{1}{\pi} \int_0^c \frac{\gamma(\xi)\,d\xi}{x - \xi}
\tag{11.231}
$$

This is an integral equation for the vortex sheet strength, $\gamma(\xi)$, that is easier to solve than the more complex Equation (11.200) discussed earlier.

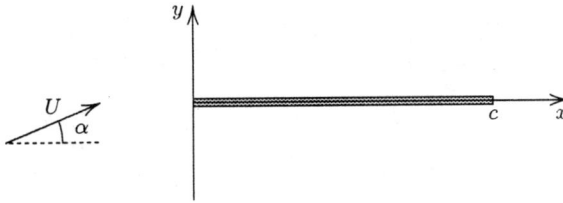

Figure 11.29: *Flat plate at angle of attack.*

As an example, consider a flat plate at angle of attack α as illustrated in Figure 11.29. The camber line is straight and parallel to the x axis, i.e., $C(x) = 0$, so that

$$\frac{dC}{dx} = 0 \quad \Longrightarrow \quad \int_0^c \frac{\gamma(\xi)\,d\xi}{x - \xi} = 2\pi U \sin\alpha \qquad (11.232)$$

This is a classical problem first solved by Poisson in the 1800's. There are two solutions for the vortex sheet, i.e.,

$$\gamma(\xi) = 2U \sin\alpha \sqrt{\frac{c - \xi}{\xi}} \quad \text{and} \quad \gamma(\xi) = 2U \sin\alpha \sqrt{\frac{\xi}{c - \xi}} \qquad (11.233)$$

We must reject the second solution because it yields $\gamma(c) \to \infty$. Then, the first of Equations (11.215) tells us it corresponds to infinite velocity at the trailing edge. By contrast, the first solution has $\gamma(c) = 0$, so that the velocity on top and bottom surfaces are finite and equal, as required by the Kutta condition. The lift then follows from Equation (11.227), wherefore

$$L = 2\rho U^2 \sin\alpha \int_0^c \sqrt{\frac{c - \xi}{\xi}}\,d\xi \qquad (11.234)$$

This integral can evaluated in closed form by changing the integration variable according to $\xi = c \sin^2\phi$. The value of the integral is $\pi c / 2$ so that the lift is

$$L = \pi\rho U^2 c \sin\alpha \qquad (11.235)$$

Recasting this result in dimensionless form, the lift coefficient, C_L, is

$$C_L = \frac{L}{\frac{1}{2}\rho U^2 c} = 2\pi \sin\alpha \qquad (11.236)$$

As it turns out, none of the linearization approximations are necessary to obtain the solution for a flat plate. Thus, this is actually an exact potential-flow solution for a flat plate at angle of attack.

Concluding Remarks. The mathematics of airfoil theory can be challenging, to say the least. A detailed explanation of the theory is available in many advanced texts such as those by Glauert (1948), Milne-Thompson (1960), Robertson (1965) and Valentine (1967). There are, nevertheless, some general results that can be used to estimate the lift of thin airfoils, including effects of camber and thickness. Using complex-variable methods, the lift coefficient for Kutta-Joukowski airfoils [see Panton (1996) or Anderson (1989)] with camber and thickness turns out to be

$$C_L = 2\pi \left(1 + 0.77\frac{T_{max}}{c} \right) \sin(\alpha + \beta), \qquad \beta = \tan^{-1}\left(\frac{2C_{max}}{c} \right) \qquad (11.237)$$

where T_{max} is the maximum thickness, and C_{max} is the maximum distance of the camber line from the chord line. This prediction shows that airfoil thickness has a small effect on lift (because $T_{max} \ll c$). While measurements show that thickness effects on lift are indeed small, the Kutta-Joukowski thickness term is somewhat misleading. This is true because viscous effects tend to change the effective shape of the airfoil due to what is known as the **displacement effect**. We will discuss this phenomenon in Chapter 14. On the one hand, thickness can sometimes increase lift, which is consistent with the Kutta-Joukowski trend. On the other hand, it can sometimes have the opposite effect. Thus, for purposes of estimating lift on an airfoil, the thickness effect is usually ignored, and the lift coefficient is estimated as

$$C_L = 2\pi \sin(\alpha + \beta), \qquad \beta = \tan^{-1}\left(\frac{2C_{max}}{c} \right) \qquad (11.238)$$

Note that this tells us a cambered airfoil has nonzero lift at zero angle of attack. This should come as no great surprise as the airfoil curvature should be expected to alter the pressure variation and hence the lift. The lift predicted by Equation (11.238) is accurate to within about 10% for most airfoils, provided the angle of attack is no larger than about 6°.

11.9 Computational Fluid Dynamics

For flow problems of interest in practical engineering applications, we must deal with irregular boundaries and perhaps even unusual farfield conditions. It should be obvious from some of the problems addressed in this chapter that closed-form analytical solutions are possible only for very simple geometries. Even a cylinder moving at constant velocity near a planar surface, for example, qualifies as a non-simple geometry (see Figure 11.23). To analyze such problems, more general solution methods are clearly needed. As the power of digital computers has increased during the second half of the twentieth century, a whole new branch of the subject known as **Computational Fluid Dynamics**, or **CFD** for short, has evolved to provide the required methods. The purpose of this concluding section is to introduce some of the basic concepts underlying CFD, and to illustrate one of the methods in the context of potential-flow theory.

In the earliest efforts aimed at solving for potential flow past complex geometries, source and vortex sheets were distributed and the attending integral equation solved numerically. The singularities were distributed either inside or on the surface of the body.[4] Using surface source and vortex sheets, A. M. O. Smith created the famous *Douglas-Neumann program* that solved for three-dimensional potential flow about arbitrary geometries. The program and its theoretical foundation are described by Hess and Smith (1966) and Hess (1975). The program has been used extensively for both aerodynamic (e.g., airplanes) and hydrodynamic (e.g., torpedoes) applications. The method was even extended to oscillating flows [Hess and Wilcox (1969)] to simulate the motion of a moored object such as an oil-drilling platform.

---

[4]Although we developed source and vortex sheets for a planar surface with the singularities within a body, they are by no means limited to such geometries.

While this method is very powerful and computationally efficient, it is restricted to inviscid flows. As the power of computers has increased, viscous effects are routinely included. Thus, the use of singular sheets to solve general fluid-flow problems has lost its appeal.

A more general numerical approach to fluid-flow problems is known as the **finite-element method** [cf. Huebner and Thornton (1983), Baker and Pepper (1991)]. In using the finite-element method, the region of interest is subdivided into a finite number of small regions called finite elements. The elements can be triangles or quadrilaterals in two dimensions, and tetrahedra, pentahedra or hexahedra in three dimensions. The dependent variables such as velocity and pressure are represented with algebraic forms involving unknown coefficients, assumed valid in each finite element. The coefficients are then determined by minimizing the error between the finite-element solution and the exact continuum solution (i.e., the solution to the exact differential equations). The finite-element method was originally devised for aircraft structural systems, with the first paper on the method published by Turner et al. (1956). The finite-element method has only recently been applied to fluid-flow problems, mainly because the form of the equations of motion makes it difficult to develop the finite-element solution algorithm [Baker and Pepper (1991)].

By far, the most common numerical technique implemented for fluid mechanical problems is the **finite-difference method**. In this approach, physical space is subdivided into finite-difference cells. In what are known as **structured grids**, the cells are quadrilaterals in two dimensions and hexahedra in three dimensions. In **unstructured grids**, the cells are typically triangles in two dimensions and tetrahedra in three dimensions. When there is a possibility of discontinuous behavior of flow properties such as shock waves in compressible flows, the most common approach is to apply the integral conservation laws to each finite-difference cell. The resulting equations, with appropriate linearization, are then solved iteratively. When there is no possibility of discontinuities, such as in potential-flow problems, we can develop approximations to the differential equations that lead to a set of coupled, linear algebraic equations, which we call the approximating finite-difference equations.

The finite-difference method has advanced by leaps and bounds since the 1960's. The first comprehensive book on these methods was written by Roache (1972). Significant advances in accuracy and efficiency have occurred since 1972. Simultaneously, computers have increased in speed dramatically. The combination has resulted in a powerful new tool for fluid mechanics. Whereas, prior to 1950, the fluid-mechanics researcher relied upon theoretical and experimental methods to aid in design studies, the researcher of today has three tools. Specifically, we now have analytical methods, experimental measurements, and CFD.

Several excellent books have been written since the pioneering effort by Roache. For the interested reader, the introductory text by Anderson (1995) is an excellent starting point. Some of the best advanced texts are those by Anderson, Tannehill, and Pletcher (1984), Ferziger and Perić (1996), Fletcher (1988a), Fletcher (1988b), Minkowycz et al. (1988), and Peyret and Taylor (1983). We will discuss various aspects of Computational Fluid Dynamics here and at the ends of Chapters 12, 13, 14 and 15. In each chapter, a new aspect of CFD is demonstrated based on a straightforward computer program. Source code for the programs is included on the diskette that accompanies this text.

11.9.1 Discretization Approximations

As noted above, we can develop finite-difference approximations for potential-flow problems by direct appeal to the differential equations. To understand how this is done, we must first introduce the concept of a **finite-difference grid**, or **finite-difference mesh**. Consider the square portion of the first quadrant in the xy plane, with $0 \leq x \leq h$ and $0 \leq y \leq h$ (see

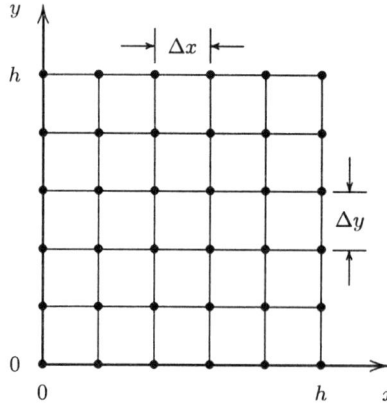

Figure 11.30: *Rectangular finite-difference grid.*

Figure 11.30). We construct a finite-difference grid by passing a series of equidistant horizontal and vertical lines as shown. If we use a total of N lines in each direction, then the spacing is

$$\Delta x = \Delta y = \frac{h}{N-1} \tag{11.239}$$

The intersections of these lines, which we call **grid lines**, define a set of points that are the **grid points**, or **nodes**, in our finite-difference grid. The idea is that, in our proposed finite-difference solution, we will determine flow properties relative to this *finite* set of grid points. The squares whose vertices are the grid points are referred to as **finite-difference cells**. Flow properties, such as velocity and pressure, will be defined at the grid points.

Of course, to tackle complicated flow problems, we will need more general finite-difference grids, preferably conforming to the shape of a complex body. The science of grid generation is a specialized branch of CFD [cf. Thompson et al. (1985), Knupp and Steinberg (1993)] that specializes in handling complex geometries. We confine our attention at this point to this simple grid so we can introduce the basic concepts of CFD without the distraction of complications attending complex grid geometry. We will discuss more general grids briefly in Chapter 14.

Describing flow properties in terms of a finite-difference grid differs from the true continuum solution. That is, the continuum solution, which is the exact solution to the differential equations of motion, provides values throughout the flow domain, *even within the finite-difference cells*. The finite-difference solution provides information only at the finite number, or **discrete number** of grid points, and thus no information about what happens within the cells. If we require such information (e.g., to compute the acceleration at the center of mass of a cell), we must interpolate from the values at the grid points.

For this simple finite-difference grid, the values of x and y are given by

$$\left. \begin{array}{l} x_i = (i-1)\Delta x, \quad 1 \le i \le N \\[2mm] y_j = (j-1)\Delta y, \quad 1 \le j \le N \end{array} \right\} \tag{11.240}$$

To simplify our notation, we use subscripts to denote the value of a given property at a given grid point. Thus, we define

$$\phi_{i,j} \equiv \phi(x_i, y_j) \tag{11.241}$$

In order to approximate a differential equation, we must develop approximations for derivatives based on the values of properties at the discrete grid points. This process is called **discretization**. To understand how we accomplish this end, consider the average value of ϕ along a horizontal grid line, i.e.,

$$\phi_{avg}(x, y) = \frac{1}{2\Delta x} \int_{x-\Delta x}^{x+\Delta x} \phi(\xi, y) \, d\xi \tag{11.242}$$

Differentiating with respect to x, we thus have[5]

$$\frac{\partial \phi_{avg}}{\partial x} = \frac{\phi(x + \Delta x, y) - \phi(x - \Delta x, y)}{2\Delta x} \tag{11.243}$$

This says that we can compute the derivative of the average value of ϕ by simply taking the difference between values at two grid points and dividing by the spacing between the points. Since ϕ_{avg} is defined at (x, y), we also see that our derivative must correspond to the point midway between $x - \Delta x$ and $x + \Delta x$. Thus, in terms of our subscript notation (dropping the subscript avg), we can rewrite Equation (11.243) as

$$\left(\frac{\partial \phi}{\partial x}\right)_{i,j} \approx \frac{\phi_{i+1,j} - \phi_{i-1,j}}{2\Delta x} \tag{11.244}$$

In a similar way, the partial derivative with respect to y is

$$\left(\frac{\partial \phi}{\partial y}\right)_{i,j} \approx \frac{\phi_{i,j+1} - \phi_{i,j-1}}{2\Delta y} \tag{11.245}$$

As an alternative, we can expand the function ϕ in Taylor series at the various grid points as follows.

$$\phi_{i+1,j} = \phi_{i,j} + \left(\frac{\partial \phi}{\partial x}\right)_{i,j} \Delta x + \frac{1}{2}\left(\frac{\partial^2 \phi}{\partial x^2}\right)_{i,j} (\Delta x)^2 + \frac{1}{6}\left(\frac{\partial^3 \phi}{\partial x^3}\right)_{i,j} (\Delta x)^3 + \cdots \tag{11.246}$$

and

$$\phi_{i-1,j} = \phi_{i,j} - \left(\frac{\partial \phi}{\partial x}\right)_{i,j} \Delta x + \frac{1}{2}\left(\frac{\partial^2 \phi}{\partial x^2}\right)_{i,j} (\Delta x)^2 - \frac{1}{6}\left(\frac{\partial^3 \phi}{\partial x^3}\right)_{i,j} (\Delta x)^3 + \cdots \tag{11.247}$$

Subtracting the series for $\phi_{i-1,j}$ from that for $\phi_{i+1,j}$, and dividing by $2\Delta x$, we find

$$\frac{\phi_{i+1,j} - \phi_{i-1,j}}{2\Delta x} = \left(\frac{\partial \phi}{\partial x}\right)_{i,j} + \frac{1}{6}\left(\frac{\partial^3 \phi}{\partial x^3}\right)_{i,j} (\Delta x)^2 + \cdots \tag{11.248}$$

If we truncate the series after the first term, we see that our approximation will be accurate provided Δx is small. Furthermore, the difference between our discrete approximation and the continuum value of $\partial \phi / \partial x$ goes to zero as $(\Delta x)^2$. We refer to this approximation as being **second-order accurate**, since the first truncated term is of second order in Δx. If the first truncated term varied as $(\Delta x)^3$, we would have third-order accuracy, and so forth.

---

[5]Recall from Section 4.7 that Leibnitz's Theorem tells us $\frac{d}{dx} \int_{a(x)}^{b(x)} f(\xi) \, d\xi = f(b)\frac{db}{dx} - f(a)\frac{da}{dx}$.

Since the equation of interest in potential-flow theory is Laplace's equation, we must develop discrete approximations for the second derivatives of ϕ. As can be easily be verified from the Taylor-series expansions,

$$\frac{\phi_{i+1,j} - 2\phi_{i,j} + \phi_{i-1,j}}{(\Delta x)^2} = \left(\frac{\partial^2 \phi}{\partial x^2}\right)_{i,j} + \frac{1}{12}\left(\frac{\partial^4 \phi}{\partial x^4}\right)_{i,j}(\Delta x)^2 + \cdots \qquad (11.249)$$

Therefore, we have the following second-order discrete approximation to the second derivative with respect to x.

$$\left(\frac{\partial^2 \phi}{\partial x^2}\right)_{i,j} \approx \frac{\phi_{i+1,j} - 2\phi_{i,j} + \phi_{i-1,j}}{(\Delta x)^2} \qquad (11.250)$$

In a completely analogous manner, the second derivative with respect to y is

$$\left(\frac{\partial^2 \phi}{\partial y^2}\right)_{i,j} \approx \frac{\phi_{i,j+1} - 2\phi_{i,j} + \phi_{i,j-1}}{(\Delta y)^2} \qquad (11.251)$$

We now have sufficient information to develop an appropriate discretized form of Laplace's equation. Working with the velocity potential, we begin with

$$\frac{\partial^2 \phi}{\partial x^2} + \frac{\partial^2 \phi}{\partial y^2} = 0 \qquad (11.252)$$

Substituting from Equations (11.250) and (11.251), we have the following finite-difference, or **difference equation**.

$$\frac{\phi_{i+1,j} - 2\phi_{i,j} + \phi_{i-1,j}}{(\Delta x)^2} + \frac{\phi_{i,j+1} - 2\phi_{i,j} + \phi_{i,j-1}}{(\Delta y)^2} = 0 \qquad (11.253)$$

For simplicity, we will confine our attention here to the square grid discussed above so that $\Delta x = \Delta y$. After a little rearrangement of terms, we conclude that

$$\phi_{i,j} = \frac{1}{4}\left(\phi_{i+1,j} + \phi_{i-1,j} + \phi_{i,j+1} + \phi_{i,j-1}\right) \qquad (11.254)$$

Therefore, according to Laplace's equation, the value of ϕ at any grid point, (x_i, y_j), is the average of the values at the four closest grid points as illustrated in Figure 11.31. We refer to this interrelation as the **finite-difference molecule** or **finite-difference stencil**.

As a final comment on discretization, Equations (11.244), (11.245), (11.250) and (11.251) are known as second-order **central-difference** approximations. They are used extensively in CFD work.

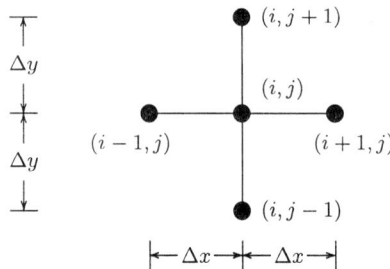

Figure 11.31: *Finite-difference molecule for Laplace's equation.*

11.9.2 Relaxation Methods

There are many methods for solving Equation (11.254), which are described in any good book on finite-difference methods. Often, when we solve fluid-flow problems we use an iterative method. Although this is not necessary in principle for linear equations like Laplace's equation, but only for nonlinear equations like Euler's equation, it is usually the most efficient approach. In an iterative method, we execute a sequence of steps to achieve a solution as follows:

1. Make an initial guess for the initial iterate;

2. Determine the error, or **residual**, that the current iterate implies from substitution into the difference equation;

3. If the residual exceeds a prescribed tolerance, ϵ, adjust the solution and repeat Step 2 — if the residual is less than ϵ, a **converged solution** has been attained.

The most common iteration method for solving Laplace's equation is called **systematic overrelaxation** or **SOR**, for short. To understand how SOR is implemented, it is worthwhile to first discuss the general procedure known as relaxation. When we use relaxation to solve iteratively, we introduce a **relaxation parameter**, λ, such that

$$\phi_{i,j}^{n+1} = (1 - \lambda)\phi_{i,j}^n + \lambda\phi_{i,j}^{\tilde{n}} \qquad (11.255)$$

where superscript n denotes the nth iterate and, in the current context, $\phi_{i,j}^{\tilde{n}}$ is given by substitution into Equation (11.254), i.e.,

$$\phi_{i,j}^{\tilde{n}} = \frac{1}{4}\left(\phi_{i+1,j}^n + \phi_{i-1,j}^n + \phi_{i,j+1}^n + \phi_{i,j-1}^n\right) \qquad (11.256)$$

We, in effect, *predict* the next iterate by direct substitution into Laplace's equation, and then use Equation (11.255) to *correct* our predicted value in some sense.

If we use a value for λ less than 1, the procedure is called **underrelaxation**. We often do this when the equation proves to be a bit difficult to solve, as evidenced by growing oscillations as the iterative procedure progresses. The example of the next subsection uses underrelaxation because of the presence of nearly singular behavior at one point in the flowfield. The underrelaxation stabilizes the computation.

When the relaxation parameter is greater than 1, we are using **overrelaxation**. To see the connection between λ and the residual, substitute Equation (11.256) into Equation (11.255). Grouping all of the terms proportional to λ together, we find

$$\phi_{i,j}^{n+1} = \phi_{i,j}^n + \lambda\underbrace{\left[\frac{1}{4}\left(\phi_{i+1,j}^n + \phi_{i-1,j}^n + \phi_{i,j+1}^n + \phi_{i,j-1}^n\right) - \phi_{i,j}^n\right]}_{\text{Residual}} \qquad (11.257)$$

Using SOR accelerates the approach to convergence. The optimum value of λ is problem dependent, and is usually determined by trial and error. For Laplace's equation, optimum values of λ between 1.5 and 1.7 are common. Using values for $\lambda > 2$ always results in a failure to converge.

To implement SOR, we begin with our initial guess, or iterate, so that $\phi_{i,j}^n$ is known at all points. We than sweep through the entire solution domain for all values of i and j, using Equation (11.257) to determine the next iterate. As discussed above, the iterative procedure continues until the residual is reduced below some prescribed tolerance.

In generating the solution, we must, of course, enforce appropriate boundary conditions. For the velocity potential, we require the velocity normal to a solid boundary to vanish, i.e., $\partial \phi / \partial n = 0$, where n is the direction normal to the surface. This kind of boundary condition is of the **Neumann type**. As with the differential equation, it must be recast in finite-difference form. For the streamfunction, we can simply specify its value at a solid boundary. This kind of boundary condition is of the **Dirichlet type**. Clearly, no discretization approximations are needed for Dirichlet-type boundary conditions.

11.9.3   Flow Past a Vertical Plate

To demonstrate the utility of numerical solutions to potential-flow problems, we now consider flow past a plate aligned at $90°$ to the freestream direction. Furthermore, we will include end walls as shown in Figure 11.32. As illustrated, the plate is $2a$ in length, while the end walls are a distance $10a$ from each other. Because the plate is centered between the end walls, the flow is symmetric about the x axis.

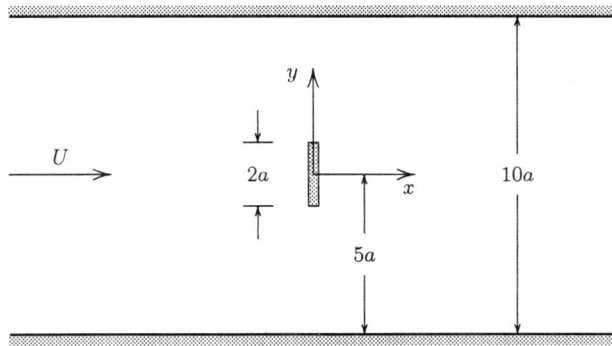

Figure 11.32: *Potential flow past a thin plate aligned normal to the flow with end walls.*

Although an exact, closed-form solution to this problem is unavailable, there is an analytical solution when the end walls are absent. This is a common occurrence in design studies. That is, we often have a theory for an object moving in a domain whose boundaries are very far from the object relative to the object's size. A missile or an airplane flying several miles above the Earth are examples. Thus, for practical purposes, the domain is effectively infinite in extent. By contrast, wind tunnel tests include end walls that are located at a finite distance from the object. It is useful to know what effect the end walls have on the solution, relative to flow in an infinite domain. Using a numerical solution, we can easily include the end walls and, hence, determine their effect. We illustrate this process in the present application by comparing our numerical solution with the exact result for an unbounded fluid.

In all of the discussion to follow, we will restrict the flow domain to the region above the symmetry plane at $y = 0$. In the absence of end walls, the potential-flow solution for flow past the plate is [see Hildebrand (1976)]:

$$\psi(x, y) = U \sqrt{\frac{1}{2} \left[\sqrt{\left(x^2 + a^2 - y^2\right)^2 + 4x^2 y^2} - \left(x^2 + a^2 - y^2\right) \right]} \qquad (11.258)$$

To compare numerical and analytical solutions, we will focus on the velocity at the plate surface and along the y axis above the plate. Differentiating the streamfunction to determine the velocity components shows that the velocities on the upstream ($x = 0^-$) and downstream ($x = 0^+$) sides of the plate are

$$u(0^\pm, y) = 0, \qquad v(0^\pm, y) = \pm \frac{Uy}{\sqrt{a^2 - y^2}} \qquad (0 \le y < a) \qquad (11.259)$$

Also, the velocity along the y axis above the plate is

$$u(0, y) = \frac{Uy}{\sqrt{y^2 - a^2}}, \qquad v(0, y) = 0 \qquad (y > a) \qquad (11.260)$$

Because the analytical solution shows that there is a singularity at the top of the plate, $(x, y) = (0, a)$, we should anticipate numerical difficulties at this point, even with end walls. We can mitigate the effects of this singularity to some extent by considering a thin plate of finite width. Thus, we let the plate have a thickness of $0.2a$. Also, we select upstream and downstream boundaries that lie $5a$ from the plate. Inspection of the exact solution without end walls suggests that this is sufficiently distant that any upstream disturbances due to the presence of the plate are small. We can thus assume the velocity is uniform at the inlet plane, $x = -5a$. Figure 11.33 is a schematic of the finite-difference grid used. As shown, the points are equally spaced, and the region within the plate is external to the computational domain

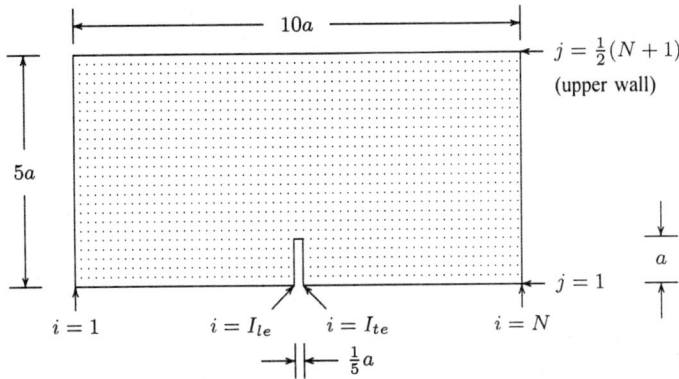

Figure 11.33: *Finite-difference grid with 51 x 26 points* ($N = 51$); *each dot represents a grid point.*

This problem is most conveniently solved in terms of the streamfunction. At the inlet, we impose uniform flow so that $\psi(-5a, y) = Uy$. On the upper wall, $v = 0$, so that the streamfunction is constant and equal to Ua. On the lower wall and plate surface, we set $\psi = 0$. Finally, we assume the outflow is everywhere parallel to the walls so that $\partial\psi/\partial x = 0$. We implement the latter condition by saying that

$$\psi(5a, y) = \psi(5a - \Delta x, y) \qquad \Longrightarrow \qquad \psi_{N,j} = \psi_{N-1,j} \qquad (11.261)$$

As mentioned in the last subsection, this computation becomes unstable if we use over-relaxation. That is, computed flow properties grow with each iteration and ultimately grow without bound. The difficulty stems from the relatively large velocities at the top of the plate. To stabilize the computation, we must use underrelaxation. The solution converges most rapidly for $\lambda = 0.9$.

Figure 11.34 shows the computed streamlines, velocity variation on the front side of the plate $[v(0^+, y)]$ and velocity on the centerline above the plate $[u(0, y)]$. The program used to do the computations, **POTFLOW**, is described in Appendix E. The source code is included on the diskette that accompanies this book. As shown, the effect of the end walls is to increase the horizontal velocity above the plate and to decrease the vertical velocity along its surface. The velocity above must increase because of the blockage offered by the plate. In an unbounded flow, the streamlines are spread out more so that $u = \partial \psi / \partial y$ is smaller. Equivalently, we can say that the constraining effect of the upper wall forces the same mass flow through a smaller area, which causes the velocity to increase. The vertical velocity along the plate is smaller than in the theoretical solution, mainly because the finite plate width removes the singularity.

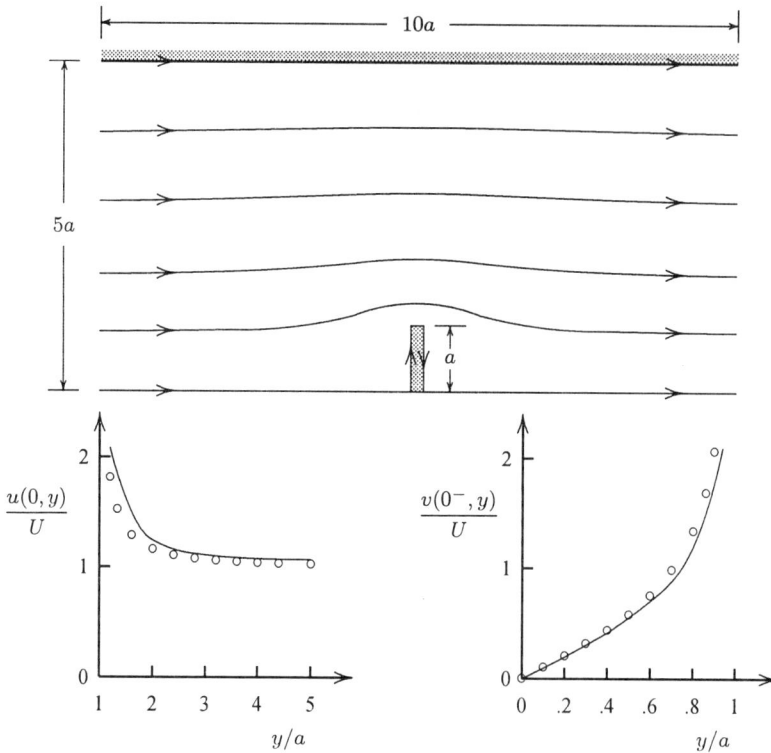

Figure 11.34: *Potential flow past a thin plate aligned normal to the freestream flow direction; —— numerical solution with end wall; ∘ exact solution for an unbounded fluid.*

11.9.4   Solution Convergence and Grid Sensitivity

All of the comments in the preceding subsection assume the numerical solution is accurate. An important question to pose is, since we have no exact solution to compare to, how can we know what the error bounds might be? This leads us to the extremely important issue of solution accuracy in CFD work. Regardless of the application, there is a need for control of numerical accuracy in CFD [Roache (1990)]. This need is just as critical in CFD work as it is in experiments where the experimenter is expected to provide estimates for the accuracy of his or her measurements. All CFD texts of any value stress this need.

One key issue determining numerical accuracy of a numerical computation is **iteration convergence**. Most numerical methods used in CFD applications require many iterations to converge. The iteration convergence error is defined as the difference between the current iterate and the exact solution to the *difference equations*. Often, the difference between successive iterates is used as a measure of the error in the converged solution, although this in itself is inadequate. A small relaxation parameter, for example, can always give a false indication of convergence [Roache (1972), Anderson et al. (1984)]. Whatever the algorithm is, we should always be careful to check that a converged solution has been obtained. This can be done by trying a stricter than usual convergence criterion, and demonstrating that there is a negligible effect on the solution. Monitoring the residual is a far more dependable indicator of solution convergence. For the vertical-plate computation described here, reducing the residual to 10^{-6} is sufficient to guarantee iteration convergence.

A second key issue is **grid convergence** or **grid insensitivity**. Because of the finite size of finite-difference cells, discretization errors exist that represent the difference between the solution to the difference equations and the *exact (continuum) solution to the differential equations*. It is important to know the magnitude of these discretization errors and to insure that a fine enough grid has been used to reduce the error to an acceptable level.

As with iteration convergence, all CFD work should demonstrate grid convergence. The most common way to demonstrate grid convergence is to repeat a computation on a grid with twice as many grid points, and compare the two solutions. If computer resources are unavailable to facilitate a grid doubling, a grid halving is also appropriate, although the error bounds will not be as sharp.

Using results for two different grids, techniques such as **Richardson extrapolation** [see Roache (1972)] can be used to determine discretization error. This method is very simple to implement, and should be used whenever possible. For a second-order accurate method with central differences, Richardson extrapolation assumes the error, $E_h \equiv \phi_{exact} - \phi_h$, where ϕ_h denotes the solution when the grid-point spacing is h, can be expanded as a Taylor series in h, wherefore

$$E_h = e_2 h^2 + e_4 h^4 + e_6 h^6 + \cdots \qquad (11.262)$$

By hypothesis, the e_i are, at worst, functions of the coordinates, but are nevertheless independent of h. Now, if we halve the number of grid points so that h is doubled, the error is given by

$$E_{2h} = 4 e_2 h^2 + 16 e_4 h^4 + 64 e_6 h^6 + \cdots \qquad (11.263)$$

For small values of h, we can drop all but the leading terms, whence the discretization error is given by

$$E_h \approx \frac{1}{3}(\phi_h - \phi_{2h}) \qquad (11.264)$$

In applying Equation (11.264), care must be taken to use numerical-solution values from the same locations. Remember that part of the foundation for the method rests on the assumption that the error may vary with the coordinates, but depends only upon grid-point spacing, h, at each grid point. So, if we have twice as many grid points in our fine grid, we use every other point and match up coordinates with the coarser grid.

The computations for the vertical plate have been done on grids with $N = 51$, $N = 101$ and $N = 201$. For example, consider the solution at $x = 0$ and $y = 2a$. The values of the dimensionless streamwise velocity, $u(0, 2a)/U$ for the 51-point, 101-point and 201-point computations are 1.247879, 1.235551 and 1.230664, respectively. Hence, the estimated error for the 101-point mesh, E_{101}, is determined from the 51-point and 101-point values to yield

$$E_{101} = \frac{1}{3}(1.235551 - 1.247879) = -4.1093 \cdot 10^{-3} \qquad (11.265)$$

Similarly, by comparing the 101-point and 201-point values, we find

$$E_{201} = \frac{1}{3}(1.230664 - 1.235551) = -1.6290 \cdot 10^{-3} \qquad (11.266)$$

Table 11.5 summarizes results of the computations at several points, including the estimated error for the 201-point computation. As shown, for the points listed, the error in the 201-point computation nowhere exceeds 0.132%.

Table 11.5: *Computed Velocities for Three Grids*

	$N = 51$	$N = 101$	$N = 201$	Error$_{N=201}$
$u(0, 2a)/U$	1.247879	1.235551	1.230664	-0.001629 (.132%)
$u(0, 3a)/U$	1.113288	1.108373	1.106067	-0.000769 (.070%)
$u(0, 4a)/U$	1.078557	1.075304	1.072984	-0.000773 (.072%)
$v(0^-, .2a)/U$	0.195446	0.194529	0.194980	0.000150 (.077%)
$v(0^-, .4a)/U$	0.415418	0.413557	0.414237	0.000227 (.055%)
$v(0^-, .6a)/U$	0.700689	0.701896	0.701834	-0.000021 (.003%)
$v(0^-, .8a)/U$	1.146758	1.202046	1.204958	0.000971 (.081%)

Problems

11.1 Determine the dimensions of the following quantities.

(a) The velocity potential, ϕ, and the streamfunction, ψ.

(b) The strength of a source, Q.

(c) The strength of a potential vortex, Γ.

(d) The strength of a doublet, $\mathcal{D}$.

11.2 Compute the circulation of a source.

11.3 Compute the mass flux from a potential vortex.

11.4 Compute the mass flux from a doublet.

11.5 Compute the circulation of a doublet.

11.6 We can construct another fundamental solution to Laplace's equation, known as the dipole, using two potential vortices as indicated below. Assume the circulation, Γ, is such that the dipole strength, $\mathcal{D}_p$, is given by

$$\mathcal{D}_p = \lim_{a \to 0} 2a\Gamma$$

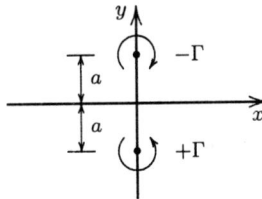

Problem 11.6

(a) Determine the velocity potential.

(b) Determine the streamfunction.

(c) Compare your results with those of a doublet.

11.7 Show that for an incompressible, irrotational flow, the equation of motion

$$\nabla \left[\frac{\partial \phi}{\partial t} + \frac{p}{\rho} + \frac{1}{2} \left(u^2 + v^2 \right) \right]$$

is identical to the Euler equation in two dimensions. **HINT:** This problem is most conveniently done in component form.

11.8 For steady, compressible, two-dimensional flows, we can define a "streamfunction" according to

$$u = \frac{1}{\rho} \frac{\partial \tilde{\psi}}{\partial y}, \qquad v = -\frac{1}{\rho} \frac{\partial \tilde{\psi}}{\partial x}$$

where ρ is the (variable) density.

(a) What are the dimensions of $\tilde{\psi}$?

(b) Verify that conservation of mass is satisfied.

(c) If the flow is irrotational, verify that

$$\nabla^2 \vec{\psi} = \mathbf{u} \times \nabla\rho \qquad \text{where} \qquad \vec{\psi} \equiv \tilde{\psi}\mathbf{k}$$

11.9 The velocity potential for a three-dimensional flow can be written as

$$\phi(x, y, z) = X(x)Y(y)z$$

where $X(x)$ and $Y(y)$ are functions to be determined.

(a) If $Y(y)$ is dimensionless, what are the dimensions of $X(x)$?

(b) What constraints must be imposed on $X(x)$ and/or $Y(y)$ to insure that the flow is irrotational?

(c) If $X(x) \to 0$ as $x \to \infty$, what is the most general form of the functions $X(x)$ and $Y(y)$?

11.10 Verify by direct substitution that the velocity potential and streamfunction for a doublet, Equations (11.89) and (11.92), satisfy Laplace's equation.

11.11 *Complex variables* are especially useful in potential-flow theory. We define the *complex potential*, $F(z)$, as a function of the complex variable, z, according to

$$F(z) = \phi + i\psi, \qquad z = x + iy = re^{i\theta}$$

where $i = \sqrt{-1}$ is the imaginary number. The quantities ϕ and ψ are the standard velocity potential and streamfunction, respectively. Determine the complex potential as a function of z for uniform flow, a point source, a potential vortex and a doublet.

11.12 *Complex variables* are especially useful in potential-flow theory. We define the *complex potential*, $F(z)$, as a function of the complex variable, z, according to

$$F(z) = \phi + i\psi, \qquad z = x + iy = re^{i\theta}$$

where $i = \sqrt{-1}$ is the imaginary number. The quantities ϕ and ψ are the standard velocity potential and streamfunction, respectively.

(a) Verify that the complex velocity is $w(z) \equiv dF/dz = u - iv$. **HINT:** Assume $F(z)$ is an *analytic function* so that, according to complex-variable theory, the derivative at any point in the complex plane is independent of the direction in which we differentiate.

(b) Letting w^* denote the *complex conjugate* of w, rewrite Bernoulli's equation in terms of the complex velocity. Assume the flow is steady and that $|\mathbf{u}| \to U$ as $r \to \infty$.

11.13 *Complex variables* are especially useful in potential-flow theory. We define the *complex potential*, $F(z)$, as a function of the complex variable, z, according to

$$F(z) = \phi + i\psi, \qquad z = x + iy = re^{i\theta}$$

where $i = \sqrt{-1}$ is the imaginary number. The quantities ϕ and ψ are the standard velocity potential and streamfunction, respectively. We want to determine the complex potential for flow past a wedge. First, beginning with the streamfunction, $\psi(r, \theta) = Ar^n \sin n\theta$, where A is a constant, find the conventional velocity potential, $\phi(r, \theta)$. Now, determine the complex potential as a function of z.

11.14 *Complex variables* are especially useful in potential-flow theory. We define the *complex potential*, $F(z)$, as a function of the complex variable, z, according to

$$F(z) = \phi + i\psi, \qquad z = x + iy = re^{i\theta}$$

where $i = \sqrt{-1}$ is the imaginary number. The quantities ϕ and ψ are the standard velocity potential and streamfunction, respectively.

(a) If the complex potential is $F(z) = U[z + R^2/z]$, where R is a constant, determine $\phi(r, \theta)$ and $\psi(r, \theta)$. What flow does this correspond to? **HINT:** Do your work in cylindrical coordinates.

(b) If we add a point vortex of strength Γ, how does $F(z)$ change?

11.15 *Complex variables* are especially useful in potential-flow theory. We define the *complex potential*, $F(z)$, as a function of the complex variable, z, according to

$$F(z) = \phi + i\psi, \qquad z = x + iy = re^{i\theta}$$

where $i = \sqrt{-1}$ is the imaginary number. The quantities ϕ and ψ are the standard velocity potential and streamfunction, respectively. If the complex potential is $F(z) = Ue^{-i\alpha}z$, where U and α are constants, determine $\phi(x, y)$ and $\psi(x, y)$. What flow does this correspond to?

11.16 Determine the streamlines and geometry for wedge flow with $n = \frac{1}{2}$.

11.17 Investigate the streamfunction $\psi(x, y) = K(x^2 - y^2)$, where K is constant. Plot the streamlines in the full xy plane, find any stagnation points, and interpret what the flow could represent.

11.18 Consider the streamfunction $\psi(x, y) = x^2 + 4y^2$. Plot the streamlines in the full xy plane, and find any stagnation points. Be sure to indicate flow direction on the streamlines. Is this flow irrotational?

11.19 Consider the streamfunction $\psi(x, y) = 3x^2y - y^3$. Plot the streamlines in the full xy plane, and find any stagnation points. Be sure to indicate flow direction on the streamlines. Is this flow irrotational?

11.20 Consider the streamfunction $\psi(x, y) = x^2y - 2y^2$. Plot the streamlines in the first quadrant for $\psi = $ -2, 0, 2, and find any stagnation points. Be sure to indicate flow direction on the streamlines. Is this flow irrotational?

11.21 Determine the velocity potential and streamfunction for a uniform flow inclined at an angle α to the x axis, i.e., for $\mathbf{u} = U(\mathbf{i}\cos\alpha + \mathbf{j}\sin\alpha)$.

11.22 Determine the velocity potential for a flow whose velocity vector is

$$\mathbf{u} = U\left[\mathbf{i} + \frac{y}{a}\mathbf{j}\right]$$

where U and a are constant velocity and length scales, respectively. Is this flow incompressible?

11.23 Determine the velocity potential for a flow whose (three-dimensional) velocity vector is

$$\mathbf{u} = \frac{U}{H}(x\mathbf{i} + y\mathbf{j} - 2z\mathbf{k})$$

where U and H are constant velocity and length scales, respectively. Is this flow incompressible?

11.24 Determine the velocity potential for an unsteady, three-dimensional flow whose velocity vector is

$$\mathbf{u} = Ue^{-\kappa x}\cos\omega t\,(\kappa z\cos\kappa y\,\mathbf{i} + \kappa z\sin\kappa y\,\mathbf{j} - \cos\kappa y\,\mathbf{k})$$

where U, ω and κ are constant velocity, frequency and wavenumber (dimensions $1/L$), respectively.

11.25 Consider stagnation-point flow for which the Cartesian velocity components are given by $u = Ax$ and $v = -Ay$ where A is a constant.

(a) Determine the streamfunction, ψ.

(b) Determine the velocity potential, ϕ.

(c) Now add a source of strength Q. What must Q be in order to move the stagnation point to $x = 0$, $y = y_o$?

(d) In general, ψ and ϕ are unique to within an additive constant. What must the constant for ψ be if the dividing streamline passes through the displaced stagnation point and is defined by $\psi = 0$?

11.26 Compute the velocity potential for flow past a wedge beginning with $\psi = r^n \sin n\theta$.

11.27 Compute the velocity potential for flow in the vicinity of a stagnation point beginning with the streamfunction, $\psi = 2\psi_o xy$. Plot three streamlines and equipotential lines in each of the upper quadrants.

11.28 The streamfunction for a two-dimensional potential flow is

$$\psi = Ar^4 \sin 4\theta$$

where A is a constant. The quantities r and θ are cylindrical coordinates.

(a) Determine the velocity components u_r and u_θ.

(b) Determine the corresponding velocity potential.

(c) Locate any stagnation points. Convert to Cartesian coordinates for the sake of clarity.

(d) The body shape is given by $\psi = 0$. If we confine our interest to the first quadrant excluding the y axis, i.e., $0 \le \theta < \pi/2$, what is the body shape?

(e) Sketch a few streamlines. Include any stagnation points, body contours and flow direction along the streamlines. **HINT:** Find the flow direction on the body surface.

11.29 The streamfunction for a two-dimensional potential flow is $\psi = -Ar^2 \cos 2\theta$ where A is a constant. The quantities r and θ are cylindrical coordinates.

(a) Determine the velocity components u_r and u_θ.

(b) Determine the corresponding velocity potential.

(c) Locate any stagnation points. Convert your answer to Cartesian coordinates for the sake of clarity.

(d) The body shape is given by $\psi = 0$. If we confine our interest to the first three quadrants, i.e., $0 \le \theta < 3\pi/2$, what is the body shape? Note that the body contours of interest lie in the first ($0 \le \theta \le \pi/2$) and third ($\pi \le \theta \le 3\pi/2$) quadrants, and that there is a dividing streamline in the second ($\pi/2 \le \theta \le \pi$).

(e) Sketch a few streamlines. Include any stagnation points, body contours and flow direction along the streamlines. **HINT:** Find the flow direction on the body surface and on the dividing streamline.

11.30 The streamfunction for a potential flow is of the form

$$\psi(x, y) = f(x) + g(y)$$

What is the most general form of the functions $f(x)$ and $g(y)$?

11.31 Determine the pressure induced by a point sink in an otherwise motionless fluid, assuming $p \to p_\infty$ as $r \to \infty$. If $p_\infty = 101$ kPa, $\rho = 998$ kg/m^3 and the sink strength is $Q = -0.9$ m^3/sec, what is the minimum value of r for which the solution is physically meaningful?

11.32 Determine the pressure induced by a doublet in an otherwise motionless fluid, assuming $p \to p_\infty$ as $r \to \infty$. If $p_\infty = 14.7$ psi, $\rho = 1.94$ slug/ft^3 and the sink strength is $\mathcal{D} = 250$ ft^3/sec, what is the minimum value of r for which the solution is physically meaningful? **NOTE:** This doublet strength would be appropriate for 40 ft/sec flow past a cylinder of radius 1 ft.

11.33 Very far from a two-dimensional lifting body, we can approximate the velocity potential as

$$\phi \approx Ur \cos\theta + \frac{\Gamma}{2\pi}\theta + \frac{\mathcal{D}}{2\pi r}\cos\theta \qquad (r \to \infty)$$

where U is freestream velocity, Γ is circulation and $\mathcal{D}$ is an effective doublet strength. Verify that as $r \to \infty$, the pressure coefficient is independent of doublet strength and is given by

$$C_p \approx \frac{C_L}{\pi}\left(\frac{R}{r}\right)\sin\theta$$

where $C_L \equiv \Gamma/(UR)$ is the *lift coefficient* and R is a length characteristic of the body size.

11.34 A simplified model of a hurricane approximates the flow as a rigid-body rotation in the inner core and a *potential vortex* outside of the core (cf. Section 4.3). That is, the circumferential velocity, u_θ, is

$$
u_\theta \approx \begin{cases} \dfrac{\Gamma r}{2\pi R^2}, & r \le R \\[2mm] \dfrac{\Gamma}{2\pi r}, & r \ge R \end{cases}
$$

where r is radial distance, R is core radius, and Γ is the circulation.

(a) Determine the velocity potential.

(b) Determine the pressure for $p \to p_\infty$ as $r \to \infty$.

(c) What is the minimum pressure if the air density is $\rho = .00234$ slug/ft^3, $p_\infty = 14.7$ psi and the peak velocity is 200 mph.

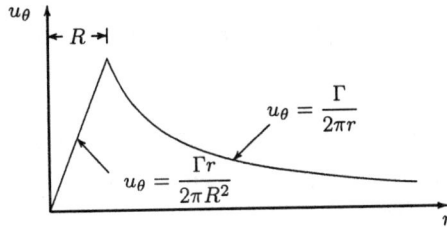

Problem 11.34

11.35 As an airplane taxis down the runway in preparation for takeoff, the velocity on the fuselage is

$$
|\mathbf{u}| = U \cdot \frac{s}{L} \cdot \frac{t}{\tau}
$$

where s is arc length and t is time. The quantities U, L and τ are reference velocity, length and time scales related by $L = U\tau$. In this problem, we examine the error in using a Pitot-static tube to infer velocity if effects of the acceleration (unsteadiness) are ignored (this problem had to be corrected to get the DC-10 certified by the FAA).

(a) Neglecting body forces and assuming the flow is incompressible, compute the pressure coefficient, C_p, defined by $C_p \equiv (p - p_s)/(\frac{1}{2}\rho U^2)$ where p_s is the pressure at the stagnation point ($s = 0$). Note that the velocity potential near the stagnation point is $\phi = \frac{1}{2}U(s^2/L)(t/\tau)$.

(b) Starting with the value for C_p computed in Part (a), use the steady-flow version of Bernoulli's Equation to infer the velocity at $s = L$ as a function of t/τ.

(c) Compare your result of Part (b) with the actual velocity given above by making a graph of the percentage error for $0.5 < t/\tau < 4$.

11.36 We can approximate a tornado by combining a potential vortex of strength Γ with a point sink of strength $-Q$, both centered at the origin. If the pressure is given by

$$
p = p_\infty - \frac{\rho Q^2}{2\pi r^2}
$$

where p_∞ is the farfield pressure, what must Γ be? Also, what is the angle between the velocity vector and the radial direction?

11.37 Determine the streamlines for a *spiral vortex*, which is the superposition of a point source of strength Q and a potential vortex of strength Γ. Plot a streamline for $\Gamma = 2Q$.

11.38 Find the stagnation-point location for a *spiral vortex* in a uniform freestream of velocity U. A spiral vortex is the superposition of a point source of strength Q and a potential vortex of strength Γ.

11.39 A uniform stream of velocity $\mathbf{u} = U\mathbf{i}$ flows past sources of strength Q at $x = \pm a, y = 0$ and a sink of strength $-2Q$ at $x = y = 0$. What is the thickness of the resulting closed body at $y = 0$?

11.40 Consider superposition of a sink of strength $Q = -Uh$ and a uniform flow of velocity U from left to right.

   (a) What is the complete streamfunction for this flow? You may choose any extra constant in the streamfunction to be zero.

   (b) Locate any stagnation points that might be present.

   (c) Noting that $y = r \sin \theta$, show that the body shape ($\psi = 0$) is such that $y \to \pm h/2$ as $\theta \to \pm \pi$.

   (d) Make a qualitative sketch of what the body and streamlines look like. Be sure to include some streamlines both inside and outside the body.

11.41 Consider superposition of a source of strength $Q = Uh$ and a uniform flow of velocity U from left to right.

   (a) What is the complete streamfunction for this flow? Include an appropriate constant in the streamfunction to insure that the negative x axis is a dividing streamline along which $\psi = 0$.

   (b) Locate any stagnation points that might be present.

   (c) Noting that $y = r \sin \theta$, show that the body shape ($\psi = 0$) is such that $y \to \pm h/2$ as $x \to \infty$.

   (d) Make a qualitative sketch of what the body and streamlines look like. Be sure to include some streamlines both inside and outside the body.

11.42 A uniform stream flows downward so that $u = 0$ and $v = -V$ where V is constant.

   (a) What are the streamfunction and velocity potential for this flow in Cartesian coordinates?

   (b) What are the streamfunction and velocity potential for this flow in cylindrical coordinates?

   (c) Now add a source of strength Q located at the origin. What are the streamfunction and velocity potential in cylindrical coordinates?

   (d) For the streamfunction of Part (c), locate any stagnation points in the flow. Express your answer in Cartesian coordinates. (Note that most of the work is most conveniently done in cylindrical coordinates — just convert when you're done.)

   (e) Now add a sink of strength Q located at $x = a$, $y = -a$ to the flow of Part (c). What are the streamfunction and velocity potential in Cartesian coordinates?

   (f) Sketch what you think the flow of Part (e) looks like, i.e., draw streamlines and indicate flow direction on the streamlines. Don't worry about precise location of stagnation point(s), etc., just indicate what the body shape might look like and give a rough idea of where the streamlines and stagnation point(s) might be.

11.43 Consider flow past a wall with a sink of strength $-Q$ at the origin. At infinity, the flow is parallel to the x axis with uniform velocity U. The freestream pressure is p_∞ and the density is ρ.

   (a) Determine the streamfunction and velocity potential for this flow. Be sure to verify that $y = 0$ is a streamline to show that you have solved the right problem. **NOTE:** $y = 0$ may be a different streamline for $x > 0$ and $x < 0$.

   (b) Determine the pressure, p, along the wall.

   (c) Find the value of x at which the pressure along the wall is a maximum.

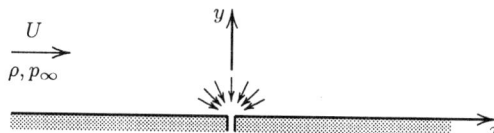

Problem 11.43

11.44 Consider flow past a wall with a source of strength Q at the origin. At infinity, the flow is parallel to the x axis with uniform velocity U. The freestream pressure is p_∞ and the density is ρ.

(a) Determine the streamfunction and velocity potential for this flow. Be sure to verify that $y = 0$ is a streamline to show that you have solved the right problem. **NOTE:** $y = 0$ may be a different streamline for $x > 0$ and $x < 0$.

(b) Determine the pressure, p, along the wall.

(c) Find the value of x at which the pressure along the wall is a maximum.

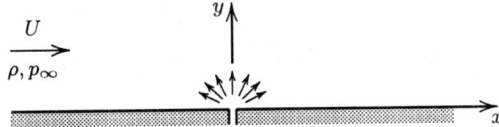

Problem 11.44

11.45 Far from a non-lifting body that is emitting fluid (e.g., from a jet engine), we can approximate the velocity potential as

$$\phi \approx Ur\cos\theta + \frac{Q}{2\pi}\ell nr + \frac{D}{2\pi r}\cos\theta \qquad (r \to \infty)$$

where U is freestream velocity, Q is the mass-injection rate and D is an effective doublet strength.

(a) Verify that

$$p - p_\infty \approx -\frac{\rho QU}{2\pi r}\cos\theta \qquad \text{as} \qquad r \to \infty$$

(b) Using a circular control volume of radius r, show that

$$\oiint_S \rho \mathbf{u}(\mathbf{u} \cdot \mathbf{n})dS = \frac{3}{2}\rho QU\,\mathbf{i} \qquad \text{and} \qquad \oiint_S (p - p_\infty)\mathbf{n}\,dS = -\frac{1}{2}\rho QU\,\mathbf{i}$$

(c) Use the results of Part (b) to determine the force on the body.

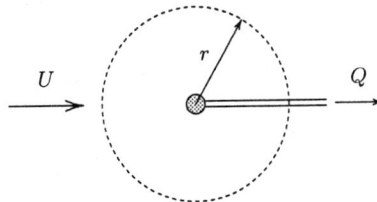

Problem 11.45

11.46 The wind is blowing with velocity U and freestream pressure p_∞ past a quonset hut, which is a half cylinder of radius R and length L (out of the page). The internal pressure is p_i. Using potential-flow theory, derive an expression for the vertical force on the hut due to the pressure difference on the hut.

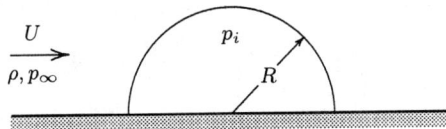

Problems 11.46, 11.47

11.47 The wind is blowing with velocity $U = 50$ mph, freestream pressure $p_\infty = 14.7$ psi and density $\rho = .00234$ slug/ft^3 past a quonset hut, which is a half cylinder of radius $R = 10$ ft and length $L = 40$ ft (out of the page). The internal pressure, p_i, is regulated so that there is zero net vertical force on the hut. Using potential-flow theory, determine the internal pressure.

11.48 An arctic hut in the shape of a half-circular cylinder has radius R. A wind of velocity U is blowing and creates a substantial aerodynamic force on the hut. This force is due to the difference between the external pressure and the pressure inside the hut, p_i. Making a thin slit at an angle θ_o from the ground level causes the net force on the hut to vanish. Determine $\sin \theta_o$. **NOTE:** Assume the opening is very small compared to R so that it has a negligible effect on the flow about the hut. Assume incompressible potential flow and note that the pressure inside the hut, p_i, depends upon θ_o.

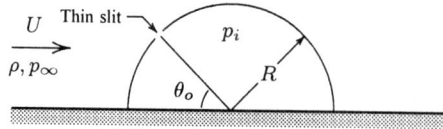

Problem 11.48

11.49 Consider flow past a semi-circular body of radius R with a blunted base as shown below. As a first approximation, assume the base pressure, p_b, is equal to the pressure at the shoulder and that the flow on the circular part of the body follows the potential-flow solution for flow past a circle.

(a) Sketch the pressure coefficient, C_p, as a function of θ for the assumed flow, with $0 \le \theta \le 2\pi$. The angle θ is zero on the positive x axis. Explain the basis of your sketch, e.g., maxima and minima, etc.

(b) Compute the drag on the body. (It is not zero because this is not an exact solution.)

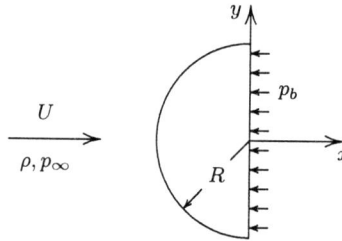

Problem 11.49

11.50 Consider flow past a circular body of radius R with a blunted base as shown below. As a first approximation, assume the base pressure, p_b, is constant across the base and that the flow on the circular part of the body follows potential-flow theory.

(a) Sketch the pressure coefficient, C_p, for the assumed flow. The angle θ is zero on the positive x axis. Explain the basis of your sketch, e.g., maxima and minima, etc.

(b) Compute the drag on the body. (It is not zero because this is not an exact solution.)

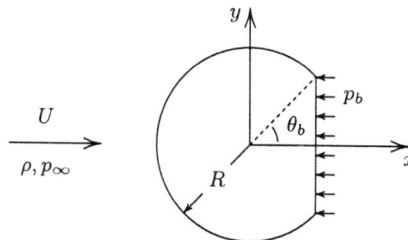

Problem 11.50

11.51 Solve for potential flow past a cylinder using separation of variables. Assume the velocity potential can be expressed as $\phi(r, \theta) = \mathcal{R}(r)\Theta(\theta)$. Substitute into Laplace's equation and the boundary conditions to solve for $\mathcal{R}(r)$ and $\Theta(\theta)$.

11.52 An interesting way to generate an exact solution for flow past a two-dimensional bump is to use a streamline for flow past a cylinder as shown. The desired height of the bump is $R/2$, where R is the cylinder radius. Determine the distance h and the maximum velocity on the bump, U_{max}.

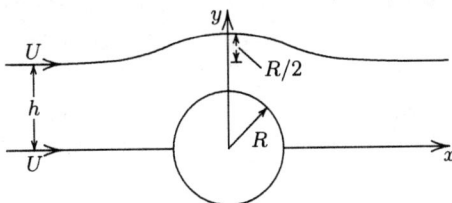

Problem 11.52

11.53 The *Flettner-Rotor* ship was designed with two rotating cylinders of height H acting as sails. The cylinders have diameter D and rotate with angular velocity Ω. The ship moves with velocity $\mathbf{U}_s = U\,\mathbf{i}$ and the wind velocity is $\mathbf{U}_w = -V\,\mathbf{j}$.

(a) Using potential-flow theory, estimate the force on the ship as a function of Ω, D, H, U, V and the density of air, ρ.

(b) Determine the thrust for $\Omega = 750$ rpm, $D = 2.75$ m, $H = 15$ m, $U = 3$ knots, $V = 12$ knots and $\rho = 1.2$ kg/m^3.

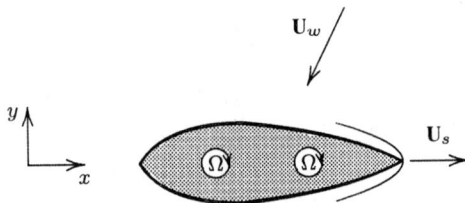

Problems 11.53, 11.54

11.54 The *Flettner-Rotor* ship was designed with two rotating cylinders of height H acting as sails. The cylinders have diameter D and rotate with angular velocity Ω. The ship moves with velocity $\mathbf{U}_s = U\,\mathbf{i}$ and the wind velocity is $\mathbf{U}_w = -V[\mathbf{i}\cos\alpha + \mathbf{j}\sin\alpha]$.

(a) Using potential-flow theory, estimate the force on the ship as a function of Ω, D, H, U, V and the density of air, ρ.

(b) If the side force vanishes, what is the thrust?

(b) Determine the thrust, in tons, for $\Omega = 10$ Hz, $D = 9$ ft, $H = 50$ ft, $U = 4$ knots, $V = 15$ knots and $\rho = .00234$ slug/ft^3.

11.55 Recall the home experiment discussed in this chapter. We would like to estimate the minimum speed, U_{min}, and rotation rate, Ω_{min}, needed to cause the cylinder to rise above the table. Assume that the angular velocity is given by $\Omega = U/R$, where U is the translation velocity and R is the cylinder radius. Let ρ, H and W denote air density, cylinder height and cylinder weight, respectively.

(a) Determine U_{min} and Ω_{min} as functions of W, ρ, R and H.

(b) Compute U_{min} (in ft/sec) and Ω_{min} (in Hz) for $W = 0.005$ lb, $\rho = .00234$ slug/ft^3, $R = \frac{1}{2}$ in and $H = 11$ in.

11.56 Consider potential flow past a cylinder of radius R where the freestream velocity, $U(t)$, varies with time, t.

(a) Compute the pressure coefficient, C_p, on the surface of the cylinder. Express your answer in terms of R, θ, U and dU/dt. **HINT:** Note that $\partial\phi/\partial t$ is always zero on the y axis.

(b) Compute the lift, L, and drag, D, on the cylinder. You may need to use the following trigonometric identity: $\cos^2\theta = \frac{1}{2}(1 + \cos2\theta)$.

11.57 Consider *steady potential flow* past a cylindrical tube with three radially drilled holes as shown below. Whenever the pressure on the two side holes is equal, the center hole (located halfway between the side holes) will point in the direction of flow. The pressure at the center hole is then the stagnation pressure. This device is called a *direction-finding Pitot tube*.

(a) Since the side holes must measure the freestream static pressure, where must they be located, i.e., what is the angle α?

(b) Now assume the freestream flow is uniform but changes with time, i.e., $U = U(t)$. Taking account of this unsteadiness, determine the difference between the pressure at the centrally located hole, p_1, and the pressure at either one of the side holes, p_2. Express your answer in terms of ρ, R, α and $U(t)$.

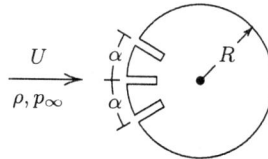

Problem 11.57

11.58 To simulate the effect of incoming and reflected waves on a pier, imagine a cylinder of radius R immersed in a fluid with a sinusoidally-varying freestream velocity, viz.,

$$U = U_o \sin \omega t$$

where U_o and ω are characteristic velocity and frequency scales, respectively.

(a) Using the potential-flow solution for flow past a cylinder, determine the pressure coefficient on the surface of the cylinder, $C_p \equiv (p - p_\infty)/(\frac{1}{2}\rho U_o^2)$. Your answer should involve a *Strouhal number*.

(b) Verify that the average value of C_p defined by

$$\overline{C}_p \equiv \frac{\omega}{2\pi} \int_0^{2\pi/\omega} C_p \, dt$$

is half the value of that for a steady flow with $U = u_o$, and is independent of Strouhal number.

11.59 The concept of *virtual mass* is valid for both two- and three-dimensional flows. The virtual mass of a sphere of radius R is $m_h = \frac{2}{3}\rho\pi R^3$, where ρ is the fluid density. Ignoring viscous drag, compute the acceleration of a sphere with density $\rho_s = \frac{1}{4}\rho$ as it rises through a fluid in a gravitational fluid.

11.60 To approximate the motion of a slender object launched from a submarine as it rises to the surface, we approximate the object as an ellipsoid of revolution of volume V. If the object's density is $\rho_m = \frac{1}{2}\rho$, where ρ is the density of water, compute the acceleration, dW/dt, according to potential-flow theory, including effects of virtual mass. Express your answer as a function of $m_h/(\rho V)$ and the acceleration of gravity, g. Make a graph of the error we would make as a function of t/c if we were to ignore the virtual mass. The virtual mass for longitudinal motion of an ellipsoid of revolution is listed below.

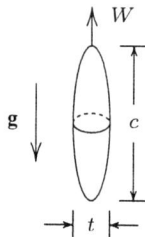

t/c	$m_h/(\rho V)$
0.0	0
0.1	0.0207
0.2	0.0591
0.3	0.1054
0.4	0.1563

Problem 11.60

11.61 Using the method of images, indicate how you would distribute sources to simulate walls along $y = 0$ and along $x = 0$, i.e., to bound the first quadrant $(x > 0, y > 0)$.

11.62 Using the method of images, indicate how you would distribute sources to simulate walls along $y = 0$ and along $x = 0$, i.e., to bound the first quadrant $(x > 0, y > 0)$.

11.63 Use the method of images to compute the flow past a cylinder a distance $4R$ from a wall as shown in Figure 11.22. Compute and compare flow velocities to the solution in an infinite domain at the four points $(x, y) = (0, 0), (0, 4R), (0, 6R), (0, 10R)$.

11.64 Using the method of images, compute the force exerted by a doublet on a planar wall. The doublet strength is $\mathcal{D}$ and it is located a distance h above the wall. **HINT:** The following integrals may be of use in deriving your answer.

$$\int_{-\pi/2}^{\pi/2} \cos^2 \phi \, d\phi = \frac{1}{2}\pi, \qquad \int_{-\pi/2}^{\pi/2} \cos^4 \phi \, d\phi = \frac{3}{8}\pi, \qquad \int_{-\pi/2}^{\pi/2} \cos^6 \phi \, d\phi = \frac{5}{16}\pi$$

11.65 As an airplane approaches a runway, its lift is altered by the so-called *ground effect*. To determine the effect, approximate the wing and the ground plane by a potential vortex of strength Γ and its image. If the approach velocity is U (parallel to the ground), determine the lift per unit wingspan when the wing is a distance h above the ground. Write your final result in terms of *aerodynamic circulation*, $\Gamma_a = -\Gamma$. Does your answer make sense for $h \to \infty$? Does the lift increase or decrease due to the *ground effect*?

11.66 Using linearized airfoil theory, evaluate the velocity components for flow past the diamond shaped airfoil shown below. Using $\epsilon = .05$, plot $u(x, 0)$ for the interval $-10 \leq x \leq 10$, and $u(1, y)$ for the interval $0 \leq y \leq 1$. Be sure to include details of what happens near the points $x = 0, 1, 2$ and $y = 0$.

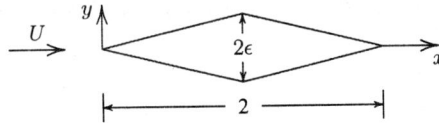

Problem 11.66

11.67 Consider a vortex sheet with strength

$$\gamma(\xi) = \gamma_o \sqrt{\frac{\xi}{c}\left(1 - \frac{\xi}{c}\right)}, \qquad \gamma_o = \text{constant}$$

The object of this problem is to determine the shape of the camber line for an airfoil simulated by superposing this vortex sheet with a uniform flow of velocity $\mathbf{U} = \mathbf{i}\, U \cos\alpha + \mathbf{j}\, U \sin\alpha$.

(a) Making the change of variables defined by

$$\xi = \frac{c}{2}(1 + \cos\theta), \qquad x = \frac{c}{2}(1 + \cos\theta')$$

show that the equation for the camber line, $C(x)$, is

$$\frac{dC}{dx} = \tan\alpha + \frac{\gamma_o}{8\pi U \cos\alpha} \int_0^\pi \frac{(1 - \cos 2\theta)d\theta}{\cos\theta - \cos\theta'}$$

(b) Using the fact that for any integer $n \geq 0$,

$$\int_0^\pi \frac{\cos n\theta \, d\theta}{\cos\theta - \cos\theta'} = \pi \frac{\sin n\theta'}{\sin\theta'}$$

determine the camber line. Set $C(0) = 0$ to eliminate any constant of integration.

(c) For the special case $\gamma_o = 4U \sin\alpha$, compute the camber line. Make a graph of the camber line, and include the chord line in your graph.

11.68 Consider a vortex sheet with strength

$$\gamma(\xi) = \gamma_o \left[\frac{\xi}{c} \left(1 - \frac{\xi}{c} \right) \right]^{3/2} , \qquad \gamma_o = \text{constant}$$

The object of this problem is to determine the shape of the camber line for an airfoil simulated by superposing this vortex sheet with a uniform flow of velocity $\mathbf{U} = \mathbf{i}\, U \cos \alpha + \mathbf{j}\, U \sin \alpha$.

(a) Making the change of variables defined by

$$\xi = \frac{c}{2}(1 + \cos \theta), \qquad x = \frac{c}{2}(1 + \cos \theta')$$

show that the equation for the camber line, $C(x)$, is

$$\frac{dC}{dx} = \tan \alpha + \frac{\gamma_o}{128\pi U \cos \alpha} \int_0^\pi \frac{(3 - 4\cos 2\theta + \cos 4\theta)d\theta}{\cos \theta - \cos \theta'}$$

Note that $\sin^4 \theta = \frac{1}{8}(3 - 4\cos 2\theta + \cos 4\theta)$.

(b) Using the fact that for any integer $n \geq 0$,

$$\int_0^\pi \frac{\cos n\theta \, d\theta}{\cos \theta - \cos \theta'} = \pi \frac{\sin n\theta'}{\sin \theta'}$$

determine the camber line. Set $C(0) = 0$ to eliminate any constant of integration.

(c) For the special case $\gamma_o = 24U \sin \alpha$, compute the camber line. Make a graph of the camber line, and include the chord line in your graph.

11.69 An airfoil has 4% camber and 10% thickness, i.e., $C_{max}/c = 0.04$ and $2T_{max}/c = 0.10$. Estimate its lift coefficient if the angle of attack is $4°$.

11.70 A thin, cambered airfoil has a lift coefficient of 0.40 at zero angle of attack and 1.09 at $6°$. Estimate the maximum thickness, $2T_{max}/c$, and maximum camber, C_{max}/c, (in percent) for this airfoil. **HINT:** To simplify your solution, use the fact that $\sin \epsilon \approx \tan \epsilon \approx \epsilon$ for small ϵ.

11.71 Using a rectangular finite-difference grid with different grid-point spacing in the x and y directions, i.e., $\Delta x \neq \Delta y$, develop a second-order accurate replacement for Equation (11.254). Write the equation in terms of $\beta \equiv \Delta y/\Delta x$.

11.72 Verify that, for a rectangular grid with $\Delta x \neq \Delta y$, the following discretization is second-order accurate in both Δx and Δy.

$$\left(\frac{\partial^2 \phi}{\partial x \partial y} \right)_{i,j} \approx \frac{\phi_{i+1,j+1} - \phi_{i+1,j-1} - \phi_{i-1,j+1} + \phi_{i-1,j-1}}{4\Delta x \Delta y}$$

HINT: The Taylor-series expansion of $\phi_{i+1,j+1}$ is given by

$$\phi_{i+1,j+1} = \phi_{i,j} + \phi_x \Delta x + \phi_y \Delta y + \frac{1}{2}\phi_{xx}(\Delta x)^2 + \phi_{xy}\Delta x \Delta y + \frac{1}{2}\phi_{yy}(\Delta y)^2 + \cdots$$

where subscripts x and y denote the partial derivative with respect to x and y, respectively.

11.73 To accurately compute the velocity at a mesh boundary from the streamfunction, ψ, we must use a second-order accurate discretization approximation based on points entirely within the computational domain. Assuming that

$$\left(\frac{\partial \psi}{\partial y} \right)_{i,1} \approx A\psi_{i,1} + B\psi_{i,2} + C\psi_{i,3}$$

expand in Taylor series for equally spaced points in the y direction. Find the values of A, B and C as functions of grid-point spacing, Δy, that yield second-order accuracy.

11.74 Using Taylor-series expansions, derive a fourth-order accurate discretization approximation for $\partial\phi/\partial x$, i.e., solve for A, B, C, D and E such that

$$\left(\frac{\partial\phi}{\partial x}\right)_{i,j} \approx A\phi_{i+2,j} + B\phi_{i+1,j} + C\phi_{i,j} + D\phi_{i-1,j} + E\phi_{i-2,j}$$

Solve for A, B, C, D and E as functions of the (constant) grid-point differential, Δx.

11.75 Consider the function $\phi(y) = y^4$. Use the second-order accurate finite-difference approximations to compute $d\phi/dy$ and $d^2\phi/dy^2$ at $y = 2$ for $\Delta y = 1.0$, 0.5, 0.25 and 0.125. Compare the numerical results with the exact values in tabular form, and use Richardson extrapolation to estimate the error for $\Delta y = 0.125$.

11.76 Suppose a finite-difference method is only first-order accurate. When this is true, Richardson's estimate of the error must be revised. Assuming

$$E_h = e_1 h + e_2 h^2 + \cdots$$

propose an alternative to Equation (11.264).

11.77 The following table represents partial results for one-dimensional finite-difference computations using a second-order accurate, time-marching method. The computations have been done on grids with 50, 100 and 200 points. Use Richardson extrapolation to estimate the discretization error at each point for the two finest grids. Based on your results, make a table of the results below and add a column with your best estimate of the continuum solution (grid-point spacing $\rightarrow$ 0) to the differential equation.

j	ϕ_{50}	j	ϕ_{100}	j	ϕ_{200}
1	3.00361	1	2.96624	1	2.95443
2	3.07446	3	3.06157	5	3.05965
3	3.09224	5	3.06523	9	3.07557
4	3.54523	7	3.53756	13	3.52365

11.78 The following table represents partial results for one-dimensional finite-difference computations using a second-order accurate, time-marching method. The computation have been done on grids with 50, 100 and 200 points. Use Richardson extrapolation to estimate the discretization error at each point for the two finest grids. Based on your results, make a table of the results below and add a column with your best estimate of the continuum solution (grid-point spacing $\rightarrow$ 0) to the differential equation.

j	ϕ_{50}	j	ϕ_{100}	j	ϕ_{200}
1	0.5592	1	0.5628	1	0.5607
2	0.5700	3	0.5740	5	0.5726
3	0.5737	5	0.5748	9	0.5745
4	0.5615	7	0.5557	13	0.5573

11.79 Using Program **POTFLOW** (see Appendix E), find the number of iterations required for the solution to converge for $\lambda = 0.85$, 0.90, 0.95 and 1.00. What is the optimum value of λ for this grid?

11.80 The purpose of this problem is to examine the issue of iteration convergence. First, examine the source code for Program **POTFLOW** (see Appendix E), and determine how to change the value of the residual that specifies solution convergence. Then make the appropriate change and run computations with the convergence criterion set to $2.0 \cdot 10^{-6}$, $1.0 \cdot 10^{-6}$ and $0.5 \cdot 10^{-6}$. (Using a smaller value than $0.5 \cdot 10^{-6}$ requires double precision to achieve a converged solution.) Use 51 grid points and a relaxation parameter of $\lambda = 0.9$. By examining $v(0^-, y)/U$ at $y/a = 0.2$, 0.4, 0.6, 0.8 and 1.0, verify that solution changes are less than 0.00001 as the residual is halved from $1.0 \cdot 10^{-6}$ to $0.5 \cdot 10^{-6}$. Also, show that these changes are less than half those that occur as the residual is reduced from $2.0 \cdot 10^{-6}$ to $1.0 \cdot 10^{-6}$.

11.81 Using a finite-difference formulation, solve for flow approaching a stagnation point. Generate a solution for the computational domain shown using equally-spaced grid points. Proceed as follows.

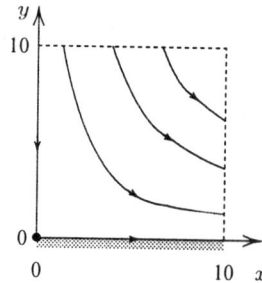

Problem 11.81

(a) Write the appropriate differential equations for ψ, u and v in terms of dimensionless variables, so that the exact solution is $\psi = xy$.

(b) Use the exact solution to set the boundary condition along $y = 10$, and assume $\partial\psi/\partial x = 0$ at $x = 10$. Express these conditions in finite-difference form. Also, state the boundary conditions along $x = 0$ and $y = 0$.

(c) Using $\psi = 1$ at all interior points for the initial condition, solve the problem using standard overrelaxation (SOR). Let the step size be $\Delta x = \Delta y = 0.10$. Find the optimum relaxation parameter, λ. Plot a few streamlines to display results of the computation.

(d) Solve the problem again with $\Delta x = \Delta y = 0.05$, and use Richardson extrapolation to estimate solution errors in ψ along $x = 2$ and also along $y = 2$. Compare with the exact solution as well.

11.82 Using a finite-difference formulation, solve for flow in a two-dimensional channel with a *forward-facing step*. Generate a solution for the computational domain shown using equally-spaced grid points. Proceed as follows.

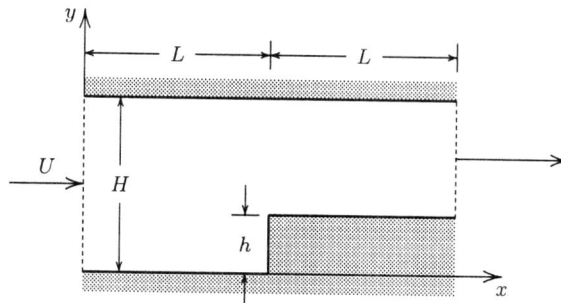

Problem 11.82

(a) Write the appropriate differential equations for ψ, u and v in terms of dimensionless variables. Use U and H as the characteristic velocity and length scales.

(b) Let $u = U$, $v = 0$ at the inlet ($x = 0$), and assume $\partial\psi/\partial x = 0$ at the outlet ($x = 2L$). Express these conditions in finite-difference form. Also, state the boundary conditions along the lower and upper walls.

(c) Using $\psi = 1$ at all interior points for the initial condition, solve the problem with $h/H = \frac{1}{3}$ and $L/H = 5$ using standard overrelaxation (SOR). Let the step size be $\Delta x = \Delta y = 0.10$. Find the optimum relaxation parameter, λ. Plot a few streamlines to display results of the computation.

(d) Solve the problem again with $\Delta x = \Delta y = 0.05$, and use Richardson extrapolation to estimate solution errors in ψ along $x = L$ and also along $y = H/2$.

11.83 Using a finite-difference formulation, solve for flow in a two-dimensional channel with a *backward-facing step*. Generate a solution for the computational domain shown using equally-spaced grid points. Proceed as follows.

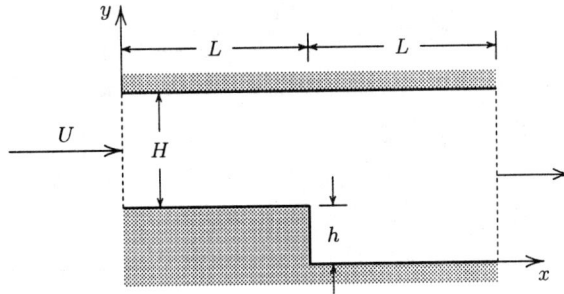

Problem 11.83

(a) Write the appropriate differential equations for ψ, u and v in terms of dimensionless variables. Use U and H as the characteristic velocity and length scales.

(b) Let $u = U$, $v = 0$ at the inlet ($x = 0$), and assume $\partial \psi / \partial x = 0$ at the outlet ($x = 2L$). Express these conditions in finite-difference form. Also, state the boundary conditions along the lower and upper walls.

(c) Using $\psi = 1$ at all interior points for the initial condition, solve the problem with $h/H = \frac{1}{4}$ and $L/H = 4$ using standard overrelaxation (SOR). Let the step size be $\Delta x = \Delta y = 0.10$. Find the optimum relaxation parameter, λ. Plot a few streamlines to display results of the computation.

(d) Solve the problem again with $\Delta x = \Delta y = 0.05$, and use Richardson extrapolation to estimate solution errors in ψ along $x = L$ and also along $y = H/2$.

11.84 Using a finite-difference formulation, solve for flow in a two-dimensional channel with a $45°$ contraction. Generate a solution for the computational domain shown using equally-spaced grid points. Proceed as follows.

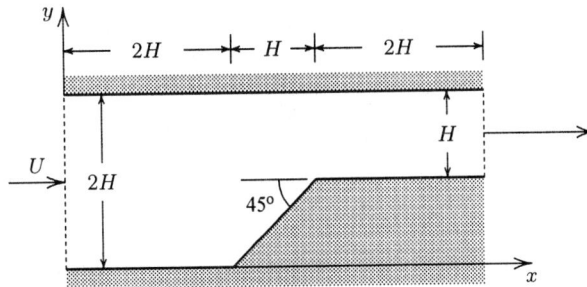

Problem 11.84

(a) Write the appropriate differential equations for ψ, u and v in terms of dimensionless variables. Use U and H as the characteristic velocity and length scales.

(b) Let $u = U$, $v = 0$ at the inlet ($x = 0$), and assume $\partial \psi / \partial x = 0$ at the outlet ($x = 5H$). Express these conditions in finite-difference form. Also, state the boundary conditions along the lower and upper walls.

(c) Using $\psi = 1$ at all interior points for the initial condition, solve the problem using standard overrelaxation (SOR). Be sure to select the grid points so that points lie precisely on the slanted part of the lower wall. Let the step size be $\Delta x = \Delta y = 0.10$. Find the optimum relaxation parameter, λ. Plot a few streamlines to display results of the computation.

(d) Solve the problem again with $\Delta x = \Delta y = 0.05$, and use Richardson extrapolation to estimate solution errors in ψ along $x = 4H$ and also along $y = \frac{5}{4}H$.

11.85 Using a finite-difference formulation, solve for flow in a two-dimensional channel with a 45°
expansion. Generate a solution for the computational domain shown using equally-spaced grid points.
Proceed as follows.

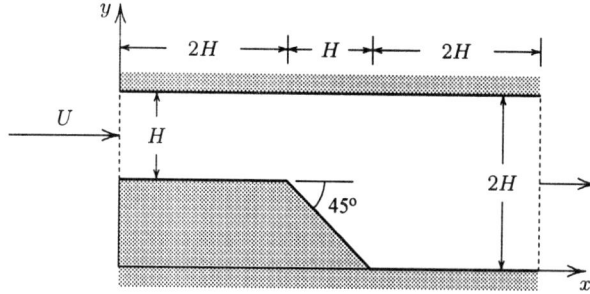

Problem 11.85

(a) Write the appropriate differential equations for ψ, u and v in terms of dimensionless variables.
Use U and H as the characteristic velocity and length scales.

(b) Let $u = U$, $v = 0$ at the inlet ($x = 0$), and assume $\partial\psi/\partial x = 0$ at the outlet ($x = 5H$). Express
these conditions in finite-difference form. Also, state the boundary conditions along the lower
and upper walls.

(c) Using $\psi = 1$ at all interior points for the initial condition, solve the problem using standard
overrelaxation (SOR). Be sure to select the grid points so that points lie precisely on the slanted
part of the lower wall. Let the step size be $\Delta x = \Delta y = 0.10$. Find the optimum relaxation
parameter, λ. Plot a few streamlines to display results of the computation.

(d) Solve the problem again with $\Delta x = \Delta y = 0.05$, and use Richardson extrapolation to estimate
solution errors in ψ along $x = 4H$ and also along $y = \frac{5}{4}H$.

Chapter 12

Viscous Effects

As we have discussed in preceding chapters, viscous effects play a small role through most of the flowfield when we consider flow past thin, streamlined bodies. This is true because the Reynolds number, $Re = \rho U L/\mu$, is very large for virtually all practical flow problems. The large value of Re is a consequence of the fact that $\mu \ll \rho U L$ for fluids such as air and water. As we will see in this chapter, the reciprocal of the Reynolds number multiplies the viscous terms in the equations of motion when we include friction, which explains why viscous effects can often be neglected.

However, viscous effects cannot be completely ignored because they give rise to a nontrivial drag force that is absent in the theory of inviscid fluids. Additionally, because of frictional effects, flows subjected to increasing pressure behave quite differently from the predictions of inviscid theory. Flow past a cylinder, for example, is very complicated due to viscous effects, and over the rear half of the cylinder it bears little relation to the potential-flow solution (see Figures 11.10 and 11.12). Hence, to develop a complete theory of fluid motion, we must reformulate the basic conservation principles with viscous effects included.

To accomplish this end, this chapter focuses on the way viscous effects manifest themselves in a fluid. We begin by examining what happens at the molecular level in a simple two-dimensional flow. A straightforward argument shows how the viscous shear stress, τ, turns out to be the product of the fluid viscosity, μ, and the velocity gradient, $\partial u/\partial y$. To generalize this result to arbitrary three-dimensional flows, we first reexamine the kinematics of a fluid particle. This provides a completely general description of how a fluid particle moves in terms of the velocity and its derivatives. As a side benefit of examining kinematics at this level, we gain insight into the nature of incompressible and irrotational fluid motion by examining the way a fluid particle deforms while moving in such flows.

After examining the kinematical aspects of fluid motion, we turn to the physical nature of the viscous force acting on a fluid particle. We will find that it is a surface force that is far more complicated than the pressure force. On the one hand, the pressure force is simply the product of the *scalar* pressure, p, the unit normal to the surface on which it acts, **n**, and the area of the surface. On the other hand, the friction force is not aligned with the unit normal, and must be represented by a more complicated relation. Specifically, we show that the viscous force acting on a surface element is the product of a *matrix* known as the **viscous stress tensor**, $[\tau]$, the unit normal, **n**, and the area. Given this representation, we reformulate the momentum-conservation principle in both integral and differential forms. The differential form is known as **Navier's equation**.

Having introduced the viscous stress tensor, we find that we need several additional equations to close our system. That is, we must develop a **constitutive relation**, i.e., a relation between $[\tau]$ and the velocity field. **Stokes' postulate** provides the required equations. Substitution into Navier's equation yields the **Navier-Stokes equation**. We then derive an equation for the vorticity, underscoring the intimate connection between vorticity and viscosity that we touched upon at the end of Chapter 10. We conclude the chapter with a discussion of Computational Fluid Dynamics in the context of viscous fluids.

12.1   Molecular Transport of Momentum

To understand how frictional forces appear in a fluid, it is instructive to discuss momentum transport at the molecular level. We begin by considering a shear flow in which the velocity is given by

$$\mathbf{U} = U(y)\,\mathbf{i} \tag{12.1}$$

where $\mathbf{i}$ is a unit vector in the x direction. Figure 12.1 depicts such a flow. We consider the flux of momentum across the plane $y = 0$, noting that molecular motion is random in both magnitude and direction. Molecules migrating across $y = 0$ are **typical of where they come from**. That is, on average, molecules moving up bring a momentum deficit and vice versa. This transfer of momentum across $y = 0$ tends to decelerate the surrounding fluid when a molecule arrives with a momentum deficit, for example, manifested as a negative force parallel to the plane. Similarly, a molecule with a momentum surplus imparts a positive x-directed force. As discussed in Section 1.8, we refer to this force (per unit area) as the shear stress, τ.

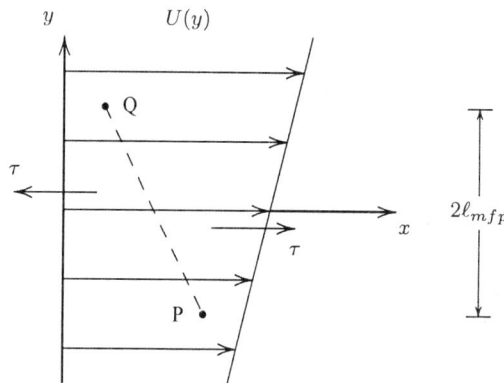

Figure 12.1: *Shear-flow schematic.*

We can appeal to arguments from the kinetic theory of gases [e.g., Jeans (1962)] to determine τ in terms of $U(y)$ and the fluid viscosity, μ. First, consider the average number of molecules moving across unit area in the positive y direction. For a perfect gas, molecules move with equal probability in all directions. The average molecular velocity is the thermal velocity, v_{th}, which is approximately 4/3 times the speed of sound in air. On average, half of the molecules move in the positive y direction while the other half move in the negative y direction. Also, the average vertical component of the velocity is $v_{th} \cos \phi$ where ϕ is the angle from the vertical. Integrating over a hemispherical shell, the average vertical speed is $v_{th}/2$. Thus, the average number of molecules moving across unit area in the positive y direction is $nv_{th}/4$, where n is the number of molecules per unit volume.

Now consider the transfer of momentum that occurs when molecules starting from point P cross the $y = 0$ plane. As noted above, we assume molecules are typical of where they come from which, on the molecular scale, is (by definition) one **mean free path** away, the mean free path being the average distance a molecule travels between collisions with other molecules. Each molecule starting from a point P below $y = 0$ brings a momentum deficit of $m[U(0) - U(-\ell_{mfp})]$, where m is the molecular mass and ℓ_{mfp} is the mean free path. Hence, the momentum flux from below is

$$\Delta P_- = \frac{1}{4}\rho v_{th}[U(0) - U(-\ell_{mfp})] \approx \frac{1}{4}\rho v_{th}\ell_{mfp}\frac{dU}{dy} \tag{12.2}$$

We have replaced $U(-\ell_{mfp})$ by the first two terms of its Taylor-series expansion in Equation (12.2) and used the fact that $\rho = mn$. Similarly, molecules moving from a point Q above $y = 0$ bring a momentum surplus of $m[U(\ell_{mfp}) - U(0)]$, and the momentum flux from above is

$$\Delta P_+ = \frac{1}{4}\rho v_{th}[U(\ell_{mfp}) - U(0)] \approx \frac{1}{4}\rho v_{th}\ell_{mfp}\frac{dU}{dy} \tag{12.3}$$

The net shearing stress is the sum of ΔP_- and ΔP_+, wherefore

$$\tau = \Delta P_- + \Delta P_+ \approx \frac{1}{2}\rho v_{th}\ell_{mfp}\frac{dU}{dy} \tag{12.4}$$

Hence, we conclude that the shear stress resulting from molecular transport of momentum in a perfect gas is given by

$$\tau = \mu\frac{dU}{dy} \tag{12.5}$$

where μ is the molecular viscosity defined by

$$\mu = \frac{1}{2}\rho v_{th}\ell_{mfp} \tag{12.6}$$

Equation (12.5) was first proposed by Isaac Newton. Fluids that satisfy this relation are thus known as **Newtonian fluids**.

The arguments leading to Equations (12.5) and (12.6) are approximate and only roughly represent the true statistical nature of molecular motion. Interestingly, Jeans (1962) indicates that a precise analysis yields $\mu = 0.499\rho v_{th}\ell_{mfp}$, so that our approximate analysis happens to give a remarkably accurate result! However, we have made two implicit assumptions in our analysis that require justification.

First, we have truncated the Taylor series appearing in Equations (12.2) and (12.3) at the linear terms. For this approximation to be valid, we must have $\ell_{mfp}|d^2U/dy^2| \ll |dU/dy|$. The length scale, L, defined by

$$L \equiv \frac{|dU/dy|}{|d^2U/dy^2|} \tag{12.7}$$

is a length scale characteristic of the mean flow. Thus, the linear relation between stress and strain-rate implied by Equation (12.5) is valid provided the *Knudsen number*, Kn, is very small, i.e.,

$$Kn = \ell_{mfp}/L \ll 1 \tag{12.8}$$

For most practical flow conditions, the mean free path is several orders of magnitude smaller than any characteristic length scale of the mean flow. Thus, Equation (12.8) is satisfied for virtually all engineering problems.

Second, in computing the rate of momentum transport across $y = 0$, we assumed that the random molecular motion is unaffected by the mean motion. This will be true if the time of flight of a molecule between collisions is very small compared to the time scale of the mean flow. Now, the average time between collisions is ℓ_{mfp}/v_{th}. The characteristic time scale for the mean flow is $|dU/dy|^{-1}$. Thus, we also require

$$\frac{\ell_{mfp}}{v_{th}} \ll \frac{1}{|dU/dy|} \quad \implies \quad \ell_{mfp} \ll \frac{v_{th}}{|dU/dy|} \tag{12.9}$$

Since v_{th} is of the same order of magnitude as the speed of sound, the right-hand side of Equation (12.9) defines yet another mean-flow length scale. As above, the mean free path is several orders smaller than this length scale for virtually all flows of engineering interest.

As a final comment, observe that the viscous stress in this simple geometry is tangential to the surface on which it acts. In more complicated flows, we find that the viscous force can have a component normal to the surface as well. In other words, viscous stresses, in general, appear as oblique stresses. This tells us that viscous stresses are fundamentally more complicated than pressure, p, which is a normal stress that can be represented simply as the product of p and the unit normal, **n**.

12.2   Kinematics of a Fluid Particle

As a fluid particle moves through a given flowfield it can experience translation, rotation and distortion of its shape. Figure 12.2 illustrates the four basic motions and distortions that a fluid particle can undergo in any given plane. General motion will be a combination of these basic components in all three coordinate directions. Regardless of the forces governing the fluid particle's motion, we can express the distortions and rotations in terms of the velocity and its partial derivatives. Developing such representations is one of the central issues of kinematics.

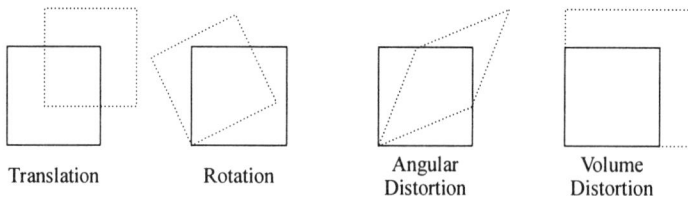

| Translation | Rotation | Angular Distortion | Volume Distortion |

Figure 12.2: *Basic motions and distortions of a fluid particle.*

12.2.1   Basic Formulation

Consider the fluid-particle cross section in the xy plane shown in Figure 12.3. At the initial time, $t = 0$, the cross section with vertices AOB is bounded by perpendicular lines of length ξ and η. In order to simplify our calculations, the coordinate axes are chosen to be coincident with the initial position of the fluid particle's lower left corner. The figure shows the same fluid particle at a time $t = \Delta t$. The fluid-particle vertices after it has moved are A′O′B′.

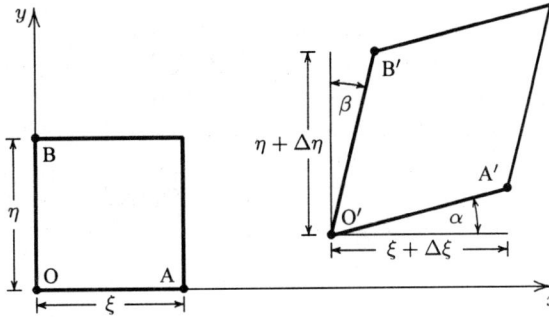

Figure 12.3: *Coordinate system for fluid-particle motion.*

As mentioned above, the fluid particle's motion consists of a linear combination of the four basic motions and distortions illustrated in Figure 12.2. For example, the displacement of the fluid particle's lower left corner from Point O to Point O' can be accounted for by a simple **translation**. Any change in length of the fluid particle's sides, such as the initially horizontal line's length changing from ξ to $\xi + \Delta \xi$, is attributable to **volume distortion**. The changes in the angles of the sides is a combination of a **rotation** through an angle $\frac{1}{2}(\alpha - \beta)$ and an **angular distortion** of $\frac{1}{2}(\alpha + \beta)$, where α and β are the angles indicated in Figure 12.3.

The point regarding the angular motions is a bit subtle and requires further explanation. As shown in Figure 12.2, in a pure rotation, both sides of a fluid particle rotate in the counterclockwise direction. By contrast, in an angular distortion, the horizontal side rotates counterclockwise while the vertical side rotates clockwise. Hence, in general motion, the angle between the horizontal side and the x axis is

$$\underbrace{\frac{1}{2}(\alpha - \beta)}_{Rotation} + \underbrace{\frac{1}{2}(\alpha + \beta)}_{\substack{Angular \\ Distortion}} = \alpha \qquad (12.10)$$

Similarly, the angle between the vertical side and the y axis is

$$\underbrace{\frac{1}{2}(\alpha - \beta)}_{Rotation} - \underbrace{\frac{1}{2}(\alpha + \beta)}_{\substack{Angular \\ Distortion}} = -\beta \qquad (12.11)$$

Our goal is to relate the four basic motions and distortions to the velocity components and their derivatives. *Keep in mind that in discussing the kinematics of a fluid particle, what we are doing is strictly geometrical in nature.* Whatever force happens to be acting on the fluid particle is of no consequence in the following analysis. We are simply developing a mathematical description of the motion in terms of the particle's velocity, rate of deformation, etc. that holds for an arbitrary force acting on the fluid particle.

We begin by determining the velocity components at the fluid particle vertices O, A and B. For simplicity, we confine our analysis to two dimensions — we will generalize our results to three dimensions later. For arbitrary two-dimensional motion, the velocity components are functions of x and y. The velocities at the three vertices prior to motion of the fluid particle can be written symbolically as follows.

$$\mathbf{u} = \begin{cases} \mathbf{u}(0,0), & \text{Point O} \\ \mathbf{u}(\xi,0), & \text{Point A} \\ \mathbf{u}(0,\eta), & \text{Point B} \end{cases} \tag{12.12}$$

Our fluid particle is assumed to be sufficiently small that we can expand the velocity in its Taylor series and truncate beyond terms linear in ξ and η. Hence, the velocities at Points A and B are

$$\mathbf{u}(\xi,0) = \mathbf{u}(0,0) + \xi \left(\frac{\partial \mathbf{u}}{\partial x} \right)_{(0,0)} + \cdots \tag{12.13}$$

$$\mathbf{u}(0,\eta) = \mathbf{u}(0,0) + \eta \left(\frac{\partial \mathbf{u}}{\partial y} \right)_{(0,0)} + \cdots \tag{12.14}$$

Thus, dropping subscripts, the velocities at the three vertices are

$$\mathbf{u} \approx \begin{cases} u\,\mathbf{i} + v\,\mathbf{j}, & \text{Point O} \\[2mm] \left(u + \dfrac{\partial u}{\partial x}\xi \right)\mathbf{i} + \left(v + \dfrac{\partial v}{\partial x}\xi \right)\mathbf{j}, & \text{Point A} \\[2mm] \left(u + \dfrac{\partial u}{\partial y}\eta \right)\mathbf{i} + \left(v + \dfrac{\partial v}{\partial y}\eta \right)\mathbf{j}, & \text{Point B} \end{cases} \tag{12.15}$$

Now, consider the movement of each vertex through an infinitesimally small time increment, Δt. In the limit $\Delta t \to 0$, we can treat the velocity at each vertex as being constant with respect to time. Hence, the locations of the three vertices after time Δt, O', A' and B', are as follows.

$$\mathbf{r} \approx \begin{cases} u\Delta t\,\mathbf{i} + v\Delta t\,\mathbf{j}, & \text{Point O}' \\[2mm] \left[\xi + \left(u + \dfrac{\partial u}{\partial x}\xi \right)\Delta t \right]\mathbf{i} + \left(v + \dfrac{\partial v}{\partial x}\xi \right)\Delta t\,\mathbf{j}, & \text{Point A}' \\[2mm] \left(u + \dfrac{\partial u}{\partial y}\eta \right)\Delta t\,\mathbf{i} + \left[\eta + \left(v + \dfrac{\partial v}{\partial y}\eta \right)\Delta t \right]\mathbf{j}, & \text{Point B}' \end{cases} \tag{12.16}$$

12.2.2   Volume Distortion

Turning first to volume distortion, we now have sufficient information to compute the new lengths of the sides of the fluid particle. Because Δt is infinitesimal by hypothesis, we may reasonably assume that the angles are also infinitesimal. Hence, referring to Figure 12.3, the distance between vertices O' and A' is $(\xi + \Delta\xi)/\cos\alpha \approx (\xi + \Delta\xi)$.[1] Similarly, the distance between O' and B' is $(\eta + \Delta\eta)/\cos\beta \approx (\eta + \Delta\eta)$. Thus, the distance between Points A' and O' is approximately equal to the difference between their x coordinates. Likewise, the distance between Points B' and O' is approximately equal to the difference between their y coordinates. From Equation (12.16), we find

---

[1] For small angles, expanding in Taylor series shows that $\cos\alpha = 1 - \frac{1}{2}\alpha^2 + \cdots \approx 1$.

$$x_{A'} - x_{O'} \approx \left[\xi + \left(u + \frac{\partial u}{\partial x} \xi \right) \Delta t \right] - u \Delta t \approx \xi + \Delta \xi \left. \right\}$$

$$y_{B'} - y_{O'} \approx \left[\eta + \left(v + \frac{\partial v}{\partial y} \eta \right) \Delta t \right] - v \Delta t \approx \eta + \Delta \eta \left. \right\} \qquad (12.17)$$

Therefore, the changes in ξ and η are

$$\Delta \xi \approx \frac{\partial u}{\partial x} \xi \Delta t, \quad \text{and} \quad \Delta \eta \approx \frac{\partial v}{\partial y} \eta \Delta t \qquad (12.18)$$

If our material were a solid, we would now compute the normal strain components, i.e., the ratio of the change in length of each side to the original length. The strain is not a particularly useful quantity for a fluid as it is proportional to Δt, and continues to increase as time passes. However, the *rate* at which straining occurs is independent of Δt, and is thus a more meaningful measure of straining. In other words, we prefer to compute the **normal strain rates**, S_{xx} and S_{yy}, defined by

$$S_{xx} \equiv \lim_{\Delta t \to 0} \frac{1}{\xi} \frac{\Delta \xi}{\Delta t} \quad \text{and} \quad S_{yy} \equiv \lim_{\Delta t \to 0} \frac{1}{\eta} \frac{\Delta \eta}{\Delta t} \qquad (12.19)$$

In a similar way, by considering either the yz or zx plane, we can derive an analogous expression for the normal strain rate along the z axis, S_{zz}. Combining Equations (12.18) and (12.19), we conclude that

$$S_{xx} = \frac{\partial u}{\partial x}, \quad S_{yy} = \frac{\partial v}{\partial y}, \quad S_{zz} = \frac{\partial w}{\partial z} \qquad (12.20)$$

where w is the velocity component in the z direction.

12.2.3   Angular Distortion and Rotation

Next, we consider angular distortion and rotation. As shown in Figure 12.3, the general motion of the fluid particle indicated has the horizontal side rotating counterclockwise by an angle α and the vertical side rotating clockwise by an angle β. As noted above, in an angular distortion, the two sides rotate in opposite directions through the same angle. By contrast, in a rotation, the two sides rotate in the same direction through the same angle (see Figure 12.2). Thus, the general motion depicted in Figure 12.3 can be achieved by having a combination of an angular distortion through an angle $\frac{1}{2}(\alpha + \beta)$ followed by a rotation through an angle $\frac{1}{2}(\alpha - \beta)$ [see Equations (12.10) and (12.11)].

We can determine the angles α and β as functions of velocity derivatives through a little trigonometry. Inspection of Figure 12.3 shows that

$$\tan \alpha = \frac{y_{A'} - y_{O'}}{\xi + \Delta \xi} \quad \text{and} \quad \tan \beta = \frac{x_{B'} - x_{O'}}{\eta + \Delta \eta} \qquad (12.21)$$

Now, from Equation (12.16), we find

$$y_{A'} - y_{O'} \approx \left(v + \frac{\partial v}{\partial x} \xi \right) \Delta t - v \Delta t \approx \frac{\partial v}{\partial x} \xi \Delta t \left. \right\}$$

$$x_{B'} - x_{O'} \approx \left(u + \frac{\partial u}{\partial y} \eta \right) \Delta t - u \Delta t \approx \frac{\partial u}{\partial y} \eta \Delta t \left. \right\} \qquad (12.22)$$

Because α and β are infinitesimally small, we can use the fact that for small angles, $\tan \alpha \approx \alpha$ and similarly for $\tan \beta$. Additionally, in the limit $\Delta t \to 0$, we can use a Taylor series expansion and Equations (12.18) to show that[2]

$$\frac{1}{\xi + \Delta \xi} \approx \frac{1}{\xi} \left(1 - \frac{\partial u}{\partial x} \Delta t \right) \qquad \text{and} \qquad \frac{1}{\eta + \Delta \eta} \approx \frac{1}{\eta} \left(1 - \frac{\partial v}{\partial y} \Delta t \right) \qquad (12.23)$$

Combining Equations (12.21) through (12.23) and dropping terms quadratic in Δt, we conclude that the angles α and β are

$$\alpha \approx \frac{\partial v}{\partial x} \Delta t \qquad \text{and} \qquad \beta \approx \frac{\partial u}{\partial y} \Delta t \qquad (12.24)$$

As with normal strain, the distortion and rotation angles increase with Δt, and are of limited value in describing fluid motion. Again, we find the rate of distortion and rotation to be independent of Δt. This motivates defining the **shear strain rate**, S_{xy}, and **rotation rate** (about the z axis), Ω_z, as follows.

$$S_{xy} = \frac{1}{2} \lim_{\Delta t \to 0} \frac{\alpha + \beta}{\Delta t} \qquad \text{and} \qquad \Omega_z = \frac{1}{2} \lim_{\Delta t \to 0} \frac{\alpha - \beta}{\Delta t} \qquad (12.25)$$

Thus, using Equations (12.24) to specify α and β, we arrive at the desired relationship for shear strain rate and rotation rate:

$$S_{xy} = \frac{1}{2} \left(\frac{\partial v}{\partial x} + \frac{\partial u}{\partial y} \right) \qquad \text{and} \qquad \Omega_z = \frac{1}{2} \left(\frac{\partial v}{\partial x} - \frac{\partial u}{\partial y} \right) \qquad (12.26)$$

All of the analysis has been confined to motion in the xy plane. By repeating the analysis in the yz and zx planes, we can derive expressions for the shear strain rates S_{yz} and S_{zx}, and the rotation rates Ω_x and Ω_y. There follows:

$$S_{yz} = \frac{1}{2} \left(\frac{\partial w}{\partial y} + \frac{\partial v}{\partial z} \right) \qquad \text{and} \qquad \Omega_x = \frac{1}{2} \left(\frac{\partial w}{\partial y} - \frac{\partial v}{\partial z} \right) \qquad (12.27)$$

$$S_{zx} = \frac{1}{2} \left(\frac{\partial u}{\partial z} + \frac{\partial w}{\partial x} \right) \qquad \text{and} \qquad \Omega_y = \frac{1}{2} \left(\frac{\partial u}{\partial z} - \frac{\partial w}{\partial x} \right) \qquad (12.28)$$

12.2.4   Strain-Rate Tensor and Vorticity Vector

At this point, we have derived 6 components of what is known as the **strain-rate tensor**. The complete strain-rate tensor has 9 components. The 3 we have not derived are S_{yx}, S_{zy} and S_{xz}. We can derive the remaining strain rates by interchanging indices, so that, for example,

$$S_{yx} = \frac{1}{2} \left(\frac{\partial u}{\partial y} + \frac{\partial v}{\partial x} \right) = S_{xy} \qquad (12.29)$$

and similarly for S_{zy} and S_{xz}. In other words, the strain-rate tensor is symmetric.

For the sake of the reader unfamiliar with the concept of a **tensor**, it is helpful to pause and define this new terminology. All of the dependent variables we normally deal with in mathematical physics are considered to be tensors. They are distinguished by what is known as their **rank**. The lowest rank tensor is rank zero which corresponds to a scalar, i.e., a

---

[2]The Taylor series expansion of $(1 + x)^{-1}$ is $(1 - x + x^2 - x^3 + \cdots)$ for $|x| < 1$.

quantity that has magnitude only. Thermodynamic properties such as pressure and density are scalar quantities. Vectors such as velocity, vorticity and pressure gradient are tensors of rank one. They have both magnitude and direction. Matrices are rank two tensors. The stress tensor, which we will introduce in the next section, is a good example for illustrating physical interpretation of a second rank tensor. It defines a force per unit area that has a magnitude and two associated directions, the direction of the force and the direction of the perpendicular to the plane on which the force acts. For a normal stress, these two directions are the same; for a shear stress, they are (by convention) perpendicular to each other. As we move to tensors of rank three and beyond, the physical interpretation becomes more difficult to ascertain. This is rarely an issue of great concern since virtually all physically relevant tensors are of rank two or less. Wilcox (1993) provides an elementary introduction to tensors and their properties. Panton (1996) provides a more advanced presentation in the context of fluid mechanics. The solid mechanics text by Fung (1965) gives a concise and comprehensive treatment of tensor calculus.

We represent the strain-rate tensor, $[\mathbf{S}]$, as the following 3 by 3 matrix.

$$[\mathbf{S}] = \begin{bmatrix} \frac{\partial u}{\partial x} & \frac{1}{2}\left(\frac{\partial v}{\partial x} + \frac{\partial u}{\partial y}\right) & \frac{1}{2}\left(\frac{\partial w}{\partial x} + \frac{\partial u}{\partial z}\right) \\ \frac{1}{2}\left(\frac{\partial u}{\partial y} + \frac{\partial v}{\partial x}\right) & \frac{\partial v}{\partial y} & \frac{1}{2}\left(\frac{\partial w}{\partial y} + \frac{\partial v}{\partial z}\right) \\ \frac{1}{2}\left(\frac{\partial u}{\partial z} + \frac{\partial w}{\partial x}\right) & \frac{1}{2}\left(\frac{\partial v}{\partial z} + \frac{\partial w}{\partial y}\right) & \frac{\partial w}{\partial z} \end{bmatrix} \tag{12.30}$$

As we saw in our study of potential-flow theory in Chapter 11, it is convenient to cast some problems in terms of cylindrical coordinates. This is certainly true for viscous flows as well, so it would be helpful to write the strain-rate tensor in terms of cylindrical coordinates, viz.,

$$[\mathbf{S}] = \begin{bmatrix} \frac{\partial u_r}{\partial r} & \frac{1}{2}\left(\frac{\partial u_\theta}{\partial r} - \frac{u_\theta}{r} + \frac{1}{r}\frac{\partial u_r}{\partial \theta}\right) & \frac{1}{2}\left(\frac{\partial w}{\partial r} + \frac{\partial u_r}{\partial z}\right) \\ \frac{1}{2}\left(\frac{1}{r}\frac{\partial u_r}{\partial \theta} + \frac{\partial u_\theta}{\partial r} - \frac{u_\theta}{r}\right) & \frac{1}{r}\frac{\partial u_\theta}{\partial \theta} + \frac{u_r}{r} & \frac{1}{2}\left(\frac{1}{r}\frac{\partial w}{\partial \theta} + \frac{\partial u_\theta}{\partial z}\right) \\ \frac{1}{2}\left(\frac{\partial u_r}{\partial z} + \frac{\partial w}{\partial r}\right) & \frac{1}{2}\left(\frac{\partial u_\theta}{\partial z} + \frac{1}{r}\frac{\partial w}{\partial \theta}\right) & \frac{\partial w}{\partial z} \end{bmatrix} \tag{12.31}$$

The computations above also determine the three components of the **local rotation vector**, i.e.,

$$\Omega = \Omega_x \mathbf{i} + \Omega_y \mathbf{j} + \Omega_z \mathbf{k} \tag{12.32}$$

Inspection of the values of Ω_x, Ω_y and Ω_z shows that this vector is proportional to the curl of the velocity. Thus, we find that the local rotation vector is given (in Cartesian and cylindrical coordinates) by

$$\Omega = \frac{1}{2}\nabla \times \mathbf{u} = \frac{1}{2}\begin{vmatrix} \mathbf{i} & \mathbf{j} & \mathbf{k} \\ \frac{\partial}{\partial x} & \frac{\partial}{\partial y} & \frac{\partial}{\partial z} \\ u & v & w \end{vmatrix} = \frac{1}{2}\begin{vmatrix} \frac{1}{r}\mathbf{e}_r & \mathbf{e}_\theta & \frac{1}{r}\mathbf{k} \\ \frac{\partial}{\partial r} & \frac{\partial}{\partial \theta} & \frac{\partial}{\partial z} \\ u_r & ru_\theta & w \end{vmatrix} \tag{12.33}$$

Summarizing, our analysis of the kinematical behavior of a fluid particle has shown the following regarding the four basic motions depicted in Figure 12.2.

- **Translation** can be represented strictly in terms of the velocity vector in the obvious way, i.e., $\Delta \mathbf{r} = \mathbf{u}\,\Delta t$.

- **Rotation**, Ω, is exactly one half the vorticity vector, $\omega = \nabla \times \mathbf{u}$.

- **Volume distortion** is quantified in terms of the normal strain rates, which are the diagonal elements of the strain-rate tensor, S_{xx}, S_{yy} and S_{zz}.

- **Angular distortion** is computed in terms of the shear-strain rates, which are the off-diagonal elements of the strain-rate tensor, S_{xy}, S_{xz}, S_{yz}, S_{yx}, S_{zx} and S_{zy}.

12.2.5   Incompressibility and Irrotationality

At this point, it is worthwhile to pause and discuss the implications of the relationships we have developed in this section. We can apply what we have learned to both incompressible and irrotational flows, and gain insight into what these constraints imply regarding the motion of a fluid particle.

Incompressibility. First, let's determine how the volume of a fluid particle varies in general. We know that at the initial instant, the volume is $V(0) = \xi \eta \zeta$. For small times, Δt, the fluid particle's shape is very nearly that of a rectangular parallelepiped, wherefore $V(\Delta t) \approx (\xi + \Delta \xi)(\eta + \Delta \eta)(\zeta + \Delta \zeta)$. Thus, the rate of change of the fluid particle's volume, relative to its initial volume is

$$\frac{1}{V}\frac{dV}{dt} = \lim_{\Delta t \to 0} \frac{(\xi + \Delta \xi)(\eta + \Delta \eta)(\zeta + \Delta \zeta) - \xi \eta \zeta}{\xi \eta \zeta \Delta t} \tag{12.34}$$

Then, from Equation (12.19) and the obvious generalization for S_{zz}, we have

$$\begin{aligned} \frac{1}{V}\frac{dV}{dt} &= \lim_{\Delta t \to 0} \frac{\xi(1 + S_{xx}\Delta t)\eta(1 + S_{yy}\Delta t)\zeta(1 + S_{zz}\Delta t) - \xi \eta \zeta}{\xi \eta \zeta \Delta t} \\ &= \lim_{\Delta t \to 0} \frac{(1 + S_{xx}\Delta t + S_{yy}\Delta t + S_{zz}\Delta t + \cdots) - 1}{\Delta t} \end{aligned} \tag{12.35}$$

Finally, taking the limit, we conclude that

$$\frac{1}{V}\frac{dV}{dt} = S_{xx} + S_{yy} + S_{zz} \tag{12.36}$$

Thus, we see that the rate of change of volume, per unit volume, of a fluid particle is the sum of the three normal strain-rate components. This sum is known as the **trace** of the tensor. We can now express the rate of change of volume in terms of the velocity by using Equation (12.20), wherefore

$$\frac{1}{V}\frac{dV}{dt} = \frac{\partial u}{\partial x} + \frac{\partial v}{\partial y} + \frac{\partial w}{\partial x} = \nabla \cdot \mathbf{u} \tag{12.37}$$

This result tells us that when we have incompressible flow, for which $\nabla \cdot \mathbf{u}$ vanishes, the volume of a fluid particle remains constant.

Figure 12.4 illustrates two examples of volume distortions that are possible in an incompressible flow. The initial shape of the fluid particle [Figure 12.4(a)] is assumed to be a cube. If the particle undergoes compression in the xy plane, it must expand in the z direction. Figure 12.4(b), drawn exactly to scale, shows what happens when the x side is reduced by 10% and the y side by 40%. To have the same volume as the original cube, the z side must increase by 85% ($.9 \times .6 \times 1.85 = 1$). Similarly, for the expansion case shown in Figure 12.4(c), doubling the length of the x side with the y side unchanged requires halving the z side to maintain constant volume.

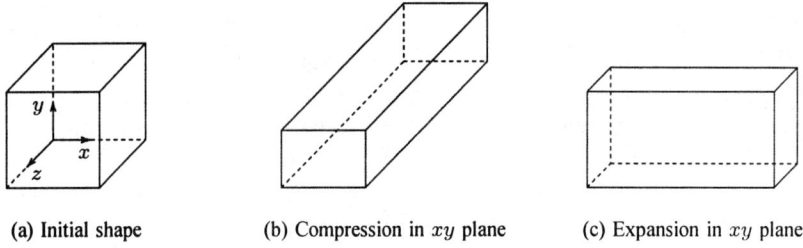

(a) Initial shape   (b) Compression in xy plane   (c) Expansion in xy plane

Figure 12.4: *Incompressible volume-deformation examples.*

Irrotationality. Turning now to the irrotationality condition, Equation (12.33) tells us the vorticity is twice the rotation rate of a fluid particle. Thus, in an irrotational flow, for which $\nabla \times \mathbf{u} = \mathbf{0}$, fluid particles always retain their original orientation. For example, consider the potential vortex that we discussed in Section 4.3 and Subsection 11.5.2. The velocity is

$$u_r = 0 \quad \text{and} \quad u_\theta = \frac{\Gamma}{2\pi r} \tag{12.38}$$

where Γ is the circulation. Reference to Equations (12.31) and (12.33) shows that both the strain-rate tensor, $[\mathbf{S}]$, and the vorticity vector, $\nabla \times \mathbf{u}$, vanish for this velocity field. Thus, because the strain-rate tensor is zero, there is no volume or angular distortion. This means fluid particles move through the flow, retaining their original shape for all time. Also, because the vorticity is zero, fluid particles experience no local rotation, but rather, always retain their original orientation. Figure 12.5(a) shows a fluid particle at several points as it moves on a streamline in potential-vortex flow.

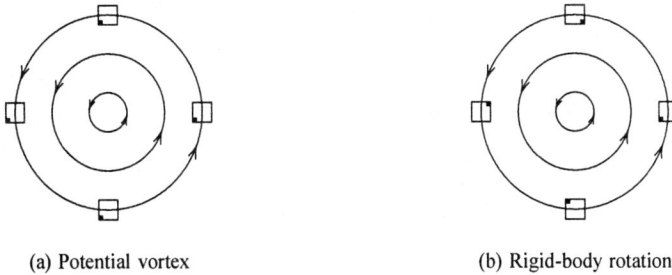

(a) Potential vortex   (b) Rigid-body rotation

Figure 12.5: *Motion of a fluid particle for a potential vortex and rigid-body rotation; one corner is marked to indicate the fluid particle's orientation.*

By contrast, we can examine the corresponding motion for a rigid-body rotation, in which the velocity is

$$u_r = 0 \quad \text{and} \quad u_\theta = \Omega r \tag{12.39}$$

By definition, in the context of fluid motion, a rigid-body rotation is a flow in which all of the fluid rotates as a unit similar to a solid body. We noted earlier (Section 4.3) that the motion in the core of a hurricane is well approximated as a rigid-body rotation. As with the potential vortex, the strain-rate tensor is zero so that fluid particles do not change in shape. However, the vorticity is

$$\nabla \times \mathbf{u} = 2\Omega \mathbf{k} \tag{12.40}$$

This means the fluid particles rotate at half this rate, which is exactly the rotation rate of the overall motion. Figure 12.5(b) illustrates the motion of a fluid particle in a rigid-body rotation. Similar to the potential vortex, the streamlines are concentric circles. In complete contrast, each fluid particle changes its orientation continuously (by rotating) as it moves along a streamline.

These two contrasting examples, irrotational potential-vortex flow and rotational rigid-body rotation, underscore the importance of vorticity. The constraint of irrotationality restricts the type of motion possible, causing fluid particles to always have the same orientation. Rotational flows, by contrast, involve continuous "tumbling" of fluid particles. The accelerations attending these fundamentally different types of motion are significantly different, and require quite different pressure fields (and other forces) to maintain.

12.3   The Viscous Stress Tensor

For an **ideal** fluid, the only surface force acting is normal to the surface of a fluid particle, viz., the pressure. Hence, the surface force on an area dS with unit normal $\mathbf{n}$ is

$$d\mathbf{F}_s = -p\,\mathbf{n}\,dS \tag{12.41}$$

so that the stress, $d\mathbf{F}_s/dS$, is

$$\frac{d\mathbf{F}_s}{dS} = -p\,\mathbf{n} \qquad \text{(Perfect fluid)} \tag{12.42}$$

By contrast, for a **real** fluid, we obtain tangential (shear) stresses as well. Thus, in a real fluid, we have

$$\frac{d\mathbf{F}_s}{dS} = \underbrace{-p\,\mathbf{n}}_{Pressure} + \underbrace{\mathbf{f}_v}_{Viscous} \qquad \text{(Viscous fluid)} \tag{12.43}$$

The surface force is a vector because it has both magnitude and direction. Pressure having only magnitude, is a **scalar** quantity, whose force acts in a direction normal to the surface. The frictional stress, $\mathbf{f}_v$, requires even more information because, in addition to having magnitude and direction, it also varies according to the orientation of the surface. To understand how we represent such a force mathematically, it is instructive to consider the tetrahedral fluid particle shown in Figure 12.6.

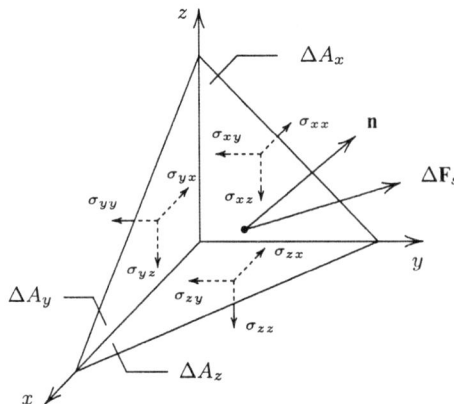

Figure 12.6: *Force balance on a tetrahedral fluid particle.*

We assume the fluid particle is of differential size with the areas of its four faces identified as follows.

- ΔA_x = Area of the face normal to the x axis (lies in the yz plane)

- ΔA_y = Area of the face normal to the y axis (lies in the xz plane)

- ΔA_z = Area of the face normal to the z axis (lies in the xy plane)

- ΔS = Area of the inclined face

Clearly, the areas ΔA_x, ΔA_y and ΔA_z are the projections of ΔS on the yz, xz and xy planes, respectively. We can express this in terms of the appropriate unit normals so that

$$\left. \begin{aligned} \Delta A_x &= (\mathbf{i} \cdot \mathbf{n}) \, \Delta S \\ \Delta A_y &= (\mathbf{j} \cdot \mathbf{n}) \, \Delta S \\ \Delta A_z &= (\mathbf{k} \cdot \mathbf{n}) \, \Delta S \end{aligned} \right\} \tag{12.44}$$

Now, in general, the force acting on any of the tetrahedron's four faces has three components. This is shown in Figure 12.6 where, for example, the force on the face normal to the x axis has components σ_{xx}, σ_{xy} and σ_{xz}. The first subscript corresponds to the unit-normal direction, and the second to the direction in which the force component acts. Similar notation holds for the other two faces. Note that, at this point, we make no distinction between pressure and viscous forces. We will separate them later.

Our goal is to examine the balance of forces in the limit $\Delta S \rightarrow 0$. By doing this, we will arrive at a representation that holds at every point in the flowfield. From Newton's second law of motion, we know that

$$(\text{Mass}) \times (\text{Acceleration}) = \sum (\text{Body Forces}) + \sum (\text{Surface Forces}) \tag{12.45}$$

As argued by Batchelor (1967), the surface forces dominate all other terms in this equation in the limit $\Delta S \rightarrow 0$. This is true because the net body force is proportional to the volume of the tetrahedron, ΔV, which is proportional to the product of the lengths of the three sides of the tetrahedron, i.e., $\Delta V = \frac{1}{6} \Delta x \Delta y \Delta z$. By contrast, the areas are all quadratic in the lengths of the sides, while the volume is cubic, so that $\Delta V \ll \Delta S$. For the same reason, we can also neglect the mass-times-acceleration contribution to the force balance since the mass of the tetrahedron is the product of fluid density and ΔV. So, we need only consider the surface forces as we reduce our tetrahedron to a single point.

First, we focus on the x direction. The x component of $\Delta \mathbf{F}_s$ balances the x components of the forces on the three faces. That is, we have

$$\begin{aligned} (\Delta \mathbf{F}_s)_x &= \sigma_{xx} \Delta A_x + \sigma_{yx} \Delta A_y + \sigma_{zx} \Delta A_z \\ &= \sigma_{xx} (\mathbf{i} \cdot \mathbf{n}) \, \Delta S + \sigma_{yx} (\mathbf{j} \cdot \mathbf{n}) \, \Delta S + \sigma_{zx} (\mathbf{k} \cdot \mathbf{n}) \, \Delta S \\ &= \mathbf{n} \cdot (\sigma_{xx} \mathbf{i} + \sigma_{yx} \mathbf{j} + \sigma_{zx} \mathbf{k}) \, \Delta S \end{aligned} \tag{12.46}$$

where we make use of Equation (12.44) to eliminate ΔA_x, ΔA_y and ΔA_z. Similarly, balancing forces in the y and z directions, we find

$$(\Delta \mathbf{F}_s)_y = \mathbf{n} \cdot (\sigma_{xy} \mathbf{i} + \sigma_{yy} \mathbf{j} + \sigma_{zy} \mathbf{k}) \, \Delta S \tag{12.47}$$

$$(\Delta \mathbf{F}_s)_z = \mathbf{n} \cdot (\sigma_{xz} \mathbf{i} + \sigma_{yz} \mathbf{j} + \sigma_{zz} \mathbf{k}) \, \Delta S \tag{12.48}$$

We can simplify these results further by observing that, in general, the unit normal to the inclined face is written as

$$\mathbf{n} = n_x\mathbf{i} + n_y\mathbf{j} + n_z\mathbf{k} \tag{12.49}$$

Hence, we have

$$\mathbf{n} \cdot \mathbf{i} = n_x, \quad \mathbf{n} \cdot \mathbf{j} = n_y, \quad \mathbf{n} \cdot \mathbf{k} = n_z \tag{12.50}$$

so that Equations (12.46), (12.47) and (12.48) can be rewritten in the following streamlined form.

$$\left.\begin{array}{l} \dfrac{(\Delta \mathbf{F}_s)_x}{\Delta S} = n_x\sigma_{xx} + n_y\sigma_{yx} + n_z\sigma_{zx} \\[2mm] \dfrac{(\Delta \mathbf{F}_s)_y}{\Delta S} = n_x\sigma_{xy} + n_y\sigma_{yy} + n_z\sigma_{zy} \\[2mm] \dfrac{(\Delta \mathbf{F}_s)_z}{\Delta S} = n_x\sigma_{xz} + n_y\sigma_{yz} + n_z\sigma_{zz} \end{array}\right\} \tag{12.51}$$

Finally, inspection of Equations (12.51) shows that they can be expressed as a matrix multiplication. Specifically, we have a 3 x 3 matrix whose components are σ_{xx}, σ_{xy}, etc., premultiplied by the unit normal, $\mathbf{n}$, which is a 1 x 3 row vector. Hence, taking the limit $\Delta S \to 0$, we have

$$\frac{d\mathbf{F}_s}{dS} = \begin{bmatrix} n_x & n_y & n_z \end{bmatrix} \begin{bmatrix} \sigma_{xx} & \sigma_{xy} & \sigma_{xz} \\ \sigma_{yx} & \sigma_{yy} & \sigma_{yz} \\ \sigma_{zx} & \sigma_{zy} & \sigma_{zz} \end{bmatrix} \tag{12.52}$$

We can write this in symbolic form as

$$\frac{d\mathbf{F}_s}{dS} = \mathbf{n} \cdot [\sigma] \tag{12.53}$$

where $[\sigma]$ is the **stress tensor**, defined by

$$[\sigma] = \begin{bmatrix} \sigma_{xx} & \sigma_{xy} & \sigma_{xz} \\ \sigma_{yx} & \sigma_{yy} & \sigma_{yz} \\ \sigma_{zx} & \sigma_{zy} & \sigma_{zz} \end{bmatrix} \tag{12.54}$$

Up to this point, everything we have done to arrive at the stress tensor applies to both fluids and solids. In fluid mechanics, it is customary to separate the pressure and viscous stresses. To do this, we rewrite $[\sigma]$ as

$$[\sigma] = \begin{bmatrix} -p + \tau_{xx} & \tau_{xy} & \tau_{xz} \\ \tau_{yx} & -p + \tau_{yy} & \tau_{yz} \\ \tau_{zx} & \tau_{zy} & -p + \tau_{zz} \end{bmatrix} \tag{12.55}$$

or,

$$[\sigma] = -p \underbrace{\begin{bmatrix} 1 & 0 & 0 \\ 0 & 1 & 0 \\ 0 & 0 & 1 \end{bmatrix}}_{[\delta]} + \underbrace{\begin{bmatrix} \tau_{xx} & \tau_{xy} & \tau_{xz} \\ \tau_{yx} & \tau_{yy} & \tau_{yz} \\ \tau_{zx} & \tau_{zy} & \tau_{zz} \end{bmatrix}}_{[\tau]} = -p[\delta] + [\tau] \tag{12.56}$$

The matrix $[\delta]$ is known as the **identity matrix** or **identity tensor**, while the matrix $[\tau]$ is the **viscous stress tensor**. Thus, in terms of our original notation [Equation (12.43)], the proper mathematical representation of the viscous stress, $\mathbf{f}_v$, is

$$\mathbf{f}_v = \mathbf{n} \cdot [\tau] \tag{12.57}$$

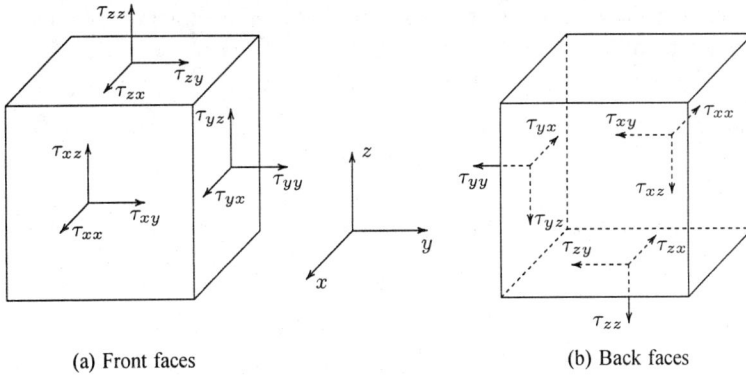

(a) Front faces   (b) Back faces

Figure 12.7: *Stress-tensor index conventions.*

We have a total of nine stress-tensor components. This corresponds to having three stress components per face with three linearly independent unit normals. The fact that we require this much information to describe the viscous stress accounts for the need to introduce a second-rank tensor.

Figure 12.7 illustrates the naming convention for the stress tensor components. As noted earlier, the first index denotes the face on which the stress acts, while the second index denotes the direction in which the force acts. For example, τ_{xy} is the stress in the y direction on an x-oriented face (i.e., a face whose unit normal is **i**).

12.4 Integral Form of the Momentum Equation

12.4.1 Derivation

Recall from Section 5.2 that application of Newton's second law of motion to a system, and use of the Reynolds Transport Theorem tells us that for a control volume (see Figure 12.8):

$$\iiint_V \frac{\partial}{\partial t}(\rho\,\mathbf{u})\,dV + \oiint_S \rho\,\mathbf{u}(\mathbf{u}\cdot\mathbf{n})dS = \mathbf{F}_s + \mathbf{F}_b \qquad (12.58)$$

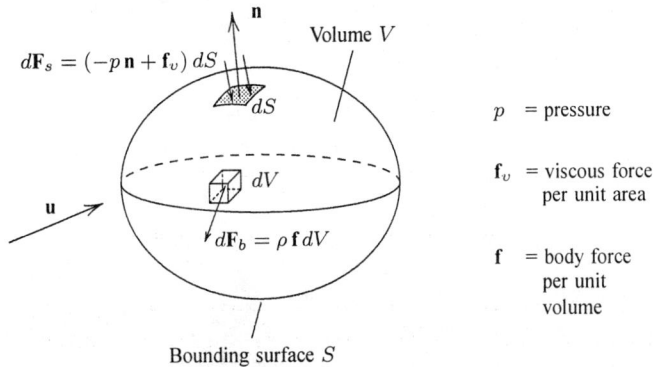

p = pressure

$\mathbf{f}_v$ = viscous force per unit area

$\mathbf{f}$ = body force per unit volume

Figure 12.8: *A general control volume for momentum conservation.*

The vectors $\mathbf{F}_s$ and $\mathbf{F}_b$ are the surface force and body force exerted by the surroundings on the control volume. As before, we introduce the *specific body force vector*, $\mathbf{f}$, so that the net body force acting on the control volume is

$$\mathbf{F}_b = \iiint_V \rho \, \mathbf{f} \, dV \tag{12.59}$$

All that remains to reformulate the momentum conservation principle for a viscous fluid is to specify the surface force.

For a viscous fluid, the surface force is given by

$$\mathbf{F}_s = \oiint_S (-p \, \mathbf{n} + \mathbf{f}_v) \, dS \tag{12.60}$$

However, we prefer to rewrite the surface force in terms of the pressure and the viscous stress tensor, $[\tau]$. So, appealing to Equation (12.57), we have

$$\mathbf{F}_s = \oiint_S (-p \, \mathbf{n} + \mathbf{n} \cdot [\tau]) \, dS \tag{12.61}$$

Combining Equations (12.58), (12.59) and (12.61), the viscous-flow replacement for Equation (5.16) is as follows.

$$\iiint_V \frac{\partial}{\partial t}(\rho \, \mathbf{u}) \, dV + \oiint_S \rho \, \mathbf{u}(\mathbf{u} \cdot \mathbf{n}) dS = - \oiint_S p \, \mathbf{n} \, dS + \iiint_V \rho \, \mathbf{f} \, dV + \oiint_S \mathbf{n} \cdot [\tau] \, dS \tag{12.62}$$

The last term is the net viscous force exerted on the control volume.

12.4.2   Control-Volume Example

The viscous-flow momentum-conservation principle in integral form applies to control volumes in the same manner detailed in Chapter 6. Its implementation is more complicated because of the viscous term, and an example at this point is instructive to underscore how friction manifests itself in a context we have studied in great detail, viz., the control volume method.

We will consider flow near a channel inlet with constant channel height, H, as sketched in Figure 12.9. As shown, the flow at the inlet is uniform, and we select a control volume that is indicated by the dashed contour. To simplify the analysis, we assume the flow is steady

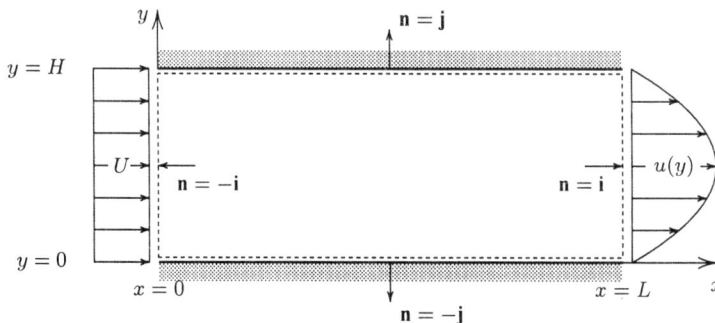

Figure 12.9: *Control volume for viscous flow in a channel.*

and incompressible with constant pressure. We also neglect body forces. Thus, the integral form of the momentum conservation principle simplifies to

$$\underbrace{\oiint_S \rho \, \mathbf{u}(\mathbf{u} \cdot \mathbf{n}) dS}_{Net \ Momentum \ Flux} = \underbrace{\oiint_S \mathbf{n} \cdot [\boldsymbol{\tau}] \, dS}_{Net \ Viscous \ Force} \qquad (12.63)$$

First, let's consider the net momentum-flux integral. At the inlet, the velocity and unit normal are $\mathbf{u} = U\,\mathbf{i}$ and $\mathbf{n} = -\mathbf{i}$, respectively. Thus,

$$\rho \, \mathbf{u}(\mathbf{u} \cdot \mathbf{n}) = -\rho \, U^2 \mathbf{i} \qquad \text{(Inlet)} \qquad (12.64)$$

At the outlet, the velocity and unit normal are $\mathbf{u} = u(y)\mathbf{i}$ and $\mathbf{n} = \mathbf{i}$, respectively. Hence, we have

$$\rho \, \mathbf{u}(\mathbf{u} \cdot \mathbf{n}) = \rho \, u^2(y)\mathbf{i} \qquad \text{(Outlet)} \qquad (12.65)$$

Because there is no flow through either of the channel walls, the only contributions to the net momentum-flux integral come from the inlet and the outlet. So, assuming unit channel width (out of the page) so that $dS = 1 \cdot dy$, there follows

$$\begin{aligned}
\oiint_S \rho \, \mathbf{u}(\mathbf{u} \cdot \mathbf{n}) dS &= \mathbf{i} \int_0^H \left[-\rho \, U^2 + \rho \, u^2(y) \right] dy \\
&= -\mathbf{i} \rho \, U^2 \int_0^H \left[1 - (u/U)^2 \right] dy \qquad (12.66)
\end{aligned}$$

Turning to the net viscous-force integral, we will confine our attention to the x component. This is the most interesting component since the momentum flux has only an x component. Also, the mathematical operations for the y and z directions are virtually identical to those for the x direction, so little additional information would be gained by including the details. The first step is to expand the net viscous-force integral into integrals over the four surfaces of the control volume. So,

$$\begin{aligned}
\oiint_S \mathbf{n} \cdot [\boldsymbol{\tau}] \, dS &= \underbrace{\int_0^H (-\mathbf{i}) \cdot [\boldsymbol{\tau}] \, dy}_{Inlet} + \underbrace{\int_0^H \mathbf{i} \cdot [\boldsymbol{\tau}] \, dy}_{Outlet} \\
&+ \underbrace{\int_0^L (-\mathbf{j}) \cdot [\boldsymbol{\tau}] \, dx}_{Lower \ Wall} + \underbrace{\int_0^L \mathbf{j} \cdot [\boldsymbol{\tau}] \, dx}_{Upper \ Wall} \qquad (12.67)
\end{aligned}$$

Next, we must compute the dot products of the unit normals with the viscous stress tensor. There follows:

$$\begin{aligned}
\mathbf{i} \cdot [\boldsymbol{\tau}] &= \begin{bmatrix} 1 & 0 & 0 \end{bmatrix} \begin{bmatrix} \tau_{xx} & \tau_{xy} & \tau_{xz} \\ \tau_{yx} & \tau_{yy} & \tau_{yz} \\ \tau_{zx} & \tau_{zy} & \tau_{zz} \end{bmatrix} \\
&= \begin{bmatrix} \tau_{xx} & \tau_{xy} & \tau_{xz} \end{bmatrix} = \tau_{xx} \mathbf{i} + \tau_{xy} \mathbf{j} + \tau_{xz} \mathbf{k} \qquad (12.68)
\end{aligned}$$

and

$$\begin{aligned}
\mathbf{j} \cdot [\boldsymbol{\tau}] &= \begin{bmatrix} 0 & 1 & 0 \end{bmatrix} \begin{bmatrix} \tau_{xx} & \tau_{xy} & \tau_{xz} \\ \tau_{yx} & \tau_{yy} & \tau_{yz} \\ \tau_{zx} & \tau_{zy} & \tau_{zz} \end{bmatrix} \\
&= \begin{bmatrix} \tau_{yx} & \tau_{yy} & \tau_{yz} \end{bmatrix} = \tau_{yx} \mathbf{i} + \tau_{yy} \mathbf{j} + \tau_{yz} \mathbf{k} \qquad (12.69)
\end{aligned}$$

Note that the dot product of a vector and a second-rank tensor (i.e., a 3 x 3 matrix) is another vector and not a scalar. The dot product reduces the order of a tensor by one. Therefore, the x component of $\mathbf{i} \cdot [\tau]$ is τ_{xx} and the x component of $\mathbf{j} \cdot [\tau]$ is τ_{yx}. The x component of the net viscous-force integral is

$$\left[\oiint_S \mathbf{n} \cdot [\tau]\, dS \right]_x = -\int_0^H \tau_{xx}(0,y)\, dy + \int_0^H \tau_{xx}(L,y)\, dy$$
$$-\int_0^L \tau_{yx}(x,0)\, dx + \int_0^L \tau_{yx}(x,H)\, dx \qquad (12.70)$$

If we further assume that the channel is of sufficient length that fully-developed flow exists at the outlet, i.e., that the flow is independent of x (see Figure 7.5), then $\tau_{xx}(L,y) = 0$. This is true because, as we will discover when we introduce Stokes' postulate in Section 12.6, τ_{xx} is proportional to $\partial u/\partial x$. Of course, since the inlet flow is uniform, $\tau_{xx}(0,y)$ also vanishes. Therefore, we conclude that

$$\left[\oiint_S \mathbf{n} \cdot [\tau]\, dS \right]_x = -\int_0^L [\tau_{yx}(x,0)\, dx - \tau_{yx}(x,H)]\, dx \qquad (12.71)$$

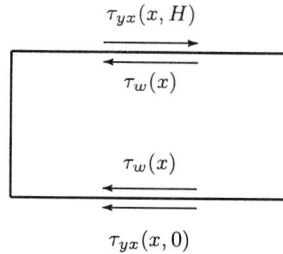

Figure 12.10: *Shear-stress sign conventions at the channel walls.*

The final step illustrates the sign conventions for the stress tensor (see Figure 12.7). On the upper channel wall, τ_{yx} is positive in the positive x direction. By contrast, it is positive in the negative x direction on the lower channel wall. Now, by symmetry about the channel centerline, we expect the viscous shearing force to point in the negative x direction on both channel walls (see Figure 12.10). Denoting the magnitude of the shear stress at the wall by $\tau_w(x)$, we must have

$$\tau_{yx}(x,0) = \tau_w(x) \qquad \text{and} \qquad \tau_{yx}(x,H) = -\tau_w(x) \qquad (12.72)$$

Therefore, the final form of the net viscous-force integral is

$$\left[\oiint_S \mathbf{n} \cdot [\tau]\, dS \right]_x = -2\int_0^L \tau_w(x)\, dx \qquad (12.73)$$

Substituting Equations (12.66) and (12.73) into Equation (12.63) yields the following equation relating the momentum flux and the viscous force.

$$\rho U^2 \int_0^H [1 - (u/U)^2]\, dy = 2\int_0^L \tau_w(x)\, dx \qquad (12.74)$$

This equation tells us that the momentum flux is reduced between outlet and inlet by an amount equal to the sum of the viscous forces on the channel walls.

12.5 Navier's Equation

12.5.1 Derivation

In Section 5.2, we showed that in the absence of viscous forces, the differential form of the momentum-conservation principle is Euler's equation. The steps required to deduce the corresponding equation for a viscous fluid are nearly identical, except that now we must include the viscous term. As in the inviscid case, it is convenient to work with the z component of Equation (12.62), i.e.,

$$\iiint_V \frac{\partial}{\partial t}(\rho w)\, dV + \oiint_S \rho w(\mathbf{u}\cdot\mathbf{n})dS = -\mathbf{k}\cdot\oiint_S p\,\mathbf{n}\,dS + \iiint_V \rho f_z\, dV$$
$$+ \mathbf{k}\cdot\oiint_S \mathbf{n}\cdot[\tau]\,dS \qquad (12.75)$$

which can be rearranged to read as follows.

$$\iiint_V \frac{\partial}{\partial t}(\rho w)\, dV + \oiint_S (\rho w\mathbf{u})\cdot\mathbf{n}\,dS + \oiint_S p\,\mathbf{k}\cdot\mathbf{n}\,dS - \iiint_V \rho f_z\, dV$$
$$- \oiint_S (\mathbf{n}\cdot[\tau])_z\, dS = 0 \qquad (12.76)$$

Before applying the divergence theorem, we must examine the viscous term more closely. First, observe that by definition, $\mathbf{n}\cdot[\tau]$ is a row vector that is formed by premultiplying the 3 x 3 viscous stress tensor, $[\tau]$, by the unit normal, $\mathbf{n}$, which is a 1 x 3 row vector, i.e.,

$$\mathbf{n}\cdot[\tau] = \begin{bmatrix} n_x & n_y & n_z \end{bmatrix} \begin{bmatrix} \tau_{xx} & \tau_{xy} & \tau_{xz} \\ \tau_{yx} & \tau_{yy} & \tau_{yz} \\ \tau_{zx} & \tau_{zy} & \tau_{zz} \end{bmatrix} \qquad (12.77)$$

Performing the indicated matrix multiplication, the z component of $\mathbf{n}\cdot[\tau]$ is

$$(\mathbf{n}\cdot[\tau])_z = n_x\tau_{xz} + n_y\tau_{yz} + n_z\tau_{zz} = (\tau_{xz}\mathbf{i} + \tau_{yz}\mathbf{j} + \tau_{zz}\mathbf{k})\cdot\mathbf{n} \qquad (12.78)$$

Note that this tells us the z component of $\mathbf{n}\cdot[\tau]$ is equal to the dot product of $\mathbf{n}$ and the vector whose components lie in the third column of $[\tau]$. Hence, we can rewrite Equation (12.76) in a form for which application of the divergence theorem is obvious, viz.,

$$\iiint_V \frac{\partial}{\partial t}(\rho w)\, dV + \oiint_S (\rho w\mathbf{u})\cdot\mathbf{n}\,dS + \oiint_S p\,\mathbf{k}\cdot\mathbf{n}\,dS - \iiint_V \rho f_z\, dV$$
$$- \oiint_S (\tau_{xz}\mathbf{i} + \tau_{yz}\mathbf{j} + \tau_{zz}\mathbf{k})\cdot\mathbf{n}\,dS = 0 \qquad (12.79)$$

Thus, using the divergence theorem, we find

$$\iiint_V \left[\frac{\partial}{\partial t}(\rho w) + \nabla\cdot(\rho w\mathbf{u}) + \nabla\cdot(p\mathbf{k}) - \rho f_z - \nabla\cdot(\tau_{xz}\mathbf{i} + \tau_{yz}\mathbf{j} + \tau_{zz}\mathbf{k})\right] dV = 0 \quad (12.80)$$

Because the volume is arbitrary, necessarily the integrand must vanish. Hence, the conservation form of the differential equation for z-momentum is

$$\frac{\partial}{\partial t}(\rho w) + \nabla\cdot(\rho w\mathbf{u}) = -\nabla\cdot(p\mathbf{k}) + \rho f_z + \nabla\cdot(\tau_{xz}\mathbf{i} + \tau_{yz}\mathbf{j} + \tau_{zz}\mathbf{k}) \qquad (12.81)$$

As in Section 5.2, we can rewrite this equation in primitive-variable form by noting that

$$\frac{\partial}{\partial t}(\rho w) + \nabla \cdot (\rho w \mathbf{u}) = \rho \frac{dw}{dt} \quad \text{and} \quad \nabla \cdot (p\mathbf{k}) = \frac{\partial p}{\partial z} \tag{12.82}$$

Thus, we arrive at

$$\rho \frac{dw}{dt} = -\frac{\partial p}{\partial z} + \rho f_z + \nabla \cdot (\tau_{xz}\mathbf{i} + \tau_{yz}\mathbf{j} + \tau_{zz}\mathbf{k}) \tag{12.83}$$

Finally, recall the origin of the viscous-force vector. It is the third column of the viscous stress tensor. In a similar way, for the y and x directions, the rules of matrix multiplication tell us the viscous force vectors are the second and first columns, respectively, of $[\tau]$. Hence, it should be clear that the differential equations for momentum conservation in the y and x directions are:

$$\rho \frac{dv}{dt} = -\frac{\partial p}{\partial y} + \rho f_y + \nabla \cdot (\tau_{xy}\mathbf{i} + \tau_{yy}\mathbf{j} + \tau_{zy}\mathbf{k}) \tag{12.84}$$

$$\rho \frac{du}{dt} = -\frac{\partial p}{\partial x} + \rho f_x + \nabla \cdot (\tau_{xx}\mathbf{i} + \tau_{yx}\mathbf{j} + \tau_{zx}\mathbf{k}) \tag{12.85}$$

We can rewrite Equations (12.83), (12.84) and (12.85) in more compact form just as we did for Euler's equation in Section 5.2. Combining these three equations into a single vector equation, we arrive at the final desired form of momentum conservation for a viscous fluid, viz.,

$$\rho \frac{d\mathbf{u}}{dt} = -\nabla p + \rho \mathbf{f} + \nabla \cdot [\tau] \tag{12.86}$$

where the divergence of the viscous stress tensor is defined by premultiplying $[\tau]$ by ∇ (represented as a row vector), i.e.,

$$\nabla \cdot [\tau] = \begin{bmatrix} \frac{\partial}{\partial x} & \frac{\partial}{\partial y} & \frac{\partial}{\partial z} \end{bmatrix} \begin{bmatrix} \tau_{xx} & \tau_{xy} & \tau_{xz} \\ \tau_{yx} & \tau_{yy} & \tau_{yz} \\ \tau_{zx} & \tau_{zy} & \tau_{zz} \end{bmatrix} \tag{12.87}$$

This equation, derived by the French mathematician Louis Navier early in the nineteenth century, is the viscous-flow replacement for Euler's equation. In his honor, it is generally referred to as **Navier's equation**. To write Navier's equation in component form in Cartesian coordinates, we simply evaluate the divergence of the stress-force terms in Equations (12.83) through (12.85). The resulting equations are

$$\rho \frac{du}{dt} = -\frac{\partial p}{\partial x} + \rho f_x + \frac{\partial \tau_{xx}}{\partial x} + \frac{\partial \tau_{yx}}{\partial y} + \frac{\partial \tau_{zx}}{\partial z} \tag{12.88}$$

$$\rho \frac{dv}{dt} = -\frac{\partial p}{\partial y} + \rho f_y + \frac{\partial \tau_{xy}}{\partial x} + \frac{\partial \tau_{yy}}{\partial y} + \frac{\partial \tau_{zy}}{\partial z} \tag{12.89}$$

$$\rho \frac{dw}{dt} = -\frac{\partial p}{\partial z} + \rho f_z + \frac{\partial \tau_{xz}}{\partial x} + \frac{\partial \tau_{yz}}{\partial y} + \frac{\partial \tau_{zz}}{\partial z} \tag{12.90}$$

12.5.2 Symmetry of the Stress Tensor

In deriving Navier's equation, we have not only arrived at a more complicated equation, we have created a closure problem. That is, we have added nine additional unknowns to our set of equations, i.e., the nine components of the viscous stress tensor. Conservation of angular momentum shows that $[\tau]$ is symmetric, which provides the following three equations.

$$\tau_{xy} = \tau_{yx}, \quad \tau_{xz} = \tau_{zx}, \quad \tau_{yz} = \tau_{zy} \tag{12.91}$$

We omit the proof here for the sake of brevity — details can be found in a more advanced text such as Batchelor (1967) or Panton (1996). Also, the Problems section outlines the proof. This simplifies our closure problem somewhat. To close our set of equations for general fluid motion, we must still find six additional equations defining the stress-tensor components.

12.5.3 Understanding Surface-Force Balances

Navier's equation is clearly quite a bit more complicated than Euler's equation. The presence of the viscous term, $\nabla \cdot [\tau]$, adds three additional terms on the right-hand sides of each component in a full three-dimensional geometry. It is thus worthwhile to pause at this point to discuss the physical meaning of $\nabla \cdot [\tau]$.

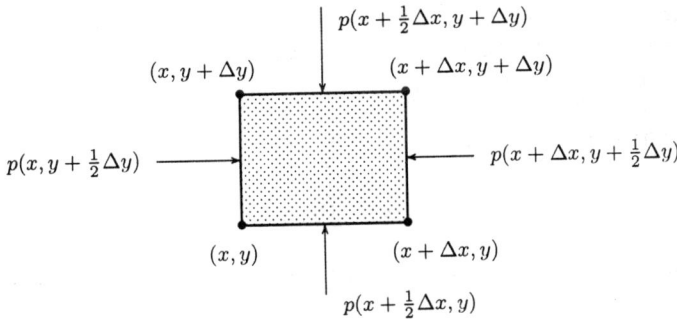

Figure 12.11: *Pressure forces acting on a fluid particle.*

To explain the way the viscous force acts on a given fluid particle, we begin by analyzing a two-dimensional geometry shown in Figure 12.11. To further simplify our construction, we examine the force balance due to the pressure. Once we have this relatively simple case in hand, we can tackle the more complicated viscous force. Consider a differential fluid particle with sides Δx and Δy as shown in Figure 12.11. We assume the pressure acts at the center of each side. The actual point of application is of no consequence as we will eventually take the limit $\Delta x \to 0$ and $\Delta y \to 0$. The net pressure force in the x direction, $\sum F_x^{(p)}$, is the difference between the products of the pressure and the area of the side (Δy in this two-dimensional geometry) acting on each side.

$$\sum F_x^{(p)} = \left[p\left(x, y + \tfrac{1}{2}\Delta y\right) - p\left(x + \Delta x, y + \tfrac{1}{2}\Delta y\right) \right] \Delta y \tag{12.92}$$

Because the volume of our fluid particle is $\Delta x \Delta y$, we conclude that the force per unit volume, in the limit $\Delta x \to 0$ and $\Delta y \to 0$, is

$$\frac{\sum F_x^{(p)}}{\Delta x \Delta y} = -\frac{p\left(x + \Delta x, y + \tfrac{1}{2}\Delta y\right) - p\left(x, y + \tfrac{1}{2}\Delta y\right)}{\Delta x} \to -\frac{\partial p}{\partial x} \tag{12.93}$$

Similarly, the net pressure force in the y direction is

$$\sum F_y^{(p)} = \left[p\left(x + \tfrac{1}{2}\Delta x, y\right) - p\left(x + \tfrac{1}{2}\Delta x, y + \Delta y\right) \right] \Delta x \tag{12.94}$$

so that the force per unit volume in the limit $\Delta x \to 0$ and $\Delta y \to 0$ is

$$\frac{\sum F_y^{(p)}}{\Delta x \Delta y} = -\frac{p\left(x + \frac{1}{2}\Delta x, y + \Delta y\right) - p\left(x + \frac{1}{2}\Delta x, y\right)}{\Delta y} \to -\frac{\partial p}{\partial y} \qquad (12.95)$$

Equations (12.93) and (12.95) show that the x and y components of the force per unit volume on the fluid particle are simply the negative of the x and y components of the pressure gradient, ∇p, respectively. Clearly, the same result holds when we consider the z direction as well. Therefore, denoting the pressure force per unit volume by $\mathbf{f}^{(p)}$, we can summarize our results as follows.

$$\mathbf{f}^{(p)} = -\nabla p \qquad (12.96)$$

In words, *the pressure gradient is the net pressure force per unit volume acting on a fluid particle.*

Having successfully completed our warm-up exercise on the pressure forces, we are now prepared to turn to the viscous-force balance. We again focus on a two-dimensional fluid particle. Although not included to eliminate clutter in Figure 12.12, the coordinates of the four vertices are the same as in Figure 12.11. As shown, each face has both a normal and a tangential force acting. Had we included the z direction, there would be a second tangential component on each face (in the z direction). The directions indicated are consistent with the conventions specified in Figure 12.7.

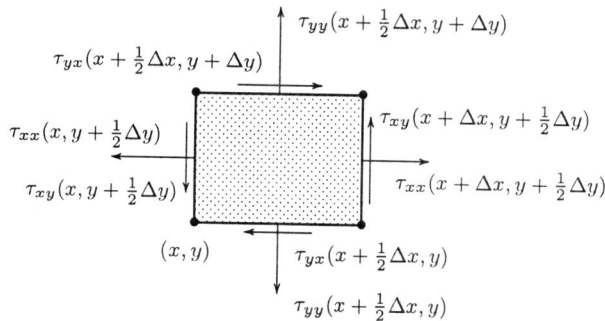

Figure 12.12: *Viscous forces acting on a fluid particle.*

Balancing forces in the x direction to determine the net viscous force, $\sum F_x^{(v)}$, we find a contribution on all four faces as follows.

$$\sum F_x^{(v)} = \underbrace{\left[\tau_{xx}\left(x + \Delta x, y + \frac{1}{2}\Delta y\right) - \tau_{xx}\left(x, y + \frac{1}{2}\Delta y\right)\right]\Delta y}_{Normal\ stress\ from\ x\ faces}$$

$$+ \underbrace{\left[\tau_{yx}\left(x + \frac{1}{2}\Delta x, y + \Delta y\right) - \tau_{yx}\left(x + \frac{1}{2}\Delta x, y\right)\right]\Delta x}_{Shear\ stress\ from\ y\ faces} \qquad (12.97)$$

Thus, the net viscous force per unit volume is

$$\frac{\sum F_x^{(v)}}{\Delta x \Delta y} = \frac{\tau_{xx}\left(x + \Delta x, y + \frac{1}{2}\Delta y\right) - \tau_{xx}\left(x, y + \frac{1}{2}\Delta y\right)}{\Delta x}$$

$$+ \frac{\tau_{yx}\left(x + \frac{1}{2}\Delta x, y + \Delta y\right) - \tau_{yx}\left(x + \frac{1}{2}\Delta x, y\right)}{\Delta y}$$

$$\to \frac{\partial \tau_{xx}}{\partial x} + \frac{\partial \tau_{yx}}{\partial y} \qquad \text{as} \qquad \Delta x, \Delta y \to 0 \qquad (12.98)$$

Turning to the y direction, the net viscous force, $\sum F_y^{(v)}$, again involves contributions from all four faces, viz.,

$$\sum F_y^{(v)} + \underbrace{\left[\tau_{xy}\left(x + \Delta x, y + \tfrac{1}{2}\Delta y\right) - \tau_{xy}\left(x, y + \tfrac{1}{2}\Delta y\right)\right]\Delta y}_{\text{Shear stress from } x \text{ faces}}$$

$$= \underbrace{\left[\tau_{yy}\left(x + \tfrac{1}{2}\Delta x, y + \Delta y\right) - \tau_{yy}\left(x + \tfrac{1}{2}\Delta x, y\right)\right]\Delta x}_{\text{Normal stress from } y \text{ faces}} \qquad (12.99)$$

Hence, the net viscous force per unit volume becomes

$$\frac{\sum F_y^{(v)}}{\Delta x \Delta y} + \frac{\tau_{xy}\left(x + \Delta x, y + \tfrac{1}{2}\Delta y\right) - \tau_{xy}\left(x, y + \tfrac{1}{2}\Delta y\right)}{\Delta x}$$

$$= \frac{\tau_{yy}\left(x + \tfrac{1}{2}\Delta x, y + \Delta y\right) - \tau_{yy}\left(x + \tfrac{1}{2}\Delta x, y\right)}{\Delta y}$$

$$\rightarrow \frac{\partial \tau_{xy}}{\partial x} + \frac{\partial \tau_{yy}}{\partial y} \qquad \text{as} \qquad \Delta x, \Delta y \rightarrow 0 \qquad (12.100)$$

Had we analyzed a three-dimensional fluid particle, Equations (12.98) and (12.100) would each have an additional term added, i.e., $\partial \tau_{zx}/\partial z$ and $\partial \tau_{zy}/\partial z$, respectively. Comparison with Equation (12.87) shows that we have arrived at the x and y components of $\nabla \cdot [\tau]$. A similar result holds for the z direction. Therefore, if we call the viscous force per unit volume $\mathbf{f}^{(v)}$, we have shown that

$$\mathbf{f}^{(v)} = \nabla \cdot [\tau] \qquad (12.101)$$

In words, *the divergence of the viscous stress tensor is the net viscous force per unit volume acting on a fluid particle.*

In summary, using this simple construction, we have shown that Navier's equation states that the sum of the forces equals mass times acceleration, per unit volume. This is identical to what we found when we derived Euler's equation for inviscid fluids.

12.6 Stokes' Postulate

As discussed at the beginning of this chapter, the shear stress in a simple shear flow is given by the Newtonian relation, Equation (12.5). In terms of the notation developed in the preceding section, we can rewrite the Newtonian relation as

$$\tau_{yx} = \mu \frac{dU}{dy} \qquad (12.102)$$

In separate research efforts, both Navier and the English mathematician Sir George Stokes addressed the issue of relating the viscous stress tensor to other flow properties, i.e., the issue of developing what we call a **constitutive relationship**. The goal of their research efforts was to generalize the Newtonian relation for arbitrary flows. The generally accepted constitutive relationship resulting from this research is referred to as **Stokes' postulate**. There are three basic premises underlying Stokes' postulate:

1. As we have already noted, conservation of angular momentum imposes a constraint on the viscous stress tensor that requires its off-diagonal terms to be equal, i.e., it is a symmetric tensor.

2. Most fluids are **isotropic**, i.e., they have no preferred directions. By contrast, a crystalline substance, with molecules arranged in a regular manner, is anisotropic.

3. For no motion and for rigid-body rotation, the viscous stress tensor must vanish so that the entire normal stress on any surface is exactly equal to the thermodynamic pressure.

These three postulates are sufficient to conclude mathematically that the viscous stress tensor is given by

$$[\tau] = 2\mu[\mathbf{S}] + \zeta \, \nabla \cdot \mathbf{u} \, [\delta] \tag{12.103}$$

where μ is molecular viscosity, ζ is the **second viscosity**, $[\mathbf{S}]$ is the strain-rate tensor defined in Equation (12.30) and $[\delta]$ is the identity tensor defined in Equation (12.56). For a monatomic gas, the second viscosity is related to μ by

$$\zeta = -\frac{2}{3}\mu \tag{12.104}$$

Although Equation (12.104) is formally valid only for monatomic gases, it is generally assumed valid in most analytical and computational research.

The mathematical steps required to arrive at Equation (12.103) are straightforward, but are most conveniently done in terms of tensor analysis. The derivation can be found in any advanced fluid mechanics text such as Batchelor (1967) or Panton (1996). In component form, the six additional equations needed to close Navier's equation are:

$$\tau_{xx} = 2\mu\frac{\partial u}{\partial x} - \frac{2}{3}\mu\nabla \cdot \mathbf{u}, \quad \tau_{yy} = 2\mu\frac{\partial v}{\partial y} - \frac{2}{3}\mu\nabla \cdot \mathbf{u}, \quad \tau_{zz} = 2\mu\frac{\partial w}{\partial z} - \frac{2}{3}\mu\nabla \cdot \mathbf{u} \tag{12.105}$$

$$\tau_{xy} = \mu\left(\frac{\partial u}{\partial y} + \frac{\partial v}{\partial x}\right), \quad \tau_{xz} = \mu\left(\frac{\partial u}{\partial z} + \frac{\partial w}{\partial x}\right), \quad \tau_{yz} = \mu\left(\frac{\partial v}{\partial z} + \frac{\partial w}{\partial y}\right) \tag{12.106}$$

The physical meaning of Stokes' postulate is clear. What it says, in words, is the following (in the case of incompressible flow for which $\nabla \cdot \mathbf{u} = 0$).

The viscous stress is proportional to the rate of strain in a fluid.

By contrast, for an elastic solid, stress is, by definition, proportional to strain. As explained in Subsections 12.2.2 and 12.2.3, strain varies continuously with time in a fluid, never reaching an equilibrium state, wherefore the rate of strain is a far more sensible quantity for use in describing deformation of a fluid particle. Hence, it is certainly reasonable that stress should be proportional to strain rate for a fluid.

Given its plausibility from the microscopic level as discussed in Section 12.1, we would certainly like to recover the Newtonian relation from our new constitutive relation. In checking for consistency, we are testing our new postulate in a simple limiting case for which we have great confidence. Assuming the velocity is given by Equation (12.1), so that

$$u = U(y), \quad v = 0, \quad w = 0 \tag{12.107}$$

clearly the divergence of the velocity is zero. Substitution of this velocity field into Equations (12.105) through (12.106) shows that

$$\tau_{yx} = \tau_{xy} = \mu\frac{dU}{dy}, \quad \text{(all other stresses vanish)} \tag{12.108}$$

Therefore, Stokes' postulate is consistent with the Newtonian relation.

12.7 Navier-Stokes Equation

By substituting Stokes' constitutive relationship [Equation (12.103)] into Navier's equation [Equation (12.86)], we arrive at the **Navier-Stokes equation**. In vector form, the equation is as follows.

$$\rho \frac{d\mathbf{u}}{dt} = -\nabla p + \rho \mathbf{f} + \nabla \cdot (2\mu[\mathbf{S}] + \zeta \nabla \cdot \mathbf{u}[\boldsymbol{\delta}]) \qquad (12.109)$$

This equation, solved in conjunction with mass and energy conservation equations, applies to general fluid motion including viscous effects. Since viscous forces have no effect on mass conservation, we still use the continuity equation derived in Section 5.1, i.e.,

$$\frac{d\rho}{dt} + \rho \nabla \cdot \mathbf{u} = 0 \qquad (12.110)$$

By contrast, viscous effects have a direct effect on energy conservation [see Equation (7.33)]. We defer development of the viscous-flow energy conservation principle to Chapter 15, and concentrate for the time being on incompressible flows with no heat transfer. This is similar to the approach we took in developing the control volume method in Chapters 1 through 6. It permits us to study effects of friction without the additional complications of compressibility and/or heat transfer.

For incompressible flow, we know that $\nabla \cdot \mathbf{u} = 0$. Assuming that the viscosity is constant (this is exactly true if temperature is constant), the Navier-Stokes Equation (12.109) simplifies to

$$\rho \frac{d\mathbf{u}}{dt} = -\nabla p + \rho \mathbf{f} + 2\mu \nabla \cdot [\mathbf{S}] \qquad (12.111)$$

where

$$\nabla \cdot [\mathbf{S}] = \begin{bmatrix} \frac{\partial}{\partial x} & \frac{\partial}{\partial y} & \frac{\partial}{\partial z} \end{bmatrix} \begin{bmatrix} \frac{\partial u}{\partial x} & \frac{1}{2}\left(\frac{\partial v}{\partial x} + \frac{\partial u}{\partial y}\right) & \frac{1}{2}\left(\frac{\partial w}{\partial x} + \frac{\partial u}{\partial z}\right) \\ \frac{1}{2}\left(\frac{\partial u}{\partial y} + \frac{\partial v}{\partial x}\right) & \frac{\partial v}{\partial y} & \frac{1}{2}\left(\frac{\partial w}{\partial y} + \frac{\partial v}{\partial z}\right) \\ \frac{1}{2}\left(\frac{\partial u}{\partial z} + \frac{\partial w}{\partial x}\right) & \frac{1}{2}\left(\frac{\partial v}{\partial z} + \frac{\partial w}{\partial y}\right) & \frac{\partial w}{\partial z} \end{bmatrix} \qquad (12.112)$$

We can simplify the divergence of the strain-rate tensor for incompressible flows. For example, consider the x component. Multiplying ∇ by the first column of the strain-rate tensor yields the x component as follows.

$$
\begin{aligned}
(\nabla \cdot [\mathbf{S}])_x &= \frac{\partial}{\partial x}\left[\frac{\partial u}{\partial x}\right] + \frac{\partial}{\partial y}\left[\frac{1}{2}\left(\frac{\partial u}{\partial y} + \frac{\partial v}{\partial x}\right)\right] + \frac{\partial}{\partial z}\left[\frac{1}{2}\left(\frac{\partial u}{\partial z} + \frac{\partial w}{\partial x}\right)\right] \\
&= \frac{\partial^2 u}{\partial x^2} + \frac{1}{2}\frac{\partial^2 u}{\partial y^2} + \frac{1}{2}\frac{\partial^2 v}{\partial y \partial x} + \frac{1}{2}\frac{\partial^2 u}{\partial z^2} + \frac{1}{2}\frac{\partial^2 w}{\partial z \partial x} \\
&= \frac{1}{2}\left(\frac{\partial^2 u}{\partial x^2} + \frac{\partial^2 u}{\partial y^2} + \frac{\partial^2 u}{\partial z^2}\right) + \frac{1}{2}\frac{\partial}{\partial x}\left(\frac{\partial u}{\partial x} + \frac{\partial v}{\partial y} + \frac{\partial w}{\partial z}\right) \\
&= \frac{1}{2}\nabla^2 u + \frac{1}{2}\frac{\partial}{\partial x}(\nabla \cdot \mathbf{u})
\end{aligned} \qquad (12.113)
$$

where ∇^2 is the Laplacian operator. Since the divergence of the velocity vanishes,

$$(\nabla \cdot [\mathbf{S}])_x = \frac{1}{2}\nabla^2 u \qquad (12.114)$$

Similar calculations for the y and z components show that

$$(\nabla \cdot [\mathbf{S}])_y = \frac{1}{2}\nabla^2 v \quad \text{and} \quad (\nabla \cdot [\mathbf{S}])_z = \frac{1}{2}\nabla^2 w \tag{12.115}$$

Summing the three components of $\nabla \cdot [\mathbf{S}]$ in vector form, we conclude that

$$\nabla \cdot [\mathbf{S}] = \frac{1}{2}\left(\mathbf{i}\,\nabla^2 u + \mathbf{j}\,\nabla^2 v + \mathbf{k}\,\nabla^2 w\right) = \frac{1}{2}\nabla^2 \mathbf{u} \tag{12.116}$$

Hence, we see that for incompressible flow,

$$2\mu\nabla \cdot [\mathbf{S}] = \mu\nabla^2\mathbf{u} \tag{12.117}$$

Thus, combining Equations (12.111) and (12.117), the incompressible Navier-Stokes equation for constant μ is

$$\rho\frac{d\mathbf{u}}{dt} = -\nabla p + \rho\,\mathbf{f} + \mu\nabla^2\mathbf{u} \tag{12.118}$$

There are two immediate observations we can make regarding the Navier-Stokes equation. First, because of the presence of second derivatives, the equation is of higher order than Euler's equation. This means, amongst other things, we can satisfy (and will require) more boundary conditions than in an inviscid fluid. We have already discussed the no-slip boundary condition at a solid boundary in Section 10.3, including its implications regarding the generation of vorticity.

Second, if we introduce dimensionless quantities according to

$$\tilde{\mathbf{u}} \equiv \frac{\mathbf{u}}{U}, \quad \tilde{p} \equiv \frac{p}{\rho\,U^2}, \quad \tilde{\mathbf{r}} \equiv \frac{\mathbf{r}}{L}, \quad \tilde{t} \equiv \frac{Ut}{L} \tag{12.119}$$

where $\mathbf{r}$ is position vector, U is a characteristic velocity and L is a characteristic length, then the dimensionless form of the incompressible Navier-Stokes equation is

$$\frac{d\tilde{\mathbf{u}}}{d\tilde{t}} = -\tilde{\nabla}\tilde{p} + \frac{1}{Re_L}\tilde{\nabla}^2\tilde{\mathbf{u}}, \quad Re_L = \frac{\rho\,U L}{\mu} \tag{12.120}$$

For simplicity, we have dropped the body force, and introduced the dimensionless operator, $\tilde{\nabla} \equiv L\nabla$. The important point to note is the coefficient of the dimensionless viscous term. Because the Reynolds number, Re_L, is very large for most practical flows, the coefficient of the viscous term is generally very small. Since the Navier-Stokes equation reduces formally to Euler's equation in the limit $Re_L \to \infty$, this explains why inviscid-flow theory provides an excellent description for many flows.

However, the limiting case $Re_L \to \infty$ yields what is referred to as a **nonuniformly valid approximation** [see Van Dyke (1975) or Wilcox (1995)]. That is, while Euler's equation provides a good description of most of the flowfield for slender objects, it cannot satisfy the no-slip boundary condition. Because of this, the Euler equation will fail to provide a meaningful solution close to solid boundaries. This is a common problem for differential equations whose highest order derivative has a small coefficient. Solutions to such equations exhibit extremely rapid variations near boundaries that, after Prandtl, are known as boundary layers. We will focus on this feature of the Navier-Stokes equation in Chapter 14.

Finally, we conclude this section by writing the incompressible Navier-Stokes equation in component form for Cartesian coordinates. The three components are:

$$\rho \frac{du}{dt} = -\frac{\partial p}{\partial x} + \rho f_x + \mu \left(\frac{\partial^2 u}{\partial x^2} + \frac{\partial^2 u}{\partial y^2} + \frac{\partial^2 u}{\partial z^2} \right) \qquad (12.121)$$

$$\rho \frac{dv}{dt} = -\frac{\partial p}{\partial y} + \rho f_y + \mu \left(\frac{\partial^2 v}{\partial x^2} + \frac{\partial^2 v}{\partial y^2} + \frac{\partial^2 v}{\partial z^2} \right) \qquad (12.122)$$

$$\rho \frac{dw}{dt} = -\frac{\partial p}{\partial z} + \rho f_z + \mu \left(\frac{\partial^2 w}{\partial x^2} + \frac{\partial^2 w}{\partial y^2} + \frac{\partial^2 w}{\partial z^2} \right) \qquad (12.123)$$

Appendix D includes the incompressible continuity and Navier-Stokes equations in other common coordinate systems.

12.8 The Vorticity Equation

In Chapter 10, we derived a differential equation for the vorticity in an inviscid fluid to help explain the role vorticity plays in developing forces in a fluid. We also noted how viscous effects change the equation, and offered an explanation for how vorticity is generated at solid boundaries. Of necessity, our discussion of the vorticity equation was brief. A more thorough examination of the vorticity equation's subtleties requires the detailed knowledge of kinematics and viscous effects presented in this chapter. Thus, we revisit the vorticity equation, armed with our new understanding of these concepts.

Since we derived the vorticity equation earlier, there is no need to do so again. By taking the curl of the Navier-Stokes equation with a conservative body force, we arrive at Equation (10.25), which we repeat here for convenience.

$$\frac{d\omega}{dt} = \omega \cdot \nabla \mathbf{u} + \nu \nabla^2 \omega \qquad (12.124)$$

In this equation, ω is the vorticity vector and $\nu = \mu/\rho$ is the kinematic viscosity.

12.8.1 Interaction With Forces on a Fluid Particle

In many ways, the vorticity equation is as important as the momentum equation. Understanding the way the various forces affect the vorticity, for example, provides insight into details of fluid motion that are not apparent from Newton's laws of motion. Observe that neither the pressure nor the body force appear in the equation, while viscous effects do. This reflects the fact that pressure and conservative body forces such as gravity act through the center of mass of a fluid particle and thus have no effect on the particle's rotation rate. By contrast, shear stresses act tangentially at the surface of the particle, and quite naturally can affect its angular motion.

To further establish the intimate connection between viscous forces and vorticity, we can make use of the following vector identity [see Appendix C].

$$\nabla \times (\nabla \times \mathbf{u}) = \nabla (\nabla \cdot \mathbf{u}) - \nabla^2 \mathbf{u} \qquad (12.125)$$

Hence, since $\nabla \cdot \mathbf{u}$ for incompressible flow, we have

$$\nabla^2 \mathbf{u} = -\nabla \times \omega \qquad (12.126)$$

Therefore, we can rewrite the incompressible Navier-Stokes equation as

$$\rho \frac{d\mathbf{u}}{dt} = -\nabla p + \rho\,\mathbf{f} - \mu \nabla \times \boldsymbol{\omega} \tag{12.127}$$

Thus, the existence of nontrivial viscous forces implies the existence of vorticity in the flow and vice versa. This is certainly the case close to solid boundaries. Conversely, if a flow has zero vorticity, the viscous term is exactly zero so that potential-flow solutions are also exact solutions to the Navier-Stokes equation. However, because potential-flow solutions fail to satisfy the no-slip boundary condition, this fact is of no particular value in solving practical flow problems.

The difference between the inviscid vorticity equation and the viscous form is the second term on the right-hand side of Equation (12.124). This term represents the diffusion process. Vorticity in a viscous fluid is transported through a fluid by molecular processes in the same way that momentum diffuses, i.e., in the manner discussed in Section 12.1. As discussed in Section 10.5, because of the no-slip boundary condition, vorticity is continuously generated at a solid boundary and carried into the flow through diffusion. For complete details on this process, see Stokes' First Problem in Section 13.4.

12.8.2   Vortex Stretching

The first term on the right-hand side of Equation (12.124) represents the **vortex-stretching** mechanism. To understand what vortex stretching means, we must introduce the concepts of **vortex lines** and **vortex tubes**. By definition, a vortex line is a line in the fluid that is everywhere tangent to the vorticity vector. A vortex tube is a tube whose sides are vortex lines. The differential equation of a vortex line follows from the fact that the cross product of the vorticity and a differential element tangent to the vortex line, $d\mathbf{r}$, is zero. Hence, similar to the derivation for a streamline [see Section 4.4], we find

$$\frac{dx}{\omega_x} = \frac{dy}{\omega_y} = \frac{dz}{\omega_z} \tag{12.128}$$

Consider a vortex tube of cross-sectional area A as shown in Figure 12.13(a). Let Γ denote the circulation computed along the circumference of the vortex tube. We demonstrated in Section 4.3 that the circulation about any curve is given by [see Equation (4.39)]

$$\Gamma = \iint_A \boldsymbol{\omega} \cdot \mathbf{n}\, dA \tag{12.129}$$

where A is, in the present context, the cross-sectional area of the vortex tube, and $\mathbf{n}$ is a unit vector normal to the cross section. Here, $\mathbf{n}$ is parallel to $\boldsymbol{\omega}$.

Now, suppose we elongate the vortex tube. In order to contain the same fluid particles, mass conservation tells us the cross-sectional area must decrease. Suppose further that its

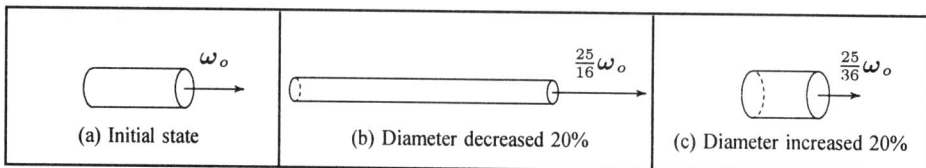

Figure 12.13: *Vortex-tube schematics.*

diameter decreases by 20% [Figure 12.13(b)]. In the absence of significant viscous effects, i.e., away from solid boundaries, the Helmholtz Theorem [Section 10.2] tells us the circulation remains constant. Thus, since A is $\frac{16}{25}$ of its initial value, the average vorticity of the vortex tube must increase to $\frac{25}{16}$ of its initial value, ω_o. Similarly, if we shorten the vortex tube [Figure 12.13(c)] so that its diameter decreases by 20% (wherefore the area is $\frac{36}{25}A$), the average vorticity decreases to $\frac{25}{36}\omega_o$.

This effect is essentially a conservation of angular momentum principle. The effect is similar to the ice skater who begins spinning with arms extended and increases his spin rate by contracting his arms. We observe vortex stretching in a draining bathtub. Away from the drain, the rotation is barely evident. It becomes visible to the naked eye just above the drain where the radius of the vortex achieves its minimum.

Now, let's consider the motion of a vortex-tube segment for an infinitesimally short time, Δt. Referring to Figure 12.14, we assume the initial length of the vortex tube is η. Also, the vorticity is $\boldsymbol{\omega} = \omega\mathbf{n}$, where $\mathbf{n}$ is normal to the cross-sectional area of the vortex tube at the initial time, $t = 0$.

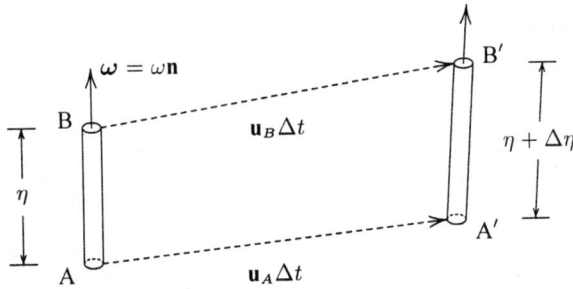

Figure 12.14: *Motion of a vortex-tube segment.*

First, assuming we are working with an infinitesimally short segment of a vortex tube, the velocity at Point B will be given by

$$\mathbf{u}_B = \mathbf{u}_A + \frac{\partial \mathbf{u}}{\partial n}\eta + \cdots \qquad (12.130)$$

where $\partial/\partial n$ indicates differentiation in the direction of the vorticity vector. Hence, truncating the Taylor series and noting that $\partial/\partial n = \mathbf{n} \cdot \nabla$, we have

$$\mathbf{u}_B \approx \mathbf{u}_A + (\mathbf{n} \cdot \nabla \mathbf{u})\,\eta \qquad (12.131)$$

After a time Δt, the ends of the vortex-tube segment are located at

$$\mathbf{r}_{A'} \approx \mathbf{r}_A + \mathbf{u}_A \Delta t \qquad (12.132)$$

$$\mathbf{r}_{B'} \approx \mathbf{r}_B + \mathbf{u}_B \Delta t \approx \mathbf{r}_B + [\mathbf{u}_A + (\mathbf{n} \cdot \nabla \mathbf{u})\,\eta]\,\Delta t \qquad (12.133)$$

This tells us the difference between the vectors connecting the endpoints of the vortex-tube segment are related by

$$\mathbf{r}_{B'} - \mathbf{r}_{A'} \approx \mathbf{r}_B - \mathbf{r}_A + (\mathbf{n} \cdot \nabla \mathbf{u})\,\eta\Delta t \qquad (12.134)$$

Rearranging terms, we arrive at the following

$$\mathbf{n} \cdot \nabla \mathbf{u} = \frac{(\mathbf{r}_{B'} - \mathbf{r}_{A'}) - (\mathbf{r}_B - \mathbf{r}_A)}{\eta\Delta t} \qquad (12.135)$$

Thus, noting that the orientation of the unit normal $\mathbf{n}$ can change with time, taking the limit $\Delta t \to 0$, the difference between $(\mathbf{r}_{B'} - \mathbf{r}_{A'})$ and $(\mathbf{r}_B - \mathbf{r}_A)$ represents the change in $\eta \mathbf{n}$, so that

$$\mathbf{n} \cdot \nabla \mathbf{u} = \frac{1}{\eta} \frac{d}{dt} (\eta \mathbf{n}) \tag{12.136}$$

Finally, recall that $\boldsymbol{\omega} = \omega \mathbf{n}$. Thus, the vortex stretching term can be written as

$$\boldsymbol{\omega} \cdot \nabla \mathbf{u} = \frac{\omega}{\eta} \frac{d}{dt} (\eta \mathbf{n}) \tag{12.137}$$

The physical meaning of this result is the following. The vortex stretching term is the product of the magnitude of the vorticity and the rate of deformation of the vortex-tube segment defined by the right-hand side of Equation (12.136). Geometric interpretation of this equation is possible, but it requires an extended analysis. Nevertheless, the origin of the terminology *vortex stretching* is evident in the special limiting case where only the velocity component parallel to the vorticity varies with distance along the tube. In this case, the other components produce a simple translation so that $\mathbf{n}$ remains constant. Accepting this limitation to help get a handle on a rather complicated phenomenon, we have

$$\boldsymbol{\omega} \cdot \nabla \mathbf{u} = \omega S_n, \qquad S_n \equiv \frac{1}{\eta} \frac{d\eta}{dt} \tag{12.138}$$

The quantity S_n is very similar to the normal strain rates $[S_{xx}, S_{yy}, S_{zz}]$, and thus represents the rate of stretching of the vortex tube, per unit length. For general three-dimensional flows, the vorticity will not remain aligned with a single unit normal $\mathbf{n}$, however. Consistent with Equation (12.137), vortex tubes will undergo rotation and translation along with extension and contraction.

12.9   Computational Fluid Dynamics

One of the most effective procedures for solving complex flowfields is the use of time-marching methods. If the desired solution is unsteady, time-marching solutions yield a true time history. Time-marching methods can also be used for steady-flow problems by letting the solution evolve in time until temporal variations become negligibly small. That is, as with our potential-flow example of Section 11.9, we begin with an initial approximation and update the solution at each timestep until the residuals differ by less than a prescribed tolerance level. In this section, we will explore elementary time-marching methods.

12.9.1   Explicit Time-Marching Methods

The simplest time-marching schemes are known as **explicit methods**. Most explicit schemes were developed prior to 1970. In an explicit scheme, the solution at time t^{n+1} depends only on the past history, i.e., the solution at time t^n. For example, consider the one-dimensional linear **Burgers' equation**:

$$\frac{\partial u}{\partial t} + U \frac{\partial u}{\partial x} = \nu \frac{\partial^2 u}{\partial x^2} \tag{12.139}$$

where u is velocity, U is an effective speed of sound, t is time and x is streamwise direction. This equation, with its obvious similarity to the one-dimensional Navier-Stokes equation, is often used in CFD research to test and validate numerical algorithms.

As in our example of Section 11.9, we will work with equally-spaced grid points. Letting u_j^n denote $u(x_j, t^n)$, we approximate $\partial u / \partial t$ with the following discretization approximation[3]

$$\frac{\partial u}{\partial t} \approx \frac{u_j^{n+1} - u_j^n}{\Delta t} \tag{12.140}$$

where $\Delta t = t^{n+1} - t^n$. For simplicity, consider simple **upwind differencing** in which we approximate $\partial u / \partial x$ according to

$$\frac{\partial u}{\partial x} \approx \frac{u_j^n - u_{j-1}^n}{\Delta x} \tag{12.141}$$

For the diffusion term, we use a **central-difference** approximation so that

$$\frac{\partial^2 u}{\partial x^2} \approx \frac{u_{j+1}^n - 2u_j^n + u_{j-1}^n}{(\Delta x)^2} \tag{12.142}$$

Using these discretization approximations, we arrive at the following first-order accurate difference equation that approximates Equation (12.139).

$$u_j^{n+1} = u_j^n - \frac{U \Delta t}{\Delta x} \left(u_j^n - u_{j-1}^n \right) + \frac{\nu \Delta t}{(\Delta x)^2} \left(u_{j+1}^n - 2u_j^n + u_{j-1}^n \right) \tag{12.143}$$

This is not a particularly accurate method, but nevertheless illustrates the general nature of explicit schemes. Note that all terms on the right-hand side of Equation (12.143) are known from time t^n. Hence, u_j^{n+1} is obtained from simple algebraic operations. Because only algebraic operations are needed (as opposed to inversion of a large matrix), explicit methods are easy to implement.

12.9.2   von Neumann Stability Analysis

The primary shortcoming of explicit schemes is a limit on the timestep that can be used. For too large a timestep, solution errors will grow with increasing iterations and the computation becomes unstable. The most commonly used method for determining the stability properties of a time-marching finite-difference scheme is von Neumann stability analysis [see Roache (1972) or Anderson et al. (1984)]. In this method, we introduce a discrete Fourier-series solution to the finite-difference equation under study, and determine the growth rate of each mode. If all Fourier modes decay as we march in time, the scheme is stable. However, if even a single mode grows, the scheme is unstable. We write each Fourier component as

$$u_j^n = G^n e^{i(j\kappa\Delta x)} \tag{12.144}$$

where G is called the **amplitude factor**, $i = \sqrt{-1}$ and κ is wavenumber. The stability of a scheme is determined as follows:

$$\left. \begin{array}{ll} |G| < 1, & \text{Stable} \\ |G| = 1, & \text{Neutrally Stable} \\ |G| > 1, & \text{Unstable} \end{array} \right\} \tag{12.145}$$

In general, G is complex, and the notation G^n means G raised to the power n. The amplitude factor for Equation (12.143) is too complicated a function for the immediate purpose, which

---

[3]We use j as the grid-point index rather than i to avoid confusion with the imaginary number i below.

is to explain how von Neumann stability analysis works. Hence, to simplify our analysis, we drop the viscous term for now by setting $\nu = 0$. We will indicate what happens with $\nu \neq 0$ after explaining the basic method.

So, for example, the velocities at time level n and grid-point numbers $j - 1$ and $j + 1$ are

$$u_{j-1}^n = G^n e^{i(j-1)\kappa\Delta x} \quad \text{and} \quad u_{j+1}^n = G^n e^{i(j+1)\kappa\Delta x} \qquad (12.146)$$

Then, after a little algebra, the amplitude factor for Equation (12.143) with $\nu = 0$ is

$$G = 1 - \frac{U\Delta t}{\Delta x}\left(1 - e^{-i\theta}\right) \quad \text{where} \quad \theta = \kappa\Delta x \qquad (12.147)$$

In terms of real and imaginary parts, the amplitude factor becomes

$$G = \left[1 - \frac{U\Delta t}{\Delta x}(1 - \cos\theta)\right] - i\frac{U\Delta t}{\Delta x}\sin\theta \qquad (12.148)$$

Thus,

$$\begin{aligned}
|G|^2 &= \left[1 - \frac{U\Delta t}{\Delta x}(1 - \cos\theta)\right]^2 + \left(\frac{U\Delta t}{\Delta x}\right)^2 \sin^2\theta \\
&= 1 - 2\frac{U\Delta t}{\Delta x}(1 - \cos\theta) + \left(\frac{U\Delta t}{\Delta x}\right)^2\left[(1 - \cos\theta)^2 + \sin^2\theta\right] \\
&= \left[1 - 2\frac{U\Delta t}{\Delta x}(1 - \cos\theta) + 2\left(\frac{U\Delta t}{\Delta x}\right)^2(1 - \cos\theta)\right] \\
&= 1 + 2(1 - \cos\theta)\frac{U\Delta t}{\Delta x}\left(\frac{U\Delta t}{\Delta x} - 1\right) \qquad (12.149)
\end{aligned}$$

In order to have a stable scheme, $|G|$ must be less than or equal to 1 for all possible values of θ. Observing that $(1 - \cos\theta) \geq 0$ for all values of θ, we conclude that for the upwind-difference scheme, errors will not grow provided the condition

$$\Delta t < \left|\frac{\Delta x}{U}\right| \quad \text{or} \quad N_{CFL} = \frac{U\Delta t}{\Delta x} < 1 \qquad (12.150)$$

is satisfied. This is the Courant-Friedrichs-Lewy (1967), or CFL condition, which means a wave cannot propagate a distance exceeding Δx in a single timestep. N_{CFL} is known as the **CFL Number**, or just the **Courant Number**. In the Euler and Navier-Stokes equations, the velocity U is replaced by $u \pm a$, where u is local velocity and a is the speed of sound. Since waves travel at very high speed for incompressible flows, the CFL condition implies smaller and smaller timesteps as Mach number, M, approaches zero. In practice, time-marching methods are acceptably efficient for Mach numbers down to about 0.1. At this low a Mach number, the flow is effectively incompressible.

If we repeat the stability analysis with $\nu \neq 0$ and $U = 0$, a straightforward computation shows that the scheme of Equation (12.143) is conditionally stable, subject to the following constraint.

$$\Delta t \leq \frac{(\Delta x)^2}{2\nu} \qquad (12.151)$$

This condition tells us Δt must not exceed the time required for u to diffuse through a finite-difference cell in a single timestep.

Explicit methods are of interest in modern CFD applications mainly for time-dependent flows. There has been renewed interest in these methods recently because of their simplicity for massively-parallel computers. In summary, their algebraic simplicity makes them especially easy to implement on any computer. Their primary drawback is their conditional stability, and thousands of timesteps are often needed to achieve steady-flow conditions.

12.9.3 MacCormack's Method

Although we could use the upwind scheme discussed in Subsection 12.9.1 to illustrate the time-marching method, it is preferable (and standard practice) to use a more accurate method. While the upwind scheme led to an algebraically simple amplification factor (which is the reason we introduced it), the method is only first-order accurate in both space and time. Many (but unfortunately not all) modern finite-difference methods in common usage are second-order accurate in space and time. One of the earliest second-order accurate, explicit methods was introduced by MacCormack (1969).

MacCormack's method proceeds in two steps, the first called the **predictor step** and the second the **corrector step**. To explain the method, it is convenient to rewrite Equation (12.139) in conservation form, i.e.,

$$\frac{\partial u}{\partial t} + \frac{\partial F}{\partial x} = 0 \quad \text{where} \quad F \equiv Uu - \nu \frac{\partial u}{\partial x} \tag{12.152}$$

Also, to understand how MacCormack's method works, we must introduce the standard, first-order accurate approximations to the first derivative. In terms of the flux, F, we define the **backward-difference** approximation by[4]

$$\left(\frac{\partial F}{\partial x}\right)_j \approx \frac{F_j^n - F_{j-1}^n}{\Delta x} \tag{12.153}$$

and the **forward-difference** approximation by

$$\left(\frac{\partial F}{\partial x}\right)_j \approx \frac{F_{j+1}^n - F_j^n}{\Delta x} \tag{12.154}$$

The predictor step uses a backward difference to advance the velocity in time, i.e.,

$$\tilde{u}_j^{n+1} = u_j^n - \frac{\Delta t}{\Delta x} \underbrace{\left(F_j^n - F_{j-1}^n\right)}_{\text{Backward Difference}} \tag{12.155}$$

The flux term is computed by using a forward difference for the diffusion term, wherefore

$$F_j^n = Uu_j^n - \frac{\nu}{\Delta x} \underbrace{\left(u_{j+1}^n - u_j^n\right)}_{\text{Forward Difference}} \tag{12.156}$$

The corrector step then uses a forward difference to complete the update of the velocity:

$$u_j^{n+1} = \frac{1}{2}\left[u_j^n + \tilde{u}_j^{n+1} - \frac{\Delta t}{\Delta x}\underbrace{\left(\tilde{F}_{j+1}^{n+1} - \tilde{F}_j^{n+1}\right)}_{\text{Forward Difference}}\right] \tag{12.157}$$

---

[4]This is the same as the upwind-difference approximation introduced earlier if $U > 0$.

combined with a backward difference for the flux term, so that

$$\tilde{F}_j^{n+1} = U\tilde{u}_j^{n+1} - \frac{\nu}{\Delta x} \underbrace{\left(\tilde{u}_j^{n+1} - \tilde{u}_{j-1}^{n+1}\right)}_{Backward\ Difference} \tag{12.158}$$

The combination of these two first-order accurate steps turns out to be second-order accurate in both time and space. In usual practice, the differencing employed in predictor and corrector steps alternates from timestep to timestep. That is, after using backward differences in the main predictor step [Equation (12.155)] and forward differences in the main corrector step [Equation (12.157)], the next iteration would use forward differences for the predictor step and backward differences for the corrector step. The differencing used in these steps for the flux term would also be reversed.

We can combine Equations (12.152) and (12.155) through (12.158) to yield the following algorithm.

Predictor:

$$\tilde{u}_j^{n+1} = u_j^n - \frac{U\Delta t}{\Delta x}\left(u_{j+1}^n - u_j^n\right) + \frac{\nu\Delta t}{(\Delta x)^2}\left(u_{j+1}^n - 2u_j^n + u_{j-1}^n\right) \tag{12.159}$$

Corrector:

$$u_n^{n+1} = \frac{1}{2}\left[u_j^n + \tilde{u}_j^n - \frac{U\Delta t}{\Delta x}\left(\tilde{u}_j^n - \tilde{u}_{j-1}^n\right) + \frac{\nu\Delta t}{(\Delta x)^2}\left(\tilde{u}_{j+1}^{n+1} - 2\tilde{u}_j^{n+1} + \tilde{u}_{j-1}^{n+1}\right)\right] \tag{12.160}$$

Like the upwind method discussed earlier, MacCormack's method is conditionally stable. The timestep limitation [Anderson (1995)] is:

$$\Delta t \le \left[\frac{|U|}{\Delta x} + \frac{2\nu}{(\Delta x)^2}\right]^{-1} \tag{12.161}$$

12.9.4   Flow Over a Plate with Uniform Suction

To illustrate a time-marching solution, we consider an idealized problem for which an exact incompressible Navier-Stokes solution is known. The problem we will address is steady flow over an infinite porous flat plate with uniform mass removal, or suction. We further assume the pressure is uniform. The flow is depicted in Figure 12.15.

Because the plate is infinite in extent, the flow depends only upon distance above the plate, y. Conservation of mass tells us that for this exactly one-dimensional flow that

$$\frac{dv}{dy} = 0 \quad\Longrightarrow\quad v = \text{constant} = -v_w \tag{12.162}$$

The x component of the Navier-Stokes equation thus simplifies to

$$-v_w\frac{du}{dy} = \nu\frac{d^2u}{dy^2} \tag{12.163}$$

The velocity must approach the freestream velocity, U_∞, far above the plate, while the no-slip boundary condition holds at $y = 0$ so that

$$u(0) = 0, \quad u(y) \to U_\infty \text{ as } y \to \infty \tag{12.164}$$

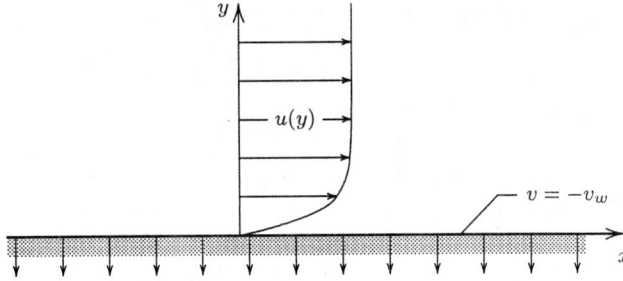

Figure 12.15: *Flow over a porous plate with uniform suction; freestream velocity is U_∞ so that $u(y) \to U_\infty$ as $y \to \infty$.*

The exact solution to Equation (12.163) subject to Equations (12.164) is

$$u(y) = U_\infty \left(1 - e^{-v_w y / \nu}\right) \tag{12.165}$$

In order to solve this steady-flow problem with the time-marching technique, we make an initial guess. Then, we replace Equation (12.163) with the unsteady one-dimensional x component of the Navier-Stokes equation, viz.,

$$\frac{\partial u}{\partial t} - v_w \frac{\partial u}{\partial y} = \nu \frac{\partial^2 u}{\partial y^2} \tag{12.166}$$

We solve Equation (12.166) numerically using MacCormack's method, seeking a long-time solution in which variations with time become vanishingly small. When this condition is achieved, what remains is the solution to Equation (12.163). Noting the exponential approach of $u(y)$ to the freestream velocity, U_∞, we select our computational domain to be the region $0 \leq v_w y / \nu \leq 10$. For initial conditions, we assume the initial value of the time-dependent velocity, $u(y, t)$, is

$$u(y, 0) = \begin{cases} U_\infty \left(\dfrac{y}{3}\right)^2, & 0 \leq \dfrac{v_w y}{\nu} \leq 3 \\[3mm] U_\infty, & 3 \leq \dfrac{v_w y}{\nu} \leq 10 \end{cases} \tag{12.167}$$

Aside from satisfying the boundary conditions, the choice of initial velocity is arbitrary for this problem. This is one of the conveniences of using the time-marching method to solve a steady-flow problem. While the initial guess does not satisfy the steady-flow equation, it does satisfy the time-dependent equation regardless of our choice since the unsteady term will accommodate the discrepancy between the exact solution and the initial guess.

Reference to Equation (12.161) tells us the theoretical maximum timestep possible with MacCormack's method is

$$\Delta t_{max} = \left[\frac{v_w}{\Delta y} + \frac{2\nu}{(\Delta y)^2}\right]^{-1} \tag{12.168}$$

In the discussion to follow, it is convenient to define an effective CFL number, $\tilde{N}_{CFL}$, as follows.

$$\tilde{N}_{CFL} = \frac{\Delta t}{\Delta t_{max}} \tag{12.169}$$

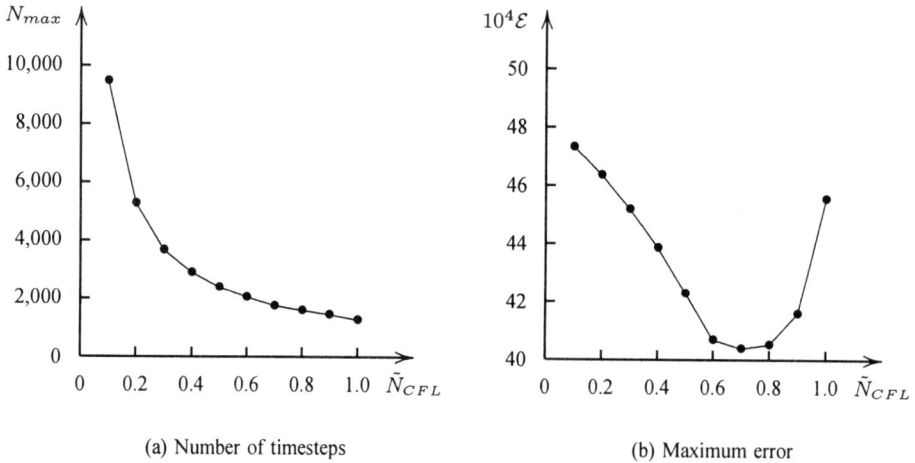

(a) Number of timesteps   (b) Maximum error

Figure 12.16: *Number of timesteps to attain steady state and the maximum error for 51 equally-spaced grid points.*

Figure 12.16(a) shows the number of timesteps required to converge to a steady solution, N_{max}, as a function of $\tilde{N}_{CFL}$. The finite-difference grid includes 51 equally-spaced points with $v_w \Delta y / \nu = 0.2$. In each case, the computations have been continued until the difference between the velocity at successive timesteps changes by less than 10^{-6}. As shown, the number of timesteps decreases monotonically from almost 10,000 when $\tilde{N}_{CFL} = 0.1$ to just under 1,300 for $\tilde{N}_{CFL} = 1.0$. The computation is unstable for $\tilde{N}_{CFL}$ in excess of 1.0, terminating with a math overflow after a few hundred steps.

Figure 12.16(b) shows the maximum difference between the computed and exact values of u/U_∞, i.e., the maximum error, $\mathcal{E}$. For this problem, we find that the error is smallest for $\tilde{N}_{CFL} = 0.7$. However, the error is only 3% larger than its minimum value when $\tilde{N}_{CFL}$ increases to 0.9, while the number if timesteps required to achieve steady flow decreases by 17%. A CFL number of 0.9 is typical for most numerical solutions generated using MacCormack's method.

Because we have an exact solution for this problem, we have an opportunity to determine how accurate Richardson extrapolation (see Subsection 11.9.4) is in estimating solution error. Figure 12.17(a) compares the exact error in u/U_∞ with the Richardson extrapolation error based on solutions with 101 and 51 grid points and with $\tilde{N}_{CFL} = 0.7$. The figure clearly shows that Richardson extrapolation provides an excellent estimate of solution error.

Figure 12.17(b) presents maximum solution error for computations with 51, 101, 201 and 401 grid points, corresponding to $v_w \Delta y / \nu$ decreasing from 0.2 to 0.025. As shown, the numerical error, $\mathcal{E}$, initially decreases as Δy decreases, but it then reverses direction and increases as $\Delta y \to 0$. This graph shows that there are two primary sources of numerical error, viz., **truncation error** and **roundoff error**. On the one hand, our computations are subject to truncation error associated with dropping terms in the Taylor-series representations of derivatives when formulating discretization approximations. Because our numerical algorithm is second-order accurate, these errors go to zero as $(\Delta y)^2$ for $\Delta y \to 0$, and this is exhibited in the figure as Δy decreases down to 0.05. On the other hand, the trend then reverses and $\mathcal{E}$ begins to increase as $\Delta y \to 0$. This reversal is caused by what is known as roundoff error.

Roundoff error occurs because a computer can represent numbers to only a finite number of significant figures. This type of error is readily observed on a standard hand-held

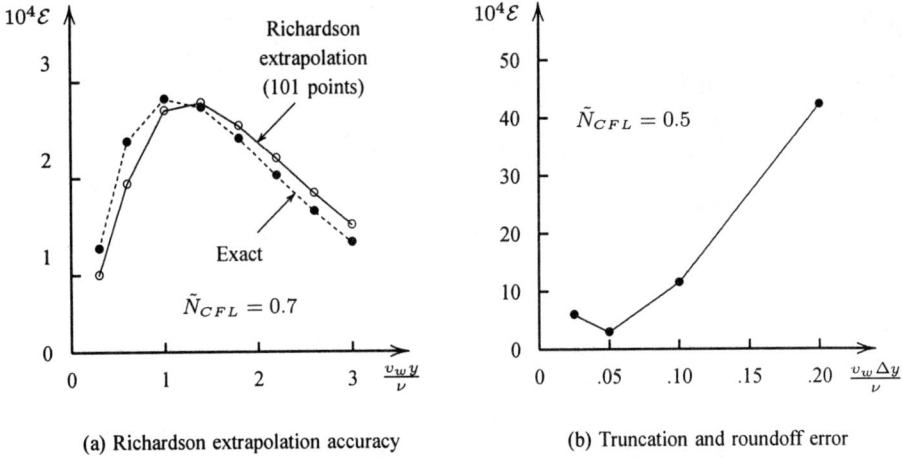

(a) Richardson extrapolation accuracy   (b) Truncation and roundoff error

Figure 12.17: *Effect of grid-point spacing on solution accuracy.*

scientific calculator. For example, some inexpensive calculators return exponentiated results accurate to 6 significant figures. The calculator tells us $2^{0.25} = 1.18921$. It also tells us $(1.18921)^4 = 2.00002$. By contrast, the square root function on the same calculator is often accurate to 9 significant figures. Taking the square root of 2 twice in succession, we find $2^{1/4} = 1.1892071$. Squaring this result twice returns the original value of 2. The difference between the two results is caused by roundoff error. Digital computers have the same shortcoming. Using single-precision arithmetic, a standard desktop or laptop personal computer represents most numbers and functions to 7 significant figures[5] (plus an exponent), while mainframe super computers typically provide 14 significant-figure accuracy in single precision.

Hence, there is a lower bound on Δy below which a computer fails to distinguish from zero. As noted above, because truncation error goes to zero as $(\Delta y)^2$, the error initially decreases with decreasing Δy. However, roundoff error eventually dominates, and further decreases in Δy eventually cause the overall accuracy to diminish. This is true because smaller increments in y require more arithmetic operations to integrate over the same range, each of which is subject to roundoff error. The cumulative effect adds to the overall numerical error. For the present problem, the minimum total numerical error occurs for 201 grid points, for which $v_w \Delta y / \nu = 0.05$.

The program used to do these computations, **SUCTION**, is described in Appendix E. Source code is included on the diskette that accompanies this book.

---

[5]There are exceptions such as the square root function, which is accurate to 16 or 17 significant figures on such machines.

Problems

12.1 If the thermal velocity in a gas is given by

$$v_{th} = \frac{4}{3}a$$

where a is the speed of sound, what is the mean free path as a function of kinematic viscosity, ν, and a? What is the size of the mean free path for air at $20°$ C?

12.2 For flow in a channel, the velocity is given by

$$U(y) = u_m \left[1 - 4\left(\frac{y}{h}\right)^2 \right]$$

Compute the *Knudsen number*, Kn for channel flow. What conclusion can you draw regarding appropriateness of regarding the fluid as being Newtonian in this situation?

12.3 The mean free path in a gas, ℓ_{mfp}, is given by

$$\ell_{mfp} = \frac{3}{2}\frac{\nu}{a}$$

where ν is kinematic viscosity and a is the speed of sound. The maximum velocity gradient for flow past an automobile is

$$\left.\frac{\partial u}{\partial y}\right|_{max} \approx \frac{.032}{Re_x^{1/5}}\frac{U^2}{\nu}$$

where Re_x is Reynolds number based on distance from the front of the automobile.

(a) Compute $\ell_{mfp}|\partial u/\partial y|_{max}/v_{th}$, where v_{th} is the thermal velocity, as a function of Re_x and Mach number, $M = U/a$.

(b) Express your answer to Part (a) as a function of x for an automobile moving at 60 mph if the temperature is $68°$ F.

(c) How small must x be to have $\ell_{mfp}|\partial u/\partial y|_{max}/v_{th} = 1$ so that Equation (12.9) is not satisfied? What is x/ℓ_{mfp} when this is true?

12.4 The exact solution for Couette flow is $u = Uy/H$ and $v = 0$, where U is the plate velocity at $y = H$, y is distance normal to the plates and H is distance between the plates.

(a) Determine the streamfunction, $\psi(x, y)$.

(b) Show that no velocity potential, $\phi(x, y)$, exists and explain why.

12.5 The exact solution for a rigid-body rotation is $u_r = 0$ and $u_\theta = \Omega r$, where Ω is the angular-rotation rate and r is radial distance.

(a) Determine the streamfunction, $\psi(r, \theta)$.

(b) Show that no velocity potential, $\phi(r, \theta)$, exists and explain why.

12.6 The velocity for incompressible flow in a converging nozzle can be approximated by

$$\mathbf{u} = U\left(1 + 2\frac{x}{L}\right)\mathbf{i} - 2U\frac{y}{L}\mathbf{j}$$

where U and L are constant velocity and length scales, respectively. Compute the vorticity vector and the strain-rate tensor.

12.7 Consider the incompressible flow whose velocity components are given by $u = Ux/L$ and $v = -Uy/L$, where U and L are constant velocity and length scales, respectively. Compute the vorticity vector and the strain-rate tensor.

12.8 Suppose the velocity components for a two-dimensional, incompressible flowfield are given by

$$u = f(y) \quad \text{and} \quad v = g(x)$$

where $f(y)$ and $g(x)$ are arbitrary functions.

(a) What must these two functions be in order to have irrotational flow?

(b) Compute the strain-rate tensor if the flow is irrotational.

12.9 Consider the flowfield whose velocity is given by $\mathbf{u} = 4Uy/a\,\mathbf{i} - Ux/a\,\mathbf{j}$ where U and a are constant velocity and length scales, respectively. Compute the vorticity vector and the strain-rate tensor.

12.10 The potential-flow solution for flow past a rotating cylinder of radius R is

$$u_r = U\left(1 - R^2/r^2\right)\cos\theta \quad \text{and} \quad u_\theta = -U\left(1 + R^2/r^2\right)\sin\theta + \frac{\Gamma}{2\pi r}$$

where Γ is circulation and r is radial distance. Compute the strain-rate tensor.

12.11 Consider the axisymmetric flowfield whose velocity components are

$$u_r = U(r/a)^3 \cos 4\theta \quad \text{and} \quad u_\theta = -U(r/a)^3 \sin 4\theta$$

where U and a are constant velocity and length scales, respectively, while r is radial distance. Compute the strain-rate tensor.

12.12 The velocity components for *Hill's spherical vortex* are as follows:

$$u_r = U\frac{rz}{R^2}, \quad u_\theta = 0, \quad w = U\left[1 - \left(\frac{z}{R}\right)^2 - 2\left(\frac{r}{R}\right)^2\right]$$

where U and R are constant velocity and length scales, respectively, while r and z denote radial and axial distance. Compute the vorticity vector and the strain-rate tensor.

12.13 An incompressible flow has the following velocity vector:

$$\mathbf{u} = \frac{U}{H}\left[x\,\mathbf{i} + y\,\mathbf{j} - 2z\,\mathbf{k}\right]$$

where U and H are constant velocity and length scales, respectively. Determine the vorticity vector and the strain-rate tensor.

12.14 An incompressible flow has the following velocity vector:

$$\mathbf{u} = \frac{U}{L^2}\left[x^2\mathbf{i} + y^2\mathbf{j} - 2(x+y)z\mathbf{k}\right]$$

where U and L are constant velocity and length scales, respectively. Determine the vorticity vector and the strain-rate tensor.

12.15 Close to a solid boundary, the velocity for an incompressible, viscous flow is given by

$$\mathbf{u} = U\frac{y}{\delta}\,\mathbf{i}$$

where U and δ are constant velocity and length scales, respectively. If an initially square fluid particle has one side parallel to the x axis, what is the shape of the fluid particle after a time Δt, where $\Delta t \ll \delta/U$?

12.16 What is the trace of the viscous-stress tensor, $tr[\boldsymbol{\tau}] \equiv \tau_{xx} + \tau_{yy} + \tau_{zz}$, for a compressible fluid with first and second viscosity coefficients μ and ζ? How does your result simplify for a monatomic gas?

12.17 For compressible viscous flow, we cannot ignore the second viscosity, and for a monatomic gas, the total stress in a fluid is

$$[\sigma] = -p[\delta] + 2\mu[S] - \frac{2}{3}\mu(\nabla \cdot \mathbf{u})[\delta]$$

Determine the average normal stress, σ_n, defined by

$$\sigma_n \equiv \frac{1}{3}\left(\sigma_{xx} + \sigma_{yy} + \sigma_{zz}\right)$$

12.18 The *bulk viscosity*, μ_v, of a gas is defined by

$$p_{\text{mechanical}} \equiv p - \mu_v \nabla \cdot \mathbf{u}$$

where p is the thermodynamic pressure and $p_{\text{mechanical}}$ is minus the average normal stress, i.e.,

$$p_{\text{mechanical}} = -\frac{1}{3}\left(\sigma_{xx} + \sigma_{yy} + \sigma_{zz}\right)$$

(a) Determine μ_v as a function of first and second viscosity coefficients, μ and ζ.

(b) What is μ_v for a monatomic gas?

12.19 Consider an incompressible flow for which

$$\mathbf{u} = A\left(x^2\mathbf{i} + z^2\mathbf{j} - 2xz\,\mathbf{k}\right)$$

where A is a constant. Determine the viscous-stress tensor, $[\tau]$, and its divergence, $\nabla \cdot [\tau]$.

12.20 Consider an incompressible flow for which

$$\mathbf{u} = U\left[\frac{x}{H}\mathbf{i} - \frac{y}{H}\mathbf{j} + \frac{z}{H}\mathbf{k}\right]\left(\frac{y}{H}\right)$$

where U and H are constant velocity and length scales, respectively. Determine the viscous-stress tensor, $[\tau]$, and its divergence, $\nabla \cdot [\tau]$.

12.21 The initial motion in a *shock tube* involves a shock wave traveling in one direction and an expansion wave in the other. The velocity in the expansion wave is very nearly one-dimensional and is given by

$$u = \frac{2}{\gamma + 1}\left(a_o + \frac{x}{t}\right), \qquad v = 0, \qquad w = 0$$

where x is streamwise distance, t is time, γ is specific-heat ratio and a_o is the speed of sound upstream of the wave. Assuming the velocity is as given above and the gas is monatomic, what is the viscous-stress tensor for the expansion wave?

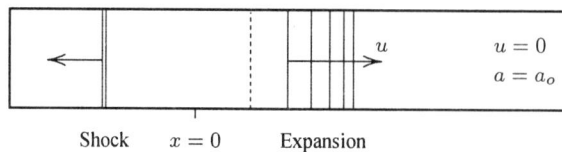

Shock $x = 0$ Expansion

u $u = 0$ $a = a_o$

Problem 12.21

12.22 For transonic flow close to an airplane wing, the velocity can be approximated as

$$\mathbf{u} \approx U(x)f(y)\,\mathbf{i} + \frac{1}{3}U(x)f(y)\frac{z}{c}\,\mathbf{k}$$

where $U(x)$ is a velocity, $f(y)$ is a dimensionless function and c is a length scale. The coordinates x, y and z lie in the streamwise, normal and spanwise directions, respectively. Assuming the second viscosity is $\zeta = -\frac{2}{3}\mu$ and that μ is constant, determine the viscous-stress tensor, $[\tau]$, and its divergence, $\nabla \cdot [\tau]$.

12.23 The vorticity vector and viscous-stress tensor for an incompressible flow are

$$\omega = 2\frac{U}{H}\left(2 - \frac{y}{H}\right)\mathbf{k} \quad \text{and} \quad [\tau] = 2\frac{\mu U}{H}\begin{bmatrix} 2x/H & -(2+y/H) & 0 \\ -(2+y/H) & -2x/H & 0 \\ 0 & 0 & 0 \end{bmatrix}$$

where U and H are constant velocity and length scales, respectively. Also, the flow has a stagnation point at $x = y = 0$. Determine the velocity components, u and v for this flow.

12.24 The vorticity vector and viscous-stress tensor for an incompressible flow are

$$\omega = -5A\,\mathbf{k} \quad \text{and} \quad [\tau] = 3\mu A\begin{bmatrix} 0 & 1 & 0 \\ 1 & 0 & 0 \\ 0 & 0 & 0 \end{bmatrix}$$

where A is a constant of dimensions $1/L$. Also, the flow has a stagnation point at $x = y = 0$. Determine the velocity components, u and v, for this flow.

12.25 The vorticity vector and viscous-stress tensor for a two-dimensional, incompressible flow are

$$\omega = \frac{K}{r^2}\sin\theta\,\mathbf{k} \quad \text{and} \quad [\tau] = -\frac{\mu K}{3r^2}\begin{bmatrix} 8\cos\theta & 9\sin\theta & 0 \\ 9\sin\theta & -10\cos\theta & 0 \\ 0 & 0 & 2\cos\theta \end{bmatrix}$$

where K is a constant of dimensions L^2/T. Also, the flow has a stagnation point at $r = R$, $\theta = \pi/2$. Determine the velocity components, u_r and u_θ for this flow as follows.

(a) One of the viscous-stress tensor components permits solving directly for $\nabla \cdot \mathbf{u}$. Identify the component and solve.

(b) Using the value of $\nabla \cdot \mathbf{u}$ obtained in Part (a), identify the stress component that permits solving directly for $\partial u_r/\partial r$. Integrate over r and call the *function of integration* $f'(\theta)$.

(c) Combining the results of Parts (a) and (b) with the stress component $\tau_{\theta\theta}$, derive an equation for $\partial u_\theta/\partial\theta$. Integrate over θ and call this *function of integration* $g(r)$.

(d) Use the remaining stress component and the results of Parts (b) and (c) to derive a differential equation involving $f(\theta)$ and $g(r)$.

(e) Use the equation for the vorticity and the results of Parts (b) and (c) to derive a second differential equation involving $f(\theta)$ and $g(r)$.

(f) Solve the coupled differential equations derived in Parts (d) and (e).

(g) Complete the solution by requiring a stagnation point at the point indicated above.

12.26 Compute the viscous-stress tensor for a *spiral vortex* with velocity components

$$u_r = u_\theta = A/r$$

where A is a constant and r is radial distance. Also, compute the viscous stress on surfaces A and B of the differential-sized fluid particle shown below. **HINT:** For the second part of this problem, rewrite your answers in terms of unit vectors in Cartesian coordinates, **i** and **j**.

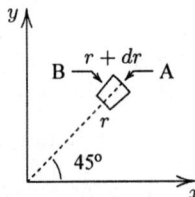

Problem 12.26

12.27 Consider a two-dimensional, viscous flow. If we rotate the coordinate system by an angle θ, the components of the viscous-stress tensor in the xy plane transform according to

$$[\tau'] = \left[\begin{array}{cc} \cos\theta & \sin\theta \\ -\sin\theta & \cos\theta \end{array} \right] \left[\begin{array}{cc} \tau_{xx} & \tau_{xy} \\ \tau_{yx} & \tau_{yy} \end{array} \right] \left[\begin{array}{cc} \cos\theta & -\sin\theta \\ \sin\theta & \cos\theta \end{array} \right]$$

(a) Taking account of the symmetry of $[\tau]$, perform the matrix multiplications to determine the components of $[\tau']$. Express your answers in terms of $\cos 2\theta$ and $\sin 2\theta$.

(b) For what angle θ does $\tau'_{xy} = 0$? The coordinate axes corresponding to this angle are called the *principal axes*.

(c) In a boundary layer, we can approximate the viscous-stress tensor by

$$[\tau] = \left[\begin{array}{cc} 0 & \tau \\ \tau & 0 \end{array} \right]$$

What are θ and $[\tau]$ for a boundary layer?

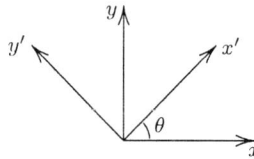

Problem 12.27

12.28 Consider the viscous stresses acting on a two-dimensional, rectangular fluid particle as shown. The flow is incompressible so that the centroid and center of mass of the fluid particle are coincident.

(a) Verify that the net torque about the centroid of the fluid particle per unit volume is given by

$$\frac{\Delta\tau_z}{\Delta x \Delta y} \to \tau_{xy} - \tau_{yx} \quad \text{as} \quad \Delta x, \Delta y \to 0$$

(b) Explain (with a sketch) why the pressure yields no net torque as $\Delta x, \Delta y \to 0$.

(c) Explain why the angular momentum of the fluid particle per unit volume is negligible compared to $\Delta\tau_z/(\Delta x \Delta y)$.

(d) If no body forces are acting, what can you conclude about τ_{xy} and τ_{yx}?

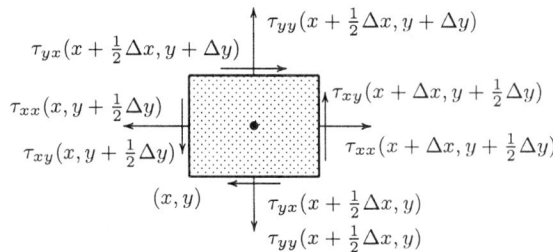

Problem 12.28

12.29 The limiting case of very small Reynolds number is called *Stokes flow* or *creeping flow*. In this limit the acceleration terms are negligible so that the continuity and Navier-Stokes equations simplify to:

$$\nabla \cdot \mathbf{u} = 0 \quad \text{and} \quad \mathbf{0} \approx -\nabla p + \mu \nabla^2 \mathbf{u}$$

If the relevant velocity and length scales for a given problem are U and L, what is the appropriate dimensionless form of the pressure in Stokes flow?

12.30 The limiting case of very small Reynolds number is called *Stokes flow* or *creeping flow*. In this limit the acceleration terms are negligible so that the continuity and Navier-Stokes equations simplify to:

$$\nabla \cdot \mathbf{u} = 0 \quad \text{and} \quad 0 \approx -\nabla p + \mu \nabla^2 \mathbf{u}$$

Verify that the pressure satisfies Laplace's equation and that the velocity satisfies the *biharmonic equation*, viz.,

$$\nabla^2 p = 0 \quad \text{and} \quad \nabla^4 \mathbf{u} = 0$$

where, by definition, $\nabla^4 \mathbf{u} = \nabla^2(\nabla^2 \mathbf{u})$.

12.31 Show that the incompressible flowfield for the *potential vortex*, i.e.,

$$u_r = 0, \quad u_\theta = \frac{\Gamma}{2\pi r}, \quad w = 0$$

with constant circulation, Γ, is an exact solution to the Navier-Stokes and continuity equations in cylindrical coordinates. Compute the pressure, $p(r, \theta, z)$, in terms of density, ρ, as well as Γ, r, and the freestream pressure, p_∞. Why is this solution unsuitable as an "exact" solution for a viscous flow?

12.32 Show that the incompressible flowfield for the *spiral vortex*, i.e.,

$$u_r = -\frac{Q}{2\pi r}, \quad u_\theta = \frac{\Gamma}{2\pi r}, \quad w = 0$$

with constant sink strength, $-Q$, and circulation, Γ, is an exact solution to the Navier-Stokes and continuity equations in cylindrical coordinates. Compute the pressure, $p(r, \theta, z)$, in terms of density, ρ, as well as Q, Γ, r, and the freestream pressure, p_∞. Why is this solution unsuitable as an "exact" solution for a viscous flow?

12.33 Show that the incompressible flowfield for *Hill's spherical vortex*, i.e.,

$$u_r = U \frac{rz}{R^2}, \quad u_\theta = 0, \quad w = U \left[1 - \left(\frac{z}{R} \right)^2 - 2 \left(\frac{r}{R} \right)^2 \right]$$

with constant velocity, U, and radius, R, is an exact solution to the Navier-Stokes and continuity equations in cylindrical coordinates. Compute the pressure, $p(r, \theta, z)$, in terms of density, ρ, as well as U, R, r and z. What value can p assume at the origin, $r = z = 0$, and in the farfield, $r, |z| \to \infty$? Why is this solution unsuitable as an "exact" solution for a viscous flow?

12.34 Consider an axisymmetric, incompressible flow for which

$$u_r = \frac{f(\theta)}{r}, \quad u_\theta = 0, \quad w = 0$$

where $f(\theta)$ is a function to be determined.

(a) Ignoring body forces, substitute this velocity field into the continuity and Navier-Stokes equations in cylindrical coordinates (Appendix D, Section D.2) and simplify. **HINT:** The circumferential component of momentum *does not* reduce to $\partial p / \partial \theta = 0$.

(b) Derive a second-order, ordinary differential equation for $f(\theta)$.

12.35 Consider a spherically-symmetric, incompressible flow for which

$$u_R = \frac{C}{R^2}, \quad u_\theta = 0, \quad u_\phi = 0$$

where C is a constant.

(a) Ignoring body forces, substitute this velocity field into the continuity and Navier-Stokes equations in spherical coordinates (Appendix D, Section D.3) and simplify.

(b) Solve for the pressure, p, subject to $p \to p_\infty$ as $R \to \infty$.

12.36 A thin layer of viscous fluid flows down an inclined plane. The velocity components are $u(y) = Cy(2h - y)$, $v = w = 0$, where C is a constant. The upper surface, $y = h$, is a free surface so that the appropriate boundary condition is $p = p_o$, where p_o is atmospheric pressure. Determine the constant C as a function of gravitational acceleration, g, kinematic viscosity, ν, and the angle θ. Also, compute the volume-flow rate, Q.

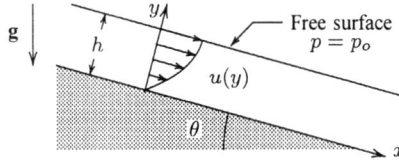

Problem 12.36

12.37 When an aircraft takes off, it leaves behind a *trailing vortex*, which can be quite strong and can take several minutes to decay. This presents a hazard to other aircraft and requires waiting periods between takeoffs and landings. To analyze the problem, we can model the motion as an *Oseen vortex* for which

$$u_r = 0, \qquad u_\theta = \frac{\Gamma}{2\pi r}\left[1 - \exp\left(-\frac{r^2}{4\nu t}\right)\right], \qquad w = 0$$

where Γ is circulation, r is radial distance, ν is kinematic viscosity and t is time.

   (a) Verify that this is an exact solution to the incompressible Navier-Stokes and continuity equations. **HINT:** By symmetry, necessarily $\partial p/\partial \theta = 0$.

   (b) How long does it take (in minutes) for u_θ to decay to 10% of its initial value at $r = 1$ ft? Assume $\nu = 1.6 \cdot 10^{-4}$ ft^2/sec.

   (c) Suppose that, to account for the fact that the flow is turbulent, we replace ν by an effective viscosity, $\nu_{\text{eff}} = 6.6 \cdot 10^{-3}$ ft^2/sec. Recompute the decay time of Part (b).

12.38 When an aircraft takes off, it leaves behind a *trailing vortex*, which can be quite strong and can take several minutes to decay. This presents a hazard to other aircraft and requires waiting periods between takeoffs and landings. To analyze the problem, we can model the motion as a *Taylor vortex* for which

$$u_r = 0, \qquad u_\theta = \frac{Hr}{8\pi\nu t^2}\exp\left(-\frac{r^2}{4\nu t}\right), \qquad w = 0$$

where H is a constant, r is radial distance, ν is kinematic viscosity and t is time.

   (a) What are the dimensions of the constant H?

   (b) Verify that this is an exact solution to the incompressible Navier-Stokes and continuity equations. **HINT:** By symmetry, necessarily $\partial p/\partial \theta = 0$.

12.39 For flow in a rotating coordinate frame, the steady, incompressible Navier-Stokes equation is

$$\mathbf{u} \cdot \nabla \mathbf{u} + 2\mathbf{\Omega} \times \mathbf{u} = -\frac{1}{\rho}\nabla p + \nu\nabla^2 \mathbf{u}$$

where $\mathbf{\Omega} = \Omega\mathbf{k}$ is the (constant) angular velocity of the coordinate system and p is the reduced pressure that accounts for centrifugal acceleration (see Section D.4 of Appendix D). In what are termed *rapidly-rotating* flows, the convective acceleration is negligible relative to the Coriolis acceleration. The Navier-Stokes equation then simplifies to

$$2\mathbf{\Omega} \times \mathbf{u} \approx -\frac{1}{\rho}\nabla p + \nu\nabla^2 \mathbf{u}$$

Show that for two-dimensional, *rapidly-rotating* flow, the z component of the vorticity, ω_z, satisfies

$$\omega_z = \frac{\nabla^2 p}{2\rho\Omega}$$

12.40 Beginning with the continuity and Navier-Stokes equations, show for two-dimensional, incompressible, viscous flow that the vorticity, ω_z, satisfies

$$\frac{d\omega_z}{dt} = \nu \nabla^2 \omega_z$$

12.41 Verify the viscous timestep limit for the upwind-differencing scheme in Equation (12.151).

12.42 Combine the CFL-condition timestep with the viscous timestep for explicit time-marching methods and determine the CFL number, N_{CFL}, as a function of the cell Reynolds number defined by $Re_{\Delta x} \equiv U \Delta x / \nu$.

12.43 Using von Neumann stability analysis, determine G and any condition required for stability of *Euler's method* applied to the inviscid Burgers' equation, $u_t + U u_x = 0$:

$$u_j^{n+1} = u_j^n - \frac{U \Delta t}{2 \Delta x} \left(u_{j+1}^{n+1} - u_{j-1}^{n+1} \right), \qquad U > 0$$

12.44 Using von Neumann stability analysis, determine G and any condition required for stability of *Richardson's method* applied to the one-dimensional diffusion equation, $u_t = \nu u_{xx}$:

$$u_j^{n+1} = u_j^{n-1} + \frac{2\nu \Delta t}{(\Delta x)^2} \left(u_{j+1}^n - 2u_j^n + u_{j-1}^n \right), \qquad \nu > 0$$

12.45 Using von Neumann stability analysis, determine G and any condition required for stability of *Crank and Nicolson's method* applied to the inviscid Burgers' equation with a sink term included, $u_t + U u_x = Su$:

$$u_j^{n+1} = u_j^n - \frac{U \Delta t}{4 \Delta x} \left(u_{j+1}^{n+1} + u_{j+1}^n - u_{j-1}^{n+1} - u_{j-1}^n \right) + \frac{1}{2} S \Delta t \left(u_j^{n+1} + u_j^n \right)$$

Assume $U > 0$ and $S < 0$. **HINT:** For a complex variable of the form $G = (a + ib)/(c + id)$, we have $|G|^2 = (a^2 + b^2)/(c^2 + d^2)$.

12.46 Using von Neumann stability analysis, determine G and any condition required for stability of *Lax's method* applied to the inviscid Burgers' equation, $u_t + U u_x = 0$:

$$\frac{u_j^{n+1} - \frac{1}{2} \left(u_{j+1}^n + u_{j-1}^n \right)}{\Delta t} + U \frac{u_{j+1}^n - u_{j-1}^n}{2 \Delta x} = 0, \qquad U > 0$$

12.47 Using von Neumann stability analysis, determine G and any condition required for stability of the following first-order accurate scheme applied to the inviscid Burgers' equation, $u_t + U u_x = 0$:

$$u_j^{n+1} = u_j^n - \frac{U \Delta t}{2 \Delta x} \left(u_{j+1}^{n+1} - u_{j-1}^n \right), \qquad U > 0$$

12.48 Using von Neumann stability analysis, determine G and any condition required for stability of the following difference scheme applied to the inviscid Burgers' equation, $u_t + U u_x = 0$:

$$u_j^{n+1} = u_j^n - \frac{U \Delta t}{2 \Delta x} \left[\beta \left(u_{j+1}^{n+1} - u_{j-1}^{n+1} \right) + (1 - \beta) \left(u_{j+1}^n - u_{j-1}^n \right) \right]$$

Assume $U > 0$ and that $\frac{1}{2} \leq \beta \leq 1$.

12.49 The purpose of this problem is to demonstrate the conditional stability of an explicit finite-difference method. The theoretical limit for MacCormack's method is $\tilde{N}_{CFL} \leq 1$. It will actually run at slightly larger values of $\tilde{N}_{CFL}$, but not very much larger. Run Program **SUCTION** (see Appendix E) with $\tilde{N}_{CFL}$ in excess of 1 for 51, 101, 201 and 401 grid points. Determine the maximum value of $\tilde{N}_{CFL}$, to the nearest hundredth, at which stable computation is possible.

12.50 The purpose of this problem is to determine the difference between first- and second-order accurate finite-difference schemes on solution accuracy. Run Program **SUCTION** (see Appendix E) and record the values of the velocity at $y(j) = 0.2$ and 1.0 for 51, 101 and 201 grid points. Do your computations using an effective CFL number, $\tilde{N}_{CFL}$, of 0.7. Now, modify the program so that only the predictor step is used. This corresponds to the scheme of Equation (12.143). **NOTE:** Be sure to add a loop that sets $u(j,1) = u(j,2)$ after advancing the solution in time. Compare the solution errors with those of MacCormack's method as a function of grid-point number.

12.51 The purpose of this problem is to examine the effect of initial conditions on how many iterations are needed to attain a steady-state solution with MacCormack's method. First, modify Program **SUCTION** (see Appendix E) so that computation terminates when the maximum difference between the computed and exact solutions is reduced to 0.005. Run the program and record the number of timesteps required to attain steady state conditions. Do your computations using 51 grid points for effective CFL numbers, $\tilde{N}_{CFL}$, of 0.2, 0.4, 0.6, 0.8 and 1.0. Now modify the program so that the initial velocity, $u(y,0)$, is U_∞ everywhere except at the wall where the no-slip condition must be enforced. Compare the number of iterations needed with those obtained from the original initial conditions.

12.52 Using MacCormack's method, determine the solution for *channel flow*, using equally-spaced grid points. Note first that the vertical velocity, v, is zero while the horizontal velocity, u, depends only upon y. The equations of motion simplify to

$$0 = -\frac{1}{\rho}\frac{dp}{dx} + \nu\frac{d^2u}{dy^2}, \qquad u(0) = 0, \quad u'(h/2) = 0$$

The exact solution is

$$u(y) = u_m\left[1 - \left(\frac{y}{h/2}\right)^2\right] \qquad \text{where} \qquad u_m \equiv -\frac{h^2}{8\mu}\frac{dp}{dx}$$

To solve numerically, proceed as follows.

(a) Rewrite the momentum equation with the unsteady term appearing on the left-hand side. Now, recast the unsteady momentum equation and the boundary conditions in dimensionless form, i.e., in terms of u/u_m, $y/(h/2)$ and $\nu t/(h/2)^2$.

(b) What is the maximum stepsize for stable computation in terms of dimensionless variables?

(c) Using $u(y,0) = u_m$ at all interior points for the initial condition, solve using MacCormack's method. Take advantage of the problem's symmetry and confine the computation to $0 \leq y \leq h/2$. Let the step size be $\Delta y = 0.01$, and use an effective CFL number, $\tilde{N}_{CFL} = 0.9$. Run the computation up to a dimensionless time, $\nu t/(h/2)^2$, of 0.10. If the Reynolds number is 1000, how many channel heights will a fluid particle travel in this amount of time? Plot the velocity profile, including data points for the exact solution, to display results of the computation. **NOTE:** Be sure to account for the pressure-gradient when you cast the equation in finite-difference form.

(d) Solve the problem again with $\Delta y = 0.005$, and use Richardson extrapolation to estimate solution errors at $y/(h/2) = 0.2$, 0.4, 0.6 and 0.8. Compare with the exact solution as well.

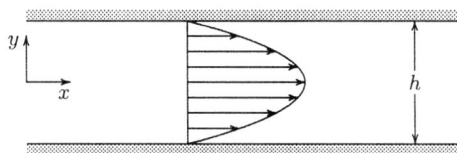

Problem 12.52

12.53 Using MacCormack's method, determine the solution for *Couette flow*, using equally-spaced grid points. Note first that the vertical velocity, v, is zero while the horizontal velocity, u, depends only upon y. The equations of motion simplify to

$$0 = \nu \frac{d^2 u}{dy^2}, \qquad u(0) = 0, \quad u(h) = U$$

The exact solution is

$$u(y) = U \frac{y}{h}$$

To solve numerically, proceed as follows.

(a) Rewrite the momentum equation with the unsteady term appearing on the left-hand side. Now, recast the unsteady momentum equation and the boundary conditions in dimensionless form, i.e., in terms of u/U, y/h and $\nu t/h^2$.

(b) What is the maximum stepsize for stable computation in terms of dimensionless variables?

(c) Using $u(y, 0) = U$ at all interior points for the initial condition, solve using MacCormack's method. Let the step size be $\Delta y = 0.01$, and use an effective CFL number, $\tilde{N}_{CFL} = 0.9$. Run the computation up to a dimensionless time, $\nu t/h^2$, of 0.10. If the Reynolds number is 1000, how many channel heights will a fluid particle travel in this amount of time? Plot the velocity profile, including data points for the exact solution, to display results of the computation.

(d) Solve the problem again with $\Delta y = 0.005$, and use Richardson extrapolation to estimate solution errors at $y/h = 0.2$, 0.4, 0.6 and 0.8. Compare with the exact solution as well.

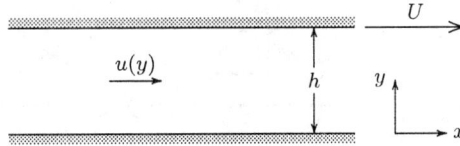

Problems 12.53, 12.54

12.54 Using MacCormack's method, determine the solution for *Couette-Poiseuille flow*, using equally-spaced grid points. Note first that the vertical velocity, v, is zero while the horizontal velocity, u, depends only upon y. The equations of motion simplify to

$$0 = -\frac{1}{\rho} \frac{dp}{dx} + \nu \frac{d^2 u}{dy^2}, \qquad u(0) = 0, \quad u(h) = U$$

The exact solution is

$$u(y) = U \left[1 - \chi \left(1 - \frac{y}{h} \right) \right] \frac{y}{h}, \qquad \chi \equiv \frac{h^2}{2\mu U} \frac{dp}{dx}$$

To solve numerically, proceed as follows.

(a) Rewrite the momentum equation with the unsteady term appearing on the left-hand side. Now, recast the unsteady momentum equation and the boundary conditions in dimensionless form, i.e., in terms of u/U, y/h, $\nu t/h^2$ and χ.

(b) What is the maximum stepsize for stable computation in terms of dimensionless variables?

(c) Using $u(y, 0) = U$ at all interior points for the initial condition, solve for the case $\chi = 4$ using MacCormack's method. Let the step size be $\Delta y = 0.01$, and use an effective CFL number, $\tilde{N}_{CFL} = 0.9$. Run the computation up to a dimensionless time, $\nu t/h^2$, of 0.10. If the Reynolds number is 1000, how many channel heights will a fluid particle travel in this amount of time? Plot the velocity profile, including data points for the exact solution, to display results of the computation. **NOTE:** Be sure to account for the pressure-gradient when you cast the equation in finite-difference form.

(d) Solve the problem again with $\Delta y = 0.005$, and use Richardson extrapolation to estimate solution errors at $y/h = 0.2$, 0.4, 0.6 and 0.8. Compare with the exact solution as well.

12.55 Using MacCormack's method, determine the solution for *Stokes' First Problem*, i.e., flow above an impulsively accelerated flat plate. Note first that the vertical velocity, v, is zero while the horizontal velocity, u, depends only upon y and t. The equations of motion simplify to

$$\frac{\partial u}{\partial t} = \nu \frac{\partial^2 u}{\partial y^2}, \qquad u(0,t) = U, \quad u(y,t) \to 0 \;\text{ as }\; y \to \infty$$

To solve numerically, proceed as follows.

(a) Recast the unsteady momentum equation and the boundary conditions in dimensionless form, i.e., in terms of u/U, y/L and $\nu t/L^2$, where L is an arbitrary length.

(b) What is the maximum stepsize for stable computation in terms of dimensionless variables?

(c) Using $u(y,0) = 0$ at all interior points for the initial condition, solve using MacCormack's method. Let the top of the grid lie at $y = 10L$ and let the step size be $\Delta y = 0.01L$. Also, use an effective CFL number, $\tilde{N}_{CFL} = 0.9$. Run the computation until $\nu t/L^2 = 4$, saving the computed velocity profile when the dimensionless time, $\nu t/L^2$, is 1, 2, 3 and 4. Plot the dimensionless velocity profiles, u/U, on a single graph as functions of $\eta \equiv y/\sqrt{\nu t}$.

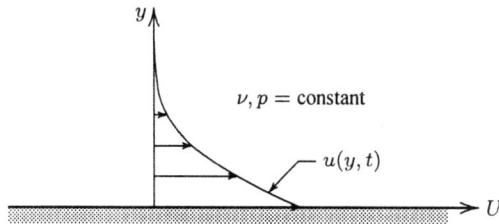

Problem 12.55

Chapter 13

Navier-Stokes Solutions

In general, the Navier-Stokes equation is a second-order, quasi-linear, vector, partial differential equation. Viewed as a group of three equations for the velocity components and pressure, coupled with the continuity equation, it is a seventh-order system in the general three-dimensional case. As such, exact analytical solutions for complex geometries are virtually nonexistent. However, exact solutions can be obtained for special geometries, and examination of these solutions serves several useful purposes. For example, we can:

- discover interesting features common to viscous-flow problems;

- derive an expression for head loss in a straight pipe;

- quantitatively describe the production and diffusion of vorticity from a solid boundary into the main flow;

- explore the connection between pressure and velocity when the assumptions underlying Bernoulli's equation do not hold;

- develop solutions that can be used in limiting cases to test complex CFD programs.

Common features of Navier-Stokes solutions for viscous flows will be evident in every flow we examine in this chapter, most notably the presence of vorticity and the passive nature of pressure (a noteworthy feature of incompressible flow). In each case, we quote the full equations and then simplify according to the problem under consideration. As we will see, all but one of these exact solutions have one thing in common. Specifically, the Navier-Stokes equation's convective term, $\mathbf{u} \cdot \nabla \mathbf{u}$, is exactly zero. The resulting simplified equation is thus linear and readily amenable to solution by straightforward algebraic or numerical methods.

Deriving an expression for head loss in a pipe is relevant because we only briefly touched on the origin of head losses in Chapter 7. While alluding to their physical connection with viscous effects, the discussion was qualitative. By manipulating the exact solution (see Section 13.3), we can show their direct quantitative relation to effects of friction.

Computing the production and diffusion of vorticity fills a quantitative gap left by our qualitative discussion in Chapter 10 of the way circulation develops in slightly-viscous fluids. Analysis of Stokes' First Problem in Section 13.4 is especially illuminating in this respect. Solution of this problem also demonstrates a valuable analytical technique known as the *similarity-solution method*.

Viscous effects on pressure are exhibited in flow near a stagnation point (Section 13.6). On the one hand, our results will show why the terminology "total pressure" and "total temperature" are preferred over "stagnation pressure" and "stagnation temperature." We mentioned this point in Section 8.6, but offered no quantitative results to support our claims. On the other hand, we will also see that the effect on pressure is very small for most practical flows.

Using exact solutions to validate CFD programs is critically important for several reasons. First, even Richardson extrapolation can be misleading if a programming error results in an error in the equations being solved. Such errors are easily detected when an exact solution is available. Second, when an exact solution is available, we have the best of all possible yardsticks for determining the accuracy of a computer program. Third, interpretation of numerical results is very difficult without a knowledge of basic fluid-mechanical principles. Since most modern CFD research focuses on full Navier-Stokes solutions, an understanding of the physics of viscous flows is essential.

13.1   Couette Flow

Description. Our first exact Navier-Stokes solution will be for the general case of **Couette-Poiseuille flow**, or **Couette flow** for short. Strictly speaking, Couette flow, named in honor of M. Couette (1858-1943), corresponds to flow with constant pressure. We discussed Couette flow briefly in Section 1.9. Without derivation, we quoted the solution, noting that velocity varies linearly between the two bounding parallel plates. Here, we include effects of pressure gradient.

Figure 13.1 shows how Couette flow, including pressure gradient, can be realized in a laboratory. In the configuration shown, we have two large reservoirs filled to different levels, connected by a thin slot, or channel. The bottom wall of the channel is stationary, while the upper wall is a moving belt. Additional piping that is not shown would be used to maintain constant depth in each reservoir.

The first object of the experiment is to simulate two infinite parallel plates, one stationary and one moving, separated by a constant distance, h. The height of the channel must be very small relative to the distance between the reservoirs, L. Measurements show that the

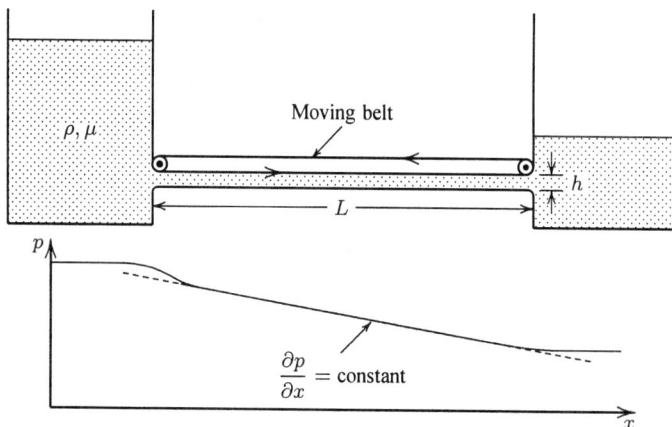

Figure 13.1: *Couette-flow schematic; clockwise belt rotation yields $\partial p / \partial x > 0$ relative to the belt velocity; counterclockwise rotation gives $\partial p / \partial x < 0$.*

flow becomes fully developed, i.e., independent of x within about 50 channel heights from the reservoirs.[1] Thus, a good simulation of Couette flow would require $L/h > 100$. Also the width of the channel (out of the page) should be at least $10h$ to avoid three-dimensional effects caused by the end walls bounding the channel (parallel to the page).

The second object is to have a linear variation of pressure along the channel, so that the pressure gradient is independent of x. By maintaining different depths in the two reservoirs, the pressures at the ends of the channel are different due to the potential-energy difference. As we will see below, once fully-developed flow is attained, the pressure must vary linearly with x. Figure 13.1 includes typical pressure variation for the experimental arrangement. It departs from linear variation only near the ends of the channel, where the flow is not fully developed.

Thus, we focus on the flow in the channel sufficiently distant from the reservoir junctions that the plates appear infinite in the x and z directions, where z is distance normal to the page. Figure 13.2 shows the geometry for Couette flow, where we denote distance normal to the channel walls as y. The gravitational force thus acts in the negative y direction.

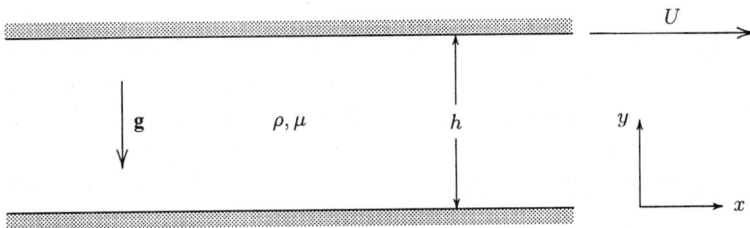

Figure 13.2: *Idealized Couette-flow geometry for fully-developed flow.*

Equations of Motion. Since the plates are infinite in the z direction, we can begin with the steady, two-dimensional form of the continuity and Navier-Stokes equations. Including the gravitational body force, we have

$$\frac{\partial u}{\partial x} + \frac{\partial v}{\partial y} = 0 \tag{13.1}$$

$$u\frac{\partial u}{\partial x} + v\frac{\partial u}{\partial y} = -\frac{1}{\rho}\frac{\partial p}{\partial x} + \nu\left(\frac{\partial^2 u}{\partial x^2} + \frac{\partial^2 u}{\partial y^2}\right) \tag{13.2}$$

$$u\frac{\partial v}{\partial x} + v\frac{\partial v}{\partial y} = -\frac{1}{\rho}\frac{\partial p}{\partial y} + \nu\left(\frac{\partial^2 v}{\partial x^2} + \frac{\partial^2 v}{\partial y^2}\right) - g \tag{13.3}$$

These equations must be solved subject to the no-slip boundary condition. Hence, the velocity components must satisfy

$$u = 0, \quad v = 0 \quad \text{at} \quad y = 0 \tag{13.4}$$

$$u = U, \quad v = 0 \quad \text{at} \quad y = h \tag{13.5}$$

In light of the proposed experimental configuration of Figure 13.1, we include the gravitational force to determine any effect it might have in the channel. As we will see, while it establishes the overall pressure gradient, it plays a completely passive role in the dynamics of the flow between the horizontal plates.

---

[1]Schlichting (1979) gives a numerical solution for developing internal flows that indicates the length required to achieve full-developed two-dimensional flow, ℓ_e, is given by $\ell_e = 0.06h(u_o h/\nu)$, where u_o is the entrance velocity.

Mass Conservation. We begin by observing that when the flow becomes fully developed, the flow is independent of x. Thus, the continuity Equation (13.1) tells us

$$\frac{dv}{dy} = 0 \quad \Longrightarrow \quad v(y) = \text{constant} \tag{13.6}$$

But, the no-slip boundary condition requires $v = 0$ at both walls, so that

$$v(y) = 0 \tag{13.7}$$

As an immediate consequence, we see that the convective acceleration is zero:

$$\underbrace{u\frac{\partial \mathbf{u}}{\partial x}}_{\substack{Zero\ because \\ \partial/\partial x \to 0}} + \underbrace{v\frac{\partial \mathbf{u}}{\partial y}}_{\substack{Zero\ because \\ v=0}} = \mathbf{0} \tag{13.8}$$

y-Momentum Conservation. Because the vertical velocity vanishes, the vertical momentum Equation (13.3) simplifies to

$$0 = -\frac{1}{\rho}\frac{\partial p}{\partial y} - g \tag{13.9}$$

Integrating with respect to y yields

$$p(x,y) = \rho f(x) - \rho g y \tag{13.10}$$

where $f(x)$ is a function of integration that will be determined by substitution into the x-momentum equation.

x-Momentum Conservation. Because we know the convective acceleration is exactly zero and $\partial^2 u/\partial x^2 = 0$ for fully-developed flow, all that remains of the x-momentum Equation (13.2) is

$$0 = -\frac{1}{\rho}\frac{\partial p}{\partial x} + \nu\frac{d^2 u}{dy^2} \tag{13.11}$$

Substituting for p from Equation (13.10), we see that

$$\frac{df(x)}{dx} = \nu\frac{d^2 u(y)}{dy^2} \tag{13.12}$$

Because the left-hand side of Equation (13.12) is a function only of x and the right-hand side a function only of y, both must be a pure constant. Therefore, once fully-developed flow is established, the only way the equation can be satisfied is to have

$$\frac{d^2 u(y)}{dy^2} = \frac{1}{\rho\nu}\frac{\partial p}{\partial x} = \frac{1}{\mu}\frac{\partial p}{\partial x} = \text{constant} \tag{13.13}$$

Integrating twice over y, we find

$$u(y) = A + By + \frac{y^2}{2\mu}\frac{\partial p}{\partial x} \tag{13.14}$$

where A and B are constants of integration. Application of the no-slip boundary conditions on $u(y)$ at both walls yields

$$\left. \begin{array}{l} u(h) = U = A + Bh + \dfrac{h^2}{2\mu}\dfrac{\partial p}{\partial x} \\[3mm] u(0) = 0 = A \end{array} \right\} \quad \Longrightarrow \quad A = 0, \quad B = \dfrac{U}{h} - \dfrac{h}{2\mu}\dfrac{\partial p}{\partial x} \qquad (13.15)$$

Therefore, the solution for the horizontal velocity is

$$u(y) = U\left[1 - \frac{h^2}{2\mu U}\frac{\partial p}{\partial x}\left(1 - \frac{y}{h}\right)\right]\left(\frac{y}{h}\right) \qquad (13.16)$$

This completes the solution for Couette flow with u and v given by Equations (13.16) and (13.7), respectively.

Discussion. Clearly, for the constant-pressure case, Equation (13.16) simplifies to the linear variation discussed in Chapter 1. The solution is a good approximation for any flow through a thin gap between two parallel plates moving at different velocities. Figure 13.3 shows solutions for three cases corresponding to:

$\partial p/\partial x < 0 \cdots$ Favorable pressure gradient
$\partial p/\partial x = 0 \cdots$ Constant pressure
$\partial p/\partial x > 0 \cdots$ Adverse pressure gradient

As shown, the fluid accelerates for a favorable pressure gradient since pressure decreases in the direction the upper wall is moving. By contrast, an adverse pressure gradient decelerates the flow. If $U = 0$, favorable pressure gradient corresponds to what we usually refer to as plane Poiseuille flow (see Section 13.2) going to the right, while adverse pressure gradient corresponds to plane Poiseuille flow going to the left.

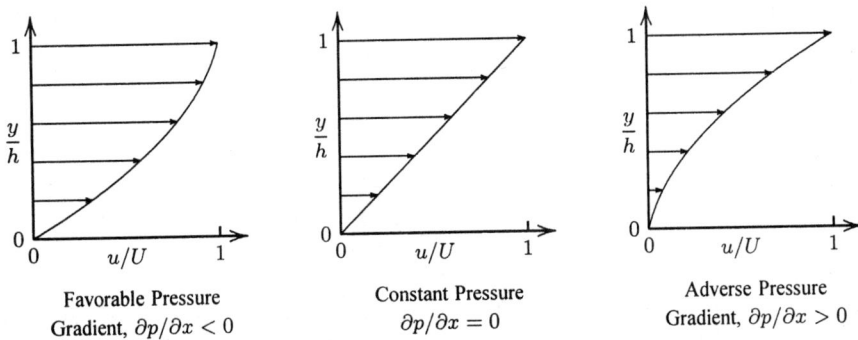

Favorable Pressure Gradient, $\partial p/\partial x < 0$ Constant Pressure $\partial p/\partial x = 0$ Adverse Pressure Gradient, $\partial p/\partial x > 0$

Figure 13.3: *Couette-flow velocity profiles.*

The shear stress for Couette flow is given by

$$\tau_{yx}(y) = \mu\frac{\partial u}{\partial y} = \frac{\mu U}{h} - \frac{h}{2}\frac{\partial p}{\partial x}\left(1 - 2\frac{y}{h}\right) \qquad (13.17)$$

Thus, the stress varies linearly with distance across the channel except in the constant-pressure case for which it is constant. The skin friction at the lower wall, c_f, is an interesting quantity that brings out another feature of Couette flow, and general viscous-flow physics. By definition,

$$c_f \equiv \frac{\tau_{yx}(0)}{\frac{1}{2}\rho U^2} = \frac{2\mu}{\rho U h}\left(1 - \frac{h^2}{2\mu U}\frac{\partial p}{\partial x}\right) \tag{13.18}$$

Therefore, introducing the Reynolds number based on channel height, Re_h, we find

$$c_f = \frac{2}{Re_h}\left(1 - \frac{h^2}{2\mu U}\frac{\partial p}{\partial x}\right), \qquad Re_h = \frac{\rho U h}{\mu} \tag{13.19}$$

Inspection of Equation (13.19) shows that for large adverse pressure gradient, the skin friction changes sign at the stationary wall. The critical value of the pressure gradient for which the skin friction is exactly zero is

$$\left(\frac{\partial p}{\partial x}\right)_{crit} = \frac{2\mu U}{h^2} \tag{13.20}$$

For larger values of $\partial p/\partial x$, skin friction is negative, corresponding to fluid near the stationary wall moving in the opposite direction of the wall at $y = h$. Figure 13.4 shows velocity profiles for the critical case and the reverse-flow case.

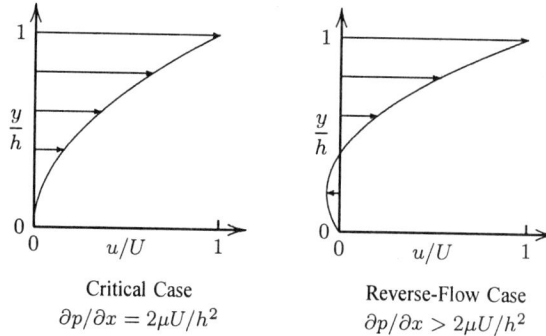

Critical Case
$\partial p/\partial x = 2\mu U/h^2$

Reverse-Flow Case
$\partial p/\partial x > 2\mu U/h^2$

Figure 13.4: *More Couette-flow velocity profiles.*

The reverse-flow case is different from flow separation observed on objects such as a cylinder (see Figure 11.12). Couette flow has zero vertical velocity, which means there is no exchange of fluid in the reverse-flow region with fluid in the forward-flow region. In other words, Couette flow features two streams flowing in opposite directions, separated by a stagnation surface, i.e., a surface on which the velocity vanishes. By contrast, a separated flow involves recirculating motion in which vertical velocities of the same order of magnitude as the horizontal velocities occur, and a given fluid particle can move from the forward-flow region to the reverse-flow region and back again.

As suggested above, the gravitational force plays an unimportant role in the dynamics of the flow within the channel. While it creates a pressure gradient, $\partial p/\partial y = -\rho g$, the absence of vertical motion means this gradient is of no consequence in a dynamic sense. Also, Equation (13.11) shows that the primary force balance is between the streamwise pressure gradient, $\partial p/\partial x$, and the shear-stress gradient, $\nu \partial^2 u/\partial y^2 = \partial \tau_{yx}/\partial y$. We can view the pressure as playing a passive role in the following sense. While the potential-energy difference of the fluid in the two reservoirs is directly responsible for the pressure difference, it plays no role in the functional form of $\partial p/\partial x$. Rather, the vorticity that exists in the channel is

$$\omega_z = \left(\frac{\partial v}{\partial x} - \frac{\partial u}{\partial y}\right) = -\frac{\partial u}{\partial y} \tag{13.21}$$

Hence, combining Equations (13.17) and (13.21), a little rearrangement of terms yields

$$\omega_z = -\frac{U}{h}\left(1 - \frac{h^2}{2\mu U}\frac{\partial p}{\partial x}\right) - \left(\frac{\partial p/\partial x}{\mu}\right)y \tag{13.22}$$

Now, since the shear stress is $\tau_{yx} = \mu\partial u/\partial y = -\mu\omega_z$, the vorticity's linear variation with y produces a constant shear-stress gradient. Hence, since the x-momentum equation tells us $\partial p/\partial x = \partial\tau_{yx}/\partial y$ for this flow [see Equation (13.11)], $\partial p/\partial x$ is necessarily constant.

13.2 Channel Flow

Description. Our next exact Navier-Stokes solution is for **channel flow**, which is also called **plane Poiseuille flow** or **duct flow**. This is again a two-dimensional flow between parallel plates with a constant pressure gradient. Unlike Couette flow, however, both plates are stationary. The flow is thus driven by a negative pressure gradient, i.e., $\partial p/\partial x < 0$ if flow is from left to right.

Channel flow can be simulated in a laboratory with the same experimental setup depicted in Figure 13.1, without the moving belt. As with Couette flow, we achieve fully-developed conditions within about 50 channel heights of the inlet under typical laboratory conditions. The actual length varies roughly as $0.06hRe_h$ [Schlichting (1979)]. Figure 13.5 defines channel-flow geometry.

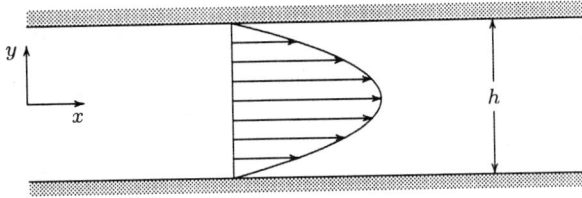

Figure 13.5: *Idealized channel-flow geometry for fully-developed flow.*

Equations of Motion. We saw from the Couette-flow solution that, aside from establishing the overall pressure drop along the channel, the gravitational force plays no direct role in the dynamics of the flow. The same is true for channel flow, wherefore the gravitational force can be omitted from the analysis. Alternatively, we can accomplish the same end by defining a **reduced pressure**, $\hat{p}$, according to

$$\hat{p} = p + \rho g y \tag{13.23}$$

where p denotes the physical pressure. What we are doing here is taking advantage of the fact that gravity is a conservative body force, with a potential function $\mathcal{V} = -gy$. Equation (13.23) defines the reduced pressure as the sum of the physical pressure and the potential energy, the reference level for potential energy being arbitrary. Because the flow is two dimensional, the continuity and Navier-Stokes equations are as follows.

$$\frac{\partial u}{\partial x} + \frac{\partial v}{\partial y} = 0 \tag{13.24}$$

$$u\frac{\partial u}{\partial x} + v\frac{\partial u}{\partial y} = -\frac{1}{\rho}\frac{\partial\hat{p}}{\partial x} + \nu\left(\frac{\partial^2 u}{\partial x^2} + \frac{\partial^2 u}{\partial y^2}\right) \tag{13.25}$$

$$u\frac{\partial v}{\partial x} + v\frac{\partial v}{\partial y} = -\frac{1}{\rho}\frac{\partial \hat{p}}{\partial y} + \nu\left(\frac{\partial^2 v}{\partial x^2} + \frac{\partial^2 v}{\partial y^2}\right) \tag{13.26}$$

For channel flow, the no-slip boundary condition becomes

$$u = 0, \quad v = 0 \quad \text{at} \quad y = \pm\frac{h}{2} \tag{13.27}$$

Mass and y-Momentum Conservation. We seek a solution for fully-developed flow, in which the velocity depends only upon y. Again, as with Couette flow, continuity simplifies to

$$\frac{dv}{dy} = 0 \quad \Longrightarrow \quad v = \text{constant} = 0 \tag{13.28}$$

where we take advantage of the no-slip boundary condition to determine the constant. Since the vertical velocity vanishes, the y-momentum equation tells us

$$0 = -\frac{1}{\rho}\frac{\partial \hat{p}}{\partial y} \quad \Longrightarrow \quad \hat{p} = \hat{p}(x) \tag{13.29}$$

x-Momentum Conservation. Observing that the reduced pressure is a function only of x, that fully-developed flow means $\partial u/\partial x = \partial^2 u/\partial x^2 = 0$, and that mass conservation requires $v = 0$, most of the terms in the x component of the Navier-Stokes equation [Equation (13.25)] are zero. What remains is

$$0 = -\frac{1}{\rho}\frac{d\hat{p}(x)}{dx} + \nu\frac{d^2u(y)}{dy^2} \tag{13.30}$$

which must be solved subject to Equation (13.27). Since $\hat{p}$ is a function only of x and u is a function only of y, we conclude immediately that

$$\frac{d^2u}{dy^2} = \frac{1}{\mu}\frac{d\hat{p}}{dx} = \text{constant} \quad \Longrightarrow \quad u(y) = A + By + \frac{y^2}{2\mu}\frac{d\hat{p}}{dx} \tag{13.31}$$

where A and B are constants of integration. From the boundary conditions at the channel walls, $y = \pm h/2$, we have

$$u(\pm h/2) = 0 = A \pm B\frac{h}{2} + \frac{h^2}{8\mu}\frac{d\hat{p}}{dx} \tag{13.32}$$

Hence, we obtain

$$A = -\frac{h^2}{8\mu}\frac{d\hat{p}}{dx}, \quad B = 0 \tag{13.33}$$

so that the horizontal velocity, $u(y)$, is

$$u(y) = -\frac{h^2}{8\mu}\frac{d\hat{p}}{dx}\left[1 - 4\left(\frac{y}{h}\right)^2\right] \tag{13.34}$$

Finally, we can cast this equation in a more illuminating form by rewriting it in terms of the centerline velocity, which is also the maximum velocity, viz.,

$$u(y) = u_m\left[1 - \left(\frac{y}{h/2}\right)^2\right], \quad u_m \equiv -\frac{h^2}{8\mu}\frac{\partial p}{\partial x} \tag{13.35}$$

We have written the final result in terms of the physical pressure, p, noting from Equation (13.23) that $d\hat{p}/dx = \partial p/\partial x$.

Discussion. Channel flow is a simple shear flow with a favorable pressure gradient. By differentiating the velocity in Equation (13.35), the vorticity, ω_z, is readily shown to be a linear function of distance across the channel. Since the shear stress, τ_{yx}, for channel flow is again given by $\tau_{yx} = \mu \partial u/\partial y = -\mu \omega_z$, the flow has a constant shear-stress gradient, $\partial \tau_{yx}/\partial y$. This, in turn, dictates that the pressure drop along the channel must occur with a linear variation in pressure, so that $\partial p/\partial x$ is constant. This is exactly what we found for Couette flow, viz., that the existing vorticity field required to maintain the flow dictates the functional form of the pressure gradient.

The shear stress at the bottom wall is given by

$$\tau_{yx}(-h/2) = \mu \frac{du}{dy}\bigg|_{y=-h/2} = -\frac{8\mu u_m}{h^2}\left(-\frac{h}{2}\right) = \frac{4\mu u_m}{h} \tag{13.36}$$

Hence, the skin friction is given as a function of Reynolds number, Re_h, by

$$c_f = \frac{4\mu u_m/h}{\frac{1}{2}\rho u_m^2} = \frac{8}{\rho u_m h/\mu} = \frac{8}{Re_h} \tag{13.37}$$

Often, it is more convenient to define Reynolds number and skin friction in terms of the average velocity, $\bar{u}$, in the channel. A simple computation, left as an exercise for the reader, shows that

$$\bar{u} = \frac{2}{3}u_m \tag{13.38}$$

Thus, the skin friction based on average velocity is

$$c_f = \frac{4\mu u_m/h}{\frac{1}{2}\rho \bar{u}^2} = \frac{6\mu \bar{u}/h}{\frac{1}{2}\rho \bar{u}^2} = \frac{12}{\overline{Re}_h} \tag{13.39}$$

where $\overline{Re}_h = \rho \bar{u} h/\mu$. In summary, the skin friction for channel flow is

$$c_f = \begin{cases} \dfrac{8}{Re_h}, & Re_h = \dfrac{\rho u_m h}{\mu} \\[2ex] \dfrac{12}{\overline{Re}_h}, & \overline{Re}_h = \dfrac{\rho \bar{u} h}{\mu} \end{cases} \tag{13.40}$$

13.3 Pipe Flow

Description. If we connect the reservoirs shown in Figure 13.1 with a cylindrical pipe, we have the axisymmetric equivalent of channel flow. This is the classical motion described as **pipe flow** or **Hagen-Poiseuille flow**. We discussed pipe flow in Subsection 1.9.2, deriving the velocity profile by balancing the pressure difference and the viscous stress. We also discussed pipe flow in Section 7.7, introducing the concept of head loss. In both of these earlier analyses, our focus was on fully-developed flow.

We again confine our attention to the fully-developed case which, like Couette and channel flow, is attained about 50 pipe diameters from the inlet reservoir under typical laboratory conditions. Correlation of measurements shows that the entrance length varies approximately

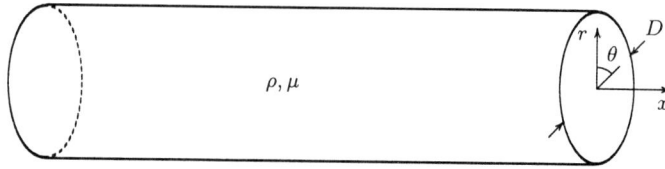

Figure 13.6: *Idealized pipe-flow geometry; pipe radius is $R = D/2$.*

as $0.06 D Re_D$ [Rosenhead (1963]. Figure 13.6 shows the notation and geometry we will use to analyze pipe flow.

Equations of Motion. For consistency with Couette-flow and channel-flow notation, we denote axial distance and velocity by x and u, respectively. Assuming the flow is axisymmetric, the azimuthal velocity, u_θ, is zero and all flow properties are independent of θ. The axisymmetric continuity and Navier-Stokes equations are:

$$\frac{\partial u}{\partial x} + \frac{1}{r} \frac{\partial}{\partial r}(r u_r) = 0 \tag{13.41}$$

$$u \frac{\partial u_r}{\partial x} + u_r \frac{\partial u_r}{\partial r} = -\frac{1}{\rho} \frac{\partial \hat{p}}{\partial r} + \nu \left(\frac{\partial^2 u_r}{\partial x^2} + \frac{\partial^2 u_r}{\partial r^2} + \frac{1}{r} \frac{\partial u_r}{\partial r} - \frac{u_r}{r^2} \right) \tag{13.42}$$

$$u \frac{\partial u}{\partial x} + u_r \frac{\partial u}{\partial r} = -\frac{1}{\rho} \frac{\partial \hat{p}}{\partial x} + \nu \left(\frac{\partial^2 u}{\partial x^2} + \frac{\partial^2 u}{\partial r^2} + \frac{1}{r} \frac{\partial u}{\partial r} \right) \tag{13.43}$$

where we again work with the reduced pressure, $\hat{p}$, [see Equation (13.23)] to effectively eliminate the gravitational force. The no-slip boundary condition applies at the pipe wall so that

$$u = 0, \quad u_r = 0 \quad \text{at} \quad r = R \tag{13.44}$$

where $R = D/2$ is the pipe radius. Additionally, the flow must be symmetric about the pipe centerline. This means $\partial u / \partial r$ must be zero at $r = 0$. It also means $u_r = 0$ at the centerline since a nonzero radial velocity on the axis would correspond to a source or a sink. Hence, the axial symmetry of the flow tells us

$$\frac{\partial u}{\partial r} = 0, \quad u_r = 0 \quad \text{at} \quad r = 0 \tag{13.45}$$

Mass and r-Momentum Conservation. Because we are seeking the solution for fully-developed flow, necessarily $\partial u / \partial x = 0$. Hence, the continuity Equation (13.41) becomes

$$\frac{1}{r} \frac{d}{dr}(r u_r) = 0 \quad \Longrightarrow \quad r u_r = \text{constant} \tag{13.46}$$

Both the symmetry condition and the no-slip condition at the pipe wall are satisfied only if the constant is zero. Therefore,

$$u_r(r) = 0 \tag{13.47}$$

Since the radial velocity, u_r, is zero, all that remains of the radial-momentum Equation (13.42) is

$$0 = -\frac{1}{\rho} \frac{\partial \hat{p}}{\partial r} \quad \Longrightarrow \quad \hat{p} = \hat{p}(x) \tag{13.48}$$

x-Momentum Conservation. Because pressure depends only upon x, all derivatives with respect to x vanish for fully-developed flow. Also, $u_r = 0$ so that $\mathbf{u} \cdot \nabla u_r = 0$. Thus, the x component of the Navier-Stokes Equation (13.43) assumes the following streamlined form.

$$0 = -\frac{1}{\rho}\frac{d\hat{p}}{dx} + \nu\left(\frac{d^2 u}{dr^2} + \frac{1}{r}\frac{du}{dr}\right) = -\frac{1}{\rho}\frac{d\hat{p}}{dx} + \frac{\nu}{r}\frac{d}{dr}\left(r\frac{du}{dr}\right) \qquad (13.49)$$

Therefore, dividing through by ν and noting that by definition $\mu = \rho\nu$, we have a pure function of x, the pressure gradient, equal to a pure function of r, as follows.

$$\frac{1}{r}\frac{d}{dr}\left(r\frac{du}{dr}\right) = \frac{1}{\mu}\frac{d\hat{p}}{dx} = \text{constant} \qquad (13.50)$$

Multiplying through by r and integrating once with respect to r, the axial velocity gradient, du/dr, is given by

$$r\frac{du}{dr} = A + \frac{r^2}{2\mu}\frac{d\hat{p}}{dx} \qquad \Longrightarrow \qquad \frac{du}{dr} = \frac{A}{r} + \frac{r}{2\mu}\frac{d\hat{p}}{dx} \qquad (13.51)$$

where A is an integration constant. Appealing to the symmetry condition on the pipe centerline [Equation (13.45)], necessarily

$$A = 0 \qquad (13.52)$$

Integrating again, the axial velocity is

$$u(r) = B + \frac{r^2}{4\mu}\frac{d\hat{p}}{dx} \qquad (13.53)$$

Finally, the no-slip condition [Equation (13.44)] tells us

$$B = -\frac{R^2}{4\mu}\frac{d\hat{p}}{dx} \qquad (13.54)$$

Thus, the axial velocity for fully-developed pipe flow varies quadratically with distance from the centerline. In terms of the maximum velocity, u_m, which occurs at $r = 0$, we find

$$u(r) = u_m\left[1 - \left(\frac{r}{R}\right)^2\right], \qquad u_m \equiv -\frac{R^2}{4\mu}\frac{\partial p}{\partial x} \qquad (13.55)$$

where we replace $d\hat{p}/dx$ by $\partial p/\partial x$ since the gravitational force acts normal to the direction of motion, so that $\partial p/\partial y$ is nonzero. Note, of course, that $\partial p/\partial x < 0$ so that u_m is positive.

Discussion. Like channel flow, pipe flow is a shear flow with favorable pressure gradient. Because the flow is axisymmetric, the vorticity is given by [see Equation (12.33)]

$$\boldsymbol{\omega} = \begin{vmatrix} \frac{1}{r}\mathbf{e}_r & \mathbf{e}_\theta & \frac{1}{r}\mathbf{k} \\ \frac{\partial}{\partial r} & 0 & \frac{\partial}{\partial x} \\ u_r & 0 & u \end{vmatrix} = \mathbf{e}_\theta\left(\frac{\partial u_r}{\partial x} - \frac{\partial u}{\partial r}\right) \qquad (13.56)$$

Thus, the vorticity vector has only a circumferential component which, for fully-developed flow, is

$$\omega_\theta = -\frac{du}{dr} \qquad (13.57)$$

Also, in this axisymmetric geometry, reference to Equation (12.31) shows that the strain-rate component S_{xr} becomes (with x and u replacing z and w):

$$S_{xr} = \frac{1}{2} \left(\frac{\partial u}{\partial r} + \frac{\partial u_r}{\partial x} \right) = \frac{1}{2} \frac{du}{dr} \tag{13.58}$$

Finally Stokes' postulate tells us the shear stress is

$$\tau_{xr} = 2\mu S_{xr} = \mu \frac{du}{dr} = -\mu\omega_\theta \tag{13.59}$$

Therefore, we again see that the shear stress is proportional to the vorticity. Since the velocity is quadratic in r, the vorticity is a linear function of r. Hence, the shear-stress gradient, $\partial\tau_{xr}/\partial r$, is constant so that $\partial p/\partial x$ must also be constant. This shows that, in exactly the same manner as for Couette and channel flow, the vorticity that develops in pipe flow forces the pressure gradient, $\partial p/\partial x$, to be constant.

The skin friction at the pipe wall follows from first computing the shear stress at $r = R$. Thus, we have

$$\tau_{xr} = \mu \left. \frac{du}{dr} \right|_{r=R} = \mu u_m \left. \left(-\frac{2r}{R^2} \right) \right|_{r=R} = -\frac{2\mu u_m}{R} \tag{13.60}$$

Reference to Figure 13.7 shows that positive shear stress at the pipe wall is in the positive x direction. This is an abbreviated version of Figure 12.7 redrawn in terms of a cylindrical geometry. As indicated, the pipe wall corresponds to the upper surface, where τ_{rx} is positive in the streamwise direction. The same is true for channel flow on the upper wall (see Figure 12.10). The skin friction is thus proportional to $-\tau_{xr}(R)$, and turns out to be a function of Reynolds number based on radius, Re_R, viz.,

$$c_f = \frac{-\tau_{xr}(R)}{\frac{1}{2}\rho u_m^2} = \frac{2\mu u_m/R}{\frac{1}{2}\rho u_m^2} = \frac{4}{\rho u_m R/\mu} = \frac{4}{Re_R} \tag{13.61}$$

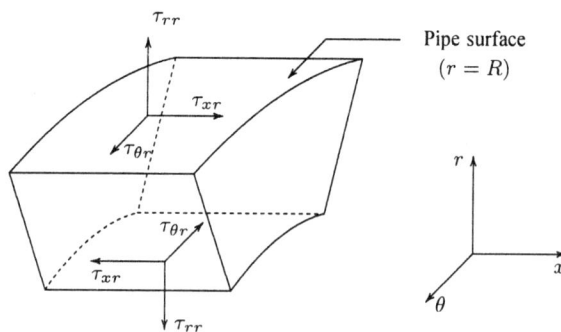

Figure 13.7: *Sign conventions for stress in a cylindrical geometry.*

The average velocity for pipe flow requires a little more attention than channel flow. We must observe the way cylindrical coordinates come into play in computing the average. Most importantly, the differential surface area, dS, is

$$dS = r\,dr\,d\theta \tag{13.62}$$

and we must integrate over θ from 0 to 2π and r from 0 to R. Therefore, the average velocity over a pipe cross section is

$$
\begin{aligned}
\bar{u} &= \frac{1}{\pi R^2} \int_0^{2\pi} \int_0^R u(r)\, r\, dr\, d\theta = \frac{2\pi u_m}{\pi R^2} \int_0^R \left[1 - \left(\frac{r}{R}\right)^2\right] r\, dr \\
&= 2u_m \int_0^R \left[1 - \left(\frac{r}{R}\right)^2\right] \left(\frac{r}{R}\right) d\left(\frac{r}{R}\right) = 2u_m \int_0^1 \left[1 - \xi^2\right] \xi\, d\xi \\
&= 2u_m \left[\frac{1}{2}\xi^2 - \frac{1}{4}\xi^4\right]_{\xi=0}^{\xi=1} = \frac{1}{2} u_m
\end{aligned}
\tag{13.63}
$$

So, the skin friction based on the average velocity becomes

$$
c_f = \frac{2\mu u_m/R}{\frac{1}{2}\rho \bar{u}^2} = \frac{4\mu \bar{u}/R}{\frac{1}{2}\rho \bar{u}^2} = \frac{8}{\overline{Re}_R}
\tag{13.64}
$$

where $\overline{Re}_R = \rho \bar{u} R/\mu$. Hence, the skin friction for fully-developed pipe flow is

$$
c_f = \begin{cases} \dfrac{4}{Re_R}, & Re_R = \dfrac{\rho u_m R}{\mu} \\[2ex] \dfrac{8}{\overline{Re}_R}, & \overline{Re}_R = \dfrac{\rho \bar{u} R}{\mu} \end{cases}
\tag{13.65}
$$

We have made use of the fact that, as shown in Equation (13.63), the average velocity is half the maximum velocity.

In Section 7.7 we discussed pipe flow and introduced the concept of head loss, h_L. We also quoted the **Darcy-Weisbach equation** [see Equation (7.101)] for the head loss in a straight pipe. Our goal is to derive this relationship for pipe flow from the exact Navier-Stokes solution.

We can do this by first recalling the x component of the Navier-Stokes equation [Equation (13.50)]. Differentiating the velocity profile given in Equation (13.55), we find

$$
\frac{d\hat{p}}{dx} = \frac{\mu}{r} \frac{d}{dr}\left(r \frac{du}{dr}\right) = -\frac{4\mu u_m}{R^2} = -\frac{8\mu \bar{u}}{R^2}
\tag{13.66}
$$

Now, for consistency with the notation of Section 7.7, we let z denote the vertical direction (see Figure 13.8), wherefore

$$
\frac{d}{dx}(p + \rho g z) = -\frac{8\mu \bar{u}}{R^2}
\tag{13.67}
$$

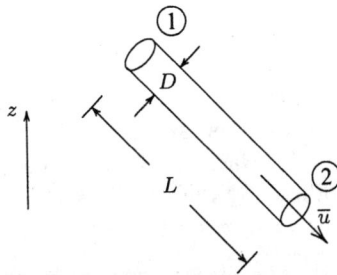

Figure 13.8: *Fully-developed flow in a pipe segment.*

Because the velocity is independent of x in fully-developed pipe flow, we can always add a kinetic-energy term, $\frac{1}{2}\alpha\rho\overline{u}^2$, for later use, as follows.

$$\frac{d}{dx}\left(p + \frac{1}{2}\alpha\rho\overline{u}^2 + \rho g z\right) = -\frac{8\mu\overline{u}}{R^2} \tag{13.68}$$

The constant α is the **kinetic-energy correction factor**, which is defined in Equation (7.82). As shown in Equation (7.97), it is equal to 2 for pipe flow.

Since the derivative in Equation (13.68) is constant, we can replace it by the difference between the quantity in parentheses at the pipe-segment endpoints divided by the length of the pipe. Therefore, we have

$$\frac{\left(p_2 + \frac{1}{2}\alpha_2\rho\overline{u}_2^2 + \rho g z_2\right) - \left(p_1 + \frac{1}{2}\alpha_1\rho\overline{u}_1^2 + \rho g z_1\right)}{L} = -\frac{8\mu\overline{u}}{R^2} \tag{13.69}$$

Hence, multiplying through by $L/(\rho g)$, and rearranging terms, our equation simplifies to

$$\frac{p_1}{\rho g} + \alpha_1\frac{\overline{u}_1^2}{2g} + z_1 = \frac{p_2}{\rho g} + \alpha_2\frac{\overline{u}_2^2}{2g} + z_2 + \frac{8\mu\overline{u}}{\rho g R^2}L \tag{13.70}$$

Comparison of this result with the approximate energy Equation (7.89) derived in Section 7.6 with pump (h_p) and turbine (h_t) contributions lumped into the head loss,[2] we conclude that

$$h_L = \frac{8\mu\overline{u}}{\rho g R^2}L \tag{13.71}$$

To show that this is identical to the Darcy-Weisbach equation quoted in Equation (7.101), we first replace the pipe radius, R, by the diameter, $D = 2R$. Rearranging terms a bit yields

$$h_L = \frac{32}{\rho\overline{u}R/\mu}\frac{\overline{u}^2}{2g}\frac{L}{D} \tag{13.72}$$

or, in terms of the skin friction,

$$h_L = 4c_f\frac{\overline{u}^2}{2g}\frac{L}{D} \tag{13.73}$$

Finally, by definition, the friction factor, f, is

$$f \equiv \frac{\tau_w}{\frac{1}{8}\rho\overline{u}^2} = 4c_f \tag{13.74}$$

Therefore, the head loss for fully-developed pipe flow according to our Navier-Stokes solution assumes the following form.

$$h_L = f\frac{\overline{u}^2}{2g}\frac{L}{D} \tag{13.75}$$

Comparison with Equation (7.101) shows we have derived the classical formula of Darcy and Weisbach. We have also shown that the approximate energy Equation (7.89) holds exactly for the highly idealized conditions of our Navier-Stokes solution.

This serves as an example of how we can use an exact Navier-Stokes solution for testing more specialized results. That is, we have developed a useful check on the Darcy-Weisbach

---

[2]Being precise, the standard definition of h_L given in Equation (7.88) involves only heat-transfer effects, while purely viscous effects appear in the definitions of h_p and h_t. Thus, what is traditionally called h_L for fully-developed, laminar pipe flow is actually $h_t - h_p$.

formula in a limiting case for which an exact solution can be obtained. Such agreement is generally regarded as a minimal requirement for establishing viability of an empirical formula. By contrast, if disagreement with a known limiting case is found, there can be little confidence in extrapolating on the basis of the formula to more complex applications.

As a final comment, retracing the steps leading to the head loss, we can relate it directly to viscous effects. Specifically, the head loss is given by

$$h_L = \frac{L}{\rho g} \frac{1}{r} \frac{d}{dr} \left(r\mu \frac{du}{dr} \right) = \frac{L}{\rho g} \left(\nabla \cdot [\tau] \right)_x \tag{13.76}$$

Thus, the head loss is directly proportional to the viscous shearing force per unit volume in the pipe.

13.4 Stokes' First Problem

Description. One of the best exact Navier-Stokes solutions for illustrating how vorticity is created in a viscous fluid is known as **Stokes' First Problem**.[3] This flow is the idealized case of an infinite flat plate impulsively accelerated to a constant velocity, U, in the direction parallel to the plate. We consider the ensuing motion in the fluid above the plate for $t > 0$, when the plate velocity instantaneously increases from 0 to U.

Our focus is on the fluid above the plate as shown in Figure 13.9, so that $y \geq 0$. The fluid far above the plate is at rest, and a typical velocity profile is indicated. As we will see below, a continuously increasing amount of fluid is set in motion as time passes. We assume that the fluid is incompressible with constant kinematic viscosity, ν, and that the pressure is constant.

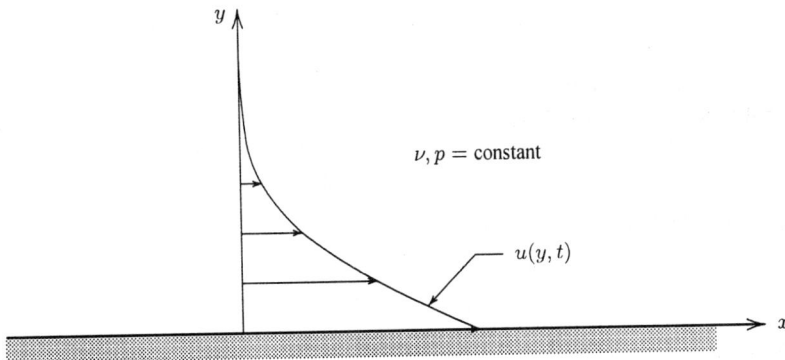

Figure 13.9: *Geometry for Stokes' First Problem; the plate moves at constant velocity U to the right.*

Before proceeding to the solution, it is worthwhile to pause and discuss the physical relevance of this idealized problem. Clearly, it is impossible to instantly attain a constant velocity in a laboratory, as this would require infinite acceleration for an instant. Since this is not physically possible, the problem posed is an idealization of a physically-realizable flow.

The idealization is in making our observations on a time scale long compared to the finite time the plate is brought from rest to a constant velocity. As with Couette flow, the plate

---

[3]Some authors refer to Stokes' First Problem as **Rayleigh Flow**.

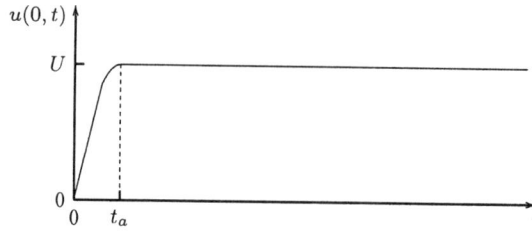

Figure 13.10: *Typical transient velocity for an accelerating plate. Stokes' First Problem solution is valid for $t \gg t_a$.*

surface could again be a moving belt. Hence, if the transient period required for the plate to achieve constant velocity is t_a, our solution will apply for $t \gg t_a$ (see Figure 13.10).

Equations of Motion. Assuming the fluid is a gas, buoyancy effects will be negligible so that we can ignore the gravity force. Since the plate is assumed infinite in the z direction (out of the page), the motion is at most two-dimensional. The unsteady continuity and Navier-Stokes equations with constant pressure are:

$$\frac{\partial u}{\partial x} + \frac{\partial v}{\partial y} = 0 \tag{13.77}$$

$$\frac{\partial u}{\partial t} + u\frac{\partial u}{\partial x} + v\frac{\partial u}{\partial y} = \nu\left(\frac{\partial^2 u}{\partial x^2} + \frac{\partial^2 u}{\partial y^2}\right) \tag{13.78}$$

$$\frac{\partial v}{\partial t} + u\frac{\partial v}{\partial x} + v\frac{\partial v}{\partial y} = \nu\left(\frac{\partial^2 v}{\partial x^2} + \frac{\partial^2 v}{\partial y^2}\right) \tag{13.79}$$

Because this is an unsteady flow, we must pose both initial and boundary conditions. The latter are provided by the no-slip boundary condition at the wall and the fact that the fluid far above the plate is at rest. The initial conditions are

$$u = 0, \quad v = 0 \quad \text{for all } y, \quad t \leq 0 \tag{13.80}$$

and the boundary conditions after the plate begins moving are

$$\left.\begin{array}{ll} u = U, \quad v = 0 & \text{for } y = 0 \\ u \to 0 & \text{for } y \to \infty \end{array}\right\} \quad t > 0 \tag{13.81}$$

Mass and y-Momentum Conservation. As in all of the examples considered thus far in this chapter, the infinite extent of the flow in the x direction tells us all horizontal locations are indistinguishable. Hence, flow properties depend only upon y and t, so that continuity simplifies to

$$\frac{\partial v}{\partial y} = 0 \quad \Longrightarrow \quad v(y, t) = f(t) \tag{13.82}$$

where $f(t)$ is a function of integration. Then, since v vanishes at the plate ($y = 0$) and as $y \to \infty$, the function $f(t)$ must be zero. Therefore,

$$v(y, t) = 0 \tag{13.83}$$

Consequently, the vertical momentum Equation (13.79) is trivially satisfied.[4]

x-Momentum Conservation. Using the facts that $v = 0$ and $\partial/\partial x \to 0$, all that remains in the x component of the Navier-Stokes equation, i.e., Equation (13.78), is

$$\frac{\partial u}{\partial t} = \nu \frac{\partial^2 u}{\partial y^2} \tag{13.84}$$

Note that this equation is similar in form to the vorticity Equation (10.25) without the "vortex-stretching" term. Both equations are similar to the classical **heat-conduction equation**. This reflects the fact that momentum (and vorticity) diffusion is identical to the process by which the fluid next to a heated surface warms.

Unlike the preceding examples, we have arrived at a partial differential equation (PDE), that can be simplified no further. By contrast, we were able to reduce the continuity and Navier-Stokes equations to a much simpler ordinary differential equation (ODE) for Couette, channel and pipe flow. This added complication is due, of course, to the unsteadiness of the flow in the present application.

To solve this equation, we will use a classical procedure known as the **similarity-solution method** [cf. Emanuel (1994), Schlichting (1979), or Wilcox (1993)]. Using this method, we will seek a change of independent variables that permits transforming the PDE in Equation (13.84) to an equivalent ODE. We begin by supposing that, on dimensional grounds, the velocity must be proportional to U. Hence, we seek a solution of the form

$$u(y, t) = U F(\eta) \tag{13.85}$$

where $F(\eta)$ is a dimensionless function and η is a dimensionless combination of y, t and any other relevant dimensional parameters. For the problem at hand the only significant dimensional parameter is the kinematic viscosity, which controls viscous processes. By contrast, other dimensional quantities such as pressure and gravitational acceleration have no effect on the dynamics of the flow, and should be excluded. Recall from dimensional analysis (Chapter 2) that we first establish the dimensions of each parameter, which are

$$[y] = L, \quad [t] = T, \quad [\nu] = L^2 T^{-1} \tag{13.86}$$

with L and T denoting length and time, respectively. Using the standard dimensional-analysis approach, a simple calculation shows that a plausible dimensionless parameter, proportional to y for convenience, is

$$\eta \propto \frac{y}{\sqrt{\nu t}} \tag{13.87}$$

We can always multiply by a constant, and for Stokes' First Problem a constant of $\frac{1}{2}$ turns out to be convenient. Thus, we seek a solution of the form

$$u(y, t) = U F(\eta), \qquad \eta \equiv \frac{y}{2\sqrt{\nu t}} \tag{13.88}$$

There is no guarantee that a solution of this form exists for the problem at hand. We are attempting to make a change of independent variables that yields an ordinary differential equation in terms of the **similarity variable**, η. Additionally, the similarity solution's existence depends upon our being able to transform the boundary conditions into forms that

---

[4]Had we included gravity, we would conclude at this point that it is balanced by $\partial p/\partial y$, and has no other effect on the dynamics of this flow.

are expressible strictly in terms of η. If y or t appears explicitly in either the transformed differential equation or boundary conditions, no similarity solution exists.

To develop the similarity solution, we must first transform Equation (13.84). We begin by using the chain rule, which tells us the first derivatives of u become:

$$\frac{\partial u}{\partial t} = \frac{du}{d\eta}\frac{\partial \eta}{\partial t} = U\frac{dF}{d\eta}\left(-\frac{1}{2}\frac{y}{2\nu^{1/2}t^{3/2}}\right) = -\frac{U}{2t}\eta\frac{dF}{d\eta} \qquad (13.89)$$

$$\frac{\partial u}{\partial y} = \frac{du}{d\eta}\frac{\partial \eta}{\partial y} = U\frac{dF}{d\eta}\left(\frac{1}{2\sqrt{\nu t}}\right) = -\frac{U}{2\sqrt{\nu t}}\frac{dF}{d\eta} \qquad (13.90)$$

Hence, the diffusion term of Equation (13.84) transforms as follows.

$$
\begin{aligned}
\nu\frac{\partial^2 u}{\partial y^2} &= \nu\frac{\partial}{\partial y}\left(\frac{U}{2\sqrt{\nu t}}\frac{dF}{d\eta}\right) = \frac{\nu U}{2\sqrt{\nu t}}\frac{\partial}{\partial y}\left(\frac{dF}{d\eta}\right) \\
&= \frac{\nu U}{2\sqrt{\nu t}}\frac{d^2 F}{d\eta^2}\frac{\partial \eta}{\partial y} = \frac{\nu U}{2\sqrt{\nu t}}\left(\frac{1}{2\sqrt{\nu t}}\right)\frac{d^2 F}{d\eta^2} \\
&= \frac{U}{4t}\frac{d^2 F}{d\eta^2}
\end{aligned}
\qquad (13.91)
$$

Substituting Equations (13.89) and (13.91) into the momentum Equation (13.84), we have

$$-\frac{U}{2t}\eta\frac{dF}{d\eta} = \frac{U}{4t}\frac{d^2 F}{d\eta^2} \qquad (13.92)$$

wherefore the transformed equation is

$$\frac{d^2 F}{d\eta^2} + 2\eta\frac{dF}{d\eta} = 0 \qquad (13.93)$$

It is a simple matter to show that the boundary conditions on u, viz., Equations (13.81), transform to

$$F(0) = 1 \quad\text{and}\quad F(\eta) \to 0 \text{ as } \eta \to \infty \qquad (13.94)$$

Therefore, because we have been able to express the equation of motion and the boundary conditions in terms of η, a similarity solution indeed exists. As a general rule, when such a solution exists mathematically, it can be observed in a laboratory (provided, of course, that the flow conditions can be simulated in the lab).

Equation (13.93) can be solved in closed form, albeit in terms of a tabulated function known as the complementary error function. We begin by rewriting the equation as follows.

$$\frac{d^2 F/d\eta^2}{dF/d\eta} = -2\eta \quad\Longrightarrow\quad \frac{d}{d\eta}\left[\ell n\frac{dF}{d\eta}\right] = -2\eta \qquad (13.95)$$

Integrating once, we find

$$\ell n\frac{dF}{d\eta} = C - \eta^2 \qquad (13.96)$$

where C is a constant of integration. Then, exponentiating both sides of this equation yields

$$\frac{dF}{d\eta} = e^{C-\eta^2} = C_1 e^{-\eta^2} \qquad (C_1 \equiv e^C) \qquad (13.97)$$

Integrating again, we arrive at

$$F(\eta) = C_1 \int_0^\eta e^{-\tilde{\eta}^2} d\tilde{\eta} + C_2 \tag{13.98}$$

where C_2 is another integration constant. Application of the boundary condition at the surface, which corresponds to $\eta = 0$, determines the value of C_2 immediately, i.e.,

$$F(0) = C_2 = 1 \tag{13.99}$$

Also, the boundary condition far above the plate ($\eta \to \infty$) tells us

$$C_1 \int_0^\infty e^{-\tilde{\eta}^2} d\tilde{\eta} + C_2 = 0 \tag{13.100}$$

Reference to a standard table of integrals shows that

$$\int_0^\infty e^{-\tilde{\eta}^2} d\tilde{\eta} = \frac{\sqrt{\pi}}{2} \tag{13.101}$$

Collecting the results given in Equations (13.99), (13.100) and (13.101), the values of C_1 and C_2 are

$$C_1 = -\frac{2}{\sqrt{\pi}}, \qquad C_2 = 1 \tag{13.102}$$

so that the solution for the function $F(\eta)$ is

$$F(\eta) = 1 - \frac{2}{\sqrt{\pi}} \int_0^\eta e^{-\tilde{\eta}^2} d\tilde{\eta} \tag{13.103}$$

We can combine Equations (13.101) and (13.103) to arrive at

$$F(\eta) = \frac{2}{\sqrt{\pi}} \left[\int_0^\infty e^{-\tilde{\eta}^2} d\tilde{\eta} - \int_0^\eta e^{-\tilde{\eta}^2} d\tilde{\eta} \right] = \frac{2}{\sqrt{\pi}} \int_\eta^\infty e^{-\tilde{\eta}^2} d\tilde{\eta} \tag{13.104}$$

By definition, the function on the right-hand side of Equation (13.104) is known as the **complementary error function**,[5] and is denoted by erfc(η), so that we can write the solution as follows.

$$F(\eta) = \text{erfc}(\eta) \tag{13.105}$$

Finally, returning to physical variables, the velocity for Stokes' First Problem is

$$u(y, t) = U \,\text{erfc}\left(\frac{y}{2\sqrt{\nu t}} \right) \mathcal{H}(t) \tag{13.106}$$

where $\mathcal{H}(t)$ is the **Heaviside step function**, defined as

$$\mathcal{H}(t) = \begin{cases} 0, & t < 0 \\ 1, & t > 0 \end{cases} \tag{13.107}$$

We must append the step function in order to satisfy the initial condition that there is no motion prior to $t = 0$, as stated in Equation (13.80). Using the step function, with its discontinuity

---

[5]Tabulations of the complementary error function are sometimes factored so that erfc$(0) = 1$, as here, but sometimes so that erfc$(0) = 1/2$. Also, the argument of the exponential may appear as $(-\eta^2)$ or $(-\eta^2/2)$ in some books.

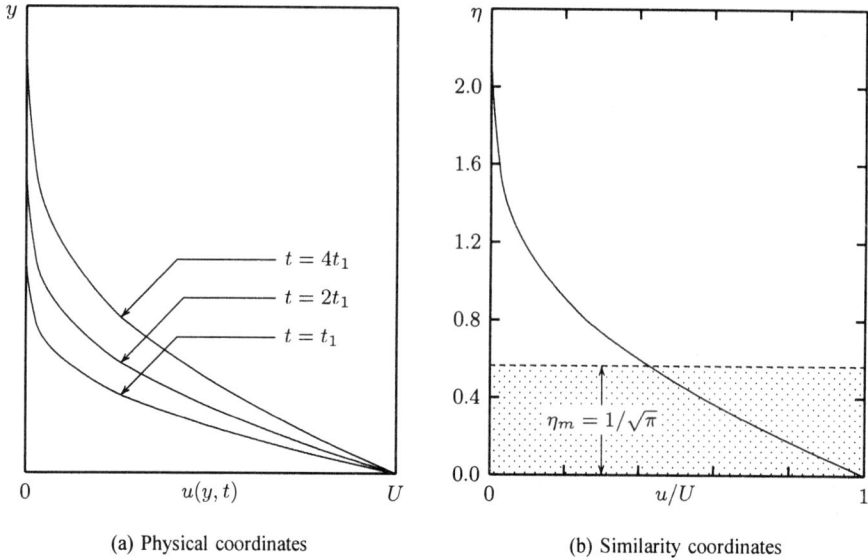

(a) Physical coordinates   (b) Similarity coordinates

Figure 13.11: *Velocity for Stokes' First Problem.*

at $t = 0$, is consistent with the idealized nature of the problem in which the plate impulsively begins moving at constant velocity. We include it for the sake of completeness — it has no other bearing on the discussion to follow.

Discussion. Figure 13.11(a) shows the solution in physical variables for three different times. As shown, as time passes, an increasing amount of fluid is set in motion. The physical process by which this occurs is diffusion, and the viscosity is often referred to as the coefficient of diffusivity. Molecules closest to the surface are typical of where they come from, and at the surface they have high x momentum. Momentum is transported into the region above the plate in the manner described in Section 12.1, and the rate of transport is proportional to ν.

We can obtain a quantitative measure of how much fluid is moving as a function of time by integrating the velocity profile from 0 to ∞. That is, the volume flux across a semi-infinite plane normal to the plate (per unit width out of the page) is

$$Q = \int_0^\infty u(y,t)\,dy = 2U\sqrt{\frac{\nu t}{\pi}} \qquad (13.108)$$

where reference to a table of integrals has been made to evaluate the integral. If we define an effective thickness, δ_m, such that [see Figure 13.11(b)]

$$Q = U\delta_m \qquad (13.109)$$

then the thickness of what is generally referred to as the **Rayleigh layer** is

$$\delta_m = 2\sqrt{\frac{\nu t}{\pi}} \qquad (13.110)$$

In other words, the thickness of the Rayleigh layer grows as $\sqrt{t}$.

Because the thickness of the Rayleigh layer is growing in time, the velocity profile has a different shape at each instant. However, as shown in Figure 13.11(b), a remarkable thing happens if we replot in terms of the similarity solution. All of the velocity profiles collapse onto the single curve defined by Equation (13.105). In terms of the similarity variable, η, the thickness of the Rayleigh layer is constant and given by

$$\eta_m = \frac{\delta_m}{2\sqrt{\nu t}} = \frac{1}{\sqrt{\pi}} \tag{13.111}$$

As should be obvious from the discussion of Q above, the shaded area in Figure 13.11(b) is equal to the area under the velocity profile, and thus has the same volume flux. When properties such as the velocity evolve in this manner, retaining the same variation in terms of scaled variables (u/U and η for this problem), we say the flow is **self similar**, or **self preserving**. If we performed a Rayleigh-flow experiment, making measurements at several times, all of our data points ought to fall on the curve defined in Equation (13.105).

Stokes' First Problem is the quintessential example of the process through which circulation develops about solid objects moving through a viscous fluid. Thus, it is worthwhile to pause and discuss the vorticity in this flow. Because this flow is two dimensional with $v = 0$, the vorticity vector has one component normal to the plane of motion given by

$$\omega = -\frac{\partial u}{\partial y}\mathbf{k} \tag{13.112}$$

We derived the differential equation for the vorticity vector in general flows in Chapter 10 [see Equation (10.25)]. We could simplify the equation for the problem at hand. However, it is simpler to differentiate the x-momentum Equation (13.84) with respect to y. Doing this, we conclude immediately that for Stokes' First Problem,

$$\frac{\partial \omega}{\partial t} = \nu \frac{\partial^2 \omega}{\partial y^2} \tag{13.113}$$

Therefore, the vorticity satisfies the classical heat equation, and diffuses from the plate surface into the main flow. Differentiating the velocity profile [Equation (13.106)], the vorticity vector is

$$\omega(y, t) = \frac{U}{\sqrt{\pi \nu t}} \exp\left(-\frac{y^2}{4\nu t}\right) \mathcal{H}(t)\mathbf{k} \tag{13.114}$$

Thus, the surface value of ω is

$$\omega(0, t) = \frac{U}{\sqrt{\pi \nu t}}\mathcal{H}(t)\mathbf{k} \tag{13.115}$$

Hence, as the plate begins its motion, the no-slip boundary condition sets up nonzero vorticity at the plate surface. At the beginning of the motion, the Rayleigh layer is very thin. Recalling that its thickness grows as $\sqrt{t}$ [see Equation (13.110)], clearly the value of $\partial u/\partial y$, and hence the vorticity, is inversely proportional to $\sqrt{t}$. The vorticity then diffuses into the fluid above the plate. This is precisely the mechanism described in Section 10.5 that accounts for the way vorticity enters a flow about an object such as an airfoil. Of course, Stokes' First Problem is a **parallel flow** (i.e., a flow with $v = w = 0$), and the convective acceleration, $\mathbf{u} \cdot \nabla \mathbf{u}$, is zero. In a general flow the operator $\mathbf{u} \cdot \nabla$ is not zero, and has the effect of convecting the vorticity throughout the flow.

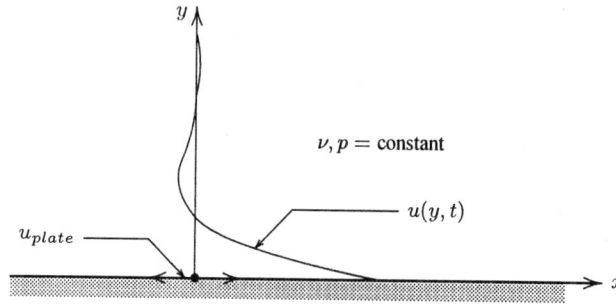

Figure 13.12: *Geometry for Stokes' Second Problem. The plate oscillates with velocity $u_{plate} = u_o \sin \Omega t$. The velocity profile is exaggerated slightly to display oscillations at large values of y.*

13.5 Stokes' Second Problem

Description. Another unsteady exact Navier-Stokes solution was first investigated by Stokes in 1845. In what is known as **Stokes' Second Problem**, we again have an infinite flat plate experiencing motion in its own plane. The plate motion for the Second Problem is oscillatory with amplitude u_o and frequency Ω, i.e., the plate velocity is

$$u_{plate} = u_o \sin \Omega t \tag{13.116}$$

As with Stokes' First Problem, the fluid is incompressible with constant kinematic viscosity, ν, and pressure, p. We again omit body forces, and consider motion in the region $y \geq 0$. Also, we assume the motion has persisted long enough that pure sinusoidal motion of the plate has been established. Figure 13.12 shows the geometry, including a typical velocity profile.

Equations of Motion. For precisely the same reasons as in Stokes' First Problem, the vertical velocity vanishes, i.e.,

$$v(y, t) = 0 \tag{13.117}$$

and the horizontal momentum equation simplifies to

$$\frac{\partial u}{\partial t} = \nu \frac{\partial^2 u}{\partial t^2} \tag{13.118}$$

The difference between Stokes' First and Second Problems is in the boundary conditions. At the plate surface, no-slip tells us the fluid velocity matches the plate velocity. Also, the fluid far above the plate is at rest. Therefore, the velocity satisfies the following boundary conditions.

$$u(0, t) = u_o \sin \Omega t, \qquad u(y, t) \to 0 \ \ \text{as} \ \ y \to \infty \tag{13.119}$$

x-Momentum Conservation. To solve Equation (13.118) subject to the boundary conditions in Equation (13.119), we try a separation of variables. That is, we assume the solution for $u(y, t)$ is the product of a function of y, that we will call $U(y)$, and a function of t whose functional form will be sinusoidal (motivated by the surface boundary condition). To determine the appropriate sinusoidal function of t, it is most convenient to work with the **Euler representation**, viz., we use the fact that

$$e^{i\Omega t} = \cos \Omega t + i \sin \Omega t \tag{13.120}$$

where $i = \sqrt{-1}$ is the imaginary number. Thus, we assume a solution of the following form.

$$u(y, t) = Im\left\{U(y)e^{i\Omega t}\right\} \tag{13.121}$$

We choose the imaginary part, of course, as our surface boundary condition tells us

$$u(0, t) = u_o \sin \Omega t = Im\left\{u_o e^{i\Omega t}\right\} \tag{13.122}$$

Note that with this choice, we have $U(0) = u_o$. So, omitting the imaginary-part notation, Im, for convenience (we will ultimately include it to complete our analysis), the appropriate derivatives of $u(y, t)$ are

$$\left.\begin{array}{rcl}\dfrac{\partial u}{\partial t} & = & i\Omega U e^{i\Omega t} \\[4mm] \dfrac{\partial^2 u}{\partial y^2} & = & \dfrac{d^2 U}{dy^2} e^{i\Omega t}\end{array}\right\} \implies i\Omega U = \nu \dfrac{d^2 U}{dy^2} \tag{13.123}$$

Therefore, our assumed form of the solution permits us to replace the original partial differential equation for $u(y, t)$ [Equation (13.118)] by the following ordinary differential equation for $U(y)$:

$$\frac{d^2 U}{dy^2} = i\frac{\Omega}{\nu}U \tag{13.124}$$

The boundary conditions [see Equation (13.119)] become

$$U(0) = u_o, \qquad U(y) \to 0 \quad \text{as} \quad y \to \infty \tag{13.125}$$

Using elementary techniques for ordinary differential equations, the solution is of the form

$$U(y) = A \exp\left(\sqrt{i\frac{\Omega}{\nu}}\,y\right) + B \exp\left(-\sqrt{i\frac{\Omega}{\nu}}\,y\right) \tag{13.126}$$

where A and B are coefficients that will be determined by imposing the boundary conditions. First, however, note that

$$\sqrt{i} = \sqrt{e^{i\pi/2}} = e^{i\pi/4} = (1 + i)/\sqrt{2} \tag{13.127}$$

Thus, we can rewrite the solution as

$$U(y) = A \exp\left[(1 + i)\sqrt{\frac{\Omega}{2\nu}}\,y\right] + B \exp\left[-(1 + i)\sqrt{\frac{\Omega}{2\nu}}\,y\right] \tag{13.128}$$

Clearly, to have $U(y) \to 0$ as $y \to \infty$, we must select $A = 0$ to prevent exponential growth. Then, to have $U(0) = u_o$, necessarily $B = u_o$. Hence, the solution for the function $U(y)$ is

$$U(y) = u_o \exp\left[-(1 + i)\sqrt{\frac{\Omega}{2\nu}}\,y\right] \tag{13.129}$$

Finally, recalling Equation (13.121), the solution for the velocity is

$$u(y, t) = Im\left\{u_o \exp\left[-(1 + i)\sqrt{\frac{\Omega}{2\nu}}\,y + i\Omega t\right]\right\} \tag{13.130}$$

We complete the solution by taking the imaginary part. The final form of $u(y,t)$ for Stokes' Second Problem becomes

$$u(y,t) = u_o \exp\left(-\sqrt{\frac{\Omega}{2\nu}}\, y\right) \sin\left(\Omega t - \sqrt{\frac{\Omega}{2\nu}}\, y\right) \qquad (13.131)$$

Discussion. To simplify our notation, we define the dimensionless distance above the plate, η, as follows.

$$\eta = \sqrt{\frac{\Omega}{2\nu}}\, y \qquad (13.132)$$

In terms of this independent variable, the velocity profile is

$$u(y,t) = u_o e^{-\eta} \sin\left(\Omega t - \eta\right) \qquad (13.133)$$

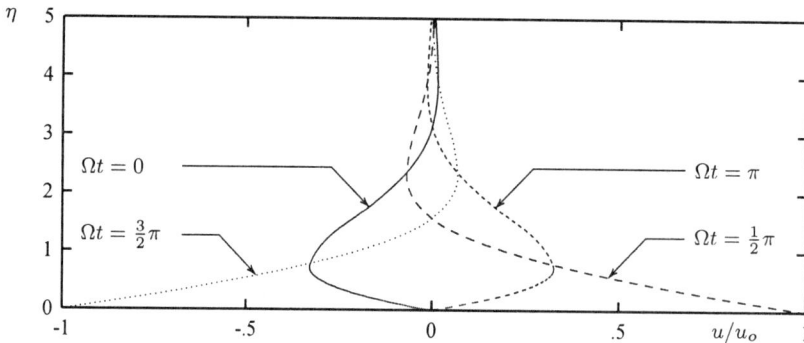

Figure 13.13: *Velocity profiles for Stokes' Second Problem.*

Figure 13.13 shows the velocity profile at four times through a complete cycle. As indicated, the velocity behaves in a rapidly-damped oscillatory manner. Unlike Stokes' First Problem, the thickness of the layer near the surface to which the motion is confined does not change with time. This reflects the fact that the vorticity, like the velocity, oscillates with time. A straightforward computation, left as an exercise for the reader, shows that the vorticity is

$$\boldsymbol{\omega} = u_o \sqrt{\frac{\Omega}{\nu}} e^{-\eta} \sin\left(\Omega t - \eta + \frac{\pi}{4}\right) \mathbf{k} \qquad (13.134)$$

Clearly, the average value of the vorticity taken over a complete cycle, i.e., for time lying in the range $0 \leq \Omega t \leq 2\pi$, is zero. While vorticity is continually diffusing from the surface, it is positive vorticity through half of the cycle and negative vorticity through the other half. Thus, there is no net vorticity transferred from the surface to the fluid above, wherefore the thickness of the layer remains constant. By contrast, there is a continuous, monotonic, source of vorticity in Stokes' First Problem, which causes the viscous layer to grow.

We can quantify the thickness of the viscous layer, which is known as the **Stokes layer**, by first noting from Equation (13.133) that

$$\left|\frac{u}{u_o}\right| \leq e^{-\eta} \qquad (13.135)$$

Thus, we can say that $|u| < 0.01u_o$ when $\eta > 4.5$, regardless of the value of t. Using this threshold to define the thickness of the Stokes layer, δ_s, we can choose

$$\delta_s \approx 4.5\sqrt{\frac{2\nu}{\Omega}} \qquad (13.136)$$

so that $e^{-\eta}$ is about 0.01 at $y = \delta_s$.

13.6 Stagnation-Point Flow

Description. Flow in the immediate vicinity of a stagnation point is another problem for which an exact Navier-Stokes solution exists. This solution differs from all the other solutions discussed in this chapter in one important respect, namely, the convective acceleration terms are nonvanishing. While this means the equation we must solve is nonlinear, we will demonstrate a separation of variables that permits reducing the equations of motion to a pair of ordinary differential equations. We use the same geometry (see Figure 13.14) and notation introduced in Subsection 11.7.4.

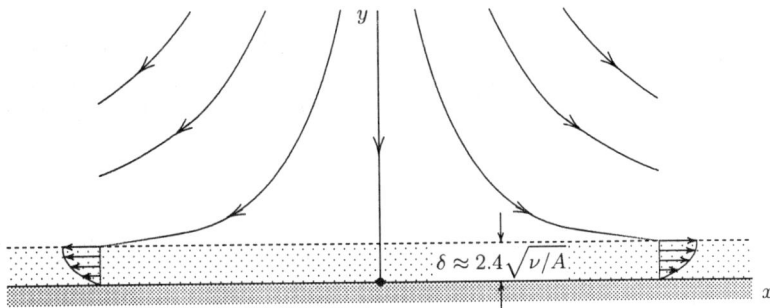

Figure 13.14: *Stagnation-point flow.*

Equations of Motion. This problem is most conveniently solved by introducing the stream-function, $\psi(x,y)$. In developing the viscous-flow equation for the streamfunction, we will discover some interesting features of the Navier-Stokes equation. Most importantly, we will see that in a two-dimensional flow, the second-order, vector form of the Navier-Stokes equation yields a fourth-order equation for ψ. We will also see that we can solve this equation with no knowledge of the pressure field, so that even in a viscous flow, the pressure plays a passive role. It can be determined in terms of the velocity field by solving a second-order partial differential equation, known as Poisson's equation.

The easiest way to arrive at the equation for the streamfunction is to begin with the differential equation for the vorticity that we derived and discussed in Chapters 10 and 12, viz., Equation (12.124).

$$\frac{d\omega}{dt} = \omega \cdot \nabla \mathbf{u} + \nu \nabla^2 \omega \qquad (13.137)$$

For two-dimensional flow we know that the vortex-stretching term, $\omega \cdot \nabla \mathbf{u}$, is exactly zero. Also, for steady flow, the Eulerian derivative of ω simplifies to $d\omega/dt = \mathbf{u} \cdot \nabla \omega$. Therefore, the steady, two-dimensional form of the vorticity equation is

$$\mathbf{u} \cdot \nabla \omega = \nu \nabla^2 \omega \qquad (13.138)$$

Now, introducing the streamfunction, $\psi(x,y)$, the velocity components are

$$u = \frac{\partial \psi}{\partial y}, \qquad v = -\frac{\partial \psi}{\partial x} \tag{13.139}$$

The vorticity vector then becomes

$$\boldsymbol{\omega} = \left(\frac{\partial v}{\partial x} - \frac{\partial u}{\partial y} \right) \mathbf{k} = -\left(\frac{\partial^2 \psi}{\partial x^2} + \frac{\partial^2 \psi}{\partial x^2} \right) \mathbf{k} \tag{13.140}$$

We can write the relationship between the vorticity and the streamfunction in a more compact form as follows.

$$\boldsymbol{\omega} = -\nabla^2 \psi \, \mathbf{k} \tag{13.141}$$

Substituting this result into Equation (13.138) yields the desired equation for the streamfunction, viz.,

$$\mathbf{u} \cdot \nabla \left(\nabla^2 \psi \right) = \nu \nabla^2 \left(\nabla^2 \psi \right) = \nu \nabla^4 \psi \tag{13.142}$$

Performing the indicated operations, we arrive at the following fourth-order, partial differential equation for the streamfunction.

$$\nu \left(\frac{\partial^4 \psi}{\partial x^4} + 2\frac{\partial^4 \psi}{\partial x^2 \partial y^2} + \frac{\partial^4 \psi}{\partial y^4} \right) = \frac{\partial \psi}{\partial y} \left(\frac{\partial^3 \psi}{\partial x^3} + \frac{\partial^3 \psi}{\partial x \partial y^2} \right) - \frac{\partial \psi}{\partial x} \left(\frac{\partial^3 \psi}{\partial y^3} + \frac{\partial^3 \psi}{\partial x^2 \partial y} \right) \tag{13.143}$$

Two boundary conditions on ψ follow from the no-slip constraint at the surface. Far above the surface, we expect the viscous-flow solution to approach the potential-flow solution of Subsection 11.7.4. Replacing the constant ψ_o in the inviscid solution by $A = 2\psi_o$ [see Equation (11.176)], we have the following four boundary conditions for the streamfunction.

$$\left. \begin{array}{ll} \dfrac{\partial \psi}{\partial y} \to Ax, & \dfrac{\partial \psi}{\partial x} \to Ay \quad \text{as } y \to \infty \\[3mm] \dfrac{\partial \psi}{\partial y} = 0, & \dfrac{\partial \psi}{\partial x} = 0 \quad \text{at } y = 0 \end{array} \right\} \tag{13.144}$$

Streamfunction. We can solve Equation (13.143) subject to the boundary conditions of Equation (13.144) using the **separation of variables** method. That is, we begin by assuming a solution of the form

$$\psi(x,y) = \Psi(x)F(y) \tag{13.145}$$

Substituting this assumed form for $\psi(x,y)$ into Equation (13.143), a straightforward but rather lengthy calculation, whose details are outlined in the Problems section, shows that

$$\Psi(x) = Ax \tag{13.146}$$

while $F(y)$ satisfies the following ordinary differential equation and boundary conditions.

$$\frac{\nu}{A} \frac{d^3 F}{dy^3} + F\frac{d^2 F}{dy^2} - \left(\frac{dF}{dy} \right)^2 + 1 = 0 \tag{13.147}$$

$$F(0) = 0, \quad F'(0) = 0, \quad F'(y) \to 1 \text{ as } y \to \infty \tag{13.148}$$

It is worthwhile at this point to recast our solution in terms of dimensionless variables, noting that $\sqrt{\nu/A}$ has dimensions of length. Thus, we can make our independent variable, i.e., distance from the surface, dimensionless by defining

$$\eta \equiv y\sqrt{\frac{A}{\nu}} \tag{13.149}$$

Also, a straightforward calculation shows that $F(y) = \psi(x, y)/(Ax)$ has dimensions of length, wherefore the quantity $G(\eta)$ defined by

$$G(\eta) \equiv \sqrt{\frac{A}{\nu}}F(y) \tag{13.150}$$

is a dimensionless function. Therefore, substituting Equations (13.149) and (13.150) into Equations (13.147) and (13.148) produces the desired dimensionless form of the stagnation-point equation and boundary conditions for the dimensionless streamfunction, viz.,

$$\frac{d^3G}{d\eta^3} + G\frac{d^2G}{d\eta^2} - \left(\frac{dG}{d\eta}\right)^2 + 1 = 0 \tag{13.151}$$

$$G(0) = 0, \quad G'(0) = 0, \quad G'(\eta) \to 1 \text{ as } \eta \to \infty \tag{13.152}$$

Pressure. Now that we have developed an equation from which we can determine the velocity, we can focus our attention on the pressure. We derived an equation for the streamfunction by beginning with the equation for vorticity, and noting that the vorticity is proportional to the Laplacian of the streamfunction. The pressure does not appear explicitly in the vorticity equation. Mathematically, when we derive the vorticity equation, we take the curl of the momentum equation, and the pressure term is eliminated because $\nabla \times \nabla p = \mathbf{0}$.

In an analogous manner, we can derive an equation for the pressure by taking the divergence of the momentum equation. Because mass conservation tells us the divergence of the velocity vanishes for incompressible flow, we will discover that the equation assumes a relatively simple form, an example of Poisson's equation, that does not explicitly involve viscosity. We begin with the x and y components of the Navier-Stokes equation, which are

$$\left.\begin{array}{l} u\dfrac{\partial u}{\partial x} + v\dfrac{\partial u}{\partial y} = -\dfrac{1}{\rho}\dfrac{\partial p}{\partial x} + \nu\nabla^2 u \\[3mm] u\dfrac{\partial v}{\partial x} + v\dfrac{\partial v}{\partial y} = -\dfrac{1}{\rho}\dfrac{\partial p}{\partial y} + \nu\nabla^2 v \end{array}\right\} \tag{13.153}$$

Next, we differentiate the first of Equations (13.153) with respect to x and the second with respect to y. There follows

$$\left.\begin{array}{l} u\dfrac{\partial^2 u}{\partial x^2} + \left(\dfrac{\partial u}{\partial x}\right)^2 + v\dfrac{\partial^2 u}{\partial x \partial y} + \dfrac{\partial v}{\partial x}\dfrac{\partial u}{\partial y} = -\dfrac{1}{\rho}\dfrac{\partial^2 p}{\partial x^2} + \nu\nabla^2\left(\dfrac{\partial u}{\partial x}\right) \\[3mm] u\dfrac{\partial^2 v}{\partial y \partial x} + \dfrac{\partial u}{\partial y}\dfrac{\partial v}{\partial x} + v\dfrac{\partial^2 v}{\partial y^2} + \left(\dfrac{\partial v}{\partial y}\right)^2 = -\dfrac{1}{\rho}\dfrac{\partial^2 p}{\partial y^2} + \nu\nabla^2\left(\dfrac{\partial v}{\partial y}\right) \end{array}\right\} \tag{13.154}$$

Adding the first and second of Equations (13.154), and rearranging terms a little yields the following equation involving the Laplacian of p.

$$u \left(\frac{\partial^2 u}{\partial x^2} + \frac{\partial^2 v}{\partial x \partial y} \right) + v \left(\frac{\partial^2 u}{\partial x \partial y} + \frac{\partial^2 v}{\partial y^2} \right) + \left(\frac{\partial u}{\partial x} \right)^2 + \left(\frac{\partial v}{\partial y} \right)^2 + 2 \frac{\partial u}{\partial y} \frac{\partial v}{\partial x}$$

$$= -\frac{1}{\rho} \nabla^2 p + \nu \nabla^2 \left(\frac{\partial u}{\partial x} + \frac{\partial v}{\partial y} \right) \qquad (13.155)$$

This equation can be rewritten as

$$u \frac{\partial}{\partial x} \left(\frac{\partial u}{\partial x} + \frac{\partial v}{\partial y} \right) + v \frac{\partial}{\partial y} \left(\frac{\partial u}{\partial x} + \frac{\partial v}{\partial y} \right) + \left(\frac{\partial u}{\partial x} \right)^2 + \left(\frac{\partial v}{\partial y} \right)^2 + 2 \frac{\partial u}{\partial y} \frac{\partial v}{\partial x}$$

$$= -\frac{1}{\rho} \nabla^2 p + \nu \nabla^2 \left(\frac{\partial u}{\partial x} + \frac{\partial v}{\partial y} \right) \qquad (13.156)$$

Now, continuity tells us the divergence of the velocity is zero, i.e., that $\partial u / \partial x + \partial v / \partial y = 0$. Thus, the first two terms on the left-hand side and the last term on the right-hand side of Equation (13.156) vanish, and all that remains is

$$\nabla^2 p = -\rho \left[\left(\frac{\partial u}{\partial x} \right)^2 + \left(\frac{\partial v}{\partial y} \right)^2 + 2 \frac{\partial u}{\partial y} \frac{\partial v}{\partial x} \right] \qquad (13.157)$$

This is a **Poisson equation** for the pressure in a two-dimensional, incompressible flow. In general, once we know the velocity, we can solve this equation for the pressure to complete the solution.

For stagnation-point flow, the pressure equation assumes a particularly simple form. Recall from the inviscid stagnation-point flow solution that the pressure is given by Equation (11.185). As with the velocity, we expect the pressure to approach the inviscid solution in the limit as $y \rightarrow \infty$. In terms of our current notation ($A = 2\psi_o$), the pressure is expected to satisfy

$$p \rightarrow p_o - \frac{1}{2} \rho A^2 \left(x^2 + y^2 \right) \qquad \text{as} \qquad y \rightarrow \infty \qquad (13.158)$$

Inspection of the farfield pressure suggests trying a solution of the following form:

$$p(x, y) = p_o - \frac{1}{2} \rho A^2 \left[X(x) + Y(y) \right] \qquad (13.159)$$

Substitution of Equation (13.159) along with the solution for the streamfunction obtained above into Equation (13.157), and doing another rather long derivation that is outlined in the Problems section, we find

$$X(x) = x^2 \qquad (13.160)$$

while the function $Y(y)$ satisfies

$$\frac{d^2 Y}{dy^2} = 4 \left(\frac{dF}{dy} \right)^2 - 2 \qquad (13.161)$$

$$Y(0) = 0 \qquad \text{and} \qquad \frac{dY}{dy} \rightarrow 2y \text{ as } y \rightarrow \infty \qquad (13.162)$$

Note that the boundary conditions for the pressure follow from requiring the pressure gradient to match the inviscid value far above the surface, and that p_o be the true stagnation pressure.

To complete our formulation of the reduced form of the equation for the pressure, it is most convenient to recast Equation (13.161) in dimensionless form. Introducing the dimensionless distance η defined in Equation (13.149), the function $G(\eta)$ given by Equation (13.150), and defining

$$P(\eta) \equiv \frac{A}{\nu} Y(y) \tag{13.163}$$

the final forms of the pressure equation and boundary conditions are

$$\frac{d^2 P}{d\eta^2} = 4 \left(\frac{dG}{d\eta} \right)^2 - 2 \tag{13.164}$$

$$P(0) = 0 \quad \text{and} \quad \frac{dP}{d\eta} \to 2\eta \text{ as } \eta \to \infty \tag{13.165}$$

Although it is not obvious, we can actually solve for the pressure in closed form. First, we can eliminate $(dG/d\eta)^2$ from Equation (13.164) by appealing to the streamfunction Equation (13.151). That is,

$$
\begin{aligned}
\frac{d^2 P}{d\eta^2} &= 4 \left[\frac{d^3 G}{d\eta^3} + G \frac{d^2 G}{d\eta^2} + 1 \right] - 2 \\
&= 4 \frac{d^3 G}{d\eta^3} + 4 \frac{d}{d\eta} \left(G \frac{dG}{d\eta} \right) - 4 \left(\frac{dG}{d\eta} \right)^2 + 2 \\
&= 4 \frac{d^3 G}{d\eta^3} + 4 \frac{d}{d\eta} \left(G \frac{dG}{d\eta} \right) - \frac{d^2 P}{d\eta^2}
\end{aligned}
\tag{13.166}
$$

Thus, rearranging terms, the pressure equation becomes

$$\frac{d^2 P}{d\eta^2} = 2 \frac{d^3 G}{d\eta^3} + 2 \frac{d}{d\eta} \left(G \frac{dG}{d\eta} \right) = 2 \frac{d^3 G}{d\eta^3} + \frac{d^2 G^2}{d\eta^2} \tag{13.167}$$

Integrating twice, and using the boundary conditions in Equation (13.165), the pressure function, $P(\eta)$, is given by

$$P(\eta) = 2 \frac{dG}{d\eta} + G^2(\eta) \tag{13.168}$$

As a final comment, we could have arrived at a third order differential equation for $G(\eta)$ and a first order differential equation for $P(\eta)$ by simply substituting the assumed form of the solution directly into the Navier-Stokes equation. This is the approach taken, for example, by Schlichting (1979). However, the reduced orders of these equations results from the special symmetries of stagnation-point flow, and cannot be expected to occur for general flows. The intent here has been twofold. One goal has been to develop the solution in a deductive manner, rather than simply verifying a known solution. A second goal has been to demonstrate that the most general form of the streamfunction equation is of fourth order [see Equation (13.143)], while the pressure satisfies Poisson's equation [see Equation (13.157)].

Discussion. To complete the solution, we must first solve for the streamfunction from Equation (13.151) subject to the boundary conditions of Equation (13.152). Because we have used the streamfunction, the continuity equation is automatically satisfied. Equation (13.151) follows from the Navier-Stokes equation. It is the viscous-flow replacement for Laplace's equation for the streamfunction, which would be appropriate for an inviscid-flow solution. As already noted, the original equation for the streamfunction was of fourth order. We have been able to integrate once so that we must deal with a third-order equation. For reference below, note that the streamfunction and velocity components are

$$\left.\begin{array}{rcl} \psi(x, y) & = & \sqrt{\nu A}\, x G(\eta) \\[4pt] u(x, y) & = & A x G'(\eta) \\[4pt] v(x, y) & = & -\sqrt{\nu A}\, G(\eta) \end{array}\right\} \tag{13.169}$$

where the dimensionless distance from the surface is

$$\eta \equiv y \sqrt{\frac{A}{\nu}} \tag{13.170}$$

Once we have determined the streamfunction, we have sufficient information to determine the pressure from Equation (13.168). This is very similar to the situation for potential-flow theory, where the velocity can be determined independent of the pressure. This suggests that even in a viscous flow, the role of the pressure is again passive. Local accelerations are dictated by the vorticity field, and the pressure adjusts as required to establish a balance of the pressure and viscous forces with the product of mass and acceleration. Again, for reference in the discussion to follow, the pressure is

$$p(x, y) = p_o - \frac{1}{2}\rho A^2 \left[x^2 + \frac{\nu}{A}P(\eta)\right] \tag{13.171}$$

Because some of the boundary conditions are specified at one of the two boundaries ($\eta = 0$) and some at the other boundary ($\eta \to \infty$), the problem we must solve is a **two-point boundary-value problem**. This type of problem is fundamentally more difficult to solve than an **initial-value problem**, where all boundary conditions are given at the same boundary. Also complicating our task of obtaining a solution, the differential equation for the streamfunction is quasi-linear and cannot be solved with elementary methods. This is a direct consequence of the nonvanishing convective acceleration terms in the Navier-Stokes equation.

A variety of techniques, both analytical (e.g., power-series solution) and numerical, is available to solve this type of problem. The first solution for stagnation-point flow was presented by Hiemenz [see Schlichting (1979)]. Consequently, stagnation-point flow is sometimes referred to as **Hiemenz flow**. We will use time-marching numerical methods to solve this problem in Section 13.7, and further discussion of solution techniques for this problem will not be given here. Rather, the rest of this section focuses on details of the solution.

Figure 13.15 shows the dimensionless streamfunction, $G(\eta)$, and horizontal velocity, $G'(\eta)$. Both functions are zero at the surface, and the horizontal velocity asymptotically approaches the inviscid value far above the surface. We can define the effective thickness of the viscous region, δ, as the point where the horizontal velocity is 99% of its inviscid-flow value. This occurs when the similarity variable $\eta = 2.4$, so that the viscous-layer thickness is

$$\delta \approx 2.4 \sqrt{\frac{\nu}{A}} \tag{13.172}$$

Combining Equations (13.168), (13.169) and (13.171), the pressure for flow near a stagnation point is

$$\begin{array}{rcl} p & = & p_o - \dfrac{1}{2}\rho A^2 \left(x^2 + 2\dfrac{\nu}{A}\dfrac{u}{Ax} + \dfrac{v^2}{A^2}\right) \\[10pt] & = & p_o - \dfrac{1}{2}\rho \left(A^2 x^2 + v^2\right) - \mu A \dfrac{u}{Ax} \end{array} \tag{13.173}$$

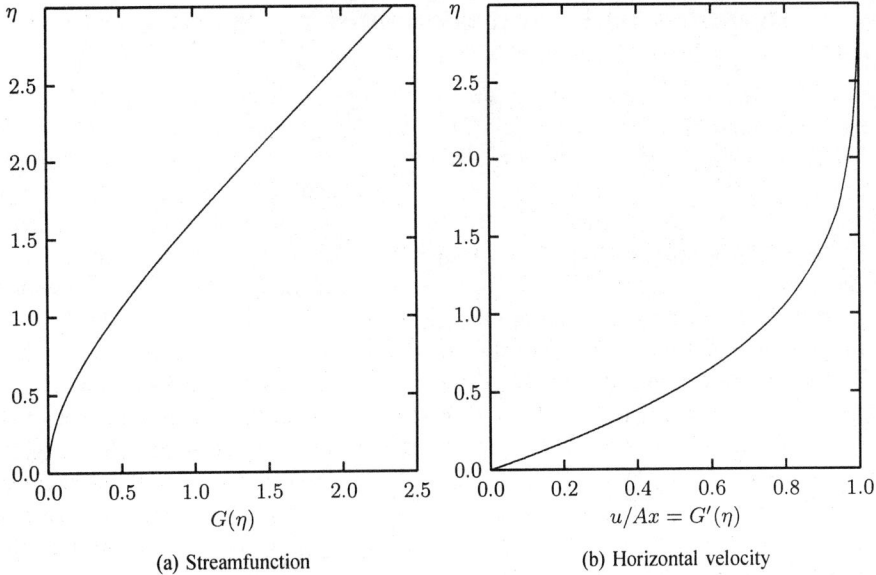

(a) Streamfunction   (b) Horizontal velocity

Figure 13.15: *Streamfunction and velocity for stagnation-point flow.*

For $\eta > 2.4$, we know that $u \to Ax$, so that the pressure will satisfy the following equation.

$$p + \frac{1}{2}\rho\left(u^2 + v^2\right) = p_o - \mu A \qquad (\eta > 2.4) \qquad (13.174)$$

Hence, the effective stagnation pressure in the inviscid flow above a stagnation point differs from the true stagnation pressure by an amount μA. This difference is very small, especially for incompressible, high Reynolds number flow. To see how small it is, recall from Subsection 11.7.4 that for flow past a cylinder

$$A = 2\psi_o = 2\frac{U}{R} \qquad (13.175)$$

Assuming the fluid is a perfect gas whose speed of sound is a_o when the pressure is p_o, we know that $p_o = \gamma\rho a_o^2$, wherefore

$$\frac{\mu A}{p_o} = 2\frac{\mu U/R}{\gamma\rho a_o^2} = \frac{2}{\gamma}\frac{U^2}{a_o^2}\frac{\mu}{\rho U R} = \frac{2}{\gamma}\frac{M^2}{Re_R} \qquad (13.176)$$

where M is the Mach number and Re_R is Reynolds number based on the cylinder radius.

For example, consider a cylindrical pipe of diameter 3 inches accidentally dropped from a skyscraper by a construction worker. Assuming the pipe's velocity reaches 100 mph, the Mach number will be 0.13 while the Reynolds number will be $Re_R \approx 10^5$. Thus, for flow past the pipe, we would have

$$\frac{\mu A}{p_o} = \frac{2}{1.4}\frac{(.13)^2}{10^5} = 2.4 \cdot 10^{-7} \qquad (13.177)$$

13.7 Computational Fluid Dynamics

At the end of Chapter 12, we examined time-marching numerical methods. Our focus was on the earliest schemes, which were explicit. In these schemes, the solution at each grid point at a given time is determined from properties at an earlier time. While explicit methods are very simple to understand and to program, they are only conditionally stable. As a consequence, very small timesteps must be taken to maintain stability with very fine grids, and a large number of steps is needed to achieve a steady-state solution. In this section, we consider **implicit schemes**, in which the solution at each grid point at a given time depends upon properties both at earlier times and at other grid points at the current time level. Although requiring the complication of having to invert a matrix, implicit schemes can be unconditionally stable. As a result, a smaller number of timesteps is needed to achieve steady state.

13.7.1 Implicit Time-Marching Methods

Implicit methods date back to 1947 when the Crank-Nicolson (1947) method first appeared. Other methods such as the Euler [Lilly (1965)] and Alternating Direction Implicit (ADI) schemes [Peaceman and Rachford (1955)] are implicit. The solution at time t^{n+1} and location x_j in this type of scheme depends not only upon the solution at the earlier timestep, but upon the solution at other spatial locations at time t^{n+1} as well. Figure 13.16 contrasts finite-difference molecules for explicit and implicit schemes.

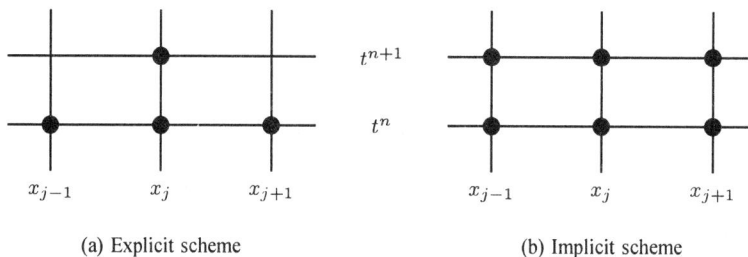

           (a) Explicit scheme   (b) Implicit scheme

Figure 13.16: *Finite-difference molecules for explicit and implicit schemes.*

Recall that the CFL condition for explicit schemes corresponds to the physical constraint that a wave cannot propagate beyond a finite-difference cell boundary in a single timestep. Intuitively, it should be plausible that using an implicit scheme removes this constraint since all points are coupled to one another. That is, rather than solving sequentially through the spatial domain as we must in an explicit method, we solve simultaneously in an implicit method. Thus, information should "travel" through the solution domain more rapidly. This is indeed what occurs, and implicit schemes thus permit a stepsize in excess of the CFL limit.

Implicit schemes have proven to be especially useful for steady-flow computations. As a rule of thumb, the CFL limit can be exceeded by factors as large as 5. There are exceptions though, and accurate results have been obtained for CFL numbers in excess of 1000 for some low Mach number flows. While these schemes will run at larger CFL numbers than explicit methods, using larger values of Δt sometimes introduces significant truncation errors. The number of timesteps required, relative to explicit methods, to achieve steady-flow conditions typically is reduced, although the factor is N_{CFL}^{-n} where $n < 1$ [cf. Roache (1972)].

13.7.2 Crank-Nicolson Method

One of the most popular implicit methods is the scheme developed by Crank and Nicolson (1947). This finite-difference scheme is second-order accurate in both space and time. For example, we consider the one-dimensional linear Burgers' equation introduced in Subsection 12.9.1, viz.,

$$\frac{\partial u}{\partial t} + U\frac{\partial u}{\partial x} = \nu\frac{\partial^2 u}{\partial x^2} \tag{13.178}$$

To implement the Crank-Nicolson method, we expand about the point $x = x_j$ and a time level midway between t^n and t^{n+1}. Denoting this point with subscript j and superscript $n+\frac{1}{2}$, the following central difference approximations are all second-order accurate in time and space.

$$\left.\left(\frac{\partial u}{\partial t}\right)^{n+\frac{1}{2}}_j = \frac{u_j^{n+1} - u_j^n}{\Delta t}\right.$$

$$\left.\left(\frac{\partial u}{\partial x}\right)^{n+\frac{1}{2}}_j = \frac{1}{2}\left[\frac{u_{j+1}^{n+1} - u_{j-1}^{n+1}}{2\Delta x} + \frac{u_{j+1}^n - u_{j-1}^n}{2\Delta x}\right]\right\} \tag{13.179}$$

$$\left.\left(\frac{\partial^2 u}{\partial x^2}\right)^{n+\frac{1}{2}}_j = \frac{1}{2}\left[\frac{u_{j+1}^{n+1} - 2u_j^{n+1} + u_{j-1}^{n+1}}{(\Delta x)^2} + \frac{u_{j+1}^n - 2u_j^n + u_{j-1}^n}{(\Delta x)^2}\right]\right.$$

Combining Equations (13.178) and (13.179) yields the following difference equation for u.

$$-\left[\frac{\nu\Delta t}{2(\Delta x)^2} + \frac{\Delta t}{4\Delta x}\right]u_{j-1}^{n+1} + \left[1 + \frac{\nu\Delta t}{(\Delta x)^2}\right]u_j^{n+1} - \left[\frac{\nu\Delta t}{2(\Delta x)^2} - \frac{\Delta t}{4\Delta x}\right]u_{j+1}^{n+1}$$

$$= \left[\frac{\nu\Delta t}{2(\Delta x)^2} + \frac{\Delta t}{4\Delta x}\right]u_{j-1}^n + \left[1 - \frac{\nu\Delta t}{(\Delta x)^2}\right]u_j^n + \left[\frac{\nu\Delta t}{2(\Delta x)^2} - \frac{\Delta t}{4\Delta x}\right]u_{j+1}^n \tag{13.180}$$

Thus, we see that at each grid point and time, (x_j, t^{n+1}), the solution for u_j^{n+1} depends upon values at the five surrounding points depicted in Figure 13.16(b). We will discover how to solve this equation in the next subsection. Before proceeding, however, it is instructive to show that this scheme is unconditionally stable for the linear Burgers' equation. Using von Neumann stability analysis [see Subsection 12.9.2], we examine a Fourier mode, i.e.,

$$u_j^n = G^n e^{i(j\kappa\Delta x)} = G^n e^{ij\theta} \tag{13.181}$$

where G is the amplitude factor, κ is wavenumber and $\theta = \kappa\Delta x$. Substituting into Equation (13.180) yields

$$\left[-\left(\frac{\nu\Delta t}{2(\Delta x)^2} + \frac{\Delta t}{4\Delta x}\right)e^{-i\theta} + \left(1 + \frac{\nu\Delta t}{(\Delta x)^2}\right) - \left(\frac{\nu\Delta t}{2(\Delta x)^2} - \frac{\Delta t}{4\Delta x}\right)e^{i\theta}\right]G$$

$$= \left(\frac{\nu\Delta t}{2(\Delta x)^2} + \frac{\Delta t}{4\Delta x}\right)e^{-i\theta} + \left(1 - \frac{\nu\Delta t}{(\Delta x)^2}\right) + \left(\frac{\nu\Delta t}{2(\Delta x)^2} - \frac{\Delta t}{4\Delta x}\right)e^{i\theta} \tag{13.182}$$

A little rearrangement of terms shows that the amplitude factor is

$$G = \frac{1 - \dfrac{\nu\Delta t}{(\Delta x)^2}(1 - \cos\theta) - \dfrac{i}{2}\dfrac{\Delta t}{\Delta x}\sin\theta}{1 + \dfrac{\nu\Delta t}{(\Delta x)^2}(1 - \cos\theta) + \dfrac{i}{2}\dfrac{\Delta t}{\Delta x}\sin\theta} \tag{13.183}$$

To have a stable scheme, the magnitude of G must be less than or equal to unity. In general, the amplitude of the ratio of two complex variables is equal to the ratio of their absolute values. Therefore, the condition for stability is

$$|G|^2 = \frac{\left[1 - \dfrac{\nu \Delta t}{(\Delta x)^2}(1 - \cos\theta)\right]^2 + \left(\dfrac{\Delta t}{2\Delta x}\right)^2 \sin^2\theta}{\left[1 + \dfrac{\nu \Delta t}{(\Delta x)^2}(1 - \cos\theta)\right]^2 + \left(\dfrac{\Delta t}{2\Delta x}\right)^2 \sin^2\theta} \leq 1 \qquad (13.184)$$

Multiplying through by the denominator, expanding the terms in brackets, and canceling common terms, all that remains is the following inequality.

$$-2\frac{\nu \Delta t}{(\Delta x)^2}(1 - \cos\theta) \leq 2\frac{\nu \Delta t}{(\Delta x)^2}(1 - \cos\theta) \qquad (13.185)$$

Since the factor $(1 - \cos\theta)$ is greater than or equal to 0 regardless of the value of θ, this inequality is satisfied provided only that $\Delta t > 0$. Thus, the Crank-Nicolson method is, by definition, unconditionally stable.

13.7.3 Thomas' Algorithm

Equation (13.180) is a coupled set of linear algebraic equations for u_j^{n+1}. This becomes evident when we rewrite Equation (13.180) in the more compact form:

$$A_j u_{j-1}^{n+1} + B_j u_j^{n+1} + C_j u_{j+1}^{n+1} = D_j \qquad (13.186)$$

where

$$\left.\begin{aligned}
A_j &= -\left[\frac{\nu \Delta t}{2(\Delta x)^2} + \frac{\Delta t}{4\Delta x}\right] \\[2mm]
B_j &= \left[1 + \frac{\nu \Delta t}{(\Delta x)^2}\right] \\[2mm]
C_j &= -\left[\frac{\nu \Delta t}{2(\Delta x)^2} - \frac{\Delta t}{4\Delta x}\right] \\[2mm]
D_j &= -A_j u_{j-1}^n - \left[B_j - 2\frac{\nu \Delta t}{(\Delta x)^2}\right] u_j^n - C_j u_{j+1}^n
\end{aligned}\right\} \qquad (13.187)$$

Because we have used equally-spaced grid points, all of the A_j, B_j and C_j are independent of grid-point index, j. For more general grid-point spacing, these coefficients will vary with j. Assuming we have a total of J grid points, we can express these equations in matrix form as follows.

$$\begin{bmatrix}
B_1 & C_1 & 0 & 0 & \cdots & 0 & 0 & 0 \\
A_2 & B_2 & C_2 & 0 & \cdots & 0 & 0 & 0 \\
0 & A_3 & B_3 & C_3 & \cdots & 0 & 0 & 0 \\
\vdots & \vdots & \vdots & \vdots & \ddots & \vdots & \vdots & \vdots \\
0 & 0 & 0 & 0 & \cdots & A_{J-1} & B_{J-1} & C_{J-1} \\
0 & 0 & 0 & 0 & \cdots & 0 & A_J & B_J
\end{bmatrix}
\begin{Bmatrix}
u_1^{n+1} \\
u_2^{n+1} \\
u_3^{n+1} \\
\vdots \\
u_{J-1}^{n+1} \\
u_J^{n+1}
\end{Bmatrix}
=
\begin{Bmatrix}
D_1 \\
D_2 \\
D_3 \\
\vdots \\
D_{J-1} \\
D_J
\end{Bmatrix}$$

$$(13.188)$$

Inspection of the matrix shows that it has nonzero elements only along the main diagonal $(B_1, B_2, B_3, \ldots, B_J)$, and along the sub-diagonals below $(A_2, A_3, A_4, \ldots, A_J)$ and above $(C_1, C_2, C_3, \ldots, C_{J-1})$ the main diagonal. All other elements are zero. The matrix thus consists of three nonzero diagonals, and is referred to as a **tridiagonal matrix**.

Because of its relatively simple structure, a tridiagonal matrix can be inverted using straightforward algebraic operations. The standard method used is known as **Thomas' algorithm** [Roache (1972)]. The algorithm can be derived by assuming the solution is of the form

$$u_j^{n+1} = P_j + Q_j u_{j+1}^{n+1} \tag{13.189}$$

where P_j and Q_j are coefficients to be determined. We can do this by substituting Equation (13.189) into Equation (13.186), wherefore

$$A_j \left[P_{j-1} + Q_{j-1} u_j^{n+1} \right] + B_j u_j^{n+1} + C_j u_{j+1}^{n+1} = D_j \tag{13.190}$$

Then, solving for u_j^{n+1}, we arrive at

$$u_j^{n+1} = \left[\frac{D_j - A_j P_{j-1}}{B_j + A_j Q_{j-1}} \right] + \left[\frac{-C_j}{B_j + A_j Q_{j-1}} \right] u_{j+1}^{n+1} \tag{13.191}$$

Comparing Equations (13.189) and (13.191) shows that the coefficients P_j and Q_j must satisfy

$$P_j = \frac{D_j - A_j P_{j-1}}{B_j + A_j Q_{j-1}} \quad \text{and} \quad Q_j = \frac{-C_j}{B_j + A_j Q_{j-1}} \tag{13.192}$$

We use these equations in the following way. First, we appeal to the boundary condition on u_j^{n+1} at $j = 1$ to set the values of P_1 and Q_1. For example, if u is given at $j = 1$ so that $u_1^{n+1} = u_o$, we would say

$$P_1 = u_o, \quad Q_1 = 0 \quad \text{for} \quad u_1^{n+1} = u_o \tag{13.193}$$

If the slope of u is given, the values of P_1 and Q_1 depend on the accuracy of the approximation used to compute the slope at the boundary. Using the first-order accurate approximation that $(u_x)_o \approx (u_2^{n+1} - u_1^{n+1})/\Delta x$, for example, yields

$$P_1 = -(u_x)_o \Delta x, \quad Q_1 = 1 \quad \text{for} \quad (u_2^{n+1} - u_1^{n+1})/\Delta x \approx (u_x)_o \tag{13.194}$$

Starting with these values for P_1 and Q_1, we can use Equations (13.192) to sequentially compute $P_2, P_3, \ldots, P_J$ and $Q_2, Q_3, \ldots, Q_J$. This is known as the **forward sweep** as the computations are done in the direction of increasing grid-point number.

Once, the coefficients P_j and Q_j are known at all points, we can use Equation (13.189) to solve for u_j^{n+1}, beginning at $j = J - 1$. This is called the **backward sweep** since computations proceed in descending order. For example, if the value of u at $j = J$ is u_f, then

$$u_{J-1}^{n+1} = P_{J-1} + Q_{J-1} u_f \quad \text{for} \quad u_J^{n+1} = u_f \tag{13.195}$$

This sweep continues for $j = J - 2, J - 3, \ldots, 2, 1$.

As a final comment, Thomas' algorithm provides an efficient method for inverting a tridiagonal matrix. It is used extensively in CFD methods. There is a constraint on the matrix that must be satisfied in order for this algorithm to work. Specifically, the matrix must be **diagonally dominant**. A sufficient condition for diagonal dominance is

$$B_j \geq -(A_j + C_j) \tag{13.196}$$

For Burgers' equation, substitution of Equations (13.187) into Equation (13.196) yields

$$1 + \frac{\nu \Delta t}{(\Delta x)^2} \geq \frac{\nu \Delta t}{(\Delta x)^2} \tag{13.197}$$

which is obviously true regardless of the values of ν, Δt and Δx.

13.7.4   Stagnation-Point Flow Revisited

Recall that for stagnation-point flow, we arrived at the dimensionless streamfunction Equation (13.151), and boundary conditions given in Equation (13.152). Also, recall that the dimensionless horizontal velocity, $U(\eta)$, is related to the dimensionless streamfunction, $G(\eta)$, according to

$$U(\eta) = \frac{dG}{d\eta} \tag{13.198}$$

Hence, we can rewrite the streamfunction equation and boundary conditions as

$$-G\frac{dU}{d\eta} = 1 - U^2 + \frac{d^2U}{d\eta^2} \tag{13.199}$$

$$G(0) = 0, \quad U(0) = 0, \quad U(\eta) \to 1 \ \text{ as } \ \eta \to \infty \tag{13.200}$$

We can solve this nonlinear two-point boundary-value problem using time-marching methods by replacing the ordinary differential equation with the following partial differential equation.

$$\frac{\partial U}{\partial t} - G\frac{\partial U}{\partial \eta} = 1 - U^2 + \frac{\partial^2 U}{\partial \eta^2} \tag{13.201}$$

The general approach is to make an initial guess at the solution and advance the solution to Equation (13.201) in time until temporal variations vanish. The steady-state solution satisfies Equation (13.199). As the solution for $U(\eta)$ proceeds, we compute the streamfunction from

$$G(\eta) = \int_0^\eta U(\tilde{\eta})d\tilde{\eta} \tag{13.202}$$

which we evaluate numerically.

To cast the equation for $U(\eta)$ in discretized form, we use the Crank-Nicolson method so that $\partial U/\partial t$, $\partial U/\partial \eta$ and $\partial^2 U/\partial \eta^2$ are approximated according to Equations (13.179), with η replacing x. Also, we represent the nonlinear term, U^2, on the right-hand side of Equation (13.201) as

$$\left(U^2\right)_j^{n+\frac{1}{2}} = U_j^n U_j^{n+1} \tag{13.203}$$

which is second-order accurate in time and space. The discretized and linearized equation is the standard tridiagonal form

$$A_j U_{j-1}^{n+1} + B_j U_j^{n+1} + C_j U_{j+1}^{n+1} = D_j \tag{13.204}$$

where

$$A_j = -\left[\frac{1}{2(\Delta\eta)^2} + \frac{G_j}{4\Delta\eta}\right]$$

$$B_j = \left[\frac{1}{\Delta t} + \frac{1}{(\Delta\eta)^2} + U_j^n\right]$$

$$C_j = -\left[\frac{1}{2(\Delta\eta)^2} - \frac{G_j}{4\Delta\eta}\right]$$

$$D_j = 1 - A_j U_{j-1}^n - \left[\frac{1}{\Delta t} - \frac{1}{(\Delta\eta)^2}\right] U_j^n - C_j U_{j+1}^n$$

(13.205)

To determine the streamfunction as the computation advances in time, we can use a number of methods to evaluate the integral of $U(\eta)$ quoted in Equation (13.202). One approach that is of comparable accuracy to our discretization approach for Equation (13.201) is to use a central-difference approximation for Equation (13.198), i.e.,

$$U_j^{n+1} = \frac{G_{j+1} - G_{j-1}}{2\Delta\eta} \implies G_{j+1} = G_{j-1} + 2U_j^{n+1}\Delta\eta \qquad (13.206)$$

We know from the boundary conditions that $G_1 = 0$. Then, given the velocity at the most current time, U_j^{n+1}, we can use Equation (13.206) to determine G_j for values of $j \geq 3$. To set the value of G_2, we observe from the boundary conditions that $G'(0) = U(0) = 0$. As can be readily verified, the following approximation for $G'(0)$ is of second-order accuracy.

$$G'(0) \approx -\frac{3G_1 - 4G_2 + G_3}{2\Delta\eta} \qquad (13.207)$$

Hence, since $G_1 = 0$, we can enforce the vanishing of $G'(0)$ by requiring

$$G_2 = \frac{1}{4}G_3 \qquad (13.208)$$

To start the computation, we assume a quadratic variation of $U(\eta)$ near the surface, i.e.,

$$U(\eta) = \begin{cases} \left(\frac{\eta}{3}\right)^2, & 0 \leq \eta < 3 \\ 1, & 3 \leq \eta \leq 5 \end{cases} \qquad (13.209)$$

Note that we have placed the upper boundary of the finite-difference grid at $\eta = 5$, with the hope that it is sufficiently distance from the surface to appear infinitely distant. A bit of numerical experimentation shows that placing the top of the grid at $\eta = 5$ is indeed far enough from the surface to have a negligible effect on the solution.

To quantify the effect of the timestep, we can use the theoretical timestep for MacCormack's explicit scheme, Δt_{max}. In terms of the notation for the present problem, the value of Δt_{max} is

$$\Delta t_{max} = \left[\frac{G_J}{\Delta\eta} + \frac{2}{(\Delta\eta)^2}\right]^{-1} \qquad (13.210)$$

and the effective CFL number is

$$\tilde{N}_{CFL} = \frac{\Delta t}{\Delta t_{max}} \qquad (13.211)$$

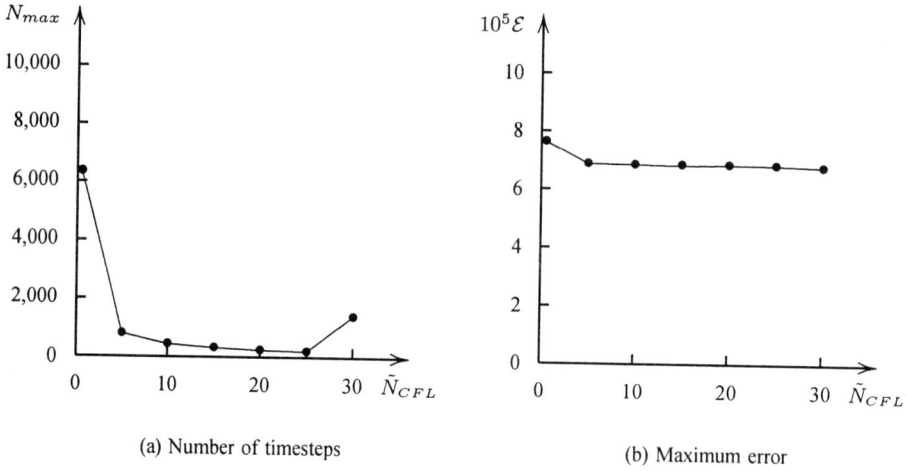

(a) Number of timesteps   (b) Maximum error

Figure 13.17: *Number of timesteps to attain steady state and maximum error for 101 equally-spaced grid points.*

We have already discussed the solution in Section 13.6. In particular, Figure 13.15 shows the computed dimensionless streamfunction, $G(\eta)$, and horizontal velocity, $U(\eta) = G'(\eta)$. The following discussion focuses on numerical aspects of the solution.

Figure 13.17(a) displays the number of timesteps, N_{max}, needed to achieve a steady solution. The computations shown were done with 101 equally-spaced grid points, so that $\Delta\eta = 0.05$. As shown, the computation is stable for large values of the effective CFL number. The number of timesteps required to achieve steady-state drops rapidly for $\tilde{N}_{CFL} > 1$. The optimum CFL number is between 20 and 25, with approximately 200 timesteps required for a converged solution. Recall that the explicit MacCormack method was optimum for a CFL number of 0.9, and required about 1400 timesteps for a similar problem (cf. Section 12.9 – Figure 12.16). For values of $\tilde{N}_{CFL}$ in excess of 25, the approach to convergence becomes erratic, as evidenced by inspection of solution residuals, ϵ_j^{n+1}, defined by

$$\epsilon_j^{n+1} \equiv \left(G\frac{dU}{d\eta} + 1 - U^2 + \frac{d^2U}{d\eta^2} \right)_j^{n+1} \tag{13.212}$$

For smaller values of $\tilde{N}_{CFL}$, the maximum residual decreases monotonically to machine accuracy as the solution proceeds. For larger values, the maximum residual oscillates as the solution approaches steady state, and the number of timesteps begins to increase as $\tilde{N}_{CFL}$ increases.

As shown in Figure 13.17(b), there is no penalty in accuracy for increasing CFL number, at least for the range of CFL numbers considered here. The error shown in the figure has been obtained from Richardson extrapolation. Eventually, for larger values of the CFL number, solution error will increase and distort the numerical solution.

The program used to perform the computations described in this section is called **STAGPT**. It is described in Appendix E, and the source code is included on the diskette supplied with this book.

Problems

13.1 Compute the vorticity vector, strain-rate tensor and viscous stress tensor for *Couette flow*.

13.2 Compute the vorticity vector, strain-rate tensor and viscous stress tensor for *channel flow*.

13.3 Compute the vorticity vector, strain-rate tensor and viscous stress tensor for *pipe flow*.

13.4 Compute the strain-rate tensor and viscous stress tensor for *Stokes' First Problem*.

13.5 Compute the strain-rate tensor and viscous stress tensor for *Stokes' Second Problem*. **HINT:** Note that $(1+i)/\sqrt{2} = e^{i\pi/4}$.

13.6 Compute the vorticity vector, strain-rate tensor and viscous stress tensor for *stagnation-point flow*. Express your answers in terms of the dimensionless streamfunction, $G(\eta)$, and its derivatives.

13.7 Determine the streamfunction for *Couette flow* and *channel flow*, assuming $\psi = 0$ when $y = 0$.

13.8 Determine the streamfunction for *pipe flow*, assuming $\psi = 0$ when $r = 0$. **HINT:** Begin with the definition of the streamfunction, $\mathbf{u} \equiv \nabla \times \boldsymbol{\Psi}$, and show that we should expect to have

$$u = \frac{1}{r}\frac{d}{dr}(r\psi) \qquad \text{where} \qquad \boldsymbol{\Psi} = \psi \, \mathbf{e}_\theta$$

13.9 Determine the streamfunction for *Stokes' First Problem*, assuming $\psi = 0$ when $y = 0$. **HINT:** Take advantage of the fact that

$$\int_0^\eta \text{erfc}(\bar{\eta})\,d\bar{\eta} = \eta\,\text{erfc}(\eta) + \frac{1}{\sqrt{\pi}}\left(1 - e^{-\eta^2}\right)$$

13.10 Determine the streamfunction for *Stokes' Second Problem*, assuming $\psi = 0$ when $y = 0$.

13.11 When the pressure gradient parameter, χ, defined by

$$\chi \equiv \frac{h^2}{2\mu U}\frac{\partial p}{\partial x}$$

exceeds 1, *Couette flow* has a reverse-flow region near the stationary plate. Determine the location and magnitude of the maximum reverse-flow velocity as a function of χ. For what value of χ is the maximum reverse-flow velocity 1/3 of the upper-plate velocity, U?

13.12 Determine the mass flux, $\dot{m}$, and momentum flux, $\dot{P}$, per unit width (out of the page), for *Couette flow* as a function of the pressure gradient parameter, χ, defined by

$$\chi \equiv \frac{h^2}{2\mu U}\frac{\partial p}{\partial x}$$

For what value of χ is the mass flux zero, and what is the corresponding momentum flux?

13.13 We want to perform laminar channel-flow experiments. To prevent transition to turbulence, we must require $Re_h \leq 2000$. If our working fluid is air at $20°$ C, and the channel height is $h = 3$ cm, what is the maximum average velocity, $(u_m)_{max}$, we can have? What is the corresponding entrance length, ℓ_e, required to achieve fully-developed flow?

13.14 We want to perform laminar pipe-flow experiments. Because of careful conditioning of the inlet flow, we can maintain laminar flow up to a Reynolds number, Re_D, of 15000. If our working fluid is water at $68°$ F, and we have sufficient room in a laboratory to accommodate an entrance length required to achieve fully-developed flow, ℓ_e, of 12 ft, what is the maximum pipe diameter (in inches) that we can use? What is the corresponding velocity, u_m?

13.15 Suppose we envision the flow in a pipe to be a *plug flow* with constant velocity, u_m, over a central core displaced a distance δ^* from the pipe walls as shown. What must δ^* be in order to have the same volume-flow rate, Q, as the exact laminar-flow solution?

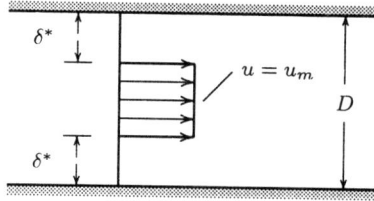

Problem 13.15

13.16 In *Stokes' First Problem*, the thickness of the viscous layer near the wall in an unspecified fluid is 1 cm at a time 1 sec after the wall is impulsively set in motion. How thick will the layer be after 1 hour?

13.17 In *Stokes' Second Problem*, the thickness of the viscous layer near the wall in an unspecified fluid is 1 cm at a time 1 sec after the wall is impulsively set in motion. How thick will the layer be after 1 hour?

13.18 Verify that, for *Stokes' Second Problem*, the shear stress at the plate surface, τ_w, is 45° out of phase with the plate velocity, $u(0,t)$. **HINT:** Note that $(1+i)/\sqrt{2} = e^{i\pi/4}$.

13.19 The purpose of this problem is to derive the separation-of-variables solution for the streamfunction in *stagnation-point flow*. We begin with the streamfunction equation and boundary conditions, which are

$$\nu\left(\frac{\partial^4\psi}{\partial x^4} + 2\frac{\partial^4\psi}{\partial x^2\partial y^2} + \frac{\partial^4\psi}{\partial y^4}\right) = \frac{\partial\psi}{\partial y}\left(\frac{\partial^3\psi}{\partial x^3} + \frac{\partial^3\psi}{\partial x\partial y^2}\right) - \frac{\partial\psi}{\partial x}\left(\frac{\partial^3\psi}{\partial y^3} + \frac{\partial^3\psi}{\partial x^2\partial y}\right)$$

$$\frac{\partial\psi}{\partial y} \to Ax, \quad \frac{\partial\psi}{\partial x} \to Ay \quad \text{as } y \to \infty$$

$$\frac{\partial\psi}{\partial y} = 0, \quad \frac{\partial\psi}{\partial x} = 0 \quad \text{at } y = 0$$

(a) Assuming a solution of the form $\psi(x,y) = \Psi(x)F(y)$, verify that

$$\Psi F'\left[\Psi'''F + \Psi'F''\right] - \Psi'F\left[\Psi F''' + \Psi''F'\right] = \nu\left[\Psi''''F + 2\Psi''F'' + \Psi F''''\right]$$

where a prime denotes differentiation, e.g., in the case of the first derivatives, $F' \equiv dF(y)/dy$ and $\Psi' \equiv d\Psi(x)/dx$.

(b) Divide through by $\Psi(x)$ and rearrange terms to show that

$$\nu\left[F'''' + 2\frac{\Psi''}{\Psi}F'' + \frac{\Psi''''}{\Psi}F\right] = \Psi'\left[F'F'' - FF'''\right] + \Psi'''FF' - \frac{\Psi'\Psi''}{\Psi}FF'$$

(c) The equation derived in Part (b) simplifies to an ordinary differential equation for $F(y)$ provided all the coefficients in the equation are independent of x. For example, the coefficient of the first term on the right-hand side is constant provided

$$\Psi' = C \quad \Longrightarrow \quad \Psi(x) = Cx + D$$

where C and D are constants of integration. Substituting $\psi(x,y) = (Cx+D)F(y)$ into the boundary conditions for $y \to \infty$ yields

$$\frac{\partial\psi}{\partial y} = (Cx+D)\frac{dF}{dy} \to Ax \quad \text{as} \quad y \to \infty$$

Using the boundary conditions, verify that

$$\Psi(x) = Ax$$

(d) Using results of Part (c), show that $F(y)$ satisfies the following ordinary differential equation and boundary conditions.

$$\frac{\nu}{A}\frac{d^4 F}{dy^4} = \frac{dF}{dy}\frac{d^2 F}{dy^2} - F\frac{d^3 F}{dy^3}$$

$$\left. \begin{array}{ll} F'(y) \to 1, & F(y) \to y \quad \text{as } y \to \infty \\ F'(0) = 0, & F(0) = 0 \quad \text{at } y = 0 \end{array} \right\}$$

(e) Take advantage of the fact that the chain rule tells us

$$F\frac{d^3 F}{dy^3} = \frac{d}{dy}\left(F\frac{d^2 F}{dy^2}\right) - \frac{dF}{dy}\frac{d^2 F}{dy^2}$$

to show that the equation for $F(y)$ can be written as

$$\frac{d}{dy}\left[\frac{\nu}{A}\frac{d^3 F}{dy^3} + F\frac{d^2 F}{dy^2} - \left(\frac{dF}{dy}\right)^2\right] = 0$$

(f) Use the boundary condition on $F(y)$ for $y \to \infty$ (note that two boundary conditions for $y \to \infty$ are redundant) to conclude that

$$\frac{\nu}{A}\frac{d^3 F}{dy^3} + F\frac{d^2 F}{dy^2} - \left(\frac{dF}{dy}\right)^2 + 1 = 0$$

$$F(0) = 0, \quad F'(0) = 0, \quad F'(y) \to 1 \text{ as } y \to \infty$$

13.20 The purpose of this problem is to derive the separation-of-variables solution for the pressure in *stagnation-point flow*. We begin with the Poisson equation for the pressure, which is

$$\nabla^2 p = -\rho\left[\left(\frac{\partial u}{\partial x}\right)^2 + \left(\frac{\partial v}{\partial y}\right)^2 + 2\frac{\partial u}{\partial y}\frac{\partial v}{\partial x}\right]$$

(a) In terms of dimensional parameters, the velocity components for stagnation-point flow are

$$u(x, y) = AxF'(y) \quad \text{and} \quad v(x, y) = -AF(y)$$

Verify that the pressure equation simplifies to

$$\nabla^2 p = -2\rho A^2\left(\frac{dF}{dy}\right)^2$$

(b) Now, assume a solution of the form $p(x, y) = p_o - \frac{1}{2}\rho A^2[X(x) + Y(y)]$. Show that the pressure equation simplifies to

$$\frac{d^2 Y}{dy^2} - 4\left(\frac{dF}{dy}\right)^2 = -\frac{d^2 X}{dx^2} = \text{constant}$$

(c) Clearly, choosing $X(x) = x^2$ satisfies the equation of Part (b) provided we select the value of the constant to be -2. This choice is also consistent with the inviscid solution. While we could add a constant term and a term linear in x to the function $X(x)$, there is no loss of generality in selecting the constants of integration that come with these terms equal to be zero. Show that we are left with the following equation for the function $Y(y)$.

$$\frac{d^2 Y}{dy^2} = 4\left(\frac{dF}{dy}\right)^2 - 2$$

13.21 A thin layer of viscous fluid flows down an inclined plane. The kinematic viscosity of the fluid is ν and the gravitational acceleration is g.

(a) State the exact conservation equations governing this flow.

(b) The appropriate boundary conditions are

$$u = v = 0 \text{ at } y = 0, \quad p = p_o \text{ at } y = h, \quad u = U \text{ at } y = h$$

where p_o is atmospheric pressure. Assuming parallel flow, i.e., that $\partial/\partial x \to 0$, make all appropriate simplifications to the equations of motion.

(c) Assuming further that $\partial p/\partial x = 0$, solve for the pressure.

(d) Solve for the velocity.

(e) If the shear stress at the wall is $\tau_w = \rho g h \sin\theta$, what is U?

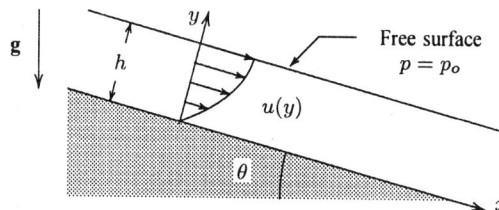

Problem 13.21

13.22 We wish to solve for fully-developed, incompressible, viscous flow between two long concentric cylinders. Both cylinders are at rest and *the flow is parallel to the centerline*. The radii of the cylinders are R_i and R_o as shown.

(a) Simplify the continuity and Navier-Stokes equations taking advantage of the axial symmetry of the geometry. Also, state the appropriate boundary conditions.

(b) Simplify the equations of Part (a) further for fully-developed flow. Solve for radial velocity, show that $\partial p/\partial r = 0$ and derive an ordinary differential equation for streamwise velocity.

(c) Solve for the streamwise velocity.

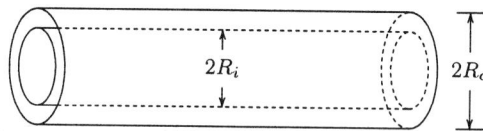

Problems 13.22, 13.23

13.23 We wish to solve for incompressible, viscous flow between two long concentric rotating cylinders. The inner cylinder rotates at an angular velocity of Ω_i, while the outer cylinder has angular velocity Ω_o. The radii of the cylinders are R_i and R_o as shown. *There is no flow in the direction parallel to the centerline, x — all motion is in the circumferential direction.*

(a) Simplify the continuity and Navier-Stokes equations taking advantage of the axial symmetry of the geometry, and the fact that the flow is independent of x. Also, state the appropriate boundary conditions.

(b) Verify that the radial velocity is zero.

(c) Solve for the circumferential velocity, u_θ. **HINT:** Assume a solution of the form $u_\theta = r^m$.

(d) Determine the limiting form of the solution for an unbounded fluid, $R_o \to \infty$, with $\Omega_o = 0$. What inviscid flow does this resemble?

13.24 Consider fully-developed, incompressible, viscous flow in a long pipe that rotates about its centerline with an angular rotation rate of Ω.

  (a) Simplify the continuity and Navier-Stokes equations taking advantage of the axial symmetry of the geometry. Also, state the appropriate boundary conditions.

  (b) Simplify the equations of Part (a) further for fully-developed flow, i.e., solve for the radial velocity, show that $\partial p/\partial r \neq 0$ and derive ordinary differential equations for the streamwise and circumferential velocities.

  (c) Solve for the streamwise and circumferential velocities. **HINT:** For u_θ, assume a solution of the form $u_\theta = r^m$.

  (d) Solve for $p(x, r) - p(x, 0)$.

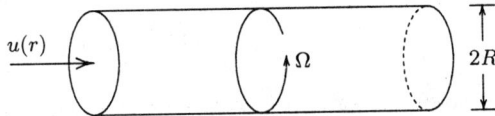

Problem 13.24

13.25 Consider incompressible, viscous flow external to a rotating cylinder of radius R in an otherwise motionless fluid. The angular-rotation rate is Ω.

  (a) Simplify the continuity and Navier-Stokes equations taking advantage of the axial symmetry of the geometry. Also, state the appropriate boundary conditions. **NOTE:** There is no reason to have $\partial p/\partial \theta \neq 0$.

  (b) Solve for u_r and u_θ. **HINT:** Try a solution of the form $u_\theta \propto r^m$.

  (c) Compute the circulation, Γ, on a circular contour of radius $r > R$. Rewrite your solution of Part (b) in terms of Γ. Is this solution familiar?

  (d) Solve for the pressure in terms of ρ, Ω, R, r and the farfield pressure, p_∞. Is there a limit on how fast the cylinder can rotate for this solution to be valid?

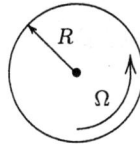

Problem 13.25

13.26 Consider fully-developed, incompressible, viscous flow in a long pipe of elliptical cross section. The lengths of the semimajor and semiminor axes are a and b, respectively.

  (a) Assume a solution of the form

$$u = A + By^2 + Cz^2, \qquad v = 0, \qquad w = 0$$

  where A, B and C are constants. How do the continuity and Navier-Stokes equations simplify?

  (b) What boundary conditions must u satisfy?

  (c) Solve for the constants A, B and C.

  (d) Does your answer make sense for the limiting case $a = b = R$?

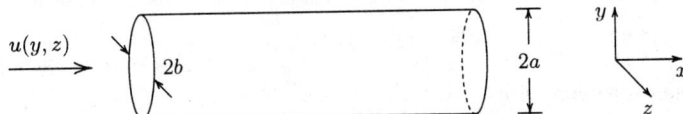

Problem 13.26

13.27 The *Ekman layer* is the viscous region above a rotating surface as observed in the rotating coordinate frame.

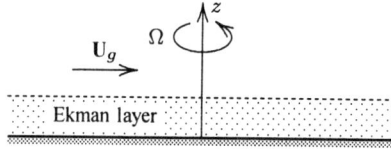

Problem 13.27

(a) Show that the incompressible continuity and Navier-Stokes equations, in rotating coordinates [see Appendix D, Section D.4], admit a solution of the form $\mathbf{u} = u(z)\,\mathbf{i} + v(z)\,\mathbf{j}$ subject to the boundary conditions $\mathbf{u}(0) = \mathbf{0}$ and $\mathbf{u}(z) \to \mathbf{U}_g$ as $z \to \infty$, where the *geostrophic* (freestream) velocity vector is $\mathbf{U}_g = U\,\mathbf{i}$ with U = constant. Assume that for all points in the flow,

$$2\rho\boldsymbol{\Omega} \times \mathbf{U}_g = -\nabla\left(p - \tfrac{1}{2}\Omega^2 r^2\right)$$

(b) Solve the resulting equations for the velocity as seen by a rotating observer, $\mathbf{u}$. **HINT:** The four roots of $(-1)^{1/4}$ are $\pm(1+i)/2, \pm(1-i)/2$.

13.28 Consider fully-developed, incompressible flow between parallel walls. The lower wall is stationary and the upper wall oscillates in its own plane with amplitude u_o and frequency Ω.

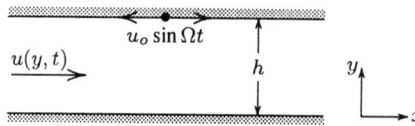

Problem 13.28

(a) Simplify the continuity and Navier-Stokes equations for this flow. Assume the pressure is constant.

(b) What are the boundary conditions?

(c) Solve for the velocity. In developing your solution, make use of the fact that, for a complex variable, z, we have $\sinh z \equiv (e^z - e^{-z})/2$.

(d) Simplify your answer for oscillations sufficiently slow that $\Omega h^2/\nu \ll 1$. **HINT:** Note that $\sinh z \approx z$ for $|z| \ll 1$.

13.29 A semi-infinite viscous fluid is bounded by a flat porous plate at $y = 0$. At time $t = 0$, the plate is impulsively set in motion at a constant velocity U in the x direction. At the same time, injection of fluid through the plate starts. The injection velocity, v_w, is in the y direction, i.e., normal to the plate. You may assume the pressure is constant for all time and at all locations in the fluid.

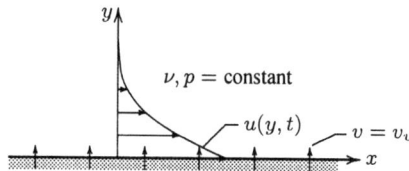

Problem 13.29

(a) Focusing upon mass conservation, what is the most general form of any solution for the velocity component in the y direction, v?

(b) Write the differential equation from which the velocity component u may be obtained.

(c) Now suppose that we seek a similarity solution. Assuming the similarity variable η is the same as in Stokes' First Problem, determine the condition that v must satisfy in order for the similarity solution in terms of η to exist.

(d) Comparing results of Parts (a) and (c), what conclusions can you draw?

13.30 We can analyze the decay of the *trailing vortex* left by an airplane that has just taken off as follows. We assume the vortex forms quickly and initially appears as a potential vortex so that

$$u_\theta(r, t) = \frac{\Gamma}{2\pi r}, \qquad u_r = 0, \qquad w = 0$$

where Γ is the initial circulation and r is radial distance. We seek a similarity solution in which the similarity variable is

$$\eta = \frac{r}{2\sqrt{\nu t}}$$

where ν is kinematic viscosity and t is time.

(a) Simplify the continuity and Navier-Stokes equations assuming the motion is axisymmetric and independent of z.

(b) What boundary and initial conditions must the velocity components satisfy?

(c) Solve for u_r and rewrite the equation for u_θ in terms of the quantity $\gamma \equiv 2\pi r u_\theta$.

(d) Transform the equations of motion and boundary conditions into similarity form.

(e) Solve for the circumferential velocity, casting the final answer in terms of Γ, r, ν and t.

13.31 The purpose of this problem is to generate the Navier-Stokes solution for flow near a three-dimensional stagnation point, i.e., for axisymmetric flow.

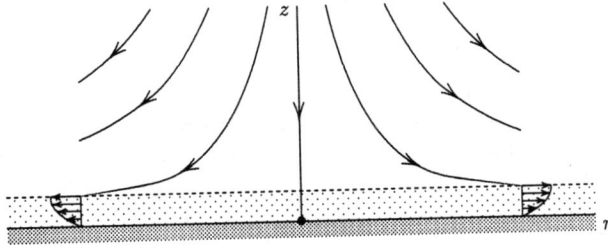

Problem 13.31

(a) Simplify the axisymmetric continuity and Navier-Stokes equations for this flow.

(b) The inviscid solution, valid in the limit $z \to \infty$, is

$$u_r = Ar \qquad \text{and} \qquad w = -2Az$$

where A is a constant. State the boundary conditions on u_r and w.

(c) Assume a solution of the form

$$u_r = ArG'(\eta), \quad w = -2\sqrt{\nu A}\, G(\eta), \quad p = p_o - \frac{1}{2}\rho A^2 \left[r^2 + 4\frac{\nu}{A} P(\eta)\right]$$

where p_o is the stagnation pressure, and η is

$$\eta \equiv z\sqrt{\frac{A}{\nu}}$$

Verify that $G(\eta)$ and $P(\eta)$ satisfy the following ordinary differential equations.

$$\frac{d^3 G}{d\eta^3} + 2G\frac{d^2 G}{d\eta^2} - \left(\frac{dG}{d\eta}\right)^2 + 1 = 0$$

$$\frac{dP}{d\eta} = \frac{d^2 G}{d\eta^2} + 2G\frac{dG}{d\eta}$$

Elaborate on how these equations differ from their two-dimensional counterparts.

13.32 *Jeffery-Hamel flows* are a family of exact incompressible, Navier-Stokes solutions for convergent and divergent channels. As illustrated, the streamlines are straight lines emanating from a point beyond the channels as indicated. The velocity is purely radial and is given by

$$u_r = \frac{\nu}{r} f(\theta), \qquad u_\theta = 0, \qquad w = 0$$

where ν is the kinematic viscosity of the fluid and $f(\theta)$ is a function to be determined.

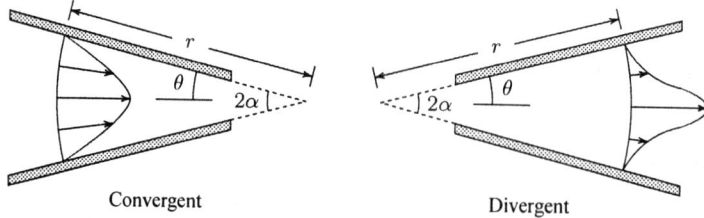

Convergent   Divergent

Problem 13.32

(a) Simplify the continuity and Navier-Stokes equations as appropriate. **HINT:** The circumferential component of the Navier-Stokes equation does not reduce to $\partial p / \partial \theta = 0$.

(b) What are the boundary conditions on u_r?

(c) Verify from the results of Parts (a) and (b) that the pressure gradient *at the walls* satisfies

$$-\frac{1}{\rho} \frac{\partial p}{\partial r} = \frac{\nu^2 K}{r^3}, \qquad K = \text{constant}$$

(d) Verify that $f(\theta)$ satisfies the following ordinary differential equation.

$$\frac{d^2 f}{d\theta^2} + 4f + f^2 + K = 0$$

13.33 Consider flow above an infinite disk rotating with angular velocity Ω about an axis perpendicular to its plane. An exact solution to the incompressible continuity and Navier-Stokes equations exists and has the following form:

$$u_r = \Omega r F(\zeta), \quad u_\theta = \Omega r G(\zeta), \quad w = \sqrt{\nu \Omega} H(\zeta), \quad p = p_o + \mu \Omega P(\zeta), \qquad \zeta \equiv z \sqrt{\frac{\Omega}{\nu}}$$

where μ is viscosity, ν is kinematic viscosity, ρ is density and p_o is stagnation pressure.

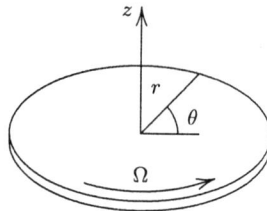

Problem 13.33

(a) Show that the axisymmetric equations of motion yield the following four ordinary differential equations for F, G, H and P.

$$\frac{dH}{d\zeta} + 2F = 0, \quad H \frac{dF}{d\zeta} + F^2 - G^2 = \frac{d^2 F}{d\zeta^2}, \quad H \frac{dG}{d\zeta} + 2FG = \frac{d^2 G}{d\zeta^2}, \quad H \frac{dH}{d\zeta} = -\frac{dP}{d\zeta} + \frac{d^2 H}{d\zeta^2}$$

(b) What are the boundary conditions on F, G, H and P?

13.34 Verify that Equation (13.203) is of second-order accuracy.

13.35 Verify that Equation (13.207) is of second-order accuracy.

13.36 Using von Neumann stability analysis, determine the amplification factor, G, for the *Euler forward time and central space* (FTCS) algorithm applied to the diffusion equation, viz.,

$$\frac{u_j^{n+1} - u_j^n}{\Delta t} = \nu \frac{u_{j+1}^{n+1} - 2u_j^{n+1} + u_{j-1}^{n+1}}{(\Delta x)^2}$$

Is the scheme unconditionally stable?

13.37 Using von Neumann stability analysis, determine the amplification factor, G, for the *Crank-Nicolson* algorithm applied to the diffusion equation with a sink term, i.e., $u_t = -Su + \nu u_{xx}$, viz.,

$$\frac{u_j^{n+1} - u_j^n}{\Delta t} = -S \frac{u_j^{n+1} + u_j^n}{2} + \nu \frac{u_{j+1}^{n+1} - 2u_j^{n+1} + u_{j-1}^{n+1} + u_{j+1}^n - 2u_j^n + u_{j-1}^n}{2(\Delta x)^2}$$

Is the scheme still unconditionally stable provided $S > 0$?

13.38 Determine whether or not the following matrices are diagonally dominant.

$$[A] = \begin{bmatrix} 10 & -5 & 0 & 0 & 0 \\ -2 & 10 & -2 & 0 & 0 \\ 0 & -3 & 10 & -3 & 0 \\ 0 & 0 & -4 & 10 & -4 \\ 0 & 0 & 0 & 0 & 10 \end{bmatrix}, \quad [B] = \begin{bmatrix} 2 & 1 & 0 & 0 & 0 \\ -1 & 3 & -1 & 0 & 0 \\ 0 & -2 & 3 & -2 & 0 \\ 0 & 0 & -1 & 3 & -1 \\ 0 & 0 & 0 & 1 & 2 \end{bmatrix}$$

$$[C] = \begin{bmatrix} 10 & 3 & 0 & 0 & 0 \\ -1 & 10 & -1 & 0 & 0 \\ 0 & -2 & 10 & -6 & 0 \\ 0 & 0 & -3 & 10 & -7 \\ 0 & 0 & 0 & 6 & 10 \end{bmatrix}$$

13.39 Under what conditions is the following matrix diagonally dominant?

$$[A] = \begin{bmatrix} x^2 & 5x & 0 & 0 & 0 \\ -x & x^2 & -x & 0 & 0 \\ 0 & -1 & 2x^2 & -1 & 0 \\ 0 & 0 & -2x & x^2 & -x \\ 0 & 0 & 0 & -x & x^2 \end{bmatrix}$$

13.40 Solve the following matrix equation using Thomas' algorithm.

$$\begin{bmatrix} 1 & 0 & 0 & 0 & 0 \\ -1 & 2 & -1 & 0 & 0 \\ 0 & -1 & 2 & -1 & 0 \\ 0 & 0 & -1 & 2 & -1 \\ 0 & 0 & 0 & 0 & 1 \end{bmatrix} \begin{Bmatrix} x_1 \\ x_2 \\ x_3 \\ x_4 \\ x_5 \end{Bmatrix} = \begin{Bmatrix} 1 \\ 2 \\ 2 \\ 2 \\ 2 \end{Bmatrix}$$

13.41 Solve the following matrix equation using Thomas' algorithm.

$$\begin{bmatrix} 100 & 0 & 0 & 0 & 0 \\ -1 & 50 & -1 & 0 & 0 \\ 0 & -1 & 20 & -1 & 0 \\ 0 & 0 & -1 & 10 & -1 \\ 0 & 0 & 0 & 0 & 1 \end{bmatrix} \begin{Bmatrix} x_1 \\ x_2 \\ x_3 \\ x_4 \\ x_5 \end{Bmatrix} = \begin{Bmatrix} 100 \\ 50 \\ 20 \\ 10 \\ 1 \end{Bmatrix}$$

13.42 The solution for a three-dimensional stagnation point is similar to the two-dimensional solution (see Problem 13.31). The velocity and pressure are

$$u_r = ArG'(\eta), \quad w = -2\sqrt{\nu A}\, G(\eta), \quad p = p_o - \frac{1}{2}\rho A^2 \left[r^2 + 4\frac{\nu}{A} P(\eta) \right], \quad \eta \equiv z\sqrt{\frac{A}{\nu}}$$

The functions $G(\eta)$ and $P(\eta)$ satisfy the following ordinary differential equations and boundary conditions.

$$\frac{d^3 G}{d\eta^3} + 2G\frac{d^2 G}{d\eta^2} - \left(\frac{dG}{d\eta}\right)^2 + 1 = 0$$

$$\frac{dP}{d\eta} = \frac{d^2 G}{d\eta^2} + 2G\frac{dG}{d\eta}$$

$$G(0) = G'(0) = P(0) = 0, \quad G'(\eta) \to 1 \quad \text{as} \quad \eta \to \infty$$

Modify Program **STAGPT** (see Appendix E) as needed to solve the three-dimensional stagnation-point equations. Make graphs to compare the two-dimensional and three-dimensional $G(\eta)$, $G'(\eta)$ and $P(\eta)$. Also, compare the two-dimensional and three-dimensional stagnation-layer thicknesses, defined as the distance above the surface where the horizontal velocity is 99% of its inviscid value.

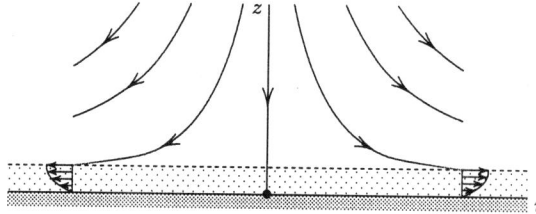

Problem 13.42

13.43 In the exact incompressible, Navier-Stokes solution for convergent and divergent channel flow (see Problem 13.32), the velocity is purely radial and is given by

$$u_r = \frac{\nu}{r}f(\theta), \quad u_\theta = 0, \quad w = 0$$

where ν is the kinematic viscosity of the fluid and $f(\theta)$ is a function to be determined. Substitution into the equations of motion shows that $f(\theta)$ satisfies the following ordinary differential equation and boundary conditions.

$$\frac{d^2 f}{d\theta^2} + 4f + f^2 + K = 0, \quad f(\pm\alpha) = 0$$

The quantity K is the (constant) dimensionless pressure-gradient parameter. Modify Program **STAGPT** (see Appendix E) as needed to solve the convergent ($K > 0$) and divergent ($K < 0$) channel-flow equations. Generate solutions for several values of K, including the case $K = 0$, and determine the maximum and minimum values of K needed to obtain $f(0) = 5000$. Make a graph of $f(\theta)/f(0)$ for several values of K. Be sure to use Richardson extrapolation on one of your computations to establish numerical error bounds.

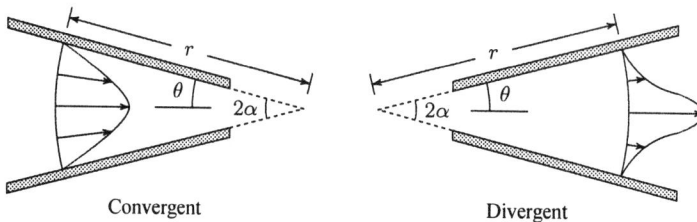

Convergent   Divergent

Problem 13.43

13.44 Using the Crank-Nicolson implicit time-marching method, determine the solution for an *Ekman layer* (see Problem 13.27), using equally-spaced grid points. Note first that the equations of motion and boundary conditions assume the following form.

$$-2\Omega v = \nu \frac{d^2 u}{dz^2}, \quad 2\Omega(u - U) = \nu \frac{d^2 v}{dz^2}$$

$$u(0) = v(0) = 0, \quad u(z) \rightarrow U, \quad v(z) \rightarrow 0 \quad \text{as} \quad z \rightarrow \infty$$

The exact solution is

$$u = U \left[1 - e^{-\zeta} \cos \zeta \right], \quad v = U e^{-\zeta} \sin \zeta, \quad \zeta \equiv z \sqrt{\Omega/\nu}$$

To solve numerically, proceed as follows.

(a) Rewrite the momentum equations with unsteady terms appearing on their left-hand sides. Now, recast the unsteady momentum equations and the boundary conditions in dimensionless form, i.e., in terms of u/U, v/U, ζ and Ωt.

(b) Using $u(z, 0) = U$ and $v(z, 0) = 0$ at all interior points for the initial condition, solve using the Crank-Nicolson method. Let the step size be $\Delta\zeta = 0.1$, and let the maximum distance above the surface lie at $\zeta = 10$. Run the computation until the maximum difference between the computed and exact solutions differs by a prescribed tolerance. Be sure to justify your choice of the tolerance level. Plot the velocity components, including data points for the exact solution, to display results of the computation.

(c) Solve the problem again with $\Delta\zeta = 0.05$, and use Richardson extrapolation to estimate solution errors at $\zeta = 0.2$, 0.5, 1.0 and 2.0. Compare with the exact solution as well.

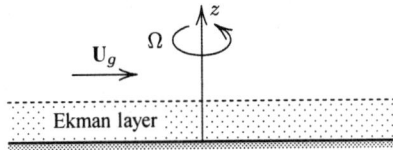

Problem 13.44

13.45 Using the Crank-Nicolson implicit time-marching method, determine the solution for a rotating disk (see Problem 13.33), using equally-spaced grid points. Note first that the transformed equations of motion and boundary conditions assume the following form.

$$\frac{dH}{d\zeta} + 2F = 0, \quad H\frac{dF}{d\zeta} + F^2 - G^2 = \frac{d^2 F}{d\zeta^2}, \quad H\frac{dG}{d\zeta} + 2FG = \frac{d^2 G}{d\zeta^2}, \quad H\frac{dH}{d\zeta} = -\frac{dP}{d\zeta} + \frac{d^2 H}{d\zeta^2}$$

where $\zeta \equiv z\sqrt{\Omega/\nu}$. To solve numerically, proceed as follows.

(a) Rewrite the momentum equations with unsteady terms appearing on their left-hand sides.

(b) Using $F(\zeta, 0) = G(\zeta, 0) = H(\zeta, 0) = P(\zeta, 0) = 0$ at all interior points for the initial condition, solve using the Crank-Nicolson method. Let the step size be $\Delta\zeta = 0.1$, and let the maximum distance above the surface lie at $\zeta = 10$. Run the computation until the maximum residual is reduced to a prescribed value. Be sure to justify your choice of the maximum residual. Plot the dimensionless velocity components, F, G and H to display results of the computation.

(c) Solve the problem again with $\Delta\zeta = 0.05$, and use Richardson extrapolation to estimate solution errors in F, G, H and P at $\zeta = 0.2$, 0.5, 1.0 and 2.0.

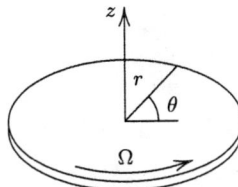

Problem 13.45

Chapter 14

Boundary Layers

The preceding chapter provides several examples of how frictional effects appear in flow over or within solid objects. Confined or **internal flows** such as Couette, channel and pipe flow are examples that are completely dominated by viscous effects. By contrast, unconfined or **external flows** such as Stokes' Problems and stagnation-point flow have viscous effects confined to layers whose thickness is proportional to $\sqrt{\nu}$. In nondimensional terms, this corresponds to having viscous effects confined to a region whose thickness is proportional to $1/\sqrt{Re}$, where Re is an appropriate Reynolds number. Even in an internal flow, if we include the inlet region, we find thin viscous layers that eventually merge on the centerline as the flow becomes fully developed.

As we will see in this chapter, effects of friction are important where velocity gradients are very large, i.e., in regions close to solid surfaces known as boundary layers. Prandtl discovered the boundary layer in 1904, and provided an explanation of how vorticity diffuses from a solid boundary and is swept into the surrounding inviscid flow. This discovery resolved d'Alembert's Paradox, and brought an end to the dominance of the empirically-based era of hydraulics. His explanation for the effect of viscosity and its relation to vorticity established the foundation for theoretical fluid mechanics as we know it today.

Because of the presence of very thin viscous layers, there is a substantial difference in computer resources required for viscous flows as compared to inviscid flows. This has not been apparent from the CFD examples in Chapters 11, 12 and 13, each requiring minuscule amounts of computing time. The examples have been deliberately chosen for their simplicity to illustrate basic concepts of CFD methods, rather than to underscore computational difficulties. In practical applications, because of the need to use finer finite-difference grids and the attendant smaller timesteps for stability and/or accuracy, a viscous-flow computation often requires 10 to 100 times as much computing time as a corresponding inviscid-flow computation.

This chapter shows that by using the boundary-layer or **thin shear layer** approximation, it is possible to arrive at a very accurate numerical solution for viscous flow past a slender body with computation times comparable to those of an inviscid-flow computation. Using this approximation, we first compute inviscid flow past the object. Then, using the inviscid-flow solution to provide the pressure and one boundary condition, we compute the flow in a thin boundary layer. If the boundary layer is thick enough, its effect on the inviscid flow may have to be accounted for by a process of successive approximation. Although complete (time-averaged) Navier-Stokes solutions are becoming more and more common for airplane geometries, this method constitutes one of the most important theoretical tools for airplane design, and continues to be used.

In addition to exploring Prandtl's boundary layer, this chapter includes an introduction to two especially complicated phenomena commonly observed in viscous fluid flows, viz., **separation** and **turbulence**. The section on turbulence includes an interesting home experiment.

14.1   Importance of Friction

Recall that for steady viscous flow of an incompressible fluid, the equations of motion are the continuity equation and the Navier-Stokes equation, i.e.,

$$\nabla \cdot \mathbf{u} = 0 \tag{14.1}$$

$$\mathbf{u} \cdot \nabla \mathbf{u} = -\frac{1}{\rho}\nabla p + \nu \nabla^2 \mathbf{u} \tag{14.2}$$

For a body of characteristic dimension L in a flow of velocity U (Figure 14.1), we can nondimensionalize the velocity according to:

$$\tilde{u} = \frac{u}{U}, \quad \tilde{v} = \frac{v}{U} \quad \Longrightarrow \quad \mathbf{u} = U\tilde{\mathbf{u}} \tag{14.3}$$

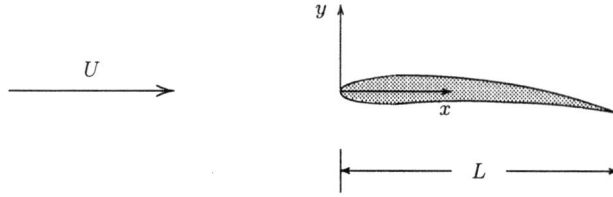

Figure 14.1: *Flow past an airfoil.*

The spatial coordinates and derivatives, in dimensionless form, are

$$\tilde{x} = \frac{x}{L}, \quad \tilde{y} = \frac{y}{L} \quad \Longrightarrow \quad \frac{\partial}{\partial x} = \frac{1}{L}\frac{\partial}{\partial \tilde{x}}, \quad \frac{\partial}{\partial y} = \frac{1}{L}\frac{\partial}{\partial \tilde{y}}, \quad \nabla = \frac{1}{L}\tilde{\nabla} \tag{14.4}$$

Also, let

$$\tilde{p} = \frac{p}{\rho U^2} \tag{14.5}$$

Then, the dimensionless form of the equations of motion for steady, incompressible, viscous flow are as follows.

$$\tilde{\nabla} \cdot \tilde{\mathbf{u}} = 0 \tag{14.6}$$

$$\tilde{\mathbf{u}} \cdot \tilde{\nabla}\tilde{\mathbf{u}} = -\tilde{\nabla}\tilde{p} + \frac{1}{Re_L}\tilde{\nabla}^2\tilde{\mathbf{u}}, \qquad Re_L = \frac{UL}{\nu} \tag{14.7}$$

Since $Re_L \gg 1$ for most flows, we can often neglect viscous effects and use Euler's equation or potential-flow techniques for *most of the flowfield*. However, as discussed in Section 10.2, inviscid solutions fail to predict any drag force resisting the body's motion. Physical observations show that drag exists and has a nontrivial effect, so that effects of friction must enter at some point.

As an example of the importance of drag, consider Figure 14.2. It depicts the flight of a baseball, which has a diameter, D, of 3 inches and a weight, mg, of 5 ounces (the quantity m is the mass of the ball, and g is the acceleration of gravity). The ball is released at a $45°$ angle to the horizontal at a speed of 80 miles per hour.

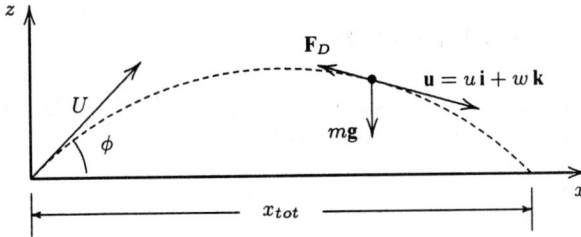

Figure 14.2: *Trajectory of a baseball.*

If we ignore friction, we can predict the motion of the baseball using classical physics. Since the only force acting on the ball is gravity, the equations of motion are

$$\left. \begin{array}{c} m\dfrac{d^2x}{dt^2} = 0 \\[2mm] m\dfrac{d^2z}{dt^2} = -mg \end{array} \right\} \tag{14.8}$$

We must solve subject to the following initial conditions.

$$x(0) = 0, \quad z(0) = 0, \quad u(0) = U\cos\phi, \quad w(0) = U\sin\phi \tag{14.9}$$

where $u = dx/dt$ and $w = dz/dt$ are the velocity components. The solution is readily found to be

$$\left. \begin{array}{c} x = Ut\cos\phi \\[2mm] z = Ut\sin\phi - \dfrac{1}{2}gt^2 \end{array} \right\} \tag{14.10}$$

Now, let's include effects of friction. Figure 14.3 shows the **drag coefficient**, C_D, for a sphere as determined from wind-tunnel experiments. By definition, the drag coefficient for a sphere is

$$C_D = \frac{F_D}{\frac{1}{2}\rho(u^2 + w^2)A} \tag{14.11}$$

where F_D is the drag force and $A = \frac{\pi}{4}D^2$ is cross-sectional area. Hence, the drag on the baseball is

$$F_D = \frac{\pi}{8}\rho D^2 C_D \left(u^2 + w^2\right) \tag{14.12}$$

Now, when the ball is traveling at an angle θ to the horizontal, the drag force in vector form, $\mathbf{F}_D$, is

$$\mathbf{F}_D = -\mathbf{i}\,F_D\cos\theta - \mathbf{k}\,F_D\sin\theta \tag{14.13}$$

However, the angle to the horizontal can be determined from the velocity components. Referring to the insert in Figure 14.3, clearly

$$\cos\theta = \frac{u}{\sqrt{u^2 + w^2}} \quad \text{and} \quad \sin\theta = \frac{w}{\sqrt{u^2 + w^2}} \tag{14.14}$$

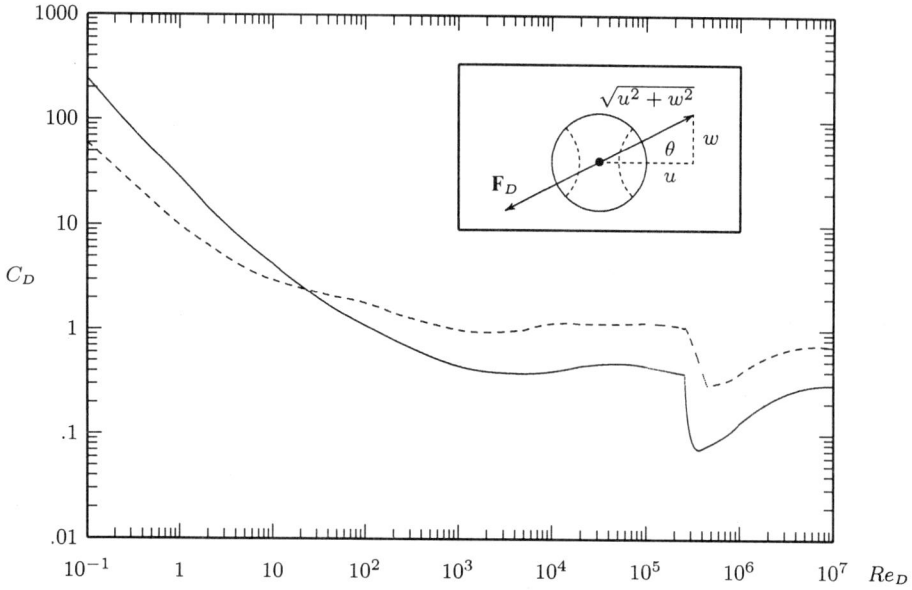

Figure 14.3: *Drag on spheres (——) and cylinders (- - -).*

Therefore, when we include viscous effects, the equations of motion for the baseball become

$$\left. \begin{array}{l} m\dfrac{du}{dt} = -\dfrac{\pi}{8}\rho D^2 C_D u\sqrt{u^2 + w^2} \\[4mm] m\dfrac{dw}{dt} = -\dfrac{\pi}{8}\rho D^2 C_D w\sqrt{u^2 + w^2} - mg \end{array} \right\} \qquad (14.15)$$

The initial conditions [Equation (14.9)] are unchanged.

There are several quantities of interest that can be gleaned from the inviscid solution. In particular, the distance traveled, x_{tot}, time of flight, t_{tot}, and average horizontal velocity, u_{avg}, provide a useful description of the ball's flight. First, to determine how far the ball travels, we note that $z = 0$ when the ball's flight begins and again when it first strikes the ground. From our solution, we have

$$z = 0 \quad \text{when} \quad t = 0 \quad \text{and} \quad t = \frac{2U \sin\phi}{g} \qquad (14.16)$$

Hence, since the second equation for t is the time of flight, the horizontal distance traveled follows from substituting into the solution for x, viz.,

$$x = U\frac{2U \sin\phi}{g}\cos\phi = \frac{U^2}{g}\sin 2\phi \qquad (14.17)$$

The average velocity is the ratio of the total distance traveled to the time of flight, i.e.,

$$u_{avg} = \frac{U^2 \sin 2\phi / g}{2U \sin\phi / g} = U \cos\phi \qquad (14.18)$$

For the conditions given above, we have the following.

$$\left. \begin{array}{ll} x_{tot} & = 428 \text{ feet} \\ t_{tot} & = 5.2 \text{ seconds} \\ u_{avg} & = 57 \text{ mph} \end{array} \right\} \qquad (14.19)$$

With viscous effects included, the initial-value problem cannot be solved in closed form. However, the solution can be generated numerically with standard integration methods. For the specified conditions, we find:

$$\left. \begin{array}{ll} x_{tot} & = 317 \text{ feet} \\ t_{tot} & = 4.8 \text{ seconds} \\ u_{avg} & = 45 \text{ mph} \end{array} \right\} \qquad (14.20)$$

Thus, friction reduces the distance traveled by 26%, the time of flight by 8% and the average velocity by 21%. These are significant reductions, especially the distance traveled, even though the Reynolds number based on the initial velocity and the ball's diameter is $Re_D = 1.8 \cdot 10^5$. To state the reduction in distance traveled in practical terms, consider the following. The inviscid prediction would indicate a major-league baseball player standing at home plate in most modern stadiums could throw a baseball over the center-field fence. By contrast, the viscous computation shows that the same player would not be able to throw the ball over the fence in many stadiums[1] even if he aimed directly down the left-field line!

14.2 Prandtl's Boundary Layer

The example of the preceding section shows that, even when the Reynolds number is very large, viscous effects can have a nontrivial cumulative effect. In 1904, Prandtl found that viscous effects in a high Reynolds number flow are confined to a thin layer near the surface of an object moving through the fluid. He originally called this the **transition layer**. Blasius renamed it the **boundary layer** in his 1908 doctoral thesis.

14.2.1 High Reynolds Number Flow Near a Surface

As pointed out by Schlichting (1979), to illustrate how viscous effects can have such a profound effect at very large Reynolds number, Prandtl discussed the solution to a differential equation of the following form.

$$\epsilon \frac{d^2 u}{dy^2} + (1 + \epsilon) \frac{du}{dy} + y = 0, \qquad 0 \le y \le 1 \qquad (\epsilon \ll 1) \qquad (14.21)$$

This equation is to be solved subject to the boundary conditions

$$u(0) = 0 \quad \text{and} \quad u(1) = 1 \qquad (14.22)$$

Prandtl offered this two-point boundary value problem as a simplified analog of the Navier-Stokes equation.[2] The second-derivative term has a small coefficient just as the second-derivative term in the Navier-Stokes equation, in nondimensional form, has the reciprocal of the Reynolds number as its coefficient. An immediate consequence is that only one boundary condition can be satisfied if we set $\epsilon = 0$. This is similar to setting viscosity to zero in the Navier-Stokes equation, which yields Euler's equation, and the attendant consequence that only the normal-velocity surface boundary condition can be satisfied. That is, we cannot enforce the no-slip boundary condition for Euler-equation solutions.

---

[1]Dodger Stadium is a counter example, measuring 295 feet down the foul lines.
[2]Strictly speaking, Prandtl's model equation did not include the $(1 + \epsilon)$ factor – it has been included here to simplify the algebra.

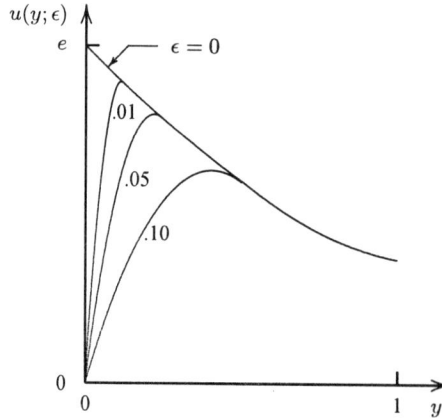

Figure 14.4: *Solutions to the model equation for several values of ϵ.*

The exact solution to this equation is

$$u(y;\epsilon) = \frac{e^{1-y} - e^{1-y/\epsilon}}{1 - e^{1-1/\epsilon}} \tag{14.23}$$

If we set $\epsilon = 0$ in Equation (14.21), we have the following first-order equation:

$$\frac{du}{dy} + y = 0 \tag{14.24}$$

and the solution, $u(y;0)$, is

$$u(y;0) = e^{1-y} \tag{14.25}$$

where we use the boundary condition at $y = 1$. Provided $y \gg \epsilon$, this solution is the limiting form of Equation (14.23). To see this, simply note that $e^{-y/\epsilon} \to 0$ as $\epsilon \to 0$. However, the solution obviously fails to satisfy the boundary condition at $y = 0$ because the solution gives $u(0;0) = e = 2.71828\cdots$. Figure 14.4 displays the solution to our simplified equation for several values of ϵ.

As shown, the smaller the value of ϵ, the more closely $u(y;0)$ represents the solution throughout the region $0 < y \leq 1$. Only in the immediate vicinity of $y = 0$ is the solution inaccurate. We describe the thin layer where $u(y;0)$ departs from the exact solution a **boundary layer**.

This simple analogy displays a key feature of solutions to the Navier-Stokes equation for *slightly-viscous* fluids, flowing over a solid boundary. We call a fluid slightly viscous whenever its kinematic viscosity, ν, is very small compared to UL, so that the Reynolds number is very large. In such a fluid, we indeed find in practice that viscous effects are confined to thin layers near the boundary.

14.2.2   Boundary-Layer Equations

In a boundary layer, Prandtl found that flow properties vary much more rapidly in the direction normal to the surface, y, than in the direction parallel to the surface, x (Figure 14.5). In fact, variations are so rapid that the convective acceleration, $u\partial u/\partial x$, and the viscous term, $\nu\partial^2 u/\partial y^2$, are of the same order of magnitude. To understand Prandtl's reasoning, we will consider flow past the flat plate depicted in the figure. The dashed line delineates the "edge" of the viscous region, which is assumed to have thickness $\delta(x)$. We do not have to define $\delta(x)$ exactly at this point.

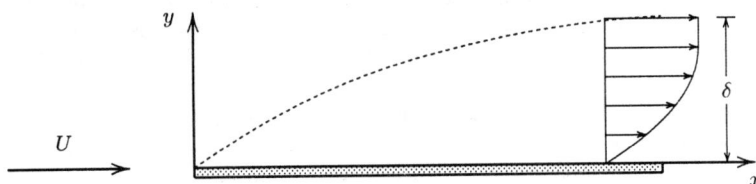

Figure 14.5: *Viscous flow over a solid surface.*

We begin with the continuity and Navier-Stokes equations. Confining our attention to steady, incompressible, two-dimensional flow, we have

$$\frac{\partial u}{\partial x} + \frac{\partial v}{\partial y} = 0 \tag{14.26}$$

$$u\frac{\partial u}{\partial x} + v\frac{\partial u}{\partial y} = -\frac{1}{\rho}\frac{\partial p}{\partial x} + \nu\left(\frac{\partial^2 u}{\partial x^2} + \frac{\partial^2 u}{\partial y^2}\right) \tag{14.27}$$

$$u\frac{\partial v}{\partial x} + v\frac{\partial v}{\partial y} = -\frac{1}{\rho}\frac{\partial p}{\partial y} + \nu\left(\frac{\partial^2 v}{\partial x^2} + \frac{\partial^2 v}{\partial y^2}\right) \tag{14.28}$$

As discussed in Section 13.6, the two-dimensional Navier-Stokes equation is of fourth order, so that four boundary conditions are needed to have a well-posed problem. Two boundary conditions follow from the no-slip boundary condition at the surface, which tells us u and v vanish at $y = 0$. Additionally, the solution must approach the freestream velocity, $u = U$ and $v = 0$ as $y \to \infty$. Thus,

$$\left.\begin{array}{l} u = 0, \quad v = 0, \quad y = 0 \\[2mm] u \to U, v \to 0, \quad y \to \infty \end{array}\right\} \tag{14.29}$$

Prandtl used order-of-magnitude estimates to develop an approximate set of equations valid in the boundary layer. His approach was heuristic, and involved excellent engineering judgment guided by experimental observations. We will follow his arguments here, noting that they are approximate and lacking in rigor. Nevertheless, the equations are well founded, and represent an excellent approximation for **attached**, or unseparated, flows. We will discuss what happens when the flow undergoes separation in Subsection 14.2.6. It is possible to deduce these equations with advanced mathematical techniques. Modern developments in applied mathematics, most notably in **singular perturbation theory** [cf. Van Dyke (1975), Kevorkian and Cole (1981), Neyfeh (1981) or Wilcox (1995)], provide a solid foundation for development of the **boundary-layer equations**.

We assume that the viscous layer is very thin, so that we can say $\delta(x) \ll x$. Anticipating that properties change over a much shorter distance in the vertical direction, δ, than in the horizontal direction, x, we expect to have $\partial f/\partial y \gg \partial f/\partial x$ for any property f. Additionally, we estimate the order of magnitude of such derivatives to be given by

$$\frac{\partial f}{\partial x} \sim \frac{F}{x} \quad \text{and} \quad \frac{\partial f}{\partial y} \sim \frac{F}{\delta} \tag{14.30}$$

where F is a measure of the change in f. The notation "$\sim$" means order of magnitude, and does not imply equality.[3] The idea behind these estimates is that if a quantity changes by an amount F over a given distance (x or δ here), then its derivative is proportional to the ratio of F to that distance. The closer the function is to being linear, the closer the proportionality coefficient is to one. Although the factor will differ from unity for general functions, these estimates still provide a reasonable estimate of the order of magnitude of a derivative.

Mass Conservation. Since the horizontal velocity, u, changes from 0 to U as we move from the plate surface to the freestream, we can plausibly estimate the order of magnitude of $\partial u/\partial x$ as

$$\frac{\partial u}{\partial x} \sim \frac{U}{x} \tag{14.31}$$

Similarly, if we denote the change in vertical velocity by V, an estimate of its derivative across the boundary layer is given by

$$\frac{\partial v}{\partial y} \sim \frac{V}{\delta} \tag{14.32}$$

So, we can estimate the order of magnitude of the terms in the continuity Equation (14.26) as follows.

$$\underbrace{\frac{\partial u}{\partial x}}_{U/x} + \underbrace{\frac{\partial v}{\partial y}}_{V/\delta} = 0 \tag{14.33}$$

Therefore, the order of magnitude of the vertical velocity is

$$\frac{V}{\delta} \sim \frac{U}{x} \quad \Longrightarrow \quad V \sim \frac{\delta}{x}U \ll U \tag{14.34}$$

Hence, the vertical velocity is very small relative to the horizontal velocity in a boundary layer. For typical boundary layers, we find that v is positive at the boundary-layer edge, corresponding to fluid being displaced from the surface as fluid moves over the plate in the streamwise direction.[4] Thus, having $v \ll u$ is consistent with having $\delta \ll x$. By contrast, if v were comparable to u, the edge of the boundary layer would spread at a finite angle, which would correspond to $\delta \sim x$. This would contradict our basic assumption that the boundary layer is very thin, which tells us it is plausible to have $v \ll u$.

Horizontal Momentum Conservation. Using Equation (14.34) for the vertical velocity, and denoting the order of magnitude of the change in pressure by P, the various terms in the x-momentum Equation (14.27) are

$$\underbrace{u\frac{\partial u}{\partial x}}_{U^2/x} + \underbrace{v\frac{\partial u}{\partial y}}_{U^2/x} = \underbrace{-\frac{1}{\rho}\frac{\partial p}{\partial x}}_{P/\rho x} + \underbrace{\nu\frac{\partial^2 u}{\partial x^2}}_{\nu U/x^2} + \underbrace{\nu\frac{\partial^2 u}{\partial y^2}}_{\nu U/\delta^2} \tag{14.35}$$

We can make three immediate observations regarding the relative importance of the terms in this equation.

1. $\nu U/x^2 \ll \nu U/\delta^2$ because $\delta \ll x$. Thus, we can ignore the term $\nu \partial^2 u/\partial x^2$.

---

[3]In the spirit of singular perturbation theory, the $\sim$ symbol means the ratio of the terms on each side of the $\sim$ approach a finite, nonzero value in the limit $\delta \to 0$, i.e., their ratio is bounded.

[4]In fact, we will find that at the edge of the boundary layer, necessarily $v < u d\delta/dx$ when the layer is growing in thickness. Even when it decreases in thickness, we have $v \sim u d\delta/dx$ at the boundary-layer edge.

2. Since we know from experiments that pressure gradient has a significant effect on a boundary layer, we expect the pressure gradient term to be of the same order of magnitude as the acceleration terms. Hence, $U^2/x \sim P/\rho x$, so that $P \sim \rho U^2$. This is consistent with Bernoulli's equation, which is valid above the boundary layer.

3. In order for viscous effects to be significant, the term $\nu \partial^2 u/\partial y^2$ must be of the same order of magnitude as the convective terms. Hence, we must have $U^2/x \sim \nu U/\delta^2$, wherefore the thickness of the boundary layer is given by

$$\delta(x) \sim \sqrt{\frac{\nu x}{U}} \qquad (14.36)$$

This result tells us that the smaller the viscosity, ν, the thinner the boundary layer will be. To cast this in quantitative terms, we first rewrite the boundary-layer thickness in terms of Reynolds number based on x, i.e., $Re_x = Ux/\nu$. There follows:

$$\delta \sim \frac{x}{\sqrt{Re_x}} \qquad (14.37)$$

Therefore, if the Reynolds number is $Re_x = 10^7$, then the boundary-layer thickness is given by $\delta \sim x/3000$. According to this estimate, the boundary layer on a 20 foot automobile moving at 55 mph would grow to about a tenth of an inch. In practice, the boundary layer on a 20 foot automobile grows to about 3 inches at this speed. However, this is not a failure of our estimate in Equation (14.37). The boundary layer is thicker because it is turbulent for a Reynolds number this large (see Section 14.3).

Vertical Momentum Conservation. Using the estimates for vertical velocity, $V \sim U\delta/x$, and pressure, $P \sim \rho U^2$, the orders of magnitude of the terms in the y-momentum Equation (14.28) are

$$\underbrace{u\frac{\partial v}{\partial x}}_{U^2\delta/x^2} + \underbrace{v\frac{\partial v}{\partial y}}_{U^2\delta/x^2} = \underbrace{-\frac{1}{\rho}\frac{\partial p}{\partial y}}_{U^2/\delta} + \underbrace{\nu\frac{\partial^2 v}{\partial x^2}}_{\nu U\delta/x^3} + \underbrace{\nu\frac{\partial^2 v}{\partial y^2}}_{\nu U/x\delta} \qquad (14.38)$$

As with horizontal momentum, we can make three immediate observations regarding the relative importance of the terms in this equation.

1. $\nu U\delta/x^3 \ll \nu U/x\delta$ because $\delta \ll x$. Thus, we can ignore the viscous term $\nu \partial^2 v/\partial x^2$.

2. Since the boundary-layer thickness is given by Equation (14.37), necessarily

$$u\frac{\partial u}{\partial x} \sim v\frac{\partial v}{\partial y} \sim \nu\frac{\partial^2 v}{\partial y^2} \sim \frac{U^2\delta}{x^2}$$

3. Because $\delta \ll x$, necessarily $U^2\delta/x^2 \ll U^2/\delta$. This means the acceleration and viscous terms are of a much smaller order of magnitude than the pressure gradient across the boundary layer. All but the normal pressure-gradient term, $\partial p/\partial y$, are unimportant in Equation (14.28), so that the vertical momentum equation simplifies to

$$0 \approx -\frac{1}{\rho}\frac{\partial p}{\partial y}$$

On reflection, the third point may seem a little odd. After all, we are, in effect, saying that $\partial p/\partial y$ is enormous compared to all other terms in the y-momentum equation. Using this information, we then conclude that it must be zero! There is no contradiction, however, and the physical meaning of this result is the following. On the one hand, the magnitude of the pressure, $p \sim \rho U^2$, is established from the balance of forces in the horizontal-momentum equation. On the other hand, any change in pressure across the boundary layer must be comparable in magnitude to the convective and viscous terms in the vertical momentum equation, which are substantially smaller than U^2/δ. Therefore, as a first approximation, the pressure must be constant across a boundary layer, so that $\partial p/\partial y \approx 0$.

Since mass conservation [Equation (14.26)] remains unaltered, the following approximate set of equations, known as **Prandtl's boundary-layer equations**, can be used to describe the flow in a boundary layer.

$$\frac{\partial u}{\partial x} + \frac{\partial v}{\partial y} = 0 \tag{14.39}$$

$$u\frac{\partial u}{\partial x} + v\frac{\partial u}{\partial y} = -\frac{1}{\rho}\frac{\partial p}{\partial x} + \nu\frac{\partial^2 u}{\partial y^2} \tag{14.40}$$

$$0 = -\frac{1}{\rho}\frac{\partial p}{\partial y} \tag{14.41}$$

Although we have developed the boundary-layer equations in a planar geometry, they hold as written in general body-oriented coordinates. That is, these equations are valid along the surface of a general geometry where x becomes arc length and y becomes distance normal to the surface (see Figure 14.6). This will be true in general because the boundary layer is very thin for most applications relative to the radius of curvature, $\mathcal{R}$, of the surface. There are exceptions, of course, and when the boundary-layer thickness, δ, is comparable in magnitude to $\mathcal{R}$, the boundary layer is said to be "thick." The boundary-layer equations as quoted above must be modified for thick boundary-layer applications [cf. Van Dyke (1975) or Emanuel (1994)]. Equations (14.39) through (14.41) are restricted to the classical case of a **thin shear layer**, which covers the vast majority of everyday applications.

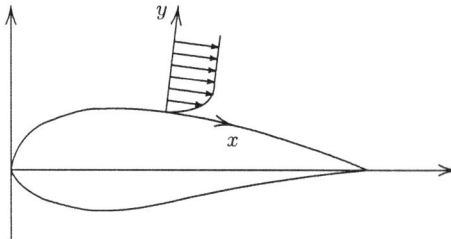

Figure 14.6: *Boundary-layer coordinates; x is arc length and y is normal distance.*

Because Equation (14.41) tells us the pressure is independent of y, necessarily the pressure is constant across a boundary layer. Consequently, the pressure can be determined from an inviscid solution. That is, we can solve Euler's equation or use potential-flow methods to determine the inviscid solution, including the pressure at all points in the flow. Then, the pressure in the boundary layer is given by the surface value of p according to the inviscid solution. The methodology for computing flow past a slender body whose surface can be described in conventional Cartesian coordinates by $y = y_b(x)$ is as follows.

1. Solve Euler's equation for flow past the body. Part of the solution is $p_{Euler}(x, y)$.

2. Compute the boundary-layer pressure from $p(x) = p_{Euler}(x, y_b)$.

3. Solve the boundary-layer equations using $p(x)$.

This is a great simplification from a computational point of view. Inviscid solutions are far less computationally intensive than Navier-Stokes solutions, mainly because of the need to resolve thin boundary layers for the latter. Similarly, because the pressure is known in advance (from the inviscid solution), the boundary-layer equations are also much easier to solve numerically than the full Navier-Stokes equation. The boundary-layer equations are accurate[5] provided the pressure gradient, dp/dx, is not strong enough to cause separation of the boundary layer. As noted earlier, the boundary-layer equations are strictly valid only for **attached flows**, i.e., flows with no separation. When separation occurs, only a complete solution to the Navier-Stokes equation is sufficient.

Since the pressure, $p(x)$, is a known function in the boundary-layer equations, we can arrive at an equation for the streamfunction, ψ, directly from Equation (14.40). The resulting equation is

$$\nu \frac{\partial^3 \psi}{\partial y^3} = \frac{1}{\rho} \frac{dp}{dx} + \frac{\partial \psi}{\partial y} \frac{\partial^2 \psi}{\partial x \partial y} - \frac{\partial \psi}{\partial x} \frac{\partial^2 \psi}{\partial y^2} \qquad (14.42)$$

The most remarkable feature of this equation is that it is of third order. This stands in contrast to the equation satisfied by ψ for the full, two-dimensional, Navier-Stokes equation, which is of fourth order [see Equation (13.143)]. Consequently, we must give up one of the boundary conditions of Equation (14.29). We retain the no-slip condition so that u and v still vanish at the surface. Also, we still insist that $u \to U$ as $y \to \infty$. Thus, we impose no condition on the vertical velocity in the freestream, and the appropriate boundary conditions for the boundary-layer equations are

$$\left. \begin{array}{ll} u = 0, \quad v = 0, \quad y = 0 \\[2mm] u \to U, \qquad\qquad y \to \infty \end{array} \right\} \qquad (14.43)$$

Note that this means we have no control over the vertical velocity as $y \to \infty$ in a boundary-layer computation — it is part of the solution.

Prior to the advent of very fast computers, great advances in aerodynamics and aviation occurred using this procedure, i.e., combined inviscid solutions and boundary-layer theory. Because most aerodynamic and hydrodynamic devices are designed to move smoothly through the fluid with minimal resistance, flow separation is usually avoided at the designated operating condition. Thus, for flow about a wing, the classical inviscid-flow/boundary-layer approach is generally sufficient for design purposes, and is still used at aircraft companies.

14.2.3 Momentum Integral Equation

In this subsection, we will use the control volume method to analyze flow in a boundary layer. While returning to the control volume methodology may appear to run counter to the aims of the differential approach that is now our primary focus, there is method in our madness. The result we obtain, known as the **momentum integral equation**, is particularly useful in both experimental and theoretical work. Additionally, the equation has played a central role

---

[5]Although this method typically overestimates the lift of an airfoil by about 10%, accounting for the displacement effect (discussed in the next subsection), removes most of the difference between theory and experiment.

in the development of boundary-layer theory, and helps illuminate some aspects of the nature of boundary layers.

Figure 14.7 shows a segment of a growing boundary layer and suitably chosen control-volume boundaries. We consider two streamwise locations, $x_1 = x$ and $x_2 = x + \Delta x$, where Δx is a small distance. The boundary-layer thickness is $\delta_1 \equiv \delta(x)$ at the upstream boundary of the control volume and $\delta_2 \equiv \delta(x + \Delta x)$ at the downstream boundary. The quantities u_e and v_e are the velocity components at the edge of the boundary layer, and are defined by

$$u_e \equiv u(x, \delta) \quad \text{and} \quad v_e \equiv v(x, \delta) \tag{14.44}$$

The surface shear stress is denoted by τ_w, and will, in general, vary with x. For simplicity, we will consider the constant-pressure case. We also assume the flow is steady, incompressible and two dimensional.

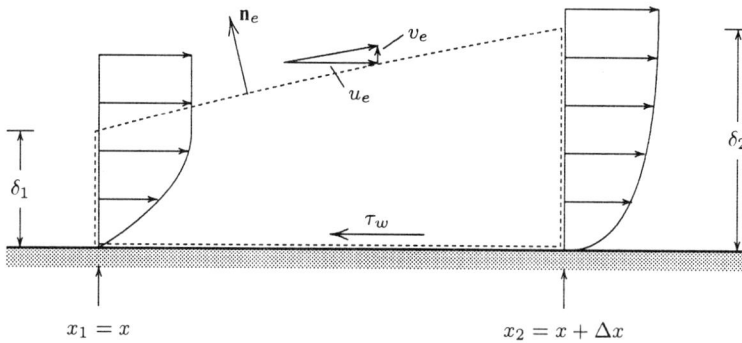

Figure 14.7: *A two-dimensional boundary layer.*

Mass Conservation. Beginning with mass conservation, we know that

$$\oiint_S \rho \mathbf{u} \cdot \mathbf{n} \, dS = 0 \tag{14.45}$$

Noting that there is no flow through the solid boundary, the closed surface integral consists of contributions from the upstream boundary ($x = x_1$), the downstream boundary ($x = x_2$), and the boundary-layer edge. At the upstream boundary, we have $\mathbf{u} \cdot \mathbf{n} = -u$, and at the downstream boundary, $\mathbf{u} \cdot \mathbf{n} = u$. At the boundary-layer edge, the unit normal is $\mathbf{n}_e$, and the velocity is equal to the freestream value. Consistent with the boundary-layer approximation, $d\delta/dx \ll 1$, the unit normal to the upper boundary is

$$\mathbf{n}_e \approx \mathbf{j} - \mathbf{i}\frac{d\delta}{dx} \tag{14.46}$$

Therefore, the value of $\mathbf{u} \cdot \mathbf{n}$ on the upper boundary of the control volume is

$$(\mathbf{u} \cdot \mathbf{n})_e \approx v_e - u_e\frac{d\delta}{dx} \tag{14.47}$$

This is the **entrainment velocity**, and it is negative when the boundary layer is growing. While v_e is generally positive, the quantity $u_e d\delta/dx$ is larger than v_e so that fluid is drawn into the boundary layer. The opposite is true, of course, if the boundary layer is decreasing in thickness. So, expanding the closed surface integral into its three contributions, we find

$$-\int_0^{\delta_1} \rho u\, dy + \int_0^{\delta_2} \rho u\, dy + \int_{x_1}^{x_2} \rho\left(v_e - u_e \frac{d\delta}{dx}\right) dx = 0 \tag{14.48}$$

which can be rewritten as

$$\frac{\int_0^{\delta(x+\Delta x)} \rho u\, dy - \int_0^{\delta(x)} \rho u\, dy}{\Delta x} + \frac{1}{\Delta x}\int_x^{x+\Delta x} \rho\left(v_e - u_e \frac{d\delta}{dx}\right) dx = 0 \tag{14.49}$$

Taking the limit as $\Delta x \to 0$, the final form of the mass-conservation principle for the boundary layer is:

$$\frac{d}{dx}\int_0^\delta \rho u\, dy + \rho\left(v_e - u_e \frac{d\delta}{dx}\right) = 0 \tag{14.50}$$

The physical interpretation of this equation is obvious. It says the rate of change of mass flux through the boundary layer is balanced by the rate at which fluid is entrained from the freestream. Clearly, when the entrainment velocity, $(\mathbf{u} \cdot \mathbf{n})_e$, is negative, Equation (14.50) tells us the mass flux increases, so that the boundary layer grows in thickness. We will use this equation to simplify the momentum equation below.

Momentum Conservation. The x component of the momentum equation is

$$\oiint_S \rho u(\mathbf{u} \cdot \mathbf{n})dS = -\mathbf{i} \cdot \oiint_S (p - p_\infty)\mathbf{n}\, dS - \mathbf{i} \cdot \oiint_S \mathbf{n} \cdot [\tau]\, dS \tag{14.51}$$

where p_∞ is the freestream pressure. Because pressure is constant across a boundary layer, necessarily $p = p_\infty$ throughout the control volume and on the entire bounding surface. Therefore, the net pressure integral vanishes. Since the entrainment velocity is generally nonvanishing so that $\rho u(\mathbf{u} \cdot \mathbf{n}) = \rho u_e(\mathbf{u} \cdot \mathbf{n})_e$ at the boundary-layer edge, the momentum-flux and viscous-stress integrals expand to yield the following equation.

$$-\int_0^{\delta_1} \rho u^2 dy + \int_0^{\delta_2} \rho u^2 dy + \int_{x_1}^{x_2} \rho u_e\left(v_e - u_e \frac{d\delta}{dx}\right) dx = -\int_{x_1}^{x_2} \tau_w dx \tag{14.52}$$

Hence, regrouping terms, we have

$$\frac{\int_0^{\delta(x+\Delta x)} \rho u^2 dy - \int_0^{\delta(x)} \rho u^2 dy}{\Delta x} + \frac{1}{\Delta x}\int_x^{x+\Delta x} \rho u_e\left(v_e - u_e \frac{d\delta}{dx}\right) dx$$
$$= -\frac{1}{\Delta x}\int_x^{x+\Delta x} \tau_w\, dx \tag{14.53}$$

Thus, taking the limit $\Delta x \to 0$, the momentum equation becomes:

$$\frac{d}{dx}\int_0^\delta \rho u^2 dy + \rho u_e\left(v_e - u_e \frac{d\delta}{dx}\right) = -\tau_w \tag{14.54}$$

Again, the physical interpretation of this equation is straightforward. The rate of change of the momentum flux through the boundary layer is balanced by the rate at which momentum is entrained from the freestream, $-\rho u_e(v_e - u_e d\delta/dx)$, and the surface shear stress.

We can rearrange Equation (14.54) to arrive at the classical result known as the momentum integral equation. To accomplish this end, we begin by substituting for the entrainment velocity from Equation (14.50).

$$\frac{d}{dx} \int_0^\delta \rho u^2 \, dy - u_e \frac{d}{dx} \int_0^\delta \rho u \, dy = -\tau_w \tag{14.55}$$

Now, recall that Euler's equation holds at the edge of the boundary layer, wherefore

$$\rho u_e \frac{du_e}{dx} = -\frac{dp}{dx} \tag{14.56}$$

Hence, since the pressure is constant, necessarily the freestream velocity is also constant. Consequently, we can rewrite Equation (14.55) as

$$\frac{d}{dx} \int_0^\delta \rho \left(u^2 - u_e u \right) dy = -\tau_w \tag{14.57}$$

Dividing Equation (14.57) through by ρu_e^2, there follows:

$$\frac{d}{dx} \int_0^\delta \frac{u}{u_e} \left(1 - \frac{u}{u_e} \right) dy = \frac{\tau_w}{\rho u_e^2} = \frac{c_f}{2} \tag{14.58}$$

Thus, we arrive at the momentum integral equation for a constant-pressure boundary layer, viz.,

$$\frac{d\theta}{dx} = \frac{c_f}{2} \tag{14.59}$$

where the quantity θ is the **momentum thickness** defined by

$$\theta \equiv \int_0^\delta \frac{u}{u_e} \left(1 - \frac{u}{u_e} \right) dy \tag{14.60}$$

Had we allowed for variable pressure, a more general version of the equation results. The complete incompressible, two-dimensional form of the equation is [see Schlichting (1979)]

$$\frac{d\theta}{dx} + (2 + H) \frac{\theta}{u_e} \frac{du_e}{dx} = \frac{c_f}{2} \tag{14.61}$$

where H is the **shape factor** defined by

$$H = \frac{\delta^*}{\theta} \tag{14.62}$$

and δ^* is the **displacement thickness**, i.e.,

$$\delta^* \equiv \int_0^\delta \left(1 - \frac{u}{u_e} \right) dy \tag{14.63}$$

The quantities c_f, θ, δ^* and H are collectively referred to as **integral parameters**. The displacement thickness is of especial interest in boundary-layer theory. It has a straightforward physical interpretation, reminiscent of the effective Rayleigh-layer thickness discussed in Section 13.4 [see Equation (13.110)]. Figure 14.8 illustrates the physical meaning of the displacement thickness.

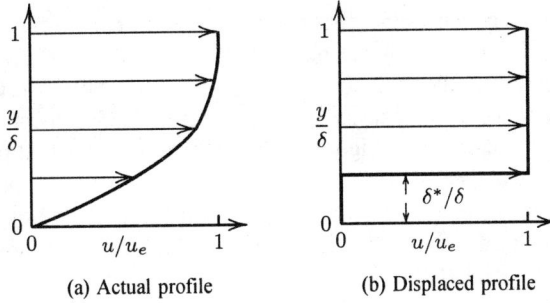

(a) Actual profile   (b) Displaced profile

Figure 14.8: *Physical meaning of displacement thickness, δ^*.*

Figure 14.8(a) shows a typical velocity profile. For incompressible flow, the mass flux per unit width (out of the page) for this profile, $\dot{m}_a$, is

$$\dot{m}_a = \rho \int_0^\delta u \, dy \tag{14.64}$$

Now, consider the velocity profile of Figure 14.8(b), for which the velocity is constant and equal to the freestream value above $y = \delta^*$, and zero below. The mass flux per unit width for this profile, $\dot{m}_b$, is

$$\dot{m}_b = \rho u_e \left(\delta - \delta^* \right) = \rho \int_0^\delta u_e \, dy - \rho u_e \delta^* \tag{14.65}$$

In order to have the same mass flux for both profiles, we conclude that

$$\dot{m}_a = \dot{m}_b \quad \Longrightarrow \quad \rho \int_0^\delta u_e \, dy - \rho u_e \delta^* = \rho \int_0^\delta u \, dy \tag{14.66}$$

With a little rearrangement of terms, the value of δ^* is

$$\delta^* \equiv \int_0^\delta \left(1 - \frac{u}{u_e} \right) dy \tag{14.67}$$

which is identical to the definition of displacement thickness given in Equation (14.63). Therefore, the displacement thickness is the distance viscous effects appear to displace the inviscid flow from the surface. This phenomenon is generally referred to as the **displacement effect** (Figure 14.9). When the displacement thickness becomes large, e.g., near a separation point, the effective shape of the body seen by the inviscid flow is quite different from the actual shape. This causes significant changes in the pressure at the surface of the body, and can change the entire flowfield as we have seen for flow past a cylinder (see Figure 11.12).

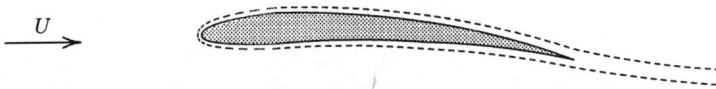

Figure 14.9: *Illustration of the displacement effect; the dashed contour shows the effect of adding δ^* to the actual airfoil shape.*

We saw another ramification of the displacement effect in our discussion of airfoil theory. Recall that at the end of Section 11.8, we noted that the theoretical effect of airfoil thickness according to Kutta-Joukowski theory is an increase in the lift coefficient, which is at variance with measured effects of airfoil thickness. Because of the displacement effect, the effective airfoil shape changes as illustrated in Figure 14.9. As a result, the actual pressure distribution differs from potential-flow theory in a way that is strongly dependent on airfoil shape. So, the actual Kutta-Joukowski shape is not realized in a real fluid, thus mitigating the theoretical results. As a general rule of thumb, the displacement effect usually **reduces** the lift, relative to the inviscid value, by about 10%.

The momentum thickness defined in Equation (14.60) does not lend itself to quite so straightforward a physical interpretation. However, it should be clear that it is a quantitative measure of the momentum lost to friction.

Equation (14.61) served as the foundation for boundary-layer theory until the early 1970's. It has given way to numerical solution of the boundary-layer equations, which provides more detailed information. The equation nevertheless still provides some interesting insight into the dynamics of the boundary layer. Also, it is useful in both experimental and computational work. On the one hand, the experimenter can measure velocity profiles at several values of x, compute θ for each profile, and use Equation (14.61) to infer the skin friction. Alternatively, if the skin friction has been measured also, Equation (14.61) can be used as a consistency check. Similarly, a computational fluid dynamicist can use this equation to check for numerical accuracy.

As an example of the various integral parameters and the momentum integral equation, consider stagnation-point flow, for which $u_e = Ax$. Let's assume an approximation to the velocity profile given by

$$\frac{u}{u_e} = \left(2 - \frac{y}{\delta}\right)\frac{y}{\delta} \tag{14.68}$$

Also, assume we have no information regarding the boundary-layer thickness, $\delta(x)$. We can use the momentum integral equation to set up a differential equation for $\delta(x)$, that can be solved in closed form. From the velocity profile, we can compute the skin friction directly. A short calculation shows that

$$c_f = \frac{\mu(\partial u/\partial y)_{y=0}}{\frac{1}{2}\rho u_e^2} = \frac{4\nu}{Ax\delta} \tag{14.69}$$

Using the definitions of δ^*, θ and H, we also find

$$\delta^* = \frac{1}{3}\delta, \qquad \theta = \frac{2}{15}\delta, \qquad H = \frac{5}{2} \tag{14.70}$$

Substitution of these integral parameters into the general form of the momentum integral equation [Equation (14.61)], we obtain

$$\frac{2}{15}\frac{d\delta}{dx} + \frac{9}{2}\frac{2}{15}\frac{\delta}{x} = \frac{2\nu}{Ax\delta} \tag{14.71}$$

Rearranging terms a bit, this equation can be rewritten as

$$x\frac{d\delta^2}{dx} + 9\delta^2 = \frac{30\nu}{A} \tag{14.72}$$

As can be verified by direct substitution, the general solution to this equation is

$$\delta^2(x) = \frac{10}{3}\frac{\nu}{A} + Cx^{-9} \tag{14.73}$$

where C is an integration constant. Far downstream, the term proportional to x^{-9} will be negligible so that we can ignore it. Thus, the thickness of this boundary layer is

$$\delta = \sqrt{\frac{10}{3}\frac{\nu}{A}} \approx 1.83\sqrt{\frac{\nu}{A}} \tag{14.74}$$

Comparison with the exact solution given in Section 13.6 shows that our inferred thickness is 24% smaller than the exact value [cf. Equation (13.172)]. Had we used a little more complicated velocity profile, say a cubic variation of u/u_e with y/δ in place of the assumed quadratic, the coefficients could be adjusted to exactly match the exact thickness, in the case where we might know what it is.

14.2.4 Blasius Solution

The boundary-layer Equations (14.39) - (14.41) have an exact solution for constant pressure, known as the **Blasius solution**. Constant pressure corresponds to a flat plate that is aligned with the freestream velocity vector (see Figure 14.5). We solve the continuity and boundary-layer equations with zero pressure gradient subject to the no-slip boundary condition and the relevant freestream condition, so that the problem is

$$\frac{\partial u}{\partial x} + \frac{\partial v}{\partial y} = 0 \tag{14.75}$$

$$u\frac{\partial u}{\partial x} + v\frac{\partial u}{\partial y} = \nu\frac{\partial^2 u}{\partial y^2} \tag{14.76}$$

$$u = v = 0 \quad \text{at} \quad y = 0 \quad \text{and} \quad u \to U \quad \text{as} \quad y \to \infty \tag{14.77}$$

This problem was the first example demonstrating Prandtl's boundary-layer theory, and was solved by H. Blasius in his Ph.D. thesis. Blasius found that, like Stokes' First Problem, the constant-pressure boundary layer is self similar. The solution is most conveniently obtained in terms of the streamfunction, which assumes the following form.

$$\psi(x, y) = \sqrt{2U\nu x}\, G(\eta), \qquad \eta \equiv y\sqrt{\frac{U}{2\nu x}} \tag{14.78}$$

Substitution into the equations of motion and the boundary conditions yields the following two-point boundary value problem for the dimensionless streamfunction, $G(\eta)$.

$$\frac{d^3 G}{d\eta^3} + G\frac{d^2 G}{d\eta^2} = 0 \tag{14.79}$$

$$\left.\begin{array}{ll} G = 0, \quad G' = 0, \quad \eta = 0 \\[2mm] G' \to 1, \qquad\qquad \eta \to \infty \end{array}\right\} \tag{14.80}$$

Although the solution for $G(\eta)$ cannot be written in closed form, it is readily obtained by numerical methods such as those used for stagnation-point flow in Section 13.7. Figure 14.10 shows $G(\eta)$ and $G'(\eta) = u/u_e$ as functions of dimensionless distance, η. Defining the boundary-layer edge as the point where $u = 0.99u_e$, we find $\eta \approx 3.5$ at this point (which corresponds to $y = \delta$), so that

$$\delta \approx 5\sqrt{\frac{\nu x}{u_e}} = \frac{5x}{\sqrt{Re_x}} \tag{14.81}$$

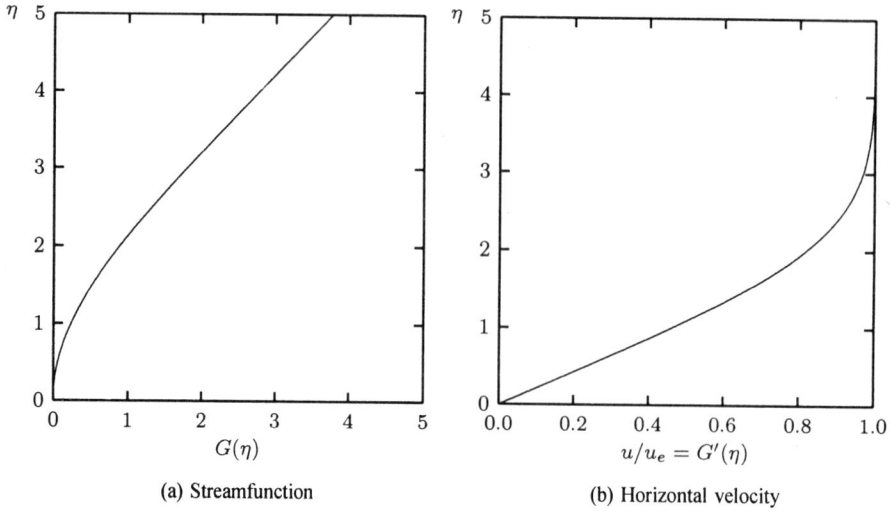

Figure 14.10: *Blasius solution.*

This confirms our earlier estimate [Equation (14.37)] of the boundary-layer thickness. Note in passing that this definition is a bit arbitrary, and therefore not worth quoting to high accuracy. Furthermore, it would be quite difficult to determine δ to such precision experimentally. Consequently, δ is less reliable than other boundary-layer length scales such as the momentum and displacement thickness.

Forming the appropriate integrals, we can compute the momentum thickness, displacement thickness and shape factor. We find

$$\delta^* = \frac{1.721x}{\sqrt{Re_x}}, \qquad \theta = \frac{0.664x}{\sqrt{Re_x}}, \qquad H = 2.59 \tag{14.82}$$

Hence, for the constant-pressure boundary layer, the displacement thickness is about a third of the boundary-layer thickness, while momentum thickness is approximately an eighth of δ. Because both δ^* and θ grow at the same rate, the shape factor remains constant.

Another quantity of interest is the surface shear stress, τ_w. From the Blasius solution, there follows

$$\tau_w = \mu \left(\frac{\partial u}{\partial y} \right)_{y=0} = \frac{\mu U G''(0)}{\sqrt{2\nu x/U}} = 0.332 \frac{\rho U^2}{\sqrt{Re_x}} \tag{14.83}$$

where we use the fact that the numerical solution gives $G''(0) = 0.332\sqrt{2}$. We can recast this in dimensionless form by introducing the **skin-friction coefficient**, c_f, defined as

$$c_f \equiv \frac{\tau_w}{\frac{1}{2}\rho U^2} \tag{14.84}$$

Thus, for the Blasius boundary layer, the skin friction is

$$c_f = \frac{0.664}{\sqrt{Re_x}} \tag{14.85}$$

To determine the drag force on one side of a plate of length L, we can integrate the surface shear stress [Equation (14.83)] from the leading to trailing edges of the plate. Denoting the drag force by F_D, we have

$$
\begin{aligned}
F_D &= \int_0^L \tau_w \, dx = 0.332 \rho \, U^2 \int_0^L \frac{dx}{\sqrt{Re_x}} \\
&= 0.332 \rho \, U^2 \frac{\nu}{U} \int_0^{Re_L} \frac{dRe_x}{\sqrt{Re_x}} = 0.664 \frac{\rho \, U^2 L}{Re_L} \sqrt{Re_L} \\
&= 0.664 \frac{\rho \, U^2 L}{\sqrt{Re_L}}
\end{aligned}
\tag{14.86}
$$

Thus, the drag coefficient for a plate of length L is

$$
C_D = \frac{F_D}{\frac{1}{2} \rho \, U^2 L} = \frac{1.328}{\sqrt{Re_L}}
\tag{14.87}
$$

14.2.5 Effects of Pressure Gradient

In addressing effects of pressure gradient on a boundary layer, Falkner and Skan (1931) found that similarity solutions exist for the boundary-layer equations when the freestream velocity is given by

$$
u_e(x) = U_o \left(\frac{x}{L} \right)^m
\tag{14.88}
$$

where U_o is a characteristic velocity, L is a characteristic length and m is a dimensionless constant. For inviscid flow past a wedge, we found in Subsection 11.7.3 that the velocity on the surface of the wedge varies with distance along the surface raised to a power [Equation (11.174)]. When m lies between 0 and 1, the freestream velocity given in Equation (14.88) corresponds to the surface velocity for flow past a wedge of angle $2m\pi/(m+1)$ as depicted in Figure 14.11.

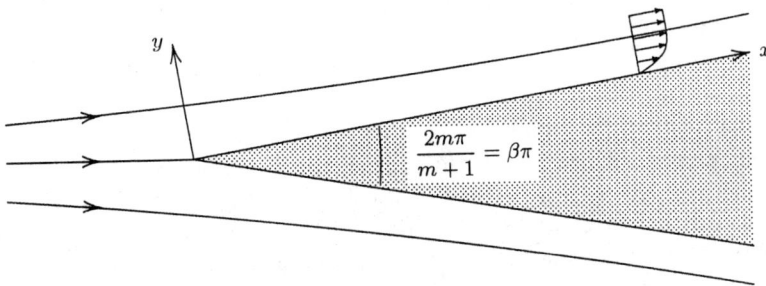

Figure 14.11: *Viscous flow past a wedge.*

The similarity solution for this flow is again most conveniently formulated in terms of the streamfunction. The Falkner-Skan solution is of the form

$$
\psi(x, y) = \sqrt{\frac{2\nu U_o L}{m+1}} \left(\frac{x}{L} \right)^{\frac{m+1}{2}} G(\eta), \qquad \eta \equiv y \sqrt{\frac{(m+1)U_o}{2\nu L}} \left(\frac{x}{L} \right)^{\frac{m-1}{2}}
\tag{14.89}
$$

Substitution into the boundary-layer equations yields

$$\frac{d^3G}{d\eta^3} + G\frac{d^2G}{d\eta^2} + \beta\left[1 - \left(\frac{dG}{d\eta}\right)^2\right] = 0 \qquad (14.90)$$

$$\left.\begin{array}{ll} G = 0, & G' = 0, \quad \eta = 0 \\ G' \to 1, & \eta \to \infty \end{array}\right\} \qquad (14.91)$$

The constant coefficient β is related to the constant m by the simple algebraic relation

$$\beta = \frac{2m}{m+1} \qquad (14.92)$$

Equation (14.90) is known as the **Falkner-Skan equation**. Hartree (1937) investigated solutions to the Falkner-Skan equation in detail. His analysis showed that physically realistic solutions exist for $-0.0905 \leq m \leq 2$. There are solutions beyond this range as well that are summarized by Stewartson (1954).

We can check the Falkner-Skan equation for consistency with two limiting cases we have already studied. The case $m = 0$ corresponds to a wedge angle of zero, which is simply a flat plate. Setting $\beta = 0$ in the Falkner-Skan equation yields the Blasius Equation (14.79). Because the boundary conditions are also identical, the Falkner-Skan solution for $\beta = 0$ is the Blasius solution. The case $m = 1$ corresponds to a wedge angle of π radians, or 180°, which is stagnation-point flow. Setting $\beta = 1$ in the Falkner-Skan equation yields the equation for stagnation-point flow [Equation (13.151)].[6] Again, since the boundary conditions are also identical, we find consistency between the Falkner-Skan solution and the solution generated in Section 13.6 when $m = 1$.

The solutions for $0 < m < 1$ correspond to the wedge shown in Figure 14.11. The solution is physically relevant for flow near the leading edge of a body with a pointed leading edge.

When $1 < m < 2$, the wedge angle exceeds 180°, so that we have what can be described as flow into a corner. As noted by Panton (1996), this type of flow can be very difficult to simulate experimentally. There are also solutions for $m > 2$, but none correspond to a simple ideal flow.

All of the solutions for $m > 0$ have one thing in common. Specifically, they correspond to decreasing pressure, which we refer to as a **favorable pressure gradient**. The description of decreasing pressure as being favorable has its origin in the effect varying pressure has on a boundary layer. Because decreasing pressure accelerates the flow, especially the low-momentum portion of the boundary layer closest to the surface, favorable pressure gradient energizes the flow, and poses no threat to the boundary layer.

By contrast, for $m < 0$, the pressure increases in the streamwise direction. This condition is described as an **adverse pressure gradient**. An adverse pressure gradient opposes the motion, and most strongly affects the low-momentum fluid near the surface. If the adverse gradient is too strong, it can cause a reversal in flow direction. When this occurs, we say that the boundary layer **separates** from the surface.

---

[6]While we obtained the stagnation-point solution for the complete Navier-Stokes equation, the solution has $\partial^2 u/\partial x^2 = \partial^2 v/\partial x^2 = 0$. Thus, the Navier-Stokes and boundary-layer solutions for the velocity, and hence the streamfunction, are identical.

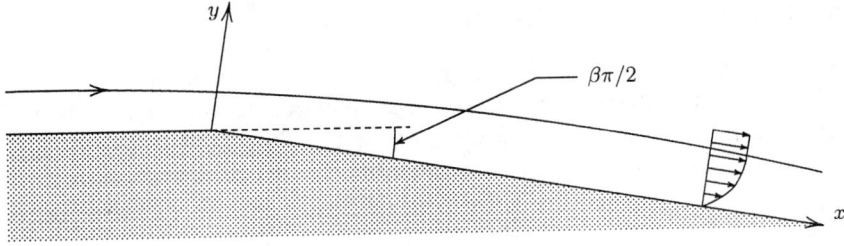

Figure 14.12: *Falkner-Skan flow past a convex corner.*

The Falkner-Skan equation has solutions for adverse pressure gradients with m in the range $-0.0905 < m < 0$. This corresponds to flow past a convex corner (Figure 14.12) with an angle as large as 18°. Hartree originally found **attached-flow** solutions for m in this range, i.e., solutions with positive velocity throughout the boundary layer. When $m = -0.0905$, corresponding to $\beta = -0.199$ (so that $\beta\pi/2 = 18°$), there is a solution with $\partial u/\partial y = 0$ at the surface, which means the skin friction is exactly zero. This corresponds to the point of separation.

Stewartson (1954) later discovered that for $-0.0905 < m < 0$, **reverse-flow** solutions also exist. These solutions have a small region close to the surface with negative velocities. It is doubtful that the Stewartson solutions bear any relation to the physical phenomenon of separation as they still entail nearly **parallel flow**, i.e., a flow that has $v \ll u$. This feature is reminiscent of the reverse-flow velocity profiles we found for Couette flow (see Figure 13.4). By contrast, when boundary-layer separation occurs, we find the vertical velocity to be comparable in magnitude to the horizontal velocity, and the boundary-layer approximation is not valid. Figure 14.13 shows four velocity profiles corresponding to some of the most interesting values of the parameter m. The results shown were computed with Program **FALKSKAN** described in Appendix E.

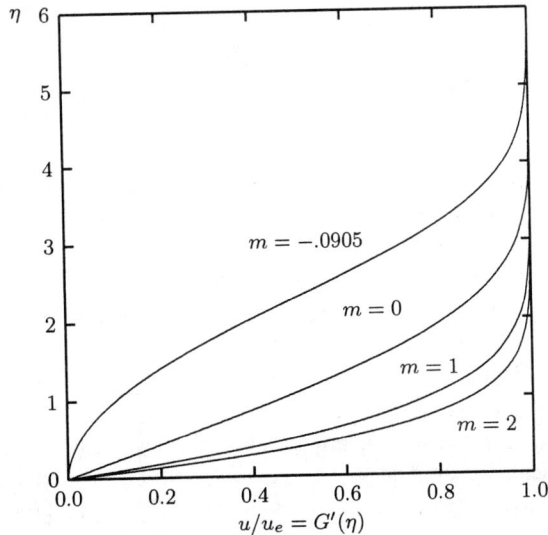

Figure 14.13: *Falkner-Skan solutions.*

14.2.6   Boundary-Layer Separation

On the one hand, because of the retarding effect of the viscous stress, the flow within a boundary layer has a smaller velocity than in the inviscid freestream. The stress is largest near the surface where the velocity gradient is largest, so that fluid particles close to the surface have far less momentum than those in the freestream. On the other hand, since the pressure is constant through a boundary layer, the pressure force acts with the same magnitude on all fluid particles within the layer. When the pressure acts to decelerate the flow, i.e., when an adverse pressure gradient exists, the low-momentum fluid particles nearest the surface are most easily stopped. If the adverse pressure gradient is strong enough, the flow will reverse direction. As discussed in the previous subsection, when this occurs we say the boundary layer has **separated**. By definition, the **separation point** in a steady, two-dimensional flow[7] is the point where $\partial u/\partial y = 0$ at the surface.

We can use a simple geometrical argument to demonstrate why having an adverse pressure gradient is a necessary (but not a sufficient) condition for separation. We begin by noting that at a solid boundary, the convective terms in the x-momentum equation vanish (because $u = v = 0$), wherefore

$$\mu \left(\frac{\partial^2 u}{\partial y^2} \right)_{y=0} = \frac{dp}{dx} \tag{14.93}$$

For a favorable pressure gradient we have $dp/dx < 0$, so that the second derivative of u is negative at the surface. Since we expect the maximum velocity to occur at the boundary-layer edge, necessarily $\partial^2 u/\partial y^2 < 0$ at the edge as well. Thus, in a favorable pressure gradient, the curvature of the velocity profile, $\partial^2 u/\partial y^2$, is everywhere negative. The Falkner-Skan profiles for $m = 1$ and $m = 2$ shown in Figure 14.13 are examples of uninflected velocity profiles with negative curvature. Similarly, for zero pressure gradient (the $m = 0$ case in Figure 14.13), there is an inflection point at the surface, i.e., a point where $\partial^2 u/\partial y^2 = 0$, while the curvature is negative at all other points in the boundary layer.

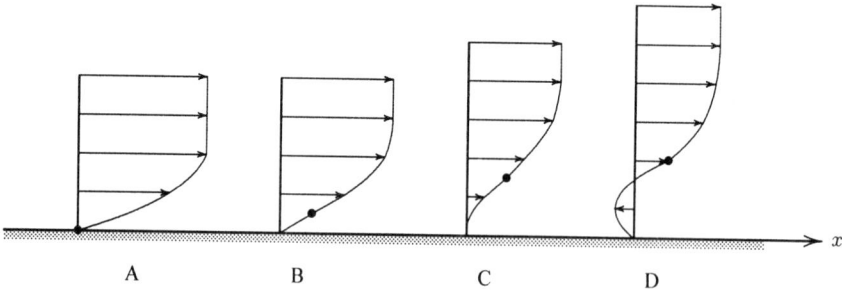

Figure 14.14: *Boundary-layer velocity profiles near separation;* • *denotes an inflection point.*

By contrast, in an adverse pressure gradient we have $dp/dx > 0$, wherefore the second derivative of u is positive at the surface. Consequently, since $\partial^2 u/\partial y^2 < 0$ at the boundary-layer edge, there must be an inflection point somewhere in the boundary layer. When an inflection point is present, the boundary-layer velocity profile shape changes as illustrated in Figure 14.14. Profile A corresponds to constant pressure. The boundary layer depicted then encounters an adverse pressure gradient and develops an inflection point above the surface as shown in Profile B. A bit farther downstream, the boundary layer separates and assumes the

---

[7]Separation in three dimensions and in unsteady flows is more complex and beyond the scope of this text.

shape of Profile C. Beyond this point, there is reverse flow near the surface corresponding to Profile D. Clearly the separated-flow profile must have an inflection point, thus demonstrating that adverse pressure gradient is a necessary condition for separation. It is not a sufficient condition, i.e., it does not guarantee separation, as a boundary layer can remain attached for mild adverse gradients (e.g., the Falkner-Skan cases with $-0.0905 < m < 0$).

Separation is generally undesirable as it leads to increased drag for devices such as wings. Separation occurs on a wing for too large an angle of attack, a condition referred to as **stall**. It causes not only an increase in drag, but a loss of lift as well. This condition is a catastrophic failure, and poses a limit on the angle of attack and load carrying capacity of an airplane.

Figure 14.15 schematically contrasts unseparated, or **attached flow** past an airfoil with the massively separated flow corresponding to stall. For attached flow [Figure 14.15(a)], the flow leaves the trailing edge smoothly and enters a thin viscous layer known as a wake. When the airfoil stalls [Figure 14.15(b)], the flow separates on the upper surface, and the boundary layer abruptly breaks away. A large region of complicated eddying motions forms in the wake of the airfoil.

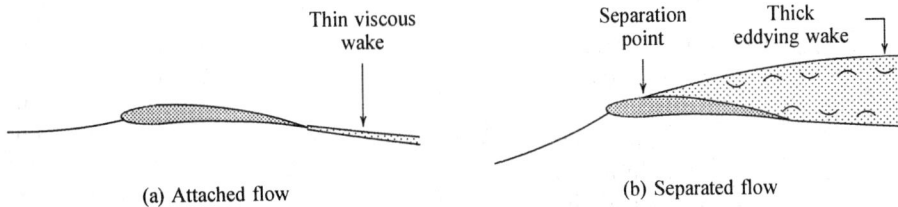

Figure 14.15: *Attached and separated flow past an airfoil.*

The loss of lift is caused by the grossly altered shape presented to the inviscid flow. The effective upper surface is the edge of the wake, which corresponds to a much thicker airfoil. Clearly, the effective angle of attack seen by the upper surface is greatly reduced. This increases the low pressure that prevails on the upper, or **suction surface** of an airfoil that provides most of the lift.[8] Since lift is proportional to the angle of attack, the reduced angle accounts for the dramatic loss of lift. This is an extreme example of the displacement effect discussed at the end of Subsection 14.2.3.

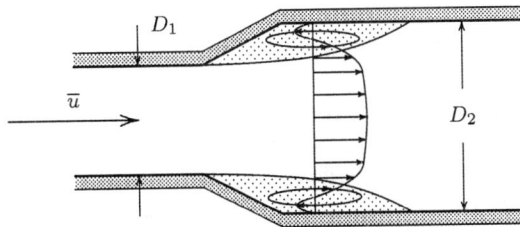

Figure 14.16: *Separation in duct or pipe flow at a sudden expansion; the flow is separated in the shaded region. As illustrated, a streamline in the separation bubble is a closed contour.*

Separation occurs for internal flows also, such as in a rapidly expanding duct or pipe. This is the cause of the head loss for a sudden expansion in ducts and pipes discussed in Subsection 7.7.2. Figure 14.16 illustrates how separation alters the flow. The adverse pressure gradient encountered as the flow expands causes an extended region of separated flow. The

---

[8]The high pressure that develops on the lower or **pressure surface** typically provides 20% or less of the lift.

presence of the boundary layer, and its susceptibility to separation, thus brings about a severe distortion in the overall flow, including strong viscous losses. This is the source of the head loss.

The contour bounding the separated region in Figure 14.16 is the locus of points, y_o, below which the net mass flux in the streamwise direction is zero, i.e.,

$$\int_{y_w}^{y_o} \rho\, u\, dy = 0 \qquad (14.94)$$

where y_w is the value of y on the duct wall. It is called the **dividing streamline**, and separates the recirculating flow from the flow in the central portion of the duct or pipe. The point where the dividing streamline intersects the surface downstream of the separation point is called the **reattachment point**. The flow downstream of reattachment has no reverse-flow region, and gradually approaches a new equilibrium state. The flow eventually reattaches when the adverse pressure gradient is removed. In our sudden-expansion example, the pressure approaches a nominally constant value downstream of the corner, so that $dp/dx \to 0$.

A boundary-layer computation that uses a measured pressure distribution provides an accurate prediction for the location of separation. However, the boundary-layer equations are singular at a separation point, with the vertical velocity and displacement thickness becoming infinite. This is true provided we insist upon specifying the pressure. The singularity is not an inherent property of the boundary-layer equations, but rather is a consequence of specifying dp/dx. Methods have been developed that specify either the displacement thickness or the surface shear stress and determine the pressure as part of the solution [cf. Panton (1996)]. These are called **inverse boundary-layer methods**, and require iteration between the inviscid and boundary-layer solutions to achieve compatible pressure distributions. These methods are limited to small separated regions.

There is a useful approximation devised by Reyhner and Flugge-Lotz (1968) that permits integration through a small separation bubble using a measured pressure distribution and the boundary-layer equations. Integration through separation can be accomplished by computing the streamwise convective term in the momentum equation according to

$$u\frac{\partial u}{\partial x} = \max[0, u] \cdot \frac{\Delta u}{\Delta x} \qquad (14.95)$$

where $\Delta u/\Delta x$ is the finite-difference approximation to $\partial u/\partial x$. In other words, on regions of reverse flow we set the streamwise convective term to zero. Wilcox (1988b), for example, has used this procedure to compute unsteady separated boundary layers.

The boundary-layer equations fail completely for massively-separated flows. A stalled airfoil, for example, cannot be treated with a boundary-layer computation. The same is true for flow past a **bluff body** such as an automobile, which has a more-or-less blunt geometry at its "trailing edge." Neither the inverse boundary-layer method nor the Reyhner and Flugge-Lotz approximation correct the deficiencies of the boundary-layer equations for massive separation.

With the increasing power of computers, approximate procedures have largely given way to full Navier-Stokes solutions. This is appropriate from a theoretical point of view as the boundary-layer equations are unsuitable for describing separation. Recalling the basic premises underlying the boundary-layer approximation, they are inappropriate for two key reasons. First, the velocity components in separated regions are both of the same order of magnitude, while the boundary-layer approximation assumes $v \ll u$. Second, the pressure varies normal to the surface in a separated region wherefore the boundary-layer approximation that $\partial p/\partial y = 0$ is violated.

14.3 Turbulence

For "low enough" velocities, in the sense that the Reynolds number is not too large, the equations of motion for a viscous fluid have well-behaved, steady solutions. We have seen several examples in Chapter 13 and in Section 14.2. Even the unsteady Stokes' problems included smoothly varying properties, primarily controlled by viscous diffusion of vorticity and momentum. For all of these solutions, the motion is termed **laminar** and can be observed experimentally and in nature.

At larger Reynolds numbers, the viscous stresses are completely overwhelmed by the fluid's inertia, and the laminar motion becomes unstable. Rapid velocity and pressure fluctuations appear and the motion becomes inherently unsteady. When this occurs, we describe the motion as being **turbulent**. In the cases of fully-developed Couette flow and pipe flow, for example, laminar flow is assured only if the Reynolds number based on channel height and pipe radius is less than 1500 and 2300, respectively.

Virtually all flows of practical engineering interest are turbulent. Turbulent flows always occur when the Reynolds number is large. For slightly viscous fluids such as water and air, large Reynolds number corresponds to anything stronger than a small breeze or a puff of wind. Thus, to analyze fluid motion for general applications, we must deal with turbulence. In this section, we will explore some of the most important aspects of this phenomenon.

14.3.1 Laminar, Transitional and Turbulent Flow

Prior to 1930, experimentalists lacked instrumentation capable of sensing rapid velocity and pressure fluctuations that are always present in turbulent flows. Rather, they determined mean values of flow properties. All flow properties undergo significant changes between laminar and turbulent regimes. In terms of mean measurements, early experimenters found quite different velocity profiles in pipe flow. Figure 14.17 schematically compares laminar and mean turbulent velocity profiles. As shown in Subsection 13.3, the laminar profile varies quadratically with pipe radius. By contrast, the turbulent-flow profile is much more filled out, appearing nearly uniform across most of the pipe. The slope of the velocity profile is much steeper at the pipe surface for the turbulent profile, corresponding to a much larger friction factor. Similar differences in velocity profiles are present for boundary layers.

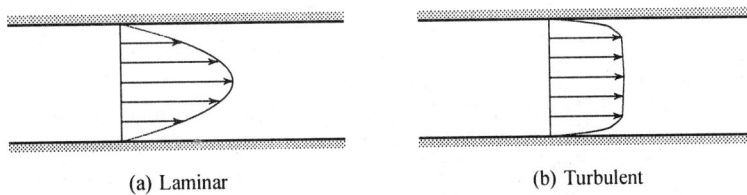

(a) Laminar   (b) Turbulent

Figure 14.17: *Pipe-flow velocity profiles.*

When measurements cover a wide enough range of Reynolds numbers to include both laminar and turbulent flow, dramatic changes in measured properties are observed as the motion changes from laminar to turbulent. The change in the character of the motion is called **transition**, and generally occurs over a finite streamwise distance.

In the case of the flat-plate boundary layer, for example, the laminar-flow (Blasius) solution is observed experimentally for Reynolds number based on plate length, Re_x, of at least $9 \cdot 10^4$. The precise point where transition begins depends on a variety of factors including the level of turbulence in the freestream and how rough the surface is. Figure 14.18 displays the skin friction for a flat-plate boundary layer with Re_x ranging from 10^3 to 10^8. As shown, for large Reynolds numbers, the skin friction is significantly larger than the Blasius value $(c_f = 0.664 \, Re_x^{-1/2})$ that we determined in Subsection 14.2.4. Also, the skin friction falls off roughly as $Re_x^{-1/5}$ (the exact origin of x depending on the transition point), which is much more gradual than the Blasius variation.

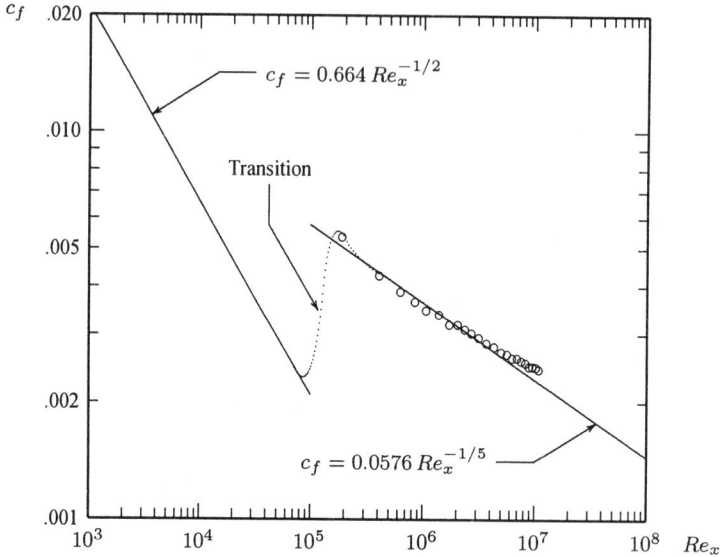

Figure 14.18: *Skin friction for a flat-plate boundary layer;* $\circ$ *measurements of Wieghardt [see Coles and Hirst (1969)].*

The same kind of abrupt change in properties is observed for internal flows. Examination of the Moody diagram for flow in a pipe, Figure 7.6, shows a similar variation of the friction factor, f, with Reynolds number. At very low Reynolds numbers, the laminar-flow solution matches measured values. As Reynolds number increases, the flow undergoes transition, and the friction factor asymptotes to values significantly above the laminar prediction. As with the boundary layer, the turbulent-flow rate of decrease of f with Reynolds number is less rapid than the laminar rate.

Flow past a cylinder is a particularly interesting example illustrating aspects of laminar, transitional and turbulent flow. Figures 14.19 and 14.20 schematically illustrate how the flow develops as Reynolds number increases. For very small Reynolds number, the flow not only remains laminar but remains attached as well. This is the limiting case known as **Stokes flow** or **creeping flow**. Figure 14.19(a) sketches typical streamlines for $Re_D < 4$. The flow is symmetric fore and aft, and has a very large drag coefficient (see Figure 14.3).

As Re_D increases to about 40, the boundary layer remains laminar over the entire surface of the cylinder and separates on the downstream side. A steady separation bubble remains fixed to the cylinder as sketched in Figure 14.19(b).

When the Reynolds number lies between 40 and 100, a remarkable phenomenon, known as the **Kármán vortex street**, appears. The laminar flow ceases being steady, and vortices are shed into the wake of the cylinder, with a regular spacing [Figure 14.19(c)]. The vortices travel downstream at a velocity slightly less than the freestream velocity, and are shed at a

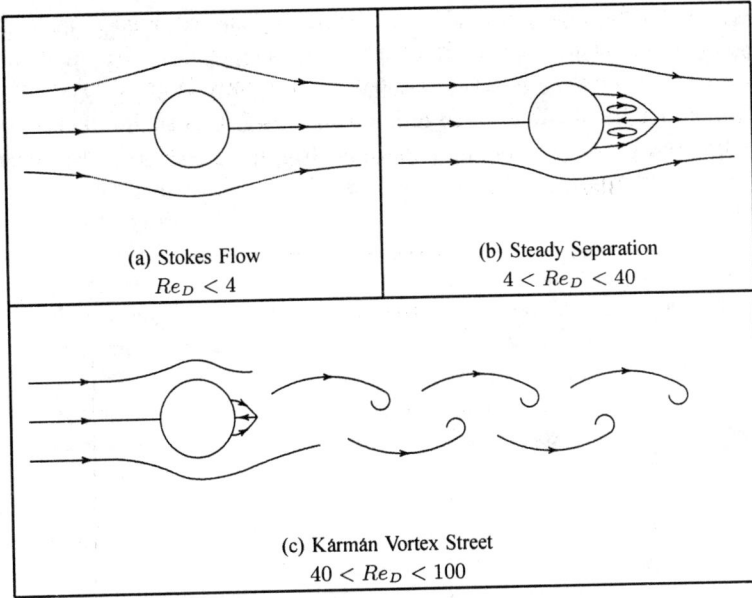

Figure 14.19: *Laminar flow past a cylinder.*

fixed frequency, ω. Measurements show that the Strouhal number for flow past a cylinder is

$$St \equiv \frac{\omega D}{U} \approx 0.2 \tag{14.96}$$

While $St \approx 0.2$ for a wide range of Reynolds numbers, it varies slightly with Re_D. The vortex street is not turbulent, and a small, steady separation bubble remains attached to the cylinder. Similar **vortex-shedding** patterns are observed on non-cylindrical bodies, even at large Reynolds numbers. A flexible body can experience resonance if the vortex-shedding frequency is near the body's structural-vibration frequency. Audible sounds emanate from electric transmission lines as a result of this phenomenon. Mooring lines that hold floating objects in place will experience significant oscillations at certain current speeds. The most startling example of how powerful such resonance can be is provided by the collapse of the **Tacoma Narrows bridge** in 1940. Resonance between a torsional oscillation mode and the vortex shedding destroyed the bridge.

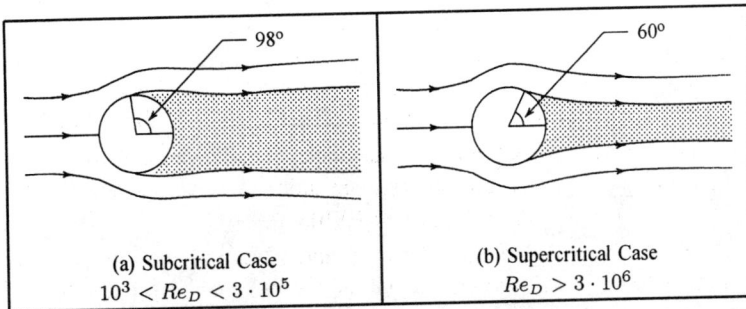

Figure 14.20: *Transitional and turbulent flow past a cylinder.*

For Reynolds numbers in excess of 100, the Kármán vortex street becomes unstable and the wake becomes turbulent. As the Reynolds number increases to $3 \cdot 10^5$, the boundary layer remains laminar, and the separation point moves upstream of the equator. Separation occurs as far forward as 98° from the axis of symmetry measured from the downstream face of the cylinder [see Figure 14.20(a)]. Separation triggers transition and a turbulent wake forms downstream of the cylinder. This is referred to as the **subcritical case**. Reference to Figure 14.3 shows that the drag coefficient for a cylinder is nearly constant for Reynolds number in the range $10^3 < Re_D < 3 \cdot 10^5$.

For Reynolds numbers in the range $3 \cdot 10^5 < Re_D < 3 \cdot 10^6$, another interesting change occurs for flow past a cylinder. Specifically, the boundary layer undergoes transition, and causes the separation point to move downstream of the equator. In general, a turbulent boundary layer can negotiate a stronger adverse pressure gradient. This is true because the fluid closest to the surface has increased momentum relative to the laminar case. As shown in Figure 14.20(b), when the Reynolds number exceeds $3 \cdot 10^6$, the separation point lies 60° from the axis. This change in flow structure has a dramatic effect on the drag coefficient. Reference to Figure 14.3 shows that there is a sudden decrease of nearly 70%, followed by a gradual increase back to the level prevailing at lower Reynolds numbers. This sudden change in drag coefficient is known as the **drag crisis**. The drag coefficient decreases because, coincident with the delay of separation, the pressure on the downstream surface of the cylinder increases, thus reducing the drag. We refer to the flow that develops for $Re_D > 3 \cdot 10^6$ as the **supercritical case**.

Van Dyke (1982) provides an excellent collection of photographs showing the various flow patterns for flow past cylinders and spheres. The drag on a sphere exhibits the same sudden decrease at a critical Reynolds number, Re_{crit}, of about $3 \cdot 10^5$. If the surface of the sphere is rough, Re_{crit} can decrease to a value as low as $2 \cdot 10^5$. This phenomenon has immediate relevance in both tennis and baseball. In both sports, the motion occurs at Reynolds numbers close to the critical range. In the case of a baseball, for example, its stitches appear as roughness elements that can trigger transition. If $Re_{crit} = 2 \cdot 10^5$, and the temperature is 90° F, then the velocity required to reach the critical Reynolds number is

$$U = 2 \cdot 10^5 \, \frac{\nu}{D} = 2 \cdot 10^5 \, \frac{1.67 \cdot 10^{-4} \, \text{ft}^2/\text{sec}}{0.25 \, \text{ft}} = 133.6 \, \frac{\text{ft}}{\text{sec}} \approx 91 \text{ mph} \qquad (14.97)$$

A pitcher who throws at speeds in excess of 90 mph is generally regarded as having a **blazing fastball**. The ball reaches the batter faster not only because of the initial speed, but also because it decelerates less due to the reduced drag attending transition to turbulence.

In 1883 Reynolds performed a simple experiment (see Figure 14.21) that demonstrates the existence of the markedly different behavior of laminar and turbulent flow. Using a transparent

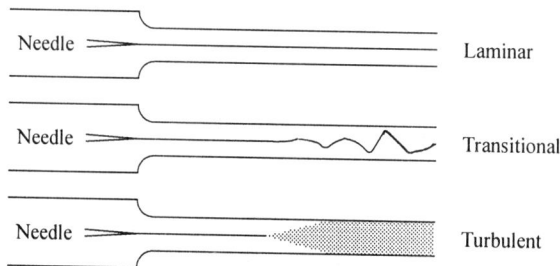

Figure 14.21: *Transition from laminar to turbulent flow in a pipe as revealed by continuous injection of dye.*

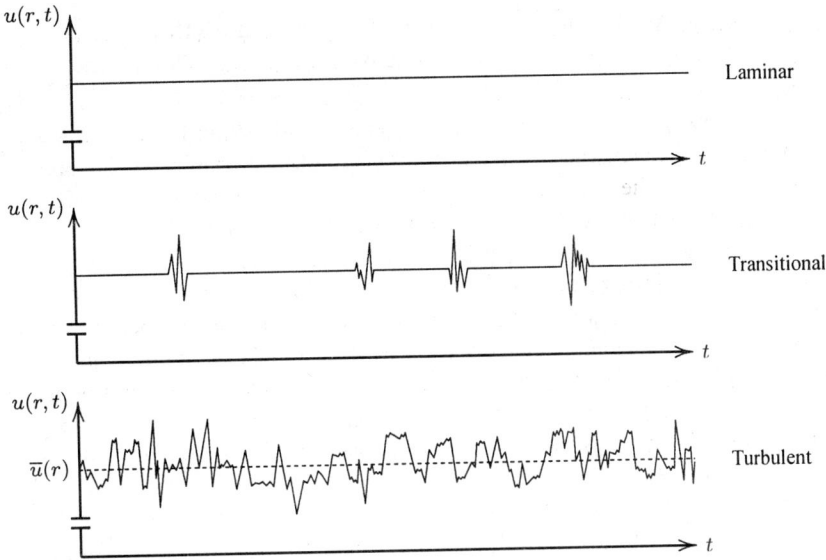

Figure 14.22: *Velocity versus time for laminar, transitional and turbulent pipe flow. Fluctuations are typically less than 10% of the mean value.*

pipe, he injected dye along the centerline. He adjusted the flow rate to control the Reynolds number, Re_D, in the pipe. For $Re_D < 2300$, the dye stream is straight and, aside from a minor amount of diffusion, remains unmixed with the surrounding fluid. As Re_D increases beyond 2300, the dye stream suddenly becomes wavy. When the Reynolds number increases still further, the distinct dye stream disappears completely and the dye spreads uniformly through the pipe. Figure 14.21 illustrates this process in sketches similar to those given by Reynolds (1883).

Since 1930, instrumentation has been developed that permits detailed measurement of flow properties revealing important aspects of turbulence. Devices such as the *hot-wire anemometer* and, more recently, the *laser-Doppler anemometer* provide instantaneous measurements sensitive enough to record a wide range of frequencies. This is important because turbulence includes oscillatory motions of a very wide range of frequencies, and the range increases with Reynolds number. If we place one of these devices in the pipe flow analyzed by Reynolds, for example, we can measure the velocity as a function of time.

Suppose we make our observations at a point a distance r from the pipe centerline near the downstream end of the pipe shown in Figure 14.21. For the three cases — laminar, transitional and turbulent — the measured velocity as a function of time, t, would be as illustrated in Figure 14.22. For laminar conditions, the velocity is constant at all points across the pipe. The flow is steady and completely consistent with the laminar-flow solution developed in Section 13.3. In the transitional case, the velocity shows sudden bursts of unsteadiness separated by finite intervals during which the velocity is constant. When the flow becomes turbulent, the velocity displays rapid fluctuations in time. The fluctuations include frequencies over a wide range, and are quite irregular and aperiodic. The magnitude of the fluctuations is typically within 10% of the average value, $\bar{u}(r)$. As mentioned above, prior to development of modern, high-frequency, measuring devices, experimenters were able to determine only averaged values such as $\bar{u}$ (e.g., Figure 14.17).

AN EXPERIMENT YOU CAN DO AT HOME
Contributed by Prof. Peter Bradshaw, Stanford University

The curious reader can perform a simple experiment to observe how fast turbulent mixing occurs. Fill a transparent cup or glass with water and allow it to settle. Next, very gently add a few drops of milk or thin cream to the water. Note that non-fat milk won't work — the drops tend to break up. Initially, the milk will sink to the bottom and spread over the bottom of the vessel. After a few seconds, a steady state will be reached in which no motion is evident, and most of the water will be clear. Although not evident to the naked eye, mixing of the milk and water proceeds at the molecular level. If the water is three inches in depth, you would have to wait approximately three weeks for the mixing to be complete. The time required for laminar mixing is proportional to the square of the size of the container, so that adding the milk to a six-inch glass of water would require just over three months to mix.

Now, dip a spoon into the cup or glass and remove it. Make no attempt to stir the mixture. You will observe a rather remarkable phenomenon. Within a few seconds, the milk and water will be completely mixed! The mixing occurs as a result of the turbulence in the wake of the spoon. To observe the turbulent-mixing process in more detail, use a larger container such as a fish bowl, being sure to remove the fish first. The mixing will take longer, and the turbulent eddies should be readily observable.

14.3.2   General Properties of Turbulence

As noted at the beginning of this section, most practical flows in nature are turbulent, so the phenomenon cannot be ignored. Although vigorous research has been conducted to help discover the mysteries of turbulence, it remains the most noteworthy unsolved scientific problem! This and the following subsections briefly outline some aspects of the analytical methods used to analyze turbulent flows. For a more-complete introduction to the physics of turbulence, refer to a basic text on the physics of turbulence such as those by Tennekes and Lumley (1983) or Landahl and Mollo-Christensen (1992).

- **Basic Definition.** In 1937, von Kármán defined turbulence in a presentation at the Twenty-Fifth Wilbur Wright Memorial Lecture entitled "Turbulence." He quoted G. I. Taylor as follows [see von Kármán (1937)]:

  "Turbulence is an irregular motion which in general makes its appearance in fluids, gaseous or liquid, when they flow past solid surfaces or even when neighboring streams of the same fluid flow past or over one another."

As the understanding of turbulence has progressed, researchers have found the term "irregular motion" to be too imprecise. Simply stated, an irregular motion is one that is typically aperiodic and that cannot be described as a straightforward function of time and space coordinates. An irregular motion might also depend strongly upon initial conditions. The problem with the Taylor-von Kármán definition of turbulence lies in the fact that there are non-turbulent flows that can be described as irregular.

Turbulent motion is irregular in the sense that it can be described by the laws of probability. Even though instantaneous properties in a turbulent flow are extremely sensitive to initial conditions, statistical averages of the instantaneous properties are not. To provide a sharper definition of turbulence, Hinze (1975) offers the following revised definition:

"Turbulent fluid motion is an irregular condition of flow in which the various quantities show a random variation with time and space coordinates, so that statistically distinct average values can be discerned."

To complete the definition of turbulence, Bradshaw [cf. Cebeci and Smith (1974)] adds the statement that *turbulence has a wide range of scales*. Time and length scales of turbulence are represented by frequencies and wavelengths that are revealed by a Fourier analysis of a turbulent-flow time history.

The irregular nature of turbulence stands in contrast to laminar motion, so called historically, because the fluid was imagined to flow in smooth laminae, or layers. In describing turbulence, many researchers refer to **eddying motion**, which is a local swirling motion where the vorticity can often be very intense. **Turbulent eddies** of a wide range of sizes appear and give rise to vigorous mixing and effective turbulent stresses that can be enormous compared to laminar values.

- **Instability and Nonlinearity.** Careful analysis of solutions to the Navier-Stokes equation, or more typically to its boundary-layer form, shows that turbulence develops as an instability of laminar flow. To analyze the stability of laminar flows, virtually all methods begin by linearizing the equations of motion. Although some degree of success can be achieved in predicting the onset of instabilities that ultimately lead to turbulence with linear theories, the inherent nonlinearity of the Navier-Stokes equation precludes a complete analytical description of the actual transition process, let alone the fully-turbulent state. For a real (i.e., viscous) fluid, the instabilities result from interaction between the Navier-Stokes equation's nonlinear inertial terms and viscous terms. The interaction is very complex because it is rotational, fully three dimensional and time dependent.

- **Vortex Stretching.** The strongly rotational nature of turbulence goes hand-in-hand with its three dimensionality. Vigorous stretching of vortex lines maintains the ever-present fluctuating vorticity in a turbulent flow. Vortex stretching is absent in two-dimensional flows (recall that $\omega \cdot \nabla \mathbf{u} = 0$ in a two-dimensional flow) so that turbulence must be three dimensional. This inherent three dimensionality means there are no satisfactory two-dimensional approximations for determining fine details of turbulent flows. This is true even when the average motion is two dimensional as the fluctuating velocities are always three dimensional.

- **Statistical Aspects.** The time-dependent nature of turbulence also contributes to its intractability. The additional complexity goes beyond the introduction of an additional dimension. Turbulence is characterized by random fluctuations thus mandating the use of statistical methods to analyze it. On the one hand, this aspect is not really a problem from the engineer's viewpoint. Even if we had a complete time history of a turbulent flow, we would usually integrate the flow properties of interest over time to extract time-averages. On the other hand, as we will see in the following subsections, time-averaging operations lead to terms in the equations of motion that cannot be determined a priori.

- **Turbulence is a Continuum Phenomenon.** In principle, we know that the time-dependent, three-dimensional Navier-Stokes equation contains all of the physics of a given turbulent flow. That this is true follows from the fact that turbulence is a continuum phenomenon. As noted by Tennekes and Lumley (1983),

    "Even the smallest scales occurring in a turbulent flow are ordinarily far larger than any molecular length scale."

    Nevertheless, the smallest scales of turbulence are still extremely small. They are generally many orders of magnitude smaller than the largest scales of turbulence, the latter often being of the same order of magnitude as the dimension of the object about which the fluid is flowing. Furthermore, the ratio of smallest to largest scales decreases rapidly as the Reynolds number increases. To make an accurate numerical simulation (i.e., a fully time-dependent three-dimensional solution) of a turbulent flow, all physically relevant scales must be resolved. While more and more progress is being made with such simulations, computers of the 1990's have insufficient memory and speed to solve any turbulent-flow problem of practical interest. To underscore the magnitude of the problem, Speziale (1985) notes that a numerical simulation of turbulent pipe flow at a Reynolds number of 500,000 would require a computer 10 million times faster than a Cray Y/MP. However, the results are very useful in developing and testing approximate methods.

- **Large Eddies and Turbulent Mixing.** An especially striking feature of a turbulent flow is the way large eddies migrate across the flow, carrying smaller-scale disturbances with them. The arrival of these large eddies near the interface between the turbulent region and nonturbulent fluid distorts the interface into a highly convoluted shape (Figure 14.23). In addition to migrating across the flow, they have a lifetime so long that they persist for distances as much as 30 times the width of the flow [Bradshaw (1972)]. Hence, the state of a turbulent flow at a given position depends upon upstream history and cannot be uniquely specified in terms of local flow properties as can be done in laminar flow.

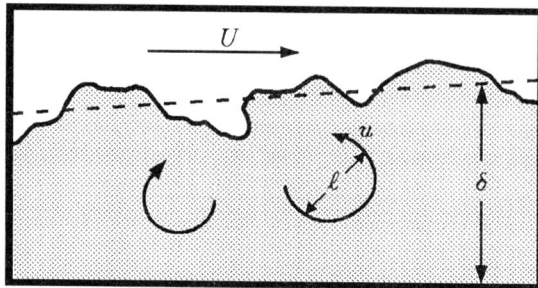

Figure 14.23: *Large eddies in a turbulent boundary layer. The flow above the boundary layer has a steady velocity U; the eddies move at randomly-fluctuating velocities of the order of a tenth of U. The largest eddy size (ℓ) is comparable to the boundary-layer thickness (δ). The interface and the flow above the boundary is quite sharp [Corrsin and Kistler (1954)].*

14.3.3 Reynolds Averaging

Because turbulence consists of random fluctuations of the various flow properties, the classical approach is statistical. Our purposes are best served by using the procedure introduced by Reynolds (1895). There are many forms of averaging that we describe as **Reynolds averaging**, each involving either an integral or a summation. The most commonly used form of Reynolds averaging in engineering applications is **time averaging**. For incompressible flow, we write the velocity and pressure as the sum of mean and fluctuating parts.

$$\left.\begin{array}{c} \mathbf{u}(x, y, z, t) = \overline{\mathbf{u}}(x, y, z) + \mathbf{u}'(x, y, z, t) \\[2mm] p(x, y, z, t) = \overline{p}(x, y, z) + p'(x, y, z, t) \end{array}\right\} \tag{14.98}$$

The quantities $\overline{\mathbf{u}}$ and $\overline{p}$ are the **mean** velocity and pressure, respectively, while $\mathbf{u}'$ and p' are the **fluctuating** parts. The complete velocity and pressure, $\mathbf{u}$ and p, are referred to as the **instantaneous** velocity and pressure, respectively. We compute the mean velocity, for example, as follows.

$$\overline{\mathbf{u}}(x, y, z) = \lim_{T \to \infty} \frac{1}{T} \int_0^T \mathbf{u}(x, y, z, t)\, dt \tag{14.99}$$

While Equation (14.99) is mathematically well defined, we can never truly realize infinite T in any physical flow. This is not a serious problem in practice however. In forming our time average, as illustrated in Figure 14.24, we just select a time T that is very long relative to the maximum period of the velocity fluctuations, T_1. In other words, rather than formally taking the limit $T \to \infty$, we do the indicated integration in Equation (14.99) with $T \gg T_1$. As an example, for flow at 10 m/sec in a 5 cm diameter pipe, an integration time of 20 seconds would probably be adequate. In this time the flow moves 4,000 pipe diameters.

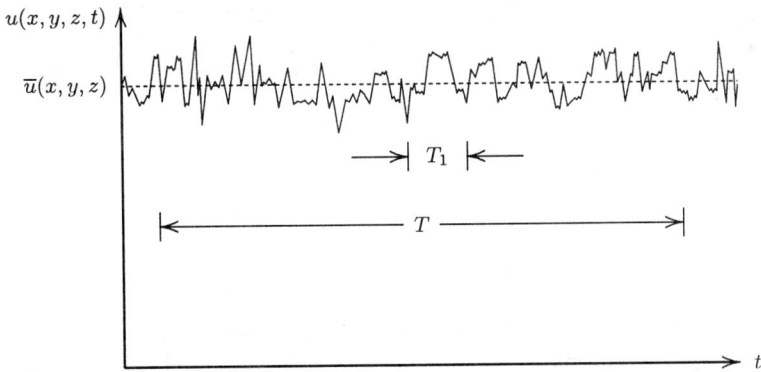

Figure 14.24: *Time averaging for turbulent flow.*

The time average of the mean velocity is again the same time-averaged value, i.e.,

$$\overline{\overline{\mathbf{u}}}(x, y, z) = \lim_{T \to \infty} \frac{1}{T} \int_0^T \overline{\mathbf{u}}(x, y, z)\, dt = \overline{\mathbf{u}}(x, y, z) \tag{14.100}$$

The time average of the fluctuating part of the velocity is zero. That is, using Equation (14.100),

$$\overline{\mathbf{u}'} = \lim_{T \to \infty} \frac{1}{T} \int_0^T [\mathbf{u}(x, y, z, t) - \overline{\mathbf{u}}(x, y, z)]\, dt = \overline{\mathbf{u}}(x, y, z) - \overline{\overline{\mathbf{u}}}(x, y, z) = \mathbf{0} \tag{14.101}$$

Clearly the time-averaging process, involving integrals over time, commutes with spatial differentiation. Thus, for p and $\mathbf{u}$,

$$\overline{\frac{\partial p}{\partial x}} = \frac{\partial \overline{p}}{\partial x} \quad \text{and} \quad \overline{\frac{\partial \mathbf{u}}{\partial x}} = \frac{\partial \overline{\mathbf{u}}}{\partial x} \tag{14.102}$$

Because we are dealing with definite integrals, time averaging is a linear operation. Thus, if c_1 and c_2 are constants while ϕ and ψ denote any two flow properties, then

$$\overline{c_1\phi + c_2\psi} = c_1\overline{\phi} + c_2\overline{\psi} \tag{14.103}$$

Up to this point we have considered time averages of linear quantities. When we time average the product of two properties, ϕ and ψ, we have:

$$\overline{\phi\psi} = \overline{(\overline{\phi} + \phi')(\overline{\psi} + \psi')} = \overline{\overline{\phi}\,\overline{\psi} + \overline{\phi}\psi' + \overline{\psi}\phi' + \phi'\psi'} = \overline{\phi}\,\overline{\psi} + \overline{\phi'\psi'} \tag{14.104}$$

where we take advantage of the fact that the product of a mean quantity and a fluctuating quantity has zero mean. There is no a priori reason for the time average of the product of two fluctuating quantities to vanish. Thus, Equation (14.104) tells us the mean value of a product, $\overline{\phi\psi}$, differs from the product of the mean values, $\overline{\phi}\,\overline{\psi}$. The quantities ϕ' and ψ' are said to be **correlated** if $\overline{\phi'\psi'} \neq 0$. They are **uncorrelated** if $\overline{\phi'\psi'} = 0$.

14.3.4   Reynolds-Averaged Equations

We now form the Reynolds-averaged continuity and boundary-layer equations. For the sake of simplicity, we again focus on steady, incompressible, two-dimensional (mean) flow. The instantaneous boundary-layer equations for conservation of mass and momentum are

$$\frac{\partial u}{\partial x} + \frac{\partial v}{\partial y} + \frac{\partial w}{\partial z} = 0 \tag{14.105}$$

$$\frac{\partial u}{\partial t} + u\frac{\partial u}{\partial x} + v\frac{\partial u}{\partial y} + w\frac{\partial u}{\partial z} = -\frac{1}{\rho}\frac{\partial p}{\partial x} + \nu\frac{\partial^2 u}{\partial y^2} \tag{14.106}$$

$$0 = -\frac{1}{\rho}\frac{\partial p}{\partial y} \tag{14.107}$$

We include all three velocity components and spatial coordinates because, while the mean velocity has $\overline{w} = 0$ and no dependence upon z, the fluctuating velocity and pressure are fully three dimensional.[9]

To begin the Reynolds-averaging process, we assume the velocity components and the pressure can be decomposed according to

$$\left. \begin{array}{rcccl} u(x,y,z,t) & = & \overline{u}(x,y) & + & u'(x,y,z,t) \\ v(x,y,z,t) & = & \overline{v}(x,y) & + & v'(x,y,z,t) \\ w(x,y,z,t) & = & & & w'(x,y,z,t) \\ p(x,y,z,t) & = & \overline{p}(x,y) & + & p'(x,y,z,t) \end{array} \right\} \tag{14.108}$$

---

[9]We could also include the z component of the momentum equation and the convective terms in the y component. However, these terms, after time averaging, produce correlations that can be ignored in a two-dimensional mean flow.

To simplify the algebra, it is convenient to rewrite the convective terms in conservation form, i.e.,

$$
u\frac{\partial u}{\partial x} + v\frac{\partial u}{\partial y} + w\frac{\partial u}{\partial z} = \frac{\partial(u^2)}{\partial x} + \frac{\partial(uv)}{\partial y} + \frac{\partial(uw)}{\partial z} - u\left(\frac{\partial u}{\partial x} + \frac{\partial v}{\partial y} + \frac{\partial w}{\partial z}\right)
$$
$$
= \frac{\partial(u^2)}{\partial x} + \frac{\partial(uv)}{\partial y} + \frac{\partial(uw)}{\partial z} \tag{14.109}
$$

where we take advantage of Equation (14.105) in order to drop the term proportional to the divergence of the velocity. Combining Equations (14.106) and (14.109) yields the x component of the momentum equation in conservation form.

$$
\frac{\partial u}{\partial t} + \frac{\partial(u^2)}{\partial x} + \frac{\partial(uv)}{\partial y} + \frac{\partial(uw)}{\partial z} = -\frac{1}{\rho}\frac{\partial p}{\partial x} + \nu\frac{\partial^2 u}{\partial y^2} \tag{14.110}
$$

Time averaging the continuity Equation (14.105) is very simple as the equation is linear. Thus, we arrive immediately at the following time-averaged form of continuity, viz.,

$$
\frac{\partial \overline{u}}{\partial x} + \frac{\partial \overline{v}}{\partial y} = 0 \tag{14.111}
$$

The term $\partial \overline{w}/\partial z$ does not appear because the mean flow is assumed to be two dimensional. Note that subtracting Equation (14.111) from Equation (14.105) shows that the fluctuating velocity vector also has zero divergence.

Time averaging the x-momentum Equation (14.110) yields

$$
\frac{\partial}{\partial x}\left(\overline{u}^2 + \overline{u'^2}\right) + \frac{\partial}{\partial y}\left(\overline{u}\,\overline{v} + \overline{u'v'}\right) = -\frac{1}{\rho}\frac{\partial \overline{p}}{\partial x} + \nu\frac{\partial^2 \overline{u}}{\partial y^2} \tag{14.112}
$$

which can be rearranged to produce the following equation.

$$
\overline{u}\frac{\partial \overline{u}}{\partial x} + \overline{v}\frac{\partial \overline{u}}{\partial y} = -\frac{1}{\rho}\frac{\partial}{\partial x}\left(\overline{p} + \rho\overline{u'^2}\right) + \frac{\partial}{\partial y}\left(\nu\frac{\partial \overline{u}}{\partial y} - \overline{u'v'}\right) \tag{14.113}
$$

Note that we have rewritten the convective terms by using Equation (14.109) in reverse. In practice, we find that $|\partial(\rho\overline{u'^2})/\partial x| \ll |\partial\overline{p}/\partial x|$ for an incompressible boundary layer, so that the final form of the boundary-layer equations for a turbulent, two-dimensional flow is as follows.

$$
\frac{\partial \overline{u}}{\partial x} + \frac{\partial \overline{v}}{\partial y} = 0 \tag{14.114}
$$

$$
\overline{u}\frac{\partial \overline{u}}{\partial x} + \overline{v}\frac{\partial \overline{u}}{\partial y} = -\frac{1}{\rho}\frac{\partial \overline{p}}{\partial x} + \frac{\partial}{\partial y}\left(\nu\frac{\partial \overline{u}}{\partial y} - \overline{u'v'}\right) \tag{14.115}
$$

$$
0 = -\frac{1}{\rho}\frac{\partial \overline{p}}{\partial y} \tag{14.116}
$$

The time-averaged conservation of mass, Equation (14.114), is identical to the laminar-flow Equation (14.39), with the mean velocity replacing the instantaneous velocity. Aside from replacement of instantaneous values by mean values, the only difference between the two-dimensional, time-averaged momentum Equations (14.115) and (14.116), and the laminar-flow momentum Equations (14.40) and (14.41), is the appearance of the correlation $\overline{u'v'}$.

Herein lies the fundamental problem of turbulence for the engineer. In order to compute all mean-flow properties of the turbulent flow under consideration, we need a prescription for computing $\overline{u'v'}$. This term is so important in turbulent flows that it has a special name. The quantity

$$\tau_t \equiv -\rho\overline{u'v'} \tag{14.117}$$

is called the **Reynolds shear stress**.[10] Except for very small distances from a solid boundary, τ_t is much larger than the laminar stress, i.e.,

$$\tau_t \gg \mu\frac{\partial\overline{u}}{\partial y} \tag{14.118}$$

14.3.5   Turbulent Boundary-Layer Structure

Experimental observations for boundary layers in external flows and for flows in pipes and channels show a common structure. According to these observations, a turbulent boundary layer can be regarded as consisting of two layers. The layers differ in their response to changes in shear and pressure gradient, with the inner layer closest to the surface adjusting much more rapidly to changes than the outer layer. The difference reflects the inner layer containing only small eddies whose lifetime is relatively short. The constraint imposed by the no-slip condition at the surface precludes larger turbulent structures near the surface. Larger eddies exist farther from the wall, and their lifetime is much longer. Thus, when the flow changes abruptly, the outer layer takes longer to adjust to the change than the inner layer. Of course there is no sharp boundary between the two layers — the lifetime increases gradually with y.

If we denote the value of the shear stress at the surface by τ_w, simple dimensional arguments show that the quantity

$$u_\tau \equiv \sqrt{\frac{\tau_w}{\rho}} \tag{14.119}$$

has dimensions of velocity. We call u_τ the **friction velocity**. Using this velocity, we can introduce dimensionless velocity, u^+, and distance, y^+, as follows.

$$u^+ \equiv \frac{\overline{u}}{u_\tau} \quad\text{and}\quad y^+ \equiv \frac{u_\tau y}{\nu} \tag{14.120}$$

Figure 14.25 shows a typical velocity profile for a turbulent boundary layer. As shown, three distinct regions are discernible, viz., the **viscous sublayer**, the **log layer** and the **defect layer**. By definition, the log layer, sometimes referred to as the "fully turbulent wall layer," is the portion of the boundary layer where the sublayer and defect layer merge. It is not really a distinct layer, but is simply a buffer region between the inner and outer parts of the boundary layer.

Dimensional arguments combined with correlation of measurements indicate that the velocity in the viscous sublayer should depend only upon u_τ, ν and y. Hence, we expect to have a relationship of the form

$$u^+ = f\left(y^+\right) \tag{14.121}$$

where $f(y^+)$ is a dimensionless function. This general functional form is often referred to as the **law of the wall**. By contrast, in the defect layer, numerous experimenters including Darcy, von Kármán and Clauser found that velocity data correlate reasonably well with the so-called **velocity-defect law**:

$$\frac{u_e - \overline{u}}{u_\tau} = g\left(\frac{y}{\Delta}\right) \tag{14.122}$$

---

[10]Although they are zero in two dimensions, the correlations $-\rho\overline{v'w'}$ and $-\rho\overline{u'w'}$ are Reynolds shear stresses as well.

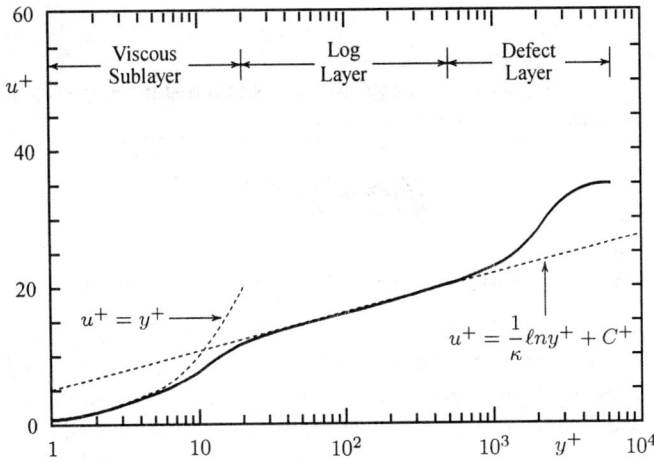

Figure 14.25: *Typical velocity profile for a turbulent boundary layer.*

where $g(y/\Delta)$ is another dimensionless function and Δ is a thickness characteristic of the outer portion of the boundary layer. The boundary-layer thickness is the scale chosen by Darcy. Clauser (1956) chose the length given by $\Delta = u_e \delta^*/u_\tau$.

Viscous Sublayer. The viscous sublayer is the region between the surface and the log layer. Since the no-slip boundary condition applies to u' and v', necessarily the Reynolds shear stress, $-\overline{\rho u'v'}$, is zero at $y = 0$. Hence, as with a laminar boundary layer, the horizontal velocity varies linearly with y close to the surface. In terms of our dimensionless velocity and distance, u^+ varies linearly with y^+ close to the surface, and gradually asymptotes to the law of the wall for large values of y^+. For $y^+ < 7$ we find

$$u^+ = y^+ \tag{14.123}$$

Log Layer. The log layer typically lies between $y^+ = 30$ and $y = 0.1\delta$. Since $\delta^+ = u_\tau \delta/\nu$ increases with Reynolds number, the location of $y/\delta = 0.1$ occurs at an increasingly larger value of y^+ as Reynolds number increases. Of particular interest to the present discussion, the velocity varies logarithmically with y in the log layer. Correlation of measurements shows that the velocity profile can be represented for a wide range of flows as

$$u^+ = \frac{1}{\kappa}\ell n y^+ + C^+ \tag{14.124}$$

where κ is the Kármán constant and C^+ is a constant that depends on surface roughness. Measurements indicate the value of κ is

$$\kappa \approx 0.41 \tag{14.125}$$

The value of C^+ for a perfectly-smooth surface is [see Kline et al. (1969)]:

$$C^+ \approx 5.0 \tag{14.126}$$

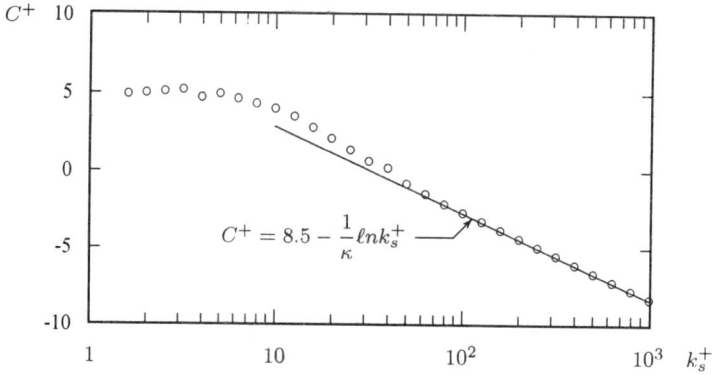

Figure 14.26: *Constant in the law of the wall, C^+, as a function of surface roughness; $\circ$ based on measurements of Nikuradse [Schlichting (1979)].*

The variation of u^+ with y^+ given in Equation (14.124) is also known as the **law of the wall**. It is of the same functional form as Equation (14.121), and corresponds to the limiting form of the function $f(y^+)$ for large y^+.

For surfaces with roughness elements of average height k_s, the law of the wall still holds, although C^+ is a function of k_s. Figure 14.26 illustrates how C^+ varies as a function of the dimensionless roughness height given by

$$k_s^+ \equiv \frac{u_\tau k_s}{\nu} \tag{14.127}$$

As shown, as k_s increases, the value of C^+ decreases. For large roughness height, measurements of Nikuradse [Schlichting (1979)] show that

$$C^+ \to 8.5 - \frac{1}{\kappa}\ell n k_s^+, \qquad k_s^+ \gg 1 \tag{14.128}$$

Substituting this value of C^+ into the law of the wall as represented in Equation (14.124) yields:

$$u^+ = \frac{1}{\kappa}\ell n \left(\frac{y}{k_s}\right) + 8.5 \quad \text{(completely-rough wall)} \tag{14.129}$$

Defect Layer. The defect layer lies between the log layer and the edge of the boundary layer. The velocity asymptotes to the law of the wall as $y/\delta \to 0$, and makes a noticeable departure from logarithmic behavior approaching the freestream. Again, from correlation of measurements, the velocity behaves as

$$u^+ = \frac{1}{\kappa}\ell n y^+ + C^+ + \frac{2\tilde{\Pi}}{\kappa}\sin^2\left(\frac{\pi}{2}\frac{y}{\delta}\right) \tag{14.130}$$

where $\tilde{\Pi}$ is **Coles' wake-strength parameter**. It varies with pressure gradient, and for constant pressure, $\tilde{\Pi} \approx 0.6$. Equation (14.130) is often referred to as the **composite law of the wall** and **law of the wake** profile.

Figure 14.27 shows how $\tilde{\Pi}$ varies with pressure gradient for the so-called **equilibrium turbulent boundary layer**. As demonstrated by Clauser (1956) experimentally and justified

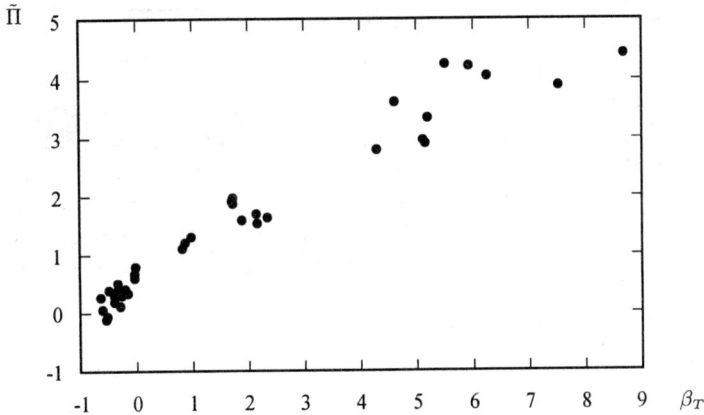

Figure 14.27: *Coles' wake-strength parameter, $\tilde{\Pi}$, as a function of pressure gradient;* • *based on data tabulated by Coles and Hirst (1969).*

by others analytically [cf. Wilcox (1993)], the velocity in the defect layer varies in a self-similar manner provided the **equilibrium parameter** defined by

$$\beta_T \equiv \frac{\delta^*}{\tau_w}\frac{d\overline{p}}{dx} \qquad (14.131)$$

is constant. Even when β_T is not constant, if it is not changing too rapidly, the value for $\tilde{\Pi}$ is close to the value shown in Figure 14.27.

Often, as an approximation, turbulent boundary-layer profiles are represented by a **power-law** relationship. That is, we sometimes say

$$\frac{\overline{u}}{u_e} = \left(\frac{y}{\delta}\right)^{1/n} \qquad (14.132)$$

where n is typically an integer between 6 and 8. A value of $n = 7$ yields a good approximation at high Reynolds number for the flat-plate boundary layer. Figure 14.28 compares a 1/7 power-law profile with measured values. As shown, the agreement between measured values and

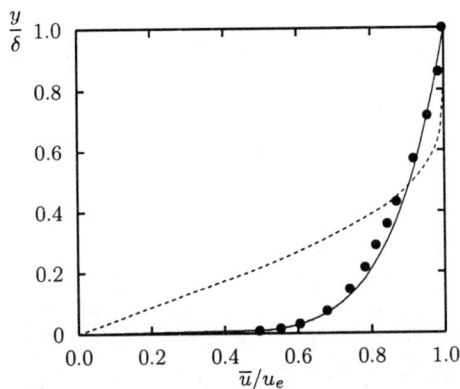

Figure 14.28: *Power-law velocity profile;* —— $\overline{u}/u_e = (y/\delta)^{1/7}$; • *Wieghardt data at* $Re_x = 1.09 \cdot 10^7$ *[Coles and Hirst (1969)]; - - - Blasius (laminar).*

the approximate profile is excellent with differences everywhere less than 3%. The laminar Blasius profile is shown to indicate the contrast between laminar- and turbulent-flow velocity profiles.

Finally, for a constant-pressure boundary layer, the measured boundary-layer thickness, δ, is

$$\delta \approx 0.37 x Re_x^{-1/5} \tag{14.133}$$

Using the 1/7 power-law profile, it is a straightforward operation to compute the displacement thickness, δ^*, momentum thickness, θ, and shape factor, H. Table 14.1 lists values for constant-pressure laminar and turbulent boundary layers. Based on these results, we can see immediately how much more filled out a turbulent boundary layer is relative to the laminar profile. Using values from the table, the ratio of displacement thickness to boundary-layer thickness is

$$\frac{\delta^*}{\delta} = \begin{cases} 0.344, & \text{Laminar} \\ 0.124, & \text{Turbulent} \end{cases} \tag{14.134}$$

Hence, the effective displacement is reduced from a little more than 1/3 of δ for a laminar boundary layer to 1/8 of δ for the turbulent case.

Table 14.1: *Integral Parameters for Laminar and Turbulent Flat-Plate Boundary Layers*

Property	Laminar Flow	Turbulent Flow
δ	$5.0 x Re_x^{-1/2}$	$0.37 x Re_x^{-1/5}$
δ^*	$1.721 x Re_x^{-1/2}$	$0.046 x Re_x^{-1/5}$
θ	$0.664 x Re_x^{-1/2}$	$0.036 x Re_x^{-1/5}$
H	2.59	1.28
c_f	$0.664 Re_x^{-1/2}$	$0.0576 Re_x^{-1/5}$

Table 14.1 also includes the skin friction, which follows from the constant-pressure version of the momentum integral equation [Equation (14.59)]. As shown in Figure 14.18, the power-law inferred skin-friction formula matches measured values for plate-length Reynolds number, Re_x, only up to about 10^6. A more accurate representation of the skin friction is provided by the Kármán-Schoenherr correlation [see Hopkins and Inouye (1971)]:

$$\frac{1}{c_f} = 17.08 \left(\log_{10} Re_\theta\right)^2 + 25.11 \log_{10} Re_\theta + 6.012 \tag{14.135}$$

This correlation more-closely follows the trend of measured skin friction, which exhibits a somewhat reduced rate of decrease with increasing Reynolds number.

14.3.6 Prandtl's Mixing-Length Hypothesis

At the end of Subsection 14.3.4, we noted that the fundamental problem of turbulence for the engineer is to establish a method for computing the Reynolds shear stress. One of the earliest attempts at formulating an engineering theory of turbulent motion is the mixing-length hypothesis proposed by Prandtl (1925). He visualized a simplified model for turbulent fluid motion in which fluid particles coalesce into lumps that cling together and move as a unit. He further visualized that in a shear flow such as that depicted in Figure 14.29, the lumps retain

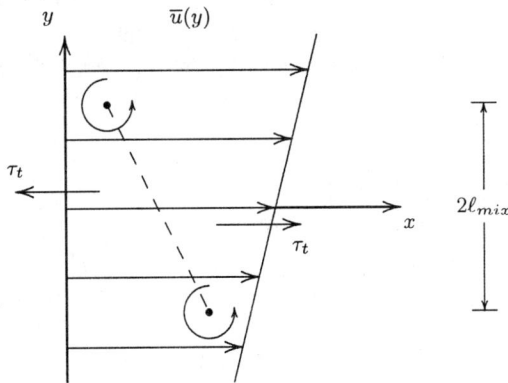

Figure 14.29: *Turbulent shear flow schematic.*

their x-directed momentum for a distance in the y direction, ℓ_{mix}, that he called the **mixing length**. In analogy to the molecular momentum-transport process with Prandtl's lump of fluid replacing the molecule and ℓ_{mix} replacing ℓ_{mfp}, we can say that similar to Equation (12.4),

$$\tau_t \equiv -\rho \overline{u'v'} = \frac{1}{2}\rho v_{mix}\ell_{mix}\frac{d\overline{u}}{dy} \tag{14.136}$$

The formulation is not yet complete because the **mixing velocity**, v_{mix}, has not been specified. Prandtl further postulated that

$$v_{mix} = \text{constant} \cdot \ell_{mix}\left|\frac{d\overline{u}}{dy}\right| \tag{14.137}$$

which makes sense on dimensional grounds. Because ℓ_{mix} is not a physical property of the fluid, we can always absorb the constant in Equation (14.137) and the factor 1/2 in Equation (14.136) in the mixing length. Thus, in analogy to Equations (12.5) and (12.6), Prandtl's mixing-length hypothesis leads to

$$\tau_t = \mu_t\frac{d\overline{u}}{dy} \tag{14.138}$$

where μ_t is the **eddy viscosity** given by

$$\mu_t = \rho\ell_{mix}^2\left|\frac{d\overline{u}}{dy}\right| \tag{14.139}$$

Our formulation still remains incomplete since we have replaced the unknown Reynolds shear stress, τ_t, with Prandtl's empirical mixing length, ℓ_{mix}. Prandtl postulated further that for flows near solid boundaries the mixing length is proportional to distance from the surface. This turns out to be a reasonably good approximation over a limited portion of a turbulent boundary layer.

Note that Equation (14.139) can be deduced directly from dimensional analysis. Assuming molecular transport of momentum is unimportant relative to turbulent transport, we expect molecular viscosity has no significance in a dimensional analysis. The only other dimensional parameters available in a shear flow are the fluid density, ρ, our assumed mixing length, ℓ_{mix}, and the velocity gradient, $d\overline{u}/dy$. (The eddy viscosity cannot depend upon $\overline{u}$ since that would violate Galilean invariance.) A straightforward dimensional analysis yields Equation (14.139).

Note also that the theoretical foundation for the mixing-length hypothesis is actually very weak. While Prandtl's approach of making an analogy between the motion of turbulent eddies and molecules provides a plausible rationale for the mixing length, it cannot be rigorously justified. As argued by Tennekes and Lumley (1983) and by Wilcox (1993), for example, the dynamics of turbulent motion are fundamentally different from that of molecules. In this spirit, the dimensional-analysis argument is as strong a theoretical case as can be made.

The mixing length is a property of the turbulent fluid motion as opposed to an intrinsic property of the fluid itself such as the mean free path. In general, ℓ_{mix} is affected by the history of the flow, the nature of the surface, the distance from the surface, the pressure gradient, etc. In general, it must be calibrated for each application. While such calibration is generally fairly easy, the model applies only to flows very similar to those for which ℓ_{mix} has been optimized. Despite these limitations, computational methods based on the mixing-length hypothesis have proven very useful in engineering applications.

Consider a constant-pressure turbulent boundary layer, for which the equations of motion simplify to

$$\frac{\partial \overline{u}}{\partial x} + \frac{\partial \overline{v}}{\partial y} = 0 \tag{14.140}$$

$$\rho \overline{u} \frac{\partial \overline{u}}{\partial x} + \rho \overline{v} \frac{\partial \overline{u}}{\partial y} = \frac{\partial}{\partial y} \left[\mu \frac{\partial \overline{u}}{\partial y} + \tau_t \right] \tag{14.141}$$

It turns out that the convective terms are negligible in the log layer, so that the sum of the viscous and Reynolds shear stress must be constant. Hence, we can say

$$\mu \frac{\partial \overline{u}}{\partial y} + \tau_t \approx \mu \left(\frac{\partial \overline{u}}{\partial y} \right)_w = \tau_w = \rho u_\tau^2 \tag{14.142}$$

where subscript w denotes value at the wall and u_τ is the friction velocity [Equation (14.119)]. As we move from the viscous sublayer into the log layer, the Reynolds stress becomes much larger than the viscous stress. Consequently, neglecting $\mu \partial \overline{u}/\partial y$ in Equation (14.142), and using the mixing-layer model to specify τ_t,

$$\tau_t = \rho \ell_{mix}^2 \left(\frac{\partial \overline{u}}{\partial y} \right)^2 \approx \rho u_\tau^2 \tag{14.143}$$

Since the velocity is given by the law of the wall, differentiation of Equation (14.124) tells us

$$\frac{\partial \overline{u}}{\partial y} = \frac{u_\tau}{\kappa y} \tag{14.144}$$

Substitution into Equation (14.143) shows that, to be consistent with the law of the wall, the mixing length must be

$$\ell_{mix} = \kappa y \tag{14.145}$$

14.3.7   Modern Turbulence Theories

Numerous attempts have been made during the twentieth century aimed at developing an accurate method for computing the Reynolds stresses in a turbulent flow. In the context of practical engineering applications, these attempts have been based on a combination of physical reasoning, dimensional analysis and empiricism. This approach has led to what are commonly referred to as **turbulence models**. Wilcox (1993) provides a comprehensive overview of the development of these models.

The earliest attempts focused on the mixing-length hypothesis. Although a few simple flows were analyzed based on mixing-length models, the development of computers was necessary to make computation of turbulent boundary layers possible. Cebeci and Smith [see Smith and Cebeci (1967)] presented what has served as one of the most widely used versions of such models. For boundary layers, the eddy viscosity is computed according to

$$\mu_t = \min\{\mu_{t_i}, \mu_{t_o}\} \tag{14.146}$$

where μ_{t_i} and μ_{t_o} are referred to as the inner and outer layer viscosities, respectively. The mixing-length formula is used in the inner layer so that

$$\mu_{t_i} = \rho \ell_{mix}^2 \left| \frac{\partial \overline{u}}{\partial y} \right|, \qquad \ell_{mix} = \kappa y \left[1 - e^{-y^+/26} \right] \tag{14.147}$$

with the Kármán constant assumed to be $\kappa = 0.40$. The outer layer viscosity is

$$\mu_{t_o} = 0.0168 \rho u_e \delta^* \left[1 + 5.5 \left(\frac{y}{\delta} \right)^6 \right]^{-1} \tag{14.148}$$

The **Cebeci-Smith model** is especially elegant and easy to implement. Most of the computational effort, relative to a laminar case, goes into computing the displacement thickness. This quantity is readily available in boundary-layer computations so that a laminar-flow program can usually be converted to a turbulent-flow program with just a few extra lines of instructions. Figure 14.30 illustrates a typical eddy viscosity profile for the Cebeci-Smith model. At Reynolds numbers representative of most applications, matching between the inner and outer layers occurs well into the log layer.

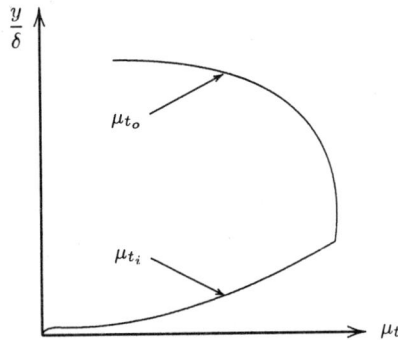

Figure 14.30: *Eddy viscosity for the Cebeci-Smith model.*

Baldwin and Lomax (1978) have formulated a similar model for use in computations where boundary-layer properties such as δ, δ^* and u_e are difficult to determine. This situation often arises in numerical simulation of separated flows, especially for flows with shock waves. The primary difference between the Cebeci-Smith and Baldwin-Lomax models is in the way μ_{t_o} is computed, which is a bit more complicated for the Baldwin-Lomax formulation. There is little difference in accuracy between the two models. While the Baldwin-Lomax model is more commonly used in modern research, the Cebeci-Smith model is detailed above because of its simplicity.

Mixing-length models adequately reproduce measured skin friction and velocity profiles for attached boundary layers. However, such models fail to accurately predict properties of separated flows. They also require modification for flows away from solid boundaries, and for flow over surfaces with mass transfer, roughness, curvature, etc. In a quest for a more universally applicable model, more sophisticated models have been developed, generally involving additional differential equations to be solved along with the basic conservation equations for mass, momentum and energy.

For example, there are **one-equation models** that involve a differential equation for the eddy viscosity, μ_t. The one-equation model formulated by Spalart and Allmaras (1992) has proven to be quite accurate for flow past airfoils, which is the primary application it has been designed for. It appears to apply to a variety of other flows reasonably well.

Considerable effort has gone into development of **two-equation models,** for which two differential equations must be solved to determine the eddy viscosity. Two such models are the k-ϵ model of Launder and Sharma (1974) and the k-ω model of Wilcox (1988a). In both models, the quantity k is the **turbulence kinetic energy** defined by

$$k \equiv \frac{1}{2} \left(\overline{u'^2} + \overline{v'^2} + \overline{w'^2} \right) \tag{14.149}$$

An equation is postulated from which k can be determined. The second equation in the k-ϵ model is the quantity ϵ, defined as the rate at which k decays. For the k-ω model, the second equation is for $\omega = \epsilon/k$. The eddy viscosity is given by

$$\mu_t = \begin{cases} C_\mu \rho k^2 / \epsilon, & k\text{-}\epsilon \text{ model} \\[2mm] \rho k / \omega, & k\text{-}\omega \text{ model} \end{cases} \tag{14.150}$$

where C_μ is a constant. The k-ϵ model has enjoyed widespread usage, although it is very inaccurate for flows with adverse pressure gradient and is completely unreliable for separated flows. By contrast, the k-ω model provides very accurate predictions for turbulent boundary layers with adverse pressure gradient. The model also provides accurate predictions for flows with modest separated regions.

Figure 14.31 compares computed and measured wake-strength, $\tilde{\Pi}$, as a function of the equilibrium parameter, β_T [see Equation (14.131)]. Computed results are shown for the Cebeci-Smith, Spalart-Allmaras, k-ω and k-ϵ models. The k-ω and Spalart-Allmaras model predictions are closest to measured values, while the Cebeci-Smith prediction is about 15% below that of the k-ω model. For the largest adverse pressure gradient included ($\beta_T = 9$), the k-ϵ model's predicted wake strength is only 60% of the k-ω value, far below the measurements.

Even more complicated formulations, known as **second-order closure models** [Launder, Reece and Rodi (1975), Wilcox (1988b)], have been developed that involve five or more differential equations (for two-dimensional flows). These models compute the Reynolds stress directly, without assuming existence of an eddy viscosity. On the one hand, they permit representation of turbulent-flow physics that cannot be properly addressed using an eddy-viscosity model. On the other hand, their complexity tends to discourage widespread use, both from a conceptual and a computational point of view.

With the increasing speed and memory of computers, solution of the full three-dimensional, time-dependent Navier-Stokes and continuity equations has become feasible, albeit at relatively low Reynolds numbers. The value of such "simulations" is obvious: they give a complete

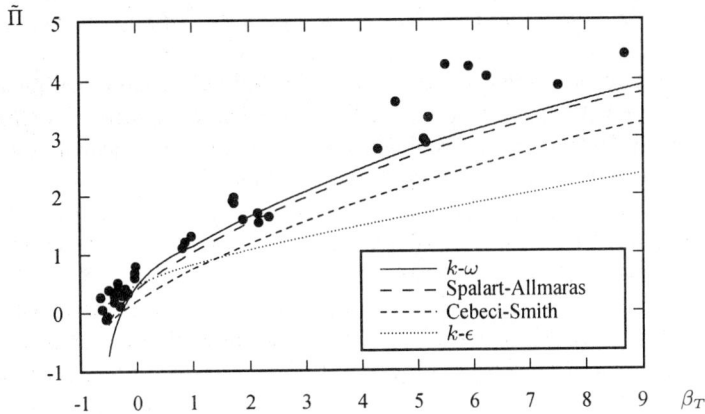

Figure 14.31: *Comparison of computed and measured wake strength,* $\tilde{\Pi}$*, as a function of pressure gradient;* • *based on data of Coles and Hirst (1969).*

description of the turbulence. From a practical standpoint, computed statistics can be used to test engineering models. At the most fundamental level, they can be used to obtain understanding of turbulence structure and processes that can be of value in developing turbulence control methods (e.g., drag reduction) or prediction methods. They can also be viewed as an additional source of experimental data that have been taken with unobtrusive measuring techniques.

Two techniques have evolved, viz., **Direct Numerical Simulation** (DNS) and **Large Eddy Simulation** (LES). A Direct Numerical Simulation is a complete time-dependent solution of the Navier-Stokes and continuity equations. A Large Eddy Simulation is a computation in which the large eddies are computed and the smallest eddies are modeled. While LES involves approximations, it also requires much less computing time than DNS, and holds promise for practical applications in the not too distant future. For more detail at an introductory level, see the excellent review articles by Rogallo and Moin (1984) and Piomelli (1994).

14.4 Computational Fluid Dynamics

In aerodynamic theory, the classic approach has been to first compute inviscid flow past a given body. Then, using the computed surface pressure, the boundary-layer equations are solved to determine viscous effects. This method is very accurate for thin, streamlined bodies such as airfoils provided the flow remains attached (although iteration to account for effects of δ^* is sometimes needed). The method is also very efficient from a computational point of view because the boundary-layer computation requires only minor computer resources. This section illustrates the classical aerodynamic approach, including details of one of the most popular numerical methods for solving the boundary-layer equations.

14.4.1 Parabolic Marching Methods

The primary reason the boundary-layer equations can be solved with great computational ease results from a key mathematical feature. Specifically, the solution at a given streamwise location, x, is completely unaffected by conditions at positions downstream of x. The boundary layer on an airfoil, for example, is oblivious to the presence of a flap or aileron until it reaches the appendage. In the theory of partial differential equations, the boundary-layer equations are classified as **parabolic**. Since the solution is unaffected by downstream conditions, beginning with suitable initial conditions at a point on a surface, we can solve by "marching" along the surface in the streamwise direction. We thus refer to computational methods used to solve the boundary-layer equations as **parabolic marching methods**.

The success of parabolic marching methods depends upon the flow remaining attached. In this case, the inviscid flow sees a slightly thicker object due to the displacement effect [Figures 14.9 and 14.15(a)] that has only a minor effect on the flowfield. The interaction between viscous and inviscid regions can be said to be weak, so that uncoupling the viscous and inviscid regions is justified. With care, small separated regions can be accommodated using inverse boundary-layer methods. Inverse methods achieve a coupling between viscous and inviscid regions through an iterative procedure. However, at each step of the iteration, the boundary-layer equations are still solved with a parabolic marching procedure.

Flows with large regions of separation include strong interaction between the inviscid and viscous flow [Figures 14.15(b) and 14.16]. These flows cannot be computed using parabolic marching methods because the basic boundary-layer approximations are not valid, and the reversed flow means that downstream conditions now affect what happens upstream. Only a solution of the full Navier-Stokes and continuity equations will suffice.

There are two particularly noteworthy methods for solving the boundary-layer equations that have enjoyed widespread use. One is the **Keller box method**, which is described in detail by Keller (1970) and by Cebeci and Smith (1974). The other is the **Blottner variable-grid method** devised by Blottner (1974). The latter scheme is the focus of this section.

14.4.2   Stretched Finite-Difference Grids

Solving the boundary-layer equations with equally-spaced grid points is impractical, especially for turbulent boundary layers. To understand why this is true, consider the following. The most rapid variations in the velocity occur in the viscous sublayer, which extends from the surface to about $y^+ = 30$. To accurately compute the velocity profile and its slope at the surface (from which the skin friction is determined), clearly we must resolve the sublayer. Experience has shown that the first grid point above the surface should lie below $y^+ = 1$. If we were to use equally-spaced grid points, the total number of points required to resolve the entire boundary layer would be prohibitively large.

To demonstrate just how many points would be required, we can use Equation (14.133) to estimate the thickness of a constant-pressure turbulent boundary layer. The dimensionless boundary-layer thickness, δ^+, is

$$\delta^+ \equiv \frac{u_\tau \delta}{\nu} \approx 0.37 \frac{u_\tau x}{\nu} Re_x^{-1/5}$$

$$= 0.37 \frac{u_\tau}{u_e} Re_x^{4/5} = 0.37 \sqrt{\frac{c_f}{2}} Re_x^{4/5} \qquad (14.151)$$

Then, using the approximate skin friction formula for a flat-plate boundary layer from Table 14.1, $c_f = 0.0576 Re_x^{-1/5}$, we find

$$\delta^+ = 0.063 Re_x^{7/10} \qquad (14.152)$$

The number of equally-spaced points, with the first at $y^+ < 1$, must be larger than this. Therefore, the dimensionless boundary-layer thickness for Reynolds numbers of 10^6 and 10^7

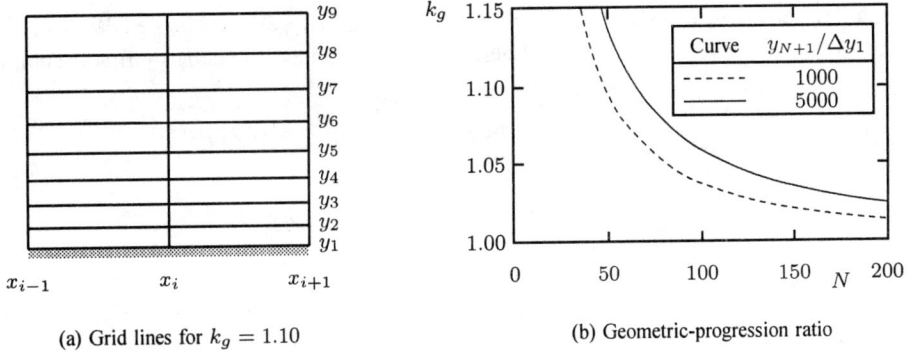

(a) Grid lines for $k_g = 1.10$

(b) Geometric-progression ratio

Figure 14.32: *Geometrically-stretched finite-difference grid.*

will be

$$\delta^+ \approx \begin{cases} 1000, & Re_x = 10^6 \\ 5000, & Re_x = 10^7 \end{cases} \qquad (14.153)$$

Affordable Navier-Stokes computations must resolve turbulent boundary layers with 40 or 50 grid points. Even boundary-layer computations rarely use more than 200 grid points.

To get by with far fewer grid points in general CFD work, we must use **non-uniform grids**, i.e., grids with variable distance between points. We adopt a strategy by which points are concentrated in regions where properties vary most rapidly. The spacing between grid points is smallest in such regions, and, of necessity, increases as we move away from where the action is. In other words, we "stretch" the grid. The use of **stretched grids** is common in viscous-flow computations because of the need to resolve the thin boundary layers near a surface.

In the case of a turbulent boundary layer, the most rapid changes occur near the surface.[11] One popular way to distribute grid points in a turbulent boundary layer is to use grid-point spacing that increases in a geometric progression. That is, the increments between grid points are given by

$$\Delta y_j = k_g \Delta y_{j-1} \qquad (14.154)$$

where k_g is the **geometric-progression ratio**. Figure 14.32(a) illustrates the vertical placement of mesh lines for a geometrically-stretched grid. If we use $N + 1$ grid points, the point most distant from the surface lies at

$$y_{N+1} = \Delta y_1 \frac{(k_g - 1)^N}{(k_g - 1)} \qquad (14.155)$$

Equation (14.155) is the sum of the first N terms in a geometric progression. Figure 14.32(b) shows how k_g varies with N for the two cases discussed above, i.e., for a turbulent boundary layer at Reynolds numbers of 10^6 and 10^7, with the first grid point at $y^+ = 1$.

If we choose to use a finite-difference grid with non-uniform spacing, the standard central-difference formulas such as Equations (11.245) and (11.251) fail to provide second-order accuracy. With a little extra effort, however, we can derive more appropriate discretization approximations for stretched grids. Consider any three consecutive grid points, y_{j-1}, y_j and y_{j+1}. Let the spacing between two adjacent grid points be defined by

$$\Delta y_j \equiv y_{j+1} - y_j \qquad (14.156)$$

---

[11] As discussed by Wilcox (1993), there is also a need to cluster grid points near the boundary-layer edge for some turbulence models.

This definition holds for all j so that, for example, $\Delta y_{j-1} = y_j - y_{j-1}$. We can derive approximations for the first and second derivatives of a function $u(x, y)$ by first assuming

$$\left. \begin{array}{rcl} \left(\dfrac{\partial u}{\partial y}\right)_{i,j} & \approx & \mathcal{D}_1^{(1)} u_{i,j+1} + \mathcal{D}_2^{(1)} u_{i,j} + \mathcal{D}_3^{(1)} u_{i,j-1} \\[3mm] \left(\dfrac{\partial^2 u}{\partial y^2}\right)_{i,j} & \approx & \mathcal{D}_1^{(2)} u_{i,j+1} + \mathcal{D}_2^{(2)} u_{i,j} + \mathcal{D}_3^{(2)} u_{i,j-1} \end{array} \right\} \qquad (14.157)$$

where $\mathcal{D}_1^{(1)}$, $\mathcal{D}_2^{(1)}$, $\mathcal{D}_3^{(1)}$, $\mathcal{D}_1^{(2)}$, $\mathcal{D}_2^{(2)}$ and $\mathcal{D}_3^{(2)}$ are coefficients to be determined. As in earlier discussion of finite-difference methods, we use $u_{i,j}$ as a shorthand for $u(x_i, y_j)$. We can uniquely solve for the coefficients in Equations (14.157) and estimate the truncation errors by using Taylor series expansions for $u_{i,j+1}$ and $u_{i,j-1}$ about the point (x_i, y_j). This is the procedure we used in Subsection 11.9.1 to determine the truncation error for the central-difference formulas. A straightforward algebraic exercise shows that if

$$\left. \begin{array}{rcl} \mathcal{D}_1^{(1)} & = & \dfrac{\Delta y_{j-1}}{\Delta y_j \left(\Delta y_{j-1} + \Delta y_j\right)} \\[4mm] \mathcal{D}_2^{(1)} & = & \dfrac{\Delta y_j - \Delta y_{j-1}}{\Delta y_j \Delta y_{j-1}} \\[4mm] \mathcal{D}_3^{(1)} & = & \dfrac{-\Delta y_j}{\Delta y_{j-1} \left(\Delta y_{j-1} + \Delta y_j\right)} \end{array} \right\} \qquad (14.158)$$

then, our discretization approximation for the first derivative becomes

$$\mathcal{D}_1^{(1)} u_{i,j+1} + \mathcal{D}_2^{(1)} u_{i,j} + \mathcal{D}_3^{(1)} u_{i,j-1} = \left(\frac{\partial u}{\partial y}\right)_{i,j} + \frac{1}{6}\left(\frac{\partial^3 u}{\partial y^3}\right)_{i,j} \Delta y_j \Delta y_{j-1} + \cdots \quad (14.159)$$

Therefore, for a non-uniform grid, we see that combining Equations (14.157) and (14.158) provides a second-order accurate approximation for the first derivative of u.

A similar exercise shows that, in a non-uniform grid, if the coefficients $\mathcal{D}_1^{(2)}$, $\mathcal{D}_2^{(2)}$ and $\mathcal{D}_3^{(2)}$ are defined as follows:

$$\left. \begin{array}{rcl} \mathcal{D}_1^{(2)} & = & \dfrac{1}{\Delta y_j \left(\Delta y_{j-1} + \Delta y_j\right)} \\[4mm] \mathcal{D}_2^{(2)} & = & \dfrac{-1}{\Delta y_j \Delta y_{j-1}} \\[4mm] \mathcal{D}_3^{(2)} & = & \dfrac{1}{\Delta y_{j-1} \left(\Delta y_{j-1} + \Delta y_j\right)} \end{array} \right\} \qquad (14.160)$$

then the second derivative is given by

$$\mathcal{D}_1^{(2)} u_{i,j+1} + \mathcal{D}_2^{(2)} u_{i,j} + \mathcal{D}_3^{(2)} u_{i,j-1} = \left(\frac{\partial^2 u}{\partial y^2}\right)_{i,j} + \frac{1}{6}\left(\frac{\partial^3 u}{\partial y^3}\right)_{i,j} \left(\Delta y_j - \Delta y_{j-1}\right) + \cdots \quad (14.161)$$

In contrast to the first derivative, we can't quite achieve second-order accuracy for the second derivative in a non-uniform grid.

For reasons that will become clear when we introduce the Blottner variable-grid method in the next subsection, we require the first derivative with respect to x to be centered at x_{i+1} [see Figure 14.32(a)]. In forming our discretization approximation, the only information available will be from the upstream grid points at x_{i-1} and x_i. For equally-spaced points in the streamwise direction, a commonly used second-order accurate approximation is the three-point forward-difference formula attributed to Adams and Bashforth [see Roache (1972) or Ferziger and Perić (1996)], viz.,

$$\left(\frac{\partial u}{\partial x}\right)_{i+1,j} \approx \frac{3u_{i+1,j} - 4u_{i,j} + u_{i-1,j}}{2\Delta x} \tag{14.162}$$

In practice, non-uniform steps are used in the streamwise direction as well as in the direction normal to the surface. This permits clustering points in regions where the boundary layer might experience rapid streamwise changes caused by, for example, a sudden change in pressure gradient. For a non-uniform streamwise grid, we replace the Adams-Bashforth formula by the following:

$$\left(\frac{\partial u}{\partial x}\right)_{i+1,j} \approx Z_1\, u_{i+1,j} - Z_2\, u_{i,j} + Z_3\, u_{i-1,j} \tag{14.163}$$

where

$$\left.\begin{aligned}
Z_1 &= \frac{\Delta x_{i-1} + 2\Delta x_i}{\Delta x_i\,(\Delta x_{i-1} + \Delta x_i)} \\[2mm]
Z_2 &= \frac{\Delta x_{i-1} + \Delta x_i}{\Delta x_i \Delta x_{i-1}} \\[2mm]
Z_3 &= \frac{\Delta x_i}{\Delta x_{i-1}\,(\Delta x_{i-1} + \Delta x_i)}
\end{aligned}\right\} \tag{14.164}$$

The interested reader can readily verify that Equations (14.157) reduce to the central-difference formulas in the limit of equally-spaced grid points in the y direction. Also, Equation (14.163) reduces to the Adams-Bashforth formula [Equation (14.162)] in the limit of constant Δx.

14.4.3 Blottner's Variable-Grid Method

Blottner (1974) has developed an especially accurate and efficient method for solving the boundary-layer equations. The power of the Blottner algorithm lies in its ability to handle stretched grids with a minimum number of grid points. For equally-spaced points, the method reduces to a combination of the Adams-Bashforth formula in the x direction and central differences in the y direction. In this limiting case, the method is second-order accurate in both the x and y directions.

To define the algorithm, consider the continuity equation and the x component of the momentum equation for a turbulent boundary layer. If we represent the Reynolds shear stress with an eddy viscosity approximation, we have

$$\frac{\partial \overline{u}}{\partial x} + \frac{\partial \overline{v}}{\partial y} = 0 \tag{14.165}$$

$$\overline{u}\frac{\partial \overline{u}}{\partial x} + \overline{v}\frac{\partial \overline{u}}{\partial y} = -\frac{1}{\rho}\frac{d\overline{p}}{dx} + \frac{\partial}{\partial y}\left[(\nu + \nu_t)\frac{\partial \overline{u}}{\partial y}\right] \tag{14.166}$$

We use the finite-difference molecule shown in Figure 14.33. The computation proceeds in the streamwise direction, i.e., in the direction of increasing subscript i. To solve along the grid line at $x = x_{i+1}$ we require information from the previous two lines at $x = x_{i-1}$ and $x = x_i$. As the computation progresses, we solve along the grid line at $x = x_{i+1}$ for flow properties at all values of y_j for $j = 1$ to $j = j_{max}$.

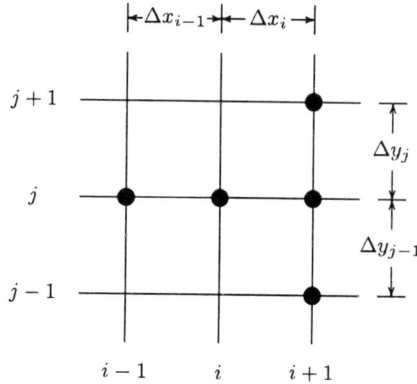

Figure 14.33: *Finite-difference molecule for Blottner's variable-grid method.*

The Blottner variable-grid method uses the following discretization approximations:

$$\overline{u}\frac{\partial \overline{u}}{\partial x} \approx \hat{u}_j \left(Z_1\,\overline{u}_{i+1,j} - Z_2\,\overline{u}_{i,j} + Z_3\,\overline{u}_{i-1,j} \right) \tag{14.167}$$

$$\overline{v}\frac{\partial \overline{u}}{\partial y} \approx \hat{v}_j \left(\frac{\overline{u}_{i+1,j+1} - \overline{u}_{i+1,j-1}}{\Delta y_{j-1} + \Delta y_j} \right) \tag{14.168}$$

$$\frac{\partial}{\partial y}\left[(\nu + \nu_t)\frac{\partial \overline{u}}{\partial y} \right] \approx \frac{\hat{\nu}_{j+\frac{1}{2}}\left(\dfrac{\overline{u}_{i+1,j+1} - \overline{u}_{i+1,j}}{\Delta y_j} \right) - \hat{\nu}_{j-\frac{1}{2}}\left(\dfrac{\overline{u}_{i+1,j} - \overline{u}_{i+1,j-1}}{\Delta y_{j-1}} \right)}{\frac{1}{2}\left(\Delta y_{j-1} + \Delta y_j\right)} \tag{14.169}$$

The quantities Z_1, Z_2 and Z_3 appearing in Equation (14.167) have been defined above in Equations (14.164). The two velocities $\hat{u}_j$ and $\hat{v}_j$ are determined by one of two possible methods. On the one hand, if an iterative solution is being done on each grid line, they are the values of $u_{i+1,j}$ and $v_{i+1,j}$ from the last iteration. On the other hand, for the first iterate, or in the case that no iteration is needed, we determine $\hat{u}_j$ and $\hat{v}_j$ by extrapolating from the two previous x stations. To extrapolate, we in effect pass a straight line through the values at x_{i-1} and x_i and compute the velocity according to

$$\hat{u}_j = Z_4\,\overline{u}_{i,j} - Z_5\,\overline{u}_{i-1,j}, \qquad \hat{v}_j = Z_4\,\overline{v}_{i,j} - Z_5\,\overline{v}_{i-1,j} \tag{14.170}$$

where the coefficients Z_4 and Z_5 are defined by

$$Z_4 = \frac{\Delta x_{i-1} + \Delta x_i}{\Delta x_{i-1}}, \qquad Z_5 = \frac{\Delta x_i}{\Delta x_{i-1}} \tag{14.171}$$

To achieve proper centering for the diffusion term in Equation (14.169), we define $\hat{\nu}_{j+\frac{1}{2}}$ and $\hat{\nu}_{j-\frac{1}{2}}$ as follows.

$$\left.\begin{aligned}
\hat{\nu}_{j+\frac{1}{2}} &= \frac{1}{2}\left[(\nu+\nu_t)_{i+1,j} + (\nu+\nu_t)_{i+1,j+1}\right] \\
\hat{\nu}_{j-\frac{1}{2}} &= \frac{1}{2}\left[(\nu+\nu_t)_{i+1,j} + (\nu+\nu_t)_{i+1,j-1}\right]
\end{aligned}\right\} \tag{14.172}$$

Collecting these discretization approximations for the various terms in the momentum Equation (14.166) yields a classical tridiagonal-matrix equation similar to the equation we encountered with Crank-Nicolson differencing (see Subsection 13.7.4). The equation is

$$A_j\overline{u}_{i+1,j-1} + B_j\overline{u}_{i+1,j} + C_j\overline{u}_{i+1,j+1} = D_j \tag{14.173}$$

where

$$\left.\begin{aligned}
A_j &= -\left[Y_3\,\hat{\nu}_{j-\frac{1}{2}} + Y_2\,\hat{v}_j\right] \\[4pt]
B_j &= \left[Z_1\,\hat{u}_j + Y_1\,\hat{\nu}_{j+\frac{1}{2}} + Y_3\,\hat{\nu}_{j-\frac{1}{2}}\right] \\[4pt]
C_j &= -\left[Y_1\,\hat{\nu}_{j+\frac{1}{2}} - Y_2\,\hat{v}_j\right] \\[4pt]
D_j &= \hat{u}_j\left[Z_2\,\overline{u}_{i,j} - Z_3\,\overline{u}_{i-1,j}\right] - \frac{1}{\rho}\frac{d\overline{p}}{dx}
\end{aligned}\right\} \tag{14.174}$$

The differencing coefficients Y_1, Y_2 and Y_3 appearing in the definitions of A_j, B_j, C_j and D_j are defined by

$$\left.\begin{aligned}
Y_1 &= \frac{1}{\Delta y_j\,(\Delta y_{j-1}+\Delta y_j)} \\[6pt]
Y_2 &= \frac{1}{\Delta y_{j-1}+\Delta y_j} \\[6pt]
Y_3 &= \frac{1}{\Delta y_{j-1}\,(\Delta y_{j-1}+\Delta y_j)}
\end{aligned}\right\} \tag{14.175}$$

Finally, the discretized form of the continuity Equation (14.165) becomes

$$Z_1\,\overline{u}_{i+1,j} - Z_2\,\overline{u}_{i,j} + Z_3\,\overline{u}_{i-1,j} + Y_2\,(\overline{v}_{i+1,j+1} - \overline{v}_{i+1,j-1}) = 0 \tag{14.176}$$

This can be solved directly for the vertical velocity, viz.,

$$\overline{v}_{i+1,j+1} = \overline{v}_{i+1,j-1} - \frac{Z_1\,\overline{u}_{i+1,j} - Z_2\,\overline{u}_{i,j} + Z_3\,\overline{u}_{i-1,j}}{Y_2} \tag{14.177}$$

The vertical velocity is computed after the momentum equation has been updated. In an iterative solution, typically accomplished with underrelaxation, the most recent iterate is used for $\overline{u}_{i+1,j}$.

14.4.4 Predicting Separation on a Rankine Oval

Armed with appropriate computational tools, we can now turn to an example of the classical aerodynamic approach in which we use inviscid theory to compute flow past a body followed

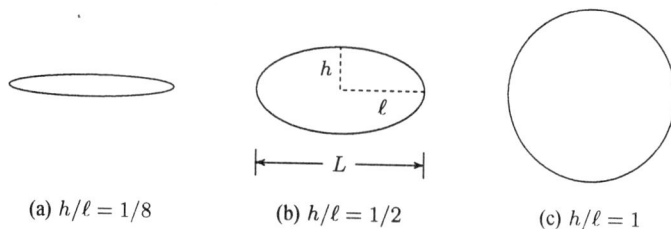

(a) $h/\ell = 1/8$ (b) $h/\ell = 1/2$ (c) $h/\ell = 1$

Figure 14.34: *Shape of Rankine ovals for several height to length ratios.*

by a boundary-layer computation. We consider Rankine ovals over the complete range of thickness ratios from $h/\ell = 0$, corresponding to a flat plate, to $h/\ell = 1$, corresponding to a circular cylinder. Figure 14.34 includes sketches of three Rankine ovals.

Except for very slender ovals, we expect the flow to separate from the downstream surface of the body. To achieve a physically realistic prediction for the separation point, a boundary-layer computation would require the measured pressure distribution as input. Since we will be using the inviscid pressure distribution, our computed separation point will be in error. Nevertheless, we can obtain some interesting information on relative trends of various turbulence models, which is one of the goals of this computational exercise. Beginning with the solution developed in Subsection 11.7.2, the inviscid surface pressure can be obtained. This is most conveniently done by writing a simple program, as the algebraic expressions for the pressure are rather complicated. Figure 14.35 shows the surface pressure for the three Rankine ovals depicted in Figure 14.34.

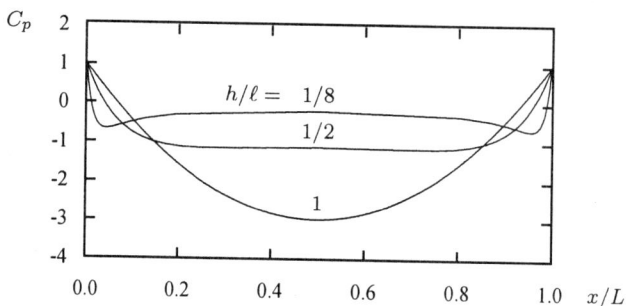

Figure 14.35: *Surface pressure coefficient for Rankine ovals.*

Figure 14.36 shows the computed separation point, x_{sep}/L, on the Rankine oval as a function of thickness ratio. Results are included for both laminar and turbulent flow. The turbulent-flow results correspond to the k-ω and k-ϵ models. All computations were done with Program **EDDYBL** [see Appendix F)], a boundary-layer program that uses the Blottner variable-grid method. The laminar computations were initiated at the stagnation point. The turbulent computations have been initiated at an axial distance $x/L = 0.15$ from approximate velocity and turbulence-property profiles.

Average computing times on a 100 MHz 80486-based microcomputer are 2 seconds for laminar flow, 7 seconds for the k-ω model and 15 seconds for the k-ϵ model. The laminar computing times are very short as no iteration is required. In addition to requiring iteration, the turbulence-model computations require solution of two additional differential equations. The k-ϵ model run times are double those of the k-ω model because stable and accurate computation requires smaller streamwise steps with the k-ϵ model.

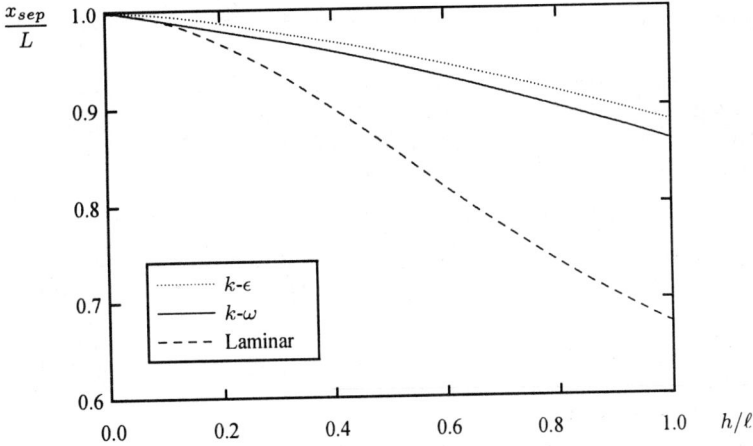

Figure 14.36: *Computed separation points for Rankine ovals.*

As shown, separation occurs much closer to the trailing edge of the oval when the flow is turbulent. The computations also show that separation occurs later according to the k-ϵ model, relative to the predictions of the k-ω model. This is completely consistent with the k-ϵ model's weak response to adverse pressure gradient. In general applications, many researchers have found that the k-ϵ model predicts flow separation at locations downstream of measured separation. By contrast, the k-ω model predicts separation in closer agreement with measurements for incompressible, separated flows. Thus, our computations are consistent with general observations.

In the limiting case where the oval becomes a cylinder, the computed laminar separation point occurs at an angle of 78° measured from the downstream symmetry axis. This is quite a bit smaller than the measured angle of 98° for subcritical flow past a cylinder [Figure 11.12(a)] in which the boundary layer remains laminar. The difference is attributable to the significant difference between the measured and potential-flow surface pressure [see Figure 11.10]. The turbulent-flow separation angle for the cylindrical case is 43° for the k-ω model and 40° for the k-ϵ model. These angles are smaller than the measured value of 60° for the supercritical case [Figure 11.12(b)] where the boundary layer on the cylinder is turbulent.

The Problems section includes an exercise that uses the measured pressure distribution for flow past a cylinder. The k-ω model predicts separation within 10% of the measured location. By contrast, the Cebeci-Smith, Spalart-Allmaras and k-ϵ models all fail to predict separation.

The program used to compute the surface-pressure distributions on the oval is called **RANKIN**, and is described in Appendix E. The program generates input-data in the format required by the boundary-layer program used to perform these computations, **EDDYBL**. The diskette supplied with this book includes source code for both programs, and several input-data files.

Problems

14.1 Compute and compare the Reynolds numbers for: (a) a 17-foot long automobile in a traffic jam moving at 7 mph on a day when the temperature is 90° F; and (b) a 30-cm long toy car traveling at 10 m/sec on a day when the temperature is 21° C. Refer to the tables in Chapter 1 for properties of air. Assume the pressure is 1 atm, and use *Sutherland's equation* to determine the viscosity.

14.2 Compute and compare the Reynolds numbers for: (a) a 26-foot long sailboat drifting at 1 knot on a day when the water temperature is 50° F; and a 50-cm long toy motorboat traveling at 8 m/sec on a day when the Jamaican water temperature is 30° C. Refer to the tables in Chapter 1 for properties of water.

14.3 At what speed must an object of characteristic size 1 foot move to have a Reynolds number of just 10^5 in air and in water? Assume pressure is atmospheric and the temperature is 68° F. Refer to Chapter 1 for properties of air and water.

14.4 Variations in viscosity with temperature are much more rapid in water than in air. To appreciate just how much of an effect there is and how it might affect viscous forces on an object, compute the Reynolds number on a 100-meter long tanker moving at 15 knots near Hawaii (water temperature 30° C) and near Alaska (water temperature 0° C – but not frozen). Refer to Table 1.7 for the kinematic viscosity of water. Repeat your computations with the viscosity of air at the same temperatures. Assume the pressure is 1 atm, and use *Sutherland's equation* to determine the viscosity.

14.5 To appreciate how viscosity affects the thickness of a boundary layer, imagine that we observe flow past a thin flat plate using fluids of widely varying viscosities. The freestream velocity, U, is the same for both fluids, and we measure the boundary-layer thickness, δ, at the same plate length, x, in both experiments. First, we use a *slightly-viscous* liquid, water ($\nu = 10^{-6}$ m²/sec), and find $\delta = 3$ mm. Next, we use a *highly-viscous* liquid, glycerin ($\nu = 1.19 \cdot 10^{-3}$ m²/sec). What is δ when glycerin is used?

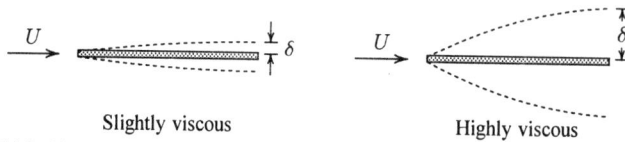

Slightly viscous   Highly viscous

Problems 14.5, 14.6

14.6 To appreciate how viscosity affects the thickness of a boundary layer, imagine that we observe flow past a thin flat plate using fluids of widely varying viscosities. The freestream velocity, U, is the same for both fluids, and we measure the boundary-layer thickness, δ, at the same plate length, x, in both experiments. First, we use a *highly-viscous* gas, helium ($\nu = 1.23 \cdot 10^{-3}$ ft²/sec), and find $\delta = 4$ in. Next, we use a *slightly-viscous* gas, carbon dioxide ($\nu = 8.44 \cdot 10^{-5}$ ft²/sec). What is δ when carbon dioxide is used?

14.7 A baseball is dropped from the Goodyear blimp, which is hovering over Yankee Stadium. Assume, for simplicity, that the ball's drag coefficient, C_D, is constant and equal to 0.4.

(a) Compute the terminal velocity, w_f, as a function of C_D, the mass of the ball, m, the diameter of the ball, D, the density of air, ρ, and gravitational acceleration, g.

(b) Determine the velocity, w, as a function of time, t. Express your answer in terms of g, t and w_f.
HINT: The following integral may be of use in arriving at your answer.

$$\int \frac{dw}{w_f^2 - w^2} = \frac{1}{w_f} \tanh^{-1}\left(\frac{w}{w_f}\right)$$

(c) Make a graph of your solution for $mg = 5$ ounces, $\rho = 0.00234$ slug/ft³ and $D = 0.25$ ft.

(d) For Reynolds number, Re_D, less than 1000, C_D is greater than 0.4. According to your solution, at what time does the Reynolds number reach 1000 (so that your solution is valid)?

14.8 We can obtain an approximate, closed-form solution for the motion of the baseball considered in Section 14.1 if we say: (a) the drag acts entirely in the horizontal direction; and (b) the drag coefficient, C_D, is constant and equal to 0.4. The equations for the ball's motion simplify to

$$m\frac{du}{dt} = -\frac{\pi}{8}\rho D^2 C_D u^2, \qquad m\frac{du}{dt} = -mg$$

Using the same initial conditions as in Section 14.1, solve for the total horizontal distance traveled, x_{tot}, and compare to the exact solution.

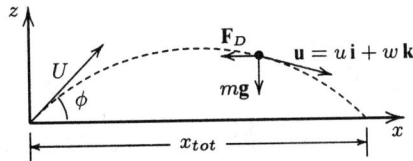

Problem 14.8

14.9 In fluid mechanics, a classical approximation is named after Oseen. In the *Oseen approximation*, we linearize the equations by writing

$$\mathbf{u} = U\mathbf{i} + \mathbf{u}'$$

where U is the freestream velocity, and $|\mathbf{u}'| \ll U$.

(a) Beginning with the two-dimensional laminar boundary-layer equations, use the Oseen approximation to linearize them, i.e., neglect all terms that involve *products of u' and v'* or their derivatives.

(b) With a straightforward change of notation, you can obtain the solution to the equation derived in Part (a) for the constant-pressure case from a Navier-Stokes solution of Chapter 13. Identify the appropriate Navier-Stokes solution, make the change of notation, and quote the solution.

14.10 Consider the differential equation similar to the one used by Prandtl to explain the boundary-layer concept, viz.,

$$\epsilon\frac{d^2u}{dy^2} + (1+\epsilon)\frac{du}{dy} + u = 0$$

The equation is to be solved subject to $u(0) = 0$ and $u(1) = 1$. Also, the parameter ϵ is assumed to be very small compared to unity.

(a) Rewrite this equation in terms of the scaled variable $\eta = y/\epsilon$.

(b) Solve the equation developed in Part (a) by setting $\epsilon = 0$ and applying the boundary condition at $\eta = 0$. If you have done everything correctly up to this point, your solution has an undetermined integration constant. We call the solution you have just generated the *inner solution* and we denote it as $u_{inner}(\eta)$.

(c) As can be easily verified, the so-called *outer solution* (the solution to the unscaled equation with $\epsilon = 0$) is $u_{outer}(y) = e^{1-y}$. We now *match* the two solutions by insisting that

$$\lim_{y \to 0} u_{outer}(y) = \lim_{\eta \to \infty} u_{inner}(\eta)$$

Perform this limiting process to determine the integration constant of Part (b).

(d) On graph paper, compare the resulting *inner solution* determined above to the exact solution for $0 < \eta < 3$ when $\epsilon = .01, .05$ and $.10$. Note that, in terms of η, the exact solution is

$$u_{exact}(\eta) = \frac{e^{1-\epsilon\eta} - e^{1-\eta}}{1 - e^{1-1/\epsilon}}$$

14.11 This and the next two problems examine the foundation of *Lubrication Theory* developed by Reynolds in 1886. The figure shows a *slipper bearing*, which is an idealized model of two machine parts in relative motion. Oil is forced into the thin gap between the slipper block and the bearing guide, with a very large pressure being supported entirely by viscous forces.

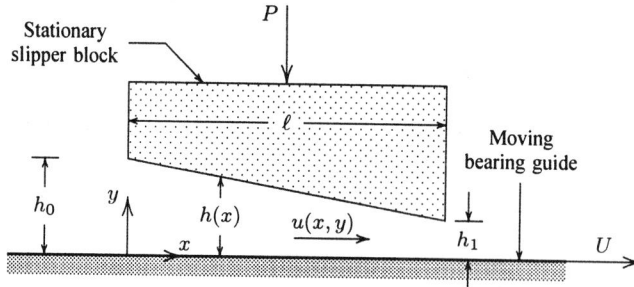

Problems 14.11, 14.12, 14.13

(a) Assuming $\alpha \equiv (h_0 - h_1)/\ell \ll 1$ and $h \ll \ell$, use order-of-magnitude estimates to conclude that

$$\nu \frac{\partial^2 u}{\partial x^2} \ll \nu \frac{\partial^2 u}{\partial y^2}, \quad u \frac{\partial u}{\partial x} \sim v \frac{\partial u}{\partial y} \quad \text{and} \quad \frac{u \partial u / \partial x}{\nu \partial^2 u / \partial y^2} \sim \left(\frac{h}{\ell}\right)^2 Re_\ell$$

(b) Based on results of Part (a), if $(h/\ell)^2 Re_\ell \ll 1$, the equations of motion are the same as for Couette flow, i.e.,

$$\frac{\partial u}{\partial x} + \frac{\partial v}{\partial y} = 0, \quad \mu \frac{\partial^2 u}{\partial y^2} = \frac{\partial p}{\partial x} \quad \text{and} \quad 0 = \frac{\partial p}{\partial y}$$

Solving for $u(x,y)$ subject to $u(x,0) = U$ and $u[x,h(x)] = 0$ for a general gap width $h(x)$, verify that

$$u(x,y) = \left[U - \frac{h^2}{2\mu} \frac{dp}{dx} \frac{y}{h}\right] \left[1 - \frac{y}{h}\right]$$

(c) Compute the volume-flow rate per unit width (out of the page), Q.

(d) Noting that $dQ/dx = 0$, verify that the pressure satisfies

$$\frac{d}{dx}\left(\frac{h^3}{\mu} \frac{dp}{dx}\right) = -6U \frac{dh}{dx}$$

14.12 Consider a *slipper bearing* with a trapezoidal gap (see figure) so that

$$h(x) = h_0 + (h_1 - h_0)\frac{x}{\ell}$$

From Reynolds *Lubrication Theory*, the pressure beneath the bearing is found from solving the following differential equation:

$$\frac{d}{dx}\left(\frac{h^3}{\mu} \frac{dp}{dx}\right) = -6U \frac{dh}{dx}, \quad p(0) = p(\ell) = p_0$$

where p_0 is the surrounding pressure. Show that, for this geometry, the pressure between the slipper block and the wall is

$$p(x) = p_0 + \frac{6\mu U \ell (h - h_0)(h_1 - h)}{(h_1^2 - h_0^2) h^2}$$

14.13 We know from Reynolds' *Lubrication Theory* applied to a *slipper bearing* with a trapezoidal gap (see figure) that the velocity and pressure are

$$u(x,y) = \left[U - \frac{h^2}{2\mu} \frac{dp}{dx} \frac{y}{h} \right] \left[1 - \frac{y}{h} \right] \quad \text{and} \quad p(x) = p_0 + \frac{6\mu U \ell \, (h - h_0)(h_1 - h)}{(h_1^2 - h_0^2) h^2}$$

(a) Compute the pressure force normal to the surface, P, defined by $P = \int_0^\ell (p - p_0) \, dx$.

(b) Compute the viscous resistance force on the bearing guide, F, defined by $F = -\int_0^\ell \tau_w dx$.

(c) Considering P to be a function of $k = h_1/h_0$, a detailed calculation shows that the maximum value of P occurs for $k = 2.2$. For this value of k, determine the effective coefficient of sliding friction, $\mu_{\text{eff}} = F/P$. Note that μ_{eff} is independent of μ.

(d) Compare μ_{eff} for a slipper bearing with $\ell/h_0 = 800$, to a typical coefficient of sliding friction, $\mu_{\text{sliding}} \approx 0.25$.

14.14 Consider steady, incompressible, viscous flow in a rotating coordinate system. Appendix D, Section D.4 includes the exact continuity and Navier-Stokes equations in such a system.

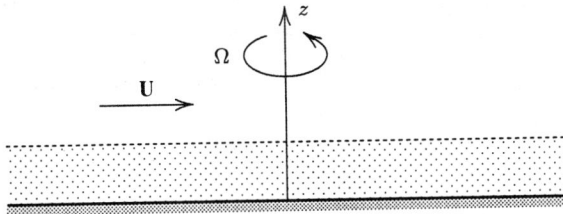

Problem 14.14

(a) Expand the continuity and Navier-Stokes equations into component form.

(b) If the velocity characteristic of an object moving in the region of interest is U and its characteristic dimension is L, use order-of-magnitude estimates in the continuity equation to show that, at least away from solid boundaries, u, v and w are all of the same order of magnitude. Also, show that all of the convective terms in the equations of motion are of the same order of magnitude.

(c) With order-of-magnitude estimates, deduce the ratio of $u\partial u/\partial x$ to Ωv. You should arrive at a dimensionless parameter known as Rossby number, Ro. Show that $u\partial v/\partial x$ and Ωu stand in the same ratio.

(d) Again, using order-of-magnitude estimates, deduce the ratio of $v\partial^2 u/\partial y^2$ to Ωv. This dimensionless parameter is the Ekman number, E.

(e) Assuming the pressure gradient is of the same order of magnitude as the Coriolis acceleration term, $2\Omega \times \mathbf{u}$, rewrite the exact equations of motion omitting all terms that are negligible in the limit of small Rossby number.

(f) Now rewrite the equations, omitting all terms that are negligible in the limit of small Ekman number.

(g) Based on your analysis, what are the appropriate equations of motion for small Rossby number *away from all solid boundaries*.

(h) Assuming there is a boundary layer near the plane $z = 0$, write the approximate equations of motion appropriate *in the boundary layer*, again assuming small Rossby number.

14.15 For viscous flow over axisymmetric bodies, it is sometimes necessary to account for an effect known as *transverse curvature*. If the boundary-layer thickness is such that $\delta \ll R$, where R is the local body radius, the boundary-layer is essentially planar, and the standard two-dimensional form of the equations is appropriate. However, when δ and R are comparable, the transverse curvature must be accounted for. Beginning with the exact continuity and Navier-Stokes equations in cylindrical coordinates (Appendix D, Section D.2), use order-of-magnitude estimates to determine the form of the boundary-layer equations appropriate for flow over an axisymmetric body when δ and R are comparable. **HINT:** For consistent notation relative to the two-dimensional case, rename w as u and z as x in the exact equations.

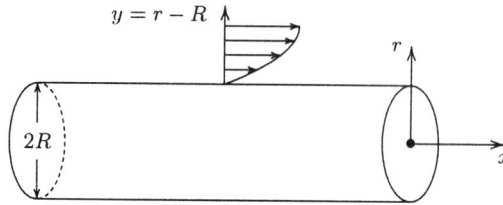

Problem 14.15

14.16 You want to use the integral technique to determine the thickness of a laminar, flat-plate boundary layer. Assume that the velocity profile can be approximated by $u/u_e = (y/\delta)^{1/2}$. Then, note that experimental data indicate $\tau_w = 1.66 u_e \mu/\delta$. Use the integral momentum equation and these relations to obtain a differential equation for δ, and solve for δ as a function of x. Compare your result with the exact laminar (Blasius) result.

14.17 Consider a simple linear approximation to the velocity profile for an incompressible flat-plate boundary layer.

$$\frac{u}{U} = \frac{y}{\delta}$$

(a) Compute the skin friction, c_f, displacement thickness, δ^*, momentum thickness, θ, and shape factor, H.

(b) Using the momentum integral equation, determine the boundary-layer thickness, δ.

(c) Quantify the differences between the results of Parts (a) and (b) and the Blasius solution.

14.18 Consider the following approximation to the velocity profile for an incompressible flat-plate boundary layer.

$$\frac{u}{U} = \sin\left(\frac{\pi}{2}\frac{y}{\delta}\right)$$

(a) Compute the skin friction, c_f, displacement thickness, δ^*, momentum thickness, θ, and shape factor, H.

(b) Using the momentum integral equation, determine the boundary-layer thickness, δ.

(c) Quantify the differences between the results of Parts (a) and (b) and the Blasius solution.

14.19 Consider the following approximation to the velocity profile for an incompressible flat-plate boundary layer.

$$\frac{u}{U} = \frac{3}{2}\left(\frac{y}{\delta}\right) - \frac{1}{2}\left(\frac{y}{\delta}\right)^3$$

(a) Compute the skin friction, c_f, displacement thickness, δ^*, momentum thickness, θ, and shape factor, H.

(b) Using the momentum integral equation, determine the boundary-layer thickness, δ.

(c) Quantify the differences between the results of Parts (a) and (b) and the Blasius solution.

14.20 Consider the following approximation to the velocity profile for an incompressible flat-plate boundary layer.

$$\frac{u}{U} = \frac{3}{2}\left(\frac{y}{\delta}\right) - \frac{1}{2}\left(\frac{y}{\delta}\right)^{5/2}$$

(a) Compute the skin friction, c_f, displacement thickness, δ^*, momentum thickness, θ, and shape factor, H.

(b) Using the momentum integral equation, determine the boundary-layer thickness, δ.

(c) Quantify the differences between the results of Parts (a) and (b) and the Blasius solution.

14.21 Consider the following approximation to the velocity profile for an incompressible flat-plate boundary layer.

$$\frac{u}{U} = 2\left(\frac{y}{\delta}\right) - 2\left(\frac{y}{\delta}\right)^3 + \left(\frac{y}{\delta}\right)^4$$

(a) Compute the skin friction, c_f, displacement thickness, δ^*, momentum thickness, θ, and shape factor, H.

(b) Using the momentum integral equation, determine the boundary-layer thickness, δ.

(c) Quantify the differences between the results of Parts (a) and (b) and the Blasius solution.

14.22 For an incompressible boundary layer with surface mass removal (suction), the momentum integral equation is modified slightly and becomes

$$\frac{d\theta}{dx} + (H+2)\frac{\theta}{u_e}\frac{du_e}{dx} = \frac{c_f}{2} - C_Q$$

where $C_Q = -v_w/u_e$ is the (dimensionless) suction coefficient, and v_w is the suction velocity.

(a) Assuming the horizontal velocity is given by $u(x,y) = u_e(x)\{1 - \exp[-y/a(x)]\}$, compute skin friction, c_f, displacement thickness, δ^*, momentum thickness, θ, and shape factor, H. **HINT:** Do all integrals from $y = 0$ to $y \to \infty$.

(b) If $C_Q Re_\theta = 1/2$, solve for $\theta(x)$ in terms of $u_e(x)$, assuming $\theta(x_o) = \theta_o$ and $u_e(x_o) = u_o$ where x_o is a reference point.

14.23 For a turbulent flat-plate boundary layer, we can estimate the boundary-layer thickness, δ, as follows. Assume the wall shear stress is the same as for pipe flow, i.e.,

$$\tau_w \approx 0.0225\rho U^2 \left(\frac{\nu}{U\delta}\right)^{1/4}$$

Then, assume that the ratio of momentum thickness, θ, to boundary layer thickness is given by the 1/7-power law velocity profile so that $\theta/\delta = 7/72$. Using the momentum integral equation, determine the ratio of δ to distance from the plate leading edge, x. Also, compute the skin friction, c_f, and the drag on a plate of length L, taking account of both sides of the plate.

14.24 A special class of turbulent boundary layers exists for which the *equilibrium parameter*, β_T, is constant, where [see Equation (14.131)]:

$$\beta_T \equiv \frac{\delta^*}{\tau_w}\frac{d\overline{p}}{dx}$$

(a) Beginning with the momentum integral equation, show that for this class of boundary layers:

$$\frac{d\theta}{dx} = \left(1 + \frac{2+H}{H}\beta_T\right)\frac{c_f}{2}$$

(b) Suppose now that the velocity profile is $u = u_e y/\delta$, where u_e is boundary-layer edge velocity and δ is boundary-layer thickness. Derive a differential equation for δ of the form:

$$\frac{d\delta}{dx} = F(\beta_T, Re_\delta)$$

14.25 *Thwaites' method* is a useful empirical procedure for predicting laminar boundary-layer properties with variable freestream pressure. To develop the method, proceed as follows.

(a) Multiply both sides of the momentum integral equation by $2\theta/\nu$, and write the equation as

$$\frac{d}{dx}\left(\frac{\theta^2}{\nu}\right) = \frac{F(K)}{u_e} \quad \text{where} \quad K \equiv \frac{\theta^2}{\nu}\frac{du_e}{dx}$$

(b) Now, observe that for laminar boundary layers, correlation of exact results shows that the function $F(K)$ is given by $F(K) \approx 0.450 - 6K$. Substitute this correlation into the equation derived in Part (a), and verify that

$$\frac{u_e\theta^2}{\nu} \approx \frac{0.450}{u_e^5}\int_0^x u_e^5(x)dx$$

(c) Use *Thwaites' method*, i.e., the equation developed in Part (b), to compute θ for constant pressure. Compare your result with the Blasius solution.

14.26 *Thwaites' method* is a useful empirical procedure for predicting laminar boundary-layer properties with variable freestream pressure, including separation point. With this method, the momentum thickness, θ, and skin friction, c_f, are given by

$$\theta^2 \approx \frac{0.450\nu}{u_e^6}\int_0^x u_e^5(x)dx, \quad c_f \approx \frac{2(K+0.09)^{0.62}}{Re_\theta} \quad \text{where} \quad K \equiv \frac{\theta^2}{\nu}\frac{du_e}{dx}$$

Also, Re_θ is Reynolds number based on θ. According to this method, boundary-layer separation occurs when $K = -0.09$. Consider a linearly-decreasing velocity, i.e.,

$$u_e(x) = U(1 - x/L)$$

where U and L are constant velocity and length scales, respectively. At what value of x/L, according to *Thwaites' method*, does the boundary layer separate? Compare your answer with the exact value of $x_{sep} = 0.120$.

14.27 *Thwaites' method* is a useful empirical procedure for predicting laminar boundary-layer properties with variable freestream pressure, including separation point. With this method, the momentum thickness, θ, and skin friction, c_f, are given by

$$\theta^2 \approx \frac{0.450\nu}{u_e^6}\int_0^x u_e^5(x)dx, \quad c_f \approx \frac{2(K+0.09)^{0.62}}{Re_\theta} \quad \text{where} \quad K \equiv \frac{\theta^2}{\nu}\frac{du_e}{dx}$$

Also, Re_θ is Reynolds number based on θ. According to this method, boundary-layer separation occurs when $K = -0.09$. Using the potential-flow solution for $u_e(x)$, determine the angle, ϕ_{sep}, at which laminar separation occurs on a circular cylinder according to *Thwaites' method*. Solve the equation you obtain by trial and error to the nearest tenth of a degree. Note that, in the spirit of boundary-layer approximations, x, is distance along the cylinder surface measured from the leading stagnation point. **HINT:** The following integral should prove helpful.

$$\int \sin^5\phi\, d\phi = -\frac{1}{15}\left[8 + 4\sin^2\phi + 3\sin^4\phi\right]\cos\phi$$

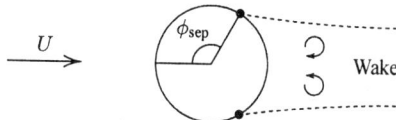

Problem 14.27

14.28 For the *Falkner-Skan* family of velocity profiles $[u_e(x) = U_o(x/L)^m]$, show that the profile corresponding to incipient separation, $m \approx -1/11$, has a shape factor of $H = 4$. **HINT:** Since the similarity variable, η, is proportional to $y/(x/L)^{(1-m)/2}$, we expect all boundary-layer thickness parameters for this profile to be proportional to $(x/L)^{(1-m)/2}$.

14.29 For a boundary layer with sufficient surface mass removal (suction), transition to turbulence can be prevented. For such a boundary layer, the laminarized velocity profile asymptotes to:

$$u = U \left[1 - e^{-C_Q U y / \nu} \right]$$

where $C_Q > 0$ is the (dimensionless) suction coefficient. Compute skin friction, c_f, displacement thickness, δ^*, momentum thickness, θ, and shape factor, H, for such a boundary layer. **HINT:** Do all integrals from $y = 0$ to $y \to \infty$.

14.30 For a turbulent boundary layer, the velocity profile can be approximated as

$$\overline{u} = u_e \left(\frac{y}{\delta} \right)^{1/n}$$

where u_e is the velocity at the boundary-layer edge and n is an integer typically between 6 and 8. Compute displacement thickness, δ^*, momentum thickness, θ, and shape factor, H.

14.31 It is possible to derive the momentum integral equation by integrating the boundary-layer equations directly. To do so, proceed as follows.

(a) Show that the momentum equation in *conservation form* is

$$\frac{\partial}{\partial x} \left(u^2 \right) + \frac{\partial}{\partial y} \left(uv \right) = u_e \frac{du_e}{dx} + \nu \frac{\partial^2 u}{\partial y^2}$$

(b) Integrate the continuity and momentum equations from $y = 0$ to $y = \delta$, taking advantage of *Leibnitz's theorem* (see Section 4.7), which tells us

$$\frac{d}{dx} \int_0^{\delta(x)} f(x,y) \, dy = \int_0^{\delta(x)} \frac{\partial f}{\partial x} dy + f(x, \delta) \frac{d\delta}{dx}$$

Verify that the continuity and momentum equations become

$$\frac{d}{dx} \int_0^{\delta} u \, dy = u_e \frac{d\delta}{dx} - v_e$$

$$\frac{d}{dx} \int_0^{\delta} u^2 dy = u_e \left[u_e \frac{d\delta}{dx} - v_e \right] + u_e \delta \frac{du_e}{dx} - \frac{\tau_w}{\rho}$$

(c) Now, introduce δ^*, θ and H, and complete the derivation. Your result should be identical to Equation (14.61).

14.32 *Wieghardt* proposed an *energy integral equation* for incompressible, laminar boundary layers, viz.,

$$\frac{d}{dx} \left(u_e^3 \theta_E \right) = 2\nu \int_0^{\delta} \left(\frac{\partial u}{\partial y} \right)^2 dy$$

The quantity θ_E is the energy thickness defined by

$$\theta_E \equiv \int_0^{\delta} \frac{u}{u_e} \left[1 - \left(\frac{u}{u_e} \right)^2 \right] dy$$

Use this equation to determine θ_E and δ for a flat-plate boundary layer assuming

$$\frac{u}{u_e} = \frac{y}{\delta}$$

Compare your result for δ with the Blasius thickness, i.e., $\delta = 5.0 \sqrt{\nu x / u_e}$.

14.33 *Wieghardt* proposed an *energy integral equation* for incompressible, laminar boundary layers, viz.,

$$\frac{d}{dx}\left(u_e^3 \theta_E\right) = 2\nu \int_0^\delta \left(\frac{\partial u}{\partial y}\right)^2 dy$$

The quantity θ_E is the energy thickness defined by

$$\theta_E \equiv \int_0^\delta \frac{u}{u_e}\left[1 - \left(\frac{u}{u_e}\right)^2\right] dy$$

Use this equation to determine θ_E and δ for a flat-plate boundary layer assuming

$$\frac{u}{u_e} = \sin\left(\frac{\pi}{2}\frac{y}{\delta}\right)$$

Compare your result for δ with the Blasius thickness, i.e., $\delta = 5.0\sqrt{\nu x/u_e}$.

14.34 The Falkner-Skan velocity profile for $m = 2$ can be approximated as

$$\frac{u}{u_e} = 1 - e^{-5y/\delta}$$

Using the momentum integral equation, solve for δ and c_f. **HINT:** Do all integrals from $y = 0$ to $y = \infty$. Compare your computed c_f with the exact value, viz.,

$$c_f = \frac{4.13}{\sqrt{Re_x}}$$

14.35 We seek an approximation to the Falkner-Skan velocity profile for $m = \frac{1}{2}$ of the form

$$\frac{u}{u_e} = 1 - e^{-\lambda y/\delta}, \qquad \lambda = \text{constant}$$

Using the momentum integral equation, solve for δ and c_f. **HINT:** Do all integrals from $y = 0$ to $y = \infty$. Compare your computed c_f with the exact value, viz.,

$$c_f = \frac{1.61}{\sqrt{Re_x}}$$

Comment on the optimum choice of λ if the goal is to compute an accurate value for c_f.

14.36 Compute the Reynolds number based on length of a 1-foot long model airplane moving at 10 ft/sec when the temperature is 68° F. Assuming the boundary layer remains attached along the fuselage and that it experiences a negligible pressure gradient, estimate the thickness of the boundary layer near the tail.

14.37 Q has designed a flying mirror to permit James Bond to reflect a laser beam around a corner. The mirror has width, $w = 10$ cm, and length, $\ell = 5$ cm. Q's design guarantees laminar flow over the mirror when it flies at $U = 50$ m/sec with its long side normal to the motion. This design feature assures that the mirror can be quickly and silently moved from one place to another. Use Blasius formulas in the computations below, and assume pressure is atmospheric while the temperature is 20° C.

(a) Assuming that transition to turbulence (a source of noise) will not occur on the mirror provided $Re_x < 4 \cdot 10^5$, will the mirror still have laminar flow if, in his haste to avoid capture, 007 makes it fly with its short side normal to the mirror's motion?

(b) What is the total drag on the mirror in the design case and when 007 rotates it as in Part (a)?

14.38 Consider the flow of kerosene ($\nu = 2.55 \cdot 10^{-5}$ ft^2/sec) past a thin flat plate. The freestream velocity is 2 ft/sec. Assuming sufficient care has been taken to maintain laminar flow, determine the boundary-layer thickness and skin friction at distances $x = 1$ ft and $x = 5$ ft from the plate leading edge.

14.39 For flow near the entrance to a wide rectangular duct, boundary layers on the lower and upper walls grow and merge on the centerline. Assuming the flow remains laminar, estimate the distance, ℓ_m, at which the boundary layers merge. Compare your answer with the distance needed to achieve fully-developed flow, $\ell_e \approx 0.06 H \, Re_H$, where $Re_H \equiv U H / \nu$.

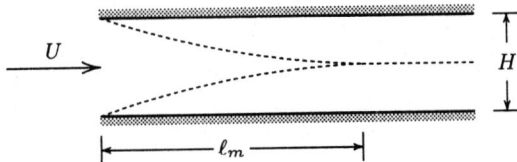

Problem 14.39

14.40 Assuming the Blasius solution applies, compute the drag coefficient, $C_D = D/(\frac{1}{2}\rho U^2 A)$, where $A = \frac{1}{2}L^2$, for the triangular flat plate shown. **HINT:** Compute the differential drag dD on a strip of width dz and integrate the result over z.

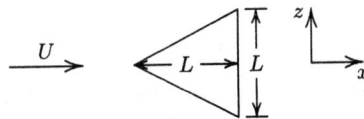

Problem 14.40

14.41 Assuming the Blasius solution applies, compute the drag coefficient, $C_D = D/(\frac{1}{2}\rho U^2 A)$, where $A = \frac{\pi}{4}d^2$, for the circular flat plate shown. **HINT:** Compute the differential drag dD on a strip of width dz and integrate the result over z. The following integral will help in arriving at an answer.

$$\int_0^{\pi/2} \cos^{3/2} \phi \, d\phi \approx 1.57$$

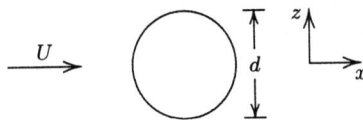

Problem 14.41

14.42 By combining Equations (14.47) and (14.50), the entrainment velocity for an incompressible boundary layer, $(\mathbf{u} \cdot \mathbf{n})_e$, is given by

$$(\mathbf{u} \cdot \mathbf{n})_e = -\frac{d}{dx} \int_0^\delta u \, dy$$

Using results obtained in the Blasius solution, determine the entrainment velocity for a flat-plate boundary layer. Express your answer in terms of freestream velocity, U, and plate-length Reynolds number, Re_x.

14.43 Show that, according to the Falkner-Skan solution,

$$c_f Re_\theta = \text{constant}$$

where the skin friction and momentum-thickness Reynolds number are defined by $c_f \equiv \tau_w/(\frac{1}{2}\rho u_e^2)$ and $Re_\theta \equiv u_e \theta / \nu$, respectively.

14.44 Beginning with the streamfunction defined in Equation (14.78), derive the similarity Equations (14.79) and (14.80) for the Blasius boundary layer. **HINT:** The chain rule for transforming derivatives in this problem is as follows.

$$\left(\frac{\partial}{\partial x}\right)_y = \left(\frac{\partial x}{\partial x}\right)_y \left(\frac{\partial}{\partial x}\right)_\eta + \left(\frac{\partial \eta}{\partial x}\right)_y \left(\frac{\partial}{\partial \eta}\right)_x = \left(\frac{\partial}{\partial x}\right)_\eta + \left(\frac{\partial \eta}{\partial x}\right)_y \frac{d}{d\eta}$$

$$\left(\frac{\partial}{\partial y}\right)_x = \left(\frac{\partial x}{\partial y}\right)_x \left(\frac{\partial}{\partial x}\right)_\eta + \left(\frac{\partial \eta}{\partial y}\right)_x \left(\frac{\partial}{\partial \eta}\right)_x = \left(\frac{\partial \eta}{\partial y}\right)_x \frac{d}{d\eta}$$

14.45 *Free shear flows* are thin viscous shear layers that form far from solid boundaries. The three classic examples are the *far wake*, the *mixing layer* and the *jet*. The governing equations are identical to the boundary-layer equations, and all three examples approach self similarity as the flow moves far downstream of the origin.

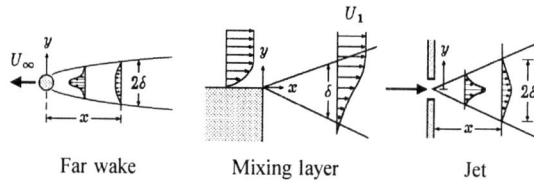

Far wake   Mixing layer   Jet
Problems 14.45, 14.46, 14.47

The equations appropriate for the laminar, two-dimensional jet are exactly the same as those for a constant-pressure boundary layer. However, the boundary conditions are quite different. In place of no-slip, we impose symmetry conditions at the centerline and, in the freestream, the velocity is zero. That is, we have

$$v = \frac{\partial u}{\partial y} = 0 \quad \text{at} \quad y = 0, \qquad u \to 0 \quad \text{as} \quad y \to \infty$$

(a) Assuming a streamfunction of the form

$$\psi(x,y) = \nu^{1/2} x^{1/3} f(\eta), \qquad \eta \equiv \frac{y}{3\nu^{1/2} x^{2/3}}$$

show that the boundary conditions on $f(\eta)$ are

$$f(0) = f''(0) = 0 \quad \text{and} \quad f'(\eta) \to 0 \quad \text{as} \quad \eta \to \infty$$

(b) Show that $f(\eta)$ satisfies the following equation:

$$\frac{d^3 f}{d\eta^3} + f \frac{d^2 f}{d\eta^2} + \left(\frac{df}{d\eta}\right)^2 = 0$$

and, noting that $(ff')' = ff'' + (f')^2$, integrate once to show that

$$\frac{d^2 f}{d\eta^2} + f \frac{df}{d\eta} = 0$$

HINT: The chain rule for transforming derivatives in this problem is as follows.

$$\left(\frac{\partial}{\partial x}\right)_y = \left(\frac{\partial}{\partial x}\right)_\eta + \left(\frac{\partial \eta}{\partial x}\right)_y \frac{d}{d\eta} \quad \text{and} \quad \left(\frac{\partial}{\partial y}\right)_x = \left(\frac{\partial \eta}{\partial y}\right)_x \frac{d}{d\eta}$$

(c) Verify that $f(\eta) = 2\alpha \tanh(\alpha\eta)$ is a solution where α is a constant. Don't forget to verify that the boundary conditions are satisfied since the purpose of this problem is to stress how much difference changing boundary conditions can make!

14.46 *Free shear flows* are thin viscous shear layers that form far from solid boundaries. The three classic examples are the *far wake*, the *mixing layer* and the *jet* (see figure). For the two-dimensional far wake, we can linearize the momentum equation far downstream of the body. The equation and boundary conditions are as follows:

$$U_\infty \frac{\partial \hat{u}}{\partial x} = \nu \frac{\partial^2 \hat{u}}{\partial y^2}, \qquad \frac{\partial \hat{u}}{\partial y} = 0 \text{ at } y = 0, \qquad \hat{u} \to 0 \text{ as } y \to \infty$$

where the velocity vector is $\mathbf{u} = U_\infty \mathbf{i} - \hat{\mathbf{u}}$, and we assume $|\hat{\mathbf{u}}| \ll U_\infty$. If the drag (per unit width) is D, the solution must satisfy the following *integral constraint*, which guarantees that momentum is conserved in the wake.

$$2\rho U_\infty \int_0^\infty \hat{u}(x, y)\, dy = D$$

(a) Assuming a similarity solution of the form

$$\hat{u}(x, y) = \frac{D}{\rho U_\infty} \sqrt{\frac{U_\infty}{\nu x}} f(\eta), \qquad \eta = y\sqrt{\frac{U_\infty}{\nu x}}$$

transform the momentum equation, boundary conditions and integral constraint to similarity form. **HINT:** The chain rule for transforming derivatives in this problem is as follows.

$$\left(\frac{\partial}{\partial x}\right)_y = \left(\frac{\partial}{\partial x}\right)_\eta + \left(\frac{\partial \eta}{\partial x}\right)_y \frac{d}{d\eta} \qquad \text{and} \qquad \left(\frac{\partial}{\partial y}\right)_x = \left(\frac{\partial \eta}{\partial y}\right)_x \frac{d}{d\eta}$$

(b) Solve for $\hat{u}(x, y)$ taking account of the integral constraint. **HINT:** The following definite integral might prove helpful: $\int_0^\infty \exp(-\frac{1}{4}\eta^2)\, d\eta = \sqrt{\pi}$.

14.47 *Free shear flows* are thin viscous shear layers that form far from solid boundaries. The three classic examples are the *far wake*, the *mixing layer* and the *jet*. For the mixing layer the equations are identical to those for a constant-pressure boundary layer. If the upper stream moves with velocity U_1 and the lower with U_2 (the figure corresponds to $U_2 = 0$), boundary conditions are as follows:

$$u(x, y) \to U_1 \text{ as } y \to +\infty \qquad \text{and} \qquad u(x, y) \to U_2 \text{ as } y \to -\infty$$

(a) The equations admit a similarity solution of the same form as the Blasius solution, viz., Equation (14.78). In addition to the farfield boundary conditions above, we define the *dividing streamline* by $\psi(x, y) = 0$. State the three boundary conditions on $G(\eta)$.

(b) We can generate an approximate solution to Equation (14.79) that is satisfactory when U_1 and U_2 are not too far apart, i.e., by assuming

$$G(\eta) = \eta + g(\eta), \qquad |g(\eta)| \ll |\eta|$$

Substitute this expression for $G(\eta)$ into Equation (14.79), and neglect the term $g d^2 g/d\eta^2$. State the resulting linearized equation and corresponding boundary conditions.

(c) Solve the equation developed in Part (b) for $u(x, y)$. **HINT:** You need only solve for $dg/d\eta$, and reference to the solution for Stokes' First Problem (Section 13.4) should prove quite helpful.

14.48 Compute the Reynolds number based on length of a 20-foot long automobile moving at 75 mph when the temperature is 68° F. Assuming the boundary layer remains attached along the side of the car and that it experiences a negligible pressure gradient, estimate the thickness of the boundary layer near the tail light.

14.49 Compute the Reynolds number based on length of a 100-meter long ship moving at 12 knots when the water temperature is 10° C. Assuming the boundary layer remains attached along the side of the hull and that it experiences a negligible pressure gradient, estimate the thickness of the boundary layer near the stern.

14.50 Assuming the flow is turbulent from the leading edge and that the skin friction is

$$c_f \approx 0.0576 Re_x^{-1/5}$$

compute the drag coefficient, $C_D = D/(\frac{1}{2}\rho U^2 A)$, where $A = bL$, for the rectangular flat plate shown. For $Re_L = 10^5$, 10^6 and 10^7, compare your answer with the laminar drag coefficient.

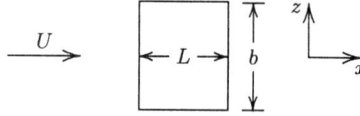

Problem 14.50

14.51 Assuming the flow is turbulent from the leading edge and that the skin friction is

$$c_f \approx 0.0576 Re_x^{-1/5}$$

compute the drag coefficient, $C_D = D/(\frac{1}{2}\rho U^2 A)$, where $A = \frac{1}{2}L^2$, for the triangular flat plate shown. For $Re_L = 10^5$, 10^6 and 10^7, compare your answer with the laminar drag coefficient, $C_D = 3.54/\sqrt{Re_L}$. **HINT:** Compute the differential drag dD on a strip of width dz and integrate the result over z.

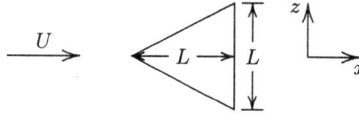

Problem 14.51

14.52 The point at which a boundary layer experiences transition from laminar to turbulent flow is very sensitive to the freestream turbulence level. Transition has been observed for the flat-plate boundary layer at plate-length Reynolds number, Re_x, ranging from about 10^5 to $3 \cdot 10^6$. Also, measurements show that the length of the transition region, $Re_{\Delta x}$, is

$$Re_{\Delta x} \approx 25 Re_{x_t}^{2/3}$$

where Re_{x_t} is the Reynolds number at which transition begins. To appreciate how abrupt transition is and how much flow properties change, consider the following two situations. In each case, compute the transition-point location, x_t, the width of the transition region, Δx, the skin friction at the beginning of transition, and the skin friction at the end of transition (i.e., at $x_f = x_t + \Delta x$). Assume that, for a turbulent flat-plate boundary layer, $c_f = 0.0576 Re_x^{-1/5}$.

  (a) Air ($\nu = 1.62 \cdot 10^{-4}$ ft^2/sec) flowing over a plate at $U = 100$ ft/sec with transition occurring at $Re_{x_t} = 2 \cdot 10^6$.

  (b) Water ($\nu = 1.08 \cdot 10^{-5}$ ft^2/sec) flowing over a plate at $U = 30$ ft/sec with transition occurring at $Re_{x_t} = 5 \cdot 10^5$.

  (c) Make a log-log plot of c_f versus Re_x based on your results of Parts (a) and (b) similar to Figure 14.18.

14.53 To appreciate why laminar flow is of minimal importance in many engineering applications, compute the percent of the vehicle over which laminar flow exists for the following situations. In each case, consider the boundary layer on a flat portion of the vehicle and assume transition occurs at a (very high) Reynolds number of $Re_{x_t} = 5 \cdot 10^5$.

  (a) A 20-foot automobile moving at 25 mph ($\nu = 1.62 \cdot 10^{-4}$ ft^2/sec).

  (b) A 20-foot automobile moving at 65 mph ($\nu = 1.62 \cdot 10^{-4}$ ft^2/sec).

  (c) A small aircraft with a wing chord length of 8 feet moving at 150 mph ($\nu = 1.67 \cdot 10^{-4}$ ft^2/sec).

  (d) A *Boeing 747* with a wing chord length of 30 feet moving at 570 mph ($\nu = 4.27 \cdot 10^{-4}$ ft^2/sec).

14.54 To appreciate why laminar flow is of minimal importance in many engineering applications, compute the percent of the vehicle over which laminar flow exists for the following situations. In each case, consider the boundary layer on a flat portion of the vehicle and assume transition occurs at a (very high) Reynolds number of $Re_{x_t} = 10^6$.

(a) A 10-meter sailboat moving at 3 knots ($\nu = 1.00 \cdot 10^{-6}$ m^2/sec).

(b) A 10-meter sailboat moving at 7.5 knots ($\nu = 1.00 \cdot 10^{-6}$ m^2/sec).

(c) A 25-meter yacht moving at 12 knots ($\nu = 0.80 \cdot 10^{-6}$ m^2/sec).

(d) A 100-meter tanker moving at 15 knots ($\nu = 1.50 \cdot 10^{-6}$ m^2/sec).

14.55 Strong Santa Ana winds of 75 mph are blowing from the desert into the Los Angeles area. They bring warm air that raises the temperature to 95° F. Determine the vortex-shedding frequency, in Hz, for a $2\frac{1}{2}$-inch diameter flagpole and a $\frac{1}{4}$-inch diameter telephone line. Noting that humans can hear sounds with frequencies between 16 Hz and 20 kHz, is it possible for the vortex shedding to be audible in either case?

14.56 A massive hurricane with 250 km/hr winds is bearing down on Miami. Compute the vortex-shedding frequency, in Hz, for a 2-centimeter diameter mooring line and a 1-meter diameter piling on a pier, assuming the winds have not yet wrought their ultimate destruction. Noting that humans can hear sounds with frequencies between 16 Hz and 20 kHz, is it possible for the vortex shedding to be audible in either case?

14.57 A small wind tunnel is being designed with a 20-cm square cross section, and a test-section length of 1.2 m. The working gas will be helium at 15° C, for which $\nu = 1.14 \cdot 10^{-4}$ m^2/sec. In order to maintain constant cross-section velocity, the test-section walls will be slanted slightly to accommodate the displacement effect of the growing boundary layers on all four walls. Assume transition occurs when $Re_x = 1.2 \cdot 10^5$, and that appropriate integral parameters for the boundary layers are given in Table 14.1. Also, assume, for simplicity, that the boundary layers have zero thickness at the entrance to the test section.

(a) If the test-section velocity is 10 m/sec, at what angle should the walls be slanted?

(b) If the test-section velocity is 200 m/sec, at what angle should the walls be slanted?

(c) Compute the test-section velocity at the end of the test section for Parts (a) and (b) if the walls are not slanted.

14.58 The viscous sublayer of a turbulent boundary layer extends from the surface to $y^+ \approx 30$. To appreciate how thin this layer is, consider the boundary layer on the hull of a large tanker. Assuming the boundary layer has negligible pressure gradient over most of the hull, you can assume

$$\delta \approx 0.37 x Re_x^{-1/5} \quad \text{and} \quad c_f \approx 0.0576 Re_x^{-1/5}$$

(a) Verify that the sublayer thickness, $\delta_{sl} = 30\nu/u_\tau$, is given by

$$\delta_{sl} \approx \frac{478}{Re_x^{7/10}} \delta$$

(b) Compute δ_{sl} at points on the hull where $Re_x = 2.8 \cdot 10^7$ and $\delta = 2.5$ in, and where $Re_x = 5.0 \cdot 10^8$ and $\delta = 25$ in. Express your answer in terms of h_δ/δ_{sl}, to the nearest integer, where $h_\delta = 1/10$ inch is the height of the symbol δ_{sl} on this page.

14.59 The viscous sublayer of a turbulent boundary layer extends from the surface to $y^+ \approx 30$. To appreciate how thin this layer is, consider the boundary layer on the side of your freshly washed and waxed (and therefore smooth) automobile. When you are moving at 65 mph, the skin friction, c_f, just below your rear-view mirror is 0.0028. Estimate the sublayer thickness and compare to the diameter of the head of a pin, $d_{pin} = 0.05$ inch. Assume the kinematic viscosity of air is $\nu = 1.62 \cdot 10^{-4}$ ft^2/sec.

14.60 A turbulent boundary layer is considered *fully developed* when the Reynolds number based on momentum thickness, Re_θ, exceeds 10^4. Using the formulas in Table 14.1, determine the skin friction for $Re_\theta = 10^4$. Compare this value to that given by the more-accurate *Kármán-Schoenherr* formula, Equation (14.135).

14.61 A surface is called *hydraulically smooth* when the surface roughness height, k_s, is such that

$$k_s^+ \equiv \frac{u_\tau k_s}{\nu} < 5$$

where u_τ is friction velocity and ν is kinematic viscosity. Consider the flow of air over a flat plate of length 1 m. For the following plate materials, what is the maximum freestream velocity, U, at which the surface will be hydraulically smooth? Use Table 14.1 as needed and assume $\nu = 1.51 \cdot 10^{-5}$ m²/sec.

Plate Material	k_s (mm)
Copper	0.0015
Galvanized iron	0.15
Concrete	1.50

Problem 14.61

14.62 A surface is called *completely rough* when the surface roughness height, k_s, is such that

$$k_s^+ \equiv \frac{u_\tau k_s}{\nu} > 70$$

where u_τ is friction velocity and ν is kinematic viscosity. Consider the flow of water over a flat plate. For the following plate materials, what is the minimum freestream velocity, U, at which the surface will be completely rough at $x = 5$ ft? Use Table 14.1 as needed and assume ($\nu = 1.08 \cdot 10^{-5}$ ft²/sec).

Plate Material	k_s (ft)
Steel	$1.5 \cdot 10^{-4}$
Cast iron	$8.5 \cdot 10^{-4}$
Concrete	$5.0 \cdot 10^{-2}$

Problem 14.62

14.63 The *atmospheric boundary layer* over a smooth beach is a very large scale turbulent, flat-plate boundary layer, and its integral parameters are quite accurately represented by Table 14.1. Suppose you are enjoying a day on the beach contemplating the A you earned in fluid mechanics. On this 85° F day (so that $\nu = 1.72 \cdot 10^{-4}$ ft²/sec), the atmospheric boundary layer is 250 ft thick and the velocity at that altitude is 20 mph. Your forehead is about 6 inches above the ground level. Is your forehead in the sublayer, log layer or defect layer? What is the wind velocity over your forehead?

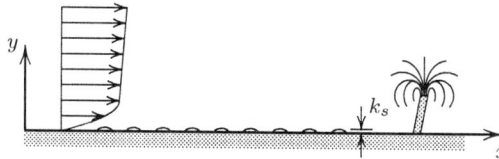

Problems 14.63, 14.64

14.64 Sunbathers are enjoying a day on the beach. They are lying on the sand with essentially uniform spacing, and their bodies appear as roughness elements of height $k_s = 30$ cm to the *atmospheric boundary layer*. One of the sunbathers is an eager fluid-mechanics student who decides to use what he learned in this chapter in a practical situation. First, just downstream of a cluster of sunbathers, he measures the wind velocity at head level, $y_1 \approx 1.8$ m, and finds $u_1 = 2.9$ m/sec. He then climbs a palm tree of height $y_2 \approx 5.0$ m and observes a wind velocity of $u_2 = 3.5$ m/sec. Assuming the beach surface is a *completely-rough* surface, what is the friction velocity according to his measurements? To verify the hypothesis that the surface is completely rough, check to see if $u_\tau k_s/\nu > 70$. Assume that $\nu = 1.60 \cdot 10^{-5}$ m²/sec.

14.65 For *completely-rough* flat plates (i.e., plate surfaces on which $u_\tau k_s/\nu > 70$), *Schlichting* developed the following correlations for skin-friction, c_f, and (one-sided) drag, C_D, coefficients:

$$c_f \approx \left(2.87 + 1.58 \log_{10} \frac{x}{k_s}\right)^{-2.5} \quad \text{and} \quad C_D \approx \left(1.89 + 1.62 \log_{10} \frac{x}{k_s}\right)^{-2.5}$$

where x is distance along the plate, L is plate length and k_s is surface-roughness height. For the rudder of a motorboat, compute c_f at the trailing edge and C_D using the smooth-surface formulas of Table 14.1 and the Schlichting formulas. The boat is moving at 16 knots, and $\nu = 10^{-6}$ m^2/sec. Approximate the rudder as a rectangular flat plate of length 0.6 m and of width (depth in the water) 1.2 m. Also, assume the surface roughness is 0.12 mm. How much larger are c_f and C_D when we account for k_s?

14.66 For a turbulent boundary layer, the velocity is given by $u^+ = y^+$ in the sublayer and by the law of the wall, Equation (14.124), in the log layer. Determine by trial and error (or Newton's iterations if you are familiar with the method) the value of y^+ (to the nearest 1/10) at which the sublayer and log-layer velocity profiles are equal.

14.67 Consider a turbulent boundary layer on a rough flat plate. An approximation proposed for the velocity profile is

$$\frac{u}{u_\tau} \approx \lambda \left(\frac{y}{k_s}\right)^{1/10}$$

where λ is a dimensionless constant, y is distance from the surface, u_τ is friction velocity and k_s is roughness height.

(a) Using the momentum integral equation, solve for the boundary-layer thickness, δ, and the skin friction, c_f.

(b) Compute the drag coefficient, C_D, which is defined by

$$C_D = \frac{1}{L} \int_0^L c_f \, dx$$

(c) Determine the coefficient λ by comparing the result of Part (b) with the von Kármán correlation:

$$C_D \approx 0.024 \left(\frac{k_s}{L}\right)^{1/6}$$

14.68 Exact theoretical considerations show that approaching a solid surface, i.e., for $y^+ \to 0$, the Reynolds shear stress goes to zero according to $\tau_t = -\overline{\rho u' v'} \propto y^3$. What does the *Cebeci-Smith model* predict as $y^+ \to 0$?

14.69 For a turbulent boundary layer with surface mass transfer, the momentum equation in the sublayer and log layer simplifies to:

$$v_w \frac{d\overline{u}}{dy} = \frac{d}{dy}\left[(\nu + \nu_t)\frac{d\overline{u}}{dy}\right]$$

where v_w is the (constant) vertical velocity at the surface.

(a) Integrate once using the appropriate surface boundary conditions. Introduce the friction velocity, u_τ, in stating your integrated equation.

(b) Focusing now upon the log layer where $\nu_t \gg \nu$, what is the approximate form of the equation derived in Part (a) if we use the mixing-length model with $\ell_{mix} = \kappa y$?

(c) Verify that the solution to the simplified equation of Part (b) is

$$2\frac{u_\tau}{v_w}\sqrt{1 + v_w\overline{u}/u_\tau^2} = \frac{1}{\kappa}\ell ny + \text{constant}$$

14.70 *Free shear flows* are thin viscous shear layers that form far from solid boundaries. The three classic examples are the *far wake*, the *mixing layer* and the *jet*. The governing equations are identical to the boundary-layer equations, and all three examples approach self similarity as the flow moves far downstream of the origin.

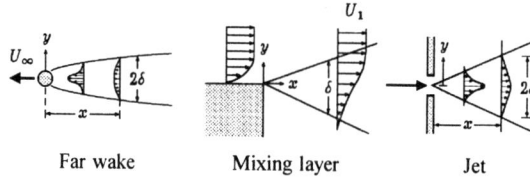

Far wake   Mixing layer   Jet

Problems 14.70, 14.71, 14.72

The equations appropriate for the turbulent, two-dimensional jet are exactly the same as those for a constant-pressure boundary layer. However, the boundary conditions are quite different. In place of no-slip, we impose symmetry conditions at the centerline and, in the freestream, the velocity is zero. The equations appropriate for a turbulent jet in terms of an eddy viscosity, ν_t, are

$$\frac{\partial u}{\partial x} + \frac{\partial v}{\partial y} = 0$$

$$u\frac{\partial u}{\partial x} + v\frac{\partial u}{\partial y} = \frac{\partial}{\partial y}\left[\nu_t\frac{\partial u}{\partial y}\right]$$

which must be solved subject to the following boundary conditions.

$$v = \frac{\partial u}{\partial y} = 0 \quad \text{at} \quad y = 0, \qquad u \to 0 \quad \text{as} \quad y \to \infty$$

The solution must also satisfy the *integral constraint*, which guarantees that the momentum flux is constant in the jet:

$$\int_0^\infty u^2 dy = J$$

To solve, assume a similarity solution of the form

$$\psi = \sqrt{Jx}\,f(\eta), \qquad \nu_t = \nu_{to}\sqrt{Jx}, \qquad \eta \equiv \frac{y}{x}$$

where ν_{to} is a constant.

(a) Show that the velocity components are given by

$$u = \sqrt{\frac{J}{x}}f'(\eta) \quad \text{and} \quad v = \sqrt{\frac{J}{x}}\left[\eta f'(\eta) - \frac{1}{2}f(\eta)\right]$$

HINT: The chain rule for transforming derivatives in this problem is as follows.

$$\left(\frac{\partial}{\partial x}\right)_y = \left(\frac{\partial}{\partial x}\right)_\eta + \left(\frac{\partial \eta}{\partial x}\right)_y\frac{d}{d\eta} \quad \text{and} \quad \left(\frac{\partial}{\partial y}\right)_x = \left(\frac{\partial \eta}{\partial y}\right)_x\frac{d}{d\eta}$$

(b) Show that the equation of motion transforms to

$$2\nu_{to}\frac{d^3 f}{d\eta^3} + f\frac{d^2 f}{d\eta^2} + \left(\frac{df}{d\eta}\right)^2 = 0$$

(c) Determine the boundary conditions and integral constraint on f and its derivatives.

14.71 *Free shear flows* are thin viscous shear layers that form far from solid boundaries. The three classic examples are the *far wake*, the *mixing layer* and the *jet* (see figure). For the two-dimensional far wake, we can linearize the momentum equation far downstream of the body. The equation and boundary conditions appropriate for a turbulent wake in terms of an eddy viscosity, ν_t, are

$$U_\infty \frac{\partial \hat{u}}{\partial x} = \frac{\partial}{\partial y}\left(\nu_t \frac{\partial \hat{u}}{\partial y}\right), \qquad \frac{\partial \hat{u}}{\partial y} = 0 \text{ at } y = 0, \qquad \hat{u} \to 0 \text{ as } y \to \infty$$

where the velocity vector is $\mathbf{u} = U_\infty \mathbf{i} - \hat{\mathbf{u}}$, and we assume $|\hat{\mathbf{u}}| \ll U_\infty$. If the drag (per unit width) is D, the solution must satisfy the following *integral constraint*, which guarantees that momentum is conserved in the wake.

$$2\rho U_\infty \int_0^\infty \hat{u}(x,y)\,dy = D$$

Assuming a similarity solution of the form

$$\hat{u}(x,y) = \sqrt{\frac{D}{\rho x}} f(\eta), \qquad \nu_t(x,y) = n(x)N(\eta), \qquad \eta = y\sqrt{\frac{\rho U_\infty^2}{Dx}}$$

transform the momentum equation, boundary conditions and integral constraint to similarity form. **HINT:** The chain rule for transforming derivatives in this problem is as follows.

$$\left(\frac{\partial}{\partial x}\right)_y = \left(\frac{\partial}{\partial x}\right)_\eta + \left(\frac{\partial \eta}{\partial x}\right)_y \frac{d}{d\eta} \qquad \text{and} \qquad \left(\frac{\partial}{\partial y}\right)_x = \left(\frac{\partial \eta}{\partial y}\right)_x \frac{d}{d\eta}$$

What must $n(x)$ be in order for a similarity solution to exist for this problem?

14.72 *Free shear flows* are thin viscous shear layers that form far from solid boundaries. The three classic examples are the *far wake*, the *mixing layer* and the *jet* (see figure). For the mixing layer the equations are identical to those for a constant-pressure boundary layer. The equations appropriate for a turbulent mixing layer in terms of an eddy viscosity, ν_t, are

$$\frac{\partial u}{\partial x} + \frac{\partial v}{\partial y} = 0$$

$$u\frac{\partial u}{\partial x} + v\frac{\partial u}{\partial y} = \frac{\partial}{\partial y}\left[\nu_t \frac{\partial u}{\partial y}\right]$$

If the upper stream moves with velocity U_1 and the lower is at rest, boundary conditions are as follows:

$$u(x,y) \to U_1 \text{ as } y \to +\infty \qquad \text{and} \qquad u(x,y) \to 0 \text{ as } y \to -\infty$$

In addition to the farfield boundary conditions above, we define the *dividing streamline* by $\psi(x,y) = 0$. The equations admit a similarity solution of the form

$$\psi(x,y) = U_1 x f(\eta), \qquad \nu_t(x,y) = n(x)N(\eta), \qquad \eta = \frac{y}{x}$$

Show that the equation of motion transforms to

$$\frac{n(x)}{Ux}\frac{d}{d\eta}\left(N\frac{d^2 f}{d\eta^2}\right) + f\frac{d^2 f}{d\eta^2} = 0$$

State the three boundary conditions on $f(\eta)$. **HINT:** The chain rule for transforming derivatives in this problem is as follows.

$$\left(\frac{\partial}{\partial x}\right)_y = \left(\frac{\partial}{\partial x}\right)_\eta + \left(\frac{\partial \eta}{\partial x}\right)_y \frac{d}{d\eta} \qquad \text{and} \qquad \left(\frac{\partial}{\partial y}\right)_x = \left(\frac{\partial \eta}{\partial y}\right)_x \frac{d}{d\eta}$$

What must $n(x)$ be in order for a similarity solution to exist for this problem?

14.73 Verify that the Adams-Bashforth formula for equally-spaced grid points, Equation (14.162), is second-order accurate. **HINT:** Be sure to expand in Taylor series about (x_{i+1}, y_j).

14.74 Verify that the Adams-Bashforth formula for nonuniformly-spaced grid points, Equation (14.163), is second-order accurate. **HINT:** Be sure to expand in Taylor series about (x_{i+1}, y_j).

14.75 Under what conditions will the tridiagonal matrix system defined in Equations (14.173) and (14.174) be diagonally dominant? What does this condition mean physically? **HINT:** Recall from Chapter 13 that a sufficient condition for diagonal dominance is $B_j \geq -(A_j + C_j)$.

14.76 Using Program **EDDYBL** (see Appendix F), determine the skin friction, c_f, momentum thickness Reynolds number, Re_θ, and shape factor, H, for an incompressible, turbulent, flat-plate boundary layer. Reynolds number based on plate length, Re_s, should range from $2 \cdot 10^6$ to $1 \cdot 10^7$. Do your computations with any turbulence model you wish. **HINT:** Aside from the choice of turbulence model, the default values in Program **SETEBL** are suitable for this problem.

    (a) To show the contrast between laminar and turbulent boundary layers, make a graph of each variable and compare, on the graph, to the corresponding Blasius value.

    (b) Verify that your numerical solution satisfies the momentum integral equation at several values of Re_s. Use a central-difference approximation, i.e., $(d\theta/ds)_i \approx (\theta_{i+1} - \theta_{i-1}) / (s_{i+1} - s_{i-1})$.

    (c) Graphically compare your computed velocity profile with the law of the wall at $Re_s = 1 \cdot 10^7$.

14.77 Using Program **EDDYBL** (see Appendix F), determine the skin friction, c_f, momentum thickness Reynolds number, Re_θ, and shape factor, H, for an incompressible, laminar, flat-plate boundary layer. Reynolds number based on plate length, Re_s, should range from 10^2 to 10^7. **HINT:** Use Program **SETEBL** to set the initial stepsize and arc length to $\Delta s = 1.5 \cdot 10^{-4}$ ft and $s_i = 0$ ft, respectively — their program variable names are DS and SI. Also, be sure to set MODEL $= -1$ to have a laminar-flow computation. All other default values in Program **SETEBL** are suitable for this problem.

    (a) Using **SETEBL**, change the appropriate input parameters to accomplish the following: use SI units (IUTYPE); set the freestream conditions to $p_{t_\infty} = 1.01858 \cdot 10^5$ N/m^2, $T_{t_\infty} = 294$ K, $M_\infty = 0.08656$ (PT1, TT1, XMA); use an initial stepsize $\Delta s = 0.01$ m (DS); set the initial boundary-layer properties so that $c_f = 0.00292$, $\delta = 0.0224$ m, $H = 1.36$, $Re_\theta = 5454$, $s_i = 0.75$ m (CF, DELTA, H, RETHET, SI); set the maximum arc length to $s_f = 2.782$ m (SSTOP); and, set up for $N = 11$ points to define the pressure distribution (NUMBER).

    (a) Make a graph of each variable and compare, on the graph, to the corresponding Blasius value.

    (b) Graphically compare computed velocity profiles at $Re_s = 10^5$, 10^6 and 10^7 to verify that the flow is self similar.

14.78 Using Program **EDDYBL** (see Appendix F), we can compute the transition from laminar to turbulent flow using the k-ω model. To do this, proceed as follows.

    (a) Using **SETEBL**, change the appropriate input parameters to accomplish the following: use SI units (IUTYPE = 0); use an initial stepsize DS $= 1.5 \cdot 10^{-4}$ ft and initial arc length SI $= 0$ ft; set the initial boundary-layer properties so that exact laminar values are used, i.e., set IBOUND = 0; select the k-ω model with viscous corrections included by setting MODEL = 0 and NVISC = 1.

    (b) Make three computations corresponding to freestream turbulence levels, T', of 0.01%, 0.1% and 1.0%. The value of T' is related to input parameter ZIOTAE $= k_e/u_e^2$ by

$$T' = 100\sqrt{\frac{2}{3}\frac{k_e}{u_e^2}} \qquad (T' \text{ in percent})$$

    (c) Determine the plate-length Reynolds numbers at the beginning (Re_{s_t}) and end (Re_{s_f}) of transition by locating the points of minimum and maximum skin friction. Make a graph, similar to Figure 14.18, based on computed results, with all three cases on the same graph. Also, make a graph of the shape factor, including results for all three computations.

14.79 The object of this problem is to compare predictions of modern turbulence models with measured properties of a turbulent boundary layer with adverse pressure gradient. The experiment to be simulated was conducted by Ludwieg and Tillman [see Coles and Hirst (1969) – Flow 1200]. Use Program **EDDYBL** and its menu-driven setup utility, Program **SETEBL**, to do the computations (see Appendix F).

(a) Using **SETEBL**, change the appropriate input parameters to accomplish the following: use SI units (IUTYPE); set the freestream conditions to $p_{t_\infty} = 1.01858 \cdot 10^5$ N/m^2, $T_{t_\infty} = 294$ K, $M_\infty = 0.08656$ (PT1, TT1, XMA); use an initial stepsize $\Delta s = 0.01$ m (DS); set the initial boundary-layer properties so that $c_f = 0.00292$, $\delta = 0.0224$ m, $H = 1.36$, $Re_\theta = 5454$, $s_i = 0.75$ m (CF, DELTA, H, RETHET, SI); set the maximum arc length to $s_f = 2.782$ m (SSTOP); and, set up for $N = 11$ points to define the pressure (NUMBER).

(b) Use the following data to define the pressure distribution in a file named **presur.dat**. The initial and final pressure gradients are $(dp_e/dx)_i = 180.9$ N/m^3 and $(dp_e/dx)_f = -15.97$ N/m^3, respectively. Also, prepare a file **heater.dat** with constant wall temperature, $T_w = 294$ K, and zero heat flux.

s (m)	p_e (N/m^2)	s (m)	p_e (N/m^2)	s (m)	p_e (N/m^2)	s (m)	p_e (N/m^2)
0.000	$1.01067 \cdot 10^5$	1.782	$1.01358 \cdot 10^5$	3.132	$1.01526 \cdot 10^5$	3.732	$1.01563 \cdot 10^5$
0.782	$1.01201 \cdot 10^5$	2.282	$1.01415 \cdot 10^5$	3.332	$1.01541 \cdot 10^5$	3.932	$1.01562 \cdot 10^5$
1.282	$1.01271 \cdot 10^5$	2.782	$1.01491 \cdot 10^5$	3.532	$1.01554 \cdot 10^5$		

(c) Do three computations using the k-ω model, one of the k-ϵ models and any other model.

(d) Compare computed skin friction with the following measured values.

s (m)	c_f	s (m)	c_f
0.782	$2.92 \cdot 10^{-3}$	2.282	$1.94 \cdot 10^{-3}$
1.282	$2.49 \cdot 10^{-3}$	2.782	$1.55 \cdot 10^{-3}$
1.782	$2.05 \cdot 10^{-3}$		

14.80 The object of this problem is to compare predictions of modern turbulence models with measured properties of a turbulent boundary layer with adverse pressure gradient. The experiment to be simulated was conducted by Schubauer and Spanganberg [see Coles and Hirst (1969) – Flow 4400]. Use Program **EDDYBL** and its menu-driven setup utility, Program **SETEBL**, to do the computations (see Appendix F).

(a) Using **SETEBL**, change the appropriate input parameters to accomplish the following: use USCS units (IUTYPE); set the freestream conditions to $p_{t_\infty} = 2052$ lb/ft^2, $T_{t_\infty} = 528.54°$ R, $M_\infty = 0.0728$ (PT1, TT1, XMA); use an initial stepsize $\Delta s = 0.006$ ft (DS); set the initial boundary-layer properties so that $c_f = 0.00340$, $\delta = 0.063$ ft, $H = 1.351$, $Re_\theta = 3066$, $s_i = 1.167$ ft (CF, DELTA, H, RETHET, SI); set the maximum arc length to $s_f = 6.167$ ft and maximum number of steps to 500 (SSTOP, IEND1); and, set up for $N = 15$ points to define the pressure (NUMBER).

(b) Use the following data to define the pressure distribution in a file named **presur.dat**. The initial and final pressure gradients are $(dp_e/dx)_i = 0$ lb/ft^3 and $(dp_e/dx)_f = 0.2365$ lb/ft^3, respectively. Also, prepare a file **heater.dat** with constant wall temperature, $T_w = 528.54°$ R, and zero heat flux.

s (ft)	p_e (lb/ft^2)	s (ft)	p_e (lb/ft^2)	s (ft)	p_e (lb/ft^2)	s (ft)	p_e (lb/ft^2)
1.167	$2.04441 \cdot 10^3$	4.500	$2.04581 \cdot 10^3$	7.833	$2.04854 \cdot 10^3$	11.167	$2.04881 \cdot 10^3$
2.000	$2.04441 \cdot 10^3$	5.333	$2.04665 \cdot 10^3$	8.667	$2.04862 \cdot 10^3$	12.000	$2.04892 \cdot 10^3$
2.833	$2.04475 \cdot 10^3$	6.167	$2.04758 \cdot 10^3$	9.500	$2.04864 \cdot 10^3$	12.833	$2.04911 \cdot 10^3$
3.667	$2.04516 \cdot 10^3$	7.000	$2.04826 \cdot 10^3$	10.333	$2.04870 \cdot 10^3$		

(c) Do three computations using the k-ω model, one of the k-ϵ models and any other model.

(d) Compare computed skin friction with the following measured values.

s (ft)	c_f	s (ft)	c_f	s (ft)	c_f
1.167	$3.40 \cdot 10^{-3}$	3.667	$2.86 \cdot 10^{-3}$	6.167	$1.33 \cdot 10^{-3}$
2.000	$3.17 \cdot 10^{-3}$	4.500	$2.38 \cdot 10^{-3}$		
2.833	$3.10 \cdot 10^{-3}$	5.333	$1.97 \cdot 10^{-3}$		

14.81 The object of this problem is to compare predictions of modern turbulence models with measured properties of a turbulent boundary layer with adverse pressure gradient. The experiment to be simulated was conducted by Bradshaw [see Coles and Hirst (1969) – Flow 3300]. Use Program **EDDYBL** and its menu-driven setup utility, Program **SETEBL**, to do the computations (see Appendix F).

(a) Using **SETEBL**, change the appropriate input parameters to accomplish the following: use USCS units (IUTYPE); set the freestream conditions to p_{t_∞} = 2148 lb/ft^2, T_{t_∞} = 537.6° R, M_∞ = 0.106 (PT1, TT1, XMA); use an initial stepsize Δs = 0.05 ft and a geometric-progression ratio k_g = 1.08 (DS, XK); set initial boundary-layer properties so that c_f = 0.00225, δ = 0.125 ft, H = 1.40, Re_θ = 9216, s_i = 2.55 ft (CF, DELTA, H, RETHET, SI); set the maximum arc length to s_f = 7.0 ft (SSTOP); and, set up for N = 11 points to define the pressure (NUMBER).

(b) Use the following data to define the pressure distribution in a file named **presur.dat**. The initial and final pressure gradients are $(dp_e/dx)_i$ = 3.410939 lb/ft^3 and $(dp_e/dx)_f$ = 0.64929 lb/ft^3, respectively. Also, prepare a file **heater.dat** with constant wall temperature, T_w = 537.6° R, and zero heat flux.

s (ft)	p_e (lb/ft^2)	s (ft)	p_e (lb/ft^2)	s (ft)	p_e (lb/ft^2)	s (ft)	p_e (lb/ft^2)
2.5	2.13123·10^3	4.0	2.13480·10^3	5.5	2.13677·10^3	7.0	2.13806·10^3
3.0	2.13272·10^3	4.5	2.13556·10^3	6.0	2.13726·10^3	7.5	2.13841·10^3
3.5	2.13387·10^3	5.0	2.13621·10^3	6.5	2.13768·10^3		

(c) Do three computations using the k-ω model, one of the k-ϵ models and any other model.

(d) Compare computed skin friction with the following measured values.

s (ft)	c_f	s (ft)	c_f	s (ft)	c_f
2.5	2.45·10^{-3}	4.00	1.91·10^{-3}	7.00	1.56·10^{-3}
3.0	2.17·10^{-3}	5.00	1.74·10^{-3}		
3.5	2.00·10^{-3}	6.00	1.61·10^{-3}		

14.82 The object of this problem is to compare predictions of modern turbulence models with measured properties of a turbulent boundary layer with adverse pressure gradient. The experiment to be simulated was conducted by Stratford [see Coles and Hirst (1969) – Flow 5300]. Use Program **EDDYBL** and its menu-driven setup utility, Program **SETEBL**, to do the computations (see Appendix F).

(a) Using **SETEBL**, change the appropriate input parameters to accomplish the following: use USCS units (IUTYPE); set the freestream conditions to p_{t_∞} = 2015 lb/ft^2, T_{t_∞} = 544° R, M_∞ = 0.05 (PT1, TT1, XMA); use an initial stepsize Δs = 0.0001 ft, a geometric-progression ratio k_g = 1.06, and near-wall boundary condition distance y^+ = 1 (DS, XK, USTOP); set the initial boundary-layer properties so that Λ = 4, c_f = 0.00368, δ = 0.0609 ft, H = 1.3616, Re_θ = 2295, s_i = 2.9075 ft (ALAMM, CF, DELTA, H, RETHET, SI); set the maximum arc length to s_f = 4.103 ft and maximum number of steps to 500 (SSTOP, IEND1); and, set up for N = 8 points to define the pressure (NUMBER).

(b) Use the following data to define the pressure distribution in a file named **presur.dat**. The initial and final pressure gradients are $(dp_e/dx)_i$ = 0.406006 lb/ft^3 and $(dp_e/dx)_f$ = 1.38916 lb/ft^3, respectively. Also, prepare a file **heater.dat** with constant wall temperature, T_w = 504° R, and zero heat flux.

s (ft)	p_e (lb/ft^2)	s (ft)	p_e (lb/ft^2)	s (ft)	p_e (lb/ft^2)
1.4308	2.01132·10^3	2.9976	2.01188·10^3	4.4496	2.01337·10^3
2.9075	2.01153·10^3	3.0580	2.01225·10^3	6.4407	2.01391·10^3
2.9903	2.01181·10^3	3.1619	2.01249·10^3		

(c) Do three computations using the k-ω model with viscous corrections (MODEL = 0, NVISC = 1), one of the k-ϵ models and any other model.

(d) Compare computed skin friction with the following measured values.

s (ft)	c_f	s (ft)	c_f
2.907	3.68·10^{-3}	3.531	0.55·10^{-3}
2.999	2.07·10^{-3}	4.103	0.53·10^{-3}
3.038	0.99·10^{-3}		

14.83 *Thwaites' method* is a useful empirical procedure for predicting laminar boundary-layer properties with variable freestream pressure. The central approximation underlying the method is the following.

$$c_f Re_\theta - 2(2 + H)\frac{\theta^2}{\nu}\frac{du_e}{dx} \approx 0.450 - 6\frac{\theta^2}{\nu}\frac{du_e}{dx}$$

All notation is that used in the context of the momentum integral equation, i.e., Equation (14.61).

(a) Modify Program **FALKSKAN** (see Appendix E) as required to generate the information needed to check the accuracy of this approximation. Note that, in terms of the Falkner-Skan solution, the flow properties pertinent to this equation are:

$$u_e(x) = U_o\left(\frac{x}{L}\right)^m, \qquad c_f = \sqrt{\frac{2(m+1)\nu}{U_o L}}\frac{G''(0)}{(x/L)^{(m+1)/2}}$$

$$\theta = \sqrt{\frac{2\nu L}{(m+1)U_o}}\left(\frac{x}{L}\right)^{\frac{1-m}{2}}\int_0^\infty G'(\eta)\left[1 - G'(\eta)\right]d\eta$$

HINT: Use the (second-order accurate) *trapezoidal rule* to evaluate integrals, viz., for the integral of a function $f(x)$ from $x = a$ to $x = b$,

$$\int_a^b f(x)dx \approx \frac{\Delta x}{2}\left[f(a) + f(b) + 2\sum_{n=1}^N f(a + n\Delta x)\right], \qquad \Delta x = \frac{b - a}{N + 1}$$

(b) Exercise your modified program for the complete range of pressure gradients for which it is valid. Make a graph comparing exact results with Thwaites' correlation.

14.84 The solution for a laminar *mixing layer* is similar to the Blasius solution (see Problem 14.47). The velocity and streamfunction are

$$u(x, y) = U_1 G'(\eta), \quad \psi(x, y) = \sqrt{2U_1\nu x}\, G(\eta), \qquad \eta \equiv y\sqrt{\frac{U_1}{2\nu x}}$$

where U_1 is the velocity of the moving stream of fluid and the lower stream is at rest. The function $G(\eta)$ satisfies the following ordinary differential equation and boundary conditions.

$$\frac{d^3 G}{d\eta^3} + G\frac{d^2 G}{d\eta^2} = 0$$

$$G(0) = 0, \qquad G'(\eta) \to 1 \quad \text{as} \quad \eta \to +\infty, \qquad G'(\eta) \to 0 \quad \text{as} \quad \eta \to -\infty$$

Modify Program **FALKSKAN** (see Appendix E) as needed to solve the laminar mixing-layer equations. Use equally-spaced points with $-10 \le \eta \le 10$. Modify FALKSKAN's initial conditions so that $G = G' = 0$ for $\eta < 0$, and be sure to enforce the boundary condition on the dividing streamline, i.e., at $\eta = 0$. Make a graph of the computed velocity profile, $G'(\eta)$. Do computations with 101 and 201 grid points, and use Richardson extrapolation to estimate numerical-solution errors.

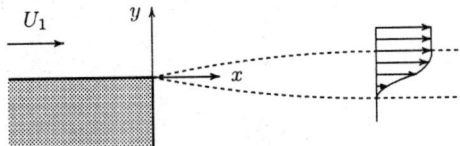

Problem 14.84

14.85 The object of this problem is to predict the separation point for flow past a circular cylinder with the boundary-layer equations, using the measured pressure distribution. The experiment to be simulated was conducted by Patel (1968), and the measured pressure distribution is shown in Figure 11.10. Use Program **EDDYBL** and its menu-driven setup utility, Program **SETEBL**, to do the computations (see Appendix F).

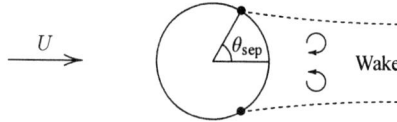

Problem 14.85

(a) Using **SETEBL**, change the appropriate input parameters to accomplish the following: use USCS units (IUTYPE); set the freestream conditions to $p_{t_\infty} = 2147.7$ lb/ft^2, $T_{t_\infty} = 529.6°$ R, $M_\infty = 0.144$ (PT1, TT1, XMA); use an initial stepsize $\Delta s = 0.001$ ft (DS); set the initial boundary-layer properties so that $c_f = 0.00600$, $\delta = 0.006$ ft, $H = 1.40$, $Re_\theta = 929$, $s_i = 0.262$ ft (CF, DELTA, H, RETHET, SI); set the maximum arc length to $s_f = 0.785$ ft and maximum number of steps to 300 (SSTOP, IEND1); set up for $N = 47$ points to define the pressure (NUMBER); and, make provision for specifying body curvature (NFLAG).

(b) Use the following data to define the pressure distribution in a file named **presur.dat**. The initial and final pressure gradients are both zero. Then, prepare a file **heater.dat** with constant wall temperature, $T_w = 529.4°$ R, and zero heat flux. Finally, prepare a file named **blocrv.dat** with constant curvature, $\mathcal{R}^{-1} = 4$ ft^{-1}.

s (ft)	p_e (lb/ft^2)	s (ft)	p_e (lb/ft^2)	s (ft)	p_e (lb/ft^2)	s (ft)	p_e (lb/ft^2)
.0000	$2.147540 \cdot 10^3$	.1000	$2.130641 \cdot 10^3$	.2500	$2.075334 \cdot 10^3$	.4000	$2.066423 \cdot 10^3$
.0025	$2.147528 \cdot 10^3$	.1125	$2.127261 \cdot 10^3$	.2625	$2.069189 \cdot 10^3$	.4125	$2.071954 \cdot 10^3$
.0050	$2.147491 \cdot 10^3$	.1250	$2.123881 \cdot 10^3$	.2750	$2.064580 \cdot 10^3$	.4250	$2.079021 \cdot 10^3$
.0075	$2.147429 \cdot 10^3$	.1375	$2.120194 \cdot 10^3$	.2875	$2.060893 \cdot 10^3$	.4375	$2.085473 \cdot 10^3$
.0100	$2.147343 \cdot 10^3$	.1500	$2.116199 \cdot 10^3$	.3000	$2.058588 \cdot 10^3$	.4500	$2.089161 \cdot 10^3$
.0125	$2.147233 \cdot 10^3$	.1625	$2.112205 \cdot 10^3$	.3125	$2.056898 \cdot 10^3$	.4625	$2.091004 \cdot 10^3$
.0250	$2.146314 \cdot 10^3$	.1750	$2.107903 \cdot 10^3$	.3250	$2.055823 \cdot 10^3$	.4750	$2.092080 \cdot 10^3$
.0375	$2.144796 \cdot 10^3$	.1875	$2.103448 \cdot 10^3$	.3375	$2.055362 \cdot 10^3$	.4875	$2.092230 \cdot 10^3$
.0500	$2.142688 \cdot 10^3$	.2000	$2.098378 \cdot 10^3$	.3500	$2.055516 \cdot 10^3$	.5000	$2.092230 \cdot 10^3$
.0625	$2.140018 \cdot 10^3$	.2125	$2.093155 \cdot 10^3$	.3625	$2.056591 \cdot 10^3$	.6500	$2.092230 \cdot 10^3$
.0750	$2.136807 \cdot 10^3$	.2250	$2.087317 \cdot 10^3$	.3750	$2.058435 \cdot 10^3$	.7850	$2.092230 \cdot 10^3$
.0875	$2.134021 \cdot 10^3$	.2375	$2.081325 \cdot 10^3$	.3875	$2.061661 \cdot 10^3$		

(c) Do three computations using the k-ω model, the Launder-Sharma k-ϵ model and any other model. (Avoid the Lam-Bremhorst, Chien and Yang-Shih k-ϵ models — they are very difficult to implement numerically for this type of flow.)

(d) Compare computed separation angle measured from the downstream symmetry axis with the measured value of $\theta_{sep} = 70°$. The radius of the cylinder is $R = 0.25$ ft, so that separation arc length, s_{sep}, is related to this angle by $\theta_{sep} = \pi - s_{sep}/R$.

Chapter 15

Viscous and 2-D Compressible Flow

We have discussed effects of compressibility in the context of one-dimensional, inviscid flows in Chapter 8. In this chapter we extend the scope of our discussion of compressibility to both viscous and two-dimensional flows. As in our earlier studies, we will find that interesting insight into compressible-flow processes can be gleaned from relatively simple mathematical analyses.

To begin the chapter, we complete formulation of the energy-conservation law for a viscous fluid. While the integral form of the energy equation developed in Chapter 7 [Equation (7.38)] is completely general, we had insufficient information at our disposal to evaluate two contributions to the energy balance. Specifically, we lacked information relating flow properties such as velocity and temperature with heat-transfer and viscous-work contributions. We acquired the information needed to specify the viscous-work term when we derived the Navier-Stokes equation. In this chapter, we will specify the heat-flux vector according to **Fourier's Law**, from which the heat-transfer term can be evaluated. From the integral-conservation form, we then deduce the differential form of the energy equation. This permits reexamining the entropy equation, which demonstrates the mechanisms by which entropy changes in a given flow.

To provide straightforward examples of how viscous and heat-transfer effects affect compressible flows, we will consider two classical one-dimensional problems. The first problem is **Fanno flow**, which is viscous compressible flow in a pipe or duct in the absence of heat transfer. The second problem is **Rayleigh flow**, which is frictionless compressible flow in a pipe or duct with heat transfer.

Next, we turn our focus to inviscid two-dimensional compressible flows by analyzing an **oblique shock wave** and the **Prandtl-Meyer expansion**. An oblique shock lies at an angle to the freestream flow direction, and is the most common type of shock wave observed in aerodynamic applications. By contrast, the normal shock occurs on an axis of symmetry and in very simple geometries. The Prandtl-Meyer expansion is an isentropic flow around a sharp corner for which pressure, temperature and density decrease, the opposite of what occurs for a shock.

We briefly discuss the effects of compressibility on boundary layers for both laminar- and turbulent-flow conditions. The chapter concludes with a discussion of some of aspects of Computational Fluid Dynamics pertinent to flows in which effects of compressibility play a significant role.

15.1   The Energy Equation

Recall that in Chapter 7, we included the appropriate heat-transfer and viscous terms in arriving at the integral form of energy conservation for an arbitrary control volume (see Figure 15.1). The result is Equation (7.33), which can be rewritten more compactly as

$$\iiint_V \frac{\partial}{\partial t}(\rho\mathcal{E})\,dV + \iint_S \rho\mathcal{H}(\mathbf{u}\cdot\mathbf{n})\,dS = \dot{Q} + \iint_S \mathbf{f}_v\cdot\mathbf{u}\,dS \qquad (15.1)$$

where $\mathbf{f}_v$ is the viscous force per unit area and the quantities $\mathcal{E}$ and $\mathcal{H}$ are the **specific total energy** and **specific total enthalpy**, respectively, defined in Equation (7.45), viz.,

$$\mathcal{E} = e + \frac{1}{2}\mathbf{u}\cdot\mathbf{u} + \mathcal{V} \quad \text{and} \quad \mathcal{H} = h + \frac{1}{2}\mathbf{u}\cdot\mathbf{u} + \mathcal{V} \qquad (15.2)$$

Note that, for simplicity, we assume the body force, $\mathbf{f}$, is conservative so that

$$\mathbf{f} = -\nabla\mathcal{V} \qquad (15.3)$$

where $\mathcal{V}$ is the body-force potential. This assumption permits representing the body force through the definitions of the specific total energy and enthalpy. When a nonconservative body force is present, a volume integral of $\mathbf{f}\cdot\mathbf{u}$ appears on the right-hand side of Equation (15.1) [cf. Equation (7.38)].

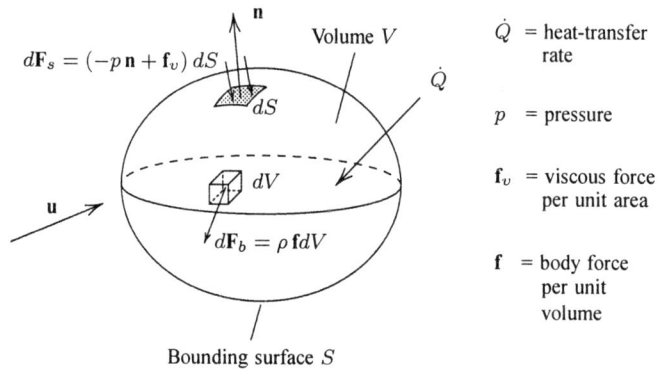

Figure 15.1: *A general control volume for energy conservation.*

15.1.1   Integral Form

The volume integral on the left-hand side of Equation (15.1) is the instantaneous rate of change of total energy in the control volume. The closed surface integral on the left-hand side is the net flux of total enthalpy out of the control volume. The integrands involve the velocity, $\mathbf{u}$, the body-force potential, $\mathcal{V}$, and standard thermodynamic properties (density, ρ, specific internal energy, e, and specific enthalpy, h). In principle, both integrals on the left-hand side can be evaluated in a control-volume computation. Additionally, the surface integral can be readily transformed to a volume integral with the divergence theorem to facilitate derivation of a differential equation for energy conservation.

The first term on the right-hand side of Equation (15.1), $\dot{Q}$, is the rate at which heat is transferred to the control volume. For our purposes, we will assume it is an integral over the closed surface bounding the control volume. We thus exclude processes such as laser irradiation or chemical heat release that would more appropriately be represented as volume integrals. Although Equation (15.1) applies to these modes of heat transfer, we choose to focus on a process that is easier to express analytically, viz., conduction. Now, the net rate of heat *transferred to the control volume* is usually defined in terms of the **heat-flux vector, q,** such that

$$\dot{Q} = -\oiint_S \mathbf{q} \cdot \mathbf{n} \, dS \tag{15.4}$$

This term has a minus sign because **n** is an outer unit normal and $\dot{Q}$ is the heat-transfer rate from the surroundings to the control volume.

The second term on the right-hand side of Equation (15.1) is the friction work done by the control volume on the surroundings. The explicit form of the integrand for the viscous-work integral requires specification of the viscous force, $\mathbf{f}_v$, acting on the surface of a fluid element. We developed an expression for this force when we derived the Navier-Stokes equation in Chapter 12 [see Equation (12.57)], viz.,

$$\mathbf{f}_v = \mathbf{n} \cdot [\tau] \tag{15.5}$$

where $[\tau]$ is the viscous-stress tensor. Thus, the rate at which the control volume does friction work on its surroundings becomes

$$\oiint_S \mathbf{f}_v \cdot \mathbf{u} \, dS = \oiint_S \mathbf{n} \cdot [\tau] \cdot \mathbf{u} \, dS \tag{15.6}$$

Substituting Equations (15.4) and (15.6) into Equation (15.1), we arrive at the following.

$$\iiint_V \frac{\partial}{\partial t} (\rho \mathcal{E}) \, dV + \oiint_S \rho \mathcal{H} (\mathbf{u} \cdot \mathbf{n}) \, dS = -\oiint_S \mathbf{q} \cdot \mathbf{n} \, dS + \oiint_S \mathbf{n} \cdot [\tau] \cdot \mathbf{u} \, dS \tag{15.7}$$

This is the energy conservation principle, in integral form, for a viscous, heat-conducting fluid.

15.1.2 Constitutive Relation for the Heat-Flux Vector

The heat-flux vector, **q**, is usually written as

$$\mathbf{q} = -k\nabla T \tag{15.8}$$

where k is the **thermal conductivity**. This relationship is known as **Fourier's law of conduction.** The minus sign is consistent with the fact that heat flows from regions of high temperature to low temperature. Furthermore, k is related to viscosity by

$$k = \frac{c_p \mu}{Pr} \tag{15.9}$$

where Pr is the **Prandtl number**. Thus, for a calorically-perfect gas (so that c_p is constant), we can write the heat-flux vector as

$$\mathbf{q} = -\frac{c_p \mu}{Pr} \nabla T = -\frac{\mu}{Pr} \nabla h \tag{15.10}$$

Typical values of the Prandtl number for gases are close to one, while liquids have Prandtl number much larger than one. At STP ($0°$ C and 1 atm), the values of Pr for air and water are:

$$Pr = \begin{cases} 0.72, & \text{Air} \\ 7, & \text{Water} \end{cases} \tag{15.11}$$

Prandtl number is essentially constant for a gas. By contrast, it falls rapidly with increasing temperature for most liquids.

15.1.3 Differential Form

To derive the differential form of the energy-conservation equation, we make use of the divergence theorem to convert the surface integrals to volume integrals.

$$\oiint_S \rho \mathcal{H} \left(\mathbf{u} \cdot \mathbf{n} \right) dS = \iiint_V \nabla \cdot \left(\rho \, \mathbf{u} \mathcal{H} \right) dV \tag{15.12}$$

$$\oiint_S \mathbf{q} \cdot \mathbf{n} \, dS = \iiint_V \nabla \cdot \mathbf{q} \, dV \tag{15.13}$$

$$\oiint_S \mathbf{n} \cdot [\tau] \cdot \mathbf{u} \, dS = \iiint_V \nabla \cdot ([\tau] \cdot \mathbf{u}) \, dV \tag{15.14}$$

While the first two integrals above are in a familiar form for application of the divergence theorem, the friction-work integral [Equation (15.14)] may appear less recognizable. The key observation to make is that the integrand is the dot product of the unit normal, $\mathbf{n}$, with the vector $[\tau] \cdot \mathbf{u}$. The integrand on the right-hand side of Equation (15.14) is very complicated, and involves a large number of terms for general flows (vector notation sometimes conceals a great deal of complexity!). In particular, the term requires first post multiplying the stress tensor by the velocity vector and then taking the divergence. In matrix notation,

$$\nabla \cdot ([\tau] \cdot \mathbf{u}) = \begin{bmatrix} \dfrac{\partial}{\partial x} & \dfrac{\partial}{\partial y} & \dfrac{\partial}{\partial z} \end{bmatrix} \begin{bmatrix} \tau_{xx} & \tau_{xy} & \tau_{xz} \\ \tau_{yx} & \tau_{yy} & \tau_{yz} \\ \tau_{zx} & \tau_{zy} & \tau_{zz} \end{bmatrix} \begin{Bmatrix} u \\ v \\ w \end{Bmatrix} \tag{15.15}$$

Bringing all terms to the left-hand side, and then grouping all of the terms in the energy equation under a single volume integral, we have

$$\iiint_V \left[\frac{\partial}{\partial t} \left(\rho \mathcal{E} \right) + \nabla \cdot \left(\rho \, \mathbf{u} \mathcal{H} \right) + \nabla \cdot \mathbf{q} - \nabla \cdot ([\tau] \cdot \mathbf{u}) \right] dV = 0 \tag{15.16}$$

wherefore the conservation form of the differential equation governing energy conservation is

$$\frac{\partial}{\partial t} \left(\rho \mathcal{E} \right) + \nabla \cdot \left[\rho \, \mathbf{u} \mathcal{H} + \mathbf{q} - [\tau] \cdot \mathbf{u} \right] = 0 \tag{15.17}$$

We can recast this equation in primitive-variable form using the same sequence of operations that we used for the adiabatic, inviscid energy equation in Section 7.3. First, we note that total enthalpy and total energy are related by

$$\mathcal{H} = \mathcal{E} + \frac{p}{\rho} \quad \Longrightarrow \quad \rho \mathcal{H} = \rho \mathcal{E} + p \tag{15.18}$$

so that the instantaneous rate of change of total energy is

$$\frac{\partial}{\partial t}(\rho\mathcal{E}) = \frac{\partial}{\partial t}(\rho\mathcal{H}) - \frac{\partial p}{\partial t} \tag{15.19}$$

Thus, the energy equation can be rewritten as

$$\frac{\partial}{\partial t}(\rho\mathcal{H}) + \nabla\cdot(\rho\,\mathbf{u}\mathcal{H}) = \frac{\partial p}{\partial t} - \nabla\cdot\mathbf{q} + \nabla\cdot([\tau]\cdot\mathbf{u}) \tag{15.20}$$

Using the chain rule and a little rearrangement of terms, we find

$$\rho\left[\frac{\partial\mathcal{H}}{\partial t} + \mathbf{u}\cdot\nabla\mathcal{H}\right] + \mathcal{H}\left[\frac{\partial\rho}{\partial t} + \nabla\cdot(\rho\,\mathbf{u})\right] = \frac{\partial p}{\partial t} - \nabla\cdot\mathbf{q} + \nabla\cdot([\tau]\cdot\mathbf{u}) \tag{15.21}$$

The first term in brackets is the Eulerian derivative of $\mathcal{H}$. The second term in brackets vanishes by virtue of the continuity equation. Hence, the primitive-variable form of the energy equation is as follows.

$$\rho\frac{d\mathcal{H}}{dt} = \frac{\partial p}{\partial t} - \nabla\cdot\mathbf{q} + \nabla\cdot([\tau]\cdot\mathbf{u}) \tag{15.22}$$

This shows that the total enthalpy will change in a flow due to the action of viscous forces and heat transfer. This equation is actually much more complicated than is evident from the compact vector notation. Expanding the terms on the right-hand side [see Equation (15.15)] and using Equation (15.10) to determine the heat-flux vector, we arrive at the following.

$$\begin{aligned}
\rho\frac{d\mathcal{H}}{dt} &= \frac{\partial p}{\partial t} + \frac{\partial}{\partial x}\left(\frac{\mu}{Pr}\frac{\partial h}{\partial x}\right) + \frac{\partial}{\partial y}\left(\frac{\mu}{Pr}\frac{\partial h}{\partial y}\right) + \frac{\partial}{\partial z}\left(\frac{\mu}{Pr}\frac{\partial h}{\partial z}\right) \\
&+ \frac{\partial}{\partial x}(\tau_{xx}u + \tau_{xy}v + \tau_{xz}w) + \frac{\partial}{\partial y}(\tau_{yx}u + \tau_{yy}v + \tau_{yz}w) \\
&+ \frac{\partial}{\partial z}(\tau_{zx}u + \tau_{zy}v + \tau_{zz}w)
\end{aligned} \tag{15.23}$$

Needless to say, Equation (15.23) is far more complex in appearance than Equation (15.22) — and we haven't even expanded $d\mathcal{H}/dt$ or used the stress/strain-rate relationship to express the stress tensor in terms of velocity derivatives! This is a self-explanatory argument for the use of vector notation throughout this text.

There are two interesting observations we can make regarding energy conservation for a viscous fluid by examining the physical processes in a boundary layer. The energy conservation equation assumes a much simpler form in a boundary layer where changes in the y direction dominate. In the two-dimensional, steady case , for example, Equation (15.23) simplifies to

$$\rho u\frac{\partial\mathcal{H}}{\partial x} + \rho v\frac{\partial\mathcal{H}}{\partial y} = \frac{\partial}{\partial y}\left(\frac{\mu}{Pr}\frac{\partial h}{\partial y}\right) + \frac{\partial}{\partial y}(\tau_{yx}u) \tag{15.24}$$

Now, the x component of the momentum equation in a compressible boundary layer is the same as for an incompressible boundary layer provided we write the viscous contribution in terms of the stress tensor, τ_{xy}. That is, we arrived at the incompressible momentum Equation (14.40) assuming constant viscosity. Specifically, the viscous term, $\nu\partial^2 u/\partial y^2$, is the limiting form of $(1/\rho)\partial\tau_{yx}/\partial y$ when μ is constant. This is the only change required to rewrite the momentum equation for a compressible boundary layer, so that

$$\rho u\frac{\partial u}{\partial x} + \rho v\frac{\partial u}{\partial y} = -\frac{dp}{dx} + \frac{\partial\tau_{yx}}{\partial y} - \underbrace{\rho\frac{\partial\mathcal{V}}{\partial x}}_{=-\rho f_x} \tag{15.25}$$

where we include any body force that might act in the streamwise direction. We can separate the kinetic energy and body-force contributions from the total enthalpy by first multiplying Equation (15.25) by u and then subtracting the result from Equation (15.24). Additionally, we can use the fact that $\tau_{yx} = \mu \partial u / \partial y$ in a boundary layer. The resulting equations governing momentum and energy for a steady, two-dimensional boundary layer assume the following form.

$$\rho u \frac{\partial u}{\partial x} + \rho v \frac{\partial u}{\partial y} = -\frac{dp}{dx} + \underbrace{\frac{\partial}{\partial y}\left(\mu \frac{\partial u}{\partial y}\right)}_{\text{Diffusion}} + \rho f_x \qquad (15.26)$$

$$\rho u \frac{\partial h}{\partial x} + \rho v \frac{\partial h}{\partial y} = u\frac{dp}{dx} + \underbrace{\frac{\partial}{\partial y}\left(\frac{\mu}{Pr} \frac{\partial h}{\partial y}\right)}_{\text{Diffusion}} + \underbrace{\mu\left(\frac{\partial u}{\partial y}\right)^2}_{\text{Dissipation}} \qquad (15.27)$$

Comparison of Equations (15.26) and (15.27) shows that both include a diffusion term. The energy equation's diffusion term is simply the heat-conduction effect. The diffusion coefficient is μ for momentum and μ / Pr for enthalpy. Thus, if the Prandtl number is close to one, as it is for most gases, the diffusion process is as important in the energy equation as it is in the momentum equation. By contrast, if Pr is large, as it is for many liquids, diffusion will not be as significant in the overall energy balance as it is for momentum. Consequently, we make our first observation based on the boundary-layer form of the energy equation.

- **For $Pr \approx 1$,** viscous and heat-conduction effects are of comparable importance. If we either include or exclude one effect in our analysis, we must do the same for the other.

- **For $Pr \gg 1$,** heat-conduction effects are less significant than viscous effects. We can perform a meaningful viscous-flow analysis for fluids with large Prandtl number without considering heat conduction.

A second noteworthy observation pertains to the **dissipation** term in Equation (15.27). First, note that for a perfect gas with constant specific heats, reference to Equations (8.5) and (8.26) shows that the enthalpy is

$$h = c_p T = \frac{\gamma}{\gamma - 1}RT = \frac{a^2}{\gamma - 1} \qquad (15.28)$$

where a is the speed of sound. Hence, we can estimate the relative orders of magnitude of the energy equation's dissipation and diffusion terms in a boundary layer of thickness δ by saying

$$\frac{\mu\left(\dfrac{\partial u}{\partial y}\right)^2}{\dfrac{\partial}{\partial y}\left(\dfrac{\mu}{Pr} \dfrac{\partial h}{\partial y}\right)} \sim \frac{\mu \dfrac{u^2}{\delta^2}}{\dfrac{\mu}{Pr} \dfrac{a^2}{\delta^2}} \sim \frac{M^2}{Pr} \qquad (15.29)$$

where M is Mach number. Therefore, we conclude that:

- **For $M \to 0$,** dissipation can be neglected relative to heat-conduction.

- **For $M \gg 1$,** dissipation is much larger than heat conduction.

15.1.4 Entropy Generation Revisited

When we analyzed energy conservation for an adiabatic, inviscid fluid subject to conservative body forces, we found that the rate of change of entropy, s, following a fluid particle is zero. This is no longer true when we include effects of friction and heat transfer. With a great deal of manipulation, we can use **Gibbs' equation**, i.e., Equation (7.11), which tells us

$$T ds = de + p\, d\left(\frac{1}{\rho}\right) \tag{15.30}$$

and show that the energy equation also takes the form

$$\rho T \frac{ds}{dt} = \underbrace{-\nabla \cdot \mathbf{q}}_{\substack{\text{Heat} \\ \text{Addition}}} + \underbrace{([\tau] \cdot \nabla)^T \cdot \mathbf{u}}_{\substack{\text{Viscous} \\ \text{Dissipation}}} \tag{15.31}$$

where superscript T denotes the transpose operation. This equation shows that the two primary mechanisms for changing the entropy in a fluid flow are heat addition and viscous dissipation. The viscous dissipation is always a positive quantity, and thus causes entropy to increase. This is very easy to demonstrate for a laminar boundary layer, where variations in the y direction are much larger than those in the x and z directions. When this is true, we can approximate the differential operator $[\tau] \cdot \nabla$ as follows.

$$[\tau] \cdot \nabla \approx \begin{bmatrix} \tau_{xx} & \tau_{xy} & \tau_{xz} \\ \tau_{yx} & \tau_{yy} & \tau_{yz} \\ \tau_{zx} & \tau_{zy} & \tau_{zz} \end{bmatrix} \begin{Bmatrix} 0 \\ \frac{\partial}{\partial y} \\ 0 \end{Bmatrix} \approx \begin{Bmatrix} \tau_{xy} \frac{\partial}{\partial y} \\ \tau_{yy} \frac{\partial}{\partial y} \\ \tau_{zy} \frac{\partial}{\partial y} \end{Bmatrix} \tag{15.32}$$

Hence, the viscous dissipation is

$$([\tau] \cdot \nabla)^T \cdot \mathbf{u} \approx \begin{bmatrix} \tau_{xy} \frac{\partial}{\partial y} & \tau_{yy} \frac{\partial}{\partial y} & \tau_{zy} \frac{\partial}{\partial y} \end{bmatrix} \begin{Bmatrix} u \\ v \\ w \end{Bmatrix} \approx \tau_{xy} \frac{\partial u}{\partial y} + \tau_{yy} \frac{\partial v}{\partial y} + \tau_{zy} \frac{\partial w}{\partial y} \tag{15.33}$$

Then, since $v \ll u$ and $v \ll w$ while $\tau_{xy} \approx \mu \partial u/\partial y$ and $\tau_{zy} \approx \mu \partial w/\partial y$ for a boundary layer, the viscous dissipation in a laminar boundary layer is

$$([\tau] \cdot \nabla)^T \cdot \mathbf{u} \approx \mu \left[\left(\frac{\partial u}{\partial y}\right)^2 + \left(\frac{\partial w}{\partial y}\right)^2\right] \tag{15.34}$$

Although the proof is more lengthy (see Problems section), the viscous dissipation is positive for all flows, not just for boundary layers.

Demonstrating that heat addition causes entropy to increase is much more subtle. In fact, there is no guarantee that it is positive at all points in a given flow. However, consistent with the second law of thermodynamics, Landau and Lifshitz (1966) have shown that Equation (15.31) implies that entropy always increases for an entire system.

15.1.5 Surface Boundary Conditions

With the addition of the energy equation to our basic set of equations governing fluid motion, we have a need to specify two more boundary conditions. We need two additional conditions because the energy equation is a second-order, scalar differential equation. Usually one of the boundary conditions is imposed far from any solid boundaries, i.e., in the freestream. In most applications, we specify the freestream temperature. The other condition is usually specified at a solid boundary.

Two distinct types of boundary conditions are typically used, corresponding to the physical configuration under study. Often, we know the temperature at a solid boundary. This is common in long gas-transmission lines, for example, where the temperature at the pipe walls is equal to that of the surroundings. Many supersonic wind tunnel tests are done with the tunnel walls maintained at a nearly constant temperature. Corresponding to such applications, we specify the temperature at the surface, i.e.,

$$T = T_w \qquad \text{at a solid boundary} \qquad (15.35)$$

If T_w is constant over the entire boundary, we would refer to the surface as being **isothermal**. In this case, we refer to Equation (15.35) as the **isothermal-wall boundary condition**.

The second type of surface boundary condition is specification of the heat-flux at the surface, q_w. From Equation (15.8), if $\mathbf{n}$ is the unit normal to the surface, we say

$$\mathbf{n} \cdot \nabla T = -\frac{q_w}{k} \qquad \text{at a solid boundary} \qquad (15.36)$$

where $k = c_p \mu / Pr$ is the thermal conductivity. In the special case where the surface heat flux is zero, we have an **adiabatic wall**. The **adiabatic-wall boundary condition** [Equation (15.36) with $q_w = 0$] is appropriate for flow over a perfectly-insulated surface. For such a surface, the heat generated by the fluid through viscous effects heats the wall until the condition $\mathbf{n} \cdot \nabla T = 0$ is satisfied. The surface temperature exceeds that of the fluid above as a result.

This effect is quite strong near a stagnation point, for example. In the absence of viscous effects, the temperature at a stagnation point would be equal to the total temperature, with all of the fluid's kinetic energy being converted to enthalpy. However, viscous losses and heat conducted away reduce the energy available to heat the surface. As a result, the temperature at a stagnation point with an adiabatic wall is less than the total temperature, T_t. We express this fact by introducing the **recovery factor**, r, so that the **adiabatic-wall temperature**, T_{aw}, is written as

$$T_{aw} = T_\infty \left[1 + \frac{\gamma - 1}{2} r M_\infty^2 \right] \qquad (15.37)$$

As noted by Schlichting (1979), a good approximation for r when Mach number lies between 0 and 5 is given by

$$r \approx \sqrt{Pr} = 0.85 \qquad (15.38)$$

which holds for stagnation-point flow and for laminar boundary layers. Thus, for example, if air flows past a body with a freestream Mach number of 3, the temperature at a stagnation point is

$$\frac{T_{aw}}{T_t} = \frac{1 + \dfrac{\gamma - 1}{2} r M_\infty^2}{1 + \dfrac{\gamma - 1}{2} M_\infty^2} = 0.904 \qquad (15.39)$$

Hence, the actual temperature at the stagnation point is 90% of the total temperature. This is the reason the term *total temperature* has replaced the older designation of *stagnation temperature* (see Section 8.6).

Clearly, we could also impose a boundary condition of **mixed type** in which a linear combination of Equations (15.35) and (15.36) is imposed. This is rare in fluid mechanics, however.

15.2 Fanno Flow

Description. As a simple example of a compressible flow in which viscous effects play a nontrivial role, we consider steady, adiabatic flow in a pipe or duct with constant area, A. This is known as **Fanno flow**, and is depicted schematically in Figure 15.2. To formulate a simplified one-dimensional set of differential equations for averaged properties on cross sections, it is most convenient to begin with the integral form of the conservation laws. The boundary of the control volume we will use is shown as the dashed rectangle in the figure.

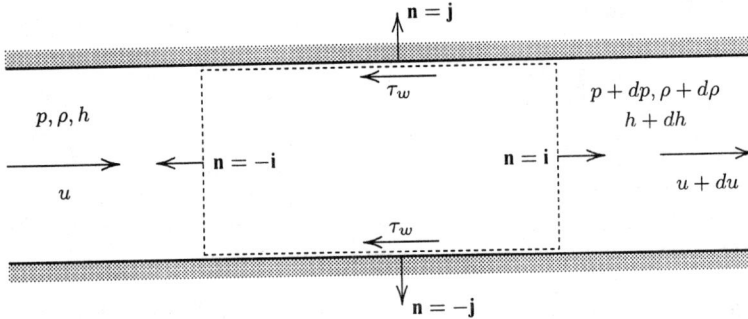

Figure 15.2: *Control volume for Fanno flow; cross-sectional area, A, is constant.*

Mass Conservation. We consider steady flow so that the mass-conservation principle tells us the net flux of fluid out of the control volume is zero. Hence,

$$\oiint_S \rho\, \mathbf{u} \cdot \mathbf{n}\, dS = 0 \tag{15.40}$$

Since there is no fluid flowing through the walls of the pipe or duct, the mass flux into the control volume balances the mass flux out. So,

$$- \rho u A + (\rho + d\rho)(u + du)A = 0 \quad \Longrightarrow \quad \rho u = (\rho + d\rho)(u + du) \tag{15.41}$$

Expanding the right-hand side and dropping the higher-order term, $d\rho du$, we find

$$\rho du + u d\rho = d(\rho u) = 0 \tag{15.42}$$

Thus, the differential form of the mass-conservation equation assumes the following form.

$$\frac{d}{dx}(\rho u) = 0 \tag{15.43}$$

Momentum Conservation. Since we have one-dimensional motion, only the x component of the momentum equation is of interest. Again appealing to the integral-conservation principle with body forces omitted, we have

$$\oiint_S \rho\, u\, (\mathbf{u} \cdot \mathbf{n})\, dS = -\mathbf{i} \cdot \oiint_S p\, \mathbf{n}\, dS + \mathbf{i} \cdot \oiint_S \mathbf{n} \cdot [\tau]\, dS \tag{15.44}$$

It is most convenient to proceed term by term in evaluating the various surface integrals in this equation. The net flux of momentum is

$$\oiint_S \rho u \left(\mathbf{u} \cdot \mathbf{n}\right) dS = \rho u (-u) A + \underbrace{(\rho + d\rho)(u + du)}_{=\rho u}(u + du) A$$

$$= -\rho u^2 A + \rho u (u + du) A = \rho u\, du\, A \tag{15.45}$$

For the net-pressure integral, only the contributions from the inlet and outlet affect momentum conservation in the x direction. Hence,

$$-\mathbf{i} \cdot \oiint_S p\, \mathbf{n}\, dS = -\mathbf{i} \cdot [-p\, \mathbf{i} A] - \mathbf{i} \cdot [(p + dp)\mathbf{i} A]$$

$$= pA - (p + dp) A = -A\, dp \tag{15.46}$$

Finally, turning to the viscous term, note that at the inlet and outlet we have $\mathbf{n} \cdot [\tau] = \tau_{xx}$. In general, the normal viscous stress is very small compared to the pressure in a viscous flow. That is, if the overall length of the pipe or duct is L, then

$$\frac{\tau_{xx}}{p} \sim \frac{\mu u / L}{\rho a^2} \sim M^2 \frac{\mu}{\rho u L} \sim \frac{M^2}{Re_L} \tag{15.47}$$

Thus, except for low Reynolds number flow, we can ignore the normal viscous stress. Turning to the perimeter of the cross section, we have $\mathbf{n} \cdot [\tau] = -\tau_w$. Hence, if P is the distance measured around the perimeter, we have

$$\mathbf{i} \cdot \oiint_S \mathbf{n} \cdot [\tau]\, dS \approx -\tau_w P\, dx \tag{15.48}$$

So, substituting Equations (15.45), (15.46) and (15.48) into Equation (15.44) yields

$$\rho u\, du\, A = -A\, dp - \tau_w P\, dx \tag{15.49}$$

Dividing through by $A\, dx$ yields the momentum equation for Fanno flow, viz.,

$$\rho u \frac{du}{dx} + \frac{dp}{dx} = -(P/A)\tau_w \tag{15.50}$$

Energy Conservation. In the absence of body forces, the energy-conservation Equation (15.7) simplifies to

$$\oiint_S \rho \left(h + \frac{1}{2}u^2\right)(\mathbf{u} \cdot \mathbf{n})\, dS = -\oiint_S \mathbf{q} \cdot \mathbf{n}\, dS + \oiint_S \mathbf{n} \cdot [\tau] \cdot \mathbf{u}\, dS \tag{15.51}$$

Assuming large Reynolds number, an order-of-magnitude estimate shows that the viscous and heat-transfer contributions on cross sections are negligible relative to the cross-sectional flux of total enthalpy, so that only contributions from the pipe or duct walls are important. Because we assume the flow is adiabatic, the heat flux is exactly zero on the perimeter, wherefore

$$\oiint_S \mathbf{q} \cdot \mathbf{n}\, dS \approx 0 \tag{15.52}$$

Also, because the no-slip boundary condition holds on the perimeter so that $\mathbf{u} = \mathbf{0}$, the friction forces do no work at the pipe or duct walls. Therefore,

$$\oiint_S \mathbf{n} \cdot [\tau] \cdot \mathbf{u} \, dS \approx 0 \tag{15.53}$$

This means the net flux of total enthalpy out of the control volume is zero, i.e.,

$$\rho u \left(h + \frac{1}{2} u^2 \right) = \text{constant} \quad \Longrightarrow \quad h + \frac{1}{2} u^2 = \text{constant} \tag{15.54}$$

where we make use of Equation (15.43) to conclude that ρu is also constant. In differential form, energy conservation becomes

$$\frac{dh}{dx} + u \frac{du}{dx} = 0 \tag{15.55}$$

We can arrive at a more illuminating form of the energy equation by appealing to Gibbs' equation [Equation (15.30)]. This will permit us to derive an equation for the entropy as follows.

$$\left.\begin{aligned}
T \frac{ds}{dx} &= \frac{de}{dx} + p \frac{d(1/\rho)}{dx} && \text{[from Gibbs' equation]} \\[2mm]
&= \frac{dh}{dx} - \frac{1}{\rho} \frac{dp}{dx} && \text{[since } h = e + p/\rho] \\[2mm]
&= -\frac{1}{\rho} \left[\rho u \frac{du}{dx} + \frac{dp}{dx} \right] && \text{[from Equation (15.55)]} \\[2mm]
&= -\frac{1}{\rho} [-(P/A)\tau_w] && \text{[from Equation (15.50)]}
\end{aligned}\right\} \tag{15.56}$$

Therefore, the entropy equation for Fanno flow is

$$\rho T \frac{ds}{dx} = (P/A)\tau_w \tag{15.57}$$

Equations (15.55) and (15.57) provide an immediate glimpse of the physical process occurring in Fanno flow. Since the right-hand side of the entropy equation is positive, the entropy increases in a pipe or duct flow with friction. However, friction changes the entropy without changing the total enthalpy.[1] The primary reason total enthalpy is conserved rests in the fact that the viscous stresses do no work at the pipe/duct surface where the stresses are largest.

To close our system of equations, we must determine the shear stress, τ_w. Following our analysis of pipe flow in Chapter 7, we introduce the conventional **friction factor**, f, and Darcy's **hydraulic diameter**, D_h, viz.,

$$f = \frac{\tau_w}{\frac{1}{8}\rho u^2} \quad \text{and} \quad D_h = 4A/P \tag{15.58}$$

Combining these relations, the right-hand side of Equation (15.57) can be written as

$$(P/A)\tau_w = \frac{1}{2}\rho u^2 f / D_h \tag{15.59}$$

---

[1] Although the physical processes are different, this also occurs for flow through a normal shock wave.

Summarizing, the equations for conservation of mass, momentum and energy appropriate for Fanno flow are as follows.

$$\left. \begin{array}{c} \dfrac{d}{dx}(\rho u) = 0 \\[2ex] \dfrac{dp}{dx} + \rho u \dfrac{du}{dx} = -\dfrac{1}{2}\rho u^2 f/D_h \\[2ex] \rho T \dfrac{ds}{dx} = \dfrac{1}{2}\rho u^2 f/D_h \end{array} \right\} \qquad (15.60)$$

Discussion. Because we have derived these equations by integrating over cross sections, flow properties such as ρ, u, h, etc. represent average conditions on each cross section. Since we have not explicitly introduced the constitutive relation for τ_w, our solution applies to either laminar or turbulent flow. Of course, the type of flow we have determines the value we use for the friction factor.

Since we are considering compressible flow, we must raise the question of the effect Mach number has on f. There is indeed an effect, and it is stronger for turbulent flow than for laminar flow. At Mach 10, for example, Schlichting (1979) indicates that the skin friction on a flat plate is about 2/3 of the incompressible value for laminar flow and 1/4 for turbulent flow. In both cases, there is only a minor reduction for subsonic flow. Similar trends should be expected for flow in pipes and ducts.

Based on these observations, for subsonic Mach numbers, the value of f can be determined from either the Moody diagram (Figure 7.6) or Equations (7.104) and (7.105). When the flow is supersonic, a compressibility correction can be applied to the incompressible friction factor. Based on measurements presented by Schlichting (1979), for turbulent flow we can compute f according to

$$f \approx \frac{f_{inc}}{\sqrt{1 + \frac{\gamma-1}{2}rM^2}} \qquad (15.61)$$

where f_{inc} is the incompressible friction factor, and r is the recovery factor. Note that for turbulent flow, r is only weakly dependent on the fluid and has been found from measurements to be

$$r = 0.89 \qquad \text{(turbulent flow)} \qquad (15.62)$$

With a bit more algebra, we can obtain a closed-form solution to Equations (15.60). Assuming a perfect, calorically-perfect gas, we proceed in a manner similar to what we did for the Laval nozzle in Section 8.9. That is, we recast the problem in terms of Mach number and reference all conditions to the sonic point. We can solve for the thermodynamic properties of the flow using basic definitions along with conservation of energy and mass, as given in Equations (15.55) and the first of (15.60). The primary relations we need to accomplish this are $a^2 = \gamma p/\rho$, $M^2 = u^2/a^2$, $p = \rho RT$. Also, the mass flux and total enthalpy are constant so that we have $\rho u = \text{constant}$ and $h + u^2/2 = \text{constant}$.

While the algebra is straightforward, it is a bit tedious and is addressed in the Problems section. Table B.3 in Appendix B includes all pertinent flow properties as a function of Mach number. Letting superscript $*$ denote sonic-flow ($M = 1$) reference conditions, the primary thermodynamic properties are given by

$$
\begin{aligned}
\frac{p}{p^*} &= \frac{1}{M}\left[\frac{\gamma+1}{2+(\gamma-1)M^2}\right]^{1/2} \\[2mm]
\frac{\rho}{\rho^*} &= \frac{1}{M}\left[\frac{2+(\gamma-1)M^2}{\gamma+1}\right]^{1/2} \\[2mm]
\frac{T}{T^*} &= \frac{\gamma+1}{2+(\gamma-1)M^2} \\[2mm]
\frac{p_t}{p_t^*} &= \frac{1}{M}\left[\frac{2+(\gamma-1)M^2}{\gamma+1}\right]^{(\gamma+1)/[2(\gamma-1)]}
\end{aligned}
\right\} \tag{15.63}
$$

Turning to the momentum equation, rearrangement in terms of Mach number yields the following differential equation.

$$
\frac{dM^2}{dx} = \gamma M^4\left[\frac{1+\frac{\gamma-1}{2}M^2}{1-M^2}\right]\frac{f}{D_h} \tag{15.64}
$$

Hence, integrating from $x = 0$ to $x = L^*$, where L^* corresponds to the sonic point, we have

$$
\begin{aligned}
\int_0^{L^*}\frac{f}{D_h}dx &= \int_{M^2}^1 \frac{\left(1-M^2\right)dM^2}{\gamma M^4\left[1+\frac{\gamma-1}{2}M^2\right]} \\[2mm]
&= \frac{1-M^2}{\gamma M^2} + \frac{\gamma+1}{2\gamma}\ell n\left[\frac{(\gamma+1)M^2}{2+(\gamma-1)M^2}\right]
\end{aligned} \tag{15.65}
$$

Finally, we define an average friction factor, $\overline{f}$, by

$$
\overline{f} \equiv \frac{1}{L^*}\int_0^{L^*} f\,dx \tag{15.66}
$$

Then, substituting Equation (15.66) into Equation (15.65), we arrive at the following equation, which relates the local Mach number and the distance from the point of interest to a reference point at which sonic conditions occur.

$$
\frac{\overline{f}L^*}{D_h} = \frac{1-M^2}{\gamma M^2} + \frac{\gamma+1}{2\gamma}\ell n\left[\frac{(\gamma+1)M^2}{2+(\gamma-1)M^2}\right] \tag{15.67}
$$

To understand how we use this relationship, it is instructive to consider two cases distinguished by the nature of the inlet flow. The two possibilities are a subsonic inlet and a supersonic inlet. For simplicity, the primary difference between the two cases we will consider is in the inlet Mach number. The temperature is the same for both cases, so that the difference in Mach number results from changing the pressure.

Subsonic Inlet. We consider the flow of air in a smooth cylindrical pipe of diameter $D = 0.2$ m and length $L = 20$ m. At the inlet, the Mach number is $M_1 = 0.3$, the pressure is $p_1 = 1$ atm and the temperature is $T_1 = 293$ K. Using our Fanno-flow solution, we would like to determine the Mach number, M_2, pressure, p_2, and temperature, T_2, at the exit. Because we have referenced all flow properties to sonic-flow conditions, we imagine that our pipe extends beyond the exit as shown in Figure 15.3.

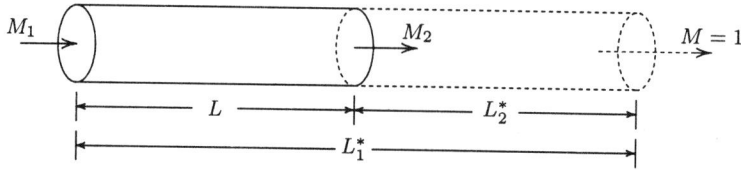

Figure 15.3: *Adiabatic flow in a pipe with friction.*

As our first step, we must compute the Reynolds number and determine the friction factor. For the inlet conditions, the speed of sound, a_1, velocity, u_1, and density, ρ_1 are:

$$
\left.
\begin{aligned}
a_1 &= \sqrt{\gamma R T_1} &&= 343 \text{ m/sec} \\[2mm]
u_1 &= M_1 a_1 &&= 103 \text{ m/sec} \\[2mm]
\rho_1 &= \frac{p_1}{R T_1} &&= 1.201 \text{ kg/m}^3
\end{aligned}
\right\}
\tag{15.68}
$$

The viscosity, μ_1, follows from Sutherland's law, Equation (1.29), which gives

$$
\mu_1 = \frac{1.46 \cdot 10^{-6} \, T^{3/2}}{T + 110.3} = 1.816 \cdot 10^{-5} \text{ kg/(m} \cdot \text{sec)}
\tag{15.69}
$$

Thus, the Reynolds number at the inlet, Re_D, is

$$
Re_D = \frac{\rho_1 u_1 D}{\mu_1} = 1.36 \cdot 10^6
\tag{15.70}
$$

Reference to the Moody diagram in Figure 7.6 shows that the flow is turbulent and the friction factor is

$$
f \approx 0.011
\tag{15.71}
$$

Now, the friction factor will vary in our pipe, and conceptually we must use the average value in Equation (15.67). For the sake of simplicity, it is customary to use the inlet friction factor as an approximation to $\overline{f}$. Also, note that the perimeter of the pipe is $P = \pi D$ and the cross-sectional area is $A = \pi D^2/4$. Hence, the hydraulic diameter is

$$
D_h = 4A/P = D
\tag{15.72}
$$

At this point, we refer to Table B.3 in Appendix B to determine the inlet properties in relation to the sonic-flow reference conditions. For Mach 0.3, we find

$$
\frac{\overline{f} L_1^*}{D_h} = 5.2993, \quad \frac{p_1}{p^*} = 3.6191, \quad \frac{T_1}{T^*} = 1.1788
\tag{15.73}
$$

Now, as shown in Figure 15.3, the length from the exit to the sonic point is $L_2^* = L_1^* - L$, wherefore

$$
\frac{\overline{f} L_2^*}{D_h} = \frac{\overline{f} L_1^*}{D_h} - \frac{\overline{f} L}{D_h} = 5.2993 - \frac{(.011)(20 \text{ m})}{(.2 \text{ m})} = 4.1993
\tag{15.74}
$$

Again referring to Table B.3, this value of $\overline{f}L_2^*/D_h$ lies 60% of the way between the values listed for Mach numbers of 0.30 and 0.35. Hence, interpolation tells us the exit Mach number, pressure and temperature are

$$M_2 \approx 0.33, \quad \frac{p_2}{p^*} \approx 3.3030, \quad \frac{T_2}{T^*} \approx 1.1743 \qquad (15.75)$$

Finally, we can compute the actual values of p_2 and T_2 according to

$$p_2 = p_1 \cdot \frac{p_2/p^*}{p_1/p^*} = 0.91 \text{ atm} \quad \text{and} \quad T_2 = T_1 \cdot \frac{T_2/T^*}{T_1/T^*} = 292 \text{ K} \qquad (15.76)$$

Supersonic Inlet. We consider the same pipe as in the subsonic case with one change. We now increase the inlet Mach number to 3. The inlet speed of sound, density and viscosity are still given by Equations (15.68) and (15.69). However, the velocity and Reynolds number both increase by a factor of 10, so that

$$u_1 = 1030 \text{ m/sec} \quad \text{and} \quad Re_D = 1.36 \cdot 10^7 \qquad (15.77)$$

We cannot use the Moody diagram to determine f_{inc} as the appropriate value is off scale for a smooth pipe. Rather, we use the turbulent-flow value for f given in Equation (7.104) to show that

$$f_{inc} \approx 0.00775 \qquad (15.78)$$

Then, using Equation (15.61) to correct for effects of compressibility, we obtain

$$f = \frac{f_{inc}}{\sqrt{1 + \frac{\gamma-1}{2}rM^2}} \approx 0.0048 \qquad (15.79)$$

Turning to the Fanno-flow tables, at Mach 3 the relevant properties for our problem are

$$\frac{\overline{f}L_1^*}{D_h} = 0.5222, \quad \frac{p_1}{p^*} = 0.2182, \quad \frac{T_1}{T^*} = 0.4286 \qquad (15.80)$$

Hence, the effective distance of the exit from the sonic point is

$$\frac{\overline{f}L_2^*}{D_h} = \frac{\overline{f}L_1^*}{D_h} - \frac{\overline{f}L}{D_h} = 0.5222 - \frac{(.0048)(20 \text{ m})}{(.2 \text{ m})} = 0.0422 \qquad (15.81)$$

Reference to Table B.3, shows that this value of $\overline{f}L_2^*/D_h$ lies 57% of the way between the values listed for Mach numbers of 1.20 and 1.25. Hence, interpolation tells us the exit Mach number, pressure and temperature for our supersonic inlet case are

$$M_2 \approx 1.23, \quad \frac{p_2}{p^*} \approx 0.7819, \quad \frac{T_2}{T^*} \approx 0.9218 \qquad (15.82)$$

and the dimensional values of p_2 and T_2 become

$$p_2 = p_1 \cdot \frac{p_2/p^*}{p_1/p^*} = 3.58 \text{ atm} \quad \text{and} \quad T_2 = T_1 \cdot \frac{T_2/T^*}{T_1/T^*} = 630 \text{ K} \qquad (15.83)$$

Mollier Diagram. There is an additional equation for Fanno flow that is particularly illuminating, and reveals the primary effect of friction on the flow. As a direct consequence of the fact that Fanno flow has constant total enthalpy, a straightforward derivation shows that the rate of change of the enthalpy is given by

$$\frac{dh}{dx} = \frac{\gamma M^2}{M^2 - 1} T \frac{ds}{dx} \tag{15.84}$$

Since the second law of thermodynamics tells us the entropy must always increase, enthalpy must decrease with x when the flow is subsonic and vice versa. Since total enthalpy, $h + \frac{1}{2}u^2$, is constant, necessarily u varies in the opposite direction to h. Thus, the cooling associated with a decrease in h for subsonic flow results in an increase in the flow's kinetic energy per unit mass, $\frac{1}{2}u^2$. Conversely, under supersonic conditions, friction heats the flow at the expense of the kinetic energy. Inspection of Equation (15.84) shows that when the Mach number is unity, necessarily $ds/dx = 0$ (to preclude the physically impossible condition that $dh/dx \to \infty$). This means s achieves its maximum value at the sonic point.

These facts are illustrated in Figure 15.4, which shows a typical variation of h with s. This type of graph is called a **Mollier diagram**. The curve is known as the **Fanno line** or, more appropriately, the **Fanno curve**. Entropy is a maximum at point a where the flow is sonic. All states with higher enthalpy lie on the subsonic, or upper, branch of the curve, while states with lower enthalpy lie on the supersonic, or lower, branch.

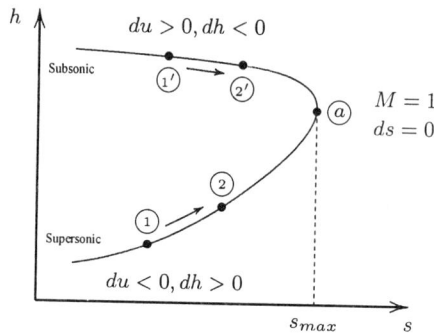

Figure 15.4: *Mollier diagram for Fanno flow.*

As indicated in the figure, the flow is always driven toward the sonic point. When the inlet flow is supersonic, for example, we begin at point 1. The flow downstream of the inlet approaches point a, with an attendant decrease in Mach number towards one. Each point, say 2, on the supersonic branch corresponds to a specific pipe/duct length, L. Increasing the length L causes point 2 to move closer to point a. When the length is selected so that the flow just becomes sonic at the exit, points 2 and a are coincident and we have $L = L^*$. The flow is then said to be **choked**, and no further increase in L can occur without modifying the inlet conditions. That is, we must move point 1 further back on the branch to a position corresponding to the desired length. If this is not done, a shock will form, causing the flow to jump from point 2 to 2′.

Similarly, when the inlet flow is subsonic, we begin at point 1′ and the flow is driven toward point a. The pipe/duct length can be increased until the flow becomes choked. Further increases in L cause a reduction in the mass-flow rate and the inlet Mach number.

For the two examples above, it is a straightforward matter to compute the length of pipe required to choke the flow. Specifically, the length is given by

$$L^* = \left(\frac{\overline{f}L^*}{D_h}\right)\left(\frac{D_h}{\overline{f}}\right) = \begin{cases} 96.4 \text{ m}, & \text{Subsonic inlet} \\ 21.8 \text{ m}, & \text{Supersonic inlet} \end{cases} \tag{15.85}$$

As a final comment, it is worthwhile to summarize how flow properties vary in Fanno flow. The direction in which most flow properties change depends upon whether the inlet flow is subsonic or supersonic. Table 15.1 lists the various flow properties and the directions in which they vary.

Table 15.1: *Variation of Fanno-Flow Properties*

Subsonic Inlet	Supersonic Inlet
M and u increase	p, ρ and T increase
T_t is constant	T_t is constant
p, ρ, T, p_t and ρ_t decrease	M, u, p_t and ρ_t decrease

15.3 Rayleigh Flow

Description. Rayleigh flow is a straightforward, one-dimensional example that illustrates effects of heat transfer in a compressible medium. We again consider steady flow of a gas in a duct or pipe with constant cross-sectional area A. In this application, we ignore viscous effects and focus rather on heat transfer as indicated in Figure 15.5. This analysis is valid for flows heated by processes such as combustion, nuclear radiation and laser irradiation. Because we are considering frictionless flow of a gas, necessarily the analysis does not apply to heat conducted from the walls of the duct or pipe. That is, consistent with the discussion at the end of Subsection 15.1.3, neglecting viscous effects and heat conduction go hand in hand since the Prandtl number is close to one for a gas. We cannot ignore one without ignoring the other.

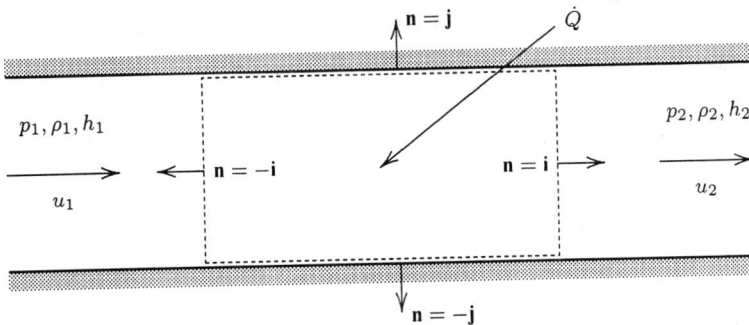

Figure 15.5: *Control volume for Rayleigh flow; cross-sectional area, A, is constant.*

Mass and Momentum Conservation. It is most convenient to apply the conservation principles to a control volume relating the inlet and a specific point downstream. Since, we are neglecting effects of friction, the integral conservation principles for mass [Equation (15.40)] and momentum [Equation (15.44) with $[\tau] = 0$] simplify immediately to

$$\rho_1 u_1 = \rho_2 u_2 \tag{15.86}$$

$$p_1 + \rho_1 u_1^2 = p_2 + \rho_2 u_2^2 \tag{15.87}$$

Energy Conservation. Again ignoring body forces, the basic energy-conservation principle in integral form is

$$\oiint_S \rho \left(h + \frac{1}{2} u^2 \right) (\mathbf{u} \cdot \mathbf{n}) \, dS = \dot{Q} \tag{15.88}$$

where $\dot{Q}$ is the heat transfer rate to the control volume. We define the **heat added per unit mass**, q, according to

$$\dot{Q} = \dot{m} q = \rho u A q \tag{15.89}$$

So, evaluating the total enthalpy-flux integral yields the following equation for energy conservation.

$$h_1 + \frac{1}{2} u_1^2 + q = h_2 + \frac{1}{2} u_2^2 \tag{15.90}$$

Discussion. The first important observation we can make regarding Rayleigh flow concerns the total temperature. Noting that the enthalpy is given by $h = c_p T$ and that $c_p = \gamma R / (\gamma - 1)$, we know that

$$h + \frac{1}{2} u^2 = c_p T + \frac{1}{2} u^2 = c_p T \left(1 + \frac{\gamma - 1}{2} M^2 \right) = c_p T_t \tag{15.91}$$

where T_t is total temperature. Therefore, we can rewrite Equation (15.90) as

$$T_{t2} = T_{t1} + \frac{q}{c_p} \tag{15.92}$$

This tells us the heat added to the duct or pipe serves to directly increase the total temperature of the flow.

Using the perfect-gas law, $p = \rho R T$, along with Equations (15.86), (15.87) and (15.90), we can solve for all flow properties as a function of Mach number. As with Fanno flow, we select our reference state as the sonic point. The solution for the primary thermodynamic variables and total conditions is tabulated in Table B.4 of Appendix B. Letting superscript * denote sonic conditions, the algebraic form of the solution is as follows.

$$\left. \begin{aligned}
\frac{p}{p^*} &= \frac{\gamma + 1}{1 + \gamma M^2} \\[2mm]
\frac{\rho}{\rho^*} &= \frac{1}{M^2} \left[\frac{1 + \gamma M^2}{\gamma + 1} \right] \\[2mm]
\frac{T}{T^*} &= M^2 \left[\frac{\gamma + 1}{1 + \gamma M^2} \right]^2 \\[2mm]
\frac{p_t}{p_t^*} &= \frac{\gamma + 1}{1 + \gamma M^2} \left[\frac{2 + (\gamma - 1) M^2}{\gamma + 1} \right]^{\gamma/(\gamma - 1)} \\[2mm]
\frac{T_t}{T_t^*} &= \frac{(\gamma + 1) M^2}{(1 + \gamma M^2)^2} \left[2 + (\gamma - 1) M^2 \right]
\end{aligned} \right\} \tag{15.93}$$

As with Fanno flow, the solution depends upon whether the flow is subsonic or supersonic at the inlet. Furthermore, our solution is valid for both heating and cooling, and the direction of changes in flow properties differs for the four possible conditions, i.e., subsonic inlet with heating, subsonic inlet with cooling, supersonic inlet with heating and supersonic inlet with cooling.

Subsonic Inlet With Heating. Consider a duct with air entering at a Mach number, $M_1 = 0.3$. The static pressure and temperature are $p_1 = 1$ atm and $T_1 = 293$ K, respectively. The heat added per unit mass is $q = 3 \cdot 10^5$ J/kg. We would like to determine M_2, p_2 and p_{t2}. Figure 15.6 schematically illustrates such a flow, including an imaginary extension to a length at which sonic flow exists. Note that beginning with the conditions at the duct exit, the heat required to bring the flow to sonic conditions is q_2^*. Thus, we infer that the heat required to take the flow from the inlet conditions to sonic conditions, q_1^*, must be

$$q_1^* = q_1 + q_2^* \tag{15.94}$$

This is different from Fanno flow in which sonic conditions correspond to an imaginary point that can be reached *adiabatically*. Here, we have *non-adiabatic* flow, so that p^*, T^*, etc. are the conditions that would exist if sufficient heat were added to reach Mach 1.

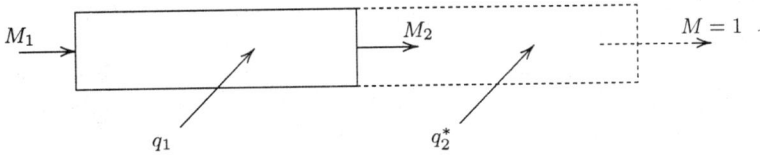

Figure 15.6: *Frictionless flow in a duct with heat addition.*

As our first step in solving this problem, we must determine the total pressure and total temperature at the inlet. We need these quantities in order to compute the magnitude of p_t and T_t at the exit. To find the inlet total conditions, we can use the isentropic flow Table B.1 in Appendix B. For an inlet Mach number, $M_1 = 0.3$,

$$\frac{p_1}{p_{t1}} = 0.9395 \quad \text{and} \quad \frac{T_1}{T_{t1}} = 0.9823 \tag{15.95}$$

Also, we need the specific heat, c_p, to compute T_{t2} from Equation (15.92).

$$c_p = \frac{\gamma R}{\gamma - 1} = \frac{(1.4)[287 \text{ J}/(\text{kg} \cdot \text{K})]}{0.4} = 1004.5 \quad \text{J}/(\text{kg} \cdot \text{K}) \tag{15.96}$$

Therefore, the inlet total pressure and total temperature are

$$\left. \begin{array}{rclcl} T_{t1} & = & \dfrac{T_1}{T_1/T_{t1}} & = & \dfrac{293 \text{ K}}{0.9823} & = & 298.3 \text{ K} \\[3mm] p_{t1} & = & \dfrac{p_1}{p_1/p_{t1}} & = & \dfrac{1 \text{ atm}}{0.9395} & = & 1.0644 \text{ atm} \end{array} \right\} \tag{15.97}$$

From Equation (15.92), the total temperature at the exit is

$$T_{t2} = T_{t1} + \frac{q}{c_p} = 298.3 \text{ K} + \frac{3 \cdot 10^5 \text{ J/kg}}{1004.5 \text{ J}/(\text{kg} \cdot \text{K})} = 298.3 \text{ K} + 298.7 \text{ K} = 597.0 \text{ K} \tag{15.98}$$

Now, we turn to the Rayleigh-flow Table B.4 in Appendix B. We require the total temperature ratio, T_{t1}/T_t^*, which we will use to determine the Mach number at the exit. We also require the static and total pressure ratios, p_1/p^* and p_{t1}/p_t^*, which will eventually be needed to determine the magnitudes of p_2 and p_{t2}. Reference to the tables shows that for $M_1 = 0.3$,

$$\frac{T_{t1}}{T_t^*} = 0.3469, \quad \frac{p_1}{p^*} = 2.1314, \quad \frac{p_{t1}}{p_t^*} = 1.1985 \tag{15.99}$$

In order to find the Mach number at the exit, M_2, we first compute T_{t2}/T_t^* as follows.

$$\frac{T_{t2}}{T_t^*} = \frac{T_{t2}}{T_{t1}}\frac{T_{t1}}{T_t^*} = \left(\frac{597.0 \text{ K}}{298.3 \text{ K}}\right)(0.3469) = 0.6943 \tag{15.100}$$

Again referring to Table B.4, we find that $T_t/T_t^* = 0.6943$ lies 4% of the way between the values listed for Mach numbers of 0.50 and 0.55. So, linear interpolation tells us that

$$M_2 \approx 0.50, \quad \frac{p_2}{p^*} \approx 1.7739, \quad \frac{p_{t2}}{p_t^*} \approx 1.1132 \tag{15.101}$$

Finally, the static and total pressure at the exit are

$$p_2 = p_1 \cdot \frac{p_2/p^*}{p_1/p^*} = 0.83 \text{ atm} \quad \text{and} \quad p_{t2} = p_{t1} \cdot \frac{p_{t2}/p_t^*}{p_{t1}/p_t^*} = 0.99 \text{ atm} \tag{15.102}$$

Supersonic Inlet With Heating. We now increase the inlet Mach number to $M_1 = 3$, leaving all other conditions the same as in the subsonic case. Reference to the isentropic flow Table B.1 shows that

$$\frac{p_1}{p_{t1}} = 0.02722 \quad \text{and} \quad \frac{T_1}{T_{t1}} = 0.3571 \tag{15.103}$$

So, the inlet total pressure and total temperature are

$$T_{t1} = \frac{T_1}{T_1/T_{t1}} = 820.5 \text{ K} \quad \text{and} \quad p_{t1} = \frac{p_1}{p_1/p_{t1}} = 36.74 \text{ atm} \tag{15.104}$$

The total temperature at the exit follows from Equation (15.92), viz.,

$$T_{t2} = T_{t1} + \frac{q}{c_p} = 820.5 \text{ K} + \frac{3 \cdot 10^5 \text{ J/kg}}{1004.5 \text{ J/(kg} \cdot \text{K)}} = 820.5 \text{ K} + 298.7 \text{ K} = 1119.2 \text{ K} \tag{15.105}$$

From the Rayleigh flow Table B.4 for an inlet Mach number, $M_1 = 3$,

$$\frac{T_{t1}}{T_t^*} = 0.6540, \quad \frac{p_1}{p^*} = 0.1765, \quad \frac{p_{t1}}{p_t^*} = 3.4245 \tag{15.106}$$

Therefore, the total temperature ratio at the duct exit is

$$\frac{T_{t2}}{T_t^*} = \frac{T_{t2}}{T_{t1}}\frac{T_{t1}}{T_t^*} = \left(\frac{1119.2 \text{ K}}{820.5 \text{ K}}\right)(0.6540) = 0.8921 \tag{15.107}$$

Referring again to the Rayleigh flow Table B.4, we see that $T_t/T_t^* = 0.8921$ lies 37% of the way between values for Mach numbers of 1.55 and 1.60. Using interpolation there follows:

$$M_2 \approx 1.57, \quad \frac{p_2}{p^*} \approx 0.5403, \quad \frac{p_{t2}}{p_t^*} \approx 1.1577 \tag{15.108}$$

Thus, the static and total pressure at the exit are

$$p_2 = p_1 \cdot \frac{p_2/p^*}{p_1/p^*} = 3.06 \text{ atm} \quad \text{and} \quad p_{t2} = p_{t1} \cdot \frac{p_{t2}/p_t^*}{p_{t1}/p_t^*} = 12.42 \text{ atm} \tag{15.109}$$

Mollier Diagram. The two examples above show that Rayleigh flow has some similarity to Fanno flow on a couple of counts. It appears that, at least in the case of heat addition, the flow is driven toward sonic conditions. This raises the question of the possibility that sufficient heating can choke the flow. As with Fanno flow, the Mollier diagram is useful for illustrating these features, along with another curious feature of this flow. The entropy for a perfect gas is given by Equation (8.7), i.e.,

$$s = c_v \, \ell n \left(\frac{p}{\rho^\gamma} \right) + \text{constant} \tag{15.110}$$

Using the Rayleigh-flow solution for p and ρ [Equations (15.93)], we can express the entropy as a function of Mach number.

$$s = c_v \, \ell n \left[M^{2\gamma} \left(\frac{\gamma + 1}{1 + \gamma M^2} \right)^{\gamma+1} \right] + \text{constant} \tag{15.111}$$

Since the enthalpy is $h = c_p T$, clearly $h/h^* = T/T^*$, which is given by the third of Equations (15.93). So, differentiating s and h with respect to x, a bit of algebra leads to the following equation relating dh/dx and ds/dx.

$$\frac{dh}{dx} = \frac{\gamma M^2 - 1}{M^2 - 1} T \frac{ds}{dx} \tag{15.112}$$

This equation defines the **Rayleigh line** or **Rayleigh curve**, which is shown in Figure 15.7.

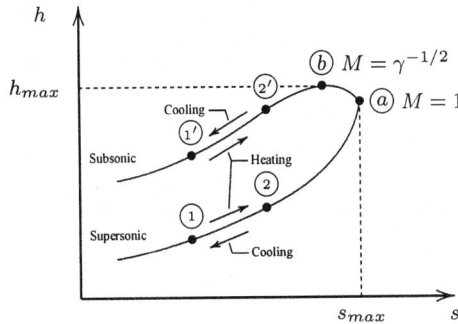

Figure 15.7: *Mollier diagram for Rayleigh flow.*

There are several similarities between the Mollier diagrams for Fanno flow (Figure 15.4) and Rayleigh flow. Both have supersonic and subsonic branches. When Rayleigh flow has heat addition, similar to Fanno flow, the flow is driven toward the sonic point a, where the entropy is a maximum. As discussed earlier, when we add the critical amount of heat, q^*, the flow will be choked. If we add heat in excess of q^*, a shock will form if the inlet is supersonic. If the inlet is subsonic, excess heating will cause the inlet conditions to change. Again, this behavior is analogous to what we found for Fanno flow.

There is an interesting feature of Rayleigh flow that has no analog in Fanno flow. Inspection of Equation (15.112) shows that on the supersonic branch, the enthalpy, and hence the temperature, increases when we heat the flow. This is obvious since $ds/dx > 0$ corresponds to $dh/dx > 0$ when $M > 1$. However, on the subsonic branch, the factor $(\gamma M^2 - 1)/(M^2 - 1)$ changes sign at $M = 1/\sqrt{\gamma}$. As a result, the enthalpy increases for Mach numbers less than $1/\sqrt{\gamma}$ and decreases for $M > 1/\sqrt{\gamma}$. In other words, for Mach numbers between $1/\sqrt{\gamma}$ (0.845 for air) and 1, the temperature decreases as a result of heating the flow! Because total temperature increases, this means the kinetic energy increases so rapidly in this range that the fluid must cool slightly to conserve energy.

When we have heat extraction, all of these trends are reversed. While the entropy of the duct or pipe decreases because of the heat extracted, the energy of the surroundings must increase so that the entropy of the entire system increases. This is dictated by the second law of thermodynamics. When heat is extracted, the flow is driven away from the sonic point on both the subsonic and supersonic branches of the Rayleigh curve. Table 15.2 summarizes the directions in which flow properties change in Rayleigh flow.

For the two Rayleigh-flow examples above, we can compute the amount of heat added per unit mass that will choke the flow. For the subsonic and supersonic cases, the inlet total temperature is $T_{t1} = 298.3$ K and 820.5 K, respectively. Reference to Table B.4 shows that T_{t1}/T_t^* is 0.3469 for subsonic flow and 0.6540 for the supersonic case. Hence, the reference total temperature, T_t^*, is 860 K for the subsonic inlet and 1255 K for the supersonic inlet. Noting that the flow is choked when $T_{t2} = T_t^*$, we use Equation (15.92) to compute the required heat added per unit mass, which we call q^*. The results are

$$q^* = \begin{cases} 5.64 \cdot 10^5 \text{ J/kg}, & \text{Subsonic inlet} \\ 4.36 \cdot 10^5 \text{ J/kg}, & \text{Supersonic inlet} \end{cases} \qquad (15.113)$$

Table 15.2: *Variation of Rayleigh-Flow Properties*

Subsonic Inlet $0 < M < \gamma^{-1/2}$	Subsonic Inlet $\gamma^{-1/2} < M < 1$	Supersonic Inlet $M > 1$
With Heat Addition:		
M, u, T and T_t increase	M, u and T_t increase	p, ρ, T and T_t increase
p, ρ, p_t and ρ_t decrease	p, ρ, T, p_t and ρ_t decrease	M, u, p_t and ρ_t decrease
With Heat Extraction:		
p, ρ, p_t and ρ_t increase	p, ρ, T, p_t and ρ_t increase	M, u, p_t and ρ_t increase
M, u, T and T_t decrease	M, u and T_t decrease	p, ρ, T and T_t decrease

15.4   Oblique Shock Waves

When a blunt-nosed object moves supersonically, it creates a shock wave that stands a finite distance ahead of the object. As sketched in Figure 15.8(a), in the immediate vicinity of the body, the shock wave is normal to the freestream flow direction. By contrast, an object with a pointed leading edge will have a shock wave that is attached to the body, provided the Mach number lies above a predictable critical value. As illustrated in Figure 15.8(b), the shock is oblique to the freestream flow direction. The shock also lies at an oblique angle far from a blunt-nosed body.

The flow patterns indicated in Figure 15.8, corresponding to objects of finite dimensions, are similar to those attending supersonic motion of an infinitesimally small object in some respects. Recall that when we discussed supersonic motion of a small body in Section 8.4, we found that the disturbance in the surrounding fluid is confined to a conical region called the **Mach cone**. We also found from the simple geometrical construction shown in Figure 8.3 that the cone half angle is

$$\mu = \sin^{-1}\left(\frac{1}{M}\right) \qquad (15.114)$$

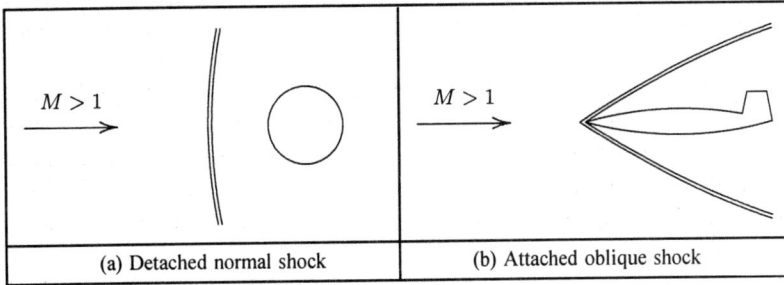

Figure 15.8: *Attached and detached shock waves.*

where μ is the **Mach angle**. The motion of a very small object is isentropic, and the wave it creates is known as a **Mach wave**. As we will see in this section, a Mach wave is the limiting case of a very weak shock wave.

We have studied normal shock waves in Sections 8.7 and 8.8, where we developed the normal shock relations. One of the key observations is that the flow behind the shock is subsonic. The flow is compressed and heated by the shock, and the pressure abruptly increases. We will find that the normal shock is the limiting case of the strongest shock possible.

Oblique shocks lie between these two extremes, and are the most commonly encountered type of shock wave. To analyze an oblique shock, it is most convenient to consider flow past a wedge. Note that we will focus on two-dimensional flows as axisymmetric and three-dimensional shock waves are inherently more difficult to analyze. In general, differential equations must be solved even for the axisymmetric case [cf. Anderson (1990)]. By contrast, we need only solve algebraic equations for the two-dimensional shock. Figure 15.9 shows supersonic flow past a wedge of half angle θ. We assume the shock wave is attached, and makes an angle β with the freestream. The shock angle is always greater than the Mach angle, and the latter is shown for reference.

To analyze an oblique shock, we can make use of a simple Galilean transformation that will reduce our analysis to a problem we have already solved. That is, if we begin with a normal shock as shown in Figure 15.10(a), the normal-shock relations are completely unaffected if we choose to observe the motion in a coordinate frame translating at a constant velocity, v, parallel to the shock. We must, of course, be careful to use the velocity component normal to the shock, u, for any computations in which the velocity appears. Figure 15.10(b) shows the same shock wave with the coordinate system aligned with the incident velocity, w_1, defined

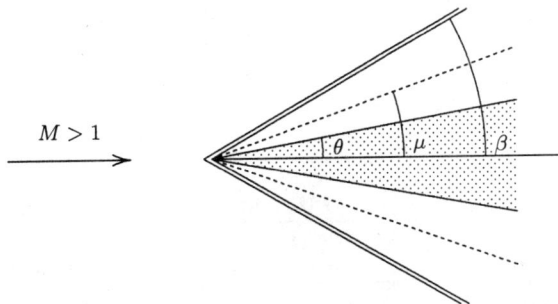

Figure 15.9: *Supersonic flow past a wedge.*

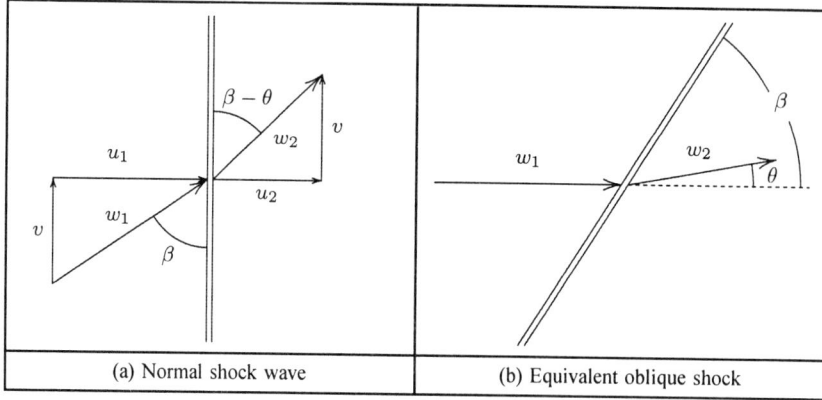

(a) Normal shock wave | (b) Equivalent oblique shock

Figure 15.10: *Using a Galilean transformation to analyze an oblique shock.*

by the following equation.

$$w_1 = \sqrt{u_1^2 + v^2} \tag{15.115}$$

Clearly, we can select our translation velocity, v, to achieve any shock angle, β, that we desire by noting from the geometry that

$$\beta = \tan^{-1}\left(\frac{u_1}{v}\right) \tag{15.116}$$

The key to understanding why we can use a Galilean transformation to analyze an oblique shock wave lies in the phrase, "we choose to observe the motion." All we have changed is the way we observe the flow through a normal shock. As observers, we do nothing to affect the flow so that its thermodynamic properties must remain the same. While it is true that we observe a different absolute velocity, a shock wave changes only the velocity component normal to the shock. As long as we take care to exclude the velocity component parallel to the shock (i.e., our translation velocity), we can use the normal shock relations, thus obviating a new control-volume analysis. So, we define the normal Mach number, M_{n1}, as

$$M_{n1} \equiv \frac{u_1}{a_1} = M_1 \sin\beta, \quad \text{where} \quad M_1 = \frac{w_1}{a_1} \tag{15.117}$$

Based on the arguments above, we arrive at the oblique-shock relations by replacing M_1 with $M_1 \sin\beta$ in Equations (8.66), (8.67) and (8.68). Therefore,

$$\frac{\rho_2}{\rho_1} = \frac{(\gamma+1)M_1^2 \sin^2\beta}{2 + (\gamma-1)M_1^2 \sin^2\beta} \tag{15.118}$$

$$\frac{p_2}{p_1} = 1 + \frac{2\gamma}{\gamma+1}\left(M_1^2 \sin^2\beta - 1\right) \tag{15.119}$$

$$\frac{T_2}{T_1} = 1 + \frac{2(\gamma-1)}{(\gamma+1)^2}\left(\frac{1 + \gamma M_1^2 \sin^2\beta}{M_1^2 \sin^2\beta}\right)\left(M_1^2 \sin^2\beta - 1\right) \tag{15.120}$$

Also, the Mach number behind the shock follows from noting that

$$M_2 = \frac{w_2}{a_2} \quad \text{and} \quad M_{n2} = \frac{u_2}{a_2} = M_2 \sin(\beta - \theta) \tag{15.121}$$

Substitution into the normal shock relation between Mach numbers ahead of and behind the shock, Equation (8.65), yields the following equation for M_2.

$$M_2^2 \sin^2(\beta - \theta) = \frac{1 + \frac{\gamma - 1}{2} M_1^2 \sin^2 \beta}{\gamma M_1^2 \sin^2 \beta - \frac{\gamma - 1}{2}} \quad (15.122)$$

We're not quite finished as we still need an equation relating the shock angle, β, and the flow deflection angle, θ. Inspection of Figure 15.10(a) provides the required equation from a simple geometric argument. Consistent with Equation (15.116), we already know that conditions ahead of the shock give

$$\frac{u_1}{v} = \tan \beta \quad (15.123)$$

Also, behind the shock, the velocity components satisfy

$$\frac{u_2}{v} = \tan(\beta - \theta) \quad (15.124)$$

Hence, we can combine these two geometric constraints to yield

$$\frac{\tan(\beta - \theta)}{\tan \beta} = \frac{u_2}{u_1} = \frac{\rho_1}{\rho_2} \quad (15.125)$$

where the second equality follows from mass conservation. Thus, using Equation (15.118) in place of the density ratio, we arrive at the following implicit relation amongst β, θ and M_1.

$$\frac{\tan(\beta - \theta)}{\tan \beta} = \frac{2 + (\gamma - 1) M_1^2 \sin^2 \beta}{(\gamma + 1) M_1^2 \sin^2 \beta} \quad (15.126)$$

Equation (15.126) is not particularly useful as it stands because it provides neither β as an explicit function of θ and M_1 nor vice versa. With a bit of manipulation, we can develop a direct relation from which θ can be obtained as a function of β and M_1, viz.,

$$\tan \theta = 2 \cot \beta \left[\frac{M_1^2 \sin^2 \beta - 1}{2 + M_1^2(\gamma + \cos 2\beta)} \right] \quad (15.127)$$

Equation (15.127) is known as the θ-β-M relation, and is the most interesting of all the oblique-shock relations. In addition to revealing several interesting properties of oblique shocks, it is a central component needed to compute their properties. Figures 15.11 and 15.12 provide detailed plots of β as a function of θ, with M_1 as a parameter. Mach numbers range from 1 to ∞. The plots are reproduced from the famous NACA[2] Report 1135, which includes extensive tables, equations and graphs for compressible flows [Ames Research Staff (1953)]. The figures give a value for β which can be used directly, or as an initial guess for an iterative solution to Equation (15.127) for a more accurate value.

The first thing to note about the oblique-shock solution of Equations (15.118), (15.119), (15.120), (15.122) and (15.127) is that all flow properties are constant after the flow passes through the shock. Thus, an oblique shock turns the flow through an angle θ, and it remains uniform downstream of the shock, so that our solution corresponds to flow past a wedge of the type shown in Figure 15.9. Since the flow is constant ahead of the shock as well, we can confine our solution to the upper half plane, which is usually referred to as flow into a compression corner. That is, we replace the dividing streamline upstream of the shock by a solid boundary (see Figure 15.13).

---

[2] The National Advisory Committee for Aeronautics (NACA) became the National Aeronautics and Space Administration (NASA) in 1958.

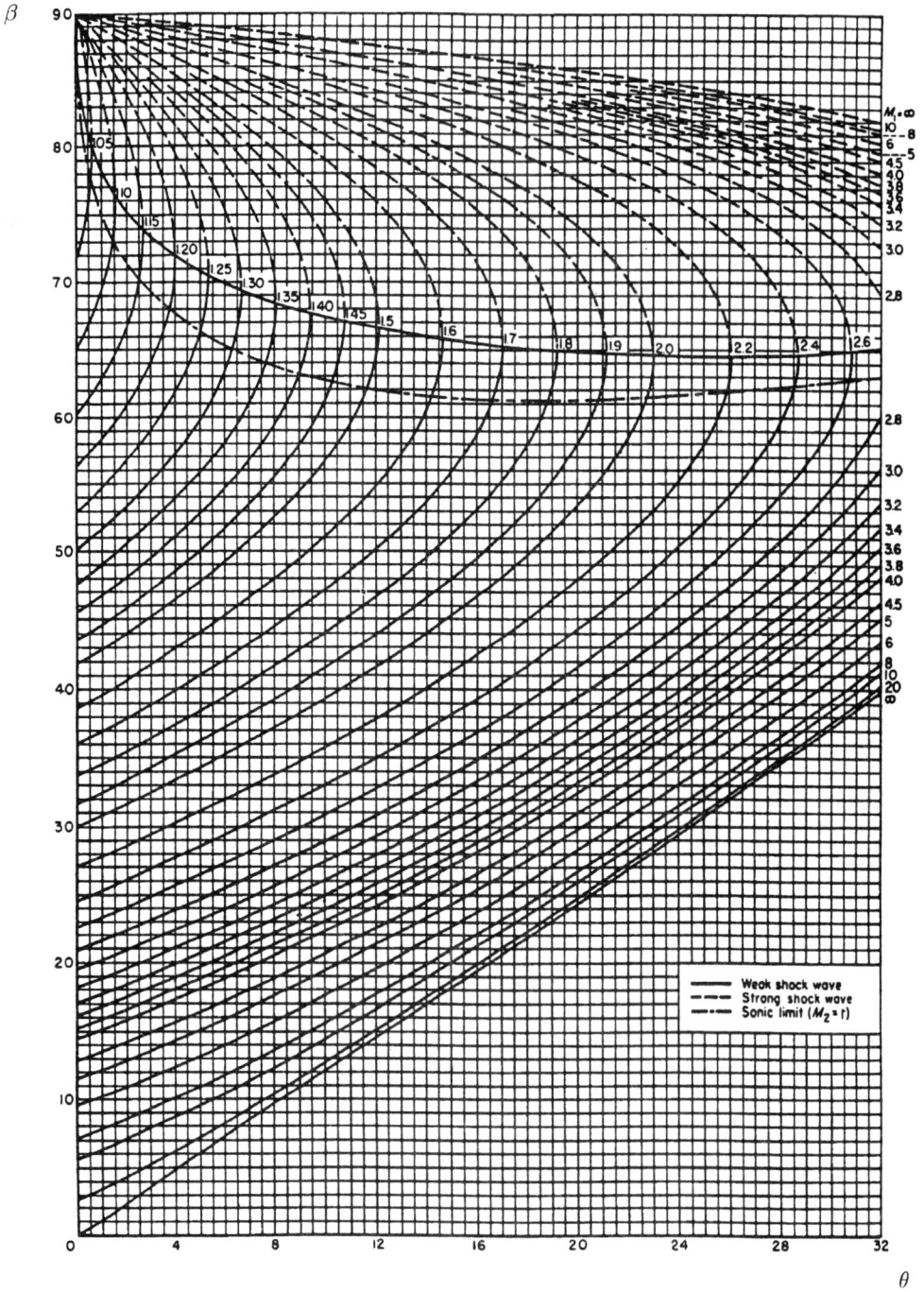

Figure 15.11: *Variation of β with θ for $\gamma = 1.4$ [From NACA-1135].*

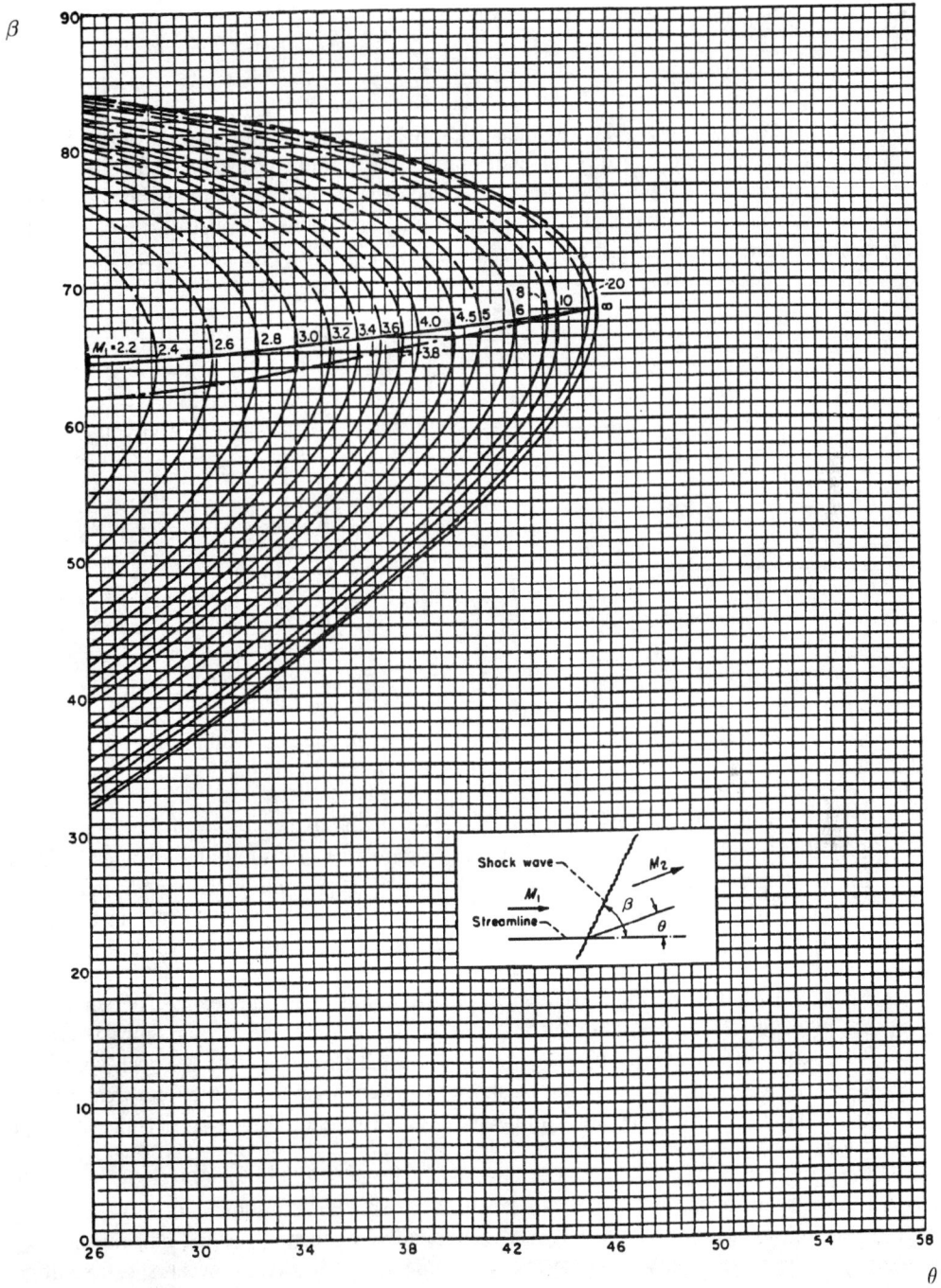

Figure 15.12: *Variation of β with θ for $\gamma = 1.4$ [From NACA-1135].*

We can use the normal-shock Table B.1 in Appendix B for oblique shocks with the understanding that we must work with the normal Mach numbers defined by Equations (15.117) and (15.121). Repeating the formulas for convenience, we have

$$M_{n1} = M_1 \sin \beta \quad \text{and} \quad M_{n2} = M_2 \sin(\beta - \theta) \tag{15.128}$$

This means that, for a given Mach number, M_1, we must first determine the normal Mach number, M_{n1}. Then, we use the normal-shock tables to determine properties behind the shock. The Mach number behind the shock listed in the tables is the normal Mach number M_{n2}, and we use the second of Equations (15.128) to determine M_2. However, before we can use the normal-shock tables, we must use either Equation (15.127) or Figures 15.11 and 15.12 to find β (or θ).

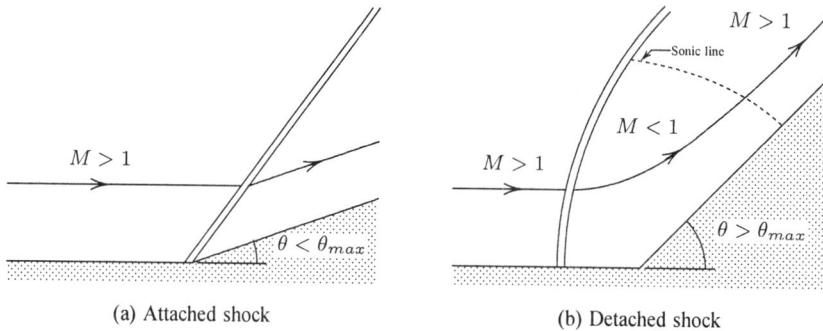

(a) Attached shock (b) Detached shock

Figure 15.13: *Supersonic flow into a compression corner.*

Focusing on the θ-β-M relation, there are several interesting features that are clearly illustrated in Figures 15.11 and 15.12. In particular, note the following:

- For each value of M_1, there is a maximum deflection angle, θ_{max}. If the geometry is such that $\theta > \theta_{max}$, the shock is **detached**. Figure 15.13 illustrates the two possibilities for a compression corner. Inspection of Figures 15.11 and 15.12 shows that θ_{max} increases with Mach number, with a maximum value of just under $46°$ in the limit $M_1 \to \infty$. The locus of θ_{max} is included as a solid curve in Figures 15.11 and 15.12.

- For any angle $\theta < \theta_{max}$, there are two values of the shock angle, β. The smaller of the two angles is the **weak-shock solution** and the larger is the **strong-shock solution**. The curve defining the locus of θ_{max} separates the strong- and weak-shock solutions. Changes in properties across a strong shock are much larger than those for a weak shock. We usually find the weak shock in nature. The strong-shock solution occurs whenever the shock is detached, as shown in Figure 15.13(b).

- For a strong shock, the flow downstream of the shock is subsonic. For a weak shock, the flow after the shock is usually supersonic, although there is a small range near θ_{max} where the flow is subsonic. The locus of angles below which the weak-shock solution has supersonic flow after the shock is indicated in Figures 15.11 and 15.12 as a dashed curve.

- For all Mach numbers, the curves join at $\theta = 0°$ and $\beta = 90°$. This means a detached shock will always begin as a normal shock, independent of M_1.

- For all Mach numbers, the weak-shock solution asymptotes to $\beta = \mu$ (the Mach angle) as $\theta \to 0$. This means that for an infinitesimal flow-deflection angle, a shock wave asymptotes to a Mach wave.

Before proceeding to examples, there are two additional observations regarding the strong-shock solution that are worthy of mention. The first regards flow past a finite-width body with a detached shock. Consider supersonic flow past a wedge with a base height L and half angle $\theta > \theta_{max}$ as shown in Figure 15.14. Beginning with a right angle on the dividing streamline, the shock is curved and lies at a decreasing oblique angle to the freestream. The flow very close to the wedge will be as shown in Figure 15.13(b), and accomplishes the required turning in a subsonic region. The shock actually assumes every angle on the β versus θ curve from the normal shock on the dividing streamline to the Mach angle far from the wedge. It must ultimately asymptote to a Mach wave at distances large compared to L so that the flow remains parallel in the freestream. Between the shock and the wedge face, the flow accelerates and becomes supersonic beyond the sonic line. The strong-shock solution holds up to the sonic line, while the weak-shock solution prevails beyond the sonic line.

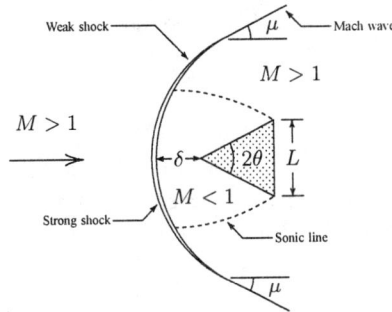

Figure 15.14: *Supersonic flow past a wedge with $\theta > \theta_{max}$; not to scale.*

Now consider the limiting case of an infinite wedge, i.e., $L \to \infty$. We pose the question as to what the flowfield must be. In particular, what is the limiting value of β far above the dividing streamline? Clearly, it cannot be a constant value as the required turning through an angle $\theta > \theta_{max}$ cannot be accomplished with a linear shock. We can appeal to dimensional analysis to determine the answer. Denote the **shock stand-off distance**, i.e., the distance between the shock and the wedge leading edge, by δ. We expect the solution to depend on Mach number, M, specific-heat ratio, γ, and the wedge base height, which is the only length in the problem. Clearly, we must have

$$\delta = L\mathcal{F}(M, \gamma) \tag{15.129}$$

where $\mathcal{F}$ is a function that can be determined either by computation or experiment. In the limit $L \to \infty$, the standoff distance must also become infinite. Thus, the solution for flow past an infinite wedge with $\theta > \theta_{max}$ is a normal shock infinitely far upstream of the wedge.

The detached-shock problem is very complicated, and not amenable to a closed-form solution for even the simplest geometry. The precise shape of a detached shock, the standoff distance, δ, and the flow between the shock and the body depend upon Mach number and body shape. The problem was a major research topic in the 1950's and 1960's, motivated by the need to understand the flow physics for blunt-nosed missiles and reentry vehicles. Numerical solution of this problem in the 1960's was one of the first major successes of CFD.

Mach 4 Compression Corner. As our first example of using the shock tables and formulas to predict oblique-shock properties, consider the compression corner shown in Figure 15.15. This configuration occurs at the intersection of a wing and a control surface on a supersonic aircraft, for example. The incident Mach number is 4, and the corner angle is 15°. We would like to determine the Mach number for the flow behind the shock, M_2, and the pressure rise, p_2/p_1.

Figure 15.15: *Mach 4 flow into a compression corner.*

As our first step, we must determine the shock angle, β. Reference to Figure 15.11 shows that for a deflection angle, $\theta = 15°$, as a first approximation

$$\beta \approx 27° \tag{15.130}$$

To obtain a more accurate value for β, we can appeal to Equation (15.127). For our purposes, one extra significant figure will suffice, i.e., we would like to know the shock angle to within a tenth of a degree. This level of accuracy is consistent with the shock tables. Using a calculator, for example, we find the following.

β	θ
27.0°	14.9°
27.1°	15.0°
27.2°	15.1°

Therefore, to three significant figures, the shock angle is

$$\beta = 27.1° \tag{15.131}$$

Next, we must compute the Mach number normal to the shock, M_{n1}. From Equation (15.128), we have

$$M_{n1} = M_1 \sin\beta = 1.82 \tag{15.132}$$

Using the normal-shock tables, Table B.1 of Appendix B, we find one row of values corresponds to $M_{n1} = 1.82$. Hence, from the table, we have

$$M_{n2} = 0.6121 \quad \text{and} \quad \frac{p_2}{p_1} = 3.6978 \tag{15.133}$$

Since no further adjustment is required for thermodynamic properties, this tells us the pressure rises by a factor of 3.70. All that remains to complete the solution is to compute the Mach number behind the shock, M_2. Again referring to Equation (15.128), we have

$$M_2 = \frac{M_{n2}}{\sin(\beta - \theta)} = 2.92 \tag{15.134}$$

Shock Reflecting from a Wall. Our next example is for Mach 4 flow in a channel with a contraction as illustrated in Figure 15.16. The ramp on the lower wall creates a shock wave identical to the preceding example. There is no difference because, for supersonic flow, disturbances propagate only in the streamwise direction. The initial shock is completely unaffected by the upper wall until it strikes it. Now, the flow behind the shock is moving parallel to the ramp. This cannot be a valid solution at the wall since the normal component must vanish there. Hence, In order to satisfy the boundary condition at the upper wall, the flow must turn back by an angle θ. This is illustrated for the streamline shown. The turning is accomplished with another shock wave that is simply the reflection of the incident shock. Our objective is to compute the angle of the reflected shock, Φ, the Mach number after the reflected shock, M_r, and the overall pressure ratio, p_r/p_1.

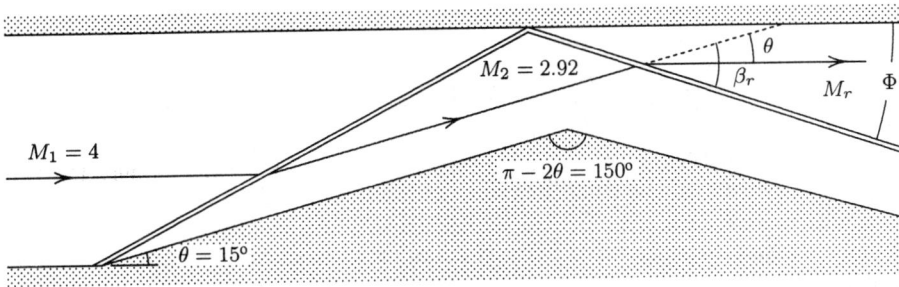

Figure 15.16: *Reflection of an oblique shock wave from a solid boundary.*

We already know that the Mach number is $M_2 = 2.92$ behind the incident shock. Reference to Figure 15.11 shows that the angle the reflected shock makes with the velocity behind the incident shock, β_r, is

$$\beta_r \approx 33° \tag{15.135}$$

Using Equation (15.127) to sharpen our estimate, we find

β_r	θ
32.8°	14.9°
32.9°	15.0°
33.0°	15.1°

Hence, the reflected shock angle is

$$\beta_r = 32.9° \tag{15.136}$$

So, the Mach number normal to the reflected shock, $\tilde{M}_{n1}$, is

$$\tilde{M}_{n1} = M_2 \sin\beta_r = 1.59 \tag{15.137}$$

Using the normal-shock Table B.1, we must interpolate between values tabulated for Mach numbers of 1.58 and 1.60. Doing the required arithmetic yields

$$\tilde{M}_{n2} = 0.6715 \quad \text{and} \quad \frac{p_r}{p_2} = 2.7829 \tag{15.138}$$

The Mach number behind the reflected shock, M_r is given by

$$M_r = \frac{\tilde{M}_{n2}}{\sin(\beta_r - \theta)} = 2.18 \tag{15.139}$$

The overall pressure ratio for the flow as it passes through both the incident and reflected shocks is

$$\frac{p_r}{p_1} = \frac{p_r}{p_2}\frac{p_2}{p_1} = (3.6978)(2.7829) = 10.29 \tag{15.140}$$

Finally, the angle the reflected shock wave makes with the upper wall is the difference between β_r and θ, i.e.,

$$\Phi = \beta_r - \theta = 17.9° \tag{15.141}$$

The astute reader will observe that our solution to this problem is incomplete. Specifically, the solution does not satisfy the boundary condition on the lower wall at points downstream of the end of the ramp. Another type of wave is needed to turn the flow around the corner, and this is the topic of the next section.

15.5 Prandtl-Meyer Expansion

Our example of the preceding section leads naturally to discussion of what must occur when a supersonic flow makes a convex turn. That is, we have seen that an oblique shock wave accommodates a concave turn,[3] which is commonly described as the flow being deflected *into itself*. We must now consider the opposite situation of a convex turn like the one encountered at the end of the ramp in Figure 15.16. In a convex turn, the flow is deflected *away* from the oncoming stream. Rather than being compressed, the flow will expand. Since the second law of thermodynamics precludes the existence of expansion shocks, the turning is accomplished through some mechanism other than a shock wave.

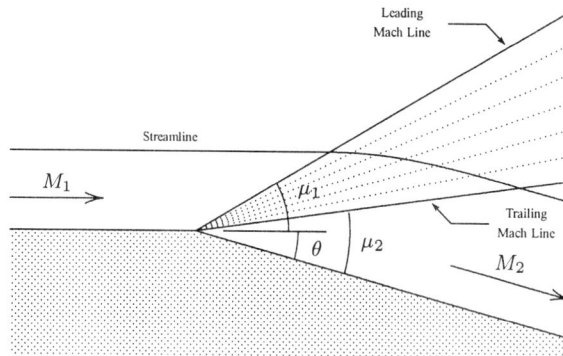

Figure 15.17: *Prandtl-Meyer expansion wave.*

For a sharp corner, Prandtl (in 1907) and Meyer (in 1908) developed the theory for what is known as a **centered expansion**. In their honor, it has come to be known as the **Prandtl-Meyer expansion wave**. As shown in Figure 15.17, it consists of a continuous succession of Mach waves, each of which turns the flow through a small angle. Because the complete turning is accomplished through a series of isentropic Mach waves, the overall process is also isentropic.

The expansion wave, sometimes referred to as an **expansion fan**, can be envisioned as an infinite number of Mach waves of infinitesimal strength. That is, each Mach wave turns

---

[3]Flow into a compression corner is the limiting case of flow into a rounded concave surface with radius of curvature approaching zero.

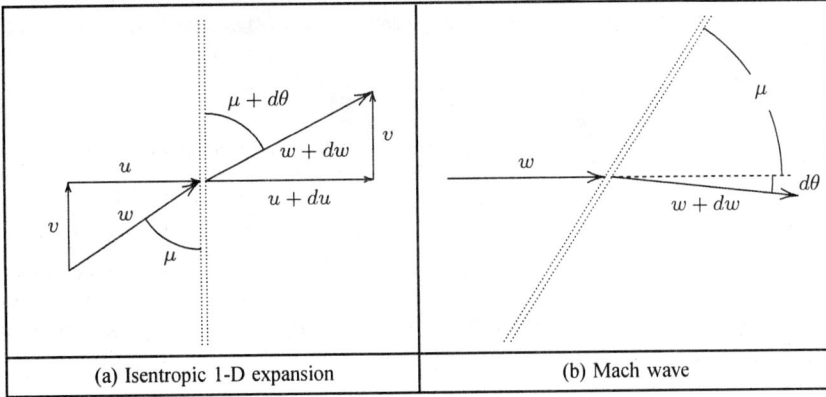

(a) Isentropic 1-D expansion   (b) Mach wave

Figure 15.18: *Using a Galilean transformation to analyze a Mach wave.*

the flow through a differential angle, $d\theta$. Since the turning occurs in this manner, streamlines through an expansion wave are smooth curves. This stands in contrast to a shock wave for which streamlines have discontinuous slope. The region over which the complete turning occurs is bounded upstream by a Mach line with angle $\mu_1 = \sin^{-1}(1/M_1)$ and downstream by a Mach line with angle $\mu_2 = \sin^{-1}(1/M_2)$.

To develop the equations describing the Prandtl-Meyer expansion, consider a continuous, one-dimensional isentropic expansion. If we focus on a point in the flow where the velocity is u, then over a differential distance it will increase to $u + du$. This is illustrated in Figure 15.18(a). As with our approach to the oblique-shock problem, we now make a Galilean transformation so that we observe the isentropic expansion in a frame translating parallel to the wave front at a constant velocity, v. Since the only thing we are doing is observing flow through the expansion front from a different coordinate frame, clearly we do not change the fact that the flow is isentropic and expanding. Obviously, we can choose v to make the resultant velocity, w, lie at any angle to the wave front that we wish. As shown in Figure 15.18(b), the angle we select is the Mach angle given by

$$\mu = \sin^{-1}\left(\frac{1}{M}\right), \qquad M \equiv \frac{w}{a} \tag{15.142}$$

where $w = \sqrt{u^2 + v^2}$ and a is the speed of sound. Aligning our coordinate system with the velocity w yields the configuration corresponding to a Mach wave.

From Figure 15.18(a), the velocities before and after the Mach wave are related according to

$$v = w\cos\mu = (w + dw)\cos(\mu + d\theta) \tag{15.143}$$

We can expand $\cos(\mu + d\theta)$ through a familiar trigonometric identity, i.e.,

$$\cos(\mu + d\theta) = \cos\mu\cos(d\theta) - \sin\mu\sin(d\theta) \approx \cos\mu - \sin\mu\, d\theta \tag{15.144}$$

where we use the facts that $\cos(d\theta) \approx 1$ and $\sin(d\theta) \approx d\theta$ in the limit $d\theta \to 0$. Substituting Equation (15.144) into (15.143), we have

$$w\cos\mu \approx (w + dw)(\cos\mu - \sin\mu\, d\theta) \approx w\cos\mu + \cos\mu\, dw - w\sin\mu\, d\theta \tag{15.145}$$

Simplifying, we arrive at the following relationship between differential changes in flow angle and flow velocity.[4]

$$d\theta = \frac{dw/w}{\tan\mu} = \sqrt{M^2 - 1}\,\frac{dw}{w} \tag{15.146}$$

Although we won't pursue the details here, it is worthwhile to pause and discuss the significance of Equation (15.146). First, while our immediate focus is on isentropic expansions, the equation applies to isentropic compressions as well. This is consistent with the notion that a Mach wave is the limiting case of a very weak shock. Liepmann and Roshko (1963), for example, derive Equation (15.146) by linearizing the oblique-shock relations in the limit of infinitesimal flow deflection. Second, this equation also serves as the foundation for a linearized supersonic airfoil theory that is discussed in most compressible-flow texts such as Shapiro (1953), Liepmann and Roshko (1963) or Anderson (1990).

We can cast Equation (15.146) in a more useful form by eliminating w in favor of Mach number, M. First, we note that M and w are related by

$$w = Ma \quad\Longrightarrow\quad \frac{dw}{w} = \frac{dM}{M} + \frac{da}{a} \tag{15.147}$$

In order to eliminate sound speed, we note from Equation (8.26) that $a = \sqrt{\gamma RT}$. Then, in terms of total temperature, T_t, we find

$$a = \sqrt{\frac{\gamma RT_t}{T_t/T}} = \sqrt{\frac{\gamma RT_t}{1 + \frac{1}{2}(\gamma - 1)M^2}} \tag{15.148}$$

Differentiation shows that the differential change in a is

$$\frac{da}{a} = -\frac{\dfrac{\gamma - 1}{2}M^2}{1 + \dfrac{\gamma - 1}{2}M^2}\,\frac{dM}{M} \tag{15.149}$$

Substituting Equation (15.149) into Equation (15.147) yields the desired relationship between differential changes in w and M, viz.,

$$\frac{dw}{w} = \frac{1}{1 + \dfrac{\gamma - 1}{2}M^2}\,\frac{dM}{M} \tag{15.150}$$

Finally, substituting Equation (15.150) into Equation (15.146) gives the following differential equation that defines a Prandtl-Meyer expansion.

$$d\theta = \frac{\sqrt{M^2 - 1}}{1 + \dfrac{\gamma - 1}{2}M^2}\,\frac{dM}{M} \tag{15.151}$$

Although not obvious by inspection, the quantity appearing on the right-hand side of Equation (15.151) *is a perfect differential*! Specifically, as can be verified by differentiation:

$$d\theta = d\nu(M) \tag{15.152}$$

---

[4]We make use of the fact that $\sin\mu = 1/M$ implies $\tan\mu = 1/\sqrt{M^2 - 1}$ to eliminate μ.

where the quantity $\nu(M)$ is the **Prandtl-Meyer function**, defined by

$$\nu(M) \equiv \sqrt{\frac{\gamma+1}{\gamma-1}}\tan^{-1}\sqrt{\frac{\gamma-1}{\gamma+1}(M^2-1)} - \tan^{-1}\sqrt{M^2-1} \qquad (15.153)$$

The Prandtl-Meyer function is tabulated as a function of Mach number in Table B.2 of Appendix B. The Mach angle is also included in the tables for convenience. There are two noteworthy limiting values for $\nu(M)$, namely, for $M \to 1$ and for $M \to \infty$. Examination of Equation (15.153) shows that for $\gamma = 1.4$,

$$\nu(M) \to \begin{cases} 0^\circ, & M \to 1 \\ 130^\circ, & M \to \infty \end{cases} \qquad (15.154)$$

Integrating Equation (15.152), we conclude that the end states for a Prandtl-Meyer expansion are related by

$$\theta_2 - \theta_1 = \nu(M_2) - \nu(M_1) \qquad (15.155)$$

The way we use this relation is as follows.

1. For a given upstream Mach number, M_1, use Table B.2 to determine $\nu(M_1)$.

2. For a given change in flow deflection angle, $\theta_2 - \theta_1$, use Equation (15.155) and the value of $\nu(M_1)$ found in Step 1 to compute $\nu(M_2)$.

3. By interpolation in Table B.2, find the value of M_2 corresponding to the value of $\nu(M_2)$ computed in Step 2.

4. Use the isentropic flow part of Table B.1 to compute other flow properties. Alternatively, we can use the isentropic relations as given by Equations (8.58), (8.60) and (8.61). So, for example, the pressure ratio across a Prandtl-Meyer expansion is

$$\frac{p_2}{p_1} = \left[\frac{1+\dfrac{\gamma-1}{2}M_1^2}{1+\dfrac{\gamma-1}{2}M_2^2}\right]^{\gamma/(\gamma-1)} \qquad (15.156)$$

Shock Reflecting from a Wall. We now have sufficient information to complete the solution for the reflecting shock-wave problem discussed in the preceding section. At the end of the ramp on the lower wall that generates the shock, the flow is turned by a Prandtl-Meyer expansion through an angle of 30°. Since the Mach number at the beginning of the expansion is 2.92, reference to Table B.2 shows the Prandtl-Meyer function and Mach angle are

$$\nu(2.92) = 48.19^\circ \quad \text{and} \quad \mu = 20.0^\circ \qquad (15.157)$$

Hence, at the end of the expansion, using Equation (15.155) gives

$$\nu(M) = 78.19^\circ \qquad (15.158)$$

Again referring to Table B.2, we find that this value of $\nu(M)$ lies 78% of the way between the values listed for Mach numbers of 5.10 and 5.15. Thus, the Mach number and Mach angle after the expansion are

$$M_e = 5.14 \quad \text{and} \quad \mu = 11.2^\circ \qquad (15.159)$$

Figure 15.19 shows the solution with the Prandtl-Meyer expansion added. There is actually still more work to be done to continue the solution downstream of the intersection of the reflected shock and the expansion fan. Once the expansion reaches the shock, the two waves interact, and the shock no longer remains straight. Although not shown in the figure, it must be curved as the upstream flow conditions seen by the shock are changing due to the expansion fan. In general, such interactions require a numerical solution, and we will not pursue the solution any further beyond noting the added complexity when waves interact.

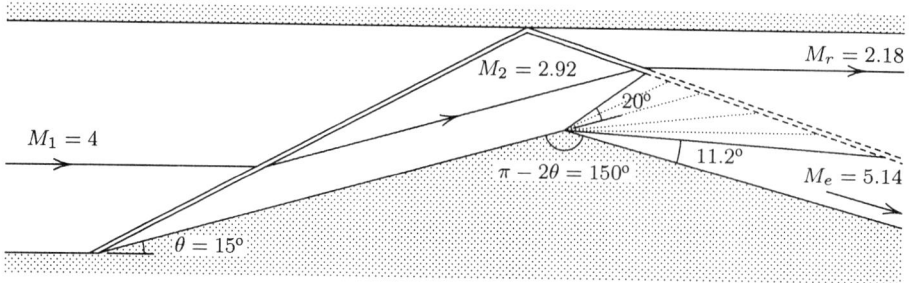

Figure 15.19: *Reflection of an oblique shock wave from a solid boundary.*

Flow Past a Diamond-Shaped Airfoil. As a second example, we consider flow past the diamond-shaped airfoil shown in Figure 15.20. In addition to illustrating use of the Prandtl-Meyer function, this example reveals an interesting aspect of compressible flow, namely, the concept of **wave drag**. The flow is turned at the leading edge of the airfoil by an oblique shock wave. At midchord, the flow turns in the opposite direction through a Prandtl-Meyer expansion. At the trailing edge, the flow must turn back to the freestream flow direction by symmetry, and this is accomplished by another oblique shock wave. Since the airfoil has zero angle of attack, the symmetry of the flow tells us there can be no lift. However, there is a possibility that the drag is nonzero, even though we are considering inviscid flow.

As noted in our discussion of oblique shock waves, flow properties after the shock are constant. The same is true of the centered expansion. Thus, the pressure is constant along both the forward and aft surfaces of the airfoil. The drag on the airfoil is given by

$$D = -\mathbf{i} \cdot \oiint_S p\,\mathbf{n}\,dS \qquad (15.160)$$

By symmetry, we can integrate over the upper surface and multiply by 2 to obtain the drag. On the forward and aft surfaces, we know that

$$\mathbf{n} = \begin{cases} -\mathbf{i}\sin\theta + \mathbf{j}\cos\theta, & \text{Forward surface} \\ \ \ \mathbf{i}\sin\theta + \mathbf{j}\cos\theta, & \text{Aft surface} \end{cases} \qquad (15.161)$$

Thus, the drag per unit width out of the page is

$$\begin{aligned} D &= 2\int_0^{\frac{1}{2}L/\cos\theta} p_2 \sin\theta\,ds - 2\int_{\frac{1}{2}L/\cos\theta}^{L/\cos\theta} p_3 \sin\theta\,ds \\ &= 2\,(p_2 - p_3)\sin\theta\,L/(2\cos\theta) \\ &= (p_2 - p_3)\,L\tan\theta = (p_2 - p_3)\,T \end{aligned} \qquad (15.162)$$

where s is tangent to the airfoil and T is the airfoil thickness. Hence, as speculated above, the airfoil has nonzero drag. This is reasonable as work must be done to push a wave through

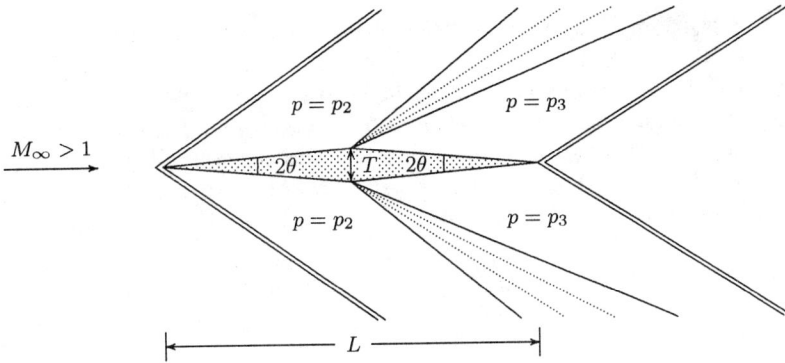

Figure 15.20: *Supersonic flow past a diamond-shaped airfoil.*

a fluid. This type of drag can be predicted by the inviscid-flow equations, and is known as **wave drag**. The drag coefficient, C_D, is given by

$$C_D \equiv \frac{D}{\frac{1}{2}\rho_\infty U_\infty^2 T} = \frac{2}{\gamma M_\infty^2}\left(\frac{p_2}{p_1} - \frac{p_3}{p_1}\right) \tag{15.163}$$

where we use the fact that $a_\infty^2 = \gamma p_\infty/\rho_\infty$ and let $p_1 = p_\infty$ for consistency with the shock tables.

As a specific example, consider Mach 2 flow past a diamond-shaped airfoil with a half angle, $\theta = 5°$. Thus, for the oblique shock at the leading edge, Figure 15.11 shows that the shock angle is $\beta = 34.3°$. The normal Mach number is $M_{n1} = M_1\sin\beta = 1.127$, and Table B.1 tells us that $M_{n2} = 0.8915$. To provide initial conditions for the expansion, note that the Mach number behind the shock is $M_2 = M_{n2}/\sin(\beta - \theta) = 1.82$. The tables also provide the pressure ratio,

$$p_2/p_1 = 1.3152 \tag{15.164}$$

Now, using Table B.2, we find $\nu(1.82) = 21.30°$. The flow turns through an angle of $10°$ at midchord, wherefore the Prandtl-Meyer function after the expansion is $\nu(M_3) = 31.30°$. A second reference to Table B.2 shows that $M_3 = 2.18$. Turning to the isentropic portion of Table B.1, the ratios of static to total pressure are $p_2/p_t = 0.1688$ and $p_3/p_t = 0.09649$. Therefore, we conclude that $p_3/p_2 = 0.5716$. Finally, using p_2/p_1 from the oblique-shock solution above, we have

$$p_3/p_1 = (p_3/p_2)(p_2/p_1) = 0.7518 \tag{15.165}$$

We now have sufficient information to compute the drag coefficient for the airfoil. Using p_2/p_1 for the forward part of the airfoil and p_3/p_1 from the Prandtl-Meyer solution, we have

$$C_D = \frac{2}{(1.4)2^2}(1.3152 - 0.7518) = 0.201 \tag{15.166}$$

As a final comment, this problem is an example of a classical method known as **shock-expansion theory**. The theory is exact for inviscid flows, and is useful for simple geometries. However, it is practical only for the portion of a flowfield that excludes interaction of intersecting waves. Because of these limitations, the method is mainly of historical interest, and has been superseded by CFD.

15.6   Compressible Boundary Layers

Laminar Flow. As discussed in Subsection 15.1.3, for a steady, two-dimensional, compressible boundary layer the viscous-flow equations of motion simplify in the same manner as for incompressible flow. The vertical momentum equation again simplifies to $\partial p/\partial y = 0$, so that the pressure is a function only of distance along the surface. Conservation of mass, momentum and energy for laminar flow are as follows.

$$\frac{\partial}{\partial x}(\rho u) + \frac{\partial}{\partial y}(\rho v) = 0 \tag{15.167}$$

$$\rho u \frac{\partial u}{\partial x} + \rho v \frac{\partial u}{\partial y} = -\frac{dp}{dx} + \frac{\partial}{\partial y}\left(\mu \frac{\partial u}{\partial y}\right) \tag{15.168}$$

$$\rho u \frac{\partial h}{\partial x} + \rho v \frac{\partial h}{\partial y} = u \frac{dp}{dx} + \frac{\partial}{\partial y}\left(\frac{\mu}{Pr}\frac{\partial h}{\partial y}\right) + \mu \left(\frac{\partial u}{\partial y}\right)^2 \tag{15.169}$$

Assuming we have a perfect gas that is also calorically perfect, we use the following thermodynamic relationships.

$$p = \rho R T, \quad h = e + \frac{p}{\rho}, \quad h = c_p T, \quad e = c_v T \tag{15.170}$$

For compressible flows, we must account for the fact that the viscosity is temperature dependent. As discussed in Section 1.8 [see Equation (1.29)], Sutherland's law provides an excellent approximation for air and is given by

$$\mu = \frac{A T^{3/2}}{T + S} \tag{15.171}$$

The empirical coefficients A and S, in USCS units, are

$$A = 2.27 \cdot 10^{-8} \; \frac{\text{slug}}{\text{ft} \cdot \text{sec} \cdot (^{\circ}\text{R})^{1/2}}, \quad S = 198.6 \; ^{\circ}\text{R} \tag{15.172}$$

In SI units their values are

$$A = 1.46 \cdot 10^{-6} \; \frac{\text{kg}}{\text{m} \cdot \text{sec} \cdot \text{K}^{1/2}}, \quad S = 110.3 \text{ K} \tag{15.173}$$

Another approximation used for gases assumes the viscosity varies as temperature raised to a power, ω. That is, the viscosity is assumed to follow a **power law**, viz.,

$$\mu = \mu_o \left(\frac{T}{T_o}\right)^{\omega} \tag{15.174}$$

where μ_o, T_o and ω are determined from correlation of measurements. In the case of air, $\omega = 0.7$ is a fairly good approximation for temperatures up to about 1000° R (about 800 K).

We impose boundary conditions at the surface, $y = 0$, and in the freestream, $y \to \infty$. In the freestream, the velocity and temperature approach their so-called **boundary-layer edge** values so that

$$u \to u_e, \quad T \to T_e \quad \text{as} \quad y \to \infty \tag{15.175}$$

At the surface, we impose the no-slip velocity boundary condition. Also, as discussed in Subsection 15.1.5, we specify either the surface temperature, T_w, or the surface heat flux, q_w. Thus,

$$\left. \begin{array}{l} u = v = 0 \\[2mm] T = T_w \quad \text{or} \quad \dfrac{\partial T}{\partial y} = -\dfrac{q_w}{k} \end{array} \right\} \quad \text{at} \quad y = 0 \qquad (15.176)$$

where k is the thermal conductivity given by Equation (15.9), which we repeat here for convenience.

$$k = \frac{c_p \mu}{Pr} \qquad (15.177)$$

In the special case of an adiabatic wall, $q_w = 0$, so that

$$\frac{\partial T}{\partial y} = 0 \quad \text{at} \quad y = 0 \qquad \text{(Adiabatic wall)} \qquad (15.178)$$

We can no longer use Bernoulli's equation to determine the pressure, of course. Rather, except for certain hypersonic-flow applications, we normally use the isentropic relations to determine freestream flow properties. If the freestream flow is not isentropic, edge conditions must be determined from a solution to Euler's equation. Nevertheless, just as in the incompressible boundary layer, the pressure (and all edge conditions) are the surface values of the Euler-equation solution.

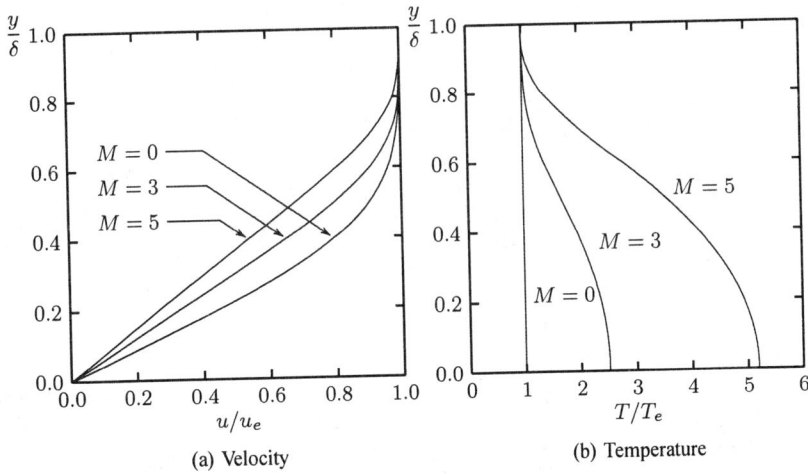

Figure 15.21: *Laminar velocity and temperature profiles for a constant-pressure, adiabatic-wall boundary layer.*

Laminar solutions to Equations (15.167) - (15.176) are a bit sensitive to the viscosity-temperature relationship, especially in the case of skin friction. Figure 15.21 shows laminar velocity and temperature profiles for a flat adiabatic-wall in air flow with Mach numbers of 0, 3 and 5. The computations have been done with Program **EDDYBL** [Appendix F] using a power-law relation for the viscosity, Equation (15.174), with $\omega = 0.7$. As shown, the velocity profile is affected somewhat by increasing Mach number. It is a little less filled out and has a reduced slope at the surface. The temperature is very large near the surface for the Mach 3 and Mach 5 cases, and approaches the freestream value at $y = \delta$. Examination of the solutions shows that at the surface, we have the adiabatic-wall temperature [see Equation (15.37)], i.e.,

$$\frac{T_{aw}}{T_e} = 1 + \frac{\gamma - 1}{2} r M_e^2, \qquad r \approx \sqrt{Pr} \tag{15.179}$$

The computations have been done with a Prandtl number, $Pr = 0.72$, so that the recovery factor, r, is 0.85.

As mentioned above, the skin friction, c_f, is affected by the viscosity law. Note that we define the skin friction for compressible flow in terms of the local boundary-layer-edge values, viz.,

$$c_f = \frac{\tau_w}{\frac{1}{2}\rho_e u_e^2} \tag{15.180}$$

A monotone variation of skin friction attends our use of a power-law viscosity, with c_f decreasing slightly as Mach number increases. Figure 15.22 shows the skin friction corresponding to the velocity and temperature profiles in Figure 15.21. Interestingly, when Sutherland's law is used, relative to the incompressible value, c_f decreases with Mach number up to about Mach 3 and then begins to increase slightly for larger Mach numbers.

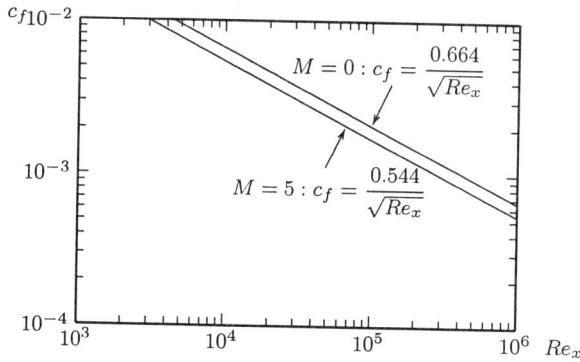

Figure 15.22: *Laminar skin friction for adiabatic-wall boundary layers with constant pressure and a power-law viscosity,* $\mu \propto T^{0.7}$.

Turbulent Flow. Turning to turbulent boundary layers, we must again use the Reynolds-averaging concepts introduced in Subsection 14.3.3 to arrive at a set of equations suitable for computation. Because the density fluctuates in time, the use of conventional time averages leads to correlations between density and velocity in the mean-flow equations. For example, time averaging the continuity equation for a steady, two-dimensional flow yields

$$\frac{\partial}{\partial x}\left(\overline{\rho}\,\overline{u} + \overline{\rho'u'}\right) + \frac{\partial}{\partial y}\left(\overline{\rho}\,\overline{v} + \overline{\rho'v'}\right) = 0 \tag{15.181}$$

This is rather inconvenient as we require information about the density-velocity correlations, $\overline{\rho'u'}$ and $\overline{\rho'v'}$, along with information about the Reynolds shear stress. There is an elegant type of Reynolds averaging known as **mass averaging** or **Favre averaging** [Favre (1965)] that simplifies the compressible-flow equations, and obviates the need to specify the density-velocity correlations in the mean conservation equations. Specifically, we use the density as a weighting function in forming our averages as follows:

$$\tilde{\mathbf{u}}(x,y,z) = \frac{1}{\overline{\rho}} \lim_{T\to\infty} \frac{1}{T} \int_0^T \rho(x,y,z,t)\mathbf{u}(x,y,z,t)\,dt \tag{15.182}$$

where $\bar{\rho}$ is the conventional time-averaged density. The quantity $\tilde{u}$ is referred to as the **mass-averaged velocity**. For the sake of brevity, we omit details of the averaging process for the compressible boundary-layer equations. Wilcox (1993) gives complete details of Favre averaging, including derivation of the Favre-averaged Navier-Stokes and continuity equations. In terms of mass-averaged quantities, the compressible boundary-layer equations are

$$\frac{\partial}{\partial x}(\bar{\rho}\tilde{u}) + \frac{\partial}{\partial y}(\bar{\rho}\tilde{v}) = 0 \tag{15.183}$$

$$\bar{\rho}\tilde{u}\frac{\partial\tilde{u}}{\partial x} + \bar{\rho}\tilde{v}\frac{\partial\tilde{u}}{\partial y} = -\frac{d\bar{p}}{dx} + \frac{\partial}{\partial y}\left[(\bar{\mu} + \mu_t)\frac{\partial\tilde{u}}{\partial y}\right] \tag{15.184}$$

$$\bar{\rho}\tilde{u}\frac{\partial\tilde{h}}{\partial x} + \bar{\rho}\tilde{v}\frac{\partial\tilde{h}}{\partial y} = u\frac{d\bar{p}}{dx} + \frac{\partial}{\partial y}\left[\left(\frac{\bar{\mu}}{Pr} + \frac{\mu_t}{Pr_t}\right)\frac{\partial\tilde{h}}{\partial y}\right] + (\bar{\mu} + \mu_t)\left(\frac{\partial\tilde{u}}{\partial y}\right)^2 \tag{15.185}$$

In arriving at Equations (15.184) and (15.185), we have introduced closure approximations for the Reynolds shear stress, τ_t, and the turbulent heat flux, q_t, that appear as a result of the mass-averaging process. Similar to what we did for the incompressible case, we assume

$$\tau_t = \mu_t\frac{\partial\tilde{u}}{\partial y} \quad \text{and} \quad q_t = -\frac{\mu_t}{Pr_t}\frac{\partial\tilde{h}}{\partial y} \tag{15.186}$$

where the quantity Pr_t is the **turbulent Prandtl number**. Measurements indicate that for boundary layers,

$$Pr_t \approx 0.9 \tag{15.187}$$

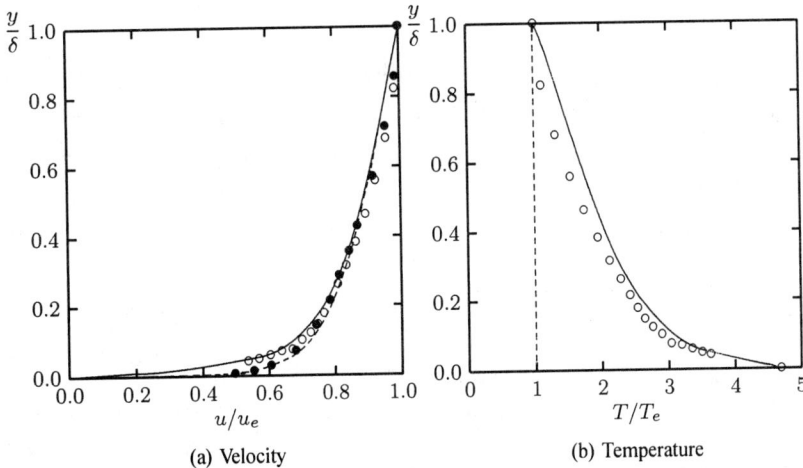

(a) Velocity   (b) Temperature

Figure 15.23: *Turbulent velocity and temperature profiles for a constant-pressure, adiabatic-wall boundary layer. Measured:* • *M = 0.1 Wieghardt [Coles and Hirst (1969)],* ○ *M = 4.5 Coles [Fernholz and Finley (1981)]. Computed: - - - M = 0.1; —— M = 4.5.*

Figure 15.23 shows measured velocity and temperature profiles for constant-pressure, adiabatic-wall turbulent boundary layers at Mach 0.1 and 4.5. Profiles computed using the k-ω model [Wilcox (1988a)] are also shown. The velocity profile is affected by compressibility mainly close to the surface, where it has reduced slope. This corresponds to reduced skin friction. Even though the slope vanishes at the surface, the temperature profile has much steeper slope than the laminar case approaching the surface. This means the temperature and density vary much more rapidly than in laminar flow in a very thin region near the surface.

Figure 15.24 shows the variation of skin friction with freestream Mach number, M_e, for constant-pressure, adiabatic-wall turbulent boundary layers. Because the turbulent stress dominates, the results are insensitive to the molecular-viscosity law used. Results shown assume the Reynolds number based on momentum thickness, $Re_\theta = 10,000$. The ratio of c_f to the incompressible value, c_{f_o}, is shown, where c_{f_o} can be obtained from the Kármán-Schoenherr correlation [Equation (14.135)]. For reference, results of computations using the k-ω model are included. The symbols are for a correlation of measurements developed by Van Driest (1951). Predictions with most turbulence models are similar for the constant-pressure case, with the exception that many versions of the k-ϵ model are quite inaccurate.

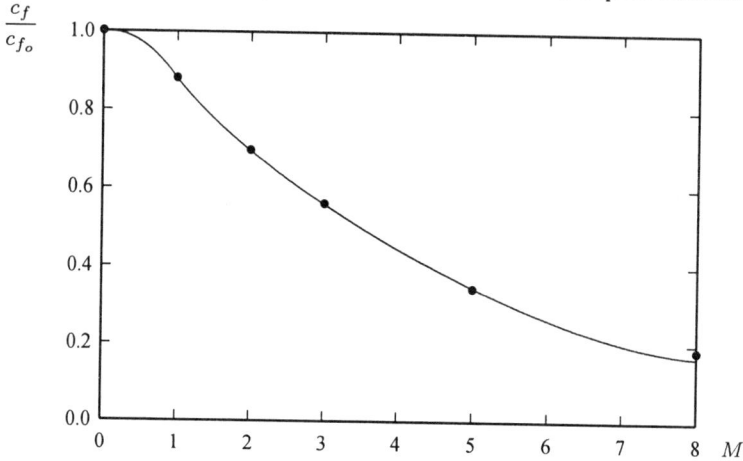

Figure 15.24: Computed and measured skin friction for a compressible flat-plate boundary layer; —— k-ω model; • Van Driest correlation.

Figure 15.25 shows laminar and turbulent skin friction as a function of plate-length Reynolds number, Re_x, for incompressible flow and for Mach 5. The turbulent-flow values shown are from computations based on the k-ω model, which are everywhere within 10% of measured values. The laminar values discussed above are included for contrast.

Compressible Law of the Wall. Since Figure 15.23 shows that compressibility has a relatively small effect on the velocity profile, the question arises about whether the law of the wall, Equation (14.124), still holds. The answer is, not quite, because the rapidly varying density close to the surface causes deviations from the law of the wall. However, as shown by Van Driest (1951), if we account for the variable density, we can deduce the **compressible law of the wall** that correlates measurements quite accurately, especially for adiabatic walls.

On dimensional grounds, we expect the velocity gradient, $\partial \tilde{u}/\partial y$, to depend upon the shear stress, τ_w, the density, $\bar{\rho}$, and distance from the surface, y. Note that because the density varies so rapidly, we use the local value rather than the surface value. So, we expect to have

$$\frac{\partial \tilde{u}}{\partial y} \propto \frac{\sqrt{\tau_w/\bar{\rho}}}{y} = \sqrt{\frac{\rho_w}{\bar{\rho}}} \frac{u_\tau}{y} \qquad (15.188)$$

where $u_\tau = \sqrt{\tau_w/\rho_w}$ is the conventional friction velocity. Measurements are consistent with this dimensional form, and the coefficient of proportionality approaching the surface is the

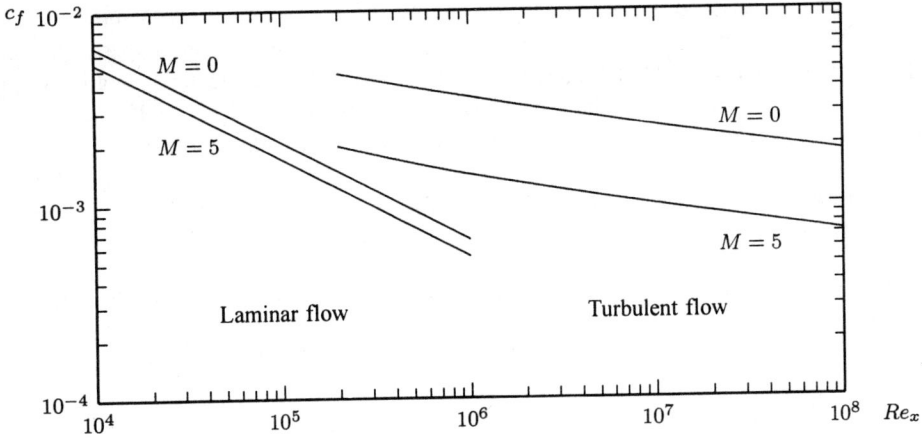

Figure 15.25: *Skin friction for adiabatic-wall boundary layers.*

Kármán constant, κ, so that

$$\frac{\partial \tilde{u}}{\partial y} = \sqrt{\frac{\rho_w}{\bar{\rho}}} \frac{u_\tau}{\kappa y} \qquad (15.189)$$

By examining the limiting form of the energy equation close to the surface, a straightforward but tedious exercise left for the Problems section shows that

$$\frac{\rho_w}{\bar{\rho}} \approx 1 - A^2 \left(\frac{\tilde{u}}{u_e}\right)^2, \qquad A^2 \equiv \frac{\gamma - 1}{2} Pr_t \frac{T_e}{T_w} M_e^2 \qquad (15.190)$$

Substituting this result into Equation (15.188), we have

$$\frac{d\tilde{u}}{\sqrt{1 - A^2 (\tilde{u}/u_e)^2}} = \frac{u_\tau}{\kappa} \frac{dy}{y} \qquad (15.191)$$

Using the change of variables defined by $\sin \phi = A\tilde{u}/u_e$, we can integrate Equation (15.191) to obtain

$$\frac{u_e}{A} \sin^{-1}\left(A\frac{\tilde{u}}{u_e}\right) = \frac{u_\tau}{\kappa} \ell n y + \text{constant} \qquad (15.192)$$

Finally, in terms of the dimensionless distance, $u_\tau y/\nu_w$, where ν_w is kinematic viscosity evaluated in terms of wall temperature and density, we have

$$\frac{u^*}{u_\tau} = \frac{1}{\kappa} \ell n \left(\frac{u_\tau y}{\nu_w}\right) + C^+ \qquad (15.193)$$

where u^* is the transformed velocity defined by

$$u^* \equiv \frac{u_e}{A} \sin^{-1}\left(A\frac{\tilde{u}}{u_e}\right), \qquad A^2 = \frac{\gamma - 1}{2} Pr_t \frac{T_e}{T_w} M_e^2 \qquad (15.194)$$

and C^+ is a constant. Equation (15.193) is the **compressible law of the wall**.

Excellent correlation of experimental data can be achieved with $\kappa = 0.41$ and $C^+ = 5.0$, the same values appropriate for incompressible flow. Figure 15.26 compares experimental data for four different Mach numbers ranging from 0 to 10 with Equation (15.193). In all cases, the data are for an adiabatic wall, and have been scaled according to Equation (15.194). As shown, while each set of data has a defect-layer component over a different range of $y^+ = u_\tau y/\nu_w$ values, the data all asymptote to the compressible law of the wall for $y^+ < 100$. For the higher Reynolds number cases (Mach 0.1 and 2.2), the data match the compressible law of the wall all the way up to $y^+ = 1000$.

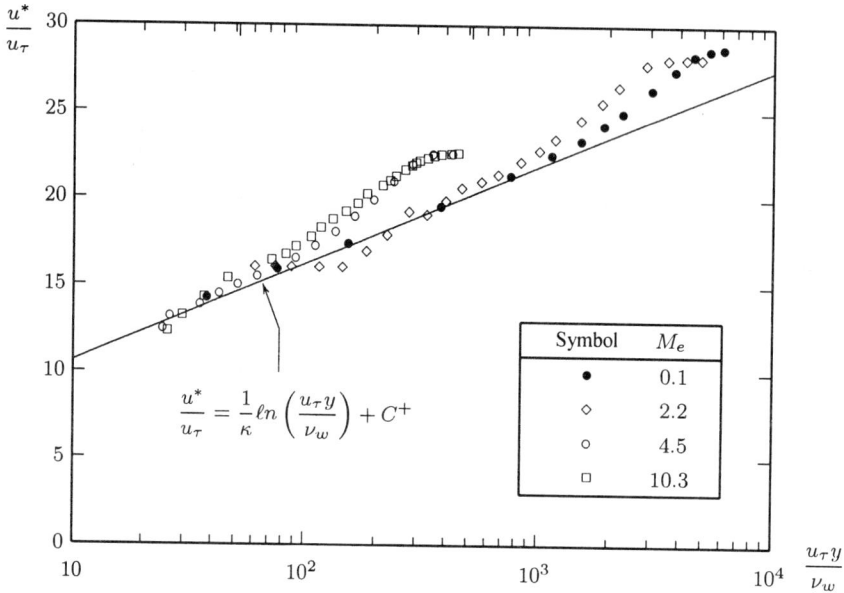

Figure 15.26: *Compressible law of the wall.*

15.7 Computational Fluid Dynamics

We conclude our discussion of compressible flows with comments on key issues concerning CFD. Because waves play such a significant role in a compressible medium, accurate prediction of wave properties is very important for achieving a satisfactory solution. There are many issues that must be considered. For example, the speed of and the pressure change induced by a wave have an important impact on a flow, and must be accurately predicted. We must also exercise caution to make sure that waves do not artificially reflect from finite-difference grid boundaries through which the physical wave would pass without reflection. Sharp interfaces such as a shock wave and even the leading and trailing edges of an expansion fan must be predicted with a minimum of numerical distortion.

These, and many other, issues all play a role in determining how faithfully a numerical solution represents physical reality. Some, such as passing waves through the boundaries of a finite-difference grid, are common to all Mach-number ranges. Even in the limit of incompressible flow where the sound speed approaches ∞, we must make sure our numerical solution doesn't include a non-physical reflected wave that would distort our solution. Other issues, such as finite wave speed and the proper handling of shock and expansion waves are specific to compressible-flow computations. The focus in this section is on the latter.

We begin by illustrating why it is so important in dealing with compressible flows to develop numerical algorithms based on the conservation form of the equations of motion. Then we explore the concepts of **numerical dissipation** and **numerical dispersion**, including their origin. These are effects that are present in all numerical methods, and can cause significant error, especially in compressible flows. We conclude with an example of numerical dissipation and dispersion in the context of MacCormack's (1969) method.

15.7.1 Conservative Differencing

Consider a normal shock wave. On the one hand, as shown in Figures 15.27(a) and (b), the density, ρ, and velocity, u, both have step discontinuities across the shock. By contrast, as shown in part (c) of the figure, their product, ρu, is constant across the shock. Another way of saying this is that the *mass flux*, ρu, is conserved across a shock wave, while the *volume flux*, u, is not conserved. Similarly, while pressure, p, and enthalpy, h, have jump discontinuities across a shock, the normal-shock relations tell us $p + \rho u^2$ and $h + \frac{1}{2}u^2$ are conserved.

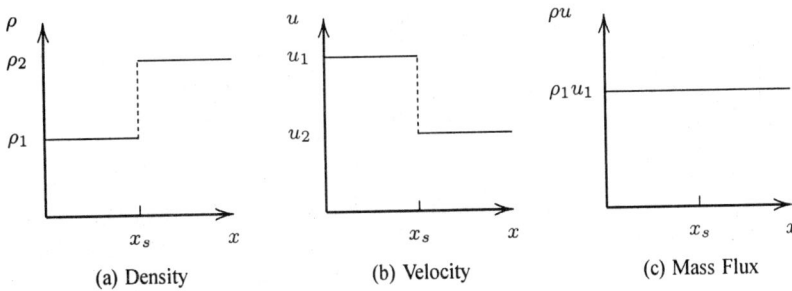

| (a) Density | (b) Velocity | (c) Mass Flux |

Figure 15.27: *Variation of properties across a normal shock.*

The implication of these ideas regarding a finite-difference formulation should be clear. In the case of a two-dimensional, inviscid flow, for example, the conservation form of the equations of motion can be written compactly as

$$\frac{\partial \mathbf{U}}{\partial t} + \frac{\partial \mathbf{F}}{\partial x} + \frac{\partial \mathbf{G}}{\partial y} = \mathbf{0} \qquad (15.195)$$

The vectors $\mathbf{U}$, $\mathbf{F}$ and $\mathbf{G}$ are

$$\mathbf{U} = \left\{ \begin{array}{c} \rho \\ \rho u \\ \rho v \\ \rho \mathcal{E} \end{array} \right\}, \qquad \mathbf{F} = \left\{ \begin{array}{c} \rho u \\ \rho u^2 + p \\ \rho u v \\ \rho u \mathcal{H} \end{array} \right\}, \qquad \mathbf{G} = \left\{ \begin{array}{c} \rho v \\ \rho u v \\ \rho v^2 + p \\ \rho v \mathcal{H} \end{array} \right\} \qquad (15.196)$$

where $\mathcal{E}$ and $\mathcal{H}$ are total energy and total enthalpy, respectively [see Equation (15.2)]. Clearly, if we base our discretization approximations on the conservation forms, we will be dealing with well-behaved quantities, even when a shock wave lies within our finite-difference cell. Clearly, no finite-difference approximation of any order of accuracy can resolve a step discontinuity within a cell where the derivative becomes infinite. By contrast, $\partial(\rho u)/\partial x = 0$ in our simple example of a normal shock. Even a first-order accurate difference formula can handle this special limiting case! When we develop our scheme based on the conservation form of the equations, we effectively treat each finite-difference cell as a control volume. This is quite reasonable as a shock is a discontinuous wave front that cannot be resolved by any grid, no matter how closely spaced the points are.[5] Since all of the conserved flux variables will be computed with minimal error, the shock relations will be preserved. That is, using the conservation form of the equations of motion as the basis of our numerical algorithm guarantees that the jumps in quantities across a shock will be correctly computed, even if local properties are distorted.

By contrast, if we develop our discretization approximations for the conservation principles in primitive-variable form, the resulting difference equations will be ill behaved. For example, in our one-dimensional example for a normal shock, the continuity equation in primitive-variable form is

$$\frac{\partial \rho}{\partial t} + u\frac{\partial \rho}{\partial x} + \rho\frac{\partial u}{\partial x} = 0 \tag{15.197}$$

In principle, both $u\partial\rho/\partial x$ and $\rho\partial u/\partial x$ are singular, while their sum is not. In practice, neither term will be accurately represented in a finite-difference scheme, and their sum will simply compound the discretization error.

Thus, for compressible flows, the conservation form of the equations of motion is clearly preferred for CFD work. The underlying reason is the presence of discontinuities in flow properties that are possible at supersonic speeds. Now, the conservation forms are an immediate consequence of the integral-conservation principles, which do not require mathematical continuity. In complete contrast, the primitive-variable forms require mathematical continuity, and thus pose an extreme (and unreasonable) constraint on a solution with shock waves.

Having a shock wave within a finite-difference cell in the manner discussed above is known as **shock capturing**. This method is helpful when shock locations must be determined as part of the solution. CFD researchers have devoted considerable effort to developing accurate shock-capturing methods.

15.7.2 Numerical Dissipation and Dispersion

Recall that when we first introduced the concepts of discretization approximations and numerical accuracy in Section 11.9, our point of view was the following. The numerical solution we seek is an approximation to a specific partial differential equation (PDE), and the numerical solution will include some degree of error. In estimating the error with Richardson extrapolation (Subsection 11.9.4), our point of reference is the continuum solution to the same specific PDE. In this subsection, we take a different point of view. Rather than viewing the PDE as our foundation for analysis and error estimation, we shift our emphasis to the approximating difference equation.

Consider the following one-dimensional, wave equation for a flow property u with a constant, positive wave speed, a.

$$\frac{\partial u}{\partial t} + a\frac{\partial u}{\partial x} = 0 \tag{15.198}$$

For purposes of illustration, we choose the following finite-difference approximation to this equation.

$$\frac{u_i^{n+1} - u_i^n}{\Delta t} + a\frac{u_i^n - u_{i-1}^n}{\Delta x} = 0 \tag{15.199}$$

---

[5]Even if we do a viscous computation, the shock thickness is of order $1/Re$, which is much thinner than any practical finite-difference cell.

Using Taylor series expansions about $x = x_i$ and $t = t^n$ similar to what we did in Section 11.9, it is a simple matter to show that this scheme is only first-order accurate in both time and space relative to Equation (15.198).

Suppose now that we pose a somewhat different question. Specifically, given the difference Equation (15.199), what PDE is implied? So, we expand in Taylor series, retaining terms up to $(\Delta x)^3$ and $(\Delta t)^3$. Thus,

$$u_i^{n+1} = u_i^n + \frac{\partial u}{\partial t}\Delta t + \frac{\partial^2 u}{\partial t^2}\frac{(\Delta t)^2}{2} + \frac{\partial^3 u}{\partial t^3}\frac{(\Delta t)^3}{6} + \cdots \qquad (15.200)$$

$$u_{i-1}^n = u_i^n - \frac{\partial \phi}{\partial x}\Delta x + \frac{\partial^2 u}{\partial x^2}\frac{(\Delta x)^2}{2} - \frac{\partial^3 u}{\partial x^3}\frac{(\Delta x)^3}{6} + \cdots \qquad (15.201)$$

where the derivatives are understood to be evaluated at the reference point. Substituting Equations (15.200) and (15.201) into Equation (15.199), and regrouping terms yields

$$\frac{\partial u}{\partial t} + a\frac{\partial u}{\partial x} = -\frac{\partial^2 u}{\partial t^2}\frac{\Delta t}{2} + a\frac{\partial^2 u}{\partial x^2}\frac{\Delta x}{2}$$
$$-\frac{\partial^3 u}{\partial t^3}\frac{(\Delta t)^2}{6} - a\frac{\partial^3 u}{\partial x^3}\frac{(\Delta x)^2}{6} + \cdots \qquad (15.202)$$

Next, we replace the temporal derivatives, $\partial^2 u/\partial t^2$ and $\partial^3 u/\partial t^3$, by spatial derivatives. We can develop the required relations by manipulating Equation (15.202). Details can be found in most CFD texts — Anderson (1995) gives a particularly lucid development. The derivatives turn out to be

$$\left.\begin{aligned} \frac{\partial^2 u}{\partial t^2} &= a^2\frac{\partial^2 u}{\partial x^2} + \left[a^3\Delta t - a^2\Delta x\right]\frac{\partial^3 u}{\partial x^3} + \cdots \\ \frac{\partial^3 u}{\partial t^3} &= -a^3\frac{\partial^3 u}{\partial x^3} + \cdots \end{aligned}\right\} \qquad (15.203)$$

Thus, substituting Equations (15.203) into Equation (15.202), shows that our chosen finite-difference equation implies the following partial differential equation for u.

$$\frac{\partial u}{\partial t} + a\frac{\partial u}{\partial x} = \nu_a\frac{\partial^2 u}{\partial x^2} + \mathcal{D}_a\frac{\partial^3 u}{\partial x^3} + \cdots \qquad (15.204)$$

The coefficients ν_a and $\mathcal{D}_a$ are effective diffusion and dispersion coefficients defined by

$$\nu_a \equiv \frac{a\Delta x}{2}(1 - \lambda) \quad \text{and} \quad \mathcal{D}_a \equiv \frac{a(\Delta x)^2}{6}\left(3\lambda - 2\lambda^2 - 1\right) \qquad (15.205)$$

where the coefficient λ is

$$\lambda = \frac{a\Delta t}{\Delta x} \qquad (15.206)$$

Equation (15.204) is referred to as the **modified equation**, corresponding to the original PDE of Equation (15.198). The meaning of this equation is the following. The exact solution to the difference Equation (15.199) is an exact solution (i.e., free of truncation error) to the modified equation. Of course, there are additional terms beyond the two appearing on the right-hand side of Equation (15.204).

The term proportional to $\partial^2 u/\partial x^2$ is a diffusion term. The coefficient ν_a has dimensions of kinematic viscosity (L^2/T), and is called **artificial viscosity**. The effect of this term is especially noticeable at wave fronts that are, in reality, very sharp. It tends to smooth out discontinuities, so that shock waves have a nonphysical finite thickness, similar to the behavior shown in Figure 15.28. In a shock wave, kinetic energy is converted to internal energy through a combination of viscous dissipation and heat conduction. In analogy to the true physical process, the effect of $\nu_a \partial^2 u/\partial x^2$ is generally referred to as **numerical dissipation**.

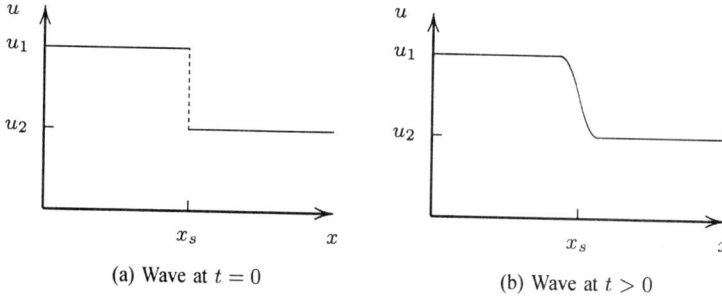

(a) Wave at $t = 0$ (b) Wave at $t > 0$

Figure 15.28: *Effect of numerical dissipation.*

The term proportional to $\partial^3 u/\partial x^3$ is a dispersion term. Dispersion is a phenomenon that has a nonuniform affect on the various frequencies in a wave, and its effect is to cause oscillations at the leading and trailing edges of the wave. The oscillations are usually bounded, and have an appearance like that shown schematically in Figure 15.29. This is strictly a numerical artifact, and is called **numerical dispersion**.

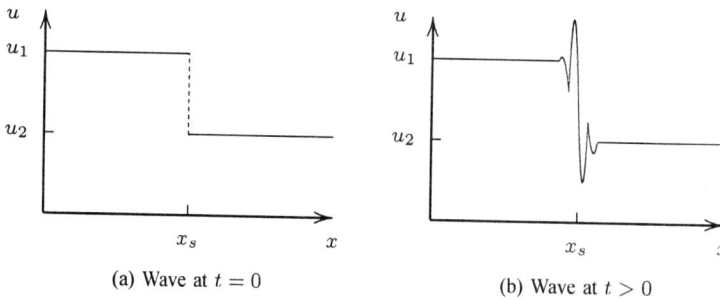

(a) Wave at $t = 0$ (b) Wave at $t > 0$

Figure 15.29: *Effect of numerical dispersion.*

With perseverance, we could continue developing additional terms in the modified equation. However, there is no need to do this as the effect of even-order derivatives, $\partial^4/\partial x^4$, $\partial^6/\partial x^6$, etc. is dissipative in nature. Similarly, odd-order derivatives, $\partial^5/\partial x^5$, $\partial^7/\partial x^7$, etc. are dispersive in nature. Thus, since the right-hand side of the modified equation is the truncation error, the leading term determines the predominate type of error for a given numerical algorithm. In particular, second-order accurate schemes will have vanishing artificial viscosity, so that the dominant type of error for such schemes is dispersion.

As a final comment, while numerical dissipation degrades numerical accuracy, it also has the desirable effect of enhancing stability. This tradeoff is so important in complex CFD applications that explicit artificial viscosity is often added to a non-dissipative scheme to improve its stability.

15.7.3 Propagating Shocks and Expansions

To illustrate dissipation and dispersion effects on a numerical scheme, we consider the inviscid, nonlinear Burgers' equation written in conservation form, viz.,

$$\frac{\partial u}{\partial t} + \frac{\partial}{\partial x}\left(\frac{1}{2}u^2\right) = 0 \tag{15.207}$$

We will solve this equation for initial conditions given by

$$u(x,0) = \begin{cases} u_1, & x \le 0 \\ u_2, & x > 0 \end{cases} \tag{15.208}$$

The exact solution to this equation depends upon the ratio of u_1 to u_2. When $u_1 > u_2$, we have the analog of a shock wave, while $u_1 < u_2$ produces a wave that spreads as time passes and thus resembles an expansion wave. The exact solutions are as follows.

Shock ($u_1 > u_2$):

$$u(x,t) = \begin{cases} u_1, & x \le u_s t \\ u_2, & x > u_s t \end{cases} \tag{15.209}$$

where the shock velocity, u_s, is given by $u_s = \frac{1}{2}(u_1 + u_2)$.

Expansion ($u_1 < u_2$):

$$u(x,t) = \begin{cases} u_1, & x \le u_1 t \\ x/t, & u_1 t \le x \le u_2 t \\ u_2, & x > u_2 t \end{cases} \tag{15.210}$$

We use MacCormack's method (see Subsection 12.9.3) to numerically simulate the solution to Burgers' equation. In the computations, we select

$$\text{Shock}: \ u_1 = 2, u_2 = 1 \quad \text{and} \quad \text{Expansion}: \ u_1 = 1, u_2 = 2 \tag{15.211}$$

Figure 15.30 compares the numerical solutions with the exact solutions after the waves have propagated for a dimensionless time, $t = 1$. As shown, both waves show evidence of dispersion. Also, a small amount of smearing has occurred for the shock and the expansion. Because MacCormack's method is second-order accurate in both time and space, the small amount of dissipation originates from the term proportional to $\partial^4 u/\partial x^4$ in the modified equation.

As a final comment on these results, this exercise is not intended to emphasize shortcomings of MacCormack's method. On the contrary, because it is second-order accurate, its artificial viscosity is very small. It is, in fact, one of the best behaved of the various explicit methods available.

The program used for these computations, **BURGER**, is described in Appendix E. Source code is included on the diskette included with this book.

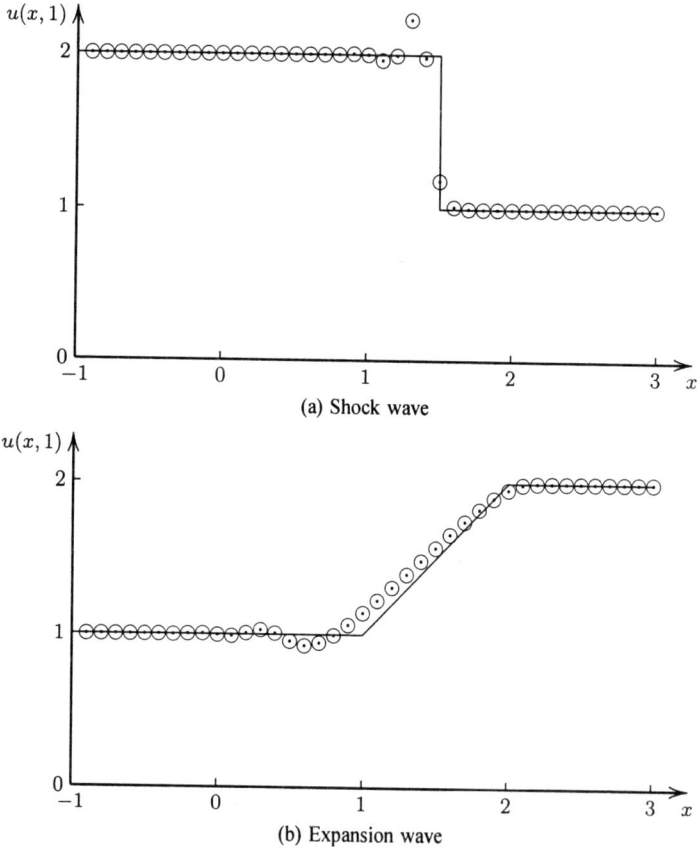

Figure 15.30: *Comparison of numerical and exact solutions to the inviscid, nonlinear Burgers'
equation; —— Exact solution; ⊙ MacCormack's method.*

Problems

15.1 The table below lists measured viscosity, μ, thermal conductivity, k, and specific heat, c_p, for air. Compute the Prandtl number and make a graph of your results.

T (°F)	$\mu \left(\frac{\text{slug}}{\text{ft·sec}} \right)$	$k \left(\frac{\text{lb}}{\text{sec·°R}} \right)$	$c_p \left(\frac{\text{ft·lb}}{\text{slug·°R}} \right)$
-50	$3.11 \cdot 10^{-7}$	$2.59 \cdot 10^{-3}$	5998
32	$3.59 \cdot 10^{-7}$	$3.03 \cdot 10^{-3}$	5998
100	$4.00 \cdot 10^{-7}$	$3.37 \cdot 10^{-3}$	5998
200	$4.53 \cdot 10^{-7}$	$3.83 \cdot 10^{-3}$	6023
400	$5.42 \cdot 10^{-7}$	$4.60 \cdot 10^{-3}$	6123
600	$6.30 \cdot 10^{-7}$	$5.47 \cdot 10^{-3}$	6273

Problem 15.1

15.2 The table below lists measured viscosity, μ, thermal conductivity, k, and specific heat, c_p, for water. Compute the Prandtl number and make a graph of your results.

T (°C)	$\mu \left(\frac{\text{kg}}{\text{m·sec}} \right)$	$k \left(\frac{\text{J}}{\text{m·sec·K}} \right)$	$c_p \left(\frac{\text{J}}{\text{kg·K}} \right)$
0	$1.79 \cdot 10^{-3}$	0.524	4010
20	$9.98 \cdot 10^{-4}$	0.605	4174
40	$6.54 \cdot 10^{-4}$	0.598	4170
60	$4.62 \cdot 10^{-4}$	0.651	4174
80	$3.50 \cdot 10^{-4}$	0.689	4196
100	$2.78 \cdot 10^{-4}$	0.684	4208

Problem 15.2

15.3 Consider flow through a normal shock wave with viscous and heat-transfer effects included.

(a) Treating the flow as steady and one dimensional, simplify the continuity, Navier-Stokes and energy equations as appropriate.

(b) What are the boundary conditions ahead of $(x \to -\infty)$ and behind $(x \to +\infty)$ the shock? **HINT:** There are 5, and the flow is essentially inviscid for $|x| \to \infty$.

(c) Integrate all three equations once, starting with continuity. Verify that the following pair of coupled first-order equations results. Note that subscript 1 denotes condition ahead of the shock.

$$\frac{4}{3}\mu \frac{du}{dx} = p + \rho u^2 - \left(p_1 + \rho_1 u_1^2 \right)$$

$$\frac{\mu}{Pr} \frac{dh}{dx} + \frac{4}{3}\mu u \frac{du}{dx} = \rho u \left(h + \frac{1}{2}u^2 \right) - \rho_1 u_1 \left(h_1 + \frac{1}{2}u_1^2 \right)$$

15.4 Consider flow through a normal shock wave with viscous and heat-transfer effects included. A straightforward derivation shows that the Navier-Stokes equation simplifies to

$$\frac{4}{3}\mu \frac{du}{dx} = p + \rho u^2 - \left(p_1 + \rho_1 u_1^2 \right)$$

(a) There is no natural length scale in this problem. To determine the appropriate characteristic length, λ, for viscous flow through a normal shock, cast this equation in dimensionless form, i.e., in terms of

$$\bar{x} \equiv \frac{x}{\lambda}, \quad \bar{u} \equiv \frac{u}{a_1}, \quad \bar{p} \equiv \frac{p}{\rho_1 a_1^2}, \quad \bar{\mu} \equiv \frac{\mu}{\mu_1}$$

where ρ_1, a_1 and μ_1 are density, speed of sound and viscosity ahead of the shock, respectively. What must λ be if all terms in the dimensionless Navier-Stokes equation are of equal importance?

(b) The mean-free path of a gas is $\ell_{mfp} \approx \frac{3}{2}\nu/a$, where ν is kinematic viscosity. How does λ compare to ℓ_{mfp}? Comment on the suitability of the Navier-Stokes equation for describing flow within a shock wave.

15.5 Verify that for two-dimensional, compressible flow of a monatomic gas, the viscous dissipation, Φ, defined by

$$\Phi \equiv \left([\tau] \cdot \nabla \right)^T \cdot \mathbf{u}$$

is always positive. **HINT:** If you can explain why, for any pair of real variables a and b, the polynomial $f(a,b) = a^2 + b^2 - ab$ is always positive, the proof will follow.

15.6 For low-speed flow of a gas induced by motion of an impulsively-accelerated flat plate that is heated, the equations of motion simplify to

$$\frac{\partial u}{\partial t} = \nu \frac{\partial^2 u}{\partial y^2}$$

$$\frac{\partial h}{\partial t} = \frac{\nu}{Pr} \frac{\partial^2 h}{\partial y^2} + \nu \left(\frac{\partial u}{\partial y} \right)^2$$

You may assume kinematic viscosity, ν, and Prandtl number, Pr, are constant. For $t > 0$, the boundary conditions are

$$u(0,t) = U, \qquad h(0,t) = h_w$$

$$u(y,t) \to 0 \quad \text{and} \quad h(y,t) \to h_\infty \quad \text{as} \quad y \to \infty$$

Since the momentum equation is not coupled to the energy equation, clearly the velocity is the same as in *Stokes' First Problem* (see Section 13.4).

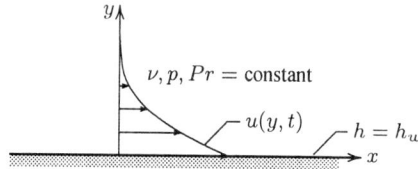

Problem 15.6

(a) Assume a solution for the enthalpy of the form

$$h(y,t) = h_\infty + (h_w - h_\infty) H(\eta), \qquad \eta \equiv \frac{y}{2\sqrt{\nu t}}$$

Verify that $H(\eta)$ satisfies the following ordinary differential equation.

$$\frac{1}{Pr} \frac{d^2 H}{d\eta^2} + 2\eta \frac{dH}{d\eta} = -\frac{4}{\pi}(\gamma - 1)M_w^2 e^{-2\eta^2}$$

where $M_w \equiv U/a_w$ and a_w is the speed of sound based on the wall temperature. Also, state the boundary conditions for $H(\eta)$.

(b) Solve for $H(\eta)$ in the limiting case $M_w \to 0$. **HINT:** Try substituting $\tilde{\eta} = \sqrt{Pr}\,\eta$.

15.7 For compressible channel flow, the Navier-Stokes and energy equations simplify to

$$0 = -\frac{dp}{dx} + \frac{d}{dy}\left(\mu \frac{du}{dy} \right)$$

$$0 = u\frac{dp}{dx} + \frac{d}{dy}\left(\frac{\mu}{Pr} \frac{dh}{dy} \right) + \mu \left(\frac{du}{dy} \right)^2$$

Assume the fluid is calorically perfect so that $h = c_p T$. Combine these equations and use the change of independent variable defined by $dy = \mu\,d\eta \implies d/d\eta = \mu\,d/dy$, to verify that, for adiabatic walls,

$$h + \frac{Pr}{2}u^2 = \text{constant}$$

15.8 For constant-pressure, compressible Couette flow, the Navier-Stokes and energy equations become

$$0 = \frac{d}{dy}\left(\mu\frac{du}{dy}\right)$$

$$0 = \frac{d}{dy}\left(\frac{\mu}{Pr}\frac{dh}{dy}\right) + \mu\left(\frac{du}{dy}\right)^2$$

Integrate the momentum equation once and verify that by regarding u as the independent variable,

$$\mu\frac{d}{dy} = \tau_w\frac{d}{du}$$

where τ_w is the surface shear stress. Use this change of variables in the energy equation and verify that

$$h + \frac{Pr}{2}u^2 = A + Bu \qquad (A, B = \text{constant})$$

15.9 Consider steady, compressible Couette flow with constant pressure, i.e., flow between two parallel plates separated by a distance h, the lower at rest and the upper moving with constant velocity U.

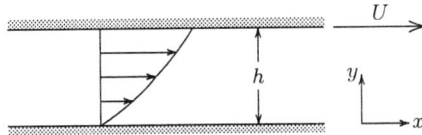

Problem 15.9

The equations of motion for this flow are:

$$\frac{\partial\rho}{\partial t} + \frac{\partial}{\partial x}(\rho u) + \frac{\partial}{\partial y}(\rho v) = 0$$

$$\rho\frac{du}{dt} = -\frac{\partial p}{\partial x} + \frac{\partial\tau_{xx}}{\partial x} + \frac{\partial\tau_{xy}}{\partial y}, \qquad \rho\frac{dv}{dt} = -\frac{\partial p}{\partial y} + \frac{\partial\tau_{xy}}{\partial x} + \frac{\partial\tau_{yy}}{\partial y}$$

$$\rho\frac{dH}{dt} = \frac{\partial p}{\partial t} - \frac{\partial q_x}{\partial x} - \frac{\partial q_y}{\partial y} + \frac{\partial}{\partial x}(u\tau_{xx} + v\tau_{xy}) + \frac{\partial}{\partial y}(u\tau_{xy} + v\tau_{yy})$$

where

$$\mathcal{H} = c_p T + \frac{1}{2}(u^2 + v^2), \qquad q_x = -\frac{\mu c_p}{Pr}\frac{\partial T}{\partial x}, \qquad q_y = -\frac{\mu c_p}{Pr}\frac{\partial T}{\partial y}$$

$$\tau_{xx} = 2\mu\frac{\partial u}{\partial x} - \frac{2}{3}\mu\left(\frac{\partial u}{\partial x} + \frac{\partial v}{\partial y}\right), \qquad \tau_{yy} = 2\mu\frac{\partial v}{\partial y} - \frac{2}{3}\mu\left(\frac{\partial u}{\partial x} + \frac{\partial v}{\partial y}\right), \qquad \tau_{xy} = \mu\left(\frac{\partial u}{\partial y} + \frac{\partial v}{\partial x}\right)$$

(a) State the four boundary conditions that must be imposed upon u and v.

(b) Assuming the plates are infinite in extent, simplify the continuity equation and solve for v.

(c) Explain why the Eulerian derivative of any quantity Υ, i.e., $d\Upsilon/dt$, is zero.

(d) Simplify the conservation of momentum and energy equations using everything you have learned from Parts (b) and (c).

(e) Simplify the viscous stresses and the heat-flux vector using everything you have learned from Parts (b) and (c).

(f) Combining results of Parts (d) and (e), write the *ordinary differential equations* satisfied by u and T.

(g) Assuming the walls are adiabatic so that $dT/dy = 0$ at $y = 0$ and $y = h$, verify that the temperatures of the lower and upper walls are related by

$$T(0) = T(h)\left[1 + \frac{\gamma - 1}{2}Pr M^2\right]$$

where M is Mach number based on U and the speed of sound at the upper wall.

15.10 Consider the motion of water waves in a shallow channel. The following one-dimensional equations provide a good approximation to the motion of waves at the surface of the fluid:

$$\frac{\partial h}{\partial t} + \frac{\partial}{\partial x}(uh) = 0 \quad \text{and} \quad \frac{\partial u}{\partial t} + u\frac{\partial u}{\partial x} = -g\frac{\partial h}{\partial x}$$

where x is horizontal distance, t is time, u is horizontal velocity, h is depth and g is gravitational acceleration. Also, the local pressure is $p = \rho g(h - z)$, where z is vertical distance and ρ is the density of water.

(a) Letting $\bar{\rho} = \rho h$ and $\bar{p} = \int_0^h p\,dz$, show that these equations can be rewritten as

$$\frac{\partial \bar{\rho}}{\partial t} + \frac{\partial}{\partial x}(\bar{\rho}u) = 0 \quad \text{and} \quad \frac{\partial u}{\partial t} + u\frac{\partial u}{\partial x} = -\frac{1}{\bar{\rho}}\frac{\partial \bar{p}}{\partial x}$$

(b) What is the effective speed of sound, a, and specific-heat ratio, γ? Express your answer for a in terms of g and h.

15.11 Derive the Fanno-flow Equations (15.63).

15.12 Show that for Fanno flow

$$\frac{dh}{dx} = \frac{\gamma M^2}{M^2 - 1}T\frac{ds}{dx}$$

15.13 Find the maximum length, in the absence of choking, for adiabatic flow of air in a 20-cm diameter pipe if the friction factor is $f = 0.022$. The inlet-flow conditions are $\bar{u} = 250$ m/sec, $T = 44°$ C and $p = 2.5$ atm. Also, determine the temperature at the exit.

15.14 Find the maximum length, in the absence of choking, for adiabatic flow of air in a 1-in square duct if the friction factor is $f = 0.0145$. The inlet-flow conditions are $\bar{u} = 955$ ft/sec, $T = 9°$ F and $p = 18.1$ psi. Also, determine the pressure at the exit.

15.15 Air enters a 5 cm by 5 cm square duct at a velocity $\bar{u} = 900$ m/sec and a temperature $T = 300$ K. The friction factor, f, is 0.018.

(a) For what duct length will the flow exactly decelerate to Mach 1?

(b) If the duct length is 2 m, will there be a normal shock in the duct?

15.16 An experiment is designed so that gas flows into a 5 cm diameter pipe with $p = 600$ kPa, $\bar{u} = 300$ m/sec and $T = 20°$ C. The friction factor is $f = 0.025$. How far from the entrance, L^*, is the flow choked for hydrogen and for helium?

15.17 Air flows adiabatically through a short pipe of diameter $D = 12$ cm. The flow enters with $p = 500$ kPa, $\bar{u} = 500$ m/sec and $T = -20°$ C. If the pipe length is $L = 1$ m and the exit pressure is $p_e = 720$ kPa, what is the average friction factor, $\bar{f}$, and the exit Mach number?

15.18 Air flows adiabatically through a short duct whose cross section is an equilateral triangle of side $s = 4$ in. The flow enters with $p = 20$ psi, $\bar{u} = 2000$ ft/sec and $T = 100°$ F. If the duct length is $L = 2$ ft and the exit temperature is $T_e = 210°$ F, what is the average friction factor, $\bar{f}$, and the exit Mach number?

15.19 Carbon dioxide flows adiabatically through a smooth cylindrical pipe of diameter $D = 10$ cm and length $L = 10$ m. At the inlet, the pressure is $p = 1$ atm, $\bar{u} = 798$ m/sec and $T = 15°$ C. Determine the Mach number, pressure and temperature at the exit. Refer to Tables 1.3 and 1.6 for properties of carbon dioxide.

15.20 Helium flows adiabatically through a smooth rectangular channel of sides 1 in by 2 in, and length $L = 4$ ft. At the inlet, the pressure is $p = 14.7$ psi, $\bar{u} = 8175$ ft/sec and $T = 59°$ F. Determine the Mach number, pressure and temperature at the exit. Refer to Tables 1.3 and 1.6 for properties of helium.

15.21 Derive the Rayleigh-flow Equations (15.93).

15.22 Show that for Rayleigh flow

$$\frac{dh}{dx} = \frac{\gamma M^2 - 1}{M^2 - 1} T \frac{ds}{dx}$$

15.23 Air with velocity $\bar{u} = 250$ ft/sec enters a circular pipe at a pressure of $p = 30$ psi and a temperature of $T = 30°$ F. How much heat per unit mass (in Btu/slug) must be added to achieve sonic conditions at the exit? What is the exit pressure?

15.24 Air with velocity $\bar{u} = 330$ m/sec enters a 6 cm x 12 cm rectangular duct at a pressure of $p = 300$ kPa and a temperature of $T = 300$ K. How much heat per unit mass (in J/kg) must be added to achieve sonic conditions at the exit? What is the exit temperature?

15.25 Air enters a frictionless duct at a Mach number of 0.3 and a total temperature of $T = 293$ K. How much heat per unit mass (in J/kg) must be added to achieve Mach numbers of 0.5, 0.7 and 0.9?

15.26 Hydrogen enters a frictionless duct at a Mach number of 4 and a total temperature of $T = 12°$ F. How much heat per unit mass (in Btu/slug) must be extracted to achieve Mach numbers of 3, 2 and 1?

15.27 An experiment is set up with gas entering a frictionless duct with $\bar{u} = 500$ m/sec, $p = 3$ atm and $T = 350$ K. How much heat per unit mass (in J/kg) must be added in order to choke the flow for air and for hydrogen? What is the exit temperature in each case?

15.28 Helium enters a frictionless duct at a Mach number of 0.8 and a total temperature of $T = -50°$ C. How much heat per unit mass (in Btu/slug) must be extracted to reduce the Mach number to 0.2?

15.29 Methane enters a frictionless duct at a Mach number of 1.3 and a total temperature of $T = 200°$ F. How much heat per unit mass (in Btu/slug) must be extracted to reduce the total temperature to $0°$ F, and what is the Mach number when this condition is reached?

15.30 Gas is flowing through a frictionless pipe, the exit Mach number is 2 and the static temperature is $20°$ C. If the heat extracted from the pipe is 100 kJ/kg, determine the inlet Mach number and static temperature for hydrogen and for air.

15.31 Determine the angle of the trailing shock in the example problem of Figure 15.20.

15.32 Consider Mach 3 flow into a compression corner with $\theta = 5°$. If the freestream static pressure is $p_1 = 100$ kPa, what is the shock angle, β, the Mach number behind the shock, M_2, and the pressure behind the shock, p_2?

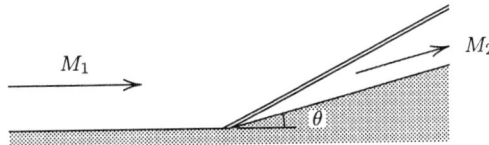

Problems 15.32, 15.33

15.33 In the flow past a compression corner, the upstream Mach number and pressure are 3.5 and 1 atm, respectively. Downstream of the corner, the pressure is 5.48 atm. Calculate the deflection angle, θ.

15.34 Consider a wedge with a half angle of $10°$ flying at Mach 2. Calculate the ratio of total pressures across the shock wave created by the wedge.

15.35 Consider a $20°$ half-angle wedge in a supersonic flow at Mach 3 at standard sea-level conditions ($p_1 = 2116$ lb/ft^2 and $T_1 = 519°$R). Calculate the wave angle, β, and the surface pressure, temperature and Mach number.

15.36 Consider an oblique shock with wave angle $\beta = 35°$. Upstream of the wave, $p_1 = 2000$ lb/ft^2, $T_1 = 520°$ R and $u_1 = 3355$ ft/sec. Calculate p_2, T_2, u_2 and the flow deflection angle, θ.

15.37 For three different supersonic wind tunnels, the reservoir pressure is 10 atm. A Pitot tube inserted in each measures pressures of (a) 1 atm, (b) 2 atm and (c) 8 atm. Calculate the test section Mach number and area ratio for each tunnel. **HINT:** Be sure to account for the *detached shock* wave standing ahead of the pitot tube.

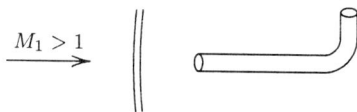

Problem 15.37

15.38 Consider Mach 3 flow past a wedge of total angle $2\theta = 16°$. The freestream static pressure is $p_1 = 20$ psi. What is the pressure at Point A when the pointed edge is forward? What is the pressure at point B if the base is forward?

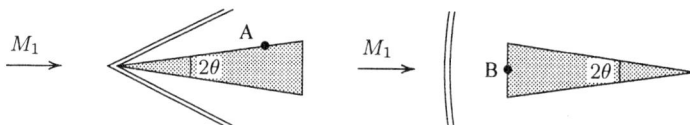

Problem 15.38

15.39 In an experimental program designed to illustrate the differences between *strong* and *weak* shocks, Mach 2 flow through a nozzle is observed with variable exit pressure. In one experiment, the exit pressure is atmospheric. A small separation bubble is present that causes an effective flow deflection of 10°. In a second experiment, the exit pressure is increased to a much higher value. The flow has a huge separation bubble that causes massive blockage. The effective flow turning angle is 10° for this case also. The separation shock in the first experiment is a *weak* shock while the shock in the second experiment is a *strong* shock.

(a) What are the shock angles for the two cases? Express your answers to within a tenth of a degree.

(b) What is the static pressure rise across the shock for the two experiments?

(c) Compute the Mach number downstream of the shock for each case.

15.40 Verify that in hypersonic flow, for an oblique shock wave, the shock angle, β, and the flow deflection angle, θ, are related by

$$\beta \approx \tfrac{1}{2}(\gamma + 1)\theta$$

You may assume that $\theta \ll 1$, $\beta \ll 1$ and M_1 is sufficiently large that $M_1 \sin \beta \gg 1$.

15.41 Compute the pressure rise, p_2/p_1, for Mach 20 flow into a 5° compression corner for air and for helium. To simplify your computations, use the fact that in hypersonic flow, $\beta \approx \tfrac{1}{2}(\gamma + 1)\theta$.

15.42 Compute the pressure rise, p_2/p_1, for Mach 30 flow into a 6° compression corner for air and for helium. To simplify your computations, use the fact that in hypersonic flow, $\beta \approx \tfrac{1}{2}(\gamma + 1)\theta$.

15.43 A cruise missile is traveling at Mach 2.13 and altitude 1200 ft. How long does it take to hear the disturbance from the missile after it passes directly overhead if the speed of sound is 1120 ft/sec?

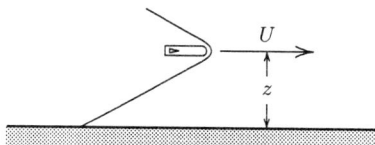

Problem 15.43

15.44 Consider reflection of an oblique shock from a solid boundary. The gas is air, the Mach number ahead of the shock is $M_1 = 2.5$, the incident shock angle is $\beta_i = 40°$, and the upstream pressure is $p_1 = 100$ kPa. Compute M_3, p_3 and the reflected shock angle, β_r.

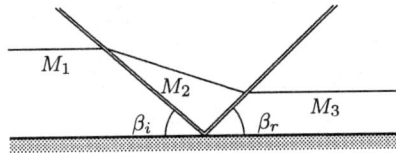

Problems 15.44, 15.45

15.45 An incident shock wave with a wave angle of $\beta = 30°$ impinges on a straight wall. If the upstream flow properties are $M_1 = 2.8$, $p_1 = 1$ atm and $T_1 = 300$ K, calculate the pressure, temperature, Mach number and total pressure downstream of the reflected wave.

15.46 A supersonic stream at $M_1 = 3.6$ flows past a compression corner with a deflection angle of $20°$. The incident shock wave is reflected from an opposite wall which is parallel to the upstream supersonic flow. Calculate the angle of the reflected shock relative to the straight wall.

15.47 Noting that expansion shocks do not exist in nature for a perfect gas, appeal to one of the oblique shock relations to show that the shock angle, β, is always larger than the Mach angle, $\mu = \sin^{-1}(1/M_1)$, where M_1 is the Mach number ahead of the shock.

15.48 Beginning with Equation (15.126), derive the θ-β-M relation as quoted in Equation (15.127). HINT: Use the following trigonometric identity.

$$\tan(\beta - \theta) = \frac{\tan \beta - \tan \theta}{1 + \tan \beta \tan \theta}$$

15.49 A Mach 2 flow of air at $p_\infty = 101$ kPa encounters a compression corner with $\theta = 13°$. A Pitot tube is located downstream of the corner. What is the pressure measured by the Pitot tube?

15.50 A Mach 1.8 flow of air at $p_\infty = 3$ psi encounters a compression corner with $\theta = 19°$. A Pitot tube is located downstream of the corner. What is the pressure measured by the Pitot tube?

15.51 For a given Prandtl-Meyer expansion, the upstream Mach number is 3 and the pressure ratio across the wave is $p_2/p_1 = 0.4$. Calculate the angles of the forward and rearward Mach lines of the expansion fan relative to the freestream direction. The gas is air.

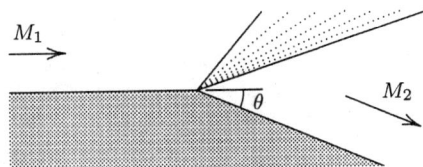

Problems 15.51, 15.52, 15.53, 15.54

15.52 The Mach number upstream of a Prandtl-Meyer expansion is 6 and the pressure ratio across the wave is 0.5. Compute the angles of the leading and trailing Mach lines of the expansion fan relative to the incident freestream direction. The gas is helium ($\gamma = 5/3$).

15.53 The Mach numbers upstream and downstream of a Prandtl-Meyer expansion are 3 and 5, respectively. What is the expansion angle, θ, and the pressure ratio across the wave. The gas is air.

15.54 The Mach number downstream of a Prandtl-Meyer expansion is 3. If the temperature ratio across the wave is 0.75, what is the Mach number upstream of the expansion? What is the expansion angle, θ? The gas is air.

15.55 Recall that the Prandtl-Meyer function, $\nu(M)$, depends on Mach number, M, as follows.

$$\nu(M) = \sqrt{\frac{\gamma + 1}{\gamma - 1}}\, \tan^{-1} \sqrt{\frac{\gamma - 1}{\gamma + 1}(M^2 - 1)} - \tan^{-1} \sqrt{M^2 - 1}$$

What is the limiting value of $\nu(M)$ as $M \to \infty$? Your answer should involve only γ. On the basis of this result, how would you propose to isentropically expand sonic flow ($M = 1$) to achieve an infinite Mach number, in principle? Assuming the incident sonic flow moves from left to right horizontally, sketch your proposed method, indicating any turning angles in degrees for carbon dioxide ($\gamma = 9/7$), air ($\gamma = 7/5$) and helium ($\gamma = 5/3$).

15.56 For Mach 4 flow of air past the body shown, we would like to determine the point where the shock and expansion fan at the corner first begin to interact. The wedge angle is $\theta = 12°$. Compute the values of x/L and y/L where the shock and leading Mach line of the expansion wave intersect.

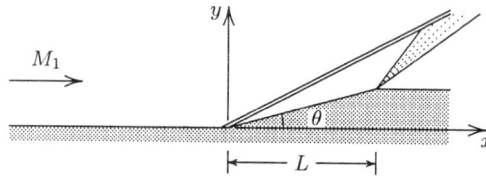

Problems 15.56, 15.57

15.57 For Mach 1.5 flow of air past the body shown, we would like to determine the point where the shock and expansion fan at the corner first begin to interact. The wedge angle is $\theta = 10°$. Compute the values of x/L and y/L where the shock and leading Mach line of the expansion wave intersect.

15.58 A Mach 3 stream of air with static pressure $p_1 = 120$ kPa is first expanded around a $20°$ corner, and then compressed through a second corner of the same angle as shown. Determine the Mach number and static pressure downstream of the compression corner.

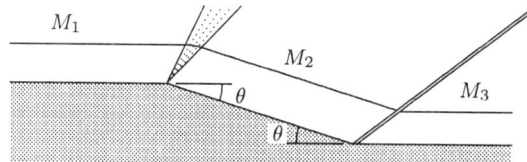

Problems 15.58, 15.59, 15.60

15.59 A Mach 4 stream of air with static pressure $p_1 = 12$ psi is first expanded around a $15°$ corner, and then compressed through a second corner of the same angle as shown. Determine the Mach number and static pressure downstream of the compression corner.

15.60 A Mach 3 stream of helium ($\gamma = 5/3$) with static pressure $p_1 = 1$ atm is first expanded around a $30°$ corner, and then compressed through a second corner of the same angle as shown. Determine the Mach number and static pressure downstream of the compression corner.

15.61 Using order-of-magnitude estimates in the x-momentum equation, we have shown that the thickness of an incompressible laminar boundary layer varies as $\delta \sim \sqrt{\nu x/U}$, where ν is kinematic viscosity, x is distance along the surface and U is freestream velocity. In a compressible boundary layer, this estimate is valid provided Mach number is not too large. However, we also have a thermal boundary layer whose thickness, δ_{th}, depends upon ν, x, U and the Prandtl number, Pr.

(a) Use order-of-magnitude estimates on the energy equation to determine δ_{th}.

(b) Compute δ_{th}/δ for air ($Pr = 0.72$), water ($Pr = 7$) and a cheap grade of oil ($Pr = 100$).

15.62 Compare a power-law representation of the viscosity of air, Equation (15.174) with $\omega = 0.7$, with *Sutherland's law*. Determine μ_o so that the formulas match when $T = 32°$ F. Display your results for $-200°$ F $\leq T \leq 4000°$ F.

15.63 For a low-speed, compressible, flat-plate boundary layer, the momentum and energy equations simplify to

$$\rho u \frac{\partial u}{\partial x} + \rho v \frac{\partial u}{\partial y} = \frac{\partial}{\partial y}\left(\mu \frac{\partial u}{\partial y}\right)$$

$$\rho u \frac{\partial h}{\partial x} + \rho v \frac{\partial h}{\partial y} = \frac{\partial}{\partial y}\left(\frac{\mu}{Pr} \frac{\partial h}{\partial y}\right)$$

If we assume the Prandtl number, Pr, is equal to 1, the enthalpy satisfies the same equation as the velocity. If T_w and T_e denote temperature at the surface and in the freestream, respectively, and if $h = c_p T$, then the temperature is given by

$$\frac{T - T_w}{T_e - T_w} = \frac{u}{u_e}$$

where u_e is freestream velocity. The *Nusselt number*, Nu_x, is defined by

$$Nu_x \equiv \frac{q_w x}{k(T_w - T_e)}$$

where q_w is surface heat flux and k is thermal conductivity. Verify that, under the assumptions above, with Re_x and c_f denoting Reynolds number based on x and skin friction, respectively,

$$Nu_x = \frac{1}{2} Re_x c_f$$

15.64 By definition, the *Nusselt number*, Nu_x, is

$$Nu_x \equiv \frac{q_w x}{k(T_w - T_e)}$$

where q_w is surface heat flux, x is distance along a surface and k is thermal conductivity. Also, T_w and T_e denote temperature at the surface and in the freestream, respectively. For low-speed, compressible, flat-plate boundary layers, correlation of measurements for laminar and turbulent boundary layers shows that

$$Nu_x = \begin{cases} 0.3320 \sqrt[3]{Pr}\, Re_x^{1/2}, & \text{laminar} \\ 0.0296 \sqrt[3]{Pr}\, Re_x^{4/5}, & \text{turbulent} \end{cases}$$

(a) Develop an equation for the total heat transfer (per unit width), Q, to a flat plate of length L for laminar and turbulent boundary layers, where

$$Q = \int_0^L q_w\, dx$$

For simplicity, assume the boundary is turbulent from the leading edge in the latter case.

(b) Determine Q for laminar and turbulent flow past a flat plate when $Re_L = 10^6$, $T_w - T_e = 10°$ F and each of the following fluids is used. Express your answers in Btu/(ft·sec).

Fluid	$k \left(\frac{\text{lb}}{\text{sec·}°\text{R}}\right)$	Pr
Mercury	1.166	0.023
Air	$3.20 \cdot 10^{-3}$	0.72
Water	$7.47 \cdot 10^{-2}$	6.88
Aircraft Engine Oil	$1.81 \cdot 10^{-2}$	9884

Problem 15.64

15.65 Estimate the laminar friction drag, C_{D_f}, for a slender, diamond-shaped airfoil moving at Mach 2. You can approximate the airfoil surface as a flat plate of length L. Use linear interpolation in Figure 15.22 to determine the variation of c_f with Re_x for a Mach 2 laminar boundary layer. Compare your result with the wave drag for $\theta = 5°$ [see Equation (15.166)] when $Re_L = 10^4$, 10^5 and 10^6.

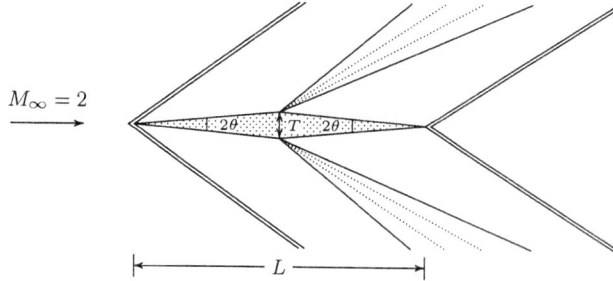

Problems 15.65, 15.66

15.66 Estimate the turbulent friction drag, C_{D_f}, for a slender, diamond-shaped airfoil moving at Mach 2. You can approximate the airfoil surface as a flat plate of length L. Use Figure 15.24 combined with $c_{f_o} \approx 0.0576 Re_x^{-1/5}$ to determine the skin friction for a Mach 2 turbulent boundary layer. Compare your result with the wave drag for $\theta = 5°$ [see Equation (15.166)] when $Re_L = 10^6$, 10^7 and 10^8.

15.67 For compressible boundary layers, the momentum integral equation is solved along with the *energy integral equation*, as follows.

$$\frac{d\theta}{dx} + \left(2 + H - M_e^2\right)\frac{\theta}{u_e}\frac{du_e}{dx} = \frac{c_f}{2}$$

$$\frac{d\theta_E}{dx} + \left[3 - (2 - \gamma)M_e^2\right]\frac{\theta_E}{u_e}\frac{du_e}{dx} = \frac{2}{\rho_e u_e^3}\int_0^\delta \mu\left(\frac{\partial u}{\partial y}\right)^2 dy$$

where the momentum thickness, θ, energy thickness, θ_E, and skin friction, c_f, are defined by

$$\theta \equiv \int_0^\delta \frac{\rho u}{\rho_e u_e}\left[1 - \frac{u}{u_e}\right]dy, \quad \theta_E \equiv \int_0^\delta \frac{\rho u}{\rho_e u_e}\left[1 - \left(\frac{u}{u_e}\right)^2\right]dy, \quad c_f \equiv \frac{2\mu_w}{\rho_e u_e^2}\left(\frac{\partial u}{\partial y}\right)_{y=0}$$

Subscripts w and e denote surface (wall) and freestream (edge), respectively. M_e is freestream Mach number and all other notation is as in the incompressible momentum integral equation. Analytical progress can be made with these equations by making the following transformation.

$$d\eta = \frac{\rho}{\rho_e}\frac{dy}{\delta_o}, \quad \delta_o = \int_0^\delta \frac{\rho}{\rho_e}dy, \quad \frac{\partial}{\partial y} = \frac{1}{\delta_o}\frac{\rho}{\rho_e}\frac{\partial}{\partial \eta}$$

Using this definition of the thickness function δ_o, $\eta = 1$ when $y = \delta$.

  (a) For a flat-plate boundary layer, assume a velocity profile of the form

$$u/u_e = \eta$$

  Determine the thickness function, δ_o, in terms of ρ_e, u_e, μ_e, x and $L_w \equiv \rho_w\mu_w/(\rho_e\mu_e)$.

  (b) Verify that for this solution to be self consistent, necessarily

$$\int_0^1 \frac{\rho\mu}{\rho_w\mu_w}d\eta = \frac{3}{4}$$

15.68 For compressible boundary layers, the momentum integral equation is solved along with the *energy integral equation*, as follows.

$$\frac{d\theta}{dx} + \left(2 + H - M_e^2\right) \frac{\theta}{u_e} \frac{du_e}{dx} = \frac{c_f}{2}$$

$$\frac{d\theta_E}{dx} + \left[3 - (2 - \gamma)M_e^2\right] \frac{\theta_E}{u_e} \frac{du_e}{dx} = \frac{2}{\rho_e u_e^3} \int_0^\delta \mu \left(\frac{\partial u}{\partial y}\right)^2 dy$$

where the momentum thickness, θ, energy thickness, θ_E, and skin friction, c_f, are defined by

$$\theta \equiv \int_0^\delta \frac{\rho u}{\rho_e u_e} \left[1 - \frac{u}{u_e}\right] dy, \quad \theta_E \equiv \int_0^\delta \frac{\rho u}{\rho_e u_e} \left[1 - \left(\frac{u}{u_e}\right)^2\right] dy, \quad c_f \equiv \frac{2\mu_w}{\rho_e u_e^2} \left(\frac{\partial u}{\partial y}\right)_{y=0}$$

Subscripts w and e denote surface (wall) and freestream (edge), respectively. M_e is freestream Mach number and all other notation is as in the incompressible momentum integral equation. Analytical progress can be made with these equations by making the following transformation.

$$d\eta = \frac{\rho}{\rho_e} \frac{dy}{\delta_o}, \quad \delta_o = \int_0^\delta \frac{\rho}{\rho_e} dy, \quad \frac{\partial}{\partial y} = \frac{1}{\delta_o} \frac{\rho}{\rho_e} \frac{\partial}{\partial \eta}$$

Using this definition of the thickness function δ_o, $\eta = 1$ when $y = \delta$.

(a) For a flat-plate boundary layer, assume a velocity profile of the form

$$u/u_e = 2\eta - \eta^2$$

Determine the thickness function, δ_o, in terms of ρ_e, u_e, μ_e, x and $L_w \equiv \rho_w \mu_w/(\rho_e \mu_e)$.

(b) Verify that for this solution to be self consistent, necessarily

$$\int_0^1 \frac{\rho\mu}{\rho_w \mu_w} (1 - \eta)^2 \, d\eta = \frac{11}{28}$$

15.69 Beginning with Equation (15.190) and the fact that $\mu_t d\tilde{u}/dy = \rho_w u_\tau^2$ in the log layer, derive the compressible law of the wall implied by the Cebeci-Smith turbulence model [see Equation (14.147)].

15.70 The object of this problem is to verify Equation (15.190). We begin with the fact that in the log layer of a compressible boundary layer, the convective terms are negligible so that the momentum and energy equations simplify to

$$0 \approx -\frac{d\overline{p}}{dx} + \frac{\partial}{\partial y} \left[(\overline{\mu} + \mu_t) \frac{\partial \tilde{u}}{\partial y}\right]$$

$$0 \approx u \frac{d\overline{p}}{dx} + \frac{\partial}{\partial y} \left[\left(\frac{\overline{\mu}}{Pr} + \frac{\mu_t}{Pr_t}\right) \frac{\partial \tilde{h}}{\partial y}\right] + (\overline{\mu} + \mu_t) \left(\frac{\partial \tilde{u}}{\partial y}\right)^2$$

(a) Multiply the momentum equation by $\tilde{u}$, add the resulting equation to the energy equation and integrate once to show that for an adiabatic wall

$$\left(\frac{\overline{\mu}}{Pr} + \frac{\mu_t}{Pr_t}\right) \frac{\partial \tilde{h}}{\partial y} + (\overline{\mu} + \mu_t) \frac{\partial}{\partial y} \left(\frac{\tilde{u}^2}{2}\right) \approx 0$$

(b) Now, use the fact that in the log layer, $\mu_t \gg \overline{\mu}$, to conclude that

$$\tilde{h} \approx \tilde{h}_w - \frac{Pr_t}{2} \tilde{u}^2$$

(c) Using appropriate thermodynamic relationships, verify Equation (15.190).

15.71 Determine the artificial viscosity, ν_a, and numerical-dispersion coefficient, $\mathcal{D}_a$, for the *Euler forward time and central space* (FTCS) algorithm applied to the one-dimensional wave equation,

$$\frac{u_i^{n+1} - u_i^n}{\Delta t} + a\,\frac{u_{i+1}^{n+1} - u_{i-1}^{n+1}}{2\Delta x} = 0$$

where a is a positive constant. **HINT:** Expand about $(i, n+1)$ so that, for example

$$u_i^n = u_i^{n+1} - u_t \Delta t + \frac{1}{2} u_{tt}(\Delta t)^2 - \frac{1}{6} u_{ttt}(\Delta t)^3 + \cdots$$

15.72 Determine the artificial viscosity, ν_a, and numerical-dispersion coefficient, $\mathcal{D}_a$, for *Lax's method* applied to the one-dimensional wave equation,

$$\frac{u_i^{n+1} - \frac{1}{2}\left(u_{i+1}^n - u_{i-1}^n\right)}{\Delta t} + a\,\frac{u_{i+1}^n - u_{i-1}^n}{2\Delta x} = 0$$

where a is a positive constant.

15.73 Determine the artificial viscosity, ν_a, and numerical-dispersion coefficient, $\mathcal{D}_a$, for the *Crank-Nicolson* algorithm applied to the one-dimensional wave equation,

$$\frac{u_i^{n+1} - u_i^n}{\Delta t} + a\,\frac{u_{i+1}^{n+1} + u_{i+1}^n - u_{i-1}^{n+1} - u_{i-1}^n}{4\Delta x} = 0$$

where a is a positive constant. **HINT:** Expand about $(i, n+1/2)$ and note that the Taylor-series expansion for $u_{i\pm1}^{n+1}$ is

$$
\begin{aligned}
u_{i\pm1}^{n+1} &= u_i^{n+1/2} + u_t(\Delta t/2) \pm u_x \Delta x + \frac{1}{2}\left[u_{tt}(\Delta t/2)^2 \pm 2u_{tx}\Delta x(\Delta t/2) + u_{xx}(\Delta x)^2\right] \\
&+ \frac{1}{6}\left[u_{ttt}(\Delta t/2)^3 \pm 3u_{ttx}\Delta x(\Delta t/2)^2 + 3u_{txx}(\Delta x)^2(\Delta t/2) \pm u_{xxx}(\Delta x)^3\right] + \cdots
\end{aligned}
$$

15.74 For isentropic flow with $p_t = 2130.6$ lb/ft^2 and $T_t = 529.39°$ R, make a graph of T/T_t as a function of Mach number, M, with specific-heat ratio, γ, as a parameter. M should vary from 0 to 20 and γ should assume values of 1.2, 1.4, 1.6 and 1.8. To make this problem less tedious, you might want to check out the Freestream Properties and Gas Properties menus in Program **SETEBL**, the input-data preparation utility for Program **EDDYBL** (see Appendix F).

15.75 The purpose of this problem is to determine the difference between first- and second-order accurate finite-difference schemes regarding solution accuracy. Run Program **BURGER** (see Appendix E) for a propagating shock and record the values of the velocity after 33 timesteps corresponding to a dimensionless time $t = 1.35$. Now, modify the program so that only the predictor step is used. This corresponds to the first-order accurate scheme of Equation (15.199). **NOTE:** Be sure to add a loop that sets $u(i,1) = u(i,2)$ after advancing the solution in time. Compare the solution after 30 timesteps, again corresponding to a dimensionless time $t = 1.35$ with that of MacCormack's method. How has the first-order method's artificial viscosity affected the solution?

15.76 The purpose of this problem is to determine the difference between first- and second-order accurate finite-difference schemes regarding solution accuracy. Run Program **BURGER** (see Appendix E) for a propagating expansion and record the values of the velocity after 30 timesteps corresponding to a dimensionless time $t = 1.35$. Now, modify the program so that only the predictor step is used. This corresponds to the first-order accurate scheme of Equation (15.199). **NOTE:** Be sure to add a loop that sets $u(i,1) = u(i,2)$ after advancing the solution in time. Compare the solution after 30 timesteps, again corresponding to a dimensionless time $t = 1.35$ with that of MacCormack's method. How has the first-order method's artificial viscosity affected the solution?

15.77 Use Program **EDDYBL** (see Appendix F) to determine the effect of Mach number on laminar, flat-plate boundary-layer skin friction. Assume the viscosity of air follows a power law of the form $\mu = AT^{0.75}$. Proceed as follows.

(a) Determine the value of A by insisting that the viscosity match the value given by Sutherland's law when $T = 68°$ F.

(b) Do computations for the following Mach numbers, using the listed total pressure. The values all correspond to a freestream static pressure of 2116.9 lb/ft^2. This permits using most of the default values and the files **presur.dat** and **heater.dat** supplied on the diskette accompanying this book.

M_∞	p_{t_∞} (lb/ft^2)
.096	$2.1306 \cdot 10^3$
1	$4.0071 \cdot 10^3$
2	$1.6564 \cdot 10^4$
5	$1.1200 \cdot 10^5$

(c) Using **SETEBL**, change the appropriate input parameters to accomplish the following: use USCS units (IUTYPE); use an initial stepsize $\Delta s = 0.0001$ ft (DS); set the viscosity law according to the value of A computed in Part (a) (SU, VISCON, VISPOW) — note that VISPOW = 1.75; select laminar flow by setting MODEL = -1; set the initial arc length $s_i = 0$ m (SI).

(d) From the values of Re_x and c_f at the final station in each computation, compute $c_f\sqrt{Re_x}$. Make a graph of your results as a function of M_∞.

15.78 The object of this problem is to compare predictions of modern turbulence models with measured properties of a Mach 2.65 turbulent boundary layer with adverse pressure gradient and surface heat transfer. The experiment to be simulated was conducted by Fernando and Smits [see Fernholz and Finley (1981)]. Use Program **EDDYBL** and its menu-driven setup utility, Program **SETEBL**, to do the computations (see Appendix F).

(a) Using **SETEBL**, change the appropriate input parameters to accomplish the following: use SI units (IUTYPE); set the freestream conditions to $p_{t_\infty} = 6.7221 \cdot 10^5$ N/m^2, $T_{t_\infty} = 270.74$ K, $M_\infty = 2.653$ (PT1, TT1, XMA); use an initial stepsize $\Delta s = 0.02$ m (DS); set the initial boundary-layer properties so that $c_f = 0.000992$, $\delta = 0.02249$ m, $H = 5.3056$, $Re_\theta = 88483$, $s_i = 1.151$ m (CF, DELTA, H, RETHET, SI); set the maximum arc length to $s_f = 1.361$ m and maximum number of steps to 200 (SSTOP, IEND1); set up for $N = 12$ points to define the pressure and set up for prescribed surface temperature (NUMBER, KODWAL).

(b) Use the following data to define the pressure distribution in a file named **presur.dat**. The initial and final pressure gradients are both zero.

s (m)	p_e (N/m^2)	s (m)	p_e (N/m^2)	s (m)	p_e (N/m^2)	s (m)	p_e (N/m^2)
0.000	$3.1039 \cdot 10^4$	1.172	$3.1500 \cdot 10^4$	1.248	$4.2870 \cdot 10^4$	1.324	$4.1000 \cdot 10^4$
1.000	$3.1039 \cdot 10^4$	1.197	$3.5839 \cdot 10^4$	1.273	$4.3650 \cdot 10^4$	1.349	$3.9630 \cdot 10^4$
1.151	$3.0490 \cdot 10^4$	1.222	$3.9890 \cdot 10^4$	1.299	$4.2510 \cdot 10^4$	1.361	$4.0170 \cdot 10^4$

(c) Use the following data to define the surface-temperature distribution in a file named **heater.dat** along with zero heat flux. The initial and final temperature gradients are both zero.

s (m)	T_w (K)	s (m)	T_w (K)	s (m)	T_w (K)	s (m)	T_w (K)
0.000	281.56	1.172	285.76	1.248	276.94	1.324	281.49
1.000	281.56	1.197	281.97	1.273	279.32	1.349	284.48
1.151	281.56	1.222	279.99	1.299	280.53	1.361	270.92

(d) Do three computations using the k-ω model, one of the k-ϵ models and any other model.

(e) Compare computed skin friction with the following measured values.

s (m)	c_f	s (m)	c_f	s (m)	c_f	s (m)	c_f
1.151	$9.92 \cdot 10^{-4}$	1.222	$9.43 \cdot 10^{-4}$	1.299	$1.01 \cdot 10^{-3}$	1.361	$1.04 \cdot 10^{-3}$
1.172	$9.96 \cdot 10^{-4}$	1.248	$9.46 \cdot 10^{-4}$	1.324	$1.07 \cdot 10^{-3}$		
1.197	$9.67 \cdot 10^{-4}$	1.273	$9.41 \cdot 10^{-4}$	1.349	$1.08 \cdot 10^{-3}$		

15.79 The object of this problem is to compare predictions of modern turbulence models with measured properties of a Mach 2.2 flat-plate turbulent boundary layer. The experiment to be simulated was conducted by Shutts [see Fernholz and Finley (1981)]. Use Program **EDDYBL** and its menu-driven setup utility, Program **SETEBL**, to do the computations (see Appendix F).

(a) Using **SETEBL**, change the appropriate input parameters to accomplish the following: use USCS units (IUTYPE); set the freestream conditions to p_{t_∞} = 5297.382 lb/ft^2, T_{t_∞} = 609.354° R, M_∞ = 2.244 (PT1, TT1, XMA); use an initial stepsize Δs = 0.001 ft (DS); start from laminar conditions by setting IBOUND = 0 and SI = 0 ft; set the maximum arc length to s_f = 3.02 ft and maximum number of steps to 1000 (SSTOP, IEND1).

(b) Prepare a file named **presur.dat** with constant freestream pressure, p_e = 462.4495 lb/ft^2. Also, prepare a file **heater.dat** with constant wall temperature, T_w = 575.7° R, and zero heat flux.

(c) Do three computations using the k-ω model, the Launder-Sharma k-ϵ model and any other model. (Avoid the Lam-Bremhorst, Chien, Yang-Shih and Fan et al. k-ϵ models — they are very difficult to implement numerically for transitional flows.)

(d) Compare computed velocity profiles with the following measured values. Also, compare to the measured skin friction at s = 3.02 ft, which is c_f = 0.00162.

y^+	u^*/u_τ	y^+	u^*/u_τ	y^+	u^*/u_τ	y^+	u^*/u_τ
$6.1100 \cdot 10^1$	16.056	$2.2402 \cdot 10^2$	17.894	$6.7206 \cdot 10^2$	21.360	$2.1995 \cdot 10^3$	26.445
$7.4670 \cdot 10^1$	16.069	$2.7900 \cdot 10^2$	19.218	$8.4178 \cdot 10^2$	22.098	$2.8776 \cdot 10^3$	27.749
$8.7570 \cdot 10^1$	16.030	$3.3197 \cdot 10^2$	19.064	$1.0115 \cdot 10^3$	22.764	$3.5573 \cdot 10^3$	28.056
$1.1540 \cdot 10^2$	16.030	$4.0052 \cdot 10^2$	19.838	$1.1812 \cdot 10^3$	23.423	$4.2367 \cdot 10^3$	28.081
$1.4420 \cdot 10^2$	16.030	$4.6841 \cdot 10^2$	20.580	$1.5200 \cdot 10^3$	24.527	$4.9150 \cdot 10^3$	28.105
$1.8261 \cdot 10^2$	16.961	$5.7090 \cdot 10^2$	20.962	$1.8607 \cdot 10^3$	25.544		

15.80 The object of this problem is to compare predictions of modern turbulence models with measured properties of a Mach 4.5 flat-plate turbulent boundary layer. The experiment to be simulated was conducted by Coles [see Fernholz and Finley (1981)]. Use Program **EDDYBL** and its menu-driven setup utility, Program **SETEBL**, to do the computations (see Appendix F).

(a) Using **SETEBL**, change the appropriate input parameters to accomplish the following: use USCS units (IUTYPE); set the freestream conditions to p_{t_∞} = 8409.111 lb/ft^2, T_{t_∞} = 563° R, M_∞ = 4.544 (PT1, TT1, XMA); use an initial stepsize Δs = 0.001 ft (DS); start from laminar conditions by setting IBOUND = 0 and SI = 0 ft; set the maximum arc length to s_f = 1.90 ft and maximum number of steps to 1000 (SSTOP, IEND1); and, set the freestream turbulence energy to k_e = 5 · 10^{-4} (ZIOTAE).

(b) Prepare a file named **presur.dat** with constant freestream pressure, p_e = 27.50813 lb/ft^2. Also, prepare a file **heater.dat** with constant wall temperature, T_w = 513.1° R, and zero heat flux.

(c) Do three computations using the k-ω model, the Baldwin-Barth model any other model. (Avoid the k-ϵ models — they are very difficult to implement numerically for transitional flows.)

(d) Compare computed velocity profiles with the following measured values. Also, compare to the measured skin friction at s = 1.90 ft, which is c_f = 0.00126.

y^+	u^*/u_τ	y^+	u^*/u_τ	y^+	u^*/u_τ	y^+	u^*/u_τ
$1.4420 \cdot 10^1$	10.295	$3.5590 \cdot 10^1$	13.848	$9.1470 \cdot 10^1$	16.559	$2.3909 \cdot 10^3$	20.951
$1.7100 \cdot 10^1$	10.972	$4.2930 \cdot 10^1$	14.465	$1.1099 \cdot 10^2$	17.258	$2.8953 \cdot 10^3$	21.951
$2.0380 \cdot 10^1$	11.713	$5.1570 \cdot 10^1$	14.990	$1.3466 \cdot 10^2$	18.052	$3.5196 \cdot 10^3$	22.523
$2.4440 \cdot 10^1$	12.456	$6.2450 \cdot 10^1$	15.472	$1.6282 \cdot 10^2$	18.943	$4.2800 \cdot 10^3$	22.540
$2.6230 \cdot 10^1$	13.182	$7.5510 \cdot 10^1$	15.968	$1.9650 \cdot 10^2$	19.893		

Appendix A

Dimensions and Units

To describe physical properties in quantitative terms, we must select a system of units for expressing their dimensions. There are several systems of units used in general engineering practice, and this appendix describes four of the commonly used systems. Of these four, this text almost exclusively uses Standard International units and one of the subsets of the U. S. Customary System of units. In the text, we refer to these two sets of units as SI and USCS, respectively. Because of their inherent simplicity, SI units are becoming the standard in many branches of engineering. Several engineering journals, for example, require authors who use USCS units to parenthetically include corresponding values in SI units.

A.1 Independent Dimensions

The first thing we must decide upon is what dimensions are independent of all others, i.e., which are most basic in some sense. Just as the spatial coordinates x, y and z are independent variables in a three-dimensional rectangular Cartesian coordinate system, so we must choose a set of **independent dimensions**. Any other dimensions are then expressed in terms of the independent dimensions, and are called **secondary dimensions**.

For engineering work, we select either mass, length, time and temperature or force, length, time and temperature as the independent dimensions. In either case, Newton's second law of motion provides the defining relationship between mass and force, regardless of which is chosen as the independent dimension.

A.2 Common Systems of Units

For the sake of uniform standards, various groups and countries have established systems of units and associated independent dimensions. Part of the objective in establishing these standards has been to promote use of **consistent units**. This means there are no special conversion factors required in a physically-based equation to make it dimensionally homogeneous, i.e., to make all terms in the equation have the same dimensions. Three of the four systems discussed in this section satisfy this constraint. As an interesting counter example, one of the systems, viz., the English Engineering system, violates this objective with regard to Newton's second law of motion.

The two most prevalent types of units are **Metric Units** and the **U. S. Customary System (USCS)**. Within each type of units, there are two sub-types. In the case of Metric Units, there is the **Standard International (SI)** system and the **Centimeter-Gram-Second (CGS)** system. Table A.1 lists mass, length, time, temperature and force in the SI and CGS systems of units.

Table A.1: *Metric Units*

Dimension	Standard International (SI)	Centimeter-Gram-Second (CGS)
Mass	kilogram (kg)	gram (g)
Length	meter (m)	centimeter (cm)
Time	second (sec)	second (sec)
Temperature	Kelvin (K)	Kelvin (K)
Force	Newton (N)	dyne (dyne)

For the USCS, the sub-types are known as the **British Gravitational (BG)** system and the **English Engineering (EE)** system. Table A.2 lists mass, length, time, temperature and force in the BG and EE systems of units. This text exclusively uses the SI system and the BG subset of the U. S. Customary System. Because of its unusual nature, the text completely avoids the EE system.

Table A.2: *U. S. Customary System of Units (USCS)*

Dimension	British Gravitational (BG)	English Engineering (EE)
Mass	slug (slug)	pound-mass (lbm)
Length	foot (ft)	foot (ft)
Time	second (sec)	second (sec)
Temperature	° Rankine (°R)	° Rankine (°R)
Force	pound (lb)	pound-force (lbf)

A.2.1   Standard International Units

The independent dimensions in the SI system[1] are as follows. The unit of mass is the kilogram (kg), the unit of length is the meter (m), the unit of time is the second (sec) and the unit of temperature is the Kelvin (K). The temperature expressed in Kelvins is absolute temperature, and is related to the commonly used Celsius, or centigrade, scale (°C) as follows.

$$K = {}^\circ C + 273.16 \tag{A.1}$$

The most important secondary dimension for applications using the laws of physics is force. The unit of force is called the Newton (N), which is defined as the force required to give a mass of 1 kilogram an acceleration of 1 meter per second per second. In dimensional terms, we say

$$1 \text{ N} = (1 \text{ kg}) \cdot (1 \text{ m/sec}^2) = 1 \text{ kg} \cdot \text{m/sec}^2 \tag{A.2}$$

---

[1]In older texts, this system is often referred to as the MKS system, which is an acronym for meter-kilogram-second. It's modern name is also Système Internationale.

Note that, since the acceleration of gravity is 9.807 m/sec^2, this means the weight of a 1 kilogram mass is 9.807 Newtons. Note also that when the 'weight' of an object is expressed in kilograms, what is actually being quantified is the object's mass.

Two other important secondary dimensions are those of work and power. In the SI system, the units of work and power are the Joule (J) and the Watt (W), respectively. A Joule is the amount of work done when a 1 Newton force produces a displacement of 1 meter in the direction of the force. A Watt is 1 Joule per second. Algebraically, we say

$$\left.\begin{array}{l} 1\ \text{J} = 1\ \text{N} \cdot \text{m} \\[2mm] 1\ \text{W} = 1\ \text{J/sec} = 1\ \text{N} \cdot \text{m/sec} \end{array}\right\} \tag{A.3}$$

A.2.2 Centimeter-Gram-Second Units

In the CGS system, we select the units of mass, length, time and temperature to be the gram (g), the centimeter (cm), the second (sec) and the Kelvin (K). The unit of force is called the dyne (dyne), which is defined as the force required to give a mass of 1 gram an acceleration of 1 centimeter per second per second. Hence,

$$1\ \text{dyne} = (1\ \text{g}) \cdot (1\ \text{cm/sec}^2) = 1\ \text{g} \cdot \text{cm/sec}^2 = 10^{-5}\ \text{N} \tag{A.4}$$

Because the acceleration of gravity is 980.7 cm/sec^2, this means the weight of a 1 gram mass is 980.7 dynes.

In the CGS system, the unit of work is the erg (erg), while power is expressed in terms of ergs per second (erg/sec). An erg is defined as the amount of work done when a 1 dyne force produces a displacement of 1 centimeter in the direction of the force.

Table A.3 lists the prefixes used with metric units.

Table A.3: *Metric-Unit Prefixes*

Factor			Prefix	Symbol
1,000,000,000,000	=	10^{12}	tera	T
1,000,000,000	=	10^{9}	giga	G
1,000,000	=	10^{6}	mega	M
1,000	=	10^{3}	kilo	k
100	=	10^{2}	hecto	h
10	=	10^{1}	deka	da
0.1	=	10^{-1}	deci	d
0.01	=	10^{-2}	centi	c
0.001	=	10^{-3}	milli	m
0.000001	=	10^{-6}	micro	μ
0.000000001	=	10^{-9}	nano	n
0.000000000001	=	10^{-12}	pico	p

A.2.3   British Gravitational Units

In the BG system, the independent dimensions are force, length, time and temperature. The unit of force is the pound (lb), the unit of length is the foot (ft), the unit of time is the second (sec) and the unit of temperature is the degree Rankine (°R). The temperature expressed in degrees Rankine is absolute temperature, and is related to the Fahrenheit scale (°F) according to

$$°R = °F + 459.67°$$ (A.5)

The unit of mass is called the slug (slug), which is defined as the mass upon which application of a 1 pound force will cause an acceleration of 1 foot per second per second. Thus,

$$1 \text{ lb} = (1 \text{ slug}) \cdot (1 \text{ ft/sec}^2) = 1 \text{ slug} \cdot \text{ft/sec}^2$$ (A.6)

The acceleration of gravity in BG units is 32.174 ft/sec^2, so that a mass of 1 slug weighs 32.174 pounds.

In the BG system, the units of work and power are the British thermal unit (Btu) and the horsepower, respectively. A Btu is the amount of heat required to raise the temperature of one pound of water (at its maximum density) by 1 degree Rankine. It is equal to 1,055 Joules. A horsepower is numerically equal to the rate of 33,000 foot-pounds of work per minute (= 550 foot-pounds per second).[2]

A.2.4   English Engineering Units

In the EE system, the units for mass and force are defined separately. The unit of force is the pound-force (lbf), while the unit of mass is the pound-mass (lbm). As with the BG system, the units of length, time and temperature are the foot, the second and the degree Rankine, respectively. This adds the complication that Newton's second law of motion must be rewritten as

$$\mathbf{F} = \frac{m\mathbf{a}}{g_c}$$ (A.7)

where $\mathbf{F}$ is force, m is mass, $\mathbf{a}$ is acceleration and g_c is a dimensional constant determined as follows. We define the pound-force and the pound-mass such that application of a 1 lbf force to a 1 lbm mass yields an acceleration equal to the acceleration of gravity on Earth, 32.174 ft/sec^2. Hence,

$$1 \text{ lbf} = \frac{1 \text{ lbm} \cdot 32.174 \text{ ft/sec}^2}{g_c} \quad \Longrightarrow \quad g_c = 32.174 \frac{\text{ft} \cdot \text{lbm}}{\text{lbf} \cdot \text{sec}^2}$$ (A.8)

Because of the inconvenience of having a dimensional parameter, g_c, appearing in a basic law of physics, EE units are inconvenient, and find use mainly when convention dictates.

---

[2]The term horsepower was originated by Boulton and Watt to state the power of their steam engines. In a practical test, it was found that the average horse could work constantly at the rate of 22,000 foot-pounds per minute. This was increased by one half in defining this arbitrary, and now universal, unit of power.

A.3 Converting Dimensional Quantities

For the discussion to follow, it is convenient to identify the independent dimensions in algebraic terms. We will refer to mass as M, length as L, time as T, temperature as Θ, and force as F. As noted above, secondary dimensions are related by law or by definition to the independent dimensions. For instance, the dimensional representation of velocity, U, is

$$[U] = \frac{L}{T} \qquad (A.9)$$

where $[U]$ means dimensions of U. Similarly, pressure has dimensions F/L^2 and acceleration is expressed dimensionally as L/T^2.

Converting from one system of units to another is straightforward if some simple algebraic rules are followed. For example, Table A.4 tells us that 1 pound is equal to 4.448 Newtons. Also, 1 foot is 0.3048 meters. Hence, to convert from pressure expressed in lb/ft^2 to N/m^2, we proceed as follows.

$$[p] = \frac{F}{L^2} \quad \Longrightarrow \quad 1 \, \frac{\text{lb}}{\text{ft}^2} = \frac{4.448 \, \text{N}}{(0.3048 \, \text{m})^2} = 47.88 \, \text{N/m}^2 \qquad (A.10)$$

Thus, we conclude that

$$1 \, \text{lb/ft}^2 = 47.88 \, \text{N/m}^2 \qquad (A.11)$$

Alternatively, we can form the ratios of independent or secondary dimensions, and use them as dimensional multiplicative factors with a magnitude of unity. That is, we can say

$$\left(\frac{4.448 \, \text{N}}{1 \, \text{lb}} \right) = 1 \qquad (A.12)$$

and

$$\left(\frac{0.3048 \, \text{m}}{1 \, \text{ft}} \right) = 1 \qquad (A.13)$$

We can now use these factors in a strict algebraic sense to convert from the USCS system to the SI system as follows.

$$1 \, \frac{\text{lb}}{\text{ft}^2} = \frac{\text{lb} \cdot \left(4.448 \, \frac{\text{N}}{\text{lb}} \right)}{\text{ft}^2 \cdot \left(0.3048 \, \frac{\text{m}}{\text{ft}} \right)^2} = 47.88 \, \frac{\text{N}}{\text{m}^2} \qquad (A.14)$$

Table A.4 lists conversion factors for properties commonly encountered in fluid mechanics.

Table A.4: *Conversion Table*

Mass				Velocity			
	1 lbm	=	0.454 kg		1 mph	=	1.467 ft/sec
	1 slug	=	14.594 kg		1 mph	=	0.447 m/sec
	1 oz	=	28.35 g		1 knot	=	1.688 ft/sec
	1 kg	=	2.205 lbm		1 knot	=	0.514 m/sec
Length				Work			
	1 ft	=	0.3048 m		1 Btu	=	778 ft·lb
	1 in	=	25.4 mm		1 ft·lb	=	1.3558 J
	1 mi	=	5280 ft		1 Btu	=	1055 J
Temperature	°C	=	$\frac{5}{9}(°F-32°)$	Power			
	K	=	°C + 273.16		1 hp	=	550 ft·lb/sec
	°R	=	$\frac{9}{5}K$		1 hp	=	2545 Btu/hr
	°R	=	°F + 459.67°		1 kW	=	1.341 hp
					1 J/sec	=	1 W
Force				Volume			
	1 lb	=	4.448 N		1 gal	=	0.0037854 m^3
	1 kN	=	224.8 lb		1 gal	=	231 in^3
	1 ton	=	2000 lb		1 gal	=	0.134 ft^3
	1 ton	=	8.897 kN		1 L	=	0.001 m^3
Pressure				Rotation			
	1 psi	=	6.895 kPa		1 rpm	=	0.1047 sec^{-1}
	1 psf	=	47.88 Pa		1 Hz	=	2π sec^{-1}
	1 atm	=	101 kPa		1 cps	=	2π sec^{-1}
	1 atm	=	2116.8 psf		1 rev	=	2π radians
	1 atm	=	760 mmHg				
Gravity	g	=	32.174 ft/sec^2				
	g	=	9.807 m/sec^2				

Appendix B

Compressible Flow Tables

This appendix includes several tables that are valid for gases with a specific-heat ratio, $\gamma = 1.4$.

B.1 Isentropic Flow and Normal-Shock Relations

Notation used in the following tables is as follows.

Isentropic flow:

M	=	local Mach number
p/p_t	=	ratio of static to total pressure
ρ/ρ_t	=	ratio of static to total density
T/T_t	=	ratio of static to total temperature
A/A^*	=	ratio of local area to reference sonic-flow area

Normal shock waves:

M_1	=	Mach number ahead of the shock wave
M_2	=	Mach number behind the shock wave
p_2/p_1	=	pressure ratio across the shock wave
T_2/T_1	=	temperature ratio across the shock wave
p_{t2}/p_{t1}	=	total pressure ratio across the shock wave

Subsonic Flow of a Perfect Gas with $\gamma = 1.4$

M	p/p_t	ρ/ρ_t	T/T_t	A/A^*	M	p/p_t	ρ/ρ_t	T/T_t	A/A^*
0.00	1.0000	1.0000	1.0000	∞	0.50	0.8430	0.8852	0.9524	1.3398
0.02	0.9997	0.9998	0.9999	28.9421	0.52	0.8317	0.8766	0.9487	1.3034
0.04	0.9989	0.9992	0.9997	14.4815	0.54	0.8201	0.8679	0.9449	1.2703
0.06	0.9975	0.9982	0.9993	9.6659	0.56	0.8082	0.8589	0.9410	1.2403
0.08	0.9955	0.9968	0.9987	7.2616	0.58	0.7962	0.8498	0.9370	1.2130
0.10	0.9930	0.9950	0.9980	5.8218	0.60	0.7840	0.8405	0.9328	1.1882
0.12	0.9900	0.9928	0.9971	4.8643	0.62	0.7716	0.8310	0.9286	1.1656
0.14	0.9864	0.9903	0.9961	4.1824	0.64	0.7591	0.8213	0.9243	1.1451
0.16	0.9823	0.9873	0.9949	3.6727	0.66	0.7465	0.8115	0.9199	1.1265
0.18	0.9776	0.9840	0.9936	3.2779	0.68	0.7338	0.8016	0.9153	1.1097
0.20	0.9725	0.9803	0.9921	2.9635	0.70	0.7209	0.7916	0.9107	1.0944
0.22	0.9668	0.9762	0.9904	2.7076	0.72	0.7080	0.7814	0.9061	1.0806
0.24	0.9607	0.9718	0.9886	2.4956	0.74	0.6951	0.7712	0.9013	1.0681
0.26	0.9541	0.9670	0.9867	2.3173	0.76	0.6821	0.7609	0.8964	1.0570
0.28	0.9470	0.9619	0.9846	2.1656	0.78	0.6691	0.7505	0.8915	1.0471
0.30	0.9395	0.9564	0.9823	2.0351	0.80	0.6560	0.7400	0.8865	1.0382
0.32	0.9315	0.9506	0.9799	1.9219	0.82	0.6430	0.7295	0.8815	1.0305
0.34	0.9231	0.9445	0.9774	1.8229	0.84	0.6300	0.7189	0.8763	1.0237
0.36	0.9143	0.9380	0.9747	1.7358	0.86	0.6170	0.7083	0.8711	1.0179
0.38	0.9052	0.9313	0.9719	1.6587	0.88	0.6041	0.6977	0.8659	1.0129
0.40	0.8956	0.9243	0.9690	1.5901	0.90	0.5913	0.6870	0.8606	1.0089
0.42	0.8857	0.9170	0.9659	1.5289	0.92	0.5785	0.6764	0.8552	1.0056
0.44	0.8755	0.9094	0.9627	1.4740	0.94	0.5658	0.6658	0.8498	1.0031
0.46	0.8650	0.9016	0.9594	1.4246	0.96	0.5532	0.6551	0.8444	1.0014
0.48	0.8541	0.8935	0.9559	1.3801	0.98	0.5407	0.6445	0.8389	1.0003
0.50	0.8430	0.8852	0.9524	1.3398	1.00	0.5283	0.6339	0.8333	1.0000

Supersonic Flow of a Perfect Gas with $\gamma = 1.4$

	Isentropic Flow				Normal Shock Wave				
M_1	p/p_t	ρ/ρ_t	T/T_t	A/A^*	M_2	p_2/p_1	ρ_2/ρ_1	T_2/T_1	p_{t2}/p_{t1}
1.00	0.5283	0.6339	0.8333	1.0000	1.0000	1.0000	1.0000	1.0000	1.0000
1.02	0.5160	0.6234	0.8278	1.0003	0.9805	1.0471	1.0334	1.0132	1.0000
1.04	0.5039	0.6129	0.8222	1.0013	0.9620	1.0952	1.0671	1.0263	0.9999
1.06	0.4919	0.6024	0.8165	1.0029	0.9444	1.1442	1.1009	1.0393	0.9998
1.08	0.4800	0.5920	0.8108	1.0051	0.9277	1.1941	1.1349	1.0522	0.9994
1.10	0.4684	0.5817	0.8052	1.0079	0.9118	1.2450	1.1691	1.0649	0.9989
1.12	0.4568	0.5714	0.7994	1.0113	0.8966	1.2968	1.2034	1.0776	0.9982
1.14	0.4455	0.5612	0.7937	1.0153	0.8820	1.3495	1.2378	1.0903	0.9973
1.16	0.4343	0.5511	0.7879	1.0198	0.8682	1.4032	1.2723	1.1029	0.9961
1.18	0.4232	0.5411	0.7822	1.0248	0.8549	1.4578	1.3069	1.1154	0.9946
1.20	0.4124	0.5311	0.7764	1.0304	0.8422	1.5133	1.3416	1.1280	0.9928
1.22	0.4017	0.5213	0.7706	1.0366	0.8300	1.5698	1.3764	1.1405	0.9907
1.24	0.3912	0.5115	0.7648	1.0432	0.8183	1.6272	1.4112	1.1531	0.9884
1.26	0.3809	0.5019	0.7590	1.0504	0.8071	1.6855	1.4460	1.1657	0.9857
1.28	0.3708	0.4923	0.7532	1.0581	0.7963	1.7448	1.4808	1.1783	0.9827
1.30	0.3609	0.4829	0.7474	1.0663	0.7860	1.8050	1.5157	1.1909	0.9794
1.32	0.3512	0.4736	0.7416	1.0750	0.7760	1.8661	1.5505	1.2035	0.9758
1.34	0.3417	0.4644	0.7358	1.0842	0.7664	1.9282	1.5854	1.2162	0.9718
1.36	0.3323	0.4553	0.7300	1.0940	0.7572	1.9912	1.6202	1.2290	0.9676
1.38	0.3232	0.4463	0.7242	1.1042	0.7483	2.0551	1.6549	1.2418	0.9630
1.40	0.3142	0.4374	0.7184	1.1149	0.7397	2.1200	1.6897	1.2547	0.9582
1.42	0.3055	0.4287	0.7126	1.1262	0.7314	2.1858	1.7243	1.2676	0.9531
1.44	0.2969	0.4201	0.7069	1.1379	0.7235	2.2525	1.7589	1.2807	0.9476
1.46	0.2886	0.4116	0.7011	1.1501	0.7157	2.3202	1.7934	1.2938	0.9420
1.48	0.2804	0.4032	0.6954	1.1629	0.7083	2.3888	1.8278	1.3069	0.9360
1.50	0.2724	0.3950	0.6897	1.1762	0.7011	2.4583	1.8621	1.3202	0.9298
1.52	0.2646	0.3869	0.6840	1.1899	0.6941	2.5288	1.8963	1.3336	0.9233
1.54	0.2570	0.3789	0.6783	1.2042	0.6874	2.6002	1.9303	1.3470	0.9166
1.56	0.2496	0.3710	0.6726	1.2190	0.6809	2.6725	1.9643	1.3606	0.9097
1.58	0.2423	0.3633	0.6670	1.2344	0.6746	2.7458	1.9981	1.3742	0.9026
1.60	0.2353	0.3557	0.6614	1.2502	0.6684	2.8200	2.0317	1.3880	0.8952
1.62	0.2284	0.3483	0.6558	1.2666	0.6625	2.8951	2.0653	1.4018	0.8877
1.64	0.2217	0.3409	0.6502	1.2836	0.6568	2.9712	2.0986	1.4158	0.8799
1.66	0.2151	0.3337	0.6447	1.3010	0.6512	3.0482	2.1318	1.4299	0.8720
1.68	0.2088	0.3266	0.6392	1.3190	0.6458	3.1261	2.1649	1.4440	0.8639
1.70	0.2026	0.3197	0.6337	1.3376	0.6405	3.2050	2.1977	1.4583	0.8557
1.72	0.1966	0.3129	0.6283	1.3567	0.6355	3.2848	2.2304	1.4727	0.8474
1.74	0.1907	0.3062	0.6229	1.3764	0.6305	3.3655	2.2629	1.4873	0.8389
1.76	0.1850	0.2996	0.6175	1.3967	0.6257	3.4472	2.2952	1.5019	0.8302
1.78	0.1794	0.2931	0.6121	1.4175	0.6210	3.5298	2.3273	1.5167	0.8215
1.80	0.1740	0.2868	0.6068	1.4390	0.6165	3.6133	2.3592	1.5316	0.8127

Supersonic Flow of a Perfect Gas with $\gamma = 1.4$ (continued)

	Isentropic Flow				Normal Shock Wave				
M_1	p/p_t	ρ/ρ_t	T/T_t	A/A^*	M_2	p_2/p_1	ρ_2/ρ_1	T_2/T_1	p_{t2}/p_{t1}
1.80	0.1740	0.2868	0.6068	1.4390	0.6165	3.6133	2.3592	1.5316	0.8127
1.82	0.1688	0.2806	0.6015	1.4610	0.6121	3.6978	2.3909	1.5466	0.8038
1.84	0.1637	0.2745	0.5963	1.4836	0.6078	3.7832	2.4224	1.5617	0.7948
1.86	0.1587	0.2686	0.5910	1.5069	0.6036	3.8695	2.4537	1.5770	0.7857
1.88	0.1539	0.2627	0.5859	1.5308	0.5996	3.9568	2.4848	1.5924	0.7765
1.90	0.1492	0.2570	0.5807	1.5553	0.5956	4.0450	2.5157	1.6079	0.7674
1.92	0.1447	0.2514	0.5756	1.5804	0.5918	4.1341	2.5463	1.6236	0.7581
1.94	0.1403	0.2459	0.5705	1.6062	0.5880	4.2242	2.5767	1.6394	0.7488
1.96	0.1360	0.2405	0.5655	1.6326	0.5844	4.3152	2.6069	1.6553	0.7395
1.98	0.1318	0.2352	0.5605	1.6597	0.5808	4.4071	2.6369	1.6713	0.7302
2.00	0.1278	0.2300	0.5556	1.6875	0.5774	4.5000	2.6667	1.6875	0.7209
2.02	0.1239	0.2250	0.5506	1.7160	0.5740	4.5938	2.6962	1.7038	0.7115
2.04	0.1201	0.2200	0.5458	1.7451	0.5707	4.6885	2.7255	1.7203	0.7022
2.06	0.1164	0.2152	0.5409	1.7750	0.5675	4.7842	2.7545	1.7369	0.6928
2.08	0.1128	0.2104	0.5361	1.8056	0.5643	4.8808	2.7833	1.7536	0.6835
2.10	0.1094	0.2058	0.5313	1.8369	0.5613	4.9783	2.8119	1.7705	0.6742
2.12	0.1060	0.2013	0.5266	1.8690	0.5583	5.0768	2.8402	1.7875	0.6649
2.14	0.1027	0.1968	0.5219	1.9018	0.5554	5.1762	2.8683	1.8046	0.6557
2.16	0.9956^{-1}	0.1925	0.5173	1.9354	0.5525	5.2765	2.8962	1.8219	0.6464
2.18	0.9649^{-1}	0.1882	0.5127	1.9698	0.5498	5.3778	2.9238	1.8393	0.6373
2.20	0.9352^{-1}	0.1841	0.5081	2.0050	0.5471	5.4800	2.9512	1.8569	0.6281
2.22	0.9064^{-1}	0.1800	0.5036	2.0409	0.5444	5.5831	2.9784	1.8746	0.6191
2.24	0.8785^{-1}	0.1760	0.4991	2.0777	0.5418	5.6872	3.0053	1.8924	0.6100
2.26	0.8514^{-1}	0.1721	0.4947	2.1153	0.5393	5.7922	3.0319	1.9104	0.6011
2.28	0.8251^{-1}	0.1683	0.4903	2.1538	0.5368	5.8981	3.0584	1.9285	0.5921
2.30	0.7997^{-1}	0.1646	0.4859	2.1931	0.5344	6.0050	3.0845	1.9468	0.5833
2.32	0.7751^{-1}	0.1609	0.4816	2.2333	0.5321	6.1128	3.1105	1.9652	0.5745
2.34	0.7512^{-1}	0.1574	0.4773	2.2744	0.5297	6.2215	3.1362	1.9838	0.5658
2.36	0.7281^{-1}	0.1539	0.4731	2.3164	0.5275	6.3312	3.1617	2.0025	0.5572
2.38	0.7057^{-1}	0.1505	0.4688	2.3593	0.5253	6.4418	3.1869	2.0213	0.5486
2.40	0.6840^{-1}	0.1472	0.4647	2.4031	0.5231	6.5533	3.2119	2.0403	0.5401
2.42	0.6630^{-1}	0.1439	0.4606	2.4479	0.5210	6.6658	3.2367	2.0595	0.5317
2.44	0.6426^{-1}	0.1408	0.4565	2.4936	0.5189	6.7792	3.2612	2.0788	0.5234
2.46	0.6229^{-1}	0.1377	0.4524	2.5403	0.5169	6.8935	3.2855	2.0982	0.5152
2.48	0.6038^{-1}	0.1346	0.4484	2.5880	0.5149	7.0088	3.3095	2.1178	0.5071
2.50	0.5853^{-1}	0.1317	0.4444	2.6367	0.5130	7.1250	3.3333	2.1375	0.4990
2.52	0.5674^{-1}	0.1288	0.4405	2.6865	0.5111	7.2421	3.3569	2.1574	0.4911
2.54	0.5500^{-1}	0.1260	0.4366	2.7372	0.5092	7.3602	3.3803	2.1774	0.4832
2.56	0.5332^{-1}	0.1232	0.4328	2.7891	0.5074	7.4792	3.4034	2.1976	0.4754
2.58	0.5169^{-1}	0.1205	0.4289	2.8420	0.5056	7.5991	3.4263	2.2179	0.4677
2.60	0.5012^{-1}	0.1179	0.4252	2.8960	0.5039	7.7200	3.4490	2.2383	0.4601

n^{-p} is shorthand for $n \cdot 10^{-p}$

Supersonic Flow of a Perfect Gas with $\gamma = 1.4$ (continued)

	Isentropic Flow				Normal Shock Wave				
M_1	p/p_t	ρ/ρ_t	T/T_t	A/A^*	M_2	p_2/p_1	ρ_2/ρ_1	T_2/T_1	p_{t2}/p_{t1}
2.60	0.5012^{-1}	0.1179	0.4252	2.8960	0.5039	7.720	3.449	2.238	0.4601
2.62	0.4859^{-1}	0.1153	0.4214	2.9511	0.5022	7.842	3.471	2.259	0.4526
2.64	0.4711^{-1}	0.1128	0.4177	3.0073	0.5005	7.965	3.494	2.280	0.4452
2.66	0.4568^{-1}	0.1103	0.4141	3.0647	0.4988	8.088	3.516	2.301	0.4379
2.68	0.4429^{-1}	0.1079	0.4104	3.1233	0.4972	8.213	3.537	2.322	0.4307
2.70	0.4295^{-1}	0.1056	0.4068	3.1830	0.4956	8.338	3.559	2.343	0.4236
2.72	0.4165^{-1}	0.1033	0.4033	3.2440	0.4941	8.465	3.580	2.364	0.4166
2.74	0.4039^{-1}	0.1010	0.3998	3.3061	0.4926	8.592	3.601	2.386	0.4097
2.76	0.3917^{-1}	0.9885^{-1}	0.3963	3.3695	0.4911	8.721	3.622	2.407	0.4028
2.78	0.3799^{-1}	0.9671^{-1}	0.3928	3.4342	0.4896	8.850	3.643	2.429	0.3961
2.80	0.3685^{-1}	0.9463^{-1}	0.3894	3.5001	0.4882	8.980	3.664	2.451	0.3895
2.82	0.3574^{-1}	0.9259^{-1}	0.3860	3.5674	0.4868	9.111	3.684	2.473	0.3829
2.84	0.3467^{-1}	0.9059^{-1}	0.3827	3.6359	0.4854	9.243	3.704	2.496	0.3765
2.86	0.3363^{-1}	0.8865^{-1}	0.3794	3.7058	0.4840	9.376	3.724	2.518	0.3701
2.88	0.3263^{-1}	0.8675^{-1}	0.3761	3.7771	0.4827	9.510	3.743	2.540	0.3639
2.90	0.3165^{-1}	0.8489^{-1}	0.3729	3.8498	0.4814	9.645	3.763	2.563	0.3577
2.92	0.3071^{-1}	0.8307^{-1}	0.3696	3.9238	0.4801	9.781	3.782	2.586	0.3517
2.94	0.2980^{-1}	0.8130^{-1}	0.3665	3.9993	0.4788	9.918	3.801	2.609	0.3457
2.96	0.2891^{-1}	0.7957^{-1}	0.3633	4.0763	0.4776	10.055	3.820	2.632	0.3398
2.98	0.2805^{-1}	0.7788^{-1}	0.3602	4.1547	0.4764	10.194	3.839	2.656	0.3340
3.00	0.2722^{-1}	0.7623^{-1}	0.3571	4.2346	0.4752	10.333	3.857	2.679	0.3283
3.02	0.2642^{-1}	0.7461^{-1}	0.3541	4.3160	0.4740	10.474	3.875	2.703	0.3227
3.04	0.2564^{-1}	0.7303^{-1}	0.3511	4.3989	0.4729	10.615	3.893	2.726	0.3172
3.06	0.2489^{-1}	0.7149^{-1}	0.3481	4.4835	0.4717	10.758	3.911	2.750	0.3118
3.08	0.2416^{-1}	0.6999^{-1}	0.3452	4.5696	0.4706	10.901	3.929	2.774	0.3065
3.10	0.2345^{-1}	0.6852^{-1}	0.3422	4.6573	0.4695	11.045	3.947	2.799	0.3012
3.12	0.2276^{-1}	0.6708^{-1}	0.3393	4.7467	0.4685	11.190	3.964	2.823	0.2960
3.14	0.2210^{-1}	0.6568^{-1}	0.3365	4.8377	0.4674	11.336	3.981	2.848	0.2910
3.16	0.2146^{-1}	0.6430^{-1}	0.3337	4.9304	0.4664	11.483	3.998	2.872	0.2860
3.18	0.2083^{-1}	0.6296^{-1}	0.3309	5.0248	0.4654	11.631	4.015	2.897	0.2811
3.20	0.2023^{-1}	0.6165^{-1}	0.3281	5.1210	0.4643	11.780	4.031	2.922	0.2762
3.22	0.1964^{-1}	0.6037^{-1}	0.3253	5.2189	0.4634	11.930	4.048	2.947	0.2715
3.24	0.1908^{-1}	0.5912^{-1}	0.3226	5.3186	0.4624	12.081	4.064	2.972	0.2668
3.26	0.1853^{-1}	0.5790^{-1}	0.3199	5.4201	0.4614	12.232	4.080	2.998	0.2622
3.28	0.1799^{-1}	0.5671^{-1}	0.3173	5.5234	0.4605	12.385	4.096	3.023	0.2577
3.30	0.1748^{-1}	0.5554^{-1}	0.3147	5.6286	0.4596	12.538	4.112	3.049	0.2533
3.32	0.1698^{-1}	0.5440^{-1}	0.3121	5.7358	0.4587	12.693	4.128	3.075	0.2489
3.34	0.1649^{-1}	0.5329^{-1}	0.3095	5.8448	0.4578	12.848	4.143	3.101	0.2446
3.36	0.1602^{-1}	0.5220^{-1}	0.3069	5.9558	0.4569	13.005	4.158	3.127	0.2404
3.38	0.1557^{-1}	0.5113^{-1}	0.3044	6.0687	0.4560	13.162	4.173	3.154	0.2363
3.40	0.1512^{-1}	0.5009^{-1}	0.3019	6.1837	0.4552	13.320	4.188	3.180	0.2322

n^{-p} is shorthand for $n \cdot 10^{-p}$

Supersonic Flow of a Perfect Gas with $\gamma = 1.4$ (continued)

	Isentropic Flow				Normal Shock Wave				
M_1	p/p_t	ρ/ρ_t	T/T_t	A/A^*	M_2	p_2/p_1	ρ_2/ρ_1	T_2/T_1	p_{t2}/p_{t1}
3.4	0.1512^{-1}	0.5009^{-1}	0.3019	6.18	0.4552	13.32	4.188	3.180	0.2322
3.5	0.1311^{-1}	0.4523^{-1}	0.2899	6.79	0.4512	14.13	4.261	3.315	0.2129
3.6	0.1138^{-1}	0.4089^{-1}	0.2784	7.45	0.4474	14.95	4.330	3.454	0.1953
3.7	0.9903^{-2}	0.3702^{-1}	0.2675	8.17	0.4439	15.81	4.395	3.596	0.1792
3.8	0.8629^{-2}	0.3355^{-1}	0.2572	8.95	0.4407	16.68	4.457	3.743	0.1645
3.9	0.7532^{-2}	0.3044^{-1}	0.2474	9.80	0.4377	17.58	4.516	3.893	0.1510
4.0	0.6586^{-2}	0.2766^{-1}	0.2381	10.72	0.4350	18.50	4.571	4.047	0.1388
4.1	0.5769^{-2}	0.2516^{-1}	0.2293	11.71	0.4324	19.44	4.624	4.205	0.1276
4.2	0.5062^{-2}	0.2292^{-1}	0.2208	12.79	0.4299	20.41	4.675	4.367	0.1173
4.3	0.4449^{-2}	0.2090^{-1}	0.2129	13.95	0.4277	21.41	4.723	4.532	0.1080
4.4	0.3918^{-2}	0.1909^{-1}	0.2053	15.21	0.4255	22.42	4.768	4.702	0.9948^{-1}
4.5	0.3455^{-2}	0.1745^{-1}	0.1980	16.56	0.4236	23.46	4.812	4.875	0.9170^{-1}
4.6	0.3053^{-2}	0.1597^{-1}	0.1911	18.02	0.4217	24.52	4.853	5.052	0.8459^{-1}
4.7	0.2701^{-2}	0.1464^{-1}	0.1846	19.58	0.4199	25.60	4.893	5.233	0.7809^{-1}
4.8	0.2394^{-2}	0.1343^{-1}	0.1783	21.26	0.4183	26.71	4.930	5.418	0.7214^{-1}
4.9	0.2126^{-2}	0.1233^{-1}	0.1724	23.07	0.4167	27.84	4.966	5.607	0.6670^{-1}
5.0	0.1890^{-2}	0.1134^{-1}	0.1667	25.00	0.4152	29.00	5.000	5.800	0.6172^{-1}
5.1	0.1683^{-2}	0.1044^{-1}	0.1612	27.07	0.4138	30.18	5.033	5.997	0.5715^{-1}
5.2	0.1501^{-2}	0.9620^{-2}	0.1561	29.28	0.4125	31.38	5.064	6.197	0.5297^{-1}
5.3	0.1341^{-2}	0.8875^{-2}	0.1511	31.65	0.4113	32.60	5.093	6.401	0.4913^{-1}
5.4	0.1200^{-2}	0.8197^{-2}	0.1464	34.17	0.4101	33.85	5.122	6.610	0.4560^{-1}
5.5	0.1075^{-2}	0.7578^{-2}	0.1418	36.87	0.4090	35.13	5.149	6.822	0.4236^{-1}
5.6	0.9643^{-3}	0.7012^{-2}	0.1375	39.74	0.4079	36.42	5.175	7.038	0.3938^{-1}
5.7	0.8663^{-3}	0.6496^{-2}	0.1334	42.80	0.4069	37.74	5.200	7.258	0.3664^{-1}
5.8	0.7794^{-3}	0.6023^{-2}	0.1294	46.05	0.4059	39.08	5.224	7.481	0.3412^{-1}
5.9	0.7021^{-3}	0.5590^{-2}	0.1256	49.51	0.4050	40.44	5.246	7.709	0.3179^{-1}
6.0	0.6334^{-3}	0.5194^{-2}	0.1220	53.18	0.4042	41.83	5.268	7.941	0.2965^{-1}
6.1	0.5721^{-3}	0.4829^{-2}	0.1185	57.08	0.4033	43.24	5.289	8.176	0.2767^{-1}
6.2	0.5173^{-3}	0.4495^{-2}	0.1151	61.21	0.4025	44.68	5.309	8.415	0.2584^{-1}
6.3	0.4684^{-3}	0.4187^{-2}	0.1119	65.59	0.4018	46.14	5.329	8.658	0.2416^{-1}
6.4	0.4247^{-3}	0.3904^{-2}	0.1088	70.23	0.4011	47.62	5.347	8.905	0.2259^{-1}
6.5	0.3855^{-3}	0.3643^{-2}	0.1058	75.13	0.4004	49.13	5.365	9.156	0.2115^{-1}
6.6	0.3503^{-3}	0.3402^{-2}	0.1030	80.32	0.3997	50.65	5.382	9.411	0.1981^{-1}
6.7	0.3187^{-3}	0.3180^{-2}	0.1002	85.80	0.3991	52.20	5.399	9.670	0.1857^{-1}
6.8	0.2902^{-3}	0.2974^{-2}	0.0976	91.59	0.3985	53.78	5.415	9.933	0.1741^{-1}
6.9	0.2646^{-3}	0.2785^{-2}	0.0950	97.70	0.3979	55.38	5.430	10.199	0.1634^{-1}
7.0	0.2416^{-3}	0.2609^{-2}	0.0926	104.14	0.3974	57.00	5.444	10.469	0.1535^{-1}
7.1	0.2207^{-3}	0.2446^{-2}	0.0902	110.93	0.3968	58.64	5.459	10.744	0.1443^{-1}
7.2	0.2019^{-3}	0.2295^{-2}	0.0880	118.08	0.3963	60.31	5.472	11.022	0.1357^{-1}
7.3	0.1848^{-3}	0.2155^{-2}	0.0858	125.60	0.3958	62.00	5.485	11.304	0.1277^{-1}
7.4	0.1694^{-3}	0.2025^{-2}	0.0837	133.52	0.3954	63.72	5.498	11.590	0.1202^{-1}
7.5	0.1554^{-3}	0.1904^{-2}	0.0816	141.84	0.3949	65.46	5.510	11.879	0.1133^{-1}

n^{-p} is shorthand for $n \cdot 10^{-p}$

Supersonic Flow of a Perfect Gas with $\gamma = 1.4$ (continued)

	Isentropic Flow				Normal Shock Wave				
M_1	p/p_t	ρ/ρ_t	T/T_t	A/A^*	M_2	p_2/p_1	ρ_2/ρ_1	T_2/T_1	p_{t2}/p_{t1}
7.5	0.1554^{-3}	0.1904^{-2}	0.8163^{-1}	141.8	0.3949	65.46	5.510	11.88	0.1133^{-1}
8.0	0.1024^{-3}	0.1414^{-2}	0.7246^{-1}	190.1	0.3929	74.50	5.565	13.39	0.8488^{-2}
8.5	0.6898^{-4}	0.1066^{-2}	0.6472^{-1}	251.1	0.3912	84.13	5.612	14.99	0.6449^{-2}
9.0	0.4739^{-4}	0.8150^{-3}	0.5814^{-1}	327.2	0.3898	94.33	5.651	16.69	0.4964^{-2}
9.5	0.3314^{-4}	0.6313^{-3}	0.5249^{-1}	421.1	0.3886	105.13	5.685	18.49	0.3866^{-2}
10.0	0.2356^{-4}	0.4948^{-3}	0.4762^{-1}	535.9	0.3876	116.50	5.714	20.39	0.3045^{-2}

n^{-p} is shorthand for $n \cdot 10^{-p}$

B.2 Prandtl-Meyer Function and Mach Angle

Prandtl-Meyer Function, $\nu(M)$, and Mach angle, $\mu(M)$, for $\gamma = 1.4$

M	$\mu(M)$	$\nu(M)$	M	$\mu(M)$	$\nu(M)$	M	$\mu(M)$	$\nu(M)$
1.00	90.00	0.00	2.00	30.00	26.38	3.00	19.47	49.76
1.05	72.25	0.49	2.05	29.20	27.75	3.05	19.14	50.71
1.10	65.38	1.34	2.10	28.44	29.10	3.10	18.82	51.65
1.15	60.41	2.38	2.15	27.72	30.43	3.15	18.51	52.57
1.20	56.44	3.56	2.20	27.04	31.73	3.20	18.21	53.47
1.25	53.13	4.83	2.25	26.39	33.02	3.25	17.92	54.35
1.30	50.28	6.17	2.30	25.77	34.28	3.30	17.64	55.22
1.35	47.79	7.56	2.35	25.18	35.53	3.35	17.37	56.07
1.40	45.58	8.99	2.40	24.62	36.75	3.40	17.10	56.91
1.45	43.60	10.44	2.45	24.09	37.95	3.45	16.85	57.73
1.50	41.81	11.91	2.50	23.58	39.12	3.50	16.60	58.53
1.55	40.18	13.38	2.55	23.09	40.28	3.55	16.36	59.32
1.60	38.68	14.86	2.60	22.62	41.41	3.60	16.13	60.09
1.65	37.31	16.34	2.65	22.17	42.53	3.65	15.90	60.85
1.70	36.03	17.81	2.70	21.74	43.62	3.70	15.68	61.60
1.75	34.85	19.27	2.75	21.32	44.69	3.75	15.47	62.33
1.80	33.75	20.73	2.80	20.92	45.75	3.80	15.26	63.04
1.85	32.72	22.16	2.85	20.54	46.78	3.85	15.05	63.75
1.90	31.76	23.59	2.90	20.17	47.79	3.90	14.86	64.44
1.95	30.85	24.99	2.95	19.81	48.78	3.95	14.66	65.12
2.00	30.00	26.38	3.00	19.47	49.76	4.00	14.48	65.78

Values for $\mu(M)$ and $\nu(M)$ are in degrees

Prandtl-Meyer Function, $\nu(M)$, and Mach angle, $\mu(M)$, for γ = 1.4 (continued)

M	$\mu(M)$	$\nu(M)$	M	$\mu(M)$	$\nu(M)$	M	$\mu(M)$	$\nu(M)$
4.00	14.48	65.78	6.00	9.59	84.96	8.00	7.18	95.62
4.05	14.29	66.44	6.05	9.51	85.30	8.05	7.14	95.83
4.10	14.12	67.08	6.10	9.44	85.63	8.10	7.09	96.03
4.15	13.94	67.71	6.15	9.36	85.97	8.15	7.05	96.23
4.20	13.77	68.33	6.20	9.28	86.29	8.20	7.00	96.43
4.25	13.61	68.94	6.25	9.21	86.62	8.25	6.96	96.63
4.30	13.45	69.54	6.30	9.13	86.94	8.30	6.92	96.82
4.35	13.29	70.13	6.35	9.06	87.25	8.35	6.88	97.01
4.40	13.14	70.71	6.40	8.99	87.56	8.40	6.84	97.20
4.45	12.99	71.27	6.45	8.92	87.87	8.45	6.80	97.39
4.50	12.84	71.83	6.50	8.85	88.17	8.50	6.76	97.57
4.55	12.70	72.38	6.55	8.78	88.47	8.55	6.72	97.76
4.60	12.56	72.92	6.60	8.71	88.76	8.60	6.68	97.94
4.65	12.42	73.45	6.65	8.65	89.05	8.65	6.64	98.12
4.70	12.28	73.97	6.70	8.58	89.33	8.70	6.60	98.29
4.75	12.15	74.48	6.75	8.52	89.62	8.75	6.56	98.47
4.80	12.02	74.99	6.80	8.46	89.89	8.80	6.52	98.64
4.85	11.90	75.48	6.85	8.39	90.17	8.85	6.49	98.81
4.90	11.78	75.97	6.90	8.33	90.44	8.90	6.45	98.98
4.95	11.66	76.45	6.95	8.27	90.71	8.95	6.42	99.15
5.00	11.54	76.92	7.00	8.21	90.97	9.00	6.38	99.32
5.05	11.42	77.38	7.05	8.15	91.23	9.05	6.34	99.48
5.10	11.31	77.84	7.10	8.10	91.49	9.10	6.31	99.65
5.15	11.20	78.29	7.15	8.04	91.75	9.15	6.27	99.81
5.20	11.09	78.73	7.20	7.98	92.00	9.20	6.24	99.97
5.25	10.98	79.17	7.25	7.93	92.24	9.25	6.21	100.12
5.30	10.88	79.60	7.30	7.87	92.49	9.30	6.17	100.28
5.35	10.77	80.02	7.35	7.82	92.73	9.35	6.14	100.44
5.40	10.67	80.43	7.40	7.77	92.97	9.40	6.11	100.59
5.45	10.57	80.84	7.45	7.71	93.21	9.45	6.07	100.74
5.50	10.48	81.24	7.50	7.66	93.44	9.50	6.04	100.89
5.55	10.38	81.64	7.55	7.61	93.67	9.55	6.01	101.04
5.60	10.29	82.03	7.60	7.56	93.90	9.60	5.98	101.19
5.65	10.19	82.42	7.65	7.51	94.12	9.65	5.95	101.33
5.70	10.10	82.80	7.70	7.46	94.34	9.70	5.92	101.48
5.75	10.02	83.17	7.75	7.41	94.56	9.75	5.89	101.62
5.80	9.93	83.54	7.80	7.37	94.78	9.80	5.86	101.76
5.85	9.84	83.90	7.85	7.32	95.00	9.85	5.83	101.90
5.90	9.76	84.26	7.90	7.27	95.21	9.90	5.80	102.04
5.95	9.68	84.61	7.95	7.23	95.42	9.95	5.77	102.18
6.00	9.59	84.96	8.00	7.18	95.62	10.00	5.74	102.32

Values for $\mu(M)$ and $\nu(M)$ are in degrees

B.3 Fanno Flow

Notation used in the Fanno-flow tables of this section is as follows.

M	=	local Mach number
fL^*/D	=	dimensionless length to sonic point
p/p^*	=	ratio of static to sonic pressure
ρ/ρ^*	=	ratio of static to sonic density
T/T^*	=	ratio of static to sonic temperature
p_t/p_t^*	=	ratio of local total to sonic total pressure

Fanno Flow for $\gamma = 1.4$

M	fL^*/D	p/p^*	ρ/ρ^*	T/T^*	p_t/p_t^*
0.00	∞	∞	∞	1.2000	∞
0.05	280.0203	21.9034	18.2620	1.1994	11.5914
0.10	66.9216	10.9435	9.1378	1.1976	5.8218
0.15	27.9320	7.2866	6.0995	1.1946	3.9103
0.20	14.5333	5.4554	4.5826	1.1905	2.9635
0.25	8.4834	4.3546	3.6742	1.1852	2.4027
0.30	5.2993	3.6191	3.0702	1.1788	2.0351
0.35	3.4525	3.0922	2.6400	1.1713	1.7780
0.40	2.3085	2.6958	2.3184	1.1628	1.5901
0.45	1.5664	2.3865	2.0693	1.1533	1.4487
0.50	1.0691	2.1381	1.8708	1.1429	1.3398
0.55	0.7281	1.9341	1.7092	1.1315	1.2549
0.60	0.4908	1.7634	1.5753	1.1194	1.1882
0.65	0.3246	1.6183	1.4626	1.1065	1.1356
0.70	0.2081	1.4935	1.3665	1.0929	1.0944
0.75	0.1273	1.3848	1.2838	1.0787	1.0624
0.80	0.0723	1.2893	1.2119	1.0638	1.0382
0.85	0.0363	1.2047	1.1489	1.0485	1.0207
0.90	0.0145	1.1291	1.0934	1.0327	1.0089
0.95	0.0033	1.0613	1.0440	1.0165	1.0021
1.00	0.0000	1.0000	1.0000	1.0000	1.0000

Fanno Flow for $\gamma = 1.4$ (continued)

M	fL^*/D	p/p^*	ρ/ρ^*	T/T^*	p_t/p_t^*
1.00	0.0000	1.0000	1.0000	1.0000	1.0000
1.05	0.0027	0.9443	0.9605	0.9832	1.0020
1.10	0.0099	0.8936	0.9249	0.9662	1.0079
1.15	0.0205	0.8471	0.8926	0.9490	1.0175
1.20	0.0336	0.8044	0.8633	0.9317	1.0304
1.25	0.0486	0.7649	0.8367	0.9143	1.0468
1.30	0.0648	0.7285	0.8123	0.8969	1.0663
1.35	0.0820	0.6947	0.7899	0.8794	1.0890
1.40	0.0997	0.6632	0.7693	0.8621	1.1149
1.45	0.1178	0.6339	0.7503	0.8448	1.1440
1.50	0.1361	0.6065	0.7328	0.8276	1.1762
1.55	0.1543	0.5808	0.7166	0.8105	1.2116
1.60	0.1724	0.5568	0.7016	0.7937	1.2502
1.65	0.1902	0.5342	0.6876	0.7770	1.2922
1.70	0.2078	0.5130	0.6745	0.7605	1.3376
1.75	0.2250	0.4929	0.6624	0.7442	1.3865
1.80	0.2419	0.4741	0.6511	0.7282	1.4390
1.85	0.2583	0.4562	0.6404	0.7124	1.4952
1.90	0.2743	0.4394	0.6305	0.6969	1.5553
1.95	0.2899	0.4234	0.6211	0.6816	1.6193
2.00	0.3050	0.4082	0.6124	0.6667	1.6875
2.05	0.3197	0.3939	0.6041	0.6520	1.7600
2.10	0.3339	0.3802	0.5963	0.6376	1.8369
2.15	0.3476	0.3673	0.5890	0.6235	1.9185
2.20	0.3609	0.3549	0.5821	0.6098	2.0050
2.25	0.3738	0.3432	0.5756	0.5963	2.0964
2.30	0.3862	0.3320	0.5694	0.5831	2.1931
2.35	0.3983	0.3213	0.5635	0.5702	2.2953
2.40	0.4099	0.3111	0.5580	0.5576	2.4031
2.45	0.4211	0.3014	0.5527	0.5453	2.5168
2.50	0.4320	0.2921	0.5477	0.5333	2.6367
2.55	0.4425	0.2832	0.5430	0.5216	2.7630
2.60	0.4526	0.2747	0.5385	0.5102	2.8960
2.65	0.4624	0.2666	0.5342	0.4991	3.0359
2.70	0.4718	0.2588	0.5301	0.4882	3.1830
2.75	0.4809	0.2513	0.5262	0.4776	3.3377
2.80	0.4898	0.2441	0.5225	0.4673	3.5001
2.85	0.4983	0.2373	0.5189	0.4572	3.6707
2.90	0.5065	0.2307	0.5155	0.4474	3.8498
2.95	0.5145	0.2243	0.5123	0.4379	4.0376
3.00	0.5222	0.2182	0.5092	0.4286	4.2346

Fanno Flow for $\gamma = 1.4$ (continued)

M	fL^*/D	p/p^*	ρ/ρ^*	T/T^*	p_t/p_t^*
3.00	0.5222	0.2182	0.5092	0.4286	4.2346
3.05	0.5296	0.2124	0.5062	0.4195	4.4410
3.10	0.5368	0.2067	0.5034	0.4107	4.6573
3.15	0.5437	0.2013	0.5007	0.4021	4.8838
3.20	0.5504	0.1961	0.4980	0.3937	5.1210
3.25	0.5569	0.1911	0.4955	0.3855	5.3691
3.30	0.5632	0.1862	0.4931	0.3776	5.6286
3.35	0.5693	0.1815	0.4908	0.3699	5.9000
3.40	0.5752	0.1770	0.4886	0.3623	6.1837
3.45	0.5809	0.1727	0.4865	0.3550	6.4801
3.50	0.5864	0.1685	0.4845	0.3478	6.7896
3.55	0.5918	0.1645	0.4825	0.3409	7.1128
3.60	0.5970	0.1606	0.4806	0.3341	7.4501
3.65	0.6020	0.1568	0.4788	0.3275	7.8020
3.70	0.6068	0.1531	0.4770	0.3210	8.1691
3.75	0.6115	0.1496	0.4753	0.3148	8.5517
3.80	0.6161	0.1462	0.4737	0.3086	8.9506
3.85	0.6206	0.1429	0.4721	0.3027	9.3661
3.90	0.6248	0.1397	0.4706	0.2969	9.7990
3.95	0.6290	0.1366	0.4691	0.2912	10.2496
4.00	0.6331	0.1336	0.4677	0.2857	10.7188
4.05	0.6370	0.1307	0.4663	0.2803	11.2069
4.10	0.6408	0.1279	0.4650	0.2751	11.7147
4.15	0.6445	0.1252	0.4637	0.2700	12.2427
4.20	0.6481	0.1226	0.4625	0.2650	12.7916
4.25	0.6516	0.1200	0.4613	0.2602	13.3622
4.30	0.6550	0.1175	0.4601	0.2554	13.9549
4.35	0.6583	0.1151	0.4590	0.2508	14.5706
4.40	0.6615	0.1128	0.4579	0.2463	15.2099
4.45	0.6646	0.1105	0.4569	0.2419	15.8735
4.50	0.6676	0.1083	0.4559	0.2376	16.5622
4.55	0.6706	0.1062	0.4549	0.2334	17.2767
4.60	0.6734	0.1041	0.4539	0.2294	18.0178
4.65	0.6762	0.1021	0.4530	0.2254	18.7862
4.70	0.6790	0.1001	0.4521	0.2215	19.5828
4.75	0.6816	0.0982	0.4512	0.2177	20.4084
4.80	0.6842	0.0964	0.4504	0.2140	21.2637
4.85	0.6867	0.0946	0.4495	0.2104	22.1497
4.90	0.6891	0.0928	0.4487	0.2068	23.0671
4.95	0.6915	0.0911	0.4480	0.2034	24.0169
5.00	0.6938	0.0894	0.4472	0.2000	25.0000

B.4 Rayleigh Flow

Notation used in the Rayleigh-flow tables of this section is as follows.

M = local Mach number
T_t/T_t^* = ratio of local total to sonic total temperature
p/p^* = ratio of static to sonic pressure
ρ/ρ^* = ratio of static to sonic density
T/T^* = ratio of static to sonic temperature
p_t/p_t^* = ratio of local total to sonic total pressure

Rayleigh Flow for $\gamma = 1.4$

M	T_t/T_t^*	p/p^*	ρ/ρ^*	T/T^*	p_t/p_t^*
0.00	0.0000	2.4000	∞	0.0000	1.2679
0.05	0.0119	2.3916	167.2500	0.0143	1.2657
0.10	0.0468	2.3669	42.2500	0.0560	1.2591
0.15	0.1020	2.3267	19.1019	0.1218	1.2486
0.20	0.1736	2.2727	11.0000	0.2066	1.2346
0.25	0.2568	2.2069	7.2500	0.3044	1.2177
0.30	0.3469	2.1314	5.2130	0.4089	1.1985
0.35	0.4389	2.0487	3.9847	0.5141	1.1779
0.40	0.5290	1.9608	3.1875	0.6151	1.1566
0.45	0.6139	1.8699	2.6409	0.7080	1.1351
0.50	0.6914	1.7778	2.2500	0.7901	1.1141
0.55	0.7599	1.6860	1.9607	0.8599	1.0940
0.60	0.8189	1.5957	1.7407	0.9167	1.0753
0.65	0.8683	1.5080	1.5695	0.9608	1.0582
0.70	0.9085	1.4235	1.4337	0.9929	1.0431
0.75	0.9401	1.3427	1.3241	1.0140	1.0301
0.80	0.9639	1.2658	1.2344	1.0255	1.0193
0.85	0.9810	1.1931	1.1600	1.0285	1.0109
0.90	0.9921	1.1246	1.0977	1.0245	1.0049
0.95	0.9981	1.0603	1.0450	1.0146	1.0012
1.00	1.0000	1.0000	1.0000	1.0000	1.0000

Rayleigh Flow for $\gamma = 1.4$ (continued)

M	T_t/T_t^*	p/p^*	ρ/ρ^*	T/T^*	p_t/p_t^*
1.00	1.0000	1.0000	1.0000	1.0000	1.0000
1.05	0.9984	0.9436	0.9613	0.9816	1.0012
1.10	0.9939	0.8909	0.9277	0.9603	1.0049
1.15	0.9872	0.8417	0.8984	0.9369	1.0109
1.20	0.9787	0.7958	0.8727	0.9118	1.0194
1.25	0.9689	0.7529	0.8500	0.8858	1.0303
1.30	0.9580	0.7130	0.8299	0.8592	1.0437
1.35	0.9464	0.6758	0.8120	0.8323	1.0594
1.40	0.9343	0.6410	0.7959	0.8054	1.0777
1.45	0.9218	0.6086	0.7815	0.7787	1.0983
1.50	0.9093	0.5783	0.7685	0.7525	1.1215
1.55	0.8967	0.5500	0.7568	0.7268	1.1473
1.60	0.8842	0.5236	0.7461	0.7017	1.1756
1.65	0.8718	0.4988	0.7364	0.6774	1.2066
1.70	0.8597	0.4756	0.7275	0.6538	1.2402
1.75	0.8478	0.4539	0.7194	0.6310	1.2767
1.80	0.8363	0.4335	0.7119	0.6089	1.3159
1.85	0.8250	0.4144	0.7051	0.5877	1.3581
1.90	0.8141	0.3964	0.6988	0.5673	1.4033
1.95	0.8036	0.3795	0.6929	0.5477	1.4516
2.00	0.7934	0.3636	0.6875	0.5289	1.5031
2.05	0.7835	0.3487	0.6825	0.5109	1.5579
2.10	0.7741	0.3345	0.6778	0.4936	1.6162
2.15	0.7649	0.3212	0.6735	0.4770	1.6780
2.20	0.7561	0.3086	0.6694	0.4611	1.7434
2.25	0.7477	0.2968	0.6656	0.4458	1.8128
2.30	0.7395	0.2855	0.6621	0.4312	1.8860
2.35	0.7317	0.2749	0.6588	0.4172	1.9634
2.40	0.7242	0.2648	0.6557	0.4038	2.0451
2.45	0.7170	0.2552	0.6527	0.3910	2.1311
2.50	0.7101	0.2462	0.6500	0.3787	2.2218
2.55	0.7034	0.2375	0.6474	0.3669	2.3173
2.60	0.6970	0.2294	0.6450	0.3556	2.4177
2.65	0.6908	0.2216	0.6427	0.3448	2.5233
2.70	0.6849	0.2142	0.6405	0.3344	2.6343
2.75	0.6793	0.2071	0.6384	0.3244	2.7508
2.80	0.6738	0.2004	0.6365	0.3149	2.8731
2.85	0.6685	0.1940	0.6346	0.3057	3.0014
2.90	0.6635	0.1879	0.6329	0.2969	3.1359
2.95	0.6586	0.1820	0.6312	0.2884	3.2768
3.00	0.6540	0.1765	0.6296	0.2803	3.4245

Rayleigh Flow for $\gamma = 1.4$ (continued)

M	T_t/T_t^*	p/p^*	ρ/ρ^*	T/T^*	p_t/p_t^*
3.00	0.6540	0.1765	0.6296	0.2803	3.4245
3.05	0.6495	0.1711	0.6281	0.2725	3.5790
3.10	0.6452	0.1660	0.6267	0.2650	3.7408
3.15	0.6410	0.1612	0.6253	0.2577	3.9101
3.20	0.6370	0.1565	0.6240	0.2508	4.0871
3.25	0.6331	0.1520	0.6228	0.2441	4.2721
3.30	0.6294	0.1477	0.6216	0.2377	4.4655
3.35	0.6258	0.1436	0.6205	0.2315	4.6674
3.40	0.6224	0.1397	0.6194	0.2255	4.8783
3.45	0.6190	0.1359	0.6183	0.2197	5.0984
3.50	0.6158	0.1322	0.6173	0.2142	5.3280
3.55	0.6127	0.1287	0.6164	0.2088	5.5676
3.60	0.6097	0.1254	0.6155	0.2037	5.8173
3.65	0.6068	0.1221	0.6146	0.1987	6.0776
3.70	0.6040	0.1190	0.6138	0.1939	6.3488
3.75	0.6013	0.1160	0.6130	0.1893	6.6314
3.80	0.5987	0.1131	0.6122	0.1848	6.9256
3.85	0.5962	0.1103	0.6114	0.1805	7.2318
3.90	0.5937	0.1077	0.6107	0.1763	7.5505
3.95	0.5914	0.1051	0.6100	0.1722	7.8820
4.00	0.5891	0.1026	0.6094	0.1683	8.2268
4.05	0.5869	0.1002	0.6087	0.1645	8.5853
4.10	0.5847	0.0978	0.6081	0.1609	8.9579
4.15	0.5827	0.0956	0.6075	0.1573	9.3451
4.20	0.5807	0.0934	0.6070	0.1539	9.7473
4.25	0.5787	0.0913	0.6064	0.1506	10.1649
4.30	0.5768	0.0893	0.6059	0.1473	10.5985
4.35	0.5750	0.0873	0.6054	0.1442	11.0486
4.40	0.5732	0.0854	0.6049	0.1412	11.5155
4.45	0.5715	0.0836	0.6044	0.1383	11.9999
4.50	0.5698	0.0818	0.6039	0.1354	12.5023
4.55	0.5682	0.0800	0.6035	0.1326	13.0231
4.60	0.5666	0.0784	0.6030	0.1300	13.5629
4.65	0.5651	0.0767	0.6026	0.1274	14.1223
4.70	0.5636	0.0752	0.6022	0.1248	14.7017
4.75	0.5622	0.0736	0.6018	0.1224	15.3019
4.80	0.5608	0.0722	0.6014	0.1200	15.9234
4.85	0.5594	0.0707	0.6010	0.1177	16.5667
4.90	0.5581	0.0693	0.6007	0.1154	17.2325
4.95	0.5568	0.0680	0.6003	0.1132	17.9213
5.00	0.5556	0.0667	0.6000	0.1111	18.6339

Appendix C

Useful Mathematical Theorems

This Appendix includes several mathematical theorems and relationships from elementary calculus that are used repeatedly throughout the text.

C.1 Mean Value Theorem of Calculus

For a function $f(x)$ that is continuous on the interval $a \leq x \leq b$, there exists a point ξ within the interval (i.e., $a < \xi < b$) such that

$$\int_a^b f(x)\, dx = (b - a) f(\xi) \tag{C.1}$$

C.2 The Vector Differential Operator ∇

Much of the analysis in the main text uses rectangular Cartesian coordinates x, y, and z with corresponding unit vectors $\mathbf{i}$, $\mathbf{j}$ and $\mathbf{k}$ as shown in Figure C.1. The vector differential operator ∇ is defined by

$$\nabla = \mathbf{i}\,\frac{\partial}{\partial x} + \mathbf{j}\,\frac{\partial}{\partial y} + \mathbf{k}\,\frac{\partial}{\partial z} \tag{C.2}$$

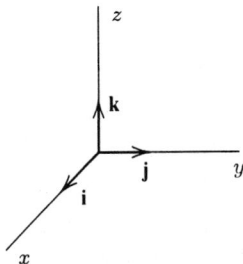

Velocity vector:
$$\mathbf{u} = u\,\mathbf{i} + v\,\mathbf{j} + w\,\mathbf{k}$$

Position vector:
$$\mathbf{r} = x\,\mathbf{i} + y\,\mathbf{j} + z\,\mathbf{k}$$

Figure C.1: *Rectangular Cartesian coordinates.*

665

The primary ways ∇ can operate on a vector $\mathbf{F} = F_x\mathbf{i} + F_y\mathbf{j} + F_z\mathbf{k}$ or a scalar ϕ are as follows.

$$\nabla \cdot \mathbf{F} = div\ \mathbf{F} = \frac{\partial F_x}{\partial x} + \frac{\partial F_y}{\partial y} + \frac{\partial F_z}{\partial z} \tag{C.3}$$

$$\nabla \phi = grad\ \phi = \mathbf{i}\frac{\partial \phi}{\partial x} + \mathbf{j}\frac{\partial \phi}{\partial y} + \mathbf{k}\frac{\partial \phi}{\partial z} \tag{C.4}$$

$$\nabla \times \mathbf{F} = curl\ \mathbf{F} = \mathbf{i}\left[\frac{\partial F_z}{\partial y} - \frac{\partial F_y}{\partial z}\right] + \mathbf{j}\left[\frac{\partial F_x}{\partial z} - \frac{\partial F_z}{\partial x}\right] + \mathbf{k}\left[\frac{\partial F_y}{\partial x} - \frac{\partial F_x}{\partial y}\right] \tag{C.5}$$

There is a particularly illuminating geometrical interpretation of the differential operator ∇. Specifically, if $\mathbf{n}$ is a unit vector, then

$$\mathbf{n} \cdot \nabla = n_x\frac{\partial}{\partial x} + n_y\frac{\partial}{\partial y} + n_z\frac{\partial}{\partial z} = \frac{\partial}{\partial n} \tag{C.6}$$

That is, $\mathbf{n} \cdot \nabla$ is the **directional derivative** with $\partial/\partial n$ representing the rate of change in the direction of $\mathbf{n}$. Finally, the following relations hold for any vectors $\mathbf{u}$ and $\mathbf{v}$ and scalars ρ and ϕ. Note that the **Laplacian operator**, ∇^2, is the dot product of ∇ with itself, i.e., $\nabla^2 = \nabla \cdot \nabla$.

$$\nabla \times \nabla \phi = \mathbf{0} \tag{C.7}$$

$$\nabla \cdot (\nabla \times \mathbf{u}) = 0 \tag{C.8}$$

$$\nabla \times (\nabla \times \mathbf{u}) = \nabla(\nabla \cdot \mathbf{u}) - \nabla^2\mathbf{u} \tag{C.9}$$

$$\nabla \cdot (\rho\,\mathbf{u}) = \mathbf{u} \cdot \nabla\rho + \rho\nabla \cdot \mathbf{u} \tag{C.10}$$

$$\nabla \times (\rho\,\mathbf{u}) = \rho\nabla \times \mathbf{u} + \nabla\rho \times \mathbf{u} \tag{C.11}$$

$$\mathbf{u} \cdot \nabla\mathbf{u} = \nabla\left(\frac{1}{2}\mathbf{u} \cdot \mathbf{u}\right) - \mathbf{u} \times (\nabla \times \mathbf{u}) \tag{C.12}$$

$$\nabla \times (\mathbf{u} \times \mathbf{v}) = (\mathbf{v} \cdot \nabla)\mathbf{u} + \mathbf{u}(\nabla \cdot \mathbf{v}) - (\mathbf{u} \cdot \nabla)\mathbf{v} - \mathbf{v}(\nabla \cdot \mathbf{u}) \tag{C.13}$$

C.3   Divergence Theorem

For any volume V bounded by a surface S with **outer** unit normal $\mathbf{n}$ (see Figure C.2), if $\mathbf{F}$ is a continuously differentiable vector, then

$$\oiint_S \mathbf{F} \cdot \mathbf{n}\,dS = \iiint_V \nabla \cdot \mathbf{F}\,dV \tag{C.14}$$

A corollary result holds for any continuously differentiable scalar, ϕ, i.e.,

$$\oiint_S \phi\,\mathbf{n}\,dS = \iiint_V \nabla\phi\,dV \tag{C.15}$$

Note that the indicated surface integrals are taken over the entire bounding surface.

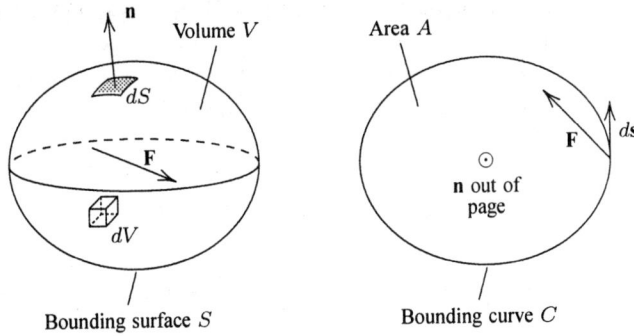

Figure C.2: *Arbitrary volume V bounded by surface S and area A bounded by curve C.*

C.4 Gauss' Theorem

For any volume V bounded by a surface S with outer unit normal $\mathbf{n}$, if $\mathbf{F}$ is a continuously differentiable vector, then

$$\oiint_S \mathbf{F} \times \mathbf{n}\, dS = - \iiint_V \nabla \times \mathbf{F}\, dV \qquad (C.16)$$

C.5 Stokes' Theorem

For any planar area A bounded by a curve C with a unit normal $\mathbf{n}$ to the plane of area A (see Figure C.2), if $\mathbf{F}$ is a continuously differentiable vector, then

$$\oint_C \mathbf{F} \cdot d\mathbf{s} = \iint_A \nabla \times \mathbf{F} \cdot \mathbf{n}\, dA \qquad (C.17)$$

C.6 A Useful Operation

At several points in the text, differential equations are deduced from integral conservation laws. This is done by first using the divergence theorem to replace all surface integrals by volume integrals. For example, mass conservation in integral form is

$$\iiint_V \frac{\partial \rho}{\partial t}\, dV + \oiint_S \rho\, \mathbf{u} \cdot \mathbf{n}\, dS = 0 \qquad (C.18)$$

Using the divergence theorem, Equation (C.18) can be rewritten as follows.

$$\iiint_V \frac{\partial \rho}{\partial t}\, dV + \iiint_V \nabla \cdot (\rho\, \mathbf{u})\, dV = 0 \qquad (C.19)$$

Because integration is a linear operation, both terms can be included under the same volume integral, wherefore

$$\iiint_V \left[\frac{\partial \rho}{\partial t} + \nabla \cdot (\rho\, \mathbf{u}) \right] dV = 0 \qquad (C.20)$$

This result holds for all volumes, which is the same as saying the limits of all three integrals are arbitrary. The only way possible to have the triple integral of an integrand yield zero for arbitrary limits of integration is for the integrand to be zero for all values of x, y and z. Hence, we conclude that

$$\frac{\partial \rho}{\partial t} + \nabla \cdot (\rho \, \mathbf{u}) = 0 \tag{C.21}$$

and this is the desired differential equation.

Appendix D

Equations of Motion in Various Coordinate Systems

Often, taking advantage of special coordinate systems can simplify the problem we wish to solve. For example, problems with cylindrical geometries such as flow in a pipe are most conveniently solved using cylindrical polar coordinates. The purpose of this Appendix is to list the incompressible continuity and Navier-Stokes equations for several useful coordinate systems.

For a general body force per unit mass, $\mathbf{f}$, the Navier-Stokes equation is

$$\frac{d\mathbf{u}}{dt} = -\frac{1}{\rho}\nabla p + \mathbf{f} + \nu\nabla^2\mathbf{u} \tag{D.1}$$

where the acceleration vector, $d\mathbf{u}/dt$, is computed as the Eulerian derivative of velocity, $\mathbf{u}$, defined as follows.

$$\frac{d\mathbf{u}}{dt} = \frac{\partial\mathbf{u}}{\partial t} + \mathbf{u}\cdot\nabla\mathbf{u} = \frac{\partial\mathbf{u}}{\partial t} + \left(u\frac{\partial}{\partial x} + v\frac{\partial}{\partial y} + w\frac{\partial}{\partial z}\right)\mathbf{u} \tag{D.2}$$

The convective part of the acceleration, $\mathbf{u}\cdot\nabla\mathbf{u}$, can be expressed in terms of the more familiar *grad* and *curl* operations. Specifically,

$$\mathbf{u}\cdot\nabla\mathbf{u} = \nabla\left(\frac{1}{2}\mathbf{u}\cdot\mathbf{u}\right) - \mathbf{u}\times(\nabla\times\mathbf{u}) \tag{D.3}$$

We can make use of a theorem from tensor calculus that tells us the following. Any vector identity that can be proven in Cartesian coordinates and written strictly in terms of *grad*, *div* and *curl* operations holds in all coordinate systems. Hence, we can rewrite the continuity equation and the Navier-Stokes equation as follows.

$$\nabla\cdot\mathbf{u} = 0 \tag{D.4}$$

$$\frac{\partial\mathbf{u}}{\partial t} + \nabla\left(\frac{1}{2}\mathbf{u}\cdot\mathbf{u}\right) - \mathbf{u}\times(\nabla\times\mathbf{u}) = -\frac{1}{\rho}\nabla p + \mathbf{f} + \nu\nabla^2\mathbf{u} \tag{D.5}$$

Beginning with Equations (D.4) and (D.5), we can write the continuity and Navier-Stokes equations in any coordinate system for which we know how to express *grad*, *div* and *curl*. The following pages list the equations, in component form, for several useful coordinate systems.

D.1   Rectangular Cartesian Coordinates

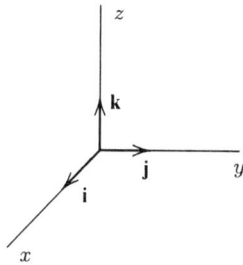

Velocity vector:
$$\mathbf{u} = u\,\mathbf{i} + v\,\mathbf{j} + w\,\mathbf{k}$$

Position vector:
$$\mathbf{r} = x\,\mathbf{i} + y\,\mathbf{j} + z\,\mathbf{k}$$

Figure D.1: *Rectangular Cartesian coordinates.*

We express the velocity vector, $\mathbf{u}$, in terms of its three components u, v and w as shown in Figure D.1. Typical *grad*, *div*, *curl* and Laplacian operations are as follows.

$$\nabla p = \mathbf{i}\,\frac{\partial p}{\partial x} + \mathbf{j}\,\frac{\partial p}{\partial y} + \mathbf{k}\,\frac{\partial p}{\partial z} \tag{D.6}$$

$$\nabla \cdot \mathbf{u} = \frac{\partial u}{\partial x} + \frac{\partial v}{\partial y} + \frac{\partial w}{\partial z} \tag{D.7}$$

$$\nabla \times \mathbf{u} = \mathbf{i}\left[\frac{\partial w}{\partial y} - \frac{\partial v}{\partial z}\right] + \mathbf{j}\left[\frac{\partial u}{\partial z} - \frac{\partial w}{\partial x}\right] + \mathbf{k}\left[\frac{\partial v}{\partial x} - \frac{\partial u}{\partial y}\right] \tag{D.8}$$

$$\nabla^2 \mathbf{u} = \left[\frac{\partial^2}{\partial x^2} + \frac{\partial^2}{\partial y^2} + \frac{\partial^2}{\partial z^2}\right](u\,\mathbf{i} + v\,\mathbf{j} + w\,\mathbf{k}) \tag{D.9}$$

The continuity equation and the three components of the Navier-Stokes equation with a body force $\mathbf{f} = f_x\mathbf{i} + f_y\mathbf{j} + f_z\mathbf{k}$ are:

$$\frac{\partial u}{\partial x} + \frac{\partial v}{\partial y} + \frac{\partial w}{\partial z} = 0 \tag{D.10}$$

$$\frac{\partial u}{\partial t} + u\frac{\partial u}{\partial x} + v\frac{\partial u}{\partial y} + w\frac{\partial u}{\partial z} = -\frac{1}{\rho}\frac{\partial p}{\partial x} + f_x + \nu\left(\frac{\partial^2 u}{\partial x^2} + \frac{\partial^2 u}{\partial y^2} + \frac{\partial^2 u}{\partial z^2}\right) \tag{D.11}$$

$$\frac{\partial v}{\partial t} + u\frac{\partial v}{\partial x} + v\frac{\partial v}{\partial y} + w\frac{\partial v}{\partial z} = -\frac{1}{\rho}\frac{\partial p}{\partial y} + f_y + \nu\left(\frac{\partial^2 v}{\partial x^2} + \frac{\partial^2 v}{\partial y^2} + \frac{\partial^2 v}{\partial z^2}\right) \tag{D.12}$$

$$\frac{\partial w}{\partial t} + u\frac{\partial w}{\partial x} + v\frac{\partial w}{\partial y} + w\frac{\partial w}{\partial z} = -\frac{1}{\rho}\frac{\partial p}{\partial z} + f_z + \nu\left(\frac{\partial^2 w}{\partial x^2} + \frac{\partial^2 w}{\partial y^2} + \frac{\partial^2 w}{\partial z^2}\right) \tag{D.13}$$

D.2 Cylindrical Polar Coordinates

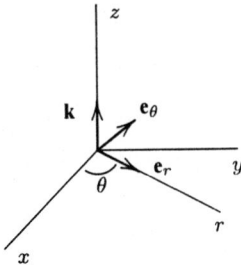

Velocity vector:
$$\mathbf{u} = u_r \mathbf{e}_r + u_\theta \mathbf{e}_\theta + w\,\mathbf{k}$$

Position vector:
$$\mathbf{r} = r\,\mathbf{e}_r + z\,\mathbf{k}$$

Unit vectors:
$$\mathbf{e}_r = \mathbf{i}\cos\theta + \mathbf{j}\sin\theta$$
$$\mathbf{e}_\theta = -\mathbf{i}\sin\theta + \mathbf{j}\cos\theta$$

Figure D.2: *Cylindrical polar coordinates.*

We express the velocity vector, $\mathbf{u}$, in terms of its three components u_r, u_θ and w as shown in Figure D.2. Typical *grad*, *div*, *curl* and Laplacian operations are as follows

$$\nabla p = \mathbf{e}_r \frac{\partial p}{\partial r} + \mathbf{e}_\theta \frac{1}{r}\frac{\partial p}{\partial \theta} + \mathbf{k}\frac{\partial p}{\partial z} \tag{D.14}$$

$$\nabla \cdot \mathbf{u} = \frac{1}{r}\frac{\partial}{\partial r}(r u_r) + \frac{1}{r}\frac{\partial u_\theta}{\partial \theta} + \frac{\partial w}{\partial z} \tag{D.15}$$

$$\nabla \times \mathbf{u} = \mathbf{e}_r\left[\frac{1}{r}\frac{\partial w}{\partial \theta} - \frac{\partial u_\theta}{\partial z}\right] + \mathbf{e}_\theta\left[\frac{\partial u_r}{\partial z} - \frac{\partial w}{\partial r}\right] + \mathbf{k}\left[\frac{1}{r}\frac{\partial}{\partial r}(r u_\theta) - \frac{1}{r}\frac{\partial u_r}{\partial \theta}\right] \tag{D.16}$$

$$\nabla^2 \mathbf{u} = \left[\frac{1}{r}\frac{\partial}{\partial r}\left(r\frac{\partial}{\partial r}\right) + \frac{1}{r^2}\frac{\partial^2}{\partial \theta^2} + \frac{\partial^2}{\partial z^2}\right](u_r\mathbf{e}_r + u_\theta\mathbf{e}_\theta + w\,\mathbf{k}) \tag{D.17}$$

The continuity equation and the three components of the Navier-Stokes equation with a body force $\mathbf{f} = f_r\mathbf{e}_r + f_\theta\mathbf{e}_\theta + f_z\mathbf{k}$ are:

$$\frac{1}{r}\frac{\partial}{\partial r}(r u_r) + \frac{1}{r}\frac{\partial u_\theta}{\partial \theta} + \frac{\partial w}{\partial z} = 0 \tag{D.18}$$

$$\frac{\partial u_r}{\partial t} + u_r\frac{\partial u_r}{\partial r} + \frac{u_\theta}{r}\frac{\partial u_r}{\partial \theta} + w\frac{\partial u_r}{\partial z} - \frac{u_\theta^2}{r} = -\frac{1}{\rho}\frac{\partial p}{\partial r} + f_r$$
$$+\nu\left(\frac{\partial^2 u_r}{\partial r^2} + \frac{1}{r^2}\frac{\partial^2 u_r}{\partial \theta^2} + \frac{\partial^2 u_r}{\partial z^2} + \frac{1}{r}\frac{\partial u_r}{\partial r} - \frac{2}{r^2}\frac{\partial u_\theta}{\partial \theta} - \frac{u_r}{r^2}\right) \tag{D.19}$$

$$\frac{\partial u_\theta}{\partial t} + u_r\frac{\partial u_\theta}{\partial r} + \frac{u_\theta}{r}\frac{\partial u_\theta}{\partial \theta} + w\frac{\partial u_\theta}{\partial z} + \frac{u_r u_\theta}{r} = -\frac{1}{\rho r}\frac{\partial p}{\partial \theta} + f_\theta$$
$$+\nu\left(\frac{\partial^2 u_\theta}{\partial r^2} + \frac{1}{r^2}\frac{\partial^2 u_\theta}{\partial \theta^2} + \frac{\partial^2 u_\theta}{\partial z^2} + \frac{1}{r}\frac{\partial u_\theta}{\partial r} + \frac{2}{r^2}\frac{\partial u_r}{\partial \theta} - \frac{u_\theta}{r^2}\right) \tag{D.20}$$

$$\frac{\partial w}{\partial t} + u_r\frac{\partial w}{\partial r} + \frac{u_\theta}{r}\frac{\partial w}{\partial \theta} + w\frac{\partial w}{\partial z} = -\frac{1}{\rho}\frac{\partial p}{\partial z} + f_z$$
$$+\nu\left(\frac{\partial^2 w}{\partial r^2} + \frac{1}{r^2}\frac{\partial^2 w}{\partial \theta^2} + \frac{\partial^2 w}{\partial z^2} + \frac{1}{r}\frac{\partial w}{\partial r}\right) \tag{D.21}$$

D.3　Spherical Coordinates

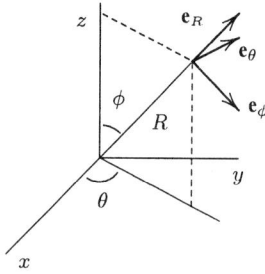

Velocity vector:
$$\mathbf{u} = u_R \mathbf{e}_R + u_\theta \mathbf{e}_\theta + u_\phi \mathbf{e}_\phi$$

Position vector:
$$\mathbf{r} = R\,\mathbf{e}_R$$

Unit vectors:
$$\mathbf{e}_R = \mathbf{i}\cos\theta\sin\phi + \mathbf{j}\sin\theta\sin\phi + \mathbf{k}\cos\phi$$
$$\mathbf{e}_\theta = -\mathbf{i}\sin\theta + \mathbf{j}\cos\theta$$
$$\mathbf{e}_\phi = \mathbf{i}\cos\theta\cos\phi + \mathbf{j}\sin\theta\cos\phi - \mathbf{k}\sin\phi$$

Figure D.3: *Spherical coordinates.*

We express the velocity vector, **u**, in terms of its three components u_R, u_θ and u_ϕ as shown in Figure D.3. Typical *grad*, *div*, *curl* and Laplacian operations are as follows.

$$\nabla p = \mathbf{e}_R \frac{\partial p}{\partial R} + \mathbf{e}_\theta \frac{1}{R\sin\phi}\frac{\partial p}{\partial\theta} + \mathbf{e}_\phi \frac{1}{R}\frac{\partial p}{\partial\phi} \tag{D.22}$$

$$\nabla\cdot\mathbf{u} = \frac{1}{R^2}\frac{\partial}{\partial R}(R^2 u_R) + \frac{1}{R\sin\phi}\frac{\partial u_\theta}{\partial\theta} + \frac{1}{R\sin\phi}\frac{\partial}{\partial\phi}(u_\phi\sin\phi) \tag{D.23}$$

$$\begin{aligned}
\nabla\times\mathbf{u} =\ & \frac{\mathbf{e}_R}{R\sin\theta}\left[\frac{\partial u_\phi}{\partial\theta} - \frac{\partial}{\partial\phi}(u_\theta\sin\phi)\right] \\
&+ \frac{\mathbf{e}_\theta}{R}\left[\frac{\partial u_R}{\partial\phi} - \frac{\partial}{\partial R}(Ru_\phi)\right] \\
&+ \frac{\mathbf{e}_\phi}{R\sin\phi}\left[\frac{\partial}{\partial R}(Ru_\theta\sin\phi) - \frac{\partial u_R}{\partial\theta}\right]
\end{aligned} \tag{D.24}$$

$$\begin{aligned}
\nabla^2\mathbf{u} =\ & \left[\frac{1}{R^2}\frac{\partial}{\partial R}\left(R^2\frac{\partial}{\partial R}\right) + \frac{1}{R^2\sin^2\phi}\frac{\partial^2}{\partial\theta^2}\right. \\
&+ \left.\frac{1}{R^2\sin^2\phi}\frac{\partial}{\partial\phi}\left(\sin\phi\frac{\partial}{\partial\phi}\right)\right](u_R\mathbf{e}_R + u_\theta\mathbf{e}_\theta + u_\phi\mathbf{e}_\phi)
\end{aligned} \tag{D.25}$$

The continuity equation and the three components of the Navier-Stokes equation with a body force $\mathbf{f} = f_R\mathbf{e}_R + f_\theta\mathbf{e}_\theta + f_\phi\mathbf{e}_\phi$ are:

$$\frac{1}{R^2}\frac{\partial}{\partial R}(R^2 u_R) + \frac{1}{R\sin\phi}\frac{\partial u_\theta}{\partial\theta} + \frac{1}{R\sin\phi}\frac{\partial}{\partial\phi}(u_\phi\sin\phi) = 0 \tag{D.26}$$

$$\begin{aligned}
&\frac{\partial u_R}{\partial t} + u_R\frac{\partial u_R}{\partial R} + \frac{u_\theta}{R\sin\phi}\frac{\partial u_R}{\partial\theta} + \frac{u_\phi}{R}\frac{\partial u_R}{\partial\phi} - \frac{u_\phi^2 + u_\theta^2}{R} \\
&= -\frac{1}{\rho}\frac{\partial p}{\partial R} + f_R + \nu\left[\frac{1}{R}\frac{\partial^2(Ru_R)}{\partial R^2} + \frac{1}{R^2\sin^2\phi}\frac{\partial^2 u_R}{\partial\theta^2} + \frac{1}{R^2}\frac{\partial^2 u_R}{\partial\phi^2}\right. \\
&\left.\quad + \frac{\cot\phi}{R^2}\frac{\partial u_R}{\partial\phi} - \frac{2}{R^2}\frac{\partial u_\phi}{\partial\phi} - \frac{2}{R^2\sin\phi}\frac{\partial u_\theta}{\partial\theta} - \frac{2u_R}{R^2} - \frac{2u_\phi\cot\phi}{R^2}\right]
\end{aligned} \tag{D.27}$$

$$\frac{\partial u_\theta}{\partial t} + u_R \frac{\partial u_\theta}{\partial R} + \frac{u_\theta}{R \sin \phi} \frac{\partial u_\theta}{\partial \theta} + \frac{u_\phi}{R} \frac{\partial u_\theta}{\partial \phi} + \frac{u_R u_\theta}{R} + \frac{u_\phi u_\theta \cot \phi}{R}$$

$$= -\frac{1}{\rho R \sin \phi} \frac{\partial p}{\partial \theta} + f_\theta + \nu \left[\frac{1}{R} \frac{\partial^2 (R u_\theta)}{\partial R^2} + \frac{1}{R^2 \sin^2 \phi} \frac{\partial^2 u_\theta}{\partial \theta^2} + \frac{1}{R^2} \frac{\partial^2 u_\theta}{\partial \phi^2} \right.$$

$$\left. + \frac{\cot \phi}{R^2} \frac{\partial u_\theta}{\partial \phi} + \frac{2}{R^2 \sin \phi} \frac{\partial u_R}{\partial \theta} + \frac{2 \cos \phi}{R^2 \sin^2 \phi} \frac{\partial u_\phi}{\partial \theta} - \frac{u_\theta}{R^2 \sin^2 \phi} \right] \qquad \text{(D.28)}$$

$$\frac{\partial u_\phi}{\partial t} + u_R \frac{\partial u_\phi}{\partial R} + \frac{u_\theta}{R \sin \phi} \frac{\partial u_\phi}{\partial \theta} + \frac{u_\phi}{R} \frac{\partial u_\phi}{\partial \phi} + \frac{u_R u_\phi}{R} - \frac{u_\theta^2 \cot \phi}{R}$$

$$= -\frac{1}{\rho R} \frac{\partial p}{\partial \phi} + f_\phi + \nu \left[\frac{1}{R} \frac{\partial^2 (R u_\phi)}{\partial R^2} + \frac{1}{R^2 \sin^2 \phi} \frac{\partial^2 u_\phi}{\partial \theta^2} + \frac{1}{R^2} \frac{\partial^2 u_\phi}{\partial \phi^2} \right.$$

$$\left. + \frac{\cot \phi}{R^2} \frac{\partial u_\phi}{\partial \phi} - \frac{2 \cos \phi}{R^2 \sin^2 \phi} \frac{\partial u_\theta}{\partial \theta} + \frac{2}{R^2} \frac{\partial u_R}{\partial \phi} - \frac{u_\phi}{R^2 \sin^2 \phi} \right] \qquad \text{(D.29)}$$

D.4 Rotating Coordinate System

For flow in a coordinate system rotating with angular velocity Ω, the absolute velocity (velocity seen by a stationary observer), $\mathbf{u}_{abs}$, is related to the velocity seen by an observer in the rotating frame, $\mathbf{u}$, according to

$$\mathbf{u}_{abs} = \mathbf{u} + \Omega \times \mathbf{r} \qquad \text{(D.30)}$$

where $\mathbf{r}$ is position vector. Figure D.4 shows a coordinate system rotating about the z axis.

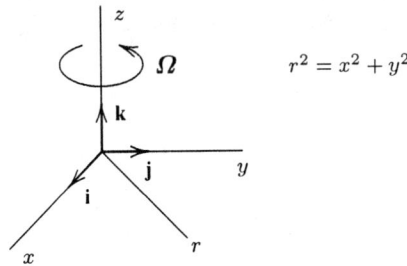

Figure D.4: *Rotating coordinate system.*

A key difference from stationary coordinates is the presence of the Coriolis and centrifugal accelerations, which are usually treated as body forces. For convenience, denote the sum of these two forces by $\mathbf{f}_{Extra}$. Then, we write

$$\mathbf{f}_{Extra} = \mathbf{f}_{Coriolis} + \mathbf{f}_{Centrifugal} \qquad \text{(D.31)}$$

where

$$\mathbf{f}_{Coriolis} = -2\Omega \times \mathbf{u} \qquad \text{(D.32)}$$

$$\mathbf{f}_{Centrifugal} = -\Omega \times \Omega \times \mathbf{r} \qquad \text{(D.33)}$$

The Coriolis force is non-conservative and cannot be simplified. By contrast, the centrifugal force is conservative as can be seen from the following operations.

$$- \boldsymbol{\Omega} \times \boldsymbol{\Omega} \times \mathbf{r} = -\Omega \mathbf{k} \times \Omega \mathbf{k} \times (x\,\mathbf{i} + y\,\mathbf{j} + z\,\mathbf{k}) = \Omega^2(x\,\mathbf{i} + y\,\mathbf{j}) \tag{D.34}$$

However, the vector $(x\,\mathbf{i} + y\,\mathbf{j})$ can be written as the gradient of $\frac{1}{2}r^2$, where r is the radial distance from the origin in the xy plane, viz,

$$x\,\mathbf{i} + y\,\mathbf{j} = \frac{1}{2}\nabla\left(x^2 + y^2\right) = \frac{1}{2}\nabla r^2 \tag{D.35}$$

Therefore, we can rewrite the centrifugal force as follows.

$$\mathbf{f}_{Centrifugal} = \nabla\left(\frac{1}{2}\Omega^2 r^2\right) \tag{D.36}$$

Thus, the Navier-Stokes equation in a rotating coordinate system assumes the following form.

$$\frac{d\mathbf{u}}{dt} + 2\boldsymbol{\Omega} \times \mathbf{u} = -\frac{1}{\rho}\nabla p + \nabla\left(\frac{1}{2}\Omega^2 r^2\right) + \mathbf{f} + \nu\nabla^2\mathbf{u} \tag{D.37}$$

The continuity equation is unaffected. As a final comment, absolute vorticity is

$$\nabla \times \mathbf{u}_{abs} = \nabla \times \mathbf{u} + \nabla \times (\boldsymbol{\Omega} \times \mathbf{r}) = \nabla \times \mathbf{u} + 2\boldsymbol{\Omega} \tag{D.38}$$

D.5   Natural (Streamline) Coordinates

Natural, or streamline, coordinates are sometimes useful as a conceptual tool for illustrating nuances of fluid motion. For two-dimensional, steady flow we envision one of the coordinates, s, being everywhere parallel to the fluid velocity, i.e., parallel to a streamline. The other coordinate, n, is chosen to be normal to streamlines (see Figure D.5).

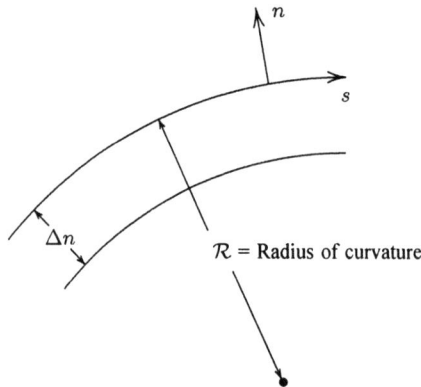

Figure D.5: *Natural coordinates.*

Since there is no flow normal to a streamline, the velocity has one component, u. An additional consequence is that the amount of fluid flowing between two streamlines is constant. Thus, denoting the distance between two streamlines by Δn, conservation of mass becomes

$$\rho u \Delta n = \text{constant} \tag{D.39}$$

The streamwise and normal components of the Euler equation are

$$u\frac{\partial u}{\partial s} = -\frac{1}{\rho}\frac{\partial p}{\partial s} + f_s \quad \text{and} \quad \frac{u^2}{\mathcal{R}} = -\frac{1}{\rho}\frac{\partial p}{\partial n} + f_n \tag{D.40}$$

where f_s and f_n are the body-force components tangent to and normal to the streamline, respectively. If the body force is conservative so that $\mathbf{f} = -\nabla\mathcal{V}$, then

$$u\frac{\partial u}{\partial s} = -\frac{1}{\rho}\frac{\partial p}{\partial s} - \frac{\partial \mathcal{V}}{\partial s} \quad \text{and} \quad \frac{u^2}{\mathcal{R}} = -\frac{1}{\rho}\frac{\partial p}{\partial n} - \frac{\partial \mathcal{V}}{\partial n} \tag{D.41}$$

The vorticity in natural coordinates is normal to the plane of motion, and its value is

$$\omega = \frac{u}{\mathcal{R}} - \frac{\partial u}{\partial n} \tag{D.42}$$

Appendix E

Companion Software

The software described in this appendix corresponds to the Computational Fluid Dynamics sections in Chapters 11 through 15. The companion diskette includes the FORTRAN source code for all of the programs. In all cases, accurate algorithms are used that guarantee grid-independent solutions on any computer from an IBM PC to a Cray Y/MP. All of the programs are short and essentially self explanatory, so only a brief description is required. Each program is described, including a discussion of input and approximate computing time. The concluding section lists the contents of the diskette provided with this book.

E.1 Program Descriptions

All key program variables are identified at the beginning of each program. To the greatest extent possible, variable names match notation used in the text. These programs can serve as the foundation for additional exploration into the world of CFD, and the reader is encouraged to experiment with the programs.

E.1.1 Program POTFLOW

Program **POTFLOW** solves for potential flow past a vertical plate, the application discussed in Subsection 11.9.3. The program prompts for two input parameters that are entered from the keyboard. The first is the relaxation parameter, λ. The program converges only for values of $\lambda < 1$. The second is an index that sets the number of grid points, $imax$. The total number of grid points in the streamwise direction, N, is as follows.

$$N = \begin{cases} 51, & imax = 0 \\ 101, & imax = 1 \\ 201, & imax = 2 \end{cases} \tag{E.1}$$

Entering any value of $imax$ other than those listed causes N to be set equal to 51.

Upon termination, the program displays the vertical velocity on the front face of the plate, $v(0^-, y)/U$, on the video display. The program pauses to permit inspection of the velocity. Upon pressing any key, the program then displays the horizontal velocity above the plate, $u(0, y)/U$, pausing each time the screen fills with new values.

As the run progresses, the program displays the iteration number and the maximum residual every 50 iterations. The solution terminates when the maximum residual is reduced to 10^{-6}. Table E.1 lists the number of iterations and CPU time for $\lambda = 0.9$ on a 100 MHz 80486-based microcomputer.

Table E.1: *Iterations and CPU Time for Program POTFLOW*

N	Iterations	CPU Time (sec)
51	1,550	5
101	5,267	66
201	14,976	750

E.1.2   Program SUCTION

Program **SUCTION** solves for viscous flow over a plate with uniform suction using MacCormack's explicit time-marching method. This application is discussed in Subsection 12.9.4. The program prompts for two input parameters that are entered from the keyboard. The first is the CFL number, $\tilde{N}_{CFL}$. The program converges for values of $\tilde{N}_{CFL} < 1$. The second is an index that sets the number of grid points, *jmax*. The total number of grid points normal to the surface, N, is as follows.

$$N = \begin{cases} 51, & jmax = 0 \\ 101, & jmax = 1 \\ 201, & jmax = 2 \\ 401, & jmax = 4 \end{cases} \qquad (E.2)$$

Entering any value of *jmax* other than those listed causes N to be set equal to 51.

Upon completion of the run, the program displays the computed and exact velocities, u/U_∞, as a function of $v_w y/\nu$.

As the run progresses, the program displays the timestep number and the maximum difference between the numerical and exact solutions every 50 timesteps. The run terminates when a pre-selected time, *tmax*, is reached. Table E.2 lists the number of timesteps for $\tilde{N}_{CFL} = 0.9$ and CPU time on a 100 MHz 80486-based microcomputer.

Table E.2: *Timesteps and CPU Time for Program SUCTION*

N	Timesteps	CPU Time (sec)
51	1,467	1
101	4,667	4
201	15,489	25
401	46,800	144

E.1.3   Program STAGPT

Program **STAGPT** solves for viscous flow in the vicinity of a two-dimensional stagnation point using the Crank-Nicolson implicit time-marching method. The stagnation-point flow solution is given in Subsection 13.7.4. The program prompts for two input parameters that are entered from the keyboard. The first is the CFL number, $\tilde{N}_{CFL}$. The program converges most rapidly for a CFL number of $\tilde{N}_{CFL} = 20$. The second is an index that sets the number of grid points, *jmax*. The total number of grid points normal to the surface, N, is as follows.

$$N = \begin{cases} 51, & jmax = 0 \\ 101, & jmax = 1 \\ 201, & jmax = 2 \end{cases} \tag{E.3}$$

Entering any value of *jmax* other than those listed causes N to be set equal to 51.

Upon completion of the run, the program displays the dimensionless velocity, $U(\eta)$, streamfunction, $G(\eta)$, and pressure, $P(\eta)$, as a function of η.

As the run progresses, the program displays the timestep number and the maximum residual every 20 timesteps. The solution terminates when the maximum residual is reduced to $5 \cdot 10^{-4}$. Table E.3 lists the number of timesteps and CPU time for $\tilde{N}_{CFL} = 20$ on a 100 MHz 80486-based microcomputer.

Table E.3: *Timesteps and CPU Time for Program STAGPT*

N	Timesteps	CPU Time (sec)
51	65	< 1
101	133	1
201	534	3

E.1.4   Program FALKSKAN

Program **FALKSKAN** solves the Falkner-Skan equation for an incompressible boundary layer with pressure gradient using the Crank-Nicolson implicit time-marching method. The Falkner-Skan equation is discussed in Subsection 14.2.5. The program prompts for three input parameters that are entered from the keyboard. The first is the pressure-gradient parameter, β. The program accepts values between -0.1988 and 2. The special cases $\beta = 0$ and $\beta = 1$ correspond to the Blasius (constant-pressure) boundary layer and stagnation-point flow, respectively. Entering a value outside this range causes the program to prompt for an acceptable value. The second is the CFL number, $\tilde{N}_{CFL}$. The program converges most rapidly for most values of β using a CFL number of $\tilde{N}_{CFL} = 10$. The third is an index that sets the number of grid points, *jmax*. The total number of grid points normal to the surface, N, is as follows.

$$N = \begin{cases} 51, & jmax = 0 \\ 101, & jmax = 1 \\ 201, & jmax = 2 \end{cases} \tag{E.4}$$

Entering any value of *jmax* other than those listed causes N to be set equal to 51.

Upon completion of the run, the program displays the dimensionless velocity, $U(\eta)$, and streamfunction, $G(\eta)$, as a function of η.

As the run progresses, the program displays the timestep number and the maximum residual every 20 timesteps. The solution terminates when the maximum residual is reduced to 10^{-4}. Table E.4 lists the number of timesteps and CPU time for $\beta = -0.1$ and $\tilde{N}_{CFL} = 10$ on a 100 MHz 80486-based microcomputer.

Table E.4: *Timesteps and CPU Time for Program FALKSKAN*

N	Timesteps	CPU Time (sec)
51	140	< 1
101	430	2
201	1,496	7

E.1.5 Program RANKIN

Program **RANKIN** computes the surface-pressure distribution on a Rankine oval and writes input data files required by Program **EDDYBL** (Appendix F). The Rankine-oval solution is discussed in Subsection 11.7.2. The program works with dimensionless quantities, including the source/sink strength, ϵ, defined by

$$\epsilon \equiv \frac{Q}{\pi U a} \qquad (E.5)$$

where Q is source/sink strength, U is freestream velocity and a is distance between the source and sink. Program **RANKIN** is valid for arbitrary $\epsilon > 0$, and can be used to duplicate the computations discussed in Subsection 14.4.4.

Input-parameter description: The program reads the following 8 input parameters from disk file **rankin.dat** in the order listed below. Integer quantities must be formatted according to (7x,i6) while floating-point quantities must be formatted as (7x,e13.6).

eps	Dimensionless source strength, $Q/(\pi U a)$
pinf	Freestream pressure, p_∞
rho	Freestream density, ρ
uinf	Freestream velocity, U
tw	Surface temperature, T_w
qw	Surface heat flux, q_w
c	Body length, $c = 2\ell$
nmax	Number of points along the x axis, N_{max}

The program first computes oval dimensions and body coordinates. Then it computes velocity components and the surface pressure from the potential-flow solution. It then prints the following message (for $\epsilon = 1$).

```
Thickness ratio   =    6.083477E-01
Maximum arclength =       12.927210
PRESUR.DAT and HEATER.DAT successfully written
Display x versus s ? (y/n)
```

Entering a y yields a display of axial distance, measured from the center of the Rankine oval, as a function of the arc length measured from the leading stagnation point.

The diskette supplied with this book includes input data files for Programs **RANKIN** and **EDDYBL** corresponding to $\epsilon = .01, .1, 1, 10$ and 100. This program runs in about one second on a 100 MHz 80486-based microcomputer.

E.1.6 Program BURGER

Program **BURGER** solves for a propagating one-dimensional wave using MacCormack's explicit time-marching method. This application is discussed in Subsection 15.7.3. The program prompts for three input parameters that are entered from the keyboard. The first is the total number of timesteps. The second is the number of timesteps between video displays of the solution. The third determines the type of wave. Entering 0 yields a shock wave. Any other number yields an expansion wave.

As the run progresses, the program displays the velocity, $u(x, t)$, as a function of position, x, and pauses to permit examination of data displayed on the monitor. The time, t, is also shown. Run times are of the order of a second between displays on a 100 MHz 80486-based microcomputer.

E.2 Diskette Contents

Flowfield Program Source:

burger.for	Source code for Program **BURGER**
falkskan.for	Source code for Program **FALKSKAN**
potflow.for	Source code for Program **POTFLOW**
rankin.for	Source code for Program **RANKIN**
stagpt.for	Source code for Program **STAGPT**
suction.for	Source code for Program **SUCTION**

Input Data Files:

verythin.zip	Input for Programs **RANKIN** and **EDDYBL**; $\epsilon = .01$
thin.zip	Input for Programs **RANKIN** and **EDDYBL**; $\epsilon = .1$
moderate.zip	Input for Programs **RANKIN** and **EDDYBL**; $\epsilon = 1$
fat.zip	Input for Programs **RANKIN** and **EDDYBL**; $\epsilon = 10$
veryfat.zip	Input for Programs **RANKIN** and **EDDYBL**; $\epsilon = 100$

Appendix F

Program EDDYBL

F.1   Overview

This appendix is the user's guide for Program **EDDYBL**, a two-dimensional and axisymmetric, compressible boundary-layer program for laminar, transitional and turbulent boundary layers that is included on the distribution diskette. An overview of the program's operation is given, along with instructions for installing the program and its menu-driven input-data preparation utility, **SETEBL**, on your computer. Two bench-mark runs are described that can be used to make sure the program is operating properly. The software includes a plotting utility for both video and hardcopy plots on IBM PC and compatible computers.

F.1.1   Acknowledgments

Program **EDDYBL** is a compressible, two-dimensional and axisymmetric program suitable for computing properties of laminar, transitional and turbulent boundary layers. The program embodies a wide variety of turbulence models ranging from mixing-length oriented algebraic models to a complete second-order closure model. This program has evolved over the past two and a half decades and can thus be termed a mature software package. Many U. S. Government Agencies have contributed to development of the program that is based on a computer code originally developed by Price and Harris (1972).

Additionally, important improvements have been made to this software package as a result of feedback from users, most notably from the outstanding fluid mechanics students at UCLA and USC. The author owes special thanks to Dr. G. Brereton of the University of Michigan whose personal efforts resulted in the addition of the option to use either USCS or SI units, and to J. Morrison of AS&M for adapting the software to a SUN Workstation.

F.1.2   Required Hardware and Software

Versions of the program are available for the following computers.

- Cray X-MP and Y-MP
- VAX 11 and 8600
- SUN Workstations
- Silicon Graphics Iris
- Intel 80386, 80486 and Pentium Based Microcomputers
- Definicon 68020 and 68030 Coprocessor Boards
- Definicon SPARC Coprocessor Boards
- IBM PC/XT/AT and Compatibles

The program requires at least 320 kilobytes of memory. To achieve sensible computing times, IBM PC/XT/AT and compatibles should have an 8087 or 80287 math coprocessor, and must use Microsoft Fortran Version 5.0 or higher. Intel 80386 based machines must have either an 80387 or Weitek math coprocessor.

F.2   Getting Started Quickly

Because **EDDYBL** and its input-data preparation utility, **SETEBL**, run on many different computers, installation of the software is a little different for each machine. The main difference occurs in the commands needed to compile and link the programs. To install the software on a computer other than an IBM PC or compatible, you must skip ahead to Sections F.3 and F.4. If you have an IBM PC or compatible microcomputer running under either the MS-DOS or Windows operating systems, executable versions of the software package are included on the distribution diskette. Regardless of the computer you are using, once you have executable programs, complete the installation as follows.

1. Read the contents of the file **read.me** in subdirectory **eddybl** on the distribution diskette. This file will tell you of any program revisions as well as the location of all pertinent program files on the diskette. Then, copy the following files to your working directory:

eddybl.exe	blocrv.dat	ploteb.exe
instl.exe	heater.dat	ploteb.dat
setebl.exe	presur.dat	exper.dat

   Omit the files **ploteb.exe**, **ploteb.dat** and **exper.dat** if you are using a computer other than an IBM PC or compatible microcomputer.

2. Run Program **INSTL** and answer all questions posed by the program. This program generates a file named **grafic.dat** that should be saved in your working directory.

3. If your computer is an IBM PC or compatible, install the **ansi.sys** driver supplied with your MS-DOS operating system by adding the following command to your **config.sys** file:

   device=ansi.sys

   Make sure the file **ansi.sys** is available in your path. If you have not previously had this command in your **config.sys** file, you must now re-boot your computer to install the **ansi.sys** driver.

A simple bench-mark case is built into the software package to allow you to quickly determine that everything is operating properly, and to see how easy it is to use Program **EDDYBL**. Because the input-data preparation utility, **SETEBL**, is menu driven, you will find that very little explanation of the program's operation is needed. After successfully completing the following bench-mark run, the first time user should nevertheless do the example of Section F.5 to be sure the software is properly installed and to learn some of the more subtle features of Program **SETEBL**.

1. The first step is to run **SETEBL**. If you have not installed **SETEBL**, you will be notified with a brief message after which the program will immediately terminate. If this happens, refer to Section F.3 and perform the installation procedure.

2. Assuming the program is properly installed, you will see a message informing you that file **eddybl.dat** does not exist. The message asks if you want to create a new file named **eddybl.dat**. For this sample session, you should answer yes by typing the letter *Y* or *y* followed by pressing the *ENTER* key.

3. Having performed Step 2, you are presented with the main menu (Figure F.1) on which ten options are listed. **SETEBL** has default values for all input parameters that correspond to an incompressible (Mach .096) flat-plate boundary layer. When you eventually exit **SETEBL**, these data will be written into an Ascii data file named **eddybl.dat**. For this bench-mark case, if you selected USCS units as the default when you ran Program **INSTL**, you have no need to change any data. However, this case must be done using USCS units. If you selected SI units as the default when you ran **INSTL**, you must change to USCS units. Type either a *U* or a *u* (for Units - note that the *U* is in reverse video on your display) and press the *ENTER* key to make the change. The menu will change to indicate which units are in effect. Before exiting, you must generate two other data files, viz., **table.dat** and **input.dat** that are needed in order to run **EDDYBL**.

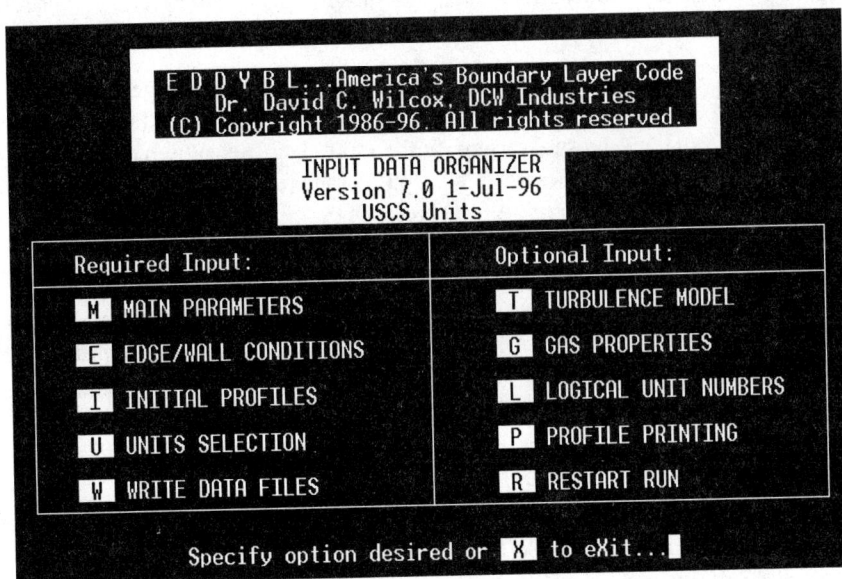

```
E D D Y B L...America's Boundary Layer Code
      Dr. David C. Wilcox, DCW Industries
  (C) Copyright 1986-96. All rights reserved.

            INPUT DATA ORGANIZER
            Version 7.0 1-Jul-96
                 USCS Units

 Required Input:                Optional Input:

   [M] MAIN PARAMETERS            [T] TURBULENCE MODEL

   [E] EDGE/WALL CONDITIONS       [G] GAS PROPERTIES

   [I] INITIAL PROFILES           [L] LOGICAL UNIT NUMBERS

   [U] UNITS SELECTION            [P] PROFILE PRINTING

   [W] WRITE DATA FILES           [R] RESTART RUN

      Specify option desired or [X] to eXit...█
```

Figure F.1: *Opening menu of Program SETEBL.*

4. To generate these files, select the **Write Data Files** option. To do this, type a *W* or a *w* followed by pressing the *ENTER* key. After a short wait, you will be notified that the binary data file **table.dat** has been successfully written. You are now presented with the following query in reverse video.

Save the profiles in Ascii form? (X=eXit, Y=Yes, ENTER=No)...

If you desire a copy of the initial profiles to be saved in a disk file named **setebl.prt** for inspection at a later time, respond with a *Y, ENTER* sequence; otherwise press the

ENTER key. After you have responded to this query, a second query will appear, viz.,

Display the profiles on the video? (X=eXit, Y=Yes, ENTER=No)...

If you want to see the profiles on your video display, respond accordingly. Otherwise, press *ENTER*. After you have responded, your screen clears again and a message appears indicating initial profiles are being generated. If you elected to display profiles on your video display, they will now be displayed, a screen at a time. Press *ENTER* to advance to the next screen. Regardless of the options you have chosen, the precise values of the integral parameters for your computed initial profiles are displayed. Finally, a message appears indicating the binary data file **input.dat** has been successfully written and that you must press *ENTER* in order to continue.

5. After you press *ENTER*, control returns to the main menu. At this point you have prepared all input-data files for the bench-mark run. Exit by typing *X*, *ENTER*.

6. All that remains now is to run Program **EDDYBL**. Program output will be directed to a disk file named **eddybl.prt**. The file **eddybl.prt** supplied on the distribution diskette contains the printout for the bench-mark run on a personal computer. Your results should agree to within several decimal places with those in the sample printout. For reference, Table F.1 summarizes approximate computing times required for several computers.

7. If you are using an IBM PC or compatible computer, you can generate a video and hardcopy plot of the computational results by running Program **PLOTEB**. Before executing this program, be sure to modify the input-data file, **ploteb.dat**, as required for your system. Section F.8 describes all input parameters in the file. If you are running under Windows, be sure to use a full-screen DOS window when you run PLOTEB.

Table F.1: *Computing Time for the Bench-Mark Case*

Computer	CPU (MHz)	FPU (MHz)	CPU Time(sec)
Cray 2	-	-	1
PC-Pentium	80586 (133)	80587 (133)	2
PC-486/DX4	80486 (100)	80487 (100)	3
CDC-7600	-	-	3
PC-486/DX2	80486 (66)	80487 (66)	5
PC-486/Weitek	80486 (25)	mW4167 (25)	7
PC-486/DX	80486 (33)	80487 (33)	9
PC-386/Weitek	80386 (33)	mW3167 (33)	12
VAX 8600	-	-	17
SPARC	7C601 (20)	8847 (20)	21
PC-386/DX	80386 (33)	80387 (33)	26
Tandy 4000	80386 (16)	mW1167 (16)	30
DSI-785+	68020 (20)	68882 (20)	42
VAX 11/785	-	-	63
PC-286/DX	80286 (12)	80287 (8)	270

F.3 Installing SETEBL

To use the supplied executable version of **SETEBL** on an IBM PC or compatible micro-computer, including the default values specified for all input-data parameters, simply copy the executable file to your working directory. Otherwise, if you wish to change some of the default values, or if you are using a computer other than an IBM PC, the first step required to install Program **SETEBL** is to compile and link the program. The main program is the file named **setebl.for**, and the various subroutines are listed in Section F.11. All routines reference three **include** files, **chars**, **comeb1** and **comeb2**. Section F.10 summarizes the commands required to compile and link Program **SETEBL**.

> **The first step required to install SETEBL for your computer is to either copy the executable file to your working directory or to compile and link Program SETEBL.**

In order to use Program **SETEBL**, you must first install it for your particular console. The program makes extensive use of reverse video, direct cursor positioning, and some graphics characters. Since no uniform standard exists for such console characteristics, the appropriate sequences used by your console must be defined for Program **SETEBL**.

F.3.1 Boot-Console Installation

In order to install **SETEBL** on your main (or boot) console, you must generate a binary data file named **grafic.dat** that contains all of the information needed by **SETEBL**. The source code for a program that generates **grafic.dat** customized for your console has been supplied as part of this software package. The program is called **INSTL**, and the source is contained in **instl.for**. If you customize Program **INSTL** or if you are using a computer other than an IBM PC, you must first compile and link Program **INSTL**. Then:

> **The second step required to install SETEBL for your console is to run Program INSTL.**

When you run Program **INSTL**, you will be given the option of specifying whether you want the default units to be USCS or SI. Make the choice best suited to your needs. You will also have to specify the type of computer you have and, in some cases, the type of console.

When you have successfully run Program **INSTL**, the required binary data file **grafic.dat** will be created and **INSTL** will print a message to that effect. Whenever you wish to run **SETEBL**, simply make sure **grafic.dat** is present in your directory. If it is not present, **SETEBL** displays a message informing you that you are attempting to run an uninstalled version of **SETEBL** and promptly terminates. If you are running Program **SETEBL** on an IBM PC based system, you must also install the **ansi.sys** driver. Thus,

> **The third step required to install SETEBL on an IBM PC based system is to install the ansi.sys driver by adding the following command to your config.sys file.**

device=ansi.sys

If you have not previously had this command in your **config.sys** file, it will not take effect until you re-boot your computer.

F.3.2   Remote-Terminal Installation

For a remote terminal whose characteristics are different from those of your boot console, you can create another **grafic.dat** by making appropriate changes to Program **INSTL**. The program is heavily commented, and customization should be straightforward.

F.4   Installing EDDYBL

To use the supplied executable version of **EDDYBL** on an IBM PC or compatible microcomputer, simply copy the executable file to your working directory. Otherwise, if you wish to make program changes, or if you are using a computer other than an IBM PC, the first step required to install Program **EDDYBL** is to compile and link the program. The main program is the file **eddybl.for** that also makes use of the include files **common** and **cpuid**. Be sure to link with the **/e** option for the Microsoft Fortran version or the **-pack** option with SVS Fortran-386/Phar Lap to reduce the size of the executable file.

> **The only step required to install EDDYBL for your computer is to either copy the executable file to your working directory or to compile and link Program EDDYBL.**

F.5   Running a General Case

This section explores, in detail, all of the salient features of the input-data preparation utility, **SETEBL**. You will be guided through the various menus and, in the process, you will set up a constant-pressure boundary-layer computation for a Mach 1 freestream. For the case you will do, freestream conditions are as follows.

$$
\begin{array}{lll}
\text{Total pressure, } p_{t_\infty} & = & 482.7 \text{ lb/ft}^2 \ (23112 \text{ N/m}^2) \\
\text{Total temperature, } T_{t_\infty} & = & 468^\circ \text{ R } (260 \text{ K}) \\
\text{Mach number, } M_\infty & = & 1
\end{array}
$$

The surface will be slightly cooled so that surface temperature is 95% of the adiabatic-wall temperature.

Your goal is to initiate the computation at a plate-length Reynolds number, Re_x, of one million and determine the point where the momentum-thickness Reynolds number, Re_θ, is 8000. You might want to do this, for example, in order to provide upstream profiles for a Navier-Stokes computation. You know from a correlation of experimental data that when $Re_x = 1.0 \cdot 10^6$, the boundary layer has the following integral properties:

$$
\begin{array}{lll}
\text{Skin friction, } c_f & = & .0038 \\
\text{Shape factor, } H & = & 1.80 \\
\text{B.L. Thickness, } \delta & = & 11.9 \ \theta \\
\text{Reynolds number, } Re_\theta & = & 1500
\end{array}
$$

Finally, the surface is perfectly smooth, there is no surface mass transfer, and the Spalart-Allmaras model will be used.

F.5.1 Preliminary Operations

To perform this exercise, delete any existing **eddybl.dat** data file that might be in your directory. Although this is not generally necessary, for the purposes of this section it will be easier if you begin with the default values.

As with the bench-mark case of Section F.2, the very first step is to run Program **SETEBL**. If you have not installed the program, you will be notified with a brief message after which the program will immediately terminate. If this happens, go back to Section F.3 and perform the installation procedure.

Assuming the program is properly installed, you will see a message informing you that file **eddybl.dat** does not exist. You will be asked if you want to create a new file named **eddybl.dat**. For this sample session, you should answer yes by typing the letter *Y* followed by pressing the *ENTER* key.

F.5.2 Units Selection

This case can be done in either USCS or SI units. Examine the main menu to determine which units are in effect. If you wish to change units, type a *U* followed by pressing *ENTER*. The menu will reflect the change in units immediately. If you change your mind and wish to go back to the original units, repeat the *U, ENTER* sequence. In the following sections, values are quoted in USCS units followed by corresponding SI values in parentheses.

F.5.3 Main Parameters

At this point, you will be presented with the main menu on which ten options are listed. Begin by entering the **Main Parameters** sub-menu. To enter this sub-menu, type an *M* followed by pressing the *ENTER* key.

Yet another sub-menu will now appear that gives you the choice of entering input data for either **Freestream Conditions** or **Body Parameters**. There is a third option that allows you to **eXit**. The latter option permits you to return to the previous menu. You will eventually do so, but first you will do some actual data preparation.

Freestream Parameters. Type an *F* followed by pressing the *ENTER* key to descend to the **Freestream Conditions** menu. You will now see a display that includes seven of the primary quantities that specify freestream flow conditions, including freestream total pressure, total temperature, Mach number, shock-wave angle, and some turbulence parameters. The bottom row provides instructions on how to proceed. Press the *ENTER* key several times, for example, and you will see the arrow move from one input variable name to the next. When you reach the last variable, pressing the *ENTER* key again will cause the arrow to move to the uppermost variable. You may make as many passes through the list of variables as you wish.

This particular menu includes a Help option to further explain the meaning of the more obscure input quantities. To display the Help menu, type an *H* followed by pressing the *ENTER* key. After reading this Help menu, pressing the *ENTER* key returns you to the Freestream Conditions data-entry menu.

Having returned from the Help menu, you will now exercise the **Change** option. First, position the arrow in front of Mach number. You accomplish this by pressing *ENTER* twice. Now, type the letter *C* (for Change) followed by pressing *ENTER*. The bottom line of the menu will now change. You are told to specify the new value, and that the FORMAT must be the standard FORTRAN floating-point format E13.6. The default value assigned to the

Mach number is .096, corresponding to essentially incompressible flow conditions. Change the Mach number to one by typing 1. (the exponent E+00 is unnecessary but the decimal point is mandatory — this is normal FORTRAN I/O). As with all commands to **SETEBL**, nothing will happen until you press the *ENTER* key. Before you do however, watch the line near the bottom of your display entitled **Static Conditions**. Keeping your eyes on the static conditions line, press the *ENTER* key. If you have done this step correctly, the new static conditions should appear in place of the old. Also, if you look at the value assigned to the Mach number you will find it has been changed to one.

At this point, you can change any of the seven input quantities. In addition to Mach number, you must change total pressure and temperature. Press the *ENTER* key five times in order to position the arrow in front of PT1, the total pressure. Using the change procedure, i.e., type a *C* followed by *ENTER*, insert the desired total pressure of 482.7 lb/ft^2 (23112. N/m^2). You may enter 4.827e+02 (2.3112e+04) or 4.827E2 (2.3112E4), etc. if you wish. Note that your keyboard's normal destructive backspace key can be used to correct typing errors. When your desired new total pressure is correctly entered, press *ENTER* and the change will be made. Verify that the new value for PT1 shown on the display is 4.827000E+02 (2.311200E+04). If you made any mistakes, repeat the change operation until you get it right.

Now press the *ENTER* key to position the arrow in front of TT1, the total temperature. Using the change procedure, change the value of TT1 to 468. (260.). Don't forget the decimal point or else your total temperature will be .000468 (.000260). Verify that the new value for TT1 shown on the display is 4.680000E+02 (2.600000E+02).

If you have changed Mach number, total pressure and total temperature correctly, the value listed below for static pressure will be very close to 255 lb/ft^2 (12209 N/m^2) and the unit Reynolds number should be approximately $1.24 \cdot 10^6$ ft^{-1} ($4.07 \cdot 10^6$ m^{-1}). Verify that the static conditions you have entered match these two values. If they do not, find and correct any errors you have made before continuing.

Jot the values of static pressure and freestream unit Reynolds number on a slip of paper for reference later. In general, knowing these values often helps expedite preparation of your input data. You can always return to this menu to find their values, of course. Later on, we will see an example of using both parameters to determine input quantities on other sub-menus.

You have now finished this sub-menu. In order to exit, simply type an *X* followed by pressing the *ENTER* key. Note that, with the exception of Help menus for which only *ENTER* is needed, you return to the previous menu by the *X*, *ENTER* sequence. Also, if you are ever in doubt about what to do, look at the last line of the display for instructions.

Body Parameters, Etc. Now you are back to the **Main Parameters** sub-menu that provides the options of altering freestream conditions, body parameters, etc. Descend to the **Body Parameters** sub-menu by typing the letter *B* followed by pressing *ENTER*. You will be presented with a menu similar to the **Freestream Conditions** sub-menu. As before, press *ENTER* several times to move the arrow. Scan the input variable definitions and default values. Examine the Help menu. In other words, begin discovering that you already know most of what is needed in order to operate **SETEBL**!

There are only two input quantities you need to change, viz., ISHORT and SSTOP. Because you are looking for the point where momentum-thickness Reynolds number is 8000, you have no need for the long printout that gives far more detail than you are interested in. Consequently, you should position the arrow next to ISHORT and change its value to 0. This is done by typing *C* and a carriage return; no value need be entered. As an experiment, you might want to try repeating this sequence. If you do, the value of ISHORT will change back to 1. Be sure you have changed ISHORT to 0 after you finish experimenting.

Turning now to SSTOP, use the *ENTER* key to position the arrow next to SSTOP. This is the maximum value of plate length to which you will permit computation to proceed. Imagine that you are certain the momentum-thickness Reynolds number will reach 8000 at a plate-length Reynolds number somewhere between three and five million. Hence, you might want to terminate your run when Reynolds number reaches five million. Referring to the unit Reynolds number of $1.24 \cdot 10^6$ ft^{-1} ($4.07 \cdot 10^6$ m^{-1}) that you jotted down earlier, a quick computation shows that a plate-length Reynolds number of five million occurs when plate length is 4.03 ft (1.23 m). Thus, change SSTOP to 4.03 (1.23).

There are no further changes you need to make at this time on this menu, so you should now exit by typing X followed by pressing the *ENTER* key. At this point, you are done with the **Main Parameters** sub-menu. In order to return to the main menu, type another X followed by pressing the *ENTER* key. Remember, nothing happens in **SETEBL** until you press the *ENTER* key.

F.5.4   Taking a Lunch Break

Before continuing setting up a new run, you are going to simulate a lunch break. Imagine that it's time to break for lunch and the systems people upstairs are notorious for causing your VAX 8600 to crash during the lunch hour. Any file you leave open will be lost as a result of a crash. In order to protect your work from such a disaster, simply exercise the exit option by typing yet another X followed by pressing the *ENTER* key.

Inspection of your directory will show that a new file named **eddybl.dat** has been created. Verify that the file exists at this time. If it does not, go back to Subsection F.5.1 and omit the mistake you made that caused you to reach this point unsuccessfully.

Now imagine you have returned from lunch, and your microcomputer (which never crashes during lunch because there are no system people to cause it to) is ready to continue serving your data-processing needs. At this point, run Program **SETEBL** again. Because the data file **eddybl.dat** exists, the program will go directly to the main menu.

F.5.5   Edge/Wall Conditions

From the main menu, you should now proceed to the **Edge/Wall Conditions** sub-menu by typing an E followed by pressing *ENTER*. This sub-menu contains five options, viz., Pressure Distribution, Heat Transfer, Mass Transfer, Body Geometry and eXit. Type a P followed by *ENTER*. The **Pressure Distribution** sub-menu explains that you must prepare a file **presur.dat** that defines the pressure distribution. You must prepare the file with an editor such as MS-DOS 5.0's EDIT, DEC's EDT, UNIX's vi, etc. All you can change in this menu is the number of points you plan on using. You will not change NUMBER because your run will have constant pressure. Thus, you need to specify pressure at two values of plate length. Note that this menu describes in detail the contents and format of **presur.dat**. Exit this menu with the usual X, *ENTER* sequence.

Now go to the **Heat Transfer** sub-menu. You are presented with a description of data file **heater.dat** that must be created with your own editor. Note that the adiabatic-wall temperature is given for your information and the value listed should be 459.4° R (255.2 K). The one parameter you can change on this menu is KODWAL which determines whether you plan on specifying surface heat flux or surface temperature. Type a C followed by *ENTER* to change KODWAL. Note that the display now indicates temperature is prescribed at the surface. Jot down the adiabatic-wall temperature for later reference. Exit this sub-menu by typing an X followed by *ENTER*.

Now go to the **Mass Transfer** sub-menu. This sub-menu describes a file, **blocrv.dat**, that must be prepared externally. You can alter the one parameter NFLAG. The default value is 0, which means **blocrv.dat** is not required to prepare your edge and surface conditions. You have no need to change its value for this application. Note that your display indicates the file **blocrv.dat** will not be required. Exit this sub-menu.

Having exited the **Mass Transfer** sub-menu, you have now made your way back to the **Edge/Wall Conditions** sub-menu. Proceed to the **Body Geometry** sub-menu. No, you didn't make a typing error. This is the same menu you just completed. It has been included for planned future enhancements to **SETEBL**. Exit back to the **Edge/Wall Conditions** sub-menu with an *X, ENTER* sequence.

The final option is to eXit. Do so by typing another *X, ENTER* sequence. You are now back at the main menu. You cannot continue until you have prepared input-data files **presur.dat** and **heater.dat** (**blocrv.dat** is not needed because NFLAG is 0). Hence, it is time to exit **SETEBL** and save all the work you have done so far.

F.5.6 Preparing Edge/Wall Condition Data Files

The easiest way to prepare data files **presur.dat** and **heater.dat** (and **blocrv.dat** as well) is to use an editor such as EDIT, EDT, vi, etc. to modify existing files from a previous run. That is why you left the files from the bench-mark run in your directory. You can delete **blocrv.dat** now if you wish as it won't be needed for this run.

If you have followed all of the instructions correctly, you have the static pressure of 255 lb/ft^2 (12209 N/m^2) and the adiabatic-wall temperature of 459.4° R (255.2 K) jotted down somewhere. Using your favorite editor, change **presur.dat** to one of the following, depending on the units you have chosen:

```
USCS Units:      0.000000E 00   2.550000E 02
                 1.000000E 01   2.550000E 02
                 0.000000E 00   0.000000E 00

SI Units:        0.000000E 00   1.220900E 04
                 1.000000E 01   1.220900E 04
                 0.000000E 00   0.000000E 00
```

As explained in the **Pressure Distribution** sub-menu, the first two lines of this file are arc-length/pressure pairs presented in format (2E14.6). The final line is the pressure gradient at the beginning and end of the interval given in (2E14.6) format also. You have specified pressure at a plate length of zero and ten feet (meters). This interval must at least cover the planned integration range. The value of the pressure is the static pressure you jotted down earlier.

Turning now to surface temperature, note that 95% of the adiabatic-wall temperature is approximately 436° R (242 K), which is the value you should use. Use your editor to modify **heater.dat** as required, noting that the values of arc length at which you specify wall conditions must match the values used for the pressure distribution. As explained on the Heat Transfer sub-menu, this file must consist of one of the following sets of four lines, depending upon which set of units (SI or USCS) you have selected:

USCS Units:	0.000000E 00	4.360000E 02	0.000000E 00
	1.000000E 01	4.360000E 02	0.000000E 00
	0.000000E 00	0.000000E 00	
	0.000000E 00	0.000000E 00	

SI Units:	0.000000E 00	2.420000E 02	0.000000E 00
	1.000000E 01	2.420000E 02	0.000000E 00
	0.000000E 00	0.000000E 00	
	0.000000E 00	0.000000E 00	

The format of the first two lines is (3E14.6), while the last two lines have format (2E14.6). The first column for the first two lines is arc length, the second is wall temperature, and the third is surface heat flux. Note that since you have chosen to specify surface temperature rather than heat flux, any value can be entered for the heat flux — it won't be used in the computation. Similarly, if you choose to specify surface heat flux, the value assigned to surface temperature is arbitrary. The third line gives surface temperature slope at the beginning and end of the interval, while the last line gives surface heat flux slope. Of course, you are not limited to constant properties in the most general case. You may prescribe as many as 50 different values for edge pressure, surface temperature, surface heat flux, etc., which should be sufficient for most boundary-layer computations. **Make sure the arc-length values match those used for the pressure distribution.**

At this point, you have prepared all of the freestream conditions, body parameters, and (from an external editor) the two data files **presur.dat** and **heater.dat**. Before reentering Program **SETEBL**, examine your directory. In addition to the input data file **eddybl.dat**, you should find another file named **eddybl.bak**. The former file is your most recent version of **eddybl.dat**. The latter file is the version you created just before taking your lunch break. Program **SETEBL** always saves your previous work in **eddybl.bak** to provide you with a little extra protection. You no longer need **eddybl.bak**, so delete it if you wish.

F.5.7 Generating Edge/Wall Conditions

Run **SETEBL** again. When the main menu appears, use a *W, ENTER* sequence to execute the Write Data Files option. A message will appear briefly indicating that edge conditions are being generated. When all computations are complete, a message appears telling you that a data file named **table.dat** has been successfully written. If you receive any other message, there are probably errors in the files you created with your editor in Subsection F.5.6, and you must correct them before you can continue setting up your run. What you are doing in this step is executing a subroutine in **SETEBL** that accomplishes two ends. First, you are generating data file **table.dat** in binary form that is used by **EDDYBL**. Second, you are computing several parameters appearing in data file **eddybl.dat** that are needed in preparing the initial profiles that are required by **EDDYBL**.

F.5.8 Initial Profiles

In addition to the message that **table.dat** has been successfully written, you also receive the message

Save the profiles in Ascii form? (X=eXit, Y=Yes, ENTER=No)...

Since you have not yet prepared the data needed to generate initial profiles, type an *X*, *ENTER* sequence. Upon returning to the main menu, you are now ready to go to the Initial Profiles sub-menu. Type an *I* followed by pressing the *ENTER* key. The sub-menu that appears has three options, viz., **Integral Parameters**, **Grid Parameters** and **eXit**.

Integral Parameters. Go to the **Integral Parameters** sub-menu first by entering another *I*, *ENTER* sequence. Press *ENTER* once to position the arrow next to skin friction. Change the value to 0.0038 in the usual manner. Press *ENTER* again to move the arrow in front of boundary-layer thickness, δ. For the conditions specified above, a quick calculation shows that δ for your unit Reynolds number is .0144 ft (.004389 m). Change the value of DELTA to 0.0144 (.004389). Now move the arrow to shape factor and change its value to 1.8. Finally, move the arrow one more time to momentum-thickness Reynolds number and change its value to 1500., being careful to remember the decimal point. Inspect your work for possible errors. When you have made all entries correctly, exit this sub-menu.

Grid Parameters. Now exercise the *G* option to enter the **Grid Parameters** sub-menu. The first quantity you should change is the initial streamwise stepsize, DS. For this constant pressure case, you can use a stepsize as big as triple the boundary-layer thickness. Hence, change the value of DS to 0.04 ft (0.0122 m). In order to start the computation at a plate-length Reynolds number of one million, the initial plate length (arc length) must be 0.806 ft (0.246 m), a fact you can deduce by using the freestream unit number of $1.24 \cdot 10^6$ ft^{-1} ($4.07 \cdot 10^6$ m^{-1}) you jotted down earlier. Hence, change SI to 0.806 (0.246). The next parameter is XK, the geometric-progression ratio. A coarser grid can be used for this case than the default grid. Change the value to 1.14, which corresponds to grid increments increasing in a geometric progression at a 14% rate. Finally, change the number of grid points normal to the surface, IEDGE, to 51. Again, inspect your work for possible errors. When your entries are error free, exit this sub-menu. Having returned to the **Initial Profiles** sub-menu, you should now exercise the exit option with the usual *X*, *ENTER* sequence to return to the main menu.

Generating Initial Profiles. As in Subsection F.5.7, exercise the **Write Data Files** option by entering a *W*, *ENTER* sequence. This will regenerate **table.dat** and you are again presented with the following message.

Save the profiles in Ascii form? (X=eXit, Y=Yes, ENTER=No)...

If you desire a copy of the profiles, in Ascii form, to be sent to a disk file named **setebl.prt** that can be printed and/or examined with an editor after exiting Program **SETEBL**, respond with a *Y*, *ENTER* sequence; otherwise simply press the *ENTER* key. After you have responded to this query, a second query will appear as follows.

Display the profiles on the video? (X=eXit, Y=Yes, ENTER=No)...

If you want to see the profiles on your video display, respond with a *Y*, *ENTER* sequence. Otherwise, press *ENTER*. After you have responded, your screen clears again and a message appears indicating initial profiles are being generated. If you elected to display profiles on your video display, they will now be displayed, a screen at a time. Press *ENTER* to view the next screen. Regardless of the options you have chosen, the precise values of the integral parameters for your computed initial profiles are displayed. Finally, a message appears indicating the binary data file **input.dat** has been successfully written and that you must press *ENTER* in order to continue.

Notice that the value of the conventional sublayer coordinate, y^+, for the point nearest the surface is printed and its value is 0.175. Subroutine START will alert you if this value ever exceeds unity as Program **EDDYBL** requires the value of y^+ nearest the surface to be less than 1 in order to remain numerically stable. If this ever happens, you must either increase XK or IEDGE. When you press *ENTER*, control returns to the main menu.

You will receive a warning if you use the k-ϵ model and the value of y^+ for the point nearest the surface is less than 0.1. Values smaller than 0.1 tend to slow the convergence rate for the k-ϵ model, and may even cause your run to crash.

F.5.9   Selecting a Turbulence Model

In order to select the Spalart-Allmaras model, go to the **Turbulence Model** sub-menu. Type a *T* followed by *ENTER*, and you will find that the fourth quantity listed is a flag called MODEL. Press *ENTER* three times to position the arrow in front of MODEL. Change its value to 4 by typing *C* followed by entering a 4 and pressing *ENTER*. Note that the highlighted bar below the menu now indicates you are using the Spalart-Allmaras model for the computation. Exit back to the main menu.

For general reference, there are 15 turbulence models implemented in Program **EDDYBL**, and the two input parameters MODEL and NVISC are used to make the selection. The choices are as follows [details and references for the turbulence models are given by Wilcox (1993)].

MODEL	NVISC	Turbulence Model
-1	-	None (Laminar Flow)
0	0	k-ω, viscous corrections excluded
0	1	k-ω, viscous corrections included
1	0	Multiscale, viscous corrections excluded
1	1	Multiscale, viscous corrections included
2	0	k-ϵ, Jones-Launder
2	1	k-ϵ, Launder-Sharma
2	2	k-ϵ, Lam-Bremhorst
2	3	k-ϵ, Chien
2	4	k-ϵ, Yang-Shih
2	5	k-ϵ, Fan-Lakshminarayana-Barnett
3	-	ν_t, Baldwin-Barth
4	-	ν_t, Spalart-Allmaras
5	-	Algebraic, Cebeci-Smith
6	-	Algebraic, Baldwin-Lomax
7	-	Half-Equation, Johnson-King

F.5.10   Logical Unit Numbers and Plotting Files

Your final input-data changes will cause printed output to go to your line printer rather than to disk file **eddybl.prt**. You will also verify that two disk files named **profil.dat** and **wall.dat** will be created that can be used as starting conditions for another program or as input to a plotting program. Go to the **Logical Unit Numbers** sub-menu by typing an *L, ENTER* sequence. The first parameter is IUNIT1 which, by default, is disk file **eddybl.prt**. For Lahey, Microsoft or SVS Fortran versions, change its value to 6. For all other versions, use your normal operating system procedure to direct the contents of **eddybl.prt** to a line printer. Verify that the value for the parameter IUPLOT is some value other than 0. If it is 0, change its value to 10 (or any other convenient value excluding unit 15 and any previously assigned unit number).

Disk file **wall.dat** is written as an unformatted file, each record of which can be read by another FORTRAN program according to the following program fragment.

```
        i=1
  10    read(iunit) s(i),res(i),cfe(i),rethet(i),
       *               h(i),che(i),anue(i),pe(i),tw(i)
        if(s(i).ne.-999.) then
          i=i+1
          go to 10
        endif
```

The various quantities saved in disk file **wall.dat** are:

Quantity	Description	Dimensions
s	s, arc length along surface	ft (m)
res	Re_s, Reynolds number based on s	None
cfe	$c_{fe} = 2\tau_w/\rho_e u_e^2$, skin friction	None
rethet	Re_θ, Reynolds number based on θ	None
h	H, shape factor	None
che	$\dot{h}/\rho_e u_e c_p$, Stanton number	None
anue	$Pr\, s\dot{h}/\mu_e c_p$, Nusselt number	None
pe	$\bar{p}$, pressure	lb/ft^2 (N/m^2)
tw	T_w, wall temperature	°R (K)

The first line of the file **profil.dat** generated by Program **EDDYBL** contains the streamwise step number, M, and the number of mesh points normal to the surface, IEDGE. The format for this line is (2I6). The remainder of the file consists of IEDGE lines of data, format (12E11.4), containing the following boundary-layer profile data, with quantities written on each line in the order listed. Note that for the k-ϵ model, the specific dissipation rate, ω, is defined by

$$\omega = \frac{\epsilon}{C_\mu k} \qquad (F.1)$$

Quantity	Description	Dimensions
y	Distance normal to surface	ft (m)
$\tilde{u}$	Horizontal velocity	ft/sec (m/sec)
$\tilde{T}$	Temperature	°R (K)
$\bar{\rho}$	Density	slug/ft^3 (kg/m^3)
k	Turbulence kinetic energy	ft^2/sec^2 (m^2/sec^2)
ω	Specific dissipation rate	sec^{-1} (sec^{-1})
$k - e$	Large eddy energy	ft^2/sec^2 (m^2/sec^2)
T_{xx}	Large eddy xx-normal stress	ft^2/sec^2 (m^2/sec^2)
T_{xy}	Large eddy shear stress	ft^2/sec^2 (m^2/sec^2)
T_{yy}	Large eddy yy-normal stress	ft^2/sec^2 (m^2/sec^2)
y^+	Compressible sublayer-scaled distance	None
u^+	Compressible sublayer-scaled velocity	None

You can now exit back to the main menu. All of your input data are prepared and you are ready to run **EDDYBL**. Exit Program **SETEBL** with a final *X, ENTER* sequence.

F.5.11   Running the Boundary-Layer Program

Run Program **EDDYBL**. Examination of program output reveals that your run didn't go far enough to determine the point where momentum-thickness Reynolds number reaches 8000. After 87 steps, the program stops at $s = 4.03$ ft (1.23 m), and Re_θ is only 7550. Linear extrapolation of your computed Re_θ indicates you needed to integrate to about $s = 4.27$ ft (1.30 m).

F.5.12   Restart Run

You could go back to **SETEBL**, increase SSTOP to, say, 1.35, and simply rerun **EDDYBL**. On your little sister's hand-me-down IBM PC/AT without an 80287, that's another 8 or 9 minutes. Since you might not really want to take another coffee break while your job runs, you might prefer a less time-consuming solution. Program **SETEBL** provides such a possibility through its **Restart** option.

Examine your directory and verify that Program **EDDYBL** has created a new file named **output.dat**. This file contains sufficient information to restart your program. Now, run Program **SETEBL** again. From the main menu, go to the Restart Run sub-menu by typing an *R* followed by pressing *ENTER*. This menu will permit you to change IEND1, the maximum streamwise step number, and SSTOP, the maximum value of arc (plate) length. Since your value for IEND1 is clearly large enough, you need only change SSTOP. With the usual procedure, change SSTOP from 1.23 to 1.35. Now type *X* followed by pressing *ENTER* in order to return to the main menu. Before returning, you will receive a message in reverse video as follows:

"Do you wish to copy OUTPUT.DAT to INPUT.DAT? (Y/N)..."

Respond Yes by typing *Y* followed by pressing *ENTER*. At this point, for all but the VAX version, **SETEBL** will inform you that it first copies **input.dat** to a new file named **input.bak**. For all versions, **SETEBL** then copies **output.dat** to **input.dat**. The point is, the final output of your original run becomes input for the restart run. Additionally, your original **input.dat** has effectively been renamed as **input.bak** (VAX/VMS creates its own backup file so this file is unnecessary). Upon completion of the copy operation, control returns to the main menu (for the VAX version, you are instructed to press *ENTER* to continue). Exit Program **SETEBL**.

At this point, data files **eddybl.dat** and **input.dat** have been modified as needed to continue your run from where you left off. The file **table.dat** requires no modification as SSTOP remains smaller than the top end of the interval for which you have defined edge and surface properties. Had we made SSTOP larger than 10, you would have to make appropriate changes to **presur.dat** and **heater.dat** to make sure edge and surface conditions are defined at least up to the new value of SSTOP. You would then have to regenerate **table.dat** via the Write Data Files option (Subsection F.5.7).

Now, run Program **EDDYBL** again. If you have made no errors, inspection of the printout combined with a little interpolation shows that Re_θ is 8000 at a plate length of approximately 4.31 ft (1.32 m).

F.5.13   Gas Properties and Profile Printing

At this point, you have seen virtually all of Program **SETEBL**'s menus and options. There are two sub-menus we didn't use in this exercise, viz., **Gas Properties** and **Profile Printing**. Both menus are self explanatory and operate in the same manner as the menus you've already explored.

The **Gas Properties** menu allows you to modify thermodynamic properties such as specific-heat ratio, perfect-gas constant, and viscosity-law coefficients. The default values are set up for air with the Sutherland viscosity law. You can implement a **power-law** viscosity relationship by setting SU = 0. Note that if you want a viscosity law of the form $\mu = \mu_o T^\omega$ then you must set input parameters VISCON = μ_o and VISPOW = $\omega + 1$.

The **Profile Printing** sub-menu permits you to print velocity, temperature, turbulent energy, etc. profiles at specified streamwise stations. Program **EDDYBL** always prints profiles at the final station. Also, whenever **EDDYBL** prints profiles, disk file **output.dat** is automatically written.

F.5.14   Selecting Laminar, Transitional or Turbulent Flow

Program **EDDYBL** can run in three different modes corresponding to (1) pure laminar flow, (2) transition from laminar to turbulent flow, and (3) pure turbulent flow. The two test cases exercise **EDDYBL** in its pure turbulent mode in which integral parameters are specified and IBOUND is set to 1.

To run in transitional mode, simply select IBOUND = 0 in the **Initial Profiles/Integral Parameters** menu. As a result, exact laminar velocity and temperature profiles will be generated in conjunction with approximate laminar profiles for the various turbulence-model parameters. The transition point is determined automatically by the model equations and depends strongly upon the freestream values of k and ω that are specified in the **Main Parameters/Freestream Conditions** menu in terms of ZIOTAE and ZIOTAL. To obtain physically realistic transition Reynolds numbers you must include low-Reynolds-number corrections in the k-ω and multiscale models by setting NVISC = 1 in the **Turbulence Model** sub-menu. Although the k-ϵ models are capable of predicting transition, extremely small streamwise steps are needed with **EDDYBL**, and stable computation is very difficult to achieve. Even if you are not interested in transition, this mode is nevertheless useful as it provides an alternate method for generating turbulent starting profiles, e.g., by starting laminar and running up to a desired value of Re_θ.

The k-ω and multiscale models are very robust and can be integrated through transition from laminar to turbulent flow with and without low-Reynolds-number corrections. By contrast, many of the other models require smaller streamwise steps than the k-ω and multiscale models, and, in general, cannot be integrated through transition unless extremely small steps are taken. If integral properties are unknown and a solution with a model other than k-ω or multiscale is desired, the optimum procedure is to start laminar with the k-ω model and integrate through transition. Then, select the desired model and use the **Restart** option to continue the run.

Finally, to run **EDDYBL** as a pure laminar boundary-layer program, the turbulence model can be suppressed by setting MODEL = -1 in the **Turbulence Model** sub-menu. When this is done, turbulence-model computations are bypassed and no transition to turbulence occurs.

F.6   Additional Technical Information

The program uses the Blottner (1974) variable-grid method augmented with an algorithm devised by Wilcox (1981) to permit large streamwise steps. Section 14.4.3 of the main text describes the algorithm. Wilcox (1993) provides details on the transformations used, equations of motion, and other details.

F.7   EDDYBL Output Parameters

Printed output from Program **EDDYBL** consists of dimensionless boundary-layer profiles, and integral parameters, some of which are dimensional. The various quantities are listed in the following two tables.

Dimensionless-profiles portion of **EDDYBL** output:

Name	Symbol/Equation	Definition
i	i	Mesh point number
y/delta	y/δ	Dimensionless normal distance
u/Ue	$\tilde{u}/u_e$	Dimensionless velocity
yplus	$y^+ = u_\tau y/\nu_w$	Compressible sublayer-scaled velocity
uplus	$u^+ = u^*/u_\tau$	Compressible sublayer-scaled velocity
k/Ue**2	k/u_e^2	Dimensionless turbulence energy
omega	$\nu_e\omega/u_e^2$	Dimensionless dissipation rate
eps/mu	μ_t/μ	Dimensionless eddy viscosity
L/delta	$\sqrt{k/\beta^*}/(\omega\delta)$	Dimensionless turbulence length scale
uv/tauw	τ/τ_w	Dimensionless Reynolds shear stress
T/Te	$\tilde{T}/T_e$	Temperature ratio

Integral-parameter portion of **EDDYBL** output:

Symbol	Meaning	USCS Units	SI Units
F	Force	pounds (lb)	Newtons (N)
L	Length	feet (ft)	meters (m)
M	Mass	slugs (sl)	kilograms (kg)
Q	Heat flux	Btu/second (Btu/sec)	Watts (W)
T	Time	seconds (sec)	seconds (sec)
Θ	Temperature	°Rankine	Kelvins

Name	Symbol/Equation	Definition	Dimensions
beta	$\beta = (2\xi/u_e)du_e/d\xi$	Pressure-gradient parameter	None
Cfe	$c_{fe} = 2\tau_w/\rho_e u_e^2$	Skin friction based on $\bar{\rho}_e$	None
Cfw	$c_{fw} = 2\tau_w/\rho_w u_e^2$	Skin friction based on $\bar{\rho}_w$	None
delta	δ	Boundary-layer thickness	L
delta*	δ^*	Displacement thickness	L
dPe/ds	$d(\bar{p}/\rho_\infty U_\infty^2)/d\bar{s}$	Dimensionless pressure gradient	None
dTe/ds	$d(T_e/T_{ref})/d\bar{s}$	Dimensionless temperature gradient	None
dUe/ds	$d(u_e/U_\infty)/d\bar{s}$	Dimensionless velocity gradient	None
H	$H = \delta^*/\theta$	Shape factor	None
hdot	$\dot{h} = q_w/(T_w - T_{aw})$	Heat-transfer coefficient	$QL^{-2}\Theta^{-1}$
Iedge	N	Total number of mesh points in B.L.	None
Itro		Number of iterations	None
kmax	$\sqrt{\beta^*(\bar{\rho}k)_{max}/\tau_w}$	Maximum turbulence energy	None
M	m	Streamwise step number	None
Me	M_e	Edge Mach number	None
Mue	μ_e	Edge molecular viscosity	$ML^{-1}T^{-1}$
Ne	N_e	Mesh point number at B.L. edge	None
Negtiv		Number of points where $k,\omega,\epsilon < 0$	None
Nerror		Number of points not converged	None
Nskip		Number of points below u^+ = USTOP	None
Nste	$\dot{h}/\rho_e u_e c_p$	Stanton number based on $\bar{\rho}_e$	None
Nstw	$\dot{h}/\rho_w u_e c_p$	Stanton number based on $\bar{\rho}_w$	None
Nue	$Pr\, s\dot{h}/\mu_e c_p$	Nusselt number based on μ_e	None
Nuw	$Pr\, s\dot{h}/\mu_w c_p$	Nusselt number based on μ_w	None
Pe	$\bar{p}$	Edge pressure	FL^{-2}
qw	q_w	Surface heat flux	QL^{-2}
radius	r_o	Body radius	L
Recov	r	Recovery factor	None
Redel*	$Re_{\delta^*} = \rho_e u_e \delta^*/\mu_e$	Reynolds number based on δ^*	None
Res	$Re_s = \rho_e u_e s/\mu_e$	Reynolds number based on s	None
Rethet	$Re_\theta = \rho_e u_e \theta/\mu_e$	Reynolds number based on θ	None
Rhoe	ρ_e	Edge density	ML^{-3}
rho*vw	$\rho_w v_w$	Surface mass flux	$ML^{-2}T^{-1}$
s	s	Arc length	L
tauw	τ_w	Surface shear stress	FL^{-2}
Te	T_e	Edge temperature	Θ
theta	θ	Momentum thickness	L
Ue	u_e	Edge velocity	LT^{-1}
utau	u_τ	Friction velocity	LT^{-1}
xi	$\xi = \int_0^s \rho_e u_e \mu_e r_o^{2j}\,ds$	Transformed streamwise coordinate	L
yplus	y_2^+	Value of y^+ nearest the surface	None
z	z	Axial distance	L

F.8 Program PLOTEB: Plotting Utility

Program **PLOTEB** creates video and hardcopy plots of skin friction, c_f, or Stanton number, St, versus arc length, s, and a u^+ versus y^+ velocity profile computed with Program **EDDYBL** on IBM PC's and compatibles.

Input-parameter description:
Program **PLOTEB** reads the following sixteen input parameters from disk file **ploteb.dat** in the order listed below. Integer quantities must be formatted according to (7x,i6) while floating-point quantities must be formatted as (7x,f6.2).

mon	Monitor type (see Appendix G)
ifore	Foreground color (see Appendix G)
iback	Background color (see Appendix G)
nprin	Printer type (see Appendix G)
mode	Graphics-mode flag for printers; number of pens for plotters (see Appendix G)
metric	Input arc-length units flag

        -1 Input arc length is in meters
         0 Input arc length is in feet
         1 Convert from feet to meters

ideccf	Number of decimal places for c_f/St scale
idecx	Number of decimal places for arc-length scale
idecup	Number of decimal places for u^+ scale
isymb	Symbol type for experimental data points

         0 Circle
         1 Triangle
         2 Square
         3 Diamond

jstart   Number of c_f/St points to skip over at beginning of computation; this is sometimes useful in order to skip over transient behavior at the beginning of a computation. The sign also determines what is plotted.

        > 0 Plot c_f versus s
        < 0 Plot St versus s

kcyccf	Increment between points to be plotted for c_f/St versus s
kcycup	Increment between points to be plotted for u^+ versus y^+
kfilt	0 to suppress data filtering; otherwise use filtering. The filtering algorithm generates a smoothed curve.
ksize	Plot scaling factor. Using 100 yields a full-size hardcopy plot. Smaller values yield a hardcopy plot reduced by *ksize* per cent. Thus, selecting *ksize* = 50 yields a half-size plot.
symsiz	Size of experimental-data symbols, in inches

Next, Program **PLOTEB** reads a single, free-formatted, line to indicate where hardcopy print is directed. This line comes immediately after the specified value for *symsiz* and defines the following five additional parameters.

devid	Device name of type character∗4; valid devices are LPT1, LPT2, LPT3, COM1, COM2, COM3, COM4
nbaud	Baud rate for a serial port; valid baud rates are 110, 150, 300, 600, 1200, 2400, 4800, 9600

parity   Parity of type character*3 or character*4 for a serial port; valid parity settings are 'even', 'odd' and 'none'

nstop   Number of stop bits for a serial port; either 1 or 2

lword   Word length for a serial port; either 7 or 8

In addition to disk file **ploteb.dat**, an optional disk file named **exper.dat** containing measured skin-friction and velocity-profile data can be included. The first line of the disk file must contain the number of input data pairs with format (i6). If no c_f or St data are available, place a zero on this line. If c_f or St data are available, this line is followed by s-c_f (or s-St) data pairs with format (2e11.4).

Next, enter the data source; as many as twenty characters can be used. The final c_f/St entry is the location of the box citing the data source. Enter a 1 for upper left, 2 for upper right, 3 for lower right, and 4 for lower left (see Figure F.2). This option is provided to help avoid having the curves pass through the citations box. The format is (7x,i6). A similar sequence of input parameters follows for velocity-profile data. The order of the data pairs is y^+ first and u^+ last. For example, the bench-mark case is an incompressible flat-plate boundary layer. Experimental data for this flow are given by Coles and Hirst (1969). The sample **exper.dat** included on the distribution diskette is as follows.

```
         19
   1.5978e 00 3.4500e-03
   2.0899e 00 3.3700e-03
   2.5820e 00 3.1700e-03
   3.0741e 00 3.1700e-03
   3.5663e 00 3.0800e-03
   4.0584e 00 3.0100e-03
   4.7146e 00 2.9300e-03
   5.5348e 00 2.8400e-03
   6.5190e 00 2.7800e-03
   7.5033e 00 2.6900e-03
   8.4875e 00 2.6600e-03
   9.4718e 00 2.6000e-03
   1.0456e 01 2.6000e-03
   1.1440e 01 2.5600e-03
   1.2425e 01 2.5300e-03
   1.3409e 01 2.4700e-03
   1.4393e 01 2.4700e-03
   1.5377e 01 2.4600e-03
   1.6362e 01 2.4300e-03
WIEGHARDT
iposcf=       2
         14
   3.8100e 01 1.4240e 01
   7.6100e 01 1.5900e 01
   1.5230e 02 1.7390e 01
   3.8070e 02 1.9520e 01
   7.6150e 02 2.1300e 01
   1.1422e 03 2.2500e 01
   1.5229e 03 2.3330e 01
   1.9037e 03 2.4220e 01
   2.2844e 03 2.4940e 01
   3.0459e 03 2.6290e 01
   3.8074e 03 2.7410e 01
   4.5688e 03 2.8270e 01
   5.3303e 03 2.8590e 01
   6.0918e 03 2.8700e 01
WIEGHARDT
iposup=      1
```

Program Output: A video plot with two graphs (see Figure F.2) is created on the screen. When the plot is complete, the following message appears:

Hardcopy output (y/n)?

Enter a *y* or a *Y* to create a hardcopy plot. Pressing any other key terminates the run without creating a hardcopy plot.

Comments:

- The following is a sample input-data file, **ploteb.dat**, for a typical Personal Computer with a standard VGA monitor and an HP LaserJet printer connected to serial port COM1:.

```
mon    =    18              (Standard VGA monitor)
ifore =    15              (Bright-white foreground)
iback =     1              (Blue background)
nprin =     5              (HP LaserJet Series III or IV)
mode   =     3              (300 dots per inch resolution)
metric=     1              (Convert feet to meters)
ideccf=     1              (One decimal place on Cf scale)
idecs =     1              (One decimal place on s scale)
idecup=    -1              (Integers on u+ scale)
isymb =     0              (Circles for experimental data)
jstart=     2              (Start plot at the second point)
kcyccf=     1              (Plot every Cf point)
kcycup=     1              (Plot every u+ point)
kfilt =     1              (Use filtering)
ksize =   100              (Full size plot)
symsiz=  .080              (.08" experimental data symbols)
'com1' , 9600 , 'none' , 1 , 8
```

The last line indicates the printer is connected to serial port COM1: and the port is set at 9600 baud, no parity, 1 stop bit and 8 data bits.

If disk file **ploteb.dat** is not available, Program **PLOTEB** uses the following set of default values :

mon = 18, *ifore* = 15, *iback* = 1, *nprin* = 5, *mode* = 3, *symsiz* = .08, *ksize* = 100, *devid* = 'LPT1'

Note that *nbaud, parity, nstop* and *lword* are not used for parallel ports.

F.9   Adapting to Other Compilers/Systems

If you change computers or compilers, the appropriate modifications may already be included in the source code provided. If your Fortran compiler is an ANSI-77 Standard compiler and supports most of the standard VAX extensions, only three categories of changes are needed.

1. You must determine the correct syntax used by your Fortran compiler for the **include** command. Then, note that the source code provided uses the VAX syntax, which is

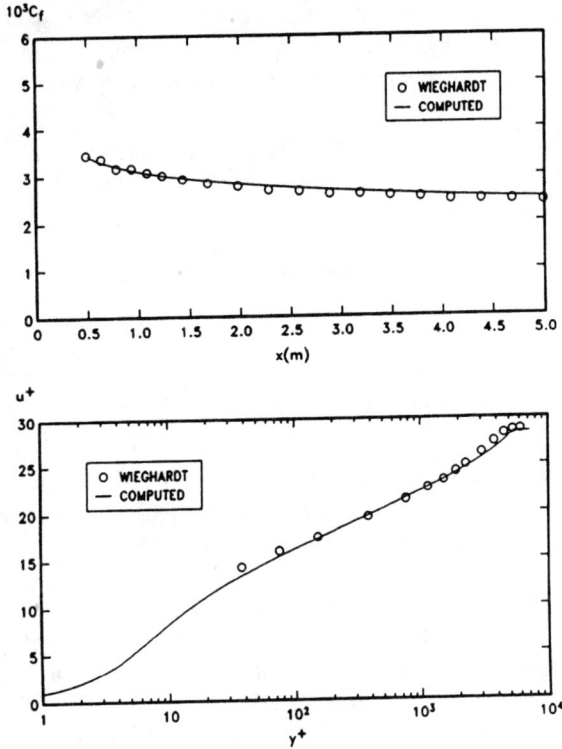

Figure F.2: *Sample plot created by Program PLOTEB.*

more-or-less standard in most modern Fortran compilers. Make the appropriate change throughout the source code for your compiler. Examples of VAX and other syntax are:

Fortran Compiler	Include Syntax
VAX, SVS, Lahey, Microsoft	`include 'filename'`
Cray (UNICOS), SUN	`include 'filename'`
Microsoft (older versions)	`$include: 'filename'`
SVS (older versions)	`$include filename`
Cray (COS)	`*CALL FILENAME`
	↑
	Column 1

2. Change the value of *icpu* defined in the include file named **cpuid**. The values currently assigned are:

icpu = 0	SVS Fortran (680x0, 80x86)
icpu = 1	Lahey/Microsoft Fortran (8088, 80x86)
icpu = 2	VAX/VMS
icpu = 3	SUN Fortran (68020, SPARC)
icpu = 4	Cray (UNICOS)
icpu = 5	Silicon Graphics Iris

3. The only other compiler-specific syntax differences are located in a subroutine called **NAMSYS** that appears in **eddybl.for**. This subroutine opens disk files depending upon the value of *icpu*. Make any changes required for your system.

4. Search the **EDDYBL** and **SETEBL** source code for occurrences of *icpu* to see if the correct action is taken for your compiler and/or operating system. Make any changes required for your system.

5. Modify Program **INSTL** as required for your video display and/or compiler-specific requirements.

F.10   Compile and Link Commands

This section describes the commands required to compile and link Programs **SETEBL**, **INSTL** and **EDDYBL** for the various Fortran compilers supported. Be sure that you have selected the appropriate value for *icpu* in the include file **cpuid**.

ICPU = 0: SVS Fortran-386 ... Phar Lap and C[3]
Special Comments: For the Phar Lap version, add the **+w1167** option to compile for a Weitek math coprocessor. Linker options for either **fastlink** or **386link** can be specified in an environment variable by including the following in your **autoexec.bat** file . . .

```
80387 version
set 386link=-l libf28 libp28 -pack -maxr ffffh -s 40000
Weitek version
set 386link=-l libf28w libp28w -pack -maxr ffffh -s 40000
```

Compile and Link:
```
svs instl.for
svs eddybl.for
svs setebl.for edge.for grafic.for initil.for ioebl.for main0.for misc.for
```

ICPU = 0: SVS Fortran-020
Special Comments: Use **pload** in place of **load** for Definicon PM-020 and PM-030 boards.

Compile and Link:
```
load fc instl -lk
load fc eddybl -lk
load fc setebl edge grafic initil ioebl main0 misc -lk
```

ICPU = 1: Lahey Fortran ... LF90
Special Comments: None.

Compile and Link:
```
lf90 instl
lf90 eddybl
lf90 setebl edge grafic initil ioebl main0 misc
```

ICPU = 1: Lahey Fortran ... F77L-EM/32
Special Comments: None.

Compile and Link:
```
f77l3 instl
386link instl
f77l3 eddybl
386link eddybl
f77l3 setebl
f77l3 edge
f77l3 grafic
f77l3 initil
f77l3 ioebl
f77l3 main0
f77l3 misc
386link setebl edge grafic initil ioebl main0 misc
```

ICPU = 1: Lahey Fortran ... F77L
Special Comments: None.

Compile and Link:
```
f77l instl
optlink instl;
f77l eddybl
optlink eddybl;
f77l setebl
f77l edge
f77l grafic
f77l initil
f77l ioebl
f77l main0
f77l misc
optlink setebl+edge+grafic+initil+ioebl+main0+misc;
```

ICPU = 1: Microsoft Fortran
Special Comments: Using the /e option reduces executable file size. You must split eddybl.for into 2 files as this compiler cannot handle large files. The commands below assume you split it into files named eddybl.for and eddybl2.for.

Compile and Link:
```
fl instl.for
fl /c eddybl.for eddybl2.for
link eddybl eddybl2,,nul, /e;
fl /c setebl.for edge.for grafic.for initil.for ioebl.for main0.for misc.for
link setebl edge grafic initil ioebl main0 misc,,nul, /e;
```

ICPU = 2: VAX Fortran
Special Comments: None.

Compile and Link:
    for instl
    link instl
    for eddybl
    link eddybl
    for setebl
    for edge
    for grafic
    for initil
    for ioebl
    for main0
    for misc
    link setebl,edge,grafic,initil,ioebl,main0,misc

ICPU = 3: SUN Fortran ... SUN/OS or MS-DOS/SP-1
Special Comments: Using the **-O3** option yields maximum optimization.

Compile and Link:
    f77 instl.f -O3 -o instl
    f77 eddybl.f -O3 -o eddybl
    f77 setebl.f edge.f grafic.f initil.f ioebl.f main0.f misc.f -O3 -o setebl

ICPU = 4: Cray Fortran ... UNICOS
Special Comments: None.

Compile and Link:
    cf77 -o instl instl.f
    cf77 -o eddybl eddybl.f
    cf77 -o setebl setebl.f edge.f grafic.f initil.f ioebl.f main0.f misc.f

ICPU = 5: Silicon Graphics Iris
Special Comments: None.

Compile and Link:
    f77 -o instl instl.f
    f77 -o eddybl eddybl.f
    f77 -o setebl setebl.f edge.f grafic.f initil.f ioebl.f main0.f misc.f

F.11 Software Package Modules

Boundary-Layer Program Source:

eddybl.for	Source code for Program **EDDYBL**
common	Include file for Program **EDDYBL**
cpuid	Include file specifying CPU type

Data-Preparation Utility Source:

setebl.for	Source code for the main program
edge.for	Source code for edge condition menus
grafic.for	Source code for reading graphics data
initil.for	Source code for initial profile menus
ioebl.for	Source code for I/O subroutines
main0.for	Source code for main input parameter menus
misc.for	Source code for miscellaneous menus
chars	Include file for Program **SETEBL**
comeb1	Include file for Program **SETEBL**
comeb2	Include file for Program **SETEBL**

Installation Program Source:

instl.for	Source code for Program **INSTL**

Bench-Mark Case Input Data:

blocrv.dat	Mass-transfer, body-curvature data file
heater.dat	Heat-transfer, surface-temperature data file
presur.dat	Pressure-distribution data file

Plotting Files:

exper.dat	Experimental data file for plotting program
ploteb.dat	Primary plotting-program data file

Bench-Mark Case Output:

eddybl.prt	Output from bench-mark test case

Executable Files for IBM PC and Compatible Microcomputers:

eddybl.exe	Program **EDDYBL**
instl.exe	Program **INSTL**
ploteb.exe	Program **PLOTEB**
setebl.exe	Program **SETEBL**

Appendix G

Plotting Program Details

The plotting program described in Appendix F, viz., Program **PLOTEB**, runs on IBM PC and compatible microcomputers with 640k of memory. It creates both video and hardcopy plots for many of the standard display devices currently available. The information in this appendix pertains to **PLOTEB**, the executable version of which is provided on the distribution diskette accompanying this book. Read the file **read.me** in the root directory of the distribution diskette to determine the location of the executable and input-data files for **PLOTEB**.

G.1 Plotting Colors

Input parameters *ifore* and *iback* set foreground and background plotting colors, respectively. The standard MS-DOS color coding scheme is used to set these colors. The MS-DOS color code is summarized below, where *icolor* is the color number and 'color' is the corresponding color.

icolor	color	*icolor*	color
0	Black	8	Dark Gray
1	Blue	9	Light Blue
2	Green	10	Light Green
3	Cyan	11	Light Cyan
4	Red	12	Light Red
5	Magenta	13	Light Magenta
6	Brown or Yellow	14	Yellow or Light Yellow
7	White	15	Bright White

G.2   Video Devices

Input parameter *mon* selects monitor type and resolution; valid devices are listed below, including the number of columns, rows and colors. The VESA modes should work on most 80486- and Pentium-based computers manufactured after 1993.

mon	Description	Columns	Rows	Colors
1	Hercules	720	348	2
2	AT&T 6300	640	400	2
3	Wyse-700	1280	800	2
4	CGA	320	200	4
5	CGA Monochrome	320	200	2
6	CGA	640	200	2
7	SVGA: Video-7	800	600	16
8	SVGA: Paradise	800	600	16
9	SVGA: ATI	800	600	16
10	SVGA: Genoa/Orchid	800	600	16
11	SVGA: Trident	800	600	16
12	SVGA: Oak	800	600	16
13	EGA	320	200	16
14	EGA	640	200	16
15	EGA Monochrome	640	350	16
16	EGA	640	350	16
17	MCGA/VGA	640	480	2
18	VGA	640	480	16
19	VGA	320	200	256
20	VESA	800	600	16
21	VESA	1024	768	16
22	VESA	1280	1024	16
23	SVGA: Unknown Card	800	600	16
24	SVGA: Everex	800	600	16
25	SVGA: Cirrus Logic	800	600	16
26	SVGA: Chips & Tech	800	600	16
27	SVGA: Genoa 5200-10	1024	768	16
28	VESA (Alternate)	1024	768	16
29	VESA (Alternate)	1280	1024	16

Important Note

To run **PLOTEB** from Windows 3.1, you must use a full-screen window. Windows 95 should switch to a full-screen window automatically.

G.3   Hardcopy Devices

Two input parameters are required for hardcopy plots, *nprin* and *mode*. The parameter *nprin* specifies the printer or plotter type. For a printer, input parameter *mode* specifies the resolution. For a plotter *mode* is the number of pens. Valid devices are as follows. If your printer or plotter is not listed, note that most popular printers and plotters used with personal computers are compatible with the Hewlett Packard or Epson printer control languages. Check your owner's manual to determine which language has been used.

nprin	Printer type
0	Hewlett-Packard LaserJet Series I
1	Hewlett-Packard LaserJet Series II
2	Hewlett-Packard DeskJet Black and White
3	Hewlett-Packard DeskJet 1-Cartridge Color
4	Hewlett-Packard DeskJet 2-Cartridge Color
5	Hewlett-Packard LaserJet Series III and IV
8	Epson with 8-pin printer head (older models)
24	Epson with 8-pin and 24-pin printer heads
25	Epson color dot-matrix printer
80	HPGL plotter without built-in circle drawing
90	HPGL plotter with built-in circle drawing
115	CGP-115 4-pen color plotter

Resolution for each printer and mode is given below as horizontal by vertical numbers of dots per inch.

nprin = 0 - 5 . . . Hewlett-Packard

mode	dots/in	*mode*	dots/in
0	75 x 75	3	300 x 300
1	100 x 100	4	600 x 600
2	150 x 150		

nprin = 8...Epson 8-pin

mode	dots/in	*mode*	dots/in
0	60 x 72	4	80 x 72
1	120 x 72	5	72 x 72
3	240 x 72	6	90 x 72

nprin = 24,25...Epson 24-pin

mode	dots/in	*mode*	dots/in
0	60 x 60	32	60 x 180
1	120 x 60	33	120 x 180
3	240 x 60	38	90 x 180
4	80 x 60	39	180 x 180
6	90 x 60	40	360 x 180

nprin = 80,90...HPGL Plotter

mode	number of pens	*mode*	number of pens
1	1	6	6
2	2	8	8

Answers to Selected Problems

SPECIAL NOTE TO STUDENTS

From a student's point of view, of all the places where typographical errors might be lurking in a book, the answers section is undoubtedly the most unwelcome place for them to hide! We have made extensive efforts to assure the accuracy of these answers. However, if you find errors in this section or any other part of the text, report them to DCW Industries' Home Page on the Worldwide Web at **http://webknx.com/dcw**. As long as this book remains in print, we will maintain an updated list of known typographical errors.

1.1 852 years

1.5 19 molecules

1.9 1225 psi

1.13 CO_2: 0.78 lb, He: 0.07 lb

1.17 Ethyl alcohol: 0.79, CCl_4: 1.59
Mercury: 13.56, Seawater: 1.03

1.21 $\mathcal{M} = 44$, yes

1.25 Air: 0.34, Water: $1.27 \cdot 10^{-5}$

1.29 (a) $\tau = [7(p + \alpha p_o)]^{-1}$
(b) $\alpha = 3041$

1.33 $\sigma = 0.0625$ N/m

1.37 $\Delta h = 7.0$ mm

1.41 $\Delta h_2 / \Delta h_1 = 1.00$

1.45 (a) $C = 4.708 \cdot 10^{-9}$ slug/[ft·sec(°R)$^{0.7}$]

1.49 $\bar{u} = U/2$

1.53 (b) $\tau_\ell = \mu U / h$, $\tau_u = -2\mu U / h$

1.57 $U = Th/(2\mu w \ell)$

1.61 $dU/dt = g \sin \alpha - \mu U / (\rho h s)$
$U \to 2.91$ ft/sec

1.65 (a) $Q = \pi(p_1 - p_2)R^4/(8\mu L)$
(b) 0.15 sec

1.69 (a) $u_m = 1.37$ ft/sec
(b) $Re_R = 810$
(c) $F = -3.14 \cdot 10^{-4}$ lb

1.3 (a) $1.75 \cdot 10^{13}$ molecules
(b) 18 molecules

1.7 $d/\ell_{mfp} = 1076$, yes

1.11 $\rho_w / \rho_a = 789$

1.15 Air: 11.77 N/m^3, Water: 9787 N/m^3

1.19 2770 ft·lb/(slug·°R)

1.23 $N = 3/8$

1.27 Air: 0.145, Ether: 888, Glycerin: 6551

1.31 $\Delta p = 4\sigma/R$

1.35 $D_a / D_w = 0.60$

1.39 $\Delta h = 4\sigma \cos \phi / (\rho g s)$

1.43 $h = \sqrt{\sigma/(\rho g)} \cot \phi$

1.47 (a) $\mu_e = 0.12/(du/dy)^{1/3}$
(b) $\mu_e / \mu_w \approx 120/(du/dy)^{1/3}$

1.51 (a) $\mu = 1.39 \cdot 10^{-3}$ slug/(ft·sec)
(b) $\rho_{max} = 2.09$ slug/ft^3

1.55 $\mu = MAd/(\pi L U D)$
$= 1.08 \cdot 10^{-3}$ kg/(m·sec)

1.59 $(M + m)dU/dt + \mu AU/h = mg$
$U \to mgh/(\mu A)$

1.63 $c_f = 4/Re_R$

1.67 $\rho u_m d / \mu = 2885$, laminar

2.5 $[\dot{h}] = MT^{-3}\Theta^{-1}$

2.9 (a) $[61.9] = M^{-0.54}L^{1.45}T^{0.08}$

(b) $[C] = L^{-0.17}T^{0.08}$

2.17 $dp/dx = \text{constant} \cdot \mu U/D^2$

2.21 $[\nu_t] = \text{constant} \cdot k/\omega$

2.25 2: $F\mu_o^2\sigma^2/\rho$, $\mu_o\sigma UL$

2.29 2: $[\rho c_p/(gk)]^{2/3}c_p\Delta T$, $c_p\mu/k$

2.33 $\Delta p = \rho\Omega^2 D^2 f[Q/(\Omega D^3)]$

2.37 $U/\sqrt{\nu\Omega}$, $\Omega h^2/\nu$, $\Omega k_s^2/\nu$

2.41 3: $\omega r^2/\Gamma_o$, $\Gamma_o t/r^2$, ν/Γ_o

2.45 3: $\Delta T/T_\infty$, $c_p\mu/k$, $\mu g/[\rho(c_p T_\infty)^{3/2}]$

2.51 4: $T/(\rho n^2 D^4)$, nD/U, UD/ν, $U/\sqrt{gD}$

2.55 (a) 46.3 m/sec

(b) $M_p = 0.205$, $M_m = 0.031$

2.59 $U_m = 687$ mph, $p_m = 3.95$ atm

2.63 (a) $U_p = 6$ m/sec, $N_p = 180$ rpm

(b) $Re_{D_m} = 6 \cdot 10^5$, $Re_{D_p} = 1.62 \cdot 10^7$

2.67 (a) $Q^2 N/\nu^3$, $\nu D/Q$

(b) $Q_m = 0.44$ m^3/sec , $N_m = 228$ rpm

2.7 2.43

2.11 $[\ell_{mix}] = L$

2.19 $c = \text{constant} \cdot \sqrt{\sigma/(\rho\lambda)}$

2.23 $R = \text{constant} \cdot (E/\rho_o)^{1/5}t^{2/5}$

2.27 2: $\rho UD/\mu$, $\rho U^2 D/\sigma$

2.31 2: UR/ν, $\mathcal{R}/R$

2.35 $\delta = \sqrt{\nu/\Omega}\, f(U/\sqrt{\nu\Omega})$

2.39 3: $\Omega^6 F/(\rho g^4)$, $\Omega U/g$, $\Omega^2 L/g$

2.43 3: $W/(\omega\mu D^2)$, ℓ/D, c/D

2.49 4: $\rho U_c D/\mu$, d/D, ρ_p/ρ, $\rho^2 g D^3/\mu^2$

2.53 $U_m/U_p = 10$

2.57 (a) 240 km/hr

(b) 750 N

(c) 16.8 hp

2.61 (a) $h_m = 0.267$ in, $U_m = 2$ ft/sec

(b) $\tau_m = 48$ min

2.65 (a) 8000 mph

(b) $T_{max} = 5380^\circ$ R $= 2716^\circ$ C

2.69 (a) 3: $\rho LF/\mu^2$, t/L, $\rho UL/\mu$

(b) $U_m = 1$ m/sec

(c) $F_p = 5000$ N/m

3.1 1.79 slug/ft^3

3.5 $\rho_u = 2.60$ slug/ft^3

3.9 $p = 12.1$ psi, $T = 40^\circ$ F

3.13 186 hp

3.17 $\tilde{\rho} = 3\rho/5$

3.21 10251 N

3.25 (b) $\lambda = 0.36$ in

3.29 $h = (p_1 - p_2)/[g(\Delta\rho + \rho d^2/D^2)]$

3.35 $F \approx 17$ kN, $h_{cp} \approx 6.7$ m

3.39 $h_p = 4h/9$

3.43 $F_{tip} = 7\rho g h^3/72$

3.47 $h = \sqrt{3}\, H$

3.51 $h = H/3$

3.55 (a) $\mathbf{F} = 4\rho g h^3(3\mathbf{i} + \mathbf{k})$

(b) $z_{cp} = 4h/3$

3.59 $h = H/4$, $F_z = -\rho g H^2/6$

3.63 $\mathbf{F} = 0$

3.67 $F_{buoy} = \rho g V/4$

3.3 Yes, since $p = 85.2$ psi at this depth

3.7 (a) $h_{H_g} = 775$ mm

(b) $h_{H_2O} = 10.5$ m

3.11 $p = 12.0$ kPa, $T = -55^\circ$ C

3.15 $\alpha = (\gamma - 1)g/(\gamma R)$

3.19 $\Delta z = 1.28$ m

3.23 $\rho = 1393$ kg/m^3

3.27 Ethyl alcohol: 14.1°, Water: 11.1°, Glycerin: 8.8°, Mercury: 0.8°

3.33 $A = h^2/2$, $I = h^4/48$

$\bar{z}$ lies at the geometric center

3.37 $H = h/3$

3.41 $H = 24.7$ ft

3.45 $F = 5\rho g H^3/3$

3.49 $F_s = \pi\rho g R^3/4$

3.53 $F = 2\rho g H^3$

3.57 $\mathbf{F} = \rho g h^3/2[\mathbf{i} + (1 + 1/n)\mathbf{k}]$

3.61 $\mathbf{F} = (15\rho g H^3/2)(\mathbf{i} + \mathbf{k})$

3.65 (a) $\mathbf{F} = -\rho g\ell^2[(3/2)\mathbf{i} + (1 + \pi/4)\mathbf{k}]$

(b) $z_{cp} = 14\ell/9$

(c) $x_{cp} = 5\ell/[6(1 + \pi/4)]$

3.69 $V = 0.025$ m^3, $\rho = 8810$ kg/m^3

3.71 2.026 lb

3.75 75%

3.79 $\rho_b = 13\rho/24$

3.85 $V = \ell^2 w/15$

3.89 $W = (1 + 3\pi/4)\rho g R^2 L$, $F_w = \rho g R^2 L/2$

3.73 6.68 ft

3.77 (b) $z = (RT/g)\ell n(\rho_o/\rho_{He})$

3.81 $F = (e - 2)\rho_o g h^2$

$z_{cp} = h/[2(e - 2)]$

3.87 (b) $N = 4/3$

4.1 (a) $\mathbf{a} = [\dot{A} + A^2]x\,\mathbf{i} - [\dot{A} - A^2]y\,\mathbf{j}$

(b) $\mathbf{a} = [\dot{A} + 2A^2x]x^2\mathbf{i} + [\dot{A} + 2A^2y]y^2\mathbf{j}$

4.5 $\mathbf{a} = (U/L)^2(x\mathbf{i} + y\mathbf{j})$

4.11 (a) $\mathbf{a} = -\Omega^2 r\,\mathbf{e}_r$

(b) $\mathbf{a} = -\Gamma^2/(4\pi r^3)\,\mathbf{e}_r$

4.17 $r = r_o$, $\theta = \theta_o + \Omega t$

4.21 (a) $\boldsymbol{\omega} = \mathbf{0}$

(b) $\boldsymbol{\omega} = \mathbf{0}$

(c) $\boldsymbol{\omega} = -(U/\delta)e^{-z/\delta}(\mathbf{i} - \mathbf{j})$

4.25 (a) $\boldsymbol{\omega} = \mathbf{0}$

(b) $\boldsymbol{\omega} = (4Ur/R^2)\cos 2\theta\,\mathbf{k}$

(c) $\boldsymbol{\omega} = \mathbf{0}$

4.29 $\Gamma = -2UL$

4.33 $xy = \text{constant}$

4.37 $y = Ce^{x^2/2}$, $C = \text{constant}$

4.41 (a) $y = x\sin\omega t + \text{constant}$

(b) $y = (U/\omega)[1 - \cos(\omega x/U)]$

4.47 $\dot{m} = \rho U R/2$

4.51 $\dot{P}_z = -\rho U^2 h^2/4$

4.55 $dN/dt = -7/72 \text{ sec}^{-1}$

4.3 (a) $\mathbf{a} = (U/H)^2(x\,\mathbf{i} + y\,\mathbf{j} + 4z\,\mathbf{k})$

(b) $\mathbf{a} = (2U^2x^2/H^4)(x\,\mathbf{i} + y\,\mathbf{j} + 3z\,\mathbf{k})$

4.7 $\mathbf{a} = \mathbf{0}$

4.13 $\mathbf{r} = x_o e^{At}\mathbf{i} + y_o e^{-At}\mathbf{j}$

4.19 $r = \sqrt{r_o^2 + Qt/\pi}$, $\theta = \theta_o$

4.23 Yes, $A = 1$

4.27 (a) $\mathbf{a} = -\mathbf{i}\omega U e^{-ky}\sin(\omega t - ky)$

(b) $\boldsymbol{\omega} = \mathbf{k}(\sqrt{2}\,kU e^{-ky})$

$\cdot \sin(\omega t - ky - \pi/4)$

4.31 $\Gamma = -UL^2/H$

4.35 $x^2 + y^2 = \text{constant}$

4.39 $r = C\sin\theta$, $C = \text{constant}$

4.43 (a) $xy = \text{constant}$

(b) $xy = x_o y_o = \text{constant}$

(c) $\mathbf{u} \times \partial\mathbf{u}/\partial t = \mathbf{0}$

4.49 $\dot{M} = 0$

$\dot{\mathbf{P}} = \pi\rho A^2 R^3/4\,\mathbf{e}_r - \rho A^2 R^3/2\,\mathbf{e}_\theta$

4.53 $\dot{m} = 0$

4.57 $dN/dt = -\dot{N}_{zap} - nUA$

5.1 Compressible, irrotational

5.5 Satisfies continuity, rotational

5.9 $u_\theta(r, \theta) = Ar^{-2}\sin\theta + Br^{-1}$

$B = \text{constant}$

5.13 (a) $u_r(r, \theta, z) = f(\theta, z)/r$

(b) $u_R(R, \theta, \phi) = g(\theta, \phi)/R^2$

5.19 $p - p_o = \rho\left(2gh - U^2\right)/2 = 26.07 \text{ kPa}$

5.23 $U_j = 10 \text{ m/sec}$, $z_{max} = 6.1 \text{ m}$

5.27 (b) $U = 6.26 \text{ m/sec}$, $p_{min} = 0.71 \text{ atm}$

5.33 $p(x, y, z, t) = -\rho g z + f(t)$

5.37 (a) $\zeta(r) = h_{min} + \Omega^2 r^2/(2g)$

5.39 $h_o = h - a\ell/(2g)$

5.41 VW Bug: $h_o = 0.92h$

Corvette: $h_o = 0.80h$

5.45 $U = 50 \text{ m/sec}$, $p_{stag} = 104.8 \text{ kPa}$

5.3 Satisfies continuity, irrotational

5.7 $A = 1$, $B = -1$

$d\mathbf{u}/dt = (t^2 + 2)x\,\mathbf{i} + t^2 y\,\mathbf{j}$

5.11 $\rho(x) = \rho_o/(1 + x/x_o)$, $x = x_o/4$

5.17 0.19 psi

5.21 $U = 3.63 \text{ ft/sec}$, $\Delta p = -1.07 \cdot 10^{-4} \text{ psi}$

5.25 $d = D[1 + \pi^2\rho^2 g D^4(\ell - z)/(8\dot{m}^2)]^{-1/4}$

5.29 (b) $p_a = 2.45 \text{ atm}$, $U = 19.8 \text{ m/sec}$

5.35 (a) $\ell_1 = L_2 - \Omega^2 L_3^2/(4g)$

$\ell_2 = L_2 + \Omega^2 L_3^2/(4g)$

(b) $\Omega = 2\sqrt{gh}/R$

5.43 $\beta = 2\alpha$

5.47 $U_{true}/U = \sqrt{1 + 2gr/U^2}$

5.51 $U_A = 2$ m/sec

5.57 $\rho \partial \phi / \partial t + \frac{1}{2} \rho \, \mathbf{u} \cdot \mathbf{u} + p + \rho \mathcal{V} = $ constant

5.53 $p_B - p_C = 18$ kPa

6.5 $\mathbf{F} = \Delta p A (\mathbf{i} - \mathbf{j})$

6.13 $dh/dt = -0.35 (d/D)^2 U$, emptying

6.17 $U_v = 3U/16$

6.21 $\tilde{u}_{avg} = 1.8U$

6.25 $z \approx 70d$

6.29 (b) $p_c = 600$ psi

6.33 $p_1 - p_2 = 1.25 \rho U^2$

6.37 (c) $U_2 = U_1 (H + 4h)/(H - 4h)$

6.41 $F_x = 3.9$ lb

6.45 $C_p \approx 0.80$

6.49 $\Delta p = \rho U^2 / 3$

6.53 $\mathbf{F} = \dot{m}[(U_1 + U_2 \cos \alpha)\mathbf{i} + U_2 \sin \alpha \, \mathbf{j}]$

6.59 $\Delta p = -0.36 \rho U^2$

6.65 (a) $V = \sqrt{2gh}$

     (b) $\mathbf{F} = (\pi/2)\rho g h d^2 \mathbf{i}$

     (c) $h = 1000d$

6.69 $\Delta p = 0.30 \rho U_1^2$, No

6.73 (b) $Mg = 12.72$ lb

6.77 $v_{min} = 2U/3$, $C_p = 7/9$

6.81 (a) $R_x = (11/16)\pi \rho U^2 D^2$

     (b) $R_x = (49/48)\pi \rho U^2 D^2$

6.87 $M(t) = M_o \exp\left(-\sqrt{\rho_e A_e} \, T t\right)$

6.11 $U = 2$ m/sec, $u = 32$ m/sec

6.15 $d = 5$, $N = 510$: $U_o = 48$ m/sec

     $d = 7$, $N = 1040$: $U_o = 42$ m/sec

6.19 $u = U/9$

6.23 (a) $u(r) = V r / h$

6.27 $dh/dt = -U/625$

6.31 $dM/dt = -2\rho_o V_o r_o h$

6.35 $\phi = \cos^{-1}(4/5)$, $\Delta p = \rho V^2 / 8$

6.39 $\mathbf{F} = (\pi/8)\rho U_i^2 D^2 \mathbf{i}$

6.43 $\tau = D/16$, $\mathbf{F} = -(9/32)\pi \rho U^2 D^2 \mathbf{i}$

6.47 $\phi = 69°$

6.51 $\mathbf{F} = (65/64)\pi \rho U^2 D^2 \mathbf{i}$

6.57 $F_x = -Mg \sin \phi / (\cos \phi + 1/5)$

6.61 $\mathbf{R} = -(9/8)\pi \rho U^2 D^2 \mathbf{i}$

6.67 $h_1 = (1 - \cos \phi)H/2$

     $h_2 = (1 + \cos \phi)H/2$

6.71 $\phi = 30°$, $F_x = -0.964 \rho U^2 A$

6.75 (b) $Mg = 1.99$ kN

6.79 $p_1 - p_2 = (31/20)\rho U_\infty^2$

6.83 (a) $R_y = 0$

     (b) $R_y = -0.077\pi \rho V^2 R^2$

6.89 (c) $U = 3.1$ knots

7.1 $\tau_T = 1/p$, smaller by $1/\gamma$

7.5 $T_2 = 81.7° C$

     $W_c = 243$ kJ

7.11 $T_2 = 18.5° C$, $u_2 = 83.3$ m/sec

7.15 (a) $\mathbf{R} = -4\sqrt{\dot{m} \dot{W}_s} \, \mathbf{i}$

     (b) $R_x = -2200$ lb

7.19 $\dot{Q} = 972$ Btu/sec, no

7.23 (a) $U_{max} = 3U$

     (c) $h_L = 0.15U^2/g$

7.27 Glass: $f = .0081$, $h_L = 8.26$ m

     Copper: $f = .0084$, $h_L = 8.57$ m

     Cast iron: $f = .0169$, $h_L = 17.24$ m

7.31 $D = 18$ cm, $Re_D = 1061 < 2300$

7.35 $\dot{W}_t = 0.80 \dot{m} g H - 0.53 \dot{m} U^2$

7.39 $\overline{u}_2 = \overline{u}_1 / 4$, $K = 0.475$

7.45 (a) $\Delta z = [1 + .0505L/D]U^2/(2g)$

     (b) $\Delta z = 17.25$ ft

     (c) $\Delta z = 17.88$ ft

7.3 $s = -R \ell n p + $ constant

7.9 (a) $n = 1.20$

     (b) $W = -8.2 \cdot 10^5$ ft·lb

     (c) $Q = -527$ Btu

7.13 (b) $P = 47.8$ kW $= 64$ hp

7.17 $T_o = 24.8° C$

7.21 $\alpha = (n + 1)^3 / (3n + 1)$

7.25 $h_p = 0.05U^2/g$

7.29 $\overline{u} = 5.74$ ft/sec

7.33 $2gh_L/U^2 = 2.8g\Delta z/U^2 - 1.05 = 0.95$

7.37 $h_p = \Delta z + 2000 \dot{m}^2/(\pi^2 \rho^2 g D^4)$

7.41 $D_h = (1 - \lambda)D$

7.47 $d/D = 0.2$: $gh_L/U^2 = 8975$

     $d/D = 0.4$: $gh_L/U^2 = 222$

     $d/D = 0.6$: $gh_L/U^2 = 28$

     $d/D = 0.8$: $gh_L/U^2 = 9$

7.49 (a) $U = \sqrt{2gh/(1.93 + 160f)}$

(b) $U = 11.6$ ft/sec

7.53 $Re_D = 672$

7.59 Chézy-Manning: $\bar{u} = 1.37$ m/sec

Colebrook: $\bar{u} = 1.38$ m/sec

7.63 $n = 0.045$

7.67 (b) $Fr = 1$

(c) $U = 7.97$ m/sec, $Fr = 3.6$

7.71 $y_c = 4.00$ ft

$S_c = 0.0119$

$\theta_c = 0.68°$

7.75 $y_2 = 4.75$ ft

$h_L = 3.23$ ft

7.51 $\mathbf{R} = \pi \rho U^2 D^2 [0.43\mathbf{i} - 0.64\mathbf{j}]$

7.57 $Q = 0.299$ ft^3/sec

7.61 Brickwork: $Fr = 0.70$

Weedy: $Fr = 0.35$

7.65 (b) $\alpha = 45°$

7.69 $\bar{u} = 1$ ft/sec

$y = 6$ in

7.73 (a) $Q = by_1^{3/2}(y_2/y_1)\sqrt{2g/(1 + y_2/y_1)}$

(b) $y_2/y_1 = 0.207$

(c) $Fr_1 = 0.27, Fr_2 = 2.83$

7.77 $y_1 = 88$ cm

$\bar{u}_1 = 3.20$ m/sec, $\bar{u}_2 = 2.80$ m/sec

$Fr_1 = 1.11, Fr_2 = 0.89$

8.1 (a) $M = 0.92$, transonic

(b) $M = 0.60$, subsonic

(c) $M = 3.00$, supersonic

(d) $M = 1.00$, sonic

8.5 197.7 m/sec

8.9 (a) $\gamma = 1 + 2/n$

8.13 $p_t = 4.25$ atm

$U = 447$ m/sec

8.21 (a) $M = 0.571$

(b) $M = 0.958$

(c) $M = 0.257$

8.29 Air: $M_1 = 1.50, T_1 = 27°$ C

He: $M_1 = 1.55, T_1 = -17°$ C

8.33 4.08

8.37 $M_\infty = \sqrt{2}/(\gamma + 1)$

8.43 $p^*/p_t = 0.487$

8.47 (a) 2.0 cm^2

(b) 17.3 cm^2

8.51 $\gamma = 1.29$

8.55 (a) $U = 1111$ ft/sec, $A^* = 9.922$ ft^2

(b) 1.046

(c) 4%

8.59 $A/A^* = \frac{1}{2}[2(2\gamma - 1)/(\gamma + 1)]^{\frac{\gamma+1}{2(\gamma-1)}}$

8.63 $M_2^2 \approx 1/M_1^2, \rho_2/\rho_1 \approx M_1^2$

$p_2/p_1 \approx M_1^2, T_2/T_1 \approx 1$

8.67 (b) $p_t = 128$ kPa, $M_e = 0.595$

(c) 298 N

8.3 $U_{air} = 229$ mph

$U_{H_2O} = 994$ mph

8.7 $a_{sea} = 1477$ m/sec, $a_{sea}/a_{fresh} = 0.997$

8.11 $T_{ta}/T_{tb} = 1$

8.15 $T_{air} = -253.0°$ C

$T_{He} = -260.7°$ C

8.23 (a) $M_\infty = 0.293$, 1.6%

(b) $M_\infty = 0.586$, 6.6%

8.31 $p_t = 2.91$ atm

8.35 86.5 kPa

8.41 Identical to original equation

8.45 $p^*/p_t = 0.546$

8.49 12 cm^2

8.53 Yes

8.57 0.112

8.61 Cadillac: -0.2%, SST: -80%

8.65 0.22

8.69 (b) 19350 ft/sec

9.1 (a) $\tau_r = 2\rho w_e A_e \ell(w_e \cos \phi - \Omega \ell)$

(c) $\phi = 83.9°$

9.5 $h_{po} = \Omega^2 r_2^2/g$

$A = \cot \beta_1 \cot \beta_2/(4\pi^2 gr_1^2 b_1 b_2)$

9.3 $\beta_2 = \tan^{-1}[(r_1/r_2)^2 \tan \beta_1]$

9.7 $Q = 0.265$ m^3/sec

$\dot{W}_p = 132$ kW

9.9 $Q = 3815$ gal/min
$\dot{W}_p = 492$ hp
9.13 (a) $Q = 3.32$ m^3/sec, $\dot{W}_p = 2.09$ kW
(b) $\alpha_2 = 52.3°$

9.19 $Q^* = 1960$ gal/min
$\eta_p = 0.81$
9.23 $[\tilde{N}_s] = L^{3/4}T^{-3/2}$
$\tilde{N}_s/N_s = 2733$ ft$^{3/4}$/sec$^{3/2}$
9.27 $C_{Q^*} = .223$, $C_{H^*} = .0435$
$C_{P^*} = .0141$, $\eta_p = 0.69$, $N_s = 4.96$
9.31 77%
9.35 (a) $(\eta_t)_{max} = C_v^2$, $\Omega r/\sqrt{2gh_p} = \frac{1}{2}C_v$
(b) $C_v = 0.94$
(c) $C_v = 0.89$

9.11 $\beta_1 = 21.2°$
$\beta_2 = 12.2°$
9.17 H$_2$O: $h_p = \quad$ 1 ft
Air: $h_p = \quad$ 829 ft
He: $h_p = 6006$ ft
9.21 (b) $\beta_2 = 127°$
$r_2 = 5.25$ in, $b_2 = 1.05$ in
9.25 $C_{Q^*} = .0124$, $C_{H^*} = .0971$
$C_{P^*} = .00133$, $\eta_p = 0.91$, $N_s = 0.64$
9.29 $D = 1.32$ m
$\Omega = 350$ rpm
9.33 $N_{sp} = \Omega(\dot{W}_m^*/\rho)^{1/2}/(gh_p^*)^{5/4}$
9.37 150 rpm

9.39 Kaplan type

10.1 $dW/dt = 0$

10.7 (a)   64 ft^2/sec
(b) 158 ft^2/sec
(c) 6835 ft^2/sec
10.13 3.2 mm

10.17 $\nu\omega_w/U^2 = -0.2/\sqrt{Re_x}$

10.3 (a) $\mathbf{f}_{vortex} = \mathbf{0}$
(b) $p + \frac{1}{2}\rho\mathbf{u}\cdot\mathbf{u} + \rho\mathcal{V} = $ constant
10.9 (a) $\Gamma = -UL$

10.11 $\Gamma = 1.11 \cdot 10^5$ ft^2/sec, $\Omega = 11.7$ rpm
10.15 Air: $\delta = 1.06$ m, $\tan^{-1}(\delta/x) = 0.6°$
H$_2$O: $\delta = 0.65$ m, $\tan^{-1}(\delta/x) = 0.4°$
10.19 (a) $\omega = -(C_Q U_\infty^2/\nu)e^{-C_Q U_\infty y/\nu}$
(b) $\delta = 4.6\nu/(C_Q U_\infty)$
(c) $\nu\omega/U_\infty^2 = -C_Q e^{-C_Q Re\,y/L}$

11.1 (a) $[\psi] = L^2 T^{-1}$
(b) $[Q] = L^2 T^{-1}$
(c) $[\Gamma] = L^2 T^{-1}$
(d) $[\mathcal{D}] = L^3 T^{-1}$
11.9 (a) $[X(x)] = LT^{-1}$
(b) None
(c) $X(x) = Ae^{-Kx}$ $(K = $ constant$)$
$Y(y) = C\cos Ky + D\sin Ky$
11.13 $\phi(r,\theta) = Ar^n\cos n\theta$
$F(z) = Az^n$

11.19 Flow is irrotational

11.23 $\phi = U(x^2 + y^2 - 2z^2)/(2H)$
Flow is incompressible

11.27 $\phi(x,y) = \psi_o(x^2 - y^2)$
11.29 (a) $u_r = 2Ar\sin 2\theta$, $u_\theta = 2Ar\cos 2\theta$
(b) $\phi(r,\theta) = Ar^2\sin 2\theta$
(c) $x = y = 0$

11.3 $\dot{m} = 0$

11.5 $\Gamma = 0$

11.11 Uniform: $F(z) = Uz$
Source: $F(z) = \frac{1}{2}(Q/\pi)\ln z$
Vortex: $F(z) = -\frac{1}{2}i(\Gamma/\pi)\ln z$
Doublet: $F(z) = \mathcal{D}/(2\pi z)$
11.15 $\phi = U(x\cos\alpha + y\sin\alpha)$
$\psi = U(y\cos\alpha - x\sin\alpha)$
Uniform flow at angle α
11.21 $\phi = U(x\cos\alpha + y\sin\alpha)$
$\psi = U(y\cos\alpha - x\sin\alpha)$
11.25 (a) $\psi = Axy + $ constant
(b) $\phi = \frac{1}{2}A(x^2 - y^2) + $ constant
(c) $Q = 2\pi Ay_o^2$
(d) $C = -Q/4$
11.31 $p = p_\infty - \frac{1}{2}\rho Q^2/(2\pi r)^2$
$r_{min} = 1.0$ cm

11.35 (a) $C_p = -(s/L)^2[1 + (t/\tau)^2]$
(b) $|\mathbf{u}| = U\sqrt{1 + (t/\tau)^2}$

11.41 (a) $\psi = Ur\sin\theta + Uh(\theta - \pi)/(2\pi)$
(b) $x = -h/(2\pi)$, $y = 0$

11.47 $p_i = 14.86$ psi

11.53 (a) $\mathbf{F} = \pi\rho\Omega D^2 H(V\mathbf{i} + U\mathbf{j})$
(b) $F_x = 207$ kN

11.57 (a) $\alpha = 30°$
(b) $p_1 - p_2 = \frac{1}{2}\rho U^2 + 0.268\rho R\dot{U}(t)$

11.65 $L = \rho U\Gamma_a[1 - \Gamma_a/(4\pi Uh)]$
Decreases

11.69 0.97

11.73 $A = -3/(2\Delta y)$, $B = 2/\Delta y$
$C = -1/(2\Delta y)$

11.39 Q/U

11.45 (c) $\mathbf{F} = -\rho QU\,\mathbf{i}$ (a thrust)

11.49 (b) $D = 8\rho U^2 R/3$

11.55 (a) $U_{min} = \sqrt{W/(2\pi\rho RH)}$
$\Omega_{min} = \sqrt{W/(2\pi\rho R^3 H)}$
(b) $U_{min} = 3$ ft/sec, $\Omega_{min} = 11.5$ Hz

11.59 $2g/3$

11.67 (b) $C(x) = x\tan\alpha$
$+ \gamma_o x(1 - x/c)/(4U\cos\alpha)$
(c) $C(x) = x\tan\alpha(2 - x/c)$

11.71 $\phi_{i,j} = \frac{1}{2}[\phi_{i,j+1} + \phi_{i,j-1}$
$+ \beta^2(\phi_{i+1,j} + \phi_{i-1,j})]/(1 + \beta^2)$

11.79 $\lambda_{opt} = 0.95$

12.1 $\ell_{mfp} = 1.5\nu/a = 0.066$ microns

12.5 (a) $\psi(r,\theta) = -\frac{1}{2}\Omega r^2 + \text{constant}$
(b) Flow is rotational

12.7 $\boldsymbol{\omega} = \mathbf{0}$, $[\mathbf{S}] = \dfrac{U}{L}\begin{bmatrix} 1 & 0 & 0 \\ 0 & -1 & 0 \\ 0 & 0 & 0 \end{bmatrix}$

12.13 $\boldsymbol{\omega} = \mathbf{0}$, $[\mathbf{S}] = \dfrac{U}{H}\begin{bmatrix} 1 & 0 & 0 \\ 0 & 1 & 0 \\ 0 & 0 & -2 \end{bmatrix}$

12.19 $[\boldsymbol{\tau}] = 2\mu A\begin{bmatrix} 2x & 0 & -z \\ 0 & 0 & z \\ -z & z & -2x \end{bmatrix}$

$\nabla\cdot[\boldsymbol{\tau}] = 2\mu A(\mathbf{i} + \mathbf{j})$

12.23 $u(x,y) = Ux^2/H^2 - 4Uy/H$
$v(x,y) = -2Uxy/H^2$

12.27 (b) $\theta = \frac{1}{2}\tan^{-1}[2\tau_{xy}/(\tau_{xx} - \tau_{yy})]$

(c) $[\boldsymbol{\tau}'] = \begin{bmatrix} \tau & 0 \\ 0 & -\tau \end{bmatrix}$

12.31 $p(r) = p_\infty - \frac{1}{2}\rho K^2/r^2$
Negative pressure for $R \to 0$

12.37 (b) 247 min
(c) 6 min

12.3 (a) $0.036M^2/Re_x^{1/5}$
(b) $1.56\cdot 10^{-5}/x^{1/5}$
(c) $x = 9.24\cdot 10^{-25}$ ft
$x/\ell_{mfp} = 4.28\cdot 10^{-18}$

12.9 $\boldsymbol{\omega} = -5\dfrac{U}{a}\mathbf{k}$, $[\mathbf{S}] = \dfrac{3}{2}\dfrac{U}{a}\begin{bmatrix} 0 & 1 & 0 \\ 1 & 0 & 0 \\ 0 & 0 & 0 \end{bmatrix}$

12.17 $\sigma_n = -p$

12.21 $[\boldsymbol{\tau}] = \dfrac{4\mu}{3(\gamma + 1)t}\begin{bmatrix} 2 & 0 & 0 \\ 0 & -1 & 0 \\ 0 & 0 & -1 \end{bmatrix}$

12.25 (g) $u_r = (K/r)(1 + r/R)\cos\theta$
$u_\theta = (K/r)(1 - r/R)\sin\theta$

12.29 $\bar{p} = pL/(\mu U)$

12.35 (b) $p(R) = p_\infty - \frac{1}{2}\rho C^2/R^4$

12.43 $G = [1 + i(U\Delta t/\Delta x)\sin k\Delta x]^{-1}$
Unconditionally stable

12.47 $G = [1 + \lambda e^{-i\theta}]/[1 + \lambda e^{i\theta}]$
$\lambda = \frac{1}{2}U\Delta t/\Delta x$, $\theta = k\Delta x$
Neutrally stable

12.49 $N = \ \ 51: \ (\tilde{N}_{CFL})_{max} = 1.11$
$N = 101: \ (\tilde{N}_{CFL})_{max} = 1.05$
$N = 201: \ (\tilde{N}_{CFL})_{max} = 1.02$
$N = 401: \ (\tilde{N}_{CFL})_{max} = 1.01$

13.1 $\boldsymbol{\omega} = -(U/h)[(1 - \chi) + 2\chi y/h]\mathbf{k}$

$$[\mathbf{S}] = -\frac{1}{2}\omega_z \begin{bmatrix} 0 & 1 & 0 \\ 1 & 0 & 0 \\ 0 & 0 & 0 \end{bmatrix}$$

$$[\boldsymbol{\tau}] = -\mu\omega_z \begin{bmatrix} 0 & 1 & 0 \\ 1 & 0 & 0 \\ 0 & 0 & 0 \end{bmatrix}$$

13.3 $\boldsymbol{\omega} = 2u_m r/R^2 \mathbf{e}_\theta$

$$[\mathbf{S}] = -\frac{u_m r}{R^2} \begin{bmatrix} 0 & 0 & 1 \\ 0 & 0 & 0 \\ 1 & 0 & 0 \end{bmatrix}$$

$$[\boldsymbol{\tau}] = -2\frac{\mu u_m r}{R^2} \begin{bmatrix} 0 & 0 & 1 \\ 0 & 0 & 0 \\ 1 & 0 & 0 \end{bmatrix}$$

13.7 Couette: $\psi = \frac{1}{2}Uh[1 - \chi(1 - \frac{2}{3}\frac{y}{h})](\frac{y}{h})^2$
Channel: $\psi = u_m h(1 - \frac{4}{3}\frac{y}{h})(\frac{y}{h})$

13.9 $\psi = 2Uy\,\mathrm{erfc}(\eta)$
$\qquad + 4U\sqrt{\nu t/\pi}(1 - e^{-\eta^2})$
$\eta \equiv y/(2\sqrt{\nu t})$

13.11 $u = -U(\chi - 1)^2/(4\chi)$
$u = -U/3$ when $\chi = 3$

13.13 $(u_m)_{max} = 1$ m/sec
$\ell_e = 3.6$ m

13.15 $\delta^* = 0.29R$
13.21 (e) $U = \frac{1}{2}(gh^2/\nu)\sin\theta$
13.25 (b) $u_r = 0$, $u_\theta = \Omega R^2/r$
(c) $\Gamma = 2\pi\Omega R^2$, potential vortex
(d) $p = p_\infty - \rho\Omega^2 R^4/r^2$
Valid for $\Omega < \sqrt{p_\infty/(\rho R^2)}$

13.17 1 cm
13.23 (d) $u_\theta \to R_i^2\Omega_i/r$, potential vortex
13.27 (b) $u(z) = U[1 - e^{-\lambda z}\cos\lambda z]$
$v(z) = Ue^{-\lambda z}\sin\lambda z$
$\lambda \equiv \sqrt{\Omega/\nu}$

13.33 (b) $F(0) = 0$, $G(0) = 1$
$H(0) = 0$, $P(0) = 0$
$F(\zeta) \to 0, G(\zeta) \to 0$ as $\zeta \to \infty$

13.39 $x \geq 3$

13.41 $\mathbf{x} = \begin{Bmatrix} 1.0000 \\ 1.0423 \\ 1.1128 \\ 1.2113 \\ 1.0000 \end{Bmatrix}$

14.1 (a) $1.0 \cdot 10^6$
(b) $2.0 \cdot 10^5$
14.5 103 mm

14.9 (b) $u(x, y) = U\,\mathrm{erf}(\eta)$
$\eta = y/(2\sqrt{\nu x/U})$
14.17 (a) $c_f = 0.577/\sqrt{Re_x}$
$\delta^* = 1.732\sqrt{\nu x/U}$
$\theta = 0.577\sqrt{\nu x/U}$, $H = 3.00$
(b) $\delta = 3.46\sqrt{\nu x/U}$

14.3 Air: 16.20 ft/sec
H$_2$O: 1.08 ft/sec
14.7 (a) $w_f = \sqrt{8mg/(\pi\rho D^2 C_D)}$
(b) $w = w_f\tanh(gt/w_f)$
(d) 0.51 sec
14.11 (c) $Q = \frac{1}{2}Uh[1 - h^2 dp/dx/(6\mu U)]$
14.19 (a) $c_f = 0.646/\sqrt{Re_x}$
$\delta^* = 1.740\sqrt{\nu x/U}$
$\theta = 0.646\sqrt{\nu x/U}$, $H = 2.69$
(b) $\delta = 4.64\sqrt{\nu x/U}$

14.21 (a) $c_f = 0.685/\sqrt{Re_x}$

$\delta^* = 1.751\sqrt{\nu x/U}$

$\theta = 0.685\sqrt{\nu x/U}$, $H = 2.55$

(b) $\delta = 5.84\sqrt{\nu x/U}$

14.25 (c) $\theta = 0.671\sqrt{\nu x/U}$

14.29 $c_f = 2C_Q$, $\delta^* = \nu/(C_Q U)$

$\theta = \nu/(2C_Q U)$, $H = 2$

14.35 $c_f = 2.12/\sqrt{Re_x}$

32% error, regardless of λ

14.39 $\ell_m = 0.01 H\, Re_H = \ell_e/6$

14.49 $\delta \approx 0.68$ m

14.53 (a) 11%

(b) 4.2%

(c) 4.7%

(d) 0.9%

14.57 (a) $0.30°$

(b) $0.14°$

(c) 10.22 m/sec for (a)

202 m/sec for (b)

14.61 Copper: 1908 m/sec

Iron: 11.4 m/sec

Concrete: 0.89 m/sec

14.65 Smooth: $c_f = .00264$, $C_D = .00330$

Rough: $c_f = .00446$, $C_D = .00573$

14.69 (a) $(\nu + \nu_t)d\overline{u}/dy - v_w\overline{u} = u_\tau^2$

(b) $\kappa^2 y^2 (d\overline{u}/dy)^2 = u_\tau^2 + v_w\overline{u}$

14.75 $Z_4\overline{u}_{i,j} - Z_5\overline{u}_{i-1,j} > 0$

Boundary layer is attached

14.23 $\delta/x = 0.37/Re_x^{1/5}$

$c_f = 0.0577/Re_x^{1/5}$

$C_D = 0.144/Re_L^{1/5}$

14.27 $\phi_{sep} = 103.1°$

14.33 $\delta = 4.82\sqrt{\nu x/u_e}$

Within 3.6% of Blasius

14.37 (a) Yes

(b) Design: 0.049 N

Rotated: 0.035 N

14.41 $C_D = 5.30/\sqrt{Re_d}$

14.51 $C_D = 0.160/Re_L^{1/5}$

14.55 Flagpole: 16.8 Hz, audible

Phone wire: 168 Hz, audible

14.59 $\delta_{sl} \approx 0.016$ in

$\delta_{sl}/d_{pin} \approx 1/3$

14.63 Log layer

7.6 mph

14.67 (a) $\delta = 10\lambda^{-5/3}k_s^{1/6}x^{5/6}$

$c_f = 1.262\lambda^{-5/3}(k_s/x)^{1/6}$

(b) $C_D = 1.514\lambda^{-5/3}(k_s/L)^{1/6}$

(c) $\lambda = 12$

14.71 $n(x) = $ constant

15.3 (b) $x \to -\infty$: $u \to u_1, h \to h_1, \rho \to \rho_1$

$x \to +\infty$: $u \to u_2, h \to h_2$

15.13 $L_{max} = 1.89$ m

$T_e = 17°\,$C

15.15 (a) 1.25 m

(b) Yes

15.19 $M_e = 1.335$, $p_e = 3.06$ atm

$T_e = 261°$ C

15.25 $M = 0.5$: $q = 2.92 \cdot 10^5$ J/kg

$M = 0.7$: $q = 4.76 \cdot 10^5$ J/kg

$M = 0.9$: $q = 5.47 \cdot 10^5$ J/kg

15.29 $q = -3543$ Btu/slug

$M_e = 2.89$

15.9 (a) $u = v = 0$ at $y = 0$

$u = U, v = 0$ at $y = h$

(b) $v(x,y) = 0$

(c) Steady $\Rightarrow$ $\partial/\partial t \to 0$

Infinite $\Rightarrow u\partial/\partial x \to 0$

$v = 0$ $\Rightarrow v\partial/\partial y \to 0$

15.17 $\overline{f} = 0.016$

$M_e = 1.18$

15.23 $q = 13265$ Btu/slug

$p_e = 13.4$ psi

15.27 Air: $q = 2.45 \cdot 10^4$ J/kg, $T_e = 415$ K

H_2: $q = 6.50 \cdot 10^6$ J/kg, $T_e = 681$ K

15.31 $31.4°$

15.33 23.6°

15.37 (a) $M = 4.39$, $A/A^* = 15.13$
(b) $M = 3.57$, $A/A^* = 7.27$
(c) $M = 1.83$, $A/A^* = 1.47$
15.41 Air: 4.930, He: 6.447
15.45 $p_3 = 4.04$ atm, $T_3 = 456$ K
$M_3 = 1.856$, $p_{t3} = 25.31$ atm
15.51 Forward: 19.5°
Rearward: 5.2°
15.55 $\nu(M) \to \frac{1}{2}\pi[\sqrt{(\gamma+1)/(\gamma-1)} - 1]$
CO_2: Turn flow 165°
Air: Turn flow 130°
He: Turn flow 90°
15.61 (a) $\delta_{th} \sim \sqrt{\nu x/(U Pr)}$
(b) Air: $\delta_{th}/\delta = 1.18$
H_2O: $\delta_{th}/\delta = 0.38$
Oil: $\delta_{th}/\delta = 0.10$
15.67 (a) $\delta_o = \sqrt{12 L_w \mu_e x/(\rho_e u_e)}$
15.71 $\nu_a = \lambda a \Delta x/2$
$\mathcal{D}_a = (4\lambda^2 - 3\lambda - 2)a(\Delta x)^2/6$
$\lambda \equiv a\Delta t/\Delta x$

15.35 $\beta = 37.8°$, $p_2 = 7996$ lb/ft^2
$T_2 = 810°$ R, $M_2 = 1.99$
15.39 (a) Weak: 39.3°, Strong: 83.7°
(b) Weak: 1.72, Strong: 4.45
(c) Weak: 1.64, Strong: 0.60
15.43 1.0 sec
15.49 706 kPa

15.53 $\theta = 27.16°$
$p_2/p_1 = 0.0694$
15.57 $x/L = 1.75$
$y/L = 2.66$

15.59 $M_3 = 3.73$, $p_3 = 11.1$ psi
15.65 $C_{Df} = 2.464/\sqrt{Re_L}$
$Re_L = 10^4$: $C_{Df}/C_{Dw} = 0.122$
$Re_L = 10^5$: $C_{Df}/C_{Dw} = 0.039$
$Re_L = 10^6$: $C_{Df}/C_{Dw} = 0.012$
15.69 Identical to Van Driest correlation
15.73 $\nu_a = 0$
$\mathcal{D}_a = -(\lambda^2 + 2)a(\Delta x)^2/12$
$\lambda \equiv a\Delta t/\Delta x$

Bibliography

Adair, R. K. (1990), *The Physics of Baseball*, Harper Perennial, Harper & Row Publishers, Inc., New York, NY.

Adamson, A. W. (1960), *Physical Chemistry of Surfaces*, Interscience, New York, NY.

Ames Research Staff (1953), "Equations, Tables, and Charts for Compressible Flow," NACA Report 1135.

Anderson, J. D. (1989), *Introduction to Flight*, Third Ed., McGraw-Hill, New York, NY.

Anderson, J. D. (1990), *Modern Compressible Flow With Historical Perspective*, Second Ed., McGraw-Hill, New York, NY.

Anderson, J. D. (1995), *Computational Fluid Dynamics: The Basics with Applications*, McGraw-Hill, New York, NY.

Anderson, D. A., Tannehill, J. C. and Pletcher, R. H. (1984), *Computational Fluid Dynamics and Heat Transfer*, Hemisphere Publishing, Washington, DC.

Baker, A. J. and Pepper, D. W. (1991), *Finite Elements 1-2-3*, McGraw-Hill, New York, NY.

Baldwin, B. S. and Lomax, H. (1978), "Thin-Layer Approximation and Algebraic Model for Separated Turbulent Flows," AIAA Paper 78-257, Huntsville, AL.

Batchelor, G. K. (1967), *An Introduction to Fluid Dynamics*, Cambridge University Press, Cambridge, England.

Bathe, W. W. (1984), *Fundamentals of Gas Turbines*, John Wiley & Sons, Inc., New York, NY.

Betz, A. (1966), *Introduction to the Theory of Flow Machines*, Pergamon Press, Oxford, England.

Blottner, F. G. (1974), "Variable Grid Scheme Applied to Turbulent Boundary Layers," *Comput. Meth. Appl. Mech. & Eng.*, Vol. 4, No. 2, pp. 179-194.

Bradshaw, P. (1972), "The Understanding and Prediction of Turbulent Flow," *The Aeronautical Journal*, Vol. 76, No. 739, pp. 403-418.

Buckingham, E. (1915), "Model Experiments and the Forms of Empirical Equations," *Transactions of the ASME*, Vol. 37, p. 263.

Carrier, G. F., Krook, M. and Pearson, C. E. (1966), *Functions of a Complex Variable*, McGraw-Hill, Inc., New York, NY.

Cebeci, T. and Smith, A. M. O. (1974), *Analysis of Turbulent Boundary Layers*, Ser. in Appl. Math. & Mech., Vol. XV, Academic Press.

Cengel, T. A. and Boles, M. A. (1994), *Thermodynamics*, McGraw-Hill, Inc., New York, NY.

Chow, V.-T. (1959), *Open Channel Hydraulics*, McGraw-Hill, Inc., New York, NY.

Churchill, R. V. and Brown, J. W. (1990), *Complex Variables and Applications*, Fifth Ed., McGraw-Hill, Inc., New York, NY.

Clauser, F. H. (1956), "The Turbulent Boundary Layer", *Advances in Applied Mechanics*, Vol. IV, Academic Press, New York, pp. 1-51.

Colebrook, C. F. (1939), "Turbulent Flow in Pipes, with Particular Reference to the Transition between the Smooth and Rough Pipe Laws," *Journal of the Institution of Civil Engineers, London*, Vol. 11, pp. 133-156.

Coles, D. E. and Hirst, E. A. (1969), *Computation of Turbulent Boundary Layers-1968 AFOSR-IFP-Stanford Conference*, Vol. II, Stanford University, CA.

Corrsin, S. and Kistler, A. L. (1954), "The Free-Stream Boundaries of Turbulent Flows," NACA TN 3133.

Courant, R., Friedrichs, K. and Lewy, H. (1967), "On the Partial Difference Equations of Mathematical Physics," *IBM Journal*, pp. 215-234.

Crank, J. and Nicolson, P. (1947), "A Practical Method for Numerical Evaluation of Solutions of Partial Differential Equations of the Heat-Conduction Type," *Proceedings of the Cambridge Philosophical Society*, Vol. 43, No. 50, pp. 50-67.

Csanady, G. T. (1964), *Theory of Turbomachines*, McGraw-Hill, New York, NY.

Dixon, S. L. (1978), *Fluid Mechanics of Turbomachinery*, Third Ed., Pergamon Press, Oxford, England.

Emanuel, G. (1994), *Analytical Fluid Dynamics*, CRC Press, Boca Raton, FL.

Eringen, A. C. (1980), *Mechanics of Continua*, Robert E. Krieger Publishing Co., Huntington, NY.

Falkner, V. M. and Skan, S. W. (1931), "Some Approximate Solutions of the Boundary Layer Equations," *Phil. Mag.*, Vol. 12, p. 865.

Favre, A. (1965), "Equations des Gaz Turbulents Compressibles," *Journal de Mecanique*, Vol. 4, No. 3, pp. 361-390.

Ferziger, J. H. and Perić, M. (1996), *Computational Methods for Fluid Dynamics*, Springer-Verlag, Berlin, Germany.

Fernholz, H. H. and Finley, P. J. (1981), "A Further Compilation of Compressible Boundary Layer Data with a Survey of Turbulence Data," AGARDograph 263.

Fletcher, C. A. (1988a), *Computational Techniques for Fluid Dynamics, Vol. I: Fundamental and General Techniques*, Springer-Verlag, Berlin, Germany.

Fletcher, C. A. (1988a), *Computational Techniques for Fluid Dynamics, Vol. II: Specific Techniques for Different Flow Categories*, Springer-Verlag, Berlin, Germany.

Fox, R. W. and McDonald, A. T. (1992), *Introduction to Fluid Mechanics*, Fourth Ed., John Wiley & Sons, Inc., New York, NY.

Fung, Y. C. (1965), *Foundations of Solid Mechanics*, Prentice-Hall, Inc., Englewood Cliffs, NJ.

Glauert, H. (1948), *The Elements of Airfoil and Airscrew Theory*, Second Ed., Cambridge University Press, Cambridge, England.

Granger, R. A. (1988), *Experiments in Fluid Mechanics*, Holt, Rinehart and Winston, Inc., New York, NY.

Hartree, D. R. (1937), "On an Equation Occurring in Falkner and Skan's Approximate Treatment of the Equations of the Boundary Layer," *Proc. Cambr. Phil. Soc.*, Vol. 33, Part II, p. 223.

Hess, J. L. and Smith, A. M. O. (1966), "Calculation of Potential Flow about Arbitrary Bodies," *Progress in Aeronautical Sciences*, Vol. 8, pp. 1-138, D. Kuchemann (ed.), Pergamon Press, Oxford, England.

Hess, J. L. and Wilcox, D. C. (1969), "Progress in the Solution of the Problem of a Three-Dimensional Body Oscillating in the Presence of a Free Surface," McDonnell Douglas Report DAC 67647.

Hess, J. L. (1975), "Review of Integral-Equation Techniques for Solving Potential Flow Problems, with Emphasis on the Surface-Source Method," *Comput. Methods Appl. Mech. Eng.*, Vol. 5, pp. 145-196.

Hildebrand, F. B. (1976), *Advanced Calculus for Applications*, Prentice-Hall, Inc., Englewood Cliffs, NJ.

Hinze, J. O. (1975), *Turbulence*, Second Ed., McGraw-Hill, New York.

Hoerner, S. F. and Borst, H. V. (1975), *Fluid-Dynamic Lift*, Hoerner Fluid Dynamics, Bricktown, NJ.

Hopkins, E. J. and Inouye, M. (1971), "An Evaluation of Theories for Predicting Turbulent Skin Friction and Heat Transfer on Flat Plates at Supersonic and Hypersonic Mach Numbers," *AIAA Journal*, Vol. 9, No. 6, pp. 993-1003.

Huebner, K. H. and Thornton, E. A. (1983), *The Finite Element Method for Engineers*, Second Ed., John Wiley & Sons, Inc., New York, NY.

Jeans, J. (1962), *An Introduction to the Kinetic Theory of Gases*, Cambridge University Press, Cambridge, England.

Jepson, R. W. (1976), *Analysis of Flow in Pipe Networks*, Ann Arbor Science Publishers, Ann Arbor, MI.

Keller, H. B. (1970), "A New Difference Scheme for Parabolic Problems," *Numerical Solutions of Partial Differential Equations, II*, J. Bramble (ed.), Academic Press, New York, NY.

Kellogg, O. D. (1953), *Foundations of Potential Theory*, Dover Publications, Inc., New York, NY.

Kevorkian, J. and Cole, J. D. (1981), *Perturbation Methods in Applied Mathematics*, Springer-Verlag, New York, NY.

Kline, S. J., Morkovin, M. V., Sovran, G. and Cockrell, D. J. (1969), *Computation of Turbulent Boundary Layers-1968 AFOSR-IFP-Stanford Conference*, Vol. I, Stanford University, CA.

Knupp, P. and Steinberg, S. (1993), *The Fundamentals of Grid Generation*, CRC Press, Boca Raton, FL.

Kolmogorov, A. N. (1941), "Local Structures of Turbulence in Incompressible Viscous Fluid for Very Large Reynolds Number," *Doklady AN. SSSR*, Vol. 30, pp. 299-303 [translated in *Proc. Roy. Soc.*, Vol. A434, pp. 9-13 (1991)].

Lakshminarayana, B. (1996), *Fluid Dynamics and Heat Transfer of Turbomachinery*, John Wiley & Sons, Inc., New York, NY.

Landahl, M. T. and Mollo-Christensen, E. (1992), *Turbulence and Random Processes in Fluid Mechanics*, Second Ed., Cambridge University Press, New York, NY.

Landau, L. D. and Lifshitz, E. M. (1966), *Fluid Mechanics*, Addison-Wesley, Reading, MA.

Launder, B. E., Reece, G. J. and Rodi, W. (1975), "Progress in the Development of a Reynolds-Stress Turbulence Closure," *Journal of Fluid Mechanics*, Vol. 68, Pt. 3, pp. 537-566.

Launder, B. E. and Sharma, B. I. (1974), "Application of the Energy Dissipation Model of Turbulence to the Calculation of Flow Near a Spinning Disc," *Letters in Heat and Mass Transfer*, Vol. 1, No. 2, pp. 131-138.

Lee, J. F. and Sears, F. W. (1963), *Thermodynamics*, Second Ed., Addison-Wesley, Reading, MA.

Liepmann, H. W. and Roshko, A. (1963), *Elements of Gasdynamics*, John Wiley & Sons, Inc., New York, NY.

Lilly, D. K. (1965), "On the Computational Stability of Numerical Solutions of Time-Dependent Non-Linear Geophysical Fluid Dynamics Problems," *Monthly Weather Review*, U. S. Weather Bureau, Vol. 93, No. 1, pp. 11-26.

Logan, E. S. (1981), *Turbomachinery: Basic Theory and Applications*, Dekker, New York, NY.

MacCormack, R. W. (1969), "The Effect of Viscosity in Hypervelocity Impact Cratering," AIAA Paper 69-354, Cincinnati, OH.

Maxworthy, T. (1968), "A Storm in a Teacup," *Journal of Applied Mechanics*, Vol. 35, No. 4.

Milne-Thompson, L. M. (1960), *Theoretical Hydrodynamics*, Fourth Ed., Macmillan Publishing Company, New York, NY.

Minkowycz, W. J., Sparrow, E. M., Schneider, G. E. and Pletcher, R. H. (1988), *Handbook of Numerical Heat Transfer*, John Wiley & Sons, Inc., New York, NY.

Moody, L. F. (1944), "Friction Factors for Pipe Flow," *Transactions of the ASME*, Vol. 66, pp. 671-684.

Munson, B. R., Young, D. F. and Okiishi, T. H. (1990), *Fundamentals of Fluid Mechanics*, John Wiley & Sons, Inc., New York, NY.

Nayfeh, A. H. (1981), *Introduction to Perturbation Techniques*, John Wiley and Sons, Inc., New York, NY.

Panton, R. L. (1996), *Incompressible Flow*, Second Ed., John Wiley & Sons, Inc., New York, NY.

Patel, V. C. (1968), "The Effects of Curvature on the Turbulent Boundary Layer," Reports and Memoranda No. 3599, Engineering Dept., Cambridge University, Cambridge, England.

Peaceman, D. W. and Rachford, H. H., Jr. (1955), "The Numerical Solution of Parabolic and Elliptic Differential Equations," *J. Soc. Indust. Applied Mathematics*, Vol. 3, No. 1, pp. 28-41. Fay, J. A. (1994), *Introduction to Fluid Mechanics*, MIT Press, Cambridge, MA.

Peyret, R. and Taylor, T. D. (1983), *Computational Methods for Fluid Flow*, Springer-Verlag, New York, NY.

Piomelli, U. (1994), "Large-Eddy Simulation of Turbulent Flows," TAM Report No. 767, UILU-ENG-94-6023, University of Illinois at Urbana-Champaign, Urbana, IL.

Prandtl, L. (1925), "Über die ausgebildete Turbulenz," *ZAMM*, Vol. 5, pp. 136-139.

Price, J. M. and Harris, J. E. (1972), "Computer Program for Solving Compressible Nonsimilar-Boundary-Layer Equations for Laminar, Transitional or Turbulent Flows of a Perfect Gas," NASA TM X-2458.

Reyhner, T. A. and Flugge-Lotz, I. (1968), "The Interaction of a Shock Wave With a Laminar Boundary Layer," *International Journal of Non-Linear Mechanics*, Vol. 3, pp. 173-199.

Reynolds, O. (1883), "An Experimental Investigation of the Circumstances which Determine whether the Motion of Water Shall Be Direct or Sinuous and of the Law of Resistance in Parallel Channels," *Philosophical Transactions of the Royal Society of London*, Vol. 174, pp. 935-982.

Reynolds, O. (1895), "On the Dynamical Theory of Incompressible Viscous Fluids and the Determination of the Criterion," *Philosophical Transactions of the Royal Society of London, Series A*, Vol. 186, pp. 123-164 [reprinted in *Proc. Roy. Soc.*, London, Vol. A451, pp. 5-47 (1995)].

Reynolds, W. C. and Perkins, H. C. (9177), *Engineering Thermodynamics*, McGraw-Hill, New York, NY.

Roache, P. J. (1972), *Computational Fluid Dynamics*, Hermosa Publishers, Albuquerque, NM.

Roache, P. J. (1990), "Need for Control of Numerical Accuracy," *Journal of Spacecraft and Rockets*, Vol. 27, No. 2, pp. 98-102.

Roberson, J. A. and Crowe, C. T. (1990), *Engineering Fluid Mechanics*, Fourth Ed., Houghton Mifflin Co., Boston, MA.

Robertson, J. M. (1965), *Hydrodynamics in Theory and Application*, Prentice-Hall, Inc., Englewood Cliffs, NJ.

Rogallo, R. S. and Moin, P. (1984), "Numerical Simulation of Turbulent Flows," *Annual Review of Fluid Mechanics*, Vol. 16, pp. 99-137.

Rosenhead, L. (1963), *Laminar Boundary Layers*, Oxford University Press, London, England.

Rouse, H. (1946), *Elementary Mechanics of Fluids*, John Wiley & Sons, Inc., New York, NY.

Sabersky, R. H., Acosta, A. J. and Hauptmann, E. G. (1989), *Fluid Flow*, Macmillan Publishing Company, New York, NY.

Saffman, P. G. (1993), *Vortex Dynamics*, Cambridge University Press, Cambridge, England.

Schlichting, H. (1979), *Boundary Layer Theory*, Seventh Ed., McGraw-Hill, New York, NY.

Shames, I. H. (1992), *Mechanics of Fluids*, Third Ed., McGraw-Hill, New York, NY.

Shapiro, A. H. (1953), *The Dynamics and Thermodynamics of Compressible Fluid Flow*, Volume 1, Ronald Press, New York, NY.

Shepherd, D. G. (1956), *Principles of Turbomachinery*, Macmillan Publishing Company, New York, NY.

Smith, A. M. O. and Cebeci, T. (1967), "Numerical Solution of the Turbulent Boundary-Layer Equations," Douglas Aircraft Division Report DAC 33735.

Spalart, P. R. and Allmaras, S. R. (1992), "A One-Equation Turbulence Model for Aerodynamic Flows," AIAA Paper 92-439, Reno, NV.

Speziale, C. G. (1985), "Modeling the Pressure-Gradient-Velocity Correlation of Turbulence," *Physics of Fluids*, Vol. 28, pp. 69-71.

Stewartson, K. (1954), "Further Solutions of the Falkner-Skan Equation," *Proc. Camb. Phil. Soc.*, Vol. 50, pp. 454-465.

Taylor, E. S. (1974), *Dimensional Analysis for Engineers*, Clarendon Press, Oxford, England.

Tennekes, H. and Lumley, J. L. (1983), *A First Course in Turbulence*, MIT Press, Cambridge, MA.

Thompson, J. F., Warsi, Z. U. A. and Mastin, C. W. (1985), *Numerical Grid Generation: Foundations and Applications*, North-Holland, Elsevier, New York, NY.

Turner, M. R., Clough, H., Martin, H. and Topp, L. (1956), "Stiffness and Deflection of Complex Structures," *Journal of Aeronautical Sciences*, Vol. 23, No. 9, pp. 805-823.

U. S. Government Printing Office (1974), *The U. S. Standard Atmosphere, 1974*, U. S. Government Printing Office, Washington, DC.

Valentine, H. R. (1967), *Applied Hydrodynamics*, Second Ed., Plenum, New York, NY.

Van Driest, E. R. (1951), "Turbulent Boundary Layer in Compressible Fluids," *Journal of the Aeronautical Sciences*, Vol. 18, pp. 145-160, 216.

Van Dyke, M. D. (1975), *Perturbation Methods in Fluid Mechanics*, Parabolic Press, Stanford, CA.

Van Dyke, M. D. (1982), *An Album of Fluid Motion*, Parabolic Press, Stanford, CA.

Van Wylen, G. J. and Sonntag, R. E. (1986), *Fundamentals of Classical Thermodynamics*, Third Ed., John Wiley & Sons, Inc., New York, NY.

von Kármán, T. (1937), "Turbulence," Twenty-Fifth Wilbur Wright Memorial Lecture, *Journal of the Aeronautical Sciences*, Vol. 41, p. 1109.

Wark, K. (1966), *Thermodynamics*, McGraw-Hill, New York, NY.

White, F. M. (1994), *Fluid Mechanics*, Third Ed., McGraw-Hill, New York, NY.

Whittaker, E. T. and Watson, G. N. (1963), *A Course of Modern Analysis*, Fourth Ed., Cambridge University Press, Cambridge, England.

Wilcox, D. C. (1981), "Algorithm for Rapid Integration of Turbulence Model Equations on Parabolic Regions," *AIAA Journal*, Vol. 19, No. 2, pp. 248-251.

Wilcox, D. C. (1988a), "Reassessment of the Scale Determining Equation for Advanced Turbulence Models," *AIAA Journal*, Vol. 26, No. 11, pp. 1299-1310.

Wilcox, D. C. (1988b), "Multiscale Model for Turbulent Flows," *AIAA Journal*, Vol. 26, No. 11, pp. 1311-1320.

Wilcox, D. C. (1993), *Turbulence Modeling for CFD*, DCW Industries, La Cañada, CA.

Wilcox, D. C. (1995), *Perturbation Methods in the Computer Age*, DCW Industries, La Cañada, CA.

Index

A

Absolute angular momentum, 287-289, 292
Absolute pressure (*defined*), 6
Absolute temperature (*defined*), 646
Absolute velocity, 156, 171-172, 290, 292-296, 306, 604, 673
Acceleration:
  advective (*defined*), 93
  centrifugal (*defined*), 673
  convective (*defined*), 93
  Coriolis (*defined*), 673
  gravitational (*defined*), 650
  in Cartesian coordinates, 670
  in cylindrical coordinates, 671
  in rotating coordinates, 673-674
  in spherical coordinates, 672-673
  in streamline coordinates, 674-675
  instantaneous (*see Acceleration: unsteady*)
  unsteady (*defined*), 93
Acoustic wave, 260-262, 269
Adams-Bashforth formula, 553
Added mass (*see Virtual mass*)
Adiabatic flow, 206-209, 261, 264, 266, 584, 587, 589, 594
  reversible (*see Isentropic flow*)
  with friction, 589-597
Adiabatic process (*defined*), 199
Adiabatic wall (*defined*), 588
  temperature, 588, 620
Adverse pressure gradient, 459-460, 524, 526-528, 532, 548, 557
Airfoil:
  camber, 376-380, 401-402
  drag, 527, 616-617
  flow past, 320-322, 332, 372-380, 475, 506, 527, 548, 616-617
  Joukowski, 373, 380, 520
  lift, 332, 358, 373-374, 377-380, 515, 520, 527
  mathematical description, 376
  pressure surface, 527
  stall, 297, 527-528
  suction surface, 527
  theory, 348, 372-380, 520, 614
  thickness, 374, 376-378, 380, 520, 527, 617
Alternate depths (*defined*), 239
Andrade's equation, 22
Anemometer:
  hot-wire, 533
  laser-Doppler, 533

Angle of attack, 373-374, 377-380, 527, 616
Angular distortion, 410-416
Angular-momentum conservation, 287-289
  differential form, 287, 427-429
  integral form, 287
Archimedes' Principle, 53, 70
Area:
  centroid of, 62
  moment of inertia of, 62
Area ratio for isentropic flow, 272-274
  tables of, 651-657
Aspect ratio, 253
Atmosphere:
  adiabatic, 73
  exponential, 57
  lapse rate, 56
  U. S. Standard, 55-57
Atmospheric pressure (*defined*), 6
Attached flow, 515
Average velocity:
  in control-volume analysis, 154, 168-169
  in pipe flow, 211-214
  in turbulent flow, 529-538, 620
Axial-flow turbomachines, 286, 300, 303-305
Axisymmetric flow:
  equations of motion for, 415, 671

B

Backward difference (*defined*), 439
Backward-facing step, 405
Barotropic fluid, 145, 325
Bearing, slipper, 560-561
Beltrami flow, 325
Bends:
  diving hazard, 55, 72
  in pipes, 218, 222
Bermuda triangle, 28
Bernoulli's equation:
  compared to energy equation, 208-210
  conditions for, 124-126, 210
  derivation from Euler's equation, 124-126
  for turbomachines, 293
  for unsteady potential flow, 338-339
BG units, 648
Blasius:
  boundary-layer solution, 521-523, 679
  equation, 521, 568
  formula for pipe friction factor, 251
  formula for skin friction, 522

Other Publications by DCW Industries

Turbulence Modeling for CFD, D. C. Wilcox (1993): A first- or second-year graduate text on modern methods for formulating and analyzing engineering models of turbulence. Presents a comprehensive discussion of algebraic, one-equation, two-equation and second-order closure models. Emphasizes an integrated balance of singularity solutions, perturbation methods and numerical integration schemes to test and formulate rational models. Includes a brief introduction to DNS, LES and Chaos theory. Accompanied by a floppy disk with a variety of useful programs, including an industrial-strength boundary layer program with a menu driven input-data preparation utility.

Perturbation Methods in the Computer Age, D. C. Wilcox (1995): Advanced undergraduate or first-year graduate text on asymptotic and perturbation methods. Discusses asymptotic expansion of integrals, including Laplace's method, stationary phase and steepest descent. Introduces the general principles of singular perturbation theory, including examples for both ODE's and PDE's. Covers multiple-scale analysis, including the method of averaging and the WKB method. Shows, through a collection of practical examples, how useful asymptotics can be when used in conjunction with computational methods.

Affordable Scientific Software: We offer several low-cost Fortran libraries compatible with Microsoft, Lahey and SVS Fortran compilers as follows.

- **Scientific Plotting Package (SPP):** Library for creating, from within a program, detailed scientific graphs on IBM/PC and compatible video and hardcopy devices.

- **Scientific Subroutine Package (SSP):** Math and statistics library compatible with the mainframe IBM Scientific Library and the Cray super computer library.

- **Interactive Fortran Library (IFL):** Utility routines to enhance the Fortran environment including enhanced string, keyboard, mouse and video handling routines.

Visit our World Wide Web Home Page for complete details about our books, software products and discounts available to purchasers of this book.

DCW Industries, Inc.

5354 Palm Drive, La Cañada, California 91011-1655 USA
Telephone: 818/790-3844   FAX: 818/952-1272
E-Mail: dcwilcox@ix.netcom.com   WWW: http://webknx.com/dcw